Subcellular Biochemistry

Volume 105

The book series SUBCELLULAR BIOCHEMISTRY is a renowned and well recognized forum for disseminating advances of emerging topics in Cell Biology and related subjects. All volumes are edited by established scientists and the individual chapters are written by experts on the relevant topic. The individual chapters of each volume are fully citable and indexed in Medline/Pubmed to ensure maximum visibility of the work.

Mauricio G. Matcu
Editor

Structure and Physics of Viruses

An Integrated Guide

Second Edition

Springer

Editor
Mauricio G. Mateu
Centro de Biología Molecular "Severo Ochoa" (CSIC-UAM), and Department of Molecular Biology
Universidad Autónoma de Madrid
Madrid, Spain

ISSN 0306-0225 ISSN 2542-8810 (electronic)
Subcellular Biochemistry
ISBN 978-3-031-65186-1 ISBN 978-3-031-65187-8 (eBook)
https://doi.org/10.1007/978-3-031-65187-8

This Springer imprint is published by the registered company Springer Nature Switzerland AG
The registered company address is: Gewerbestrasse 11, 6330 Cham, Switzerland

Foreword

The great challenges of modern science are increasingly requiring the combined efforts of different branches of knowledge. A remarkable example is our longstanding crusade to fight and understand viruses, those tiny pathogens that are a permanent companion and threat for living beings and humankind. The recent COVID-19 pandemic stands as a poignant testament to this reality, where a single viral entity swiftly orchestrated a global upheaval, drastically altering our existence. However, SARS-CoV-2 represents not an isolated incident but rather a stark reminder of the imperative to deepen our comprehension of these microscopic infectious agents to increase our odds of combatting future pandemics.

Viruses have traditionally been a subject of study for biologists, but are increasingly fascinating physicists and chemists alike, dazzled by their clever operation and their outstanding potential for promising applications. What makes viruses special is that they are comparatively simple in the biological context, yet extremely effective in replicating themselves in all kinds of hosts. Viruses have found optimal solutions to survive in very harsh conditions and are teaching us important lessons on how to work efficiently at the nanoscale, making the most of their very limited resources.

A large part of our present knowledge of viruses has been facilitated by techniques, theories, and methods developed in physics and chemistry. In biology, structure is the key to understand function, and physical techniques like X-ray diffraction or cryo-electron microscopy have opened the door to unveil structural details of viruses with nearly atomic resolution. In addition, function ultimately involves a set of processes that cannot escape physical laws and that, in fact, very often make a profitable use of them. In recent years, physics and chemistry are not only helping in providing new experimental tools to investigate viruses but are also increasingly contributing to achieve a qualitative and quantitative comprehension of the processes involved. This close collaboration between physics, chemistry, and biology has led to great advances and has boosted our present understanding of viruses.

But interdisciplinarity has its toll. It is often very hard for scientists trained in one of the traditional sciences to navigate between two worlds, overcoming the different points of view and the methodological and terminological barriers between disciplines. In this context, a book like this constitutes an invaluable resource to cover this gap for students and scientists entering the fascinating world of viruses. The book gathers a complete selection of chapters written by leading experts in the field. It has been written primarily with a non-expert audience of students and researchers in mind; however, it has been aimed also at providing practitioners in one area of structural or physical virology with updated overviews of most other areas of these broad scientific disciplines. It covers distinct but complementary aspects of the study of viruses that go from techniques required for their characterization to their potential applications, including its structure and functioning. In this second edition, this knowledge has been updated including the latest developments in this thriving field, while maintaining the didactic essence of the book.

This effort has been partially possible thanks to the *Spanish Network on the Biophysics of Viruses* (BioFiViNet), supported by the Spanish Ministry of Science. This platform coordinates the efforts of the Spanish physicists, chemists, and biologists interested in the interdisciplinary study of viruses and has been the initial spark to agglutinate most of the experts who ended up being authors of this book. It has also promoted the added value of collaborative work between apparently widely different disciplines, thanks to the support and enthusiasm of all its members.

But, without doubt, the real artificer of this great work is its editor Mauricio G. Mateu who has meticulously planned, coordinated, and assisted the efforts of all authors to accomplish this interdisciplinary and formative spirit. In the name of all authors, I would like to thank Mauricio for his relentless effort and countless hours of dedication to this book. I am sure that future physical and structural virologists will also appreciate his efforts, and that this book will help them to prepare for the myriad wonderful lessons that viruses still have to teach us.

Barcelona, Spain
February 2024

David Reguera

Preface to the Second Edition

Structural Virology is today an all-important discipline that permeates most other virological disciplines. The application of physical and physicochemical techniques has led to the determination of the high-resolution molecular structure and dynamics of a rapidly increasing number of viruses and viral components. Moreover, the interplay between biophysical, biochemical, and biological approaches is allowing the elucidation of the structural basis of viral function in unprecedented detail. In addition, a growing number of theoretical or experimental researchers are engaged in a fundamental physics-oriented approach to study different aspects of the architecture, assembly, disassembly, and material properties of virus particles, and the packaging and uncoating of viral genomes. A new term, *physical virology*, has been coined to encompass these and related studies. Structural and physical virology are rapidly merging to provide a renewed and richer view on viruses.

Since many years ago my group's research on viruses has benefitted from multidisciplinary collaborations with physicists, chemists, and biologists. As those collaborations progressed, I became aware of the difficulties we sometimes faced in finding common scientific goals and interdigitating research pathways instead of merely crossing scientific trajectories. Like many others in similar situations I found that, once we began to understand each other's scientific language and aims, we were able to tackle together fruitful combined studies on viruses. This experience and talks with colleagues and students convinced me of the need for an integrated, interdisciplinary textbook on structural and physical virology, addressed mainly to students and non-specialists. I contacted some collaborators and close colleagues belonging to the Spanish Society for Virology and/or the Spanish multidisciplinary network on the biophysics of viruses (*BioFiViNet*), which involved physicists, chemists, and biologists. All of them are internationally recognized structural or physical virologists at the cutting edge of their research areas working in Spanish research institutes or university departments. Some of them recruited other close colleagues to coauthor their chapters. Such general closeness among book authors allowed intense mutual feedback during the organization and writing of the book through frequent e-mails, phone calls, and meetings of the editor with the authors, and also between authors of different chapters.

Ten years have passed since the first edition of this book was published. The book was very well received, but the last decade has seen many important advances in our understanding of the structure and physics of viruses. While many concepts and techniques as described in the first edition are still fully valid, some parts had become outdated. In addition, the dramatic consequences of the COVID-19 pandemic caused by the coronavirus SARS-CoV-2 have provided a stark reminder that viruses are a permanent and very serious threat to humankind. As a result, a new wave of researchers coming from widely different scientific disciplines have recently converged in the study of viruses. A second edition of this book was in order to include updates on some important concepts, techniques, and developments in structural and physical virology. In response to an invitation by Springer, I asked the corresponding authors of the first edition if they were willing to participate in a new edition. Unfortunately Prof. José Antonio Melero had prematurely passed away, and a few other senior authors had retired or were unavailable. Happily, the vast majority of the authors did agree to update their chapters. Other close colleagues, also internationally recognized experts, were recruited to write new chapters for this edition. As a result of the professional and geographical closeness between authors, each chapter was written and revised based on intense feedback between authors and editor and between some authors of different chapters.

This book is aimed first at M.Sc. students, Ph.D. students, and postdoctoral researchers with a university degree in biology, chemistry, physics, or related sciences who share an interest in or are actually working on viruses. We have aimed also at providing an updated account of many important concepts, techniques, studies, and applications in structural and physical virology for established scientists working on viruses, irrespective of their physical, chemical, or biological background and their field of expertise. We have *not* attempted to provide a collection of *for-experts-only* reviews focused mainly on the latest research on specific topics; we have *not* generally assumed that the reader knows all of the jargon and all but the most recent and advanced results regarding each topic dealt with in this book. In short, we have attempted to write a book basic enough to be useful to M.Sc. and Ph. D. students, as well as advanced and current enough to be useful to senior scientists.

After an infectious virus particle (a virion) targets a host cell, it loses its integrity and releases its nucleic acid genome inside the cell. In the period of time that goes from that point until the progeny viruses are formed, the virus as a discrete physical entity ceases to exist, but the viral genetic instructions inside the infected cell may subvert the cell metabolism. Eventually, many copies of the viral genome and of the viral proteins required to form a new virus particle will be made in the cell, and a number of progeny virions will be assembled from them, closing an infectious cycle (or viral "life" cycle). The structural biology of viral metabolic processes such as replication, recombination, integration, transcription or translation and their spatial and temporal regulation, and the virus-induced alteration of cellular components and reactions are outside the scope of this book. Structure-based aspects regarding these processes are integrated in many excellent Molecular Virology books, a few of which are referred to at the end of the introductory chapter in this book (Chap. 1).

The present book has instead been focused on the "other half" of the viral cycle, which is generally less well known. Specifically, this book contemplates the structure, dynamics, and physics of virus particles: From the moment they come into existence by self-assembly from viral components produced in the infected cell, through their extracellular stage, until they recognize and infect a new host cell and cease to exist by losing their physical integrity to start a new infectious cycle. (Bio)physical techniques used to study the structure of virus particles and components and some applications of structure-based studies of viruses are also contemplated.

Inevitably, some compromises had to be made. Because of space limitations, not every possible topic has been contemplated; however, we believe most of the important general aspects of the structure and physics of virus particles as they are known have been covered. Space limitations have also prevented the authors of the different chapters to include explanations of some general concepts or terms. However, we believe the most important aspects in each chapter will be clearly understandable by those with a B.Sc.-level knowledge of physics, chemistry, biology, or related scientific areas. Consultation to *Wikipedia* or other general sources may quickly solve an occasional doubt on a specific term by a reader coming from a different area of knowledge. In any case, several teaching aids have been implemented in the book to facilitate the reading and understanding of each chapter by a newcomer to the field. These aids include:

An introductory chapter. Chapter 1 includes a brief general introduction to viruses and their structure, some basic concepts and terms in molecular and structural virology, and general descriptions of different steps in the virus cycle. That chapter is also intended as a guide to help the non-expert reader integrate in a general picture the topics treated in each monographic chapter (Chaps. 2–22).

A similar outline in different chapters in each part of the book. Each Part II chapter dealing with a specific technique includes sections on the principles of the technique, on relevant examples of contributions of the technique for understanding viruses, and on technical perspectives. Each Part III chapter dealing with a stage in the viral cycle includes sections that connect the stage described with stages described in other chapters.

Frequent cross-references between chapters. They may be useful to find in other chapters additional information on certain subjects and/or to connect particular aspects treated in different chapters.

Basic systematic information on viruses species and families mentioned in the book. This information is intended mainly to help the reader navigate among the multitude of virus names that will inevitably appear in the different chapters of this book (or of any other virology book). *Table 1.1* includes (nearly) all virus species, families and orders mentioned in this book, with indication of host and viral genome types. *Figure 1.3* gives a scheme of the general virion structure for some of the most important families of animal viruses. *Specific indexes* at the end of the book include (nearly) all virus species and families mentioned, each with reference to pages where some information regarding the species or family is mentioned.

A list of references focused largely on review articles. The list of references at the end of each chapter has purposefully been kept relatively short, in order not to overwhelm the non-expert reader; reviews have been included wherever possible, and the advanced reader is referred in many cases to those reviews for original research articles that have not been cited in the book. We apologize to the many authors whose work, however important, could not be directly cited.

A section on further reading. In each chapter, a short list under the subheading *further reading* includes references to a few books and/or reviews. These may be particularly useful either for learning or refreshing basic principles or for more detailed/advanced information on the subject.

This book is organized into four parts. Although the chapters are self-contained and may be understood individually, all chapters, especially those within each part (I to IV), are inter-related. They have been designed with the input of editor and authors to tell a complete story. Some minor overlaps between chapters have been kept to facilitate the understanding of each chapter subject without a need to read other chapters in the book. We feel these overlaps may also help in connecting the different chapter contents.

Part I, *The Viral Machine*, includes an introductory chapter to the rest of the book (Chap. 1) and another chapter on the fundamental composition and basic architecture of virus particles (Chap. 2). This knowledge is essential for understanding many subjects treated in the following chapters.

Part II, *Determination of the Structure, Dynamics and Physical Properties of Viruses*, contemplates most of the major experimental techniques used in structural and physical virology. These include different electron microscopy techniques, with emphasis on cryo-electron microscopy and cryo-electron tomography (Chap. 3); X-ray crystallography (Chap. 4); nuclear magnetic resonance spectroscopy (Chap. 5); other spectroscopic techniques (circular dichroism and fluorescence) and mass spectrometry (Chap. 6); the combination of the above and other structural biology methods (Chap. 7); 3D correlative microscopy techniques for studies on the structure and dynamics of viruses within cells (Chap. 8); and single-molecule techniques including atomic force microscopy (Chap. 9); and optical tweezers (Chap. 10). In all cases, the techniques are described not in broad, general terms (as can be found in general technical books) but in the ways they are specifically applied to study virus particles and/or their components.

Part III, *Structural Foundations of Viral Properties and Functions*, deals with the different stages in the viral cycle in which virus particles and/or their structural components are involved. Confronting the *chicken and egg* problem, we decided to start the endless viral cycle when a viral particle comes into existence by self-assembly; we then proceed along the cycle until the viral particle ceases to exist by losing its integrity and releasing its nucleic acid in a host cell. As previously noted, the metabolism of the viral nucleic acid and other viral components until new viral particles are formed (thus closing the viral cycle) does not involve virus particles and is, thus, out of the scope of this book. Chapters 11 to 17 individually consider specific stages of the infectious cycle in which virus particles are involved. In most of these chapters the authors have focused on studies on some model viruses that constitute paradigms to understand the structural bases of virus function at that stage in the cycle. In addition, some general conclusions are extracted from the cases described and other studies.

Chapter 11 deals with experimental studies on the basic assembly of structurally simple viruses; both structural aspects and the cellular environment where assembly actually occurs are contemplated. Chapter 12 builds on what has been described in Chaps. 2 and 11, to describe important specific aspects of the structure and assembly of more complex viruses, including scaffold-mediated assembly. Chapter 13 focuses in the different ways in which the viral nucleic acid is packaged inside the virus particle. Chapter 14 deals with the process of maturation of viral particles to become infectious virions. Self-assembly or assisted assembly of a virus shell (capsid), packaging of the nucleic acid, and virus particle maturation are not clear-cut steps and they frequently overlap in space and time; thus, particularly frequent cross-references between Chaps. 11 and 14 have been included.

No specific chapter in Part III is devoted to the properties of the extracellular mature virions because, for the purposes of this book, they may be sufficiently covered in Chap. 1 and/or in other chapters dealing with processes in which those properties come into play. Virion conformational stability and structural dynamics are briefly revised in Chap. 1 in the context of the complete viral cycle and are dealt with in more detail in other chapters as related with some particular stages in the infectious cycle including assembly, genome packaging, maturation, cell receptor recognition, entry into the host cell, and genome uncoating. Several aspects of the recognition of extracellular virus particles by the immune system, and of mechanisms of virus recognition and neutralization by antibodies as well as virus escape from antibody recognition, are briefly revised in Chaps. 20 and 1, respectively.

Chapters 15–17 describe the complex process of virus entry into cells and viral nucleic acid transfer. Chapter 15 deals with the specific recognition of host cells by animal viruses through interaction with receptor molecules leading to entry into the host cell, and the release (uncoating) of the viral genome of nonenveloped viruses inside the infected cell. Chapter 16 contemplates the process of membrane fusion that allows the delivery of the genetic material of enveloped viruses into the host cell. Chapter 17 describes the mechanisms that bacteriophage viruses follow for bacterial cell recognition and genome transfer into the host cell.

The last chapters in Part III (Chaps. 18 and 19) deal with salient studies on the fundamental physics of viruses. Chapter 18 describes experimental studies on the mechanical properties of virus particles; the contributions of mechanical studies on viruses to understand their biological function; and the biological relevance of the mechanical properties of viruses for their survival. Chapter 19 contemplates theoretical physics-based studies on the more fundamental aspects of virus architecture, material properties, assembly and entry into host cells. All of these are highly relevant subjects about which some experimental studies have been described in previous chapters (Chaps. 2 and 11–18). Readers of any one of the Part III chapters are strongly encouraged to read also Chap. 19 to acquire a more complete understanding of the structure and properties and functions of virions, and the synergy between experimental and theoretical, structural and physical virology studies.

Part IV, *Applied Structural and Physical Virology*, includes some important current or potential concepts, developments and applications of the structural and physical knowledge being acquired on viral particles. We chose three major applied areas. Chapter 20 contemplates the general design and structural basis of action of antivirals. In this chapter antivirals against molecules other than virus particles or their components are contemplated first, because the design principles and structural basis of action do not largely depend on the target, and some of those antivirals have been well-studied and proved remarkably successful; the chapter then continues by considering novel approaches based on the inhibition or misdirection of virus entry or morphogenesis. Chapter 21 describes novel approaches to vaccines based on virus-like particles or chimeric virions. Chapter 22 contemplates the engineering of virus particles by chemical and/or genetic manipulation, and some rapidly expanding developments in the use of engineered viral particles for different medical, biotechnological, or nanotechnological applications.

I would like to gratefully acknowledge the authors of this book for their enthusiasm, time, effort, and patience devoted not only to write their original chapters, but also to meet and discuss contents, presentation, and improvements, and for their willingness to update their chapters for this second edition; Prof. David Reguera also for writing the foreword and to him and Profs. Carmen San Martín and Luis Menéndez-Arias for critically reading many of the original book chapters; Miguel Angel Fuertes for great assistance with a number of tasks during the preparation of the first edition of this book; and José A. Pérez for formatting some figures for the same edition. The authors of this book wish to collectively express here our gratitude to those other book authors who critically read our chapter's manuscripts. We gratefully acknowledge the help by Dr. Alexis Rivas and Mr. Srinivasan Manavalan at Springer and by everyone involved in the production of the second edition of this book.

Madrid, Spain
February 2024

Mauricio G. Mateu

Contents

Part I
The Viral Machine

Chapter 1
Introduction: The Structural Basis of Virus Function

Mauricio G. Mateu

Abstract Viruses may be regarded as dynamic nucleoprotein assemblies capable of assisted multiplication within cells, and of propagation between cells and organisms. Infectious virus particles (virions) assembled in a host cell are dynamic, generally metastable particles: They are robust enough to protect the viral genome outside the cell but are also poised to undergo structural changes and execute mechanochemical actions required for infection of other cells. This chapter provides a broad introduction to the structural and physical biology of viruses and is intended mainly for virology students. It includes (i) an elementary overview on virions and on the structural basis of virus function; (ii) a concise summary on basic techniques used in structural or physical virology; and (iii) brief structure-based general descriptions of the different stages in the virus cycle, especially those in which virions and/or their components are involved. These contents may facilitate a better understanding of the specialized subjects treated in the rest of this book. This chapter is also intended as a "road map" to help the nonexpert reader interconnect and integrate in a single picture the different topics described in depth in the 21 monographic chapters in this book.

Keywords Virus · Virus cycle · Infection · Capsid · Viral genome · Capsid subunits · Capsid building blocks · Oligomerization · Self-assembly · Assisted assembly · Assembly intermediates · Scaffolding proteins · Conformational stability · Conformational dynamics · Nucleic acid packaging · Capsid-nucleic acid condensation · Virus maturation · Virus stability and dynamics · Virus-antibody recognition · Virus-receptor recognition · Virus entry · Fusion · Uncoating · Antivirals · Vaccines · Biotechnology · Nanotechnology

---

M. G. Mateu (✉)
Centro de Biología Molecular "Severo Ochoa" (CSIC-UAM), and Department of Molecular Biology, Universidad Autónoma de Madrid, Madrid, Spain
e-mail: mgarcia@cbm.csic.es

M. G. Mateu (ed.), *Structure and Physics of Viruses*, Subcellular Biochemistry 105,
https://doi.org/10.1007/978-3-031-65187-8_1

Abbreviations

AAV	adeno-associated virus
AFM	atomic force microscopy
CBB	capsid building block
CP	capsid protein
ds	double-stranded
cryo-EM	cryogenic electron microscopy
cryo-ET	cryogenic electron tomography
DNA	deoxyribonucleic acid
EMDB	Electron Microscopy Data Bank
Fab	antigen-binding antibody fragment
FMDV	foot-and-mouth disease virus
HIV-1	human immunodeficiency virus type 1
HRV	human rhinovirus
HSV-1	herpes simplex virus type 1
IAV	influenza A virus
ICTV	International Committee on Taxonomy of Viruses
MD	molecular dynamics
mRNA	messenger RNA
MS	mass spectrometry
MVM	minute virus of mice
NMR	nuclear magnetic resonance
PDB	Protein Data Bank
PV	poliovirus
RNA	ribonucleic acid
RT	reverse transcriptase
SAXS	small-angle X-ray scattering
ss	single-stranded
TMV	tobacco mosaic virus
VLP	virus-like particle

1.1 The Structure and Physics of Viruses

Viruses are biological entities capable of assisted multiplication within cells and of propagation between cells and organisms. Virus multiplication and propagation is generally a cyclic process: An infectious viral particle (*virion*) introduces its genome into a host cell, new virions are formed in the cell and released, and these, in turn, may infect other host cells. This cycle of infection is often called the *virus life cycle*. There has been a largely philosophical debate on whether viruses are alive or not. We use the term *virus life cycle* as a synonym of *infectious cycle*; we are not making the statement that viruses are "living" organisms. In this book, viruses are contemplated as macromolecular complexes that, through biological evolution, came into existence and were endowed with the capacity to make copies of themselves by using the genetic instructions they enclose and the host cell machinery.

Because of the effects many viral infections cause on humans and other living beings, viruses are frequently considered only as pathogens causing disease and human suffering, economic losses, and social problems. However, since the times of the "Phage Group" and the advent of molecular virology in the 1940s and 1950s, more and more scientists coming from different areas have become increasingly aware that viruses also provide outstanding, relatively simple models to explore biomolecular structure-function relationships using a combination of physical, chemical, and biological approaches. The knowledge thus acquired is being decisive not only to combat viral disease but also in the quest to understand in physicochemical terms the molecular machinery of life. This knowledge is also being directly applied to the development of novel biomedical or technological applications of viral components.

Some specific reasons for studying the structure and dynamics and the physical and (bio)chemical properties of virus particles are listed next:

(i) Virus particles constitute excellent models to understand and learn to manipulate molecular self-assembly.
(ii) Virus particles are paradigms to understand structure-function relationships in biomacromolecular assemblies and biological machines.
(iii) A profound knowledge of virus structure, dynamics, and properties is essential to understand the life cycles of viruses.
(iv) Virus particles, their components, and the processes in which they participate provide novel targets for the design of antiviral drugs.
(v) Understanding the structural determinants of virus stability, dynamics, and function may facilitate the rational manipulation of virus particles to develop new or improved vaccines, gene therapy vectors, and nanoparticles for drug delivery or other biomedical or bio/nanotechnological uses.

This chapter provides a general introduction to structural and physical virology. It is intended mainly for M.Sc. students, Ph.D. students, and postdoctoral researchers in physics, chemistry, biology, or related areas who are interested in viruses but who may be relatively unfamiliar with the subject. It intends also to provide a "road map" to help the nonexpert reader integrate in a general picture the topics treated in depth in the other chapters in this book. To achieve those aims, the present chapter includes the following:

(i) Brief explanations on some elementary concepts and terms in molecular, structural, and physical virology; a detailed description of the basic architecture of viruses will follow in the accompanying Chap. 2 in Part I of this book.
(ii) Some broad guidelines on the applicability of most of the different experimental techniques described in Part II, Chaps. 3–10, to characterize the structure, dynamics, and physical properties of virus particles.
(iii) A brief overview of a generic virus cycle and of the major roles in the cycle of virus particles and their components.

(iv) Brief accounts of general structural concepts and descriptions regarding each of the different stages of the viral cycle in which virions or their components are involved; and of relevant properties of virus particles. These short, elementary accounts may facilitate those readers with little background in molecular and structural virology a fullest and better integrated understanding of the contents of Part III, Chaps. 11–17; each of these chapters deals in detail with structural aspects of one of other stage of the virus cycle, including virus assembly, viral genome packaging, virus maturation, cell receptor recognition, entry into the host cell, and genome uncoating.

(v) A schematic overview of novel, physics-based approaches to study virus structure, dynamics, and properties, as a brief introduction to detailed accounts of some physical virology methods (Part II, Chaps. 9, 10), studies (Part III, Chaps. 18, 19), and applications (Part IV, Chap. 22).

(vi) A very brief overview of applied studies in structural virology, to put in a general context the particular applications described in detail in Part IV, Chaps. 20–22.

1.1.1 Structural Virology

Our knowledge of the molecular structure and function of viruses has grown spectacularly in the last decades, largely because these entities have uniquely and increasingly attracted the interest of biologists, (bio)chemists, and physicists alike. Viruses have, thus, been rediscovered as organized, dynamic complexes of biomolecules that act as minute machines. These nanomachines are continuously being modified and diversified through mutation and biological adaptation. However, they are invariably determined by the laws of physics and chemistry to blindly perform sophisticated mechanochemical actions, including penetration into host cells and self-assembly from their molecular components after these have been replicated in the cell. The application of physical and physicochemical techniques has led to the determination of the high-resolution molecular structures of many viruses; the interplay of this approach with (bio)chemical and biological approaches have allowed in many cases the elucidation of the structural basis of viral function in unprecedented detail. *Structural Virology* has permeated other virological disciplines to provide a molecular view of viruses and their biology. The detailed structural knowledge on viruses and their components has made and will surely continue to make, decisive contributions in the never-ending fight against viral disease and also on the exploiting of virus components for novel biomedical and technological developments.

1.1.2 *Physical Virology*

In the last years, the advent of nanoscience and nanotechnology and the increasing awareness on the outstanding features of virus particles as solid-state nanodevices are leading to a renewed look at viruses from the physicist's point of view. Theoretical physicists have begun to tackle at the most fundamental level different aspects of the architecture, self-assembly and physical properties of virus particles. Also, the development of techniques to study individual molecules have opened up new possibilities for the experimental study of the structure, properties, and mechanochemical actions of viruses and their components. A new term, *Physical Virology*, has been coined to encompass theoretical and experimental physics-oriented studies of viruses. This novel approach is merging with long-standing structural virology approaches based on electron microscopy, X-ray crystallography, and many other established or developing physical or physicochemical techniques. As a consequence, viruses are currently being investigated for new developments not only in biomedicine and biotechnology but also in nanotechnology, including nanomaterials and nanoelectronics.

1.2 Virions and Their Structural Components

1.2.1 *Molecular Composition of Viruses*

From a structural point of view, virions may be generally regarded as nucleoprotein assemblies. They all include a nucleic acid genome and many copies of one or more proteins. However, virions present remarkable differences in size, shape, molecular composition, structural organization, and complexity (Fig. 1.1). Detailed structural models of virions and other virus particles determined by electron microscopy (Chap. 3) are available in the *Electron Microscopy Data Bank* (EMDB; https://www.ebi.ac.uk/emdb/). Atomic coordinates and high-resolution structural models of virions, capsids, or other virus particles, and of viral components whose structures have been determined by X-ray crystallography (Chap. 4) or electron microscopy (Chap. 3), are available in the *Protein Data Bank* (PDB; http://www.rcsb.org). A large amount of structural information on viruses can be found in the *Virus Particle Explorer database* (VIPERdb; https://viperdb.org).

Considering molecular composition only, viruses are generally classified into two large groups, *non-enveloped viruses* and *enveloped viruses*, depending on the absence or presence of an outer lipid layer.

Non-enveloped Viruses

The simplest non-enveloped virions are composed of just a protein shell or *capsid* (sometimes called *coat*) made of multiple copies of one or more proteins that contains the viral nucleic acid (Fig. 1.1, top). In less simple non-enveloped virions, the capsid may contain not only the viral genome but also other proteins and/or other

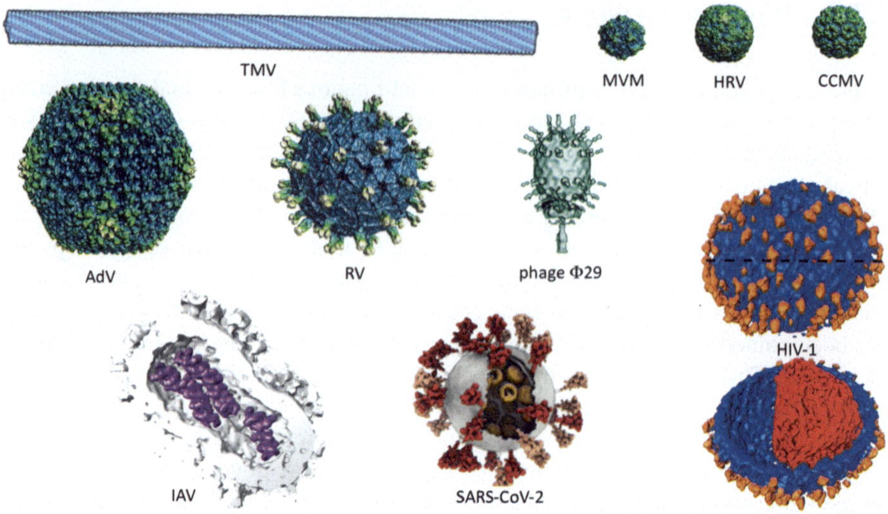

Fig. 1.1 Examples of virus major structural types. (i) Nonenveloped, structurally simple viruses (Chap. 11) include viruses with helical symmetry (TMV, tobacco mosaic virus) or icosahedral symmetry (MVM, minute virus of mice; HRV, human rhinovirus; CCMV, cowpea chlorotic mottle virus). (ii) Nonenveloped, structurally complex viruses (Chaps. 12 and 17) with icosahedral symmetry (AdV, adenovirus; RV, rotavirus; bacteriophage ϕ29). (iii) Enveloped, structurally complex viruses (Chap. 12) (IAV, influenza A virus; SARS-CoV-2, severe acute respiratory syndrome coronavirus 2; HIV-1, human immunodeficiency virus type 1). Enveloped viruses are shown in cross-sections to reveal the different organization of the virion interior (see text). All viruses are represented at approximately the same scale. TMV is 300 nm long x 18 nm wide and provides a reference for scale. The TMV model is reproduced from the PDB molecule of the month shown in https://doi.org/10.2210/rcsb_pdb/mom_2009_1. Models for MVM, HRV, CCMV, AdV, and RV have been obtained from VIPERdb (https://viperdb.org) [Montiel-Garcia D, Santoyo-Rivera N, Ho P, Carrillo-Tripp M, Johnson JE, Reddy VS (2021). Nucleic Acids Res 49: D809-D816.]. The Φ29 model is reproduced from [Wikoff WR, Johnson JE (1999) Curr Biol 9: R296-R300], with permission. IAV and SARS-CoV-2 images were taken from Fig. 3.13 Chap. 13 of this book. The HIV-1 image is reproduced from [Briggs JAG, Grünewald K, Glass B, Förster F, Kräusslich H-G, Fuller SD (2006) Structure 14:15–20], with permission

macromolecules, which may be organized in subassemblies (Fig. 1.1, middle). These additional biomolecules or subassemblies can be enclosed in the capsid shell or externally attached to its surface (Fig. 1.1, middle) (see Chaps. 2, 11, 12, 17).

Enveloped Viruses

In enveloped virions, the capsid and/or other internal structures are typically surrounded by a lipid bilayer, or *envelope*, in which some proteins are embedded (Fig. 1.1, bottom); some enveloped virions have a complex multilayer structure made of organized lipid, protein, and/or nucleoprotein layers (see Chaps. 2, 12).

1.2.2 The Virus Capsid

The capsid plays a fundamental role in both the architecture and the biological function of a virus. A virus capsid can be generally described as a hollow symmetric protein oligomer or multimer made from several tens to many hundreds of copies of one or a few different types of folded polypeptides, the *capsid protein* (CP) subunits. Most CPs (or their individual structural domains if formed by more than one domain) can be ascribed to one of a very limited number of protein architectures or *folds* that can assemble into a very limited number of quaternary structures (Chap. 2).

In each virus, oligomerization of the CPs during capsid assembly normally leads to a defined type of symmetric quaternary structure. Only very few types of capsid symmetry are frequent. The basic types are *helical* (Fig. 1.2, left) and *icosahedral* (Fig. 1.2, right). In some viruses, capsids adopt an elongated (prolate) icosahedral architecture (e.g., phage ϕ29 in Fig. 1.1). Other capsids, such as those of retroviruses (e.g., human immunodeficiency virus type 1, HIV-1, in Fig. 1.1) or poxviruses, are made of other less simple arrangements of CP subunits (see Chaps. 2, 11, 12).

Helical Capsids

Helical capsids (e.g., that of tobacco mosaic virus (TMV) in Fig. 1.2) are extremely simple, theoretically infinite multimers of CPs and (in principle) could be made as

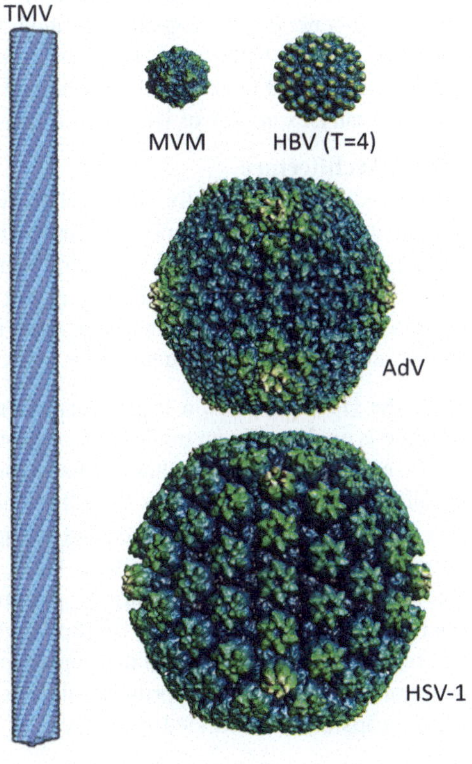

Fig. 1.2 Examples of major types of capsid symmetry. Left: helical (TMV). Right: icosahedral (MVM (triangulation number T = 1), HBV (T = 4), AdV (pseudo-T = 25), HSV-1 (T = 16)) (see Chaps. 2, 11, and 12). All capsids are represented at approximately the same scale. TMV is 300 nm long x 18 nm wide, and provides a reference for scale. The TMV model is reproduced from the PDB molecule of the month shown in https://doi.org/10.2210/rcsb_pdb/mom_2009_1. Models for MVM, HBV, AdV, and HSV-1 have been obtained from VIPERdb (https://viperdb.org) [Montiel-Garcia D, Santoyo-Rivera N, Ho P, Carrillo-Tripp M, Johnson JE, Reddy VS (2021). Nucleic Acids Res 49:D809-D816]

long as required to encapsidate a nucleic acid genome irrespective of its length; thus, there is no theoretical limitation on the amount of genetic information these capsids could store. However, physical and biological restrictions limit the length of helical capsids, which are much less frequent than icosahedral capsids; regular helical capsids are found in less than 10% of the virus families. Chapter 2 provides a description of the architecture of helical capsids; examples of helical viruses can be found in other chapters of this book.

Icosahedral Capsids
In contrast to helical capsids, strictly icosahedral capsids (e.g., that of the minute virus of mice (MVM) in Fig. 1.2) must be made of exactly 60 structurally identical components (CP monomers or oligomers) in order to fulfill intrinsic geometric constraints. This fact could severely limit the size of the nucleic acid genome and, hence, the amount of genetic information that could be enclosed (see Chap. 2). However, evolution has led to different structural solutions to make very large viral capsids with icosahedral symmetry, made of hundreds of CP subunits, that can enclose very large genomes (e.g., that of herpes simplex virus type 1 (HSV-1) in Fig. 1.2). Many large capsids containing many subunits are made of only one type of CP that is capable to adopt different *quasi-equivalent* (similar) conformations (given by the *triangulation number* T); this feature allows the CP molecule to fit into T nonequivalent positions in the icosahedral capsid (see Chap. 2 for a detailed description). This evolutionary strategy minimizes the amount of genetic information required to encode the capsid. Icosahedral capsids are extremely frequent; they occur in about half the virus families. See Chap. 2 for an in-depth description of the architectures of icosahedral viral capsids; Chaps. 2, 11, 12 and others in this book provide many examples of icosahedral capsids and viruses.

Virion Architecture
Enveloped virions typically adopt relatively flexible, frequently spheroidal shapes and are in many cases pleomorphic (individual particles may differ in size and exact shape). They can contain an icosahedral capsid (e.g., HSV-1, Fig. 1.2), a capsid with a different architecture (e.g., HIV-1 in Fig. 1.1), other types of compact (nucleo)-protein complexes (e.g., influenza A virus (IAV) in Fig. 1.1), or combinations of a capsid and a nucleocapsid inside (e.g., HIV-1).

In the HSV-1 virion, a large icosahedral capsid (Fig. 1.2) containing the viral genome is enclosed in an outer layer of proteins (*tegument*), surrounded in turn by a lipidic envelope; the virion that results is considerably larger than the capsid, and the capsid architecture determines the spherical shape of the virion. In the IAV virion, several roughly helical nucleocapsid complexes (ribonucleoprotein particles) are directly surrounded by a protein layer (the *matrix*) and the lipid envelope (Fig. 1.1), and this virus is largely pleomorphic. In the mature HIV-1 virion, a nucleocapsid complex formed by the RNA genome and the nucleocapsid protein and other proteins is contained in a capsid in the shape of a truncated cone, made of a hexagonal lattice of capsid protein subunits that includes pentagonal "defects." The HIV-1 capsid is surrounded by a layer of the matrix protein underneath the lipid envelope. Many other variations in virion architecture, some of them remarkably complex, do exist. In some cases, two or more concentric capsids are found; in others, internal lipidic envelopes are present. The structures of some complex viruses are described in detail in Chaps. 2, 12. Other examples of complex virion architectures can be found in several chapters in this book.

1.2.3 Types of Viral Nucleic Acid

The type of genomic nucleic acid used by different viruses (Table 1.1) is of prime importance to determine the mechanisms of genome replication and expression during the metabolically active, intracellular phase of the viral infectious cycle. From the perspective of this book, the type of nucleic acid, especially whether it is single stranded (ss) or double stranded (ds), is extremely relevant for the assembly of many virus capsids (Chaps. 2, 11, 12, 19), the mechanism of nucleic acid packaging (Chap. 13), the organization of the nucleic acid inside the capsid (Chap. 13), virus particle maturation (Chap. 14), and some properties and functions of the virion, including genome uncoating (Chaps. 2, 11–14, 18).

RNA Viruses
Riboviruses (*RNA viruses*) use RNA as genetic material. Some of them use ssRNA (*ssRNA viruses*); others use dsRNA (*dsRNA viruses*). In turn, two types of ssRNA viruses can be distinguished depending on the polarity of their RNA genome strand (*ssRNA(+) viruses* and *ssRNA(−) viruses*). A genomic single-stranded nucleic acid is considered of positive (+) polarity if its sequence corresponds to that in the viral messenger RNAs (mRNA) and of negative (−) polarity if its sequence corresponds to the complementary of the mRNA sequences. In some viruses, the virion encloses more than one copy of the genome (e.g., retroviruses), or the genome is *segmented*, i.e., split into several nucleic acid molecules (e.g., influenza virus) that can even be carried by different virus particles. Most of the virus species known are RNA viruses.

DNA Viruses
Deoxyviruses (*DNA viruses*) use DNA as their genetic material. Some of them use ssDNA (*ssDNA viruses*); others use dsDNA (*dsDNA viruses*).

1.2.4 Host Cells and Organisms

Viruses can be grouped also by the kind of cell they can infect (Table 1.1). Many known viruses infect eukaryotic cells (*eukaryotic viruses*). Of them, many infect animals such as vertebrates (including humans) or insects (*animal viruses*); others infect fungi (*fungal viruses* or *mycoviruses*), plants (*plant viruses*) or protista (*protist viruses*). Many other viruses infect prokaryotic cells such as bacteria

Table 1.1 Some virus families and species grouped according to host type and viral genome

Host[a]	Viral genome[b]	Virus group or *family*[c]	Virion morphology[d]	Some virus species[e]
Bacteria	dsDNA	Myovirus group	icosahedral, contractile tail	P1, P2, P4, SPO1, T4, μ, ϕ92, 8a, 44RR
		Podovirus group	icosahedral, short tail	BPP-1, K1F, P22, Sf6, T7, ε15, ϕ15, ϕ29
		Siphovirus group	icosahedral, long tail	HK97, p2, SF6, SPP1, TP901-1, T5, λ
		Aoguangviridae	icosahedral, tailed	HK97
		Autographiviridae	icosahedral, tailed	T7
		Corticoviridae	icosahedral	PM2
		Demerecviridae	icosahedral, tailed	T5
		Herelleviridae	icosahedral, tailed	SPO1
		Peduoviridae	icosahedral, tailed	P2
		Straboviridae	icosahedral, tailed	T4
		Salasmaviridae	icosahedral, tailed	ϕ29
		Tectiviridae	icosahedral	PRD1, P23-77
		Lederbergvirus (unassigned genus)	icosahedral, tailed	P22
		Uetakevirus (unassigned genus)	icosahedral, tailed	ε15
		Lambdavirus (unassigned genus)	icosahedral, tailed	λ
	ssDNA	*Inoviridae*	filamentous	fd, M13
		Microviridae	icosahedral, very small	ϕX174
	dsRNA	*Cystoviridae*	enveloped, icosahedral capsid	ϕ6, ϕ12
	ssRNA(+)	*Fiersviridae*	icosahedral	MS2, Qβ
Archea	dsDNA	*Ampullaviridae*	bottle-shaped	ABV
		Guttaviridae	enveloped, droplet-shaped	SNDV
		Rudiviridae	rod-shaped	SIRV-2
		Ungulaviridae	filamentous	AFV-1
Protista	dsDNA	*Phycodnaviridae*	pseudoicosahedral, giant	PBCV-1
Plant	ssDNA	*Geminiviridae*	icosahedral, geminated	BGYMV
	ssRNA (+)	*Alphaflexiviridae*	rod-shaped	PapMV, PVX
		Bromoviridae	icosahedral	AMV, BMV, CCMV, CMV
		Potyviridae	filamentous	BYMV
		Secoviridae	icosahedral	BPMV
		Solemoviridae	icosahedral	RYMV, SBMV
		Tombusviridae	icosahedral	CPMV, RCNMV, STNV, TBSV, TNV
		Tymoviridae	icosahedral	PhMV, TYMV
		Virgaviridae	rod-shaped	TMV
	dsDNA-RT	*Caulimoviridae*	icosahedral	CaMV
Fungi	dsRNA	*Partitiviridae*	icosahedral	*Atkinsonella hypoxylon* virus
		Chrysoviridae	icosahedral	*Penicillium chrysogenum* virus
Animal	dsDNA	*Adenoviridae*	icosahedral	(h)AdV
		Asfarviridae	icosahedral, large	ASFV
		Baculoviridae	elongated, rod-like capsid	AcNPV, cytoplasmic polyhedrosis virus, WNPV
		Iridoviridae	icosahedral	Frog virus 3
		Herpesviridae	enveloped, spherical, large, icosahedral capsid	HCMV, HSV-1, VZV
		Mimiviridae	pseudoicosahedral, giant, internal membranes	*Acanthamoeba polyphaga* mimivirus
		Papillomaviridae	icosahedral	BPV, HPV 1
		Polyomaviridae	icosahedral	SV40, HPV(human polyomavirus)
		Poxviridae	enveloped, brick-shaped, large	VACV, Variola virus
	ssDNA	*Circoviridae*	icosahedral	PCV2
		Parvoviridae	icosahedral, small	AAV-2, CPV, FPV, H1-PV, MEV, MVM, PPV

(continued)

Table 1.1 (continued)

dsRNA	*Birnaviridae*	icosahedral	IBDV
	Sedoreoviridae	icosahedral	BTV, RV
ssRNA(+)	*Alphatetraviridae*	icosahedral	NωV
	Astroviridae	icosahedral	Mamastrovirus 1
	Caliciviridae	icosahedral	NV (Norovirus), RHDV
	Coronaviridae	enveloped, pleiomorphic	MERS-CoV, SARS-CoV-1, SARS-CoV-2
	Flaviviridae	enveloped	Dengue virus, HCV, TBEV, YFV
	Hepeviridae	icosahedral	HEV
	Nodaviridae	icosahedral	FHV, PaV
	Picornaviridae	icosahedral	CV, echoV, FMDV, HAV, HRV, PV
	Togaviridae	enveloped, icosahedral capsid	Chikungunya virus, Semliki Forest virus, SINV
ssRNA(-)	*Arenaviridae*	enveloped, spheroidal, pleiomorphic	LCMV
	Filoviridae	enveloped, filamentous	Ebola virus, Marburg virus
	Hantaviridae	enveloped, spheroidal, pleiomorphic	Hantaan virus
	Kolmioviridae	enveloped, pleiomorphic	HDV
	Orthomyxoviridae	enveloped, pleiomorphic	IAV, IBV, ICV
	Paramyxoviridae	enveloped, pleiomorphic	HeV, hMPV, hPIV, measles virus, MuV, NDV, NiV, Sendai virus, RSV(Respiratory Syncytial virus)
	Periabunyaviridae	enveloped, spheroidal, pleiomorphic	Bunyamwera virus
	Phenuiviridae	enveloped, spheroidal,pleiomorphic	RVFV
	Rhabdoviridae	enveloped, bullet-shaped	Rabies virus, VSV
ssRNA(+)-RT	*Retroviridae*	enveloped, spheroidal, pleiomorphic	BLV, EIAV, HIV-1, HIV-2, HTLV-I, MLV, MMTV, MoMuLV, MPMV; RSV(Rous sarcoma virus), SIV
dsDNA-RT	*Hepadnaviridae*	enveloped, spherical, icosahedral capsid	HBV

[a]Different virus species of some families may infect organisms from different kingdoms. For example, some species of the *Sedoreoviridae* , *Bunyaviridae,* or *Rhabdoviridae* can infect animals, while other species of those same families can infect plants

[b]RT indicates that virus replication involves a reverse transcriptase

[c]Myovirus, Podovirus, and Siphovirus are polyphyletic groups related to morphology (tail structure) and are no longer recognized as families by the current ICTV classification (2022 release, https://ictv.global/taxonomy) (see text and Chap. 17). The families (or genera unassigned to a family) of the Class *Caudoviricetes* to which some representative viruses of those three morphological groups are ascribed are also listed. The current ICTV virus classification includes 264 families, plus additional subfamilies and genera that have not been ascribed to families thus far

[d]Features refer to those of protoype viruses and are intended only as a rough approximation to the general morphology of the virus group/family

[e]The virus species listed as representative examples include the vast majority of those mentioned in this book, plus a few additional ones. Virus species abbreviations: *AAV* adeno-associated virus; *ABV Acidianus* bottle-shaped virus; *AcNPV Autographa californica* nuclear polyhedrosis virus; *AdV* adenovirus; *AFV Acidianus* filamentous virus; *AMV* alfalfa mosaic virus; *ASFV* african swine fever virus; *BGYMV* bean gold yellow mosaic virus; *BLV* bovine leukemia virus; *BMV* brome mosaic virus; *BPMV* bean pod mottle virus; *BPV* bovine papillomavirus; *BTV* bluetongue virus; *BYMV* barley yellow mosaic virus; *CaMV* cauliflower mosaic virus; *CCMV* cowpea chlorotic mottle virus; *CMV* cucumber mosaic virus; *CPMV* cowpea mosaic virus; *CPV* canine parvovirus; *CV* coxsackie virus; echoV echovirus; *EIAV* equine infectious anemia virus; *FHV* flock house virus; *FMDV* foot-and-mouth disease virus; *FPV* feline parvovirus; *H1-PV* parvovirus H1; *(h)Ad* (human) adenovirus; *HAV* hepatitis A virus; *HBV* hepatitis B virus; *HCMV* human cytomegalovirus; *HCV* hepatitis C virus; *HDV* hepatitis delta virus; *HeV* Hendra virus; *HEV* hepatitis E virus; *HIV* human immunodeficiency virus; *hMPV* human metapneumovirus; *HPV* human papillomavirus; *HPV* human polyomavirus; *hPIV* human parainfluenza virus; *HRV* human rhinovirus; *HSV* human herpesvirus; *HTLV* human T-lymphotropic virus; *IAV IBV ICV* influenza A, B or C virus; *IBDV*

(continued)

Table 1.1 (continued)

infectious bursal disease virus; *LCMV* lymphocytic choriomeningitis virus; *MERS-CoV* Middle East respiratory syndrome coronavirus; *MEV* mink enteritis virus; *MLV* murine leukemia virus; *MMTV* mouse mammary tumor virus; *MPMV* Mason-Pfizer monkey virus; *MuV* mumps virus; *MVM* minute virus of mice; *NDV* Newcastle disease virus; *NiV* Nipah virus; *NV* Norwalk virus; *NωV Nudaurelia capensis* ω virus; *PapMV* papaya mosaic virus; *PaV* Pariacoto virus; *PBCV Paramecium bursaria Chlorella* virus; *PCV* porcine circovirus; *PhMV Physalis* mottle virus; *PPV* porcine parvovirus; *PV* polio virus; *PVX* potato virus X; *RCNMV* red clover necrosis mosaic virus; *RHDV* rabbit haemorrhagic disease virus; RSV respiratory syncytial virus; *RSV* Rous sarcoma virus; *RV* rotavirus; *RVFV* Rift Valley fever virus; *RYMV* rice yellow mottle virus; *SARS-CoV-2* severe acute respiratory syndrome coronavirus 2; *SBMV* southern bean mosaic virus; *SINV* Sindbis virus; *SIRV Sulfolobus islandicus* rod-shaped virus; *SIV* simian immunodeficiency virus; *SNDV Sulfolobus neozelandicus* droplet-shaped virus; *STMV* satellite tobacco mosaic virus; *STNV* satellite tobacco necrosis virus; *SV40* simian virus 40; *TBEV* tick-borne encephalitis virus; *TBSV* tomato bushy stunt virus; *TMV* tobacco mosaic virus; *TNV* tobacco necrosis virus; *TYMV* turnip yellow mosaic virus; *VACV* vaccinia virus; *VSV* vesicular stomatitis virus; *VZV* varicella-zoster virus; *WNPV Wiseana* nuclear polyhedrosis virus; *YFV* yellow fever virus

(*prokaryotic viruses*). Bacterial viruses are usually called *bacteriophages* or, simply, *phages*. Archea are also infected by viruses (*archeal viruses*). Most viruses have evolved to infect one or a few species of organisms (limited *host range*) and one or a few cell types (specific *tropism*). The type of cell a virus can infect is, in part, the consequence of the structural and functional features the virus particle has evolved to enter a particular cell. In general, infection by an animal virus or bacteriophage depends largely on the receptor molecule(s) on the cell surface that the virion can specifically bind (see Sect. 1.4.5 and Chaps. 15–17). Although some archeal, protist, and fungal viruses are providing fascinating cases for structural study (a few of which are mentioned in this book), the vast majority of viruses that have been the subject of detailed structural and molecular virology studies infect animals, plants, or bacteria (see Table 1.1). Detailed information on the molecular biology of animal viruses, plant viruses, and/or phages can be found also in the books listed at the end of this chapter [1–9].

Animal Viruses

Animal viruses are extremely diverse in terms of size, shape, presence or absence of envelope, capsid architecture, type of nucleic acid genome, and specific mechanisms employed to complete the different stages of the viral cycle as required for infection, multiplication, and propagation. Some animal viruses are among the structurally simplest viruses known. These include the parvoviruses (e.g., MVM; Fig. 1.1; or the adeno-associated viruses, AAV; see Chap. 11). Others are structurally much more complex, such as the adenoviruses (AdV; Fig. 1.1), herpesviruses (e.g., HSV-1), retroviruses (e.g., HIV-1; Fig. 1.1), or the giant mimiviruses (see Chap. 12). The study of the structure, function, biology, and pathogenicity of many animal viruses has been or is still hampered by a number of difficulties, including their structural and/or functional complexity, technical problems to grow and/or manipulate them, and/or safety issues. However, despite all the difficulties, and mainly because of the biomedical or socioeconomic importance of a large number of animal viruses that are pathogenic for humans or livestock, many of these are among the most intensively studied viruses of all. This is reflected also in the many human and animal viruses used as examples and case studies in most chapters of this book. Many particular aspects related to the molecular biology and structure of many different animal viruses can be found also in the books listed at the end of this chapter [1–4, 8–16].

Bacteriophages

Bacteriophages are extremely abundant in the biosphere. They show widely different structures and type of nucleic acid genome; they have helical or icosahedral capsids and may or may not include a lipid envelope. Phages range from very small and structurally simple non-enveloped icosahedral viruses (e.g., ϕX174) and long but simple helical viruses (some filamentous phages) to large and very complex viruses (e.g., the extremely abundant tailed DNA phages of the *Caudoviricetes* class, such as ϕ29; Fig. 1.1). Since the origins of molecular biology over three quarters of a century ago, and continuing through several decades, some bacteriophages were found to present important advantages as model systems for molecular and genetic studies compared to most animal and/or plant viruses. Their advantages include the relative ease with which they can be obtained in large amounts; their relative structural and functional simplicity and ease of handling of bacterial cells compared to eukaryotic cells; their innocuity for humans, animals and plants; and the possibility to readily obtain certain types of mutant viruses to investigate virus structure and function. These and other reasons led to the intensive use of phages as paradigms for many molecular biology studies on nucleic acid replication, gene expression, and their regulation. Several bacteriophages have also provided, and continue to provide, model systems for studying molecular recognition and self-assembly during the morphogenesis of biomolecular complexes. As a consequence, the structure and function of some bacteriophages, and most stages in their life cycles, are known in great detail (see Chaps. 12, 17). Many particular aspects of the molecular biology of different bacteriophages can be found in several books listed at the end of this chapter. For some books on bacteriophage molecular biology, see [5, 6].

Plant Viruses

Most of the very abundant plant viruses are non-enveloped ssRNA(+) viruses with a slender helical capsid or a relatively small icosahedral capsid. The structure and function of a few plant viruses, such as TMV (Fig. 1.1), have been intensively studied for many decades, in many cases because of some advantages of those viruses as model systems, for example, the facility to grow plant viruses in very large quantities by simply infecting host plants. In addition, studies on many plant viruses have been greatly stimulated because of the economically important diseases they cause in crop plants. Recently, some of the advantages of phages and plant viruses referred to above have led to their preferential use as platforms for many bio/nanotechnological developments (see Chap. 22). However, generally speaking plant viruses have been the subject of fewer studies than animal viruses or bacteriophages, and many stages of the life cycles of the former remain less well known than those of the latter. For a book on plant virus molecular biology and structure, see [7].

1.2.5 Classification of Viruses

Comparisons of the sequences of viral genes and genomes have led to the establishment of phylogenetic relationships between many viruses. In addition, comparison of the tertiary structure of viral proteins, especially CPs, has allowed the identification of very distant evolutionary relationships among different viruses, or at least between some of their genes (see Chap. 7). It must be noted here that genetic recombination and horizontal gene transfer between even very different, unrelated viruses are frequent. Thus, viruses of widely different origins could share some evolutionarily closely related genes, and of course the proteins these genes encode. It is not yet possible to solidly establish a general phylogenetic-based classification of viruses. In 1973, the *International Committee on Taxonomy of Viruses* (ICTV) was established, and a general database on viruses and its classification was created and is regularly updated (https://ictv.global).

Known viruses have been classified into seven major groups based on the type of nucleic acid genome (*Baltimore classification*). These groups are as follows: I, dsDNA viruses; II, ssDNA viruses; III, dsRNA viruses; IV, ssRNA(+) viruses; V, ssRNA(−) viruses; VI, ssRNA(+) virus whose replication involves the action of a reverse transcriptase (RT) that synthesize DNA from a RNA template; and VII, dsDNA viruses whose replication involves the action of an RT (see Table 1.1).

In addition to their classification according to type of nucleic acid genome, viruses have been classified by ICTV in a number of taxonomic groups (*taxons*) that reflect phylogenetic relationships. Such relationships are inferred mainly by comparative sequence analysis, but other features are also considered, including structural features, genome organization, gene expression programs, etc. The main viral taxons, in descending hierarchical order, includes *realm, kingdom, phylum, class, order, family, genus,* and *species*. The most useful taxon regarding virus classification may be the *family* (see Table 1.1). Viruses in a same family probably share a not too distant evolutionary relationship. In its more recent classification, the ICTV has defined six virus realms, 10 kingdoms, 17 phyla, 40 classes, 72 orders, 264 families, 2818 genera, and 11,273 recognized viral species. Many discovered viruses have not been classified, and the biosphere undoubtedly contains a huge number of viruses that have not been discovered yet.

Virus family Latin names (italicized) include the suffix *viridae*. Very frequently, the family English name (non-italicized), which include the suffix virus (plural: viruses), is used instead of the Latin name. However, this may occasionally cause confusion on whether one is referring to a viral family or genus, unless this point is specified. In the literature, virus species are usually referred to by their English names, and most have been given standard abbreviations. Most bacteriophages were originally named according to a rather arbitrary mixture of Latin letters, Greek letters, and/or numbers. The trend is now toward a binomial identification of virus species.

In this book, viruses will generally be identified by type of nucleic acid, by family (Latin or English name) and/or by species (common name). For example, in different chapters, repeated mention is made to HIV-1. HIV-1 is a virus species of the Retroviridae (retrovirus family), which, in the classification by nucleic acid type, belongs to a group of ssRNA(+) viruses whose replication involves an RT (group VI). A list of the vast majority of virus families and species (names) mentioned in this book, along with a few other important families and species, is included in Table 1.1. This table is mainly intended to help the reader navigate among the multitude of virus names that will inevitably appear in the different chapters of this book (or in any other virology book). In addition, an index of names of nearly all virus species and families mentioned in this book can be found at the end of the book.

Viruses in each family do share many features and present many structural and functional similarities, apart from the type of nucleic acid genome. Of the utmost relevance for the subject of this book is that viruses from the same family generally show a clear similarity regarding virion and capsid structures. However, it must emphatically be noted that even viruses of a same family may sometimes *drastically* differ in some or many structural features, properties, functions and underlying mechanisms, cells and organisms they can infect, pathogenic effects, etc. Substantial phenotypic differences may occur even between some individual viruses from a same clone, and some of these differences can be due to a single amino acid substitution in some viral protein. The converse is also true: some viruses from different families may share many structural and functional similarities, probably as an evolutionary consequence of similar "lifestyles," either by conservation of basic structural and/or functional features or by convergent evolution. See, for example, Chaps. 2, 7 regarding structural similarities in CPs; Chap. 17 for similarities in the tails of tailed bacteriophages (Class *Caudoviricetes*); and Chaps. 2, 19 for fundamental physics-based similarities in architecture and some properties of icosahedral virus capsids. Figure 1.3 shows very simple, general schemes of virion structures for members of some important families of animal viruses, many of which are mentioned in this book.

1.3 Techniques Used to Study the Structure and Physics of Viruses

Today, the structural virologist has at his/her disposal a vast array of sophisticated and powerful techniques to study virus structure, dynamics (including conformational rearrangements, assembly, and disassembly), and physicochemical properties. Most of the major structural and several other biophysical experimental techniques

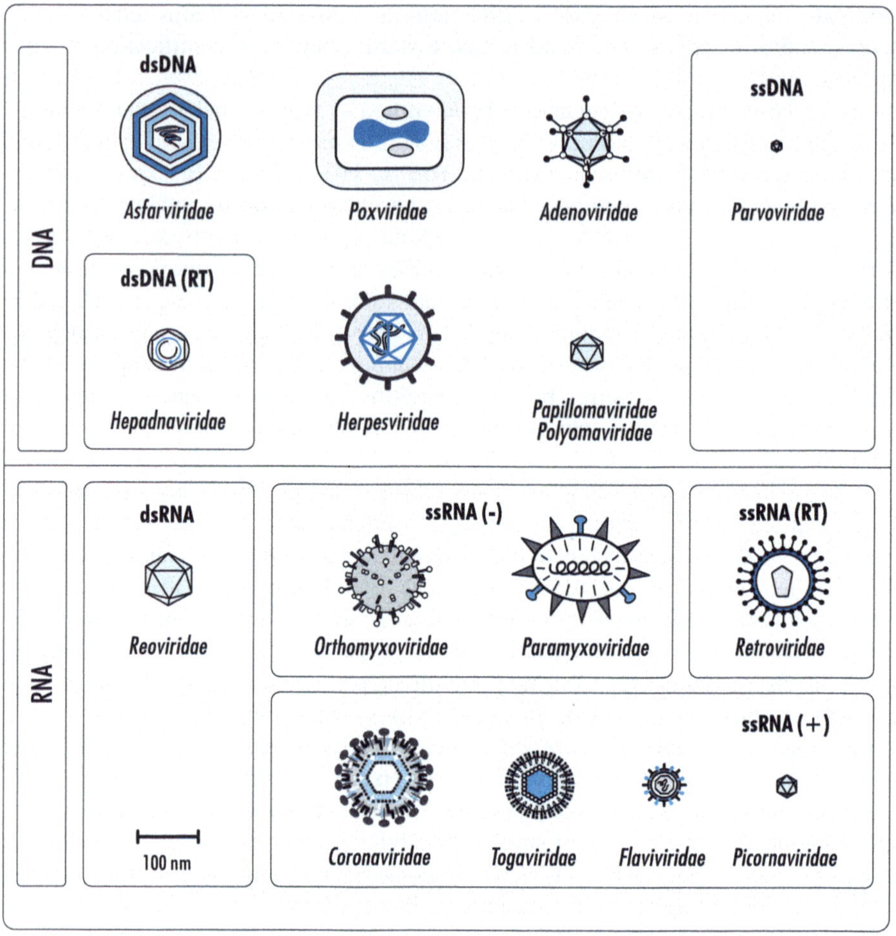

Fig. 1.3 General morphologies of virions belonging to some important animal virus families. The families are grouped according to viral DNA type. Figure reproduced from [*Carrasco L, Almendral JM (coord.) (2006) Virus Patógenos. Ed. Hélice and Fundación BBVA, Madrid, Spain*], with permission

in current use to study isolated viruses are covered in Part II (Chaps. 3–10). Several physics-based theoretical approaches to study the structure, dynamics, and properties of viruses are covered in Chap. 19.

The novice virology student may wonder how he/she could approach different specific questions related to the structural or physical study of his/her chosen viral system. There are no easy recipes to decide which one(s) of the available techniques and approaches a researcher should use for solving a specific problem in structural virology. The decision will depend, of course, on many scientific, technical, and practical circumstances and considerations. Some options may be easily discarded, but several other options may, in principle, be adequate for any intended goal, viral system, and defined situation. Asking oneself a number of questions when considering a structural virology project should greatly help in defining the technical choices. A number of relevant considerations (including a few obvious ones) are listed next:

Some scientific considerations:

(i) What are the researcher's specific scientific goals?
(ii) What information on the chosen or related virus(es) is already available?
(iii) Is the researcher interested in determining the structure of the complete virion or only of some structural component(s)? Are the virus particles asymmetrical or pleomorphic? Are they enveloped?
(iv) Is the researcher interested in studying some dynamic process in which the viral particle could be involved? Dynamic studies may be focused on fluctuations of the viral structure or conformational rearrangements induced by external factors and may include the structural characterization of intermediates of virus particle assembly, maturation, genome packaging, or disassembly.
(v) Does the researcher contemplate the study of some aspect of virus structure or dynamics in situ (inside a host cell)?
(vi) Is the researcher interested in characterizing some physicochemical property of the virus particle? which one(s)? What for?
(vii) Is the researcher interested in determining the structure or properties of a virus particle or component bound to a ligand? Which one? What for?
(viii) Is the researcher interested in trying to relate the structural information to be acquired on the viral particle or component or virus-ligand complex with some viral function? Which one(s)? Does the researcher aim at identifying the molecular mechanism involved?
(ix) Is the researcher interested in trying to relate the structural information to be acquired with the molecular biology of the virus? With pathogenic effects? With viral evolution? Any other aspect?
(x) Does the researcher intend to develop any particular application based on the structural information acquired? Which application?

Some technical considerations:

(xi) Can the virus in question be produced by infection of cultured cells? or can the viral CPs be produced in some expression system, and will they form viral capsids or virus-like particles (VLPs) in vitro? Can isolated viral components of interest be obtained in a stable form?
(xii) Can the virions, capsids, or VLPs be purified? Which level of purity must be reached? What amount of sufficiently pure viral particles or components can be reasonably obtained?
(xiii) Do the viral particles or viral structural components aggregate at high concentrations? Can they be crystallized?

(xiv) Are the viral particles or components stable enough in the conditions required for structure determination by the technique(s) being considered?
(xv) Should the viral particles or components be chemically or genetically modified? Is there an infectious DNA clone of the virus genome available to allow site-directed genetic modifications?

A few practical considerations:

(xvi) Which structural virology expert(s), if any, could be contacted for possible collaboration? What structural facilities are available to the researcher and/or his collaborators?
(xvii) What about safety issues? Will containment facilities be available, if required? Permissions needed?
(xviii) How much time and what resources will be available for the project? What partial goals should be achieved along the project? In which order should they be tackled? By whom? How long will each goal possibly take? What are the alternatives if something fails or takes (much) more time than expected? These latter considerations are indeed common to all scientific projects, but the very complex nature of many structural techniques applied to study a large molecular assembly, and the time, effort, and uncertainties involved, make such considerations particularly relevant in this case (and difficult to decide upon).

Given the very large number of considerations and uncertainties, it would be illusory to try and provide some general guidelines for choosing the "right" approach (es) and technique(s) for analyzing virus structure, dynamics, or physical properties. However, a brief comparison on the general applicability, strengths, and limitations of each of the different structural techniques contemplated in Part II of this book for generic studies in structural and physical virology is possible and is included next. Please do note that each case may be different and that many exceptions will be found. Many relevant examples/case studies on the application of different structural and physical techniques to specific problems in virology are provided in most chapters of this book.

1.3.1 Techniques for Studying the Structure of Virus Particles or Their Components

Several powerful techniques are being used to determine the high-resolution structure of virions or their structural components. The detailed structures of many different virus particles have provided both critical answers and relevant new questions in molecular virology; they have led also to a spectacular advance in our knowledge of viruses and biological processes and our options to combat viral disease.

In most cases, structural or physical studies of virus particles or viral components involve their production and purification, and in some cases their genetic or chemical modification. Brief descriptions of these initial steps are included in Chap. 22. In some cases, structural aspects of virus infection can be studied in situ (within the host cell itself). Technical aspects for the preparation of infected cells for structural studies are included in Chap. 8.

Transmission Electron Microscopy Using Negative Staining
Conventional transmission electron microscopy using negative staining (Chap. 3) remains an all-important imaging technique, especially for initial structural characterization of isolated virus particles and for many in-cell studies of virions or their components (Chap. 8). It generally requires very small amounts of virus particles that do not need to be very pure and is extremely fast. The particle's general morphology and architecture can be determined. This technique generally provides a rapid, first step for the structural characterization of different types of virus particles.

However, the sample is fixed and dehydrated, and a heavy metal layer (opaque to electrons) is used to increase contrast between sample and background. This treatment may introduce artifacts, including deformation of flexible particles, alterations of their fine structure and blurring of their images. The resolution achieved on stained viral particles is, in general, relatively limited (see Chap. 3).

Cryogenic Electron Microscopy (Cryo-EM)
Since the recent "resolution revolution" (Chap. 3), *Cryo-EM* is frequently the first choice for determining the atomic or quasi-atomic structure of virus particles and virus-ligand complexes. The sample is vitrified and kept at an extremely low temperature; the virus particles remain hydrated, and no staining is generally required. The possibility of artifacts is greatly reduced compared to conventional EM. Even more important, cryo-EM is invariably used together with three-dimensional reconstruction techniques that average the images of many thousands of virus particles visualized in different orientations and result in a volumetric (three-dimensional) structural model of the imaged particles. Because the images of so many individual particles are averaged, different artifacts affecting different individual particles make very little contribution to the averaged model, and the signal-to-noise ratio is greatly increased. Also, as no staining agent needs to be used, the structural model obtained reveals also the anatomy (internal structure) of the viral particle. Some years ago, the typical resolution achieved on virus particles using cryo-EM could reach 0.5–1 nm. Outstanding improvements in instrumentation and in the procedures involved (sample preparation, image processing, and three-dimensional reconstruction, etc.) now allow cryo-EM to reach quasi-atomic or even atomic resolution when determining the structure of many different macromolecular complexes, including viruses (Chap. 3).

Cryo-EM currently allows the determination not only of symmetry averaged viral structures but also the resolution of asymmetric components of virus particles, such as the viral nucleic acid molecule or auxilliary proteins and substructures in the viral particle. Moreover, the classification of individual virus particle images and independent image processing allows the possibility to identify at atomic level structural differences between the different states of a virus (e.g., the genome-filled virion compared to the empty viral capsid or different intermediate states in a virus morphogenetic process) (Chap.3)

Cryogenic Electron Tomography (Cryo-ET)

Cryo-ET (Chap. 3) is a variant of cryo-EM in which an individual microscopic object, such as a single virus particle, can be imaged from various angles of view. As a result, a three-dimensional reconstruction of the structure of a single virus particle, and not a reconstruction based on many different particles, can be achieved. Because of this feature, cryo-ET can be used to determine detailed structures of pleomorphic virions such as IAV, retroviruses, coronaviruses (Fig. 1.1), and many others.

Cryo-ET also allows the detailed visualization of virus particles and viral structures within infected cells (see Chap. 8). However, the current technical limitations of cryo-ET do not allow the determination of atomic or quasi-atomic structures (see Chap. 3).

X-Ray Crystallography

Until relatively recently, *X-ray crystallography* (Chap. 4) was the only available technique to determine the atomic or quasi-atomic structure of virus particles and of many other large macromolecular complexes, including asymmetric assemblies such as the ribosome. Molecular crystals contain a large proportion of water, the virus particles or proteins in them are fully hydrated, and the structures obtained are physiologically relevant. If the conditions are right, structural models of virus particles and macromolecular viral components obtained by X-ray crystallography can reach true atomic resolution.

A major weakness of X-ray crystallography is that adequate molecular crystals of the virus particles or their components must be obtained, which is not always easy, or even achieved at all. For example, filamentous, enveloped, or pleomorphic viruses will not crystallize; adequate crystals of flexible multimeric complexes between virus particles and large protein ligands such as cellular receptors or antibody fragments have very rarely been obtained. Also, relatively large amounts of very pure virus particles or proteins are most often required. Crystals must be well-ordered, of sufficient size, and stable enough under an intense X-ray beam to allow the collection of adequate data. The question of how to obtain some necessary data for macromolecular structure determination using X-ray crystallography (the *phase problem*, see Chap. 4) has been severely limiting in the past. However, this is generally no longer a critical issue, as procedures involving molecular replacement, anomalous dispersion effects, cryo-EM models, or AlphaFold structural predictions are used for initial phase estimation (Chap 4). Recent technological advances have increased the speed and ease with which crystallographic structures of macromolecules and viruses can be obtained.

Today, the relative strengths and weaknesses of high-resolution cryo-EM and X-ray crystallography strongly favor the use of cryo-EM for the determination of the structure of complete virus particles, while X-ray is generally more suitable for the determination of the structure of viral components (see Chaps. 3 and 4). In many cases, structural virology projects are best approached by a suitable combination of more than one technique from a basic pool that includes X-ray crystallography (Chap. 4), cryo-EM (Chap. 3), cryo-ET (Chap 3), RMN (Chap. 5), and atomic force microscopy (Chap. 9), among others (see Chaps. 7 and 8)

Nuclear Magnetic Resonance (NMR) Spectroscopy
Solution NMR spectroscopy complements X-ray crystallography and cryo-EM for the determination of the atomic or near atomic resolution structure of viral macromolecular components (Chap 5). The molecules remain in solution, generally in close to physiological conditions, and the uncertainties of the crystallization process are avoided. NMR is also a powerful technique for determining certain aspects of the conformational dynamics of viral components. The major disadvantage of NMR for structural determination is a rather severe limitation in the size of the biomolecule whose structure can be solved. At present, molecules for structural determination by NMR must be no larger than about 35 kDa. However, by using state-of-the-art instrumentation and sophisticated technical approaches, the structures of some larger proteins (50 kDa and more) have been solved. Unfortunately, direct structural determination of complete viral particles, most viral subassemblies, and many multidomain or oligomeric proteins are still way beyond the reach of solution NMR spectroscopy. Large amounts of purified sample and, usually, isotopic labelling of the sample (see Chap. 5) are required too, and the biomolecules must not aggregate during the slow acquisition of data at moderate temperature. *Solid-state NMR spectroscopy* has been applied for structural studies of a few viral particles (Chap. 5).

Atomic Force Microscopy
In *atomic force microscopy* (AFM), small objects are not probed by electrons or photons as in EM, X-ray crystallography or NMR spectroscopy. Instead, the objects are probed by "touching" them with a very fine stylus (tip) that is used to trace the fine topography of the object (Chap. 9). AFM has already been applied to image a number of virus particles (see Chap. 9). Its greatest advantage as a static virus imaging technique is that the structure of single particles can be determined in liquid, in close to physiological conditions. Most importantly, AFM can also be used as a dynamic imaging technique, where large-scale, relatively slow structural changes of macromolecules and macromolecular assemblies can be followed in real time (see below). A clear disadvantage is that, like conventional EM but unlike cryo-EM, cryo-ET, X-ray crystallography, and NMR spectroscopy, only the surface topography of the object (not its internal structure), can be determined. Because of basic geometric considerations regarding the tip-sample interaction, care has to be exercised in interpreting the images obtained (see Chap. 9). Also, flexible particles could be deformed by adsorption to the solid substrate. The resolution achieved on virus particles is currently limited to about 1 nm. In special conditions (ultra-high vacuum) with special instrumentation, atomic resolution of small organic molecules has been achieved using AFM.

Combined Structural Approaches

Some limitations of X-ray crystallography, NMR spectroscopy, cryo-EM, or cryo-ET for structural studies of viruses and viral components have been overcome by combining some of these techniques (see Chap. 7). For example, a comparatively low-resolution structure of an asymmetric virus can be solved by cryo-ET; atomic resolution structures of subunit proteins of the same virus can be determined by X-ray crystallography; and by combining the structural models obtained, a high-resolution pseudo-atomic model of the virus particle can be obtained. In another example, the atomic structure of a viral protein can be solved by X-ray crystallography at very high resolution and the conformational dynamics of that protein studied by NMR. The power of combining these and other structural, biophysical, or biochemical techniques for structural virology studies is fully illustrated in Chaps. 7 and 8 and can be also appreciated in examples described in other chapters of this book.

1.3.2 Biophysical Techniques for Studying the Dynamics and Physical Properties of Virus Particles

The high-resolution "photographs" obtained by applying the techniques mentioned in Sect. 1.3.1 provide invaluable structural information; however, there is a clear need to complement those static images with "movies" on the dynamics of virus particles using the highest possible spatial and temporal resolution. This is a difficult task, and the available techniques applied so far to viral particles have generally provided relatively limited insights into virus particle dynamics. However, the results already obtained have certainly revealed the critical importance of the local and global dynamics of virus particles and/or some of their components during the infection cycle, and fascinating mechanisms that underlie specific aspects of the infection cycle by different viruses (see Chaps. 5–19 for many examples).

Two types of dynamism in virus particles can be distinguished: *equilibrium dynamics* (e.g., *virus breathing*), in which the structure of the virus particle in any particular state fluctuates around a minimum free energy conformation; and *reversible* or (frequently) *irreversible conformational transitions* between two states of the viral particle, generally induced by external physical and/or chemical external factors acting in vivo and/or in vitro (see Sect. 1.4.4).

In the case of conformational rearrangements, the initial and final states and any reaction intermediates that happen to be stable enough may, in principle, be structurally characterized using the techniques already mentioned in Sect. 1.3.1 and described in detail in Chaps. 3–5, 7, 8. In some cases, unstable intermediates of dynamic processes involving virus particles have been stabilized by kinetic trapping using biochemical or genetic methods, and structurally characterized using some of the techniques referred to in Sect. 1.3.1. In addition, other biophysical or biochemical techniques can be used to detect changes in secondary, tertiary, and/or quaternary structure in virus particles or viral components. These techniques include *circular dichroism spectroscopy* and *fluorescence spectroscopy*, which have been used also to probe in solution some particular aspects of the structure and properties of virus particles, including chemical and conformational stability (see Chap. 6). Some transient intermediates of assembly/disassembly have been identified by using *mass spectrometry* (MS) techniques (Chap. 6). *Small-angle X-ray scattering* (SAXS) (Chap. 7) can be used to determine the shape of virus particles and contributes to structural determination in combination with other techniques; in addition, SAXS can be used to study virus dynamics including assembly and disassembly (see Chap. 7). *Differential scanning calorimetry* can also be used to indirectly follow conformational transitions in virus particles (see Chap. 7).

Single-molecule physical techniques such as AFM (Chap. 9) and *optical tweezers* (Chap. 10) have recently begun to be applied to study dynamic processes in viral particles. AFM offers the possibility to structurally characterize initial and final states and stable intermediates in dynamic processes and determine their mechanical properties that may be biologically relevant. In addition, high-speed AFM is starting to unveil in real time and in near-physiological physicochemical conditions some conformational rearrangements or the disassembly of viral particles (see Chaps. 9, 18). Optical tweezers have been used, for example, to follow and quantify in real time the dynamics of DNA packaging in bacteriophage capsids (see Chap. 10).

Some information on the equilibrium dynamics of virus particles and viral proteins can be obtained from data obtained by X-ray crystallography or high-resolution cryo-EM. As already mentioned, NMR spectroscopy is extremely useful to determine the conformational dynamics at equilibrium of viral components (see Chap. 5). Global or local equilibrium dynamics of complete virus particles in solution has been probed by different biophysical or biochemical techniques including NMR spectroscopy (Chap. 5), limited proteolysis/MS (Chap. 6) and hydrogen exchange/MS (Chap. 6).

Physics-based models and computational simulations, including *coarse-grain and all-atom molecular dynamics* (MD), as well as other theoretical approaches (Chap. 19), are being used to predict some aspects of virus particle dynamics. At present, virus particles in general are still too large and complex to carry out long all-atom MD simulations. All-atom MD simulations of some aspects of the dynamics of very simple viral capsids have, however been carried out with quite interesting results. In many studies, highly simplified (coarse-grained) structural models of simple viral capsids are used instead, and are proving their value to capture the essence of some aspects of virus dynamics. These and other physics-based theoretical approaches to study virus particle dynamics, including conformational changes, assembly and disassembly are described in Chap. 19.

1.3.3 Techniques for Studying Structure-Function Relationships in Viral Particles

Once the molecular structure and/or conformational dynamics of a virus particle or structural component are known, a vast array of biophysical, biochemical, genetic, cell biology, and other techniques can be used to investigate the relationship between viral structure and viral properties or function. An account of those techniques and of the ways they can be used to study virus structure, properties, and function is outside the scope of this book. However, some of these techniques and their uses in structural virology are briefly mentioned in Chaps. 7 and 8. Examples of the results obtained in structure-function analyses of virus particles or their components can be found also in other chapters of this book (see especially Chaps. 11–18).

1.4 The Roles of Virus Particles and Their Components Along the Virus Life Cycle

1.4.1 General Overview of the Virus Cycle

Viruses have evolved different alternatives for their multiplication and propagation. However, these alternatives share a number of general features and involve a number of broadly similar processes, albeit with multiple variations.

Several types of virus infection are known, which one may occur will depend on the virus species, host cell, and conditions in which the cells are infected. In the case of animal viruses, many infections lead directly to the production of progeny virions in large numbers and the death of the host cell (*lytic infection*). In contrast, some infections involve the constant production of virus particles in small numbers, without killing the infected cells which are able to divide, and both the infected cells and the virus will persist in time (*persistent infection*). Other infections lead to the indefinite perpetuation of the viral genome in the host cell, which is replicated along with the cellular genome and propagated by passing to the progeny cells (*latent infections*). In these infections, the viral genome may be propagated by becoming physically integrated in the cellular genome, as in the case of HIV-1, or by being maintained as an independent (*episomal*) nucleic acid molecule, as in the case of HSV-1. However, even latent infections can eventually lead to the production and release of virus particles that will be able to infect other host cells, closing the virus life cycle. Some bacteriophages can also produce both lytic and latent (*lysogenic*) infections, which are viewed as evolved alternative strategies for virus survival; either alternative will be followed depending on the conditions the infected cell and the virus may encounter.

A generic, simplified infectious cycle of an animal virus is illustrated in Fig. 1.4. It must be emphasized again that many variations are possible depending on the virus species, host cell, and environmental conditions. However, this simplified virus

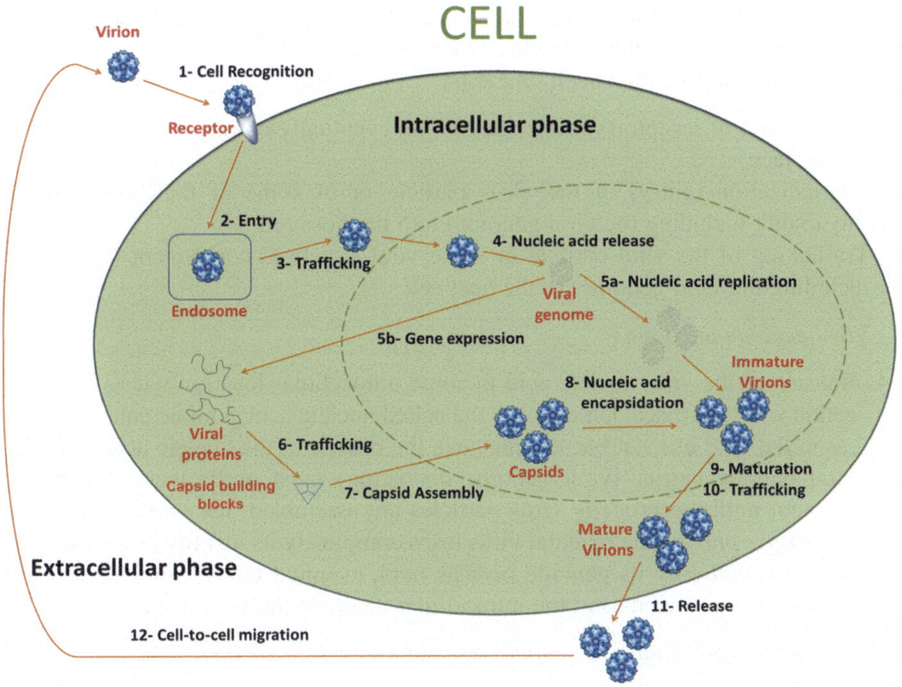

Fig. 1.4 A generic viral cycle. The following stages of the cycle are schematized: *1.* Recognition of host cell receptor(s) on the host cell surface. *2.* Entry of the virus particle (or of some components) into the host cell. *3.* Trafficking of the viral particle (or some components) in the host cell. *4.* Release of the nucleic acid. *5a.* Replication of the viral genome. *5b.* Expression of the viral genome. *6.* Trafficking of the synthesized virus components to sites where the morphogenesis of virus particles will take place. *7.* Assembly of the virus capsid from the previously synthesized CP subunits. *8.* Packaging of the nucleic acid genome inside an assembled capsid (capsid assembly and nucleic acid packaging may occur concomitantly). *9.* Maturation of the virus particle. Depending on the animal virus species, stages 4, 5a, 5b (in part), 6, 7, 8, and 9 may occur in either the cytoplasm or the nucleus of the host cell; this is indicated in the figure by encircling those steps by a discontinuous line, which represents the nuclear membrane only in those cases where the processes indicated occur in the cell nucleus. *10.* Intracellular trafficking of the virus particles and/or mature virions, which may occur between, during, and/or after steps 7, 8, and/or 9. *11.* Release of virions from the cell. *12.* Navigation of virus particles in the extracellular environment. See text for a more complete description and some important variations. Figure kindly provided by M.A. Fuertes

cycle may be useful in this book for putting into perspective the different processes in which virus particles and their molecular components participate during the infection process. The cyclic existence of a virus physical entity and its descent is traditionally contemplated in textbooks from the perspective of an infected host cell. Thus, a viral cycle is generally described by starting from a pre-existing, already formed virus particle which infects a cell and ending with the progeny virus particles that are eventually produced in, and released from the infected cell (either directly in a lytic infection, or after a number of cellular generations have gone through a latent infection). From this classic perspective, the major generic steps in the life cycle of a virus (especially adapted for animal viruses, but mostly valid also for bacteriophages) can be listed in the following order (see Fig. 1.4):

Cell recognition by the virion and entry

1. Recognition of receptor(s) molecule(s) (and eventually co-receptors) on the host cell surface
2. Internalization (entry) of the virus particle or of some of their components (invariably including the viral genome) into the host cell
3. Trafficking of the viral particle or some viral structural component (invariably including the viral genome) in the host cell

Viral genome uncoating

4. Release of the viral nucleic acid in some intracellular location where its replication will take place. Unless only the naked nucleic acid genome enters the cell (as in the case of most bacteriophages), the release of the nucleic acid from the viral particle carrier (viral genome uncoating) must occur there. From this moment until the progeny virus particles are assembled, the virus fully enters the *eclipse phase*. The parental virus loses completely its identity as a molecular complex, and only its genome persists (and, eventually, also some of its viral proteins if required during the intracellular phase of the viral cycle).

Viral gene expression and genome replication

5. Expression and replication of the viral genome. Unless a latent infection is established, e.g., by integration of the viral genome into the cellular genome, the cellular metabolism is subverted to the virus' advantage. Viral genes are transcribed into mRNAs in spatially and temporally controlled processes, and viral proteins required during the different processes of viral metabolism and structural viral proteins required to form the progeny virus particles are synthesized. In addition, the viral nucleic acid genome is replicated by viral and/or cellular polymerases, following one of a vast array of mechanisms that will depend in part on the type of viral nucleic acid genome and virus species.
6. Trafficking of the synthesized virus components to intracellular sites where the morphogenesis of virus particles will take place.

Virion morphogenesis

7. Assembly of the virus capsid from the previously synthesized CP subunits.
8. Packaging of the nucleic acid genome inside an assembled capsid. For many viruses, capsid assembly and nucleic acid packaging occur concomitantly in a single condensation process.
9. Maturation of the virus particle to become an infectious virion. For some viruses, maturation occurs in response to nucleic acid packaging, and these processes cannot be clearly separated.

Virion release

10. Intracellular trafficking of the progeny virus particles and/or mature infectious virions, which may occur between, during, and/or after steps 7, 8, and/or 9.
11. Release of virions from the cell. For some viruses, maturation may occur during or even after release.

Extracellular existence of the virion

12. Navigation of virus particles in the extracellular environment, within and between organisms. During this apparently passive phase, the virion is in fact confronted with, and must fight against, multiple external factors that can lead to its disruption or inactivation. These factors include many physical and chemical aggressions in the environment and multiple antiviral biochemical defenses mounted in the host organism: for example, virus neutralization by antibodies and/or other components of the immune system of an animal host.

This book is focused on virus particles, and a major part of it (Part III, Chaps. 11–19) deals with each of the major general stages in the virus cycle *in which virus particles or their molecular components participate* (i.e., steps 1–4 and 7–12 above). Thus, we have chosen the "beginning" of the (in fact) endless virus cycle (Fig. 1.4) from the perspective of a virus particle, and not from that of an infected cell. We start our description after the viral components are synthesized in a host cell, when the viral particle begins its existence by self-assembly from some of those components (Chaps. 11–14), and we proceed until the same viral particle ceases to exist by uncoating its nucleic acid genome in another host cell (Chaps. 15–17). A logical advantage of this approach is that during morphogenesis of a virion critical features are created that will be needed later by that same virion, for example, when it enters the cell or when its genome is uncoated. Examples of such features include metastability of many viruses or their components (Chaps. 11–16) or built-in pressure in tailed bacteriophages (Chaps. 13, 17–19).

1.4.2 Viral Metabolism: Viral Genome Expression and Replication

After a virion infects a host cell, it begins to lose its physical integrity, finally disappearing as such when its nucleic acid is released inside the cell. From this point until the progeny viruses are formed (if they are formed at all), the virus as a discrete physical entity ceases to exist. What we find instead is an infected cell that contains an additional set of (viral) genetic instructions. The viral genome and, in the case of some viruses, a few viral proteins that have entered the host cell along with the genetic material may not only start a number of new metabolic events but may also alter the cellular metabolic processes and the structural organization of the cell. Eventually, many copies of the viral genome and of the viral proteins required to form a new virus particle are made in the infected cell, and a number of progeny virions are assembled from them.

The diverse viral metabolic processes, such as viral genome replication, recombination, integration, transcription, and translation and their regulation, and some of the virus-induced alterations of cellular components and reactions, are generally much better known than most stages in which viral particles participate, such as virus morphogenesis or disassembly and uncoating. Viral metabolism, including some structural and many functional aspects, constitutes the major subject of most molecular virology textbooks and is out of the scope of the present textbook, which focuses on the structural biology and physics of virus particles instead. Thus, only a few very general basic facts on viral metabolism will be sketched here. For detailed accounts on viral metabolism, the reader is referred at the end of this chapter to a few excellent molecular virology textbooks and compendiums [1–9] and the specialized reviews and references contained therein.

Partly because of the different types of nucleic acid molecules that can act as viral genomes, different viruses follow a bewildering diversity of pathways and mechanisms for genome expression and replication. Moreover, genetic information in viral genomes is stored in a remarkable compact ("streamlined") configuration, probably because of size limitations in the viral capsid and other constraints. This fact also contributes to the very complex regulation of viral gene expression for many viruses.

A central requisite for viral gene expression (as for cellular gene expression) is the synthesis of mRNA molecules (*genome transcription*). The dsDNA genomes of herpesviruses, adenoviruses, or poliomaviruses (to cite a few examples) are replicated by the action of a DNA polymerase in the cell nucleus, and their genes are transcribed into mRNAs by an RNA polymerase much like the genes in cellular chromosomes. In hepadnaviruses, replication of the dsDNA genome involves a RNA intermediate and its reverse-transcription by a RT. The ssDNA genomes of parvoviruses are copied by a DNA polymerase in the nucleus to yield dsDNA molecules, which are transcribed into mRNAs, and also used as intermediates to yield multiple progeny ssDNA genomes that are encapsidated during virus morphogenesis in the nucleus.

ssRNA viruses are extremely abundant and also diverse. In order to be transcribed into mRNAs, viral ssRNA(+) genomes, such as those of picornaviruses or coronaviruses, are replicated in the cytoplasm using a viral RNA replicase to yield complementary ssRNA(−) molecules, which are in turn replicated/transcribed to yield complementary ssRNA(+) molecules that may act as viral genomes and/or mRNAs. Viral ssRNA(−) genomes, such as those of orthomyxoviruses (influenza virus) or paramyxoviruses, are transcribed in the cytoplasm to yield complementary ssRNA(+) molecules that may act as mRNAs and/or are in turn replicated to yield ssRNA(−) molecules that act as viral genomes. In retroviruses, the ssRNA(+) genome is reverse-transcribed by a viral RT into a complementary ssDNA(−) molecule, which is then used as template to form a viral dsDNA molecule which may be integrated in the cellular genome and/or transcribed to yield viral mRNAs. The dsRNA genomes of reoviruses are directly transcribed into mRNAs and replicated in the cytoplasm.

The transcription of viral genes into viral mRNAs inside the infected cell is a carefully regulated process, both in space and time. In many but not all viruses, during a lytic infection some viral genes are preferentially transcribed early in the infection process, and the corresponding mRNAs are translated in the cell ribosomes to yield *early viral proteins*, which will be generally needed to subvert the cell metabolism or control later stages of viral gene expression. At some point, replication of the viral genome starts, and genomic nucleic acid molecules accumulate in some location of the infected cell. In later stages, other viral mRNAs are translated to yield *late viral proteins*. These are generally synthesized in much larger amounts than early viral proteins and usually include, among others, the *structural viral proteins*, those that will form a part of the newly assembled virus particles during virion morphogenesis, the next stage of the viral cycle.

1.4.3 Virion Morphogenesis

Virion morphogenesis is a very complex and tightly regulated process and, except for a few viruses, is still poorly understood in many aspects. Several different stages that sometimes overlap can be distinguished: capsid assembly (Chaps. 11, 12), nucleic acid packaging (Chap. 13), and virus particle maturation (Chap. 14).

Capsid Assembly

Assembly of a virus capsid is a complex oligomerization (or multimerization) process that proceeds along a pathway regulated by ordered interactions between the participating macromolecules. The structural complexity of these large oligomers (or multimers) usually requires intricate assembly pathways mediated by allosteric switches, irreversible steps, and the assistance of other molecules including scaffolding proteins or viral nucleic acid. Three different general strategies for capsid assembly can be recognized (Fig. 1.5): (i) *Capsid self-assembly*, which strictly requires only CP subunits that self-associate under appropriate conditions (Fig. 1.5a); in some cases, some factors may be required to provide the right conditions for self-assembly; (ii) *scaffolding protein-assisted capsid assembly*, which invariably requires not only the CP subunits but also the assistance of scaffolding proteins to produce a precursor, immature capsid (procapsid) (Fig. 1.5b); and (iii) *viral nucleic acid-assisted capsid assembly*, which requires the simultaneous interaction of the CP subunits and the viral nucleic acid in a condensation process in which capsid assembly and nucleic acid packaging occur simultaneously (Fig. 1.5c). In this book, and from a morphogenetic perspective only, we have arbitrarily included viruses that do not require scaffolding proteins for capsid assembly within the *structurally simple viruses* (Chap. 11); and those viruses that require scaffolding proteins (and, thus, will follow a more complex capsid assembly pathway) within the *structurally complex viruses* (Chap. 12). The morphogenetic process of some structurally complex viruses involve further complications, such as acquisition of lipid envelopes (viral membranes) (enveloped viruses) or the separate assembly of different virus parts (e.g., heads and tails of bacteriophages) (Chap. 12).

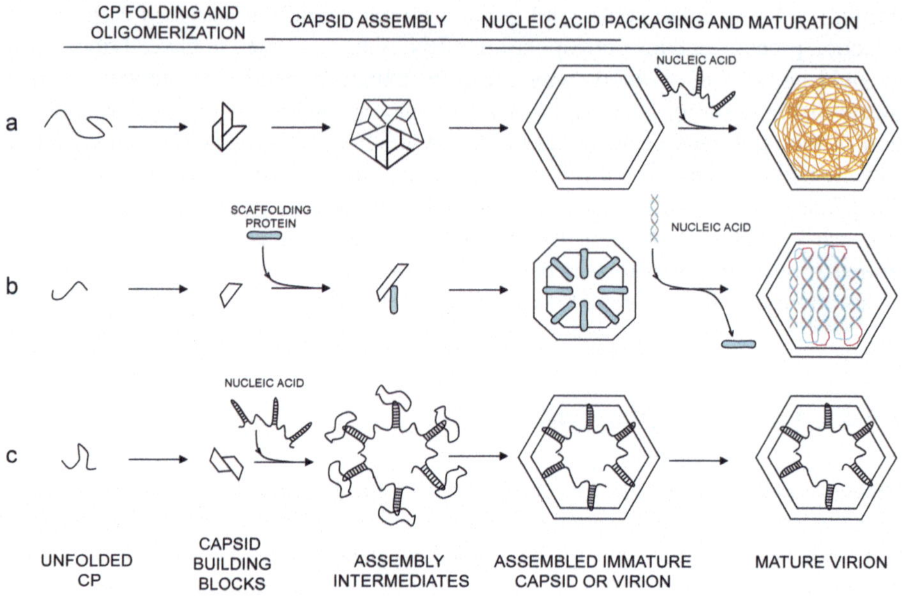

Fig. 1.5 Simplified schemes of three general strategies for assembly of virus capsids. (**a**) Unassisted self-assembly. (**b**) Scaffolding protein-assisted assembly. (**c**) Viral nucleic acid-assisted assembly (see text for details). Figure kindly provided by M.A. Fuertes and reproduced from [Mateu MG (2013). Arch Biochem Biophys 531:65–79], with permission

Virus capsids are generally assembled from soluble, stable capsid building blocks (CBB) (Fig. 1.5). Depending on the virus, CBBs may be folded CP monomers or, frequently, small oligomers that are formed by association of CPs in a previous step; stable CBBs can be, thus, indistinctly regarded as preformed starting substrates for the capsid assembly reaction itself or as early intermediates in the complete morphogenetic process from unfolded CPs to native virions (see Chaps. 11, 12). In vitro or in the cell, the CBBs may be found in an assembly-incompetent state (depending on virus and conditions), and some environmental change or additional factor may be required to activate them and/or to assist in their association.

Capsid self-asembly Several important aspects of the thermodynamics and kinetics of capsid self-assembly (Fig. 1.5a) have been experimentally determined using in vitro systems; in these systems, simple virus capsids (or virions in a few cases) are assembled from its CBBs under close-to-physiological or, sometimes, non-physiological conditions. Unfortunately, the assembly pathways from CBBs to simple icosahedral capsids have proved to be very difficult to trace experimentally, as populated (stable) intermediates (not counting the initial CBBs) are rarely observed, and transient intermediates can be difficult to detect. Despite these difficulties, scarcely populated, transient assembly intermediates have been identified and characterized during the assembly or disassembly of simple capsids, for example, using electrospray ionization-MS or ion-mobility-MS. In addition, several stable intermediates have been identified during the assembly of a few simple viruses, for example, in picornaviruses (see Chap. 11). In contrast, self-assembly and subsequent maturation of more complex icosahedral capsids frequently involve a number of intermediates, including fairly stable intermediates populated at equilibrium and/or transient intermediates, some of which have been identified using different structural or dynamic techniques (see Chaps. 12, 13).

A number of simplified theoretical models and computational simulations on the thermodynamics and/or kinetics of virus capsid self-assembly have been developed (Chap. 19) and contrasted with the results of in vitro experiments in which simple virus capsids are self-assembled from their CBBs. In turn, results obtained in these experiments have been used as constraints to develop improved or novel theoretical models. It must be noted that in most studies, the theoretical models and simulations are limited to the self-association of CPs or CBBs, although some coarse-grained simulations include also the viral nucleic acid. Thus, from a physiological perspective, the results thus obtained in most studies may be particularly relevant in those cases in which the capsid self-assembles in vivo without the assistance of scaffolding proteins or the viral nucleic acid. However, theoretical and experimental capsid-only approaches may also contribute to (qualitatively) understand the assembly of capsids that, in vivo, require scaffolding proteins or the viral nucleic acid. This possibility is supported by the fact that some of these capsids can be correctly assembled in vitro without the assistance of any other macromolecules.

A high-order reaction in which 12 pentameric, 20 trimeric, or 30 dimeric CBBs collide simultaneously and in the right orientations to form an icosahedral capsid is unlikely. In fact, theoretical and experimental analyses support the view that simple icosahedral virus capsids can self-assemble from CBBs through a cascade of second-order reactions. Those analyses have also revealed other rather general features: (i) Self-assembly may be thermodynamically considered as two-state: only the dissociated state (soluble CBBs) is populated below a certain CP concentration, and only the fully associated state (complete capsids) is populated above that concentration. (ii) Capsid assembly kinetics can be represented by a sigmoidal curve and includes a lag phase. (iii) The reaction rate is strongly dependent on protein concentration, and this and the previous features suggest that capsid assembly proceeds through nucleation and growth. (iv) At high protein concentrations, free CBBs disappear but only partially assembled capsids are produced. (v) There is hysteresis to dissociation: the capsid disassembles at much lower CP concentrations than those required for assembly. (vi) Off-pathway reactions may occur, leading to aberrant particles, capsids with non-native quaternary structure or polymorphisms. See Chap. 19 for detailed descriptions of theoretical models on capsid self-assembly that are supported by experimental results.

Scaffolding protein-assisted capsid assembly The proper assembly of many icosahedral virus capsids in vivo, and even in vitro, requires the assistance of scaffolding proteins (Fig. 1.5b; see Chap. 12). Scaffolding proteins establish specific but transient protein-protein interactions with the CP subunits (CBBs) during the assembly process and are later removed, being absent in the mature virus particle. Thus, scaffolding proteins can be regarded as assembly chaperones. Most, but not all, capsids that require scaffolding proteins for proper assembly have a complex quaternary structure. In some cases, the scaffold may self-assemble in the absence of CPs, and is subsequently used as a template for capsid assembly by recruiting CP subunits. In many cases the internal scaffold and the capsid are formed in a co-assembly process where the scaffolding protein subunits promote CP-CP interactions, and the CP subunits promote in turn interactions between the scaffolding protein subunits by eliciting conformational changes in them. Once a precursor, or immature capsid (*procapsid*) is assembled, the scaffolding proteins are removed, either before or during packaging of the viral nucleic acid. In some cases, the scaffolding proteins subunits are extruded through openings in the capsid and recycled, thus acting as assembly catalysts. In other cases, they are degraded by an encapsidated viral protease.

The use of scaffolding proteins for viral capsid assembly offer additional possibilities for controlling the process. Several functional roles for the scaffolding proteins of different viruses have been identified or suggested:

(i) Initiation of capsid assembly by nucleation.
(ii) Promotion of capsid assembly by increasing the effective concentration of CPs and/or lowering the energy barrier of a conformational transition that activates the CP or CBB.
(iii) Directing assembly, the size of the scaffold would determine capsid size by preventing incorrect CP-CP interactions, and the scaffolding protein-CP interactions would promote the proper conformational switches in the CP subunits being incorporated.
(iv) Mediating the incorporation of other viral structural proteins and subassemblies to the capsid.
(v) Preventing the capsid from being filled with some nonspecific proteins which would be difficult to remove.
(vi) Stabilization of the assembled procapsid.

It must be noted here that other viral proteins or protein assemblies in addition to scaffolding proteins can assist the assembly of some virus capsids, irrespective of its structural simplicity. See Chaps. 11 and 12 for detailed descriptions of the assembly of different virus capsids.

Nucleic Acid Packaging
Viruses have evolved two general strategies for packaging their nucleic acid genome in their capsid (Fig. 1.5): condensation of the CPs or CBBs and the viral nucleic acid or insertion of the nucleic acid in a preformed empty capsid (Chap. 13).

Nucleic acid-assisted capsid assembly and genome packaging In many viruses, the viral nucleic acid is recruited to assist the assembly process, to directly yield nucleic acid-containing virions in a combined assembly-packaging process (Fig. 1.5c). In these cases, no empty capsid intermediates are normally formed during virus particle assembly in vivo. The experimental observations suggest that, for many ssRNA viruses, the CP subunits (or CBBs) may bind the RNA and influence the acquisition of a defined tertiary structure by the latter, mainly composed of secondary structure elements but not showing a highly compacted, unique fold. In turn, the folded RNA can promote CP oligomerization and influence the geometry of protein-protein interactions between CP subunits to direct the formation of the structurally correct particle. See Chap. 13 for a detailed description of viral capsid and ssRNA co-assembly.

Viral nucleic acid packaging into a preformed capsid In many other viruses, a nucleic-acid-free capsid is assembled first, either without or with the assistance of scaffolding proteins and/or other auxiliary proteins; the viral nucleic acid genome is later packaged into the preformed capsid (Fig. 1.5a and b). This strategy is followed by many dsDNA viruses, including tailed phages, and dsRNA viruses (see Chap. 13) and also by ssDNA viruses (parvoviruses; see Chap. 11). dsDNA packaging in tailed phages has been extensively studied in exquisite detail. In essence, it occurs as an ATP-driven reaction. This reaction involves a molecular motor and a portal protein complex located at one of the vertices of the icosahedron-based capsid, through which the DNA molecule is inserted. See Chap. 13 for detailed descriptions of active dsDNA and dsRNA packaging into preformed viral capsids.

Virus Particle Maturation

Many virus particles are assembled in a noninfectious form. Maturation refers to a generally irreversible reaction, or series of reactions, that convert a noninfectious (immature) virus particle into an infectious (mature) virion (Chap. 14). Maturation reactions may occur before, during, and/or after nucleic acid encapsidation. In addition to the activation of the virion, a frequent (albeit not universal) effect of maturation is the physical stabilization of the capsid. Proper assembly of many viral capsids may require weak association energies (see Chap. 19), and the resulting capsids may be thermodynamically unstable. The observed hysteresis to dissociation may be enough, in some cases, to prevent capsid disassembly by extreme dilution during the extracellular phase. However, irreversible post-assembly reactions may have evolved as an additional, or alternative, mechanism to stabilize many virions. Irreversible maturation frequently blocks the capsid in a rather stable (actually metastable, or "spring loaded") state through covalent modification and/or conformational rearrangements, which sometimes involve also the formation of stabilizing covalent bonds and/or binding of additional, "cementing" capsid proteins. Maturation of some other virus particles, for example, those of retroviruses, may actually destabilize them (see Chap. 18). When considering maturation, or any other stage of the viral cycle, one must consider the different selective constraints that may be acting of different viruses depending on their structure and "lifestyle" (e.g., genome protection in enveloped vs. non-enveloped viruses, entry of animal viruses vs. nucleic acid injection in bacteriophages, etc.).

Probably because of different selective constraints, maturation does take several forms in different virus families, and even species from the same family. In several families of small, non-enveloped ssRNA viruses, maturation involves a proteolytic cleavage of some capsid subunits. This reaction frequently results in the release of a peptide required for interaction of the virion with cellular membranes during the infection process. In tailed dsDNA phages, maturation overlaps with DNA packaging and is a complex irreversible process that involves capsid conformational changes and achieves different functional goals: removal of the no longer needed scaffolding proteins, capsid expansion, creation of new binding sites for the attachment of additional structural proteins and subassemblies required for virion function, and capsid stabilization. As for simpler icosahedral capsids, the association energy between CPs in tailed phage capsids is low and the procapsid (*prohead*) is rather unstable. The viral dsDNA is packaged into the capsid to a very high (crystalline) density, which results in the capsid becoming pressurized. Capsid pressurization is thought to be required for assisting injection of the viral genome into the infected host bacterium but will tend to disrupt the unstable procapsid (see Chaps. 10, 17, 18, 19). Thus, different dsDNA phages have evolved different specific strategies for capsid stabilization during and after maturation. See Chap. 14 for a detailed description of these and other maturation strategies and mechanisms in different viruses. Examples of maturation in some other viruses can be found also in a few other chapters in this book (e.g., Chap. 11). Factors contributing to the stability of an already mature virion are briefly described in Sect. 1.4.4.

Host Cell Factors Involved in Virus Morphogenesis

Virion morphogenesis in the infected cell is a very complex process. The cell interior is a molecularly crowded environment, where many competing reactions take place. Macromolecular crowding has been shown to contribute to virus capsid assembly, but it can also promote the formation of aberrant assembly products. Moreover, replication of the viral genome, capsid assembly, nucleic acid packaging, and virus maturation must not only occur but need to be strictly coordinated in space and time in order to produce an infectious virion. As a result, virion morphogenesis in the host cell generally requires auxiliary viral or cellular factors, including chaperones, and frequently occurs in *virus factories,* very complex subcellular structures that are formed inside infected cells (see Chap. 8). Virus factories can be regarded as huge dynamic scaffolds that assist virion replication and assembly; they are currently being studied, for example, by electron microscopy and electron tomography of infected cells. Chapter 8 describes structural, molecular, and cell biology techniques that are being used to identify host cell factors that may participate in virus assembly or for studies on virus morphogenesis in the infected cell; the same chapter describes also some important results obtained on this complex and still poorly known process.

1.4.4 The Virion During the Extracellular Phase

Virus Stability

The many physical and chemical aggressions a virion has to withstand outside cells and organisms have led to multiple structural solutions for virion stabilization; these solutions, however, must not interfere with the capacity of the virion to release its genome into a host cell. Some of the stabilization strategies that different viruses have evolved are summarized below. Studies on the structural determinants of the stability of virus particles are providing further clues on the biological success of viruses; they are also leading to the identification of novel targets for antiviral intervention and to explore possibilities for the engineering of viral particles of increased stability for bio/nanotechnological applications.

Noncovalent interactions between capsid subunits Interfaces between CPs or CBBs can differ widely between capsids, or even within a capsid: They can be small or quite large; discontinuous or continuous; quite planar or convoluted; independent of or overlapping with other CP-CP interfaces; including a hydrophobic core and polar rim, or essentially polar; etc. (see Chap. 11 for a few examples). In general, mutational analyses have identified many specific structural elements and residues in the CPs that are relevant for capsid assembly and/or stability in different viruses. The assembly and/or stability of viral capsids may be critically altered by single amino acid substitutions; in addition to the hydrophobic effect (as estimated by buried surface area), buried electrostatic interactions (hydrogen bonds and charge-charge attractive interactions) may be most important for assembly of a stable capsid in some viruses. Conversely, charge-charge repulsions between CP subunits have been clearly involved in the destabilization of very different virus capsids. Electrostatic repulsions between CP subunits in the capsid can be modulated by protonation/deprotonation both in vitro and in vivo and may contribute to the capsid stability-instability balance and the control of biologically relevant capsid conformational transitions.

Covalent bonds between capsid subunits Disulfide bonds or other covalent bonds are formed in the capsids of very different viruses; in many cases, they have been shown to transiently or permanently contribute to capsid assembly and/or stability. The reversibility of disulfide bonding (bonds formed in the oxidizing extracellular environment, but likely disrupted in the reducing environment inside the cell) may provide a convenient strategy to meet the conflicting requirements between extracellular stability versus intracellular lability for some viruses.

Cementing proteins The interactions established between scaffolding protein subunits, and between them and the CP subunits, can help to transiently stabilize some immature capsids before they become irreversibly stabilized by maturation. The attachment of bridging or "cementing" proteins during capsid maturation constitutes an evolutionary strategy of some viruses to increase capsid stability by providing additional intersubunit interactions.

Capsid-viral nucleic acid interactions In many icosahedral ssRNA viruses, the N-terminal and/or C-terminal segments of the CPs are positively charged and normally located in the capsid interior, where they may establish ionic interactions with the negatively charged phosphates of the nucleic acid. In some viruses a fair number of positive residues are located at the capsid inner wall. Other virions contain polyamines. All of these alternatives can neutralize a substantial fraction of the RNA negative charge and contribute significantly to the nucleic acid-assisted assembly and/or virion stability. In several ssRNA plant and animal viruses and a few ssDNA viruses (parvoviruses), short stretches of the viral nucleic acid molecule conform to the icosahedral symmetry of the capsid and can be directly visualized by X-ray crystallography or cryo-EM of the virions. Non-covalent interactions between structured regions at the capsid inner wall and nucleic acid stretches may have a role in RNA-assisted capsid assembly or DNA packaging into a preformed capsid and/or in genome uncoating. They may also stabilize the assembled viral particle against a non-productive conformational change or dissociation.

Hysteresis to dissociation For non-enveloped viruses whose capsids are not irreversibly stabilized after assembly through maturation, inbuilt hysteresis to dissociation may be essential to prevent disassembly in the extracellular environment by extreme dilution.

Entropic stabilization Increased capsid breathing reduces the difference in conformational entropy between the dissociated and associated states. Thus, breathing may help to stabilize the assembled capsid.

Stabilizing ligands Some ligands (metal ions, antibodies, etc.) may increase the stability of virus particles. For example, small hydrophobic molecules (*pocket factors*) are able to bind hydrophobic pockets in the capsids of some picornaviruses (poliovirus, PV, and human rhinovirus, HRV) and can entropically stabilize the virion against conformational changes, inhibiting genome uncoating (see also below).

Covalent attachment of functional groups Some irreversible enzymatic reactions on viral capsids during maturation stabilize the viral particle, and chemical modifications of specific capsid residues may have different roles in assembly and/or stability.

Mechanical stability of virus particles Nearly every study on the molecular determinants of viral capsid stability so far has focused on analysis of the resistance of the virion against inactivation of its infectivity (as an indication of structural alterations) or of the viral particle against chemical modification (including proteolytic cleavage), conformational rearrangement or dissociation into subunits by the action of heat or (bio)chemical agents. However, viruses may also be subjected to severe mechanical stress both outside and inside cells and must be robust enough to withstand the mechanical forces acting on them. Moreover, viruses may have evolved not only to withstand but also to use those mechanical forces. Investigation on the mechanical stability of virus particles and of other aspects of virus mechanics, and their possible biological consequences, have begun using AFM (Chap. 18) and physics-based theoretical approaches (Chapt. 19).

Virus Dynamics

Virions must achieve enough stability in the extracellular environment without compromising viral genome uncoating in the host cell. This conflict between stability and instability has generally been met by many viruses by evolving a virion that is relatively robust under certain conditions but conformationally dynamic and metastable.

Breathing Biochemical and biophysical studies of non-enveloped virions and capsids in solution reveal that they are not static at equilibrium but constitute dynamic assemblies which sample a range of similar conformations around a minimum free energy state. *Virus breathing* is a term sometimes used to refer to the transient exposure to solvent of some structural elements in the particle, including functional CP peptide segments, and these events are biologically required during some steps of the infectious cycle of different viruses (e.g., picornaviruses or parvoviruses). Capsid breathing can be modulated by internal components or capsid-bound external molecules. In the picornaviruses PV and HRV and in other enteroviruses, a large hydrophobic cavity (pocket) in the virus capsid is involved both in virus breathing and in facilitating a transition of the particle from a metastable conformation to another conformation during cell receptor recognition and nucleic acid uncoating. Binding of antiviral agents to the hydrophobic pocket (see Chap. 20) or of a virus-neutralizing antibody to a region nearby, or mutations in the capsid that may partly fill the pocket, were all shown to impair viral breathing. The same factors also impair the transition between the metastable and final states of the capsid. The above and other results indicate that global or local conformational equilibrium dynamics (breathing) may be a general feature in virus particles and may be required for infection by facilitating structural transitions. AFM studies have revealed that different regions in viral particles can differ in mechanical stiffness and indicate that local regions in the particle with high mechanical flexibility may also present high conformational dynamism which may be biologically relevant (see Chap. 18).

Conformational rearrangements in mature virions Excess stability and rigidity may lead to a biologically inert virion, unable to replicate in a host cell. In addition to conformational dynamism at equilibrium, metastable virions respond to the appropriate stimuli by undergoing controlled conformational transitions required for infectivity. In non-enveloped viruses, the entire viral particle may be conformationally rearranged, although the greatest alterations may sometimes be localized (Chap. 15). In enveloped viruses, a particular structural component (e.g., an envelope glycoprotein on the surface) may change its conformation to allow membrane fusion and entry into the host cell (Chap. 16). The transitions must be controlled through the action of regulatory factors, to ensure they occur at the appropriate location and time. The same conformational rearrangements caused by physiological factors can be frequently induced in vitro, even under nonphysiological conditions, using different agents (e.g., changes in temperature or pH). These facts have facilitated the ex vivo or in vitro studies of capsid conformational rearrangements using a vast array of techniques. In some cases, the biological relevance of the transitions observed is not yet clear. In other cases, the identified capsid rearrangements have been shown to be biologically required, and the underlying mechanisms have been unveiled. These structural changes may be required, for example, for the exit of the virion from the cell compartment or the cell where it was formed, entry into another cell, intracellular transport, or uncoating of the viral genome. Specific examples of conformational rearrangements of viral particles or some of their components, and their biological implications, are described in many chapters of this book, including Chaps. 11–19.

To summarize, the end product of virus morphogenesis is a robust enough macromolecular complex able to effectively protect the viral genome during its extracellular existence. However, this complex is also a flexible, dynamic, breathing, metastable particle poised to undergo controlled conformational transitions required to perform biologically critical functions during different stages of the virus life cycle, including entry into cells, intracellular trafficking, and viral genome uncoating.

Virion Neutralization by Antibodies and Virus Escape from Neutralization
An extracellular virion must not only be stable enough to withstand physicochemical aggressions without compromising its function; it must also be able to evade the host defenses without compromising virion stability and function. Higher animals, especially mammals, fight viral infection by mounting a complex molecular and cellular immune response (see Chap. 21 for an outline; see also Chap. 10 in ref. [18] for a detailed description of the mammalian immune response against foot-and-mouth disease virus (FMDV) as a model virus).

Antibodies elicited against invasive virus particles constitute a major part of the immune defense of animals against many viruses. A number of structural studies have focused in the recognition of virions or some of their components by antibodies, the mechanisms by which antibodies neutralize virion infectivity, and the structural strategies viruses follow to escape from antibody recognition. Structural descriptions of complexes between virions or their components and antibodies can be found in Chap. 14 of ref. [13]. Chapter 4 of ref. [18] provides a review on the relationships between the structural basis of virus-antibody recognition and virus escape from neutralization, focusing on foot-and-mouth disease virus (FMDV) as a model system for small, non-enveloped viruses. From an applied perspective, the information obtained on virus-antibody recognition and virus escape has been shown to be critical for the development of improved or novel vaccines, including VLPs and chimeric viral particles (see Chap. 21).

Virus-antibody recognition The structures of complexes between individual viral proteins (e.g., envelope proteins of enveloped viruses) and monovalent antigen-binding antibody fragments (Fabs) are generally determined by X-ray crystallography (Chap. 4) or cryo-EM (Chap. 3) or by combined approaches (Chap. 7). In some respects, virus-antibody recognition is structurally similar in many respects to virus-receptor recognition and other virus-protein ligand interactions. The reader is referred to Chap. 15 for general structural principles of virus-receptor recognition, some of which can also be applied to virus-antibody interaction (but see below).

In enveloped viruses (such as IAV or HIV-1), some antibodies recognize and bind strongly to specific patches (*neutralization epitopes*) on the surface of the envelope proteins. In non-enveloped viruses, some antibodies bind to neutralization epitopes on the capsid surface. In some small non-enveloped viruses (e.g., PV and

HRV of the major receptor group), a remarkable difference between receptor and antibody recognition is that the receptor-binding site lies in a *canyon*, a depression of the virus surface, while antibodies cannot penetrate very deep into the depression and bind preferentially to capsid protrusions (see Chap. 15). This is not the case in other similar viruses, including FMDV, whose receptor-binding site is located in a fully exposed, protruding and mobile protein loop (termed the *major antigenic loop*) which also binds many neutralizing antibodies.

On a viral capsid, epitopes may be formed by residues in just one loop of a CP subunit (*continuous epitopes*; e.g., most epitopes in the *major antigenic loop* of some FMDV types), or several segments of a same CP subunit or of several neighboring CP subunits or CBBs (*discontinuous epitopes*, e.g., epitopes identified in other, equally immunodominant antigenic regions in FMDV). If segments from different CBBs are involved, the epitope will be formed only after some intermediate or the complete capsid has been assembled; some will form only after the capsid has been structurally reorganized during virus maturation (Chap. 14). Antibodies against these epitopes can be, thus, used as probes for studying capsid dynamics. It must be noted here that even if the antibody-binding residues are all located in one protein loop, not every residue in a peptide stretch may be directly involved in binding and, strictly speaking, nearly all epitopes are discontinuous. In addition, all viral antigenic epitopes are *conformation-dependent* as, like any other protein ligand, they must adopt a defined conformation either before or on binding the antibody.

About 15–20 residues in the virus capsid or envelope protein may form the *contact epitope* and interact with a similar number of residues in the antibody *paratope,* contributing to binding strength and specificity through the hydrophobic effect and multiple non-covalent interactions (van der Waals, hydrogen bonds and charge-charge interactions). However, in many viral epitopes very few residues in the contact epitope, those that form the *energetic epitope*, contribute most of the binding energy. In addition, some residues that may not form a part of the contact epitope may indirectly contribute to binding because, if substituted by mutation, the epitope conformation may be altered, preventing antibody binding. The residues whose mutation affects antibody recognition, located either within the contact epitope or elsewhere in the viral particle, may be considered a part of the *functional epitope*.

Virus neutralization by antibodies Once bound to a capsid (non-enveloped viruses) or envelope protein (enveloped viruses), antibodies can neutralize virions by one of several different mechanisms. A frequent mechanism is interference with cell receptor or co-receptor recognition. For example, antibodies that bind the major antigenic loop of FMDV overlap extensively with the receptor binding site and sterically interfere with receptor binding (see Chap. 15; [18]). Other mechanisms involve impairment of a conformational rearrangement of a protein or of the virion that would be required at some stage of the viral cycle. For example, some antibodies prevent a conformational transition of the viral fusion protein that is required during entry of enveloped viruses (e.g., HIV-1) into the host cell (see Chap. 16). Some antibodies may exert a neutralizing effect by binding bivalently to the capsid, increasing capsid stability and probably impairing uncoating; they may even act intracellularly after the virus-receptor antibody has been internalized. Other antibodies may exert their neutralizing effect by facilitating virus elimination through *opsonization*. Some mechanisms of antibody-mediated neutralization of viruses have not been elucidated or well characterized yet.

Virus escape from neutralization by antibodies In general, virus escape occurs through negative selection by circulating antibodies in the organism; it may occur also by genetic drift. RNA virus populations are *quasispecies,* a term that refers to their extreme genetic heterogeneity, being formed by multiple genetic variants in different and sometimes rapidly varying proportions [17]. This generally leads to the frequent emergence of virus variants differing in some phenotypic trait, usually but not necessarily when subjected to the appropriate selection pressure [17]. As a consequence, drug-resistant variants (see Chap. 20) or antibody-resistant variants, especially in the case of RNA viruses (such as FMDV or HIV-1, among many others) will appear most readily; this phenomenon severely hampers viral disease control.

In some RNA viruses, such as influenza A virus, drastic antigenic changes leading to virus escape from antibody recognition may occur abruptly (*antigenic shift*) by genetic reassortment that substitutes at once a glycoproteins type exposed on the viral envelope by an antigenically very different type. This sudden antigenic change poses particularly severe problems for disease control. In addition, in these and many other viruses, antigenic variation may occur more gradually (*antigenic drift*), through the accumulation of amino acid substitutions in epitopes located in several antigenic regions of the envelope proteins (in enveloped viruses) or of the viral capsid (in non-enveloped viruses). These amino acid substitutions generally eliminate some virus-antibody interactions, sterically interfere with the antibody recognition process, and/or distort the epitope conformation.

As already mentioned in PV, HRV of the major receptor group, other enteroviruses, and viruses from other families, depressions may have been selectively favored as receptor-binding sites (see above) to allow for partial hiding from antibody recognition, thus preserving to some extent receptor binding under antibody pressure (see Chap. 15). However, a clear-cut structural separation between receptor-binding sites (or other functional regions) and antibody-binding sites does not generally occur in small non-enveloped viruses. For example, some antibodies that can neutralize HRV recognize epitopes that include part of the canyon wall and also residues involved in receptor binding. In FMDV, the receptor-binding site is exposed and nearly fully overlap with a major antibody-binding site (see above; [18]). Thus, additional or alternative mechanisms must operate to preserve cell receptor recognition, and also other virion functions, in the face of a strong antibody selection pressure (see next).

Very thorough genetic mapping of epitopes in small non-enveloped viruses (e.g., FMDV) by selecting many antibody-escape mutants revealed that escape mutations tend to repeatedly occur at very few positions in the contact epitope. These residues may sometimes correspond to those in the energetic epitope, but this is not always the case at all. A paradigmatic case is that of the major antigenic loop of FMDV type C. In this loop, an RGD motif and some neighboring residues are critically involved in binding both the cell receptor and different virus-neutralizing antibodies. FMDV cannot escape from neutralization by any of these antibodies through mutation of nearly any one of the residues critically involved in antibody binding, because these are precisely the ones critically involved also in cell receptor binding. However, the virus manages to readily and repeatedly escape neutralization by adopting either of two non-exclusive alternatives: i) a drastic one (a lesser kind of antigenic "shift," albeit not due to reassortment, affecting a major antigenic site), by mutation of one of a pair of unique residues that are critical for binding most antibodies that recognize this loop, but less so for binding the receptor; and ii) a gradual one (a kind of antigenic "drift" in that same antigenic site) by accumulation of mutations of several residues that, individually, are of minor relevance for antibody binding but that, together, severely impair binding of most antibodies that recognize this viral loop. Studies with many FMDV type C field variants isolated during several decades in two continents indicate that antigenic "shifts" in the major antigenic loop involving mutation of those unique residues have occurred repeatedly and independently, but infrequently. This paucity may be related to a significant loss of fitness caused by those "balancing act" mutations. In contrast, antigenic "drifts" in this loop by gradual accumulation of less critical mutations appear to have frequently contributed to FMDV escape in the field.

In small non-enveloped virions, the overlap between antigenic regions and regions involved in virion functions may extend to the entire capsid surface. Competition between antibodies or Fabs for binding to FMDV or other viruses suggest that a large part of the capsid surface is involved in neutralization epitopes. In contrast, viral escape mutations identified in the laboratory or in virus variants isolated in the field collectively tend to cluster in very few discrete regions highly exposed on the capsid surface.

Thorough mutational studies with FMDV and other small viruses indicate that most residues in the capsid surface are involved in virus function: if mutated, they severely impair infection, or at least decrease the biological fitness of the virus. Unlike cellular proteins, the protein capsids of small, non-enveloped viruses may have been functionally *streamlined* through multiple, sometimes conflicting selective pressures imposed on a very simple structure (see Chaps. 11–19). As a consequence, only very few residues in the viral capsid can be mutated to evade the antibody response and still allow normal virus infection. Immune evasion of FMDV type C in the field appears to have repeatedly involved mainly combinations of mutations of a surprisingly restricted number of residues on the capsid surface; these capsid residues may be among the few whose mutation has no severe effect on virus function.

1.4.5 Virus-Host Cell Recognition, Entry, and Uncoating

After an often long and dangerous voyage in the extracellular environment (within and between organisms), a few virions may have succeeded in preserving their integrity and functionality while evading the host defenses, and some of them may eventually encounter other host cells/organisms. The virion may then infect the cell by anchoring to appropriate receptor molecule(s), which will trigger the penetration of the entire virion or some of its components (invariably including the viral nucleic acid) into the cell. Unless the nucleic acid enters alone and naked, viral genome replication and expression will require uncoating of the internalized viral nucleic acid. Cell recognition, entry, and uncoating frequently overlap and rely on quite different mechanisms depending on the virus species and host type. These processes are very briefly summarized here separately for animal viruses (see also Chaps. 15, 16), bacteriophages (see also Chap. 17) and plant viruses.

Cell Recognition, Entry, and Uncoating of Animal Viruses

Cell receptor recognition Like any other cell, animal cells are surrounded by a soft, lipidic plasma membrane. However, unlike bacteria or plant cells, animal cells are not surrounded by a hard, thick cell wall, and this absence facilitates virus penetration. In general, animal viruses start the infection of a host cell by specifically recognizing some *receptor* molecule embedded in the cell membrane (Chap. 15). In the case of non-enveloped viruses, a part of the capsid surface, or a specialized capsid-attached viral protein can bind some cellular molecule (typically a glycoprotein, carbohydrate, or glycolipid) that normally serves some cellular function. In the case of enveloped virions, a viral glycoprotein embedded in the viral lipid envelope can specifically interact with the cell receptor molecule. In some cases, *co-receptor* molecules in the cell membrane may be also recruited during the recognition process. Virus-receptor binding is an energetically favorable process and may trigger conformational changes in the bound capsid or viral protein that mediate the entry process.

Entry into the host cell Attachment of the virus to the receptor molecule starts a complex process in which the entire virus particle, or a part of it (usually the capsid or nucleocapsid containing the nucleic acid), enters the cell. Penetration may, in general, occur by direct entry through the cell membrane or, much more frequently, by *endocytosis* mediated by interaction with the receptor(s), which lead to the engulfment of the virion in an endosome. In this latter case, a subsequent mechanism is required for the release of the viral particle or uncoated genome in the appropriate cellular compartment.

The mechanisms of penetration of non-enveloped animal viruses are in general poorly known, but for some viruses, this process is understood in some detail (Chap. 15). For example, cell-receptor recognition by many non-enveloped viruses may activate a process of endocytosis, in which the virion enters the cell largely intact and is transported through one or several different internalization routes, ending with the release of the viral genome in the appropriate cellular compartment.

Release of the virus or some component (including the viral genome) in the appropriate compartment may occur after lysis of the endosome and diffusion of the virion or disassembled components (see below) or by permeabilization of the endosome membrane, which may include the transient opening of a channel through which the viral nucleic acid can be translocated.

The mechanism of penetration of enveloped viruses is, in general, better known. Binding of a viral glycoprotein to a cell receptor invariably triggers a conformational change in a viral *fusion protein* that leads to the physical joining (fusion) of the viral envelope and a cellular membrane (Chap. 16). In direct entry (e.g., by HIV-1), after receptor recognition, the virus envelope and the plasma membrane are fused. In endosome-mediated entry, after the virus is engulfed in the endosome, acidification leads to the fusion between the virus envelope and the endosome membrane. In both cases, the fusion event allows the diffusion of the viral capsid or nucleocapsid containing the viral genome inside the cell. This is followed by trafficking of the relevant viral component to the appropriate cellular compartment for genome expression and replication.

Genome uncoating Uncoating is, in general, a very poorly known step in the life cycles of animal viruses. In some non-enveloped virions, such as the picornaviruses PV and HRV, receptor binding and/or low pH mediate capsid conformational rearrangements that may open pores large enough for nucleic acid extrusion. In other non-enveloped virions such as the picornavirus FMDV, acidification at the endosomes dissociates the viral capsid into subunits, and the nucleic acid is released into the cytoplasm. Important aspects of these processes are still uncertain or unknown (see Chap. 15). In enveloped virions, the viral nucleic acid-containing capsid or nucleocapsid that entered the cell after fusion of the virus and cell membranes is transported to the appropriate cell compartment for viral nucleic acid uncoating. In some cases, the capsid is disassembled and the nucleic acid just released, either by dilution or by a poorly characterized mechanism that may require the participation of cellular factors (e.g., in HIV-1). In other cases, direct replication and expression of the genome may occur with the viral nucleic acid being still a part of a ribonucleoprotein particle (e.g., ininfluenza virus) or being still associated with a viral capsid core (e.g., in reoviruses). See Chaps. 15 and 16 for a detailed description of different mechanisms of cell recognition, entry, and uncoating in different non-enveloped or enveloped animal viruses.

Cell Recognition and Nucleic Acid Transfer by Bacteriophages

Bacterial cells are surrounded by thick and hard cell walls that viruses cannot penetrate. Like animal viruses, bacteriophages recognize cellular receptors on the cell surface. However, the phage particle does not actually enter the cell. The barrier posed by the cell wall is overcome in different ways. In many dsDNA phages, a tube penetrates the cell wall as if it were the needle of a hypodermic syringe, and the nucleic acid genome is injected inside the cell, driven in part by the internal pressure built-in in the capsid during packaging of the nucleic acid. See Chap. 17 for a detailed description of cell receptor recognition by different bacteriophages and phage nucleic acid delivery into the host bacterium.

Entry of Plant Viruses

Plant cells are surrounded, like bacterial cells, by a rigid cell wall a virus cannot easily penetrate. Nearly all plant viruses are ssRNA(+) viruses which have a slender helical capsid or a relatively small icosahedral capsid. Most seem to enter the plant cell in a passive way, through small breaches in the cell wall and transient openings in the plasma membrane that may be caused by mechanical injury, for example, by the action of herbivorous animals or insects. Once a cell is infected, progeny virions may propagate from cell to cell in the organism through discontinuities in the cell wall crossed by channels that establish interconnections between neighboring cells (e.g., plasmodesmata). A general description of plant virus entry and infection can be found, for example, in the book by Hull [7].

1.5 Experimental and Theoretical Developments in Physical Virology

Physical virology (see Sect. 1.1.2 for a brief introduction to the field) is becoming rapidly integrated with classical structural virology. Physical virology studies are providing novel, physics-based fundamental insights into the structure, dynamics, and properties of virus particles. In this book, we have described in detail dominant experimental (Chaps. 9, 10) and theoretical (Chap. 19) methods and several basic studies (Chaps. 18, 19) in physical virology, as well as some developing or potential applications of those studies (Chap. 22).

Two very important physics-based, single-molecule techniques that are being applied to study viruses are AFM (Chap. 9) and optical tweezers (Chap. 10). AFM can be used to image viruses and also to follow structural changes in viral particles, even in real-time using HS-AFM. In addition, AFM can be used also to probe some physical properties of virus particles. Mechanical features of virus particles such as particle stiffness, Young's modulus of the capsid material (as a measure of its intrinsic stiffness), tensile strength, or resistance to mechanical fatigue are being thus determined. A dominant procedure to study virus mechanics involves the indentation of virions or capsids by "pushing" on defined points on the particle surface with the AFM tip (Chap. 18). AFM is also being used to determine the physical forces involved in molecular recognition between virus particles and different ligands, for example, cell receptors or antibodies, even in physiologically relevant conditions, with receptors embedded in the membrane of a cell or liposome. The general approach in these studies involves attaching a virus particle or virus ligand to the AFM tip and the interacting partner to a solid substrate, and "pulling" until the interaction between them is broken. Mechanical aspects of some stages in the viral cycle, such as virus entry into cells, can also be followed by AFM (see Chap. 9). Optical tweezers have been particularly successful in determining physical forces involved in the action of molecular motors during the mechanochemical process of dsDNA encapsidation inside a preformed bacteriophage capsid (see Chap. 10).

Several theoretical physics-based approaches are being applied to understand the physical foundations of virus capsid architecture; material properties of viruses, especially virus mechanics, and different aspects of the infection process in which viral particles are involved, including capsid assembly and maturation and virus entry into cells. In this book, we have integrated theoretical studies in physical virology in a single monographic chapter written by experts in the field (Chap. 19). In that chapter, a theoretical physics-based perspective is used to contemplate many aspects of virus structure, dynamics, and properties that had been described in previous chapters (Chaps. 2, 11, 13, 15, 18) from an empiric, structural biology-based perspective.

1.6 Applied Structural and Physical Virology

Several decades of intensive research on the structure, dynamics, and properties of virus particles, part of which is described in Chaps. 1–19 of this book, have opened up many possibilities for the exploitation of the knowledge acquired, and of the viral particles themselves, in biomedicine and biotechnology. Moreover, the advent of nanotechnology as well as novel studies on the physics of viruses has led to a growing awareness by researchers of physical or chemical disciplines on the multiple possibilities virus particles present also for nanotechnological applications. In addition to their possible direct use, viruses are also providing inspiration for the design of nonviral or hybrid nanoparticles and nanomaterials. Among others, two extremely important applied areas in human and animal health in which the knowledge acquired on virus structure is having a clear impact are the development of novel antiviral drugs (Chap. 20) and virus vaccines (Chap. 21). Knowledge on the structure and physical properties of virus particles is also being applied for the development of viral-based nanoparticles, nanomaterials, and nanodevices for multiple applications (Chap. 22).

1.6.1 Virus Structure-Based Antiviral Research: Antiviral Drugs and Vaccines

The knowledge acquired from structural studies on virus morphogenesis (Chaps. 11–14) and cell receptor recognition and entry of animal viruses into cells (Chaps. 15–16) have provided excellent opportunities to develop new antiviral agents based on the inhibition or misdirection of these processes. Chapter 20 focuses on the general structural basis of antiviral action and describes principles and methods for the structure-based search and design of new antiviral agents. Of particular relevance for the subject of this book, that chapter includes several examples where the target of antiviral research is virus assembly, cell receptor recognition, or entry into the host cell.

Likewise, the structural knowledge acquired on the recognition of viral particles by the immune system is leading to the development of novel or improved vaccines, including those based on virus-like particles (VLPs) obtained by production of CPs in expression systems and self-assembly in the cell or in vitro, or chimeric VLPs or virions carrying foreign epitopes from a pathogen (Chap. 21).

1.6.2 Other Applications of Virus Particles

Virus particles have acquired through evolution outstanding properties and functionalities: to name just a few, capsid self-assembly; capsid metastability and conformational rearrangements; the mechanochemical action of viral molecular motors for nucleic acid packaging; the targeting of specific cells through precise molecular recognition; and chemical and mechanical actions for delivery of the genome into the host cells, including particle penetration and controlled particle disassembly, or nucleic acid injection. These properties have opened up many possibilities to turn viral particles into useful materials and devices in biomedicine, biotechnology, and nanotechnology. However, viruses as evolved in nature do not have all of the several properties and functionalities required, or are not optimized, for any of their contemplated applications. Thus, methodologies have to be devised to manipulate virus particles and modify their structure, properties, and/or functions to serve the intended goals. Virus particles can be modified by direct chemical means and/or by protein engineering approaches using genetic techniques (see Chap. 22). Learning to successfully modify virus particles may not only provide useful products but also lead to a deeper understanding on the structure and function of these and other biomolecular complexes.

A noncomprehensive list of current or developing applications of the knowledge acquired on virus particles, and of adequately engineered viral particles may include the following (Fig. 1.6 and Chaps. 20, 21, 22):

Biomedical applications (Chap. 22):

(i) Design and development of antiviral drugs (already mentioned in Sect. 1.6.1; see Chap. 20)
(ii) Design and development of vaccines (already mentioned in Sect. 1.6.1; see Chap. 21)
(iii) Phage therapy: developing treatments to treat bacterial infections by administration of bacteriophages
(iv) Oncolytic virotherapy: developing anticancer treatments that involves the administration to a patient of infectious but attenuated viruses to selectively infect and eliminate cancer cells

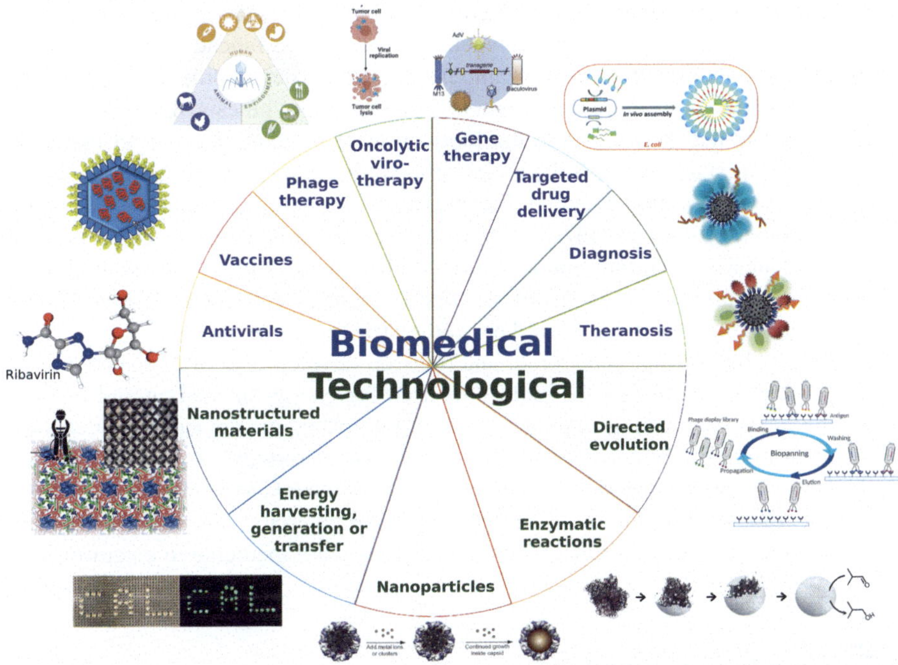

Fig. 1.6 Some applications of virus particles in biomedicine, biotechnology and nanotechnology (see text and Chap. 22). This Figure corresponds to Fig. 22.4 in Chap. 22 of this book

(v) Gene therapy: developing procedures for the introduction of genes into target cells in a patient, using a suitable vector (including engineered viral particles), in order to prevent or treat a genetic disorder by correction or replacement of the defective gene(s)

(vi) Targeted drug delivery: developing procedures to specifically direct drugs to diseased cells (e.g., cancer cells), using highly selective NP carriers or nanocontainers such as engineered viral particles

(vii) Diagnosis: developing procedures for noninvasive in vivo imaging using modified viral particles to carry and specifically deliver to target cells certain compounds that provide imaging signals. Also, the use of modified virus particles as biosensors for the detection of specific substances

(viii) Theranosis (combined diagnosis and therapy): combining noninvasive in vivo imaging methods with therapies based on targeted drug delivery

Technological applications (Chap. 22)

(ix) Directed evolution of proteins and protein assemblies. Use of *phage display* techniques for the selection from a combinatorial library of variant proteins or peptides of those with the desired properties (binding to specific ligands, improved binding, chemical or conformational stability, catalysis of specific substrates or improved catalytic efficiency, etc.).

(x) Enzymatic reactions: development of engineered virus particles containing heterologous enzymes as nanoreactors for improved biocatalysis.

(xi) Metal-containing and bioinorganic nanoparticles: development of engineered virus particles as templates or nanoreactors for the synthesis of nanostructured inorganic nanoparticles and the fabrication of bioinorganic hybrid materials for different applications. Potential applications of metal nanoparticles include their use as contrast agents for diagnostic imaging, biosensors, magnetic nanoparticles for localized magnetic hyperthermia therapy, etc. Potential applications of bioinorganic nanoparticles include their use as improved catalysts or as components of sensors, batteries, electronic or optoelectronic devices (e.g., solar cells), or data storage (memory) devices.

(xii) Energy harvesting, generation, and transfer: development of engineered virus particles as platforms for molecules used for energy harvesting and transfer (e.g., light harvesting in photovoltaic or photocatalytic devices) and development of engineered virus particles to generate electricity.

(xiii) Nanostructured materials: development of engineered virus particles as building blocks for the construction of nanomaterials, nanostructures, and nanodevices for multiple technological uses, from medicine to electronics.

1.7 Concluding Remarks

A vast amount of information on the structure, dynamics, and properties of virions and their molecular components has accumulated at an ever-increasing pace. The interplay between structural, physical, and molecular virology has led to a giant leap in our understanding of virus structure and function; a deeper and better comprehension on the workings of the molecular machinery of life; critical advances in the fight against viral disease; and promising developments in the application of virus particles to medicine and technology. However, there is still much to be learned on how viruses are made and how they work.

I would like to end this chapter with a note intended for the young virology student: Truly inspiring views of coral reefs are obtained not by snorkeling but by diving. William Beebe recommended readers of his 1928 book *Beneath Tropic Seas* (Putnam's sons, NY): "Don't die without having borrowed, stolen, purchased or made a helmet of sorts, to glimpse for yourself this new world." This chapter has provided just a quick look from the *surface* at some of what has been learned on virus structure and physics so far. By *diving deep* into the chapters that follow, the reader will be rewarded with amazing close-up views of these fascinating molecular machines.

Acknowledgments I gratefully acknowledge former and current collaborators and members of my group for their invaluable contributions to our studies on virus structure-function relationships, Miguel Angel Fuertes for providing some figures and technical assistance and Ricardo Arias-González, José R. Castón, Pedro J. de Pablo, Ignacio Fita, Julio Gómez-Herrero, José Luis Neira,

David Reguera, Wouter Roos, Alejandro Valbuena, Nuria Verdaguer, and members of my group for the fruitful discussions on the connection between the structure, physics, and biology of viruses. Work in the author's laboratory is currently funded by a grant from the Spanish Government (PID2021-126973OB-I00) and by an institutional grant from Fundación Ramón Areces. The author is an associate member of the Institute for Biocomputation and Physics of Complex Systems, Zaragoza, Spain.

References[1]

Molecular Virology Textbooks[2]

1. Flint SJ, Enquist LW, Racaniello VR, Skalka AM (2020) Principles of virology, 5th edn. John Wiley & Sons, Hoboken, NJ, USA. (*note: an intermediate level, highly recommended book on molecular virology*)
2. Dimmock N, Easton A, Leppard K (2016) Introduction to modern virology, 7th edn. Wiley-Blackwell. (*note: an introductory text on general principles in molecular virology*)
3. Carter JB, Saunders VA (2013) Virology. Principles and applications, 2nd edn. John Wiley & Sons, Chichester. (*note: introductory text on general principles in molecular virology*)
4. Cann AJ (2016) Principles of molecular virology, 6th edn. Academic Press, Waltham, MA, USA. (*note: introductory text on general principles in molecular virology*)
5. Hatfull GF (2023) Bacteriophages. In: Knipe DM, Howley PM, Cohen JL, Damania BA (eds) Fields Virology, vol 4, 7th edn. Wolters Kluwer, pp 590–623
6. Sanz-Gaitero M, Seoane-Blanco M, van Raaij MJ (2019) Structure and function of bacteriophages. In: Harper DR, Abedin ST, Burrowes BH, McConville ML (eds) Bacteriophages. Springer Nature, Switzerland, pp 1–73
7. Hull R (2014) Plant virology, 5th edn. Elsevier. (*note: a specialized book on the molecular biology of plant viruses, including many structural features*)

Advanced Reference Books on Virology

8. Knipe DM, Howley PM, Cohen JL, Damania BA (eds) (2023) Fields virology, 7th edn. Wolters Kluwer. (*note: advanced, extensive coverage of virology; a classical several-volume reference book*)

[1]Because of the elementary, student-oriented nature of this introductory chapter, we have purposefully limited the list of references to some recommended books. For further references to scientific reviews and original articles on structural and physical virology, the reader is referred to the sections on "References and Further reading" included in the rest of the chapters in this book.

[2]These books contain elementary to intermediate information on the molecular biology of viruses, with emphasis on viral genome replication and expression. Reference to some structural aspects directly related with these processes, which are out of the scope of the present book, may be also found in the books listed below.

9. Bamford DH, Zuckerman M (eds) (2021) Encyclopedia of virology, 4th edn. Academic Press. (*note: advanced, extensive coverage of virology topics based on concise reviews; a five-volume reference book*)
10. Casjens S (ed) (1985) Virus structure and assembly, Jones and Bartlett, Boston, USA. (*note: now outdated, but still very useful for understanding some fundamental principles in structural virology*)

Advanced Books on Structural or Physical Virology[3]

11. Chiu W, Burnett RM, Garcea RL (eds) (1997) Structural biology of viruses, Oxford University Press, New York, USA. (note: *now outdated in many respects, but an extremely useful book that describes clearly many principles in structural virology and contains detailed advanced monographs on a variety of structural virology topics*)
12. Stockley PG, Twarock R (eds) (2010) Emerging topics in physical virology, Imperial College Press, London, UK. (*note: An excellent (though somewhat outdated now) collection of advanced reviews on a number of specific topics in theoretical or experimental physical virology. Highly recommended for advanced, specialised reading*)
13. Agbandje-McKenna M, McKenna R (eds) (2011) Structural virology, RSC Publishing, Cambridge, UK. (*note: an excellent book containing monographs on many structural virology topics. Some chapters in that book complement others in the present book by covering topics such as evolution of viral capsid structures, structural aspects of virus-antibody recognition, or viral vectors for gene therapy. Physical virology-specific techniques (AFM, optical tweezers, physics-based models and simulations) and studies using these approaches are not contemplated. Highly recommended*)
14. Rossmann MG, Rao VB (eds) (2012) Viral molecular machines, Adv Exp Med Biol vol. 726, Springer, New York, USA. (*note: a collection of excellent, quite advanced and detailed reviews on some structural and molecular aspects of viruses considered as molecular machines, with emphasis on nucleic acid packaging. Physical virology-specific topics (see comment on ref. 15) are not contemplated. Highly recommended for advanced, specialised reading*)
15. Greber U (ed) (2019) Physical virology. Virus structure and mechanics, Springer Nature, Switzerland. (*note: a collection of updated, excellent reviews on some advanced topics in structural and physical virology. Highly recommended for advanced, specialised reading*)
16. Comas-Garcia M, Rosales-Mendoza S (eds) (2023) Physical virology, Springer Nature, Switzerland. (*note: A very recent collection of updated, excellent reviews on some advanced topics in structural and physical virology. Highly recommended for advanced, specialised reading*)

Other References

17. Domingo E, Biebricher C, Eigen M, Holland J (2001) Quasispecies and RNA virus evolution: principles and consequences, Landes Bioscience, Austin, TX, USA. (*note: a reference book on RNA virus variation, quasispecies and evolution. Important to understand the evolutionary*

[3]These specialized books contain excellent advanced monographs on many structural or physical virology topics.

concepts and genetic mechanisms behind the origin and continuous modification and adapta tion of viral machines described in the present book)
18. Sobrino F, Domingo E (eds) (2017) Foot and mouth disease virus, Caister Academic Press, Norfolk, UK. (*note: see Chap. 4 for a general account on the relationships between virus structure and biological function during the infectious cycle, virus- antibody recognition and viral escape from the immune response, using a thoroughly studied virus model, FMDV. See Chap. 10 for a detailed description on how animal virus particles are recognized by the immune system, again using FMDV as a model*)

Chapter 2
The Basic Architecture of Viruses

José R. Castón

Abstract Viruses are elegant macromolecular assemblies and constitute a paradigm of the economy of genomic resources; they must use simple general principles to complete their life cycles successfully. Viruses need only one or a few different capsid structural subunits to build an infectious particle, which is possible for two reasons: extensive use of symmetry and built-in conformational flexibility. Although viruses come in many shapes and sizes, two major symmetric assemblies are found: icosahedral and helical. The enormous diversity of virus structures appears to be derived from one or a limited number of basic schemes that became more complex by consecutive incorporation of additional structural elements. The intrinsic structural polymorphism of the viral proteins results in dynamic capsids. The study of virus structures is required to understand structure-function relationships, including those related to morphogenesis and antigenicity, among many others. These structural foundations can be extended to other macromolecular complexes that control many fundamental processes in biology.

Keywords Capsid · Nucleocapsid · Helical symmetry · Icosahedral symmetry · Triangulation number · Conformational polymorphism · Quasi-equivalence · Prolate capsid · Metastable capsid · Molecular switch

Abbreviations

BTV	bluetongue virus
cryo-EM	cryogenic electron microscopy
cryo-ET	cryogenic electron tomography
CP	capsid protein
HBV	hepatitis B virus

J. R. Castón (✉)
Department of Macromolecular Structure, Centro Nacional de Biotecnología (CNB-CSIC), Madrid, Spain
e-mail: jrcaston@cnb.csic.es

M. G. Mateu (ed.), *Structure and Physics of Viruses*, Subcellular Biochemistry 105,
https://doi.org/10.1007/978-3-031-65187-8_2

HIV	human immunodeficiency virus
T	triangulation number
TBSV	tomato bushy stunt virus
3D	three-dimensional

2.1 Introduction

Viruses are probably the most extreme example of nature's efficient use of limited coding capacity. Optimization of resources is reflected in their multiplication and dissemination strategies as well as in their construction. Viruses use simple, general principles to successfully complete their life cycles [1]. They need only one or a few proteins to build an infectious viral particle with many essential functions, including assembly to form a protective shell for the genome during extracellular transport, packaging of the genome, cell entry by binding to cell surface receptors, genome uncoating, and, in some instances, viral genome replication. All of these activities are carried out efficiently, despite a high mutation rate of capsid proteins (CP).

The solution for this optimization is dual: extensive use of symmetry and conformational polymorphism of structural proteins [2], both of which allow the virus to build three-dimensional (3D) structures of defined size and shape by using many copies of one or only a few proteins. If we do not consider complex viruses (see Chap. 12), the two main classes of symmetric virus structures are icosahedral and helical [3, 4]. The need for multiple copies of one or more protein subunit types arranged symmetrically can also be inferred from the nature of the genetic code. A base triplet (codon) has an approximate molecular weight of 1000 Da and encodes a single amino acid with an average molecular weight of 150 Da. In the best of cases, a nucleic acid can encode 15% of the virus weight as protein; nevertheless, 50–90% of the virus molecular weight is protein. This high protein-to-nucleic acid ratio is necessary to enclose the viral genome in a large enough protein capsid.

The classical vision of capsids as inert containers that protect the fragile genome during the extracellular phase of the virion is inappropriate; capsids are dynamic structures whose components have transient conformations related to specific functions in the viral cycle [5, 6]. The capsid is a metastable assembly and should be considered a compromise: it is robust enough to protect the genome and ensure its propagation during passage from one cell (or host) to another and labile enough to allow genome delivery in the host cell to initiate infection.

A virus particle is a structure consisting of a nucleic acid surrounded by a protein shell, also termed capsid or coat. Capsids are built of identical protein subunits, or protomers. Protomers cluster into capsomers, morphological units that are chemically definable oligomers of one or more viral proteins, usually identified by electron microscopy (e.g., hexamers, pentamers, trimers, or dimers). Most capsids are spherical or filamentous, although some have more complex shapes. For enveloped viruses, the capsid and the nucleic acid together are referred to as nucleocapsid or core (this concept is also used for the nucleic acid in complex with proteins, which compose the packaged genome form normally involved in replication). In those viruses, the nucleocapsid is surrounded by a lipid bilayer membrane associated with glycoproteins and other structures (fibers) of cellular and/or viral origin. The infectious (entire) virus particle is termed virion.

Interactions among protein subunits in the viral capsid are noncovalent and generally weak. There are exceptions, such as the bacteriophage HK97 protein shell, in which pentameric and hexameric subunits are covalently joined to form rings that are also topologically interlocked, creating a protein chainmail [7]. For the most part, however, as a result of viral particle symmetry, numerous weak interactions (hydrogen bonds, salt bridges, van der Waals forces, and hydrophobic interactions) cooperate to maintain an appropriate 3D structure and to facilitate dynamic structural processes.

How is a closed shell with a defined shape built? This is a complex process that must reconcile two apparently contradictory requisites: energy requirements must be minimal to ensure high assembly efficiency, which impedes aberrant assemblies in vivo, and the protein container should be stable and tough. A number of factors increase the effectiveness of assembly. The tertiary structure of capsid structural subunits has built-in conformational flexibility, which allows them to maintain the slightly different intersubunit contacts necessary to build most icosahedral capsids (see below). Capsid structural proteins also have large interaction areas and, even isolated from their natural environment, capsid subunits can self-assemble spontaneously into complex assemblies similar or identical to authentic virus capsids. This spontaneous assembly indicates that the viral particle is in a free-energy minimum state. To facilitate subunit interactions, viruses can use host resources such as molecular chaperones and membrane-specific regions, or other viral components such as scaffolding, accessory, cement, and/or proteolytic proteins.

A highly stable capsid is usually assembled after an in situ maturation process that takes place through the rearrangement of contacts among protein subunits, again illustrating the concept of viral particles as metastable structures. For bacteriophages and other dsDNA viruses, the immature assembly is termed procapsid (or prohead). Coordinated conformational changes in the procapsid lead to a functional (infectious mature) particle. No direct energetic cost is associated with the complex process of virion formation, except for DNA packaging in preformed (pro)capsids, as seen in some dsDNA viruses and bacteriophages (see Chap. 12). In addition to these built-in features of the protein subunits, the unidirectionality of the viral morphogenetic process is driven by chemical modifications (proteolysis, glycosidation, addition of phosphate groups, etc.) and, in some instances, by incorporation of accessory components.

In this chapter, we describe the principles of capsid organization and discuss numerous examples of virus architecture. X-ray crystallography (Chap. 4) and 3D cryogenic electron microscopy (cryo-EM) (Chap. 3) techniques have shown the molecular mechanisms by which identical building blocks, the CPs, make use of symmetry and structural polymorphism to assemble a multifunctional container in a biologically feasible period.

2.2 How Virus Structures Are Studied

Viruses were first visualized by electron microscopy, which provided the basis for their classification following morphological criteria. Development of electron microscopy techniques has had a central role in the study of viral structure and life cycle (see Chap. 3). The introduction of adequate fixation methods (chemical or physical) and contrast protocols (such as negative staining) for viral samples allowed initial observations of isolated viruses and virus-infected cells in thin sections. More recently, cryo-EM has become a major tool to study virus structure. Samples are frozen so rapidly that they are immobilized in an amorphous (or vitreous) solid state, without physical damage of the specimen by ice crystal growth [5]. Cryo-EM allows direct visualization of viral particles, as no staining or fixative agents are introduced during sample preparation, and native conformation is preserved, as samples remain hydrated (Fig. 2.1). Combined with image reconstruction techniques, cryo-EM is used widely to determine the 3D structure of viruses, and it has contributed notably to reveal the metastable nature of viral particles throughout the virus life cycle. Dynamic studies of viruses have shown conformational changes of structural subunits that helped to explain the molecular mechanisms underlying specific processes. 3D cryo-EM maps are currently calculated at near-atomic resolution, enabling us to infer the positions of amino acid residues in the capsid structural subunits, and to define the surface contacts among structural components of the virion [8, 9]. Enveloped virions are flexible (pleomorphic) and less symmetric than naked (nonenveloped) icosahedral or helical viruses; cryogenic electron tomography (cryo-ET) is used to study individual particles of complex enveloped viruses such as orthomyxo- and poxvirus, albeit at lower resolution than single particle cryo-EM.

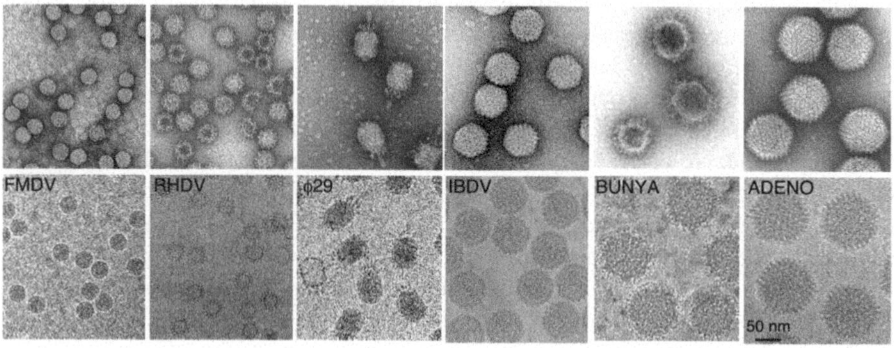

Fig. 2.1 Conventional EM and cryo-EM analysis of several viruses. A collection of representative virus preparations visualized by conventional negative staining (top), or directly vitrified without staining and/or fixing agent (bottom). FMDV, foot-and-mouth disease virus (sample provided by E. Domingo, CSIC); RHDV, rabbit hemorrhagic disease virus (provided by J. Bárcena, CSIC); ϕ29, prolate bacteriophage ϕ29; IBDV, infectious bursal disease virus; BUNYA, the enveloped bunyavirus, Bunyamwera virus (provided by C. Risco, CSIC); ADENO, adenovirus (micrographs by C. San Martín, CSIC)

Cryo-ET is used also to study the viral life cycle directly in infected cells (see Chap. 3).

X-ray crystallography is the other major procedure for determining virus structure and interactions at the atomic level. The major limitation of this approach is the need to obtain sufficiently large (milligram) quantities of highly pure (and stable) virus preparations to form suitable 3D crystals for X-ray diffraction (see Chap. 4). The first high-resolution virus structure, for the ssRNA plant virus tomato bushy stunt virus (TBSV), was determined in 1978 [10]. Structures of small human viruses were obtained in the mid-1980s for poliovirus [11] and rhinovirus [12]. Atomic resolution maps of viral capsids have been essential for establishing the mechanisms that govern the assembly and stability of viral structures.

X-ray crystallography and cryo-EM are used not only to solve atomic structures of entire viral capsids, but also to obtain detailed structural maps of isolated viral proteins such as hemagglutinin and neuraminidase of influenza virus, several viral proteases (from picornavirus, herpesvirus, and hepatitis B virus (HBV)), human immunodeficiency virus (HIV) reverse transcriptase, and several RNA polymerases. Some larger viruses and intermediate structural states are refractory to high-resolution analysis by X-ray crystallography, but they are accessible to 3D cryo-EM. Many studies have relied on the complementarity between these techniques. For example, pseudoatomic models of viral capsids have been obtained by docking atomic structures of individual capsid components determined by X-ray crystallography into lower-resolution cryo-EM density maps of the complete capsid (see Chap. 7). Nuclear magnetic resonance spectroscopy has also contributed valuable structural information by showing the structural dynamics of domains or proteins (see Chap. 5). Atomic structures have supplied valuable information for the design of antiviral drugs that alter particle stability or prevent its assembly (see Chap. 20).

2.3 Structure-Based Viral Lineages

Despite their minimal sequence similarity, many structural proteins of spherical viruses (in animals and plants) adopt a very limited number of tertiary folds indicating deep evolutionary relationships (see also Chaps. 7 and 12). Structural comparisons based on CP folds indicate that evolution led to a small number of viral lineages derived from a common ancestor [13]. Four well-established lineages have been described to date. These lineages are (i) the picornavirus-like lineage with a single β-barrel parallel to the capsid surface as the CP fold; (ii) the viruses with an upright double β-barrel CP (prototypes are phage PRD1 and adenovirus); (iii) the tailed dsDNA phages, tailed haloarchaeal viruses, and herpesviruses, which all share the Hong Kong 97 phage (HK97)-like CP fold; and (iv) the double-stranded RNA (dsRNA) or bluetongue virus (BTV)-like viruses. Viruses of the HK97- and PRD1/adenovirus-like lineages infect hosts from all kingdoms of life, *Archaea*, *Bacteria* and *Eukarya*, which indicates a likely origin before that of the current diversified

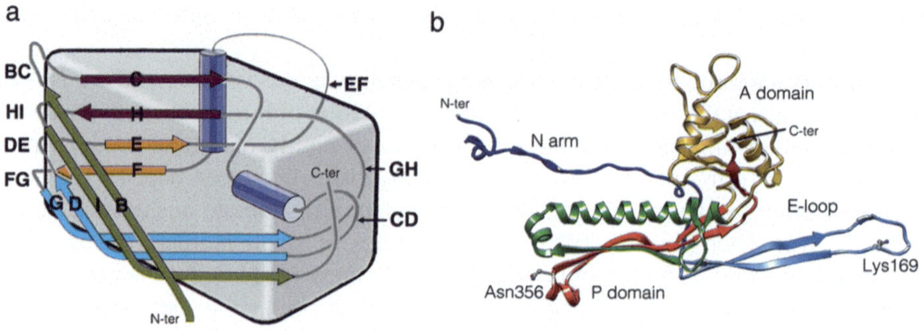

Fig. 2.2 The wedge-shaped β-barrel fold and the HK97 capsid protein fold. (**a**) Scheme showing the connecting loops between secondary structural elements (eight β-strands, arrows labeled B-I, and two α-helices, shown as cylinders). Loops CD, EF and GH generally have large insertions; loops BC, HI, DE and FG are short. This wedge-like domain has large interaction surfaces, appropriate for icosahedral symmetry. (**b**) Structure of the HK97 capsid protein (gp5), color-coded by domains (PDB entry 1OHG). The two domains A and P are indicated, as well as the extended N-arm (dark blue) and the E-loop (cyan). In the mature capsid, Lys^{169} on the E-loop forms a covalent bond with Asn^{356} on a neighboring subunit

cells; this might also be true for the other lineages. Helical and enveloped viruses are thought to fall into a limited number of lineages [14]. Recent studies described structural homology between the core domains (a four-helix bundle) of nucleoproteins from plant flexible filamentous positive-strand viruses [15] (potyviruses and alphaflexivirus) and animal segmented negative-strand viruses (bunyaviruses and orthomyxoviruses) [16].

One of the most common folds of all viruses known is the antiparallel β-barrel [17]. The β-barrel fold is also known as a jelly-roll or a Swiss-roll β-barrel. This wedge-shaped structure is formed by eight antiparallel β-strands (with two opposing BIDG and CHEF β-sheets) and two α-helices (Fig. 2.2a). Structural subunit size varies between 20 and 40 kDa, with the major differences located in N- and C-terminal ends (facing the interior and the outer surface of the capsid, respectively) and in the loop sizes between β-strands (in some cases, more than 100 amino acids that form additional domains). The β-strands usually lie roughly tangential to the capsid surface, defining a protein shell ~30 Å thick. The wedge-like shape allows the subunits to pack tightly and, due in part to this built-in flexibility, the jelly-roll motif-based CP can assemble into unrelated capsids of variable sizes (by increasing the T number, see below). The β-barrel is also found in some bacilliform viruses such as alfalfa mosaic virus (AMV). The large subunit of comoviruses, ssRNA viruses that infect plants, is also made by fusion of two parallel jelly-roll β-sandwiches [18].

Joined, perpendicular β-barrels were first observed in the adenovirus trimeric capsomer; each monomer consists of two successive jelly-roll motifs, producing a pseudo-hexameric structure [19]. Many other large dsDNA viruses have similar trimeric capsomers, such as the *Paramecium bursaria* chlorella virus 1 (PBCV1) [20], the dsDNA bacteriophage PRD1 [21], and the sulfolobus-infecting virus, STIV (sulfolobus turreted icosahedral virus) [22]. The non-icosahedral poxviruses have been linked to the PRD1/adenovirus like lineage, as the poxvirus scaffolding protein has two fused β-barrels and forms pseudohexamers that assemble into a transitory lattice on the surface of immature particles [23].

Tailed dsDNA viruses constitute the most successful self-replicating system on Earth. The HK97 CP forms an L-shaped, mixed α/β structure with two domains [24]. The peripheral (P) domain consists of a long α-helix and an elongated, three-stranded β-sheet plus extensions (N-arm and E-loop), and the axial (A) domain, composed of two α-helices and an additional β-sheet. The two residues that form covalent crosslinks with neighboring subunits are on opposite sides of the protein (Fig. 2.2b). This structure is optimal for CP self-assembly to form a metastable particle, in which subunit interactions are modulated by the CP N-terminal scaffolding domain. A proteolytic process then takes place to remove the scaffolding domain, followed by massive conformational changes and crosslinking, leading to a robust mature capsid [24]. Some bacteriophages incorporate additional domains in the main shell protein fold, resulting in different assembly mechanisms [25].

Whereas virus evolution has led to a limited number of CP folds, two general virion assembly strategies predominate, the co-assembly mechanism and the sequential mechanism [26]. In the co-assembly mechanism, a virus uses its own genome (a ssRNA or ssDNA molecule) to nucleate CP subunits around itself, a cooperative condensation driven by electrostatic, specific genome-CP interactions. This mechanism is typical of the picornavirus-like lineage, but assembly and packaging processes are probably also coupled in BTV-like viruses and in adenoviruses. The sequential assembly mechanism is exemplified by viruses of the HK-97-like lineage, which assemble as a precursor particle, the procapsid, and for which scaffolding proteins are needed. In PRD1/adenoviruses and BTV-like viruses, the scaffold function is performed by internal lipid membranes (except for adenovirus) and inner icosahedral cores, respectively. The procapsid subsequently matures during genome packaging, a process often associated with the action of a protease that cleaves scaffolding proteins. The portal (or dsDNA-packaging protein) complex, incorporated in the initial steps of procapsid assembly, is responsible for dsDNA encapsidation and includes an ATP driven packaging nanomotor; it is also found in the PRD1-adenovirus lineage.

2.4 Capsid Symmetry

Natural crystals are formed from multiple copies of single molecules that establish identical interactions with adjacent molecules and thus have equivalent environments. This is the simplest solution to form a regular or symmetric structure. There are fundamental differences between natural crystals and viruses, however; whereas a crystal is a continuous 3D network with no size limitations, viruses build an empty container of defined shape and limited cargo space for packaging of the viral nucleic acid (and, in some cases, other components as well) in a biologically feasible time frame [27]. The capsid shell is therefore seen as a two-dimensional crystal that

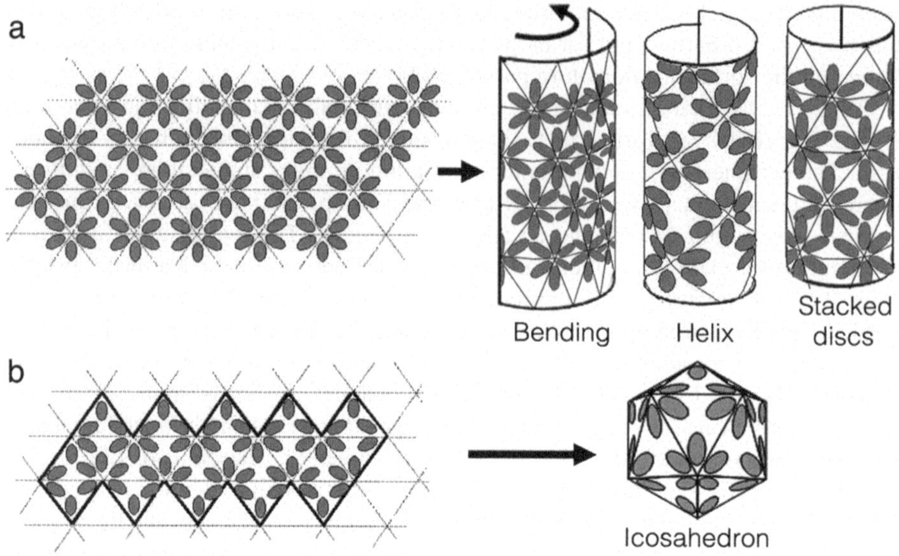

Fig. 2.3 Generation of curved structures such as a helix and an icosahedron from a planar hexagonal lattice. (**a**) Planar hexagonal lattice with subunits arranged in the triangular faces defined by the lattice vectors. (**b**) Icosahedron generated from a defined subset of faces derived from the hexagonal array

closes, resulting in a large cavity for enclosing the viral genome. Although they use basic principles of crystal growth, viruses require additional mechanisms to control their dimensions [28].

There are several alternatives for generating 2D crystals within the 17 planar crystallographic groups. Some of these groups are variants of hexagonal lattices. In a hexagonal network, identical subunits can be arranged to maintain them in identical environments. If correct distortions are introduced, two types of closed shells can be made from this planar network, helices and icosahedra. If after bending and binding the network edges are out of register, a tube is obtained in which layer lines are in register and generate a helix; if the edges are in register, a cylinder with a pile of stacked discs is obtained (Fig. 2.3a). In a helix, structural subunits have identical interactions, except those located at the ends, and the self-assembly of protomers can easily be envisaged. The diameter of the tubular structure depends on the curvature in the subunit interactions, whereas its length is limited by that of the nucleic acid and by subunit-nucleic acid interactions.

The icosahedron is a closed polyhedron with cubic symmetry, and asymmetric units can therefore be laid on the symmetric polyhedron surface with identical interactions among them. In 1956, Watson and Crick proposed the first principles for the construction of spherical viruses [29]. In theory, a limited number of regular solids (termed Platonic polyhedra) allow close packing of repeated (asymmetric) subunits such as the proteins. These polyhedra are the tetrahedron (four triangular faces and 12 equivalent positions), the cube (six square faces and 24 equivalent

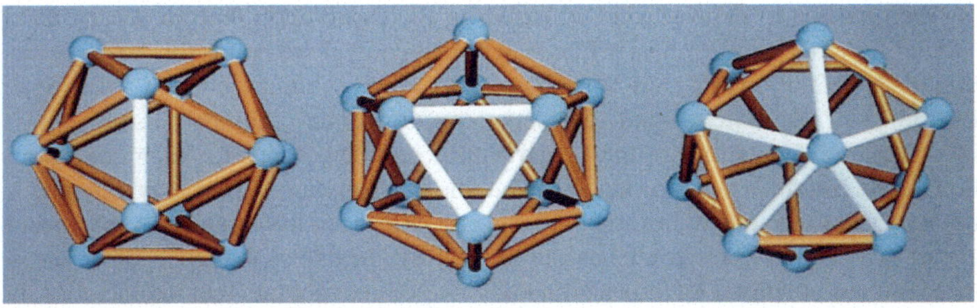

Fig. 2.4 Rotational icosahedral symmetry axes. The icosahedron viewed along two-, three- and fivefold symmetry axes (left to right). The twofold axis passes through the middle of each edge, the threefold axis passes through the center of each icosahedral face (an equilateral triangle), and the fivefold axis through the vertices of the icosahedron

positions), the octahedron (eight triangular faces and 24 equivalent positions), the dodecahedron (12 pentagonal faces and 60 equivalent positions), and the icosahedron (20 triangular faces and 60 equivalent positions). The icosahedron (and the dodecahedron) is the most economic natural structure in terms of the genetic information it requires, as a larger number of identical subunits can be laid in equivalent positions. In addition, the icosahedron is the cubic symmetrical structure in which the surface/volume ratio is lowest, involving the best component use to generate a hollow polyhedron, that is, a container. An icosahedron can be generated from a hexagonal planar lattice by conversion of planar hexagonal nodes (or hexamers) into convex pentagonal nodes (or pentamers) after removing one subunit/hexagonal node (Fig. 2.3b). Inclusion of pentamers allows curvature in the lattice, which is closed as a polyhedron with 12 pentagonal vertices and 60 identical subunits. The icosahedron has a series of characteristic views depending on its orientation relative to the rotational symmetry of the solid, known as 532 symmetry. An icosahedron has six fivefold axes through the 12 vertices, 10 threefold axes through the 20 triangular faces, and 15 twofold axes through the center of each edge (Fig. 2.4). The capsid described (with three identical subunits to form each triangular face) would be the simplest icosahedral capsid formed by 60 identical subunits in equivalent positions.

2.5 Quasi-equivalence Theory

As more virus structures were characterized, it became evident that the limitation in subunit number did not account for the molar mass and size of most capsids. Based on the conceptual model of the hexagonal planar lattice, some hexamers must be added between pentamers to build larger icosahedra. In 1962, Caspar and Klug developed the quasi-equivalence theory, which constitutes the basis of the structural studies on icosahedral viruses [30].

The quasi-equivalence theory is based on the assumption that identical capsid subunits interact, forming nearly identical (quasi-equivalent) bonds for the construction of stable icosahedral capsids with multiples of 60 subunits. There is a degree of flexibility in the intersubunit contacts (conformational polymorphism), which involves minor variations in the 3D structure of protein subunits. Each conformational state is termed a conformer. For example, the structure of pentamer subunits is not identical to the one of hexamer subunits. A number of different-sized icosahedra can be generated, because each has a unique ratio of hexamers to the 12 pentamers in the basic icosahedron. As indicated above, five or six subunits that interact closely are called pentameric or hexameric capsomers, respectively, or simply pentons and hexons. Given that the icosahedron is derived from the planar hexagonal lattice, not every combination of pentamers and hexamers is allowed; conversion of 12 nonadjacent hexamers into pentamers to form a closed shell must take place at regular, correctly separated intervals (Fig. 2.5a). The ratio that systematically defines all possible icosahedra from the hexagonal lattice is termed triangulation number (T), which matches the unit triangles of the elemental planar lattice contained in each triangular face of the icosahedron (facet) (Fig. 2.5a), i.e., the subtriangulation of the 20 facets. The smallest viruses follow T = 1 geometry. The rules of quasi-equivalence allow only certain values of T; T is given by the formula $T = h^2 + k^2 + hk$, in which h and k are positive integers that define the pentamer positions in the original hexagonal lattice (Fig. 2.5a). Icosahedral viruses larger than T = 1 have T values of 3, 4, 7, 9, 12, 13, 16, 19, 21, 25, 27, 28, and so on.

The quasi-equivalence theory, combined with the concept of T, provides a systematic nomenclature with which to describe icosahedral capsid structure. Icosahedra fulfill Euler's polyhedral formula, faces (20) + vertices (12) = edges (30) + 2. Taking T into account, this formula becomes 20 T + V = 30 T + 2. As there must always be 12 pentameric vertices, the total number of hexamers is 10 (T-1), and the total number of subunits is 60 T. T also describes the classes of conformers, or different quasi-equivalent (similar) environments occupied by a subunit, but T does not describe the basic morphological unit or repeated oligomer (other that hexamer/pentamer), which must be determined empirically for each icosahedral virus (Fig. 2.5b).

Viral capsids, as macromolecular protein assemblies, are enantiomeric structures. Handedness is defined by the arrangement of lattice singular points between neighboring pentameric positions. The T = 7 lattice is skewed, and there are two enantiomorphs, the left-handed (T = 7 *laevo* or 7 *l*) and right-handed (T = 7 *dextro* or 7*d*) configurations. Even though T = 1 and T = 3 capsids are nonskewed capsids, capsomers (dimers or pentamers) might show intrinsic handedness (Fig. 2.5a).

Paradoxically, a T = 1 capsid with only one protein subunit is the least frequently observed quaternary structure in native viruses [17]. A few small ssRNA viruses with protein subunits of 20–40 kDa, such as satellite tobacco necrosis virus (STNV; 159-residue CP), are known to have this capsid type. Other T = 1 capsids are formed by ssDNA viruses such as parvo- and microviruses (such as bacteriophage ϕX174; 426-residue CP) with CP as large as 75 kDa ([31] and references therein).

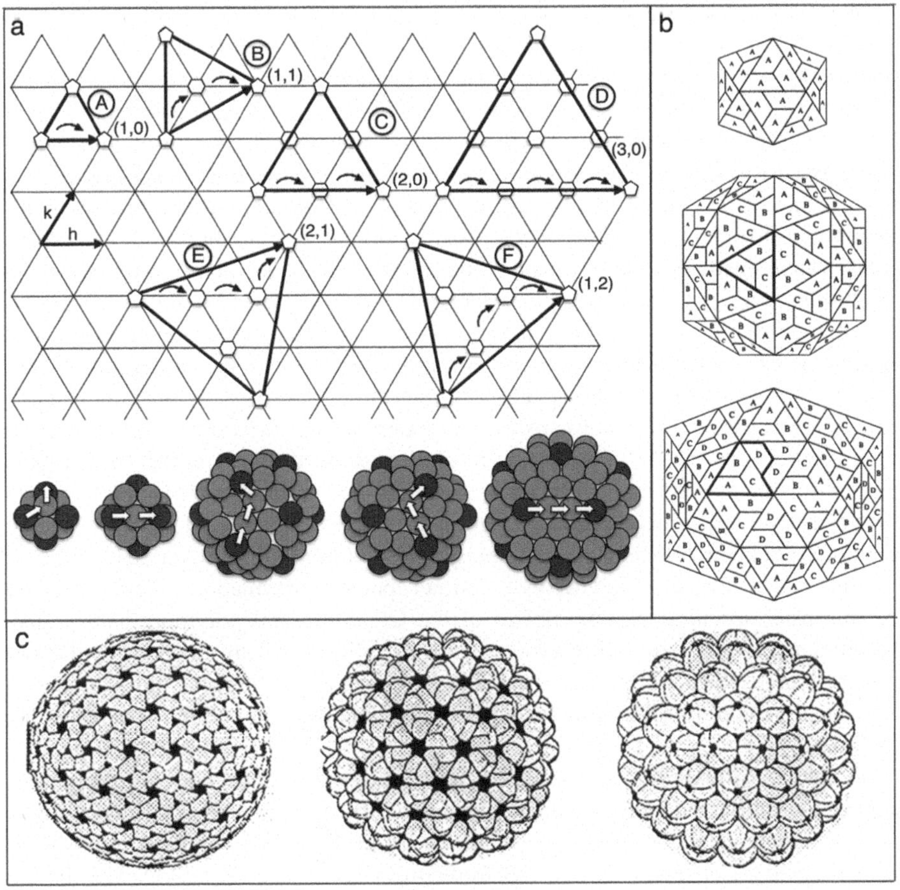

Fig. 2.5 The triangulation number T, and how it is determined in an icosahedral lattice. (**a**) The coordinate system of a planar hexagonal net with axes h and k crossing at a 60° angle is formed by equilateral triangles; each triangle is formed by three protein subunits (as in Fig. 2.3), and specific hexamer-to-pentamer conversion introduces curvature into the hexamer sheet (see main text). Capsomers are located at the lattice intersection points; hexagons represent hexamers (or planar hexameric positions) and pentagons correspond to (convex) pentamers. Several triangular facets of an icosahedron are indicated (bold outline), and a facet edge of the icosahedron (bold vector) connects the two closest neighboring pentamers. The two unit vectors of the underlying hexagonal net, h and k, define the coordinates of the vector relative to its origin (positions at which hexamers are replaced by pentamers), and determine the parameters of the icosahedral capsid. The number of steps or spacings (curved arrows) from one structural unit (a hexamer or pentamer) to the adjacent one is therefore an integer. In A), a pentamer is separated from the adjacent pentamer by one step along the h axis (i.e., pentamers are in direct contact); hence, $h = 1$, $k = 0$, which corresponds to a $T = 1$ icosahedron (in which case 12 adjacent hexamers are converted to pentamers). Larger icosahedra are built by converting 12 nonadjacent hexamers to pentamers at regular, precisely spaced intervals. In B), one pentamer is separated from another by one step along the h axis and one step along the k axis; $h = 1$, $k = 1$, and $T = 3$. Note that T indicates the number of unit triangles contained in the icosahedral facet. In C) and D), pentamers are separated by two and three steps along the direction of the h axis and are $T = 4$ and $T = 9$, respectively. A $T = 4$ lattice consists of 80 unit triangles, 240 subunits, and 4 quasi-equivalent protein subunits in the icosahedral asymmetric unit. Triangulated icosahedra might have local (or quasi) six-, three-, and/or twofold

High-resolution structural studies have provided some clues as to how identical subunits adopt different conformations [8, 9, 17, 32]. The distinct conformational states can be controlled by order/disorder of flexible regions of the protein (e.g., in loops or N and C termini), ds or ssRNA, metal ions, pH, or combinations of any of these [33]. These inherent and/or extrinsic factors that alter the conformational states are referred to as molecular switches. Removal of the molecular switching mechanism results in assembly of aberrant or nonnative structures. Molecular switches might be insufficient, especially when capsids with large T numbers and/or complex viruses are considered. These capsids require one or more auxiliary proteins that act as morphogenic factors to trigger structural changes (scaffold, minor capsid or enzymatic proteins) [2, 34]. In theory, quasi-equivalence involves small differences in subunit interactions and conformations; in practice, for capsids analyzed to date, viruses show broad variation, and the literature describes equivalent, quasi-equivalent and nonequivalent capsids. For example, the structure of cowpea chlorotic mottle virus (CCMV) at atomic resolution showed that building blocks have quasi-equivalent contacts, with allowed variations in noncovalent bond angles between residues, which stabilize oligomers, whereas the two classes of dimeric building blocks in TBSV have completely different bonding contacts [31].

Quasi-equivalence, which involves distinct spatial conformations, is supported by the intrinsic structural polymorphism of the CP. Structural polymorphism is also fundamental to the maturation process, since it allows in situ transient conformational changes throughout morphogenesis [35]. This strategy is optimal, as no additional genetic information is needed to generate these structural variants of the capsid.

Despite the flexibility and multiple possibilities that quasi-equivalence offers, capsid assembly frequently requires additional structural proteins. In such viruses, quasi-equivalence in the strict sense cannot be used, since quasi-equivalent positions are occupied by different proteins, as in picornaviruses (see Chap. 11). The mature picornavirus capsid has 60 copies each of three subunits (VP1, VP2, VP3) that are synthesized together as a precursor polyprotein. There is very little sequence

Fig. 2.5 (continued) symmetry. When the coordinates h or k are zero, or h = k, the icosahedral face is symmetric relative to the coordinate system (examples A, B, C, and D). When h differs from k, the capsid is asymmetric or skewed, as in E) and F). In a T = 7 icosahedron, the icosahedral face is turned relative to the hexagonal lattice, but the surfaces of the unit triangles gained or lost are compensated and T is an integer. The handedness of a T = 7 capsid is defined by the separation steps between the two closest pentamers, giving the maximum number of steps in the selected direction (this movement is similar to that of a knight in chess). In the T = 7 laevo (left-handed) or 7 *l* (as in E), h = 2 and k = 1 (this step is leftward, along the direction of the k axis); similarly, in the T = 7dextro (right-handed) or 7*d* (as in F), k = 2 and h = 1 (a rightward step, along the direction of the h axis). The two T = 7 lattices are enantiomorphs, i.e., mirror images of each other. The three-dimensional models (bottom) show how T number is determined; capsomers are shown as balls (black, pentamer; grey, hexamer). Note that capsid size is increased at higher T numbers, as capsomers are the same size. (**b**) Three different capsids with the same T number (T = 9) but different capsomers, viewed from a twofold axis; structural subunits are assembled without forming oligomers (top), as trimers (middle), or as pentamers and hexamers (bottom) similarity among them, but they all have the same fold, the canonical β barrel. The additional genetic cost provides advantages that will be difficult (if not impossible) to achieve only by conformational changes in a single protein; for example, the rapid acquisition of mutations to evade the immune system without altering capsid assembly [11, 12, 17]. When capsids are assembled from different gene products, T numbers are no longer valid, and pseudo T (P) numbers are used instead (e.g., P = 3 in picornaviruses). Another well-known group with P = 3 capsids are the comoviruses, whose capsid has two subunits, a small subunit with a β-barrel domain and a large subunit formed by fusion of two β-barrels [18].

There are many variations on the icosahedral theme; geminivirus particles consist of two fused T = 1 icosahedra, in which each lacks a pentameric cap [36]. In TBSV capsids, among others, particle curvature results from both pentamers and hexamers [10]. These variations are not restricted to capsids based on the tangential wedge-like β -barrel domain; the CP of HBV forms cores of varying T numbers and does not share this fold. The proportion of HBV T = 3 and T = 4 particles is affected by modifying the length of the C terminus, which is located on the inner surface [37].

There are notable exceptions [34] that do not adhere to quasi-equivalence rules, such as the all-pentamer T = 7d capsid of papovaviruses (papillomaviruses and polyomaviruses), formed by 72 pentamers (360 subunits and 6 conformers, which would correspond to a disallowed T = 6) rather than 12 pentamers and 60 hexamers (420 subunits and 7 conformers as predicted by quasi-equivalence) [38, 39]. In adenoviruses, which have a T = 25 shell (in fact, a pseudoT = 25, as pentameric subunits differ from hexameric subunits), each hexon is a trimer [19], but each subunit is formed by two similar domains with the β -barrel fold (see Chap. 12). A T = 25 capsid would have 25 conformers, but as the hexameric trimer is the basic unit, only four classes of (trimeric) conformers are needed. Another striking exception is the inner core of dsRNA viruses (formally, T = 1) formed by 12 decamers, in which dimers are the asymmetric unit; these capsids would correspond to the forbidden "T = 2" [40–42]. A number of studies have firmly established that this "T = 2" architecture is a common feature for dsRNA viruses ([43] and references therein).

Structural comparison of CP is used to establish relatedness when sequence conservation is limited [17, 44, 45] and has led to detection of relationships among icosahedral viruses that infect hosts from three different domains (bacteria, eukarya, and archaea) (see Sect. 2.2).

2.6 Variations in the Use of Icosahedral Capsids: Multiple Layers and Prolate Icosahedra

Most animal viruses are icosahedral and follow the general principles described above, although many viruses have additional nonsymmetric lipid envelopes. The simplest icosahedral viruses include RNA and DNA viruses. Examples of icosahedral RNA viruses include the Sedoreo-, Birna-, Calici-, Noda-, Tetra- and Picornaviridae families; examples of icosahedral DNA viruses include the Parvo-, Papova- and Adenoviridae families. Some icosahedral viruses have locally broken the symmetry at one of 12 fivefold vertices, due to the presence of the portal and tail complexes in place of a pentamer. To illustrate the variation in size and structural features among different viral capsids, we show 3D reconstructions of several distinct viruses calculated from cryo-EM images, from the ~30 nm-diameter capsid of picornaviruses to the ~100 nm rotavirus capsid (Fig. 2.6). 3D maps are intentionally presented at low to medium resolution (some are available at much higher resolution) to emphasize the capsomers (di-, tri-, penta- or hexamer) as well as the T numbers. Detection of specific capsomers implies stronger intra- than intercapsomeric interactions; isolated capsomers of some viruses can be purified for structural and functional studies.

Many dsRNA viruses, including important human, animal and plant pathogens (rota-, reo-, and orbiviruses) consist of a multilayered concentric capsid with icosahedral symmetry [56]. This complex organization is based on the role of each protein shell. Outer layers, usually a T = 13 shell, provide protection against chemical, physical, and enzymatic damage and participate in receptor interaction and cell entry. The innermost core, a T = 1 layer with a protein dimer as the asymmetric unit, is essential for the organization of the genome and the replicative, participates in genome transcription and replication, and isolates the virus genome from host defense mechanisms (see Chap. 12). Fungal dsRNA viruses, which lack an extracellular route, have only one shell, a T = 1 capsid.

Capsids with larger T numbers could be considered advantageous, as the formation of larger containers would in turn allow the incorporation of larger genomes; the concomitant packaging of specific viral components, such as polymerases and other enzymes, would reduce virus dependence on host cell machinery. However, such a container, although made at a low economic cost, presents enormous challenges. For example, in a quasi-equivalent T = 25 capsid that roughly emulates the adenovirus capsid, 1500 copies of an individual protein must be laid accurately into the shell. In a relatively simpler T = 13 lattice, as in the birnavirus capsid, "only" 780 VP2 proteins must be arranged on the outer capsid surface (see also Chap. 12). The VP2 subunit has three domains, an outer domain with an β-barrel fold in which β-strands are oriented radially, a shell domain with a β-barrel in tangential orientation, and an interior helix bundle contributed by the N and C termini [57]. The protein distribution problem is reduced in several ways: VP2 subunits form trimers, and five classes of these trimers are needed (rather than 13 conformers) to assemble into pentamers and hexamers. VP2 structural polymorphism is controlled by an inherent molecular switch, a C-terminal segment, that is absent in the mature VP2 form. Proteolytic processing of this switch is a finely tuned process in which viral and cellular proteases participate; these proteolytic events take place in a procapsid-like structure, in which multiple scaffolding protein copies are essential for stabilizing or promoting interactions between precursor subunit trimers within hexamers [50, 51, 58].

The largest icosahedral viruses analyzed by cryo-EM include the insect larvae iridovirus CIV (T = 147, 185 nm diameter [59]), the algal dsDNA viruses

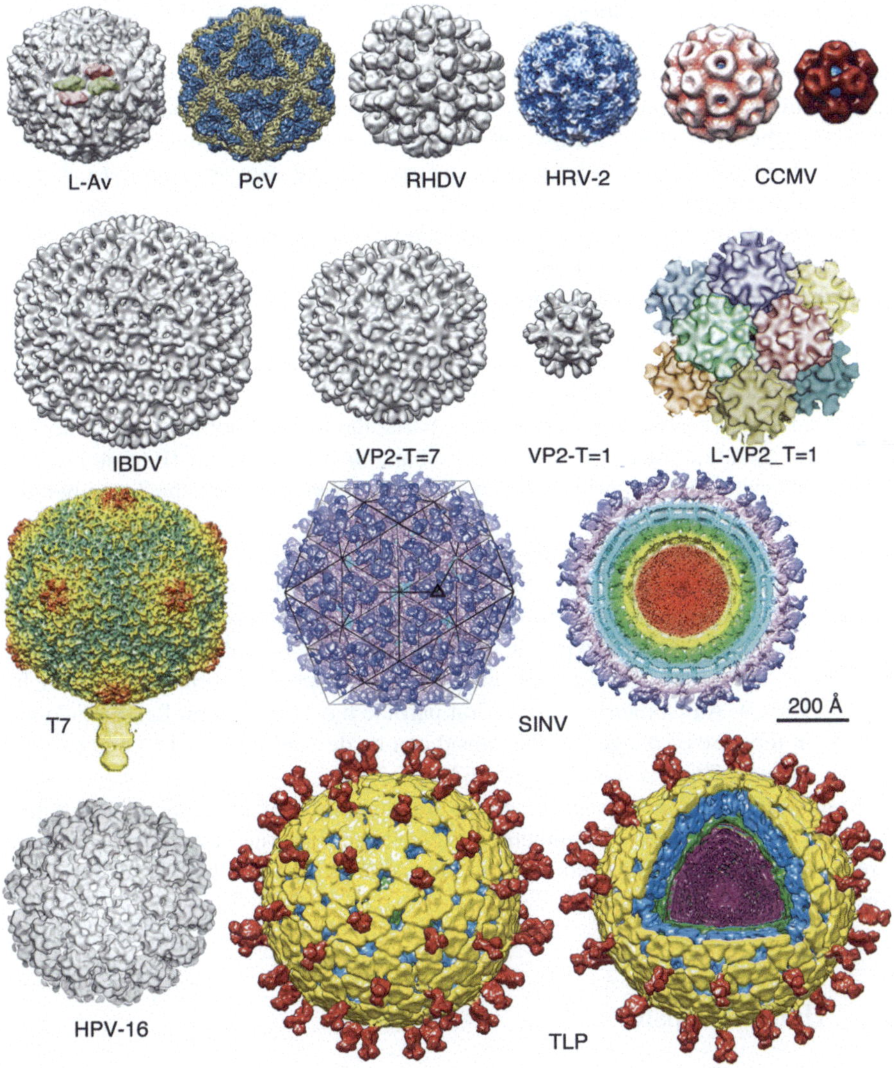

Fig. 2.6 Three-dimensional maps of icosahedral viruses calculated from two-dimensional cryo-electron microscopy images. L-Av, L-A virus of the yeast *Saccharomyces cerevisiae* with a T = 1 (or "T = 2") capsid (430 Å diameter); the basic unit is an asymmetric dimer (green and red) that assembles into decamers [46]; PcV, *Penicillium chrysogenum* virus, T = 1, 400 Å; the asymmetric basic unit has two similar domains (yellow and blue) [43]; RHDV, rabbit hemorrhagic disease virus, T = 3, 400 Å; the basic unit is a dimer of the capsid protein [47]; HRV-2, human rhinovirus 2, P = 3, 304 Å; the asymmetric unit is made of three different proteins [48]; CCMV, cowpea chlorotic mottle virus, T = 3, 300 Å (left) and T = 1 virus-like particle, T = 1, 200 Å [49]; IBDV, infectious bursal disease virus, T = 13 *l* lattice, ~700 Å; the building block is a VP2 trimer [50–52]; VP2 can assemble into T = 7 (VP2-T = 7; the left-handed form is chosen arbitrarily) and T = 1 (VP2-T = 1) capsids; T = 1 capsids self-assemble into a dodecahedron with 12 T = 1 capsids (L-VP2-T = 1); T7, T7 bacteriophage head, T = 7, ~600 Å [53]; the map shows a tail, located at the same vertex where the connector core is assembled; SINV, Sindbis virus, an alphavirus with a double-layered particle, ~700 Å: outer T = 4 capsid (blue), a lipid bilayer (cyan), and an inner T = 4 capsid (green and yellow) (Modified and reproduced with permission from Ref [54]); HPV-16,

Paramecium bursaria chlorella virus 1 (PBCV-1, T = 169d, 190 nm [60]) and *Phaeocystis pouchetii* virus (PpV01, T = 219d, 220 nm [61]), and the amoebal dsDNA mimivirus (estimated T between 972 and 1200, 750 nm, [62]); all have a lipid membrane beneath the capsid and use minor CP for stabilization and to determine their assembly. Mimivirus can be "infected" by its own virus, Sputnik (with a T = 27 icosahedral capsid), and its genome size is similar to that of intracellular parasitic bacteria (see Chap. 12).

A different strategy to increase the internal volume of the capsid is followed by a group of icosahedral virus variants, which elongate the capsid along a fivefold axis with additional subunits in the central band or cylinder. This type of prolate capsid is typical of bacteriophages such as ϕ29 [63] and T4 [64] (with a portal, the DNA packaging machine, and a tail, the DNA injecting machine, in a pentameric vertex) but is also found in several fungus, plant, and animal viruses.

Consider an icosahedron made of three parts, the top and bottom caps (with five equilateral triangular facets) and the central band or cylinder (with 10 triangles) that connect the caps. These capsids are characterized by two numbers, the T number and the elongation number Q (also referred to as T_{end} and T_{mid}, respectively). As the T number specifies the unit triangles of the triangular facet of an icosahedron (subdivisions into smaller triangles), the Q number also indicates the number of unit triangles in the stretched triangles of the central band but can be any positive integer (Fig. 2.7). For icosahedral viruses, the total number of subunits is defined by $n = 60$ T; for prolate capsids, the total number of protein subunits is given by $n = 30(T + Q)$. An isometric, icosahedral particle has $Q = T$; a prolate particle has $Q > T$. In the case of phage T4, the capsid can be defined by $T = 13\ l$ and $Q = 20$ [64]; in phage ϕ29, $T = 3$ and $Q = 5$ [63].

Assembly of prolate capsids based on the elongation of an icosahedron along a threefold or twofold axis of symmetry has also been suggested for some bacilliform viruses such as AMV; geometrical rules have been established for their construction [65].

2.7 Helical Capsids

Helical symmetry facilitates the assembly of nucleoprotein complexes without size limitations, as genomes of different lengths could, in principle, be coated by identical protein subunits, forming a cylinder with helical symmetry ([66], and references therein). The model case is the tobacco mosaic virus (TMV), the first identified virus.

Fig. 2.6 (continued) human papillomavirus 16, T = 7d, ~600 Å; this capsid is made of 72 pentameric capsomers ([55], courtesy of BL Trus); TLP, a rotavirus with a triple-layered particle, ~700 Å (without spikes): VP4 spikes (red), VP7 outermost T = 13 *l* layer (yellow), VP6 intermediate T = 13 *l* layer (blue), VP2 innermost T = 1 layer (green), and the internal density of dsRNA and polymerase complexes (violet) (courtesy of D Luque and JM Rodríguez)

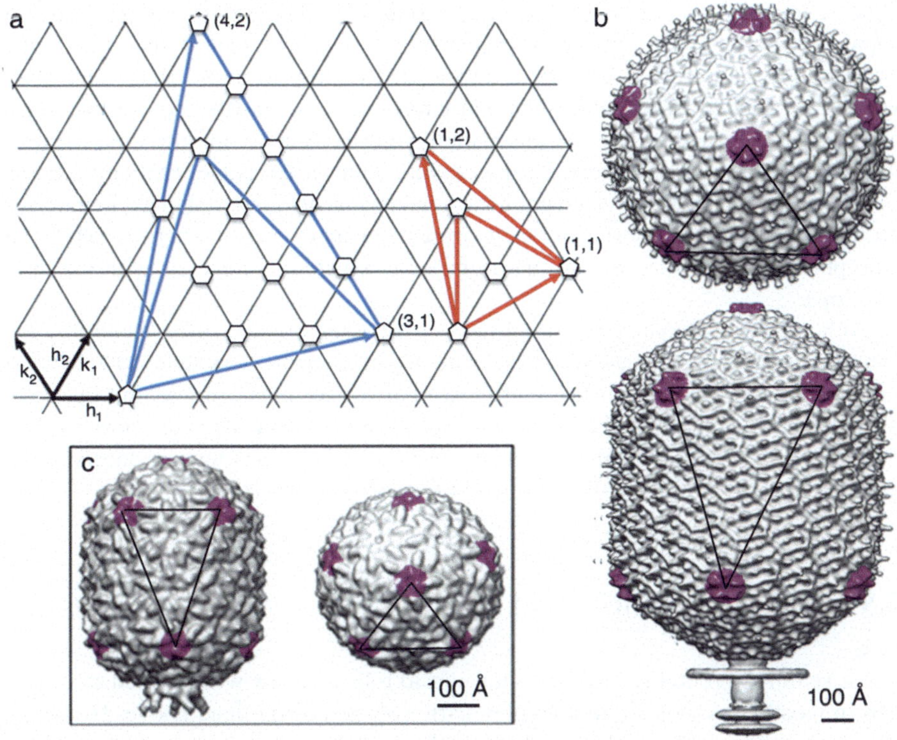

Fig. 2.7 Determination of the triangulation number (T) and the elongation number (Q) in a prolate icosahedral lattice. (**a**) A planar equilateral triangular lattice (as in Fig. 2.5) is used to show the triangulation numbers, T (or T_{end}) and Q (or T_{mid}), of prolate capsids; two unit vector systems, (h_1, k_1) and (h_2,k_2), are used to represent the equilateral triangular facets at the end caps and the elongated triangular facets at the central band (see text). The large equilateral triangle (blue) represents an end cap facet of a bacteriophage T4 head, and is defined by the (h_1, k_1) coordinate point [the (h_1, k_1) lattice point corresponds to a (h, k) point in Fig. 2.5]; $h_1 = 3$, $k_1 = 1$, and $T_{end} = h_1^2 + h_1k_1 + k_1^2 = 13\ l$. The large triangle with unequal sides (blue) represents the phage T4 midsection facets and is described by two vectors, one from the origin to the lattice points (h_1, k_1) and the other to (h_2,k_2); $h_2 = 4$, $k_2 = 2$, and the triangulation of the extended facet is $Q = T_{mid} = h_1h_2 + h_1k_2 + k_1k_2 = 20$. For bacteriophage ϕ29, the equilateral facets at the end caps (red) are defined by $h_1 = 1$, $k_1 = 1$, and $T_{end} = 3$, and the stretched facets at the midsection triangles by $h_2 = 1$, $k_2 = 2$, and $Q = 5$. (**b**) Surface-shaded representation of the 3D cryo-EM map of the T4 bacteriophage head ([64], cryo-EM map deposited at the EMDataBank; ID EMD-1075), viewed along the fivefold axis of an end cap (top) and perpendicular to the fivefold axis (bottom). The equilateral and stretched triangles show the icosahedral equilateral and elongated facets defined by pentameric vertices (pink). (**c**) 3D map of ϕ29 bacteriophage prohead ([63], EMD-1117) showing equivalent views as for T4 phage

A helix is defined by its pitch (P), which depends on two parameters: the number of subunits per helix turn (μ) and the axial rise per subunit (ρ) (P = μ x ρ). For TMV, P is 23 Å, μ = 16.3 and ρ = 1.4 Å. TMV particles are rigid, right-handed, rodlike structures, 300 nm long and 18 nm in diameter, with a central 4 nm-diameter hole.

Other helical viruses, however, are longer and flexible. The genome of many animal negative-sense ssRNA viruses is protected by a helical capsid that forms the nucleocapsid, surrounded by a lipid envelope. Contrary to many phage and plant viruses, no helical non-enveloped animal viruses are known. Rhabdoviruses (rabies and stomatitis vesicular viruses) and paramyxovirus (Sendai, mumps, and measles viruses) have left-handed helix nucleocapsids with an envelope. Orthomyxovirus (influenza virus) have multisegment helical nucleocapsids, referred to as ribonucleoproteins (RNP). As for most enveloped viruses, the nucleocapsid is surrounded by an amorphous layer formed by the matrix protein, which also interacts with the lipid envelope.

Although helical symmetry simplifies capsid assembly and offers flexibility for genome packaging, this assembly is relatively infrequent. Less than 10% of viral families have helical symmetry, and most of them are plant viruses and bacteriophages, such as those of the *Inoviridae* (phages M13 and fd). The most evident problems associated with this geometry are the disadvantageous surface/volume ratio and the structural instability inherent to long, thin structures.

2.8 The Viral Nucleic Acid Inside

The assembly of proteins to build a viral capsid follows a few well-defined strategies and in most cases yields a unique geometric solution as outlined above. However, the organization of the nucleic acid inside the capsid is more diverse. Although this topic will be reviewed in depth in Chap. 13, we will note here several examples that describe how the nucleic acid organization is related to capsid geometry.

Whereas most filamentous bacteriophages contain a cylindrical cavity for the packaged genome, in archaea, plant, and mammalian filamentous viruses, the nucleic acid is highly ordered, embedded in the capsid proteins [67]. In the rigid rod-shaped plant viruses (such as tobacco mosaic virus, TMV), ssRNA is closely assembled with the CP, integrated such that it strictly follows the geometrical characteristics imposed by the CP assembly. Other DNA-containing helical viruses (like phage M13) also show a close structural relationship between nucleic acid and protein, although the fit is less tight than in TMV-like viruses, and is based on defined contacts between the DNA and the inner walls of the helical protein shell. A different type of viral helical protein-nucleic acid arrangement is found in flexible nucleoprotein complexes characteristic of *Mononegavirales* (Borna-, Filo-. Paramyxo- and Rhabdoviridae families). These helical assemblies, which are enclosed within lipid-based envelopes, work as an active complex for replication and transcription.

Many small positive-sense ssRNA virus with icosahedral capsids co-assemble their capsids around their genomes, which contain sequence/structure elements (termed packaging signals, PS) involved in capsid assembly and genome compaction [68]. That is the case for satellite tobacco necrosis virus (STNV) and bacteriophage MS2 with a $T = 1$ and $T = 3$ capsid, respectively, in which the specific interactions between the viral genome PS and the capsid protein increase the assembly efficiency and result in a high degree of ordering. In nodaviruses, such as Pariacoto virus and flock house virus (FHV), their ssRNA genomes are organized into dodecahedral cages composed of A-type duplexes that account for ~35% of the encapsidated RNA [69]. Weak ordering of packaged ssRNA is observed when nonspecific interactions between an ensemble of RNA conformations and the capsid inner surfaces are dominant, as in brome mosaic virus (BMV) [70].

At the other extreme, dsDNA bacteriophages show very high nucleic acid density inside the capsid (>500 mg/ml). In these cases, viruses have incorporated an ATP-driven motor to package the DNA molecule as a tight coil that forms concentric shells inside the capsid, with the DNA strands arranged in a parallel pseudo-hexagonal crystalline array.

2.9 Enveloped Viruses

As already mentioned above, there are many viruses that are structurally more elaborate than naked capsids or nucleocapsids. Above we described several complex viruses in which the capsid or nucleocapsid is made of several proteins that can be ordered in concentric layers or that can include membranes beneath an icosahedral shell, such as the PRD1-like viruses. In this section, we will provide an overview of some specific structural aspects of enveloped viruses in which the outermost structural layer is a lipid membrane. For detailed descriptions of the structure, assembly, and function of enveloped virus and other structurally complex viruses, see Chaps. 12, 16, and 17.

Capsid or nucleocapsid assembly is not always the final step in virion morphogenesis, as some are surrounded by one or several membrane envelopes of cellular origin, modified by insertion of viral proteins. The lipid bilayer is acquired during the budding of the viral particle from a membrane, which can be the plasma membrane during virus escape from the host cell, or membrane-delimited structures in eukaryotic cells such as the nuclear membrane, endoplasmic reticulum, Golgi apparatus, and vacuoles. Enveloped viruses can exit the cell without causing cell death, since the budding process does not alter cell membrane integrity, and can cause either persistent or latent infections. Viral glycoproteins are inserted in the membranes via the compartmentalization routes and constant flux of vesicular trafficking. Glycoproteins have an external domain (ectodomain) with sugar chains, which is anchored in the membrane by a transmembrane hydrophobic segment (typically α-helix) that terminates in a small domain (endodomain) facing the cytoplasm. Viral glycoprotein monomers associate to form multimeric spikes (visible by electron microscopy), which are the functional state for at least four activities: identification and binding to cell receptors, membrane fusion, transcription activation, and receptor destruction. These functions can be carried out by one or several glycoproteins. Envelopes studded with glycoproteins usually have minor proteins that make ionic channels.

Finally, matrix proteins are internal proteins that mediate the association of glycoprotein endodomains with the nucleocapsid (see Chap. 12).

Membrane acquisition differs among viruses. Alphaviruses and flaviviruses, which have a positive-sense ssRNA genome, have a nucleocapsid that is assembled in the host cytoplasm and only subsequently interacts with the glycoprotein endodomains. In alphaviruses (of the Togaviridae family), lateral interactions among glycoproteins exclude cell membrane proteins, resulting in a three-layered capsid: a glycoprotein outer icosahedral layer, a lipid bilayer, and a protein inner icosahedral layer (both icosahedral shells are in register with a T = 4 geometry [71]). Flaviviruses have an outer ordered protein layer above a polygonal lipid bilayer and an inner disordered nucleocapsid; the outer capsid is formed by 90 closely packed glycoprotein dimers that lie flat on the particle surface, rather than forming protruding spikes. Dimer arrangement differs from a true T = 3 shell, however, and the capsid should be considered a T = 1 capsid with three monomers per asymmetric unit ([72] and references therein). Whereas budding takes place in the plasma membrane for alphaviruses, flavivirus buds in the endoplasmic reticulum lumen.

Contrary to the well-ordered, quasi-crystalline arrangement found in icosahedral capsids, most enveloped viruses lack regular organization in the membrane. Orthomyxoviruses (influenza virus) have an additional protein, the matrix protein, which mediates interactions between helical nucleocapsids (in eight segments) and the viral envelope. Retroviruses have a similar membrane-associated matrix protein. The Influenza virus membrane contains two glycoproteins, neuraminidase, which cleaves sialic acid groups from receptor glycoproteins, and hemagglutinin, a trimer that contains the binding site for the virus receptor and is responsible for viral and cell membrane fusion. Rhabdoviruses and paramyxoviruses are similar to orthomyxoviruses, except that the genome is a single helical nucleocapsid. Bunyaviruses have a tripartite genome and are very flexible since they lack matrix proteins; they have glycoprotein spikes, some ordered in a T = 12 capsid, some locally ordered, and some lacking any order. Herpesvirus virions comprise a T = 16 icosahedral capsid (125 nm diameter) and interact with the envelope (derived from the trans-Golgi network) through a complex asymmetric layer with many viral (and a few cellular) proteins, called the tegument (virion size, 200 nm diameter). The herpesvirus envelope has 12 different glycoproteins and about 800 spikes per virion. Poxviruses (or brick-shaped viruses) such as vaccinia virus encompass multiple structural elements, and are described in detail in Chap. 12.

Retroviruses, such as the lentivirus HIV, are enveloped pleomorphic virions. The mature HIV virion is about 100 nm in diameter and has two copies of a positive sense ssRNA genome associated with the nucleoprotein (NC) in the nucleoprotein complex or viral core. The dimeric ribonucleoprotein complex is enclosed in a protein shell made of ~1500 copies of a single protein, the capsid protein CA, which is assembled into a cone-shaped capsid in the mature HIV virion. There is a matrix protein (MA) layer that interacts with the inner leaflet of the viral envelope (MA is a peripheral membrane protein). MA, CA, and NC are synthesized as a single precursor, Gag, that is arranged radially in immature, incomplete spherical particles. Immature HIV virions bud at lipid rafts; these membranes have viral spikes formed by the surface (gp120) and transmembrane (gp41) glycoproteins, which are also initially fused. Virions mature after budding by the concomitant activation of the viral protease. CA forms hexamers on a variably curved hexagonal lattice closed by insertion of 12 pentamers, 7 at the wide and 5 at the narrow end of the fullerene cone. CA has two domains, CA^{NTD} and CA^{CTD}; in the mature capsid, CA^{NTD} forms hexameric and pentameric rings, whereas CA^{CTD} dimeric contacts, at base of the protein shell, connect neighboring rings ([73] and references therein).

Acknowledgments I thank Daniel Luque for preparing several figures used in this chapter and Catherine Mark for editorial help. This work was supported by grants PID2020-113287RB-I00) from the Ministry of Science and Innovation and P2018/NMT-4389 from the Comunidad Autónoma de Madrid to JRC.

References

1. Harrison SC (2007) Principles of virus structure. In: Knipe DM, Howley PM, Griffin DE, Lamb RA, Martin MA, Roizman B, Strauss SE (eds) Fields virology, vol 1. Lippincott Williams & Wilkins, Philadelphia, pp 59–98
2. Dokland T (2000) Freedom and restraint: themes in virus capsid assembly. Structure Fold Des 8:R157–R162
3. Flint SJ, Enquist LW, Racaniello VR, Skalka AM (2009) Principles of virology. ASM Press, Washington, DC
4. Cann AJ (2012) Principles of molecular virology. Elsevier, Amsterdam
5. Baker TS, Olson NH, Fuller SD (1999) Adding the third dimension to virus life cycles: three-dimensional reconstruction of icosahedral viruses from Cryo-electron micrographs. Microbiol Mol Biol Rev 63:862–922
6. Prasad BV, Schmid MF (2012) Principles of virus structural organization. Adv Exp Med Biol 726:17–47
7. Hendrix RW, Johnson JE (2012) Bacteriophage HK97 capsid assembly and maturation. Adv Exp Med Biol 726:351–363
8. Chang J, Liu X, Rochat RH, Baker ML, Chiu W (2012) Reconstructing virus structures from nanometer to near-atomic resolutions with cryo-electron microscopy and tomography. Adv Exp Med Biol 726:49–90
9. Zhou ZH (2011) Atomic resolution cryo electron microscopy of macromolecular complexes. Adv Protein Chem Struct Biol 82:1–35
10. Harrison SC, Olson AJ, Schutt CE, Winkler FK, Bricogne G (1978) Tomato bushy stunt virus at 2.9 Å resolution. Nature 276:368–373
11. Hogle JM, Chow M, Filman DJ (1985) Three-dimensional structure of poliovirus at 2.9 a resolution. Science 229:1358–1365
12. Rossmann MG, Arnold E, Erickson JW, Frankenberger EA, Griffith JP, Hecht HJ, Johnson JE, Kamer G, Luo M, Mosser AG et al (1985) Structure of a human common cold virus and functional relationship to other picornaviruses. Nature 317:145–153
13. Abrescia NG, Bamford DH, Grimes JM, Stuart DI (2012) Structure unifies the viral universe. Annu Rev Biochem 81:795–822
14. Nasir A, Caetano-Anolles G (2017) Identification of capsid/coat related protein folds and their utility for virus classification. Front Microbiol 8:380
15. Zamora M, Mendez-Lopez E, Agirrezabala X, Cuesta R, Lavin JL, Sanchez-Pina MA, Aranda MA, Valle M (2017) Potyvirus virion structure shows conserved protein fold and RNA binding site in ssRNA viruses. Sci Adv 3:eaao2182

16. Valle M (2018) Structural homology between nucleoproteins of ssRNA viruses. Subcell Biochem 88:129–145
17. Rossmann M, Johnson J (1989) Icosahedral RNA virus structure. Annu Rev Biochem 58:533–573
18. Lomonossoff GP, Johnson JE (1991) The synthesis and structure of comovirus capsids. Prog Biophys Mol Biol 55:107–137
19. Roberts MM, White JL, Grutter MG, Burnett RM (1986) Three-dimensional structure of the adenovirus major coat protein hexon. Science 232:1148–1151
20. Nandhagopal N, Simpson AA, Gurnon JR, Yan X, Baker TS, Graves MV, Van Etten JL, Rossmann MG (2002) The structure and evolution of the major capsid protein of a large, lipid-containing DNA virus. Proc Natl Acad Sci USA 99:14758–14763
21. Abrescia NG, Cockburn JJ, Grimes JM, Sutton GC, Diprose JM, Butcher SJ, Fuller SD, San Martin C, Burnett RM, Stuart DI, Bamford DH, Bamford JK (2004) Insights into assembly from structural analysis of bacteriophage PRD1. Nature 432:68–74
22. Fu CY, Johnson JE (2012) Structure and cell biology of archaeal virus STIV. Curr Opin Virol 2: 122–127
23. Bahar MW, Graham SC, Stuart DI, Grimes JM (2011) Insights into the evolution of a complex virus from the crystal structure of vaccinia virus D13. Structure 19:1011–1020
24. Wikoff WR, Liljas L, Duda RL, Tsuruta H, Hendrix RW, Johnson JE (2000) Topologically linked protein rings in the bacteriophage HK97 capsid. Science 289:2129–2133
25. Veesler D, Johnson JE (2012) Virus maturation. Annu Rev Biophys 41:473–496
26. Luque D, Caston JR (2020) Cryo-electron microscopy for the study of virus assembly. Nat Chem Biol 16:231–239
27. Zlotnick A (2005) Theoretical aspects of virus capsid assembly. J Mol Recognit 18:479–490
28. Johnson JE (2008) Multi-disciplinary studies of viruses: the role of structure in shaping the questions and answers. J Struct Biol 163:246–253
29. Crick FH, Watson JD (1956) Structure of small viruses. Nature 177:473–475
30. Caspar DLD, Klug A (1962) Physical principles in the construction of regular viruses. Cold Spring Harb Symp Quant Biol 27:1–24
31. Liljas L (1986) The structure of spherical viruses. Prog Biophys Mol Biol 48:1–36
32. Johnson JE, Speir JA (1997) Quasi-equivalent viruses: a paradigm for protein assemblies. J Mol Biol 269:665–675
33. Johnson JE (1996) Functional implications of protein-protein interactions in icosahedral viruses. Proc Natl Acad Sci USA 93:27–33
34. Steven A, Trus B, Booy F, Cheng N, Zlotnick A, Castón J, Conway J (1997) The making and breaking of symmetry in virus capsid assembly: glimpses of capsid biology from cryoelectron microscopy. FASEB J 11:733–742
35. Chen DH, Baker ML, Hryc CF, DiMaio F, Jakana J, Wu W, Dougherty M, Haase-Pettingell C, Schmid MF, Jiang W, Baker D, King JA, Chiu W (2011) Structural basis for scaffolding-mediated assembly and maturation of a dsDNA virus. Proc Natl Acad Sci USA 108:1355–1360
36. Zhang W, Olson NH, Baker TS, Faulkner L, Agbandje-McKenna M, Boulton MI, Davies JW, McKenna R (2001) Structure of the maize streak virus geminate particle. Virology 279:471–477
37. Zlotnick A, Cheng N, Conway JF, Booy FP, Steven AC, Stahl SJ, Wingfield PT (1996) Dimorphism of hepatitis B virus capsids is strongly influenced by the C-terminus of the capsid protein. Biochemistry 35:7412–7421
38. Rayment I, Baker TS, Caspar DL, Murakami WT (1982) Polyoma virus capsid structure at 22.5 a resolution. Nature 295:110–115
39. Liddington RC, Yan Y, Moulai J, Sahli R, Benjamin TL, Harrison SC (1991) Structure of simian virus 40 at 3.8-a resolution. Nature 354:278–284
40. Grimes JM, Burroughs JN, Gouet P, Diprose JM, Malby R, Zientara S, Mertens PP, Stuart DI (1998) The atomic structure of the bluetongue virus core. Nature 395:470–478

41. Reinisch KM, Nibert ML, Harrison SC (2000) Structure of the Reovirus core at 3.6 Å resolution. Nature 404:960–967
42. Naitow H, Tang J, Canady M, Wickner RB, Johnson JE (2002) L-A virus at 3.4 a resolution reveals particle architecture and mRNA decapping mechanism. Nat Struct Biol 9:725–728
43. Luque D, Gonzalez JM, Garriga D, Ghabrial SA, Havens WM, Trus B, Verdaguer N, Carrascosa JL, Caston JR (2010) The T=1 capsid protein of Penicillium chrysogenum virus is formed by a repeated helix-rich core indicative of gene duplication. J Virol 84:7256–7266
44. Bamford DH, Grimes JM, Stuart DI (2005) What does structure tell us about virus evolution? Curr Opin Struct Biol 15:655–663
45. Baker ML, Jiang W, Rixon FJ, Chiu W (2005) Common ancestry of herpesviruses and tailed DNA bacteriophages. J Virol 79:14967–14970
46. Castón JR, Trus BL, Booy FP, Wickner RB, Wall JS, Steven AC (1997) Structure of L-A virus: a specialized compartment for the transcription and replication of double-stranded RNA. J Cell Biol 138:975–985
47. Luque D, Gonzalez JM, Gomez-Blanco J, Marabini R, Chichon J, Mena I, Angulo I, Carrascosa JL, Verdaguer N, Trus BL, Barcena J, Caston JR (2012) Epitope insertion at the N-terminal molecular switch of the rabbit hemorrhagic disease virus T = 3 capsid protein leads to larger T= 4 capsids. J Virol 86:6470–6480
48. Pickl-Herk A, Luque D, Vives-Adrian L, Querol-Audi J, Garriga D, Trus BL, Verdaguer N, Blaas D, Caston JR (2013) Uncoating of common cold virus is preceded by RNA switching as determined by X-ray and cryo-EM analyses of the subviral A-particle. Proc Natl Acad Sci USA 110:20063–20068
49. Luque D, de la Escosura A, Snijder J, Brasch M, Burnley RJ, Koay MST, Carrascosa JL, Wuite GJL, Roos WH, Heck AJR, Cornelissen JJLM, Torres T, Caston JR (2014) Self-assembly and characterization of small and monodisperse dye nanospheres in a protein cage. Chem Sci 5:575–581
50. Saugar I, Luque D, Ona A, Rodriguez JF, Carrascosa JL, Trus BL, Caston JR (2005) Structural polymorphism of the major capsid protein of a double-stranded RNA virus: an amphipathic alpha helix as a molecular switch. Structure 13:1007–1017
51. Luque D, Saugar I, Rodriguez JF, Verdaguer N, Garriga D, Martin CS, Velazquez-Muriel JA, Trus BL, Carrascosa JL, Caston JR (2007) Infectious bursal disease virus capsid assembly and maturation by structural rearrangements of a transient molecular switch. J Virol 81:6869–6878
52. Castón JR, Martínez-Torrecuadrada JL, Maraver A, Lombardo E, Rodríguez JF, Casal JI, Carrascosa JL (2001) C terminus of infectious bursal disease virus major capsid protein VP2 is involved in definition of the T number for capsid assembly. J Virol 75:10815–10828
53. Agirrezabala X, Martin-Benito J, Caston JR, Miranda R, Valpuesta JM, Carrascosa JL (2005) Maturation of phage T7 involves structural modification of both shell and inner core components. EMBO J 24:3820–3829
54. Tang J, Jose J, Chipman P, Zhang W, Kuhn RJ, Baker TS (2011) Molecular links between the E2 envelope glycoprotein and nucleocapsid core in Sindbis virus. J Mol Biol 414:442–459
55. Buck CB, Trus BL (2012) The papillomavirus virion: a machine built to hide molecular Achilles' heels. Adv Exp Med Biol 726:403–422
56. Pesavento JB, Crawford SE, Estes MK, Prasad BV (2006) Rotavirus proteins: structure and assembly. Curr Top Microbiol Immunol 309:189–219
57. Coulibaly F, Chevalier C, Gutsche I, Pous J, Navaza J, Bressanelli S, Delmas B, Rey FA (2005) The birnavirus crystal structure reveals structural relationships among icosahedral viruses. Cell 120:761–772
58. Irigoyen N, Caston JR, Rodriguez JF (2012) Host proteolytic activity is necessary for infectious bursal disease virus capsid protein assembly. J Biol Chem 287:24473–24482
59. Yan X, Yu Z, Zhang P, Battisti AJ, Holdaway HA, Chipman PR, Bajaj C, Bergoin M, Rossmann MG, Baker TS (2009) The capsid proteins of a large, icosahedral dsDNA virus. J Mol Biol 385:1287–1299

60. Zhang X, Xiang Y, Dunigan DD, Klose T, Chipman PR, Van Etten JL, Rossmann MG (2011) Three-dimensional structure and function of the Paramecium bursaria chlorella virus capsid. Proc Natl Acad Sci USA 108:14837–14842
61. Yan X, Chipman PR, Castberg T, Bratbak G, Baker TS (2005) The marine algal virus PpV01 has an icosahedral capsid with T=219 quasisymmetry. J Virol 79:9236–9243
62. Xiao C, Kuznetsov YG, Sun S, Hafenstein SL, Kostyuchenko VA, Chipman PR, Suzan-Monti M, Raoult D, McPherson A, Rossmann MG (2009) Structural studies of the giant mimivirus. PLoS Biol 7:e92
63. Tao Y, Olson NH, Xu W, Anderson DL, Rossmann MG, Baker TS (1998) Assembly of a tailed bacterial virus and its genome release studied in three dimensions. Cell 95:431–437
64. Fokine A, Chipman PR, Leiman PG, Mesyanzhinov VV, Rao VB, Rossmann MG (2004) Molecular architecture of the prolate head of bacteriophage T4. Proc Natl Acad Sci USA 101: 6003–6008
65. Luque A, Reguera D (2010) The structure of elongated viral capsids. Biophys J 98:2993–3003
66. Stubbs G, Kendall A (2012) Helical viruses. Adv Exp Med Biol 726:631–658
67. Zanotti G, Grinzato A (2021) Structure of filamentous viruses. Curr Opin Virol 51:25–33
68. Twarock R, Stockley PG (2019) RNA-mediated virus assembly: mechanisms and consequences for viral evolution and therapy. Annu Rev Biophys 48:495–514
69. Tang L, Johnson KN, Ball LA, Lin T, Yeager M, Johnson JE (2001) The structure of pariacoto virus reveals a dodecahedral cage of duplex RNA. Nat Struct Biol 8:77–83
70. Beren C, Cui Y, Chakravarty A, Yang X, Rao ALN, Knobler CM, Zhou ZH, Gelbart WM (2020) Genome organization and interaction with capsid protein in a multipartite RNA virus. Proc Natl Acad Sci USA 117:10673–10680
71. Zhang R, Hryc CF, Cong Y, Liu X, Jakana J, Gorchakov R, Baker ML, Weaver SC, Chiu W (2011) 4.4 a cryo-EM structure of an enveloped alphavirus Venezuelan equine encephalitis virus. EMBO J 30:3854–3863
72. Mukhopadhyay S, Kuhn RJ, Rossmann MG (2005) A structural perspective of the flavivirus life cycle. Nat Rev Microbiol 3:13–22
73. Ganser-Pornillos BK, Yeager M, Pornillos O (2012) Assembly and architecture of HIV. Adv Exp Med Biol 726:441–465

Further Reading

Bamford D, Zuckerman M (eds) (2021) Encyclopedia of virology, 4th edn. Academic Press, Oxford
Steven AC, Baumeister W, Johnson LN, Perham RN (eds) (2016) Molecular biology of assemblies and machines. Garland Science, New York
Comas-García M, Rosales-Mendoza S (eds) (2023) Physical virology. Springer Nature, Cham, Switzerland
Patton JT (ed) (2008) Segmented double-stranded RNA viruses, structure and molecular biolgy. Caister Academic Press, Norfolk
Chiu W, Burnnett RM, Garcea RL (eds) (1997) Structural virology of viruses. Oxford University Press, New York
Rossmann MG, Rao VB (eds) (2012) Viral Molecular Machines. Springer, New York
Harris JR, Bhella D (eds) (2018) Virus protein and nucleoprotein complexes. Springer, New York
Chiu W, Johnson JE (eds) (2003) Virus Structure, Advances in protein chemistry, vol, vol 64. Academic Press, Amsterdam

Part II
Determination of the Structure, Dynamics, and Physical Properties of Viruses

Chapter 3
Conventional Electron Microscopy, Cryogenic Electron Microscopy, and Cryogenic Electron Tomography of Viruses

José R. Castón and Daniel Luque

Abstract Electron microscopy (EM) techniques have been crucial for understanding the structure of biological specimens such as cells, tissues and macromolecular assemblies. Viruses and related viral assemblies are ideal targets for structural studies that help to define essential biological functions. Whereas conventional EM methods use chemical fixation, dehydration, and staining of the specimens, cryogenic electron microscopy (cryo-EM) preserves the native hydrated state. Combined with image processing and three-dimensional reconstruction techniques, cryo-EM provides three-dimensional maps of these macromolecular complexes from projection images, at atomic or near-atomic resolutions. Cryo-EM is also a major technique in structural biology for dynamic studies of functional complexes, which are often unstable, flexible, scarce, or transient in their native environments. State-of-the-art techniques in structural virology now extend beyond purified symmetric capsids and focus on the asymmetric elements such as the packaged genome and minor structural proteins that were previously missed. As a tool, cryo-EM also complements high-resolution techniques such as X-ray diffraction and NMR spectroscopy; these synergistic hybrid approaches provide important new information. Three-dimensional cryogenic electron tomography (cryo-ET), a variation of cryo-EM, goes further, and allows the study of pleomorphic and complex viruses not only in their physiological state but also in their natural environment in the cell, thereby bridging structural studies at the molecular and cellular levels. Cryo-EM and

J. R. Castón (✉)
Department of Macromolecular Structure, Centro Nacional de Biotecnología (CNB-CSIC), Madrid, Spain

Nanobiotechnology Associated Unit CNB-CSIC-IMDEA, Madrid, Spain
e-mail: jrcaston@cnb.csic.es

D. Luque (✉)
School of Biomedical Sciences, The University of New South Wales, Sydney, NSW, Australia

Electron Microscope Unit, Mark Wainwright Analytical Centre, The University of New South Wales, Sydney, NSW, Australia
e-mail: d.luque@unsw.edu.au

M. G. Mateu (ed.), *Structure and Physics of Viruses*, Subcellular Biochemistry 105,
https://doi.org/10.1007/978-3-031-65187-8_3

cryo-ET have been applied successfully in basic research, shedding light on fundamental aspects of virus biology and providing insights into threatening viruses, including SARS-CoV-2, responsible for the COVID-19 pandemic.

Keywords Capsid · Cryogenic electron microscopy · Cryogenic electron tomography · Electron microscopy · Fourier transform · Image processing · Resolution · Single-particle analysis · Sub-tomogram averaging · Three-dimensional reconstruction · Viral macromolecular assembly · Virus

Abbreviations

CCD	charge-coupled device
CLEM	correlative light and electron microscopy
cryo-EM	cryogenic electron microscopy
cryo-ET	cryogenic electron tomography
CTF	contrast transfer function
DED	direct electron detector
ds	double-stranded
EM	electron microscopy, electron microscope
FEG	field emission gun
FIB	focus ion bean
FSC	Fourier shell correlation
FT	Fourier transform
NMR	nuclear magnetic resonance
RdRp	RNA-dependent RNA polymerase
SEM	scanning electron microscopy
SNR	signal-to-noise ratio
SPA	single-particle analysis
ss	single-stranded
SSE	secondary structure element
STA	subtomogram averaging
TEM	transmission electron microscopy
2D	two-dimensional two dimensions
3D	three-dimensional, three dimensions
3DR	three-dimensional reconstruction

3.1 Introduction

Microscopy is probably the technique with the greatest influence in the history of biology. Optical (or light) microscopy was introduced to biology in the seventeenth century by Antoni van Leeuwenhoek, who described bacteria and protozoa using a microscope able to enlarge image size approximately 300 times. Light microscopy uses visible light and a system of lenses to magnify images, providing intuitive information on the morphology and general structure of the biological model under study. The end of the nineteenth and beginning of the twentieth centuries saw the discovery of electrons and X-rays; both provided new alternatives for the observation of matter and stimulated the development of the electron microscope and the X-ray diffractometer. In 1931, Ernst Ruska and colleagues invented the electron microscope (Ruska obtained the Nobel Prize in Physics in 1986). One of the early micrographs (photographic films) taken in the 1930s was of bacterial viruses or bacteriophages, called "tiny hostile bacteria."

It took 300 years to improve the simple light microscope but less than 40 years to perfect the electron microscope. With electron microscopy (EM), scientists were provided with a powerful method that extended the range of observable structures far beyond the limits imposed by the physics of visible light. Our understanding of cell structure and tissues is based in numerous EM contributions. As in many other fields in biology, viruses have had a decisive role in the development of EM, as they constitute simplified tools with which to standardize preparation methods in biological applications. In turn, EM has made outstanding contributions to understand virus structure and its relationship with biological function.

Biological EM might be considered a descriptive technique in its origins. The questions addressed by EM have evolved together with our concept of the molecular reality of the cell [1]. The experience accumulated over many decades in the multidisciplinary field of biology has shown that each molecular event is more than the sum of its parts. This is reflected, for example, in the regulation and integration of physiological signals, termed "systems biology." Genomics has provided a complete list of the macromolecules in a cell, which led to structural genomics, which in turn provides the basic principles of the structure of proteins, of DNA and of RNA. The basic cellular entities that carry out fundamental biological processes are nonetheless multiprotein assemblies [2]. These macromolecular complexes (some of which may contain nucleic acids) respond to factors that determine their association, disassembly, or signal transmission, leading finally to specific molecular events.

To decipher biological processes, the organization of macromolecules and components of macromolecular complexes must be established. The three major techniques in current use for structural determination are X-ray crystallography (Chap. 4), nuclear magnetic resonance (NMR) spectroscopy (Chap. 5), and cryogenic electron microscopy (cryo-EM). Historically, X-ray crystallography and, to a lesser extent, NMR spectroscopy have been instrumental in unveiling the atomic structures of the individual modules or subunits comprising biological molecular machines. In the current landscape, three-dimensional (3D) cryo-EM provides an effective means of determining the structure of many macromolecular assemblies at atomic or near-atomic resolution [3, 4]. The ability to construct atomic models ab initio, with no prior knowledge of the structure of the subunits, is not only possible but quite routine in cryo-EM [5]. Whereas NMR is limited by the size of the macromolecules under study, X-ray crystallography is limited by the difficulty of crystallizing large, flexible complexes, and the difficulties in producing sufficient material for crystallization trials. The number of structures of large macromolecular complexes that have been determined at near-atomic resolution by X-ray crystallography is limited compared to the number of protein structures available. EM acted as a link between structural cell biology and high-resolution structural molecular biology. Over time, descriptive EM has given way to a quantitative approach [6]. Three-dimensional (3D) density maps can be synthesized from two-dimensional (2D) images, as electron images are 2D projections of the 3D (multi-)macromolecular assembly under study [7]. EM images nonetheless have deficiencies, as the electron microscope is not a perfect instrument. These problems can be reduced by subsequent image processing of the micrographs.

EM facilitates the understanding of an atomic structure in its biological context, particularly when macromolecular assemblies and the relationships of their components are involved. The frontiers between NMR spectroscopy, X-ray crystallography, and (cryo-)EM are diffuse, as they form a smooth continuum for visualizing macromolecular structures [8] (see also Chaps. 4, 5, 7, and 8). The emerging relationship among these approaches is a complementary one that can overcome the limitations of each method alone; for instance, atomic models can be fitted accurately into EM-derived maps of larger assemblies [9] (see Chap. 7). Viruses are excellent models of nanomachines; their capsids and other viral protein assemblies have provided some of the most striking images using this hybrid approach [10].

In the past few years, 3D cryo-EM has revolutionized the field of structural biology [11–13] (Fig. 3.1). The 2017 Nobel Prize for Chemistry was awarded to Jacques Dubochet (University of Lausanne), Joachim Frank (Columbia University), and Richard Henderson (MRC Laboratory of Molecular Biology), in recognition of the profound influence of cryo-EM on many scientific disciplines [14]. Viruses, owing to their large mass and inherent symmetry, played a pivotal role in advancing this methodology. 3D cryo-EM has the versatility to investigate a wide spectrum of biological specimens, ranging from individual proteins [15, 16] to large macromolecular complexes such as virus particles [17, 18]. Cell organelles, entire eukaryotic cells, bacteria, and tissue sections can be analyzed by a related method known as cryogenic electron tomography (cryo-ET) [19, 20]. A distinctive capability of cryo-EM is its ability to capture complexes in their various native states and immobilize intermediates at specific time points, also known as time-resolved cryo-EM [21].

Cryo-EM includes several imaging modes through a transmission electron microscope (TEM). Among these, single-particle analysis (SPA) and cryo-ET are the most widely used methods, as they can address a broad range of biological problems. SPA excels in determining the structures at near-atomic resolution of infectious virions spanning a wide size range, viral macromolecular assemblies that are large and dynamic, with compositional heterogeneity (encompassing states with intermediate functional conformations during assembly), soluble proteins such as viral polymerases or surface spikes, and genomes that follow an asymmetric organization. The structure of apoferritin, a 24-subunit protein nanocage resembling a closed icosahedral capsid, was recently solved at 1.2 Å resolution using SPA cryo-EM (i.e., true

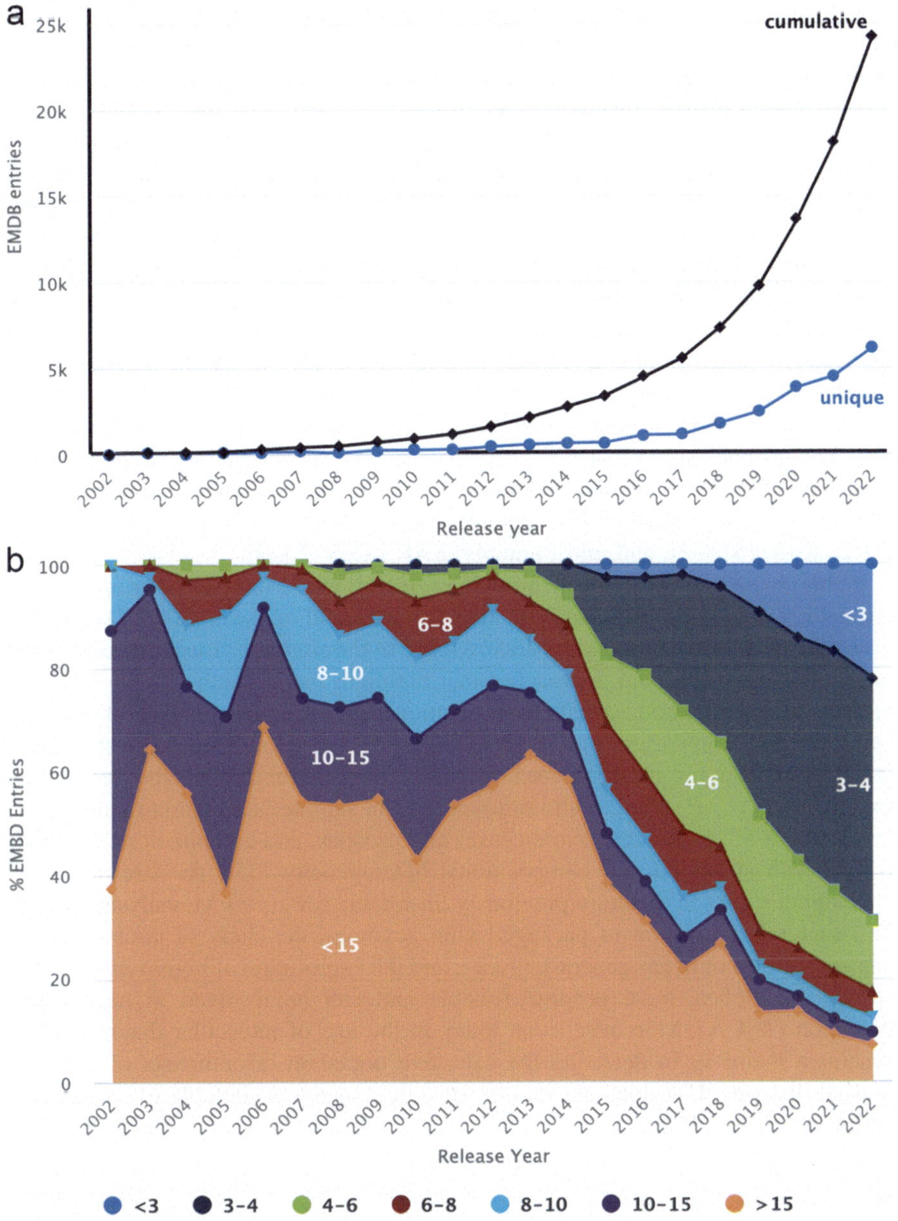

Fig. 3.1 Statistics of electron microscopy maps release. (**a**) EMDB entries released per year and cumulatively. (**b**) EMDB entries resolution in shells per year as percentage of released maps. Charts have been generated with EMDB chart builder (https://www.ebi.ac.uk/emdb/statistics)

atomic resolution), with individual hydrogen atoms clearly discerned [22, 23]. Cryo-ET enables analysis of nonsymmetric, pleomorphic, and/or complex viruses at unprecedented resolution and of viruses not only in their native physiological state but also in their natural cell environment (in situ) [17, 18] (see Chap. 8). Sub-volume or sub-tomogram averaging (STA), a method that averages repeating structures from tomographic reconstructions (or tomograms) [24], can be further exploited to resolve higher resolution detail of any repeating structures such as the viral surface glycoproteins of the severe acute respiratory syndrome coronavirus 2 (SARS-CoV-2) [25] and the internal nucleocapsid of the human immunodeficiency virus (HIV) [26].

3D cryo-EM is currently a go-to technique because of three key factors: (i) the broad molecular weight range available for study (~50–100,000 kDa), (ii) the need for relatively small sample quantities, and (iii) the ability to handle heterogeneous samples to discern multiple conformational or compositional states [17, 18, 27]. In the field of virology, it also enables rapid structural analysis of emerging viruses [28]. For instance, cryo-EM has been instrumental in combating various viruses responsible for recent deadly epidemics, including Ebola virus (2014–2016) [29], Zika virus (2015–2016) [28], dengue virus (2019–2020) [30], MERS-CoV (2012–2015) [31, 32], and, most notably, SARS-CoV-2 [33].

In this chapter, we describe the principles of EM, with examples of its application to the study of viruses. Most sections in this chapter outline the basis of 3D cryo-EM (cryo-EM plus image processing) and cryo-ET to the study of viruses, discuss some outstanding examples and approaches that highlight the strength of cryo-EM, and describe how SPA and cryo-ET are revolutionizing our understanding of virology across the resolution scale, from cells to atoms. The development of direct electron detectors (DED) for image recording was a key factor driving the "resolution revolution" [11, 34], along with improvements in sample preparation methods, the development of specialized cryo-electron microscopes, and computational methods to deal with structural and compositional heterogeneity [35]. As a result, many challenging topics in virology previously unsuitable for cryo-EM analysis, such as the structural elucidation of packaged viral genomes, the study of jumbo viruses, characterization of heterogeneous viruses, and the exploration of high-dynamic viral surface assemblies, have become feasible and can be resolved at near-atomic resolution (2–4 Å). Moreover, even today in the age of molecular diagnosis, EM remains a mainstay in detecting the causative organism in outbreaks of new and unusual diseases; EM does not require specific reagents to be able to recognize a pathogenic agent [36].

3.2 Transmission Electron Microscopy of Viruses

3.2.1 Basic Concepts and General Experimental Design

The physical basis of EM is the electron charge. Electrons are provided by an emission source, such as a thermionic tungsten or lanthanum hexaboride filament or, alternatively, by a field emission gun (FEG). A FEG generates a much brighter electron beam with better spatial and temporal coherence than a thermal ionic electron source. Electrons are accelerated by high voltage, and focused by electromagnetic lenses (condenser lenses) to make a collimated (highly parallel) beam that interacts with the specimen. The interior of the microscope column must be under a high vacuum system to reduce the frequency of electron collision with gas atoms to negligible levels in the path of the electrons; this system also prevents degradation of the coherence and monochromaticity of the electron beam (Fig. 3.2a). The sample is inserted in the column through an airlock. Cutting-edge cryo-electron microscopes feature an FEG operated at 300-keV, although 200-keV microscopes can be used to achieve sub-2 Å-resolution structures (Fig. 3.2b). Due to the substantial maintenance and operational expenses, national facilities offering general use access are granted upon application.

There are two basic types of electron microscopes, the transmission electron microscope (TEM) and the scanning electron microscope (SEM). TEM projects electrons through a very thin specimen, the beam interacts with the specimen and the transmitted electrons are focused by electromagnetic lenses (objective lens and projector lenses) onto a detector to produce a magnified 2D image of the specimen [37]. The electron beam covers a wide area of the specimen. Detectors are a fluorescent screen, a photographic film or a charge-coupled device (CCD). The brightness of a particular area of the image is proportional to the number of electrons that are transmitted through the specimen. DED (without scintillators to convert the electrons into photons) are currently extended, as they are superior recording devices for cryo-EM images. DED allow acquisition of micrograph "movies," grouped into individual frames, in which sample drift and other deficiencies can be corrected (below).

SEM produces an image that gives the impression of three dimensions. In this microscope, the electron beam is focused to a fine probe and scans the surface of the specimen, point by point, to generate secondary (reflected) electrons from the specimen; these are then detected by a sensor. The image is produced over time as the entire specimen is scanned. There are also scanning transmission electron microscopes (STEM) for imaging thicker specimens and for analytical use; a scanning beam of electrons penetrates thin samples and provides quantitative determinations of macromolecular masses, and indicates the presence and distribution of the atomic elements in the specimen. The majority of macromolecular (and viral) EM studies are based on TEM.

EM specimens must be solid and thin. Although water is by far the foremost component of living entities, it has long been neglected in EM, as it evaporates in the vacuum of the microscope. Years of research have established many useful procedures for observing biological samples with acceptable structural preservation although their water has been removed (see below). Full preservation of native conformation after complete dehydration nonetheless remains unresolved. Cryo-EM techniques, which began to emerge in the 1980s, consider water as the natural intrinsic component that it actually is. Specimens remain fully hydrated, immobilized in vitreous water (or vitreous ice). Water is cooled very rapidly in

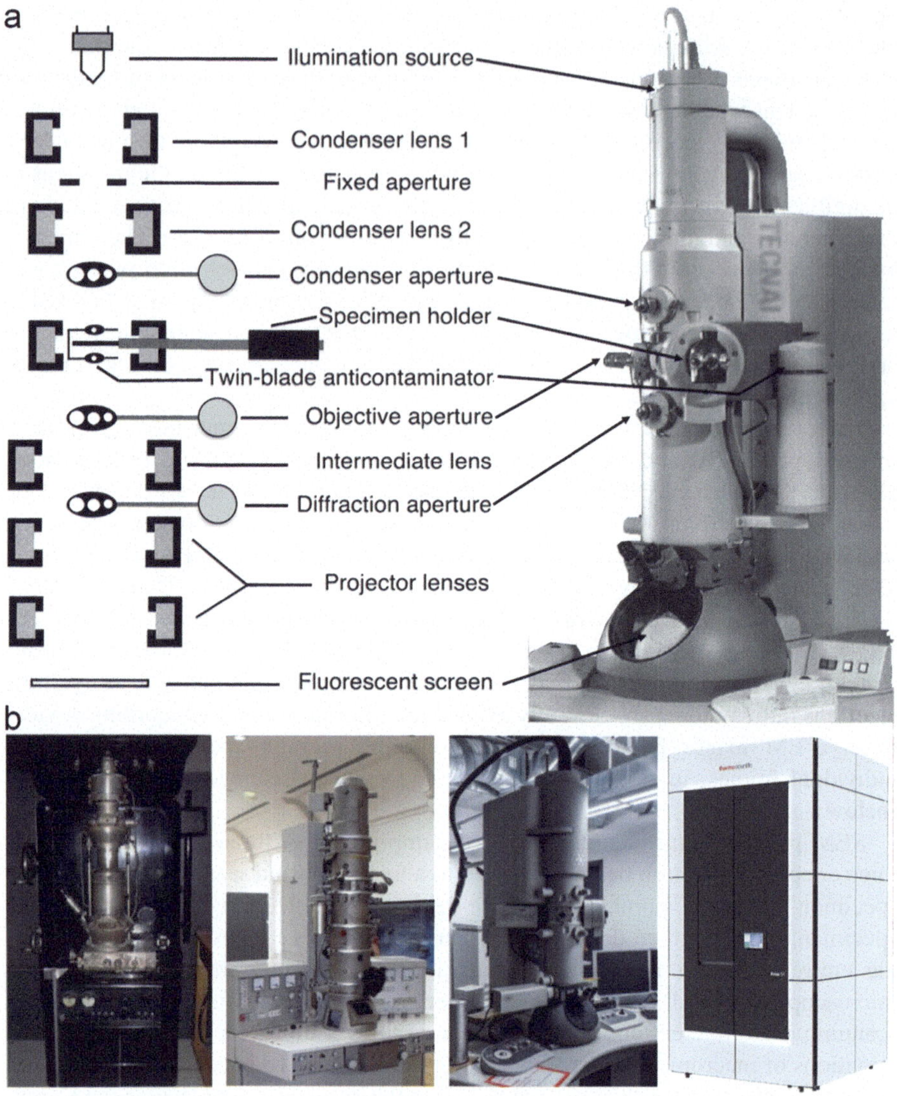

Fig. 3.2 Major components of a transmission electron microscope. (**a**) Diagram of major components in the column of a modern transmission (cryo-)electron microscope. On the right, a Philips Tecnai 12, 120 kV electron microscope. (**b**) Evolution of electron microscopes (from left to right): Ruska's electron microscope built in 1933, a Siemens ELMISKOP 102 electron microscope (1973), a FEI Tecnai G2 Twin (200 kV), and a 300 kV Thermofisher Scientific Titan Krios (2022)

such a way it is solidified without ordering, as it is in hexagonal or cubic ices; the resulting vitreous water is in an amorphous solid state. The specimen is maintained at a very low temperature (−180°C or lower) and the evaporation rate becomes negligible, preserving the hydrated state of the specimen in the microscope [38].

The amount of electron scattering is dependent on specimen density and overall thickness. The specimen must be thin enough to prevent image degradation due to the numerous elastic and inelastic scattering interactions of electrons passing through the specimen [39]. Electrons that interact with the potential field of atomic nuclei usually undergo elastic collisions and are deflected with no loss of energy. They emerge with their wavelength unchanged but experience an associated phase shift. Elastically scattered and unscattered electrons contribute to image formation. Electrons that interact with the outer electrons of atomic nuclei undergo inelastic collisions and transfer some of their energy to the specimen atoms; this process contributes to image blurring and is responsible for radiation damage to the specimen. Due to the nonconductive nature of the specimen, these inelastic interactions lead to the formation of many highly reactive ions and free radicals; this results in the breakage of covalent bonds, causing molecular structures to degrade rapidly during observation. Image blurring can be caused by specimen movement, which can include mechanical drift (stage drift), vibration, and/or image thermal drift. These adverse effects can be minimized using imaging strategies in which the electron dose is low [40].

A constant problem with biological specimens is their low contrast, defined as perceptible variation in the intensity between different regions. Two components contribute to contrast, amplitude, and phase contrast. Amplitude contrast is due to the corpuscular nature of electrons and requires the use of an objective lens aperture that traps the electrons scattered at large angles (inelastic scattering or high angle elastic scattering). Biological samples scatter electrons weakly (at small angles), and artificial methods are used to enhance scattering properties, such as the introduction of heavy atoms (e.g., negative staining). Regions of the specimen that give rise to these trapped electrons appear dark in the image. This mechanism is negligible in unstained specimens. In theory, smaller objective apertures increase the amplitude contrast but are more prone to astigmatism problems if the aperture is displaced from the center of the beam axis.

Phase contrast is due to the wavelike nature of electrons and arises from the constructive or destructive interference between the elastically scattered electrons and the unscattered, transmitted electrons. This is the dominant contrast mechanism in cryo-EM of vitrified (unstained) specimens. Elastically scattered electrons experience a phase shift (or difference in path length) as they emerge from the specimen, with no alterations in their amplitude, and this phase shift can be modified. Spherical aberration (inherent to the objective lens of the microscope) and focus setting (directly manipulated by the microscopist) are factors that generate phase contrast. No phase contrast would be seen in an aberration-free microscope when the specimen is imaged exactly in focus (see below).

3.2.2 Sample Preparation Techniques: Negative Staining, Metal Shadowing, Embedding, and Ultramicrotomy

As important as the development of the electron microscope itself is the development of adequate sample preparation techniques to be used with it. In the late 1950s, detailed images of viruses were obtained by embedding them in a high-density (electron-opaque), highly soluble stain, which replicates specimen structure. Heavy metal salts such as phosphotungstic acid, uranyl acetate, and ammonium molybdate have been used successfully as negative stains. Stains stabilize the sample in the harsh conditions necessary for observation, including the hostile atmosphere (high vacuum) and continuous electron irradiation (energy transfer processes). As a general rule, the difficulties associated with biological samples are solved by following four steps: (1) fixation (chemical or physical), (2) dehydration or drying, (3) the introduction of support (plastic and amorphous carbon layers, resin, metal shadow, or vitreous ice), and (4) increase in contrast (heavy, electron-dense elements).

For optimal results, a stable, structureless substrate is essential for specimen support; plastic films such as collodion or formvar stabilized with a thin layer of carbon are appropriate. These films are used as a support in a specimen grid, the electron microscope analog of the glass slide used in light microscopy. EM grids are fine mesh supports (~3 mm in diameter), on which a plastic/carbon film is deposited, and the biological material is adhered for transport and viewing in the electron microscope. Copper is normally used in the fabrication of grids, although other more inert metals such as gold or nickel may be needed.

There are several negative staining procedures; the flotation method is probably the most popular. A plastic/carbon-coated grid is floated on a droplet containing the specimen for 1–3 min to permit adsorption of the specimen. The grids are previously glow-discharged to render the carbon film more hydrophilic and increase specimen adhesion. The grid is then passed through 2–3 droplets of water and transferred onto a drop of negative stain for 30–60 seconds, blotted with filter paper, and air-dried. In principle, the low pH of the negative stain solution, the adsorption of the specimen to the carbon layer, the embedding in a heavy-metal salt cast, and the dehydration method can all be expected to severely disrupt the native conformation of the specimen. The technique nonetheless preserves quite well the structure of a wide range of specimens and constitutes an indispensable initial step for higher resolution studies by cryo-EM [41, 42], as resolution is limited by the granularity of the stain to ~20 Å.

The negative staining technique is a very quick and easy method for assessing the homogeneity and quality of many particulate solutions such as purified virus preparations and requires a minimum of experience and equipment. Limitations are due mainly to structural preservation; the stain depicts only the surface features of the specimen and any cavities the stain penetrates, but little or no information is derived about the partially or unstained internal features. The method is important in structural virology for fast but unrefined morphological identification. Enveloped viruses are very susceptible to disruption by negative stains and/or air-drying.

Metal shadowing techniques also use heavy metals to enhance the contrast of the low intrinsic electron scattering of biological samples. Shadowing experiments reveal topographic features of specimens by spraying particles of vaporized metal to form a thin coating on whole pre-dried specimens. Fixing and drying of the specimen should preserve the details to be studied; the simplest method is to allow the sample to air-dry, although this technique can result in excessive shrinkage, cracking, and/or collapse of fine structures. The freeze-drying method freezes the specimen rapidly, while it is still in the aqueous phase. The frozen specimen is transferred to a chamber that maintains a temperature below −80°C under vacuum, and the solid ice gradually sublimates; preservation is usually better using this process. The freeze-fracture technique replicates fractured surfaces of frozen specimens obtained with a sharp knife; it is especially suited for studies of membrane structure. A variant of this approach is freeze-etching, in which sublimation (etching of ice) is introduced after fracturing.

Platinum or tungsten/tantalum are usually melted and evaporated from a heated electrode, and a second coat of carbon is often used to improve stability of the replica coating. Specimen shadowing can be either unidirectional or rotary. In unidirectional shadowing, the specimen is immobile and is shadowed at an average angle of ~45° in a high vacuum evaporator. This method can be used to measure the height of a specimen from the length of its shadow, and these micrographs are analyzed to obtain topographic maps. In rotary shadowing, the specimen is rotated on a motorized stage and is shadowed from all directions. Metal replicas are released from the underlying biological material by careful chemical treatments such as a hypochlorite solution, and the floating replica is washed thoroughly, carefully fished on fine grids, dried, and then viewed in the microscope.

Embedding and ultramicrotomy of biological samples, including isolated virus particles or virus-infected tissue or cells, is an elaborate procedure. Specimens are immersed in a matrix that is solidified and then cut into thin slices or sections (50 to 100 nm thick). The specimen is first chemically fixed to preserve its structure; osmium tetroxide was the first fixative used for EM, but laboratories today use many variations on the fixation process, including potassium permanganate and the standard glutaraldehyde-osmium tetroxide protocol. The sample is next dehydrated, which consists of the gradual replacement of water with an organic solvent that acts as a "transition solvent" between the aqueous environment of the specimen and the hydrophobic plastic embedding medium; ethanol and acetone are common dehydrating agents. The sample is infiltrated in a resin; in this process, dehydrants are gradually replaced by (viscous) resin monomers such as epoxy mixtures (often called Epon). The resin specimens are transferred into molds or capsules and placed in an oven in which the epoxy components polymerize to form a solid. There are many types of embedding media with special uses, such as Lowicryl and LR White (acrylic media) used for immunocytochemistry. Finally, the embedded sample is thinly sectioned with a precision (diamond or glass) knife on an ultramicrotome,

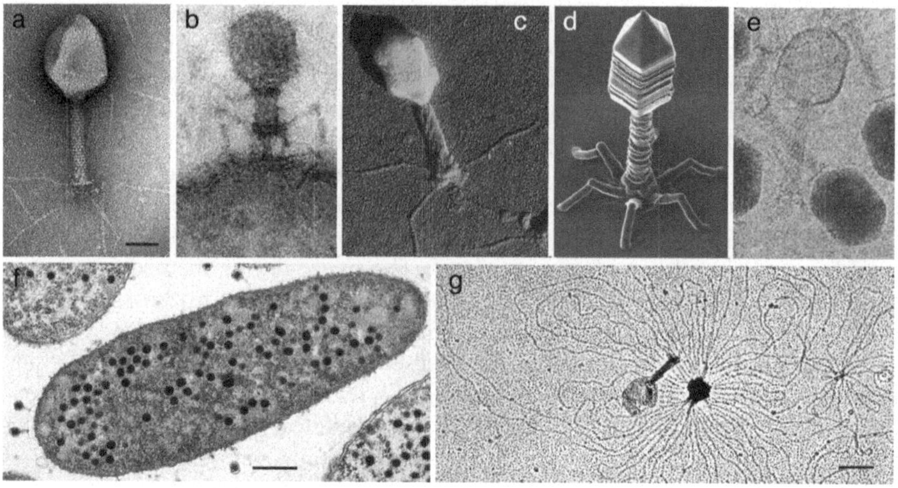

Fig. 3.3 Analysis of bacteriophage T4 by conventional EM and cryo-EM tecniques. (**a**) T4 phage negatively stained with 2% uranyl acetate; darker areas represent local concentrations of stain. The general morphology of the T4 phage, including head, tail, baseplate, and fibers can easily be distinguished. The 40.5 Å tail spacing is seen as stacked disks. Note that phage heads are collapsed to different extents, giving the impression of structural heterogeneity due to phage collapse. Bar, 500 Å. (**b**) Thin section showing phage T4 injecting its DNA genome through the bacterial wall; fibers adhere to the outer wall surface and the tail is contracted (Dr. Michel Wurtz, University of Basel, Switzerland; micrograph by J. v.d. Broek). (**c**) Freeze-dried, metal-shadowed phage T4. General morphology is seen as with negative staining, and additional structural details can be inferred: head and tail shadows indicate different heights for these structures; the helical pattern in the tail region is evident in these conditions. (**d**) SEM image of an artificial nano T4 phage fabricated from carbon by focused-ion-beam-induced chemical vapor deposition on a Si surface, about 10x larger than the real virus (micrograph by Reo Kometani and Shinji Matsui, University of Hyogo, Japan). (**e**) Vitrified phage T4; full (dark) and empty (clear) heads are seen. At this defocus, tail spacing of 40.5 Å is enhanced. (**f**) Thin section through a phage T4-infected *E. coli* bacterium (Dr. Michel Wurtz; micrograph by M. Maeder). Bar, 0.5 μm. (**g**) Air-dried and shadowed broken (by osmotic shock) phage T4 that has released its packaged DNA (Dr. Michel Wurtz; micrograph by J. Meyer). The partially extended DNA indicates high packaging density within the head. Bar, 1000 Å

sections are picked up on a naked copper grid, and stained with heavy metal salts to enhance contrast.

Bacteriophage T4 of the myovirus group is an excellent model that illustrates the complementarity of the information that can be obtained by a variety of conventional TEM procedures (Fig. 3.3). A double-stranded (ds) DNA-tailed virus that infects *Escherichia coli*, phage T4 is characterized by a large elongated (prolate icosahedral) capsid (head) (1150 Å long, 850 Å wide), a tail made of two concentric protein cylinders (1000 Å long, 210 Å wide), and a complex base plate (460 Å diameter) with six long tail fibers attached to it (1450 Å long). It is approximately 200 nm long and 80–100 nm wide. The tail is surrounded by a contractile sheath, with helical symmetry (with a pitch of 40.5 Å), which contracts during infection of the bacterium. DsDNA is densely packaged, ~24 Å interstrand spacings, inside the head [43].

3.2.3 Sample Preparation Techniques: Freeze-Substitution, High-Pressure Freezing, and Cryo-ultramicrotomy

The high vacuum required in the electron microscope column does not allow visualization of living cells in liquid water, but, as discussed above, specimen ultrastructure is altered as soon as water is removed. The approaches described thus far increase the stability of dehydrated biological samples by introducing chemical fixatives such as a negative stain agent or osmium tetroxide. As we will see below, the closest approach to the native state is the vitrification of the biological material. This approach is restricted to "thin" structures (viruses are ideal samples), which can be cryo-fixed on a specimen grid by plunge-freezing. Thick samples can be cryo-fixed by high-pressure freezing.

Freeze-substitution is a low-temperature dehydration method; it can be considered a hybrid technique that bridges the gap between cryo-EM and conventional EM. To ensure good structural preservation, biological samples are cryo-fixed, for example, by high-pressure freezing (see below). They are then freeze-substituted, which consists of dehydration at −90 °C by replacing ice with an organic solvent based on acetone, ethanol, or methanol and, if desired, chemical fixatives. After dehydration (or substitution), the temperature is raised to −40°C to −30°C, and the sample is resin-embedded. Cryo-fixation by high-pressure freezing provides excellent preservation of cellular and molecular architecture for samples between 10 and 400 μm thick. This approach has provided realistic views of the crowded eukaryotic cytoplasm, in which protein concentration is ~300 mg/ml.

High-pressure freezing is an approach that permits optimal preservation of biological ultrastructure by cryo-fixation, provided the biological material is not destroyed by water crystallization. This requires high cooling rates (higher than 10,000°K/s) at atmospheric pressure (~1 bar). The native structure is well preserved by conventional freezing (or vitrification) of specimens up to 0.5 μm. Thicker specimens (up to 30 μm) can be frozen without visible ice crystal damage using antifreezing (cryoprotectant) agents, but chemical fixatives are needed; this implies a deviation of the native structure and, therefore, loss of high-resolution details. Thicker samples (~200 μm) can be adequately frozen under high pressure (~2000 bars), in which physical properties of water are changed (high pressure is, in fact, a physical cryoprotectant that prevents ice nucleation and crystal growth and increases the depth of vitrification).

Material embedded in vitreous ice is ideal for cutting directly into ultrathin cryo-sections (with a cryo-ultramicrotome using diamond knives) and can be observed in a cryo-electron microscope at liquid nitrogen temperatures, although cutting-induced deformations and artifacts are a serious problem. This approach is known as cryo-EM of vitreous sections (CEMOVIS). Recently developed focus ion bean (FIB) milling of cell lamellas (below) overcomes this problem and allows the visualization of the cellular environment without artifacts. A three-dimensional reconstruction of the specimen in the cryo-section or lamella (150–300 nm) can be generated by computerized cryo-ET. This is the best method for preserving whole cells and tissues in their native state, and relatively high resolution is achievable [44] (see Chap. 8).

3.3 Cryogenic Electron Microscopy of Viruses

3.3.1 Basic Concepts and General Experimental Design

In cryo-EM of viruses (and other types of biological macromolecular assemblies), the specimen is not dehydrated; instead, it is included in a thin film of water (~0.2 µm or less) that is vitrified [45]. An aliquot (~3 µL) of a virus suspension is placed on a grid coated with a holey carbon support film. Alternative grids and support materials are available, such as all gold grids for reducing beam-induced motion of the sample [46], graphene substrates used as support films [47], and affinity grids, which use a grid substrate (e.g., bound antibodies [48]) to purify sample particles or immobilize them. Excess sample solution is blotted away with filter paper, leaving a very thin film (up to several thousand Å) of virus suspension in the holes, which is rapidly plunged into a cryogen (such as ethane slush) at liquid nitrogen temperature. Freezing is so rapid that water molecules are immobilized in an amorphous state, so-called vitreous or glasslike ice, preventing structural alteration in the specimen caused by ice crystal formation [38] (Fig. 3.3e). This method provides a natural water-like environment for the virus particle, with no staining or fixing artifacts, and the native structure is preserved with structural integrity near-atomic resolution. Ideally, vitreous ice will be uniform across the entire grid, only slightly thicker than the assembly or molecule of interest and will contain well-distributed particles at adequate concentration. To improve reproducibility, this is done using automated plunge-freezer devices, which are commercially available. Recent studies have indicated that macromolecules under study may exhibit a preference for interacting with the air-water interface or can undergo (full or partial) denaturation when exposed to it. The conventional vitrification process, which involves forming a thin solvent film through blotting paper followed by immersion in cryogenic fluids, typically takes a few seconds. During this time, there is a risk of the sample coming into contact with the air-water interface [49, 50]. Various methods are currently under development to enhance the quality and repeatability cryo-EM samples. These strategies include (i) direct sample deposition technique onto self-wicking grids, which employs an inkjet piezo dispenser; (ii) employing microcapillarity for sample deposition; (iii) the application of an ultrasonic humidifier to spray the sample onto the grid; (iv) pin printing followed by rapid vitrification through the use of a cryogenic stream; and (v) time-resolved cryo-EM through microfluidic spraying. These various approaches are contributing to the ongoing advancements in cryo-EM sample preparation, ultimately improving the quality and reliability of data acquisition in this field.

To avoid devitrification of the sample, the grid bearing the virus sample is maintained at liquid nitrogen temperatures, while it is manually introduced in the side-entry stage of the microscope with a specimen cryoholder, and for several hours for the recording of micrographs (or movies). Alternatively, the grids can be stored indefinitely in liquid nitrogen. Heat transfer between the cryoholder and the microscope can be a serious problem and cause the specimen to vibrate (a result of liquid nitrogen boiling in the cryoholder dewar). Modern cryogenic electron microscopes are equipped with a high-throughput sample loader that allows for robotic loading up to 12 grids at once. Mechanical drift can be another obstacle when the cryoholder is not stably inserted in the column.

At −180°C, vitrified ice does not sublime at a significant rate in the high-vacuum conditions of the electron microscope column. Transfer steps must be very rapid to minimize contamination by deposition of hoarfrost (from water vapor), and the cryoholder and the interior of the microscope column are equipped with a cryoshielding, anti-contaminator device, termed cryobox. Holey carbon film grids are used as is or can be glow-discharged, washed with acetone vapor to enhance hydrophilicity, or replaced by continuous non-perforated carbon film grids. Holey gold support foil on gold grids provides another alternative.

The concentration of the virus suspension is critical for obtaining a thin film with even specimen distribution after blotting; negative staining analysis is used to determine an appropriate range of particle concentration. When sample concentration is limited, a continuous carbon support foil is used rather than the holey film. Another limiting factor is that of the solutes used during virus purification by ultracentrifugation in gradients such as glycerol, sucrose, and cesium chloride. These chemicals bubble when exposed to the electron beam and are therefore incompatible with high-resolution imaging [35]. Therefore, these compounds must be removed by floating the grid on a drop of distilled water (routine for preparing negatively stained samples) or dialysis of the selected virus fractions.

Sample stability and homogeneity should be checked by conventional biochemical and biophysical analyses prior to cryo-EM analysis [51]. Size exclusion chromatography, which separates specimen subpopulations based on the hydrodynamic radius of their particles, serves as a suitable method for preparing homogeneous samples [52, 53]. In addition to contaminant complexes, specimen heterogeneity can result from a sample in distinct compositional or conformational states. The GraFix technique provides a means to "purify" the sample [54]. In this approach, a chemical cross-linker, such as glutaraldehyde, is used to stabilize complexes, preventing dissociation and aggregations.

3.3.2 Magnification, Calibration, and Minimal Electron Dose

The definition of absolute magnification is necessary, and real magnifications should be calibrated periodically to avoid uncertainty in image processing. The hysteresis of magnetic lenses and changes in specimen height in the objective lens cause small changes in nominal magnification that must be considered in high-resolution studies. Comparisons of the three-dimensional reconstruction (3DR) of a virus whose dimensions have been calculated by X-ray analysis is a precise approach to determining the absolute scale. Apoferritin, a 24-subunit protein complex, and β-galactosidase, a tetramer, are popular standard samples and benchmarks for use as internal magnification standards and quality control samples. The choice of working magnification depends on the size of the sample being analyzed, the aimed resolution and the physical size of the detector. In automated acquisition using DED, rather than the nominal magnification, the sampling rate or pixel scale expressed as Å/pixel is the primary descriptor. Most high-resolution cryo-EM studies typically correspond to a pixel scale below 1 Å/pixel.

The major factor limiting the resolution of the cryo-EM structure is radiation damage induced by the energy transfer from electron beams onto the sample. Vitrified specimens are highly sensitive to electron irradiation, and the act of imaging with electrons could eventually destroy the specimen. In the days of manual data collection, after insertion of the specimen stage into the microscope, the cryogrid was searched at low magnification and irradiation levels. An experienced cryomicroscopist was able to locate optimal areas by visual inspection; dark areas in the grid are too thick for electron penetration, while bright areas are probably too thin for correct virus particle embedding or can even be dried. Most viruses can be seen directly at low magnification as dark spots, allowing assessment of particle distribution and concentration. These particles are destroyed at high magnification but can be observed directly for a few seconds as a "bubbling" effect. These sacrificed specimen areas are a convenient form of direct evidence of vitrified specimen quality, assuming that intact adjacent areas in the grid have similar particle concentrations and ice thickness.

Nowadays, the strategy for image acquisition is based in this "low dose" approach and is fully automated in modern cryo-microscopes equipped with DED. Promising specimen areas are searched at low magnification and their positions in the grid are selected and stored: next, the system switches among two states, focusing at high magnification and recording at the same high magnification. All of the electrons that pass through the specimen are used to form its image. The image is focused in an adjacent sacrificed area by observing the Fresnel fringes that appear at the edge of radiation-induced bubbles (that first form within the specimen). In this step, astigmatism is corrected and other adjustments are made (Fig. 3.4). A number of automated data collection software packages are available for collecting large data sets for reliable statistical analysis.

While the use of low electron doses is essential to prevent radiation damage, it often results in the production of noisy, low-contrast images characterized by a notably poor signal-to-noise ratio (SNR). This diminished SNR can involve significant challenges in subsequent image-processing tasks, such as accurate particle alignment and the achievement of atomic-resolution reconstruction. To overcome this limitation, a common approach is to combine numerous distinct views of the same macromolecular complex through averaging, thereby improving the SNR and

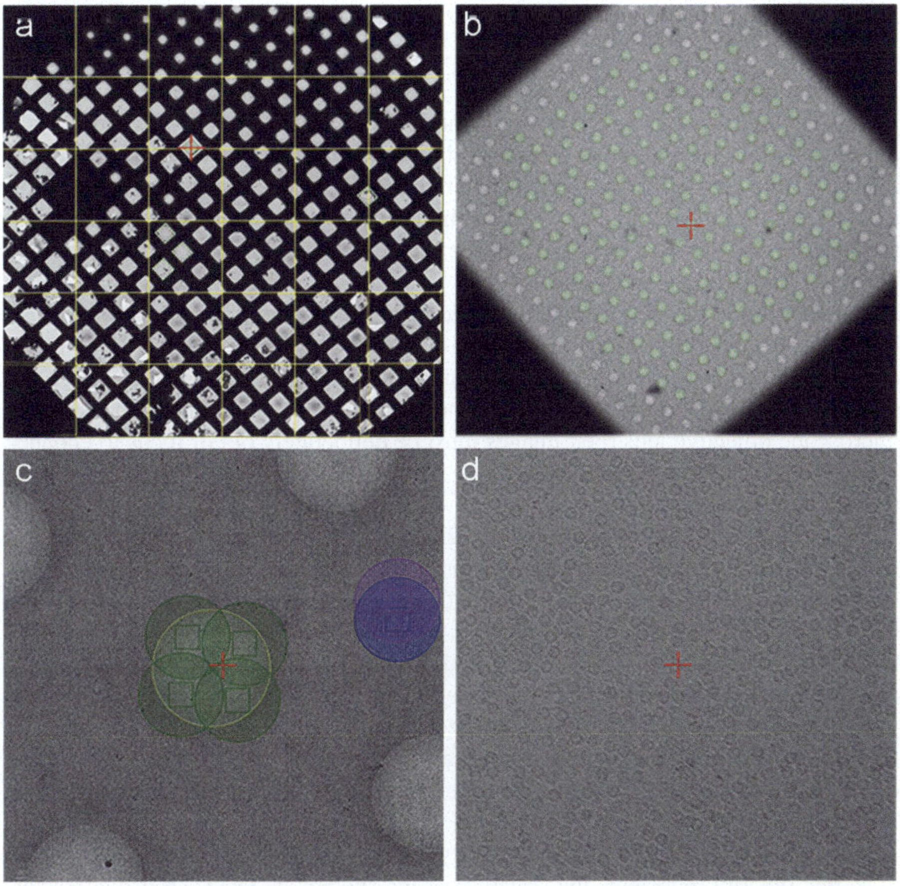

Fig. 3.4 Scheme of automatic cryo-EM data collection. (**a**) A grid atlas recorded by automatic acquisition and montage of overlapping tiles at low magnification. Grid squares with ice thickness appropriate for collection are selected as potential targets. (**b**) Grid squares checked, and automated hole finder selects appropriate holes. Holes too close to the grid bars, cracked, contaminated, or empty are excluded. (**c**) A hole image used to center it and define an acquisition template that will be repeated every hole: acquisition areas in the hole (green) plus focus (violet) and drift (purple) areas in the carbon between holes. Colored circles and squares show the beam and detector size, respectively. (**d**) A high-magnification movie is acquired at the programmed position and focus

enabling the calculation of a more reliable 3D reconstruction. Additionally, energy filters enhance the contrast of a cryo-EM image by selectively reducing inelastically scattered electrons through energy loss selection, thereby improving the visibility of specimen details and reducing background noise, which results in a clearer and more detailed representation of the sample.

3.3.3 Imaging Conditions: Contrast Transfer Function

To record high-resolution images, it is critical to work on a perfectly aligned microscope. All major components of the microscope should be aligned, with a top-down approach. Particular emphasis should be placed to the eucentricity adjustment (i.e., the area of interest stays centered as the microscope tilts around the tilt axis), correction of the objective lens astigmatism (by fine-tuning the power spectrum, or the Fourier transform [FT], of the image), and minimization of the residual beam tilt (coma-free alignment).

Another drawback in cryo-EM is related to imaging conditions. Images are noisy because there is no staining agent that enhances the scattering. The amplitude contrast of unstained biological specimens is thus quite low (~10% or less), and phase contrast is the dominant contrast mechanism, which requires defocusing the microscope objective lens (i.e., images must be deliberately out of focus to enhance contrast). Extensive computer processing is then needed to correct the images and recover structural details.

In a perfect electron microscope, the 2D image recorded on film or in a movie would be a realistic representation of the projection of the 3D scattering matter in the specimen. As in any other optical system, the image of an object is modulated by the lens characteristics such that each point in the image is the result of a convolution of the corresponding point of the object and a mathematical function, known as the contrast transfer function (CTF). Image defocus correction is necessary for accurate interpretation of specimen structure. In the case of the electron microscope, the CTF depends on the spherical aberration coefficient (Cs) of the objective lens, the effects of defocusing and inelastic electron scatter, and the partial coherence of the beam. This fundamental concept implies that all TEM images are altered by the CTF and must be corrected (deconvoluted) to retrieve the high-resolution signal and obtain a 3DR that reliably represents the structure of the specimen. The CTF is a characteristic function of each microscope (i.e., the Cs of the objective lens) and the imaging settings (defocus level of the objective lens, beam coherence, and accelerating voltage).

During data acquisition, images are recorded at a range of defocus settings (typically between −0.5 μm to −3 μm). By choosing different defocus settings (focal series are usually taken), specific structural features (or frequencies in the Fourier space) of the image can be accentuated selectively at the expense of others. In appropriate imaging conditions, a contrast-enhancing effect can thus be obtained. The CTF have a complex, oscillatory shape: CTF-modulated information implies the existence of frequency ranges in which there are no data (zeros in the CTF), regions in which the information has reversed contrast (between the first and the second zero, between the third and fourth zero, and so on), and a fall-off of useful information at higher frequencies (the envelope function), dependent mainly on beam coherence. Whereas the first zero of the CTF determines the image defocus values, the last zero defines the information limit of the image. In the same way that observing a sample by different conventional EM techniques provides complementary data, recording

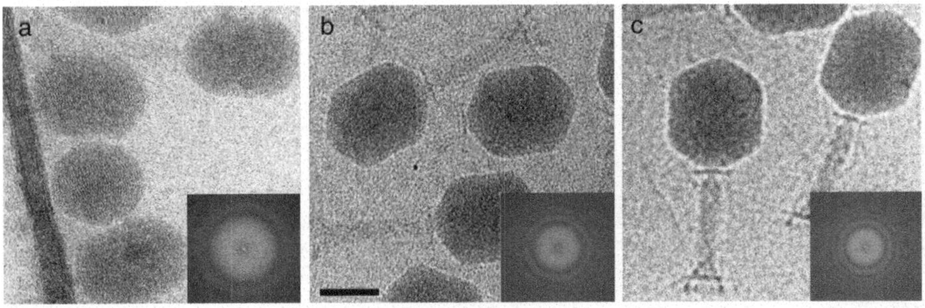

Fig. 3.5 Unstained, frozen, hydrated specimen of phage T4 in vitreous ice, prepared on holey carbon film. Focus series of phage T4 under imaging conditions that enhance dominant spatial frequencies corresponding to (**a**) interstrand spacings between double-stranded DNA filaments (~26 Å) within the head in a relatively close-to-focus image, (**b**) tail sheath spacing (40.5 Å) in a regular defocused image, and (**c**) general morphology in a highly defocused image. Note that the detail of the tail is almost lost in the close-to-focus image. Insets show FT (Fourier transform) for each cryomicrograph; the further from focus, the closer the Thon ring is to the FT center. Bar, 500 Å

several images at different focal settings to enhance distinct specimen features is a "must-do" (Fig. 3.5).

Quality of Cryo-EM micrographs or movies can be analyzed by inspection of the CTF zeros or Thon rings in the image computed diffraction patterns (the sum of the Fourier transform amplitudes is termed the average power spectrum). Precise determination of the Thon ring pattern (CTF zeros or points of minimum contrast) in the power spectrum allows reversing the phases in the appropriate regions of the FT. Multiple images with complementary defocus must be used to fill in the missing data near the zeros. This method is reliable when FEG of high-voltage (200–300 kV) data are used. An image free of aberrations has the signal in the power spectrum as a series of concentric circular rings (zeros or Thon rings) that extends outward from the center of the Fourier transform; the greater the extent of rings, the higher the resolution in the image. Images with drift or astigmatism are easily detected and excluded from further structural analysis (Fig. 3.6). Large symmetrical assemblies can be imaged at small defocus levels with good image contrast and preserved high-resolution signals. For small (100–300 kDa) and asymmetric assemblies, high-defocus settings are needed, which limit the resolution that can be achieved. As good contrast is required for the cryo-micrograph, a compromise should be reached among the factors that modulate contrast. The use of a small objective aperture increases image contrast but might introduce astigmatism, and lower acceleration voltages increase image contrast but cause more radiation damage to the specimen. Higher defocus means greater contrast, but also more rapid dampening of the envelope function and many zero transitions of the CTF; CTF of the envelope function is reduced using a brighter beam, such as that generated by FEG. Image contrast can also be improved by additional instrumentation. Energy filters remove inelastically scattered electrons, responsible for much of the noise in micrographs, thus improving the signal-to-noise ratio of the images.

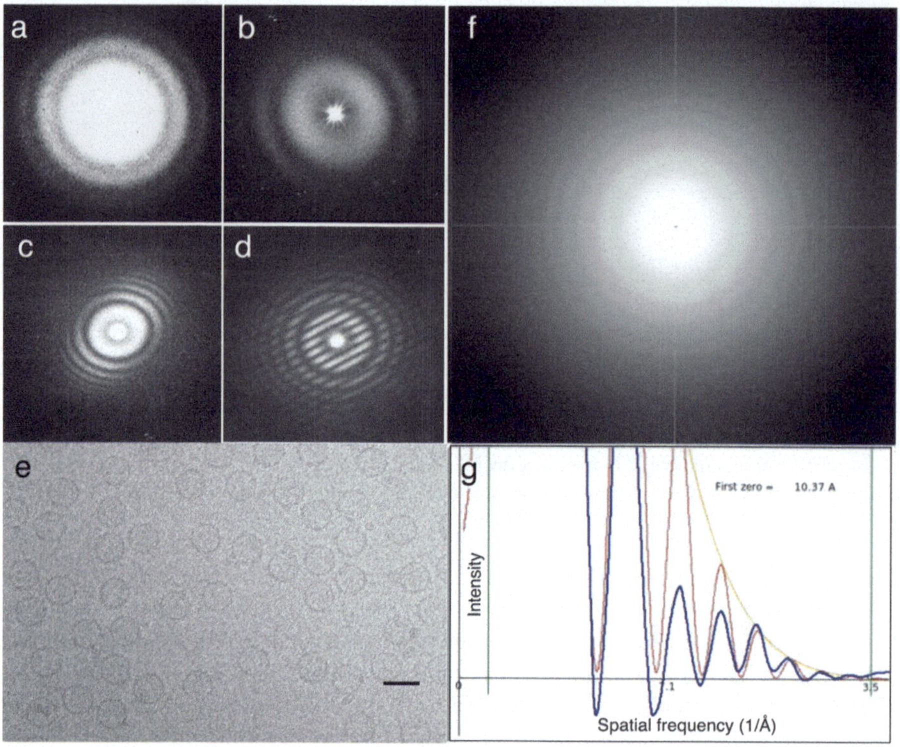

Fig. 3.6 Power spectrum (equivalent to diffraction pattern) analysis of cryomicrographs with distinct aberrations. (**a**) An aberration-free image shows circular concentric Thon rings (zeros). The exact underfocus can be calculated from the first zero position. Signal near the center of the FT corresponds to lower-spatial frequencies (low-resolution details); signal extending outward from the center corresponds to high-spatial frequencies. The last Thon ring indicates the maximum detectable frequency (i.e., resolution) that could be rescued from the 2D projection image. (**b**) Astigmatic migrographs are detected because Thon rings appear elliptical. This aberration can originate from the illumination source and is corrected by the objective lens (an objective aperture that is not well centered can also cause astigmatism). (**c**) Sample drift occurs when resolution is lost preferentially in one direction of the power spectrum. Drift can be thermal (temperature instability in the refrigeration system of the objective lens) or mechanical (stage movement during exposure). Vibrations (acoustic or physical due to liquid nitrogen bubbling in the cryo-stage dewar) limit resolution in all directions. (**d**) Double exposure to determine drift speed. Taking two exposures on the same micrograph, spaced 15–30 seconds apart, produces a pattern with these typical bands; narrow bands mean larger drift movements and vice versa. (**e**) An excellent image of the mycovirus *Rosellinia necatrix* quadrivirus 1 (RnQV1) taken in a Titan Krios 300 kV microscope, in which contrast is very low. Bar, 500 Å. There is no doubt as to the quality of this micrograph, since (**f**) the average FT calculated from many boxed particle images shows Thon rings to ~3.5 Å resolution. (**g**) Average power spectrum curve (blue line) is used to fit theoretical contrast transfer function (CTF) curves (red line). Underfocus values of 0.7 μm; first zero of the contrast transfer function at 10.37 Å

3.3.4 Direct Electron Detectors

The traditional means of recording electron microscope images relied on photographic film. However, using film was labor-intensive and severely limited data collection throughput. After developing, the film must be digitized using a high-precision film scanner to convert it to a numerical array that represent the darkness levels on the film, making it suitable for computer processing. For many EM applications (as discussed in Sects. 3.2.2 and 3.2.3), CCD cameras have become the preferred choice and digital images are available directly. CCD cameras offer other advantages, such as easy handling, high signal linearity in signal response as a function of the number of incoming electrons, and a broad dynamic range. Nonetheless, CCD detectors were worse than film in terms of sensitivity and resolution due to their coupling with an electron scintillator screen. DED were key in improving achievable resolution, as they surpassed film in both resolution and sensitivity. Additionally, the DED readout rate is much faster than that of CCD detectors [55].

DED can directly observe the transmission of a single electron through the sample to the detector as a discrete "electron event" rather than using a scintillator or a photographic film to record images. DED are based on complementary metal oxide semiconductor (CMOS) sensors to convert electrons directly into an electrical current. This conversion is accomplished with a high detective quantum efficiency (DQE), a measure of detection efficiency across spatial frequencies [55, 56], thereby increasing the SNR. DED offer two operational modes: integration and counting. In integration mode, the signal is integrated over the entire exposure time or is dose-fractionated into multiple movie frames. In counting mode, individual electron scattering events are detected. Although integrative mode was faster and counting optimized the acquisition quality, more recent detectors are able to record high-quality counting acquisitions at high speed (up to 900 exposures per hour). Some direct detectors can use so-called "super-resolution" acquisition, where advanced algorithms estimate the precise entry point of each electron into the detector with sub-pixel accuracy.

DED has a rapid readout rate and images are collected as movies composed of multiple frames. This approach allows for dose fractionation into multiple frames per second [57, 58], where each frame receives an extremely low electron dose, typically around 0.5–1 electron per pixel. More importantly, the beam-induced movement of the specimen and support foil (and/or the mechanical drift of the stage) that results in motion-induced image blurring can be tracked and corrected by computational alignment of a series of low-dose frames taken in one area before averaging [13, 59] (Fig. 3.7). Moreover, subframes with optimized doses can be selectively chosen for subsequent image processing [60], as later frames would be of lower quality due to radiation damage. Whereas damage is observable easily in burned areas, high-resolution information in the image might be lost even before visible signs of damage occur (for example, when an area is examined for too long). The combination of dose fractionation and motion correction greatly enhances data acquisition efficiency ensuring that nearly all images are of a quality suitable for

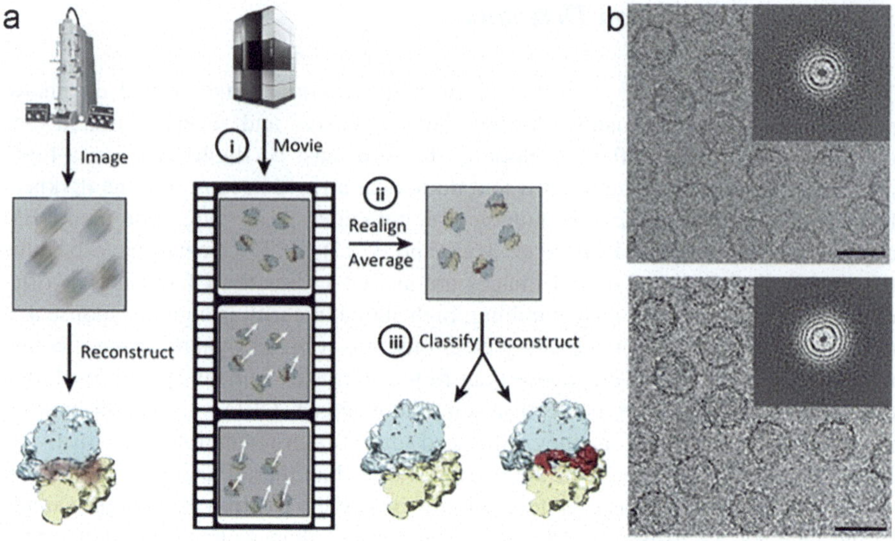

Fig. 3.7 Correction of beam-induced motion and recovery of high-resolution information. (**a**) Photographic films record noisy images in which beam-induced sample motion led to image blurring and resulting 3D reconstructions are limited in resolution (left). DED records movies, and the movie frames can be realigned to correct movements induced by the electron beam; resulting 3D reconstructions are at near- and atomic-resolution (right). Reproduced with permission from Ref. [59]. (**b**) Two-dimensional movie average after translational/rotational alignment. Cryo-EM images correspond to vitrified empty capsids of the mycovirus RnQV1 (~450-Å-diameter virus). Bars, 500 Å. Insets show power spectra calculated from the averaged movie before (top) and after (bottom) motion correction. Motion correction restores Thon rings from ~6–7 (top) til 3–4 Å (bottom) resolution

recovering high-resolution information [61]. DED provides instant feedback for image quality, as FT is calculated dynamically during image acquisition. To further mitigate beam-induced motion, the use of holey gold support foil has proven effective [45]. This support foil helps reduce specimen movement during imaging, enhancing the overall quality of data acquisition in cryo-electron microscopy [46].

3.4 Cryo-EM Image Processing and Three-Dimensional Reconstruction

Electron microscopes have a large depth of focus compared to virus size. This means that all characteristics at different heights in the virus are in focus simultaneously, and each image recorded in cryo-EM conditions is a 2D projection of 3D scattering density in the specimen (i.e., all structure is superimposed onto a single plane). The superposition of all structural information from different levels results in a garbled image. In a single projected view, a specific feature can be derived from any level in the specimen. To reconstruct the native 3D structure, many images must be recorded to obtain different views of the object; viruses tend to be randomly oriented in the vitreous specimen layer [62]. Alternative views can be obtained by tilting the specimen in the microscope. A 10° tilt is usually sufficient to give an adequate sampling of orientations. This approach is necessary, for example, to identify the absolute handedness of icosahedral viruses using cryo-EM (see Chap. 2).

In 1968, DeRosier and Klug found that a 3DR could be recovered from micrographs with 2D images [63], using the mathematical device termed the FT. Sir Aaron Klug received the Nobel Prize in chemistry in 1982. The FT separates different components of the image into all 2D spatial frequency components (or constituent waves). The FT of each image corresponds to a central plane of the 3D FT of the object density; this is the central section theorem. Distinct projected views fill in different central sections of the transform; the larger the number of views included in the 3D FT, the better the sampling and the 3D object density recovered in Fourier synthesis. It is essential to ensure that sufficient particle orientations have been provided for adequate sampling of Fourier space since interpolation errors can limit the resolution. In addition to this approach, FT are used for general quality control (evaluating the CTF of each micrograph), correcting the defocus effects (in high-resolution studies), and determining particle orientation.

The FT of an image (termed reciprocal or frequency space) is an equivalent, reversible representation of the image (real or Cartesian space). FT provides a mathematical description of the diffraction pattern when imaging an object with a lens (optical or electromagnetic). In the diffraction pattern (the object's FT) produced in the back focal plane of the lens, waves are recombined (positive and negative interference) to form an image. This concept of FT to give a diffraction pattern, followed by Fourier synthesis to produce an image, is central to image processing. In the real-space representation, the parameters that define an image are Cartesian coordinates (x, y, z) and an associated density value. In the FT representation, these parameters are reciprocal coordinates, an associated amplitude (strength) and an associated phase (which specifies position relative to origin of the wave representing spatial frequency). FT also allows filtering according to spatial frequency. The highest spatial frequencies usually have a large fraction of noise; by eliminating these frequencies in Fourier space and using the inverse FT to convert back to real space, the resulting image is less noisy [64]. The same principles used for 3D image reconstruction are applied for structure determination in X-ray crystallography (Chap. 4).

3.4.1 Determination and Refinement of Particle Orientation and Origin

After the selection of many thousands of single-particle images, the center of symmetry of each image is established using cross-correlation with a reference image or with the same image rotated 180°. To calculate the 3DR of the object, it must then be determined whether a projection is the front, side or top view, or somewhere in between. Particle orientation determination then begins, that is, the viewing direction relative to the icosahedral symmetry axes. Particle orientations are specified relative to the smallest repeating unit of the icosahedron, termed the asymmetric unit.

The first step of this process requires an initial 3D model as a reference. Many icosahedral virus density maps are available from the Electron Microscopy Data Bank (EMDB: http://www.ebi.ac.uk/pdbe/emdb). The map must be scaled to match the dataset under study and then low-pass filtered (~35–40 Å resolution) to reduce model bias. A preliminary, low-resolution 3D model can be generated ab initio from a set of averaged 2D images with a high signal-to-noise ratio. Also, the initial 3D model might be calculated from a low-resolution negative stained EM structure. The implementation of stochastic gradient descent algorithms for ab initio model generation in modern software packages has enormously simplified this initial and critical step. Due to the inherent difficulties of processing low-contrast close-to-focus images, it is often useful (and mandatory) to collect focal series with a range of defocus values to facilitate orientation determination. The next step when calculating a 3D virus map is the refinement (improvement) of the orientation and center parameters. This analysis involves comparing the particle images with 2D reprojections from a 3D model; projection matching and common lines are the method most commonly used to determine these parameters [65].

In the projection matching method, the model is projected computationally in all possible directions to generate a set of reference images that are used to match each image with one projection corresponding to the appropriate orientation. The initial low-resolution model is updated at the end of each refinement cycle during this iterative process until no further improvement is obtained. Projection matching can be carried out in real or Fourier space.

In the, almost non-used, common line method, the central section theorem provides the mathematical foundation for orientation determination of projection images from the same object. The FT of two projection images with different orientations are planes in the 3D FT of the object (a virus) that intersect one another. This intersection line is termed common line and has the same values in each plane for amplitudes and phases. For an icosahedral virus, a single image with a defined orientation in the asymmetric unit is associated with 59 symmetry-related orientations. Each of the 60 equivalent orientations defines the corresponding planes in the 3D FT. The intersection line between any of these planes is known as a self-common line, and there is a maximum of 37 pairs of self-common lines in the FT of a 2D projection image. To estimate individual particle orientation parameters, an exhaustive search is made for all possible orientations within the asymmetric unit. Cross-common lines arise from the application of icosahedral symmetry between a particle and another particle (with different orientations); there are 60 pairs of cross-common lines between any two images. The orientation angles are initially estimated by assuming an orientation and comparing the agreement of the phases between the common line pairs.

3.4.2 Resolution, Model Building, and Validation

The resolution of the 3D map depends on the total number of particles, distribution of views within the asymmetric unit, accuracy of center and orientation parameters, the quality of the data, and many other factors such as the CTF correction. The resolution in 3DR is less objectively determined than in X-ray crystallography, for which resolution has come to mean the highest spatial frequency contributing meaningfully to the map [66].

Assessment of the resolution is performed statistically, by splitting the particle set in half and computing two independent maps from each half-set. The two maps (often referred to as even and odd maps) are then correlated as a function of spatial frequency to determine the extent to which the fine structural details have been determined reliably (termed FSC method, Fourier Shell Correlation). A typical FSC curve exhibits a value close to +1 at low frequencies, indicating that low-resolution details are highly correlated, and decays to zero at higher frequencies (structures differ in high-resolution details). The frequency at which the FSC curve drops below a threshold of 0.143 determines the final resolution of the structure. The "gold standard" FSC method requires that the FSC curve is calculated from two entirely independent maps (i.e., the dataset is randomly split into two independent halves that are separately refined and reconstructed) [67, 68]. The resolution of a cryo-EM model can be non-isotropic in different regions of the map [69]. Typically, the highest resolution is achieved in well-ordered portions of the macromolecule, while some parts of the map may exhibit significantly lower resolution due to the intrinsic flexibility of the molecule [69]. Different computer programs are available to assess local variations in the resolution within the map.

Amplitude scaling is used to restore high-resolution features, using X-ray solution scattering profiles as a reference, or by determining an effective B-factor (see below). This process of map sharpening scales specific spatial frequencies (resolutions) above others. Modern deep learning techniques provide a robust solution for amplitude enhancement cryo-EM maps, effectively mitigating anisotropy in resolution and enabling a more accurate representation of complex biological structures.

Although image variability is a limiting resolution factor and might result from incomplete, truncated or damaged assemblies, it can also indicate multiple conformations. Conformational or functional states of a macromolecular complex are difficult to classify, as variability must be differentiated from viewing geometry. If these (normally evasive) conformations are assigned to a time-ordered sequence, the resulting density maps can be used to make elegant time-resolved movies of dynamic processes. As noted above, cryo-EM is the best approach for trapping and visualizing transient structures with a lifetime of a millisecond or longer, as the specimen is rapidly vitrified. Buffer conditions can also be adjusted to extend the lifetime of a transient state; this approach has been used in virus maturation analysis for systems such as herpes simplex virus and several dsDNA bacteriophages (reviewed in [70], and references therein).

Once a 2–4 Å cryo-EM map has been obtained, the subsequent step is to build and refine a model. The known protein sequence is fitted into the 3D map by positioning the bulky amino acid side chains first. This initial fitting allows for de novo structure determination. Alternatively, an existing PDB model can be adapted and modified using the cryo-EM density map as a constraint. Advanced methods like AlphaFold [71] and RoseTTAFold [72] have emerged for predicting protein structures from primary sequence information. These preliminary models can serve as a starting point for building into cryo-EM data, offering valuable insights and guidance in the structural determination process. Furthermore, machine learning techniques, such as the ModelAngelo algorithm [73], integrate data from the cryo-EM map with information derived from protein sequence and structure within a unified graph neural network (see Chap. 7). This approach yields atomic protein models of comparable quality to those traditionally crafted by human experts. The 3D atomic model created is then subjected to further refinement. In some cases, homology models of atomic structures obtained by X-ray crystallography and NMR can be fitted into a low-resolution cryo-EM map, a strategy referred to as the hybrid approach. There are several model-building and validation tools, most of them originally developed for X-ray crystallography analysis. These tools include Phenix [74], Chimera [75], and Coot [76]. As an example, the cryo-EM-derived structural models of a spherical virus, the 40 nm-diameter fungal virus *Penicillium chrysogenum* virus (PcV) are shown at different resolutions as well as their corresponding subnanometer and near-atomic models for the asymmetric unit [77–79] (Fig. 3.8).

3.4.3 Structure Determination by Cryo-EM SPA

The general workflow for SPA encompasses several key steps, including quality assessment of motion-corrected average images, particle picking, image defocus estimation and CTF correction, followed by 2D particle classification, particle orientation and refinement, image reconstruction and 3D refinement, resolution evaluation, and cryo-EM map validation [80] (Fig. 3.9). The 2D classification procedure results in the assignment of all particles to different 2D projections of the macromolecule. Particles within the same classes are then aligned and averaged to generate class averages. For handling nonsymmetric or flexible components within the analyzed complex, projection subtraction and local reconstruction are used routinely (below).

Numerous image processing packages are available to carry out these tasks. All this has become computationally feasible with improved central processing units (CPU) and, more recently, graphic processing units (GPU). Many of these software packages continuously integrate novel approaches with better automatic particle-picking algorithms, more precise and faster CTF determinations, and improvements in beam-induced motions correction, local-resolution estimation, 2D and 3D classification methods, as well as alignment and reconstruction methods. The continuous

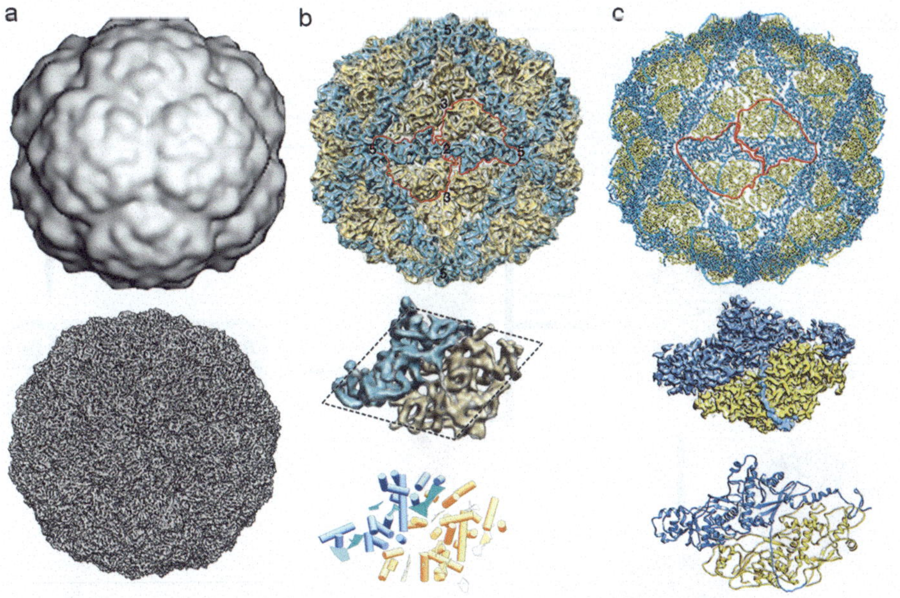

Fig. 3.8 Three-dimensional cryo-EM maps of the fungal virus *Penicillium chrysogenum* virus (PcV) at different resolutions. (**a**) Surface-shaded representation of the outer surface of the PcV T = 1 capsid (400 Å in diameter) at 26 [77] and 4.1 Å [79] resolution (top and bottom, respectively), viewed along a twofold axis of icosahedral symmetry. Cryo-EM is no longer a method of "blobology." (**b**) Structure of the PcV capsid at 8 Å resolution, viewed as in (**a**), with capsid protein elements in cyan and yellow; boundaries for two asymmetric units are outlined in red (top) [78]. Segmented asymmetric unit of PcV capsid formed by two ellipsoid-like structures with roughly similar morphology; the dashed line highlights the rhomboidal shape (middle). PcV capsid protein secondary structural elements (cylinders, α-helices; planks, β-sheets) (bottom). This subnanometer resolution analysis indicates that the two PcV capsid protein halves have a comparable structural signature. (**c**) Atomic model of the PcV capsid calculated from a 4.1 Å resolution map, shown as in (**b**) (top). Cryo-EM density, segmented asymmetric unit of PcV capsid (middle). Atomic model of a PcV capsid protein showing the N-terminal domain A (blue) and the C-terminal domain B (yellow)

development of these software tools offers multiple possibilities for combining distinct programs and their different parameters.

Single-particle cryo-EM provides the 3D structure of the specimens examined by computationally merging images of many (easily exceeding tens or hundreds of thousands) individual macromolecules in identical or highly similar conformations. Each particle image represents a randomly orientated 2D projection image that contains all the structural detail of the 3D specimen. Angular orientation parameters are determined by comparing the 2D projections with spatially defined reprojections of an initial 3D model filtered at low resolution. Subsequently, a new 3D map of the macromolecular complex is calculated from these 2D projections through a process known as "back-projection," which combines all the views to form a single

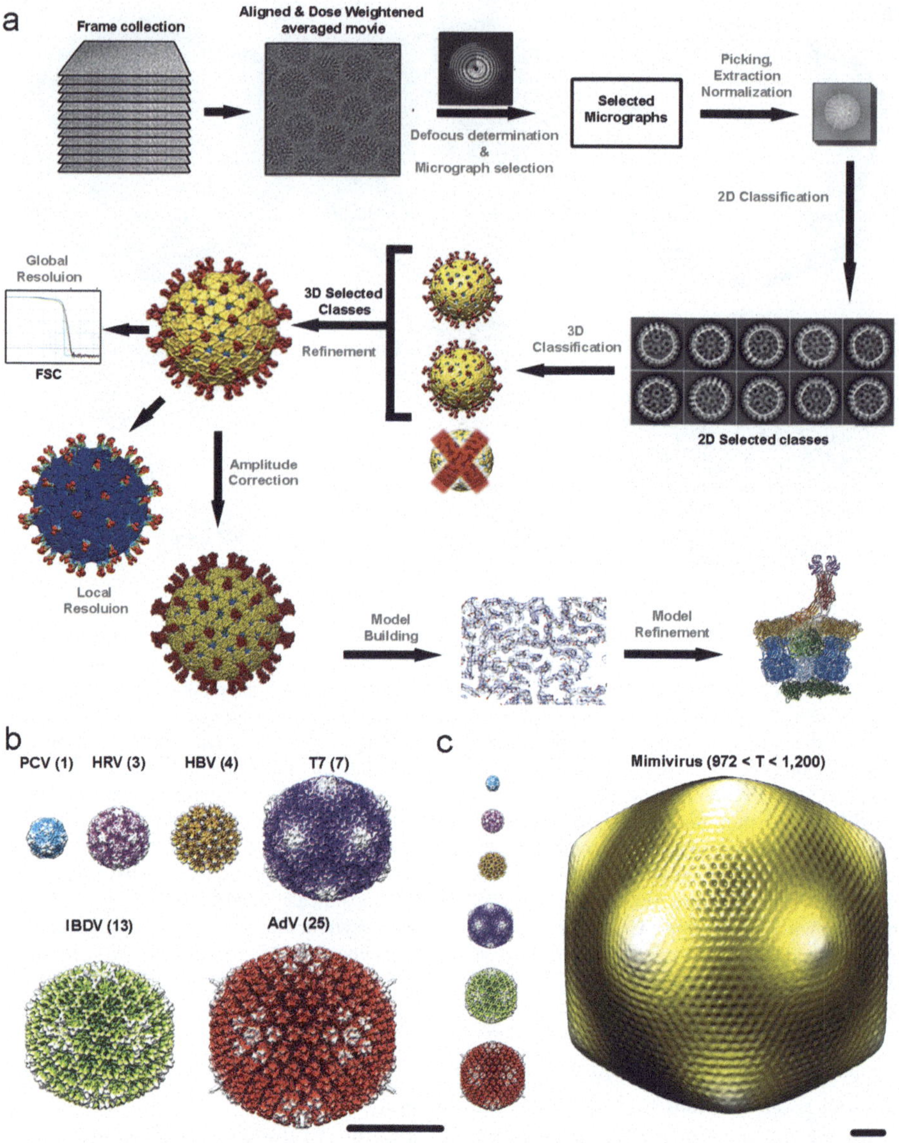

Fig. 3.9 Overview of the single-particle cryo-EM workflow, from data collection to 3D model and 3D maps of icosahedral viruses. (**a**) The vitrified virus sample is imaged by collecting movie frames that are aligned, dose weighted and averaged. Defocus is calculated in motion-corrected images, good quality micrographs are selected, and individual viral particles are picked and extracted. Next, particle images are subjected to 2D classification and averaging, to get class averages according to their similarity and clean up the data set of bad particles. Selected averaged classes can be used to obtain an ab initio 3D model. This low-resolution model is then used as a reference for 3D classification, a necessary step for identifying distinct conformations, macromolecular heterogeneities, or particle subsets with different structural integrity. Orientation refinement is iteratively done until the structure converges, as indicated in the resolution analysis by the FSC method. The final map is amplitude-corrected to reveal high-resolution details. Finally, the protein sequence is fitted

Fig. 3.9 (continued) into the 3D map to build a de novo 3D model of the protein which is refined to obtain a final model. (**b**) Cryo-EM-derived icosahedral capsids with different triangulation numbers (indicated in parenthesis): porcine circovirus (PCV, EMD-8969, [81]), human rhinovirus (HRV, EMD-2107, [82]), hepatitis B virus (HBV, EMD-2278, [83]), bacteriophage T7 (T7, EMD-1810, [84]), infectious bursal disease virus (IBDV, EMD-1238, [85]) and adenovirus (AdV, EMD-1579, [86]). (**c**) Same icosahedral capsids using the same color code as in (b), including Mimivirus (EMD-5039, [87]). Bars for b and c, 500 Å

3D map [88]. This projection-matching process is performed iteratively to yield new 3D reconstructions with improved resolution until no further improvement is possible. Finally, the known protein sequence is fitted into the 3D map, initially positioning the bulkier amino acid side chains, followed by further refinement of the 3D atomic model.

Owing to the very low electron doses used to prevent radiation damage, biological specimens yield images with low contrast and significant noise. Resolution can also be influenced by the presence of conformational and/or compositional heterogeneity within the sample, in which multiple states of a macromolecular assembly coexist. Additional methods such as 3D classification of particle conformations [89–91], nonhomogeneous refinement, multibody refinement, refinement of subparticle orientations, per-particle focus and motion refinement, and correction of Ewald sphere effects for large viruses have recently been applied to overcome limitations in the conventional data processing pipeline [92, 93]. These methods reconstruct 3D structures by combining many images of a homogeneous subset of particles that have been aligned with each other at high precision. One of the promising developments is the ability to capture short-lived states (ranging from 10 to 1000 milliseconds) through time-resolved cryo-EM [21, 94], opening new avenues for studying dynamic processes in structural biology.

3.5 Near-Atomic Resolution of Virus Structures by Cryo-EM

Near-atomic resolution of virus structures that address important biological questions has been reported for a number of viruses (reviewed in [66, 95, 96], and references therein). Cytoplasmic polyhedrosis virus (CPV) was the first example (at 3.88 Å resolution) of a complete de novo chain trace from cryo-EM maps. Due to the ordered nature, structural integrity, and homogeneity of many viral assemblies, cryo-EM is used extensively to solve their near-atomic (2–4 Å) and atomic resolution structures (Fig. 3.10). This technique is extensively utilized to elucidate structures of infectious virions of a broad size range, with helical or icosahedral symmetry, tailed phages, viruslike particles (VLP), hybrid VLP (with heterologous cargos), and viral complexes representing intermediate functional states during assembly [99]. Even isolated proteins, such as viral polymerases purified from

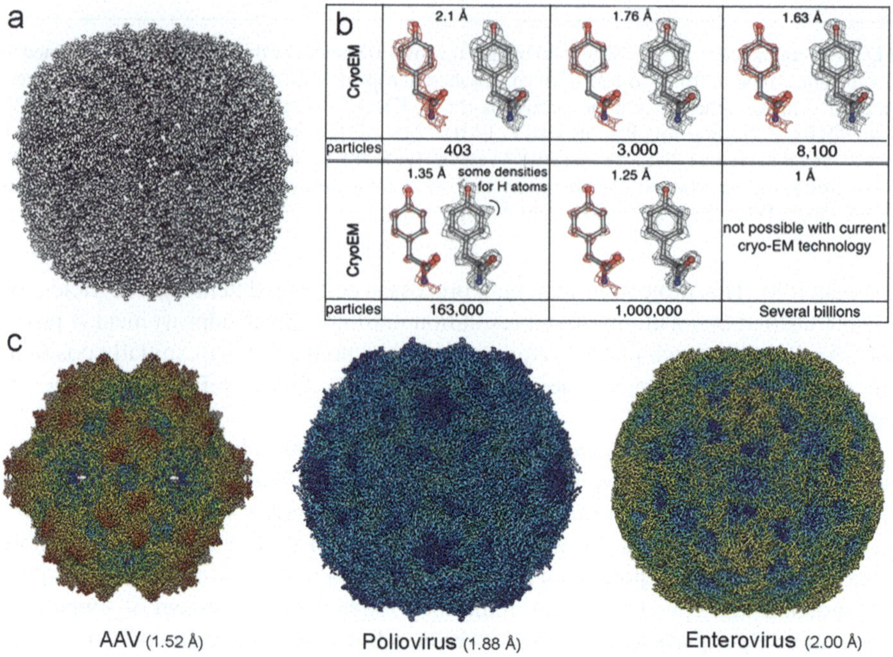

Fig. 3.10 Atomic resolution cryo-EM structures. (**a**) 1.25 Å-resolution structure of apoferritin obtained by cryo-EM (EMD-11668, [23]). (**b**) Tyrosine 32 of human apoferritin shown at high (red) and low (grey) thresholds to better judge the structural details and the map quality at different resolutions. Individual atoms can be distinguished starting at 1.35 Å resolution or better (adapted from [23] with permission). (**c**) Near-atomic structures by cryo-EM of adeno-associated virus AAV (EMD-22854, [97]), Poliovirus (EMD-15727, [98]), and Enterovirus (EMD-25765)

viruses, or non-structural proteins, are analyzed by cryo-EM (termed cryo-EM ex vivo or *ex virio*), although these structural targets require much more expertise for cryo-EM sample preparation and optimization.

Symmetry-based averaging is a common approach in cryo-EM, but it can lead to the loss of unique yet ubiquitous asymmetric features. Recent innovations have expanded the scope of cryo-EM analysis to encompass the asymmetric organization of the packaged genome and minor structural proteins, such as viral polymerases, which were previously missed (cryo-EM in situ) [100, 101]. Asymmetric structures play pivotal roles in various steps of the virus replication cycle, and many of these might be key targets for the development of new antiviral drugs [102, 103]. Some recent studies showing asymmetric structural features in viruses and the reconstruction methods used to identify these novel or unexpected components associated with essential functions. Additionally, challenging viral systems, such as giant (or jumbo) viruses, have been investigated using cryo-EM and cryo-electron tomography (cryo-ET). These techniques have played a crucial role in the battle against the COVID-19 pandemic. Moreover, the latest structural insights obtained through cryo-ET shed light on complex and pleomorphic viruses, unveiling the mechanisms underlying their entry, replication, assembly, and budding, i.e., dynamic processes that occur within the cellular context.

3.6 Reconstructing Viruses Without Imposing Symmetry

Many icosahedral symmetry-based virions present features/structures that do not obey icosahedral symmetry. Examples include the packaging complex, the tail, or the core in the capsid interior of dsDNA tailed bacteriophages [104]. These non-icosahedral components can have their own symmetry, which is obscured by the symmetry mismatch with the icosahedral capsid shell (see Chap. 12). A map without icosahedral symmetry requires at least 60 times as much data to achieve a similar resolution. Elaborate approaches during the data processing step have enabled structural resolution of complex viruses as a single entity without imposing symmetry, using projection-matching techniques to orient the particles. Probing asymmetry in icosahedral and helical viruses can be challenging when the asymmetric feature is small or has a weak signal. Tailed bacteriophages are an exception, which have a large assembly that allows correct alignment of the asymmetric assembly [105, 106]. Various image processing-based strategies have been employed to study nonsymmetric capsid features. These approaches include the following: (1) standard asymmetric refinement, in which no symmetry is applied during data processing (C1 symmetry), and orientation sampling is performed across the entire orientation space; (2) symmetry relaxation, in which the icosahedral orientations determined are relaxed and iterative sampling is tested for the 60 symmetry-related orientations; (3) symmetry expansion, where each of the 60 symmetry-related orientations is assigned to every particle before conducting focused (or masked) classification and refinement; and (4) subparticle classification/refinement, in which subregions or subvolumes are re-extracted and allowed to deviate from icosahedral symmetry [92, 93, 107].

The major components of T4 phage, head, tail, and tail fibers, are assembled by three independent pathways and then joined together to form the infectious virion [108] (see Chap. 12). The head, an elongated icosahedron ~1200 x 860 Å, is built with three capsid proteins: 930 copies of the mature major capsid protein gp23* that form the capsid hexamers, 55 copies of the vertex protein gp24* that form pentamers at eleven of the twelve vertices (see Chap. 2), and 12 copies of the portal protein gp20 that form the unique dodecameric portal vertex. The portal vertex is a ring structure with a central channel through which a dsDNA molecule ~100 MDa enters the capsid during packaging and exits during infection. The tail consists of a sheath, an internal tail tube and a baseplate, situated at the distal end of the tail. Two types of fibers attached to the baseplate, the long tail fibers and the short tail fibers, are responsible for host cell recognition and binding (Fig. 3.11a).

Analysis of the viral genome is central in understanding dynamic processes during the life cycle, such as assembly, infection, and genome replication. For

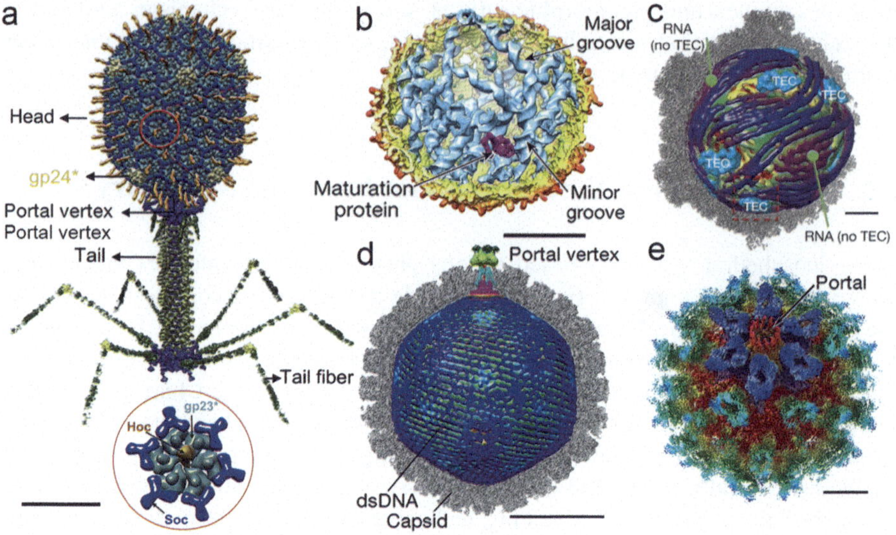

Fig. 3.11 Cryo-EM asymmetric reconstructions of prolate and icosahedral viruses for visualization of asymmetric components in situ. (**a**) Structural model of bacteriophage T4. The enlarged capsomer shows the major capsid protein gp23*, Soc, and Hoc (bottom). Yellow subunits at the five-fold vertices correspond to gp24*. The unique portal vertex connects the head to the tail. Reproduced with permission from Ref. [108]. (**b**) A cut-open view of bacteriophage MS2 to show a single copy of the maturation protein (magenta) and the ssRNA viral genome (blue) mostly organized as dsRNA (major and minor grooves of double-stranded regions are indicated). Reproduced with permission from Ref. [109]. (**c**) Cryo-EM asymmetric reconstruction of cypovirus, with the front half of the icosahedral capsid removed to show the asymmetric components, the dsRNA genome (dark blue), and the transcriptional enzyme complexes (TEC, cyan). Reproduced with permission from Ref. [110]. (**d**) Herpes simplex virus type 1 (HSV-1) with the front half of the protein shell removed to visualize the portal vertex elements (portal, magenta; portal cap, green) and ordered dsDNA coils (blue) spooled around a disordered dsDNA core. Reproduced with permission from Ref. [111]. (**e**) Radially colored full feline calicivirus (FCV) virion to highlight the receptor proteins (blue) and the VP2 portal-like assembly (red) at a unique threefold symmetry axis. Reproduced with permission from Ref. [112]. Scale bars 100 Å (**b**, **c**, and **e**) and 450 Å (**a** and **d**)

example, bacteriophage MS2 and Qβ capsids each incorporate a single copy of the maturation protein (relatively rigid), responsible for binding both viral single-stranded (ss) RNA and host receptors and the genome. Asymmetric reconstruction and refinement have revealed that ssRNA contains multiple stem loops that can act as packaging signals (PS). These PS bind cooperatively to a capsid protein recognition motif, and serve as allosteric switches, control assembly efficiency, and are responsible for in vivo packaging specificity [113, 114]. High-resolution asymmetric cryo-EM reconstructions, achieved through standard asymmetric refinements, traces 80% of the genome with 16 stem-loops for MS2, confirm its assembly model [109, 115, 116] (Fig. 3.11b).

Whereas dsDNA viruses in the extracellular life cycle remain in a quiescent state, dsRNA virus capsids in the host cytoplasm exhibit dynamic behavior, serving as nanocages for mRNA synthesis. These capsids use their genome segments as templates for transcriptional enzyme complexes (TECs), which include an RNA-dependent RNA polymerase (RdRp) and other enzymes. In the case of cypovirus, where there are 10 genomic dsRNA segments with a non-spooled organization, 10 TEC are anchored to the inner surface of the capsid shell around the fivefold axis, leaving two of the 12 possible positions vacant [110, 117] (Fig. 3.11c). RdRp, resolved to 3.3 Å resolution, adopt different conformations in transcribing and non-transcribing states. These studies relied on subtracting capsid density from the original images, followed either by standard asymmetric refinement [110] or refinement based on symmetry relaxation [117]. Similar in situ structural analyses have been conducted for rotavirus, reovirus, aquareovirus, and bluetongue virus, shedding light on the mechanism of RNA transcription and replication [118–121]. In bacteriophage ϕ6, which has a trisegmented genome, the dsRNA adopts a dsDNA-like single-spooled genome organization, and RdRp is detached from the capsid inner surface. These variations might be a consequence of the differences in transcription mechanisms between the two virus families, namely, semiconservative vs. conservative transcription [122].

Herpesvirus consists of an inner icosahedral capsid that contains a linear dsDNA genome, a middle proteinaceous tegument layer, and an outer envelope with trans-membrane glycoproteins. The viral genome is packaged and ejected by an ATP-driven terminase through the asymmetric portal complex located at a predefined fivefold symmetry axis of the capsid. Cryo-EM studies of herpes simplex virus type 1 (HSV-1) virions using sequential localized classification and symmetry relaxation have allowed for the decoupling and reconstruction of heterosymmetric and asymmetric capsid elements. The HSV-1 map shows the in situ structures of the unique portal vertex, genomic termini, and ordered dsDNA coils [111] (Fig. 3.11d). A similar approach resolved the human cytomegalovirus structures of the portal and the capsid vertex-specific components [123].

Feline calicivirus (FCV) has a positive-sense ssRNA genome that is encapsidated within a $T = 3$ icosahedral capsid. Upon binding its receptor, focused classification has led to the identification of a portal-like structure formed of 12 copies of VP2, a minor structural protein, located at a unique three-fold symmetry axis [112]. Conformational changes observed in the capsid protein VP1 result in the extrusion of the internalized VP2 proteins and the formation of a small pore in the capsid shell. This VP2 portal-like assembly may play a role in endosomal escape and genome delivery, a process not well understood in nonmembrane viruses (Fig. 3.11e).

3.7 Reconstructing Giant Viruses

Giant viruses belong to the nucleocytoplasmic large DNA viruses (NCLDV) which comprise nine families: *Ascoviridae, Asfarviridae, Iridoviridae, Marseilleviridae, Mimiviridae, Pandoraviridae, Phycodnaviridae, Pithoviridae*, and *Poxviridae*, along with some independent species. These viruses, with the dsDNA genome enclosed in a lipid bilayer [124], vary in size from ~150 nm to 2 μm. They encompass high-symmetry icosahedral viruses, such as *Paramecium bursaria* chlorella virus 1 (PBCV-1), and amphora-shaped (asymmetric) viruses like Pithovirus.

Two primary strategies are used to calculate high-resolution cryo-EM structures of large viruses [125]: (1) the use of electron microscopes with high-acceleration voltage, e.g., one megavolt cryo high-voltage electron microscopy (1MV cryo-HVEM), which enhances sample penetration and overcomes limitations associated with electro-optical physics at lower voltages [126], and (2) the "block-based reconstruction" method [127].

For thick samples, the influence of depth of field causes an internal focus shift (a focus gradient within a single virus particle), imposing limitations on achievable resolution [128]. Standard CTF correction methods, which assume a single constant defocus for the entire particle, are inadequate for large, high-resolution structures. Increasing the accelerating voltage increases depth of field, however, and improves the electron optical conditions in thick samples. For 300 kV electron microscopes, recent advances in image processing, such as Ewald sphere correction to tackle the focus gradient within a single virus particle, can significantly improve attainable resolutions. "Block-based" reconstruction [127] focuses on sub-sections ("blocks") of the virus to allow localized defocus refinement across each particle and reduces the required box size.

For high-resolution 3D SPA of giant viruses, there are currently six structures exceeding 1 nm resolution. Two of these structures are for PBCV-1 [129, 130], two for African swine fever virus (ASFV) [131, 132], one for the Marseillevirus Singapore grouper iridovirus [133], and one for Tokyovirus [126]. All were analyzed using "block-based" reconstruction (with 300 kV microscopes), except for Tokyovirus (with a maximum diameter of 250 nm), which was analyzed as a complete viral particle using 1 MV cryo-HVEM. PBCV-1 was resolved at 3.5 Å resolution [129], while ASFV achieved a resolution of 4.1 Å resolution [132] using cryo-EM SPA. The PBCV-1 and ASFV capsids consist of one "major" capsid protein (MCP) and a combination of 14 (PBCV-1) or four kinds (ASFV) of "minor" capsid proteins (mCP). The MCP of PBCV-1 and ASFV are trimeric and share the double "jelly roll" motif, each of which consists of eight β strands [134].

3.8 Reconstructing Viruses with Helical Symmetry

The first virus ever isolated, tobacco mosaic virus (TMV), is a helical polymer of a protein bound to ssRNA. The capsids of many plant viruses and bacteriophages have helical symmetry [135]. In addition, the genome of some enveloped animal viruses is surrounded by a helical symmetry-based nucleocapsid; these include filoviruses (Ebola and Marburg viruses), rhabdoviruses (rabies and vesicular stomatitis viruses (VSV)), paramyxoviruses (measles and mumps), orthomyxoviruses (influenza virus), and retroviruses (human immunodeficiency virus, HIV) (see Chap. 2). In certain nonphysiological conditions, some capsid proteins of icosahedral viruses may assemble into helical structures, revealing an intrinsic structural polymorphism [136]. X-ray fiber diffraction has led to high resolution for several rigid filamentous viruses [137], but this method is inappropriate for the more common flexible filamentous viruses. Whereas the first 3DR from a negatively stained sample of the helical tail of phage T4 was calculated at ~35 Å resolution [63], using cryo-EM has been reconstructed at 3.4 Å resolution [138].

The redundancy intrinsic to helical assemblies renders them optimal for cryo-EM averaging, since a single projection image provides different views of identical subunits needed to reconstruct the 3D structure [139]. Helical parameters are determined by measuring the pitch of the helix from layer lines in the FT of the tubular structure. A 2D classification and averaging is made due to the high degree of noise present in cryo-EM images and if structural variability or heterogeneity is detected (or suspected). The 3DR is computed using the iterative helical real space reconstruction (IHRSR) method, a single-particle type approach that does not require long high-ordered filaments [140]. The IHRSR algorithm uses helical symmetry as a constraint to impose it on asymmetric reconstructions [141, 142].

In the last years, cryo-EM has enabled the construction of a de novo atomic model of numerous filamentous viruses in the three domains of life: *Bacteria*, *Archaea*, and *Eukarya*. Two key findings have emerged from these analyses: (1) bacteriophages have a cylindrical cavity within their capsids to accommodate the packaged genome, and (2) in filamentous viruses in archaea, plant, and mammalian, the nucleic acid is highly organized, embedded within the capsid proteins [143]. In addition, the structural similarity observed in the capsid proteins of these viruses suggests that helical viruses can be grouped into a limited number of lineages. Interestingly, filamentous viruses that infect hyperthermophilic acidophile archaea have a genomic dsDNA with a conformation corresponding to the A-form DNA [144, 145].

3.9 Cryogenic Electron Tomography of Viruses

3.9.1 Basic Concepts and General Experimental Design

A preparation of icosahedral viruses observed by cryo-EM normally provides multiple views with random orientations of identical particles (without considering structural heterogeneity). Many enveloped viruses are pleiomorphic, however; examples include orthomixo-, paramyxo-, retro-, and poxviruses (see Chap. 12). Herpesviruses and other enveloped viruses are intermediate forms, with an icosahedral capsid surrounded by a pleiomorphic structure made of protein and lipid layers (see Chap. 2). The structural variability of all these viruses means that each particle has a different shape, size, or conformation, and 3DR by averaging cryo-EM images as described above is not possible. However, a reconstructed 3D image can be produced by obtaining multiple views of a single viral particle. As in medical computer tomography scanning, the 3D structure of the viral sample is reconstructed from a set of 2D projections recorded at predefined tilt angles. To do this, the

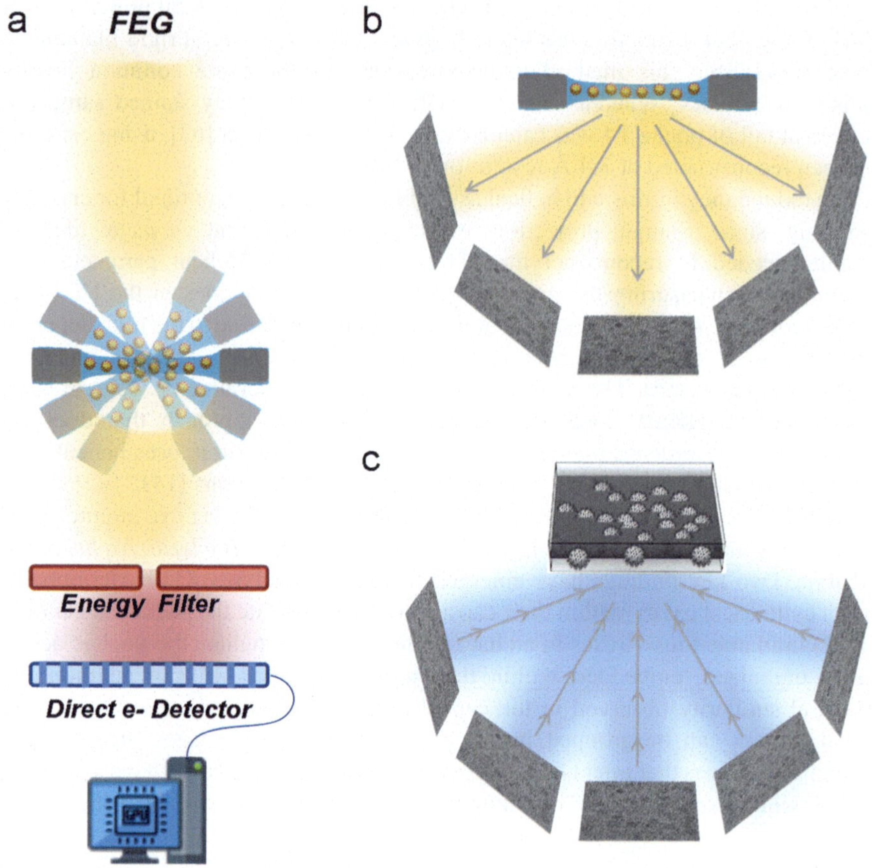

Fig. 3.12 Cryo-ET. (**a**) 2D projection images generated with minimum energy loss electron (energy filter) are acquired on a direct electron detector (or CCD camera), while the same specimen area embedded in a vitreous thin layer is tilted incrementally (~1–2°, tilt range ± 70°). (**b**) Diagram illustrating the images projected by a specimen at successive tilt angles. (**c**) After mutual alignment of all these 2D projection images, a density map (tomogram) is reconstructed by a weighted back-projection procedure

specimen is tilted around an axis perpendicular to the electron beam inside the microscope. This type of cryo-EM is termed cryo-ET [19, 146]. Tilting of the specimen provides images of the same field of view (typically more than 100), for example, over a range of ±70° at steps of 1° or 2° (limited by the design of specimen holders and because specimen thickness increases very rapidly at high tilt angles) (Fig. 3.12).

The 3D reconstruction of the field of interest, or tomogram, is calculated as a back-projection in real space; the assembly of correctly aligned projections is reverse-projected into 3D space. Cryo-ET has allowed the study of samples ranging in size from large (pleomorphic) complexes to organelles, prokaryotic cells, and eukaryotic cell lamellae (thin cell sections obtained by FIB milling [147]), that is, samples not amenable to SPA averaging method. Cryo-ET has therefore allowed the study of viruses in their natural cell environment (see Chap. 8), thereby providing a bridge between structural studies at the molecular and cellular levels (reviewed in [65, 148–150], and references therein).

The structural information within a reconstructed tomogram is limited by three notable factors: low signal in each tilt image to limit accumulated electron dose, missing information due to mechanical limitations in the microscope stage (termed the "missing wedge"), and accuracy of alignments between each successive tilt image prior to reconstructing a 3D volume [19]. To obtain a detailed, undistorted reconstruction, the tilt series must cover as wide an angular range as possible in as many increments as possible. Tomogram resolution is limited because the range of angles sampled is incomplete, and there are no data in some parts of the 3D FT (the missing wedge); in addition, data are missing in the spaces between sampled planes. At the same time, the electron dose must be kept as low as possible to avoid radiation damage, and the resulting projections are very noisy, making their alignment and CTF correction unreliable. To increase contrast, images are taken at high defocus values (at least 3–10 μm; compared to 0.5–3 μm for cryo-EM), and the resolution is somewhat lower than the first zero of the CTF (in the 4 nm range), as restoration of the true projection is difficult with these low signal-to-noise ratio micrographs. These limitations can be overcome with a double-tilt series; two single-tilt series of the objects are recorded, during which the specimen is rotated by 90° around the beam direction after the first series. Similar to SPA cryo-EM, multiple copies of homogeneous structures, such as glycoproteins, of the polymorphic viruses can be extracted from a tomogram as a volume, aligned, and averaged. This STA approach delivers structures of large macromolecular complexes at subnanometer resolution in their native cellular environment [151]. In some cases, STA can achieve high resolution by averaging 3D structures [26], as opposed to the 2D images used in SPA cryo-EM. The introduction of correlative light and electron microscopy (CLEM) and FIB milling have further enhanced the capabilities of cryo-ET. These advancements allow cryo-ET to reveal the native ultrastructures of large eukaryotic cells and even tissues with unparalleled spatiotemporal resolution [24, 152].

A contemporary cryo-ET setup consists of a high-voltage cryo-electron microscope equipped with an FEG, a CCD/DED detector, and possibly an energy filter [153]. Automation of the acquisition process is essential for acquiring tomographic data sets in low-dose conditions. High-voltage electrons can penetrate thicker samples, which enables imaging of cells or organelles. The interaction of electrons with thick (>200 nm) specimens produces many inelastic scattering events, leading to strong blurring. Energy filters allow imaging of thick specimens and are indispensable for tomography of whole ice-embedded cells or other objects of 0.5–1 μm thickness. Cryo-sectioning is a solution for very thick specimens frozen at high pressure. Without sectioning, only the peripheral areas of flat cells are accessible to cryo-ET.

3DR of an object from its projections involves two steps: first, the micrographs must be aligned to a common coordinate system, and second, the aligned micrographs are merged into a tomogram. Fiducial markers added to the specimen (typically colloidal gold), which appear as high-density dots in the images, are commonly used for alignment, although this can also be done without markers. The most common reconstruction method in cryo-ET is the weighted back-projection, to form a 3DR. Because of uneven sampling in the Fourier space, the low frequencies are artificially enhanced, which requires weighting of projections prior to back-projection to obtain a reliable reconstruction.

The interpretation of tomograms at the ultrastructural level requires the decomposition of a tomogram into its structural components, e.g., the segmentation of defined components such as membranes and glycoproteins. Manual assignment of features is commonly used. In addition, to increase the signal-to-noise ratio, so-called denoising algorithms have been developed for analysis and three-dimensional visualization.

3.9.2 *Structure of Pleiomorphic Viruses and Capture of Viral Life Cycles Using Cryo-ET*

Cryo-ET has provided a means to conduct 3D analysis of complex and pleomorphic enveloped viral particles, such as those shown in Fig. 3.13 [25, 154–159]. Many enveloped viruses are responsible for animal and human diseases [163, 164], including HIV-1 [26, 164], herpesviruses [165], and influenza virus [166]. Furthermore, cryo-ET enables the study of the complete viral replication cycle within the native environment, the cell, at an unprecedented level of resolution (Chap. 8). STA typically improves resolution to ~2 nm and in some cases was able to achieve ~4 Å resolution, marking a significant advancement in cryo-ET resolution capabilities. Several noteworthy examples are highlighted wherein cryo-ET has yielded crucial and, in some cases, unanticipated discoveries related to complex viruses and in situ viral infections.

Influenza A virus (IAV) has two types of protein spikes emerging from its envelope, hemagglutinin (HA) and neuraminidase (NA); there are about 400 densely packed spikes (85% HA, 15% NA) and NA spikes tend to cluster; in addition, HA molecules show conformational changes after fusion with liposomes. The eight genome segments are organized as ribonucleoproteins [166], and the matrix protein (M1) forms an endoskeleton beneath the virus membrane within intact virions. Cryo-ET/STA at ~8 Å resolution of M1 within purified virus particles as well as the structure of M1 oligomers reconstituted in vitro, offered insights into IAV assembly and disassembly [167] (Fig. 3.13c). Cryo-ET of intact virions of herpes simplex virus 1 (HSV-1) has provided new information about the pleiomorphic structures crucial for infection of host cells: tegument (an amorphous protein layer that includes host structures such as actin filaments) and an envelope coated with glycoprotein spikes. The icosahedral $T = 16$ capsid enclosing the viral DNA adopts an eccentric position within the approximately spherical virus. The capsid images were extracted

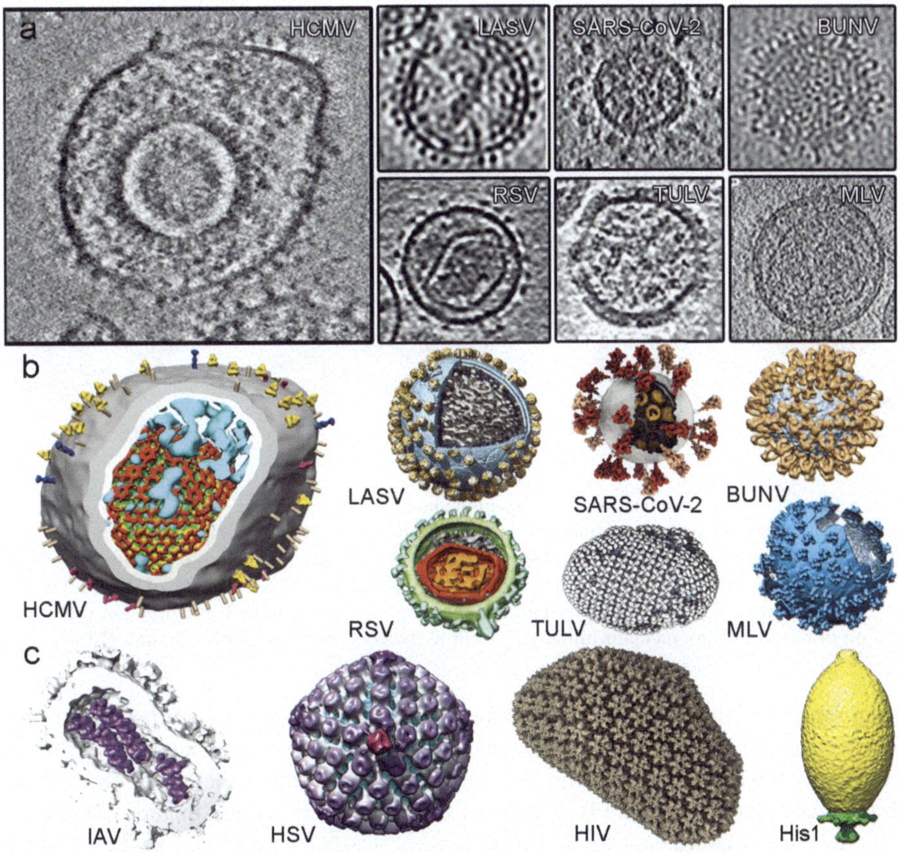

Fig. 3.13 Cryo-ET of pleiomorphic viruses. (**a, b**) Tomographic reconstructions of human cytomegalovirus (HCMV, reproduced with permission from Ref. [154]), Lassa virus (LASV, [155]), severe acute respiratory syndrome coronavirus-2 (SARS-CoV-2, reproduced with permission from Ref. [25]), Bunyamwera virus (BUNV, [156]), Rous sarcoma virus (RSV, [157]), Tula virus (TULV, [158]), and murine leukemia virus VLP (MLV, [159]). A slice of each tomogram is shown (a), and surface-shaded representations of the tomograms are cut to see the virion interior (b). Scale bar, 50 nm. (**c**) Segmented rendered surfaces of 3D reconstructed tomograms from influenza A virus (IAV, courtesy of J. Martín-Benito), HSV-1 (EMD-5452, [160]), HIV (PDB 3J3Y, [161]), and haloarchaeal virus His1 (EMD-6222, [162]). The genome of IAV is organized as ribonucleoprotein complexes (purple); the glycoproteins HA and NA are distinguishable. See Fig. 3.14 for details of HCMV, TULV, and MLV surface glycoproteins

from tomograms, averaged, and compared to the non-tegumented capsid, revealing capsid-tegument contacts. The entry portal in different herpesvirus capsids was also visualized in one of the 12 fivefold vertices [160] (Fig. 3.13c). An equivalent cryo-ET study to detect asymmetric components inside icosahedral viruses was performed with poliovirus, a ~ 30 nm icosahedral virus that lacks an envelope; isolate particles were imaged releasing their genome, and the RNA exited through channels near the twofold axes. In the study of HIV-1, a 7 Å resolution cryo-ET structure of the capsid

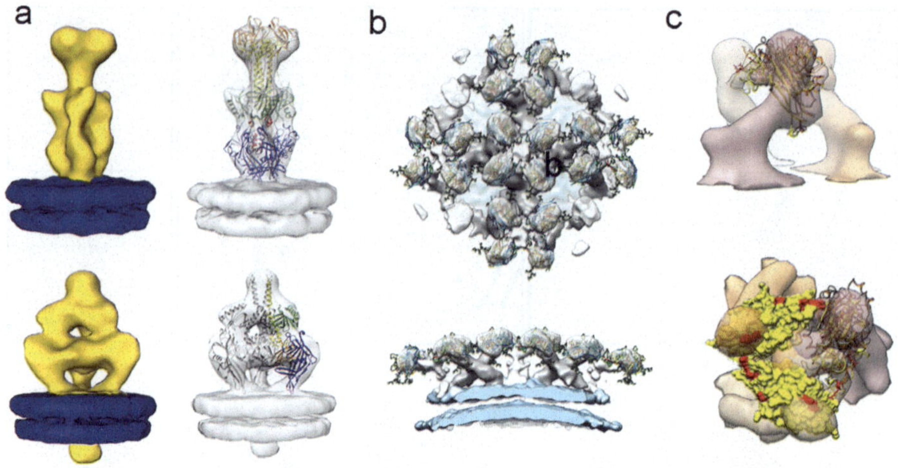

Fig. 3.14 Subtomogram averaging and hybrid methods. (**a**) Structures of HCMV gB protein in the postfusion (top) and prefusion (bottom) conformations as revealed by STA. The structures are shown as either as shaded surfaces or as semi-transparent gray fitted with the gB ectodomain trimer crystal structure (ribbon) [154]. (**b**) Top and side views of the STA structure of TULV Gn-Gc glycoprotein spike fitted with the crystallographic structure of Puumala hantavirus Gn [158]. (**c**) Fitting of the MLV Env receptor binding domain in the corresponding STA structure [159]

from intact virions confirmed the distinctive hollow cone shape of the capsid and allowed precise positioning of individual capsid hexamers and pentamers within the lattice structure [161, 168] (Fig. 3.13c). Furthermore, the structure of the capsid domain and spacer peptide 1 (CA-SP1) of Gag (the precursor that includes MA, CA, and NC) in immature capsids was elucidated at 3.9-Å resolution [26]. This level of detail was previously feasible primarily through SPA cryo-EM. Within the archaeal viruses, the icosahedral viruses belonging to the HK97 and PRD1 lineages have been extensively studied. However, other archaeal viruses, characterized by unique bottle, lemon, or spindle-shaped morphologies, offer intriguing assembly dynamics and are ideal candidates for cryo-electron tomography (cryo-ET) investigations [162, 169] (Fig. 3.13c).

Cryogenic CLEM (cryo-CLEM), which combines spatiotemporal information from fluorescence microscopy, where proteins of interest are tagged with fluorescent labels, with structural data from cryo-ET, is very useful for studying viral assembly and/or rare and short-lived events within specific regions of virus-infected cells [170, 171] (Chap. 8). One notable example involves Pseudomonas phages, which assemble a compartment for DNA replication. Prior to cell lysis and the release of phages, empty capsids assemble on the host cell membrane and migrate along PhuZ spindles to the compartment surface for DNA packaging before the release of viral progeny [172, 173]. A phase plate cryo-ET study of cyanophage Syn5, which infects the cyanobacterium *Synechococcus*, has identified distinct assembly intermediates in situ, such as procapsids with scaffolding proteins, expanded empty and full capsids, as well as complete virions with the tail [174]. Cryo-ET of the ε15 bacteriophage infecting its bacterial host has allowed visualization of structural changes in the tail machinery, including a tubular density that spans the periplasmic space of the bacterial wall that is the conduit for viral DNA entry into the cell. The life cycle of murine gammaherpesvirus, including viral attachment, entry, assembly in virus-induced nuclear inclusion bodies, and egress, has been also tracked using cryo-ET. In summary, advances in cryo-ET and associated techniques such as cryo-FIB milling are providing key insights that have expanded our knowledge of intracellular activities of viruses at unprecedented resolution [175] (see Chap. 8).

At subnanometer resolution, cryo-ET structural models (and cryo-EM to lesser extension) yield significant information about macromolecular complexes at low to medium resolution, or the folding of individual components. If more detail is required in the structural model, and higher-resolution structural models of individual subunits in the complex are available, the high-resolution structures obtained by X-ray crystallography (see Chap. 4) and by NMR (see Chap. 5) can be fit ("docked") to the cryo-EM electron density map to obtain a "pseudo-atomic" model of the complex [176–178]. Such fitting can reveal, for example, interaction interfaces between individual components in atomic detail, providing structural insight into how individual intersubunit interactions might affect complex functions. A more extensive description of combined approaches is included in Chap. 7.

As the resolution of the cryo-EM map improves, discrepancies in some parts of the individual subunits are evident once fitted into the cryo-EM map; they can be adjusted optimally by permitting flexible fitting of specific domains and/or SSE. Flexible docking can be achieved manually using molecular graphics, but the most-used methods are based on molecular dynamic simulations or on normal mode analysis (see Chap. 19 for a general description of these theoretical approaches). In addition, predicted structures using AlphaFold [71] and RosettaFold [72] (see Chap. 7) can be fit into low-resolution SPA volumes to clarify the positions of proteins within a complex.

Human cytomegalovirus (HCMV) enters host by glycoprotein B (gB)-mediated membrane fusion upon receptor-binding to gH/gL-related complexes. Using cryo-ET with energy filtering, the structures of glycoproteins in the native environment of HCMV virions were reported as well as the metastable prefusion viral fusion proteins [154] (Fig. 3.14a). *Hantaviruses*, a genus of human pathogens, initiate cell infection through the concerted action of Gn and Gc viral surface glycoproteins. Cryo-ET of intact Gn-Gc spike complexes from the non-pathogenic Tula hantavirus localized Gn to the membrane-distal surface of the virion, and the x-ray structure of Gn from Puumala hantavirus (PUUV), a causative agent of hemorrhagic fever with renal syndrome, was fitted. The fold of PUUV Gn is likely to be widely conserved across hantaviruses [158] (Fig. 3.14b). The process of attachment and fusion in retroviruses is mediated by Env, an envelope glycoprotein, and its conformational changes in the course of fusion are a focus of intense studies. In murine leukemia virus (MLV), the structure of natively membrane-anchored MLV Env and its major rearrangements upon interaction with a host cell were determined by cryo-ET and STA [159] (Fig. 3.14c).

3.10 Cryo-EM and Cryo-ET of SARS-CoV-2

The coronavirus disease 2019 (COVID-19) pandemic, caused by the severe acute respiratory syndrome coronavirus 2 (SARS-CoV-2), has led to a global crisis with over approximately 760 million cases and approximately seven million deaths. Cryo-EM and cryo-ET studies played a pivotal role in a record-breaking time frame [179]. These techniques have been instrumental in the development of therapeutics targeting the spike protein (S protein) that decorates the SARS-CoV-2 viral surface. Not only was the atomic structure of many S protein specimens established by SPA cryo-EM analysis but also its multiple conformational states, crucial for unraveling its mechanism of action, as well as for studying the spike complexes with the human angiotensin-converting enzyme-2 (ACE2) receptor and neutralizing antibodies [179, 180]. Basic research on previously limited outbreaks of closely related coronaviruses, such as SARS-CoV-1 (or SARS-CoV) in 2002 and the Middle Eastern respiratory syndrome coronavirus (MERS-CoV) in 2012, as well as the accumulated experience with other emergent viral pandemics (HIV and influenza viruses), were also fundamental to the development of mRNA-based vaccines and effective antibody therapies [31, 32]. This rapid structural analysis of SARS-CoV-2 has demonstrated that our global society is prepared to apply similar strategies to combat future emerging viruses.

The immature S protein undergoes proteolytic cleavage into two subunits, S1 and S2, which form a heterodimer. This heterodimer then assembles into a trimer that results in the characteristic spike on the virion that coats the viral surface [181] (Fig. 3.15). The S trimeric protein serves as a metastable fusion nanomachine; in situ analysis of S trimers by cryo-ET and STA of intact virions has revealed that most spikes are in the prefusion conformation, with only 3% in the postfusion conformation [25, 183]. The receptor-binding domain (RBD) within the S1 subunit can assume either a "down" or "up" conformation, corresponding to ACE2-inaccessible or -accessible conformations, respectively [184]. During infection, RBD of the prefusion form in the up conformation binds to ACE2 before the fusion of viral and host membrane, mediated by the fusion peptide in S2 (see Chap. 16). The S protein, which elicits a strong immune response, stands as the most critical target antigen for therapeutic development and current vaccines. Moderna and Pfizer mRNA vaccines are based on S protein stabilized in the prefusion conformation by two-proline substitutions [33]. Promising therapeutic agents against emergent variants include neutralizing antibodies targeting specific regions of the S protein and recombinant ACE2-derived inhibitors. Approximately 25 other viral proteins, some yet to be structurally characterized, are also being investigated as potential targets for antiviral drug development [182].

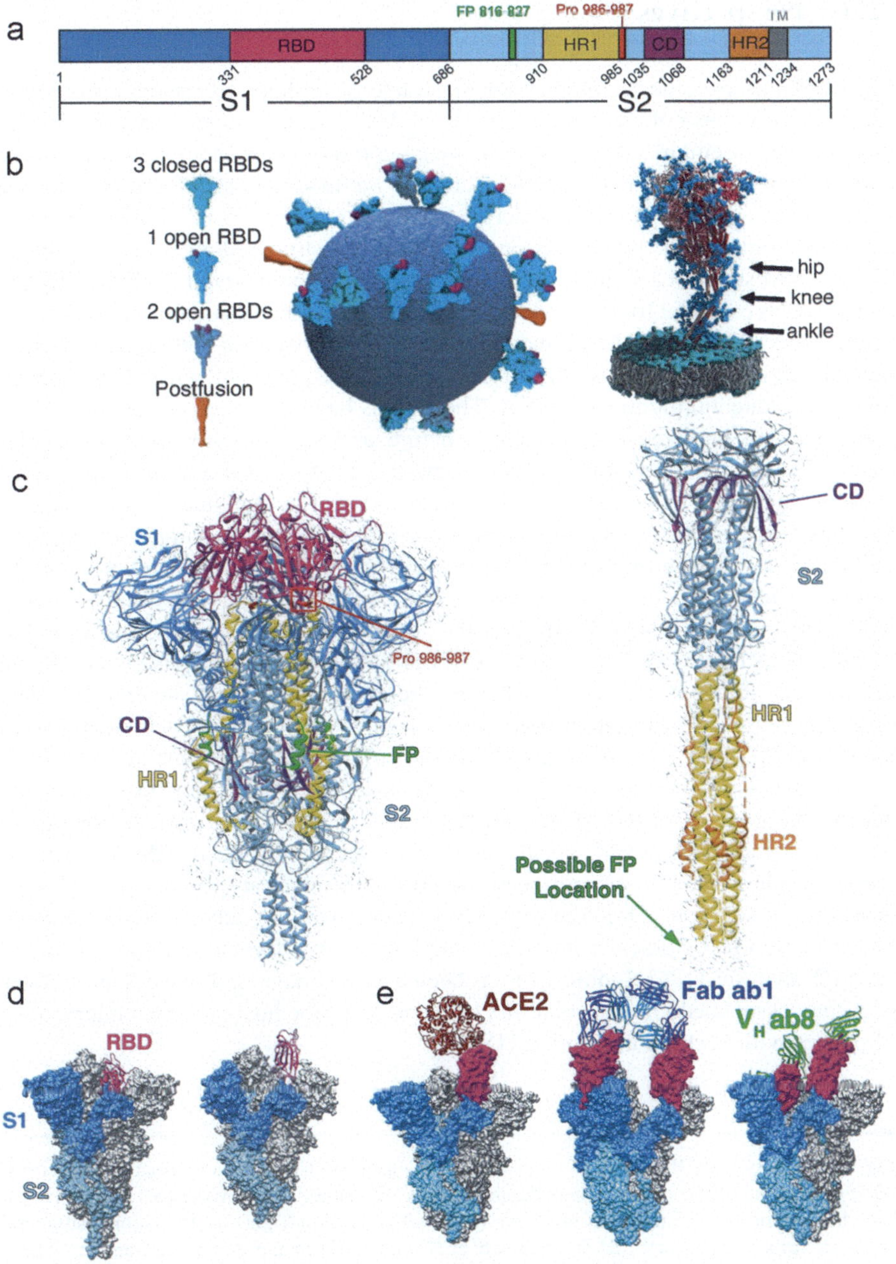

Fig. 3.15 Structures of SARS-CoV-2 spikes. (**a**) Scheme of the SARS-CoV-2 S domains showing the S1 subunit, S2 subunit, receptor-binding domain (RBD), fusion peptide (FP), heptad repeat 1 (HR1), connector domain (CD), heptad repeat 2 (HR2), and the transmembrane region (TM). (**b**)

3.11 Perspectives

Cryo-EM has become a unique tool in structural biology to obtain near-atomic resolution structures of more and more, large, flexible and dynamic, macromolecular (viral) mega-assemblies. Many of these mega-complexes are membrane proteins or may contain nucleic acids. Cryo-EM facilitates the in silico purification (sorting via image processing) of specimens derived from ex vivo sources (roughly purified fractions), which could potentially pave the way for a structural proteomic approach. 3D cryo-EM studies have unveiled that the conformational variability of these mega-complexes is essential for their correct molecular mechanisms and functions (Part III of this book). Cryo-EM, therefore, complements perfectly with cutting-edge prediction structure advanced tools like AlphaFold and RoseTTAFold, which can generate highly accurate atomic models [185]. These tools have led to a leap forward in our computational knowledge of protein structure and are currently used on regular basis. The structural biology field is expected to grow toward fully integrated computational and experimental cryo-EM methods. The combination of structural findings by cryo-EM with bioinformatic tools (and molecular dynamics simulations) will allow description of proteome interactions at an unprecedent level of detail. In addition, this wealth of cryo-EM knowledge can lead to antiviral drug discovery (Chap. 20), vaccine design (Chap. 21), and advancement of new platforms for virus-based nanotechnology (Chap. 22). Cryo-ET allows structural analysis directly in the cellular environment (Chap. 8). One of the major challenges in cryo-ET data processing is to identify target biomacromolecules accurately and efficiently from the noisy and distorted tomograms within the crowded cellular context.

Liquid electron microscopy (liquid-EM) is a relatively recent technique that permits the imaging of biological samples in their native liquid state. In contrast to traditional transmission EM, which requires samples to be fixed (EM) or vitrified (cryo-EM) in a vacuum environment, liquid-EM offers real-time data of dynamic processes in solution. Although liquid-EM results may not achieve the same resolution as cryo-EM studies, it provides complementary information to cryo-EM and cryo-ET analyses [186]. Liquid-EM has been used to image multiple dynamic states of Adeno-associated virus in solution [187] and real-time host-pathogen interactions of phages with bacteria in solution [188].

Fig. 3.15 (continued) 3D model of a SARS-CoV-2 virion showing the conformations and flexibility of S on the virion surface. Three flexible hinges are marked by arrows (right). (**c**) SPA cryo-EM structure of the S trimer in its prefusion (left) and postfusion (right) conformations [structural elements colored as in (a)]. (**d**) Comparison of S with RBD in the down (left) and up (right) conformations. (**e**) Cryo-EM structures of S bound to ACE2 with RBD in the up conformation (left), SN501Y mutant bound by Fab ab1 with RBD in the up conformation (middle), and SN501Y bound by VH ab8 in both the RBD up and down conformations (right). Reproduced from [182]

Acknowledgments We apologize to colleagues who made outstanding contributions but were not mentioned here due to space limitations. The authors thank C. Mark for editorial assistance. JRC is indebted to current and former members of its group and other colleagues for their hard work, skills, and enthusiasm that made work possible and enjoyable. This work was supported by grants from the Spanish Ministry of Science and Innovation (PID2020-113287RB-I00) and the Comunidad Autónoma de Madrid (P2018/NMT-4389) to JRC and by a grant of the Instituto de Salud Carlos III (PI20CIII-00014) to DL.

References

1. Alberts B, Johnson A, Lewis J, Raff M, Roberts K, Walter P (2007) Molecular biology of the cell. Garland Science, New York
2. Nogales E, Grigorieff N (2001) Molecular machines: putting the pieces together. J Cell Biol 152:F1–F10
3. Chua EYD, Mendez JH, Rapp M, Ilca SL, Tan YZ, Maruthi K, Kuang H, Zimanyi CM, Cheng A, Eng ET, Noble AJ, Potter CS, Carragher B (2022) Better, faster, cheaper: recent advances in Cryo-electron microscopy. Annu Rev Biochem 91:1–32
4. Chmielewski D, Chiu W (2021) Cryo-electron microscopy (CEM) structures of viruses. In: Bamford DH, Zuckerman M (eds) Encyclopedia of virology, vol 1, 4th edn. Academic Press, Oxford, pp 233–241
5. DiMaio F, Song Y, Li X, Brunner MJ, Xu C, Conticello V, Egelman E, Marlovits T, Cheng Y, Baker D (2015) Atomic-accuracy models from 4.5-a cryo-electron microscopy data with density-guided iterative local refinement. Nat Methods 12:361–365
6. Steven A, Belnap D (2005) Electron microscopy and image processing: an essential tool for structural analysis of macromolecules. Curr Protoc Protein Sci 17:Unit 17 12
7. Klug A (2010) From virus structure to chromatin: X-ray diffraction to three-dimensional electron microscopy. Annu Rev Biochem 79:1–35
8. Harrison SC (2004) Whither structural biology? Nat Struct Mol Biol 11:12–15
9. Baumeister W, Steven AC (2000) Macromolecular electron microscopy in the era of structural genomics. Trends Biochem Sci 25:624–631
10. Steven AC, Baumeister W (2008) The future is hybrid. J Struct Biol 163:186–195
11. Kuhlbrandt W (2014) Biochemistry. The resolution revolution. Science 343:1443–1444
12. Egelman EH (2016) The current revolution in Cryo-EM. Biophys J 110:1008–1012
13. Li X, Mooney P, Zheng S, Booth CR, Braunfeld MB, Gubbens S, Agard DA, Cheng Y (2013) Electron counting and beam-induced motion correction enable near-atomic-resolution single-particle cryo-EM. Nat Methods 10:584–590
14. Bertozzi C (2017) Atoms out of blobs: CryoEM takes the Nobel prize in chemistry. ACS Cent Sci 3:1056
15. Khoshouei M, Radjainia M, Baumeister W, Danev R (2017) Cryo-EM structure of haemoglobin at 3.2 Å determined with the Volta phase plate. Nat Commun 8:16099
16. Merk A, Bartesaghi A, Banerjee S, Falconieri V, Rao P, Davis MI, Pragani R, Boxer MB, Earl LA, Milne JLS, Subramaniam S (2016) Breaking Cryo-EM resolution barriers to facilitate drug discovery. Cell 165:1698–1707
17. Jiang W, Tang L (2017) Atomic cryo-EM structures of viruses. Curr Opin Struct Biol 46:122–129
18. Kaelber JT, Hryc CF, Chiu W (2017) Electron Cryomicroscopy of viruses at near-atomic resolutions. Annu Rev Virol 4:287–308
19. Lucic V, Rigort A, Baumeister W (2013) Cryo-electron tomography: the challenge of doing structural biology in situ. J Cell Biol 202:407–419

20. Wan W, Briggs JA (2016) Cryo-electron tomography and subtomogram averaging. Methods Enzymol 579:329–367
21. Klebl DP, Aspinall L, Muench SP (2023) Time resolved applications for Cryo-EM; approaches, challenges and future directions. Curr Opin Struct Biol 83:102696
22. Nakane T, Kotecha A, Sente A, McMullan G, Masiulis S, Brown P, Grigoras IT, Malinauskaite L, Malinauskas T, Miehling J, Uchanski T, Yu L, Karia D, Pechnikova EV, de Jong E, Keizer J, Bischoff M, McCormack J, Tiemeijer P, Hardwick SW, Chirgadze DY, Murshudov G, Aricescu AR, Scheres SHW (2020) Single-particle cryo-EM at atomic resolution. Nature 587:152–156
23. Yip KM, Fischer N, Paknia E, Chari A, Stark H (2020) Atomic-resolution protein structure determination by cryo-EM. Nature 587:157–161
24. Quemin ERJ, Machala EA, Vollmer B, Prazak V, Vasishtan D, Rosch R, Grange M, Franken LE, Baker LA, Grunewald K (2020) Cellular electron Cryo-tomography to study virus-host interactions. Annu Rev Virol 7:239–262
25. Yao H, Song Y, Chen Y, Wu N, Xu J, Sun C, Zhang J, Weng T, Zhang Z, Wu Z, Cheng L, Shi D, Lu X, Lei J, Crispin M, Shi Y, Li L, Li S (2020) Molecular architecture of the SARS-CoV-2 virus. Cell 183:730
26. Schur FK, Obr M, Hagen WJ, Wan W, Jakobi AJ, Kirkpatrick JM, Sachse C, Krausslich HG, Briggs JA (2016) An atomic model of HIV-1 capsid-SP1 reveals structures regulating assembly and maturation. Science 353:506–508
27. Cianfrocco MA, Kellogg EH (2020) What could go wrong? A practical guide to single-particle Cryo-EM: from biochemistry to atomic models. J Chem Inf Model 60:2458–2469
28. Sirohi D, Chen Z, Sun L, Klose T, Pierson TC, Rossmann MG, Kuhn RJ (2016) The 3.8 a resolution cryo-EM structure of Zika virus. Science 352:467–470
29. Sugita Y, Matsunami H, Kawaoka Y, Noda T, Wolf M (2018) Cryo-EM structure of the Ebola virus nucleoprotein-RNA complex at 3.6 a resolution. Nature 563:137–140
30. Zhang X, Ge P, Yu X, Brannan JM, Bi G, Zhang Q, Schein S, Zhou ZH (2013) Cryo-EM structure of the mature dengue virus at 3.5-a resolution. Nat Struct Mol Biol 20:105–110
31. Yuan Y, Cao D, Zhang Y, Ma J, Qi J, Wang Q, Lu G, Wu Y, Yan J, Shi Y, Zhang X, Gao GF (2017) Cryo-EM structures of MERS-CoV and SARS-CoV spike glycoproteins reveal the dynamic receptor binding domains. Nat Commun 8:15092
32. Pallesen J, Wang N, Corbett KS, Wrapp D, Kirchdoerfer RN, Turner HL, Cottrell CA, Becker MM, Wang L, Shi W, Kong WP, Andres EL, Kettenbach AN, Denison MR, Chappell JD, Graham BS, Ward AB, McLellan JS (2017) Immunogenicity and structures of a rationally designed prefusion MERS-CoV spike antigen. Proc Natl Acad Sci USA 114:E7348–E7357
33. Wrapp D, Wang N, Corbett KS, Goldsmith JA, Hsieh CL, Abiona O, Graham BS, McLellan JS (2020) Cryo-EM structure of the 2019-nCoV spike in the prefusion conformation. Science 367:1260–1263
34. Nogales E, Scheres SH (2015) Cryo-EM: a unique tool for the visualization of macromolecular complexity. Mol Cell 58:677–689
35. Elmlund D, Le SN, Elmlund H (2017) High-resolution cryo-EM: the nuts and bolts. Curr Opin Struct Biol 46:1–6
36. Goldsmith CS, Miller SE (2009) Modern uses of electron microscopy for detection of viruses. Clin Microbiol Rev 22:552–563
37. Bozzola JJ, Russell LD (1999) Electron microscopy. Principles and techniques for biologist. Jones and Bartlett publishers, Boston
38. Dubochet J, Adrian M, Chang JJ, Homo JC, Lepault J, McDowall AW, Schultz P (1988) Cryo-electron microscopy of vitrified specimens. Q Rev Biophys 21:129–228
39. Amos LA, Henderson R, Unwin PN (1982) Three-dimensional structure determination by electron microscopy of two-dimensional crystals. Prog Biophys Mol Biol 39:183–231
40. Henderson R (2004) Realizing the potential of electron cryo-microscopy. Q Rev Biophys 37: 3–13

41. Harris JR (1997) Negative staining and cryoelectron microscopy: the thin film techniques, vol 5. BIOS Scientific Publishers Ltd., Oxford, p 18
42. Wild P (2008) Electron microscopy of viruses and virus-cell interactions. Methods Cell Biol 88:497–524
43. Leiman PG, Kanamaru S, Mesyanzhinov VV, Arisaka F, Rossmann MG (2003) Structure and morphogenesis of bacteriophage T4. Cell Mol Life Sci 60:2356–2370
44. Cavalier A, Spehner D, Humbel BM (2008) Handbook of cryo-preparation methods for electron microscopy. CRC Press, London
45. Baker TS, Olson NH, Fuller SD (1999) Adding the third dimension to virus life cycles: three-dimensional reconstruction of icosahedral viruses from Cryo-electron micrographs. Microbiol Mol Biol Rev 63:862–922
46. Russo CJ, Passmore LA (2014) Electron microscopy: Ultrastable gold substrates for electron cryomicroscopy. Science 346:1377–1380
47. Russo CJ, Passmore LA (2014) Controlling protein adsorption on graphene for cryo-EM using low-energy hydrogen plasmas. Nat Methods 11:649–652
48. Yu G, Li K, Huang P, Jiang X, Jiang W (2016) Antibody-based affinity Cryoelectron microscopy at 2.6-a resolution. Structure 24:1984–1990
49. D'Imprima E, Floris D, Joppe M, Sanchez R, Grininger M, Kuhlbrandt W (2019) Protein denaturation at the air-water interface and how to prevent it. elife 8
50. Klebl DP, Gravett MSC, Kontziampasis D, Wright DJ, Bon RS, Monteiro DCF, Trebbin M, Sobott F, White HD, Darrow MC, Thompson RF, Muench SP (2020) Need for speed: examining protein behavior during CryoEM grid preparation at different timescales. Structure 28:1238–1248 e1234
51. Passmore LA, Russo CJ (2016) Specimen preparation for high-resolution Cryo-EM. Methods Enzymol 579:51–86
52. Duong-Ly KC, Gabelli SB (2014) Gel filtration chromatography (size exclusion chromatography) of proteins. Methods Enzymol 541:105–114
53. Skiniotis G, Southworth DR (2016) Single-particle cryo-electron microscopy of macromolecular complexes. Microscopy (Oxf) 65:9–22
54. Kastner B, Fischer N, Golas MM, Sander B, Dube P, Boehringer D, Hartmuth K, Deckert J, Hauer F, Wolf E, Uchtenhagen H, Urlaub H, Herzog F, Peters JM, Poerschke D, Luhrmann R, Stark H (2008) GraFix: sample preparation for single-particle electron cryomicroscopy. Nat Methods 5:53–55
55. McMullan G, Faruqi AR, Henderson R (2016) Direct electron detectors. Methods Enzymol 579:1–17
56. McMullan G, Faruqi AR, Clare D, Henderson R (2014) Comparison of optimal performance at 300keV of three direct electron detectors for use in low dose electron microscopy. Ultramicroscopy 147:156–163
57. Brilot AF, Chen JZ, Cheng A, Pan J, Harrison SC, Potter CS, Carragher B, Henderson R, Grigorieff N (2012) Beam-induced motion of vitrified specimen on holey carbon film. J Struct Biol 177:630–637
58. Scheres SH (2014) Beam-induced motion correction for sub-megadalton cryo-EM particles. elife 3:e03665
59. Bai XC, McMullan G, Scheres SH (2015) How cryo-EM is revolutionizing structural biology. Trends Biochem Sci 40:49–57
60. Grant T, Grigorieff N (2015) Measuring the optimal exposure for single particle cryo-EM using a 2.6 a reconstruction of rotavirus VP6. elife 4:e06980
61. Ripstein ZA, Rubinstein JL (2016) Processing of Cryo-EM movie data. Methods Enzymol 579:103–124
62. Crowther RA (2008) The Leeuwenhoek lecture 2006. Microscopy goes cold: frozen viruses reveal their structural secrets. Philos Trans R Soc Lond Ser B Biol Sci 363:2441–2451
63. De Rosier DJ, Klug A (1968) Reconstruction of three dimensional structures from electron micrographs. Nature 217:130–134

64. Moody MF (1990) Image analysis of electron micrographs. In: Hawkes PW, Valdré U (eds) Biophysical electron microscopy. Basis concepts and modern techniques. Academic Press, London, pp 145–288
65. Rochat RH, Chiu W (2012) Cryo-electron microscopy and tomography of virus particles. In: Egelman EH (ed) Comprehensive biophysics, biophysical techniques for structural characterization of macromolecules, vol 1. Academic Press, Oxford, pp 311–340
66. Grigorieff N, Harrison SC (2011) Near-atomic resolution reconstructions of icosahedral viruses from electron cryo-microscopy. Curr Opin Struct Biol 21:265–273
67. Henderson R, Sali A, Baker ML, Carragher B, Devkota B, Downing KH, Egelman EH, Feng Z, Frank J, Grigorieff N, Jiang W, Ludtke SJ, Medalia O, Penczek PA, Rosenthal PB, Rossmann MG, Schmid MF, Schroder GF, Steven AC, Stokes DL, Westbrook JD, Wriggers W, Yang H, Young J, Berman HM, Chiu W, Kleywegt GJ, Lawson CL (2012) Outcome of the first electron microscopy validation task force meeting. Structure 20:205–214
68. Scheres SH, Chen S (2012) Prevention of overfitting in cryo-EM structure determination. Nat Methods 9:853–854
69. Subramaniam S, Earl LA, Falconieri V, Milne JL, Egelman EH (2016) Resolution advances in cryo-EM enable application to drug discovery. Curr Opin Struct Biol 41:194–202
70. Steven AC, Heymann JB, Cheng N, Trus BL, Conway JF (2005) Virus maturation: dynamics and mechanism of a stabilizing structural transition that leads to infectivity. Curr Opin Struct Biol 15:227–236
71. Jumper J, Evans R, Pritzel A, Green T, Figurnov M, Ronneberger O, Tunyasuvunakool K, Bates R, Zidek A, Potapenko A, Bridgland A, Meyer C, Kohl SAA, Ballard AJ, Cowie A, Romera-Paredes B, Nikolov S, Jain R, Adler J, Back T, Petersen S, Reiman D, Clancy E, Zielinski M, Steinegger M, Pacholska M, Berghammer T, Bodenstein S, Silver D, Vinyals O, Senior AW, Kavukcuoglu K, Kohli P, Hassabis D (2021) Highly accurate protein structure prediction with AlphaFold. Nature 596:583–589
72. Baek M, DiMaio F, Anishchenko I, Dauparas J, Ovchinnikov S, Lee GR, Wang J, Cong Q, Kinch LN, Schaeffer RD, Millan C, Park H, Adams C, Glassman CR, DeGiovanni A, Pereira JH, Rodrigues AV, van Dijk AA, Ebrecht AC, Opperman DJ, Sagmeister T, Buhlheller C, Pavkov-Keller T, Rathinaswamy MK, Dalwadi U, Yip CK, Burke JE, Garcia KC, Grishin NV, Adams PD, Read RJ, Baker D (2021) Accurate prediction of protein structures and interactions using a three-track neural network. Science 373:871–876
73. Jamali K, Kall L, Zhang R, Brown A, Kimanius D, Scheres SHW (2023) Automated model building and protein identification in cryo-EM maps. bioRxiv
74. Afonine PV, Poon BK, Read RJ, Sobolev OV, Terwilliger TC, Urzhumtsev A, Adams PD (2018) Real-space refinement in PHENIX for cryo-EM and crystallography. Acta Crystallogr D Struct Biol 74:531–544
75. Pettersen EF, Goddard TD, Huang CC, Meng EC, Couch GS, Croll TI, Morris JH, Ferrin TE (2021) UCSF ChimeraX: structure visualization for researchers, educators, and developers. Protein Sci 30:70–82
76. Casanal A, Lohkamp B, Emsley P (2020) Current developments in coot for macromolecular model building of electron Cryo-microscopy and crystallographic data. Protein Sci 29:1069–1078
77. Caston JR, Ghabrial SA, Jiang D, Rivas G, Alfonso C, Roca R, Luque D, Carrascosa JL (2003) Three-dimensional structure of penicillium chrysogenum virus: a double-stranded RNA virus with a genuine T=1 capsid. J Mol Biol 331:417–431
78. Luque D, Gonzalez JM, Garriga D, Ghabrial SA, Havens WM, Trus B, Verdaguer N, Carrascosa JL, Caston JR (2010) The T=1 capsid protein of Penicillium chrysogenum virus is formed by a repeated helix-rich core indicative of gene duplication. J Virol 84:7256–7266
79. Luque D, Gomez-Blanco J, Garriga D, Brilot AF, Gonzalez JM, Havens WM, Carrascosa JL, Trus BL, Verdaguer N, Ghabrial SA, Caston JR (2014) Cryo-EM near-atomic structure of a dsRNA fungal virus shows ancient structural motifs preserved in the dsRNA viral lineage. Proc Natl Acad Sci USA 111:7641–7646

80. Cheng Y, Grigorieff N, Penczek PA, Walz T (2015) A primer to single-particle cryo-electron microscopy. Cell 161:438–449
81. Dhindwal S, Avila B, Feng S, Khayat R (2019) Porcine circovirus 2 uses a multitude of weak binding sites to interact with Heparan sulfate, and the interactions do not follow the symmetry of the capsid. J Virol 93:e02222–e02218
82. Pickl-Herk A, Luque D, Vives-Adrian L, Querol-Audi J, Garriga D, Trus BL, Verdaguer N, Blaas D, Caston JR (2013) Uncoating of common cold virus is preceded by RNA switching as determined by X-ray and cryo-EM analyses of the subviral A-particle. Proc Natl Acad Sci USA 110:20063–20068
83. Yu X, Jin L, Jih J, Shih C, Zhou ZH (2013) 3.5A cryoEM structure of hepatitis B virus core assembled from full-length core protein. PLoS One 8:e69729
84. Ionel A, Velazquez-Muriel JA, Luque D, Cuervo A, Caston JR, Valpuesta JM, Martin-Benito J, Carrascosa JL (2011) Molecular rearrangements involved in the capsid shell maturation of bacteriophage T7. J Biol Chem
85. Luque D, Saugar I, Rodriguez JF, Verdaguer N, Garriga D, Martin CS, Velazquez-Muriel JA, Trus BL, Carrascosa JL, Caston JR (2007) Infectious bursal disease virus capsid assembly and maturation by structural rearrangements of a transient molecular switch. J Virol 81:6869–6878
86. Perez-Berna AJ, Marabini R, Scheres SH, Menendez-Conejero R, Dmitriev IP, Curiel DT, Mangel WF, Flint SJ, San Martin C (2009) Structure and uncoating of immature adenovirus. J Mol Biol 392:547–557
87. Xiao C, Kuznetsov YG, Sun S, Hafenstein SL, Kostyuchenko VA, Chipman PR, Suzan-Monti M, Raoult D, McPherson A, Rossmann MG (2009) Structural studies of the giant mimivirus. PLoS Biol 7:e92
88. Penczek PA (2010) Fundamentals of three-dimensional reconstruction from projections. Methods Enzymol 482:1–33
89. Scheres SH (2012) RELION: implementation of a Bayesian approach to cryo-EM structure determination. J Struct Biol 180:519–530
90. Scheres SH, Gao H, Valle M, Herman GT, Eggermont PP, Frank J, Carazo JM (2007) Disentangling conformational states of macromolecules in 3D-EM through likelihood optimization. Nat Methods 4:27–29
91. Lyumkis D, Brilot AF, Theobald DL, Grigorieff N (2013) Likelihood-based classification of cryo-EM images using FREALIGN. J Struct Biol 183:377–388
92. Goetschius DJ, Lee H, Hafenstein S (2019) CryoEM reconstruction approaches to resolve asymmetric features. Adv Virus Res 105:73–91
93. Huiskonen JT (2018) Image processing for cryogenic transmission electron microscopy of symmetry-mismatched complexes. Biosci Rep 38
94. Chen B, Frank J (2016) Two promising future developments of cryo-EM: capturing short-lived states and mapping a continuum of states of a macromolecule. Microscopy (Oxf) 65:69–79
95. Chang J, Liu X, Rochat RH, Baker ML, Chiu W (2012) Reconstructing virus structures from nanometer to near-atomic resolutions with cryo-electron microscopy and tomography. Adv Exp Med Biol 726:49–90
96. Zhou ZH (2011) Atomic resolution cryo electron microscopy of macromolecular complexes. Adv Protein Chem Struct Biol 82:1–35
97. Xie Q, Yoshioka CK, Chapman MS (2020) Adeno-associated virus (AAV-DJ)-Cryo-EM structure at 1.56 a resolution. Viruses 12
98. Bahar MW, Nasta V, Fox H, Sherry L, Grehan K, Porta C, Macadam AJ, Stonehouse NJ, Rowlands DJ, Fry EE, Stuart DI (2022) A conserved glutathione binding site in poliovirus is a target for antivirals and vaccine stabilisation. Commun Biol 5:1293
99. Luque D, Caston JR (2020) Cryo-electron microscopy for the study of virus assembly. Nat Chem Biol 16:231–239
100. Stass R, Ilca SL, Huiskonen JT (2018) Beyond structures of highly symmetric purified viral capsids by cryo-EM. Curr Opin Struct Biol 52:25–31

101. Zhu D, Cao D, Zhang X (2023) Virus structures revealed by advanced cryoelectron microscopy methods. Structure 31:1348
102. Ceska T, Chung CW, Cooke R, Phillips C, Williams PA (2019) Cryo-EM in drug discovery. Biochem Soc Trans 47:281–293
103. Renaud JP, Chari A, Ciferri C, Liu WT, Remigy HW, Stark H, Wiesmann C (2018) Cryo-EM in drug discovery: achievements, limitations and prospects. Nat Rev Drug Discov 17:471–492
104. Briggs JA, Huiskonen JT, Fernando KV, Gilbert RJ, Scotti P, Butcher SJ, Fuller SD (2005) Classification and three-dimensional reconstruction of unevenly distributed or symmetry mismatched features of icosahedral particles. J Struct Biol 150:332–339
105. Li F, Hou CD, Lokareddy RK, Yang R, Forti F, Briani F, Cingolani G (2023) High-resolution cryo-EM structure of the pseudomonas bacteriophage E217. Nat Commun 14:4052
106. Fang Q, Tang WC, Tao P, Mahalingam M, Fokine A, Rossmann MG, Rao VB (2020) Structural morphing in a symmetry-mismatched viral vertex. Nat Commun 11:1713
107. Jose J, Hafenstein SL (2022) Asymmetry in icosahedral viruses. Curr Opin Virol 54:101230
108. Tao P, Zhu J, Mahalingam M, Batra H, Rao VB (2019) Bacteriophage T4 nanoparticles for vaccine delivery against infectious diseases. Adv Drug Deliv Rev 145:57–72
109. Dai X, Li Z, Lai M, Shu S, Du Y, Zhou ZH, Sun R (2017) In situ structures of the genome and genome-delivery apparatus in a single-stranded RNA virus. Nature 541:112–116
110. Zhang X, Ding K, Yu X, Chang W, Sun J, Zhou ZH (2015) In situ structures of the segmented genome and RNA polymerase complex inside a dsRNA virus. Nature 527:531–534
111. Liu YT, Jih J, Dai X, Bi GQ, Zhou ZH (2019) Cryo-EM structures of herpes simplex virus type 1 portal vertex and packaged genome. Nature 570:257–261
112. Conley MJ, McElwee M, Azmi L, Gabrielsen M, Byron O, Goodfellow IG, Bhella D (2019) Calicivirus VP2 forms a portal-like assembly following receptor engagement. Nature 565: 377–381
113. Twarock R, Bingham RJ, Dykeman EC, Stockley PG (2018) A modelling paradigm for RNA virus assembly. Curr Opin Virol 31:74–81
114. Twarock R, Stockley PG (2019) RNA-mediated virus assembly: mechanisms and consequences for viral evolution and therapy. Annu Rev Biophys 48:495–514
115. Koning RI, Gomez-Blanco J, Akopjana I, Vargas J, Kazaks A, Tars K, Carazo JM, Koster AJ (2016) Asymmetric cryo-EM reconstruction of phage MS2 reveals genome structure in situ. Nat Commun 7:12524
116. Gorzelnik KV, Cui Z, Reed CA, Jakana J, Young R, Zhang J (2016) Asymmetric cryo-EM structure of the canonical Allolevivirus Qbeta reveals a single maturation protein and the genomic ssRNA in situ. Proc Natl Acad Sci USA 113:11519–11524
117. Liu H, Cheng L (2015) Cryo-EM shows the polymerase structures and a nonspooled genome within a dsRNA virus. Science 349:1347–1350
118. Pan M, Alvarez-Cabrera AL, Kang JS, Wang L, Fan C, Zhou ZH (2021) Asymmetric reconstruction of mammalian reovirus reveals interactions among RNA, transcriptional factor micro2 and capsid proteins. Nat Commun 12:4176
119. Ding K, Celma CC, Zhang X, Chang T, Shen W, Atanasov I, Roy P, Zhou ZH (2019) In situ structures of rotavirus polymerase in action and mechanism of mRNA transcription and release. Nat Commun 10:2216
120. Cui Y, Zhang Y, Zhou K, Sun J, Zhou ZH (2019) Conservative transcription in three steps visualized in a double-stranded RNA virus. Nat Struct Mol Biol 26:1023–1034
121. He Y, Shivakoti S, Ding K, Cui Y, Roy P, Zhou ZH (2019) In situ structures of RNA-dependent RNA polymerase inside bluetongue virus before and after uncoating. Proc Natl Acad Sci USA 116:16535–16540
122. Ilca SL, Sun X, El Omari K, Kotecha A, de Haas F, DiMaio F, Grimes JM, Stuart DI, Poranen MM, Huiskonen JT (2019) Multiple liquid crystalline geometries of highly compacted nucleic acid in a dsRNA virus. Nature 570:252–256
123. Li Z, Pang J, Dong L, Yu X (2021) Structural basis for genome packaging, retention, and ejection in human cytomegalovirus. Nat Commun 12:4538

124. Abergel C, Legendre M, Claverie JM (2015) The rapidly expanding universe of giant viruses: Mimivirus, Pandoravirus, Pithovirus and Mollivirus. FEMS Microbiol Rev 39:779–796
125. Burton-Smith RN, Murata K (2021) Cryo-electron microscopy of the giant viruses. Microscopy (Oxf) 70:477–486
126. Chihara A, Burton-Smith RN, Kajimura N, Mitsuoka K, Okamoto K, Song C, Murata K (2022) A novel capsid protein network allows the characteristic internal membrane structure of Marseilleviridae giant viruses. Sci Rep 12:21428
127. Zhu D, Wang X, Fang Q, Van Etten JL, Rossmann MG, Rao Z, Zhang X (2018) Pushing the resolution limit by correcting the Ewald sphere effect in single-particle Cryo-EM reconstructions. Nat Commun 9:1552
128. Downing KH, Glaeser RM (2018) Estimating the effect of finite depth of field in single-particle cryo-EM. Ultramicroscopy 184:94–99
129. Fang Q, Zhu D, Agarkova I, Adhikari J, Klose T, Liu Y, Chen Z, Sun Y, Gross ML, Van Etten JL, Zhang X, Rossmann MG (2019) Near-atomic structure of a giant virus. Nat Commun 10:388
130. Zhang X, Xiang Y, Dunigan DD, Klose T, Chipman PR, Van Etten JL, Rossmann MG (2011) Three-dimensional structure and function of the Paramecium bursaria chlorella virus capsid. Proc Natl Acad Sci USA 108:14837–14842
131. Liu S, Luo Y, Wang Y, Li S, Zhao Z, Bi Y, Sun J, Peng R, Song H, Zhu D, Sun Y, Li S, Zhang L, Wang W, Sun Y, Qi J, Yan J, Shi Y, Zhang X, Wang P, Qiu HJ, Gao GF (2019) Cryo-EM structure of the African swine fever virus. Cell Host Microbe 26:836–843 e833
132. Wang N, Zhao D, Wang J, Zhang Y, Wang M, Gao Y, Li F, Wang J, Bu Z, Rao Z, Wang X (2019) Architecture of African swine fever virus and implications for viral assembly. Science 366:640–644
133. Pintilie G, Chen DH, Tran BN, Jakana J, Wu J, Hew CL, Chiu W (2019) Segmentation and comparative modeling in an 8.6-a Cryo-EM map of the Singapore grouper Iridovirus. Structure 27:1561–1569 e1564
134. Benson SD, Bamford JK, Bamford DH, Burnett RM (2004) Does common architecture reveal a viral lineage spanning all three domains of life? Mol Cell 16:673–685
135. Stubbs G, Kendall A (2012) Helical viruses. Adv Exp Med Biol 726:631–658
136. Lepault J, Petitpas I, Erk I, Navaza J, Bigot D, Dona M, Vachette P, Cohen J, Rey FA (2001) Structural polymorphism of the major capsid protein of rotavirus. EMBO J 20:1498–1507
137. Namba K, Stubbs G (1986) Structure of tobacco mosaic virus at 3.6 a resolution: implications for assembly. Science 231:1401–1406
138. Zheng W, Wang F, Taylor NMI, Guerrero-Ferreira RC, Leiman PG, Egelman EH (2017) Refined Cryo-EM structure of the T4 tail tube: exploring the lowest dose limit. Structure 25:1436–1441 e1432
139. Ge P, Zhou ZH (2011) Hydrogen-bonding networks and RNA bases revealed by cryo electron microscopy suggest a triggering mechanism for calcium switches. Proc Natl Acad Sci USA 108:9637–9642
140. Egelman EH (2015) Three-dimensional reconstruction of helical polymers. Arch Biochem Biophys 581:54–58
141. Egelman EH (2007) Single-particle reconstruction from EM images of helical filaments. Curr Opin Struct Biol 17:556–561
142. Egelman EH (2000) A robust algorithm for the reconstruction of helical filaments using single-particle methods. Ultramicroscopy 85:225–234
143. Zanotti G, Grinzato A (2021) Structure of filamentous viruses. Curr Opin Virol 51:25–33
144. Liu Y, Osinski T, Wang F, Krupovic M, Schouten S, Kasson P, Prangishvili D, Egelman EH (2018) Structural conservation in a membrane-enveloped filamentous virus infecting a hyperthermophilic acidophile. Nat Commun 9:3360
145. DiMaio F, Yu X, Rensen E, Krupovic M, Prangishvili D, Egelman EH (2015) Virology. A virus that infects a hyperthermophile encapsidates A-form DNA. Science 348:914–917

146. Zhang P, Mendonça L (2021) Analysis of viruses in the cellular context by electron tomography. In: Bamford DH, Zuckerman M (eds) Encyclopedia of virology, vol 1, 4th edn. Academic Press, Oxford, pp 242–247
147. Schaffer M, Mahamid J, Engel BD, Laugks T, Baumeister W, Plitzko JM (2017) Optimized cryo-focused ion beam sample preparation aimed at in situ structural studies of membrane proteins. J Struct Biol 197:73–82
148. Lucic V, Forster F, Baumeister W (2005) Structural studies by electron tomography: from cells to molecules. Annu Rev Biochem 74:833–865
149. Cope J, Heumann J, Hoenger A (2011) Cryo-electron tomography for structural characterization of macromolecular complexes. Curr Protoc Protein Sci 17(Unit17):13
150. Hong Y, Song Y, Zhang Z, Li S (2023) Cryo-electron tomography: the resolution revolution and a surge of in situ Virological discoveries. Annu Rev Biophys 52:339–360
151. Oikonomou CM, Jensen GJ (2017) Cellular electron Cryotomography: toward structural biology in situ. Annu Rev Biochem 86:873–896
152. Turk M, Baumeister W (2020) The promise and the challenges of cryo-electron tomography. FEBS Lett 594:3243–3261
153. Obr M, Hagen WJH, Dick RA, Yu L, Kotecha A, Schur FKM (2022) Exploring high-resolution cryo-ET and subtomogram averaging capabilities of contemporary DEDs. J Struct Biol 214:107852
154. Si Z, Zhang J, Shivakoti S, Atanasov I, Tao CL, Hui WH, Zhou K, Yu X, Li W, Luo M, Bi GQ, Zhou ZH (2018) Different functional states of fusion protein gB revealed on human cytomegalovirus by cryo electron tomography with Volta phase plate. PLoS Pathog 14:e1007452
155. Li S, Sun Z, Pryce R, Parsy ML, Fehling SK, Schlie K, Siebert CA, Garten W, Bowden TA, Strecker T, Huiskonen JT (2016) Acidic pH-induced conformations and LAMP1 binding of the Lassa virus glycoprotein spike. PLoS Pathog 12:e1005418
156. Bowden TA, Bitto D, McLees A, Yeromonahos C, Elliott RM, Huiskonen JT (2013) Orthobunyavirus ultrastructure and the curious tripodal glycoprotein spike. PLoS Pathog 9: e1003374
157. Butan C, Winkler DC, Heymann JB, Craven RC, Steven AC (2008) RSV capsid polymorphism correlates with polymerization efficiency and envelope glycoprotein content: implications that nucleation controls morphogenesis. J Mol Biol 376:1168–1181
158. Li S, Rissanen I, Zeltina A, Hepojoki J, Raghwani J, Harlos K, Pybus OG, Huiskonen JT, Bowden TA (2016) A molecular-level account of the antigenic Hantaviral surface. Cell Rep 15:959–967
159. Riedel C, Vasishtan D, Siebert CA, Whittle C, Lehmann MJ, Mothes W, Grunewald K (2017) Native structure of a retroviral envelope protein and its conformational change upon interaction with the target cell. J Struct Biol 197:172–180
160. Schmid MF, Hecksel CW, Rochat RH, Bhella D, Chiu W, Rixon FJ (2012) A tail-like assembly at the portal vertex in intact herpes simplex type-1 virions. PLoS Pathog 8:e1002961
161. Zhao G, Perilla JR, Yufenyuy EL, Meng X, Chen B, Ning J, Ahn J, Gronenborn AM, Schulten K, Aiken C, Zhang P (2013) Mature HIV-1 capsid structure by cryo-electron microscopy and all-atom molecular dynamics. Nature 497:643–646
162. Hong C, Pietila MK, Fu CJ, Schmid MF, Bamford DH, Chiu W (2015) Lemon-shaped halo archaeal virus His1 with uniform tail but variable capsid structure. Proc Natl Acad Sci USA 112:2449–2454
163. Li S (2022) Cryo-electron tomography of enveloped viruses. Trends Biochem Sci 47:173–186
164. Vankadari N, Shepherd DC, Carter SD, Ghosal D (2022) Three-dimensional insights into human enveloped viruses in vitro and in situ. Biochem Soc Trans 50:95–105
165. Zeev-Ben-Mordehai T, Vasishtan D, Hernandez Duran A, Vollmer B, White P, Prasad Pandurangan A, Siebert CA, Topf M, Grunewald K (2016) Two distinct trimeric conformations of natively membrane-anchored full-length herpes simplex virus 1 glycoprotein B. Proc Natl Acad Sci USA 113:4176–4181

166. Arranz R, Coloma R, Chichon FJ, Conesa JJ, Carrascosa JL, Valpuesta JM, Ortin J, Martin Benito J (2012) The structure of native influenza virion ribonucleoproteins. Science 338:1634–1637
167. Peukes J, Xiong X, Erlendsson S, Qu K, Wan W, Calder LJ, Schraidt O, Kummer S, Freund SMV, Krausslich HG, Briggs JAG (2020) The native structure of the assembled matrix protein 1 of influenza a virus. Nature 587:495–498
168. Mattei S, Glass B, Hagen WJ, Krausslich HG, Briggs JA (2016) The structure and flexibility of conical HIV-1 capsids determined within intact virions. Science 354:1434–1437
169. Hochstein R, Bollschweiler D, Dharmavaram S, Lintner NG, Plitzko JM, Bruinsma R, Engelhardt H, Young MJ, Klug WS, Lawrence CM (2018) Structural studies of Acidianus tailed spindle virus reveal a structural paradigm used in the assembly of spindle-shaped viruses. Proc Natl Acad Sci USA 115:2120–2125
170. Hampton CM, Strauss JD, Ke Z, Dillard RS, Hammonds JE, Alonas E, Desai TM, Marin M, Storms RE, Leon F, Melikyan GB, Santangelo PJ, Spearman PW, Wright ER (2017) Correlated fluorescence microscopy and cryo-electron tomography of virus-infected or transfected mammalian cells. Nat Protoc 12:150–167
171. Chaikeeratisak V, Nguyen K, Khanna K, Brilot AF, Erb ML, Coker JK, Vavilina A, Newton GL, Buschauer R, Pogliano K, Villa E, Agard DA, Pogliano J (2017) Assembly of a nucleus-like structure during viral replication in bacteria. Science 355:194–197
172. Chaikeeratisak V, Khanna K, Nguyen KT, Sugie J, Egan ME, Erb ML, Vavilina A, Nonejuie P, Nieweglowska E, Pogliano K, Agard DA, Villa E, Pogliano J (2019) Viral capsid trafficking along Treadmilling tubulin filaments in bacteria. Cell 177:1771–1780 e1712
173. Chaikeeratisak V, Nguyen K, Egan ME, Erb ML, Vavilina A, Pogliano J (2017) The phage nucleus and tubulin spindle are conserved among large pseudomonas phages. Cell Rep 20: 1563–1571
174. Dai W, Fu C, Raytcheva D, Flanagan J, Khant HA, Liu X, Rochat RH, Haase-Pettingell C, Piret J, Ludtke SJ, Nagayama K, Schmid MF, King JA, Chiu W (2013) Visualizing virus assembly intermediates inside marine cyanobacteria. Nature 502:707–710
175. Graham M, Zhang P (2023) Cryo-electron tomography to study viral infection. Biochem Soc Trans 51:1701–1711
176. Tang L, Johnson JE (2002) Structural biology of viruses by the combination of electron cryomicroscopy and X-ray crystallography. Biochemistry 41:11517–11524
177. Topf M, Lasker K, Webb B, Wolfson H, Chiu W, Sali A (2008) Protein structure fitting and refinement guided by cryo-EM density. Structure 16:295–307
178. Rossmann MG, Morais MC, Leiman PG, Zhang W (2005) Combining X-ray crystallography and electron microscopy. Structure 13:355–362
179. Rapp M, Shapiro L, Frank J (2022) Contributions of single-particle cryoelectron microscopy toward fighting COVID-19. Trends Biochem Sci 47:117–123
180. Xu C, Han W, Cong Y (2023) Cryo-EM and cryo-ET of the spike, virion, and antibody neutralization of SARS-CoV-2 and VOCs. Curr Opin Struct Biol 82:102664
181. Cai Y, Zhang J, Xiao T, Peng H, Sterling SM, Walsh RM Jr, Rawson S, Rits-Volloch S, Chen B (2020) Distinct conformational states of SARS-CoV-2 spike protein. Science 369:1586–1592
182. Hardenbrook NJ, Zhang P (2022) A structural view of the SARS-CoV-2 virus and its assembly. Curr Opin Virol 52:123–134
183. Ke Z, Oton J, Qu K, Cortese M, Zila V, McKeane L, Nakane T, Zivanov J, Neufeldt CJ, Cerikan B, Lu JM, Peukes J, Xiong X, Krausslich HG, Scheres SHW, Bartenschlager R, Briggs JAG (2020) Structures and distributions of SARS-CoV-2 spike proteins on intact virions. Nature 588:498–502
184. Walls AC, Park YJ, Tortorici MA, Wall A, McGuire AT, Veesler D (2020) Structure, function, and antigenicity of the SARS-CoV-2 spike glycoprotein. Cell 181:281–292 e286
185. Azinas S, Carroni M (2023) Cryo-EM uniqueness in structure determination of macromolecular complexes: a selected structural anthology. Curr Opin Struct Biol 81:102621

186. Kelly DF, DiCecco LA, Jonaid GM, Dearnaley WJ, Spilman MS, Gray JL, Dressel-Dukes MJ (2022) Liquid-EM goes viral–visualizing structure and dynamics. Curr Opin Struct Biol 75: 102426
187. Jonaid GM, Dearnaley WJ, Casasanta MA, Kaylor L, Berry S, Dukes MJ, Spilman MS, Gray JL, Kelly DF (2021) High-resolution imaging of human viruses in liquid droplets. Adv Mater 33:e2103221
188. Kennedy E, Nelson EM, Tanaka T, Damiano J, Timp G (2016) Live bacterial physiology visualized with 5 nm resolution using scanning transmission electron microscopy. ACS Nano 10:2669–2677

Further Reading

Agbandje-McKenna M, McKenna R (eds) (2011) Structural virology. RSC Publishing, Cambridge
Rossmann MG, Rao VB (eds) (2012) Viral molecular machines. Springer, New York
Kuo J (ed) (2014) Electron microscopy. Methods and protocols. Humana Press, Springer
Crowther RA (ed) (2016) Methods in enzymology Vol. 579. Cryo-EM, the resolution revolution: recent advances in CryoEM. Academic Press
Steven AC, Baumeister W, Johnson LN, Perham RN (eds) (2016) Molecular biology of assemblies and machines. Garland Science, New York
Comas-García M, Rosales-Mendoza S (eds) (2023) Physical virology. Springer Nature, Cham Switzerland
Also especially recommended for further reading are references [3, 18, 24, 33, 84, 88, 139, 170, and 176] *from those listed above*

Chapter 4
X-Ray Crystallography of Viruses

Núria Verdaguer, Cristina Ferrer-Orta, and Damià Garriga

Abstract Since the 1970s and for about 40 years, X-ray crystallography has been by far the most powerful approach for determining virus structures at close to atomic resolutions. Information provided by these studies has deeply and extensively enriched and shaped our vision of the virus world. In turn, the ever-increasing complexity and size of the virus structures being investigated have constituted a major driving force for methodological and conceptual developments in X-ray macromolecular crystallography (MX). Landmarks of the structure determination of viral particles, such as the ones from the first animal viruses or from the first membrane-containing viruses, have often been associated with methodological breakthroughs in X-ray crystallography.

In recent years, the advent of new detectors with fast frame rate, high sensitivity, and low-noise background has changed the way MX data is collected, enabling new types of studies at X-ray free-electron laser and synchrotron facilities. In parallel, a very high degree of automation has been established at most MX synchrotron beamlines, allowing the screening of large number of crystals with very high throughputs. This has proved crucial for fragment-based drug design projects, of special relevance for the identification of new antiviral drugs.

This change in the usage of X-ray crystallography is also mirrored in the recent advances in cryo-electron microscopy (cryo-EM), which can nowadays produce macromolecule structures at resolutions comparable to those obtained by MX. Since this technique is especially amenable for large protein assemblies, cryo-EM has progressively turned into the favored technique to study the structure of large viral particles at high resolution.

In this chapter, we present the common ground of proteins and virus crystallography with an emphasis in the peculiarities of virus-related studies.

N. Verdaguer (✉) · C. Ferrer-Orta
Institut de Biologia Molecular de Barcelona (CSIC), Parc Científic de Barcelona, Barcelona, Spain
e-mail: nvmcri@ibmb.csic.es

D. Garriga
ALBA Synchrotron Light Source, Cerdanyola del Vallès, Spain

M. G. Mateu (ed.), *Structure and Physics of Viruses*, Subcellular Biochemistry 105,
https://doi.org/10.1007/978-3-031-65187-8_4

Keywords Molecular replacement · Non-crystallographic symmetry · Phase problem · Structural virology · Virus capsid · Viral proteins · Virus X-ray crystallography

Abbreviations

AdV	Adenovirus
BTV	Bluetongue virus
cryo-EM	Cryogenic electron microscopy
HIV-1	Human immunodeficiency virus 1
HRV	Human rhinovirus
MR	Molecular replacement
MX	Macromolecular X-ray crystallography
PDB	Protein Data Bank
SX	Serial crystallography
SARS-CoV-2	Severe acute respiratory syndrome coronavirus 2
XFEL	X-ray free-electron laser

4.1 Introduction

"Seeing is believing": visualizing the three-dimensional structures of intact virus particles, as well as their constituent proteins and complexes, provides us a tool for a full understanding of their biological function. Macromolecular X-ray crystallography (MX) is one of the most powerful approaches for visualizing such macromolecular assemblies at atomic resolution. The first high-resolution pictures of intact virus particles were provided some 40 years ago with the crystal structures of small RNA plant viruses [1–4]. This seminal work was followed by two landmark studies in macromolecular crystallography, the structures of human rhinovirus (HRV) (Fig. 4.1) and poliovirus (PV) [8, 9]. By the end of the 1990s, virus crystallography achieved a new breakthrough, the structure of the large core particle (70 nm in diameter) of bluetongue virus (BTV) [10]. The cores of these viruses act as transcriptional machines in infected cells, holding their genomes and transcriptional enzymes within a protective shell so that the viral double-stranded RNA is never revealed to the infected cell, eliminating any chance of cellular response. The BTV structure (Fig. 4.2) showed the assembly of nearly 1000 protein subunits: the core was made from two principal protein components—a thin skin covering the genome and enzymes, made from 120 copies of a large protein VP3 (triangulation number $T = 2$), which assembles into an icosahedral arrangement via conformational switching to yield a pattern of subunits not predicted in the theory of Caspar and Klug (see Chap. 2). This was clothed in 780 copies of the protein VP7 ($T = 13$) in an arrangement that follows the theory of quasiequivalence.

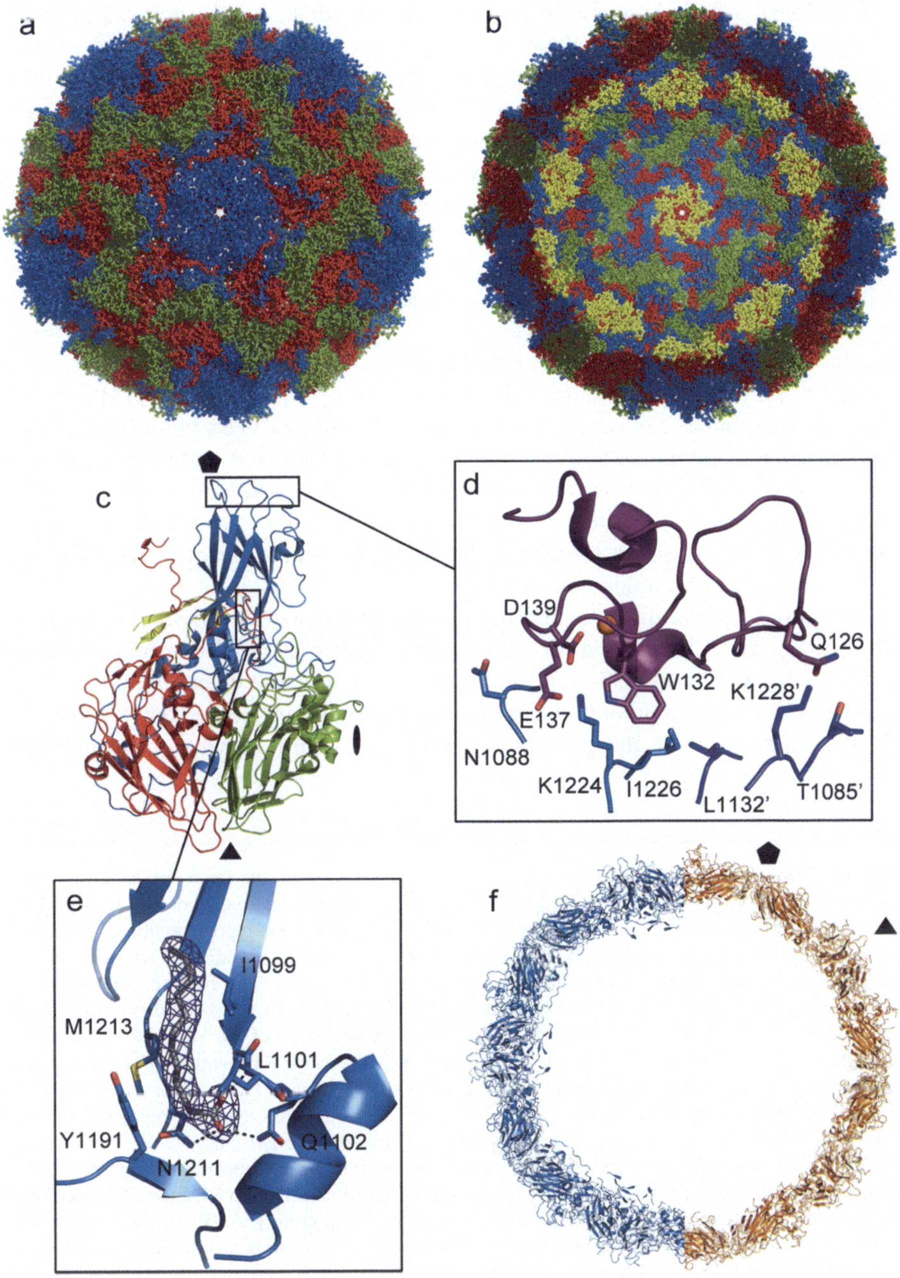

Fig. 4.1 The human rhinovirus capsid. Representation of the external (**a**) and the internal (**b**) surfaces of the HRV2 capsid (PDB code 1FPN) [5], seen along a fivefold symmetry axis. The capsid, of about 30 nm in diameter, is built by 60 protomers, arranged in a T = 1 icosahedral lattice. Each protomer is composed by four different capsid proteins: VP1 (*colored in blue*), VP2 (*green*), VP3 (*red*), and VP4 (*yellow*). VP1, VP2, and VP3 proteins constitute the capsid shell, whereas VP4

Enveloped viruses (see Chap. 2) have rarely yielded crystals; however, solubilized protein components of their membranes have been crystallized and studied with great success. Only very regular structures can form single crystals, and in order to study the molecular details of larger and more complex virus particles by MX, it is necessary to dissect them into well-defined subunits or substructures. This dissection was originally done with proteases, by disassembly or by isolation of substructures from infected cells. For example, the influenza virus hemagglutinin, the first viral glycoprotein for which atomic details were visualized [11], was obtained from crystals of protein cleaved from the surface of purified virions. In addition, the structure of the adenovirus (AdV) hexon was obtained from excess unassembled protein derived from adenovirus-infected cells [12]. Since the middle of the 1990s, this dissection has more commonly been carried out using recombinant proteins (i.e., the gp120 protein from the human immunodeficiency virus (HIV-1); [13]). So far, only one membrane-containing virus X-ray structure has been published: PRD1, possessing an internal membrane, was successfully crystallized, and its structure was solved by MX [14, 15]. Four proteins (MPC, P16, P30, and P31) out of a total of 18 structural proteins identified in PRD1 were resolved in the icosahedrally ordered capsid. A conformational switch of the MPC N- and C-terminus controls the quasiequivalent interactions within the pseudo-T = 25 lattice.

MX-based structural studies of viral proteins and assemblies (see also Chaps. 2, 11, 12, 13, 14, 15, 16, and 17) have been facilitated by the accumulation of experience in their purification, crystallization, and structure determination, as well as by the development of multidisciplinary technologies, including (i) crystallization robotics that automate the process, (ii) third and fourth generation synchrotron

Fig. 4.1 (continued) is located at the interior of the capsid, lying on the inner surface, around the fivefold symmetry axes. A prominent feature of the HRV shell is a star-shaped dome on each of the fivefold axes surrounded by a large depression or "canyon." (**c**) Ribbon representation of the capsid protomer, colored as in **a** and **b** and with the icosahedral symmetry axes indicated with an *ellipsoid* (twofold axis), a *triangle* (threefold), and a *pentamer* (fivefold). The high-resolution crystal structure of a virus is used to gather information for more than the mere spatial disposition of the capsomers; they may also provide the molecular basis of key steps of the virus life cycle. Some examples of this are illustrated in panels (**d**), (**e**), and (**f**). (**d**) Ribbon representation of the structure the HRV2 capsid interacting with the V3 module of VLDL receptor. The cognate receptor for the minor group rhinoviruses (PDB code 1V9U) [6]. One V3 module (colored in purple) interacts with two VP1 molecules from neighboring protomers (*colored in different blue tones*). The residues involved in the virus-receptor interaction are represented as sticks and labeled. (**e**) Electron density map (*dark blue wire*) corresponding to the "pocket factor," a cell derived fatty acid that often occupies the hydrophobic pocket of VP1, at the bottom of the canyon. The VP1 residues interacting with this pocket factor are represented as sticks. The pocket factor can be displaced by a number of antiviral drugs that are thought to rigidify the capsid, preventing the required structural changes needed for viral genome uncoating during cell infection. (**f**) Comparison of the native HRV2 capsid (*blue*) with the 80S subviral B-particle (*orange*), the end-stage of RNA uncoating (PDB code 3TN9) [7]. Only half of a ~ 80 Å slab of each capsid shells are represented as to illustrate the expansion in the 80S particle with respect to the native virion. The positions of one threefold and one fivefold axis of the icosahedral particle are indicated with a *triangle* and a *pentamer*, respectively

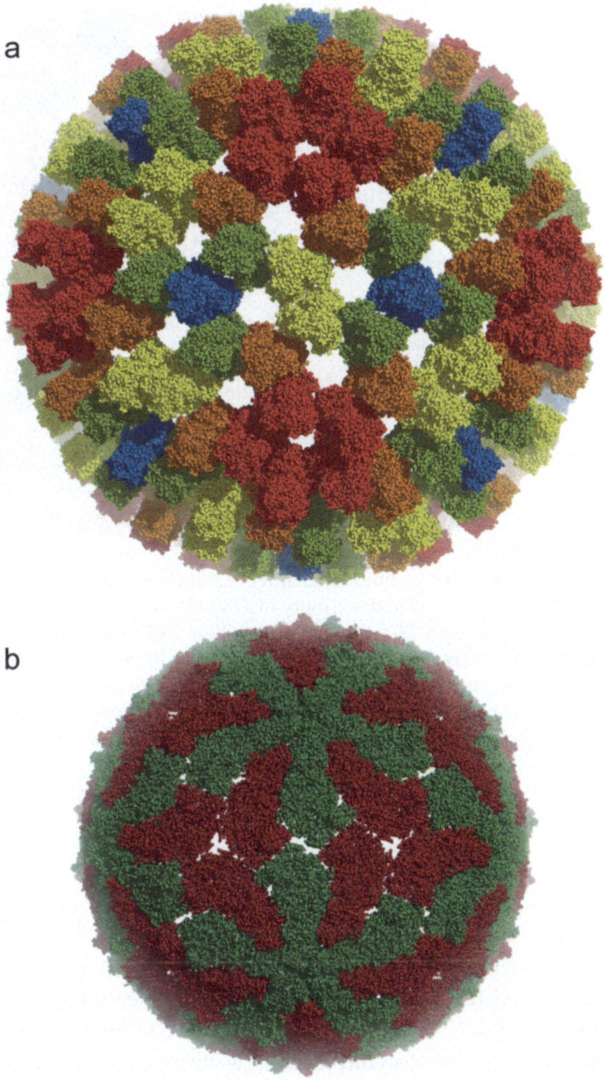

Fig. 4.2 Bluetongue virus core particle structure. Stick representation of the atomic structure of BTV1 virus (PDB code 2BTV) [10], seen along a twofold symmetry axis. The BTV core particle is composed by two differentiated protein shells. The outer shell (**a**) contains 780 copies of VP7 protein, arranged as trimers on a T = 13 quasiequivalent lattice. In the figure, these VP7 trimers are colored according to their relative position to the symmetry axes (*blue* at the threefold axes, *yellow* at the twofold axes, *red* at the fivefold axes, *green* between the two- and threefold axes trimers, *orange* between the two- and fivefold axes trimers). The inner shell (**b**) consists in a thin subcore generated by 120 copies of VP3 protein organized in a T = 2 lattice. The two structural conformations of the VP3 protein are shown in dark red and dark green. Both layers of the core are represented at scale

beamlines with high brilliance and a great degree of automation, (iii) large-area hybrid photon-counting detectors facilitating data collection from weakly diffracting crystals with large unit cells (see below), and (iv) faster computers and improved software for crystallographic data integration and for structure determination. Current frontiers include defining the architectures of very large assemblies, as exemplified by the 3.8 Å X-ray structure of a complete human adenovirus particle [16, 17]. This particle of 150 MDa and an average diameter of 91 nm represents the largest structure solved to date using MX (Fig. 4.3). Furthermore, comparative analyses of multiple structures of these assemblies, captured in different

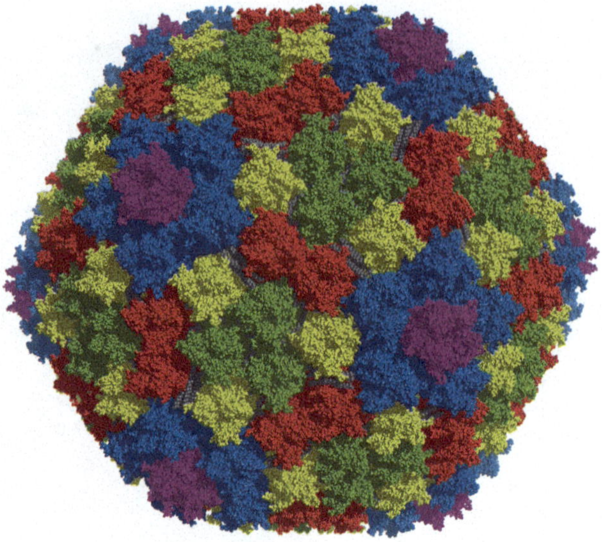

Fig. 4.3 Structure of human adenovirus. Stick representation of the crystal structure of AdV5 (PDB code 1VSZ and updated entry 6CGV) [16, 17]. The adenovirus capsid is mainly composed of hexon protein, with 720 subunits arranged as 240 trimers on a pseudo-T = 25 icosahedral lattice (12 hexon trimers per icosahedral facet). In the figure, the hexon trimers are colored according to their relative position to the symmetry axes: *cyan* around the fivefold axes, *green* around the threefold axes, *red* around the twofold axes, and *yellow* between the two- and fivefold axes. Each of the 20 vertices of the capsid is occupied by 5 copies of the penton base protein (*colored in pink*), in which the Ad fiber shafts are inserted (not shown in the figure). Additional proteins (proteins IIIa, VI, VIII, and IX, the so-called cement proteins, represented as *gray balls* in the figure) mediate and stabilize the interactions between neighboring hexon trimers

conformational states, greatly help to precisely define how they function and are regulated. The growing number of structures of virus-receptor complexes [5, 18] (Fig. 4.1b) (see Chap. 15), mature and immature capsid proteins [19] (see Chap. 14), membrane fusion intermediates [20, 21] (see Chap. 16), or RNA uncoating intermediates [7, 22] (Fig. 4.1f) (see Chap. 15) illustrate some examples. It is now clear that solving the structure of a new virus or a new viral protein at high resolution typically signifies the beginning, not the end, of a modern structural virology study [23]. An example of this is the influence in virus evolutionary studies: the accumulation of viral structures solved over the last decades allows us to classify the virus world based on their structural relationships, giving rise to a new concept in virus evolution that uses specific structural markers, such as the folding of a capsid protein, to infer a common ancestor. The structural approach in virus classification reduces the complexity of studying a large number of genomic sequences to a space of lower variability, such as protein folding. This strategy allows inferring older evolutionary relationships than those detected using methods based on sequence comparison, given that the requirement to maintain a function implies the imposition of restrictions on the protein structure. The power of classification based on structure

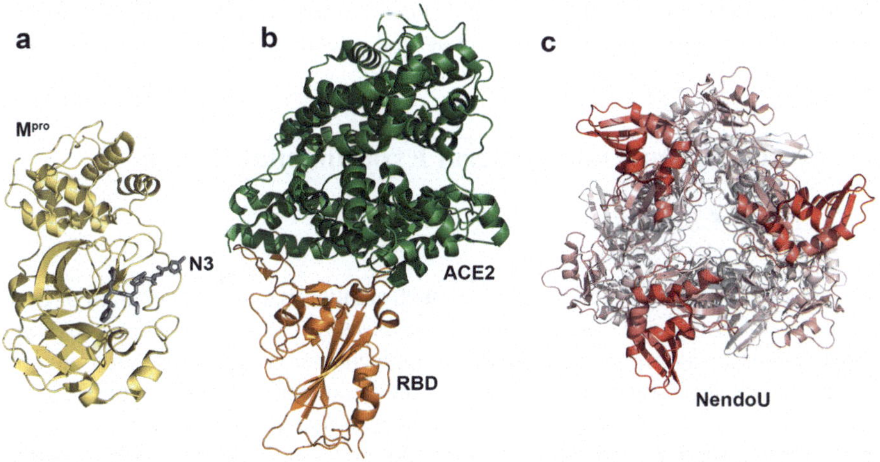

Fig. 4.4 Crystallographic structures of some antiviral targets of SARS-CoV-2. (**a**) The main protease M^{pro} (*yellow cartoon*) in complex with the covalent inhibitor N3 (*gray sticks*; PDB: 6 LU7; 26). M^{pro}, encoded by the nonstructural protein 5 (nsp5), plays a crucial role in viral replication as responsible of the proteolytic cleavage at 11 sites of the polyproteins pp1a and pp1ab. It is considered one of the main targets for drug discovery. (**b**) The structure of the complex of the receptor binding domain RBD of the SARS-CoV-2 spike protein (S; *orange cartoon*) in complex with the cell receptor ACE2 (*green cartoon*), revealing insights into the mechanisms of cell recognition and binding that have served as a basis for the development of new targeted treatments against this pathogen (PDB: 6LZG; [27]). (**c**) Structure of the hexameric nsp15 protein, also known as the uridine-specific nidoviral endoribonuclease (NendoU). The NendoU protein plays an important role in the ability of the virus to evade the host innate immune system by cleaving the polyuridine leader sequence of antisense viral RNA. This structure was solved by serial femtosecond crystallography at room temperature, with an X-ray free-electron laser (PDB: 7K9P; [28]). The NendoU hexamer is represented in cartoons, colored by B-factors (*from blue to red*). The *darkest red* highlights the highest B-factors, representing the highest flexibility of the molecule that is mainly concentrated in the RNA-binding pockets. These structural results, combined with independent cryo-EM observations [29], indicating different RNA-binding affinities to the active sites, have been interpreted such that this enzyme functions as an allosteric switch [30]

becomes evident by the fact that almost half of the currently described viral families can be organized into a relatively small number of lineages, each of which including viruses that infect different domains of life [24, 25] (see also Chap. 7).

Furthermore, MX has been one of the earliest methods used to investigate the binding of small molecule inhibitors or inhibitor fragments to viral proteins and assemblies, providing precise structural information of the protein-inhibitor interactions to guide progression from hit compounds to clinical candidate, playing a central role in high-throughput drug discovery approaches [31–33] (see Fig. 4.4 for recent examples on severe acute respiratory syndrome coronavirus 2, SARS-CoV-2). In addition, in the past decade, a powerful methodology, known as serial femtosecond X-ray crystallography, is enabling structural biologists to investigate how biological molecules transiently interact with amazing precision, shedding new light into the accurate functioning of different viral targets and guiding the development of novel optimized drugs to combat viral infections [29, 30] (see Chap. 20).

4.2 Basic Concepts and General Experimental Design

4.2.1 Basic Concepts

We cannot form images of things that are much smaller than the wavelength of light we are using. X-ray wavelengths are in the order of atom diameters and bond lengths, allowing atoms in a molecule or molecular complex to be individually resolved. However, X-ray scattering from a single molecule would be unimaginably weak. In contrast, crystals arrange a huge number of molecules (e.g., virus particles) in the same orientation, and scattered waves add up in phase, increasing the signal to a level which can be measured.

Figure 4.5 shows two examples of packing of a virus in two different crystal types. The lattice is defined as a construct dividing space into regular translationally periodic units. The combination of the unit lattice with molecular motifs (e.g., a virus particle) generates the unit cell (Fig. 4.5a) of the crystal which is built from translationally stacked unit cells. In addition, the molecules contained in each unit cell can be related by internal symmetry, generating a unit cell packed with multiple symmetry equivalent copies of the molecule (Fig. 4.5b). The number of different ways to pack molecules in space is limited. For asymmetric chiral molecules (such as proteins), there are only 65 discrete and distinct ways to assemble three-dimensional periodic crystals through combination of translational and rotational symmetry. These 65 types of arrangements form 65 space groups. Their symmetry properties and the rules for constructing each crystal structure are described in the International Tables for Crystallography, Volume A [34].

The asymmetric unit of a unit cell contains all the necessary information to generate the complete unit cell of a crystal structure by applying its symmetry operations to the asymmetric unit. Besides the strict crystallographic symmetry between molecules, additional non-crystallographic symmetry can locally exist between molecules in the asymmetric unit. In virus crystals, the molecular symmetry results in most cases from the combination of local and crystallographic symmetry (Fig. 4.5c).

In a basic diffraction experiment, a single molecular crystal is placed into a finely focused X-ray beam, and the diffraction images are recorded in a detector. The electron density representing the atomic structure of the molecules in the crystal is reconstructed by Fourier methods from the diffraction data, and an atomic model of the structure is built into the electron density (Fig. 4.6). It is important to note that the diffraction experiment is fundamentally different from imaging with an optical microscope. The crucial difference is that visible light scattered from objects can be focused through refractive lenses to create a magnified image of the object. In contrast, as no refractive lenses are available for X-rays, the X-ray diffraction

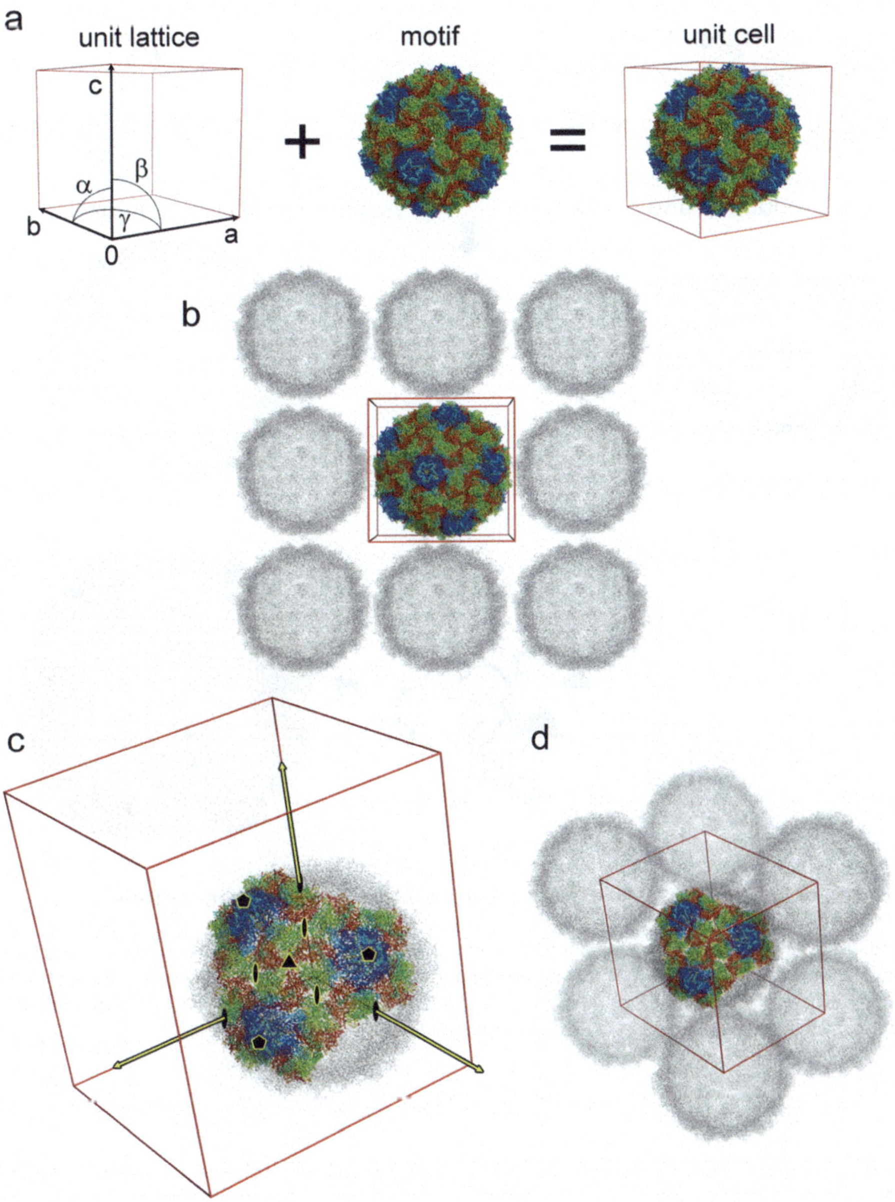

Fig. 4.5 Assembly of virus crystals. (**a**) The crystal unit cell is built by a motif (in this case, a whole virus) that fills the unit lattice. The unit lattice is defined by three basic vectors [a, b, c] that extend from a common origin [0] in a right handed system. The angle between a and b is γ, the angle between b and c is α, and the angle between a and c is β. The stacking of multiple unit cells will produce a crystal. (**b**) Crystal packing of an icosahedral virus, in which the whole virus is the asymmetric unit (*colored*). The other viruses of the crystal (*gray*) can be obtained by translation of the unit cell. For clarity, only one two-dimensional layer of the crystal is shown. (**c**) An icosahedral

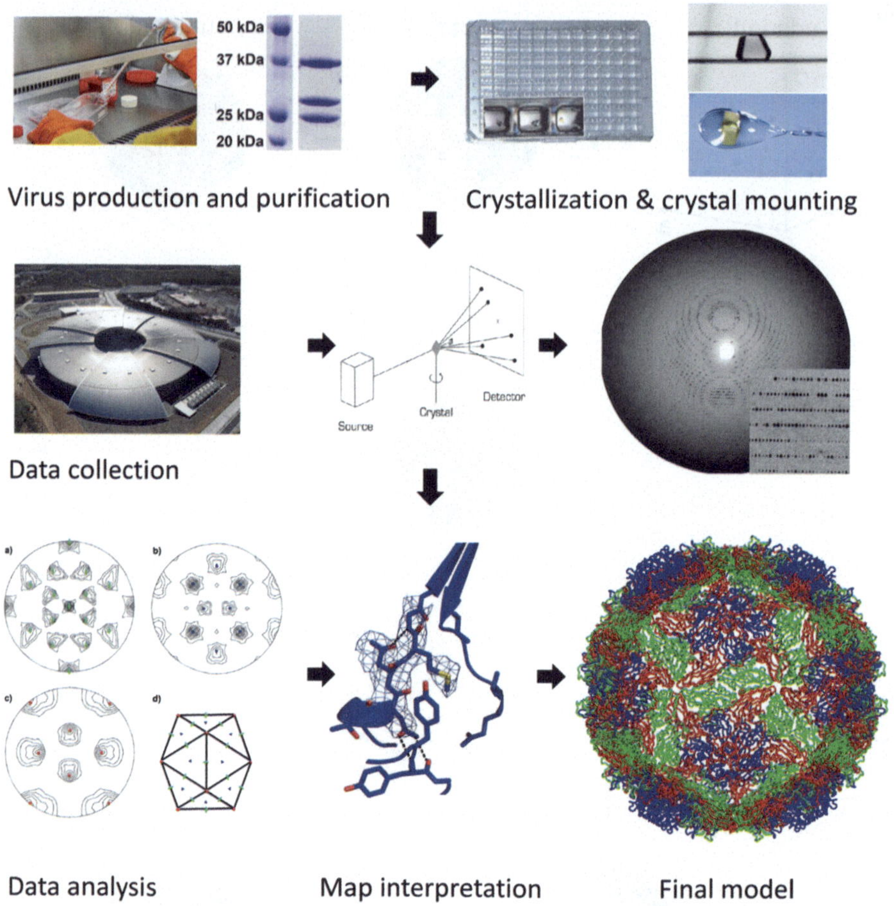

Fig. 4.6 Virus structure determination by X-ray crystallography. The solution of a virus structure involves the following steps: (i) preparation and purification of virus samples, (ii) crystallization and mounting the virus crystals, (iii) measurement of the diffraction data, (iv) phase calculation usually by molecular replacement, and (v) map interpretation and model building

Fig. 4.5 (continued) capsid has 12 fivefold symmetry axes (indicated with pentamers, in the figure), 20 threefold symmetry axes (triangles), and 30 twofold symmetry axes (ellipsoids), relating the quasi-identical subunits that build up the virus capsid. When this virus is fitted in a crystal lattice, it may happen that some of the icosahedral symmetry axes coincide with the symmetry axes defined by the unit cell (*yellow arrows*); in this case, only part of the virus forms part of the crystal asymmetric unit (*colored*). The rest of the virus particle (*gray*) can be obtained applying the crystal symmetry operators to the asymmetric unit. (**d**) The crystal packing of an icosahedral virus, in which one-fourth of the virus forms the crystal asymmetric unit patterns do not provide a direct image of the molecular structure. The electron density of the scattering molecular structure must be reconstructed by Fourier transform techniques. For this type of reconstruction, from reciprocal diffraction space back into the direct molecular space, two terms are required: (i) the structure factor amplitudes, directly accessible in the form of the square root of the intensities of each measured diffraction spot or reflection, and (ii) the relative phase angle of each reflection. These phase angles are not directly accessible and must be supplied by additional phasing experiments. The absence of directly accessible phases constitutes the so-called phase problem in macromolecular crystallography (see Sect. 4.6).

At the end of the chapter (section on Further Reading), three reference books are provided for an extended description of the general principles and techniques in X-ray crystallography of biological molecules. The first two books are aimed at non-practitioners; the last one contains a more in-depth treatment.

4.2.2 General Experimental Design

The general approach to structure determination of viral particles, proteins, and other macromolecular complexes by X-ray crystallography is schematically shown in Fig. 4.6. The procedure involves the following basic steps: (i) preparation and purification of virus samples; (ii) crystallization and crystal mounting; (iii) measurement of the X-ray diffraction data; (iv) phase calculation, usually by molecular replacement (MR); and (v) map interpretation and model building. These steps will be briefly described in the following sections in this chapter (Sects. 4.3, 4.4, 4.5, 4.6, and 4.7).

The first limitation for an X-ray experiment is the requirement of well-ordered three-dimensional crystals. This implies that we cannot determine the structure of a macromolecular complex that does not have a single, well-defined structure or that is difficult to obtain in a pure and stable state. Furthermore, the structure determination of an intact virus particle entails a significant scaling up in most respects compared to a traditional crystallographic analysis of a single protein. At a fixed resolution, the number of reflection spots to be measured is proportional to the volume of the asymmetric unit of the crystal. For a virus particle, there are several hundred times as many data to be measured than for an isolated protein. Not only are there far more X-ray reflections to be measured, but these reflections will each be, on average, several hundred times weaker. Therefore, virus crystallography makes considerable demands on the qualities of X-ray sources, the performance and sensitivity of X-ray detectors and computer hardware and software [35].

Having a well-diffracting single crystal is a first step, though not necessarily sufficient, to solve the structure. As mentioned in Sect. 4.2.1, X-ray detectors are able to measure the position and intensity of reflections without recording the phase information. Therefore, in order to compute the Fourier transform, to go from the diffraction pattern to the electron density map of the structure, we need to obtain information about the phases. In virus crystallography, molecular replacement is now the method most frequently used to estimate the phases, needed to solve crystal structures (see Sect. 4.6). Spherical viruses are made from protein shells with icosahedral (532) symmetry (Chap. 2). According to the theory of quasiequivalence, the icosahedral shell is composed of 60 × T chemically identical subunits which can then, in principle, make similar interactions to form the capsid (see Chap. 2). The presence of icosahedral symmetry leads to the presence of non-crystallographic redundancy (with a minimum of fivefold non-crystallographic symmetry, for those viruses that crystallize with the viral capsid lying on a point of 23 crystallographic symmetry). This redundancy is of enormous importance in a crystallographic analysis, providing extremely powerful constraints that facilitate many aspects of the analysis [35] (Sect. 4.6).

4.3 Production and Purification of Viral Particles and Proteins for Structural Studies

4.3.1 Growth and Purification of Intact Viruses and Viral Capsids

The first step in the route to solving the structure of a viral particle by X-ray crystallography (Fig. 4.6), and a prerequisite for crystallization of the particle, is the availability of sufficient amounts of the sample in purified form. For virus particles, the purification from native sources often relies on classical biochemical protein purification procedures (see Chap. 22). Virus yields vary, and, for animal viruses grown in a monolayer cell culture, a vast number of roller bottles or large culture plates may be required. A key step in purification is usually ultracentrifugation using sucrose and/or CsCl gradients. In addition, chromatographic FPLC techniques are also available to separate high-molecular-weight species. Many virus particles may be readily concentrated using 100-kDa micro-concentrators or precipitated by ultracentrifugation. Standard spectrophotometric concentration measurements should be corrected for the nucleic acid content; an approximate formula is as follows: virus particle concentration = $(1.55\ A_{280}\text{–}0.76\ A_{260}) \times$ dilution [36].

With a macromolecular assembly, there is a possibility of crystallizing a substructure such as an assembly intermediate (see Chaps. 11 and 12). Homogeneity and integrity of the sample used for crystallization and of the crystallized molecule should be carefully analyzed. The crystallization of intact virus particles can be verified using SDS-PAGE to demonstrate the presence of the correct complement of proteins, using tests to show recovery of infectivity from dissolved crystals and by electron microscopic analysis of the crystals.

4.3.2 Expression and Purification of Recombinant Viral Proteins

Low yield virus production and sample heterogeneity sometimes impede purification of native virions with the quality and in the quantity required for structural studies. The developments in molecular biology techniques that allow cloning the viral genomes or capsid protein genes for overexpression and assembly of recombinant virus-like particles (VLPs) in heterologous systems are instrumental for the structural studies of these viral assemblies. Recombinant techniques can further provide the means to modify complexes at the gene level, resulting in alteration of the protein complex subunits to meet the high quality requirements for structural analyses, as well as for site-directed structure-function studies. A successful example in the application of these techniques was achieved in the X-ray structure determination of human adenovirus type 5 (AdV5). High quality crystals were obtained of a genetically modified virus, in which short penton fibers from another serotype have replaced those of the wild-type AdV5 [16]. Such strategies can also be applied for the recombinant production of viral nonstructural proteins.

Molecular cloning, protein expression, and crystallization are the major preparatory steps of a structural study. All three processes are experimentally driven and its outcome cannot be predicted. A parallel approach pursuing multiple variants of the protein target significantly increases the chances of success. A number of high-throughput initiatives have been developed using affordable methods and equipment to automate molecular cloning of expression constructs, recombinant expression screenings, and purification. The advantages of automation are obvious: procedures can be carried out in parallel and scaled up accordingly to process large amounts of samples at reasonable cost. The methodology is also available for overexpression of multicomponent protein complexes, either in prokaryotic or eucaryotic cells [37, 38].

4.4 Crystallization

The biochemical and biophysical process of crystallization is still a largely empirical process. Crystals grow from an aqueous protein solution, when the solution is brought into supersaturation (Fig. 4.7). Supersaturation is achieved by varying concentrations of precipitant, protein, additives, and other parameters affecting solubility, such as pH and temperature. Crystallization proceeds in two phases, nucleation and growth. Nucleation is the conception stage of crystallization during which a molecular aggregate of critical size forms, allowing for subsequent crystal growth. Once the crystal nucleus has formed, crystal growth follows spontaneously (see [39] for a complete view).

The crystallization process can be effectively illustrated using a phase diagram (Fig. 4.7). The phase diagram indicates which state is stable under different chemical

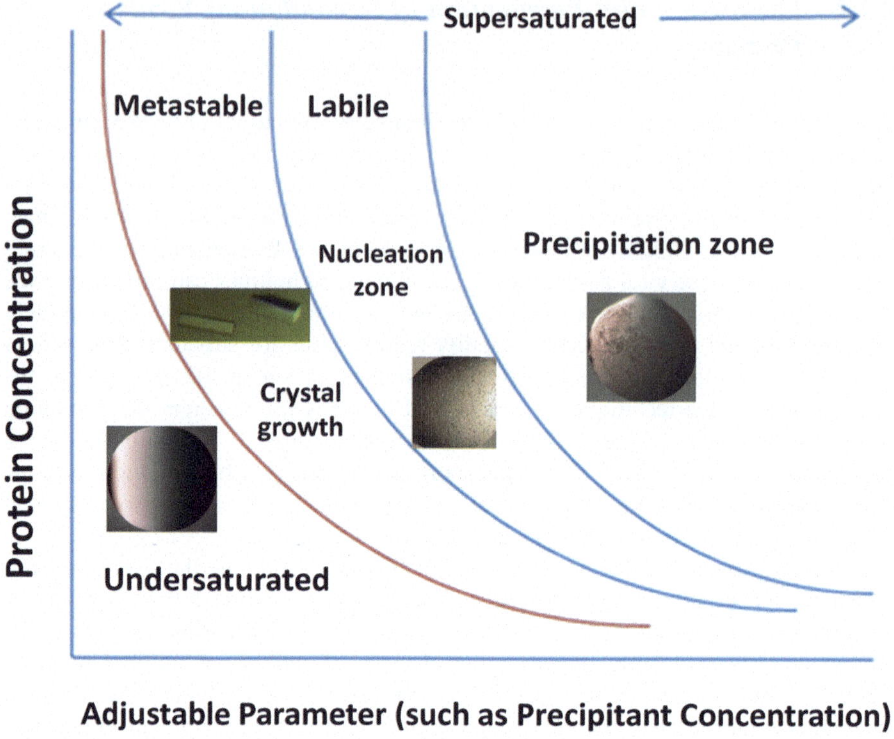

Fig. 4.7 Crystallization phase diagram. A schematic drawing of a protein crystallization phase diagram based on two of the most commonly varied parameters, protein and precipitant concentrations. The figure schematically illustrates four areas: (i) an area of very high supersaturation where the protein will precipitate; (ii) an area of moderate supersaturation where spontaneous nucleation takes place; (iii) an area of lower supersaturation, just below the nucleation zone where crystals are stable and may grow. This area, known as metastable zone, is thought to contain the best conditions for growth of large and well-ordered crystals: (iv) an undersaturated area where the protein is fully dissolved and will never crystallize

and physical conditions. It provides means of quantifying the influence of parameters such as the concentrations of protein, precipitants, additives, etc. on the production of crystals, allowing the design of crystal optimization experiments. Crystallization conditions for viruses and other complex assemblies are similar to those of single proteins.

4.4.1 Classical Methods

There are four major techniques for growing crystals of biological macromolecules. These techniques have been in use for over 40 years and include batch crystallization, dialysis, liquid-liquid diffusion (also known as free interface diffusion), and vapor diffusion [39]. The four methods work well, the latter being the most extended technique for viral particle crystallization. The typical time of crystal growth, using any one of these techniques, is 1–3 weeks. However, the time for obtaining crystals can vary from several hours to as long as a year. Standard virus concentrations for the first crystallization trials range around 10–20 mg/ml; however, if the solubility of the virus is limited to less than this, it is still worth trying at whatever concentration can be achieved.

The method of vapor diffusion begins with an aqueous crystallization drop containing the macromolecule and the crystallization agents at a concentration lower that that required for the formation of crystals. This drop (typically from 0.5 μl to 10 μl) is equilibrated against a reservoir (0.5–25 ml) containing a solution with crystallization agents at different concentrations than in the crystallization drop. Usually, the crystallization cocktail is used as reservoir solution but it is not a requirement—the reservoir solution and the crystallization cocktail can be different chemical species [40]. Equilibration proceeds by diffusion of the volatile species, typically water, until vapor pressure equilibrium is reached between the experiment drop and the reservoir. As a consequence of the equilibration process, there is an increase in the relative concentration of solutes in the crystallization drop. By decreasing the amount of water in the drop, the reservoir acts to gradually concentrate the protein and other chemicals in the crystallization drop. This gradual change and increase in the concentration may lead to supersaturation of the macromolecule and subsequent crystallization. Three commonly used vapor diffusion methods are i) hanging drop, where the drop is suspended over a reservoir from the underside of a hydrophobic surface; ii) sitting drop, where the drop sits in a small concave surface; and iii) sandwich drop, where the drop makes contact with two narrowly separated surfaces.

The first step of finding initial crystallization conditions for a new macromolecule is to set up screening trials, exposing the macromolecule to numerous different chemical agents in different drops in order to find "hits" or "leads" that point to conditions that may be conducive to crystallization. Crystals, crystalline precipitate, and phase separation are all considered leads that are worth pursuing.

One of the main phenomena preventing crystallization is particle aggregation. Electron microscopy is a useful tool for checking this, in addition to methods such as dynamic light scattering. If aggregation proves to be a problem, it may be relieved by the addition of detergents. Other useful additives include divalent cations known to interact with some viral capsids. With viruses, highly ordered crystals are often obtained once the optimum crystallization conditions are found. This may be attributable to their isometric nature and reflects the perfection with which these systems assemble. There is a tendency toward high symmetry space groups, often with some of the capsid symmetry axes coinciding with those of the crystal [35].

4.4.2 *High-Throughput Approaches*

Nowadays, structural studies of proteins and large molecular assemblies are facilitated by the development of the robotic crystallization technology. The use of robotic nanoliter screenings allows exploring a vast number of different conditions with a small amount of sample (usually just few milligrams for a whole 98-well plate) [41, 42]. High-throughput technologies render the—often repetitive—crystallization screening process, much less labor-intensive and sample consuming and with fewer errors and better recordkeeping than most humans. Many stages of the procedure are automated, starting with the mixing of concentrated chemical stock solutions into a set of crystallization cocktails, to dispensing the macromolecule plus precipitant mixture droplets into plates or even sealing the plates.

Correspondingly, automated solutions have also been developed to assist with the later steps of the process: monitoring the crystallization drops, scoring the outcomes, and even suggesting new experiments based on those outcomes. Current commercial drop imaging systems are equipped with several imaging methods, each providing different information to evaluate the potential crystals growing in the droplets, including visible light, polarized light (for birefringent imaging, which gives increased contrast for most protein crystals), ultraviolet light (UV; to determine if crystals are protein or salt), multi-fluorescence imaging (MFI; for protein-protein complexes), fluorescence recovery after photobleaching (FRAP; ideal for LCP crystal identification), and second-order nonlinear imaging of chiral crystals (SONICC; able to detect microcrystals, <1 μm) [43]. Combined with these technologies, powerful machine learning algorithms have been developed to assist classification of images from protein crystallization droplets, achieving accuracies $>90\%$ [44, 45].

All these developments have significantly raised the success rate of the crystallization process and reduced the time required to identify crystal conditions for new targets.

4.5 Data Collection and Processing

Once adequate crystals are available, they are mounted in an appropriate holder and irradiated with X-rays (Fig. 4.6). Macromolecular crystals commonly suffer rapid radiation damage during room temperature X-ray data collection. Therefore, data are now routinely collected with the sample held at around 100 K, usually resulting in higher resolution and better quality data. In addition to the cryo-protection techniques, other sophisticated post-crystallization treatments, as crystal dehydration, are crucial for the structure determination of large macromolecular assemblies.

4.5.1 Cryo-Crystallography

Macromolecular crystallographic data are nowadays routinely collected at cryogenic temperatures (100 K). The key advantage of this technique is the enormous reduction in secondary radiation damage to the sample during X-ray data collection compared to data collection at room temperature. This is because, at cryotemperatures, most of the free radicals produced by the incident radiation have much lower diffusion rates, so they do not travel as far through the crystal, thus avoiding as much damage. The prolonged crystal lifetimes usually result in better quality and higher resolution data. Nevertheless, even in samples that are cryocooled, radiation damage is often observed at undulator synchrotron beamlines. It is a limiting problem in the optimum use of these high brilliant beams, which are routinely attenuated or defocused.

Currently, mounting a crystal for cryo-crystallography involves suspending it by surface tension in cryo-buffer within a thin fiber loop (Fig. 4.6) and then flash-cooling it in gaseous or liquid cryogen. The cryo-buffer is designed to vitrify and prevent crystalline ice formation [46]. Cryo-crystallography experiments are affected by two classes of variables: first, the physical and chemical environment of the crystal in the loop, including choice and concentration of cryoprotectant agent, solvent content of the crystal, size of the crystal, and surface area/volume ratio of the crystal, and second, the external variables, including the cooling regime (the cryogen used for flash-cooling, the cryogen flow rate, and the temperature used during data collection) and the incident beam conditions of dose, dose rate, flux, and wavelength.

The most popular cryoprotectant is glycerol at concentrations between 20% and 25%. This does not mean that it is the best cryoprotectant; most experiments involve little optimization of cryoprotocols, and glycerol is often just the easiest available solution that will "do the job." The optimum scenario, however, is to grow the crystal in a solution that is already cryoprotected.

Due to faster heat transfer, small crystals are generally easier to flash-cool than larger ones. Larger crystals, however, have a greater diffracting power and will yield more useable diffracted photons before reaching their dose limit. There is thus a minimum crystal size that can be expected to give a data set at a specified resolution for a certain unit cell.

4.5.2 Other Post-crystallization Treatments

Data collection at cryogenic temperatures, which reduce radiation damage and increase crystal lifetime in the X-ray beam, is essential when dealing with crystals of large macromolecular assemblies. However, the development of protocols for cryo-crystallography in large viruses sometimes fails. Even in sufficiently cryoprotected crystals, rapid freezing can introduce lattice disorders that reduce the diffraction limit. This effect, most likely attributable to the unequal spatial contraction of the solvent and the crystal lattice, is more pronounced for larger crystals with high solvent content [47]. Furthermore, to reduce safety hazards, crystals of infectious pathogenic viruses may have to be mounted in sealed capillary tubes for data collection. In these cases, a large number of crystals will have to be examined in order to solve the structure. Alternative post-crystallization treatments may be helpful to reduce the amount of virus crystals needed for structural studies of human pathogens.

Crystal dehydration can lead to remarkable improvements of diffraction quality. The controlled reduction of the solvent content in a crystal decreases the volume of the unit cell, altering in most cases the crystal packing. These rearrangements may reinforce existing and create new crystal contacts and can thereby lead to better ordered and better diffracting crystals. A sophisticated example of this application has been found in the structural analysis of the membrane-containing bacteriophage PRD1. Crystals of PRD1 were very delicate, and the mechanical handling involved in preparing them for data collection rendered them useless for this purpose. In this case, a method for growing PRD1 crystals in thin-walled quartz capillary tubes by vapor diffusion was developed, which obviated the need for crystal handling and rendered X-ray data collection practical. These crystals grown in capillaries were further treated by the addition of small amounts of a 20% (w/v) PEG20K solution injected into the capillary prior data collection [48].

The new developments in techniques for measuring diffraction data directly from crystals in their crystallization plates allow to bypass a number of problems associated with crystal handling and cryo-protection. By eliminating crystal handling, plate-based data collection in modern synchrotron microfocus MX beamlines offers fully automated data collection of macromolecular crystals in situ, to obtain diffraction data for crystals adverse to cryogenic harvesting (e.g., viruses, membrane proteins), but also providing a safe platform for the analysis of pathogenic particles [49, 50]. This methodology also allows a rapid characterization of protein-ligand interactions by collecting large numbers of data sets from similar crystals with different compounds added in industrial drug development [50].

4.5.3 Large Unit Cells and Fragile Crystals

Large unit cell dimensions are inherent to crystals of viruses. Therefore, various crystallographic challenges have to be undertaken when starting with data collection; crystals with large unit cells diffract weaker because fewer unit cells per crystal volume contribute to scattering. Furthermore, as a consequence of long unit cell axes, the distances between reflections in reciprocal space are small, leading to closely spaced diffraction spots on the detector. Modern undulator beamlines in third generation synchrotrons offer the best instrumentation for successful data collection when working with large unit cells. Their high beam intensity compensates for the weak diffraction, while the small beam divergence and low energy bandwidth decrease spot size, reducing the problem of overlapping reflections. The increased surface area of modern detectors allows crystal to detector distances to be increased; this further reduces the problem of overlapping reflections associated to the very densely spaced Bragg reflections characteristic of diffraction patterns from crystals with large unit cells [35, 51]. It also increases the signal to noise ratio, as the background radiation is inversely proportional to the square of the detector distance. In addition to the best possible instrumentation, an optimized data collection strategy is needed for a successful experiment. Among the general considerations that should be taken into account, the orientation of unit cell axes is of special importance. If one of the unit cell axes is substantially longer than the others, it should be aligned with the spindle axis of crystal rotation to avoid overlapping reflections from neighboring lunes (groups of reflections formed from the same plane in the reciprocal lattice, appearing on the diffraction images from rotating crystals, especially those with large unit cell dimensions, as is the case of crystallized viral particles) [47].

The strong beam intensity and long exposures necessary to collect high-resolution diffraction data from crystals of large macromolecular complexes cause severe radiation damage and frequently limit the amount of data that can be collected from one crystal, even at cryogenic temperatures. In favorable cases, only a few crystals are needed to collect useful data (i.e., for frozen crystals of a virus belonging to a highly symmetric space groups). On the other hand, for other more challenging analyses, a large number of crystals can be required (e.g., in the studies on BTV and reovirus cores over 1000 crystals, mounted in sealed capillaries, had to be examined [10, 52]). Potential problems with non-isomorphism of crystals and crystal quality can be associated with data collection from many crystals. Post-crystallization treatments will reduce problems with non-isomorphism and variations in cell dimensions.

The measurement of low-resolution reflections is often overlooked in traditional protein crystallography. For virus crystallography, these reflections can be very important, for instance, in using a low-resolution cryo-electron microscopy reconstruction as a phasing model (see Sect. 4.6 and Chaps. 3 and 7) or for visualizing less well-ordered structures [53].

4.5.4 New Methods

X-ray crystallography has witnessed a massive development over the past decade, thanks to the large increase in the intensity and brightness of X-ray sources (see Sect. 4.9) but also to the advent of new types of high frame rate detectors.

Hybrid photon-counting (HPC) detectors allow counting the number of single events (i.e., *single photons* of a certain energy reaching the detector) within a given time interval. Compared to the older charge-coupled device (CCD) technology, pixel-array HPC detectors offer a much higher sensitivity and a very low-noise background, together with previously unthinkable fast frame rates.

With these detectors, diffraction spots with significantly less intensity can be recorded, which results particularly benefiting for experiments using microcrystals or in the studies with large macromolecular complexes, both typical conditions when working with virus particles. In addition, since detector sensitivity is higher, lower radiation doses can be used, meaning that much less radiation damage will be induced. In parallel, the fast frame rates (reaching over 2 kHz for large-area detectors at full bit depth [54]) and negligible dead times between frames have unlocked the possibility to collect in continuous (shutterless) mode.

This way, the traditional data collection strategy, based on high-dose, low multiplicity, and coarse φ-slicing, has been gradually replaced by new protocols based on a shutterless data collection with low-dose, high multiplicity, and fine φ-slicing strategies [55].

On the other hand, such faster frame rates enable new screening methods, such as fast X-ray based scanning (i.e., sample rastering) to detect diffracting (micro)crystals on a sample support. In turn, this capability unlocked the establishment of automated sample centering procedures and unattended data collection protocols [56] but also the possibility to collect a partial data set, or even just one single still image, from each identified crystal in a given sample support, which can later on be combined into a complete data set. This method, termed serial crystallography (SX) [57], has been rapidly implemented at most X-ray free-electron laser and synchrotron facilities and has since evolved into a myriad of versions, based on different sample delivery strategies. One of the advantages of serial data collection is that it allows to bypass the effects of radiation damage inflicted on macromolecular crystals during diffraction experiments (see Sect. 4.5.1). Since it is only using information from image (s) collected from new, unaffected crystals, before these result affected by radiation, the resulting data can maintain the high-resolution information of the crystals and prevent the appearance of artefacts in the solved structures which affect biological interpretations (e.g. broken disulfide bonds, decarboxylation of glutamate and aspartate residues).

In addition, since SX does not require cryogenic conditions, collection of still diffraction images at room temperature has been developed in recent years to study protein dynamics during reactions taking place at the molecules within the crystal. Such time-resolved crystallography studies can describe multiple time points of an enzymatic reactions with time resolutions in the order of milliseconds, when using synchrotron radiation, and femtoseconds when performed at X-ray free-electron laser (XFEL) facilities. Despite being a very recent methodology, reports on this kind of studies performed on viral proteins are starting to appear [28, 30].

4.5.5 Data Processing

The most common data collection method in macromolecular crystallography is the rotation method that involves, as the name indicates, a simple rotation of the crystal around a single axis in small increments (generally between 0.1° and 0.05° for viruses) while the crystal is exposed to X-rays and the reflections are recorded on an area detector (Fig. 4.5).

Once the crystal is centered, the first rotation image is recorded, and the raw electronic signal from the detector is corrected by the detector software for detector-specific nonuniformity defects. The corrected raw image includes a header section with relevant experimental parameters, such as detector distance, beam center position, wavelength, etc., and is displayed in a graphical display window. The image is then visually inspected for a number of qualitative key parameters of the crystal. Good diffraction implies single, resolved, and strong spots, extending far out in the diffraction angle to high resolution.

Indexing is the assignment of a consistent set of unit cell vectors matching the diffraction pattern. In order for indexing to succeed, the position of the incident primary beam on the detector and the crystal-detector distance needs to be known with precision. For crystals with large unit cells, defining the correct starting values for the detector parameters is the most important for quality indexing. Once a unit cell has been selected, its lattice symmetry determines the minimal symmetry of the reciprocal space. Based on lattice symmetry, a data collection strategy can be determined, aiming to collect at least enough reflections to cover the entire asymmetric unit of the reciprocal space, in order to obtain complete data. This, however, has become less critical with the use of the newer pixel-array detectors, which allow much lower radiation doses to be used, thus delaying the loss of diffraction of the exposed crystal. Once all image frames are collected, the data processing programs must convert the raw pixel intensity of the measured and indexed diffraction spots into a list of properly scaled intensities and their estimated standard errors. The quality of data is normally judged by the agreement of multiple recorded or symmetry related data.

Nowadays, (initial) data processing can be performed in a highly automated way at MX beamlines from most synchrotron sources. In addition to being convenient and more time efficient, the main advantage of automated data processing is that the obtained results are not as dependent on the experience of the experimenter with a particular program, but rather on the data quality itself.

4.6 Phase Determination

The large unit cell dimensions inherent to virus crystals posed also daunting challenges to the determination of phases required to compute the electron density maps from diffraction data using Fourier transformations (Fig. 4.8). However, the first animal virus crystal structures [8, 9] were landmarks confirming the power of density modification, in particular non-crystallographic symmetry averaging, to solve the phase problem, clearing the path toward more accurate and more complex virus structures [10, 15, 16]. In virus X-ray crystallography, the solution to the phase problem usually proceeds in two differentiated steps with first the determination to a

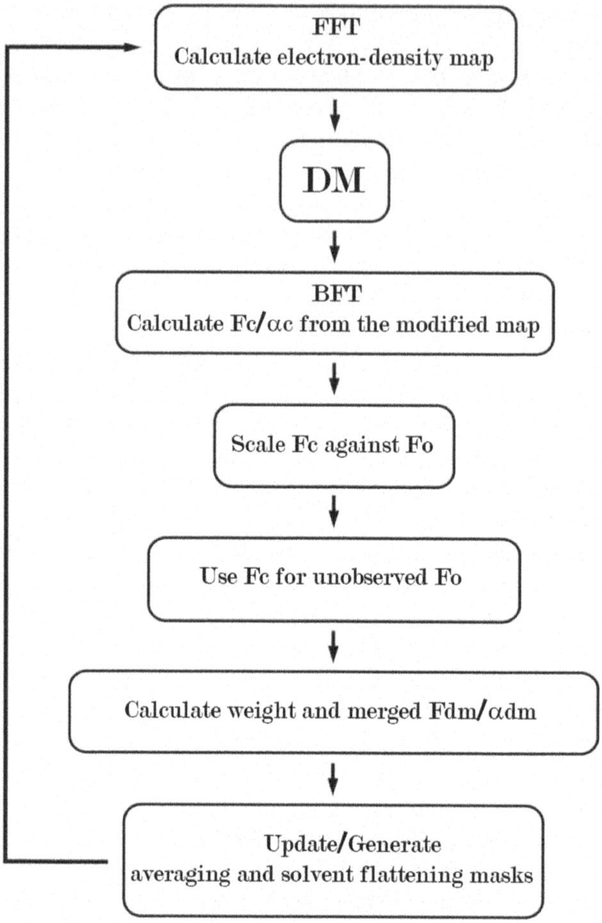

Fig. 4.8 Flowchart of the density modification iterative procedure. Electron density maps, obtained with a Fourier transformation (FFT step), are modified according to the density modification constraints available (DM step), particularly non-crystallographic symmetry averaging and solvent flattening in virus crystallography. The back Fourier transform is then applied (BFT step) to obtain the corresponding structure factor amplitudes and phases (Fc and αc, respectively). After the experimental and calculated amplitudes are scaled together, the calculated structure factors are used to fill the unmeasured reflections. Finally, new amplitudes and phases, ready to provide a new electron density map, are obtained combining and weighting reflections mainly according to differences between experimental and calculated amplitudes. At this point, new solvent flattening and averaging masks, defining the volumes where these density constraints can be applied in the next DM cycle, are calculated. The process can be repeated iteratively until reaching convergence

certain resolution of some initial phases that are later refined and extended to higher resolution.

This section contains a number of technicalities presenting the approaches used to solve the phase problem in virus crystallography. Readers interested mainly in an overall perspective should only retain the idea of the central importance of the phase problem in diffraction experiments and that some powerful procedures have been developed to solve it in virus crystallography. Interested readers can find more in-depth explanations in the references listed under *Further Reading* at the end of this chapter, in particular the last reference.

4.6.1 Initial Phase Determination in Crystals from Icosahedral Virus

To obtain a set of initial phases for an icosahedral virus crystal, it is, in general, necessary to determine first the orientations and positions of virus particles in the asymmetric unit of the crystal. Orientations can be determined using the self-rotation function (or even better the locked self-rotation: a self-rotation obeying the restraints imposed by the particle symmetry). Computation of the self-rotation, which requires information only about the structure factor amplitudes, is usually performed in polar coordinates and represented in sections at constant values of the rotation angle κ corresponding to the icosahedral symmetry axes (Fig. 4.6). Once orientations have been determined, information on the position of the center of the viral particles, the translational parameters, can generally be derived from packing considerations. In some cases, electron microscopy analysis of thin-sectioned crystals has permitted direct visualization of packing with the determination of the number and position of particles in the unit cell [58]. Often the symmetry of the crystal can constraint the possible orientations and positions of the virus particles by requiring the coincidence of crystal and viral particle symmetries (Fig. 4.5c). For example, when three orthogonal crystal twofold axes superimpose with three twofold axes from the icosahedral particle, only two alternative orientations are possible. In this particular case, the position of the virus particle is also fixed by the packing that places the center of the icosahedra at the intersection of the three crystal axes. Coincidence of crystal and icosahedral axis reduces the multiplicity of the non-crystallographic symmetry that is 60 when a complete icosahedral particle is found in the crystal asymmetric unit. Fivefold axes are not among the possible crystal symmetries, and consequently the fivefold axes from icosahedral particles will remain always as local, non-crystallographic, symmetries.

When orientations and positions of the viral particles in the unit cell are known, initial phases can be obtained using either isomorphous replacement (with heavy atom derivatives) or MR. Isomorphous replacement, historically the first method capable of solving the phase problem in protein crystallography, requires the preparation of heavy atom derivatives, which are crystals isomorphous to the ones from the native sample (ideally with the same packing and the same molecular structures), but containing also a few extra heavy atoms (with high atomic numbers) [59]. Diffraction differences between native and isomorphous crystals can then provide information about the phases. In turn, MR requires the availability of independent structural information about the sample that could be used to obtain approximate phases or, at least, some constraints on the possible phases. This independent structural information can be very diverse ranging from atomic models or electron density maps of related samples to the presence of non-crystallographic symmetries. Anomalous diffraction, an alternative phasing methodology used frequently in protein crystallography, has essentially never been applied in virus crystallography mainly due to the experimental difficulties of collecting the weak anomalous signal with enough accuracy from virus crystals.

Preparation of isomorphous heavy atom derivatives is similar to all protein crystals including virus crystals, but because there are 60 equivalent units in the icosahedral particle, at least 60 positions on the particle are expected to be modified, which results in a large number of vectors (at least 60 × 59) between heavy atoms within one particle. Information about the non-crystallographic symmetry is taken into account both during the heavy atom search, which can be restricted to an asymmetric unit of the icosahedron, and also during the refinement of the heavy atom parameters [60]. Heavy atom derivative was the method used to solve the first icosahedral virus structures such as TBSV, SBMV, STNV, or human rhinovirus (HRV) 14 [2–4, 60]. However, for about the last 30 years, the vast majority of new virus crystal structures have been solved without heavy atom derivates. Instead, a diversity of ingenious molecular replacement approaches, including the use of new in silico models generated by AlphaFold [61–63] (see Chap. 7), in combination with the presence of non-crystallographic symmetry has revolutionized virus crystallography, allowing the explosive rise of the number of crystal structures from viruses and viral proteins that have been determined.

When the atomic coordinates of an accurate molecular model are available, classical MR techniques can be employed to place the searching model in the asymmetric unit of the cell of the unknown structure. The calculated structure factors provide then approximate phases and amplitudes, with correlation coefficients between observed and calculated amplitudes often used as an estimation of the quality of phases at a given resolution. Starting at this stage, very accurate phases for the unknown structure can be derived reaching near-atomic resolutions (as described in Sect. 4.6.2) by density modification (Fig. 4.7), even in the presence of the minimal fivefold non-crystallographic symmetry and low similarity. Improved searching models have also been obtained in conjunction with the information provided by three dimensional cryo-electron microscopy (cryo-EM), where the coordinates of a starting model built by similarity have been trimmed and adjusted to fit the cryo-EM map [64] (see also Chaps. 3 and 7). True quasi-atomic models, obtained by combining the X-ray coordinates of one of the capsid components with a cryo-EM reconstruction of the complete capsid, have provided successful starting models in challenging studies.

After the initial MR solution, the quasi-atomic model is generally improved by rigid body refinement of the capsid components against the experimental diffraction data from the crystal. Using cryo-EM maps of the BTV virion, a quasi-atomic model was built for the outer shell formed by protein VP7, which had been solved by MX [10]. The quasi-atomic model was then placed in the unit cell of BTV crystals providing initial phases, which were then refined and extended to higher resolution (see below) allowing the accurate structure determination not only of the outer VP7 shell but also of the inner shell formed by 120 subunits of VP2. Similarly, initial phases were obtained by MR for the structure of the membrane-containing bacteriophage PRD1 using a quasi-atomic model of the viral capsid derived from fitting a cryo-EM reconstruction [14]. Position and orientation of the initial model were refined against the experimental diffraction data of PRD1 crystals, which was followed by rigid body refinement of the 240 P3 trimers included in the model. The PRD1 crystal solved at about 4 Å resolution allowed to determine the conformation variants adopted by P3 and also the structures of proteins P16 and P31, forming the pentameric base of the vertex spike, and of P30. Initial phases for the crystal structure of human adenovirus solved at 3.5 Å resolution were also obtained by MR using a quasi-atomic model built by fitting the known hexon and penton base structures into the cryo-EM density of the virus [16]. The 150-megadalton adenovirus capsid with nearly one million residues in the crystal asymmetric unit represents, as indicated before, the largest crystal structure ever solved, confirming the power of MR and non-crystallographic symmetry in virus crystallography.

MR has also been sufficient to solve a number of virus crystal structures when using as searching models the electron density of the virus obtained by cryo-EM. The crystal structure of human hepatitis B virus capsid, with a T = 4 triangulation icosahedral lattice, was solved using as searching model the 7 Å resolution cryo-EM map [65]. The initial phases to 8 Å resolution were then extended using density modification procedures to 3.3 Å resolution. At this point, it became evident that in the electronic density α helices were left handed, showing that the wrong hand had been selected for the EM map used. Luckily in this case, one of the icosahedral twofold axis was coincident with a crystal twofold and the phase extension protocol had not been affected except by producing the inverted hand map. Even the virus masks of cryo-EM maps, with constant density inside, have proved capable to provide searching models for good molecular replacement solutions with initial phases that have been then refined and extended to almost atomic resolution [52].

The feasibility of ab initio phase determination has been demonstrated for icosahedral viruses with initial phases calculated using spherical hollow shell models of the viruses [58, 66]. However, this fascinating possibility has seldom been fulfilled in practice because of a number of difficulties mainly related with the following: (i) the sensibility to certain parameters, such as the outer and inner radius of the shell; (ii) the availability of accurate values for intensities of reflections at low resolution (at least two-thirds of the viral radius); (iii) the packing of the virus particles, particularly when positions present a center of symmetry in the unit cell, as for example when an icosahedral twofold axis coincides with a crystallographic even-fold axis.

4.6.2 Phase Refinement and Extension: The Power of Non-crystallographic Symmetry

To improve and extend the resolution of the initial phases, obtained at resolutions lower or much lower than atomic resolution, iterative density modification procedures, in particular non-crystallographic averaging and solvent flattening, can be used (Fig. 4.8). The power of phase determination provided by an iteration of density modification is, according to the semiquantitative expression derived by Arnold and Rossmann [67], directly related to the square root of the non-crystallographic redundancy (N). Therefore, non-crystallographic symmetry averaging with high non-crystallographic multiplicities, as often available in virus crystallography, can be an extremely powerful mechanism to refine and to extend the initial phases. The accuracy of the observed amplitudes and the proportion of data measured also affect the expected phasing power. In fact, high non-crystallographic redundancy can overcome efficiently the lack of completeness of the experimental diffraction data. Reciprocally, low redundancy can be compensated with the high accuracy and completeness of the measured intensities.

Many of the original problems in virus crystallography related with the averaging algorithms had to deal with limited computer memories and are now essentially solved with the standard software of protein crystallography. It was found that filling the missing experimental amplitudes by including calculated values (suitably weighted) from the back-transformed electron density map can be critical to allow the propagation of phase information and accordingly also critical for the success of the density modification procedure. The power of density averaging is greatly reinforced when used together with solvent flattening, setting the solvent and sometimes also the capsid interior to their respective mean electron densities. In fact, for crystals with a large solvent percentage volume, solvent flattening provides a considerable amount of phasing power, in particular for new reflections added at higher resolution. Thus, the envelope or mask defining the solvent must allow the maximum volume without truncating the protein shells. Finding and optimizing the envelopes of the volumes within which the different density modification operations are applied is a critically important aspect of the density modification procedures as already mentioned for the ab initio phasing determination. A number of elaborated alternatives have been developed to refine automatically the envelopes that now are routinely used.

Density modification is mostly applied in real space although many of the concepts are better controlled in the reciprocal space formulation [68]. In particular, the shell thickness for phase extension should be limited by consideration of the spherical viral particle radius and of the unit cell size. Weighting reflections according to the likely error in their phases can hasten the process. Phases for the newly phased reflections should be allowed a few cycles to converge before the unobserved data are added in. Poor agreement of the strong low-resolution terms may inhibit phase extension. It may therefore be necessary to impose low-resolution cutoffs on the observed data if, for instance, reflections close to the backstop shadow have not been measured properly. Termination errors can be reduced by extending the reflections a little beyond the resolution limit of the data, thus allowing the calculated data to contribute to the proper phasing of the observed reflections at the highest resolution. The process of phase refinement and extension is normally monitored by reference to averaging agreement R factors and more reliably to the correlation coefficients between the observed and the calculated structure factors analyzed in terms of resolution.

4.7 Map Interpretation

The final step in the structure determination is the building of a chemically reasonable model that fits the electron density maps generated in the previous steps (Fig. 4.6). Refinement allows completing and optimizing the model maximizing the agreement with all the experimental information available.

4.7.1 Model Building

The accuracy of the phases obtained by the averaging procedure results in high quality electron density maps where the amino acid sequence is often recognized even despite the relatively low resolution (in some cases lower than 3.5 Å) of many of the virus crystal structures determined. In these maps, a model is normally constructed for a single icosahedral subunit using standard protein crystallography tools (Fig. 4.6). For viruses with triangulation numbers higher than 1, the icosahedral subunit contains multiple structurally different copies of a given protein, but once a model for the first protein copy becomes available, it can be adapted to provide an initial fit for the electron density corresponding to the remaining copies. Minimaps, skew maps, and other map interpretation techniques, widely used in virus crystallography and referred in the literature till the early 1990s, have now been abandoned by the possibilities offered by powerful graphic systems and interactive software such as COOT [69]. For some virus structures solved at low resolution, like PRD1 at about 4 Å resolution [14], averaged difference Fourier maps corresponding to selenium methionine derivatives were calculated using the refined final phases of the native virus. Peaks in these maps, corresponding do the different number of electrons between the selenium and the sulfur atoms (from the selenium and the normal methionines, respectively), flagged the positions of methionines facilitating and providing controls to the sequence assignment during model building.

4.7.2 Model Refinement

The coordinates and atomic temperature factors can be refined to minimize the differences between the calculated and observed structure factors while maintaining reasonable stereochemistry. The number of refined parameters can be sharply reduced by applying strict non-crystallographic symmetry constraints resulting in a much higher data-to-parameters ratio, though constrained refinement, available in a few refinement packages such as CNS [70] or REFMAC [71]. Since phases derived from the averaging process can be very accurate, these phases can also be used as experimental observations. In fact, due to the quality of the density maps obtained by averaging, the aim of refinement is, in general, not to improve the maps but simply to improve the atomic model. The quality of the averaged maps together with the large computational requirements of virus refinement in reciprocal space explained the successful application of real-space refinement to a number of virus structures [2, 72]. Currently, a real-space refinement procedure including constraints on internal molecular symmetry, implemented in the PHENIX package [73, 74], allows automatic refinement of virus structures solved by both X-ray crystallography and cryo-EM. In cases where high resolution and good data completeness are available, the strict non-crystallographic symmetry constraints can be relaxed to analyze deviations from perfect symmetry among the subunits found in the crystal asymmetric unit. This can be particularly important for some external parts of the virus shells involved in crystal contacts [75, 76].

4.8 Understanding Viruses: Major Contributions of X-Ray Crystallography and the Transition to Cryo-EM

The amount, diversity, and complexity of viral capsid structures now available at near-atomic resolution and stored in the PDB (Protein Data Bank; https://www.rcsb.org) or in specialized structural databases, such as VIPER (VIrus Particle ExploreR; https://viperdb.org), are the result of the outstanding achievements of structural virology, which now treasures the three-dimensional structures of over 1390 icosahedral capsids, from more than 70 viral families (July 2024). These structures provided fundamental insights into virus assembly and maturation but also on the use of viral particles as targets for antiviral drug discovery (Chap. 20), vaccine development (Chap. 21), bio-/nanotechnological developments (Chap. 22), or structure-based virus classification, among others [77]. Of the total number of icosahedral viral particle structures deposited so far in the PDB (July 2024), about a third (425) have been solved by MX. It is important to note that until 10 years ago, MX was the dominant technique, allowing to reach high resolution with large systems. However, since the so-called resolution revolution in single particle cryo-EM, the enormous potential of this technique has been demonstrated by providing near-atomic resolution for an exponentially growing number of viral particles (see Chap. 3).

Many more virus-related protein structures are also available at the PDB (about 8700 on July 2024), and a very high percentage of them, more than 75%, have been determined by MX. These include from structural proteins not forming icosahedral capsids, such as the proteins of helical viruses or the different components of tailed phages, to many viral enzymes, such as viral proteinases or RNA and DNA polymerases. Major technological advances—including improved and faster X-ray detectors, fully automated data collection, and automated software for structure solution—have dramatically increased MX performance, opening the possibility of much more efficient structure-based drug discovery projects [78, 79].

The enormous amount of information contained in the structures of virus capsids and proteins, covering many of the relevant issues in virology and allowing many other types of research, is deeply changing our view of the virus world at an accelerated pace. Many of the structures available correspond to potential targets for antiviral drugs, and structural virology has had a major role in the development of pharmaceuticals for clinically important viral diseases such as AIDS (see Chap. 20). Drugs directed against the HIV aspartic proteinase are a paradigmatic example of the relevance of structural biology in the development of viral medications, with the structure determination of thousands of crystals from the proteinase complexed with different compounds. The high degree of standardization and automation of the method has also boosted the throughput for determination of new crystal structures. A very illustrative example of the speed of response of the MX community was seen during the recent COVID-19 pandemic, when the coordinates of the first determined SARS-CoV-2 protein structure, that of the protease M in complex with inhibitor [26] (Fig. 4.4), were released in January 26, 2020, only 16 days after the release of the virus sequence. By now, there are 4406 structures of SARS-CoV-2 proteins available in the PDB, 2960 of which (>67%) are determined by MX (July 2024).

The structures of essentially all the protein types encoded by some RNA viruses, such as picornaviruses, flaviviruses, and coronaviruses, have been determined, which allows evaluating the possibilities and limitations of the structural genomics strategies. In fact, it appears clear from these examples that the structural information available for the individual proteins provides only an initial glimpse of the viral functioning as the structures of a diversity of protein complexes and of the proteins interacting with nucleic acids together with alternative protein conformations are still required.

The near-atomic resolution structure and organization of the capsids of some viruses have enriched the information about the underlying building principles expanding the quasiequivalence symmetry and icosahedral triangulation concepts toward, for example, the interpretation of departures from these principles (as in Polyomavirus or BTV) or to pseudo triangulation arrangements (as in phage PRD1, adenoviruses, and many other viruses). This type of information has also contributed to the detailed modeling of the largest spherical viruses, although they are not yet amenable to high-resolution X-ray crystal studies [80, 81].

Specific contributions of X-ray crystallography for understanding particular aspects of stages in the virus life cycle are extremely numerous. We will not deal with particular examples in this chapter, because a substantial number of those contributions have been described in many other chapters of this book, specifically Chaps. 2, 7, 11, 12, 13, 14, 15, 16, 17, 20, 21, and 22. The reader is referred to those chapters to find relevant studies where high-resolution structures of viral particles or components determined by X-ray crystallography have proved instrumental for our understanding of the molecular biology of viruses.

4.9 Perspectives

At synchrotrons, new X-ray detectors and microbeams together with "in situ" diffraction are leading to a step change that is central to virus X-ray crystallography. In parallel, several existing synchrotron facilities are upgrading their sources from a third to fourth generation electron storage rings—already a reality for the European Synchrotron Radiation Facility Extremely Brilliant Source (ESRF-EBS) and a planned process for many more facilities [82, 83], resulting in large increases, of ~5 orders of magnitude, in brilliance of their delivered photon beam. Such increase will be hugely beneficial for the typically photon-hungry studies of viral crystals, due to their intrinsically weak diffraction signal.

Even beyond this, ideas are being developed to grow viral crystals in infected cells, bypassing the requirements for purification and crystallization. Such tiny crystals (typically sub-micrometer) might be amenable to diffraction with XFEL, which offer even larger X-ray brilliances than synchrotron sources. Studies at XFELs indicate that the sample size can be reduced to just a single molecule offering the possibility of single virus particle analysis. The oversampling of the continuous scattering originated by the diffraction of a nonperiodic sample allows an alternative approach to solve the "phase problem" [84]. In these experiments, a sufficiently brief (in the femtoseconds' range) and intense X-ray pulse should provide sufficient photons for a useful pattern before radiation damage commences. There are now clear experimental evidences that this single-molecule X-ray "diffract and destroy" analysis is feasible and can be applied in particular to viruses, though at present only very low resolution has been achieved [85–88].

Since the first virus structures were determined at almost atomic resolution, about 45 years ago, virus crystallography has continued providing an always deeper and wider understanding of the virus world with a major impact in the innovation of the X-ray methodology. It is clear that X-ray crystallography will still play an important role in the field of structural virology for the years to come, in particular, for structure-based drug design projects relaying on automation and high throughput, as well as for structure determination of small or medium-sized proteins, such as viral nonstructural proteins and monomeric components of virions and nucleocapsids. Where well-diffracting crystals are available, X-ray diffraction can provide greater resolution details more quickly than any other technique, and it is expected that its speed and power will continue to increase. However, we are likely to see crystallography no longer leading large and complex virus projects as cryo-EM methods take over in determining these structures at near-atomic resolution. Structural virologists must choose the best method in each case and be prepared to combine methods (see Chap. 7). The future is integrative structural virology—many of the questions remaining in virus structure and functioning will require the use of X-ray studies in combination with complementary experimental approaches, such as high-resolution cryo-EM, cryogenic electron tomography, and single-molecule techniques (see also Chaps. 3, 7, 8, 9, and 10).

Acknowledgments We would like to dedicate this chapter to Prof. M.G. Rossmann (1930–2019) for his pioneering and continuous groundbreaking results in structural virology. We are indebted to Ignacio Fita who wrote the sections on data processing, phase determination, and refinement in the previous edition, which have been maintained in this new edition of the chapter. This work was supported by grant PID2020-117976GB-100 from the Spanish Ministry of Science and Innovation.

References

1. Holmes KC, Stubbs GJ, Mandelkow E, Gallwitz U (1975) Structure of tobacco mosaic virus at 6.7 Å resolution. Nature Mar 254:192–196
2. Harrison SC, Olson AJ, Schutt CE, Winkler FK, Bricogne G (1978) Tomato bushy stunt virus at 2.9 a resolution. Nature 276:368–373
3. Abad-Zapatero C, Abdel-Meguid SS, Johnson JE, Leslie AG, Rayment I, Rossmann MG, Suck D, Tsukihara T (1980) Structure of southern bean mosaic virus at 2.8 a resolution. Nature 286:33–39
4. Liljas L, Unge T, Jones TA, Fridborg K, Lövgren S, Skoglund U, Strandberg B (1982) Structure of satellite tobacco necrosis virus at 3.0 a resolution. J Mol Biol 159:93–108 (3)
5. Verdaguer N, Blaas D, Fita I (2000) Structure of human rhinovirus serotype 2 (HRV2). J Mol Biol 300:1179–1194
6. Verdaguer N, Fita I, Reithmayer M, Moser R, Blaas D (2004) X-ray structure of a minor group human rhinovirus bound to a fragment of its cellular receptor protein. Nat Struct Mol Biol 11: 429–434
7. Garriga D, Pickl-Herk A, Luque D, Wruss J, Castón JR, Blaas D, Verdaguer N (2012) Insights into minor group rhinovirus uncoating: the X-ray structure of the HRV2 empty capsid. PLoS Pathog 8:e1002473
8. Rossmann MG, Arnold E, Erickson JW et al (1985) Structure of a human common cold virus and functional relationship to other picornaviruses. Nature 317:145–153
9. Hogle JM, Chow M, Filman DJ (1985) Three-dimensional structure of poliovirus at 2.9 Å resolution. Science 229:1358–1136
10. Grimes JM, Burroughs JN, Gouet P, Diprose JM, Malby R, Ziéntara S, Mertens PP, Stuart DI (1998) The atomic structure of the bluetongue virus core. Nature 395:470–478
11. Wilson IA, Skehel JJ, Wiley DC (1981) Structure of the haemagglutinin membrane glycoprotein of influenza virus at 3 Å resolution. Nature 289:366–373
12. Roberts MM, White JL, Grütter MG, Burnett RM (1986) Three-dimensional structure of the adenovirus major coat protein hexon. Science 232:1148–1151
13. Kwong PD, Wyatt R, Robinson J, Sweet RW, Sodroski J, Hendrickson WA (1998) Structure of an HIV gp120 envelope glycoprotein in complex with the CD4 receptor and a neutralizing human antibody. Nature 393:648–659

14. Abrescia NG, Cockburn JJ, Grimes JM et al (2004) Insights into assembly from structural analysis of bacteriophage PRD1. Nature 432:68–74
15. Cockburn JJ, Abrescia NG, Grimes JM et al (2004) Membrane structure and interactions with protein and DNA in bacteriophage PRD1. Nature 432:122–125
16. Reddy VS, Natchiar SK, Stewart PL, Nemerow GR (2010) Crystal structure of human adenovirus at 3.5 a resolution. Science 329:1071–1075
17. Natchiar SK, Venkataraman S, Mullen TM, Nemerow GR, Reddy VS (2018) Revised crystal structure of human adenovirus reveals the limits on protein IX quasi equivalence and on analyzing large macromolecular complexes. J Mol Biol 430:4132–4141
18. Stehle T, Casasnovas JM (2009) Specificity switching in virus-receptor complexes. Curr Opin Struct Biol 19:181–188
19. Ganser-Pornillos BK, Yeager M, Sundquist WI (2008) The structural biology of HIV assembly. Curr Opin Struct Biol 18:203–217
20. Harrison SC (2005) Mechanism of membrane fusion by viral envelope proteins. Adv Virus Res 64:231–261
21. Mukhopadhyay S, Kuhn RJ, Rossmann MG (2005) A structural perspective of the flavivirus life cycle. Nat Rev Microbiol 3:13–22
22. Wang X, Peng W, Ren J et al (2012) A sensor-adaptor mechanism for enterovirus uncoating from structures of EV71. Nat Struct Mol Biol 19:424–429
23. Rossmann MG (2013) Structure of viruses: a short history. Q Rev Biophys 46:133–180
24. Abrescia NGA, Bamford DH, Grimes JM, Stuart DI (2012) Structure unifies the viral universe. Annu Rev Biochem 81:795–822
25. Simmonds P, Adriaenssens EM, Zerbini FM, Abrescia NGA, Aiewsakun P, Alfenas-Zerbini P et al (2023) Four principles to establish a universal virus taxonomy. PLoS Biol 21:e3001922
26. Jin Z, Du X, Xu Y et al (2020) Structure of Mpro from SARS-CoV-2 and discovery of its inhibitors. Nature 582:289–293
27. Wang Q, Zhang Y, Wu L et al (2020) Structural and functional basis of SARS-CoV-2 entry by using human ACE2. Cell 181:894–904.e9
28. Jernigan RJ, Logeswaran D, Doppler D (2023) Room-temperature structural studies of SARS-CoV-2 protein NendoU with an X-ray free-electron laser. Structure 31:138–151.e5
29. Frazier MN, Wilson IM, Krahn JM, Butay KJ, Dillard LB, Borgnia MJ, Stanley RE (2022) Flipped over U: structural basis for dsRNA cleavage by the SARS-CoV-2 endoribonuclease. Nucleic Acids Res 50:8290–8301
30. Botha S, Fromme P (2023) Review of serial femtosecond crystallography including the COVID-19 pandemic impact and future outlook. Structure 31:1306–1319
31. Blundell TL, Patel S (2004) High-throughput X-ray crystallography for drug discovery. Curr Opin Pharmacol 4:490–496
32. Bauman JD, Patel D, Arnold E (2012) Fragment screening and HIV therapeutics. Top Curr Chem 317:181–200
33. Maveyraud L, Mourey L (2020) Protein X-ray crystallography and drug discovery. Molecules 25:1030
34. Burzlaff H, Gruber B, Wolf PM et al (2002) Crystal lattices. In: Hahn T (ed) International tables for crystallography, vol A. Kluwer Academic Publishers, Dordrecht
35. Fry E, Grimes J, Stuart DI (1999) Virus crystallography. Mol Biotechnol 12:13–23
36. Fry E, Logan D, Stuart D (1996) Virus crystallography. In: Jones C, Mulloy B, Sanderson M (eds) Methods in molecular biology, vol 56: crystallographic methods and protocols. Humana Press, Totowa
37. Vijayachandran LS, Viola C, Garzoni F et al (2011) Robots, pipelines, polyproteins: enabling multiprotein expression in prokaryotic and eukaryotic cells. J Struct Biol 175:198–208
38. Schütz A, Bernhard F, Berrow N et al (2023) A concise guide to choosing suitable gene expression systems for recombinant protein production. STAR Protoc 4:102572
39. Chayen NE, Helliwell JR, Snell EH (2010) Macromolecular crystallization and crystal perfection. Oxford University Press, Oxford

40. Newman J (2005) Expanding screening space through the use of alternative reservoirs in vapor-diffusion experiments. Acta Crystallogr D Biol Crystallogr 61:490–493
41. Chayen NE (2005) Protein crystallization: automation robotization and miniaturization. In: Sundstrum M, Norin M, Edwards A (eds) Structural genomics and high throughput structural biology. CRC Press/Taylor & Francis, Abingdon
42. McPherson A, Gavira JA (2014) Introduction to protein crystallization. Acta Crystallogr F Struct Biol Commun 70:2–20
43. Meyer A, Betzel C, Pusey M (2015) Latest methods of fluorescence-based protein crystal identification. Acta Crystallogr F Struct Biol Commun. 71:121–131
44. Watkins CJ et al (2019) A crystal/clear pipeline for applied image processing. In: Abramson D, de Supinski B (eds) Supercomputing Frontiers. SCFA 2019. Lecture notes in computer science, vol 11416. Springer, Cham
45. Bruno AE, Charbonneau P, Newman J, Snell EH, So DR et al (2018) Classification of crystallization outcomes using deep convolutional neural networks. PLoS One 13:e0198883
46. Garman E (2003) 'Cool' crystals: macromolecular cryo-crystallography and radiation damage. Curr Opin Struct Biol 13:545–551
47. Mueller M, Jenni S, Ban N (2007) Strategies for crystallization and structure determination of very large macromolecular assemblies. Curr Opin Struct Biol 17:572–579
48. Cockburn JJ, Bamford JK, Grimes JM, Bamford DH, Stuart DI (2003) Crystallization of the membrane-containing bacteriophage PRD1 in quartz capillaries by vapour diffusion. Acta Crystallogr D Biol Crystallogr 59:538–540
49. Axford D, Owen RL, Aishima J et al (2012) In situ macromolecular crystallography using microbeams. Acta Crystallogr D Biol Crystallogr 68:592–600
50. Sanchez-Weatherby J, Sandy J, Mikolajek H, Lobley CMC, Mazzorana M, Kelly J, Preece G, Littlewood R, Sørensen TLM (2019) VMXi: a fully automated, fully remote, high-flux in situ macromolecular crystallography beamline. J Synchrotron Radiat 26:291–301
51. Rossmann MG (1999) Synchrotron radiation as a tool for investigating virus structures. J Synchrotron Radiat 6:816–821
52. Reinisch KM, Nibert ML, Harrison SC (2000) Structure of the reovirus core at 3.6 Å resolution. Nature 404:960–967
53. Gouet P, Diprose JM, Grimes JM et al (1999) The highly ordered double-stranded RNA genome of bluetongue virus revealed by crystallography. Cell 97:481–490
54. Leonarski F, Nan J, Matej Z et al (2023) Kilohertz serial crystallography with the JUNGFRAU detector at a fourth-generation synchrotron source. IUCrJ 10:729–737
55. Winter G, Gildea RJ, Paterson NG, Beale J, Gerstel M, Axford D, Vollmar M, McAuley KE, Owen RL, Flaig R, Ashton AW, Hall DR (2019) How best to use photons. Acta Crystallogr D Biol Crystallogr 75:242–261
56. Aragão D, Flaig R, Romano P, Mazzonana M (2022) Interactive and unattended data collection driven by target dose on the diamond light source I04 beamline. Acta Crystallogr A 78:e264–e265
57. Diederichs K, Wang M (2017) Serial synchrotron X-ray crystallography (SSX). Methods Mol Biol 1607:239–272.
58. Taka J, Naitow H, Yoshimura M, Miyazaki N, Nakagawa A, Tsukihara T (2005) Ab initio crystal structure determination of spherical viruses that exhibit a centrosymmetric location in the unit cell. Acta Crystallogr D Biol Crystallogr 61:1099–1106
59. Baker TS, Johnson JE (1997) Principles of virus structure determination. In: Chiu W, Burnett RM, Garcea RL (eds) Structural biology of viruses. Oxford University Press, New York
60. Arnold E, Vriend G, Luo M, Griffith JP, Kamer G, Erickson JW, Johnson JE, Rossmann MG (1987) The structure determination of a common cold virus, human rhinovirus 14. Acta Crystallogr A43:346–361
61. Gutnik D, Evseev P, Miroshnikov K, Shneider M (2023) Using AlphaFold predictions in viral research. Curr Issues Mol Biol 45:3705–3732

62. Jumper J, Evans R, Pritzel A et al (2021) Highly accurate protein structure prediction with AlphaFold. Nature 596:583–589
63. Callaway E (2022) What's next for AlphaFold and the AI protein-folding revolution. Nature 604:234–238
64. Speir JA, Munshi S, Wang G, Baker TS, Johnson JE (1995) Structures of the native and swollen forms of cowpea chlorotic mottle virus determined by X-ray crystallography and cryoelectron microscopy. Structure 3:63–78
65. Wynne SA, Crowther RA, Leslie AGW (1999) The crystal structure of the human hepatitis B virus capsid. Mol Cell 3:771–780
66. Tsao J, Chapman MS, Wu H, Agbandje M, Keller W, Rossmann MG (1992) Structure determination of monoclinic canine parvovirus. Acta Crystallogr B48:75–88
67. Arnold E, Rossmann MG (1985) Effect of errors, redundancy and solvent content in the molecular replacement procedure for the structure determination of biological macromolecules. PNAS 83:5489–5493
68. Rossmann MG (1989) The molecular replacement method. Acta Crystallogr A46:73–82
69. Emsley P, Cowtan K (2004) Coot: model-building tools for molecular graphics. Acta Crystallogr D Biol Crystallogr 60:2126–2132
70. Brunger AT, Kuriyan J, Karplus M (1987) Crystallographic R factor refinement by molecular dynamics. Science 235:458–460
71. Murshudov GN, Skubak P, Lebedev AA et al (2011) REFMAC5 for the refinement of macromolecular crystal structures. Acta Crystallogr D67:355–367
72. Coulibaly F, Chevalier C, Gutsche I, Pous J, Navaza J, Bressanelli S, Delmas B, Rey FA (2005) The birnavirus crystal structure reveals structural relationships among icosahedral viruses. Cell 120(6):761–772
73. Afonine PV, Poon BK, Read RJ, Sobolev OV, Terwilliger TC, Urzhumtsev A, Adams PD (2018) Real-space refinement in PHENIX for cryo-EM and crystallography. Acta Crystallogr D Struct Biol 74:531–544
74. Liebschner D, Afonine PV, Baker ML et al (2019) Macromolecular structure determination using x-rays, neutrons and electrons: recent developments in phenix. Acta Crystallogr D Struct Biol 75:861–877
75. Garriga D, Querol-Audí J, Abaitua F et al (2006) The 2.6-angstrom structure of infectious bursal disease virus-derived T=1 particles reveals new stabilizing elements of the virus capsid. J Virol 80:6895–6905
76. Sarker S, Terrón MC, Khandokar Y, Aragão D, Hardy JM, Radjainia M, Jiménez-Zaragoza M, de Pablo PJ, Coulibaly F, Luque D, Raidal SR, Forwood JK (2016) Structural insights into the assembly and regulation of distinct viral capsid complexes. Nat Commun 7:13014
77. Johnson JE, Olson AJ (2021) Icosahedral virus structures and the protein data bank. J Biol Chem 296:100554
78. Erlanson DA, Fesik SW, Hubbard RE, Jahnke W, Jhoti H (2016) Twenty years on: the impact of fragments on drug discovery. Nat Rev Drug Discov 15:605–619
79. Jacquemard C, Kellenberger E (2019) A bright future for fragment-based drug discovery: what does it hold? Expert Opin Drug Discov 14:413–416
80. Nandhagopal N, Simpson AA, Gurnon JR et al (2002) The structure and evolution of the major capsid protein of a large, lipid-containing DNA virus. PNAS 99:14758–14763
81. Zhanga X, Xianga Y, Duniganb DD et al (2011) Three-dimensional structure and function of the Paramecium bursaria chlorella virus capsid. PNAS 108:14837–14842
82. Raimondi P, Benabderrahmane C, Berkvens P et al (2023) The extremely brilliant source storage ring of the European Synchrotron Radiation Facility. Commun Phys 6:82
83. Chapman HN (2023) Fourth-generation light sources. IUCrJ 10:246–247
84. Raines KS, Salha S, Sanberg RL, Jiang H, Rodríguez JA, Fahimian B, Kapteyn HC, Du J, Miao J (2010) Three-dimensional structure determination from a single view. Nature 463:214–217
85. Seibert MM, Ekeberg T, Maia FR et al (2011) Single mimivirus particles intercepted and imaged with an X-ray laser. Nature 470:78–81

86. Ekeberg T, Svenda M, Abergel C et al (2015) Three-dimensional reconstruction of the Giant Mimivirus particle with an X-ray free-electron laser. Phys Rev Lett 114:098102
87. Lundholm IV, Sellberg JA, Ekeberg T et al (2018) Considerations for three-dimensional image reconstruction from experimental data in coherent diffractive imaging. IUCrJ 5:531–541
88. Rose M, Bobkov S, Ayyer K, Kurta RP et al (2018) An advanced workflow for single-particle imaging with the limited data at an X-ray free-electron laser. IUCrJ 5:727–736

Further Reading

Blow D (2002) Outline of crystallography for biologists. Oxford University Press, Oxford
Rhodes G (2006) Crystallography made crystal clear, third edn. Academic Press, London
Rupp B (2010) Macromolecular crystallography. Garland Science, Taylor & Francis group, LLC, New York

Chapter 5
Nuclear Magnetic Resonance Spectroscopy to Study Virus Structure

José L. Neira

Abstract Nuclear magnetic resonance (NMR) is a spectroscopic technique based on the absorption of radiofrequency radiation by atomic nuclei in the presence of an external magnetic field. NMR has followed a "bottom-up" approach to solve the structures of isolated domains of viral proteins, including capsid protein subunits, or to provide information about other macromolecular partners with which such proteins interact. NMR has been instrumental in describing conformational changes in viral proteins and nucleic acids, showing the presence of dynamic equilibria which are thought to be important at different stages of the virus life cycle. In this sense, NMR is also the only technique currently available to describe, in atomic detail, the conformational preferences of intrinsically disordered viral proteins. Furthermore, NMR can provide insights into the thermodynamic parameters governing binding reactions between different viral macromolecules. NMR has also complemented X-ray crystallography and has been combined with electron microscopy to obtain pseudo-atomic models of entire virus capsids. Finally, the joint use of liquid and solid-state NMR has allowed the identification of conformational changes in viral capsids upon insertion into host membranes.

Keywords Assembly · Binding · Capsid · Chemical shifts · Conformational changes · Coupling constants · Dynamics · Electron paramagnetic resonance (EPR) · Equilibrium · Flexibility · Hydrogen exchange · Lipids · Membrane · Nuclear magnetic resonance (NMR) · Nucleic acid · Protein-protein interactions · Protein-nucleic acid interactions · Solid-state · Solution structure · Unfolded protein · Virus

J. L. Neira (✉)
IDIBE, Universidad Miguel Hernández, Elche, Alicante, Spain

Instituto de Biocomputación y Física de Sistemas Complejos, Zaragoza, Spain
e-mail: jlneira@umh.es

M. G. Mateu (ed.), *Structure and Physics of Viruses*, Subcellular Biochemistry 105,
https://doi.org/10.1007/978-3-031-65187-8_5

Abbreviations

ACE2	Angiotensin-converting enzyme 2
BLV	Bovine leukemia virus
CA	Retrovirus capsid protein
CoV	Coronavirus
CTD	C-terminal domain of CA
COSY	Correlation spectroscopy
EIAV	Equine infectious anemia virus
cryo-EM	Cryogenic electron microscopy
EPR	Electron paramagnetic resonance
FD	Fusion domain
FD-SARS-CoV-2	FD of severe acute respiratory syndrome coronavirus 2
FID	Free induction decay
HIV-1	Human immunodeficiency virus type 1
hLysRS	Human lysyl-tRNA synthetase
HTLV	Human T-cell leukemia virus (human T-lymphotropic virus)
IDP	Intrinsically disordered protein
IDR	Intrinsically disordered regions
MA	Matrix protein
MMTV	Mouse mammary tumor virus
MoMuLV	Moloney murine leukemia virus
MV	Monkey virus
NC	Nucleocapsid protein
NC-SARS-CoV-2	NC of severe acute respiratory syndrome coronavirus 2
NMR	Nuclear magnetic resonance
NOE	Nuclear Overhauser effect
NOESY	Nuclear Overhauser effect spectroscopy
nsp	Non-structural protein
NTD	N-terminal domain of CA
N_{tail}	C-terminal region of the Measles virus nucleoprotein
PIP2	Phosphatidylinositol-4,5-diphosphate
PPI	Protein-protein interaction
RBD	Receptor-binding domain of the spike protein of SARS-CoV-2
RSV	Rous sarcoma virus
SARS-CoV-2	Severe acute respiratory syndrome coronavirus 2
S	Spike protein of SARS-CoV-2
SL	Stem loop
SIV	Simian immunodeficiency virus
SP	Spacer region
SSNMR	Solid-state NMR
TOCSY	Total correlation spectroscopy
VPr	Accessory viral protein R

5.1 Introduction

Although cryogenic electron microscopy (cryo-EM) (Chap. 3) and X-ray crystallography (Chap. 4) are extensively used for determining the three-dimensional structures of virus particles and viral proteins at or close to atomic resolution, not all biological macromolecules (or their complexes with other molecules) can form crystals. Additionally, many biomolecules can behave differently in the crystalline state or in a vitrified state than they do in solution. In those cases, nuclear magnetic resonance (NMR) spectroscopy has been used to determine their high-resolution structures. Furthermore, the structures of some biomolecular complexes (which cannot be crystallized), the monitoring of enzymatic reactions, the description of protein folding reactions, or the exploration of the dynamic properties of biomolecules can be addressed by NMR. Therefore, NMR spectroscopy has emerged as an important complement to cryo-EM and X-ray crystallography in structural biology in general, and in structural virology in particular.

In this chapter, I shall describe the foundations of NMR spectroscopy and some of its applications to study viral proteins and virus particles. This spectroscopy has a serious limitation when faced with the study of large proteins or macromolecular assemblies: the signals from those biomolecules become too broad to be clearly observed and resolved in a spectrum. Due to this limitation, NMR has dealt with the study of whole virus particles and large viral proteins by solving the structure of small protein subunits or isolated protein domains of larger proteins (a "bottom-up" approach). NMR has also been applied to study the conformational dynamics of viral proteins in solution, and there have been attempts to tackle the dynamics of complete viral particles. Only recently, solid-state NMR (SSNMR) techniques have been applied to the structural study of whole, intact virus particles in "quasi-solution" conditions. In this chapter, there will be a brief mention of some of the applications of the "youngest sister" of NMR, electron paramagnetic resonance (EPR) spectroscopy, to address questions concerning the dynamics and protein-protein interactions (PPIs) of some viral proteins. The answers to specific questions provided by EPR complement those provided by NMR and other spectroscopic techniques (Chap. 6).

5.2 Physical Principles of Magnetic Spectroscopies

5.2.1 The Basis of NMR

NMR is one of the few spectroscopies that can be explained from both a classical perspective and a quantum mechanics point of view. In-depth explanations of NMR theory are beyond the scope of this chapter, so there will be a brief description only on some basic aspects in the following paragraphs. There are excellent books where the interested reader can find a detailed treatment of both complementary views on the NMR foundations [1–5].

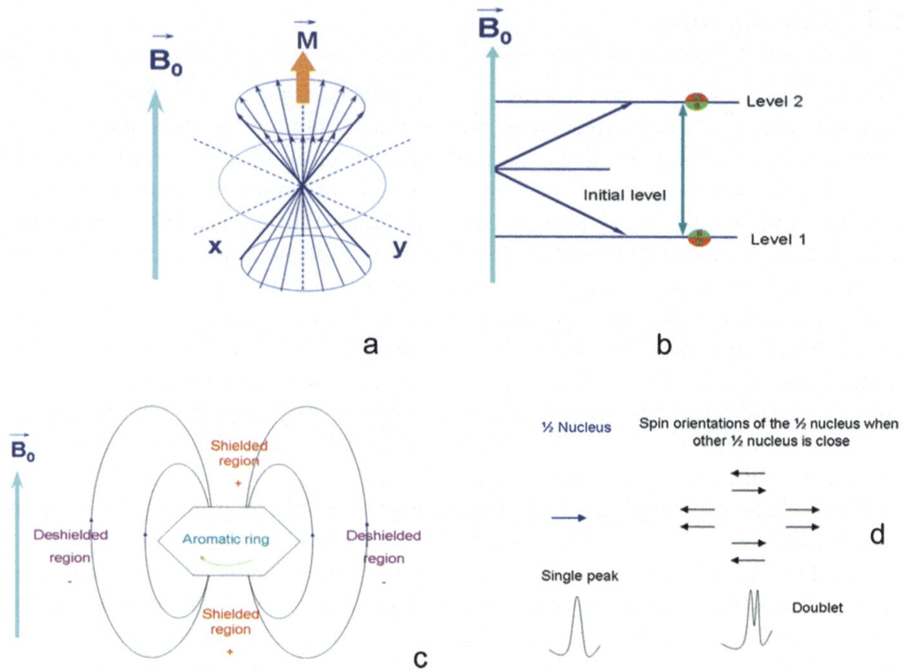

Fig. 5.1 Physical basis of NMR. (**a**) Classical view of NMR. Nuclear spins are turning around (in a precession movement) the external magnetic field (B_O, *cyan arrow*); the net magnetization of the system (the sum of all nuclear magnetic dipoles of the sample) is shown in *orange* (M). (**b**) Quantum view of NMR. Two quantum levels resulting from B_0 (*blue arrow*) are shown (assuming two possible orientations for nuclear spin, as in ^{1}H); spins are represented as *small circles* indicating polarity. The *double arrow* indicates the transition between quantum levels. (**c**) Concept of shielding in aromatic side chains. The *small blue arrows* indicate the ring-current magnetic field direction, for which two lines have also been drawn. The *green line* on the aromatic ring shows the movement of the electron rings. The *orange region* is the shielded one, and the *purple zone* is the deshielded region. The direction of the external magnetic field is indicated by a *large cyan arrow*. (**d**) Coupling constant concept for a nucleus with spin ½ (two possible orientations), with a nearby nucleus also with spin ½. From a quantum perspective (see [2–7]), some of those possible orientations are indistinguishable, and they result in two lines (a doublet) in the NMR spectrum instead of the expected single line

NMR is a spectroscopy that "measures" the interaction between matter and light based on the magnetic properties of atomic nuclei. Those nuclei with odd numbers of protons and/or neutrons have a magnetic moment or magnetic dipole (μ) due to their spinning. From a classical point of view, in the presence of an external static magnetic field (B_0), the μ of the nuclei will orient parallel to (with) or anti-parallel to (against) B_0. This interaction causes a precession of the nuclear dipoles (the matter in the end) around the external field (Fig. 5.1a), analogous to a gyroscope spinning in a gravitational field. The angular frequency (ω) of this motion depends on the strength of B_0 and the value of μ. The Boltzmann distribution dictates that the most stable state (nuclear magnetic dipoles oriented with B_0) will be the most populated at equilibrium, albeit by only a small fraction, because of the small difference in energy with the less stable state (dipoles oriented against B_0). If an appropriate amount of energy (light in the region of radiofrequency in the end) is applied, the nuclear magnetic dipoles will move away from their equilibrium position and become oriented against B_0. The corresponding absorption of energy by the nuclei is measured by NMR.

From a quantum perspective, if we consider the simplest nucleus (that of ^{1}H, with a single proton) in the absence of an external magnetic field, B_0, the two possible orientations of the nuclear spin (± ½) have equivalent energy (they are degenerate). In the presence of B_0, two different quantum levels will appear (Fig. 5.1b). The energy difference (E) between those levels is the dot product of B_0 and μ, and it matches the energy within the radiofrequency region of the electromagnetic spectrum; the small separation between both quantum levels is responsible for the low sensitivity of NMR spectroscopy. To induce a transition between those levels, it is necessary to apply electromagnetic radiation with energy given by the Planck equation [1–7]. Thus, the technique is called "nuclear" because we are observing nuclei; it is called "magnetic" because we need a magnetic field to have different non-degenerate quantum levels; and it is called "resonance" because we need to apply the exact energy (in the radiofrequency region), provided by the Planck equation corresponding to the energetic separation between the split levels.

In general, nuclei with even numbers of protons and neutrons (such as the most abundant isotope of carbon, $^{12}_{6}$C, will have zero spin and will not yield an NMR spectrum (because there are no levels to split in the presence of B_0). Those nuclei with an odd number of either protons or neutrons will have a half-integer spin (1/2, 3/2, ...), and those with odd numbers of both protons and neutrons will have an integer spin (larger than ½). If the spin of the nucleus is larger than ½, there are more than two quantum states in the presence of B_0, or, in other words, more than two different ways of spinning around B_0 in the classical view. More spectral transitions and, therefore, more complicated spectra are obtained, which make those nuclei much less convenient for NMR analyses than ^{1}H and other nuclei with spin ½. Since the total number of protons and neutrons varies from one isotope to another, it is clear that for an element with several isotopes, some of them will be NMR-active. The most interesting nuclei, from a biomolecular point of view, are ^{1}H (100% naturally abundant); ^{15}N, ^{13}C (natural abundance of 0.36% and 1.11%, respectively; biomacromolecules need to be artificially enriched in ^{15}N and/or ^{13}C if these isotopes are to be observed in NMR experiments; see below); and ^{31}P (100% naturally abundant; especially useful for studies on nucleic acids or lipids).

5.2.2 *The Chemical Shifts*

The atomic nuclei in a molecule are not isolated: they interact with their own electrons, those of other atoms, and with other atomic nuclei. Since the electronic shells have their own spin, they will create an internal magnetic field which will be "felt" in different ways by the nearby nuclei. Therefore, each nucleus in a molecule will have a different "view" (the so-called shielding) of the external magnetic field B_0, depending on its surroundings. Perhaps the clearest example of this shielding is provided by aromatic rings. In the presence of B_0, a planar aromatic ring with its delocalized π electrons "creates" its own magnetic field (Fig. 5.1c). A nucleus above or below the ring plane will be "shielded" from the external magnetic field, but it will be "unshielded" if it is located at other positions relative to the ring; thus, the magnitude of the total field "felt" by a nucleus informs on its location with respect to the ring. In the same way, hydrogen bonding, as in some proteins, decreases the electronic shell around a nucleus and reduces the shielding effect. Therefore, it is possible to say that each nucleus in a molecule resonates at a slightly different energy (frequency) relative to others, because of its specific magnetic environment (i.e., each nucleus needs a slightly different energy to be promoted to its quantum excited state).

It is impractical to measure the absolute frequency of resonance for every nucleus, so a relative frequency is provided instead. A standard small compound is introduced with the NMR sample, and the resonance frequencies of every nucleus in the molecule under study are referred to that of the standard. In a first stage, we measure the differences between the resonance frequency of each nucleus and that of the standard. Since these measurements of frequencies are provided in hertz (Hz), they are B_0-dependent, and therefore, the spectrometer magnetic field must be specified. To avoid an instrument-dependent parameter, a unit-less chemical shift (δ) is provided, given in parts *per* million (ppm). Thus, δ has the same operational meaning as the wavelength (λ) or frequency (υ) in other spectroscopies (see Chap. 6), but in a way, the δ at which a nucleus resonates is a measure of nucleus-electron cloud interactions.

5.2.3 Spin-Spin Coupling: Seeing Other Covalently Bound Nuclei

Another important parameter in NMR is the *coupling constant*, *J*, which measures the magnetic interaction between the spins of two or more covalently bound nuclei (and thus, the interaction occurs *through* the electrons of the bond). The magnetic field "felt" by a nucleus depends on whether the covalently bound nucleus, with a μ, is aligned or not with B_0 (Fig. 5.1d). Therefore, the local field "felt" will be either slightly reduced or increased (when compared to the original magnetic field "felt" when only electronic shielding is considered) due to the presence of the neighboring, covalently bound nuclei. Then, the original absorption line in the NMR spectrum is split into several lines (two for the simplest case of a single nearby nucleus with spin ½, as the hydrogen, Fig. 5.1d); the strength of the interaction, and that of the splitting, is given by *J*. An NMR-active nucleus covalently bound to nuclei with a null μ (such as protons bound to $^{12}_{6}C$ will not show a splitting of its NMR spectral lines; conversely, nuclei with a nuclear spin larger than ½ (such as ^{14}N) will tend to broaden neighboring nuclei lines in the NMR spectrum rather than splitting them.

The J is measured in Hz; it is not B_0-dependent and it depends on the nuclei involved and the number of covalent bonds between the nuclei. For instance, the values of J for protons range from 0 to about 20 Hz, but the one-bond J for the ^{15}N-^{1}H is close to 90 Hz. As J depends on the covalently bound nuclei, it provides information on the dihedral angle connecting them (for protons, e.g., this is given by a mathematical relationship termed the Karplus equation [1]). This finding has profound implications, since it allows us to obtain structural information. For instance, the J between the amide proton of an amino acid residue and its H_α proton varies depending on the type of protein secondary structure element (α-helix or β-sheet) where the residue is located: 3–5 Hz for residues in α-helices and 8–9 Hz for residues in β-sheets.

5.2.4 Relaxation Mechanisms

For simplicity, let us consider a nucleus such as ^{1}H with a nuclear spin of ½ and then two quantum levels in the presence of B_0. Electromagnetic radiation of the appropriate υ (in the radiofrequency region) causes spins to jump from the lower to the higher quantum level. As the lower level (μ aligned with the B_0, according to the classical point of view) has a larger population of spins than the higher one, there will be a net absorption of radiofrequency radiation, leading to a signal in the NMR spectrum. If the electromagnetic radiation is intense enough, the number of spins in the lower and higher levels becomes equal, and therefore, there is no net absorption of energy because there is no "room" in the higher level to allocate more spins. From the quantum point of view, we say that the signal is "saturated": as many photons are emitted as are absorbed, and no signal is observed in the NMR spectrum.

However, there are always spontaneous transitions allowing the spins in the saturated higher level to return to the equilibrium population (to the lower level). In other spectroscopic techniques, the recovery of the equilibrium population occurs by "crashing" with other molecules and losing the excess energy by heat release (see Chap. 6). In NMR, a nucleus relaxes by through-space interaction (through the μ) with a nearby nucleus (from either the same or different molecules); the *spin-lattice* relaxation time, T_1, is the time the equilibrium spin population takes to recover by exchanging energy with the surroundings (i.e., the *lattice*). For a nucleus to contribute effectively to the relaxation of another nucleus, it should be close, and it should belong to a molecule(s) which rotate(s) with an angular frequency similar to the magnetic resonance transition frequency of the studied nucleus (the quantum level separation).

There is another characteristic time, the *spin-spin* relaxation time, T_2, which also measures the recovery of the equilibrium population. The time T_2 refers to interactions *between* spins and measures the rate at which spins exchange energy with each other, without involving interactions with the environment (i.e., with the *lattice*). The time T_2 depends, like T_1, on the molecular size and the speed of rotation of the molecule (which is related to the molecular tumbling). Furthermore, T_2 is inversely proportional to the NMR resonance linewidths: the more efficient the T_2 relaxation process, the larger the linewidths (as observed in large molecules, or small ones in rigid environments, e.g., in SSNMR; see Sect. 5.4.4.). Both T relaxation times are important, as they provide information about the dynamics of the molecule.

5.2.5 The Nuclear Overhauser Effect (NOE): Seeing Other Nuclei Through Space

A very important parameter in NMR is the *nuclear Overhauser effect* (NOE), which is due to a through-space interaction between nuclear spins and their μ values (and therefore, it is related to the two T relaxation times described above). A NOE is a change of intensity of an NMR resonance corresponding to a particular nucleus when a nearby nucleus is irradiated, that is, when such nucleus is illuminated with a particular radiofrequency radiation. Those nuclei do not have to be covalently bound; it is necessary and sufficient that they are spatially close enough. Thus, determination of NOEs for different nuclei in a macromolecule allows us to determine which nuclei are near each other in space, a crucial step in determining its three-dimensional structure by NMR (see Sect. 5.3.2).

When intense radiation corresponding to the resonance transition frequency of one nucleus is applied to the sample, spin saturation occurs, and, therefore, equal populations of spins are found in the lower and higher quantum levels (as described in Sect. 5.2.4). However, that nucleus is not isolated; it can interact with other nuclei nearby (belonging to the same or different molecule). This results in a change in the equilibrium population of spins. Therefore, when we irradiate a nucleus with the proper energy to cause a transition between its levels (and then, to observe its corresponding peak in the NMR spectrum), we shall observe a variation in the intensity of the peaks of nearby nuclei compared to the intensity observed if the first nucleus had not been excited. Since the interaction between the corresponding magnetic dipoles depends on r^{-6} (r being the distance between them), a NOE is only observed for close neighbors (usually less than 5 Å). The NOE short-range limits its applications, but it provides great specificity in the determination of molecular structures by NMR (see below).

5.2.6 Pulse Techniques and Multidimensional NMR

Most spectroscopic techniques (such as fluorescence and CD; see Chap. 6) collect spectra by varying the wavelength of the excitation source. We could do the same in NMR: varying the wavelength of the radiofrequency source would sequentially induce transitions of different atomic nuclei between their quantum levels (Fig. 5.1b). However, modern NMR spectrometers use radiofrequency pulses due to improved signal-to-noise ratio and versatility. Since most of the tiny magnetic dipoles—but not all, due to the equilibrium Boltzmann distribution—are aligned with the external field (Fig. 5.1a), there is a net magnetization, M, along the direction defined by B_0 (in the classical view), and most of the spins will start turning around along that direction (see Sect. 5.2.1) (Fig. 5.2a, left). If we apply a short (microsecond) radiofrequency perturbation (a *pulse*) along the x-axis (Fig. 5.2a), B_1, with a frequency matching the separation energy between the quantum levels (or, in other words, in the classical view, the frequency of precession of the magnetization), the spins initially aligned with the z-axis will start moving away from that axis and tend to align with the y-axis (thus, in the classical view, they will start turning around along that axis). When we remove the perturbation (the pulse), over time, the different spins forming the M will begin to separate at their different precession frequencies (due to their diverse δs), returning to the z-axis (Fig. 5.2a, right). Thus, the magnetization vector will decrease with a T_1 relaxation time constant (see Sect. 5.2.4), leading to an exponentially decreasing signal termed a *free induction decay* (FID). The FID is usually observed during a time called t_2 (note that this is an *observation* time, different from the relaxation time T_2) (Fig. 5.2a, right side). Each spin (nucleus) contributes differently to the FID due to its slightly different δ; the individual contributions of the spins can be separated by means of a Fourier transformation of the time-dependent signal (which relies on the mathematical principle that any time-dependent function is equivalent to a frequency-dependent function) [2–7].

With the application of a single pulse to the sample, we obtain a one-dimensional NMR (1D-NMR) (Fig. 5.2b, top panel) spectrum, which graphically represents absorption (or emission) intensity plotted against δ (or υ), similar to other spectroscopies. As molecular size increases, the number of nuclei also increases, leading to greater overlap of δs among different nuclei in a 1D-NMR spectrum. This overlap occurs because some nuclei experience similar magnetic environments by chance, even if they are located in different parts of the molecule. Therefore, to identify (and isolate) the δ (or υ) of each specific nucleus, it is necessary to resolve those overlapping frequencies. By applying a series of pulses to the sample, overlapping frequencies can be separated into two (or more) axes, thereby increasing spectral resolution. This results in a multidimensional spectrum where absorption intensity is plotted as a function of two (in 2D-NMR) or more (n-dimensional NMR) δ (or υ) values (Fig. 5.2b, bottom panel). A multidimensional spectrum is valuable for monitoring interactions of nuclei through bonds (obtaining J values) or through space (obtaining NOEs). These interactions can be homonuclear (involving the same

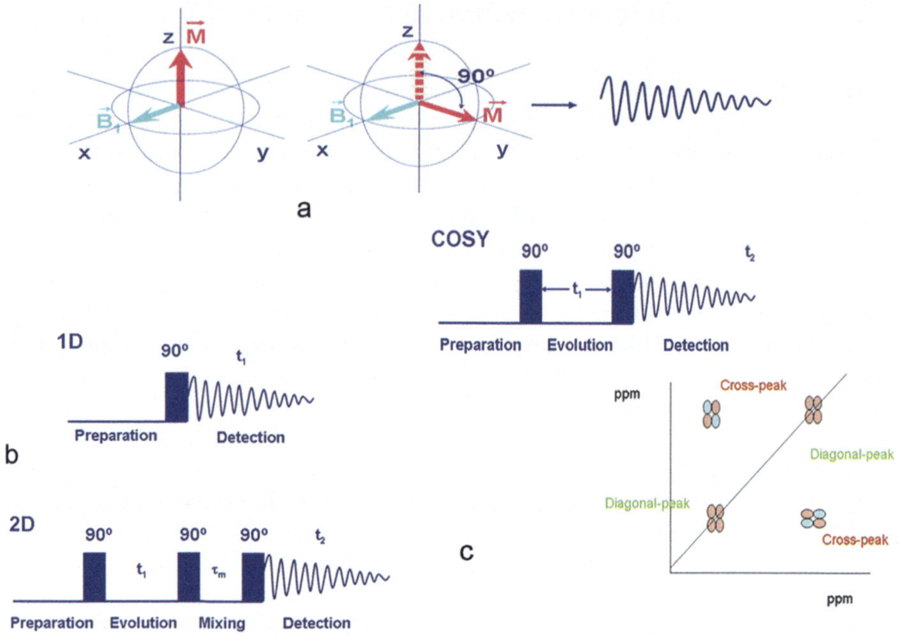

Fig. 5.2 Pulses and n-dimensional NMR. (**a**) How does a pulse work? Initially (*left image*), the system is in equilibrium in the presence of an external magnetic field (not indicated); that is, this image shows the state of the system as in Fig. 5.1a. A pulse (B_1, green) is applied along the x-axis, and its effect (*right image*) is to tilt the magnetization (M, red arrow) away from the z-axis. I have represented a pulse completely tilting the magnetization along the y-axis, but this may vary (see, e.g., [2–7] for different pulse lengths). From the evolution of this tilted magnetization over time as it returns to equilibrium, a FID is obtained (*far right*). (**b**) Series of pulse sequences (pulses indicated by *rectangles*): *top*, scheme to obtain a 1D-NMR spectrum; *bottom*, scheme to obtain a 2D-NOESY spectrum. The time periods during the experiments are labeled; conventionally, acquisition time in 1D-NMR is called t_1 (t_2 in 2D-NMR spectra). Numbers above rectangles (pulses) indicate the tilt caused by the pulse (i.e., a 90° pulse tilts magnetization completely along the y-axis). (**c**) A scheme to obtain a 2D-COSY spectrum. Pulse sequence is depicted on top, where time t_1 is linearly increased in each experiment. The resulting spectrum shows diagonal peaks and cross-peaks (*bottom*). The clusters of peaks in a COSY are four-lobed with alternated signs (positive and negative) due to the effect of the coupling constants between covalently bound nuclei

type of nuclei, e.g., ^{1}H-^{1}H) or heteronuclear (involving different types of nuclei, e.g., ^{1}H-^{15}N). In the following paragraphs, two types of homonuclear 2D-NMR experiments will be briefly described; for a clear description of key 2D-NMR experiments used in structure determination, refer to [2–8].

The simplest 2D-NMR spectrum (a homonuclear one called 2D-COSY) is obtained by applying two pulses before observing the FID during the time t_2 (Fig. 5.2c). The time t_1, which separates the two pulses (Fig. 5.2c, top panel), is varied in successive experiments. Consequently, the FID observed during t_2 changes because the spins evolve differently during each increment of t_1. Essentially, this results in *k* (the number of increments of t_1) individual 1D-NMR experiments, where each experiment varies the parameter t_1. Therefore, the FIDs obtained are a function of two times: t_1 and t_2. A Fourier transformation of this matrix leads to a 2D-NMR spectrum in which resonances are obtained as a function of two frequencies υ (or chemical shifts δ) or "dimensions" (Fig. 5.2c, bottom panel). In the end, we have a contour plot level (analogous to those used in weather forecasts) where we see different resonances of protons split across two dimensions (on two axes, akin to a Cartesian coordinate map). In the resulting spectrum, each resonance from the 1D spectrum correlates with itself, leading to the so-called diagonal peaks (because there is a diagonal through the two opposite vertices of the contour map); these peaks do not yield useful information since they provide the same information as in the 1D-NMR spectrum. On the other hand, we have off-diagonal cross-peaks, which provide useful information as they correlate nuclei (protons, in the 2D-COSY) that are connected by covalent bonds (Fig. 5.2c, bottom panel).

On the other hand, a homonuclear 2D-NOESY spectrum consists of the addition of a third pulse after the last pulse of a COSY, separated by a fixed time (mixing time) maintained throughout the experiment (Fig. 5.2b bottom panel); during this time, spatially close spins interact through the NOE (see Sect. 5.2.5). We can also design heteronuclear spectra, where instead of observing the same nucleus (^{1}H) in both dimensions, we observe ^{1}H in the t_2 dimension (horizontal axis) and another type of nuclei (such as ^{15}N or ^{13}C) in the t_1 dimension (vertical axis). With biomolecules, these heteronuclear experiments (especially one called HSQC, which typically shows correlations between ^{1}H (horizontal axis) and ^{15}N (or ^{13}C) (vertical axis), although generally any heteronucleus can be explored in the vertical axis; see Sect. 5.4.3) are useful for mapping biomolecule-biomolecule interactions. However, it must be noted that due to the low sensitivity of the NMR technique and the low natural abundance of the latter NMR-active isotopes, the biomolecule must be isotopically enriched in those nuclei (^{15}N or ^{13}C).

The key to 2D-NMR spectroscopy is having a time (t_1) which varies in each of the 1D-NMR spectra acquired (Fig. 5.2b, c), during an observation time, t_2. Therefore, if instead of varying a single time (t_1), we modify two times t among the pulses, we have 3D-NMR. In 3D-NMR, we obtain "cubes" with diagonals between two opposite edges, and there are off-diagonal cross-peaks that contain useful information. To achieve good resolution, 3D-NMR experiments require the use of labeled samples. 3D-NMR spectroscopy is widely used in the assignment (see Sect. 5.3) and structural elucidation of biomolecules. If we modify three times t among the pulses, we have 4D-NMR spectroscopy, and so on for n-NMR spectroscopy. In most examples described in the literature, the use of 2D- and 3D-NMR spectra with double labeled (^{15}N and ^{13}C) or even triply labeled (^{15}N, ^{13}C and ^{2}H) samples is enough to obtain the structure of a biomolecule.

5.2.7 *The Youngest Sister: Electron Paramagnetic Resonance (EPR)*

Not only do nuclei (the proton) have a spin, but electrons also possess spin. Therefore, the same reasoning applied above for the proton in atomic nuclei can be applied to an electron in electronic shells. The main differences with the features of nuclei are as follows: (i) the μ of an electron is larger than that of the proton (1000 times more); (ii) we know relatively well how electrons are arranged in the shells of atoms in orbitals, making it easy to predict whether an atom or molecule will yield an EPR spectrum when there are unpaired electrons; (iii) since the μ of the electron is larger, the energy required to split the levels must be higher than that in NMR, reaching into the microwave region rather than radiofrequency; and (iv) because the energy is higher, electrons, in the presence of an external B_0 field, can interact with other electrons—as nuclei do. This is similar to the spin-spin coupling described above for nuclei (see Sect. 5.2.3)—as well as with nuclei—and therefore the EPR spectrum shows, in addition to electron-electron interactions, nuclei-electron coupling (interaction). Thus, the number of lines in an EPR spectrum depends on the number of unpaired electrons (similar to NMR with the number of nearby nuclei), and these lines can also be split due to coupling from the spin of the nuclei.

Molecules with unpaired electrons are termed paramagnetic, whereas those with paired electrons are diamagnetic. Similar to NMR (see Sect. 5.2.1), the technique is called "electronic" because it observes electrons, "paramagnetic" because it requires a magnetic field to create non-degenerate quantum levels observed only in paramagnetic substances, and "resonance" because it involves applying precise energy (in the radiofrequency region) corresponding to the separation between split levels, as *per* the Planck equation. Some authors prefer the term "electronic spin resonance" instead of EPR to emphasize that electron spins are being observed.

Molecules such as proteins, which are usually diamagnetic, can be rendered paramagnetic by introducing a paramagnetic label, usually a nitroxide radical (though other paramagnetic substances are also available), at a specific cysteine residue. This modification is usually achieved through site-directed mutagenesis. In such cases, EPR spectroscopy provides insights into the dynamics of two main aspects: (i) the specific polypeptide region where the probe is attached; and (ii) the behavior of the probe itself. In the simplest scenario, EPR spectra offer information about these two different dynamics (times). In EPR, the focus is more on analyzing the dynamics and shape of the spectral lines rather than their specific frequencies. By designing mutants with paramagnetic probes at different positions along the chain (assuming the probe dynamics remain consistent across the protein), researchers can gather information on various dynamics at different locations. The EPR spectrum reflects changes in the environment surrounding the label, enabling the technique to reveal local structural changes in proteins, introduced by PPIs or changes in the solution environment where the macromolecule and probe are embedded.

5.3 Determination of Biomolecular Structures by NMR

5.3.1 Sample Preparation

Most NMR experiments are carried out in H_2O, with the addition of 5–10% ^2H_2O to facilitate monitoring of the spectrum of ^2H during long acquisition times (for details, refer to [3–5], which delves deeper into this topic beyond this chapter's scope). However, there is a notable exception for experiments aimed at describing hydrogen bond scaffolding in molecules (see Sect. 5.3.2). For studies involving peptides, protein fragments, or situations where membrane-associated proteins or lipids are involved, organic solvents like trifluoroethanol or dimethyl sulfoxide are often used. In cases involving proteins associated with membranes or other lipids, detergents such as sodium dodecyl sulfate or others (see Sect. 5.4.3 for examples with proteins from the severe acute respiratory syndrome coronavirus 2 (SARS-CoV-2)) are necessary. In some instances, conventional high-resolution NMR methods may not suffice, prompting the use of specialized techniques such as SSNMR to provide structural insights (see Sect. 5.4.4).

As NMR is a low-sensitivity spectroscopic technique, experiments aimed at solving molecular structures typically operate at concentrations around 1 mM (for current spectroscopic configurations). Therefore, the biomolecule must be soluble (i.e., monodisperse) at these concentrations and stable over long acquisition times. The sample is placed in a tube (typically 5 mm wide) made of special, non-paramagnetic materials [3–5], ensuring it is free of contaminants or other paramagnetic compounds. If the biomolecule is stable, NMR spectra are often acquired at low pH to minimize hydrogen exchange of amide protons (crucial for assignment processes; see Sect. 5.3.2) with water. Buffers used must not introduce signals that could obscure sample resonances; hence, deuterated buffers are common in homonuclear (^1H-detected) experiments. High ionic strength can reduce signal-to-noise ratio due to local magnetic fields from ions in solution. Lastly, NMR experiments are conducted at moderately elevated temperatures to decrease resonance linewidths (thus enhancing resolution), ensuring the biomolecule remains stable without unfolding or aggregating.

5.3.2 Structure Determination

The steps involved in determining the structure of a protein or nucleic acid via NMR follow a straightforward logic. Assuming we know the biomolecule's primary structure, the goal is to derive its secondary and tertiary structures. There are two main approaches to assigning a biomolecule: (i) using natural abundance ^1H (the homonuclear procedure), primarily through 2D-NMR spectroscopy; and (ii) using labeled samples (the heteronuclear approach), relying on 3D (or higher-dimensional)-NMR spectroscopy. The difference lies in the types of experiments used

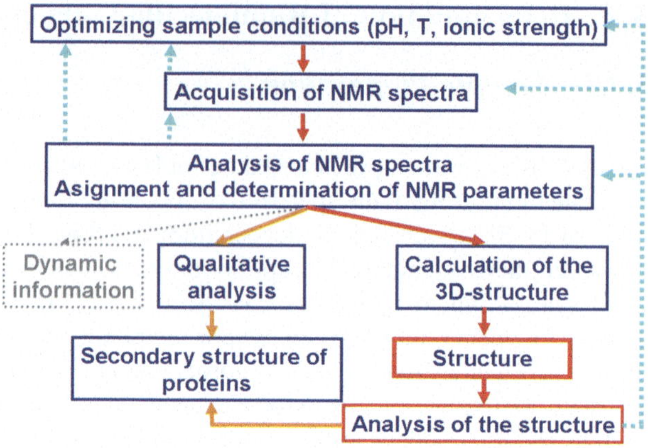

Fig. 5.3 Solving structures by NMR. Flow chart indicating the different steps in the determination of the 3D structure of a protein using NMR

for spectrum assignment, while the structural constraints required are essentially the same.

Although there are many variations in the assignment process (beyond this chapter's scope, detailed in [8, 9]), whether employing homonuclear or heteronuclear approaches, the process can be distilled into a series of steps. Initially, the multidimensional NMR spectra obtained must be *assigned*, meaning that each intensity peak at a defined δ in the spectra must be linked to a particular nucleus within the molecule.

In the homonuclear approach, the assignment process initially involves identifying different types of residues (Fig. 5.3). To illustrate this procedure further, we will focus on protein assignments in the following lines. Analogously, in this first step, we assign a *name* (e.g., Asn, or any other amino acid type) to each different set of signals associated with the same amide proton observed in a TOCSY (a more sophisticated type of COSY) experiment. However, distinguishing which specific residue (e.g., Asn25, Asn72, Asn80, or Asn100) within the 200-residue protein we are studying remains challenging at this stage. This inability to pinpoint the exact residue is analogous to not having a *surname* for each set of signals assigned to a *name* (amino acid type). In the subsequent step, we assign a *surname* to each set of signals already assigned a *name* by acquiring a NOESY spectrum. The NOESY spectrum facilitates the identification of spatial connectivity within the chain (see Sect. 5.2.5). Therefore, in the homonuclear approach, while we are "naming" and "surnaming" each residue in the polypeptide chain, we can also infer the type of secondary structure in which these assigned residues are located based on the observed NOE patterns [2, 5, 7–10] (Fig. 5.3). It is noteworthy that both the "naming" and "surnaming" steps in protein assignment predominantly rely on the presence of amide protons. Hence, the more amide proton resonances identified, the greater the likelihood of completing the entire protein assignment.

In the heteronuclear approach, the assignment step does not involve the use of NOESY spectra [5, 8–10] as a necessary stage for achieving complete assignment. This method relies instead on determining J values for interactions between ^{1}H with

^{13}C, ^{1}H with ^{15}N, ^{13}C with ^{15}N, and ^{13}C with ^{13}C nuclei. This approach avoids ambiguities by relying on information provided by covalent bonds (spin-spin coupling) rather than through-space interactions as in NOESY. Consequently, the assignment process can be relatively easily computerized [9]. NOESY spectra are later used in this approach to confirm the through-bond assignments and to provide a dataset of restraints (distances between nuclei) for the subsequent steps in the structure determination protocol [9] (see below). Similar to the homonuclear approach in proteins, resonance assignment in the heteronuclear approach is primarily based (though not exclusively) on the presence of amide protons.

Once the spectra are assigned, various parameters for the identified nuclei can be obtained (see Sect. 5.2). These include chemical shifts (δ), coupling constants (J), peak intensities, NOEs, and, for dynamic studies, relaxation times (in some cases, even for the heteronuclei). Notably, distances between atomic nuclei (known as distance restrictions) are obtained from NOEs, and dihedral angles between bonded nuclei are derived from J values. Distance restraints, derived mainly from NOESY spectra, are crucial for the pulse sequence toolkit in NMR structural determination. Currently, there is software available that can generate initial, preliminary structures based solely on chemical shift values of ^{1}H, ^{13}C, and ^{15}N backbone nuclei (interested readers can explore tools like those available at the Biological Magnetic Resonance Bank, BMRB: http://csrosetta.bmrb.io). These initial structures can be refined later using NOESY information or J values.

There is another critical parameter used in structural determination, providing information on the hydrogen bond scaffolding of biomolecules (secondary and tertiary structures in proteins and hydrogen-bonding patterns in nucleic acids). For clarity, the focus will be on proteins in the following discussion. In proteins, amide hydrogens and other hydrogens bound to nitrogens (such as in the side chains of Lys, Arg, Asn, or Gln) are weakly bound due to nitrogen's hybridization and electronegativity, similar to hydrogens in water molecules. Therefore, N-H bonds in proteins and O-H bonds in water can break due to thermal fluctuations and can immediately interact with either another water molecule's oxygen or by a protein's nitrogen, facilitating *hydrogen exchange* between the protein and water. When a protein is dissolved in $^{2}H_2O$ instead of H_2O, we observe an exchange between the labile ^{1}H of the protein and the ^{2}H of the solvent. Importantly, since the resonance frequency (υ) of ^{2}H differs from that of ^{1}H, we cannot observe ^{2}H directly in NMR while observing ^{1}H. By acquiring spectra at different times after dissolving the biomolecule in $^{2}H_2O$, we observe the disappearance of ^{1}H involved in N-H bonds in those 2D-NMR spectra. However, not all ^{1}H atoms exchange with the same ease; those involved in hydrogen bonds are less likely to exchange with the solvent. In structural determination, knowing which protons are more "reluctant" to exchange with $^{2}H_2O$ is crucial. Moreover, the rate of exchange provides insights into the stability of the exchanged N-H hydrogen bonds. Thus, the pattern of exchange between ^{1}H and ^{2}H depends on hydrogen bonding and protein stability, and the extent of incorporation of ^{2}H nuclei allows us an estimation of overall protein stability. A similar approach is used to measure the stability of virus capsids using mass spectrometry (see Chap. 6), where the mass of ^{2}H is approximately twice that of

^{1}H, and the number of incorporated ^{2}H nuclei can be precisely determined with current mass spectrometers.

It is worth noting that the most commonly used structural restrains in NMR have been described. However, there are other parameters, applicable to partially folded proteins or intrinsically disordered proteins (IDPs), that are beyond the scope of this chapter. Some of them involve the use of paramagnetic probes (similar to those used in EPR; Sect. 5.2.7). For a more comprehensive overview, interested readers can refer to [5, 7, 9].

To construct a structural model of a macromolecule using NMR, we need to identify the conformation(s) consistent with experimentally obtained restraints (distances, dihedral angles, and hydrogen bonds) (Fig. 5.3). Usually, these restraints are incorporated into potential energy functions to compute molecular dynamics and minimum energy structures [9]. Through this computationally intensive process, structures with the fewest violations of NMR restraints are identified. With sufficient restraints, only one structure (or a set of very similar structures) may be feasible. With fewer restraints, multiple structurally different models consistent with the experimental data can be found. During computation, these obtained structures are used to refine restrictions and reassign many NOEs observed in NOESY spectra, which were previously ambiguous (Fig. 5.3, dotted line). The final outcome is an ensemble of refined, similar structural models that represent the minimum-energy three-dimensional structure of the macromolecule. Sometimes these models are averaged to generate a representative structure of the ensemble. These structures typically exhibit well-defined regions (where there are typically numerous structural restrictions *per* amino acid or base) and other regions where the protein backbone or nucleic acid appears less restricted. These less defined regions may correspond to areas with higher mobility, often implicated in molecular recognition processes. Alternatively, these regions may lack local and/or long-range restrictions, contributing to their ambiguous definition.

5.4 NMR Structures of Viral Macromolecules

This section describes several examples of the use of NMR spectroscopy to solve the structure and investigate the dynamics of viral proteins and nucleic acids involved in defining the structure of virus particles (the so-called structural biomolecules). Numerous proteins from different viruses have been extensively studied using NMR. However, particular focus has been on structural proteins and nucleic acid domains from viruses such as human immunodeficiency virus type 1 (HIV-1), other retroviruses, and the coronavirus SARS-CoV-2, mainly due to their significance as targets for developing new antiviral therapies (Chap. 20). A summary of various structural studies on retroviral proteins and nucleic acids is provided as case studies to illustrate what NMR can reveal about virus structure. In addition, a few examples of proteins from other types of viruses will be provided to illustrate additional applications of NMR in structural virology. Examples will be discussed to show

how, knowing the structures of macromolecules through cryo EM (Chap. 3) or X-ray (Chap. 4), NMR can shed light on the different conformational features of macromolecules modified by changing their environment or through their interactions with others.

5.4.1 NMR Structures of Viral Proteins

The main protein components of retrovirus particles are synthesized as three polyproteins that play distinct roles in the internal structure of the virion (Gag), the viral enzymes (Pol), and the glycoproteins of the virion envelope (Env). The Gag polyprotein is essential for virus particle assembly and budding. Specifically, the Gag polyprotein is responsible for capturing viral RNA and assembling into a spherical capsid within the immature virion. This assembly occurs either in the cytosol (B- and D-type retroviruses) or at the cell membrane (C-type, human T-lymphotropic virus (HTLV) (first known as human T-cell leukemia virus), bovine leukemia virus (BLV), and lentiviruses). During virion maturation, the HIV-1 Gag polyprotein undergoes processing by the viral protease, resulting in the formation of matrix (MA), mature capsid (CA), nucleocapsid (NC), and p6 proteins.

The Capsid Protein of Retroviruses
In HIV-1, the CA domain plays a crucial role in mediating Gag-Gag interactions critical for both immature and mature capsid formation [11]. Additionally, HIV-1 CA interacts with various host cell factors, including cyclophilin A, a host restriction factor, and human lysyl- tRNA synthetase (hLysRS) [12]. CA is a two-domain protein, composed of largely helical N- and C-terminal domains (NTD and CTD, respectively) connected by a flexible linker [13, 14]. Upon proteolytic processing during virion maturation, CA is released from the Gag precursor and rearranges to form a conical capsid that surrounds the viral genome and associated proteins [13, 15]. The capsid is later disassembled after virus entry into the host cell to allow uncoating of the viral genome. Mutations that impair CA assembly or alter its stability lead to dramatic reductions in viral infectivity. Assembly and stability of CA depend on inter- and intramolecular interactions involving the NTD and/or CTD of CA [14, 16]. The NTD (CA residues 1–146), which is monomeric in solution, forms hexamers in the virion capsid; the CTD (CA residues 146–231), which is dimeric in solution, links adjacent hexamers [14]. The atomic structure of HIV-1 CA has been studied by both X-ray crystallography and NMR using a "bottom-up" approach, starting with solving the structures of the isolated domains. Recently, the NMR structure of the full-length CA has been determined, confirming previous results with the isolated domains.

(a) *The structure of the HIV-1 NTD.* The structures of the HIV-1 NTD determined by NMR and X-ray crystallography ([17] and references therein) are in good agreement. NTD consists of seven α-helices, two β-hairpins, and an exposed,

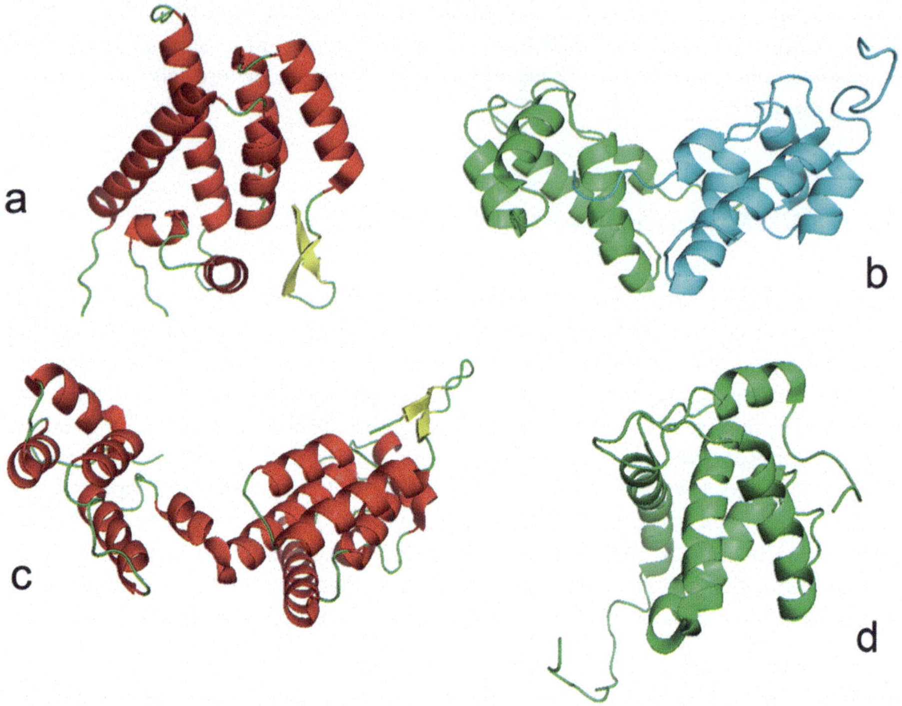

Fig. 5.4 Structures of some HIV-1 proteins solved by NMR. (**a**) Isolated NTD monomer (PDB number 1GWP) [17, 18]. (**b**) Isolated CTD dimer (monomers shown in green and cyan) (PDB 2KOD) [19]. (**c**) Full-length monomeric CA mutant (PDB number 2LF4) [20]; the NTD and CTD are respectively on the right and left sides of the figure. (**d**) Un-myristylated MA (PDB number 2H3F) [21]. Images in this and following figures were produced with PyMol [22]. In panels (**a**) and (**c**), helices are shown in red, β-sheets in yellow, and loops in green

partially ordered loop (Fig. 5.4a). This domain has an arrowhead shape, with the β-hairpins and loop extending from the trailing edge and the C-terminal helix projecting from the tip. This structure differs from those of other RNA virus coat proteins [17]. At its N-terminus, there is a terminal proline that forms a salt bridge with a conserved buried aspartate residue (Asp51). This salt bridge cannot exist in the Gag precursor polyprotein, suggesting that the β-hairpin itself plays a role in triggering core particle condensation. The binding site for cyclophilin A is located on the exposed loop and includes Pro90, which adopts both *cis* (absent in X-ray structures) and *trans* conformations in the free monomeric domain in solution due to loop flexibility. In the X-ray structure of the complex with cyclophilin A, Pro90 adopts a *trans* conformation [17].

(b) *The structure of the HIV-1 CTD.* The structure of the HIV-1 CTD has been determined by both NMR and X-ray crystallography. The CTD consists of a short 3_{10}- helix, a highly conserved sheet-turn-helix element known as the major homology region (MHR), and four helices (8–11, in the numbering of the intact

CA protein) (Fig. 5.4b), which are interconnected by short loops or turn-like structures. The dimerization interface of the CTD is formed by the mutual docking of α-helix 9 from each monomer, with the side chains of Trp184 deeply buried in the dimer interface, and hydrophobic interactions between the 3_{10}-helix of one subunit and α-helix 9 of the other. While there are minor variations between the crystal and solution structures of the dimeric CTD (discussed in the next paragraph), the fundamental interactions stabilizing the lattice of the mature virus capsid remain consistent with those observed in the solution dimer interface.

Although both X-ray and NMR structures of active dimeric CTD reveal similar structures within each monomer (NMR structure solved in 2009 and X-ray structure in 1996), slight variations occur in the structure of α-helix 9 ([14] and references therein). These variations occur at the tertiary level (packing of α-helix 9 against other helices within the same monomer) and at the quaternary level (the angle formed by α-helix 9 across the dimer dyad). Furthermore, in solution, the CTD exhibits two conformers whose populations are influenced by the protonation state of Glu175, a residue involved in the dimerization interface [23]. In addition, structurally different dimers have been characterized, including the canonical dimer described above and a domain-swapped arrangement primarily involving the MHR [24]. Different structures of mutant, monomeric species of CTD have also been determined by NMR. Taken together, these findings underscore the high structural plasticity of the CTD, which likely plays a significant biological role in capsid assembly, maturation, and function [14].

(c) *Structure of the intact retrovirus CA proteins.* The structures of full-length HTLV type I CA and Rous sarcoma virus (RSV) CA have been determined by NMR [25, 26]. Similar to HIV-1 CA, these proteins consist of two independent domains connected by a flexible linker. Minor structural differences between the CTDs may explain why HIV-1 CA forms dimers, whereas HTLV type I CA and RSV CA are monomeric. NMR studies of these proteins also indicate that the two domains are structurally independent, as evidenced by the absence of inter-domain NOEs. Furthermore, the flexibility of the entire protein, measured by T_1 and T_2 relaxation times, suggests that both regions exhibit dynamic independence.

The structure of full-length monomeric HIV-1 CA (with an inactive dimerization interface) has also been solved by NMR (Fig. 5.4c). While the overall structures of both domains resemble those of the isolated regions, significant differences are observed in the dimerization helix of the CTD [20]. In dimeric CA structures from crystallography and NMR, the dimerization helix (helix 9) is longer and shows a bend, whereas in the monomeric mutant, α-helix 9 is shorter and straight. This finding supports previous conclusions suggesting substantial conformational rearrangements during CA dimerization ([14] and references therein). The NTD and CTD are connected by a five-residue linker (residues 145–149) (Fig. 5.4c). Similar to other CA structures, this linker is highly flexible, as evidenced by random coil chemical shifts for the central three residues and T_1/T_2 relaxation data. This inter-domain flexibility likely plays a crucial role in facilitating the docking of other Gag domains during immature capsid assembly [27] and accommodating the structural variations observed between subunits in mature retroviral capsids [14].

The Matrix Protein of Retroviruses

The MA protein has similar roles in all retroviruses, directing membrane localization of the assembling viral particle and after virus maturation, forming a stable structural shell associated with the inner surface of the mature viral lipid envelope ([14] and references therein).

The three-dimensional structures of the HIV-1, BLV, and HTLV type II MA proteins have been determined by NMR ([17] and references therein); there are also X-ray structures of the simian immunodeficiency virus (SIV) and HIV-1 MA [17]. The HIV-1 MA consists of five α-helices, two short 3_{10}-helical stretches, and a three-stranded mixed β-sheet (Fig. 5.4d shows the NMR structure, where only the helices are observed). Close to the C-terminus of the second helix, there is a putative membrane-binding surface, which could anchor the protein to the membrane surface, in concert with an N-terminal myristoyl group. HIV-1 MA is a monomer in solution but forms trimers in the crystal; the HIV-1 MA trimers may also be observed in lower-resolution EM images and could serve as a fundamental building block for the formation of the MA shell within the mature virion. The individual proteins of the trimer are arranged to create a basic surface, which interacts directly with the acidic inner layer of the lipid ([17] and references therein). Comparison between the monomer in crystallographic trimers and the NMR structure of the monomer in solution shows a displacement of the 3_{10}-helix, suggesting this structural element may act as a "hinge" during virion assembly and disassembly. This structural organization resembles the rearrangements proposed for CA dimerization, highlighting the critical role of protein flexibility in retrovirus morphogenesis. Exploring the dynamics of this helix through measurements of T_1 and T_2 relaxation times by NMR could provide further insights into MA flexibility.

The Nucleocapsid Protein of Retroviruses

NC is the domain of the Gag polyprotein responsible for directing genome packaging. With the exception of spumaviruses, all retroviral NC proteins contain one or two copies of a conserved CCHC motif (Cys-X2-Cys-X4-His-X4-Cys), which serve as zinc-binding sites, known as "zinc knuckles." In HIV-1, NC does not form direct lattice stabilizing interactions but contributes to particle formation by tethering Gag to the RNA genome (see [14] and references therein).

NMR structures are available for synthetic peptides containing sequences from the HIV-1 NC (including the zinc knuckle domains) and for intact NC proteins from HIV-1, Moloney murine leukemia virus (MoMuLV) (which contains a single CCHC zinc knuckle), mouse mammary tumor virus (MMTV), and monkey virus (MV) [17, 28]. In HIV-1 and MoMuLV NC proteins, the CCHC zinc knuckle domains adopt similar folds. The N-terminal residues form a metal-coordinating reverse turn known as a "rubredoxin knuckle," resembling metal-coordinating substructures observed in the iron domain of rubredoxin. Subsequent residues form a

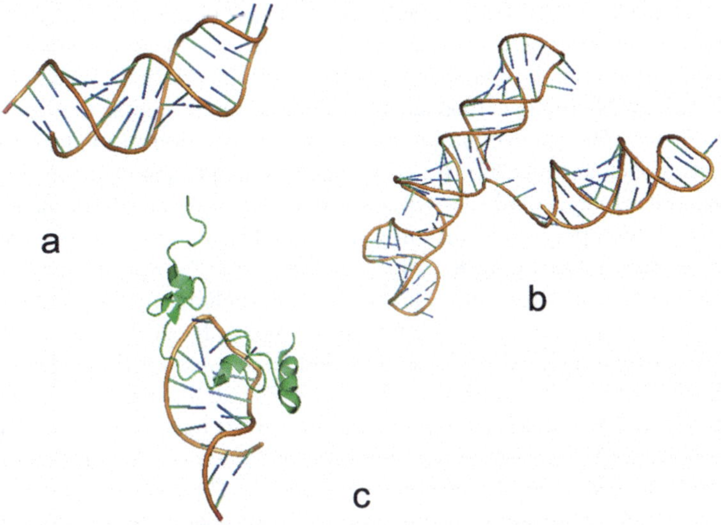

Fig. 5.5 NMR structures of isolated viral RNA or viral RNA-protein complexes. (**a**) Structure of the SL1 of HIV-1 (PDB number 1N8X) [29]. (**b**) Structure of the MLV ψ-RNA site (PDB number 1S9S) [30]. (**c**) Structure of the HIV-1 NC bound to SL2 of the RNA (PDB number 1FSU) [31]

loop that leads to a C-terminal 3_{10}-helix (see Fig. 5.5c for an example, which also illustrates the interaction of NC with nucleic acid). In HIV-1 NC, weak NOE inter-knuckle contacts exist between two bulky aromatic side chains, but dynamic studies (measured by T_1 and T_2 relaxation times) suggest that the two zinc knuckles behave independently, similar to the NTD and CTD domains in CA (as discussed earlier). Thus, NC in solution may explore different conformations with interacting and non-interacting knuckle domains [28]. On the other hand, one of the zinc knuckles in MMTV and Monkeypox virus (MPXV) undergoes a conformational transition involving an additional C-terminal β-hairpin, albeit oriented differently in each protein. Therefore, NMR studies suggest that distinct structural features of the zinc knuckles in diverse retroviruses may facilitate the packaging of diverse genomes.

Intrinsically Disordered Proteins (IDPs)

One major contribution of NMR to the field of virus structure is the identification and characterization of proteins that are nearly or completely unfolded. This topic will be briefly described. Such proteins cannot be crystallized and are not amenable to X-ray structure determination. Currently, NMR and in some cases EPR are the sole techniques providing atomic-level insights into this special class of proteins. Other examples of intrinsically disordered regions (IDRs), not entire proteins, are discussed in Sect. 5.4.3. These IDRs can be crucial in viral entry and neutralization, as recently shown for a short region in the envelope protein of HIV-1 (gp41), which is affected by a monomer-to-α-helical trimer equilibrium [32].

The p6 protein of HIV-1 serves as a docking site for several viral and cellular binding factors, facilitating virus budding [33] and regulating CA processing and virus core assembly [34]. In aqueous solution, the peptide is disordered, but under hydrophobic solution conditions (e.g., in the presence of trifluoroethanol), which may mimic a lipidic environment, p6 adopts a helix-hinge-helix structure: a short α-helix (residues 14–18) is connected by a flexible hinge region to a longer helix (residues 33–44). This conformation allows the latter region to bind to a vacuolar protein sorting factor and to the HIV-1 accessory viral protein (VPr) [35].

The Gag polyproteins of G-retroviruses contain a conserved p12 domain between MA and CA, which plays key roles in virus assembly and nuclear transcription. A MoMuLV fragment containing the p12 region and the NTD is unfolded; even the β-hairpin at the N-terminus of the NTD region is unfolded [36]. The p12 region and the hairpin may fold upon interaction with other molecules during capsid assembly.

Some non-structural viral proteins also contain substantial IDRs. For example, the protein W from bacteriophage λ is required for DNA stabilization within the capsid (head) and for attaching the tail during virion morphogenesis. The NMR structure of protein W consists of two α-helices and a two-stranded β-sheet arranged around a well-packed hydrophobic core. However, the C-terminal segment, which is essential for DNA stabilization, is unfolded [37]. Another example is the 117-residue-long bacteriophage λ FII protein, which is crucial for joining the phage head and tail during late stages of phage morphogenesis. In solution, this protein forms a 3D structure composed of seven β-strands and a short α-helix. However, two large regions remain unstructured: the N-terminus (residues 1–24) and a large loop near the middle of the protein (residues 46–62) [38]. It is suggested that both regions may fold during assembly. In yet another example, NMR studies of the Gn membrane glycoprotein of hantaviruses show that the zinc finger domain is the only folded polypeptide patch in the cytoplasmic tail of the protein. This region interacts with the ribonucleoprotein during virus assembly [39].

The C-terminal region of the Measles virus nucleoprotein (N_{tail}), comprising residues 401–525 of the whole protein, is an IDP. It interacts with the X domain of the viral phosphoprotein (residues 459–507 of the intact protein). The use of EPR (by introducing cysteine residues at selected sites along the polypeptide chain) has demonstrated an example of *folding upon binding*: residues 488–502 of N_{tail} adopt an α-helical conformation, while the rest of the polypeptide chain remains disordered [40–43]. The binding affinity of the complex is high (in the order of nM). Furthermore, in this example, both proteins were labelled with different paramagnetic probes, revealing that even though residues 488–502 fold into an α-helix upon binding, the dynamics of the probe attached to the IDP differ from those of the probe attached to the X domain [43]. Therefore, while the 488–502 region forms a folded structure, its dynamics are distinct from those of a similar secondary structural element in a properly folded protein. Chapter 6 will include a discussion on how this complex can also be studied using mass spectrometry.

5.4.2 NMR Structures of Viral Nucleic Acids

All retroviruses specifically package two copies of their genomes during virus assembly. Packaging is mediated by the Gag proteins, which in the absence of viral RNA can incorporate cellular RNAs. However, viral RNA contains specific packaging signals (Ψ-sites) that are preferentially packaged; genome selection proceeds via direct binding of NC to these sites, although not exclusively. The Ψ-sites are initially sequestered by intramolecular base pairing in monomeric RNA and become exposed upon dimerization, facilitating high-affinity binding of the RNA dimer to NC [44, 45]. A ribonucleoprotein complex, consisting of a small number of Gag molecules and two copies of the genome, is directed toward assembly sites on the plasma membrane, where thousands of additional Gag molecules assemble to form the immature virus particle.

Numerous NMR structural studies on retroviral RNA have focused on the 5′--untranslated region containing packaging signals. Studies indicate that the 120 nucleotides upstream of the Gag start codon are crucial for efficient packaging. This segment is proposed to form four stem-loops (SL1–SL4), each with different structures and affinities for NC [29, 46, 47] (Fig. 5.5a, b). For instance, the NMR structure of the SL2 stem-loop reveals a "platform motif" where the rings of sequential U14 and A15 bases are approximately coplanar, which may mediate long-range interactions stabilizing the ψ-RNA association or facilitating splicing and/or packaging [46]. A similar "platform motif" is observed in the NMR structure of the 101-nucleotide core encapsidation signal of MoMuLV, accompanied by an "A-minor K-turn" where unpaired bases of a bulge interact with the groove of a proximal stem [30].

The structure of the entire HIV-1 5′-leader RNA (712-nucleotide dimer) has been studied using NMR [48]. Residues spanning the Gag start codon (AUG) form a hairpin structure in the monomeric leader RNA and base-pair with bases in the unique 5′ region in the dimeric form. This association causes dimerization by displacing and exposing a dimer-promoting hairpin, thereby enhancing binding to NC. Taken together, these findings suggest a mechanism where translation, RNA dimerization, RNA-NC binding, and RNA packaging within the assembled capsid are triggered by a common RNA structural switch. This switch involves a shift in conformational equilibrium among different RNA conformations.

5.4.3 NMR Structures of Viral Protein-Ligand Complexes

This section presents examples of the type of information NMR spectroscopy can provide when viral proteins interact with other partners. The latest subsections focus on PPIs, emphasizing not the structures of the complexes formed (which can be unknown) but how NMR findings can propose interactions involving specific regions of each macromolecule in the complex. A specific section dedicated to viral protein-sugar complexes is not included due to the limited number of examples in the literature. Most protein-sugar complexes described focus on measuring the affinity of interactions (i.e., thermodynamic parameters) between the protein and sugar. For an example of the insights NMR can provide on such complexes, readers are encouraged to refer to a recent study discussing the interaction between the glycosylated SARS-CoV-2 spike (S) protein and two sialic acid-containing trisaccharides [49, 50].

Protein-Nucleic Acid Complexes
In RSV, the dynamics of an RNA segment in both free and bound states with a fragment of Gag (comprising CTD, NC, and the in-between spacer region (SP)) have been described. Both NC and CTD regions exhibit independent dynamic behaviors. Contacts with the RNA are distributed across these three regions: one of the zinc fingers in NC appears to make key contacts with the RNA; the SP region undergoes a slow conformational equilibrium in the absence of the RNA, which is restrained upon binding; and finally, the first and fourth helices in CTD are also affected by RNA binding. These results suggest that NC-RNA binding disrupts intramolecular interactions between SP and CTD regions, thereby enhancing accessibility of the Gag dimer contact surface and promoting intermolecular organization [51].

The structures of several stem-loop sites in the retroviral RNA complexed with the NC of HIV-1 have been determined by NMR [31, 47, 52] (Fig. 5.5c). Both zinc knuckles bind to exposed guanine bases of the tetraloop, while basic residues at the N-terminus of NC pack against the closer zinc knuckle near the RNA stem. In addition, the N-terminal zinc knuckle interacts with the platform motif as discussed earlier. Some interactions resemble those observed in SL3, whereas others are notably different; for instance, NC interacts with the major groove of SL3, and the orientations of the two zinc domains differ between the complexes [31].

However, not all retroviruses bind with high affinity to individual SL RNA loops; for instance, in MoMuLV, NC binds only to the three SLs with a nM affinity, indicating specificity toward a region present only in the intact three loops [47]. The NMR structure of the MoMuLV-NC in complex with the 101-nucleotide core encapsidation segment of the MoMuLV Ψ-site suggests a network of interactions that promote specific binding by the NC's sole zinc knuckle.

Conversely, structural studies of retroviral RNA-NC complexes reveal significant differences in genome recognition and NC-RNA binding among different viruses [44, 48]. Furthermore, these data suggest that the mechanism of genome recognition/packaging involves one or more NC domains of several Gag molecules interacting with multiple Ψ-site stem-loops [47].

Finally, it is noteworthy that NMR has been used to detect and map new interactions between viral proteins and nucleic acids without requiring the structure of the complex. We describe two examples where the abovementioned HSQC method has been exploited (Sect. 5.2.6). Both illustrate a general technique (further examples involving PPIs are discussed at the end of this section), enabling easy and rapid detection of protein-ligand interactions in high-throughput screening. In addition, the second example briefly introduces the concept of liquid-liquid phase separation (LLPS) in organelles, which is crucial in both cells and viruses. Cells organize and carry out many biochemical reactions in non-membrane-bound compartments. Biomolecular condensates are membrane-less intracellular assemblies or compartments formed through LLPS in the cytoplasm [53]. Apart from compartmentalization, these condensates can concentrate biopolymers, and many proteins crucial for cell growth and development undergo phase separation.

In the first example, recent NMR studies have detected MA-DNA interactions, although the structure of the complex remains unsolved. Instead, the DNA-interacting region was identified by changes in chemical shifts of assigned protein resonances upon nucleic acid addition, using heteronuclear HSQC experiments [54].

In the second example, it is noteworthy that the SARS-CoV-2 genome, encapsulated within a lipid membrane, consists of a single-stranded RNA approximately 30 kb in length. This RNA is packaged by the NC, which plays crucial roles in genome packaging, viral replication upon host infection, and also SARS-CoV-2 replication through RNA-driven LLPS [55–57]. The NC of SARS-CoV-2, NC-SARS-CoV-2, comprises two well-folded domains flanked by IDRs: the first domain contains the RNA-binding region, although all parts of the protein have been involved in RNA or protein binding [58]); the second domain is responsible for dimerization of NC-SARS-CoV-2. The N-terminal region of the RNA-binding domain includes a disordered region with prion-like primary sequence properties, which modulates LLPS formation in stress granules, as demonstrated by HSQC and fluorescence experiments (see Chap. 6) [59]. Acquisition of HSQC spectra in the absence and presence of polyU RNA, coupled with previous domain assignments within NC-SARS-CoV-2, facilitated identification of specific protein residues directly involved in RNA interactions during stress granule formation. In this context, the 3D structure of the granules is less relevant than identifying amino acids crucial for binding and probably triggering LLPS, which can serve as targets in designing new drugs (see Chap. 20).

Protein-Lipid Complexes

NMR has also been used to study interactions between retroviral MA proteins and lipids. Interactions of MA from equine infectious anemia virus (EIAV) or HIV-1 with phosphatidylinositol-4,5-diphosphate (PIP2) have been described [60]. The interaction between PIP2 and HIV-1 MA promotes membrane binding and induces conformational changes at the N-terminus of MA, exposing the myristate moiety. The binding region involves the loop between the first and second helices and the cleft between the second and fourth helices. In un-myristylated EIAV MA, the lipid binding site includes the loop between the second and third helices, accompanied by conformational changes detected through T_1 and T_2 relaxation times, which may trigger MA trimerization. The structure of myristylated HIV-2 MA in complex with PIP2 has also been resolved [61] (Fig. 5.6a). Similar to HIV-1 MA, the myristoyl group of HIV-2 MA is partially sequestered within a narrow hydrophobic tunnel formed by side chains of residues from helices 1, 2, 3, and 5. However, the myristate of HIV-2 MA is more tightly sequestered than that of the HIV-1 protein, and the

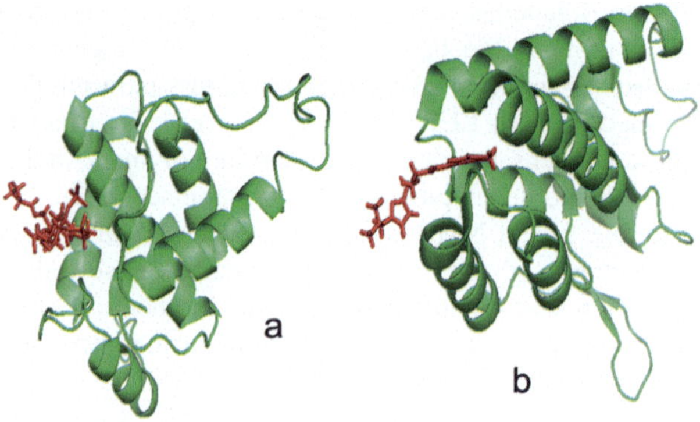

Fig. 5.6 NMR structures of the complexes of small ligands with viral proteins. (**a**) NMR structure of the myristylated HIV-2 MA in complex with the PIP2 lipid (PDB number 2K4I) [61]. The PIP2 lipid moiety is shown in *red sticks*. (**b**) NMR structure, obtained by joint refinement with X-ray data, of the CAP-1 inhibitor in complex with CA of HIV-1 (PDB number 2JPR) [62]. The CAP-1 compound is shown in *red sticks*

addition of PIP2 induces minor changes in the structure of HIV-2 MA. These NMR studies contribute to understanding MA-envelope interactions in retroviruses ([14] and references therein).

SARS-CoV-2 may enter target cells through membrane fusion occurring either at the endosomal membrane (approximately between pH 5.0 and pH 6.5) or the plasma membrane (approximately between pH 7.0 and pH 7.4). The fusion domain (FD) of the S protein is responsible for initiating fusion. Acquisition of HSQC spectra of FD-SARS-CoV-2 in the presence and absence of dodecyl phosphocholine micelles indicates that membrane fusion is more readily initiated at low pH, resulting in elongation of an α-helix within the domain. The FD's ability to mediate membrane fusion is strongly pH-dependent, with a tendency to induce fusion events at low pH [63]. These results suggest a possible preference for the endocytic pathway in virus infection.

Protein-Peptide and Protein-Organic Molecule Complexes

The CA protein of HIV-1 is emerging as a new therapeutic target (see Chap. 20). To help in understanding how certain peptides and organic molecules binding to CA may inhibit HIV-1 capsid assembly, the structures of complexes between some of these compounds and the CTD of HIV-1 have been elucidated using NMR ([64] and references therein). Next, two examples will be described.

In the first example, NMR was used to investigate the interactions of the peptide CAI and a stapled derivative (conformationally restricted by a covalent bond between non-contiguous residues in the sequence) with a monomeric mutant of CTD. Binding of these peptides to CA inhibits HIV-1 capsid assembly and virion infectivity [64]. The binding interface of CAI with CA was mapped by observing changes in the chemical shifts of ^{15}N-labeled CA in an HSQC spectrum (similar to the binding studies of DNA with MA or RNA with NC-SARS-CoV-2, as mentioned above) ([64, 65] and references therein). The peptides bind to a hydrophobic pocket within CTD, delineated by residues from the four helices of the domain, adopting a helical conformation where their hydrophobic side chains establish extensive contacts with specific hydrophobic patches in CTD. The structural results suggest that peptide binding sterically interferes with CTD-NTD interactions within the hexameric rings that form the capsid and distorts the CTD dimerization interface. These results further underscore the structural plasticity of CTD ([61] and references therein).

The molecule N-(3-chloro-4-methylphenyl)-N′-{2-[({5-[(dimethylamino)-methyl]-2-furyl}-methyl)-sulfanyl]ethyl}-urea) (CAP-1) binds to HIV-1 CA and inhibits capsid assembly and virus infectivity. In this example, both X-ray and NMR studies show that CAP-1 binds within a pocket delineated by parts of helices 1, 2, 4, and 7 in the NTD [62] (Fig. 5.6b). This region is spatially close to the groove created by α-helices 8 and 9 (the dimerization helix) of the CTD. The residue Phe32 of NTD pulls away from its position, creating a large cavity where the aromatic ring of CAP-1 inserts, making hydrophobic contacts with other residues in the protein [62]. One general conclusion from these studies is that assembly of the HIV-1 capsid can be inhibited by small molecules (peptides or organic compounds) that are hydrophobic enough to bind small pockets in CA, induce local structural distortions, and impede CA-CA interactions ([14, 64] and references therein; see also Chap. 20).

This study of CAP-1 in complex with CA also underscores an important aspect of NMR. The combined use of this spectroscopy with other biophysical techniques, such as X-ray crystallography (as illustrated in the aforementioned example) or EM, has yielded promising advances in recent years. For instance, the NMR structure of a viral capsid protein can be integrated with the lower-resolution cryo-EM structure of the entire capsid to generate a pseudo-atomic model. This approach was used in a detailed structural study of the mature HIV-1 capsid, revealing a small protein-protein interface that was previously undetected [19]. In another example, the structure of a retroviral RNA packaging element was jointly elucidated by NMR and cryo-electron tomography [66].

Protein-Protein Complexes: Detecting Interactions, Not Only Structures

The S protein of SARS-CoV-2 binds to the angiotensin-converting enzyme 2 (ACE2) receptor of the host, which is also glycosylated during the initial stages of infection. In certain viral infections like influenza or Ebola, glycan-mediated interactions play a crucial role in the initial contact between the virus and the host. Glycans not only regulate host-pathogen recognition processes but also influence cell signaling, differentiation pathways, and immune responses. The human immune system contains various glycan-binding proteins (lectins), which can specifically detect glycan epitopes and initiate innate immune responses in a glycan-dependent manner. Studies have demonstrated that human lectins interact with the recognition binding domain (RBD) of the S protein using HSQC spectroscopy, either with ^{15}N-labeled lectins or ^{15}N-labeled RBD. Analyzing changes in δs in the presence and absence of each partner helps identify the regions involved in the interaction.

Leveraging the known structures of the human lectin and its assigned residues, along with the X-ray structure of the RBD, researchers have constructed models of the complex formed between the RBD and the lectin [67].

In the previous example, HSQC spectra were used to detect interacting regions and propose a model of the RBD-lectin complex (i.e., mapping PPIs). However, such spectra can also be used to map modifications in proteins resulting from enzymatic reactions between macromolecules. The SARS-CoV-2 RNA genome has two open reading frames that encode two large polyproteins. These polyproteins are cleaved by viral proteases to yield 16 non-structural proteins (nsp1–16). Some of these proteins make up the replication-transcription complex, with nsp12 serving as the core RNA-dependent RNA polymerase targeted by antiviral drugs like remdesivir or molnupivarir in COVID-19 treatment. The nsp12 protein includes an NiRAN domain capable of transferring AMP from ATP to protein substrates (a process known as AMPylation). Its target substrates are nsp7, nsp8, and nsp9. Using HSQC to detect ^{31}P on the vertical axis (instead of ^{15}N as previously described), researchers have measured how the NiRAN domain AMPylates the N-terminus of nsp9 at Asn1 (where AMP attaches to the nitrogen backbone of asparagine via a phosphoramidate linkage) [68]. In this example, the presence of AMP indicates the interaction between nsp9 (substrate) and the NiRAN domain of nsp12 (enzyme), despite the structure of the nsp9-nsp12 complex remaining unknown.

5.4.4 Structure and Dynamics of Viral Particles via SSNMR Spectroscopy

The study of intact viral particles and complexes involving viral or cellular macromolecules poses challenges for solution-state NMR and, in some cases, even for X-ray crystallography (see Chap. 4), due to their large size, limited solubility, and difficulties in crystallization. Attempts have been made to address the structure of whole virus capsids using NMR; however, these efforts have primarily yielded information about highly dynamic regions of the capsid proteins [69, 70].

In addition to the solution-state NMR, SSNMR spectroscopy is emerging as a promising technique to study the structure and dynamics of viral particles. SSNMR is not limited by solubility, molecular size, or lack of long-range order. Challenges in obtaining NMR structures of large macromolecules and supramolecular assemblies arise from short T_2 relaxation times and broad linewidths (see Sect. 5.2.4), which are influenced by the molecule's correlation time (the time it takes for the molecule to rotate ~57°). SSNMR's ability to suppress molecular motions through radiofrequency irradiation is crucial for obtaining spectra of large molecular assemblies or aligned molecules within membrane-like environments. Detailed technical aspects of these methods are beyond the scope of this chapter, and interested readers are referred to recent literature [71–73].

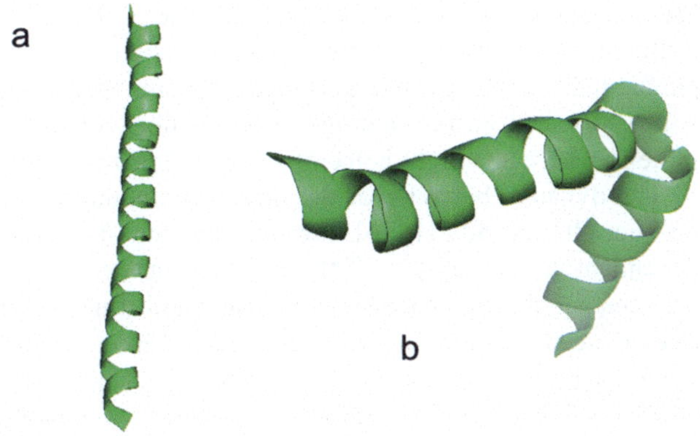

Fig. 5.7 Protein structures solved by SSNMR. (**a**) Structure of a fd capsid protein subunit in intact bacteriophage particles (PDB number 1NH4) [74]. (**b**) Structure of the membrane- bound species of the fd capsid protein (PDB number 1MZT) [75]

Filamentous bacteriophages are rod-shaped particles encasing a single-stranded circular DNA within a cylindrical capsid. This capsid is primarily composed of several thousand copies of the major capsid protein. Although these virions lack an envelope, their life cycle involves interactions with the host cell membrane, with the capsid protein also serving as a membrane protein. One of the most extensively studied filamentous bacteriophages is fd (Class I), which infects *Escherichia coli*. Its capsid protein is initially expressed as a procapsid protein with a 23-residue sequence. In vivo, the procapsid is inserted into the cell membrane, becoming the most abundant membrane protein in infected cells. It consists of a transmembrane helix connected by a short loop to an amphipathic helix. During virus assembly, the capsid is extruded from the membrane, and the DNA is packaged within the capsid. The structure of the capsid protein in filamentous bacteriophage particles was determined using SSNMR [71, 74] (Fig. 5.7a). The first five residues are unfolded and highly mobile. Following the Pro6-Ala7 patch, the capsid protein forms a continuous straight helix between residues 8 and 38, albeit not a perfectly straight helix: near residue 39 (midway along the hydrophobic helix), there is a kink that alters the helix rotation. Interestingly, the structure of the capsid protein in the virion differs from that determined by X-ray fiber diffraction. In the membrane-bound protein, the helix undergoes substantial rearrangement around residue 21 (which was part of the straight helix in the bacteriophage), a change that accompanies assembly. The rotation of the transmembrane helix (residues Gln16–Ala46) around its long axis shifts by 16° to achieve proper alignment for packing into viral particles (Fig. 5.7b) [76].

To facilitate comparison among different bacteriophages, the capsid protein of Pf1 (Class II), which infects *Pseudomonas aeruginosa*, has also been studied using SSNMR [71]. The capsid protein of Pf1 shares similarities with that of fd, including a comparable distribution of charged and hydrophobic residues. However, distinct differences have been observed between their structures. In contrast to fd, the N-terminal region of Pf1's capsid protein forms a hook with the six N-terminal residues. This structural feature contrasts with results from X-ray fiber diffraction studies, where multiple chemical shifts were observed for the side-chain and alkyl backbone atoms of Thr5, indicating a diverse conformation among the 7300 subunits in the 36 MDa virion [77]. Furthermore, residues around Gln16 (where the hydrophobic helix starts) exhibit a complex dynamic behavior, with evidence of a bulge, which differs from the findings of X-ray fiber diffraction studies on the fd capsid protein, where a straight helix was detected instead of a hinge around residue 21.

The dynamic behavior of the membrane-bound coat protein also clearly differs between Pf1 and fd [71, 76]. For instance, the Pf1 capsid undergoes a temperature-dependent transition [71, 78], resulting in structural changes at the interfaces between neighboring capsid subunits and adjustments in the hydrophobic interfaces, as evidenced by shifts in chemical signals. Moreover, SSNMR studies of ^{31}P of the DNA backbone suggest that the DNA is packaged differently: fd packages DNA randomly, whereas Pf1 shows a uniform orientation of the packaged DNA [70]. Therefore, SSNMR studies indicate structural similarities but also notable differences between these two bacteriophages.

SSNMR, combined with solution-state NMR, has also been used for structural studies of the mature HIV-1 capsid. The capsid structure resembles a fullerene cone, comprising a lattice of CA hexamers with specific arrangement of pentamers at defined positions (see [14] for details). SSNMR studies have shown that both open tubes made only of hexamers and the authentic conical capsid are conformationally homogeneous (even allowing partial assignment of chemical shifts for several residues in the conical capsid). The structure of CA does not differ between these morphologies [16, 23]. Furthermore, this structure appears similar to that observed in solution for isolated protein. Once again, the linker region connecting the NTD and CTD was found to be highly flexible (as indicated by T_1 and T_2 relaxation times) (Sect. 5.4.1) As already mentioned, this flexibility allows the protein to adopt multiple conformations essential for the assembly of pleiomorphic retroviral capsids.

5.5 Understanding Viruses: Some Major Contributions of NMR

This chapter describes how NMR spectroscopy addresses challenging task of solving the structures of large viral proteins, viral particles, and other viral complexes and assemblies using a "bottom-up" approach. This technique has successfully elucidated the structures of numerous isolated domains from different viral proteins. The results have not only offered new insights into the relationships between protein structure and viral function but have also complemented structures resolved by X-ray crystallography. NMR has been indispensable in uncovering the conformational dynamics of several viral proteins, including entire viral particles. It has highlighted the existence of alternate conformations, delineating the high intrinsic mobility of viral protein domains and how such dynamic behavior contributes to virion morphogenesis. Furthermore, NMR has underscored the inherently unfolded nature of certain viral proteins at the atomic level. The intrinsic mobility and/or unfolded character of many regions of these viral proteins appear to be defining features of their biological function. These features likely facilitate interactions with other biomolecules (not only during virus assembly but also throughout the viral cycle), as increased mobility allows residues to explore a greater variety of conformations, enhancing the likelihood of adopting the optimal active structure. The joint efforts of X-ray crystallography, cryo-EM, and NMR have provided a more precise understanding of these dynamic equilibria and have resolved structural ambiguities between these techniques. Their combined use has provided a comprehensive and detailed view of interactions during viral assembly. Lastly, advances in SSNMR have contributed significantly to understanding the structures of filamentous phage particles and the mechanisms underlying their membrane insertion.

5.6 Perspectives

A promising future for NMR spectroscopy in structural virology is evident from the techniques and studies described in this chapter, as well as from other developments in NMR that could not be detailed due to space limitations. In the near future, NMR is poised to contribute significantly to studies focusing on the following:

(i) The structure and conformational dynamics of viral proteins, including disordered regions.
(ii) The structural characterization of viral particles or details of viral proteins in solution, through the integration of cryo-EM (Chap. 3) or X-ray crystallography (Chap. 4) with both solution and solid-state NMR approaches.
(iii) The exploration of viral particle structure and dynamics using SSNMR, in conjunction with methods like X-ray fiber or neutron diffraction, to provide deeper insights into membrane-bound conformations of viral proteins. SSNMR may extend to elucidating initial infection steps, such as fusion complex formation.
(iv) Advancements in solution NMR for understanding the structure and dynamics of viral capsids.
(v) Investigating structural aspects of viral proteins within live cells using in-cell NMR, which could provide insights into various stages of the viral lifecycle in vivo.
(vi) Determining the thermodynamic parameters governing interactions involving viral macromolecules.

Acknowledgments I would like to thank the regional Valencian Government for their support (CIAICO/2021/135) in funding the research conducted in my laboratory.

References

1. Atkins PW, de Paula J (2008) Physical chemistry for the life sciences. Oxford University Press, Oxford
2. Wüthrich K (1986) NMR of proteins and nucleic acids, vol 17. Wiley, New York, p 11
3. Derome AE (1987) Modern NMR techniques for chemistry research. Pergamon Press, Oxford
4. Carbajo RJ, Neira JL (2013) NMR for chemists and biologists. Springer, Dordrecht
5. Cavanagh J, Fairbrother WJ, Palmer AG III, Skelton N (1996) Protein NMR spectroscopy. Principles and practice. Academic Press, New York
6. Keeler J (2006) Understanding NMR spectroscopy, 2nd edn. Wiley, New York
7. Evans JE (1996) Biomolecular NMR spectroscopy. Oxford University Press, Oxford
8. Bax A, Lerner L (1986) Two-dimensional NMR spectroscopy. Science 232:960–967
9. Reid DG (ed) (1997) Protein NMR techniques, Methods in molecular biology, vol, vol 60. Humana Press, Totowa
10. Dingley AJ, Lorenzen I, Grötzinger J (2008) NMR analysis of viral protein structures. In: Foster GD, Johansen IE, Hong Y, Nagy PD (eds) Plant virology protocols: from viral sequence to protein function. Humana Press, Totowa
11. Adamson CF, Freed EO (2007) Human immunodeficiency virus type 1 assembly, release and maturation. Adv Pharmacol 55:347–387
12. Mascarenhas AP, Musier-Forsyth K (2009) The capsid protein of human immunodeficiency virus: interactions of HIV-1 capsid with host protein factors. FEBS J 276:6118–6127
13. Ganser-Pornillos BK, Yeager M, Sundquist WI (2008) The structural biology of HIV assembly. Curr Opin Struct Biol 18:203–217
14. Ganser-Pornillos BK, Yeager M, Pornillos O (2012) Assembly and architecture of the HIV. Adv Exp Med Biol 726:441–465
15. Adamson CS, Salzwedel K, Freed EO (2009) Virus maturation as a new HIV-1 therapeutic target. Expert Opin Ther Targets 13:895–908
16. Pornillos O, Ganser-Pornillos BK, Kelly BN, Hua Y, Whitby FG, Stout CD, Sundquist WY, Hill CP, Yeager M (2009) X ray structures of the hexameric building block of the HIV capsid. Cell 137:182–192
17. Turner BG, Summers MF (1999) Structural biology of HIV-1. J Mol Biol 285:1–32
18. Tang C, Ndassa Y, Summers MF (2002) Structure of the N-terminal 283-residue fragment of the immature HIV-1 gag polyprotein. Nat Struct Biol 9:537–543
19. Byeon IJ, Meng X, Jung J, Jung J, Zhao G, Yang R, Ahn J, Shi J, Concel J, Aiken C, Zhang P, Gronenborn AM (2009) Structural convergence between cryo-EM and NMR reveals intersubunit interactions critical for HIV-1 capsid function. Cell 139:178–190
20. Shin R, Zhou YM, Krishna NR (2011) Structure of a monomeric mutant of the HIV-1 capsid. Biochemistry 50:9457–9467
21. Saad J, Miller J, Tai J, Kim A, Granam RH, Summers MF (2006) Structural basis for targeting HIV-1 gag proteins to the plasma membrane for virus assembly. Proc Natl Acad Sci USA 103: 11364–11369
22. DeLano WL (2002) The PyMOL molecular graphics system. DeLano Scientific, San Carlos
23. Byeon IJ, Hou G, Han Y, Suiter CL, Ahn J, Jung J, Byeon CH, Gronenborn AM, Polenova T (2012) Motions on the millisecond time scale and multiple conformations on HIV-1 capsid protein: implications for structural polymorphism of CA assemblies. J Am Chem Soc 134: 6455–6466
24. Ivanov D, Tsodikov OV, Kasanov J, Elnberger T, Wagner G, Collins T (2007) Domain-swapped dimerization of the HIV-1 capsid C-terminal domain. Proc Natl Acad Sci USA 104: 4353–4358
25. Khorasanizadeh S, Campos-Olivas R, Summers MF (1999) Solution structure of the capsid protein from the human T-cell leukemia virus type-I. J Mol Biol 291:491–505

26. Campos Olivas R, Newman L, Summers MF (2000) Solution structure and dynamics of the Rous sarcoma virus capsid protein and comparison with capsid proteins of other retroviruses. J Mol Biol 296:633–649
27. Jiang J, Ablan SD, Derebail S, Hercick K, Soheilian F, Thomas JA, Tang S, Hewlett I, Nagashima K, Gorelick RJ et al (2011) The interdomain linker region of HIV-1 capsid protein is a critical determinant of proper core assembly and stability. Virology 421:253–265
28. Klein DJ, Johnson PE, Zollars ES, De Guzman RN, Summers MF (2000) The NMR structure of the nucleocapsid protein from the mouse mammary tumor virus reveals unusual folding of the C-terminal zinc knuckle. Biochemistry 39:1604–1612
29. Lawrence DC, Stover CC, Noznitsky J, Wu Z, Summers MF (2003) Structure of the intact stem and bulge of HIV-1 psi-RNA stem-loop SL1. J Mol Biol 326:529–542
30. D'Souza V, Dey A, Habib D, Summers MF (2004) NMR structure of the 101-nucleotide core encapsidation signal of the Moloney murine leukemia virus. J Mol Biol 337:427–442
31. Amarasinghe GK, De Guzmán RN, Turner RB, Chancellor KJ, Wu ZR, Summers MF (2000) NMR structure of HIV-1 nucleocapsid protein bound to stem-loop SL2 of the psi-RNA packaging signal. Implications for genome recognition. J Mol Biol 301:491–511
32. Chiliveri SC, Louis JM, Bax A (2021) Concentration-dependent structural transition of the HIV-1 gp41 MPER peptide from disordered monomers into α-helical trimers. Angew Chem Int Ed 60:166–170
33. Yu XF, Matsuda Z, Yu QC, Lee TH, Essex M (1995) Role of the C terminus gag protein in human immunodeficiency virus type I virion assembly and maturation. J Gen Virol 76:3171–3179
34. Votteler J, Neumman L, Hahn S, Hahn F, Rauch P, Schmidt K, Studtrucker N, Solbak SM, Fossen T, Henklein P et al (2011) Highly conserved serine residue 40 in HIV-1 p6 regulates capsid processing and virus capsid assembly. Retrovirology 8:11
35. Fossen T, Wray V, Bruns K, Rachmat J, Henklein P, Tessmer U, Maczurek A, Klinger P, Schubert U (2005) Solution structure of the human immunodeficiency virus type 1 p6 protein. J Biol Chem 280:42515–42527
36. Kyere SK, Joseph PR, Summers MF (2008) The p12 domain is unstructured in a murine leukemia virus p12-CA(N) gag construct. PLoS One 3:e1902
37. Maxwell KL, Yee AA, Booth V, Arrowsmith CH, Gold M, Davidson AR (2001) The solution structure of bacteriophage lambda protein W, a small morphogenetic protein possessing a novel fold. J Mol Biol 308:9–14
38. Maxwell KL, Yee AA, Arrowsmith CH, Gold M, Davidson AR (2002) The solution structure of the bacteriophage head-tail joining protein. J Mol Biol 318:1395–1404
39. Estrada DF, Conner M, Jeor SC, Guzmán RN (2011) The structure of the hantavirus zinc finger domain is conserved and represents the only natively folded region of the Gn cytoplasmic tail. Front Microbiol 2:251–262
40. Bourhis JM, Johansson K, Receveur-Bréchot V, Oldfield CJ, Dunker KA, Canard B, Longhi S (2004) The C-terminal domain of measles virus nucleoprotein belongs to the class of intrinsically disordered proteins that fold upon binding to their physiological partner. Virus Res 99: 157–167
41. Bourhis JM, Receveur-Bréchot V, Oglesbee M, Zhang X, Buccellato M, Darbon H, Canard B, Finet S, Longhi S (2005) The intrinsically disordered C-terminal domain of the measles virus nucleoprotein interacts with the C-terminal domain of the phosphoprotein via two distinct sites and remains predominantly unfolded. Protein Sci 14:1975–1992
42. Blocquel D, Habchi J, Costanzo S, Doizy A, Oglesbee M, Longhi S (2012) Interaction between the C-terminal domains of measles virus nucleoprotein and phosphoprotein: a tight complex implying one binding site. Protein Sci 21:1577–1585
43. Le Breton N, Longhi S, Rockenbauer A, Gigliarelli B, Marque SRA, Belle V, Martinho M (2019) Probing the dynamic properties of two sites simultaneously in a protein-protein interaction process: a SDSL-EPR study. Phys Chem Chem Phys 21:22584–22588

44. D'Souza V, Summers MF (2005) How retroviruses select their genomes. Nat Rev Microbiol 3: 643–655
45. D'Souza V, Summers MF (2004) Structural basis for packaging the dimeric genome of Moloney murine leukemia virus. Nature 431:586–590
46. Amarasinghe GK, De Guzmán RN, Turner RB, Summers MF (2000) NMR structure of stem-loop SL2 of the HIV-1 psi-RNA packaging signal reveals a novel A-U-A base-triple platform. J Mol Biol 299:145–156
47. Lu K, Heng X, Summers MF (2011) Structural determinants and mechanism of HIV-1 genome packaging. J Mol Biol 410:609–633
48. Lu K, Heng X, Garyu L, Monti S, Garcia EL, Kharytonchyk S, Dorjsuren B, Kulandaivel G, Jones S, Hiremath A et al (2011) NMR detection of the structures in the HIV-1 5′ leader RNA that regulate genome packaging. Science 334:242–245
49. Unione L, Moure MJ, Lenza MP, Oyenarte I, Ereño-Orbea J, Arda A, Jiménez-Barbero J (2022) The SARS-CoV-2 spike glycoprotein directly binds exogeneous sialic acids: a NMR view. Angew Chem Int Ed 61:e202201432
50. Tomris I, Unione L, Nguyen L, Zaree P, Bouwman KM, Liu L, Li Z, Fok JA, Carrasco, MR, van der Woude R et al (2022) The SARS-CoV-2 spike N-terminal domain engages 9-O-catylated α2-8-linked sialic acids. bioRxiv. https://doi.org/10.1101/2022.09.14.507904
51. Taylor GM, Ma L, Vogt VM, Post CB (2010) NMR relaxation studies of an RNA-binding segment of the Rous sarcoma virus gag polyprotein in free and bound states: a model for autoinhibition of assembly. Biochemistry 49:4006–4017
52. De Guzmán RN, Wu ZR, Stalling CC, Pappalardo L, Borer PN, Summers MF (1998) Structure of the HIV-1 nucleocapsid protein bound to the SL3 Ψ-RNA recognition element. Science 279: 384–388
53. Hyman AA, Weber CA, Jülicher F (2014) Liquid-Liquid phase separation in biology. Annu Rev Cell Dev Biol 30:39–58
54. Cai M, Huang Y, Craigie R, Clore GM (2010) Structural basis for the association of HIV-1 matrix protein with DNA. PLoS One 5:e15675
55. Chen H, Cui Y, Han X, Hu W, Sun M, Zhang Y, Wang PH, Song G, Chen W, Lou J (2020) Liquid-liquid phase separation by SARS-CoV-2 nucleocapsid protein and RNA. Cell Res 30: 1143–1145
56. Perdikari TM, Murphy AC, Ryan VH, Watters S, Naik MT, Fawzi NL (2020) SARS-CoV-2 nucleocapsid protein phase-separates with RNA and with human hnRNPs. EMBO J 39: e106478
57. Savastano A, Inañez de Opakua A, Rankovic M, Zweckstetter M (2020) Nucleocapsid protein of SARS-co-V-2 phase separates into RNA-rich polymerase-containing condensates. Nat Commun 11:6041
58. Chang C-K, Hsu Y-L, Chang Y-H, Chao F-A, Wu M-C, Huang Y-S, Hu C-K, Huang T-H (2009) Multiple nuclei acid binding sites and intrinsic disorder of severe acute respiratory syndrome coronavirus nucleocapsid protein: implications for ribo-nucleocapsid protein packaging. J Virol 83:2255–2264
59. Zachrdla M, Savastano A, Inañez de Opakua A, Cima-Omori MS, Zweckstetter M (2022) Contribution of the N-terminally intrinsically disordered region of the severe acute respiratory syndrome coronavirus 2 nucleocapsid protein to RNA-induced phase separation. Protein Sci 31: e4409
60. Chen K, Bachtiar I, Psizcek G, Bouamr F, Carter C, Tjandra N (2008) Solution NMR characterizations of oligomerization and dynamics of equine infectious anemia virus matrix protein and its interaction with PIP2. Biochemistry 47:1928–1937
61. Saad JS, Ablan SD, Ghanam RH, Kim A, Andrews K, Nagashima K, Soheilian F, Freed EO, Summers MF (2008) Structure of the myristylated human immunodeficiency virus type 2 matrix protein and the role of the phosphatidylinositol-(4,5)-biphospahte in membrane targeting. J Mol Biol 382:434–437

62. Kelly BN, Kyere S, Kinde I, Tang C, Howard BR, Robinson H, Sundquist WI, Summers MF, Hill CP (2007) Structure of the antiviral assembly inhibitor CAP-1 complex with the HIV-1 CA protein. J Mol Biol 373:355–356
63. Birtles D, Oh AE, Lee J (2022) Exploring the pH dependence of the SARS-CoV-2 complete fusion domain and the role of its unique structural features. Protein Sci 31:e4390
64. Neira JL (2009) The capsid protein of human immunodeficiency virus: designing inhibitors of capsid assembly. FEBS J 276:6110–6117
65. Bocanegra R, Nevot M, Doménech R, López I, Abián O, Rodríguez-Huete A, Cavasotto CN, Velázquez-Campoy A, Gómez J, Martínez MÁ et al (2011) Rationally designed interfacial peptides are efficient in vitro inhibitors of HIV-1 capsid assembly with antiviral activity. PLoS One 6:e23877
66. Summers MF, Irovalieba RN, Tolbert B, Smalls-Mantey A, Iyalla K, Loeliger K, D'Souza V, Khant H, Schmid MF, García EL et al (2010) Structure of a conserved retroviral RNA packaging element by NMR spectroscopy and cryo-electron tomography. J Mol Biol 404: 751–772
67. Lenza MP, Oyenarte I, Diercks T, Quintana JI, Gimeno A, Coelho H, Diniz A, Peccati F, Delgado S, Bosch A et al (2020) Structural characterization of N-linked glycans in the receptor binding domain of the SARA-CoV-2 spike protein and their interactions with human lectins. Angew Chem Int Ed 59:23763–23771
68. Park GJ, Osinski A, Hernández G, Eitson JL, Majumdar A, Tonelli M, Henzler-Wildman K, Pawlowski K, Chen Z, Li Y et al (2022) The mechanism of RNA capping by SARS-CoV-2. Nature 609:793–800
69. Vriend G, Hemminga MA, Verduin BJM, De Wit JL, Schaafsma TJ (1981) Segmental mobility involved in protein-RNA interaction in cowpea chlorotic mottle virus. FEBS Lett 134:167–171
70. Szymczyna BR, Gan L, Johnson JE, Williamson JR (2007) Solution NMR studies of the maturation intermediates of a 13 MDa viral capsid. J Am Chem Soc 129:7867–7876
71. Opella SJ, Zeri AC, Park SH (2008) Structure, dynamics, and assembly of filamentous bacteriophages by nuclear magnetic resonance spectroscopy. Annu Rev Phys Chem 59:635–637
72. Sun S, Han Y, Paramasivam M, Yan S, Siglin AE, Williams JC, Byeon IJ, Ahn J, Gronenborn AM, Polenova T (2012) Solid-state NMR spectroscopy of protein complexes. Methods Mol Biol 831:303–331
73. Duer MJ (2004) Introduction to solid state NMR spectroscopy. Blackwell, London
74. Zeri AC, Mesleh MF, Nevzorov AA, Opella SJ (2003) Structure of the coat protein in fd filamentous bacteriophage particles determined by solid-state NMR spectroscopy. Proc Natl Acad Sci USA 100:6458–6463
75. Marassi FM, Opella SJ (2003) Simultaneous assignment and structure determination of a membrane protein from NMR orientational restrains. Protein Sci 12:403–411
76. Park SH, Marassi FM, Black D, Opella SJ (2010) Structure and dynamics of the membrane-bound form of Pf1 coat protein: implications of structural rearrangements for virus assembly. Biophys J 99:1465–1474
77. Goldbourt A, Gross B, Day LA, McDermott AE (2007) Filamentous phage studied by magic-angle spinning NMR: resonance assignment and secondary structure of the coat protein in Pf1. J Am Chem Soc 129:2388–2344
78. Goldbourt A, Day LA, McDermott AE (2010) Intersubunit hydrophobic interactions in Pf1 filamentous phage. J Biol Chem 285:37051–37059

Further Reading

For the basis of NMR:

Brey WS (ed) (1996) Magnetic resonance in perspective: highlights of a quarter of century. Academic Press, New York
Claridge TDW (1999) High-resolution NMR techniques in organic chemistry. Pergamon Press, Oxford
Homans SW (1992) A dictionary of concepts in NMR. Oxford University Press, Oxford
Neuhaus D, Williamson MP (2000) The nuclear Overhauser effect in structural and conformational analysis, 2nd edn. Wiley-VCH, New York
For dynamics by NMR:
Alderson TR, Kay LE (2021) NMR spectroscopy captures the essential role of dynamics in regulating biomolecular function. Cell 184:577–595
Sekhar A, Kay LE (2019) An NMR view of protein dynamics in health and disease. Annu Rev. Biophys 48:297–319
For IDPs:
Dyson HJ (2011) Expanding the proteome: disordered and alternatively folded proteins. Q Rev. Biophys 44:467–518
Xue B, Williams RW, Oldfield CJ, Goh GK, Dunker AK, Uversky VN (2010) Viral disorder or disordered viruses: do viral proteins possess unique features? Protein Peptide Lett 17:932–951
For LLPS:
Alberti S, Hyman AA (2021) Biomolecular condensates at the nexus of cellular stress, protein aggregation disease and ageing. Nat Rev. Mol Cell Biol 22:196–213
Banani SF, Lee HO, Hyman AA, Rosen MK (2017) Biomolecular condensates: organizers of cellular biochemistry. Nat Rev. Mol Cell Biol 18:285–298
Brangwynne CP, Eckmann CR, Courson DS, Rybarska A, Hoege C, Gharakhani J, Jülicher F, Hyman AA (2009) Germline P granules are liquid droplets that localize by controlled dissolution/condensation. Science 324:1729–1732
Nott TJ, Petsalaki E, Farber P, Jervis D, Fussner E, Plochowietz A, Craggs TD, Bazett-Jones DP, Pawson T, Forman-Kay JD, Baldwin AJ (2015) Phase transition of a disordered nuage protein generates environmentally responsive membrane-less organelles. Mol Cell 57:936–947
Also recommended for further reading are references [2–9, 73] listed above.

Chapter 6
Fluorescence, Circular Dichroism and Mass Spectrometry as Tools to Study Virus Structure

José L. Neira

Abstract Fluorescence and circular dichroism, as analytical spectroscopic techniques, and mass spectrometry, as an analytical tool to determine molecular mass, are important biophysical methods in structural virology. Although they do not provide atomic or near-atomic details as cryogenic electron microscopy, X-ray crystallography or nuclear magnetic resonance spectroscopy can, they do deliver important insights into virus particle composition, structure, conformational stability and dynamics, assembly and maturation and interactions with other viral and cellular biomolecules. They can also be used to investigate the molecular determinants of virus particle structure and properties and the changes induced in them by external factors. In this chapter, the physical foundations of these three techniques will be described, alongside examples demonstrating their contribution in understanding the structure and physicochemical properties of virus particles.

Keywords Assembly · Calorimetry · Capsid · Chemical denaturants · Chirality · Circular dichroism · Conformational changes · Dissociation · Dynamics · Flexibility · Fluorescence · Interactions · Intermediate · Ligand · Mass spectrometry · Protein stability · Refractive index · Structure · Virus

Abbreviations

AAV	Adeno-associated virus
ANS	1-anilino-8-naphtalenesulfonate
bisANS	bis(4-anilinonaphtalene-5-sulfonic acid)
BMV	Brome mosaic virus
CA-HBV	Capsid protein of hepatitis B virus
CD	Circular dichroism

J. L. Neira (✉)
IDIBE, Universidad Miguel Hernández, Elche, Alicante, Spain

Instituto de Biocomputación y Física de Sistemas Complejos, Zaragoza, Spain
e-mail: jlneira@umh.es

M. G. Mateu (ed.), *Structure and Physics of Viruses*, Subcellular Biochemistry 105,
https://doi.org/10.1007/978-3-031-65187-8_6

CDMS	Charge detection mass spectrometry
CTD-HBV	C-terminal domain of hepatitis B virus capsid
DENV-PR	Protease of the dengue virus
ds	Double-stranded
DSC	Differential scanning calorimetry
cryo-EM	Cryogenic electron microscopy
EPR	Electron paramagnetic resonance
ESI	Electrospray ionization
FHV	Flock house virus
FPOP	Photochemical oxidation of proteins
FRET	Förster resonance energy transfer
FTICR	Fourier-transform ion cyclotron resonance
GdmCl	Guanidine hydrochloride
HA	Hemagglutinin
HBV	Hepatitis B virus
HIV	Human immunodeficiency virus
HK97	Hong-Kong 97 virus
HPV	Human papilloma virus
HRF	Hydroxyl radical footprinting
HRV	Human rhinovirus
HSV	Herpes simplex virus
HX	Hydrogen exchange
IDP	Intrinsically disordered protein
IM	Ion mass
m/z	Mass-to-charge ratio
MALDI	Matrix-assisted laser desorption ionization
MS	Mass spectrometry
MVM	Minute virus of mice
MW	Molecular weight
NC	Nucleocapsid protein
NLS	Nuclear localization signal
NMR	Nuclear magnetic resonance
$NS3_{pro}$	Protease domain of DENV-PR
$NS2B_{cf}$	Cofactor of DENV-PR
PKC	Protein kinase C
RYMV	Rice yellow mottle virus
RSV	Respiratory syncytial virus
SARS-CoV-2	Severe acute respiratory syndrome coronavirus 2
SINV	Sindbis virus
SL	Stem-loop
SP	Spacer polypeptide
T_m	Thermal denaturation midpoint
TMV	Tobacco mosaic virus
TOF	Time of flight

UV	Ultraviolet
WHV	Woodchuck hepatitis virus

6.1 Introduction

In Chap. 5, nuclear magnetic resonance (NMR) spectroscopy was discussed, which is a powerful analytical technique for examining the structure and dynamics of viruses and viral proteins in solution. In the current chapter, the focus will shift to fluorescence and circular dichroism (CD) spectroscopies, as well as mass spectrometry (MS). These techniques do not provide information at or close to atomic resolution, as cryogenic electron microscopy (cryo-EM) (Chap. 3), X-ray crystallography (Chap. 4) and NMR (Chap. 5) can do, but they allow us to investigate several aspects of virus structure and properties such as stability and conformational dynamics.

The initial segment of this chapter will cover fluorescence spectroscopy and CD techniques. Both are faster (experiments can be completed in minutes) and require lower biomolecular concentrations than cryo-EM, X-ray crystallography or NMR. Thus, fluorescence and CD can be used as a first approach to obtain some information on the structure and properties of the biomolecule, which can then be studied in structural detail by using the latter techniques. Both fluorescence and CD are based on the absorption of light, which induces transitions between the electrons of the external shells of the molecules involved. Their uses to study virus structure, stability, assembly and interactions are described in this part of the chapter.

The subsequent segment of this chapter will delve into the foundations and applications of MS in structural virology. This technique is based on changes in the trajectory experienced by ions derived from the target molecule in the presence of electric and magnetic fields. Contrary to fluorescence, CD or NMR, MS does not use electromagnetic radiation to induce leaps among quantum levels, and, therefore, it is not a spectroscopic technique. From the data obtained, MS provides information about the mass of the target biomolecules, even of intact virus particles. Data acquisition is very fast, and the amount of sample required is even lower than that needed for CD or fluorescence spectroscopies. The use of MS in virus structure, stability, dynamics and assembly and virus-ligand interactions are also described in this section of the chapter.

6.2 Physical Principles of Fluorescence and Circular Dichroism Spectroscopies

6.2.1 Fluorescence

A detailed account of the general principles of fluorescence can be found in the book by Lakowicz [1], and some of its applications to biochemistry are described in the textbook by Albany [2]. Some basic principles will be briefly reviewed to explain the foundation of this spectroscopy.

The energy of a couple of bound atoms (E(total)) depends on the following:

$$\begin{aligned} \mathrm{E(total)} &= \mathrm{E(translation)} + \mathrm{E(rotational)} + \mathrm{E(vibrational)} + \mathrm{E(electronic)} \\ &\quad + \mathrm{E(electronic\ spin\ orientation)} + \mathrm{E\ (nuclear\ spin\ orientation)} \end{aligned}$$

The translational energy comprises the changes in the position of the molecule; the rotational energy involves the molecular rotation around the gravity centre of the molecule; the vibrational energy contains the deformation (compression or stretching) of bonds and angles; and the electronic energy encompasses the movement of electrons within the molecular shell. The last two terms involve the change of the spin of the electron and the change of the nuclear spin (see Chap. 5), respectively. The variations in each of those types of energy lead to different spectroscopies; the fluorescence and CD techniques are based on electronic transitions.

Each of those types of energies does not distribute continuously between the two atoms but is "quantized". For instance, the energetic separation between the "rungs" in the electronic energy ladder is 10^{-17} J; those in the vibrational one range around 10^{-20} J; those in the rotational one are 10^{-24} J; and those in the translational one are 10^{-36} J. This energetic distribution leads to two important conclusions. First, inducing an electronic transition involves also the induction of rotational, vibrational and translational transitions because they have smaller energy. And second, for an electronic level, there are vibrational ones; for each vibrational level, there are rotational ones, and so on.

The Basis of Fluorescence

Fluorescence involves photon absorption and electrons located in specific orbitals within a molecule. The absorption of photons induces leaps from a ground electronic level to an excited state. In this excited state, the molecule can return to the ground state following any of these processes (Fig. 6.1):

(a) The molecules dissipate the excess energy in the surrounding environment, usually, but not exclusively, as heat.
(b) The molecules can release the same amount of absorbed energy by the emission of a photon. This is called resonance.

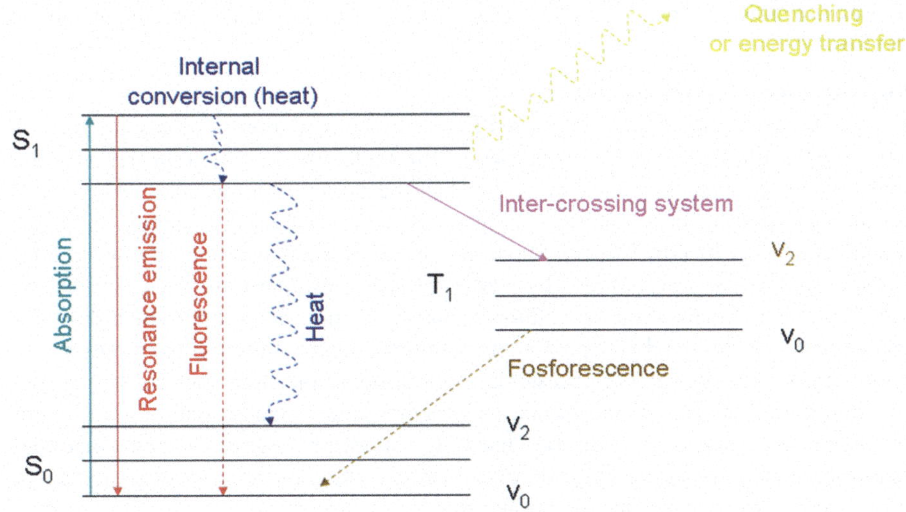

Fig. 6.1 The electronic energy levels and transitions in a molecule. Two electronic levels are shown (S_0 and S_1) with their corresponding vibrational ones (the number of vibrational levels drawn is arbitrary, labelled as $V_0 \ldots V_i$). The absorption of energy is represented by a continuous *green line* and emission by a *red* line; internal conversion processes (such as the emission of heat) are indicated by *curved dotted lines*. Fluorescence emission is represented by a *dotted red line*. The intersystem crossing, involving transition from an electronic state to another unstable state (T_1) with different spin properties, is shown in *pink* and phosphorescence emission by a *dotted brown line*. Quenching or energy transfer is denoted by a *curved dotted yellow line*. The meaning and nomenclature of electronic states S and T are beyond the scope of this book. For further details, readers can refer to [1]

(c) Part of the absorbed energy is dissipated in the medium as heat. Then, the molecules are in the lower vibrational level of the electronic excited state, and from this "rung" in the energetic ladder, they release a photon to return to the ground electronic level. The released photon will have a lower energy than the absorbed one (because some amount has been dissipated as heat), and then, its wavelength will be larger. This process is fluorescence.

(d) A transient passage occurs to an unstable electronic state, from where the molecule can pass to the ground state by photon emission. This process is phosphorescence.

(e) Some molecules in the excited state can release energy to nearby molecules (by collisional quenching or energy transfer).

It is clear that the fluorescence emission spectrum is the mirror image, shifted towards longer wavelengths, of the absorption spectrum, provided that the vibrational distribution in both ground and excited states is basically the same. However, we cannot conclude that all the absorbing molecules will have fluorescence because energy can be dissipated within the environment as collisional or thermal energy, or as energy transfer to other molecules. For instance, the heme group does absorb, but it does not show fluorescence due to transfer of energy from the porphyrin ring to the iron.

Fluorescence of Biomolecules

In proteins, tryptophan (Trp), tyrosine (Tyr) and phenylalanine residues are responsible for the ultraviolet (UV) properties of the molecule, but whereas Trp absorbs more intensely than the two others, Tyr has the largest fluorescence. When both residues are present in a protein, Tyr fluorescence is weaker, due mainly to energy transfer from Tyr to Trp. The Trp fluorescence can be quenched not only by external agents (such as oxygen) but also by nearby residues. Furthermore, the environment of a Trp affects the position of its fluorescence spectrum maximum (which does not occur for Tyr residues): hydrophobic environments lead to smaller maximum wavelengths; conversely, solvent-accessible environments produce spectra with larger maximum wavelengths. Thus, variations in fluorescence intensity and/or wavelength of maximum intensity can be used to monitor stability, conformational rearrangements, dissociation, unfolding, ligand binding and affinity of proteins and protein complexes, including viral particles.

Isolated sugars show no fluorescence, and the fluorescence work with lipids is rather limited due to the possibility of dispersion due to vesicle size. Purine and pyrimidine bases in nucleic acids show weak fluorescence (although they absorb strongly at 260 nm) at physiological pH, but they fluoresce strongly at acidic pHs and low temperature. Purine bases (especially guanine) show fluorescence at least three times more intense than pyrimidine ones. In the native state, DNA or RNA show very weak fluorescence (except transfer RNAs which contain a highly fluorescent modified base); however, extrinsic fluorescent agents capable of binding to nucleic acids (such as ethidium bromide) can be used to detect nucleic acids. In this way, for example, the structural stability of some virions or the process of genome uncoating (Chap. 15) can be followed by quantifying the amount of nucleic acid released (accessible to the external fluorophore) on capsid dissociation or conformational rearrangement.

Extrinsic fluorophores, such as ANS, capable of binding to proteins, can change the protein fluorescence emission spectrum depending on the conformation of the macromolecule and can be used to (i) monitor folding or unfolding; (ii) detect intermediate states, and (iii) monitor conformational rearrangements in proteins and protein complexes, including virus capsids. Furthermore, extrinsic fluorophores can be covalently attached to proteins—similarly to how paramagnetic probes are attached to proteins in the application of EPR (Chap. 5)—and used to map protein-ligand interactions within living systems [3, 4], after excitation at the appropriate wavelength.

6.2.2 CD

The Basis of CD

A property of many molecules is their chirality or molecular asymmetry. A description of the properties and characterization of organic chiral molecules can be found in organic chemistry texts [5]. The simplest cases result from the presence of asymmetric carbons; however, molecular chirality does not rely exclusively on the asymmetrically substituted carbon atoms but rather on any molecular feature providing "handedness". For instance, proteins in the α-helical conformation wind in a right-handed manner, although it is important to note that this helical "handedness" ultimately derives from the stereochemistry of the amino acids. In nucleic acids, the helical polynucleotide winds in either a left- or right-handed manner [6].

Chiral structures can be characterized by polarized light. In non-polarized light (Fig. 6.2a), the electric vector of the electromagnetic radiation can oscillate in any

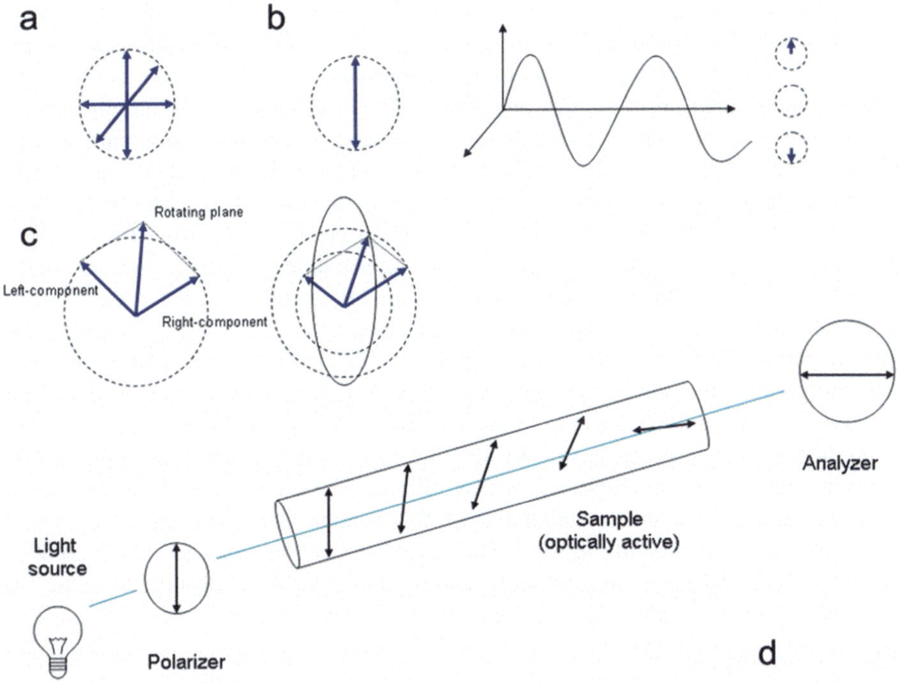

Fig. 6.2 Polarized light and CD. For simplicity, only the electric field (not the magnetic field) of the electromagnetic radiation is depicted (*double arrowhead*): (**a**) unpolarized light as seen by a stationary observer. (**b**) Linearly (or plane) polarized light: *left panel* shows it as seen by a stationary observer; *right panel* illustrates how it appears to an observer travelling with the beam. (**c**) Circularly polarized light as seen by a stationary observer, with left and right circularly polarized components (*left panel*); when combined, they generate linearly polarized radiation (*right panel*); passing through an optically active sample results in elliptically polarized light (*solid line*). (**d**) Setup of a CD spectropolarimeter. The *green line* indicates the beam

direction perpendicular to the direction of propagation (for simplicity, the following brief description will be reduced to the electric field, as Maxwell's equations allow the characterization of electromagnetic radiation solely by the electrical component). Conversely, in a plane-polarized beam, the electric field oscillates only in a plane (Fig. 6.2b); if we moved with the field, we would see the vector representing the electric field oscillating back and forth along a line (Fig. 6.2b, right panel). In addition, in a circularly polarized beam, the plane of oscillation rotates (Fig. 6.2c, left side); if we moved with the field, we would observe the electric vector moving clockwise (right-handed circularly polarized light) or counterclockwise (left-handed circularly polarized light). This rotating plane can be decomposed into two components: left and right circularly polarized vectors.

The two enantiomers of a chiral molecule have different refractive indexes, *n*; that is, the refraction of light as it moves through each enantiomer occurs differently for each of them. This phenomenon is called *birefringence*. Therefore, the variation of *n* is not the same in every direction within that material (when both enantiomers are present), and thus, it is said that the variation of *n* is anisotropic. Biological cells or crystals have different *n* depending on (i) the direction of the beam that illuminates them and (ii) the different planes of material they are made of. Chiral molecules exhibit *circular* birefringence, also known as circular dichroism (CD). A thorough and rigorous description of the physical basis of CD can be found in several textbooks [7–9]. From a quantum perspective, CD is related to a transition between electronic states in a molecule; this transition is active if, for an instant, the electric and magnetic components of both states are not perpendicular to each other (which occurs only if the molecule is chiral). The CD spectrum results from the different absorption (or phase changes of the wave) of right-handed or left-handed circularly polarized light by a chiral molecule. In addition to phase shifts relative to the incident beam, the amplitude of the emerging electric field is also modified differently for left-handed and right-handed circularly polarized lights. Thus, the resulting electric field amplitude (the sum of the left-handed and right-handed circularly polarized electric field amplitudes) does not follow a circle but forms an ellipse (Fig. 6.2c, right side). This resulting beam is said to be elliptically polarized, and the units of CD measure the degree of ellipticity of this ellipse. The changes in ellipticity for a molecule depend on the wavelengths of the incident light and result in a CD spectrum (see examples in Fig. 6.3). Specific rules exist that predict which component of polarized light is preferentially absorbed [9–11]. An experimental setup for measuring CD in an optically active sample is shown in Fig. 6.2d.

CD of Biomolecules

Traditionally, the regions of the electromagnetic spectrum useful for CD analysis of biomolecules have been divided into (i) the far-UV region (from 190 to 250 nm); and (ii) the near-UV region (from 250 to 320 nm). In proteins, the peptide bonds absorb UV light below 240 nm; the aromatic side chains absorb mainly in the 260–320 range (although they can also absorb around 222 nm [11]); and the disulphide bridges show a broad absorption band at 260 nm. In addition, the spectra of different types of secondary structure have characteristic shapes (see below). Some CD-based

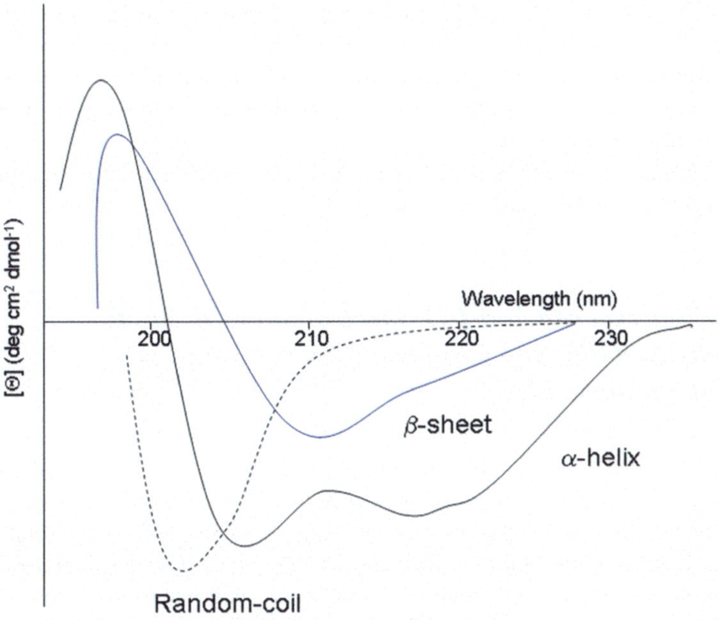

Fig. 6.3 Far-UV CD spectra of different model proteins. The pure α-helix (*black continuous line*), pure β-sheet (*blue continuous line*) or random coil (*black dashed line*)

studies use wavelengths longer than those indicated above (i.e. >400 nm), but in most cases, they involve molecules with a strong coloured chromophore.

Secondary structure elements in proteins (α-helices, β-sheets or stretches with irregular structure such as loops or random-coil polypeptides) exhibit characteristic far-UV CD spectra (Fig. 6.3); perhaps the most characteristic is that of the α-helix, with two minima at 208 and 222 nm. Obviously, proteins containing a mixture of the three types of structure will show a CD spectrum that results from the combination of the three spectra mentioned above, depending on the percentages of the corresponding types of secondary structure elements. Deconvolution of a protein's spectrum allows for an estimation of the percentage of each type of secondary structure element, and there are online programs implemented to facilitate this [12, 13]. Caution must be exercised, of course (as with any mathematical procedure applied to complex experimental data), since the presence of overlapping bands may lead to incorrect conclusions about the protein's secondary structure prediction.

The near-UV spectrum of proteins is usually rather weak, except for those with a large number of aromatic residues. Thus, large protein concentrations in a wide path-length cell are generally required to obtain a decent spectrum with a good signal-to-noise ratio [11]. Contrary to what happens in the far-UV region, the near-UV spectrum of proteins is mainly associated with the asymmetric environment of the aromatic residues. This region is only used in specific cases, since the information provided is similar to that obtained by fluorescence spectroscopy. On the other hand, the near-UV spectrum of double-stranded (ds) DNA is very intense due to the packing of the bases. Similar to proteins (Fig. 6.3), there are also "model spectra" for the different nucleic acid conformations [9, 11]. In addition, transitions among different nucleic acid conformers can be monitored by following changes in the spectrum, similar to proteins.

The CD analysis of lipids is rather limited due to the possibility of dispersion caused by the size of the vesicles formed.

6.2.3 *Fluorescence and CD Spectroscopies to Study the Stability and Conformational Dynamics of Biomolecules*

In addition to investigating some structural aspects of biomolecules and biomolecular complexes including viruses, fluorescence and CD spectroscopies can be used to detect conformational changes in proteins, nucleic acids and their complexes by following changes in their fluorescence and/or CD spectra. One particular example is the determination of the conformational stability of such macromolecules. Since the far-UV CD spectrum of a random-coil polypeptide differs significantly from those of other types of secondary structure in proteins (Fig. 6.3), it is possible to monitor variations in protein secondary structure by measuring changes in ellipticity at a selected wavelength (usually at 222 nm). Similarly, since the Trp or Tyr fluorescence in a protein is sensitive to changes in exposure to solvent, it is possible to track changes in protein tertiary (and quaternary) structure by measuring changes in fluorescence intensity at a selected wavelength (or, in the case of Trp, changes in the wavelength of maximum intensity).

For the determination of protein stability using these spectroscopies, protein structure melting can be induced by chemicals (such as concentration of H^+ (which is the pH), urea or GdmCl) or temperature. In some cases, the use of these chemicals, combined with analyses at different concentrations of viral protein, has provided information on the protein's oligomerization state during denaturation [14]. In simple cases where the melting of the viral protein is highly cooperative (an all-or-none transition), a sigmoidal curve will be obtained that can be fitted to a two-state model, allowing extraction of the relevant thermodynamic parameters [15, 16]. On the other hand, if the thermodynamic parameters governing the binding reaction of a macromolecule to a ligand need to be determined, an isothermal titration, followed by fluorescence or CD, should be performed. This involves adding increasing amounts of the ligand to a fixed concentration of the macromolecule and monitoring changes in fluorescence or CD upon ligand addition, which can then be fitted to a hyperbolic curve [2]. However, the use of these spectroscopic techniques for determining binding does not preclude the possibility of determining the thermodynamic parameters governing binding reactions using the "gold-standard" method in association studies: isothermal titration calorimetry (Chap. 7).

6.3 Fluorescence and CD Spectroscopies to Study Virus Structure

CD and fluorescence spectroscopies have been used to study structural features, stability and conformational changes of isolated viral capsid proteins, the stable oligomers they form and, occasionally, even those of entire virus particles. In addition, both techniques have been used to study the interactions of viral proteins with other biomolecules. The studies in the literature will be divided into three different aspects.

6.3.1 Conformational Changes in Viral Proteins and Capsids

A common application of far-UV CD and fluorescence in structural virology is to investigate the overall secondary and tertiary structures of virus capsid proteins and their potential changes during capsid assembly. For example, the hepatitis B virus (HBV) capsid protein was mainly helical and showed minimal conformational restructuring upon assembly [17]. Numerous examples exist where mutations in capsid proteins, resulting in impaired self-association, did not significantly alter the fluorescence or far-UV spectra compared to those of the wild-type protein. In these cases, CD and/or fluorescence analyses showed that capsid assembly could proceed without major changes in the tertiary (or secondary) structures of the capsid building blocks. They also demonstrated that subtle structural changes were enough to hinder capsid assembly, suggesting potential for developing antiviral strategies targeting this process (Chap. 20). Conversely, dimerization of the capsid protein of human immunodeficiency virus type 1 (HIV-1) involves a transient monomeric intermediate and conformational rearrangements, while maintaining the overall fold of the monomer [18]. Structural changes in capsid proteins have been observed, for example, during pressure-induced dissociation of human rhinovirus (HRV14) [19] or multimerization of the N-terminal region of Mason-Pfizer monkey virus p12 protein [20]. These and other examples using CD or fluorescence have revealed that dissociation or association of some viral capsids can involve substantial conformational rearrangements, a phenomenon extensively studied in structural detail using cryo-EM and X-ray crystallography (Chaps. 3, 4 and 14).

Another common application of far-UV CD and fluorescence in structural virology involves studying conformational rearrangements of virus proteins in response to environmental factors mimicked in vitro, which stimulate conditions within host cells or virions. For instance, the HIV-1 polyprotein Gag contains a spacer polypeptide (SP1) where the first five residues are critical for proper assembly of Gag into an immature capsid. In aqueous solution at low concentrations, isolated SP1 behaves as an intrinsically disordered protein (IDP); however, at high concentrations, it exhibits an α-helix-like far-UV CD spectrum. This suggests that Gag aggregation potentially induces secondary structure acquisition by SP1, which could mediate its function

[21]. Other intracellular or intra-virion factors affecting virus protein conformation have been mimicked in vitro using high concentrations of salts (ions), organic solvents, changes in pH or lipids. In all cases, conformational changes have been monitored using CD and fluorescence. The results have contributed to understanding substantial conformational dynamics in virus structural proteins. However, no general trends have emerged, and conclusions largely depend on the specific protein under investigation.

Not only can the presence of the reagents described in the previous paragraph trigger conformational changes, but also interactions with other viral proteins play a significant role. For instance, the protease of the dengue virus (DENV-PR) is a heterodimeric trypsin-like serine protease consisting of NS3 protease domains ($NS3_{pro}$) and the cofactor NS2B ($NS2B_{cf}$). $NS3_{pro}$ exhibits catalytic activity independently, regardless of whether $NS2B_{cf}$ is present, maintaining its spherical shape. Conversely, $NS2B_{cf}$ exists in at least two conformations, mainly differing in the relative positioning of the cofactor relative to the active centre of the $NS3_{pro}$ domain. In the closed conformation (i.e. the catalytically active state), $NS2B_{cf}$ wraps around $NS3_{pro}$, contributing to the formation of several enzymatic pockets. In contrast, in the open state, the $NS2B_{cf}$ domain is disordered and does not contribute to the formation of the substrate binding pocket. The elucidation and description of these two conformations, and thus the dynamic equilibrium between them, have been achieved using single-molecule Förster resonance energy transfer (FRET) [22]. In this technique, two fluorescent probes are introduced into DENV-PR at different positions within $NS3_{pro}$ and another within the $NS2B_{cf}$ domain. The introduction of these probes was facilitated by the engineered introduction of cysteines into both domains (similar to the use of paramagnetic probes, as described in Chap. 5). The labelling positions were strategically chosen to observe significant changes in energy transfer efficiency (FRET efficiency) between the two probes (similar to what happened between Tyr and Trp; Sect. 6.2.1), depending on the conformation adopted by the two proteins in the heterodimer [23]. Furthermore, it was shown that the presence of a competitive inhibitor strongly favours the catalytically active closed conformation of the enzyme.

Likewise, far-UV CD and, more frequently, fluorescence spectroscopy have also been used to detect conformational rearrangements in assembled virus capsids in response to factors such as heat or changes in pH that may mimic physiological conditions. For instance, fluorescence spectroscopy detected a subtle conformational rearrangement in the capsid of the minute virus of mice (MVM), which was associated with the biologically relevant peptide translocation through capsid pores [24].

6.3.2 *Stability and Association of Viral Capsid Proteins*

Both CD and fluorescence spectroscopy have been used to determine the stability of viral proteins and particles in thermal or chemical denaturation assays (Sect. 6.2.3) and, in some cases, to elucidate the dominant forces behind capsid stabilization [25]. An excellent review on assembly as a binding process can be found in the work by Katen and Zlotnick [26] (see also Chap. 19; reference [26] also contains a review on the stability of several viruses followed by different spectroscopic techniques). For example, it was found that the wild-type HBV capsid protein dimer is highly stable, showing a T_m above 95 °C [26]. The ability of this dimer to self-assemble is improved by the naturally occurring mutation F97L, but this mutant has the same structure (as monitored by both techniques) and the same stability as the wild-type [27]. The authors suggested that this mutation could affect the conformational dynamics of the dimer.

In some cases, the stability and self-association of capsid proteins are linked to the presence of the corresponding nucleic acid (Chaps. 12 and 13). For instance, the respiratory syncytial virus (RSV), belonging to the *Pneumoviridae*, is a single-stranded RNA virus containing a nucleocapsid protein wrapped around the RNA, forming the nucleocapsid (NC) with helical symmetry. Studies in vitro of NC show the presence of decameric assemblies with nucleocapsid protein-RNA interfaces. These assemblies and the protein-nucleic acid interfaces are unaffected by variations in temperature, pH or ionic strength; furthermore, the rings of proteins formed in the assembly are highly resistant to chemical denaturation. Interestingly, the release of RNA from the protein-nucleic acid complex and conformational changes in the secondary structure (as mapped by far-UV CD) occur concomitantly [28]. These results suggest that NC stability and nucleic acid binding are highly interdependent processes.

Among the comparatively few examples where the stability of whole virus particles was studied, urea denaturation analyses of HRV14 followed by far-UV CD and fluorescence showed limited stability [19]. However, the dissociated capsid proteins of HRV14, as analysed by far-UV CD, retained residual structure even at 8 M urea [19]. Interestingly, it was also possible to detect signals in the near-UV region attributed to viral RNA, and there was a decrease in the nucleic acid signal at 260 nm with increasing urea concentration. This result suggests that RNA becomes more ordered within the assembled capsid, potentially influencing the formation of membrane-less compartments (Chap. 5).

Far-UV CD and fluorescence have also been used to identify the molecular determinants of folding, stability and oligomerization of capsid proteins through mutational analysis. For example, the thermodynamic roles of each residue at the dimerization interface in the HIV-1 capsid protein were probed using far-UV CD (secondary structure), fluorescence (tertiary structure) and gel filtration (quaternary structure) [29]. Mutational studies analysing the folding, stability and/or oligomerization of viral capsid proteins have been reported for bacteriophage P22 [30], *Physalis* mottle virus [31] and MVM [32]. In all cases, thermal and chemical denaturation of mutated residues involved in inter-subunit interactions disrupted both subunit protein folding and particle assembly (in contrast to observations with F97L mutant of HBV [26]). From these studies, it can be concluded that the distribution of critical residues and binding energy in inter-subunit interfaces in virus capsids may be complex and quite different among viruses, involving, in some cases, a few key residues, interactions and allosteric effects.

Studies on the folding, stability and association of viral proteins monitored by fluorescence and CD can be complemented by thermal denaturation analysis followed by differential scanning calorimetry (DSC) (Chap. 7). In this biophysical technique, instead of following a spectroscopic property, the heat exchanged in the dissociation and/or unfolding reaction of a viral capsid or protein is monitored. For instance, the thermal stability of the empty capsid of MVM has been studied using fluorescence and DSC [24]. The DSC results indicated that the dissociation/denaturation transition involves a high enthalpy change and proceeds through one or more intermediates, which disappear in the presence of GdmCl. Furthermore, thermal dissociation of the capsid involves a partially folded species of the capsid protein [32], with high flexibility and practically devoid of tertiary structure. A similar partially folded conformation has been observed, using fluorescence and CD, for the isolated capsid protein of herpes simplex virus (HSV) [33] and during maturation of the bacteriophage HK97 capsid [34]. Temperature- and molecular partner-dependent disorder-to-order transitions in the polymerase cofactor phosphoprotein—which is a crucial tethering polypeptide among the RNA polymerase complex in RSV—have been observed using ANS fluorescence and far-UV CD [35]. All these results, taken together with the existence of viral IDPs and other IDP regions (Chap. 5), suggest that partially folded conformations (or completely disordered structures) of viral proteins may be very important in the virus life cycle [36]. Probably, flexibility may permit virus proteins to explore a wider conformational space, which facilitates the selection of the required folded conformation for self-association or interaction with other biomolecules (Chap. 5).

The thermal stability of some viral capsid proteins has been analysed at different pHs to elucidate the importance of several hydrogen bonds, which disappear (or are formed) as the pH is modified. For instance, it has been shown that the thermal stability of the capsid protein of adenovirus type 2 is increased in mildly acidic conditions [37]. The effect of covalent modifications in viral capsid proteins on their thermal stability has also been analysed. For example, PKC (Protein kinase C)-mediated phosphorylation of the HBV core protein was shown to stabilize the protein and also improve its self-association [38].

In general, the whole stability of viral capsids using fluorescence is usually determined by a thermal shift assay, in which probes become strongly fluorescent upon binding to hydrophobic patches of the protein capsid (e.g. SYPRO Orange) or to the viral RNA genome (e.g. SYTO-82) that become exposed upon heating virus particles. The amount of required viral proteins is small, and many of these results are usually robust, but the thermal shift assay with SYPRO Orange is sensitive to surfactants and EDTA [39]. Furthermore, interactions between the probe and capsid-binding antivirals, as well as mutual competition for binding sites, cannot be excluded. To overcome these caveats, it has been recently proposed that differential scanning fluorimetry (following the variations in intensity of the natural fluorescent residues of the protein), as a label-free alternative, can also be used; the results obtained with HRV-A2 were promising [39].

6.3.3 Binding of Ligands to Viral Proteins

If binding of a ligand induces a change in the secondary and/or tertiary structure of the protein, the binding reaction can be monitored by far-UV CD and/or fluorescence, respectively. In a general sense, the addition of ions or changes in pH can be considered particular examples of binding. Near- and far-UV CD, together with fluorescence, can be used to detect the changes occurring in the protein (or the nucleic acid) upon binding of DNA or RNA. Alternatively, an increase in the T_m of the thermal denaturation of a protein monitored by far-UV CD, upon addition of a ligand [18], may indicate binding [40]. However, if the dissociation constant of the complex is large, or the heat released upon binding is very small, there might not be any changes in the measured T_m.

Both CD and fluorescence can be used to monitor the binding of ions to capsid proteins, which in some cases can trigger capsid assembly. A particularly well-studied example is that of HBV, where the affinity for Zn(II) has been determined by following changes in the fluorescence spectrum upon collecting spectra after addition of increasing concentrations of the ion [26]. In some cases, it is possible to determine the affinity constants, not from the spectra obtained at different concentrations of the ion but by following the changes in the T_m-values from far-UV CD spectra at increasing concentrations of the ion [14]. The findings with different viral proteins and ions suggest that: (i) the assembly reaction is usually kinetically trapped; and (ii) the overall process of assembly triggered by the ion is allosterically regulated. Binding can also be explored by using minimalist models of the capsid protein, as described with fragments of the human papilloma virus capsid [41]. Changes in fluorescence can be monitored not only in the capsid protein but also in other molecules used as reporters (such as ANS or the SYPRO Orange, Sect. 6.3.2), as shown, for example, in the inhibition of capsid assembly of *Salmonella typhimurium* phage P22 by 1,1-bis(4-anilinonaphtalene-5-sulfonic acid) [42]: the bisANS-bound coat or scaffolding proteins are incapable of assembly into procapsids.

Binding of antivirals or other molecules that affect virus capsid assembly (Chap. 20) has also been monitored by fluorescence and/or CD [43]. In the reported cases, binding induces conformational changes in the protein, which can hamper proper assembly [44]. An excellent application of the design and use of these molecules can be found in the work by Zlotnick's group on HBV [45, 46]. A recent example involving the main protease of severe acute respiratory syndrome coronavirus 2 (SARS-CoV-2)—responsible for processing viral polyproteins synthesized from viral RNA after infection—has shown how the presence of a drug (quercetin) causes changes in the T_m value of the isolated protein, followed by fluorescence (Sect. 6.3.2) and far-UV CD. Subsequently, the enzymatic parameters of the inhibition of the protease by quercetin were determined by a FRET (Sect. 6.3.1) assay [47].

6.3.4 Understanding Viruses: Some Major Contributions of Fluorescence and CD

The experiments with various viral systems, such as those described above, have underscored the importance of spectroscopic analyses based on CD and fluorescence for analysing the folding, stability, conformational dynamics and association of viral capsid proteins. These techniques have also provided insights into the stability and conformational dynamics of viral particles, the effects of different factors and the molecular determinants of these processes have identified potential drugs capable of inhibiting or disrupting the function of viral proteins. Results obtained using these and other biophysical techniques have revealed some general underlying themes and a broad diversity of specific strategies employed by different viruses to adapt to selective pressures affecting the conformational stability and dynamics of capsids and virions.

6.4 Mass Spectrometry (MS) as a General Analytical Technique in Biochemistry and Biology: Basis and Current Spectrometers

6.4.1 The Basis of MS

MS identifies molecules based on their mass and charge, and it has become one of the most sensitive methods for studying aspects of virus structure. The first mass spectrometer was designed at the turn of the twentieth century by Sir Joseph J. Thomson. Thomson observed the deflection of a beam of positively charged ions when subjected to a combination of magnetic and electrostatic fields. Ion deflection occurred through a series of parabolic curves corresponding to particular mass/charge (*m*/*z*) ratios of the ions. The position of each curve was related to the velocity of the ion and, consequently, its mass and acceleration. Although MS has evolved since those experiments, the basic setup of the technique remains consistent: all MS instruments have three key elements: (i) an ion source; (ii) an analyser (sieve) of ion beams according to the m/z ratio; and (iii) a detector that monitors the deflected beams (Fig. 6.4, top).

In the first stage of MS, molecules are vaporized and, in some experimental setups, broken down (typically using highly energetic particles); these fragmentation events produce ions. In the second stage, the ions are accelerated in a vacuum chamber under the influence of combined electric and magnetic fields. According to Newton's second law, lighter ions deflect further than heavier ones, resulting in different trajectories depending on the ion masses and charges. The third stage involves monitoring these ion deflections to calculate the *m*/*z* ratio, which generates the mass fingerprint or mass "spectrum". Graphically, a typical mass "spectrum" plots intensity (signal) data versus the *m*/*z* ratio (Fig. 6.4, bottom). Despite the name

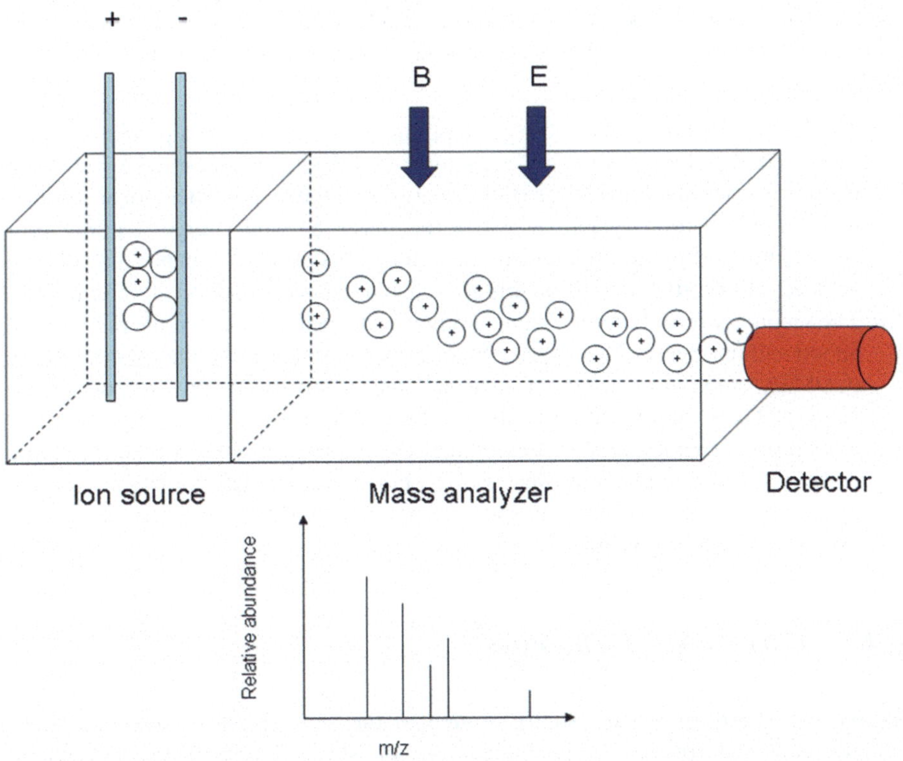

Fig. 6.4 Mass spectrometry. (*Top*) A mass spectrometer setup, showing the ion source, the ion (mass) analyser and the detector. In the ion analyser, the deflection of ion beams is depicted. Initially, the sample consists of neutral and charged molecules. (*Bottom*) Example of a mass "spectrum" of a model molecule

"spectrum", which traditionally implies light involvement, MS spectra are derived solely from ions. The nature of the mass spectrum depends on how the spectrometer handles the sample: while some MS instruments fragment molecules into multiple ions, others detect intact ions (single ionization).

It is important to note that conventional MS measures the *m/z* ratio. We obtain a "spectrum" (distribution) of *m/z* for many ions. The product of *m/z* and the charge gives us the mass, and repeating the measurement with different ions results in masses that are binned to form a mass distribution. If the ions generated are singly charged, then *m/z* directly represents the mass, and our "spectrum" shows a single mass distribution. However, larger ions, particularly relevant in biomolecules or viral capsids (and in ESI-MS, Sect. 6.4.2), result in peak distributions due to ions with different charge but the same molecular weight (MW). These distributions become more complex when analysing biological samples containing large assemblies (complexes, viral capsids, etc.) or samples with post-translational modifications, which introduce heterogeneity and broaden these distributions. One solution to handle this heterogeneity is adopting a single-particle approach that simultaneously measures *m/z* and the charge, *z*, where the product of *m/z* and *z* gives the mass. Currently, two basic approaches are used to determine the masses of individual ions by MS:

(a) The first approach involves charge stepping, where the *m/z* of a particular ion is measured, the charge of the ion is changed and the *m/z* is measured again. With two *m/z* values and knowledge of the charge difference, a mathematical solution to the two-equation system provides the mass. Several charge-stepping processes are performed to confirm the charge steps and enhance calculation accuracy. For further details on this approach, interested readers can refer to [48].
(b) The second approach involves the simultaneous measurement of the *m/z* ratio of an ion and its charge. This approach utilizes charge detection mass spectrometry (CDMS), employing detectors such as Fourier-transform ion cyclotron resonance (FTICR) and Orbitrap analysers (Sect. 6.4.2), which offer slightly lower mass resolution compared to the first method. Time-of-flight MS with cryogenic detectors is also applicable (Sect. 6.4.2). For more detailed information on the operation of both analysers and specifics about the CDMS approach, readers are encouraged to consult [48].

6.4.2 Current MS Techniques

Before the 1980s, MS was usually limited to the analysis of volatile and semi-volatile chemical species. Early applications of MS in biochemistry focused on characterizing the primary protein structure through peptide sequencing, which remains central to many current MS workflows in peptide biomarker analysis and proteomics. Previous attempts to analyse biomolecules often failed because the ionization methods used caused thermal decomposition of non-volatile species before they could be transferred into the gas phase. In 1984, electrospray ionization (ESI), an ion source that preserved weakly bound ionic solvent clusters during transfer to the gas phase, was developed. Shortly thereafter, the first ESI-MS of intact protein ions was developed [49]. Almost simultaneously, matrix-assisted laser desorption ionization time (MALDI) was developed in 1985 [50]. Comprehensive reviews of current MS methods can be found in [51, 52], and the application of MS for analysing protein complexes (with a focus on ESI-MS) is described in [53, 54]. There are currently two major "soft ionization" MS techniques for working with biomolecules, which rely on different modes of sample ionization (in some studies, both techniques can be combined [53]):

(a) *ESI-MS*: A comprehensive description of the ESI technique can be found in [49, 51, 52, 55]. In this type of MS instrument, ESI vaporizes the sample. A volatile solvent containing the molecule is pushed through a very small charged capillary, creating a spray of charged droplets. Upon evaporation, ionization occurs (Fig. 6.5a), generating singly or multiply charged ions. The sample is directed towards the analyser, which in most cases is an ion trap. The ion trap

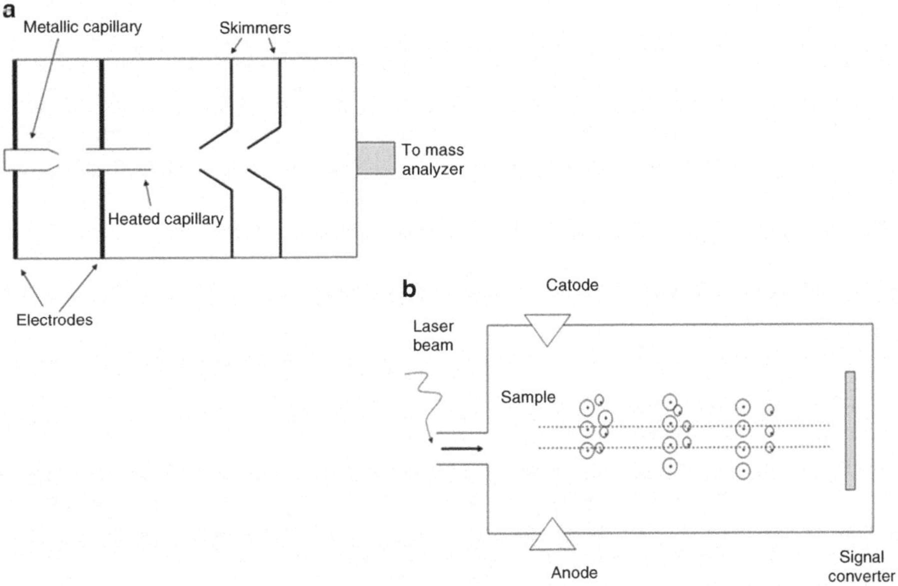

Fig. 6.5 Current types of MS in use to study biological samples. (**a**) ESI-MS schematic setup representation: The solution is passed through a metallic capillary tube to produce a fine spray of droplets; the droplets cross through a large electric field that ionizes them (*dark*, *thick lines* indicate the electrodes). The heated capillary dries the sample. The skimmers are cones with a central orifice designed to intercept the centre of the droplet spray to sample the central portion of the expansion. (**b**) MALDI-TOF MS schematic setup representation: The sample is vaporized by a laser beam and ionized by the use of an electric field; as the ions pass through the analysing chamber, they are separated according to their time of flight

uses a combination of magnetic and electric fields to trap ions in a vacuum by varying the intensity and/or direction of both fields (known as a quadrupole trap). Recently, an Orbitrap analyser has been developed for CDMS (simultaneous measurement of *m*/*z* and charge), essentially an electrostatic ion trap where ions circulate around a spindle electrode while oscillating back and forth along its length [56]. These analysers, along with FTICR analysers, can be used in the following MS technique.

(b) *Matrix-assisted laser desorption ionization time-of-flight (MALDI-TOF)-MS*: This MS method also relies on "soft ionization" of the sample. This ionization is achieved using a high-energy laser beam directed at a macromolecule immersed in a protective matrix (usually α-cyano-4-hydroxicynnaminic acid). The mixture is added to a plate and allowed to dry (crystallize) before insertion into the MS. The laser vaporizes and ionizes the crystallized mixture (hence the name, matrix-assisted laser desorption ionization) (Fig. 6.5b). The ions produced are accelerated in an electric field within a vacuum tube. The time taken for ions to reach the detector ("time of flight") is measured: larger ions reach the detector later than lighter (low *m*/*z*) ones. Because acceleration occurs in a vacuum, the velocity depends solely on mass, and ions reach the detector within 1 ms. MALDI-TOF-MS does not require sample fragmentation and can be applied to biomolecules with large masses, including viral particles (i.e. it can be used for native MS [57]). In contrast to ESI-MS, MALDI-TOF-MS primarily produces singly charged ions. Excellent reviews describing the advantages, applications and development of MALDI-TOF-MS can be found in [50, 57, 58].

6.5 MS as a Tool to Study Virus Structure

There are several advantages to using MS compared to other structural or analytical techniques described in this book (Chaps. 3, 4 and 5 and Sects. 6.2 and 6.3 of this chapter). A primary technical advantage is the minimal amount of material required—usually two or three orders of magnitude less than needed for fluorescence or CD, and several orders of magnitude less than cryo-EM, X-ray crystallography or NMR. A second technical advantage over these structural techniques is the speed of analysis, which can be completed in just a few minutes.

Currently, the term "structural MS" encompasses a set of methods used to gain insights into biomolecular features and interactions. These techniques have provided valuable information on the conformational features of proteins and nucleic acids. The most widely used methods within the "structural MS" toolkit include affinity purification-MS (AP-MS)—also known as affinity proteomics mass spectrometry (Sect. 6.5.4), hydrogen-deuterium exchange MS (HDX-MS) (Sect. 6.5.3), cross-linking MS (XL-MS) (Sect. 6.5.4), hydroxyl radical footprinting MS (HFR-MS) (Sect. 6.5.3), native MS (Sects. 6.5.1 and 6.5.2), ion mobility MS (IM-MS) (Sect. 6.5.4), and top-down MS. For further details on these methods and their applications, interested readers can refer to references [48, 54, 57]. In structural virology, MS is ideally suited for real-time monitoring of changes and interactions of macromolecules. Specifically, MS can track virus particle assembly pathways and conformational changes and provide insights into the stability, conformational dynamics and stoichiometry of viruses.

To study virus particles, MS is usually carried out at near-neutral pH, where ionization occurs to a limited extent, and under relatively high ionic strength conditions (close to physiological levels). However, "soft ionization" has two major technical drawbacks: (i) weak signals in MS; and (ii) high *m/z* ratios. Furthermore, the presence of salt and buffer ions (whether specifically or non-specifically bound to virus particles) results in heterogeneous, broad signals that make accurate mass determination challenging. This issue is exacerbated in structural studies of virions containing nucleic acids due to cation binding to the nucleic acid. Despite these drawbacks, MS is increasingly used in varied and interesting applications in virus assembly and structure, driven in part by impressive recent developments in analyser design [48, 57]. Several examples are detailed in the following subsections, which are interrelated; in fact, many aspects using the same virus model are shared and discussed from various approaches. For recent reviews and further examples on the use of MS to study the structure and dynamics of virus particles, readers can refer to [59–68].

6.5.1 Virus Particle Composition and Structure

An early study describing intact virus capsids by MS was reported in the mid-1990s [69]. The author ionized rice yellow mottle virus (RYMV) and tobacco mosaic virus (TMV) (Fig. 6.6). RYMV has a 29-nm-diameter icosahedral capsid (with a total molar mass of 6.2 MDa), whereas TMV forms a rod-shaped helical capsid with a diameter of 17.5 nm and a length of 300 nm (totalling 40.5 MDa in mass). The masses determined by MS were within ±15% from the calculated values [73]. Usually, these experiments generate highly charged states with various *m/z* rates that are difficult to assign. The authors showed that the virus samples used in MS remained infectious and retained their native morphology. These experiments provided

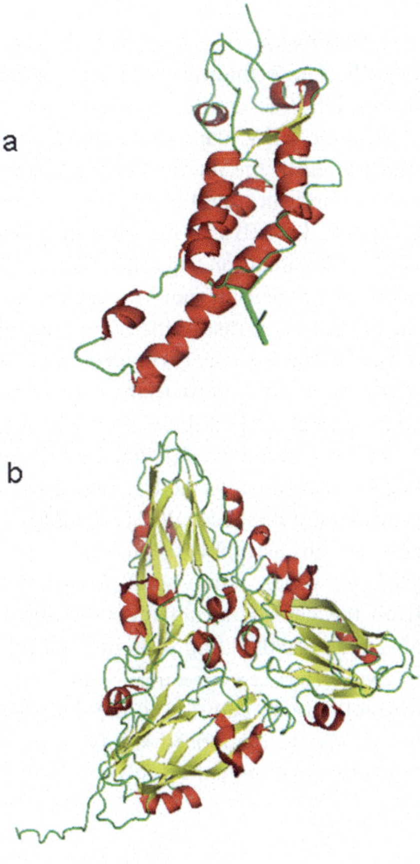

Fig. 6.6 Structures of two virus capsid proteins, whose assembly to form virus capsids has been studied by MS. (**a**) Structure of the TMV capsid protein (PDB number: 2OM3 [70]). The figure was obtained with PyMol [71]. A nucleotide-binding region is shown in *green* on the right side of the structure. (**b**) Structure of a trimer of the RYMV capsid protein (PDB number: 1F2N [72]). Helices are shown in *red*, loop regions in *green* and β-strands in *yellow*

proof-of-concept for the non-invasiveness of MS and its capability to study weak, non-covalent biological complexes; in fact, this was one of the initial examples of native MS [57].

Since then, CDMS has been used to determine the masses of viral particles with different morphologies. For instance, HBV capsids exhibit two different icosahedral geometries (with $T = 3$ and $T = 4$, Chap. 2), composed of 90 and 120 dimers of the capsid protein (CA-HBV), respectively, although assembly studies often employ a truncated capsid protein comprising the first 149 residues (lacking the RNA-binding domain) (Sect. 6.5.2). The theoretical masses of both geometries are ~3 and ~ 4 MDa, respectively. Experimentally determined masses obtained by MS were within 0.1% of the calculated values [74]. The predominance of one geometry type is salt concentration-dependent: at 210 mM ammonium acetate, capsid growth with $T = 4$ predominates over 90 minutes of monitoring, with no significant intermediate populations observed with fewer than 120 dimers. In contrast, at 510 mM ammonium acetate, trapped intermediates with capsids ranging between 3 and 4 MDa are observed, with the 3 MD species considered an intermediate resulting from a collapse of a strained nonspherical shell. These intermediates are thought to be off-pathway assembled species due to assembly errors, requiring partial disassembly for correction. An intermediate with a mass of ~3 MDa is observed during the disassembly of $T = 4$ capsids [75–78]. All these studies were carried out using CDMS. These findings indicate multiple assembly pathways with distinct kinetics during HBV assembly, involving self-correction disassembly processes.

Mature HBV capsids (with $T = 4$ capsid with packaged dsDNA) must be transported to the nucleus through the nuclear membrane using receptors (carriers) called importin α and β [69, 73], which bind to the C-terminal domain of the HBV capsid (CTD-HBV). This binding occurs through specific sequences known as nuclear localization signals (NLSs) [79, 80]. The CTD-HBV is initially located inside the capsid in mature capsids but transiently exposed to the exterior to show the NLS. In contrast to the behaviour observed with single proteins, where importin α usually binds to the cargo protein, CDMS experiments indicate that either importin α or β can bind directly to the capsids, with up to 90 importin β molecules binding to some capsids [81, 82].

P22 is a tailed dsDNA bacteriophage that infects *Salmonella*. Assembly typically begins with an icosahedral precursor exhibiting $T = 7$ symmetry, known as the procapsid, which contains 420 copies of the capsid protein. The scaffolding protein acts as an assembly chaperone, ensuring the correct structure of procapsids [83]. However, some scaffolding proteins remain trapped and are later expelled from the capsid during the viral life cycle. Procapsids can assemble even in the absence of scaffolding proteins, but this process is slow and results in the formation of two different geometries with $T = 4$ and $T = 7$. CDMS measurements show a distinct peak in addition to the one corresponding to the proper MW (~23 MDa) of the procapsid composed solely of capsid protein, indicating the presence of ~110 scaffolding proteins [84–86].

6.5.2 Virus Particle Stability, Conformational Dynamics and Assembly

MS has also been used to explore the factors influencing the correct pathways of assembly and the stability of assembly intermediates. For instance, the CTD-HBV comprises two regions: a core (residues 1–140) and the protamine domain (residues 150–183), connected by a linker whose exact length and the conditions under which assembly occurs determine the proportion of $T = 3$ and $T = 4$ capsids formed during assembly (Sect. 6.5.1). MS has enabled the determination of the rate of inter-conversion between both capsid forms and between these species and the soluble pool of capsid proteins (using a mixture of ^{15}N-labelled and unlabelled proteins); therefore, the relative stability of both forms can also be explored. Both capsids showed similar temperature-dependent stability, although the $T = 3$ form maintained its integrity better during the MS experiments [74]. Thus, in HBV, it can be concluded that there are several defined pathways for capsid assembly.

The stability and assembly of virus capsids may depend on capsid protein concentration, pH, redox conditions, the presence of disulphide bridges, ionic strength, metal ions, scaffolding proteins (Sect. 6.5.1), the viral nucleic acid and other factors (Chaps. 1, 11, 12, 13, 14 and 19). We have described how the geometry of the capsids of HBV is ammonium acetate concentration-dependent (Sect. 6.5.1). A study on Norwalk virus provides an example on the use of MS to investigate the influence of pH on capsid assembly. The fully assembled capsid is formed by 180 copies of capsid protein VP1. MS analysis showed that the capsid is very stable at acidic pHs, but at alkaline pHs, intermediate or aberrant structures formed by 2, 4, 6, 18, 40, 60 and 80 copies of VP1 were identified by MS (the latter two being the most abundant) [87]. These species could be reversibly converted into the native capsid (180 VP1 copies) at acidic pHs. The authors proposed a nucleation model of capsid assembly (Chap. 19) based on the formation of smaller oligomers, mainly dimers, prior to their association into larger 60- and 80-copy intermediate species.

Another example of how pH, and in this case metal ion removal as well, can affect assembly and disassembly is found in brome mosaic virus (BMV) [88]. Changing to basic pH with low ionic strength triggers disassembly, yielding a broad mass distribution of 3.3 and 1.4 MDa (the sum of both is close to the MW of the intact virion). The higher mass distribution is close to the expected MW of an empty capsid with $T = 3$, while the second lower mass is due to RNA associated with some capsid protein. This lowest fragment accumulates over time when Mg(II) is removed from the solution under the same pH; if the ion is removed, the higher mass distribution shifts to a larger mass value, concomitantly with a broadening of the distribution. The explanation of such behaviour can be rationalized in terms of charge-charge interactions. Removal of Mg(II) causes electrostatic repulsion among glutamic residues in the capsid protein, disrupting the protein-protein interfaces in the capsid and exposing pores through which the RNA becomes solvent-exposed. To avoid this exposure, the RNA recruits more capsid proteins, leading to an increase in the MW of the virion [88, 89]. These findings suggest two different pathways in the disassembly of BMV, consistent with HBV findings (Sect. 6.5.1), indicating that disassembly pathways of viruses are complex. Furthermore, these findings also indicate that RNA (in general, nucleic acids) can play a key role in the assembly of non-enveloped, icosahedral RNA viruses ([90] and the next paragraph in this section).

MS has also been used to study the influence of viral nucleic acid on virus particle assembly (Chaps. 12 and 13) and how it can improve the stability of the entire capsid. Apart from the example described earlier in this section, two additional examples are provided. In the first study, in vitro reassembly of bacteriophage MS2 was initiated by a sequence-specific RNA-protein interaction between the capsid protein dimer and an RNA stem-loop (SL) encompassing the Ψ-site (packaging signal) in the viral RNA [91–93]. MS analysis revealed that the protein-RNA complexes had a 1:1 stoichiometry. In the second study, concurrent dimerization and packaging of the HIV-1 genome (Chap. 5) were investigated using MS. The two copies of the genomic RNA are bound by an intermolecular base-paired region formed by self-complementarity of a sequence located in the SL of the Ψ-site. The RNA dimer is stabilized through interaction with the nucleocapsid protein, and ESI-MS studies showed that binding sites are present in two, but not all four, of the SLs of RNA [92]. Therefore, comparing assembly pathways in the presence and absence of the RNA SL for MS2 has allowed discerning the importance of different intermediate self-associated capsid protein species in achieving efficient capsid assembly [93]. In fact, intermediates where each subunit protein capsid had only a handful of contacts with nearby subunits would be favoured during capsid assembly, suggesting a nucleation mechanism.

A study on bacteriophage Qβ provides an example of using MS to analyse the influence of disulphide bridges in capsid assembly [94]. MS analysis of capsid disassembly after disulphide bridge reduction yields capsid protein dimers. However, disassembly of intact capsids (without reduction of disulphides) leads mainly to pentamers; reduction of disulphide bridges in the pentamers yields monomers. Thus, the assembly pathway of the Qβ capsid may be directed by the intermolecular formation of covalent (disulphide) bonds around the capsid fivefold axes.

Intermediates during capsid assembly are observed when mutations of residues in corresponding capsid proteins at selected sites are introduced. For instance, single mutations in the capsid protein of P22 at position Ala285 yield geometries of $T = 4$ and $T = 7$ (A285T mutant), or more dramatically $T = 3$ or $T = 4$ (for the A285Y mutant) [95]. In an extreme case of mutation, HBV and Woodchuck hepatitis virus (WHV) share 65% identity in their capsid proteins, but they exhibit very different assembly pathways. WHV does not show the self-correction pathways observed in HBV (Sect. 6.5.1), and CDMS experiments show capsids with a $T = 4$ geometry containing 150 dimers of capsid protein. In this geometry, capsids have been elongated along their fivefold axis symmetry by the addition of a ring of hexamers of proteins [96].

Thermal stability of viral capsids can also be mapped by CDMS. As described earlier, fluorescence and far-UV CD can provide clues about the thermal stability of viral capsids or proteins (Sects. 6.3.1 and 6.3.2). The combined use of CDMS with an ESI source, controlling the temperature of the solution before the electrospray step, has shown that filled adeno-associated viral vectors (AAVs) release their cargo first and form intermediate empty capsids before dissociating. pH variations also induce significant structural variations in both filled or empty AAVs [97]. Future advancements in this emerging technique may enable thermal stability measurements followed by MS, without the high concentrations required in fluorescence and far-UV CD.

Finally, an example will be mentioned where CDMS has helped identify possible determinants of virus infectivity, including modifications of the capsid protein or the lipid bilayer surrounding the nucleocapsid in enveloped viruses (Chap. 12). Sindbis virus (SINV) is an enveloped single-stranded RNA virus. Its hosts include mosquitoes and vertebrates, but SINV particles derived from mosquitoes are more infective than those from mammals. CDMS measurements revealed that SINV from mammals have a higher MW than those from mosquitoes, suggesting that differences in infectivity could be linked to different glycoside modifications of the virus spike proteins and/or variations in the lipid bilayer composition [98].

6.5.3 *Dissection of Conformational Stability and Dynamics in Virus Particles*

The study of the dynamic features of viral proteins (and biomacromolecules in general) poses challenges for most analytical techniques: X-ray (Chap. 3) and cryo-EM (Chap. 4) are biased towards well-structured elements, providing insights into viral dynamics primarily through interpolation of rigid structural snapshots. NMR (Chap. 5) can offer hints, but only if the macromolecule studied is small enough. Structural dynamics are crucial for viral infection and replications, as viral capsids are conformationally dynamic assemblies. For instance, some capsid shells must expose their internal components externally to bind receptors (Sect. 6.5.1, as seen in the example of NLS in HBV capsid) or to facilitate host membrane lysis, while others must expand by up to 25% of their diameters to enable fusion. In enveloped viruses, dynamics of surface glycoproteins play critical roles in membrane fusion and cell entry. Understanding the dynamics of viral capsids is therefore essential.

Major advances in studying specific regions of a virus particle in terms of conformational stability and dynamics have been achieved using techniques such as limited proteolysis or hydrogen exchange (HX) coupled with MS, among others. For further details, interested readers can refer to reviews available in [99, 100].

Limited Proteolysis

Even at equilibrium, a capsid may sample different conformations and transiently expose some structural elements that would otherwise be buried (the so-called capsid breathing). Limited proteolysis-MS (alongside HX-MS and hydroxyl radical footprinting, discussed below) has been used to identify dynamic regions in virus capsids. The limited proteolysis approach involves identifying peptides generated through time-dependent, limited proteolytic attacks on virus particles (an application dating back to the early days of MS, Sect. 6.4.2). Particle regions that are more conformationally dynamic tend to expose cleavage sites more readily and are cleaved earlier. The size of the fragments of the capsid proteins obtained at different times reveals the cleavage sites in the particle, thereby indicating dynamic regions [100].

Using this approach, biologically relevant transient externalization of normally internal terminal segments of capsid proteins has been detected in picornaviruses (e.g. poliovirus, HRV) [101], nodaviruses (e.g. flock house virus, FHV) and HBV. In general, the tertiary and quaternary structure of virions (i.e. their capsid assembly) can be mapped using limited proteolysis ([102] and references therein). In addition, this approach has demonstrated differences in local and global breathing among different states of viral capsids, such as during the maturation of tetraviruses (*Nudaurelia capensis* ω virus). Processes such as the binding of viral RNA genomes to the capsid inner wall in FHV [103] or of hydrophobic compounds to pockets in the capsid of HRV [104] have also been found to impair local or global breathing. Some of these effects on capsid dynamics are crucial for virus biology (Chap. 18) or the action of antiviral agents (Chap. 20): while some antivirals stabilize capsids against proteolysis, others have the opposite effect ([102] and references therein).

HX-MS

The basis of hydrogen exchange in biomolecules, including proteins and protein complexes, was explained in Chap. 5 as a part of the description of the HX-NMR technique. The pattern of exchanged protons with deuterium depends on the conformational stability of the region considered and the environmental conditions where HX is conducted. Therefore, under these conditions, the number of incorporated deuterium nuclei in each analysed region can be used to estimate its conformational stability. Since the mass of deuterium is roughly twice that of a proton, the number of incorporated deuterium nuclei can be precisely determined by current mass spectrometers [100, 105]. HX-MS provides a wealth of structural information about conformational flexibility, solution accessibility of specific protein regions and hydrogen bonding of polypeptide patches, although this information cannot be easily quantified to define structures as in cryo-EM (Chap. 3), X-ray (Chap. 4) or NMR (Chap. 5).

Several examples of the use of HX-MS to investigate the conformational stability or dynamics of virus particles and specific regions within them are described next. Additional examples beyond those mentioned below involve understanding the interfaces of neutralizing antibodies or targeted molecules capable of binding and neutralizing viral surface glycoproteins in enveloped viruses. Readers interested in applications of HX-MS in structural virology can find more examples in [106, 107].

HX-MS studies on HRV indicated that inter-subunit interfaces are conformationally rather static, while regions around the capsid fivefold axes, involved in the translocation of polypeptide segments, are more dynamic

[108]. Some aspects of the conformational changes of the phage HK97 capsid during maturation (Chap. 14) have been described by the combined use of protein engineering and HX-MS. During maturation, the HK97 capsid undergoes an expansion of roughly 20% in diameter. The number of exchanged hydrogens suggest that a pre-expanded particle undergoes significant refolding and twisting of the tertiary structure to accommodate the expansion [109, 110]. The differences in solvent accessibility observed between capsid subunits in different quasi-equivalent conformations disappear after the conformational transition that leads to the first maturation intermediate. However, other quaternary interactions close to the icosahedral threefold axes are preserved during maturation. These threefold "staples" are formed by residues Arg and Glu and a metal-binding site; mutations at these sites facilitate the conformational transition, as revealed by a larger proportion of protons being exchanged and a decrease in the in vitro expansion rate. Taken together, these and other results suggest that the HK97 capsid, before the expansion, is in a high-energy metastable state.

The dynamics of maturation of the HIV-1 capsid have also been studied by HX-MS; both the stability of the immature (either non-mutated or carrying a mutation that prevented the maturation cleavage) and the mature capsids have been described [111–113]. Among other results, it has been found that the mature capsid is more stable (i.e. it exchanges more slowly) than the immature capsid. Changes in conformational stability during maturation and expansion of the bacteriophage P22 capsid have also been analysed by HX-MS [114]. The P22 capsid protein consists of two domains connected by a stable linker. The HX pattern indicates that during maturation, the C-terminal region maintains its conformational flexibility (dynamics), whereas the N-terminal region becomes conformationally less dynamic. These experiments are consistent with the possibility that maturation increases the number of inter-subunit contacts.

HX-MS has also been used to investigate structural and dynamic aspects of nucleic acid encapsidation in dsRNA or DNA bacteriophages. These phages rely on an active packaging mechanism, in which an ATP-dependent molecular motor forces the nucleic acid into the preformed procapsid (Chap. 13). These packaging motors consist of at least two regions: the ATPase complex and the connector portal complex; the portal complex replaces the capsid proteins at one of the fivefold vertices of the capsid. The use of HX-MS reveals: (i) the inter-subunit interactions between the portal and the capsid protein and; (ii) the number of subunits forming the portal (in P22, the portal proteins assembled exclusively as a dodecamer, aided by the scaffolding proteins) [113, 115, 116]. Conformational changes related to nucleic acid packaging have also been investigated by HX-MS. For instance, the structure of the packaging motor of the dsRNA bacteriophage Φ8 is not structurally modified during ATP binding and hydrolysis (as suggested by the absence of large changes in the HX pattern); conversely, RNA binding causes changes in kinetics that allow the identification of several nucleotide-binding sites [117]. Finally, the HX-MS results suggest that in the presence of the motor, the capsid is conformationally less dynamic [118].

Biophysical characterization of viruses can be carried out by HX-MS to describe the conformational stability of viral capsids (in parallel to the studies described in Sects. 6.3.2 and 6.5.2). In the next lines, there are two examples. The first one describes the use of urea (as in the unfolding of isolated viral proteins followed by fluorescence and far-UV CD; Sects. 6.3.1 and 6.3.2) on a small icosahedral plant virus [119]. The use of increasing concentrations of urea reveals how the capsid disentangles responding to this chemical stress, which could mimic the same pathway followed under other chemical or cellular stresses. The second example explores the heat response of MVM capsid using HX-MS to see whether the results confirm previous findings using fluorescence and DSC (Sect. 6.3.2) [24, 32]. Heating of the MVM capsid triggers the externalization of the N-termini of capsid proteins, exposing an NLS, which is important for nuclear translocation (Sect. 6.5.1) and then infectivity; therefore, heating mimics a key structural transition in viral replication. Studies at different temperatures allow us to dissect the precise regions melting and which ones are firstly externalized [120]. It would be interesting to explore the same capsid by employing the methods using the CDMS with a variable temperature ionization source [97].

Allosteric effects can also be monitored by HX-MS, but care must be taken to avoid misunderstanding with the binding region (which must be clearly defined in the system under study before starting the HX-MS experiment). In the next lines, there are two examples. The first one concerns changes in receptor-binding domains, as these modifications are crucial in balancing immune evasion with host interactions. In Nipah virus, allosteric effects were observed in the ectodomain upon binding to ephrin B2, its receptor. Comparison of the MS in the absence and presence of $^{2}H_2O$ shows the binding interface becoming more protected in the presence of ephrin B2, but several sites on the opposite side of the interface also showed increased HX [121]. These allosteric effects in Nipah virus seem to be caused by the host-virus membrane fusion. The second example of allosterism is observed in some antibodies upon binding to their antigens; describing the allosterism in antibodies can give insight into how these molecules can neutralize and recognize their viral targets [122, 123]. Furthermore, the findings can provide invaluable information to optimize those antibodies in therapeutic applications.

In enveloped viruses, glycoproteins can have several conformations (pre- and post-fusion ones), which are important during fusion steps and in the host recognition process. HX-MS has been used to characterize the pH-triggered pre- to post-fusion transition of influenza A virus hemagglutinin (HA) spike protein [124], revealing a sequential kinetic pathway with intermediates, affecting distinct regions of HA.

Hydroxyl Radical Footprinting Mass Spectrometry (HRF-MS)

As in HX-MS, HRF-MS relies on covalently labelling viral macromolecules to probe protein conformation and binding to other molecules. Interested readers can find more details on general applications in [125, 126]. The method uses hydroxyl radicals ($OH^{\cdot}$) to irreversibly modify solvent-exposed residues of proteins. The $OH^{\cdot}$

can be generated by laser-based fast photochemical oxidation of proteins (FPOP), laser photolysis of hydrogen peroxide or Fenton chemistry using EDTA [125].

Infection by Paramyxoviridae family of viruses relies on the interplay of a highly variable receptor-binding protein and a more conserved fusion protein causing membrane fusion required for genomic transfer. During viral fusion, the fusion protein undergoes temperature-dependent conformational changes from a pre-fusion state to a post-fusion state. Studies at different temperatures where FPOPs have been performed allows for dissecting the dynamics of the fusion protein [127]. The pre-fusion state undergoes a greater extent of oxidation, indicating that this state is more open and flexible than the post-fusion state.

6.5.4 Virus-Ligand Interactions

Native MS preserves non-covalently bound complexes, including ligand-bound viral particles. Thus, interactions between virus particles or their components and other biomolecules can be determined using MS. Several examples will be described in the following paragraphs, concluding with the use of MS in antiviral drug and vaccine development.

In several studies, interactions between lipids or membrane proteins and viral capsids were detected by MS, and in some cases, they were confirmed by other techniques such as cryo-EM (Chap. 3). For instance, the outer membrane proteins OmpA and OmpC from *Shigella*, the bacterial host of phage Sf6, have been detected in complex with the Sf6 capsid [128]. Also, MS in conjunction with enzyme digestion has shown that most of the proteins forming the capsid of iridovirus are capable of binding to lipids [129].

Some viral capsids recruit proteins from host cells or other viral proteins required for infection. For instance, when HSV-1 nucleocapsids exit the nucleus, they become surrounded by a primary lipid envelope. The use of MS to detect macromolecular interactions has allowed the identification of most of the proteins forming those primary enveloped virions: a glycoprotein, a cellular protein and a tegument protein [130, 131].

MS has also proven useful, together with electrophoresis, in analysing the composition of viral particles and their covalent modifications. In many of these proteomic studies, XL-MS or proteolysis has been used to facilitate protein identification (see [102] for a review). These studies have helped determine the stoichiometry of complexes involving viral proteins and/or nucleic acids. For instance, a map of the interactions between HIV and human protein complexes has been obtained: as many as 497 interactions, involving 435 human proteins, have been identified [132]. A similar approach has been used to identify the entire proteome of SARS-CoV-2 (viral and host cell proteins) [133]. Both proteomic studies are also good examples of the use of MS applied to the analysis of proteome.

Native MS can also be used to map interactions between IDPs of viral proteins and their partners. The complex between N_{tail} and the phosphoprotein X domain from measles virus (Chap. 5) has been studied using this approach. A broad charge-state distribution was observed for isolated N_{tail} (in agreement with its disordered nature). When both proteins are mixed, a 1:1-stoichiometry complex is formed, but there were two different conformations (open and closed) for the complex [134]. Apart from native MS, the authors in this work also used IM-MS, a technique where ions generated are separated based on their charge, mass and physical characteristics, providing a measurement of the protein conformational envelope. This is a good example on how MS can confirm findings obtained by other techniques (EPR and NMR (Chap. 5), as well as fluorescence).

Molecular interactions among viral proteins and partners can also be addressed using XL-MS. The general XL-MS approach involves the use of a chemical cross-linking reagent on a sample of protein or proteinaceous mixture. This mixture can involve tissues, *in cellulo* samples, or in vitro ones. More information in this general structural biology approach can be found in [57, 135], or from a virus structure perspective in [64]. A comprehensive view of the host-pathogen protein interaction network has been obtained with this approach in poliovirus [136].

Small molecule inhibitor binding to the hepatitis C virus RNA-dependent RNA polymerase, the protease M of SARS-CoV-2 (in both cases involving allosteric changes of the enzyme), the HIV integrase and reverse transcriptase enzymes has also been investigated by HX-MS [137–141]. In these examples, the addition of the inhibitor does not imply an increase in the rigidity of the proteins, and in the case of the protease, one of the drugs causes dissociation of the dimeric enzyme [141] (see [102] and references therein and more examples in Sect. 6.5.3).

6.6 Understanding Viruses: Some Major Contributions of MS

MS has become a very important technique for various analyses on the structure and properties of virus particles, as demonstrated by the examples mentioned above and others. MS can be used to determine the size and composition of capsid building blocks, assembly intermediates, assembled capsids and virions and aberrant viral particles. MS can also be used to track structural changes in response to external factors during the assembly or disassembly of virus particles. It has proven instrumental in analysing viral capsid breathing, detecting conformationally dynamic regions in viral particles and understanding the role of molecules or conditions that facilitate or hinder those changes. Furthermore, MS has enabled the detection of the entire interactome of some viruses.

6.7 Perspectives

CD, fluorescence spectroscopy and MS will likely hold a critical supporting role in structural virology. Fluorescence and CD are highly valuable for initial or rapid general structural characterization of viral proteins and viral particles; their information is particularly important when proteins or viral particles cannot be crystallized or are unsuitable for liquid NMR or solid-state NMR analyses. CD and, especially, fluorescence spectroscopy serve as effective probes for the rapid detection and monitoring of conformational rearrangements in virus particles. MS has become a critically important technique across various fields, including structural virology. The use of MS in this area is expected to become widespread for identifying and studying the conformational dynamics of virus capsids and for proteomic analyses in virology, including detecting interactions between viral proteins, virus particles and cellular biomolecules involved in virus replication and assembly factories (Chaps. 8 and 12). Furthermore, similar to other structural spectroscopic techniques (fluorescence or NMR), MS is anticipated to be applied to *in cellulo* measurements and kinetic studies to provide more details on kinetic pathways. In the coming years, MS will likely be routinely used for studying membrane proteins (embedded in micelles, detergents or lipid nanodiscs).

The power of the techniques described in this chapter is maximized in those cases where they can be combined (see Chap. 7) with structural techniques including EM (Chap. 3), X-ray crystallography (Chap. 4) and/or NMR (Chap. 5). Moreover, only in rare circumstances can a single technique provide answers to all the questions at hand. Therefore, structural biology, and in our case structural virology, is increasingly moving towards integrative approaches. These approaches combine high-resolution structural data with lower-resolution information to obtain the most detailed picture of the biological system's function.

Acknowledgements I thank Mauricio G. Mateu for suggestions. I apologize to those colleagues whose works were not cited as examples due to lack of space or inadvertent omission. I also thank the regional Valencian Government for their support (CIAICO/2021/135) for carrying out the research in my laboratory.

References

1. Lakowicz JR (1999) Principles of fluorescence spectroscopy, 2nd edn. Plenum, New York
2. Albani JR (2007) Principles and application of fluorescence spectroscopy. Blackwell, Oxford
3. Cubitt AB, Heim R, Adams SR, Boyd AE, Gross LA, Tsien RY (1995) Understanding, improving and using green fluorescent proteins. TIBS 20:448–455
4. Stepanenko OV, Stepanenko OV, Scherbakova DM, Kuznetsova IM, Turoverov KK (2011) Modern fluorescence proteins: from chromophore formation to novel intracellular applications. BioTechniques 51:313–327
5. Morrison RN, Boyd RN (1992) Organic chemistry, 6th edn. Prentice Hall, Englewood Cliffs

6. Wang AHJ, Quigley GJ, Koplpak FJ, Crawford JL, van Boom JH, van der Marel G, Rich A (1979) Molecular structure of a left-handed double helical DNA fragment at atomic resolution. Nature 282:680–686
7. Schmidt W (2005) Optical spectroscopy in chemistry and life sciences. Wiley-VGH, Weinheim
8. Atkins P, de Paula J (2006) Physical chemistry for the life sciences. Oxford University Press, Oxford
9. Tinoco J, Sauer K, Wang Y (1996) Physical chemistry: principles and applications for the life sciences, 3rd edn. Prentice Hall, Englewood Cliffs
10. Crabbe P (1967) Optical rotatory dispersion and optical circular dichroism in organic chemistry. Top Stereochem 1:93–98
11. Kelly SM, Price NC (2000) The use of circular dichroism in the investigation of protein structure and function. Curr Protein Pept Sci 1:349–384
12. Whitmore L, Wallace BA (2004) DICHROWEB, an online server for protein secondary structure analyses from circular dichroism spectroscopic data. Nucleic Acids Res 32:W668–W673
13. Whitmore L, Wallace BA (2008) Protein secondary structure analyses from circular dichroism spectroscopy: methods and reference databases. Biopolymers 89:392–400
14. Esperante SA, Alvarez-Pagi D, Salgueiro M, Desimone MF, de Oliveira GAP, Aran M, García-Pardo J, Aptekmann AA, Ventura S, Alonso LG et al (2022) A finely tuned interplay between calcium binding, ionic strength and pH modulates conformational and oligomerization equilibria in the respiratory syncytial virus matrix (M) protein. Arch Biochem Biophys 30: 109424
15. Clarke J, Fersht AR (1993) Engineered disulfide bonds as probes of the folding pathway of barnase: increasing the stability of proteins against the rate of denaturation. Biochemistry 32: 4322–4329
16. Privalov PL (1992) Physical basis of the stability of the folded conformations of proteins. In: Creighton TE (ed) Protein folding. Freeman, New York, pp 83–126
17. Wingfield PT, Stahl SJ, Williams RW, Steven AC (1995) Hepatitis core antigen produced in *Escherichia coli*: subunit composition, conformational analysis, and *in vitro* capsid assembly. Biochemistry 34:4919–4932
18. Mateu MG (2002) Conformational stability of monomeric and dimeric forms of the C-terminal domain of human immunodeficiency virus-1 capsid protein. J Mol Biol 318:519–531
19. Gonçalves RB, Mendes YS, Soares MR, Katpally U, Smith TJ, Silva JL, Oliveira AC (2007) VP4 protein from human rhinovirus 14 is elapsed by pressure and locked in the capsid by the antiviral compound WIN. J Mol Biol 366:295–306
20. Kneijzlik Z, Ulbrich P, Strohalm M, Lastůvková H, Kodïcek M, Sakalian M, Ruml T (2009) Conformational changes of the N-terminal part of the mason-Pfizer monkey virus p12 protein during multimerization. Virology 393:168–176
21. Datta SA, Temeselew LG, Christ RM, Soheilian F, Kamata A, Mirro J, Harvin D, Nagashima K, Cachau RE, Rein A (2011) On the role of SP1 domain in HIV-1 particle assembly: a molecular switch? Virology 85:4111–4121
22. Förster T (1948) Zwischenmolekulare Energiewanderung und Fluoreszenz. Ann Phys 437:55–75
23. Götz C, Hinze G, Gellert A, Maus H, von Hammerstein F, Hammerschmidt SJ, Lauth UA, Schirmeister T, Basché T (2021) Conformational dynamics of the dengue virus protease revealed by fluorescence correlation and single-molecule FRET studies. J Phys Chem B 125:6837–6846
24. Carreira A, Menéndez M, Reguera J, Almendral JM, Mateu MG (2004) In vitro disassembly of a parvovirus capsid and effect on capsid stability of heterologous peptide insertions in surface loops. J Biol Chem 279:6517–6525

25. Watts NR, Palmer IW, Eren E, Steven AC, Wingfield PT (2020) Capsids of hepatitis B virus e antigen with authentic C termini are stabilized by electrostatic interactions. FEBS Lett 594: 1052–1061
26. Katen S, Zlotnick A (2009) The thermodynamics of virus capsid assembly. Methods Enzymol 455:395–417
27. Ceres P, Stray SJ, Zlotnick A (2004) Hepatitis B virus capsid assembly is enhanced by naturally occurring mutation F97L. J Virol 78:9538–9543
28. Alvarez-Pagi D, Esperante SA, Salgueiro M, Camporeale G, de Oliveira GAP, Prat Gay G (2019) A conformational switch balance viral RNA accessibility and protection in a nucleocapsid ring model. Arch Biochem Biophys 671:77–86
29. del Alamo M, Neira JL, Mateu MG (2003) Thermodynamic dissection of a low affinity protein-protein interface involved in human immunodeficiency virus assembly. J Biol Chem 278:27923–27929
30. Green B, King J (1999) Folding and stability of mutant scaffolding proteins defective in P22 capsid assembly. J Biol Chem 274:16141–16146
31. Umahankar M, Murthy MRN, Savithiri HS (2003) Mutation of interfacial residues disrupts subunit folding and particle assembly of Physalis mottle tymovirus. J Biol Chem 278:6145–6152
32. Reguera J, Carreira A, Riolobos L, Almendral JM, Mateu MG (2004) Role of interfacial amino acid residues in assembly, stability and conformation of a spherical virus sample. Proc Natl Acad Sci USA 101:2724–2729
33. Kirkitadze MD, Barlow PN, Price NC, Kelly SM, Boutell CJ, Rixon FJ, McClelland DA (1998) The herpes simplex virus protein, VP23, exists as a molten globule. J Virol 72:10066–10072
34. Lee KK, Gan L, Tsuruta H, Hendrix RW, Duda RL, Johnson JE (2004) Evidence that a local refolding event triggers maturation of HK97 bacteriophage capsid. J Mol Biol 340:419–433
35. Noval MG, Esperante SA, Molina IG, Chemes LB, Prat Gay G (2016) Intrinsic disorder to order transitions in the scaffold phosphoprotein P from the respiratory syncytial virus RNA polymerase complex. Biochemistry 55:1441–1454
36. Xue B, Willians RW, Oldfield CJ, Goh GK, Dunker AK, Uversky VN (2010) Viral disorder or disordered viruses: do viral proteins possess unique features? Protein Pept Lett 17:932–951
37. Rexroad J, Martin TT, McNeilly D, Godwin S, Middaugh CR (2006) Thermal stability of adenovirus type 2 as a function of pH. J Pharm Sci 95:1469–1479
38. Kang H, Yu J, Jung G (2008) Phosphorylation of hepatitis B virus core C-terminally truncated protein (Cp149) by PKC increases capsid assembly and stability. Biochem J 416:47–54
39. Real-Hohn A, Groznica M, Löffler N, Blaas D, Kowalski H (2020) nanoDSF: in vitro label-free method to monitor picornavirus uncoating and test compounds affecting particle stability. Front Microbiol 11:1442
40. Waldron TT, Murphy KP (2003) Stabilization of proteins by ligand binding: application to drug screening and determination of unfolding energetics. Biochemistry 42:5058–5064
41. Sun J, Yu JS, Jin S, Zha X, Wu Y, Yu Z (2010) Interactions of synthetic HPV-16 capsid peptides with heparin: thermodynamic parameters and binding mechanism. J Phys Chem B 114:9854–9861
42. Teschke CM, King J, Prevelige PE Jr (1993) Inhibition of viral capsid assembly by 1,1-bis (4-anilinonaphthalene-5-sulfonic acid). Biochemistry 32:10658–10665
43. Zlotnich A, Mukhopadhyay S (2010) Virus assembly, allostery and antivirals. Trends Microbiol 19:14–23
44. Du S, Betts L, Yang R, Shi H, Concel J, Ahn J, Aiken C, Zhang P, Yeh JL (2011) Structure of the HIV-1 full-length capsid protein in a conformational trapped unassembled state induced by small-molecule binding. J Mol Biol 406:371–386
45. Zlotnick A, Lee A, Bourne CR, Johnson JM, Domanico PL, Stray SJ (2007) In vitro screening for molecules that affect virus capsid assembly (and other protein association reaction). Nat Protoc 2:490–498

46. Stray SJ, Johnson JM, Kopek BJ, Zlotnick A (2006) An in vitro fluorescence screen to identify antivirals that disrupt hepatitis B virus capsid proteins. Nat Biotechnol 24:358–362
47. Abián O, Ortega-Alarcón D, Jiménez-Alesanco A, Ceballos-Laita L, Vega S, Reyburn HT, Rizzuti B, Velázquez-Campoy A (2020) Structural stability of SARS-CoV-2 3CLpro and identification of quercetin as an inhibitor by experimental screening. Int J Biol Mac 164:1693–1703
48. Jarrold MF (2022) Applications of charge detection mass spectrometry in molecular biology and biotechnology. Chem Rev 122:7415–7441
49. Fenn JB, Mann M, Meng CKAI, Wong SF, Whitehouse CM (1989) Electrospray ionization for mass spectrometry of large biomolecules. Science 246:64–71
50. Karas M, Hillenkamp F (1988) Laser desorption ionization of proteins with molecular masses exceeding 10,000 Da. Anal Chem 60:2299–2301
51. Johnstone RAW, Rose ME (1996) Mass spectroscopy for chemist and biochemist. Cambridge University Press, Cambridge
52. Gault VA, McClenaghan NH (2009) Understanding bioanalytical chemistry: principles and applications. Wiley and Blackwell, Oxford
53. Ashcroft AE (2005) Recent developments in electrospray ionization mass spectrometry: non-covalently bound protein complexes. Nat Prod Rep 22:452–464
54. Benett JL, Nguyen GTH, Donald WA (2022) Protein small-molecule interactions in native mass spectrometry. Chem Rev 122:7327–7385
55. Wilm MS, Mann M (1994) Electrospray and Taylor-cone theory, Dole's beam of macromolecules at last? Int J Mass Spectrom Ion Proc 136:167–180
56. Eliuk S, Makarov A (2015) Evolution of Orbitrap mass spectrometry instrumentation. Annu Rev Anal Chem 8:61–80
57. Britt HM, Cragnolini T, Thalassinos K (2022) Integration of mass spectrometry data for structural biology. Chem Rev 122:7952–7986
58. Tanaka K, Waki H, Ido Y, Akita S, Yoshida T (1988) Protein and polymer analysis up to m/z 100000 by laser ionization time-of-flight mass spectrometry. Rapid Commun Mass Spectrom 2:151–153
59. Uetrecht C, Heck AJR (2011) Modern biomolecular mass spectrometry and its role in studying virus structure, dynamics and assembly. Angew Chem Int Ed 50:8248–8262
60. Greco TM, Diner BA, Cristea IM (2014) Spectrometry-based proteomics on fundamental discoveries in virology. Annu Rev Virol 1:581–604
61. Guttman M, Lee KK (2016) Isotope labeling of biomolecules: structural analysis of viruses by HDX-MS, 1st edn. Elsevier
62. Ashcroft AE (2019) Mass spectrometry-based studies of virus assembly. Curr Op Virol 36:17–12
63. Hargett AA, Renfrow MB (2019) Glycosylation of viral surface proteins probed by mass spectrometry. Curr Opin Virol 36:56–66
64. Dülfer J, Kadek A, Kopicki JD, Krichel B, Uetrecht C (2019) Structural mass spectrometry goes viral. Adv Virus Res 105:189–238
65. Cipollo JF, Parsons LM (2020) Glycomics and glycoproteomics of viruses: mass spectrometry applications and insights toward structure-function relationships. Mass Spec Rev 39:371–409
66. Milewska A, Ner-Kluza J, Dabrowska A, Bodzon-Kulakowska A, Pyrc K, Suder P (2020) Mass spectrometry in virological sciences. Mass Spec Rev 39:499–452
67. Witkowska D (2020) Mass spectrometry and structural biology techniques in the studies of coronavirus-receptor interaction. Molecules 25:4133
68. Wörner TP, Shamorkina TM, Snijder J, Heck AJR (2021) Mass spectrometry-based structural virology. Anal Chem 93:620–640
69. Siuzdak G (1996) Mass spectrometry and viral analysis. Chem Biol 9:707–715
70. Sachse C, Chen JZ, Coureux PD, Stroupe ME, Fandrich M, Grigorieff N (2007) High-resolution electron microscopy of helical specimens: a fresh look at tobacco mosaic virus. J Mol Biol 371:812–835

71. DeLano WL (2002) The PyMOL molecular graphics system. De Lano Scientific, San Carlos
72. Qu C, Liljas L, Opalka N, Brugidou C, Yeager M, Beachy RN, Fauquet CM, Johnson JE, Lin T (2000) 3D domain swapping modulates the stability of members of an icosahedral virus group. Struct Fold Des 8:1095–1103
73. Fuesternau SD, Benner WH, Thomas JJ, Brugidou C, Bothner B, Siuzdak G (2001) Mass spectrometry of an intact virus. Angew Chem Int Ed 40:541–544
74. Uetrecht C, Versluis C, Watts NR, Roos WH, Wuite GJ, Wingfield PT, Steven AC, Heck AJ (2008) High resolution mass spectrometry of viral assemblies: molecular composition and stability of dimorphic hepatitis B virus capsids. Proc Natl Acad Sci USA 105:9216–9220
75. Lutomski CA, Lytktey NA, Pierson EE, Zhao Z, Zlotnick A, Jarrold MF (2018) Multiple pathways in capsid assembly. J Am Chem Soc 140:5784–5790
76. Lutomski CA, Lytkei NA, Zhao Z, Pierson EE, Zlotnick A, Jarrold MF (2017) HBV capsid completion occurs through error correction. J Am Chem Soc 139:16932–16938
77. Pierson EE, Keifer DZ, Selzer L, Lee LS, Contino NC, Wang JCY, Zlotnick A, Jarrold MF (2014) Detection of late intermediates in virus capsid assembly by charge detection mass spectrometry. J Am Chem Soc 136:3536–3541
78. Luque A, Reguera D, Morozov A, Rudenock J, Bruinsam R (2012) Physics of shell assembly: line tension, hole implosion, and closure catastrophe. J Chem Phys 136:184507
79. Wing CE, Fung HYJ, Chook YM (2022) Karyopherin-mediated nucleocytoplasmic transport. Nat Moll Cell Biol 23:307–328
80. Cautian B, Hill R, de Pedro N, Link W (2015) Components and regulation of nuclear transport processes. FEBS J 282:445–462
81. Kann N, Schmitz A, Rabe B (2007) Intracellular transport of hepatitis B virus. World J Gastroenterol 13:39–47
82. Chen C, Wang JCY, Pierson EE, Keifer DZ, Delaleu M, Gallucci L, Cazenave CM, Kann M, Jarrold MF, Zlotnick A (2016) Importin β can bind hepatitis B virus core protein and empty core-like particles and induce structural changes. PLoS Pathog 12:e1005802
83. Prevelige PE, Thomas D, King J (1988) Scaffolding proteins regulate the polymerization of P22 coat subunits into icosahedral shells in vitro. J Mol Biol 202:743–757
84. Keifer DZ, Pierson EE, Hogan JA, Bedwell G, Prevelige PE, Jarrold MF (2014) Charge detection mass spectrometry of bacteriophage P22 procapsid distributions above 20 MDa. Rapid Commun Mass Spectrom 28:483–488
85. Thuman-Commike PA, Greene B, Malinski JA, King J, Chiu W (1998) Role of the scaffolding protein in P22 procapsid size determination suggested by T = 4 and T = 7 procapsid structures. Biophys J 74:559–568
86. Zlotnick A, Suhanovsky MM, Teschke CM (2012) The energetic contributions of scaffolding and coat proteins to the assembly of bacteriophage procapsids. Virology 428:64–69
87. Shoemaker GK, van Dujin E, Crawford SE, Uetrecht C, Baclayon M, Roos WH, Wuite GJ, Estes MK, Prasad BV, Heck AJ (2010) Norwalk virus assembly and stability monitored by mass spectrometry. Mol Cell Proteomics 9:1742–1751
88. Bond KM, LyKtey NA, Tsvetkova IB, Dragnea B, Jarrold MF (2020) Disassembly intermediates of brome mosaic virus identified by charge detection mass spectrometry. J Phys Chem B 124:2124–2131
89. Chakravarty A, Reddy VS, Rao ALN (2020) Unraveling the stability and capsid dynamics of the three virions of brome mosaic virus assembled autonomously in vivo. J Virol 94:e01794–e01719
90. Twarock R, Stockely PG (2019) RNA-mediated virus assembly: mechanisms and consequences for viral evolution and therapy. Annu Rev Biophys 48:495–514
91. Stockley PG, Rolfsson O, Thompson GS, Basnak G, Francese S, Stonehouse NJ, Homans SW, Ashcroft AE (2007) A simple, RNA-mediated allosteric switch controls the pathway to formation of a T = 3 capsid. J Mol Biol 369:541–552

92. Fabris D, Chaudhari P, Hagan N, Turner K (2007) Functional investigation of retroviral protein ribonucleic acid complexes by nanospray Fourier transform ion cyclotron resonance mass spectrometry. Eur J Mass Spectrom 13:29–33
93. Knapman TW, Morton VL, Stonehouse NJ, Stockley PG, Ashcroft AE (2010) Determining the topology of virus assembly intermediates using ion mobility spectrometry-mass spectrometry. Rapid Commun Mass Spectrom 24:3033–3042
94. Ashcroft AE, Lago H, Macedo JM, Horn WT, Stonehouse NJ, Stockley PG (2005) Engineering thermal stability in RNA phage capsids via disulphide bonds. J Nanosci Nanotechnol 5: 2034–2041
95. Keifer DZ, Motwani T, Tescke CM, Jarrold MF (2016) Acquiring structural information on virus particles via charge detection mass spectrometry. J Am Soc Mass Spectrom 27:1028–1036
96. Pierson EE, Keifer DZ, Kukreja AA, Wang JCY, Zlotnick A, Jarrold MF (2016) Charge detection mass spectrometry identifies preferred non-icosahedral polymorphs in the self-assembly of woodchuck hepatitis virus capsids. J Mol Biol 428:292–300
97. Kostelic MM, Ryan JP, Brown LS, Jackson TW, Hsieh CC, Zak CK, Sanders HM, Liu Y, Chen VS, Byrne M et al (2022) Stability and dissociation of adeno-associated viral capsids by variable temperature-charge detection mass spectrometry. Anal Chem 94:11723–11727
98. Dunbar CA, Rayaprolu V, Wang CY, Brown CJ, Leishman E, Jones-Burrage S, Trinidad JC, Bradshaw HB, Clemmer DE, Mukhopadhyay S et al (2019) Dissecting the components of Sindbis virus from arthropod and vertebrate hosts: implication for infectivity differences. ACS Infect Dis 5:892–902
99. Johnson JE (2003) Virus particle dynamics. Adv Protein Chem 64:197–218
100. Bothner B, Hilmer JK (2011) Probing viral capsids in solution. In: Agbandje-McKenna M, McKenna R (eds) Structural virology. RSC Publishing, London, pp 41–61
101. Bothner B, Dong XF, Bibbs L, Johnson JE, Siuzdak G (1998) Evidence of viral capsid dynamics using limited proteolysis and mass spectrometry. J Biol Chem 272:672–676
102. Morton VL, Stockley PG, Stonehouse NJ, Ashcroft AE (2008) Insights into virus capsid assembly from non-covalent mass spectrometry. Mass Spectrom Rev 27:575–595
103. Bothner B, Schneemann A, Marshall D, Reddy V, Johnson JE, Siuzdak G (1999) Crystallographically identical virus capsids display different properties in solution. Nat Struct Biol 6: 114–116
104. Lewis JK, Bothner B, Smith TJ, Siuzdak G (1998) Antiviral agent blocks breathing of the common cold virus. Proc Natl Acad Sci USA 95:6774–6778
105. Tuma R, Coward LU, Kirk MC, Barnes S, Prevelige PE Jr (2001) Hydrogen-deuterium exchange as a probe of folding and assembly in viral capsids. J Mol Biol 306:389–396
106. James EI, Murphree TA, Vorauer C, Engen JR, Guttman M (2022) Advances in hydrogen/ deuterium exchange mass spectrometry and the pursuit of challenging biological systems. Chem Rev 122:7562–7623
107. Vallejo DD, Rojas-Ramírez C, Parson KF, Han Y, Gadkari VV, Ruotolo BT (2022) Mass spectrometry methods for measuring protein stability. Chem Rev 122:7690–7719
108. Wang L, Smith DL (2005) Capsid structure and dynamics of a human rhinovirus probed by hydrogen exchange mass spectrometry. Protein Sci 14:1661–1672
109. Gertsman I, Fu CY, Huang R, Kornives EA, Johnson JE (2010) Critical salt bridges guide capsid assembly, stability and maturation behavior in bacteriophage HK97. Mol Cell Proteomics 9:1752–1763
110. Gertsamn J, Komives EA, Johnson JE (2010) HK97 maturation studies by crystallography and H/2H exchange reveals the structural basis for exothermic particle transitions. J Mol Biol 397: 560–574
111. Monroe EB, Kang S, Kyere SK, Li R, Prevelige PE Jr (2010) Hydrogen/ deuterium exchange analysis of HIV-1 capsid assembly and maturation. Structure 18:1483–1491
112. Cortines JR, Monroe EB, Kang S, Prevelige PE Jr (2011) A retroviral chimeric capsid protein reveals the role of the N-terminal β-hairpin in mature core assembly. J Mol Biol 410:641–652

113. Fu CY, Uetrecht C, Kang S, Morais MC, Heck AJ, Walter MR, Prevelige PE Jr (2010) A docking model based on mass spectrometric and biochemical data describes phage packaging motor incorporation. Mol Cell Proteomics 9:1764–1773
114. Kang S, Prevelige PE Jr (2005) Domain study of bacteriophage P22 coat protein and characterization of the capsid lattice transformation by hydrogen/deuterium exchange. J Mol Biol 347:935–948
115. Motwani T, Lokareddy RK, Dunbar CA, Cortines JR, Jarrold MF, Cingolani G, Teschke CM (2017) A viral scaffolding protein triggers portal ring oligomerization and incorporation during procapsid assembly. Sci Adv 3:e1700423
116. Kang S, Poliakov A, Sexton R, Renfrow MB, Prevelige PE Jr (2008) Probing conserved helical modules of portal complexes by mass spectrometry-based hydrogen/deuterium exchange. J Mol Biol 381:772–784
117. Lisal J, Kainov DE, Lam TT, Emmett MR, Wei H, Gottlieb P, Marshall AG, Tuma R (2006) Interaction of packaging motor with the polymerase complex of dsRNA bacteriophage. Virology 351:73–79
118. Lisal J, Lam TT, Kainov DE, Emmett MR, Marshall AG, Tuma R (2005) Functional visualization of viral molecular motor by hydrogen deuterium exchange reveals transient states. Nat Struct Mol Biol 12:460–466
119. Ramesh R, Lim XX, Raghuvamsi PV, Wu C, Wong SM, Anand GS (2019) Uncovering metastability and disassembly hotspots in whole viral particles. Prog Biophys Mol Biol 143:5–12
120. Van de Waterbeemd M, Llauró A, Snijder J, Valbuena A, Rodríguez-Huete A, Fuertes MA, de Pablo PJ, Mateu MG, Heck AJR (2017) Structural analysis of a temperature-induced transition in a viral capsid probed by HDX-MS. Biophys J 112:1157–1165
121. Wong JJW, Young TA, Zhang J, Liu S, Leser GP, Komives EA, Lamb RA, Zhou ZH, Salafsky J, Jardetzky TS (2017) Monomeric Ephrinb2 binding induces allosteric changes in Nipah virus G that precede its full activation. Nat Commun 8:781
122. Meng B, Lan K, Xie J, Lerner RA, Wilson IA, Yang B (2020) Inhibitory antibodies identify unique sites of therapeutic vulnerability in rhinovirus and other enteroviruses. Proc Natl Acad Sci USA 117:13499
123. Fernandez E, Kose N, Edeling MA, Adhikari J, Sapparapu G, Lazarte SM, Nelson CA, Govero J, Gross ML, Fremont DH, Crowe JE Jr, Diamond MS (2018) Mouse and human monoclonal antibodies protect against infection by multiple genotypes of Japanese encephalitis virus. MBio 9:e00008–e00018
124. Benhaim MA, Mangala-Prasad V, Garcia NK, Guttman M, Lee KK (2020) Structural monitoring of a transient intermediate in the hemagglutinin fusion machinery on influenza virions. Sci Adv 6:eaaz8822
125. McKenzie-Coe A, Montes NS, Jones LM (2022) Hydroxyl radical protein foot-printing: a mass spectrometry-based structural method for studying the higher order structure of proteins. Chem Rev 122:7532–7561
126. Chance MR, Farquhar ER, Yang S, Lodowski DT, Kiselar J (2020) Protein footprinting: auxiliary engine to power the structure biology revolution. J Mol Biol 432:2973–2984
127. Poor TA, Jones LM, Sood A, Leser GP, Plasencia MD, Rampel DL, Jardetzky TS, Woods RJ, Gross ML, Lamb RA (2014) Probing the paramyxovirus fusion protein-refolding event from pre- to post-fusion by oxidative foot-printing. Proc Natl Acad Sci USA 111:E3596–E2605
128. Zhao H, Sequeira RD, Galeva NA, Tang L (2011) The host outer membrane proteins OmpA and OmpC are associated with the Shigella phage Sf6 virion. Virology 409:319–327
129. Wu J, Chan R, Wenk MR, Hew C-L (2010) Lipidomic study of intracellular Singapore group iridovirus. Virology 399:248–256
130. Padula ME, Sydnor ML, Wilson DW (2009) Isolation and primary characterization of herpes simplex virus 1 primary enveloped virions from the perinuclear space. J Virol 83:4757–4765

131. Harper AL, Meckes DG Jr, Marsh JA, Ward MD, Yeh PC, Baird NL, Wilson CB, Semmes OJ, Wills JW (2010) Interaction domain of the UL16 and Ul21 tegument proteins of herpes simplex virus. J Virol 84:2963–2971
132. Jäger S, Cimermancic P, Gulbahce N, Johnson JR, McGovern KE, Clarke SC, Shales M, Mercenne G, Pache L, Li K et al (2011) Global landscape of HIV-human protein complexes. Nature 481:365–370
133. Gordon DE, Jang GM, Bouhaddou M, Xu J, Obernier K, White KM, O'Meara MJ, Rezelj VV, Guo JZ, Swaney DL et al (2020) A SARS-CoV-2 protein interaction map reveals targets for drug repurposing. Nature 583:459–468
134. D'Urzo A, Konijnenberg A, Rosetti G, Habchi J, Li J, Carloni P, Sobott F, Longhi S, Grandori R (2015) Molecular basis for structural heterogeneity of an intrinsically disordered protein bound to a partner combined ESI-IM-MS and modelling. J Am Soc Mass Spectrom 26:472–481
135. Piersimoni L, Kastritis PL, Arlt C, Sinz A (2022) Cross-linking mass spectrometry for investigating protein conformations and protein-protein interactions- a method for all seasons. Chem Rev 122:7500–7531
136. DeBlasio SL, Chavez JD, Alexander MM, Ramsey J, Eng JK, Mahoney J, Gray SM, Bruce JE, Cilia M (2016) Visualization of host-poliovirus interaction topologies using protein interaction reporter technology. J Virol 90:70–75
137. Feng L, Dharmarajan V, Serrao E, Hoyte A, Larue RC, Slaughter A, Sharma A, Plumb MR, Kessl JJ, Fuchs JR et al (2016) The competitive interplay between allosteric HIV-1 integrase inhibitor BI/D and LEDGF/p75 during the early stage of HIV-1 replication adversely affects inhibitor potency. ACS Chem Biol 11:1313–1321
138. Nguyen PDM, Zheng J, Gremminger TJ, Qiu L, Zhang D, Tuske S, Lange MJ, Griffin PR, Arnold E, Chen SJ et al (2020) Binding interface and impact on protease cleavage for an RNA aptamer to HIV-1 reverse transcriptase. Nucleic Acids Res 48:2709–2722
139. Tuske S, Zheng J, Olson ED, Ruiz FX, Pascal BD, Hoang A, Bauman JD, Das K, DeStefano JJ, Musier-Forsyth K et al (2020) Integrative structural biology studies of HIV-1 reverse transcriptase binding to a high-affinity DNA aptamer. Cur Res Struct Biol 2:116–129
140. Deredge D, Li J, Johnson KA, Wintrode PL (2016) Hydrogen/deuterium exchange kinetics demonstrate long range allosteric effects of thumb site 2 inhibitors of hepatitis C viral RNA-dependent RNA polymerase. J Biol Chem 291:10078–10088
141. El-Baba TJ, Lutomski CA, Kantsadi AL, Malla TR, John T, Mikhailov V, Bolla JR, Schofield CJ, Zitzmann N, Vakonakis I et al (2020) Allosteric inhibition of the SARS-CoV-2 main protease: insights from mass spectrometry-based assays. Angew Chem Int Ed 59:23544–22354

Further Reading

For the basis of MS:

Baker M (2010) Mass spectrometry for biologists. Nat Methods 7:157–161

Burlingame AL, Carr SA, Baldwin MA (eds) (2000) Mass spectrometry in biology. Humana Press, Totowa

Fenn JB, Mann M, Meng CK, Wong SF (1990) Whitehouse CM electrospray ionization-principles and practice. Mass Spectrom Rev 9:37–70

Yamashita M, Fenn JB (1984) Electrospray ion source. Another variation on the free-jet theme. J Phys Chem 88:4451–4459

For proteins analyzed by CD:

Kelly SM, Ness TJ, Price NC (2005) How to study proteins by circular dichroism. Biochim Biophys Acta 1751:119–139

Applications and approaches of MS in structural biology in general:
Chavez JD, Bruce JE (2019) Chemical cross-linking with mass spectrometry: a tool for systems structural biology. Curr Opin Chem Biol 48:8–18
Liu XR, Rempel DL, Gross ML (2020) Protein higher-order-structure determination by fast photochemical oxidation of proteins and mass spectrometry analysis. Nat Protoc 15:3942–3970
Petrotchencko EV, Borchers CH (2022) Protein chemistry combined with mass spectrometry for protein structure determination. Chem Rev 122:7488–7499
Tamara S, A den Boer M, Heck AJR (2022) High-resolution native mass spectrometry. Chem Rev 122: 7269–7326
Applications of MS in structural virology:
See references [59–68] *listed above.*

Chapter 7
Integrative Approaches to Study Virus Structures

David I. Stuart, Hanna M. Oksanen, and Nicola G. A. Abrescia

Abstract A virus particle must work as a strongroom to protect its genome, but at the same time it must undergo dramatic conformational changes to infect the cell in order to replicate and assemble progeny. Thus, viruses are miniaturized wonders whose structural complexity requires investigation by a combination of different techniques that can tackle both static and dynamic processes. In this chapter, we will illustrate how major structural techniques such as X-ray crystallography and electron microscopy can be combined with other techniques to determine the structure of complex viruses. The power of these hybrid approaches is discussed through a number of examples.

Keywords Hybrid methods · X-ray crystallography · Electron microscopy · Cryogenic electron microscopy · Electron tomography · Cryogenic electron tomography · Small-angle X-ray scattering · Virus · Bacteriophage · Capsid · Mutagenesis · Dissociation · Crystal structure · Fitting

Abbreviations

BMV	bromegrass mosaic virus
CD	circular dichroism

D. I. Stuart
Division of Structural Biology, The Wellcome Trust Centre for Human Genetics, University of Oxford, Oxford, UK

Diamond Light Source Ltd, Diamond House, Harwell Science and Innovation Campus, Didcot, UK

H. M. Oksanen
Molecular and Integrative Biosciences Research Programme, Faculty of Biological and Environmental Sciences, University of Helsinki, Helsinki, Finland

N. G. A. Abrescia (✉)
Structure and Cell Biology of Viruses Lab, CIC bioGUNE – Basque Research and Technology Alliance, Derio, Spain

Ikerbasque, Basque Foundation for Science, Bilbao, Spain
e-mail: nabrescia@cicbiogune.es

M. G. Mateu (ed.), *Structure and Physics of Viruses*, Subcellular Biochemistry 105,
https://doi.org/10.1007/978-3-031-65187-8_7

cryo-EM	cryogenic electron microscopy
cryo-FIB	cryogenic focused ion beam
cryo-SIM	cryogenic structured illumination microscopy
cryo-ET	cryogenic electron tomography
DSC	differential scanning calorimetry
EM	electron microscopy
ET	electron tomography
EMDB	electron microscopy data bank
FMDV	foot-and-mouth disease virus
HCV	hepatitis C virus
HRV-16	human rhinovirus 16
HSV	herpes simplex virus
HIV	human immunodeficiency virus
ITC	isothermal titration calorimetry
MS	mass spectrometry
MR	molecular replacement
NCS	non-crystallographic symmetry
NMR	nuclear magnetic resonance
RVFV	Rift Valley fever virus
SANS	small-angle neutron scattering
SAXS	small-angle X-ray scattering
SBMV	Southern bean mosaic virus
SeMet	selenomethionine
SPA	single particle analysis
SIV	simian immunodeficiency virus
TBSV	tomato bushy stunt virus
TEM	transmission electron microscopy
TMV	tobacco mosaic virus
VLP	virus-like particle
WNV	West Nile virus
wwPDB	worldwide protein data bank
XFEL	X-ray free-electron laser
2D	two-dimensional
3D	three-dimensional

7.1 Introduction: The 'Multidisciplinary Approach' Concept in Structural Virology

7.1.1 Early Virus Structural Studies

In previous chapters, several independent approaches to study virus structure, including electron microscopy (EM, Chap. 3), X-ray crystallography (Chap. 4), nuclear magnetic resonance spectroscopy (NMR, Chap. 5), other spectroscopic methods (Chap. 6) and mass spectrometry (MS, Chap. 6), have been described. In this chapter, we will illustrate how some of those methods can be combined to provide further insights into virus structures. In order to help understand how these initially independent methods have converged to complement each other in the study of viruses, we will provide a brief historical perspective and recapitulation of some of these structural methods as applied to virology.

Structural virology began in 1931 when Ernst Ruska built the first electron microscope and, together with his brother and medical doctor, Helmut Ruska, initiated the imaging of viruses and other submicroscopic structures during the years 1938 and 1939. Alongside this, the principles of small-angle X-ray scattering (diffraction) (SAXS) were proposed by André Guinier (see Sect. 7.2.2). In 1935, Wendell Stanley provided the conditions for the crystallization of the first virus, tobacco mosaic virus (TMV) that infects plants, which contributed to his award, 11 years later, of the Nobel Prize in Chemistry. Ruska also received a Nobel Prize but in Physics for his fundamental contribution in electron optics; incredibly, this was 55 years after building his first microscope! In 1937, Bawden and Pirie crystallized another plant virus, tomato bushy stunt virus (TBSV), and 4 years later, crystals of TBSV and TMV produced clear X-ray diffraction patterns.

The 1950s saw the beginning of modern biology with the determination of the three-dimensional (3D) structure of DNA by James Watson and Francis Crick and the first high-resolution protein X-ray structures, myoglobin by John Kendrew and haemoglobin by Max Perutz. Interestingly, in 1956 Watson and Crick also proposed the principles of virus organization. According to their hypothesis, based on electron micrographs of plant viruses, their propensity to form crystals and concepts of genetic economy, simple viruses were formed of a capsid of multiple copies of relatively small proteins arranged in a symmetrical fashion, providing identical protein contacts among their neighbours. They suggested that cubic symmetry (including possibly icosahedral symmetry) or a cylindrical shell would satisfy the above requirements, allowing a larger volume of nucleic acid to be packaged. The case for icosahedral symmetry of viruses was made in 1956 by Donald Caspar, who provided evidence of the structure of TBSV using X-ray diffraction, and then by John Finch and Aaron Klug in 1959 when they determined the low-resolution poliomyelitis virus structure. Caspar and Klug took this idea further and formulated the quasi-equivalence theory, indicating that certain multiples of 60 subunits could also form viral capsids but the equivalent interactions among neighbour subunits could no longer exist. Klug's thinking on this was inspired by the work of Buckminster Fuller, particularly *The Dymaxion World of Buckminster Fuller* [1]. Caspar and Klug based the quasi-equivalent theory on the fact that the capsomers are adaptable molecules that organize in different ways within a highly ordered structure, thus introducing the (partially correct) concept of protein flexibility to explain these different arrangements (see also Chap. 2). Methods development for 3D reconstruction from electron micrographs began in 1964 when Aaron Klug and Jacob Berger introduced the use of Fourier transform in micrograph analysis. This analysis was extended to objects with rotational symmetry, helical symmetry (for rod-shaped viruses) and finally to icosahedral specimens (for a historical

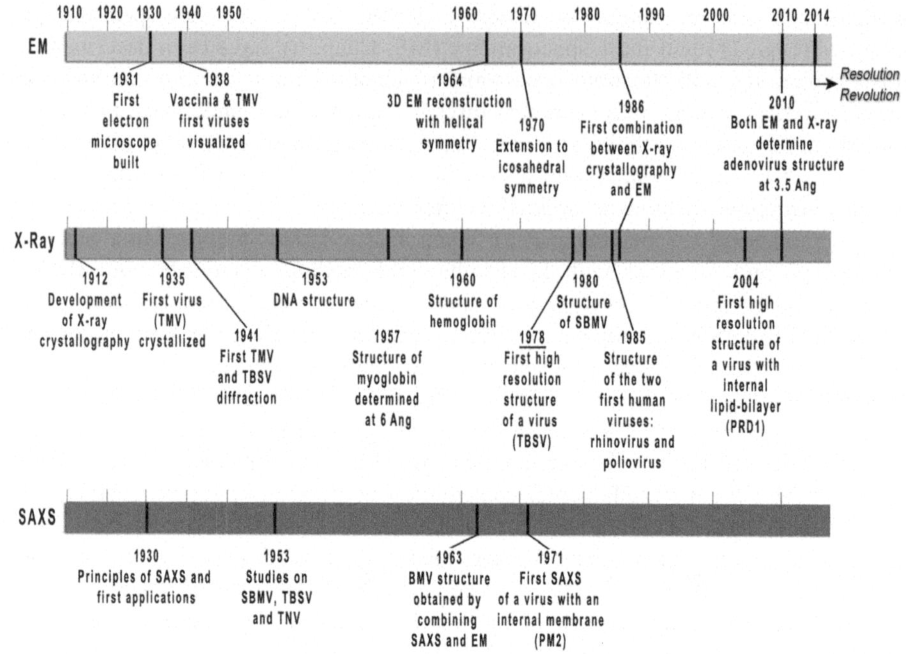

Fig. 7.1 Timeline of the major achievements in structural virology by EM, X-ray crystallography and SAXS techniques for the determination of virus structures. Due to their relevance as landmarks, structures of DNA and myoglobin have been included. Tobacco mosaic virus (TMV); tomato bushy stunt virus (TBSV); lipid-containing bacteriophage PRD1; Southern bean mosaic virus (SBMV); bromegrass mosaic virus (BMV); lipid-containing bacteriophage PM2

account, see 'Further Reading' at the end of this chapter). In the same period, Walter Hoppe, who contributed enormously to the development of methods and instrumentation in X-ray analysis, moved to the EM field, transferring the same momentum and know-how to advances in EM methods. The scale of Hoppe's contributions is more obvious now than at the time; for instance, he also originated electron tomography (ET), a specialized application of EM, which is now poised to provide transformative contributions to virology (see below and Sects. 7.5 and 7.7.1).

Together with X-ray and EM, small-angle scattering (using both X-rays [SAXS] and neutrons [small-angle neutron scattering, or SANS]) also contributed to the analysis of virus structures. In the years that followed, the impact of SAXS as a primary method became much weaker, leaving the leading role in structural virology to the other two structural techniques (Fig. 7.1). In parallel to the above developments, efforts were made to improve the diffraction of virus crystals and to devise ways to analyse symmetric structures that ultimately culminated into the structure of TBSV at 2.9 Å resolution in 1978 [2] and 2 years later, the structure of Southern bean mosaic virus (SBMV) at 2.8 Å resolution [3]. Since then, both X-ray crystallography and EM have been fundamental techniques in providing structural insights into complex viruses. The techniques are complementary and well suited to hybridization.

Since 2014, however, cryogenic EM (cryo-EM) has taken centre stage in structural virology, thanks to advances in image classification software based on Bayesian criteria, improved transmission electron microscopes and the availability of detector technology capable of directly recording scattered electrons, allowing cryo-EM density maps to routinely reveal atomic detail [4–6]. This new capability has revolutionized the field with enormous impact across different disciplines [6]. In 2017 Joachim Frank, Richard Henderson and Jacques Dubochet received the Nobel Prize in Chemistry for developing cryo-EM for high-resolution structure determination of biomolecules in solution (https://www.nobelprize.org/uploads/2018/06/advanced-chemistryprize2017.pdf). While cryo-EM is nowadays a widespread technique, at least for single particle analysis (SPA), its democratization in order to release the enormous potential of the cryo-EM revolution has been and is still 'in divenire' due to the relatively high cost of acquiring and maintaining state-of-the-art cryo-EM instrumentation and facilities [7].

7.1.2 The Emerging Concept of Hybrid Methods

To grasp the concept of hybrid methods in structural virology, one has to go back to the early culture of structural biology. The methods were developed to answer fundamental biological questions, but the X-ray diffractometers and electron microscopes were designed, built and maintained by engineers, physicists and biophysicists. The interdisciplinary expertise gathered in pioneering laboratories in, among others, Birbeck and King's College in London (JD Bernal and Rosalind Franklin), Oxford University (Dorothy Hodgkin), the Royal Institution London (Lawrence Bragg and David Phillips) and the MRC-Unit Laboratory of Molecular Biology at Cambridge University (Max Perutz) around the middle of the twentieth century forged the spirit and ambition to tackle the unexplored complexity of biology by complementary approaches (see Max Perutz's video interview and commentary on http://www.youtube.com/watch?v=5I4eq63-3_I).

Thus, on the one hand, complex biological systems were reduced to smaller pieces (proteins or protein domains), in a reductionist (or 'top-to-bottom') approach, studied separately and then, possibly, put back again in an overall framework (see also Chap. 5 regarding the intensive use of this approach in virus structure studies using NMR). On the other hand, the complexity of larger macromolecular assemblies such as viruses could be looked at head-on, thanks to the diverse expertise in those laboratories. So, hybrid methods and expertise were originally supporting each other, validating the results obtained by the individual novel emerging structural techniques.

Over the past four decades, the multidisciplinary approach has moved on from its original rationale towards the use of hybrid methods in predicting biological interactions, uncovering dynamic processes and bridging different resolution scales in molecular and cellular processes. The study of human adenovirus, a complex non-enveloped icosahedral dsDNA virus (see below and Chap. 12), provides a good example of this. Indeed, this is the first case of combining X-ray crystallography and EM data, where the crystal structure of the viral protein capsid assembled as stable capsomeric hexamers (hexons) was used to build an atomic model of the entire virion by exploiting the capsomer arrangement revealed by EM images [8]. Since then, the individual methods have become more powerful, and the combination of X-ray crystallography, cryo-EM and cryogenic electron tomography (cryo-ET) represents a strategy of choice not just for many challenging viruses but generally within the fields of structural and cellular structural biology.

In this chapter, we will mainly focus on describing how hybrid methods, with emphasis on the early combination of X-ray crystallography and EM, have shaped our understanding of virus structures. We will provide examples of their contributions to the structure solution of very large and complex viruses and briefly discuss the challenges ahead, highlighting the power and vigour of methodological interactions in structural virology.

7.2 Some Classical Methods in Structural Virology: A Brief Overview

In this section, we recapitulate the three major structural techniques that have primarily contributed to the understanding of virus structures: EM, SAXS/SANS and X-ray crystallography. EM and X-ray crystallography as applied to viruses have been described extensively in Chaps. 3 and 4, respectively. Here, we simply underline those concepts that interlink one method to another.

7.2.1 Electron Microscopy

Transmission electron microscopy (TEM) relies on an electron beam passing through a virus sample either negatively stained to enhance the contrast or (in cryo-EM) vitrified at <100 K to ameliorate the radiation damage (vitrification is usually through simple flash freezing, but freezing rates may be enhanced by a simultaneous pressure wave; see Chap. 3). Electrons, elastically scattered via the Coulomb potential of the atoms composing the virus (or any biological/material sample), are combined via magnetic lenses to generate a magnified image of the specimen. Depending on the TEM setup, data can be recorded in real space (images) or a reciprocal space (diffraction). Mathematically, what allows the scattering process to be formalized in one or the other space is the Fourier transform (see also Chaps. 3 and 4). Since the energy of the electrons is very high, their equivalent wavelength is much shorter than that for the X-rays used in crystallography. As a consequence, in image mode, the result is very close to a two-dimensional (2D) projection (real space), conceptually equivalent to the optical image obtained by cameras and visible light microscopes. In reciprocal space, this corresponds to a plane of information passing through the origin of reciprocal space (a central section), and information on both amplitude and phase of the transform of the elastically scattered electrons can be derived (Fig. 7.2). In diffraction mode, only the amplitudes can be directly calculated from the measured intensity of diffraction spots (reciprocal space) but with higher accurancy than the corresponding values extracted from real-space images, which provide the less precise phase information (Fig. 7.2).

The dual use of TEM in image and diffraction modes on the same sample constitutes the field of 2D electron crystallography. However, to collect such electron diffraction images, it is fundamental that the specimen forms a 2D ordered lattice, a requirement not needed for image mode. By combining these modes, 2D electron crystallography circumvents the 'phase problem' that X-ray crystallography must inevitably face (see Chap. 4). Each EM snapshot is equivalent to a 'still' X-ray image. Interestingly, it has proved much easier to generate 3D crystals than 2D (perhaps especially so for viruses which are often isometric), so the field of 2D crystallography remains much less developed. More recently, microcrystal electron diffraction (MicroED) is emerging as a valuable tool when small 3D crystals (<50 μm) are available. Cryogenic electron diffraction data are collected as the crystals are continuously rotated within the electron beam. So far, data collected from crystals of biological macromolecules have been phased using the molecular replacement technique.

Viruses have been investigated by EM since its early days due to their large and regular shape and relative ease of purification. Indeed, since the visualization of the first viruses in 1938–1939 [9], a head-to-head race began in the technological advances of each of the major structural techniques at the time: EM, SAXS and X-ray crystallography (Figs. 7.1 and 7.2). EM remains a workhorse method in several different aspects of virology from virus detection and identification to the 3D structure determination and analysis of virus-cell interactions. In addition recent advances in EM technology mean that nowadays this technique can provide atomic resolution structures of viruses and virus-ligand complexes [10–13] (see Chap. 3). The highest resolution analyses come from combining many projections of the object collected from different particles thrown down at random; however, in recent years, it has become possible to get insightful data from an individual virus particle by collecting a full tilt series and doing tomographic reconstruction (cryo-ET) (Chap. 3). Furthermore, the resolution of both EM and ET can be enhanced for symmetrical virus particles by applying symmetry constraints or sub-tomogram averaging techniques in the case of cryo-ET (see Sect. 7.5). X-ray and electron methods begin to almost fully overlap when we consider X-ray microscopy, in particular soft X-ray cryo-ET, a technique that is capable of visualizing specimens as thick as a eukaryotic cell in quasi-native condition to a resolution beyond 30 nm [14]. Together, this armoury of methods allows, for instance, the visualization of viruses and/or replication factories in situ within the cell (see below and Chap. 8).

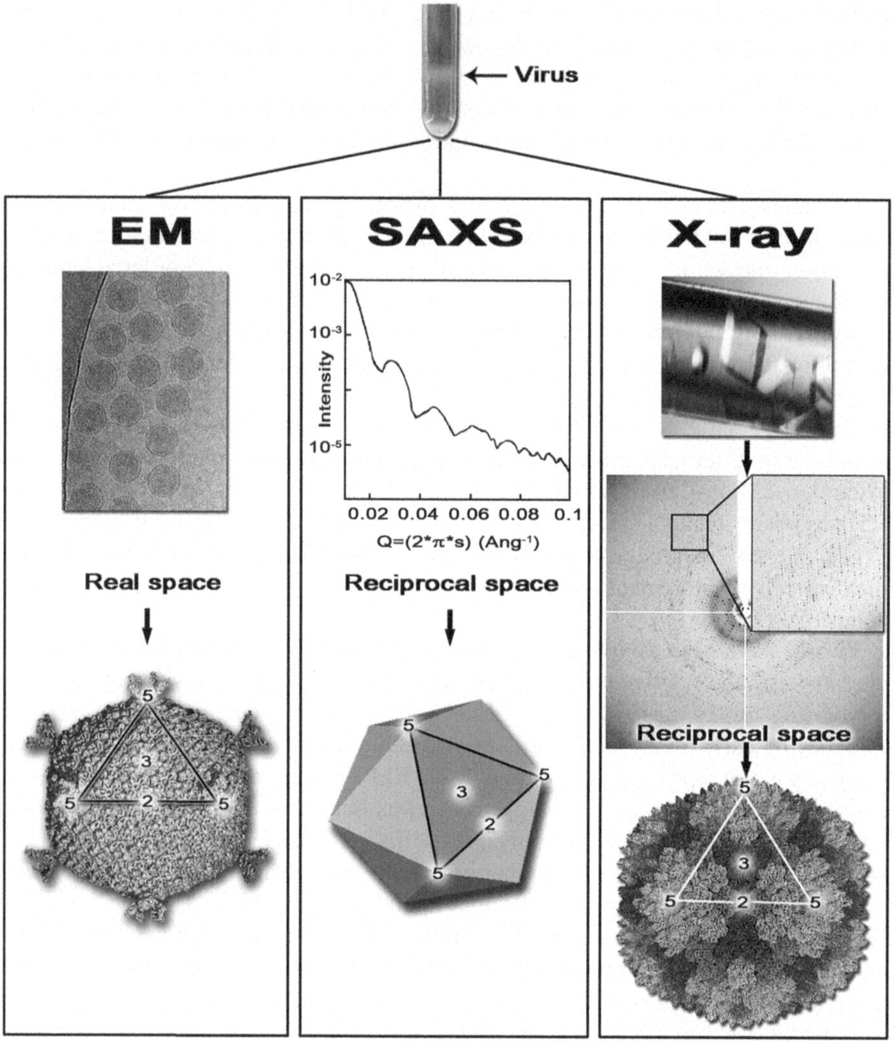

Fig. 7.2 Three major structural techniques at a glance. Schematic representation of the workflow of virus structure determination in EM, SAXS and X-ray. *Top*: All three techniques require purified virus; the black arrow indicates the virus band resulting from ultracentrifugation. (**a**) In EM, the purified virus is placed on a microscopy grid, flash-frozen, and then the 2D projections of the virus are visualized in an electron microscope, producing real-space images (*top*); post-processing of these images provides the virus structure (*bottom*). (**b**) In SAXS, the purified virus in solution is irradiated by an X-ray beam, producing a low-angle scattering curve (*top*) containing raw data in reciprocal space; post-processing of the data produces the virus's overall shape and molecular architecture. (**c**) In X-ray crystallography, the purified virus is used to obtain virus crystals (*top*) that are then diffracted, producing diffraction images (*centre*) containing raw data in reciprocal space; once data processing and the phase problem is solved, the virus structure is obtained at atomic resolution (*bottom*). The outlined region in the virus models in (**a**–**c**) delineates 1 of the 20 viral facets, and numbers indicate icosahedral symmetry axes (2, twofold; 3, threefold; 5, fivefold). Each technique provides structural information in a different resolution range; in some cases, these ranges overlap (see Fig. 7.4)

7.2.2 Small-Angle X-Ray Scattering

SAXS is a technique that provides a structural characterization of a protein, protein complex or virus by using the elastic X-ray scattering from a solution containing the system of interest. The mathematical principles of small-angle scattering were laid out in 1931 by André Guinier (Fig. 7.1) [for a review, see [15]]. As used for the analysis of virus structure, it is typical to work in the solution phase and to use an X-ray (SAXS, wavelength usually ~1.5 Å) or neutron beam (SANS) onto a homogenous solution containing the molecule of interest, and to record the one-dimensional (1D) plot of the scattered beam as a function of scattering angle (Q in the terminology of this method) close to the undeflected beam (hence 'small-angle'). This 1D scattering profile is the average of all scattering profiles deriving from the individual molecules (Fig. 7.2). Thus, in comparison to EM, where individual slices of the virus particle transform are assembled to produce as complete a version as possible of the 3D transform, in small-angle scattering the virus particle transform is spherically averaged to produce the linear scattering profile. Despite this massive loss of information, it is possible to extract information that relates to the volume, radius of gyration and shape of the biological object. Modern synchrotron radiation sources are capable of producing highly accurate curves which can support precise interpretation. Due to the development of software which uses the massive computational power available today, it is remarkably possible to gain ab-initio low-resolution structural information by the use of a randomly distributed dummy atom modelling approach that has replaced the original trial-and-error shape assignment [16]. In brief, a large number of differently shaped molecules are generated, and the scattering patterns simulated by analytical calculation are correlated and compared with the experimentally measured profile. Additional and/or external information on the studied sample, for example, disulphide bonds' presence, oligomerization state, etc., will put constraints on the generation of the different models, thereby increasing the reliability of the final shape assigned by the ab-initio procedure. Nonetheless, it is not possible to reconstruct a detailed 3D structure of a biological macromolecule from solely SAXS data due to the limited information content of the raw data recorded (Fig. 7.2).

SAXS has been used in the characterization of the shape and composition of viruses since the early days of structural virology. Leonard and co-workers in 1953 carried out SAXS studies on SBMV, TBSV and TMV, the same year that the structure of DNA was deduced from X-ray diffraction images (Fig. 7.1). At that point, no virus structure was solved at high resolution, but TBSV and TMV had already been crystallized and showed striking diffraction patterns. It is worth mentioning along this line that, though not strictly SAXS, wide-angle diffraction and the interpretation of X-ray scattering from a partially ordered specimen, often referred to as fibre diffraction, have been pivotal in the elucidation not only of the DNA structure but also of virus structures such as TMV or phage Pf1 (for an exhaustive review on fibre diffraction studies of filamentous viruses, see [17]). With the rise of electron imaging methods, the use of such ordered fibres for structure determination has declined and is nowadays almost unused.

In the early days, turnip yellow mosaic virus, cucumber mosaic virus and bromegrass mosaic virus (BMV) were also investigated by SAXS. Interestingly, in the BMV study, SAXS data were used alongside negative-stain images of the virus obtained by EM to reveal the virus dimensions and the presence of a hollow structure in the centre of the particles. The small set of plant viruses already mentioned were the initial common targets for the different structural techniques. A deviation from this theme was the structural study of a marine bacteriophage, PM2, the first phage to have been described possessing an internal lipid bilayer [18]. This painstaking work combined both SAXS and negative-stain EM methods to describe in detail the architectural organization and assembly of proteins and lipids within the phage. Finally, 40 years later, the *quasi*-atomic model of the entire PM2 virion was revealed by X-ray crystallography [19]. In retrospect, the value of the method is evident: if specific questions can be asked, such as the radius or particle shell thickness, and if it is possible to construct a unique model, the method can give highly accurate answers and distinguish between competing models. This is exemplified by the results from neutron diffraction.

SANS has provided insightful information on virus structures by exploiting the differential contrast generated by the radically different interaction of neutrons with the nuclei of water (^1H_2O) and deuterated water (^2H_2O) to spatially differentiate the genome and lipids from the protein mass. The advantages of this systematic contrast matching lie in the fact that the intensity pattern measured (see Fig. 7.2) derives from the spherically averaged Patterson function, defined as the Fourier transform of the structure factors describing the particle embedded in the solvent (for more details on Patterson function, see general references on X-ray crystallography provided in Chap. 4). Whole sets of Patterson vectors are eliminated by matching out certain components, that markedly disturb the diffraction.

In recent years, SAXS has been re-emerging as an important structural tool to study biological macromolecules, thanks to synchrotron radiation, automation in sample preparation with in-line coupling of size exclusion chromatography and the capabilities of counting detectors such as the DECTRIS-EIGER2 [20, 21]. However, from the virus point of view, major insights using SAXS have been gathered in the study of kinetic and time-resolved processes (TR-SAXS) of virus assembly and disassembly, genome packaging and the influence of ions/cations on the stability of viruses and virus-like particles [22, 23].

7.2.3 X-Ray Crystallography

In contrast to TEM and SAXS, virus X-ray crystallography (Chap. 4) relies on viruses ordering themselves into a 3D lattice, or crystal (Fig. 7.2). Illuminating the crystal with X-rays allows the recording of the Fourier transform of the unique portion of the lattice (unit cell) and thus the transform of the virus particle. The transform can only be sampled at points specified by the Bragg condition; however, this still allows for a fully detailed reconstruction if all unique Bragg spots are measured and phases can be determined. Radiation damage is mitigated since it is distributed across all particles in the illuminated portion of the X-ray beam—usually exceeding a billion particles. All viruses crystallized so far possess icosahedral symmetry (icosahedron: a Platonic solid with 20 identical equilateral triangular faces, 30 edges and 12 vertices; see Chap. 2 for details). Due to this regular isometric shape, crystallization is often not the bottleneck in a virus crystallographic project, particularly for relatively small and biochemically stable viruses. Indeed, the stability of virus crystals decreases with increasing virus size and is approximately proportional to the inverse of the square of the virus radius [24]. Limitations in the amount or homogeneity of purified virus are the most restrictive factors in a successful project. However, recent advances in nano-crystallization techniques and efficient data collection strategies have allowed the determination of virus structures using only approximately 30 μg of virus, as demonstrated in the case of Equine rhinitis A virus [25]. Once the crystals are available and show useful X-ray diffraction patterns, data recording and processing, phase determination and subsequent model building and refinement lead to the final atomic virus structure (for details of the entire protocol, see Chap. 4 and [24]).

Thus, X-ray crystallography allows the determination of virus structure at the atomic level and has substantially contributed, and continues to contribute, to the understanding of virus biology. Milestones in structural virology (Fig. 7.1) include the TBSV structure, the first virus solved at 2.9 Å resolution by Harrison's group in 1978 [2] and the structure of SBMV solved at 2.8 Å resolution by Rossmann's group in 1980 [3]. Five years later, the structures of two viruses infecting humans, poliovirus and the common cold virus, were solved by X-ray crystallography [26, 27] (see Chap. 4). Thereafter, more virus structures have been elucidated by this technique, as evidenced by the entries in the Protein Data Bank (PDB). The number of virus structures determined by crystallography peaked in 2000. Despite a subsequent decrease, a significant contribution to the field came in 2004 from the study of bacteriophage PRD1. PRD1 is the first and only virus with an internal lipid bilayer solved by X-ray crystallography at atomic detail, revealing a general principle of virus assembly scalable to very large viruses [28, 29].

Finally, in 2010, after several years of parallel advances, both X-ray crystallography and EM stunningly converged in terms of resolution achieved (~3.5 Å) with the elucidation of the structure of adenovirus (~900 Å diameter), which is so far the largest structure to be successfully tackled by X-ray crystallography [30, 31]. Meanwhile, for individual virus protein structure determination, X-ray crystallography has become much more routine, in part thanks to efforts to systematize all aspects of the process by structural genomics/proteomics activities (see Chap. 4) [32]. Methodological improvements have also impacted whole virus studies; for instance, data collection directly from crystallization plates has made structure determination more routine and safer, and continued developments at synchrotrons, including microbeams and pixel array detectors, indicate the viability of the method, although with the capability of cryo-EM, these are niche methods at the moment [33, 34]. An area currently in development is the use of nanocrystals at highly powerful pulsed X-ray sources (the so-called X-ray free-electron lasers). Continued work suggests that structure analysis might be possible from much smaller arrays of viruses than has been possible with more conventional synchrotron sources (see Sect. 7.7.1 and Chap. 4).

Since 2014, virus crystallography has been declining as a major technique to elucidate the 3D structure of icosahedral viruses, surrendering to the advances of the cryo-EM technique, which also benefits from using a liquid sample rather than a crystal for data collection. Different public databases bring together all structural information available on studied viruses (see, e.g., the Electron Microscopy Data Bank (EMDB) https://www.ebi.ac.uk/emdb/, the Worldwide Protein Data Bank (wwPDB) https://www.wwpdb.org/ and the VIPER Database https://viperdb.org/).

7.3 Combining X-Ray Crystallography and Electron Microscopy

7.3.1 General Approach: The Beginning of Quasi-Atomic Models

Since early times, structural virologists have endeavoured to combine knowledge and methods to offer reliable insights into virus structures. Prior to 1978 (when X-ray virus crystallography first revealed the atomic details of a complete virus [2]), SAXS, EM and X-ray diffraction were usually combined to cross-validate the low-resolution information gathered by each technique. Increasingly powerful electron microscopes and the definitive establishment of protein and virus X-ray crystallography as a high-resolution technique have made it possible to combine structural information at different resolution levels, yielding *quasi*-atomic virus models capable of predicting molecular interactions within the virus or between virus particles and ligands, including virus-host cell recognition processes and virus-antibody interactions.

EM and cryo-EM allow visualization and 3D reconstruction of whole viruses and virus-ligand complexes. At the same time, X-ray crystallography can frequently provide atomic snapshots of the composite pieces of the virus, and, like pieces in a 3D jigsaw puzzle, structural virologists collate these different pieces to produce the most accurate and precise representation of the virus and its interactions with other (macro)molecules from the external environment. This process defines the procedure of fitting X-ray protein structures—either viral proteins or ligands of the virus—into density maps of viruses and virus-ligand complexes obtained by cryo-EM when the achieved resolution does not allow direct structure determination.

Despite the fundamental differences in the way electrons and X-rays interact with matter, the end result of 3D analyses in each is a 'density' map that looks surprisingly similar between the two methods and can therefore be interpreted and analysed in a similar way, using similar software. It is important to mention that NMR spectroscopy, although it cannot tackle the structure of a whole virus, has provided structures of viral proteins that have been subsequently used for fitting into EM maps (for a relevant example, see Chap. 5). NMR is also increasing its role as a hybrid method, providing constraints on possible dynamic events or conformational changes of virus structural components (see Chap. 5).

After an initial manual docking aided by computer graphics programs or an optimized likelihood-based docking of the entire viral and/or ligand protein structure into the corresponding volume of the EM reconstruction, improvement and refinement of the fitting can be performed either in real space or reciprocal space using a correlation coefficient criterion and/or a least-square minimization criterion [35, 36]. This initial rigid-body refinement protocol aims to determine a total of six parameters, a rotation matrix and a translation vector that would maximize the accuracy and precision of the fitting relative to virus map (Fig. 7.3a). Naturally, successful fitting requires the linear scale of the EM-derived density map to match that of the atomic model, which may not always be the case.

It has been estimated that this procedure for low-resolution cryo-EM maps could achieve an accuracy of 2.2 Å in a density map of 22 Å and 4 Å in a density map of 24 Å [38]; the higher the resolution of the map, the better the docking. Early software to accomplish this task includes the program X-PLOR (https://atbweb.stanford.edu/atb_publications/brunger_xplor_manual_1992.pdf), the General Averaging Program (GAP; D. I. Stuart and J. M. Grimes, unpublished software available upon request from authors), EMfit (written by Michael Rossmann), Situs (http://situs.biomachina.org/index.html) and UROX (https://sites.google.com/site/xsiebert/urox) (Fig. 7.3b). In recent years, the concept of flexible fitting has also been introduced, allowing distinct domains of the fitted structure to undergo conformational changes to match the volume restrictions imposed by low- and medium-resolution cryo-EM maps, cf. Norma (http://www.igs.cnrs-mrs.fr/elnemo/NORMA/), Flex-EM (http://salilab.org/Flex-EM/) or ISOLDE (https://tristanic.github.io/isolde/). One critical aspect of flexible fitting is how to define what is fixed and what is flexible. This can leverage prior knowledge or utilize normal mode analysis (e.g. Norma), which mathematically assigns directions of motion to structural elements of the fitting structure.

However, when the resolution of the virus map reaches 7 Å, it becomes possible to identify the fold of the most ordered part of the virus, such as major capsid proteins, and model it using templates for β-barrel or α-helical motifs [39]. Template substitution methods, like those offered by Rosetta software for modelling into X-ray and EM-derived density maps of medium-resolution (4–7 Å), have also been adopted [40] (see also below for the impact of more recent software like AlphaFold, RoseTTAFold and ModelAngelo for structure prediction and model building). In recent years, software packages traditionally devoted to crystallographic refinement, such as REFMAC and PHENIX, have been enhanced with routines for real-space refinement optimized for cryo-EM [41, 42].

Nevertheless, a major challenge persists in combining different levels of information, particularly in generating quasi-atomic models of viruses. Specifically, how

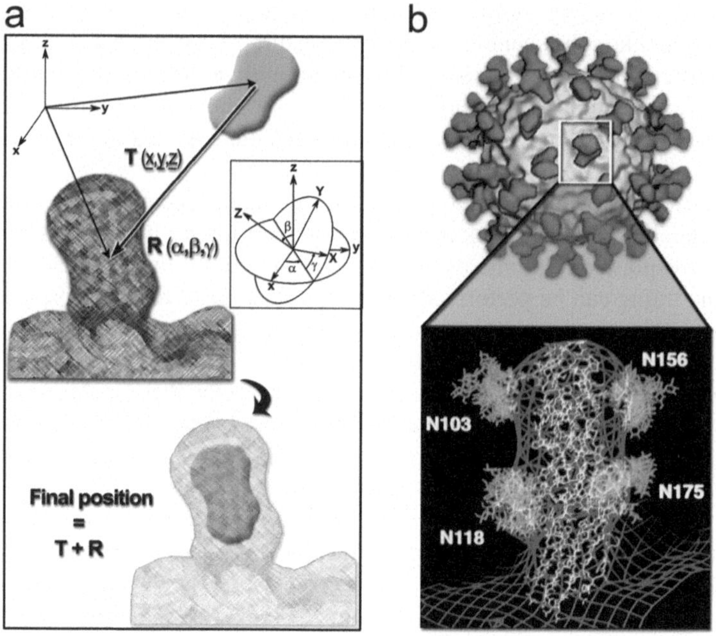

Fig. 7.3 Fitting X-ray data into cryo-EM maps. (**a**) Schematic of the fitting procedure; the crystal structure of a virus-binding molecule (*ligand*, represented here as a grey two-lobe object) is moved coarsely from its initial position into the corresponding electron density map of the virus-ligand complex obtained by cryo-EM (dark-grey mesh) by a translation (T) and a rotation (R); the final position of the object within the density is achieved by the refinement of six parameters [x, y, z and α, β, γ (Euler angles)] using either real- or reciprocal-space refinement procedures (see [35]). (**b**) *Top*, cryo-EM reconstruction of rhinovirus 16 (HRV16) (light grey) with ICAM-1 cellular receptor molecules bound (dark grey); *bottom*, the ICAM-1 X-ray structure (white stick model) fitted into the virus-receptor cryo-EM density (dark-grey mesh); the N-glycosidic groups in the ICAM-1 structure are labelled according to the asparagine residue to which each is bound. Adapted from [37] with permission

to finely balance prior knowledge relative to the achieved resolution remains a critical consideration. In Fig. 7.4, we illustrate the number of effective data points versus parameters for various methods, providing insight into what each technique and their combination can achieve. In principle, there is a need for an objective method to assess the information content of each data source and produce a model that fits the data accurately without making unfounded assumptions. We will revisit this challenge later in the discussion.

As mentioned earlier, one of the first examples of combining structural methods can be found in the study of adenovirus by Roger Burnett and his laboratory in 1986 [8]. Adenovirus is a large dsDNA virus that infects vertebrates, including humans. Burnett's interest in adenovirus stemmed from the beautiful EM images collected in the 1970s by Nick Wrigley, which showed hexamers of the major capsid protein, known as hexons, that constitute architectural elements of the virion. However,

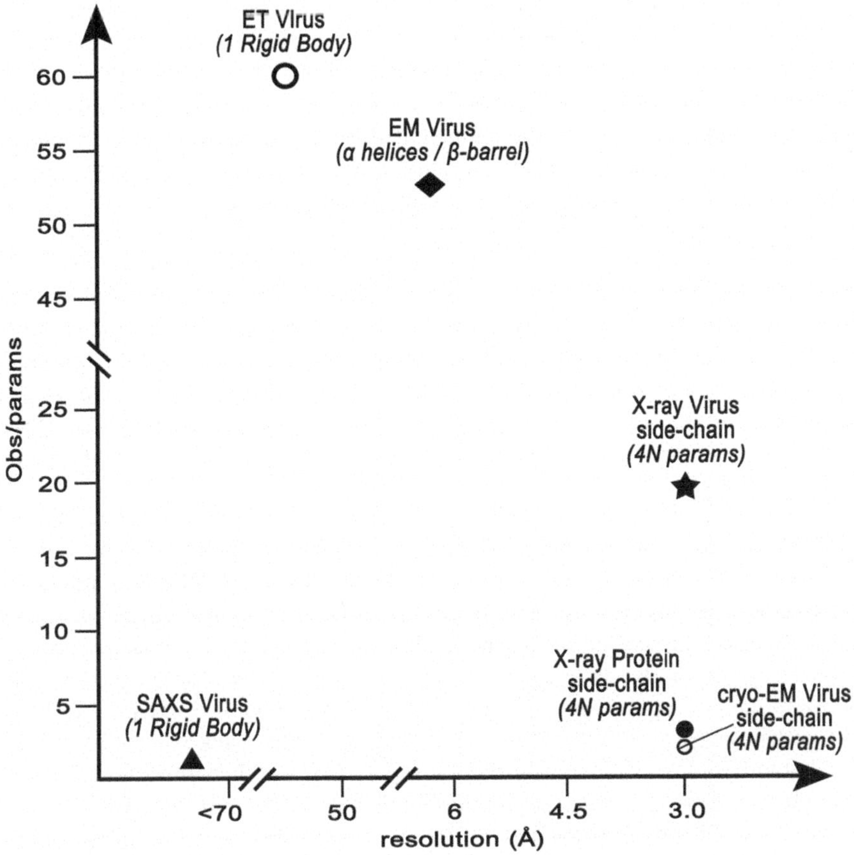

Fig. 7.4 Simplified guideline of observations-parameters ratio *per* technique. The data points were extracted by estimating the number of observations for a virus of a determined size and achievable resolution divided the number of structural elements (parameters) to be fit. For the determination of the virus structure by X-ray crystallography, we considered the lowest NCS scenario in a virus crystal. As of 2014, cryo-EM also provides virus structures at close to atomic resolution (often better than 4 Å and in some cases better than 2 Å)

X-ray structural studies on the hexon structure had already started [43], prefiguring the approach that ultimately elucidated protein-protein interactions involved in the assembly of the large icosahedral adenovirus particle. Undoubtedly, by solving the X-ray structure of the hexon at 2.9 Å resolution and integrating it into the adenovirus EM reconstruction, Roberts and co-workers pioneered the hybrid methods approach in structural virology [8]. Since then, many other examples of such hybridization of techniques have enabled the elucidation of biomedically relevant interactions, such as those governing the binding of the human cellular receptor ICAM-1 to human rhinovirus 16 (HRV-16) or those involved in the binding of neutralizing antibodies to foot-and-mouth disease virus (FMDV) (Fig. 7.3b) [37, 44].

Prior to the resolution revolution [6], *quasi*-atomic models were generated for enveloped and membrane-containing viruses. In the former case, a lipid bilayer decorated by glycoproteins surrounds a nucleocapsid, while in the latter case, the vesicle is enclosed by the proteinaceous capsid shell. One of the best structurally described enveloped virus is the human pathogenic dengue virus (*Flaviviridae* family). Years of investigation on the entire virion by EM and on its glycoproteins by X-ray crystallography have finally produced a clear picture of the dengue virus maturation, assembly pathway and cell entry mechanism [45, 46]. Additionally, by combining the crystal structure of a Fab fragment in complex with the West Nile virus (WNV) E glycoprotein and the dengue and WNV structures in complex with the Fab by cryo-EM, Cherrier and co-workers were able to establish the structural basis for antibody neutralization of infectious immature particles [47]. Similarly, *quasi*-atomic models for other biomedically relevant enveloped viruses have been generated using a similar 'fitting X-ray structures into cryo-EM maps' approach. These include viruses such as Chikungunya virus (*Togaviridae* family) and vesicular stomatitis virus (*Rhabdoviridae* family).

In the case of lipid-containing viruses, the original example in hybrid methods is PRD1 (*Tectiviridae* family), a bacteriophage that infects gram-negative bacteria and whose *quasi*-atomic model was proposed in 2001 [48, 49]. PRD1 is indeed the model system for this type of virus, having been intensively studied biochemically, genetically and structurally. Its architecture, major capsid protein fold and principle of assembly strikingly recapitulate those elucidated earlier for adenovirus, thus leading to the concept of viral lineages and virus common ancestry [8, 28, 29, 50, 51] (see below, Sect. 7.8.2) and the description of higher ranks of viral taxonomy (kingdoms) based on structural information [52, 53]. Another illustrative example, although not strictly a case of fitting X-ray structures into cryo-EM maps, is presented by the process that led to the X-ray structure solution of the marine lipid-containing bacteriophage PM2 (*Corticoviridae* family). The moderate 7 Å resolution achieved with the PM2 virus crystals (in comparison to the ~4 Å obtained for PRD1 virus) prompted the need to determine the structure of the individual components at higher resolution [19]. This 'divide et impera' approach led to high-resolution structures of the two major viral capsid components that were then docked into the low-resolution X-ray map. As a result, key aspects of viral evolution, capsid assembly and membrane morphogenesis were elucidated [19].

Furthermore, it is worth underlining that the protocol of fitting atomic structures into electron density maps can also be extended to volumes obtained by sub-tomogram averaging (see below and Chap. 3). Sub-tomogram averaging algorithms—more recently implementing a Bayesian approach—allow the classification and averaging of equivalent structures, thus improving the signal-to-noise ratio [54–56]. Early sub-tomogram application led to plausible fitting of X-ray crystal structures, as in the case of the capsid protein domain of the Gag polyprotein in HIV [57]. Nowadays, although not routinely, the sub-tomogram averaging technique can deliver density maps up to 8–4 Å resolution not only of viruses and viral components but also of cellular components such as ribosome structures in vivo [58, 59].

Finally, credit must be given to the fact that the combination of methods in structural virology has been adopted more widely in biology and defined as 'integrative modelling'. Indeed, this approach aims to combine different available sources of information—structural, biochemical, biophysical and proteomics—on a target molecule to convey a molecular picture or a unified structural view of the system under study. Good examples of this idea are embodied in the 'Integrative Modelling Platform' (IMP) (https://integrativemodeling.org/) and Haddock (https://www.bonvinlab.org/software/haddock2.4/cryoEM/), with an example being the analysis of the nuclear pore [60]. Again, how and with what weights the different sources of information are combined in an appropriate statistical framework (Bayesian, Maximum-Likelihood, etc.) that provides a built-in calibration to allow validation and statistical significance assessment of the resultant model remain an active area of research.

7.3.2 Validating the Quasi-Atomic Models

Validation of the proposed *quasi*-atomic model is a fundamental step for a reliable interpretation of molecular interactions and their usefulness to the scientific community, particularly for the rational design of drugs (Chap. 20) and vaccines (Chap. 21). Several programs (see Sect. 7.3.1) that carry out positional refinement of the initial docked structure into low-medium-resolution maps (e.g. < 4 Å resolution) provide correlation coefficients (for real-space analysis) or metrics equivalent to the crystallographic R-factor (for reciprocal space analysis), which are now compulsory for inclusion in the final publication [61]. It must be said, however, that there is no unique criterion that describes the accuracy of the model, and a multi-parameter assessment approach (atomic clashes, stereochemistry, biological meaningfulness) together with the above-mentioned statistical measures is best suited. Furthermore, if additional biochemical or biophysical information is available, such as a proven interaction between specific residues of the ligand and the virus, then it should be considered, for example, by imposing restraints on the movements of the fitted X-ray structure during refinement.

In 2010, an attempt to standardize practices in the EM community was pursued during the inaugural meeting of the Electron Microscopy Validation Task Force (https://www.wwpdb.org/task/em). A series of guidelines and recommendations were suggested with the ultimate goal of increasing the impact and significance of EM in biology and biomedicine. It is also worth mentioning the earlier efforts of the X-ray community through the X-ray Validation Task Force (https://www.wwpdb.org/task/xray) in reinforcing criteria for X-ray structure validation, where the broad principles are now accepted and required for publication. Thus, since the criteria for structure validation are continuously evolving and still in need of improvement, it is advisable, when combining X-ray with medium- or low-resolution EM-derived density maps, to maintain a conservative attitude in the biological interpretation of the resulting *quasi*-atomic model. Also, remember that the data-to-parameter ratio generally available during this hybridization process does not favour over-interpretations (see Fig. 7.4)—better safe than sorry!

Finally, nowadays, for structure determination of macromolecules (including symmetrical viruses), single-particle cryo-EM routinely achieves resolutions beyond 4 Å, necessitating the adoption of structure modelling and refinement strategies usually implemented in crystallography [62]. The 'only' difference with classical crystallographic refinement is that during refinement against a cryo-EM map, the map itself does not change while the atomic model improves and must fulfil strict model quality criteria for deposition in public repository and for publication.

7.3.3 *Phase Information 'Interplay'*

A major problem in crystallography, as previously stated, is the so-called phase problem, a methodological 'inconvenience' that prevents the retrieval of atomic positions directly from the raw data (see Chap. 4). EM images embody the phases of the object by visualizing it in real space and thus are a potential source of phase information for diffraction methods. When highly symmetrical particles such as icosahedral viruses are investigated structurally, it is possible to exploit the redundancy of information for phase refinement (see Chaps. 3 and 4). In crystallography, in the best-case scenario, when none of the symmetry axes of the virus align with the crystallographic symmetry axes of the crystal, the full 60-fold non-crystallographic symmetry (NCS) can be iteratively used to improve the starting phase assignment (in the worst-case scenario, there will be only fivefold NCS). It has been shown that the availability of 60-fold NCS can allow convergence of phase refinement starting from phases derived from a spherical object [63, 64] (for an overview on ab-initio phasing, see [65]). However, for this protocol to succeed, it is paramount to know (*i*) the exact location of the icosahedral symmetry axis relative to the asymmetric unit (AU) of the unit cell, (*ii*) the radius of the starting model [63, 64] and (*iii*) the protein region for the generation of appropriate envelopes used in the averaging and solvent flattening procedures.

In virus crystallography, apart from the first virus structures that were solved using heavy atom methods [2, 3], it has been a common practice to use as a starting phasing model a virus previously solved (unrelated or even a sphere; see above) or, if available, a low-resolution-derived EM map located within the AU of the unit cell of the crystal through the molecular replacement (MR) technique [24, 66] (see Chap. 4). It is worth noting that for the phasing of FMDV, a hybrid virus model was built from existing viruses, weighting-up the conserved structural pieces, and used as a searching model in MR—an approach that is now implemented in some standard protein MR protocols [67].

The availability of low-medium resolution cryo-EM maps has propelled the application of this phasing strategy to X-ray data obtained from crystals of biological macromolecules smaller than viruses and often less symmetrical [68–70]. This alternative approach for circumventing the crystallographic phase problem requires that there is resolution range overlap between the 3D EM data and those obtained by X-ray crystallography; pivotal are the data corresponding to the low-resolution terms and the presence of NCS. The former condition allows the object to be located within the AU of the unit cell (by molecular replacement), while the latter boosts phase refinement and phase extension procedures. To be more precise, the power of phase extension depends on the degree of oversampling of the transform of the unique portion of the object, and NCS delivers this directly, as does a very high solvent content (e.g. a thin virus shell with disordered interior).

One of the first examples in which the use of low-resolution models was successful in phasing X-ray data of a small protein (38 kDa) was inspired by studies on mengovirus and FMDV [68]. Villeret and co-workers were not only able to build the correct geometry of the oligomeric state (12-fold) using a monomeric X-ray structure of a related protein and generate low-resolution phases at 8 Å but also able to phase-extend those initial phases to the full 3 Å resolution of the X-ray data.

In principle, SAXS could also provide initial low-resolution phasing as described in the case study of nitrite reductase, a 34.4 kDa protein that forms homotrimers. However, in this case, once the SAXS envelope was correctly located in the AU of the unit cell, only phases up to 20 Å were reliable, and no further phase extension was attempted [71].

An enlightening case of phase interplay between EM and X-ray crystallography comes from the structure of the marine bacteriophage PM2 at 7 Å resolution [19] (Fig. 7.5). The initial phases for solving the entire virus X-ray crystal were derived from the cryo-EM-based PM2 structure at 8.4 Å resolution [72]. Despite the apparently minimal ~1.4 Å improvement in resolution of the X-ray over the EM data, the quality of the resulting X-ray map, thanks to the accuracy of the crystallographically measured structure factor amplitudes at the time, allowed a clearer interpretation of the different 60-fold structurally arranged components, particularly those pivotal elements located between the capsid and the lipid vesicle [19]. However, the limited 7 Å X-ray map could not unequivocally determine the fold of the major capsid protein P2 (32 kDa). Therefore, an X-ray crystallography study of isolated P2 was undertaken [74]. The failure of molecular replacement using supposedly related X-ray structures as search models and the difficulty of obtaining heavy atom-soaked crystals or suitable selenomethionine (SeMet)-derivatized X-ray data (see Sect. 7.4) required the use of the low-resolution electron density corresponding to a P2 trimer extracted from the X-ray averaged map of PM2 as the search model. The location of the two P2 trimers within the AU of the unit cell by molecular replacement led to the determination of initial phases at 7.6 Å resolution, which were then extended to 2.5 Å by means of NCS and solvent flattening [73, 75].

In conclusion, although the examples mentioned above do not yet represent routine procedures, the experience gained in virus phasing has made possible to implement an additional structure determination tool for protein crystallography.

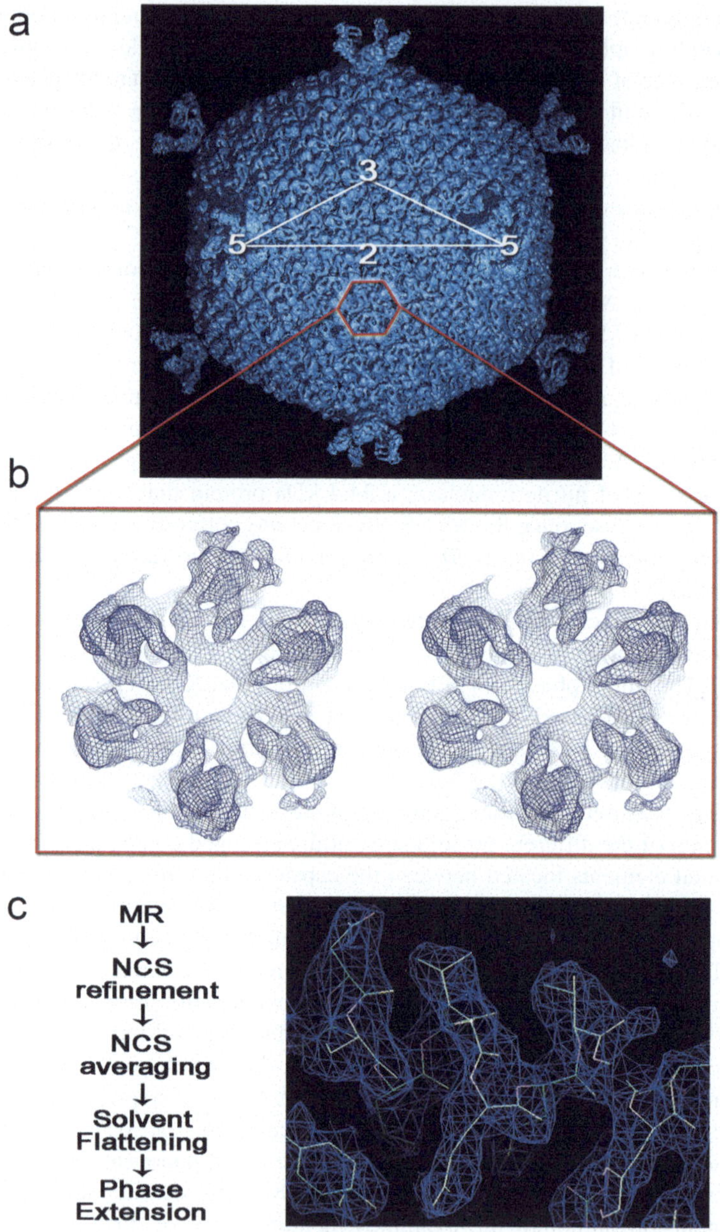

Fig. 7.5 Phasing with low-resolution phases. (**a**) X-ray-derived map of the lipid-containing bacteriophage PM2 [19] (~600 Å in diameter), which was phased using the corresponding EM map [72]; 2-fold, 3-fold and 5-fold icosahedral symmetry axes in the capsid are labelled with numbers; a trimeric capsomer of the major capsid protein P2 is outlined in red. (**b**) Enlarged stereo view of the extracted electron density at 7 Å corresponding to the P2 trimer used in the molecular replacement (MR) protocol. (**c**) *Left*, workflow for phasing X-ray data using the low-resolution

7.4 Dissecting Virus Structures by Integrating Biochemical, Genetic and Biophysical Tools

Virus structures can be tackled by complementing not only the structural techniques mentioned above but also methods that involve biochemical and genetic manipulation of the virion. It is clear that the prerequisite for high-quality structural data is the quality of the original biological material (though cryo-EM can potentially use computer-sorting methods to select high-quality particles). Traditional virus purification methods rely on their sedimentation behaviour (such as in rate zonal centrifugation) or virus particle density (equilibrium centrifugation). These methods are often combined with differential centrifugation, concentration methods (e.g. filter devices) or precipitation (e.g. using various polyethylene glycols or ammonium sulphate). New approaches have been developed to purify large macromolecules like viruses, based on flow field-flow fractionation or using large-pore-size monoliths or agarose matrix in chromatography, enabling the analysis and separation of a wide range of virus samples [76–79]. The addition of biochemical and genetic modifications allows for the production and study of different kinds of virus material, from wild-type and mutant virus infections to quantitative virus particle dissociation experiments, or using recombinant protein expression systems. In the following sections, we will illustrate some examples where the tailored use of biochemical, genetic and biophysical techniques has provided insightful views into virus structure.

7.4.1 Classical Antibody Labelling Methods

Immuno-EM serves as a powerful tool for localizing specific proteins that are asymmetrically distributed in virus particles. This method relies on the specific recognition of a target protein (antigen) exposed on the particle surface by a primary antibody. Subsequently, a secondary molecule recognizing the primary antibody is introduced, such as gold-conjugated protein A, which binds with high affinity to IgG molecules. The gold particles generate high contrast under TEM (see Sect. 7.2.1), facilitating the localization of the specific protein within the virus structure, although detailed structural information is not preserved. Using secondary binders conjugated to gold particles of different sizes, different antigens can be co-localized within the same sample. Before labelling, virus particles are usually fixed using cross-linkers, with glutaraldehyde being one of the most popular choices. The toolbox of cross-

Fig. 7.5 (continued) density map in (b) as a search model (NCS, non-crystallographic symmetry); *right*, experimental electron density map at 2.5 Å for the P2 major capsid protein resulting from MR, followed by phase refinement, extension and solvent flattening. Reproduced from [73] with permission

linkers and chemical probing protocols has expanded, enabling exploration of new approaches in virus biochemical manipulation [80]. Furthermore, virus particles can be loosened, e.g. by gently treating them with non-ionic detergents such as Triton X-100 during sample and grid preparation, to enhance the accessibility of antigens to primary antibodies (e.g. see [81]).

Typical issues in immune-EM, such as nonspecific labelling or background noise, can be mitigated by implementing appropriate controls and adjusting the amount of primary antibody. The most reliable conclusions, however, emerge when immunogold-labelling EM data is complemented with other approaches, as demonstrated by studies on the dsDNA bacteriophage PRD1. Combining immunolabelling with investigations involving PRD1 mutants verified the presence of the unique vertex structure crucial for genome packaging [81, 82]. Subsequently, using cryo-EM and cryo-ET techniques, the same unique vertex was demonstrated to (*i*) facilitate viral genome packaging into the procapsid and (*ii*) serve as the vertex from which a proteo-lipidic tube protrudes for delivering the viral genome during infection onset [83, 84].

7.4.2 Mutagenesis Studies

Classical mutagenesis techniques using shotgun mutagens, induced by ultraviolet light (UV) or N-methyl-N-nitro-N-nitrosoguanidine (NTG) and, especially, site-directed mutagenesis are powerful tools for generating virus mutants. These methods allow for the creation of specific virus mutants aimed at identifying proteins and investigating their functions. Mutants can be generated by introducing a stop codon within the target gene (nonsense mutants) or by deleting part of or the entire protein-coding region (deletion mutants). In cases where mutations are lethal, rescue systems such as suppression or complementation are required, limiting the use of virus mutants to those infecting hosts for which a genetic system has been established. Genetic engineering of obscure microorganisms isolated from the environment poses additional challenges [85]. Bacteriophage PRD1 infects gram-negative bacteria like *Escherichia coli* and *Salmonella enterica*, making it amenable to established genetic tools. PRD1 serves as a notable example where numerous suppressor-sensitive mutants have been created, proving invaluable in assigning functions to viral proteins and elucidating the architecture of the packaging complex through cryo-EM studies [84]. Furthermore, the genetic removal of the PRD1 flexible receptor-binding proteins from the virion vertices was crucial for obtaining diffracting crystals, leading to the determination of the X-ray virus structure at 4 Å resolution [28, 29].

Moreover, mutagenesis has facilitated the insertion of green-fluorescent protein (GFP) into large and complex viruses such as herpes, vaccinia or Zika viruses—an important tool for studying cell interactions or virus localization within the cells (see Chap. 8). Careful selection of multiple tags also provides structural information, such as from fluorescence resonance energy transfer (FRET) and correlation spectroscopy methods. An early example comes from research on influenza virus polymerase, revealing interactions within the cell [86].

Detailed structure-function relationships in virus particles at the amino acid residue level can be explored by individually replacing specific residues based on high-resolution structural information. Typically, residues are substituted with alanine, removing the side chain (beyond the Cβ atom) and the intra- or intermolecular interactions it establishes with other residues, thus avoiding additional groups and non-native interactions and minimizing the probability of altering the main-chain conformation of the viral proteins. For example, systematic alanine-scanning mutational analyses have elucidated the structural and functional relevance of protein-protein and protein-nucleic acid interactions in virus particles such as FMDV and the parvovirus minute virus of mice [87]. In addition, the structure determination of some single-amino acid mutants is providing high-resolution insights into the structural basis of physical properties, assembly, conformational dynamics and disassembly of virus particles. Examples include research on rhinovirus and poliovirus [88].

7.4.3 Quantitative Dissociation Studies

The controlled removal of specific proteins from virus particles is highly valuable for structural and structure-function studies. Using this biochemical quantitative dissociation approach, several viral proteins have been accurately assigned to their positions within the virion. Interactions between capsid proteins and underlying membrane components can be disrupted using treatments like urea or guanidine hydrochloride. For example, studies on bacteriophages PM2 and PRD1 employed these methods, revealing membrane-associated proteins and releasing major capsid proteins [89, 90]. The released proteins were then further purified using standard biochemical methods for protein crystallography studies [50, 73, 90]. Quantitative protease treatments of virus particles can also identify structures protruding from the virion surface. In the case of membrane-containing viruses, different detergent treatments can selectively extract various membrane-associated proteins from the virus particle. For instance, in the case of the internal membrane-containing PRD1 virus, integral and peripheral membrane proteins have been successfully separated from each other [91].

One example of a biochemical quantitative dissociation analysis involves the marine icosahedral dsDNA bacteriophage PM2 (Fig. 7.6a). For PM2, several unique subviral particles have been distinguished based on their sedimentation properties in a sucrose gradient [72, 90]. A combination of biochemical, cryo-electron microscopic and X-ray crystallographic approaches has enabled the visualization of the entire PM2 virion structure at subnanometric resolution and facilitated the determination of its major structural protein structures (Figs. 7.5 and 7.6a) [19, 72–74]. Additionally, archaeal icosahedral membrane-containing viruses such as SH1 and HHIV-2 (*Sphaerolipoviridae* family), archaeal pleomorphic viruses (*Pleolipoviridae*

a

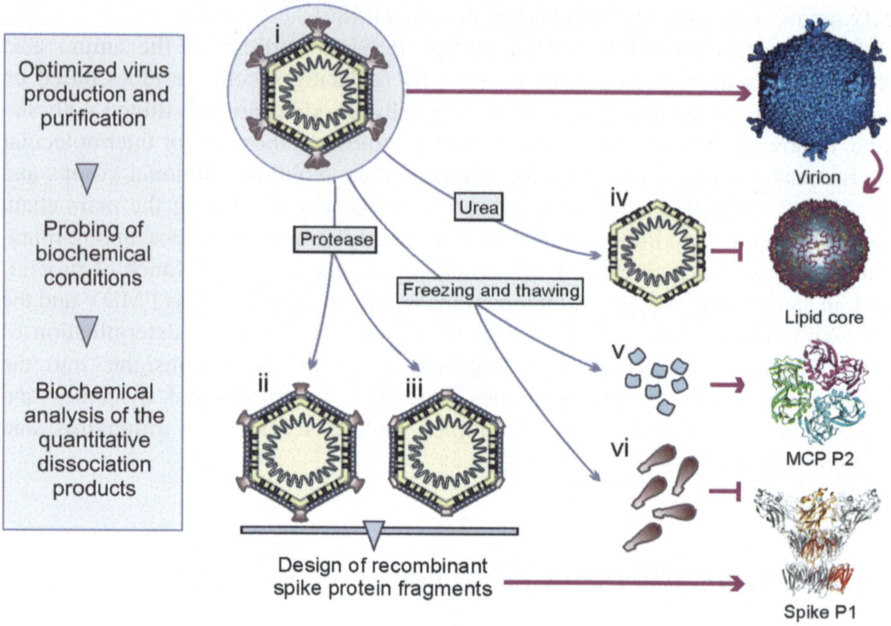

b

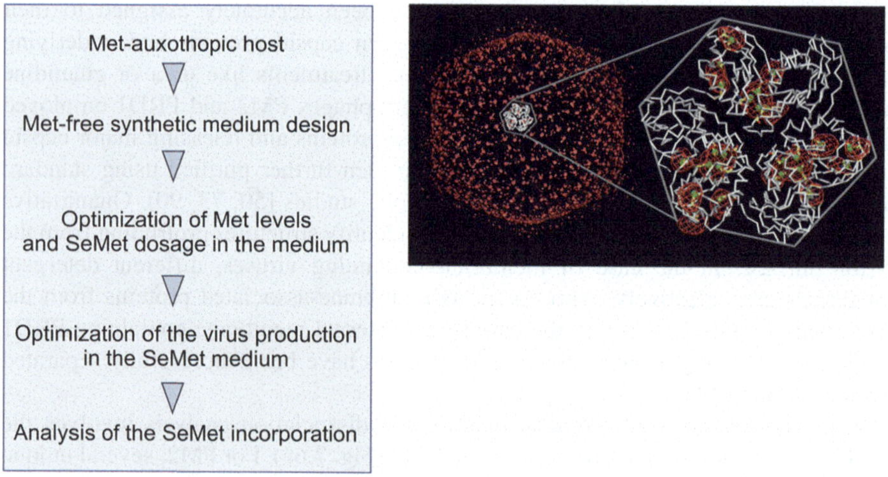

Fig. 7.6 Schematic workflow for biochemical probing. (**a**) *Left*, workflow in a study on the quantitative dissociation of virus particles. *Right*, schematic representation of the controlled dissociation of bacteriophage PM2 [92] used to solve the structures of virion components. (i) Highly purified and infectious virions used to solve the virion structures by cryo-EM [72] and X-ray crystallography [19]; (ii) proteinase K-treated virus particles devoid of the distal receptor-binding domain of the spike protein P1; (iii) bromelain-treated virus particles devoid of the two outermost domains of the spike protein P1; (iv) urea treatment exposes the internal lipid core particle

family) and *Thermus* bacteriophage P23–P77 (*Matsushitaviridac* family) are other viruses for which quantitative dissociation analysis has provided valuable information and contributed significantly to structural analysis [93–95].

7.4.4 Selenomethionine Incorporation as a Labelling Tool

Efficient selenomethionine (SeMet) labelling of protein species within bacteria is a standard technique but less so for proteins expressed in mammalian cells [96]. Even more challenging is the SeMet labelling of viruses, which, combined with X-ray crystallography, has proven valuable in structural virology. To our knowledge, only two efficient methods for SeMet labelling of complex viruses have been reported, and their roles in interpreting the electron density maps of PRD1 and PM2 were significant [19, 28, 92, 97]. The PM2 labelling method relied on a methionine auxotrophic host (Fig. 7.6b), whereas successful PRD1 SeMet incorporation was achieved with a methionine prototrophic host grown in minimal medium and pulsed with SeMet at a late stage of infection [97]. In addition to obtaining phase information in X-ray crystallography, SeMet labelling combined with difference Fourier maps can facilitate the interpretation of protein folds and even the identification of specific protein species [19, 98]. Using SeMet-labelled and native virus X-ray data, it was possible to assign electron density to all ordered protein components in the virion based on their amino acid sequence [19, 92].

7.4.5 Some Old and New Biophysical Techniques to Study Viruses

Together with the above techniques, other biophysical methods for studying virus particles in solution, including fluorescence spectroscopy, circular dichroism (CD), hydroxyl radical protein footprinting or limited proteolysis coupled to mass spectrometry (MS) (see Chap. 6), differential scanning calorimetry (DSC), isothermal

Fig. 7.6 (continued) (membrane proteins, lipids and the genome); (v) and (vi) freezing and thawing of the virions release (v) the trimeric major capsid protein (MCP) and (vi) the monomeric spike proteins. *Far right, top to bottom*: X-ray crystallographic structure of a virion and its lipid core particle, MCP P2 trimer and the pentameric vertex-spike complex of protein P1. Adapted from [19] with permission (**b**) *Left*, workflow of the SeMet labelling of bacteriophage PM2. *Right*, SeMet difference Fourier map for the entire SeMet-labelled PM2 virion contoured at 4σ (red mesh) with a single trimer of the MCP P2 outlined in white; an enlarged view of the P2 Cα backbone is shown, with the methionine residues depicted as balls-and-sticks (the incorporated selenium atoms are shown as green spheres). Reproduced from [92] with permission titration calorimetry (ITC) and mass photometry (MP) may be used to study different aspects of virus particle stability and conformational dynamics and relate these properties to virus structure. DSC and ITC, in particular, may contribute respectively to assess the thermal stability and conformational transitions of virus particles and to study virus binding to the cell, viral cell entry process and genome release. Several studies have been carried out using DSC and ITC on viruses such as phage λ [99]. Recently, although still a niche technique, MP has been shown to successfully quantify empty, partially filled and full adeno-associated virus vectors [100]. Changes in secondary structure of the capsid proteins in viral particles have been followed by far-UV CD; intrinsic fluorescence of viral particles or extrinsic fluorescence by binding dyes to the viral nucleic acid exposed on capsid disassembly has been used to follow capsid conformational rearrangements and determine the thermal or chemical stability of viral particles against dissociation (see Chap. 6).

An application of the thermofluor assay used to assess protein stability before crystallization has been adapted for high-throughput screening of thermal stability of viruses. This new application, termed the PaSTRy assay, involves the simultaneous use of two distinct fluorescent dyes, one with affinity to hydrophobic residues and the other to the genome [101]. Upon heating, hydrophobic residues become exposed, and genome release is simultaneously monitored by fluorescence intensity, rapidly assessing virus stability.

Finally, developments in MS technique have shown that it is possible to investigate very large protein complexes such as viruses by native MS (see Chap. 6). This formidable advance has opened the way to the analysis of virus disassembly processes and assembly principles traditionally considered out of range for this method [102, 103].

7.5 Combining Electron Microscopy and Electron Tomography

7.5.1 General Approach

The understanding of the architectural and molecular organization of enveloped and complex viruses is a current challenge in structural biology (Chap. 12). Traditional approaches such as EM and, more stringently, X-ray crystallography require sample homogeneity in terms of structural organization and size, a premise that fails when studying those viruses that do not possess a regular shape or composition, the so-called pleomorphic viruses. Indeed, such viruses often come in several sizes and shapes, and until recent years, the unavailability of a structural technique capable of tackling such viruses has left them on the margins of structural investigation.

The advent of ET and cryo-ET has allowed these 'difficult' viruses, which lack regular symmetry, to be tackled [104, 105]. With this technique (for details and some examples, see Chaps. 3 and 12), viruses can be studied as individual particles, and those structural elements conserved and redundant within the virus structure can be averaged, thus improving the signal-to-noise ratio and overcoming typical problems inherent to the technique such as anisotropy and data incompleteness due to the missing wedge [104]. To enable functional interpretation of cellular tomograms without sub-tomogram averaging, IsoNet, a deep learning-based software package, has been recently developed to iteratively reconstruct the missing wedge information and increase the signal-to-noise ratio, using knowledge learned from raw tomograms [106].

Thus, if a virus possesses distinct structural levels, some with no symmetry (e.g. the envelope) and others with symmetry (e.g. the nucleocapsid), it is in principle possible to combine the cryo-ET 3D reconstruction with other symmetry-imposed and/or symmetry-averaged structural elements. These latter elements are usually at higher resolution and can be obtained either by sub-tomogram averaging procedure within the same tomographic data collection (e.g. see [55, 56, 107] and websites https://www.dynamo-em.org/w/index.php?title=Main_Page; https://pytom.sites.uu.nl/) or by other structural methods such as cryo-EM, X-ray crystallography or NMR. This combination will depict the architectural complexity of the virus at different resolution levels. ET is also very powerful in studying viruses in the cellular context, particularly during viral entry and replication (see below and Chaps. 3 and 8). In the following section, we will provide a few examples of the combination of cryo-ET and SPA cryo-EM techniques and illustrate one of the earliest examples of the smart use of tomography in assessing particle symmetry [108].

7.5.2 Case Studies

One of the most elegant early examples of combining EM techniques to depict a complex virus is the cryo-ET study of herpes simplex virus (HSV) [109]. In this work, the regular icosahedral nucleocapsid was separated from the enclosing pleomorphic envelope, which was decorated with tegument and glycoproteins essential for infection. Nucleocapsids were extracted from the tomograms, 60-fold averaged (similar to the standard icosahedral 3D reconstruction approach) and reinserted into the whole virus 3D reconstructed tomogram (Fig. 7.7a), revealing contacts between the nucleocapsid and the tegument that were not visible in individual tomographic reconstructions.

The combination of ET and EM is not limited to viruses with inherent symmetry mismatches. Several situations during the infection cycle can lead to symmetry loss even in highly symmetric viruses. Picornaviruses, for instance, are small, icosahedral, non-enveloped viruses extensively studied for several viral processes such as receptor attachment and RNA translocation. By using decorated liposomes with the poliovirus receptor CD155, researchers were able to image the receptor bound to poliovirus using both single-particle reconstruction and cryo-ET [112].

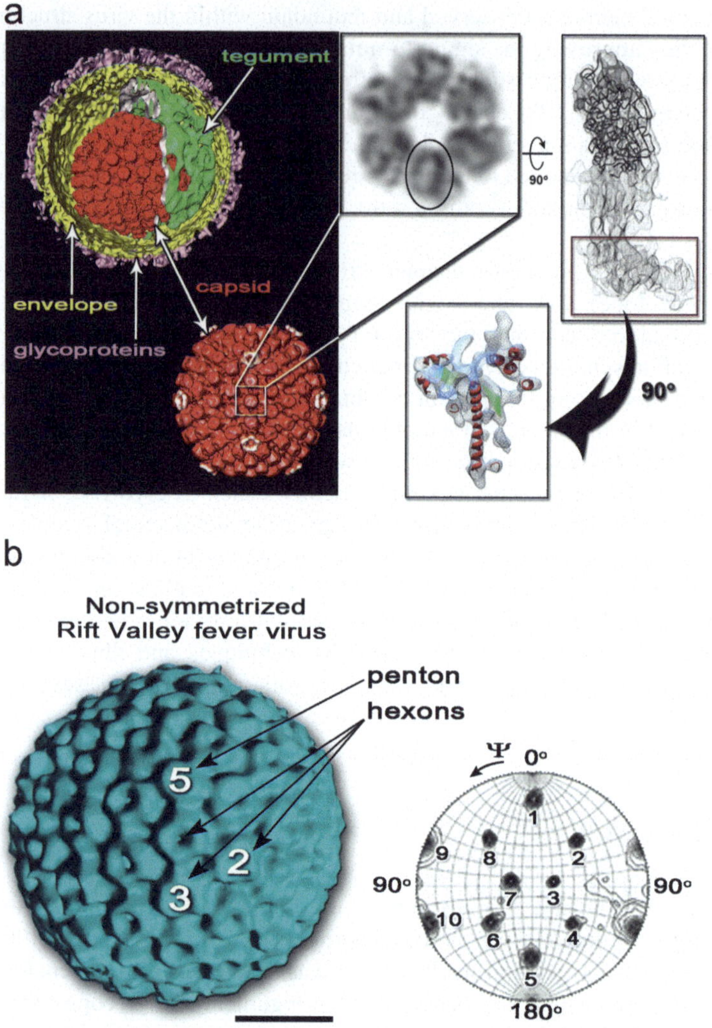

Fig. 7.7 Using X-ray tools in ET. (**a**), *Top left,* herpes virus electron cryo-tomography reconstruction with the different structural elements [109]; below, herpes virus nucleocapsid reconstruction obtained by sub-tomogram averaging combined with icosahedral averaging (courtesy of K Grünewald, CSSB Centre for Structural Systems Biology, Hamburg); one capsomer sitting on a twofold icosahedral axis and composed by six copies (top centre) of VP5 protein is outlined; *centre top,* cryo-EM-derived density corresponding to this capsomer; *top right,* a VP5 monomer composed of three domains (upper, middle and floor domain); the crystal structure of the upper domain is shown as a Cα ribbon model (dark grey) fitted into the corresponding cryo-EM density; no X-ray structure exists for the middle domain; the floor domain is delineated by a red rectangle. Figure adapted from [110] with permission. *Bottom right,* cryo-EM density corresponding to the floor domain; secondary structure templates have been fitted to suggest structural homology with the gp5 capsid protein of bacteriophage HK97. Figure adapted from [111] with permission. (**b**) *Left,* electron density reconstruction of RVFV by single-particle ET [108]; capsomers and icosahedral two-, three- and fivefold symmetry axes are marked; *right,* stereographic projection of the

For retroviruses, cryo ET has provided structural insights into the human and simian immunodeficiency viruses HIV/SIV envelope spike proteins gp120 and gp41, which are responsible for binding to the cellular receptor and fusion with cellular membranes, respectively. The gp120/gp41 complex is randomly distributed on the virion surface, making it difficult to study [113, 114]. Sub-tomogram averaging of thousands of individual spikes and imposition of threefold symmetry yielded a high-quality 3D map, allowing the structure of the SIV-gp120 core previously solved by X-ray crystallography to be docked into it [115]. Additional studies involving bound receptor and antibody have elucidated key aspects of spatial arrangement during cell and antibody recognition and conformational changes [116]. Moreover, beneath the HIV/SIV envelope, during maturation and after cleavage of the Gag polyprotein, the capsid domain forms a cone-shaped core shell organized into a hexagonal lattice. Tomographic images of this lattice, followed by sub-tomogram averaging of the hexagonal building blocks, enabled the identification of individual protein domains and the fitting of the corresponding X-ray crystal structure [57, 117].

Cryo-ET has also provided a 3D representation of influenza virus, which exhibits noticeable pleomorphism [118]. The two different surface proteins could be directly distinguished from the tomograms, and molecular shape comparison with crystal structures of haemagglutinin (HA) or neuraminidase (NA) facilitated mapping of the individual spikes in the tomograms.

During the COVID-19 pandemic, the detailed architecture of intact SARS-CoV2 virus with the native spike protein S structure was determined through sub-tomogram averaging, combined with the spike protein model derived from high-resolution SPA cryo-EM [119–122]. Moreover, in situ cellular studies through cryo-ET provided insightful information on the budding and structure of intracellular virions, along with the virus cellular egress mechanism, including analysis of the spikes on released virions using averaging techniques [123–125].

Tomography is also useful for assessing virus particle symmetry in cases where no prior evidence exists. Early analysis of negatively stained samples of several bunyaviruses revealed a wide diversity in size, morphogenesis, structure and surface characteristics. One notable member of this virus family is Rift Valley fever virus (RVFV), a major pathogen affecting both animals and humans in Africa. Cryo-ET provided the first useful 3D reconstruction of RVFV particles by reconstructing 46 individual particles with similar overall size and shape and then averaging them without imposing symmetry [108]. The averaged map underwent Fourier transformation, and the derived structure factor amplitudes were used to calculate a self-rotation function—a routine procedure used in X-ray crystallography to detect

Fig. 7.7 (continued) Kappa = 120° section for the self-rotation function of the tomographic reconstruction, revealing the presence of 10 threefold axes and thus demonstrating the icosahedral nature of the RVFV virion; scale bar represents 25 nm (original images courtesy of SJ Watowich, University of Texas Medical Branch)

redundant structural elements within the crystal's AU (see Chap. 4 and Fig. 7.7b). This analysis identified five-, three- and twofold axes, conclusively assigning icosahedral symmetry to the virus. Subsequently, imposing 60-fold icosahedral symmetry further improved the resolution of the 3D reconstruction from 75 Å to 61 Å, revealing the arrangement of capsomers and the symmetry of the lattice (triangulation number *pseudo-T* = 12). These findings laid the groundwork for higher-resolution structural studies achieved through systematic single-particle sub-tomogram averaging of 267 particles, resulting in a virus map at 20 Å resolution. This significant improvement enabled detailed analysis of the heterodimer formed by the G_N and G_C glycoproteins that decorate the envelope [126].

7.6 From Viruses in Solution to Virus in Cells

7.6.1 Virus Structures in the Cellular Context

Viruses have traditionally been observed within infected cells using either fluorescently labelled viruses and light microscopy (resolution worse than 200 nm) or through cell-sectioning techniques for subsequent imaging in an electron microscope [9] (see Chap. 8). In the latter case, different cell and sample preparation techniques have been implemented and optimized to maintain virus and cell in close to native conditions. These efforts have been complemented by advances that have made routine acquisition of tomographic data possible. Cellular tomography thus provides depth information at higher resolution compared to the serial addition of ultrathin 2D sections (for details and examples, see Chaps. 3 and 8).

One of the major difficulties in interpreting the structure of virus-infected cells is the confusion caused by macromolecular crowding within the cell. To provide an atlas of the different stages of viral infection—tracking how viruses enter the cell, replicate, assemble and exit—different time points are structurally investigated during infection. For each time point, tomograms are collected, and combined computational techniques, from segmentation to template matching, are employed to detect recurrent cellular or virus structures [107, 127]. Sub-tomogram averaging of equivalent structures increases the signal-to-noise ratio and allows the improved structure to be integrated back into the cellular landscape, as demonstrated by previously mentioned structural studies on SARS-CoV-2 [120, 125]. This approach can use virus structures obtained at higher resolution by other methods such as cryo-EM or X-ray crystallography, providing more accurate information about the geometrical arrangement and spacing between virus particles and cellular structures. This hybrid approach helps to fill in information inevitably missing from grid-based ET methods (see Chaps. 3 and 8) and bridges the resolution gap between 3D cellular tomography and other methods such as single-particle EM and X-ray crystallography.

7.6.2 Cryo-Super-Resolution Microscopy Coupled with Soft X-Ray Tomography

While pursuing the reductionist approach, the analysis of biological functions from whole organisms to cells and subsequently to single molecules has been a longstanding aspiration for structural scientists over the past decades. Structural biology and virology have increasingly focused on dynamic cellular and virus-cell interaction processes. Achieving this requires a time-resolved approach to visualize different size levels, and correlating them in time and space remains non-routine.

Cryo-structured illumination microscopy (cryo-SIM) (Chap. 8) can generate 3D images of volumetric samples with superior resolution compared to any other confocal microscopy technique, with a maximum depth penetration exceeding 10 μm [128]. This achievement is made possible by using low light doses (10–100 W/cm^2 laser power) and short image acquisition times (20–100 ms per single exposure), minimizing sample heating [14]. In essence, super-resolution is achieved employing patterned light that interacts with fluorophores (e.g. within cells and tagged proteins), producing low-frequency interference fringes (the Moiré effect) that carry structural information beyond the diffraction limit. This information is collected for each Z plane across the cellular target and computationally recombined to produce an image with up to double resolution in all three dimensions [128]. The scope of cryo-SIM at cryogenic temperatures lies in its integration with correlative ultrastructural imaging (Fig. 7.8a, top). Consequently, cryo-SIM has been implemented at the Diamond Light Source (UK) on the correlative cryo-imaging beamline B24 and is currently being set up at the ALBA synchrotron on beamline BL09-Mistral (Spain). Synchrotron facilities can generate soft X-rays in the water window (wavelength 2.3–4.4 nm), and their differential absorption by oxygen and carbon atoms (10 times lower for O than C) provides contrast for cellular tomography imaging. Virus X-ray tomography has demonstrated pioneering applications using model systems such as the virus life cycles of vaccinia virus, herpesvirus, and, more recently, reovirus and SARS-CoV-2 [14, 125, 129, 130]. These studies illustrate that soft X-rays can offer valuable 3D structural insights into virus maturation events at subnanometer resolutions (50–25 nm) in reasonably thick (5–8 μm) cells and compartments, including the cell nucleus.

7.6.3 Cryo-Focused Ion Beam Milling and Correlative Microscopy

High-resolution EM studies of biological objects within cells present several technical challenges due to the limited penetration depth of electrons (200–300 nm). Thus, one of the major challenges in cryo-TEM is the thickness of biological samples including cells. Mammalian cells vary greatly in thickness, typically ranging from 1 to 100 μm. To image thicker biological samples, techniques such as serial

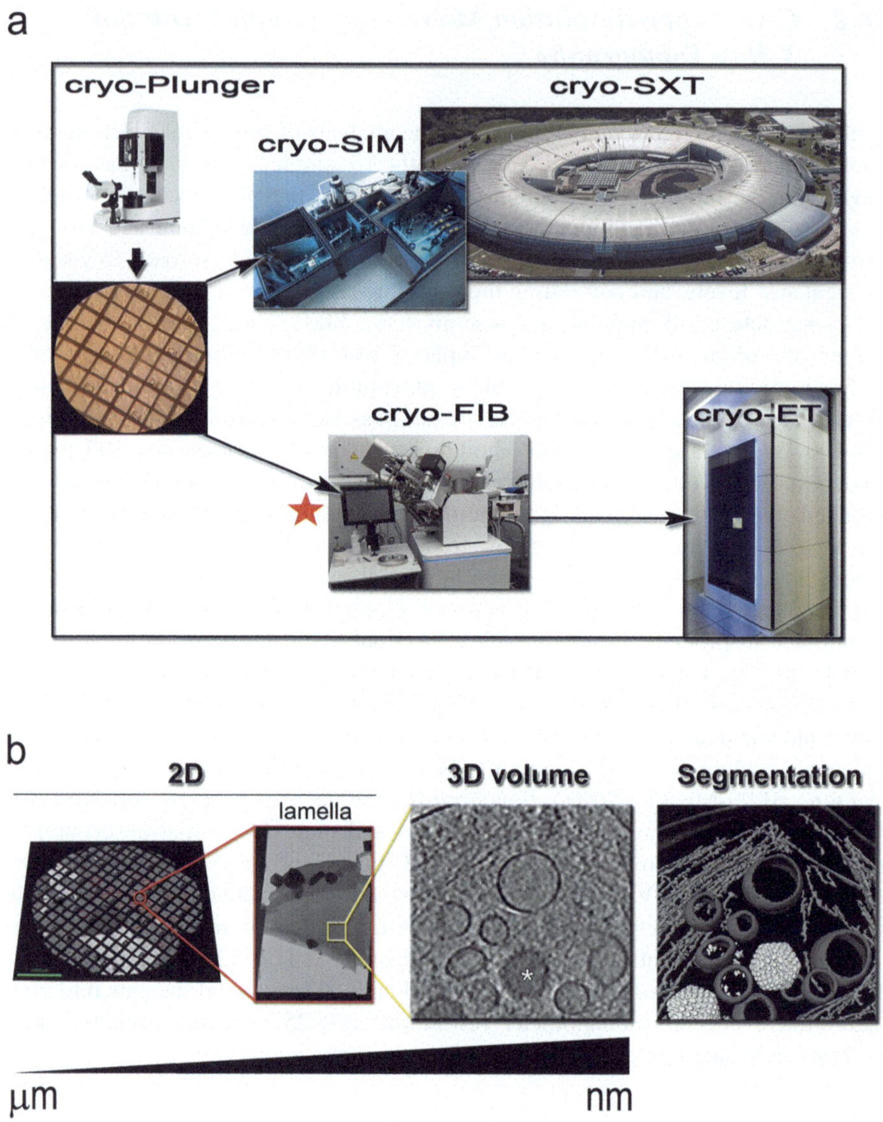

Fig. 7.8 Workflow in cellular tomography. (**a**), *Top*, schematic of the methodological steps involved in performing correlative cryo-super-resolution structural illumination microscopy (cryo-SIM) and cryo-soft X-ray tomography (cryo-SXT). Cells vitrified using a plunge freezing device (top left) are identified on a grid and targeted by cryo-SIM; then the grid is transferred to a cryo-holder, and the same cell and the corresponding region is targeted by cryo-SXT at synchrotron facilities. *Bottom*, schematic of the steps for cellular cryo-FIB/cryo-ET. Targeted cells are milled to produce lamellas in cryogenic conditions that are then transferred to a high-end electron microscope for cryo-ET data collection. The red star indicates that after vitrification and prior milling, cells can be imaged using a fluorescence microscope (offline or nowadays inline within the cryo-FIB microscope) to locate the region of interest for milling. (**b**), Left to right: detection of the cell of

cell sectioning, slice-and-view or cryo focused ion beam (cryo FIB) milling are necessary to thin down the sample (Chap. 8). FIB milling, originally developed in materials science and more recently adapted for biological sciences, involves preparing thin samples for TEM studies (Fig. 7.8a, bottom). The operation of a focused ion beam system involves ablating or milling a sample using a highly focused ion beam, usually gallium or in plasma FIB xenon, argon, oxygen or nitrogen [131]. Vitrified grids containing the cells of interest are deliberately oriented so that the angle of incidence of the ion beam is nearly parallel to the grid surface, resulting in a final lamella that is parallel to the grid and thin enough to be manipulated with conventional tilting strategies during cryo-ET studies (Fig. 7.8b).

Correlative microscopy allows the integration of data across the microns to nanometers range, using fluorescence light microscopy and EM [130, 132, 133] (see Chap. 8); however, from the point of view of combined structural strategies, it is important to underline that the aim is to identify cellular regions of interest using specific chromophore labels in the light microscope and track biological events before stopping the clock and imaging the same region at much higher resolution using EM. In this way, virus-cell interactions are reconstructed in 3D and framed in time. A critical step is finding the same cellular region once the cell is within the electron microscope. To this purpose, a common reference system between the light and electron microscopes must exist. Technological advances in robotics and microscopy cryo-holders have made possible data collection and putting forward a general workflow of the procedure, pioneered by Plitzko [132]. Correlative microscopy for 3D tomographic reconstruction is moving at a fast pace, and benefits of its routine use are enormous in studying virus-infected cells [134]. Early relevant contributions come, for example, from the study of retroviruses where it has been possible to identify HIV particles attached to cell membranes and to show the release of mutant HIV cores into the cytoplasm of host cells [135, 136]. More recent ones include the elucidation of the infectious cycle of HSV-1 within the host cell and the unveiling of SARS-CoV-2 assembly and egress [125, 137]. The integration of correlative microscopy with cryo-ET will benefit the study of those viruses that are characterized by low replication and infection efficiencies and have therefore escaped direct visualization.

Fig. 7.8 (continued) interest in 2D, e.g. using SEM and/or light microscopy (marked by a red circle on the grid); cryo-FIB milling and lamella generation and identification of the region of interest within the lamella (indicated by a yellow square). This region is then targeted for tilt-series data collection and subsequent 3D reconstruction and segmentation of the different structural elements

7.7 Emerging Hybrid Methods

7.7.1 *Developing 3D Imaging Techniques: XFEL and Ptychography*

In recent years, virus crystallography has also made strides towards in vivo structural analysis. Crystals of large viruses are extremely fragile, prompting efforts to grow intracellular crystals for structure determination—a significant challenge in crystallography (see also Chap. 4). The inspiration behind this crystallogenesis strategy stems from observations that protein crystallization can occur in vivo, alongside the formation of paracrystalline arrays of virus particles within infected cells [138–140]. However, further advances in synchrotron radiation technology, particularly in delivering high-brilliance synchrotron light, are needed to enable X-ray data collection from these in vivo crystals. Optimism for imaging is fuelled by tests of the X-ray free-electron laser (XFEL) on viruses, which, although technically successful, provided only very low-resolution information (< 200-Å) due to the limited number of incident photons [141, 142].

One of the most recent and debated methodological advancements in 3D image reconstruction algorithms, potentially applicable for initial low-resolution phasing, is ankylography [143]. Essentially, ankylography involves capturing the diffraction pattern generated by a coherent soft X-ray laser beam hitting the object of interest (not necessarily crystalline) on a curved detector. Ankylography proposes an advantage over XFEL crystallography or X-ray tomography by utilizing a 2D spherical diffraction pattern to reconstruct the full 3D structure of the object. A proof-of-principle for this reconstruction method was demonstrated with poliovirus at 20–30 Å resolution [143]. Despite its potential impact, ankylography faces a fundamental challenge: capturing a thin spherical shell of data in reciprocal space is a limited approximation for acquiring the full 3D volume needed to produce a high-resolution reconstruction [144].

As a variant of coherent X-ray scattering, ptychography—an imaging technique proposed in the 1960s by Hoppe [145] aiming to solve the phase problem through interference phenomena—has recently resurfaced as an EM technique potentially capable of achieving super-resolution (~5 Å) [146]. The principles of ptychography have also been adapted for X-ray tomography in life science experiments [147]. It will be interesting to observe the achievable resolution of ptycho-tomographic 3D reconstructions, especially with advancements in next-generation detectors and synchrotrons that are speeding up the methodology. Recently, a version of ptychography based on electrons, cryogenic electron ptychography has been developed and applied to viruses, yielding spatial information down to ~33 Å [148].

7.7.2 The Advent of AlphaFold, RoseTTAFold and ModelAngelo

A further change in paradigm in structural biology, with a knock-on effect on the study of viruses, is represented by the development of software such as AlphaFold (and AlphaFold-Multimer), RoseTTAFold and ModelAngelo [149–151]. After decades of research on fundamental science, software and algorithm developments to predict the 3D structure of proteins from their primary sequences, in 2021, two pieces of software, AlphaFold and RoseTTAFold, emerged with a striking ability to accurately predict protein structures.

These software systems, while powerful, require a high level of expertise for accurate interpretation. AlphaFold, in particular, relies on artificial intelligence and outperformed other methods in the challenging 14th Critical Assessment of Protein Structure Prediction (CASP14) (https://predictioncenter.org/casp14/) [152]. It incorporates physical and biological knowledge about protein structure, leveraging multi-sequence alignments in the design of its deep learning algorithm. These advances build upon the extensive prior structural knowledge accumulated over the years—including approximately 160,000 released PDB X-ray structures between 1976 and 2021, facilitated by the advent of synchrotron radiation sources.

RoseTTAFold produces structure predictions with accuracies approaching those of AlphaFold, but it is also capable of predicting structures of de novo-designed proteins and of using cryo-EM maps to enable modelling of large complexes. The availability of reliable 3D models facilitates the interpretation of cryo-EM density maps and drastically shortens the time required for de novo model building into density by the user. However, even when the fold is accurately predicted, supervised rebuilding and careful model refinement into density are always necessary.

In the case of predicting structures of viral proteins, some extra caution should be exercised, as the environmental and host cell-induced pressures to which viruses are subjected and the timescales for virus evolution are not comparable to those of proteins from cellular organisms—thus, a significant portion of the proteome of a complex virus may be poorly predicted [153]. However, it is undeniable that AlphaFold can impact viral research [154]. Furthermore, the advent of AlphaFold-Multimer paves the way for multimeric protein complex predictions relevant for understanding the function of many biological systems where proteins interact with each other—in a living cell, a protein is unlikely to be alone! The weakness of this approach lies mainly in the complexity of interactions that AlphaFold has to predict, considering that there are far fewer experimentally determined complex structures than individual proteins. Thus, experimental structure determination of derived complexes and interactions remains pivotal for drawing any valuable and meaningful conclusions, and these results will accelerate improvements in artificial intelligence-based methods such as AlphaFold.

A breakthrough in density map interpretation is represented by ModelAngelo software, which uses artificial intelligence to build a 3D model directly into density [151]. If the resolution of the cryo-EM density map is better than 2.5 Å,

ModelAngelo does not even need the primary sequence of the targeted protein; it can attempt to identify the residue type, thereby unveiling the unknown protein's identity. The power of ModelAngelo is still emerging, and feedback on this first version from users in real cases will shape future upgrades to increase reliability in difficult scenarios (e.g. anisotropic or low-resolution density maps). Its full potential will likely unfold when used for identifying proteins in their native cellular context. Indeed, thanks to advances in cellular cryo-ET methodologies and constant improvements in resolution achieved by sub-tomogram averaging techniques, cryo-ET density maps of proteins and protein complexes can approach 4 Å resolution. The ultimate goal is to provide a cellular atlas with identified protein players contributing to the physio-pathological cellular state. ModelAngelo, combined with AlphaFold-Multimer moving towards an AlphaFold-populated cell, for example, could start to populate a protein interactome within a reconstructed 3D cell volume.

7.8 The Biology Behind the Combined Methods

7.8.1 Implications for Biomedical Research

There are two main types of drugs—small molecule and biologics—and the role of hybrid methods in their development differs. Current methods for small-molecule drug discovery rely mainly on (i) ligand-based design and (ii) structure-based design. In the first case, the search for new compounds against a target molecule is based on previously known drugs; in the second case, the search is based on the availability of the 3D structure of the target molecule (see Chap. 20 for a comprehensive view on the translational applications of structural virology in antiviral research). Both screening strategies are computationally intensive. Structure-based design uses atomic or quasi-atomic models to rationalize the interaction of the potential candidate drug with the target biomolecule, for example, to block the binding of the virus to the cellular receptor. One of the first examples where the interdisciplinary approach of virus EM with X-ray crystallography has provided clues for the development of antiviral drugs is represented by the docking of ICAM-1 to human rhinovirus [37]. Indeed, this quasi-atomic model contributed to the development of pleconaril, a drug that inhibited the uncoating of rhinovirus upon infection by binding to the major capsid protein VP1 [155] (Fig. 7.3b). Although this drug was never licensed, failing in phase III clinical studies, the whole process reinforced the utility of the combined approach in virus structure determination [155]. Undoubtedly, the interaction of viruses with their cellular receptors is one of the major structural targets in antiviral research, facilitating the rational design of novel drugs aiming to halt cell entry.

Similarly, studies aimed at understanding the mechanism of viral evasion of antibody neutralization, for example, for HIV-1 gp120, have made use of crystal structures and negative-stain EM [156] (Fig. 7.9a). Nevertheless, there are other steps during virus morphogenesis that can be tackled by combining EM information

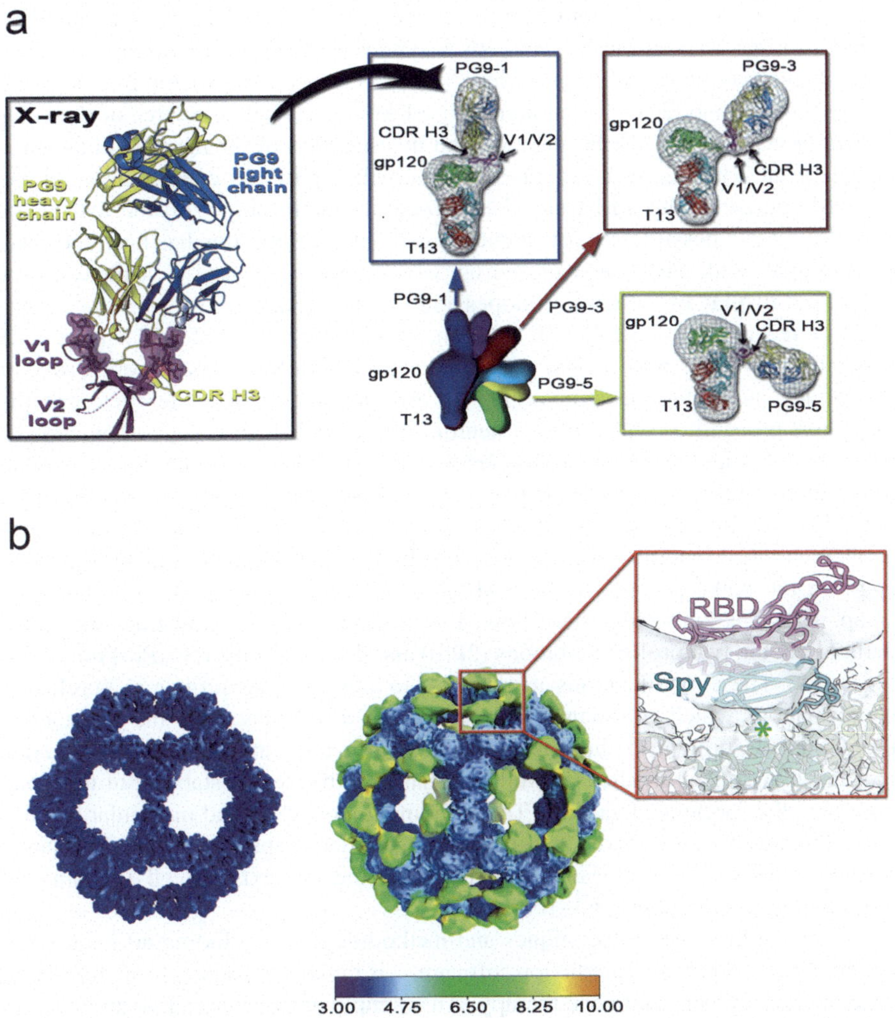

Fig. 7.9 Impact of the combination of structural methods. (**a**), *Left*, X-ray structure of HIV gp120 V1/V2 domain (in magenta) complexed with neutralizing antibody PG9 (yellow, heavy chain; blue, light chain) represented as ribbons; *right*, negative-stain EM reconstruction classes (six in total, colour-coded from blue to green) of ternary complexes composed by gp120, PG9 and CD4-binding site antibody T13 with the corresponding fitting of the gp120 V1/V2-PG9 crystal structure in three of them (insets). Adapted from [156] with permission. (**b**) 3D cryo-EM density of SpyCatcher003-mi3 virus-like particle (VLP) conjugated with the SARS-CoV-2 receptor-binding domain (RBD) (immunogenic nanoparticle-based vaccine) with the icosahedral symmetry imposed and displayed at two contour levels (left, 0.0428; right, 0.2260) coloured by resolution as per the colour scale (Å) below. In the inset, an enlarged view of the region marked by the red square on the VLP-RBD supposed to correspond to the SpyTag/SpyCatcher003 (cyan cartoon tube) and RBD (pink cartoon tube) can be seen (the green asterisk marks the N-terminus of mi3 cage). Reproduced from [157]

and X-ray crystallography, for example, targeting the ribonucleoprotein complex of influenza virus responsible for viral RNA synthesis [158]. Furthermore, the structural elucidation of virus-like particles (VLPs) as transporters of foreign antigenic epitopes (see Chap. 21) by combining cryo-EM and crystal structures of the component proteins allows the localization and investigation of the modifications introduced in the virus capsid structure. The knowledge of the conformation of the inserted epitope in the VLPs and their structural relationship with the rest of the particles directs possible improvements of construct design; this has been the case, for example, with TBSV and norovirus, which have been used as a platform for polyvalent display of antigenic epitopes and vaccine design.

More recently, with the COVID-19 pandemic, a plethora of biologics have been generated, enabling a very fast response to SARS-CoV-2. Apart from different disciplines and techniques converging on elucidating the virus life cycle and the virus-host recognition and antibody neutralization mechanisms, the combination of in silico design and biochemistry has led to the SpyCatcher-mi3 protein nanoparticle to produce a highly immunogenic nanoparticle-based vaccine (Fig. 7.9b) [157, 159, 160]. Remarkably, the success of RNA-based COVID-19 vaccines, like the Pfizer-BioNTech and Moderna vaccines, also results from the detailed structural knowledge on the spike protein of the SARS-CoV-2 virus in the prefusion state (see Chap. 16). This knowledge has allowed the introduction of stabilizing mutations, particularly the two proline mutations (2Pro) and the six proline (HexPro) mutations [161, 162]. These modifications are designed to lock the spike protein in a prefusion conformation, which is thought to improve the effectiveness of the vaccine by presenting the protein to the immune system in a form that is likely to generate more neutralizing antibodies, as well as enhancing the overall stability of the spike protein in the formulated vaccine. The combination of NMR and molecular dynamics techniques has also proven instrumental in elucidating the structural and dynamic roles of SARS-CoV-2 glycans, particularly in the contexts of cell tropism and immunological shielding [163–165].

The integrative use of techniques and methodologies—including artificial intelligence-based 3D protein prediction software— is rapidly impacting drug discovery, vaccine development and gene therapy. An example of this integrative approach in vaccinology is the structure-based design of the human-infecting respiratory syncytial virus fusion glycoprotein, stabilized in its prefusion conformation, which has proven to be a highly effective immunogen in non-human primates, representing a viable vaccine candidate [166]. The power of combining structural insights with vaccine development is also evident in the manufacturing of synthetic vaccines for FMDV and poliovirus [101, 167].

Finally, cryo-EM, apart from its democratization in academic laboratories, has proven to be one of the essential techniques in assessing the Critical Quality Attributes in the manufacturing of adeno-associated virus gene vectors [7, 168, 169]. It is worth noting that the cell and gene therapy sector is one of the fastest-growing areas in biomedicine, with already approved AAV-based drugs by the US Food and Drug Administration (e.g. Luxturna for curing the monogenetic disease hereditary retinopathy).

7.8.2 Novel Concepts in Virus Evolution

In the last two decades, the accumulation of virus and viral protein structures determined mainly by X-ray crystallography and cryo-EM, or the combination of these two techniques, has allowed us to explore the virosphere based on the structural relationships between different viruses. This has led to the emergence of a novel concept in virus evolution that uses specific structural markers, the so-called virus-self elements (e.g. the fold of the capsid protein) to infer common ancestry [51]. Although the observation that viruses infecting different species (cf. SBMV vs. rhinovirus) displayed a similar structural fold in their capsid protein (single jelly roll) dates back to 1985, it was only in 2002 that the capsid-centred view in virus classification was formalized as a precise hypothesis and validated over the years through systematic experimental structure comparisons [27, 51, 170]. This novel taxonomic approach simplifies the organization of the virosphere and elucidates relationships between viruses infecting different hosts across the three domains of life. A clear example of this is represented by adenovirus and PRD1, which infect vertebrate and bacteria, respectively. Their major capsid proteins (MCPs) share a vertical double-jelly roll fold, and the capsids share common architectural principles [28] (Fig. 7.10a). The power of this structural classification became evident when, at least until 2012, almost half of the virus families described—many of which did not belong to any order according to the International Committee on Taxonomy of Viruses (ICTV) classification scheme at the time—could be traced back to one of the four viral lineages described: (*i*) PRD1/adenovirus-like, (*ii*) blue tongue virus-like, (*iii*) HK97-like and (*iv*) picornavirus-like [170].

Eight years after the structure-based classification of the virosphere was introduced, the ICTV revised virus taxonomy by implementing a 15-rank classification. Within this framework, the PRD1-adenovirus structure-based lineage now constitutes the new kingdom *Bamfordvirae* (etymology of the taxon name available at https://ictv.global/taxonomy/taxondetails?taxnode_id=202208707&taxon_name=Bamfordvirae), which belongs to the realm *Varidnaviria* [52, 53] (Fig. 7.10b). This kingdom includes some of the largest known viruses, with very high T numbers. Central to our understanding is not only the major capsid protein but also the concept of a tape measure protein, as exemplified in PRD1, which explains how such massive structures can be accurately assembled. Examples include structures with T numbers such as $h = 7$, $k = 8$ or $k = 12$ as in Paramecium bursaria chlorella virus 1 (PBCV-1; $T = 169$) or African swine fever virus (ASFV; $T = 277$), respectively [28, 172–174] (Fig. 7.10a).

It is unlikely that many more virus lineages will be unveiled by the structure-based classification method, as this depends on the discovery of new viral protein folds. However, the complexity introduced by evolutionary processes continually raises new questions regarding assembly and viral evolution [175]. High-resolution cryo-EM structures of three archaeal viruses—SH1, Haloarcula hispanica icosahedral virus 2 (HHIV-2) and Haloarcula californiae icosahedral virus 1 (HCIV-1)—at ~3.8 Å resolution have revealed that their assembly relies on homo- and/or

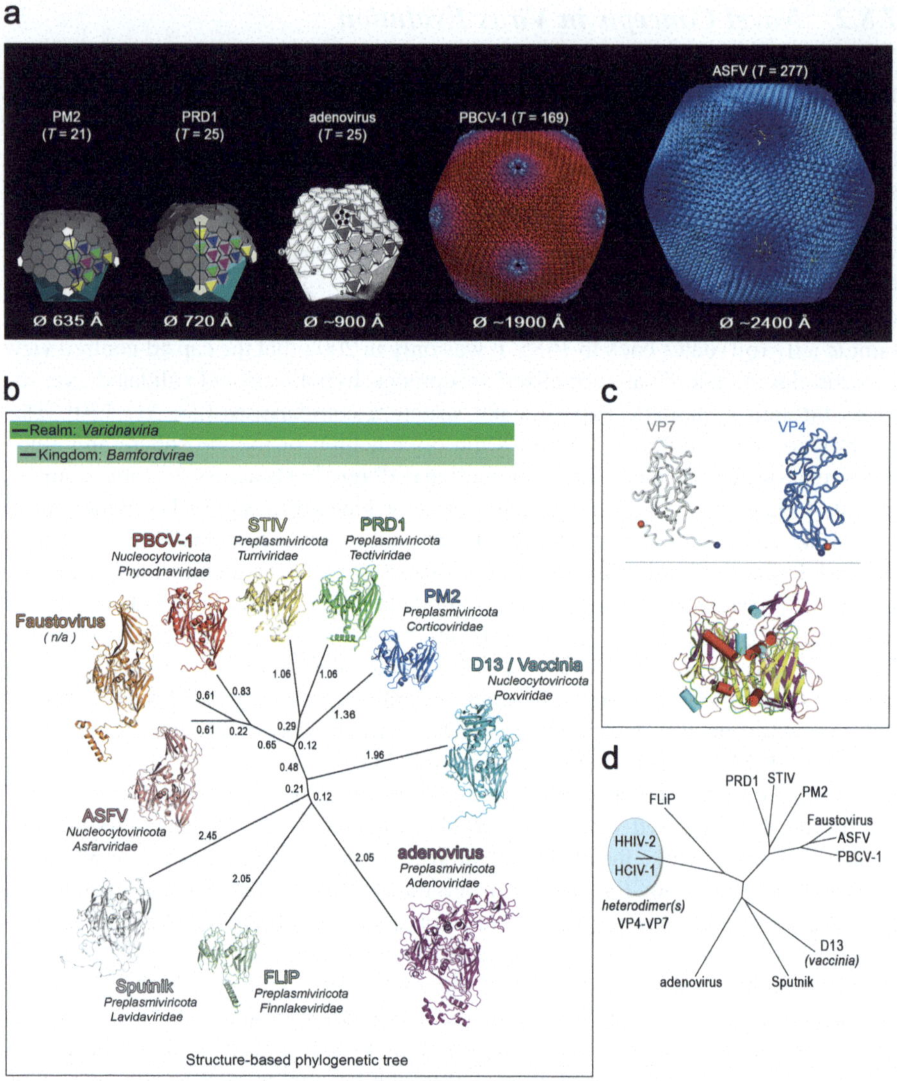

Fig. 7.10 Structure-based virus phylogeny. (**a**), schematic representation of the architecture of bacteriophages PM2 and PRD1 (grey hexagons represent the pseudo-hexameric morphology displayed by viral major capsid proteins forming trimers; the trimers composing the icosahedral asymmetric unit are represented as colour-coded triangles; black lines on the capsomers indicate the T numbers), adenovirus (adapted from [8]), the cryo-EM density of PBCV-1 virus (image taken from http://viperdb.scripps.edu/info_page.php?VDB=1m4x) and the cryo-EM density of ASFV (EMD-10346). All four represented viruses have a major capsid protein (MCP) with a vertical double β-barrel fold and are members of the PRD1-adenovirus viral lineage. (**b**) Phylogenetic tree based on the structural characteristics of the MCP within the kingdom *Bamfordvirae*, belonging to the *Varidnaviria* realm. *Bamfordvirae* members share a distinctive vertical double-jelly roll fold in their MCP, serving as a defining feature for this viral group. Each MCP structure is accompanied by the corresponding virus name (top), phylum (middle) and family (bottom). It is worth noting that *Faustovirus* awaits official classification (recently FLiP was assigned to *Preplasmiviricota*). The evolutionary distances across the described members of the originally called PRD1-adenovirus viral

heterodimerization motifs in the MCPs, along with additional molecular guiding proteins between the capsid protein shell and the outer membrane leaflet [12, 95, 176]. In viruses belonging to the kingdom *Helvetiavirae* in the realm *Varidnaviria*, the two vertical single-jelly roll MCPs form heterodimers that mimic the relative angular orientation of the individual β-barrel in the vertical double-jelly roll MCPs of the PRD1-adenovirus lineage [12, 176, 177] (Fig. 7.10c-d).

Deciphering sequence and structural homologies not only satisfies our desire to classify and understand the complexity that surrounds us but also contributes to the fight against viral infections [171].

7.9 Perspectives and Challenges Ahead

Much progress has been made in understanding virus biology, thanks to the combination of biochemical, genetic and structural techniques and more recently computational methods. Overall, technology advances have underpinned a major shift towards a physico-chemical structure-based view of viruses, driven mainly by cryo-EM and X-ray crystallography. Their hybridization has provided insights into mechanisms such as virus-host recognition, antibody-mediated virus neutralization, antigen display, virus assembly and maturation and other properties and functions of virus particles (see Chap. 1 and Part III of this book, Chaps. 11–19). Nevertheless, efforts in the past have mainly focused on virus particles displaying icosahedral morphology (about half of the known virus families), which are easier to analyse, as described above.

Structural virology still faces many challenges ahead, but its power has been clearly demonstrated by the response to the emergence of SARS-CoV-2. Data gathered through X-ray crystallography, cryo-EM and cryo-ET on viral proteins and whole native viruses, including in infected cells, have dissected several aspects of viral pathogenesis in record time. Importantly, the expertise acquired in

Fig. 7.10 (continued) lineage were calculated with the Homologous Structure Finder software and depicted with PHYLIP (https://evolution.genetics.washington.edu/phylip.html); the evolutionary distances are shown next to each branch. The Protein Data Bank identifiers (PDB id) for the structures are as follows: PRD1, PDBid 1HX6; PBCV-1, 1M3Y; adenovirus, 1P2Z; STIV, 2BBD; Vaccinia D13, 2YGB; Sputnik, 3 J26; Faustovirus, 5J7O; FLiP, 5OAC; ASFV p72, 6KU9; and PM2, 2W0C. Adapted from [53, 171]. (**c**), At the top, HCIV-1 (the kingdom *Helvetiavirae*, the realm *Varidnaviria*) vertical single-jelly rolls VP7 and VP4 MCPs represented as cartoon tubes; the N-terminus and C-terminus are indicated by blue and red spheres, respectively. At the bottom, superimposition of the chimeric vertical double β-barrel resulting from the heterodimerization of VP4–VP7 with the vertical double β-barrel PM2 P2 MCP, both displayed by secondary structural elements (VP4-VP7: β-strands magenta, α-helices cyan cylinders, loops/turns pink; P2: β-strands yellow, α-helices red cylinders, loops/turns green). (**d**) Structure-based phylogenetic tree as in Fig. 10b within the kingdom *Bamfordvirae* and *Helvetiavirae* including the chimeric double β-barrel generated by the heterodimer VP4–VP7 (HHIV-2 and HCIV-1). Reproduced from [171] responding to the urgent search for vaccines, therapeutics and technologies has not only tangibly contributed to delivering vaccines that saved millions of lives but also provided a pipeline from basic to translation research that can be used to combat emerging virus threats in the future.

Technological developments in synchrotron radiation and related technologies are transforming the landscape of structural biology, especially in virus research. Technologies like XFEL and in-situ diffraction, coupled with correlative microscopy in cryogenic EM/ET, open new avenues for detailed structural analysis. Specifically, developments like the dynamic transmission electron microscopy (DTEM) at the Lawrence Livermore National Laboratory are pushing the boundaries of our ability to capture real-time, dynamic processes at the nanoscale. It is like having a high-speed camera for incredibly tiny events! The revolution in cryo-EM depended on detector developments, and these continue with, for example, DECTRIS SINGLA® detector, providing excellent data from lower-voltage machines (potentially further enabling the democratization of the method) and future direct electron detectors likely to markedly increase the rate of data collection.

To sum up, the hybridization of methods in structural virology is a powerful approach that generates knowledge and tools to serve in the fight against viral infections. These contributions are fundamental to the health of humans and animals, and they are helping to create a "Systems Virology" vision capturing the complex biology of viruses with unprecedented sensitivity and resolution.

Acknowledgements D.I.S. is supported by the UKRI MRC and the Chinese Academy of Medical Sciences and is a Jenner investigator. H.M.O. was supported by the University of Helsinki and the Research Council of Finland by funding for FINStruct and Instruct Centre Finland, Instruct-ERIC. N.G.A.A. is supported by the Spanish Ministerio de Ciencia e Innovacion (PID2021-126130OB-I00) and by the Basque Government (COVID-19 CO-2020/00001). MCIN is also thanked for the Severo Ochoa Center of Excellence Accreditation to the CIC bioGUNE (CEX2021-001136-S).

References

1. Buckminster Fuller R, Marks R (1973) The Dymaxion world of Buckminster Fuller. Doubleday
2. Harrison SC, Olson AJ, Schutt CE, Winkler FK, Bricogne G (1978) Tomato bushy stunt virus at 2.9 Å resolution. Nature 276:368–373
3. Abad-Zapatero C, Abdel-Meguid SS, Johnson JE, Leslie AGW, Rayment I, Rossmann MG et al (1980) Structure of Southern bean mosaic virus at 2.8 Å resolution. Nature 286:33–39
4. Scheres SH (2012) RELION: implementation of a Bayesian approach to cryo-EM structure determination. J Struct Biol 180:519–530
5. McMullan G, Faruqi AR, Henderson R (2016) Direct electron detectors. Methods Enzymol 579:1–17
6. Kuhlbrandt W (2014) Biochemistry. The resolution revolution. Science 343:1443–1444
7. Stuart DI, Subramaniam S, Abrescia NG (2016) The democratization of cryo-EM. Nat Methods 13:607–608
8. Roberts MM, White JL, Grutter MG, Burnett RM (1986) Three-dimensional structure of the adenovirus major coat protein hexon. Science 232:1148–1151

9. Goldsmith CS, Miller SE (2009) Modern uses of electron microscopy for detection of viruses. Clin Microbiol Rev 22:552–563
10. Sirohi D, Chen Z, Sun L, Klose T, Pierson TC, Rossmann MG et al (2016) The 3.8 Å resolution cryo-EM structure of Zika virus. Science 352:467–470
11. Kotecha A, Wang Q, Dong X, Ilca SL, Ondiviela M, Zihe R et al (2017) Rules of engagement between alphavbeta6 integrin and foot-and-mouth disease virus. Nat Commun 8:15408
12. Santos-Perez I, Charro D, Gil-Carton D, Azkargorta M, Elortza F, Bamford DH et al (2019) Structural basis for assembly of vertical single beta-barrel viruses. Nat Commun 10:1184
13. Duyvesteyn HME, Santos-Perez I, Peccati F, Martinez-Castillo A, Walter TS, Reguera D et al (2021) Bacteriophage PRD1 as a nanoscaffold for drug loading. Nanoscale 13:19875–19883
14. Kounatidis I, Stanifer ML, Phillips MA, Paul-Gilloteaux P, Heiligenstein X, Wang H et al (2020) 3D correlative cryostructured illumination fluorescence and soft X-ray microscopy elucidates reovirus intracellular release pathway. Cell 182:515–530 e17
15. Da Vela S, Svergun DI (2020) Methods, development and applications of small-angle X-ray scattering to characterize biological macromolecules in solution. Curr Res Struct Biol 2:164–170
16. Tuukkanen AT, Kleywegt GJ, Svergun DI (2016) Resolution of ab initio shapes determined from small-angle scattering. IUCrJ 3:440–447
17. Stubbs G (2001) Fibre diffraction studies of filamentous viruses. Rep Prog Phys 64:1389–1425
18. Harrison SC, Caspar DL, Camerini-Otero RD, Franklin RM (1971) Lipid and protein arrangement in bacteriophage PM2. Nat New Biol 229:197–201
19. Abrescia NG, Grimes JM, Kivela HM, Assenberg R, Sutton GC, Butcher SJ et al (2008) Insights into virus evolution and membrane biogenesis from the structure of the marine lipid-containing bacteriophage PM2. Mol Cell 31:749–761
20. Bucciarelli S, Midtgaard SR, Nors Pedersen M, Skou S, Arleth L, Vestergaard B (2018) Size-exclusion chromatography small-angle X-ray scattering of water soluble proteins on a laboratory instrument. J Appl Crystallogr 51:1623–1632
21. Zielinski KA, Katz AM, Calvey GD, Pabit SA, Milano SK, Aplin C et al (2023) Chaotic advection mixer for capturing transient states of diverse biological macromolecular systems with time-resolved small-angle X-ray scattering. IUCrJ 10:363–375
22. Bauer DW, Li D, Huffman J, Homa FL, Wilson K, Leavitt JC et al (2015) Exploring the balance between DNA pressure and capsid stability in herpesviruses and phages. J Virol 89: 9288–9298
23. Asor R, Khaykelson D, Ben-Nun-Shaul O, Oppenheim A, Raviv U (2019) Effect of calcium ions and disulfide bonds on swelling of virus particles. ACS Omega 4:58–64
24. Fry EE, Abrescia NGA, Stuart DI (2007) Virus crystallography. In: Sanderson M, Skelly J (eds) Macromolecular crystallography: conventional and high-throughput methods—a practical approach. Oxford University Press
25. Tuthill TJ, Harlos K, Walter TS, Knowles NJ, Groppelli E, Rowlands DJ et al (2009) Equine rhinitis a virus and its low pH empty particle: clues towards an aphthovirus entry mechanism? PLoS Pathog 5:e1000620
26. Hogle JM, Chow M, Filman DJ (1985) Three-dimensional structure of poliovirus at 2.9 Å resolution. Science 229:1358–1365
27. Rossmann MG, Arnold E, Erickson JW, Frankenberger EA, Griffith JP, Hecht HJ et al (1985) Structure of a human common cold virus and functional relationship to other picornaviruses. Nature 317:145–153
28. Abrescia NG, Cockburn JJ, Grimes JM, Sutton GC, Diprose JM, Butcher SJ et al (2004) Insights into assembly from structural analysis of bacteriophage PRD1. Nature 432:68–74
29. Cockburn JJ, Abrescia NG, Grimes JM, Sutton GC, Diprose JM, Benevides JM et al (2013) Membrane structure and interactions with protein and DNA in bacteriophage PRD1. Nature 432:122–125

30. Liu H, Jin L, Koh SB, Atanasov I, Schein S, Wu L et al (2010) Atomic structure of human adenovirus by cryo-EM reveals interactions among protein networks. Science 329:1038–1043
31. Reddy VS, Natchiar SK, Stewart PL, Nemerow GR (2010) Crystal structure of human adenovirus at 3.5 Å resolution. Science 329:1071–1075
32. Perrakis A, Daenke S, Stuart DI, Sussman JL (2011) From SPINE to SPINE-2 complexes and beyond. J Struct Biol 175:105
33. Axford D, Owen RL, Aishima J, Foadi J, Morgan AW, Robinson JI et al (2012) In situ macromolecular crystallography using microbeams. Acta Crystallogr D Biol Crystallogr 68: 592–600
34. Wang X, Peng W, Ren J, Hu Z, Xu J, Lou Z et al (2012) A sensor-adaptor mechanism for enterovirus uncoating from structures of EV71. Nat Struct Mol Biol 19:424–429
35. Rossmann MG (2000) Fitting atomic models into electron-microscopy maps. Acta Crystallogr D Biol Crystallogr 56:1341–1349
36. Millan C, McCoy AJ, Terwilliger TC, Read RJ (2023) Likelihood-based docking of models into cryo-EM maps. Acta Crystallogr D Struct Biol 79:281–289
37. Kolatkar PR, Bella J, Olson NH, Bator CM, Baker TS, Rossmann MG (1999) Structural studies of two rhinovirus serotypes complexed with fragments of their cellular receptor. EMBO J 18:6249–6259
38. Gilbert RJ, Grimes JM, Stuart DI (2003) Hybrid vigor: hybrid methods in viral structure determination. Adv Protein Chem 64:37–91
39. Topf M, Sali A (2005) Combining electron microscopy and comparative protein structure modeling. Curr Opin Struct Biol 15:578–585
40. Kaufmann KW, Lemmon GH, Deluca SL, Sheehan JH, Meiler J (2010) Practically useful: what the Rosetta protein modeling suite can do for you. Biochemistry 49:2987–2998
41. Brown A, Long F, Nicholls RA, Toots J, Emsley P, Murshudov G (2015) Tools for macromolecular model building and refinement into electron cryo-microscopy reconstructions. Acta Crystallogr D Biol Crystallogr 71:136–153
42. Afonine PV, Poon BK, Read RJ, Sobolev OV, Terwilliger TC, Urzhumtsev A et al (2018) Real-space refinement in PHENIX for cryo-EM and crystallography. Acta Crystallogr D Struct Biol 74:531–544
43. Franklin RE, Harrison SC, Pettersson U, Pihilpson L, Branden CJ, Werner P (1972) Structural studies on the adenovirus Hexon. Cold Spring Harb Symp Quant Biol 36:503–510
44. Hewat EA, Verdaguer N, Fita I, Blakemore W, Brookes S, King A et al (1997) Structure of the complex of a Fab fragment of a neutralizing antibody with foot-and-mouth disease virus: positioning of a highly mobile antigenic loop. EMBO J 16:1492–1500
45. Yu IM, Zhang W, Holdaway HA, Li L, Kostyuchenko VA, Chipman PR et al (2008) Structure of the immature dengue virus at low pH primes proteolytic maturation. Science 319:1834–1837
46. Li L, Lok SM, Yu IM, Zhang Y, Kuhn RJ, Chen J et al (2008) The flavivirus precursor membrane-envelope protein complex: structure and maturation. Science 319:1830–1834
47. Cherrier MV, Kaufmann B, Nybakken GE, Lok SM, Warren JT, Chen BR et al (2009) Structural basis for the preferential recognition of immature flaviviruses by a fusion-loop antibody. EMBO J 28:3269–3276
48. Bamford DH, Caldentey J, Bamford JK (1995) Bacteriophage PRD1: a broad host range DSDNA tectivirus with an internal membrane. Adv Virus Res 45:281–319
49. San Martin C, Burnett RM, de Haas F, Heinkel R, Rutten T, Fuller SD, Butcher SJ, Bamford DH (2001) Combined EM/X-ray imaging yields a quasi-atomic model of the adenovirus-related bacteriophage PRD1 and shows key capsid and membrane interactions. Structure 9: 917–930
50. Benson SD, Bamford JK, Bamford DH, Burnett RM (1999) Viral evolution revealed by bacteriophage PRD1 and human adenovirus coat protein structures. Cell 98:825–833
51. Bamford DH, Burnett RM, Stuart DI (2002) Evolution of viral structure. Theor Popul Biol 61: 461–470

52. International Committee on Taxonomy of Viruses Executive C (2020) The new scope of virus taxonomy: partitioning the virosphere into 15 hierarchical ranks. Nat Microbiol 5:668–674
53. Simmonds P, Adriaenssens EM, Zerbini FM, Abrescia NGA, Aiewsakun P, Alfenas-Zerbini P et al (2023) Four principles to establish a universal virus taxonomy. PLoS Biol 21:e3001922
54. Nickell S, Forster F, Linaroudis A, Net WD, Beck F, Hegerl R et al (2005) TOM software toolbox: acquisition and analysis for electron tomography. J Struct Biol 149:227–234
55. Scaramuzza S, Castano-Diez D (2021) Step-by-step guide to efficient subtomogram averaging of virus-like particles with dynamo. PLoS Biol 19:e3001318
56. Zivanov J, Oton J, Ke Z, von Kugelgen A, Pyle E, Qu K et al (2022) A bayesian approach to single-particle electron cryo-tomography in RELION-4.0. elife 11:e83724
57. Briggs JA, Riches JD, Glass B, Bartonova V, Zanetti G, Krausslich HG (2009) Structure and assembly of immature HIV. Proc Natl Acad Sci USA 106:11090–11095
58. Hoffmann PC, Kreysing JP, Khusainov I, Tuijtel MW, Welsch S, Beck M (2022) Structures of the eukaryotic ribosome and its translational states in situ. Nat Commun 13:7435
59. Shah PNM, Gilchrist JB, Forsberg BO, Burt A, Howe A, Mosalaganti S et al (2023) Characterization of the rotavirus assembly pathway in situ using cryoelectron tomography. Cell Host Microbe 31:604–615 e4
60. Alber F, Dokudovskaya S, Veenhoff LM, Zhang W, Kipper J, Devos D et al (2007) The molecular architecture of the nuclear pore complex. Nature 450:695–701
61. Brunger AT (1992) Free R value: a novel statistical quantity for assessing the accuracy of crystal structures. Nature 355:472–475
62. Lawson CL, Kryshtafovych A, Adams PD, Afonine PV, Baker ML, Barad BA et al (2021) Cryo-EM model validation recommendations based on outcomes of the 2019 EMDataResource challenge. Nat Methods 18:156–164
63. Tsao J, Chapman MS, Rossmann MG (1992) Ab initio phase determination for viruses with high symmetry: a feasibility study. Acta Crystallogr A 48:293–301
64. Chapman MS, Tsao J, Rossmann MG (1992) Ab initio phase determination for spherical viruses: parameter determination for spherical-shell models. Acta Crystallogr A 48:301–312
65. Rossmann MG (1995) Ab initio phase determination and phase extension using non-crystallographic symmetry. Curr Opin Struct Biol 5:650–655
66. Rossmann MG (1990) The molecular replacement method. Acta Crystallogr A 46:73–82
67. Acharya R, Fry E, Stuart D, Fox G, Rowlands D, Brown F (1989) The three-dimensional structure of foot-and-mouth disease virus at 2.9 Å resolution. Nature 337:709–716
68. Villeret V, Tricot C, Stalon V, Dideberg O (1995) Crystal structure of Pseudomonas aeruginosa catabolic ornithine transcarbamoylase at 3.0-Å resolution: a different oligomeric organization in the transcarbamoylase family. Proc Natl Acad Sci USA 92:10762–10766
69. Dodson EJ (2001) Using electron-microscopy images as a model for molecular replacement. Acta Crystallogr D Biol Crystallogr 57:1405–1409
70. Trapani S, Schoehn G, Navaza J, Abergel C (2010) Macromolecular crystal data phased by negative-stained electron-microscopy reconstructions. Acta Crystallogr D Biol Crystallogr 66: 514–521
71. Hao Q, Dodd FE, Grossmann JG, Hasnain SS (1999) Ab initio phasing using molecular envelope from solution X-ray scattering. Acta Crystallogr D Biol Crystallogr 55:243–246
72. Huiskonen JT, Kivelä HM, Bamford DH, Butcher SJ (2004) The PM2 virion has a novel organization with an internal membrane and pentameric receptor binding spikes. Nat Struct Mol Biol 11:850–856
73. Abrescia NG, Grimes JM, Oksanen HM, Bamford JK, Bamford DH, Stuart DI (2011) The use of low-resolution phasing followed by phase extension from 7.6 to 2.5 Å resolution with noncrystallographic symmetry to solve the structure of a bacteriophage capsid protein. Acta Crystallogr D Biol Crystallogr 67:228–232
74. Abrescia NG, Kivelä HM, Grimes JM, Bamford JK, Bamford DH, Stuart DI (2005) Preliminary crystallographic analysis of the major capsid protein P2 of the lipid-containing bacteriophage PM2. Acta Crystallogr Sect F Struct Biol Cryst Commun 61:762–765

75. Stuart DI, Abrescia NG (2013) From lows to highs: using low-resolution models to phase X-ray data. Acta Crystallogr D Biol Crystallogr 69:2257–2265
76. Oksanen HM, Domanska A, Bamford DH (2012) Monolithic ion exchange chromatographic methods for virus purification. Virology 434:271–277
77. Eskelin K, Lampi M, Meier F, Moldenhauer E, Bamford DH, Oksanen HM (2016) Asymmetric flow field flow fractionation methods for virus purification. J Chromatogr A 1469:108–119
78. Eskelin K, Poranen MM, Oksanen HM (2019) Asymmetrical flow field-flow fractionation on virus and virus-like particle applications. Microorganisms 7:555
79. Schönn MA, Eskelin K, Bamford DH, Poranen MM, Unoson C, Oksanen HM (2022) Inline-tandem purification of viruses from cell lysate by agarose-based chromatography. J Chromatogr B Analyt Technol Biomed Life Sci 1192:123140
80. Piersimoni L, Kastritis PL, Arlt C, Sinz A (2022) Cross-linking mass spectrometry for investigating protein conformations and protein-protein interactions horizontal line a method for all seasons. Chem Rev 122:7500–7531
81. Gowen B, Bamford JK, Bamford DH, Fuller SD (2003) The tailless icosahedral membrane virus PRD1 localizes the proteins involved in genome packaging and injection at a unique vertex. J Virol 77:7863–7871
82. Strömsten NJ, Bamford DH, Bamford JK (2003) The unique vertex of bacterial virus PRD1 is connected to the viral internal membrane. J Virol 77:6314–6321
83. Peralta B, Gil-Carton D, Castano-Diez D, Bertin A, Boulogne C, Oksanen HM et al (2013) Mechanism of membranous tunnelling nanotube formation in viral genome delivery. PLoS Biol 11:e1001667
84. Hong C, Oksanen HM, Liu X, Jakana J, Bamford DH, Chiu W (2014) A structural model of the genome packaging process in a membrane-containing double stranded DNA virus. PLoS Biol 12:e1002024
85. Kivelä HM, Madonna S, Krupovic M, Tutino ML, Bamford JK (2008) Genetics for Pseudoalteromonas provides tools to manipulate marine bacterial virus PM2. J Bacteriol 190:1298–1307
86. Avilov SV, Moisy D, Munier S, Schraidt O, Naffakh N, Cusack S (2011) Replication-competent influenza a virus that encodes a split-green fluorescent protein-tagged PB2 polymerase subunit allows live-cell imaging of the virus life cycle. J Virol 86:1433–1448
87. Rincon V, Rodriguez-Huete A, Mateu MG (2015) Different functional sensitivity to mutation at intersubunit interfaces involved in consecutive stages of foot-and-mouth disease virus assembly. J Gen Virol 96:2595–2606
88. Marsian J, Fox H, Bahar MW, Kotecha A, Fry EE, Stuart DI et al (2017) Plant-made polio type 3 stabilized VLPs-a candidate synthetic polio vaccine. Nat Commun 8:245
89. Bamford D, Mindich L (1982) Structure of the lipid-containing bacteriophage PRD1: disruption of wild-type and nonsense mutant phage particles with guanidine hydrochloride. J Virol 44:1031–1038
90. Kivelä HM, Kalkkinen N, Bamford DH (2002) Bacteriophage PM2 has a protein capsid surrounding a spherical proteinaceous lipid core. J Virol 76:8169–8178
91. Caldentey J, Luo C, Bamford DH (1993) Dissociation of the lipid-containing bacteriophage PRD1: effects of heat, pH, and sodium dodecyl sulfate. Virology 194:557–563
92. Kivela HM, Abrescia NG, Bamford JK, Grimes JM, Stuart DI, Bamford DH (2008) Selenomethionine labeling of large biological macromolecular complexes: probing the structure of marine bacterial virus PM2. J Struct Biol 161:204–210
93. Kivelä HM, Roine E, Kukkaro P, Laurinavicius S, Somerharju P, Bamford DH (2006) Quantitative dissociation of archaeal virus SH1 reveals distinct capsid proteins and a lipid core. Virology 356:4–11
94. Rissanen I, Pawlowski A, Harlos K, Grimes JM, Stuart DI, Bamford JKH (2012) Crystallization and preliminary crystallographic analysis of the major capsid proteins VP16 and VP17 of bacteriophage P23-77. Acta Crystallogr F68:580–583

95. Gil Carton D, Jaakkola ST, Charro D, Peralta B, Castano Diez D, Oksanen HM et al (2015) Insight into the assembly of viruses with vertical single beta-barrel major capsid proteins. Structure 23:1866–1877
96. Aricescu AR, Lu W, Jones EY (2006) A time- and cost-efficient system for high-level protein production in mammalian cells. Acta Crystallogr D Biol Crystallogr 62:1243–1250
97. Benson SD, Bamford JK, Bamford DH, Burnett RM (2002) The X-ray crystal structure of P3, the major coat protein of the lipid-containing bacteriophage PRD1, at 1.65 Å resolution. Acta Crystallogr D Biol Crystallogr 58:39–59
98. Hendrickson WA (1991) Determination of macromolecular structures from anomalous diffraction of synchrotron radiation. Science 254:51–58
99. Jeembaeva M, Jonsson B, Castelnovo M, Evilevitch A (2010) DNA heats up: energetics of genome ejection from phage revealed by isothermal titration calorimetry. J Mol Biol 395: 1079–1087
100. Wagner C, Fuchsberger FF, Innthaler B, Lemmerer M, Birner-Gruenberger R (2023) Quantification of empty, partially filled and full adeno-associated virus vectors using mass photometry. Int J Mol Sci 24:11033
101. Kotecha A, Zhang F, Juleff N, Jackson T, Perez E, Stuart D et al (2016) Application of the thermofluor PaSTRy technique for improving foot-and-mouth disease virus vaccine formulation. J Gen Virol 97:1557–1565
102. Shoemaker GK, van Duijn E, Crawford SE, Uetrecht C, Baclayon M, Roos WH et al (2010) Norwalk virus assembly and stability monitored by mass spectrometry. Mol Cell Proteomics 9: 1742–1751
103. Worner TP, Bennett A, Habka S, Snijder J, Friese O, Powers T et al (2021) Adeno-associated virus capsid assembly is divergent and stochastic. Nat Commun 12:1642
104. Lucic V, Forster F, Baumeister W (2005) Structural studies by electron tomography: from cells to molecules. Annu Rev Biochem 74:833–865
105. Subramaniam S, Bartesaghi A, Liu J, Bennett AE, Sougrat R (2007) Electron tomography of viruses. Curr Opin Struct Biol 17:596–602
106. Liu YT, Zhang H, Wang H, Tao CL, Bi GQ, Zhou ZH (2022) Isotropic reconstruction for electron tomography with deep learning. Nat Commun 13:6482
107. Himes BA, Zhang P (2018) emClarity: software for high-resolution cryo-electron tomography and subtomogram averaging. Nat Methods 15:955–961
108. Freiberg AN, Sherman MB, Morais MC, Holbrook MR, Watowich SJ (2008) Three-dimensional organization of Rift Valley fever virus revealed by cryoelectron tomography. J Virol 82:10341–10348
109. Grünewald K, Desai P, Winkler DC, Heymann JB, Belnap DM, Baumeister W et al (2003) Three-dimensional structure of herpes simplex virus from cryo-electron tomography. Science 302:1396–1398
110. Bowman BR, Baker ML, Rixon FJ, Chiu W, Quiocho FA (2003) Structure of the herpesvirus major capsid protein. EMBO J 22:757–765
111. Baker ML, Jiang W, Rixon FJ, Chiu W (2005) Common ancestry of herpesviruses and tailed DNA bacteriophages. J Virol 79:14967–14970
112. Bostina M, Bubeck D, Schwartz C, Nicastro D, Filman DJ, Hogle JM (2007) Single particle cryoelectron tomography characterization of the structure and structural variability of poliovirus-receptor-membrane complex at 30 Å resolution. J Struct Biol 160:200–210
113. Zhu P, Liu J, Bess J Jr, Chertova E, Lifson JD, Grise H et al (2006) Distribution and three-dimensional structure of AIDS virus envelope spikes. Nature 441:847–852
114. Zanetti G, Briggs JA, Grunewald K, Sattentau QJ, Fuller SD (2006) Cryo-electron tomographic structure of an immunodeficiency virus envelope complex in situ. PLoS Pathog 2:e83
115. Chen B, Vogan EM, Gong H, Skehel JJ, Wiley DC, Harrison SC (2005) Structure of an unliganded simian immunodeficiency virus gp120 core. Nature 433:834–841
116. Liu J, Bartesaghi A, Borgnia MJ, Sapiro G, Subramaniam S (2008) Molecular architecture of native HIV-1 gp120 trimers. Nature 455:109–113

117. Qu K, Ke Z, Zila V, Anders-Osswein M, Glass B, Mucksch F et al (2021) Maturation of the matrix and viral membrane of HIV-1. Science 373:700–704
118. Harris A, Cardone G, Winkler DC, Heymann JB, Brecher M, White JM et al (2006) Influenza virus pleiomorphy characterized by cryoelectron tomography. Proc Natl Acad Sci USA 103: 19123–19127
119. Yao H, Song Y, Chen Y, Wu N, Xu J, Sun C et al (2020) Molecular architecture of the SARS-CoV-2 virus. Cell 183:730–738 e13
120. Ke Z, Oton J, Qu K, Cortese M, Zila V, McKeane L et al (2020) Structures and distributions of SARS-CoV-2 spike proteins on intact virions. Nature 588:498–502
121. Wrapp D, Wang N, Corbett KS, Goldsmith JA, Hsieh CL, Abiona O et al (2020) Cryo-EM structure of the 2019-nCoV spike in the prefusion conformation. Science 367:1260–1263
122. Walls AC, Park YJ, Tortorici MA, Wall A, McGuire AT, Veesler D (2020) Structure, function, and antigenicity of the SARS-CoV-2 spike glycoprotein. Cell 181:281–292 e6
123. Klein S, Cortese M, Winter SL, Wachsmuth-Melm M, Neufeldt CJ, Cerikan B et al (2020) SARS-CoV-2 structure and replication characterized by in situ cryo-electron tomography. Nat Commun 11:5885
124. Fukuhara H, Dokainish HM, Kita S, Tabata K, Takasu A, Huiskonen JT, et al (2023) Unprecedented spike flexibility revealed by BSL3 Cryo-ET of active SARS-CoV-2 virions. bioRxiv
125. Mendonça L, Howe A, Gilchrist JB, Sheng Y, Sun D, Knight ML et al (2021) Correlative multi-scale cryo-imaging unveils SARS-CoV-2 assembly and egress. Nat Commun 12:4629
126. Huiskonen JT, Overby AK, Weber F, Grunewald K (2009) Electron cryo-microscopy and single-particle averaging of Rift Valley fever virus: evidence for GN-GC glycoprotein heterodimers. J Virol 83:3762–3769
127. Pintilie GD, Zhang J, Goddard TD, Chiu W, Gossard DC (2010) Quantitative analysis of cryo-EM density map segmentation by watershed and scale-space filtering, and fitting of structures by alignment to regions. J Struct Biol 170:427–438
128. Phillips MA, Harkiolaki M, Susano Pinto DM, Parton RM, Palanca A, Garcia-Moreno M et al (2020) CryoSIM: super-resolution 3D structured illumination cryogenic fluorescence microscopy for correlated ultrastructural imaging. Optica 7:802–812
129. Carrascosa JL, Chichon FJ, Pereiro E, Rodriguez MJ, Fernandez JJ, Esteban M et al (2009) Cryo-X-ray tomography of vaccinia virus membranes and inner compartments. J Struct Biol 168:234–239
130. Hagen C, Guttmann P, Klupp B, Werner S, Rehbein S, Mettenleiter TC et al (2012) Correlative VIS-fluorescence and soft X-ray cryo-microscopy/tomography of adherent cells. J Struct Biol 177:193–201
131. Dumoux M, Glen T, Smith JLR, Ho EML, Perdigao LMA, Pennington A et al (2023) Cryo-plasma FIB/SEM volume imaging of biological specimens. elife 12:e83623
132. Plitzko JM, Rigort A, Leis A (2009) Correlative cryo-light microscopy and cryo-electron tomography: from cellular territories to molecular landscapes. Curr Opin Biotechnol 20:83–89
133. van Driel LF, Valentijn JA, Valentijn KM, Koning RI, Koster AJ (2009) Tools for correlative cryo-fluorescence microscopy and cryo-electron tomography applied to whole mitochondria in human endothelial cells. Eur J Cell Biol 88:669–684
134. Hampton CM, Strauss JD, Ke Z, Dillard RS, Hammonds JE, Alonas E et al (2017) Correlated fluorescence microscopy and cryo-electron tomography of virus-infected or transfected mammalian cells. Nat Protoc 12:150–167
135. Kukulski W, Schorb M, Welsch S, Picco A, Kaksonen M, Briggs JA (2011) Correlated fluorescence and 3D electron microscopy with high sensitivity and spatial precision. J Cell Biol 192:111–119
136. Jun S, Ke D, Debiec K, Zhao G, Meng X, Ambrose Z et al (2011) Direct visualization of HIV-1 with correlative live-cell microscopy and cryo-electron tomography. Structure 19: 1573–1581

137. Vijayakrishnan S, McElwee M, Loney C, Rixon F, Bhella D (2020) In situ structure of virus capsids within cell nuclei by correlative light and cryo-electron tomography. Sci Rep 10:17596
138. Ji X, Sutton G, Evans G, Axford D, Owen R, Stuart DI (2009) How baculovirus polyhedra fit square pegs into round holes to robustly package viruses. EMBO J 29:505–514
139. Iwasaki K, Omura T (2010) Electron tomography of the supramolecular structure of virus-infected cells. Curr Opin Struct Biol 20:632–639
140. Koopmann R, Cupelli K, Redecke L, Nass K, Deponte DP, White TA et al (2012) In vivo protein crystallization opens new routes in structural biology. Nat Methods 9:259–262
141. Seibert MM, Ekeberg T, Maia FR, Svenda M, Andreasson J, Jonsson O et al (2011) Single mimivirus particles intercepted and imaged with an X-ray laser. Nature 470:78–81
142. Meents A, Wiedorn MO (2019) Virus structures by X-ray free-electron lasers. Annu Rev Virol 6:161–176
143. Raines KS, Salha S, Sandberg RL, Jiang H, Rodriguez JA, Fahimian BP et al (2010) Three-dimensional structure determination from a single view. Nature 463:214–217
144. Reich ES (2011) Three-dimensional technique on trial. Nature 480:303
145. Hoppe W, Langer R, Frank J, Feltynowski A (1969) Image differentiation procedures in electron microscopy. Naturwissenschaften 56:267–272
146. Maiden AM, Humphry MJ, Zhang F, Rodenburg JM (2011) Superresolution imaging via ptychography. J Opt Soc Am A Opt Image Sci Vis 28:604–612
147. Dierolf M, Menzel A, Thibault P, Schneider P, Kewish CM, Wepf R et al (2010) Ptychographic X-ray computed tomography at the nanoscale. Nature 467:436–439
148. Pei X, Zhou L, Huang C, Boyce M, Kim JS, Liberti E et al (2023) Cryogenic electron ptychographic single particle analysis with wide bandwidth information transfer. Nat Commun 14:3027
149. Jumper J, Evans R, Pritzel A, Green T, Figurnov M, Ronneberger O et al (2021) Highly accurate protein structure prediction with AlphaFold. Nature 596:583–589
150. Baek M, DiMaio F, Anishchenko I, Dauparas J, Ovchinnikov S, Lee GR et al (2021) Accurate prediction of protein structures and interactions using a three-track neural network. Science 373:871–876
151. Jamali K, Kall L, Zhang R, Brown A, Kimanius D, Scheres SHW (2023) Automated model building and protein identification in cryo-EM maps. bioRxiv
152. Callaway E (2020) 'It will change everything': DeepMind's AI makes gigantic leap in solving protein structures. Nature 588:203–204
153. Mutz P, Resch W, Faure G, Senkevich TG, Koonin EV, Moss B (2023) Exaptation of inactivated host enzymes for structural roles in orthopoxviruses and novel folds of virus proteins revealed by protein structure modeling. MBio 14:e0040823
154. Gutnik D, Evseev P, Miroshnikov K, Shneider M (2023) Using AlphaFold predictions in viral research. Curr Issues Mol Biol 45:3705–3732
155. Rossmann MG (2012) Crystallography, evolution, and the structure of viruses. J Biol Chem 287:9552–9559
156. McLellan JS, Pancera M, Carrico C, Gorman J, Julien JP, Khayat R et al (2011) Structure of HIV-1 gp120 V1/V2 domain with broadly neutralizing antibody PG9. Nature 480:336–343
157. Tan TK, Rijal P, Rahikainen R, Keeble AH, Schimanski L, Hussain S et al (2021) A COVID-19 vaccine candidate using SpyCatcher multimerization of the SARS-CoV-2 spike protein receptor-binding domain induces potent neutralising antibody responses. Nat Commun 12:542
158. Arranz R, Coloma R, Chichon FJ, Conesa JJ, Carrascosa JL, Valpuesta JM et al (2012) The structure of native influenza virion ribonucleoproteins. Science 338:1634–1637
159. Bruun TUJ, Andersson AC, Draper SJ, Howarth M (2018) Engineering a rugged Nanoscaffold to enhance plug-and-display vaccination. ACS Nano 12:8855–8866
160. Cohen AA, Gnanapragasam PNP, Lee YE, Hoffman PR, Ou S, Kakutani LM et al (2021) Mosaic nanoparticles elicit cross-reactive immune responses to zoonotic coronaviruses in mice. Science 371:735–741

161. Lu M, Chamblee M, Zhang Y, Ye C, Dravid P, Park JG et al (2022) SARS-CoV-2 prefusion spike protein stabilized by six rather than two prolines is more potent for inducing antibodies that neutralize viral variants of concern. Proc Natl Acad Sci USA 119:e2110105119
162. Ni T, Mendonca L, Zhu Y, Howe A, Radecke J, Shah PM et al (2023) ChAdOx1 COVID vaccines express RBD open prefusion SARS-CoV-2 spikes on the cell surface. iScience 26: 107882
163. Lenza MP, Oyenarte I, Diercks T, Quintana JI, Gimeno A, Coelho H et al (2020) Structural characterization of N-linked Glycans in the receptor binding domain of the SARS-CoV-2 spike protein and their interactions with human lectins. Angew Chem Int Ed Engl 59:23763–23771
164. Casalino L, Gaieb Z, Goldsmith JA, Hjorth CK, Dommer AC, Harbison AM et al (2020) Beyond shielding: the roles of Glycans in the SARS-CoV-2 spike protein. ACS Cent Sci 6: 1722–1734
165. Stagnoli S, Peccati F, Connell SR, Martinez-Castillo A, Charro D, Millet O et al (2022) Assessing the mobility of severe acute respiratory syndrome Coronavirus-2 spike protein Glycans by structural and computational methods. Front Microbiol 13:870938
166. Crank MC, Ruckwardt TJ, Chen M, Morabito KM, Phung E, Costner PJ et al (2019) A proof of concept for structure-based vaccine design targeting RSV in humans. Science 365:505–509
167. Porta C, Kotecha A, Burman A, Jackson T, Ren J, Loureiro S et al (2013) Rational engineering of recombinant picornavirus capsids to produce safe, protective vaccine antigen. PLoS Pathog 9:e1003255
168. Subramanian S, Maurer AC, Bator CM, Makhov AM, Conway JF, Turner KB et al (2019) Filling adeno-associated virus capsids: estimating success by cryo-electron microscopy. Hum Gene Ther 30:1449–1460
169. Arriaga I, Navarro A, Etxabe A, Trigueros C, Samulski RJ, Moullier P et al (2022) Cellular and structural characterization of VP1 and VP2 knockout mutants of AAV3B serotype and implications for AAV manufacturing. Hum Gene Ther 33:1142–1156
170. Abrescia NG, Bamford DH, Grimes JM, Stuart DI (2012) Structure unifies the viral universe. Annu Rev Biochem 81:795–822
171. Ravantti JJ, Martinez-Castillo A, Abrescia NGA (2020) Superimposition of viral protein structures: a means to decipher the phylogenies of viruses. Viruses 12:1146
172. Andres G, Charro D, Matamoros T, Dillard RS, Abrescia NGA (2020) The cryo-EM structure of African swine fever virus unravels a unique architecture comprising two icosahedral protein capsids and two lipoprotein membranes. J Biol Chem 295:1–12
173. Wang N, Zhao D, Wang J, Zhang Y, Wang M, Gao Y et al (2019) Architecture of African swine fever virus and implications for viral assembly. Science 366:640–604
174. Fang Q, Zhu D, Agarkova I, Adhikari J, Klose T, Liu Y et al (2019) Near-atomic structure of a giant virus. Nat Commun 10:388
175. Abrescia NGA (2021) Evolution steered by structure. In: Bamford DH, Zuckerman M (eds) Encyclopedia of virology., Fourth Edition. Elsevier Ltd
176. De Colibus L, Roine E, Walter TS, Ilca SL, Wang X, Wang N et al (2019) Assembly of complex viruses exemplified by a halophilic euryarchaeal virus. Nat Commun 10:1456
177. Rissanen I, Grimes JM, Pawlowski A, Mäntynen S, Harlos K, Bamford JK et al (2013) Bacteriophage P23-77 capsid protein structures reveal the archetype of an ancient branch from a major virus lineage. Structure 21:718–726

Further Reading

https://www.nobelprize.org/prizes/chemistry/2017/dubochet/lecture/
https://www.nobelprize.org/prizes/chemistry/2017/frank/lecture/

https://www.nobelprize.org/prizes/chemistry/2017/henderson/lecture/

Henderson R, Sali A, Baker ML, Carragher B, Devkota B, Downing KH, Egelman EH, Feng Z, Frank J, Grigorieff N, Jiang W, Ludtke SJ, Medalia O, Penczek PA, Rosenthal PB, Rossmann MG, Schmid MF, Schröder GF, Steven AC, Stokes DL, Westbrook JD, Wriggers W, Yang H, Young J, Berman HM, Chiu W, Kleywegt GJ, Lawson CL (2012) Outcome of the first electron microscopy validation task force meeting. Structure 20:205–214

Perutz MF, Rossmann MG, Cullis AF, Muirhead H, Will G, North AC (1960) Structure of haemoglobin: a three-dimensional Fourier synthesis at 5.5-a. resolution, obtained by X-ray analysis. Nature 185:416–422

Read RJ, Adams PD, Arendall WB 3rd, Brunger AT, Emsley P, Joosten RP, Kleywegt GJ, Krissinel EB, Lütteke T, Otwinowski Z, Perrakis A, Richardson JS, Sheffler WH, Smith JL, Tickle IJ, Vriend G, Zwart PH (2011) A new generation of crystallographic validation tools for the protein data bank. Structure 19:1395–1412

Steven AC, Baumeister W (2008) The future is hybrid. J Struct Biol 163:186–195

Chapter 8
3D Cryo-Correlative Methods to Study Virus Structure and Dynamics Within Cells

Rocío Arranz, Francisco Javier Chichón, Ana Cuervo, and José Javier Conesa

Abstract Understanding the dynamic processes involving virus structural components within host cells is crucial for comprehending viral infection, as viruses rely entirely on host cells for replication. Viral infection involves various intracellular stages, including cell entry, genome uncoating, replication, transcription and translation, assembly of new virus particles in a complex morphogenetic process, and the release of new virions from the host cell. These events are dynamic and scarce and can be obscured by other cellular processes, necessitating novel approaches for their in situ characterization. Among these methods, correlative microscopy integrates the labeling, localization, and functional characterization of events of interest through visible light microscopy, complemented by the structural insights provided by high-resolution imaging techniques. This correlative approach enables a comprehensive exploration of subcellular events within the cellular context, including those related to viral morphogenesis. This chapter provides an introduction to correlative three-dimensional imaging methods, specifically designed to study viral morphogenesis and other intracellular stages of the viral cycle under conditions closely resembling their native environment. The integration of whole-cell imaging and high-resolution structural biology techniques is emphasized as essential for unraveling the mechanisms by which viruses generate and disseminate their progeny.

Keywords Viral factory · Virus morphogenesis · Correlative microscopy · Sample vitrification · X-ray microscopy · Electron microscopy · Visible light microscopy · Super-resolution microscopy · Cryogenic soft X-ray tomography · Cryogenic electron tomography · Lamellae preparation · Cryogenic focused ion beam scanning electron microscopy · Volume imaging · Volume electron microscopy

R. Arranz · F. J. Chichón · A. Cuervo · J. J. Conesa (✉)
Department of Macromolecular Structure, Centro Nacional de Biotecnología (CNB-CSIC), Madrid, Spain
e-mail: jj.conesa@cnb.csic.es

M. G. Mateu (ed.), *Structure and Physics of Viruses*, Subcellular Biochemistry 105,
https://doi.org/10.1007/978-3-031-65187-8_8

Abbreviations

AI	After Infection
Cherry-G3BP1	Ras-GTPase-Activating Protein-Binding Protein 1
cryo-CLEM	Cryogenic Correlative Light and Electron Microscopy
cryo-ET	Cryogenic Electron Tomography
cryo-FIB-SEM	Cryogenic Focused Ion Beam Scanning Electron Microscopy
cryo-SXT	Cryogenic Soft X-ray Tomography
cryo-vEM	Cryogenic Volume Electron Microscopy
cryo-VLM	Cryogenic Visible Light Microscopy
FIB-SEM	Focused Ion Beam Scanning Electron Microscopy
Gal3	Galectin-3
GFP	Green Fluorescent Protein
IV	Immature Virion
IEV	Intracellular Enveloped Virus
MV	Mature Virion
MuV	Mumps Virus
SARS-CoV-2	Severe Acute Respiratory Syndrome Coronavirus 2
SE	Secondary Electrons
SEM	Scanning Electron Microscopy
SIM	Structured Illumination Microscopy
SMLM	Single-Molecule Localization Microscopy
STED	Stimulated Emission Depletion
STA	Subtomogram Averaging
TEM	Transmission Electron Microscopy
VACV	Vaccinia Virus
vEM	Volume electron microscopy
VLM	Visible Light Microscopy
2D	Two-dimensional
3D	Three-dimensional

8.1 Introduction

Viruses rely entirely on target host cells for their replication. Therefore, investigating their interaction with cellular components during the intracellular stages of the infectious cycle is crucial to gain insights into how they manipulate the host to generate progeny. In this intracellular stage, viruses produce different molecules, replicate, and mature to release new viruses from the host cell. To achieve this, several viruses form specialized intracellular regions where viral components are concentrated, known as viral factories. These regions form during the infection process and are dedicated to viral replication and/or morphogenesis (including capsid assembly (Chaps. 11 and 12), viral genome packaging (Chap. 13), and virus maturation (Chap. 11)) and to evade the cellular antiviral response [1]. Viral factories are induced by different viruses and result from the remodeling of the cellular environment, their intracellular location and nature being dependent on the virus causing the infection [2]. DNA and RNA viruses replicating in the cytoplasm establish viral factories that lead to alterations in the distribution and dynamics of the cytoskeleton and the relocation of host cell organelles. Implicated organelles include the Golgi apparatus, endocytic compartment, mitochondria, or endoplasmic reticulum (e.g., vaccinia virus or severe acute respiratory syndrome coronavirus-2, SARS-CoV-2; see Sects. 8.5.2 and 8.5.3, respectively) [3, 4]. In the case of DNA viruses replicating in the nucleus, the formation of viral factories involves the reorganization of nuclear components such as chromatin and nuclear domains, including nuclear bodies or the nucleolus [5]. Additionally, membraneless subcellular compartments, known as biomolecular condensates, which compartmentalization involves a liquid-liquid phase separation mechanism [6], are also hijacked or built de novo by viruses to play a role in every step of viral infection [7]. Examples of biomolecular condensates implicated in the intracellular viral cycle include Negri bodies [8], viral inclusions of the measles virus [9], or intracellular inclusions of the mumps virus (MuV) (see Sect. 8.5.1).

While biochemical and genomic analyses play a significant role in understanding these processes, they have limitations since they mostly involve the generation of bulk data. Moreover, events within infected cells that involve viral components are characterized by a highly dynamic and rare nature, often being masked by other cellular processes. In this context, there is a need for characterizing these processes while overcoming those limitations. With this aim, advanced imaging modalities now provide a high-resolution understanding of subcellular events, allowing us to dissect viral processes at the individual event level within the cellular context. Particularly, coupling visible light microscopy (VLM), which allows us to capture the full context of viral processes, with high-resolution information provided by structural biology techniques, is key for understanding how viruses generate and disseminate their progeny.

In this chapter, we provide a comprehensive introduction to correlative three-dimensional (3D) imaging methods suitable for studying virus morphogenesis and other intracellular processes during viral infection, in conditions closest to their native environment.

8.2 A Comprehensive Study of Virus Intracellular Cycle: 3D Correlative Microscopy

Correlative microscopy is a broad term that includes the use of two or more imaging modalities applied to the same sample. Typically, it combines an imaging approach that captures the entire cellular context, such as fluorescence microscopy, with a high-resolution technique like electron microscopy, to provide detailed information at the subcellular level [10, 11]. This integration combines advantages from different approaches, resulting in information that cannot be obtained through any of the individual methodologies alone. As a result, correlative microscopy becomes a valuable tool for the unambiguous identification of different subcellular compartments, the precise location and conformational study of protein complexes (molecular mapping), and the observation of rare and dynamic events from a quantitative perspective.

There is a plethora of possible correlative workflows by combining different VLM methods to provide functional and localization information, and high-resolution techniques, to enable the characterization of the ultrastructure of the cell. That being said, it is important to choose the proper techniques to be combined to obtain the answer to a specific biological question. The appropriate combination can be identified based on the needs of spatial resolution, data throughput, sample preservation, sample volume to be analyzed, and temporal resolution. In this chapter, we will focus on high-precision correlative approaches performed on samples preserved in close-to-native conditions, producing high-resolution 3D information. In the following sections, we will illustrate the different correlative workflow steps: sample preparation in close-to-native conditions, localization of the structures of interest and acquisition of functional information, and high-resolution structural characterization. Additionally, we will describe this methodology focusing on some examples applied to the study of viral morphogenesis and other processes in virus-infected cells.

8.3 Sample Preparation for Its Characterization in Close-to-Native Conditions

Sample preparation is a fundamental step in order to obtain reliable results, especially when the experiment targets the description of high-resolution information. Besides the sample environment required for conserving structural features, in a correlative approach, sample preparation should be compatible with the chosen imaging techniques. In this context, it is also important to consider that the sample preparation method selected will determine the precision of the correlation between the results obtained by VLM and those obtained by the high-resolution technique employed. The required correlation precision depends on the spatial extent of the event of interest. When events are widely distributed within a cell, the necessary correlation precision falls within the range of micrometers. In such cases, identifying whether a specific cell contains the event of interest is generally sufficient. However, when the event of interest is confined to a specific location within the cell, a higher correlation precision in the range of nanometers becomes imperative. This higher correlation precision is essential for effectively targeting and imaging the events of interest at high resolution. A detailed description of the correlation process will be provided later in this section.

High-resolution image acquisition methods considered in this chapter cryogenic electron microscopy and cryogenic soft X-ray tomography (cryo-SXT)—must be conducted under vacuum conditions to avoid the attenuation of the source used (either electrons or X-rays) during its optical path. As previously mentioned, this would modify the sample environment, making it mandatory either to remove (dehydrate) or immobilize (freeze) the water within biological samples to prevent boiling, which damages the sample, interferes with the microscope beam, and reduces the spatial resolution.

8.3.1 Sample Preparation by Dehydration

Conventional transmission electron microscopy (TEM) typically uses dehydration as the preferred method. This process involves several steps that may modify the sample structure, such as chemical fixation with aldehydes, dehydration, resin embedding, and sectioning [12]. Sample fixation through aldehyde cross-linking can create artifacts in biological structures because the process is not instantaneous and can affect protein conformation and structure [13]. However, the primary artifacts are caused by dehydration of the sample, which leads to alterations in the sample's ultrastructure due to shrinkage, membrane reshaping, redistribution, and extraction of soluble elements [14].

8.3.2 Sample Preparation by Freezing Methods: Sample Vitrification

Sample preparation by freezing allows the retention of water molecules in the sample environment. This methodology avoids previously described chemical and dehydration artifacts. Therefore, this approach preserves the sample hydrated in its close-to-native conditions, enabling the acquisition of higher spatial resolution information. To achieve this, freezing has to be performed under specific conditions in a process called vitrification. In this methodology, the sample has to be frozen fast enough so the water contained inside acquires an amorphous ice structure rather than a crystalline one that could damage the specimen [15]. In fact, this nondestructive freezing [16] is an essential step in sample preparation, and its application to biological samples meant a breakthrough in attainable resolution [17].

Vitrification can be achieved through various approaches, either at atmospheric or high pressure, depending on the sample characteristics. At atmospheric pressure, vitrification is accomplished by rapidly immersing the sample in a cryogen (ethane, propane, or a mix of both) at close to liquid nitrogen temperature (plunge freezing) [18] or by projecting the cryogen onto the sample (jet freezing) [19]. This approach results in cooling rates on the order of 100,000 K/s for samples with a thickness of approximately 5 µm, suitable for vitrifying purified complexes, bacteria, and some eukaryotic cells [20]. Vitrifying thicker samples, like larger eukaryotic cells, tissues, or small animals, is more challenging and requires reduced cooling rates. One approach is to use cryoprotectants (like sucrose, some polymers, or glycerol) that slow down the formation of ice crystals by increasing osmotic pressure [21]. However, it is important to note that cryoprotectants may also impact the ultrastructure of the sample. Another possibility is to avoid crystal formation by increasing the pressure to 210 MPa during cooling. This methodology, named high-pressure freezing, reduces the cooling rate to a few 1000 K/s, enabling the vitrification of samples with a maximum thickness of around 200 µm [22]. Regardless of the method used, it is important to note that in order to prevent the formation of ice crystals once the sample is vitrified, it must be maintained at a temperature close to that of liquid nitrogen (−196 °C) and should not exceed the re-vitrification point of water (around −140 °C).

8.3.3 *Correlation Between VLM Imaging and High-Resolution Imaging*

The precision of the correlation between VLM and high-resolution imaging results is influenced by both the sample preparation method and the timing of VLM data acquisition within the workflow. In a correlative workflow that involves chemical fixation and dehydration of the sample, VLM data can be performed at various stages. These stages include before fixation (live imaging), after fixation but before subsequent sample preparation steps, and post-preparation, once the sample is prepared for high-resolution data acquisition. Live imaging and post-fixation imaging tend to yield poorer correlation precision due to substantial differences in the structural distribution and preservation in the sample between imaging modalities. For example, cells analyzed by in vivo VLM could move on the support surface prior to fixation, hindering their relocation in further analysis steps. The highest correlation precision is achieved in the post-preparation stage, after the sample is prepared for high-resolution imaging.

In the case of a workflow involving cryofixation, VLM imaging can be achieved either before vitrification (live imaging) or after vitrification (cryogenic VLM (cryo-VLM) imaging). Similar to conventional sample preparation workflows described, live imaging results in lower correlation precision due to the absence of sample fixation, which allows for cellular or subcellular rearrangements, whereas cryoVLM imaging, which involves sample fixation by freezing, provides a higher level of precision in the correlation between different imaging steps.

As mentioned in the introduction, this chapter will focus on high-precision correlative approaches for obtaining high-resolution 3D information in close-to-native conditions. Therefore, the next section will concentrate only on the description of cryo-VLM imaging approaches, which provide a high correlation level due to the sample vitrification technique that allows fixing the sample while preserving high-resolution features. To learn more about VLM approaches that are included in correlative workflows but do not provide close-to-native information, we recommend the works of Scher and Avinoam [23] or Sanders and collaborators [14].

8.4 Targeting the Structures of Interest: VLM Imaging

VLM provides essential insights into biological processes inside a cell, including viral morphogenesis. In a correlative workflow, VLM plays a fundamental role in localizing the structures of interest and gaining functional information. In this section, we will provide an overview of various labeling approaches and describe different VLM types that can be integrated into a correlative workflow.

8.4.1 Sample Labeling in VLM

Given that biological samples often lack contrast in transmitted light microscopy, different dyes (specific and nonspecific) can be used to label biological structures. Among the nonspecific methods, some examples include Gram staining, which allows classifying bacteria [24], and the hematoxylin and eosin staining, widely applied in histology and pathology [25]. These staining methods are relatively nonspecific. For example, hematoxylin targets acidic structures (like nucleic acids in the cell nucleus), and eosin stains proteins. More specific staining techniques are essential in correlative workflows as they allow the identification of precise cellular or viral structures relevant to the process under study. This need is frequently addressed by fluorescence staining, e.g., by labeling proteins of interest with a fluorescent tag (like green fluorescence protein, GFP) or by immunostaining using antibodies labeled with fluorophores. A fluorophore is a molecule that, when excited by light of a specific wavelength, emits light at a longer wavelength. In fluorescence microscopy, fluorophores label specific structures, enabling high-sensitivity and high-specificity imaging. Fluorescence staining can be conducted in several ways; for convenience, we will group the different procedures into two categories: live cell staining and dead cell staining.

Dead Cell Staining
In dead cell fluorescent staining, it is required that the cell membrane is permeated so that molecules can access their intracellular targets. An example of this is immunostaining, where antibodies labeled with fluorescent molecules target specific viral proteins or cellular compartments [26].

Live Cell Fluorescent Staining
In live cell fluorescent staining, fluorescent labeling must be compatible with living cells, and thus, cell membrane integrity cannot be compromised. Moreover, these

labels must be nontoxic to the cells and should not interfere with cellular processes and functions. Therefore, the fluorophores used must be either cell-permeable or synthesized by the cells themselves (genetically encoded fluorescent proteins). Cell-permeable fluorophores are typically organic small fluorescent molecules (e.g., Alexa Fluor dyes, Atto dyes, Cy dyes, and Dyomics dyes) with an affinity for biological structures, such as specific cellular compartments, nucleic acids, or proteins [27]. Genetically encoded fluorescent proteins can be employed as markers to label specific proteins. For instance, GFP was employed to label the tegument protein VP22 of herpes simplex virus 1 for the study of their internalization and egress processes [28].

8.4.2 *Cryogenic Fluorescent Microscopy*

When considering fluorescence imaging under cryogenic conditions in a correlative workflow, the goal is to provide information about the cellular environment and its functionality for high-resolution data acquisition. In this context, only live fluorescent staining is usable. Compromising cellular integrity, necessary for dead cell staining, can also impact the preservation of biological structures, thus limiting the achievable resolution.

Wide-Field Microscopy and Confocal Microscopy
Among the various fluorescence microscopy approaches, the most widely used ones are wide-field and confocal microscopy.

In wide-field microscopy, the sample is illuminated by a broad beam, resulting in rapid image acquisition (Fig. 8.1a). Wide-field microscopy provides two-dimensional (2D) images and is known for its simplicity and cost-effectiveness due to the use of fewer optical elements compared to other imaging modalities. However, it is important to note that this method illuminates the entire sample volume in each acquired image, which can lead to the inclusion of out-of-focus information, causing image blur.

In confocal microscopy, a laser beam is collimated to a spot that is raster-scanned over the sample at different depths, providing 3D information. Before reaching the detector, the emitted fluorescent light from each illuminated spot encounters a pinhole, which removes most of the out-of-focus information coming from sample depths different from the one where the excitation spot is positioned (Fig. 8.1b). This results in images containing information from a specific sample depth and with higher resolution compared to wide-field microscopy.

Wide-field and confocal microscopy can be adapted for operation under cryogenic conditions; however, these approaches exhibit certain differences when used in such conditions. The sample stage must maintain the specimen at temperatures close to those of liquid nitrogen (−196 °C) to prevent devitrification and damage of biological structures. Additionally, the objectives used need to have a long working distance to withstand the low temperatures applied to the sample. Consequently,

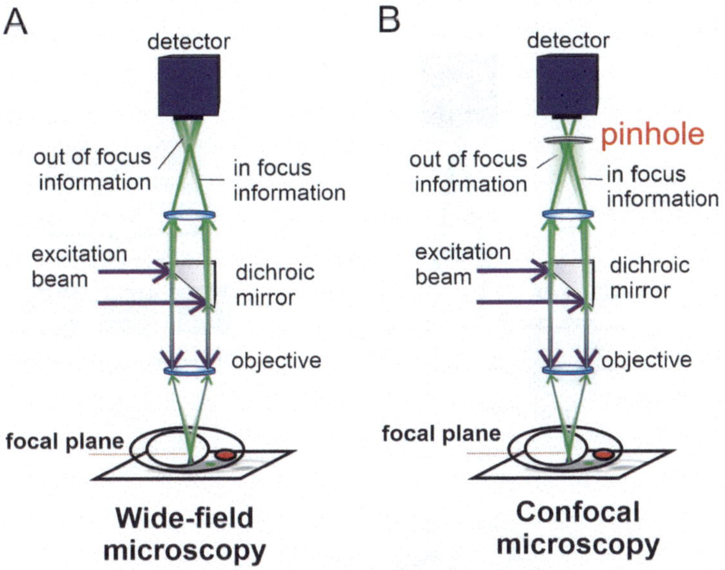

Fig. 8.1 Schematic representation of wide-field and confocal microscopes. (**a**) optical path scheme of a wide-field microscope. The excitation light (purple) illuminates the sample, and emitted light (green) reaches the detector after being selected by the dichroic mirror. (**b**) optical path of a confocal microscope. The excitation light (purple) illuminates the sample. The emitted light (green) reaches the detector after being selected by a dichroic mirror and passing through a pinhole, which blocks the out-of-focus information

these objectives often have a lower numerical aperture, which limits the achievable spatial resolution in the resulting images [29].

However, wide-field and confocal microscopy are constrained by the diffraction limit [30], restricting their lateral spatial resolution to around 250 nm [31]. VLM approaches that surpass this spatial resolution limit are referred to as super-resolution VLM techniques.

Super-Resolution Visible Light Microscopy

As explained before, these methods involve VLM approaches that are able to improve spatial resolution. These techniques could require sample immobilization, such as in stimulated emission depletion (STED) microscopy, or utilize the blinking or photoswitching of fluorescent molecules, as seen in single-molecule localization microscopy (SMLM) [32].

In cryogenic conditions, the sample is already immobilized by the vitrification process; however, maintaining the sample under these conditions faces some challenges. STED microscopy and SMLM techniques, like photoactivated localization microscopy or stochastic optical reconstruction microscopy, demand high laser intensities that are incompatible with vitrification preservation. Additionally, the physical properties of fluorophores under cryogenic conditions can differ from those

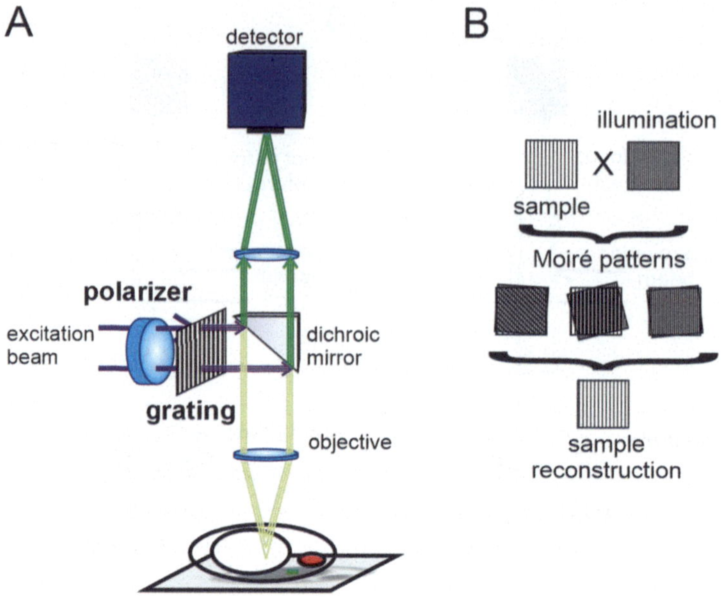

Fig. 8.2 Schematic representation of a structured illumination microscope. (**a**) optical path scheme of a structured illumination microscope. The excitation light (purple) passes through a polarizer, linearly polarizing it. Then, the excitation light travels through a grating with lines parallel to the polarization vector of light, creating a sinusoidal illumination pattern. This structured light is then projected onto the sample, and the emitted light (green) reaches the detector after being selected by the dichroic mirror. (**b**) concept of resolution enhancement by SIM. When the structured light with a known pattern is projected onto the sample, the emitted light forms a Moiré pattern. Structured light is projected out of the sample at three different defined angles and phases, producing a set of Moiré patterns. These generated Moiré patterns are then processed computationally to obtain a SIM-reconstructed image with increased resolution [35]

at room temperature [33], requiring further investigation to adapt these technologies for use in cryogenic conditions [34].

However, there are well-established VLM super-resolution approaches that work effectively in cryogenic conditions as they do not require high laser intensities, blinking, or photo-switchable fluorophores. These techniques are structured illumination microscopy (SIM) and Airyscan microscopy.

Structured Illumination Microscopy. The SIM technique is a super-resolution VLM method that can double the resolution achieved by conventional wide-field microscopy by illuminating the sample with striped patterns of light, named Moiré fringes (Fig. 8.2a). The interference between these light patterns and the emitted sample fluorescence can be used to computationally recover information beyond the diffraction limit [35] (Fig. 8.2b).

Airyscan Microscopy. Airyscan microscopy is a VLM super-resolution approach that, when combined with confocal microscopy, can provide a resolution increase similar to that offered by cryo-SIM. In this approach, the traditional pinhole used in confocal microscopy is replaced by an array detector, enabling the collection of more

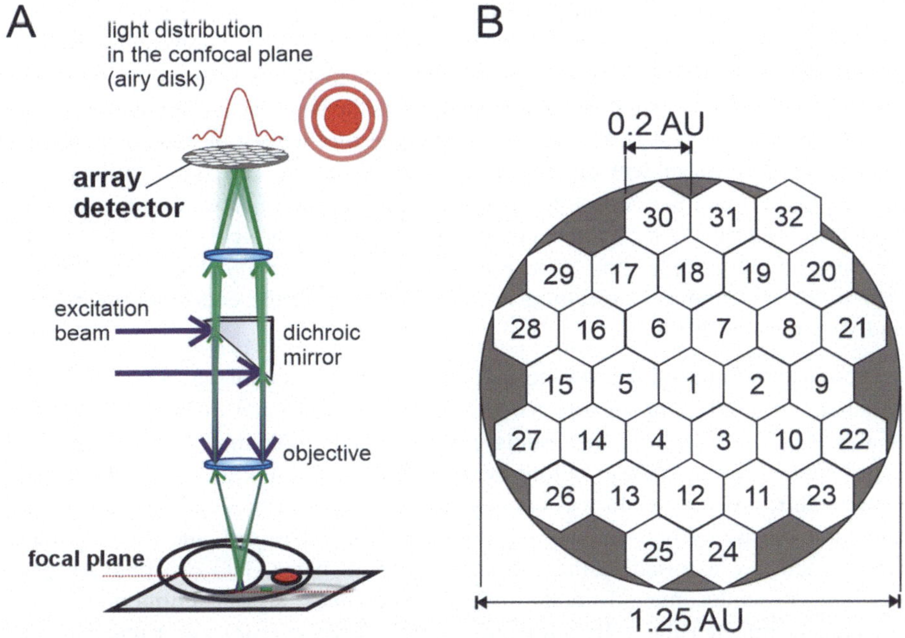

Fig. 8.3 Schematic representation of an Airyscan microscope. (**a**) optical path scheme of an Airyscan microscope. The excitation light (purple) illuminates the sample. The emitted light (green) is selected by a dichroic mirror and reaches the array detector placed in the confocal plane, where, in a confocal setup, the pinhole would be placed. Above the detector, the red plot and dot surrounded by rings depict the light distribution (Airy disk) in the confocal plane, representing the lateral and transverse direction, respectively. (**b**) layout of a 32-element array detector, each element functioning as an individual 0.2-Airy disk unit pinhole. This design enables the rejection of out-of-focus information without a pinhole and allows determining the Airy disk center displacement relative to the optical axis. The circular arrangement of elements generates images equivalent to those produced with a 1.25-Airy disk unit pinhole confocal system, resulting in enhanced SNRs. Consequently, images acquired with this array detector design, after reconstruction (pixel reassignment principle) [37], exhibit a resolution limited by the detection units (0.2-Airy disk units) and sensitivity equivalent to those acquired with a 1.25-Airy disk unit pinhole in confocal microscopy

light compared to confocal microscopy and thereby increasing the signal-to-noise ratio. In this method, each of the detection elements within the detector array acts as an independent pinhole. By knowing the dimensions and positions of each element relative to the optical axis, an improved image can be reconstructed [36] (Fig. 8.3).

Case Study of Cryogenic Super-Resolution VLM: Reovirus Escape from Endosomes

We illustrate the utility of cryogenic super-resolution VLM methods in the study of viral infection by addressing the work by Kounatidis and collaborators [38]. In this work, the authors applied cryo-SIM to investigate the infection process of reovirus, specifically focusing on the mechanism of viral escape from endocytic vesicles.

Reoviruses are non-enveloped, icosahedral virus with a double-stranded RNA genome, belonging to the *Sedoreoviridae* family. It is now widely recognized as prevalent in the human population, primarily spreading via respiratory or enteric routes [39]. Reovirus, being non-enveloped, lacks the capacity for membrane fusion to enter host cells; however, it is known that the mechanism of entry is mediated through clathrin-dependent endocytosis. Once inside the cell, viral particles are transported through the endosomal pathway. The escape of reovirus core particles from these endosomes is facilitated by low-pH-dependent proteolytic degradation of the capsid protein $\sigma3$, leading to the release of the amino-terminal domain of the $\mu1$ capsid protein, ultimately resulting in pore formation in the endosome membrane and viral release [40].

To track the intracellular vesicle trafficking of reovirus particles, Kounatidis and collaborators used a human osteosarcoma cell line expressing mCherry Galectin-3 (Gal3). Gal3 is a protein that binds extracellular carbohydrates and is initially distributed in the cytoplasm before being transported to the cell surface. In their experiments, mCherry-Gal3 was employed to monitor the endocytic compartments and viral release. When endosomal membranes are disrupted, the carbohydrates within them are exposed to the cytoplasm, and mCherry-Gal3 can bind to them resulting in endosomes decorated with fluorescence signal. Furthermore, the researchers infected the cells with fluorescently labeled (Alexa647) reovirus and transfected them with a GFP-Rab7 protein, a late endosome marker, to distinguish between different endocytic compartments.

Using cryo-SIM, they were able to establish the time frame for reovirus endosomal release. One hour after infection (AI), viral particles were confined within a specific endosomal population, as identified by GFP-Rab7. By 2 h AI, approximately 60% of these vesicles had been permeated and labeled by Gal3, leading to the release of the viral particles. Moreover, these vesicles further developed into multivesicular bodies as they progressed toward the perinuclear area within the first 3 h AI. Using fluorescence information, they could determine the 3D spatial locations of the various vesicles containing viral particles (Fig. 8.4), providing the localization information needed for further higher-resolution analysis using transmitted soft X-rays. Further details on transmitted soft X-ray microscopy will be provided in the next section.

8.5 Structural Characterization: High-Resolution Imaging Techniques

After targeting the study subject using cryo-VLM techniques and appropriately identifying the sample area of interest, the next step in a cryogenic correlative approach is the structural characterization of the identified region using high-resolution imaging techniques.

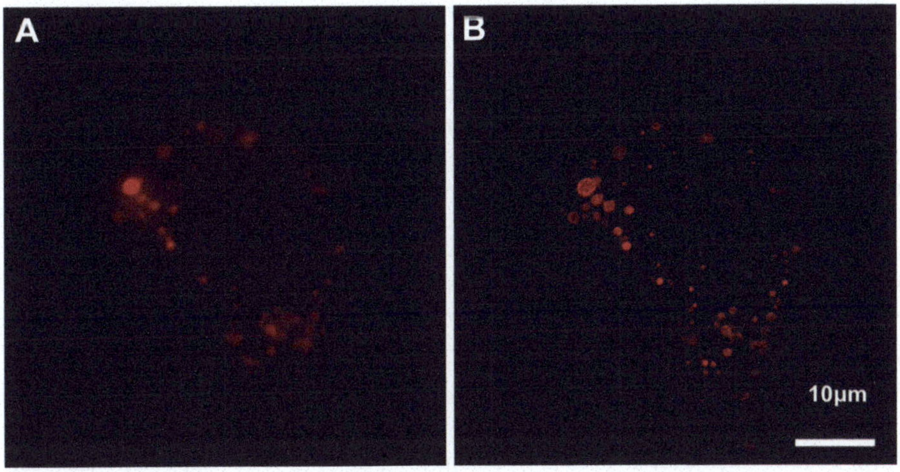

Fig. 8.4 cryo-SIM of reovirus U2OS infected cells. (**a**) wide-field image of a mammalian U2OS reovirus-infected cell. Red signal corresponds to mCherry-Gal3 protein construction. (**b**) reconstructed SIM image of the same cell shown in A. Adapted from Kounatidis and collaborators [38]

Conventional 2D imaging methods for studying cells provide information about single planes within much larger structures. While sufficient for identifying viruses and related structures, they may omit many elements and contacts. For instance, a single 2D TEM image cannot definitively determine whether a viral particle is inside a vesicle or undergoing budding into a membrane. In contrast, 3D methods overcome this limitation. 3D reconstructions can be obtained using different approaches. In TEM and cryo-SXT, image processing software combines 2D images from different views into a volume. These 2D images can be obtained from different samples in different orientations, as in single-particle cryo-electron microscopy techniques (see Chap. 3), or by tilting the sample holder, as in tomography techniques (see the following section). On the other hand, 3D ultrastructure can be obtained by serially sectioning the sample, as in vEM.

There is a plethora of technologies that provide 3D ultrastructural information. As stated previously, we will only refer to 3D imaging methods compatible with cryopreservation, capable of providing high-resolution ultrastructural information in the nanometer range and that can be correlated with VLM with high precision. These methods include those based on X-rays, such as cryo-SXT, as well as electron-based methods like cryogenic electron tomography (cryo-ET), and cryogenic volume electron microscopy (cryo-vEM).

8.5.1 Structure Determination Using cryo-ET

cryo-ET is a structural biology technique that employs electrons to image the sample of interest under cryogenic conditions. It involves obtaining projections of an object following a tomographic acquisition scheme and then computationally combining them to render the 3D structure of the object (see Chap. 3). This powerful method is capable of conducting in situ structural studies in unperturbed cellular environments, combining the ability for 3D molecular-level imaging with the best achievable structural preservation physically possible [41]. cryo-ET can be used to image a variety of samples, including complex nonsymmetric or pleomorphic viruses.

Sample Preparation for cryo-ET

In the last 20 years, cellular cryo-ET has been deployed to directly visualize cellular compartments and allow the study of the conformation of protein complexes and their cellular distribution (visual proteomics) [42, 43]. This approach has been applied extensively in cell infection research [44–48]. However, several limitations hinder its direct application to in situ studies, including the thickness of cell and tissue samples and difficulties in locating and identifying complexes of interest within them. In TEM, besides the requirements of sample vitrification for size, the sample thickness is restricted to less than 500 nm to make them transparent to electrons, and most samples require additional thinning steps.

Traditionally, this reduction of sample thickness was accomplished using a cryo-ultramicrotome, where a knife sequentially cuts the sample of interest. Although sufficiently thin sections can be obtained with a cryo-ultramicrotome, this method has seen limited use due to compression artifacts that alter the cellular ultrastructure and cause protein denaturation. To overcome this limitation, a new methodology applied to biological samples has been developed in recent years, named cryo-focused ion beam scanning electron microscopy (cryo-FIB-SEM). Originating in the material science field, this technique uses an ion beam (usually gallium) to remove part of the sample. cryo-FIB-SEM has emerged as the preferred method for compression-free specimen thinning, allowing the creation of layers approximately 100–500 nm thick (lamellae) throughout the bulk of vitrified samples from cells, tissues, or small animals [49]. Gallium ion sources offer several advantages, such as low melting point, good emission and vacuum properties, and the ability to be focused to spots as small as tens of nanometers, providing an ideal spot dimension for the precise erosion of materials. The production of lamellae has opened the possibility of visualizing intricate molecular details in thicker samples using cryo-ET, such as regions deep within eukaryotic cells, rather than just the thin cell periphery.

Nonetheless, technical challenges related to the other limiting factor, the localization of features of interest, make it difficult to ensure that the milled lamella contains the features of interest. Cryogenic correlative light and electron microscopy (cryo-CLEM) can help overcome this challenge by labeling targets with fluorophores [50] (see Sect. 8.4.2). This labeling assists in guiding the cryo-FIB-

SEM milling process [51] and selecting the most suitable imaging areas for cryo-ET experiments. The success of this workflow also relies on the accuracy of the correlation between the sample positions determined by cryo-VLM and those assigned to the FIB-SEM microscope. Such labeling is also essential at the end of the workflow to analyzed the results and assist in the interpretation of observed features.

In the study of very thick samples, cryo-FIB-SEM can be combined with cryo-compatible micromanipulators. Slabs from high-pressure frozen samples, which are not compatible with TEM, can be lifted out and placed on electron microscopy (EM) grids for further thinning. This cryo-FIB-SEM "lift-out" method significantly broadens the range of samples that can be examined using cryo-ET, including eukaryotic cells, multicellular organisms, and tissues [52]. As mentioned previously, lamellae preparation has typically been carried out using a gallium beam to erode part of the sample on top and bellow the desired region. However, this approach may present challenges when dealing with very thick samples that require substantial material removal. Achieving a significant material removal rate within a reasonable time frame requires the application of high FIB beam currents. Yet, at these elevated currents, the beam diverges, limiting the precision of the milling process. Consequently, these limitations extend the milling times required for high-precision erosion. Additionally, ion implantation can occur in any focused ion beam setup, and gallium has been shown to react with various samples, potentially altering the underlying structures [53]. An alternative to the gallium ion source involves using plasmas generated from gasses like xenon, oxygen, or argon. These sources offer a smaller focus spot at higher currents [54, 55], presenting the opportunity to enhance throughput in lamellae preparation. This capability allows for the milling of large volumes within a reasonable time frame. Recently, this technology has been developed and demonstrated its capacity to produce cryogenic lamellae of cells for in situ structure determination [55].

cryo-ET Image Acquisition

In cryo-ET image acquisition, different views of an electron-transparent vitrified sample are obtained by rotating the sample holder around a defined tilt axis inside the transmission electron microscope (Fig. 8.5). The typical acquisition scheme covers a maximal range of 140°, between ±70° [56]. This process results in a series of 2D projections, known as a tilt series (Fig. 8.5), which are collected under low electron dose conditions (typically around 100–120 $e^{-}/Å^{2}$ accumulated dose across the tilt series) to prevent radiation damage to the sample [57]. The tilt range is limited due to physical instrument constraints and sample thickness, resulting in an incomplete set of projections and missing information (missing wedge), which causes feature elongation and reduced resolution in the direction of the electron beam [56]. To obtain the 3D volume of the sample (tomogram), these tilt series images have to be combined computationally (Fig. 8.5). The different steps of this process will be detailed in the following section.

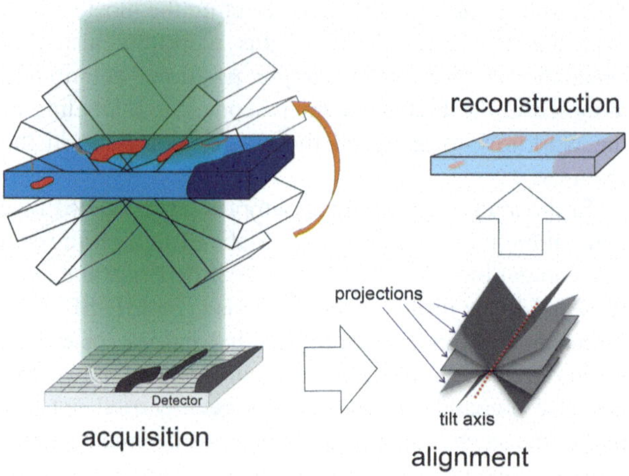

Fig. 8.5 Electron tomography working principle. 2D projections of the samples are collected at various tilting angles. These projections are then aligned to a common rotation axis and combined to generate a 3D reconstruction, known as a tomogram

Tomographic Data Processing

Tomographic data processing can be divided into three broad categories: reconstruction, segmentation, and averaging of repetitive structures. The reconstruction process involves building a volume (tomogram) in silico from the acquired tilt series of 2D projections (Fig. 8.5). This is achieved through tomographic reconstruction algorithms such as back projection, weighted back projection, or iterative reconstruction methods [58]. These algorithms are based on the inverse of the Radon transform [59], which means that a 3D object can be uniquely defined from its projections in all spatial directions. After volume reconstruction, it is essential to identity and assign electron densities to different biological structures (e.g., viral capsids, membranes, nucleus). This process enables the distinction of these structures within noisy and low-contrast data. Segmentation can be performed manually or automatically [60] and results in a color-coded representation of the data, which significantly aids in interpretation and communication. To enhance the details of repetitive components within the tomogram (e.g., viral capsids), an additional processing step called subtomogram averaging (STA) can be carried out. STA is an in silico procedure aimed at combining subtomograms (subvolumes within tomograms) containing repetitive structures (macromolecular complexes) to produce a 3D model (subtomogram average) with enhanced resolution and signal-to-noise ratio (SNR) compared to the initial tomograms. The successful application of subtomogram averaging depends on identifying the structural homogeneity of the complex within the cellular environment and its presence in various orientations. This method enables structures to be resolved with accuracy of a few Å, providing valuable functional and mechanistic information [61]. Moreover, the refined model can be placed back into the positions of the original subtomograms with their initial orientations, facilitating quantitative analysis of their abundance, distribution, and interactions with other biological structures [62].

Case Study of cryo-ET: Mumps Virus Condensates
In the context of viral infection, initial cryo-ET studies were limited to the periphery of cells and usually focused on the early stages of infection, analyzing surface proteins and virus entry. However, recent advancements in the field, including cryo-thinning techniques and the implementation of correlative approaches, have played a pivotal role in significantly expanding research capabilities. These advancements have proven to be powerful tools for unveiling the underlying molecular mechanisms of replication, regardless of their location within the cell. Combined with progress in specialized image analysis and processing methods, these developments have enabled researchers to uncover intricate details of virus assembly and release at sub-nanometer resolutions [63].

As an example, the study by Zhang and collaborators demonstrates how the combination of whole-cell proteomics and cryo-CLEM provides a powerful tool to probe molecular and structural aspects of viral replication within cells. This approach allows for the proposition of a mechanistic model for stress-mediated reactivation in persistent infection, particularly in the context of biomolecular condensates [64]. The authors investigated the basis for stress-mediated reactivation in a member of negative-stranded RNA viruses, MuV (Paramyxoviridae family), by performing in situ cryo-CLEM and cryo-ET workflows on the viral factories. For this purpose, HeLa cells infected with MuV and expressing a fluorescent protein as a marker of cellular stress (mCherry-G3BP1, Ras GTPase-activating protein-binding protein 1 labeled with the fluorophore mCherry), were seeded on grids, treated with different stress conditions, and then vitrified (Fig. 8.6a). MuV factories were found in close proximity to stress granules, necessitating the use of mCherry-G3BP1 fluorescence for targeted 3D cryo-FIB thinning and to locate unlabeled viral factories for cryo-ET (Fig. 8.6a). Tomogram analysis demonstrated that MuV viral factories and nucleocapsids undergo structural changes during stress. Notably, prolonged stress induced a morphological transition from curved to straight nucleocapsid morphology (Fig. 8.6b). The conformation of the nucleocapsids in the cellular context was examined through subtomogram averaging, revealing an RNA-accessible state under prolonged stress (Fig. 8.6c).

8.5.2 *Ultrastructural Characterization Using cryo-SXT*

cryo-SXT is a method that uses soft X-rays, which encompass low-energy X-rays ranging from a few electronvolts (eV) to approximately 10 keV, to image the sample of interest. Sample imaging with X-rays has several advantages: improved contrast in the biological structures imaged and greater penetration power, thus eliminitating the need for the sample thinning step described earlier for cryo-ET (see Sect. 8.5.1).

cryo-SXT Image Acquisition and Processing
cryo-SXT operates with incoming X-ray energies falling within the water window energy range (284–543 eV). In this energy range, carbon, a primary component of

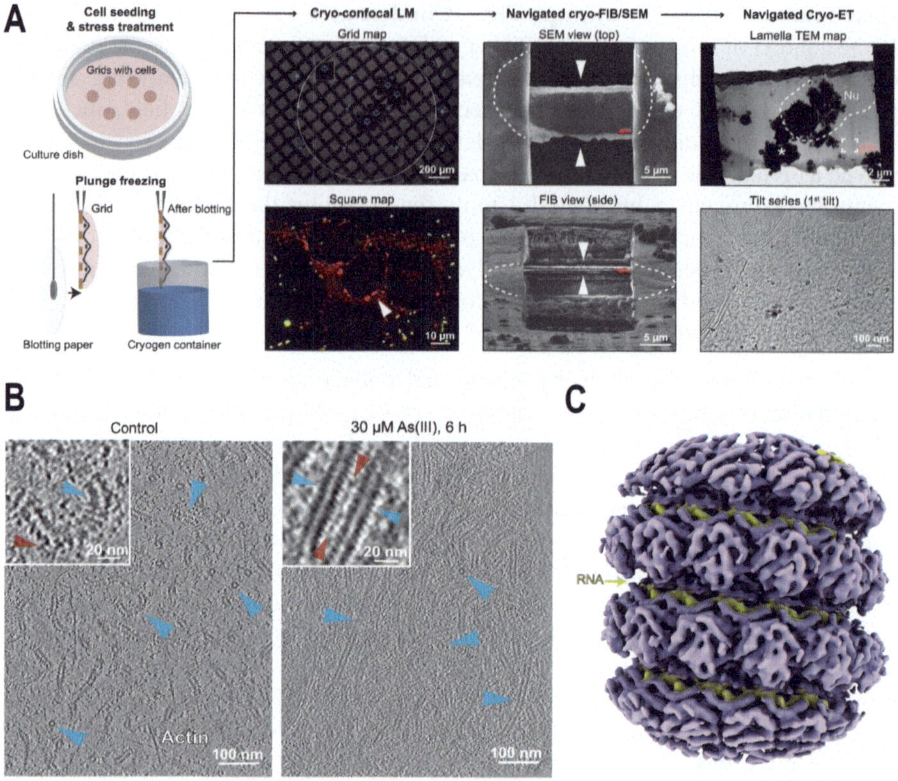

Fig. 8.6 Cryogenic correlative electron tomography and tomogram analysis of MuV condensates. (**a**) 3D cryo-CLEM workflow targeting unlabeled MuV factories in the vicinity of stress granules. Cells were seeded and treatments were applied at different time points of stress. Then, they were plunge-frozen, followed by imaging with a cryo-confocal light microscope (LM). Cells showing stress granule signals (red; indicated with white arrowhead) and fluorescent beads (yellow) as fiducials for 3D correlation between LM and SEM/FIB images. Lamellae were then prepared by cryo-FIB. Correlating SEM/FIB images with 3D LM signals after lamella preparation confirmed stress location preservation (red signal) in lamella. Overlaying the SEM image with transformed fluorescent signal onto the cryo-TEM image of the lamella. At nearby regions where filamentous structures were seen, tilt series acquisition was performed, and later, they were identified to be MuV nucleocapsids. Nu, nucleus. Asterisk: ice contaminants introduced during transfers between the cryo-microscopes. (**b**) representative tomographic slices of nucleocapsids (cyan arrowheads) and associated flexible densities (red arrowheads in insets) in viral factories at indicated control and stress condition. (**c**) subtomogram average of the in-cell MuV nucleocapsids resolved to 6.5 Å. RNA densities in yellow. Adapted from [64]

biological structures, absorbs significantly, while oxygen, present in the water within cells, exhibits poor absorption. This intrinsic contrast in biological samples eliminates the need for staining agents, enabling the analysis of fully hydrated biological specimens. This technique typically involves placing the microscope in a synchrotron, although tabletop setups are now available [65]. To maintain the samples under the vacuum conditions required by the microscope, it is necessary to immobilize them (and the water within them) through vitrification. As mentioned earlier, cryo-SXT is also characterized by its substantial penetration power, enabling the examination of samples with a thickness of approximately 10 μm. This range encompasses the typical thickness of most cellular samples, allowing the imaging of whole cells without sectioning.

The acquisition scheme used to obtain the 3D ultrastructure of the cell in cryo-SXT closely resembles that described for cryo-ET (see Sect. 8.5.1 and Fig. 8.5). It involves acquiring projections at various tilting angles around a rotation axis ranging from ±70°. Subsequently, these projections are computationally combined to reconstruct the cell's volume in a similar way as in cryo-ET (see Sect. 8.5.1 and Fig. 8.5). Because X-ray absorption is proportional to the elemental composition of the sample, the absorption values obtained for each voxel in the 3D reconstructions serve as a signature of their biochemical composition. As a result, they can be utilized to differentiate between various cellular compartments [66]. The highest lateral resolution achieved in cryo-SXT is approximately 25 nm, enabling the visualization of subcellular compartments but not the resolution of individual protein complexes [67].

Case Study of cryo-SXT: Vaccinia Virus Viral Factories

cryo-SXT has found applications in various viral studies [38, 68–73]. To illustrate the insights that can be gained through a correlative approach involving cryo-SXT, we will discuss the work of Chichón and collaborators [74]. In this early study, they examined the morphogenesis of the vaccinia virus (VACV) using cryo-SXT on infected cells previously identified through cryo-VLM.

The vaccinia virus belongs to the Poxviridae family, comprising large, enveloped, double-stranded DNA viruses that replicate in the cytoplasm and infect both vertebrate and invertebrate animals [75]. During the VACV morphogenetic cycle, a viral factory is formed in the cytoplasm. Within this factory, viroplasmic dense bodies contain viral core proteins and DNA. These dense particles incorporate membrane sheets derived from the endoplasmic reticulum, forming extended crescents that evolve into spherical immature virions (IVs). These IVs undergo maturation in a complex process involving the reorganization of viral envelopes and the formation of the viral core, resulting in mature viruses (MVs). MVs have a characteristic brick shape, with a diameter of 250 nm and a length of 350 nm. The maturation process of MVs further proceeds through interactions with the trans-Golgi network, acquiring two additional membranes and forming intracellular enveloped viruses (IEVs). IEVs can exit cells through exocytosis, losing their outermost membrane and becoming extracellular enveloped viruses (EEVs) [76].

The study of VACV maturation process in whole cells using cryo-SXT enabled the identification of IVs and MVs, facilitating the examination of their differential distribution. Viral particles tracked trough GFP labeling (Fig. 8.7a and b) were found compartmentalized within viral cytoplasmic factories based on their maturation state (Fig. 8.7c and d). While IVs tended to accumulate in perinuclear areas, MVs were predominantly located in the cellular periphery. This observation also suggested the

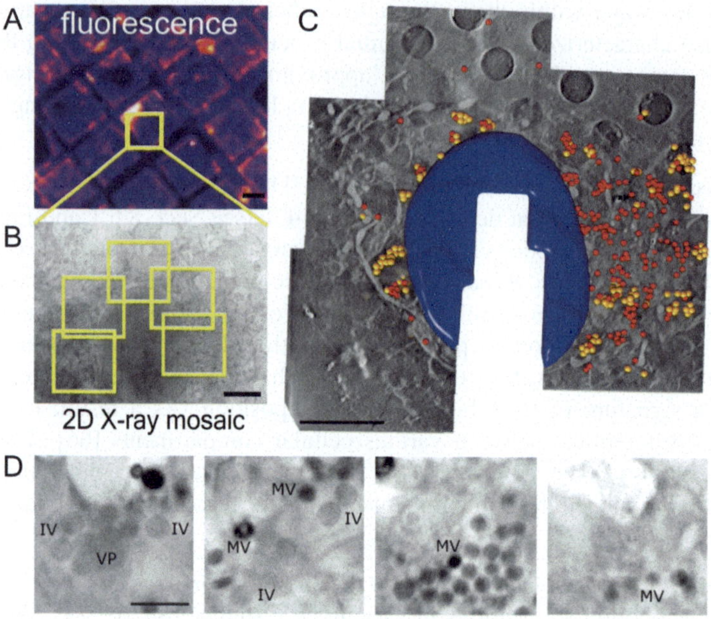

Fig. 8.7 Cryo-SXT 3D ultrastructure of a vaccinia-infected cell. (**a**) vaccinia-infected cells were located in the sample by fluorescence imaging based on the signal of a GFP-labeled vaccinia virus. (**b**) a mosaic of X-ray 2D images was collected on the same area shown in A. The mosaic was used to position the acquisition of cryo-SXT tilt series. (**c**) acquired tilt series in B were reconstructed to provide the distribution of vaccinia intracellular particles; differential distribution of the immature and mature forms (red and yellow particles, respectively) is shown. The cell nucleus is colored in blue. (**d**) tomogram sections showing different maturation states of vaccinia virus, viroplasm (VP), immature virions (IV), and mature virions (MV). Scale bars: A, 50 μm; B, 10 μm; C, 5 μm; D, 500 nm. Adapted from [74]

presence of transitional regions between IVs and MVs within the factories. Additionally, other maturation intermediates, such as viroplasm attached to endoplasmic reticulum-derived crescents and incomplete IVs, were identified in the viral factories (Fig. 8.7c and d).

8.5.3 Cryo-Volume Electron Microscopy

Volume electron microscopy (vEM) encompasses a set of nanometer-resolution imaging methodologies used to reveal the 3D structure of cells, tissues, and small model organisms, particularly those with volumes exceeding 1 μm. Within these methodologies, we include TEM and SEM approaches. In these techniques, the sample of interest is imaged and sliced sequentially, providing a complete picture of samples thicker than 1 μm.

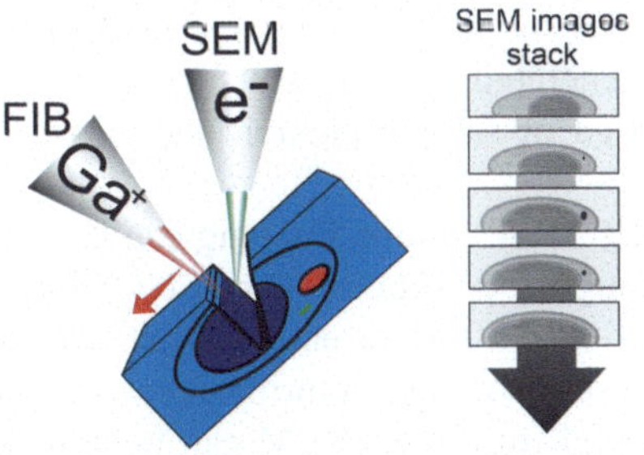

Fig. 8.8 Working principle of cryo-FIB-SEM volume imaging. The cryo-FIB-SEM data acquisition begins with the FIB gun eroding a trench in a vitrified sample, opening the perspective for SEM image acquisition. Subsequently, the FIB beam is shifted in the direction of the red arrow to remove a thin layer of material. A new surface is created and imaged using SEM. This cycle is repeated until the entire volume of interest is captured, resulting in a stack of SEM images which represents the 3D structure of the sample

Volume Imaging Techniques Involving Transmission Electron Microscopy (TEM)
As explained in Sect. 8.5.1, in TEM, the electron beam passes through a thin slice of biological material, typically up to 500 nm thick, to generate a projection. To acquire 3D information at depths beyond 1 μm, consecutive sample sections must be imaged following a single projection or tomography scheme (as detailed in Sect. 8.5.1) and then stacked to create the final volume. It is important to note that these methodologies are not performed under cryogenic conditions due to the challenges associated with cryo-ultramicrotomy and the accompanying artifacts [77]. Therefore, they fall outside the scope of this chapter but are extensively reviewed elsewhere [78].

Volume Imaging Techniques Involving Scanning Electron Microscopy
In volume imaging techniques involving scanning electron microscopy (SEM), the electron beam is raster-scanned across the sample surface, resulting in the generation of secondary (SE) and back-scattered electrons used to create sample images. Among SEM volume imaging techniques, we find different methods such as serial block-face SEM and focused ion beam SEM (FIB-SEM). Of these approaches, only FIB-SEM is compatible with the requirements of cryogenic conditions. The other techniques mentioned are beyond the scope of this book chapter; further information on these techniques can be found in this review [78].

In cryo-FIB-SEM, the cryopreserved sample is initially imaged by SEM using the SE signal to create a surface image. Subsequently, a thin layer of the sample surface (as thin as 10 nm) is eroded through FIB milling, producing a new surface that is then imaged by SEM again. This process is repeated sequentially until the entire volume of interest is completely eroded, resulting in a stack of SEM images that provides information of the complete analyzed sample. After processing and alignment, this stack of images is used to create a volume representation of the sample's ultrastructure (tomogram), achieving a resolution at the subcellular compartment level (around 10 nm resolution) and enabling the tracking of large protein complexes (Fig. 8.8)

[79]. As stated in Sect. 8.5.1 for lamellae preparation, the FIB gallium ion source can be replaced by plasma generated from gasses such as xenon, oxygen, or argon for vEM, increasing the speed of sample erosion and allowing larger volumes to be reached [54, 55].

Furthermore, cryo-FIB-SEM volume imaging can be seamlessly integrated with cryo-VLM imaging through a correlative approach. This enables the localization information collected by cryo-VLM to enhance the data acquisition in cryo-FIB-SEM, increasing the success rate in capturing areas of interest. Additionally, the functional insights obtained from cryo-VLM can be contextualized within the 3D ultrastructure, providing supplementary information [80, 81]. While this approach has not yet been applied to the study of viral morphology, it is easy to envision how this technology can significantly contribute to our understanding of viral morphogenesis. For instance, it could elucidate the specific subcellular compartments to which a particular viral intermediate is related by fluorescently labeling a specific compartment and correlating this information with the ultrastructure of infected cells.

Case Study of cryo-FIB-SEM: Viral Morphogenesis of SARS-CoV-2
The cryo-FIB-SEM approach is relatively novel, and to date, in viral morphology studies, it has only been applied to study the infection of SARS-CoV-2 [68]. Coronaviruses are small enveloped viruses with RNA-positive non-segmented genomes. Coronaviruses infect many different animals, and they can cause mild to severe respiratory infections in humans. In 2019, SARS-CoV-2 emerged rapidly spreading across the globe and causing an exceptional threat to global public health [82]. By means of cryo-FIB-SEM, Mendonça and collaborators were able to describe a series of cytopathological alterations in SARS-CoV-2-infected cells. The authors observed numerous electron-dense particles at the cellular periphery, which were consistent with the dimensions of individual viruses (Fig. 8.9a and e). Additionally, they identified tubular structures extending from the cell surface into the cytoplasm, containing virus-like particles (Fig. 8.9b and e). In the cytoplasm, the authors observed vesicles and large complex membranous compartments housing virus-like particles, along with extensive cytoplasmic vacuolization, especially in the perinuclear area (Fig. 8.9b, c, and e). Moreover, they observed severe alterations in the nuclear ultrastructure, characterized by cytoplasmic invaginations that could occupy nearly half of the cell nucleus (Fig. 8.9c, d, and e).

8.6 Perspectives

The emergence of cutting-edge tools and technologies has been crucial in driving significant progress in our understanding of how cells, tissues, and organisms operate in both health and disease. Obtaining the 3D structures of viral components in isolation does not suffice to achieve a comprehensive understanding of the numerous processes occurring during the viral life cycle. Complementing high-

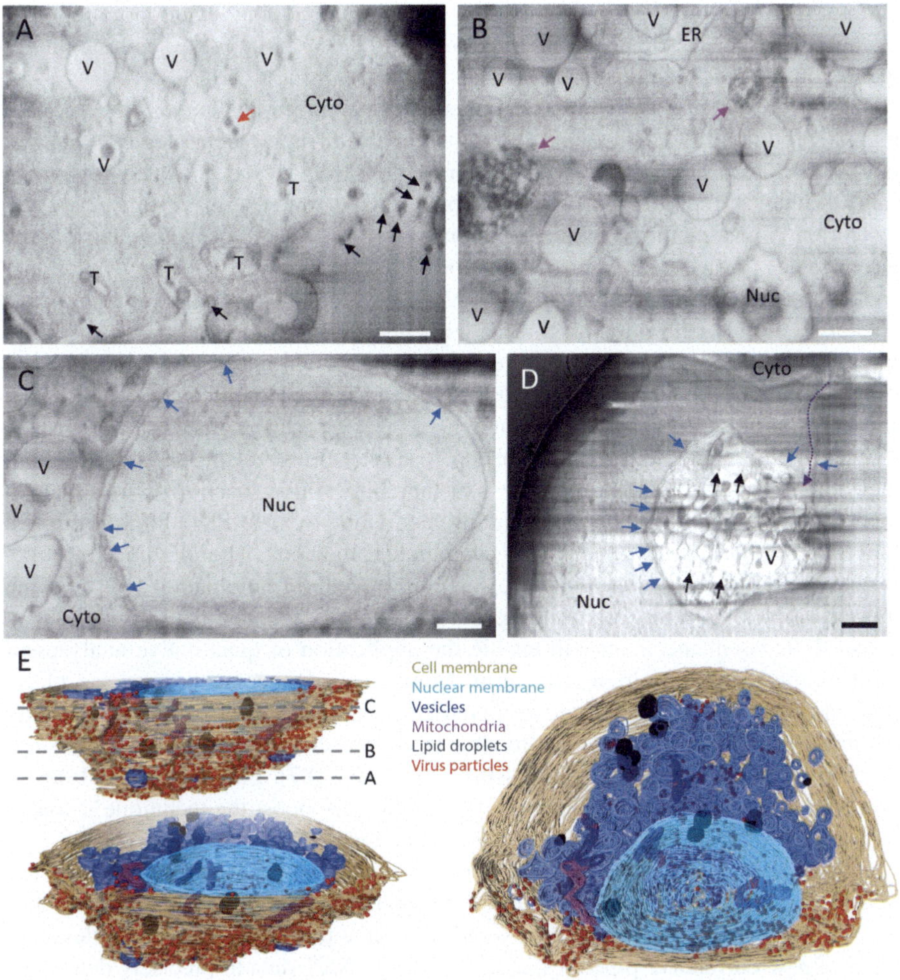

Fig. 8.9 cryo-FIB-SEM volume imaging of a SARS-CoV-2-infected cell. (**a–d**), SEM sections of the reconstructed tomogram. (**a**) cell periphery. (**b**) cytoplasmic region. (**c**) nucleus. (**d**) cytoplasm invagination in the nuclear region. Black and red arrows, extracellular and intracellular invaginations, respectively; blue arrows, nuclear pores; pink arrows, complex membrane compartments; dashed purple arrow, nuclear invagination path. V, vesicles. T, tunnels. Nuc, nucleus. Cyto, cytoplasm. ER, endoplasmic reticulum. (**e**) 3D rendering of the segmented ultrastructure shown in (**a–c**) Dashed lines in the upper left panel indicate the positions of sections shown in (**a–c**) Scale bars, (**a–c**) 500 nm. (**d**), 1 μm. Modified from [68]

resolution information from in vitro studies using structural biology approaches, such as single-particle analysis, nano-electron diffraction, X-ray crystallography, and nuclear magnetic resonance, with in situ nanometric resolution in native conditions, will lead to the development of new models and hypotheses that will be key to advancing knowledge at different stages of the viral life cycle. Improvements in the various stages of the workflow, including sample preparation, event targeting, and structural characterization (discussed below), will be essential for the routine implementation of these techniques.

Cryogenic fixation of cellular samples currently faces challenges in achieving high reproducibility. However, the emergence of new technologies such as pin-printing [83] and support micropatterning [84] devices is expected to significantly enhance reproducibility in this process. As interest in cryogenic super-resolution VLM methods increases, alongside concurrent advancements in new technologies, the development of new correlative workflows will bring functional and localization information provided by VLM closer to the resolution of nanometric resolution ultrastructural methods. This will enhance correlation precision, enabling more accurate targeting of very rare events and better integration of functional information into the ultrastructural context, aiding in the identification of biological structures in crowded cellular environments.

cryo-ET is remarkable for its capacity to unveil the structure of viruses within their native cellular context, regardless of their subcellular location. This advancement has been made possible through sample thinning by cryo-FIB milling, adding a functional dimension to genetic and biochemical methods. The ongoing improvements in cryo-FIB milling methods, such as procedural automation [85] and the incorporation of plasma FIB milling, are expected to enhance data throughput in cryo-ET. This enhancement will enable the application of quantitative analyses of viral morphogenesis at high resolution. Moreover, advancements in phase plates [86] and detectors are anticipated to increase the resolution of tomograms, allowing the analysis of smaller molecules in situ. However, challenges persist in the steps involving data collection and processing in cryo-ET [87]. Further technical developments to address these challenges will significantly boost the technique's contribution to virology research, drug development, and vaccine development.

Thanks to the high penetration power of X-rays, cryo-SXT provides nanometric resolution 3D ultrastructure of whole cells without sectioning and with high throughput. The nondestructive nature of cryo-SXT is anticipated to facilitate the development of new correlative pipelines, e.g., using cryo-SXT data to precisely determine areas for erosion via FIB milling, followed by cryo-ET analysis.

cryo-vEM enables whole-cell 3D imaging at nanometric resolution. While in genomics, proteomics, and metabolomics, the challenge has been to move from a system level to single cells, cryo-vEM faces the reverse: moving from single cell to a system level. The advancements in fast milling technologies, such as plasma FIB, the increased image acquisition speed via multi-beam SEM [88], and the automation of image acquisition pipelines, are expected to increase data throughput in cryo-vEM. This advancement will have an impact on the study of viral infections, offering valuable support for the development of novel antiviral therapeutics and vaccine research.

The efficacy of integrating methods lies in acquiring information that surpasses the capabilities of each approach, overcoming technical limitations, and mitigating artifacts inherent to singular methods. Cryo-correlative microscopy approaches discussed in this chapter complement functional and localization information provided by cryo-VLM with the structural information gained by high-resolution methods. However, integrating these methods with additional approaches, such as cryo-hard X-ray fluorescence [89], could add a chemical dimension to the study of viral infection. These advancements can only occur within a multidisciplinary environment where virologists, structural biologists, physicists, chemists, engineers, and computational scientists collaborate.

Acknowledgments RA, FJC, and JJC's work was supported by the CRIOMECORR project (ESFRI-2019-01-CSIC-16). AC's work was supported by Grant P2022/BMD-7232 TomoXliver2.

References

1. Novoa RR, Calderita G, Arranz R, Fontana J, Granzow H, Risco C (2005) Virus factories: associations of cell organelles for viral replication and morphogenesis. Biol Cell 97:147–172
2. Fernández de Castro I, Tenorio R, Risco C (2021) Virus factories. In Encyclopedia of virology. Elsevier, pp. 495–500
3. Schmid M, Speiseder T, Dobner T, Gonzalez RA (2014) DNA virus replication compartments. J Virol 88:1404–1420
4. Ketter E, Randall G (2019) Virus impact on lipids and membranes. Annu Rev Virol 6:319–340
5. Gaete-Argel A, Márquez CL, Barriga GP, Soto-Rifo R, Valiente-Echeverría F (2019) Strategies for success. Viral Infections and Membraneless Organelles. Front Cell Infect Microbiol 9:336
6. Brangwynne CP, Eckmann CR, Courson DS, Rybarska A, Hoege C, Gharakhani J, Jülicher F, Hyman AA (2009) Germline P granules are liquid droplets that localize by controlled dissolution/condensation. Science 324:1729–1732
7. Etibor TA, Yamauchi Y, Amorim MJ (2021) Liquid biomolecular condensates and viral lifecycles: review and perspectives. Viruses 13:366
8. Nikolic J, Le Bars R, Lama Z, Scrima N, Lagaudrière-Gesbert C, Gaudin Y, Blondel D (2017) Negri bodies are viral factories with properties of liquid organelles. Nat Commun 8:58
9. Pfaller CK, Radeke MJ, Cattaneo R, Samuel CE (2014) Measles virus C protein impairs production of defective copyback double-stranded viral RNA and activation of protein kinase R. J Virol 88:456–468
10. Koning RI, Celler K, Willemse J, Bos E, van Wezel GP, Koster AJ (2014) Correlative cryo-fluorescence light microscopy and cryo-electron tomography of Streptomyces. Meth Cell Biol 124:217–239
11. Briegel A, Chen S, Koster AJ, Plitzko JM, Schwartz CL, Jensen GJ (2010) Correlated light and electron Cryo-microscopy. Meth Enzymol 481:317–341
12. Davies HA (1999) General preparation of material and staining of sections. Meth Mol Biol 117: 1–11
13. Singh H, Bishen KA, Garg D, Sukhija H, Sharma D, Tomar U (2019) Fixation and fixatives: roles and functions—a short review. Dental J Adv Studies 7:051–055
14. Sanders S, Jensen Y, Reimer R, Bosse JB (2023) From the beginnings to multidimensional light and electron microscopy of virus morphogenesis. Adv Virus Res 116:45–88
15. Adrian M, Dubochet J, Lepault J, McDowall AW (1984) Cryo-electron microscopy of viruses. Nature 308:32–36
16. Hawes PC (2015) Preparation of cultured cells using high-pressure freezing and freeze substitution for subsequent 2D or 3D visualization in the transmission electron microscope. Meth Mol Biol 1282:271–282
17. Dubochet J, Lepault J, Freeman R, Berriman JA, Homo JC (1982) Electron microscopy of frozen water and aqueous solutions. J Microsc 128:219–237

18. Kasas S, Dumas G, Dietler G, Catsicas S, Adrian M (2003) Vitrification of cryoelectron microscopy specimens revealed by high-speed photographic imaging. J Microsc 211:48–53
19. Weissenberger G, Henderikx RJM, Peters PJ (2021) Understanding the invisible hands of sample preparation for cryo-EM. Nat Methods 18:463–471
20. Dobro MJ, Melanson LA, Jensen GJ, McDowall AW (2010) Plunge freezing for electron cryomicroscopy. Meth Enzymol 481:63–82
21. Studer D, Humbel BM, Chiquet M (2008) Electron microscopy of high pressure frozen samples: bridging the gap between cellular ultrastructure and atomic resolution. Histochem Cell Biol 130:877–889
22. Moor H (1987) Theory and practice of high pressure freezing. In: Steinbrecht RA, Zierold K (eds) Cryotechniques in biological electron microscopy. Springer, Berlin, pp 175–191
23. Scher N, Avinoam O (2021) 50 shades of CLEM: how to choose the right approach for you. Meth Cell Biol 162:1–11
24. Popescu A, Doyle RJ (1996) The gram stain after more than a century. Biotech Histochem 71: 145–151
25. Fischer AH, Jacobson KA, Rose J, Zeller R (2008) Hematoxylin and eosin staining of tissue and cell sections. CSH Protoc 2008, pdb.prot4986
26. McClelland RD, Culp TN, Marchant DJ (2021) Imaging flow cytometry and confocal immunofluorescence microscopy of virus-host cell interactions. Front Cell Infect Microbiol 11: 749039
27. Zhang M, Li M, Zhang W, Han Y, Zhang Y-H (2019) Simple and efficient delivery of cell-impermeable organic fluorescent probes into live cells for live-cell superresolution imaging. Light: Sci App 8:73
28. Elliott G, O'Hare P (1999) Live-cell analysis of a green fluorescent protein-tagged herpes simplex virus infection. J Virol 73:4110–4119
29. Schellenberger P, Kaufmann R, Siebert CA, Hagen C, Wodrich H, Grünewald K (2014) High-precision correlative fluorescence and electron cryo microscopy using two independent alignment markers. Ultramicroscopy 143:41–51
30. Editorial (2009) Beyond the diffraction limit. Nat Photon 3: 361
31. Schermelleh L, Ferrand A, Huser T, Eggeling C, Sauer M, Biehlmaier O, Drummen GPC (2019) Super-resolution microscopy demystified. Nat Cell Biol 21:72–84
32. Wolff G, Hagen C, Grünewald K, Kaufmann R (2016) Towards correlative super-resolution fluorescence and electron cryo-microscopy. Biol Cell 108:245–258
33. Faro AR, Adam V, Carpentier P, Darnault C, Bourgeois D, de Rosny E (2010) Low-temperature switching by photoinduced protonation in photochromic fluorescent proteins. Photochem Photobiol Sci 9:254–262
34. Mantovanelli AMR, Glushonkov O, Adam V, Wulffelé J, Thédié D, Byrdin M, Gregor I, Nevskyi O, Enderlein J, Bourgeois D (2023) Photophysical studies at cryogenic temperature reveal a novel Photoswitching mechanism of rsEGFP2. J Am Chem Soc 145:14636–14646
35. Gustafsson MGL (2000) Surpassing the lateral resolution limit by a factor of two using structured illumination microscopy. J Microsc 198:82–87
36. Korobchevskaya K, Lagerholm BC, Colin-York H, Fritzsche M (2017) Exploring the potential of Airyscan microscopy for live cell imaging. Photo-Dermatology 4:41
37. Sheppard CJ, Mehta SB, Heintzmann R (2013) Superresolution by image scanning microscopy using pixel reassignment. Opt Lett 38:2889–2892
38. Kounatidis I, Stanifer ML, Phillips MA, Paul-Gilloteaux P, Heiligenstein X, Wang H, Okolo CA, Fish TM, Spink MC, Stuart DI et al (2020) 3D correlative Cryo-structured illumination fluorescence and soft X-ray microscopy elucidates Reovirus intracellular release pathway. Cell 182:515–530.e517
39. Eledge MR, Zita MD, Boehme KW (2019) Reovirus: Friend and Foe. Curr Clin Microbiol Rep 6:132–138
40. Boulant S, Stanifer M, Kural C, Cureton DK, Massol R, Nibert ML, Kirchhausen T (2013) Similar uptake but different trafficking and escape routes of reovirus virions and infectious

subvirion particles imaged in polarized Madin–Darby canine kidney cells. Mol Biol Cell 24: 1196–1207
41. Turk M, Baumeister W (2020) The promise and the challenges of cryo-electron tomography. FEBS Lett 594:3243–3261
42. Robinson CV, Sali A, Baumeister W (2007) The molecular sociology of the cell. Nature 450: 973–982
43. Grünewald K, Medalia O, Gross A, Steven AC, Baumeister W (2003) Prospects of electron cryotomography to visualize macromolecular complexes inside cellular compartments: implications of crowding. Biophys Chem 100:577–591
44. Grünewald K, Cyrklaff M (2006) Structure of complex viruses and virus-infected cells by electron cryo tomography. Curr Opin Microbiol 9:437–442
45. Cyrklaff M, Linaroudis A, Boicu M, Chlanda P, Baumeister W, Griffiths G, Krijnse-Locker J (2007) Whole cell cryo-electron tomography reveals distinct disassembly intermediates of vaccinia virus. PLoS One 2:e420
46. Vijayakrishnan S, Loney C, Jackson D, Suphamungmee W, Rixon FJ, Bhella D (2013) Cryotomography of budding influenza a virus reveals filaments with diverse morphologies that mostly do not bear a genome at their distal end. PLoS Pathog 9:e1003413
47. Wolff G, Limpens R, Zevenhoven-Dobbe JC, Laugks U, Zheng S, de Jong AWM, Koning RI, Agard DA, Grünewald K, Koster AJ et al (2020) A molecular pore spans the double membrane of the coronavirus replication organelle. Science 369:1395–1398
48. Chmielewski D, Schmid MF, Simmons G, Jin J, Chiu W (2022) Chikungunya virus assembly and budding visualized in situ using cryogenic electron tomography. Nat Microbiol 7:1270–1279
49. Rigort A, Bäuerlein FJB, Villa E, Eibauer M, Laugks T, Baumeister W, Plitzko JM (2012) Focused ion beam micromachining of eukaryotic cells for cryoelectron tomography. Proc Natl Acad Sci USA 109:4449–4454
50. Timmermans FJ, Otto C (2015) Contributed review: review of integrated correlative light and electron microscopy. Rev Sci Instrum 86:011501
51. Sartori A, Gatz R, Beck F, Rigort A, Baumeister W, Plitzko JM (2007) Correlative microscopy: bridging the gap between fluorescence light microscopy and cryo-electron tomography. J Struct Biol 160:135–145
52. Schaffer M, Pfeffer S, Mahamid J, Kleindiek S, Laugks T, Albert S, Engel BD, Rummel A, Smith AJ, Baumeister W et al (2019) A cryo-FIB lift-out technique enables molecular-resolution cryo-ET within native Caenorhabditis elegans tissue. Nat Methods 16:757–762
53. Giannuzzi LA, Stevie FA (1999) A review of focused ion beam milling techniques for TEM specimen preparation. Micron 30:197–204
54. Smith NS, Skoczylas WP, Kellogg SM, Kinion DE, Tesch PP, Sutherland O, Aanesland A, Boswell RW (2006) High brightness inductively coupled plasma source for high current focused ion beam applications. J Vac Sci Technol B 24:2902–2906
55. Berger C, Dumoux M, Glen T, Yee NBY, Mitchels JM, Patáková Z, Darrow MC, Naismith JH, Grange M (2023) Plasma FIB milling for the determination of structures in situ. Nat Commun 14:629
56. Frank J (2006) Introduction: principles of electron tomography. In: Frank J (ed) Electron tomography: methods for three-dimensional visualization of structures in the cell. Springer, New York, pp 1–15
57. Dierksen K, Typke D, Hegerl R, Baumeister W (1993) Towards automatic electron tomography II. Implementation of autofocus and low-dose procedures. Ultramicroscopy 49:109–120
58. Fernandez JJ (2012) Computational methods for electron tomography. Micron 43:1010–1030
59. Radermacher M (1994) Three-dimensional reconstruction from random projections: orientational alignment via radon transforms. Ultramicroscopy 53:121–136
60. Chen M, Dai W, Sun SY, Jonasch D, He CY, Schmid MF, Chiu W, Ludtke SJ (2017) Convolutional neural networks for automated annotation of cellular cryo-electron tomograms. Nat Methods 14:983–985

61. Metskas LA, Wilfong R, Jensen GJ (2022) Subtomogram averaging for biophysical analysis and supramolecular context. J Struct Biol X 6:100076
62. Guo Q, Lehmer C, Martínez-Sánchez A, Rudack T, Beck F, Hartmann H, Pérez-Berlanga M, Frottin F, Hipp MS, Hartl FU et al (2018) In situ structure of neuronal C9orf72 poly-GA aggregates reveals proteasome recruitment. Cell 172:696–705
63. Quemin ERJ, Machala EA, Vollmer B, Pražák V, Vasishtan D, Rosch R, Grange M, Franken LE, Baker LA, Grünewald K (2020) Cellular electron Cryo-tomography to study virus-host interactions. Annu Rev Virol 7:239–262
64. Zhang X, Sridharan S, Zagoriy I, Eugster Oegema C, Ching C, Pflaesterer T, Fung HKH, Becher I, Poser I, Müller CW et al (2023) Molecular mechanisms of stress-induced reactivation in mumps virus condensates. Cell 186:1877–1894
65. McEnroe T, O'Reilly F, Sheridan P, Howard J, Byrne R, O'Connor A, Rogers D, Rogers C, McCarthy C et al (2017) Development of a commercial laboratory scale soft X-ray microscope. Microsc Microanal 23:982–983
66. Weiß D, Schneider G, Niemann B, Guttmann P, Rudolph D, Schmahl G (2000) Computed tomography of cryogenic biological specimens based on X-ray microscopic images. Ultramicroscopy 84:185–197
67. Groen J, Conesa JJ, Valcárcel R, Pereiro E (2019) The cellular landscape by cryo soft X-ray tomography. Biophys Rev 11:611–619
68. Mendonça L, Howe A, Gilchrist JB, Sheng Y, Sun D, Knight ML, Zanetti-Domingues LC, Bateman B, Krebs A-S, Chen L et al (2021) Correlative multi-scale cryo-imaging unveils SARS-CoV-2 assembly and egress. Nat Commun 12:4629
69. Aho V, Myllys M, Ruokolainen V, Hakanen S, Mäntylä E, Virtanen J, Hukkanen V, Kühn T, Timonen J, Mattila K et al (2017) Chromatin organization regulates viral egress dynamics. Sci Rep 7:3692
70. Aho V, Mäntylä E, Ekman A, Hakanen S, Mattola S, Chen J-H, Weinhardt V, Ruokolainen V, Sodeik B, Larabell C et al (2019) Quantitative microscopy reveals stepwise alteration of chromatin structure during herpesvirus infection. Viruses 11:935
71. Myllys M, Ruokolainen V, Aho V, Smith EA, Hakanen S, Peri P, Salvetti A, Timonen J, Hukkanen V, Larabell CA et al (2016) Herpes simplex virus 1 induces egress channels through marginalized host chromatin. Sci Rep 6:28844
72. Hagen C, Guttmann P, Klupp B, Werner S, Rehbein S, Mettenleiter TC, Schneider G, Grünewald K (2012) Correlative VIS-fluorescence and soft X-ray cryo-microscopy/tomography of adherent cells. J Struct Biol 177:193–201
73. Pérez-Berná AJ, Rodríguez MJ, Chichón FJ, Friesland MF, Sorrentino A, Carrascosa JL, Pereiro E, Gastaminza P (2016) Structural changes in cells imaged by soft X-ray Cryo-tomography during hepatitis C virus infection. ACS Nano 10:6597–6611
74. Chichón FJ, Rodríguez MJ, Pereiro E, Chiappi M, Perdiguero B, Guttmann P, Werner S, Rehbein S, Schneider G, Esteban M et al (2012) Cryo X-ray nano-tomography of vaccinia virus infected cells. J Struct Biol 177:202–211
75. Greseth MD, Traktman P (2022) The life cycle of the vaccinia virus genome. Annu Rev Virol 9: 239–259
76. Moss B (2013) Reflections on the early development of poxvirus vectors. Vaccine 31:4220–4222
77. Al-Amoudi A, Studer D, Dubochet J (2005) Cutting artefacts and cutting process in vitreous sections for cryo-electron microscopy. J Struct Biol 150:109–121
78. Peddie CJ, Genoud C, Kreshuk A, Meechan K, Micheva KD, Narayan K, Pape C, Parton RG, Schieber NL, Schwab Y et al (2022) Volume electron microscopy. Nat Rev Methods Primers 2: 51
79. Schertel A, Snaidero N, Han H-M, Ruhwedel T, Laue M, Grabenbauer M, Möbius W (2013) Cryo FIB-SEM: volume imaging of cellular ultrastructure in native frozen specimens. J Struct Biol 184:355–360

80. de Beer M, Daviran D, Roverts R, Rutten L, Macías Sánchez E, Metz JR, Sommerdijk N, Akiva A (2023) Precise targeting for 3D cryo-correlative light and electron microscopy volume imaging of tissues using a FinderTOP. Comm Biol 6:510
81. Wu GH, Mitchell PG, Galaz-Montoya JG, Hecksel CW, Sontag EM, Gangadharan V, Marshman J, Mankus D, Bisher ME, Lytton-Jean AKR et al (2020) Multi-scale 3D Cryo-correlative microscopy for vitrified cells. Structure 28:1231–1237
82. Hu B, Guo H, Zhou P, Shi Z-L (2021) Characteristics of SARS-CoV-2 and COVID-19. Nat Rev Microbiol 19:141–154
83. Ravelli RBG, Nijpels FJT, Henderikx RJM, Weissenberger G, Thewessem S, Gijsbers A, Beulen BWAMM, López-Iglesias C, Peters PJ (2020) Cryo-EM structures from sub-nl volumes using pin-printing and jet vitrification. Nat Commun 11:2563
84. Engel L, Gaietta G, Dow LP, Swif MF, Pardon G, Volkmann N, Weis WI, Hanein D, Pruitt BL (2019) Extracellular matrix micropatterning technology for whole cell cryogenic electron microscopy studies. J Micromech Microeng 29:115018
85. Klumpe S, Fung HKH, Goetz SK, Zagoriy I, Hampoelz B, Zhang X, Erdmann PS, Baumbach J, Müller CW, Beck M et al (2021) A modular platform for automated cryo-FIB workflows. elife 10:e70506
86. Schwartz O, Axelrod JJ, Campbell SL, Turnbaugh C, Glaeser RM, Müller H (2019) Laser phase plate for transmission electron microscopy. Nat Methods 16:1016–1020
87. Jiménez de la Morena J, Conesa P, Fonseca YC, de Isidro-Gómez FP, Herreros D, Fernández-Giménez E, Strelak D, Moebel E, Buchholz TO, Jug F et al (2022) ScipionTomo: towards cryo-electron tomography software integration, reproducibility, and validation. J Struct Biol 214: 107872
88. Eberle AL, Zeidler D (2018) Multi-beam scanning electron microscopy for high-throughput imaging in Connectomics research. Front Neuroanat 12:112
89. Conesa JJ, Carrasco AC, Rodríguez-Fanjul V, Yang Y, Carrascosa JL, Cloetens P, Pereiro E, Pizarro AM (2020) Unambiguous intracellular localization and quantification of a potent iridium anticancer compound by correlative 3D Cryo X-ray imaging. Angew Chem Int Ed Engl 59:1270–1278

Further Reading

Dumoux M, Glen T, Smith JLR, Ho EML, Perdigão LMA, Pennington A, Klumpe S, Yee NBY, Farmer DA, Lai PYA et al (2023) Cryo-plasma FIB/SEM volume imaging of biological specimens. elife 12:e83623

Prakash K, Diederich B, Heintzmann R, Schermelleh L (2022) Super-resolution microscopy: a brief history and new avenues. Phil Trans R Soc A 380:20210110

Navarro PP (2022) Quantitative Cryo-electron tomography. Front Mol Biosci 9:934465

Pereiro E (2019) Correlative cryo-soft X-ray tomography of cells. Biophys Rev 11:529–530

Verkade P, Collinson L (2019) Correlative imaging: focusing on the future. John Wiley & Sons

Chapter 9
Atomic Force Microscopy of Viruses

Pedro J. de Pablo

Abstract Atomic force microscopy (AFM) makes it possible to obtain images at nanometric resolution, and to accomplish the manipulation and physical characterization of specimens, including the determination of their mechanical and electrostatic properties. AFM has an ample range of applications, from materials science to biology. The specimen, supported on a solid surface, can be imaged and manipulated while working in air, ultra-high vacuum or, most importantly for virus studies, in liquid. The adaptability of AFM is also favored by the large variety of specimens of very different sizes that it can deal with, such as atoms, molecules, and molecular complexes including viruses and cells. AFM allows, in addition, the possibility to observe dynamics in real time. Indeed, AFM facilitates single molecule experiments enabling not only to *see* but also to *touch* the material under study (i.e., mechanical manipulations) and constitutes a fundamental source of information for materials characterization. In particular, the study of the mechanical properties of viruses and other biomolecular aggregates at the nanoscale is providing humongous information This helps to elaborate mechano-chemical structure/function models of complex protein aggregates, expanding and complementing the information obtained by other techniques.

Keywords Atomic force microscopy · Force · Beam deflection · Tip · Cantilever · Stylus · Topography · Mechanics · Liquids · Tip-sample dilation · Adsorption · Virus · Capsid · Virion · Deformation · Nanoindentation · Force curve · Disruption · Breakage · Physical Virology

---

P. J. de Pablo (✉)
Department of Physics of the Condensed Matter, C03 and IFIMAC (Instituto de Física de la Materia Condensada). Universidad Autónoma de Madrid, Madrid, Spain
e-mail: p.j.depablo@uam.es

M. G. Mateu (ed.), *Structure and Physics of Viruses*, Subcellular Biochemistry 105,
https://doi.org/10.1007/978-3-031-65187-8_9

Abbreviations

AdV	Adenovirus
AFM	Atomic force microscopy
CCD	Charge-coupled device
CM	Contact mode
Cryo-EM	Cryogenic electron microscopy
DM	Dynamic mode
EDL	Electrostatic double layer
EM	Electron microscopy
HIV	Human immunodeficiency virus
HOPG	Highly oriented pyrolytic graphite
HS-AFM	High-speed atomic force microscopy
JM	Jumping mode
MVM	Minute virus of mice
PBV	Picobirnavirus
SIMV1	*Sulfolobus* monocaudavirus 1
STM	Scanning tunneling microscopy
TIRFM	Total internal reflection fluorescence microscopy
UHV	Ultra-high vacuum.

9.1 Introduction

The word *microscope* is typically understood as an optical device that uses light to achieve a magnified image of a small object. Consequently, the question that arises when seeing an atomic force microscope (AFM) for the first time is: where do I have to look to see the object? In a conventional microscope (Fig. 9.1), a source emits particles, such as electrons or photons, that interact with the specimen. A detector registers and communicates this interaction to an analyzer, which processes the information received to make it comprehensible. Optical and electron microscopes fit into this source-specimen-detector-analyzer scheme.

In the optical microscope, the photons emitted by the incandescent filament are refracted by a system of lenses, both before and after the interaction with the specimen, finally arriving at the eyepiece where the detector (i.e., the eyes, film, or charge-coupled device, CCD) collects the information. A typical optical microscope using visible light can reach a resolution of $\lambda/2 \sim 200$ nm (λ being the wavelength of the light).

In the case of electron microscopy (EM) (see Chap. 3), a filament emits electrons, and a system of electromagnetic lenses are used to deflect and focus the electron beam both before and after the interaction with the specimen. Afterward the electrons are collected by a phosphorescent screen, film, or CCD. In 1933, the German engineer Ernst Ruska constructed an electron microscope that exceeded the maximum resolution of any optical microscope, reaching about 1 nm.

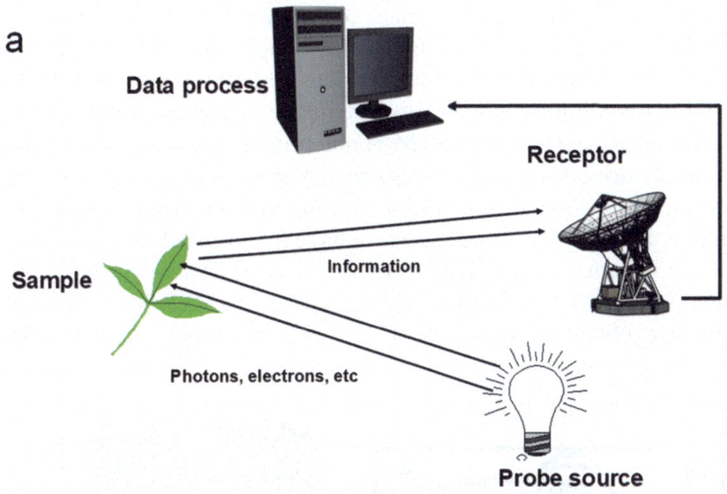

Fig. 9.1 Concept of microscopy. Any microscope requires a probe source, sample, receptor, and data analysis

AFM does not use photons or electrons as probes, the probe itself is a nanometric tip at the end of a microcantilever that mechanically palpates the object under study like the white cane used by blind people. In this way, AFM enables the construction of a topographic image based on recording the deflection angle of the microcantilever at each point of the studied sample. In this chapter, I describe the conceptual and technical bases of AFM. I refer to working methods in liquids and their applications for imaging virus particles and viral components [1]. I also comment on the combination of AFM with Total Internal Reflection Fluorescence Microscopy (TIRFM) and include a brief introduction to High-Speed Atomic Force Microscopy (HS-AFM). In a related chapter (Chap. 18), the specific application of the AFM to determine the mechanical and other physical properties of virus particles is described.

9.2 Basic Concepts

In scanning probe microscopy a sharp stylus of a few nanometers in diameter, which can be considered as a probe, approaches the surface of the sample. Gerd Binnig and Heinrich Rohrer [2] invented the Scanning Tunneling Microscope (STM) and received the Nobel Prize for Physics along with Ernst Ruska in 1986. The STM is based on a quantum effect (tunneling) that occurs when a sharp metallic tip is approached to a distance (z) of less than 1 nm from a conductive surface. This effect involves the flow of an electronic current (I) between the surface and the tip according to the formula $I \sim \exp(-\sqrt{\phi} z)$, where ϕ is the work-function of the metallic surface [3]. The strong dependence on the tip-surface distance can be used to obtain topographic and electronic maps of the sample by moving (i.e., scanning), the tip on the surface while keeping the tip-sample distance constant through a feedback algorithm. Although this tool provides true atomic resolution in ultra-high vacuum (UHV) conditions, a mandatory requisite is that both the tip and sample should be conductors. Therefore, STM is not suitable for biological samples since these are mainly insulators [4]. As a historical remark, it is convenient to remember here that the idea of using a sharp tip for imaging biological specimens was initially endeavored with STM [5]. Indeed, bacteriophage Φ29 was adsorbed on a conductive surface of highly oriented pyrolytic graphite (HOPG). Although the quality of the data of this pioneering publication [5] (Fig. 9.2a) is well below that of the AFM images that are obtained nowadays (Fig. 9.2b), the beauty of that work lies in

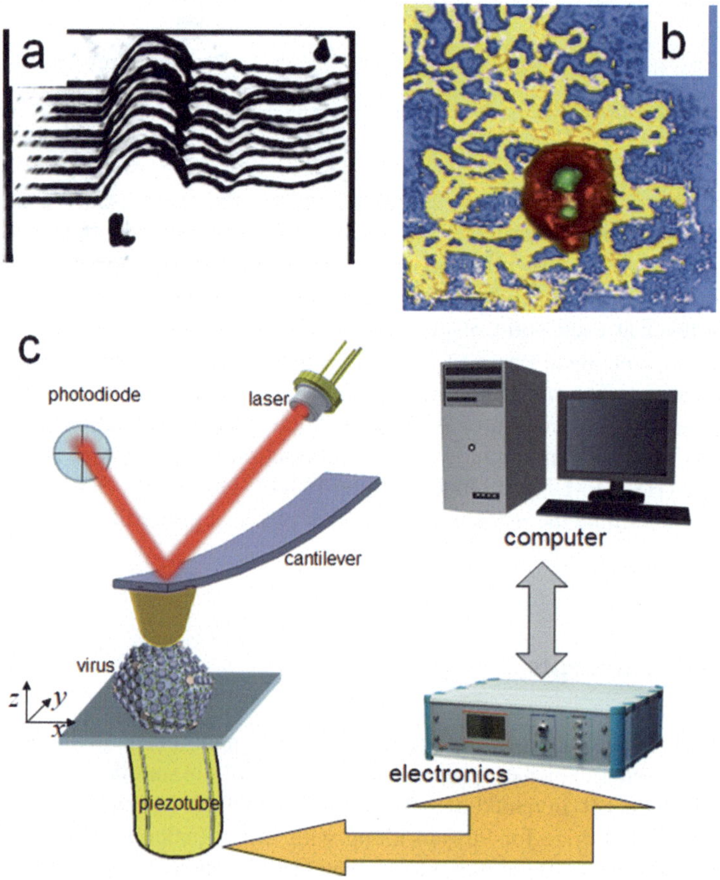

Fig. 9.2 Origins and principles of Atomic Force Microscopy. (**a**) Topographical profiles of bacteriophage Φ29 obtained with a pen engaged to a potentiometer with STM [5]. (**b**) Recent AFM image of an exploded bacteriophage Φ29 (adapted from [6]). (**c**) The software running on the computer controls the electronics, which gives and receives information to and from the AFM head

proposing the *in illo tempore* (1985) novel idea of visualizing biological samples with a sharp tip.

In 1986, Binnig, Quate, and Gerber invented the AFM [2] combining the principles of both the STM and the so-called stylus profilometer [7]. In an AFM (Fig. 9.2c), a sharp tip (~tens of nm in radius) attached to the end of a microcantilever is approached to the surface by means of a piezoelectric device. Therefore, a force appears between the tip and surface that can be attractive or repulsive (see below) causing the cantilever to bend. A feedback algorithm acting on the piezoelectric device controls the relative tip-surface distance, and a topographic map of the object is obtained by scanning the surface in a plane perpendicular to the tip. The original article presents the topographic profiles obtained for ceramics [2]. The fact that both tip and sample may be insulators is one of the main advantages of AFM, opening the door to a whole range of possibilities such as the study of biological samples including proteins, nucleic acids, membranes, cells, and biomolecular complexes such as viruses.

9.3 AFM Implementation

Although there are a variety of routes to control the deflection of the cantilever in AFM, here we will focus on the beam deflection method [8] since it is commonly used when working with biological samples. The beam deflection system involves focusing a laser beam on the end of the cantilever and collecting the reflected light with a photodiode (Fig. 9.2c). Therefore, any bending of the cantilever will affect the position of the reflected laser spot on the photodiode. A normal bending will originate a so-called normal force signal F_n on the photodiode sectors, whereas a lateral torsion will result in the lateral force F_l.

The core of an AFM is the head, where the beam deflection system is integrated along with the piezoelectric device (piezo-tube) that moves the sample in all three directions (*x, y, z*). In Fig. 9.2c, an AFM head configuration is shown where the tip stays at rest and the sample is moved by the piezo-tube. In another configuration known as "stand alone," the sample remains still, and the tip makes the scanning movement controlled by a piezo-tube that carries the cantilever. The electronic components receive the signals coming from the photodiode, mainly F_n and F_l, and provide high voltages (~100 volts) to move the piezo-tube to which the sample is attached. The computer is in charge of managing the data, calculating all the parameters required to move the piezo-tube in a convenient way by using the electronics. Details on the different components of a typical AFM and its operation are given in the following subsections.

9.3.1 Cantilevers

Integrated tip and cantilever assemblies can be fabricated from silicon or silicon nitride using photolithographic techniques. Thousands of tip-cantilver devices can be produced on a single silicon wafer. The cantilevers can be rectangular (Fig. 9.3a) or V-shaped, and they typically range from 60 to 200 μm in length, 10 μm to 40 μm in width, and 0.3 μm to 2 μm in thickness. The usual tip radius is about 30 nm although lower diameters can be obtained (Fig. 9.3b). The mechanical stiffness of the cantilever (see Chap. 18) is given by its spring constant k, which normally ranges

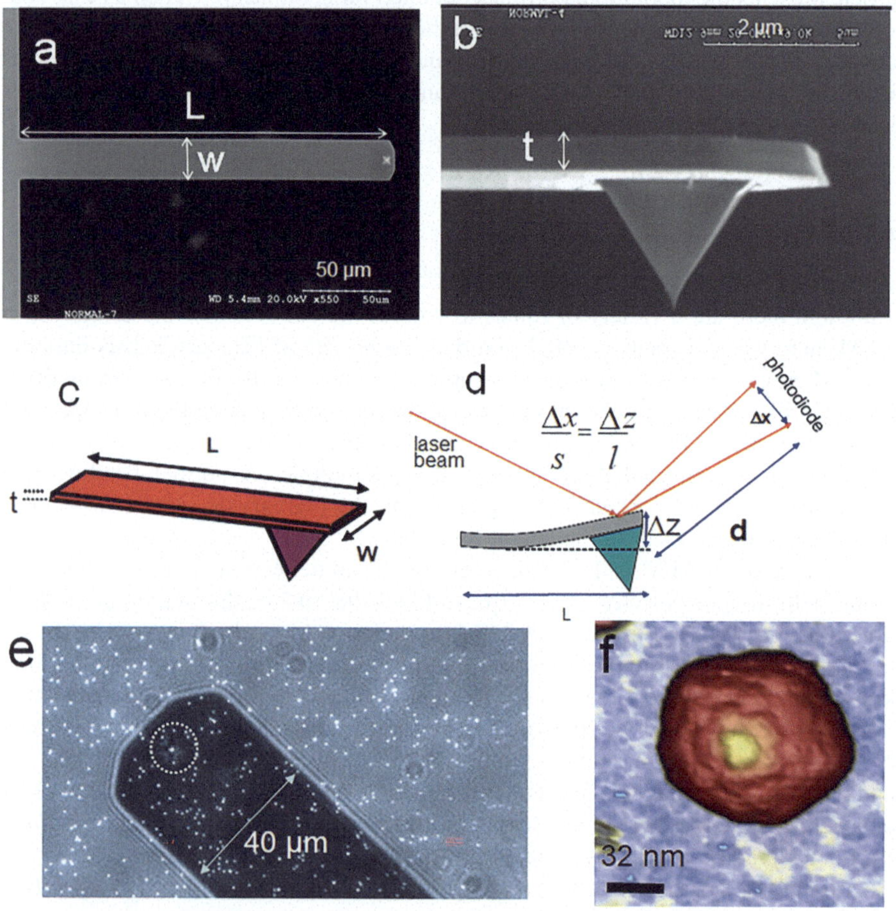

Fig. 9.3 AFM cantilevers. (**a**) and (**b**), respectively, show top and side views of a rectangular cantilever of 150 μm in length (L) and 25 μm in width (w) attached to a chip, also presenting the pyramidal tip. (**c**) sketches the geometry of a rectangular cantilever. The cartoon in (**d**) indicates the vertical resolution of the bending cantilever as a function of several geometric parameters. (**e**) demonstrates a comparison between the cantilever and virus sizes. (**f**) AFM image of a phage λ particle obtained in a liquid medium

between 0.03 N/m and 40 N/m, and strongly depends on the cantilever's dimensions. For example, for a rectangular cantilever $k = \frac{Ew}{4}\left(\frac{t}{L}\right)^3$, where E is the Young Modulus (a measurement of the intrinsic elasticity of a solid material, see Chap. 18) of the cantilever, and W, t, and L are the cantilever width, thickness, and length, respectively (Fig. 9.3c). While W and L can be precisely known, thickness t is always difficult to measure. Therefore, the manufacturers normally provide the cantilever spring constant with an error of 10%–30% and, consequently, the user should calibrate each cantilever [9]. The spring constant k is used to calculate the force applied on the cantilever as a function of bending Δz using Hooke's law, i.e., $F = -k\Delta z$, where the minus sign means that the force is opposite to Δz (Fig. 9.3d).

The vertical resolution of the cantilever, Δz, strongly depends on the noise of the photodiode, since it defines the minimum significant displacement Δx, on the detector (Fig. 9.3d). A typical cantilever of 100 µm length has about 0.1 Å resolution assuming a signal-to-noise ratio around 1 [9].

Another important parameter that can influence the vertical resolution is the thermal noise [10] Due to the equiparticion theorem, the cantilever oscillates at the resonance frequency with an amplitude $\Delta z = \sqrt{\frac{k_B T}{k}}$, where k_B is the Boltzmann constant, T is the absolute temperature and k is the cantilever spring constant. For example, at room temperature, there is a thermal noise of about 5 Å for a cantilever of spring constant 0.02 N/m.

It is instructive to compare the dimensions of the cantilever with a typical virus size. Figure 9.3e shows a cantilever of 40 um in width, with the pyramid of the tip encircled. The cantilever is on top of a glass surface where the disseminated white blobs are plastic beads of 100 nm in diameter, which is the order of magnitude of many icosahedral viruses. Due to the diffraction limit, these beads appear to have ~250 nm in diameter. In spite of these very different dimensions, AFM images of a single virus can resolve even relatively small topographic details, such as the individual protein capsomers composing the capsid of a λ phage particle (Fig. 9.3f).

9.3.2 *Interaction Between the AFM Tip and the Sample Surface*

To understand the interaction between tip and sample we shall refer to potentials rather than forces. Physicists prefer this approach as potentials are scalar and therefore easier to deal with, compared to vectors such as forces. For this purpose, let us graphically depict the Lennard-Jones potential in Jules (J) between two atoms separated by r meters (m) in vacuum [11] (dashed line in Fig. 9.4a).

$U(r) = -A/r^6 + B/r^{12}$, where r is the tip-sample distance, $A = 10^{-77}$ J m^6 and $B = 10^{-134}$ J m^{12}.

A rough approximation to the AFM tip–sample interaction is to consider the approach of two such atoms (inset, Fig. 9.4a) as a solid surface and the apex of a sharp tip at r distance. An interaction force $F(r) = -\frac{dU}{dr}$ will occur, and this force

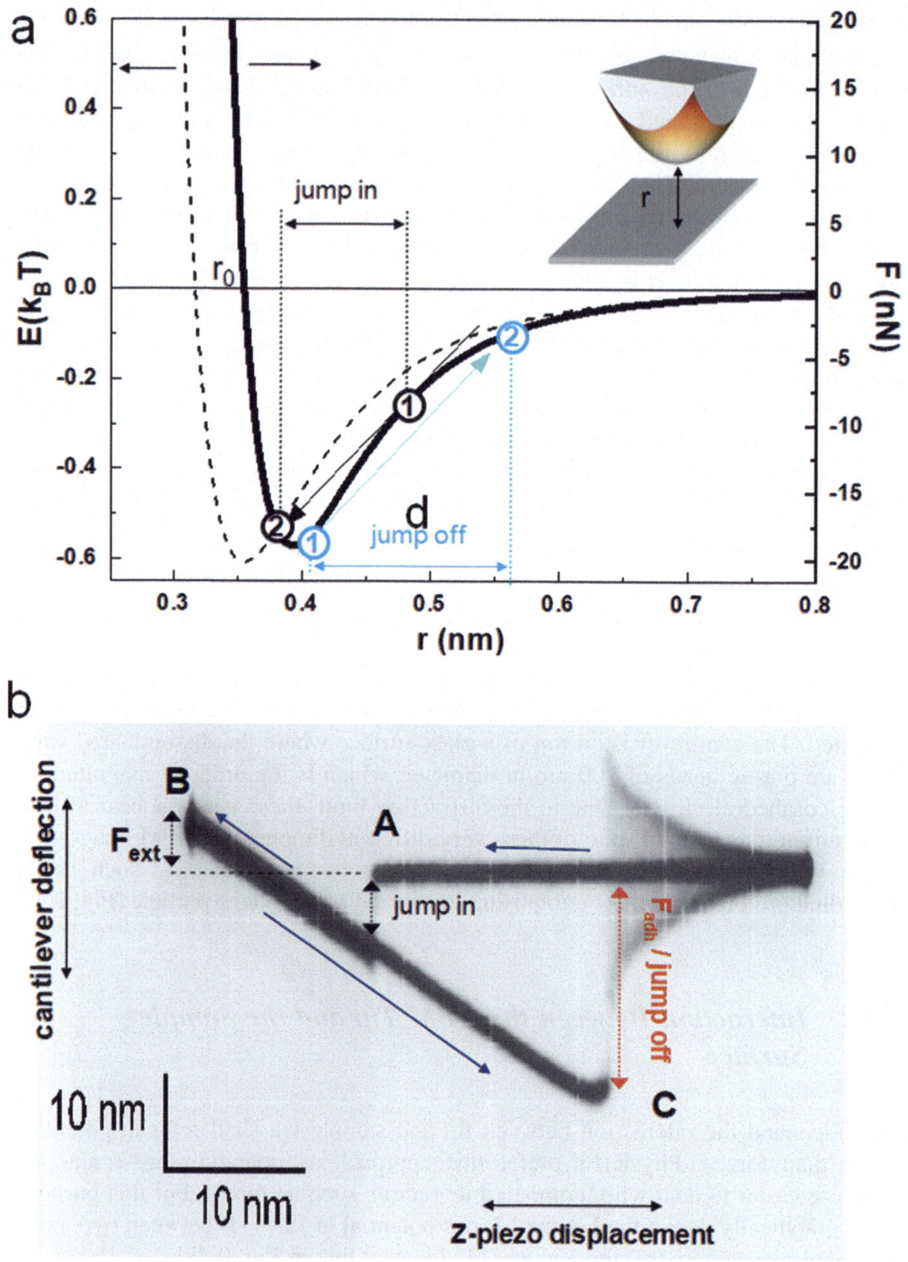

Fig. 9.4 Tip-surface interaction and force vs. distance curve. (**a**) The dashed line represents a Lennard-Jones potential as a function of the distance between atoms, *r*, mimicking the tip-surface interaction (represented in the inset). The solid line is the interaction force obtained from the potential. The numbers inside the circles indicate the cantilever instabilities (see text for a complete description). Panel (**b**) shows an oscilloscope snapshot of force vs. *z*-piezo displacement curve in air

will be attractive ($F < 0$) when $r > r_0$, or repulsive ($F > 0$) when $r < r_0$, (solid line curve in Fig. 9.4a). These regions define the attractive and the repulsive regimes of the AFM operation, respectively.

Now, let us consider the role of these regimes in a force vs. distance (*F-z*) AFM experiment that involves the tip approaching to the surface (Figs. 9.4b). The experiment starts with the tip placed far from the surface in the attractive regime. As the tip is approaching and as soon as the gradient (i.e., the slope) of the force equals the cantilever spring constant, the tip jumps to the surface from black point 1 to black point 2 in Fig. 9.4a (both connected by a downwards arrow). This effect is seen in the *F-z* curve of Fig. 9.4b at point A as a sudden jump of the cantilever deflection (vertical scale). In this way, the tip establishes mechanical contact with the surface and it rapidly enters the repulsive regime ($F > 0$). The *z*-piezo (horizontal scale) is elongated until a given *F* or deflection value is reached and then stops (point B in Fig. 9.4b). The external loading force F_{ext} can be calculated as the difference between the zero-deflection position (i.e., before the jump to contact in point A in Fig. 9.4b) and the deflection at point B in Fig. 9.4b. Since the vertical scale is about 10 nm per division and the cantilever spring constant $k = 0.1$ N/m we have $F_{ext} \sim 0.5$ div × 10 nm/div × 0.1 nN/nm = 0.5 nN.

Subsequently, the *z*-piezo retraces and the tip is released from the surface at point C in Fig. 9.4b (i.e., once more where the derivative of the tip-surface force equals the cantilever spring constant (blue arrow, Fig. 9.4a) jumping from the blue point 1 to blue point 2 in Fig. 9.4a. The cantilever deflection jumps-off to zero with a dampened oscillation. This jump-off is known as the adhesion force F_{adh} (in Fig. 9.4b $F_{adh} \sim 2$ div × 10 nm/div × 0.1 nN/nm = 2 nN). The total force at point B in Fig. 9.4b is the sum of F_{ext} and F_{adh}, (i.e., 2.5 nN). It is interesting to notice that no matter how small F_{ext} is, in air the total force applied to the surface will always be at least F_{adh}.

9.3.3 Contact Mode

Contact Mode (CM) is the simplest operational method used for AFM and it was the first to be developed [4]. Here, the tip is brought into contact with the surface until a given deflection in the cantilever (F_n,) is reached; the tip then scans a square area of the surface to obtain a topographic map. By elongating or retracting the *z*-piezo, the feedback algorithm tries to keep the cantilever deflection constant by comparing the F_n signal with a set point reference value established by the user. Topographic data are obtained by recording the *z*-piezo voltage that the feedback algorithm is applying to correct the cantilever deflection at each pixel of the surface. Since the *z*-piezo is calibrated, the voltage is transformed into height and a topographic map is obtained. Let us consider a simple example where the tip is scanning a step (black line in Fig. 9.5a) with $F_n = -k\Delta z$. When the cantilever moves to the upper part of the step, it undergoes a deflection greater than Δz (light gray, Fig. 9.5a). Therefore, the feedback algorithm retracts the z-piezo to recover the same deflection Δz as when the tip was

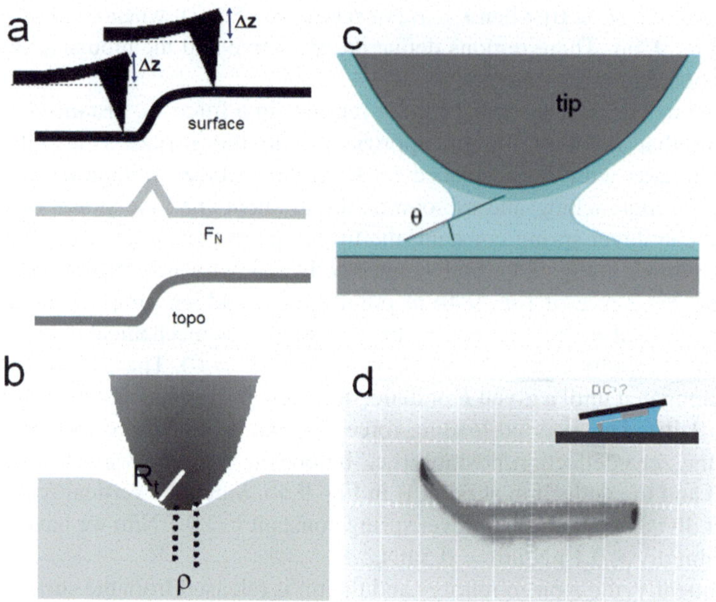

Fig. 9.5 Contact mode, feedback, lateral resolution and force vs. distance curve in liquid. (**a**) presents the contact mode showing the variation of F_n (light gray), and z-piezo tube voltage (topography, dark gray) as a function of a step (black). In (**b**) the geometric features of the tip-surface contact are depicted (see text for a complete description). (**c**) presents a cartoon of the water meniscus formed between the AFM tip and the surface in air conditions. (**d**) This meniscus does not appear in liquid conditions, since the force vs. distance curve does not show any trace of adhesion force

in the lower part of the step and, consequently, a topographic profile of the step is obtained (dark gray in Fig. 9.5a). The latter is known as the constant deflection mode. When the z-piezo elongation is not modified the method is called constant height mode and the map of the changes in F_n is the informative channel. The reader is encouraged to reproduce the profiles of Fig. 9.5a when the tip goes downwards the step.

Let now us consider the lateral resolution that can be achieved in contact mode. We can make an estimate of this parameter by applying the Hertz theory [12], which accounts for the deformation of solids in contact. Once the tip is in contact with the sample, the radius ρ of the tip-surface contact area is given by (Fig. 9.5b):

$$\rho = \left(\frac{3FR}{4E^{*}}\right)^{1/3}$$

where F is the applied force i.e. F_n; E^{*} is the effective Young modulus expressed by:

$$\frac{1}{E^*} = \frac{1 - {\nu_t}^2}{E_t} + \frac{1 - {\nu_s}^2}{E_s}$$

with E_t, ν_t, and E_s, ν_s being the Young Modulus and Poisson ratio for the tip and sample, respectively. R is the effective radius expressed as a combination of the tip R_t and sample radius R_s:

$$\frac{1}{R} = \frac{1}{R_{tip}} + \frac{1}{R_{sample}}$$

It is worth stressing that the main condition for Hertz model applicability is that the radius of contact ρ must be much less than the effective radius R ($\rho < < R$). This approximation roughly implies that the deformation is smaller than the dimensions of the tip-sample system [12]. It is important to recall that this assumption is especially suited to deformations of flat samples. In the case of metals $E = 100$ GPa, $\nu \sim 0.5$ and the mechanical thermal noise of the cantilever is 10 pN, with $R_t \sim 20$ nm, the radius of contact ρ is about 0.15 nm, which implies atomic resolution. However, the adhesion force in air (Fig. 9.4b) is about 5 nN, increasing ρ to ~1 nm. Atomic resolution can be achieved by either working in liquids [13] or in UHV conditions [14].

CM is not frequently used to image isolated biological materials such as virus particles. The high lateral forces that are applied to the surface during scanning [15], which can damage the sample, present an important problem when using this mode. This is especially important for mechanically delicate biomolecules adsorbed onto a surface.

9.3.4 Dynamic Modes

In Dynamic Modes (DM) operation [16], the cantilever is oscillated close to or at its resonance frequency $\omega_0 \approx \sqrt{\frac{E}{\rho}\frac{t}{L^2}}$ for a rectangular cantilever of density ρ, Young's modulus E, thickness t, and length L. As the tip approaches the sample the oscillating amplitude decreases until it establishes contact with the sample (Fig. 9.6a), following a similar cycle to that in Fig. 9.4b, but now with oscillation. Now the feedback loop grasps the amplitude rather than the F_n and, by keeping the oscillation amplitude constant, a topographical map of the object can be obtained. The amplitude is reduced because the resonance frequency ($\Delta\omega$) changes with the tip-sample distance (z and with the tip–sample interaction force F_{ts} according to $\frac{\Delta\omega}{\omega_0} \approx \frac{1}{2k}\frac{dF_{ts}}{dr}$ [17], thereby decreasing with attractive forces. The new resonance frequency ω, is lower than ω_0 and, since the cantilever is still oscillating at ω_0, the cantilever amplitude decreases (Fig. 9.6b).

While CM can damage isolated biological samples on surfaces, DM operated in non-contact mode does not apply high dragging forces [17] and is commonly used to

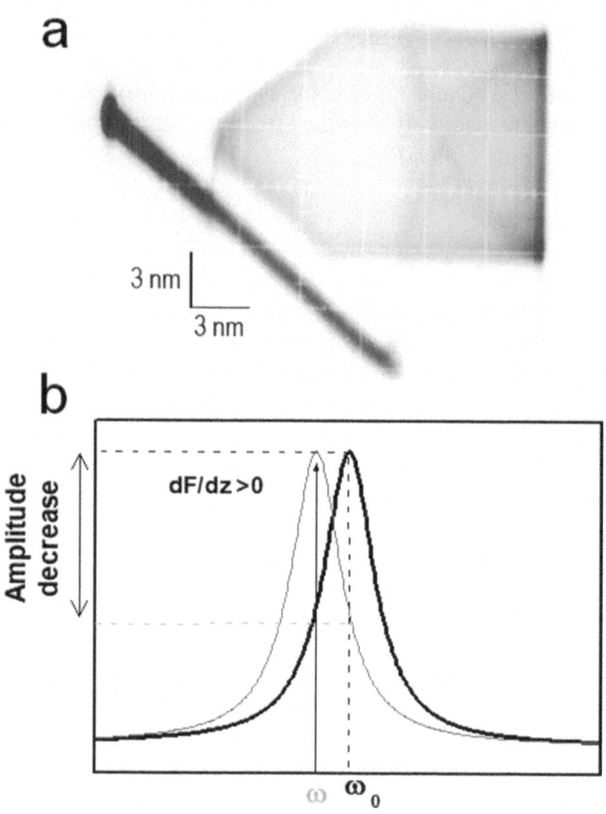

Fig. 9.6 Dynamic modes. (**a**) depicts the force vs. Piezo-tube z displacement curve of an oscillating cantilever. (**b**) illustrates the reduction of the resonance frequency of the cantilever from ω_0 to ω as the tip approaches the surface (see text)

image weakly attached molecules, including delicate biomolecules, in air. In addition to topographic maps, DM can be used to obtain other information on the sample, including a phase map (the time difference between the excitation and the response of the cantilever measured for each pixel of the sample), which in air carries information on the different materials composing the sample [17].

When DM is used in liquid the situation changes completely, as the viscosity of water reduces the resonance frequency about fourfold and the quality factor (a measure of the cantilever damping) of the oscillation is reduced from ~100 to 10. When oscillating the cantilever in liquid a mechanical contact between tip and sample is established, thereby applying lateral and normal forces that may damage the specimen, especially when isolated biological objects are imaged [18]. Thus, while DM can be advantageously used to image virus particles in air, its operation in physiological (liquid) conditions may be limited, unless fixation agents such as glutaraldehyde are used [19]. However, glutaraldehyde may induce some changes in virus particle conformation or mechanical properties.

9.3.5 Jumping Mode (JM) or Pulse Force Mode

Jumping Mode (JM), also called pulse force mode [20], is a variation of CM where lateral tip displacement occurs when the tip and sample are not in mechanical contact, thereby largely avoiding shear forces and the corresponding damage to the tip-sample system. JM performs a *F-z* curve (Fig. 9.4b) at every pixel of the image, moving the tip to the next point at the end of each cycle when the tip and sample lose contact. Feedback is engaged at point B (Fig. 9.4b), moving the z-piezo in a convenient way to maintain a constant deflection or loading force F_{ext}.

JM [21] routinely produces adhesion force maps, which provide compositional or geometrical information of the surface [22]. The adhesion force between the tip and the surface can be described as, $F_{adh} = 4\pi R\gamma_L cos\theta + 1.5\pi\Delta\gamma R$, where γ_L is the water surface tension and θ the angle of the water meniscus which is between the tip and surface; R is the effective radius (described above) and $\Delta\gamma$ the tip-surface energy difference. Since the first term accounts for the water meniscus between tip and surface (Fig. 9.5c), it mainly informs about the hydrophobicity of the sample. A rough estimation of adhesion force in air at room temperature results in ~7 nN for a tip radius of 20 nm. The second term depends mainly on the tip-surface geometry. We denote the importance of the first term by comparing the *F-z* curves taken from glass in both air (Fig. 9.4b) and water (Fig. 9.5d). In liquid, the adhesion force is almost gone, since there is no water meniscus between the tip and sample, although some hysteresis appears in the *F-z* due to the dragging force of the water on the cantilever. In our view, JM is a mode of choice for analyzing virus particles in a liquid milieu, close to physiological conditions, because it minimizes lateral forces and enables a precise control of the applied force [23].

9.3.6 Tip-Sample Geometrical Dilation

All the operational modes described above give rise to an effect that is inherent to the very nature of the AFM approach: the so-called tip–sample geometrical dilation. Because of this geometry-based effect, the topography of an object in an AFM image appears as laterally expanded (see Fig. 9.7). AFM users must understand and control this artifact as much as possible to preclude any misunderstanding of the data obtained.

When surface asperities are of dimensions comparable to the tip radius, which is very common with many samples imaged by AFM, including virus particles, the size of the tip plays an important role on image definition and the apparent image lateral expansion due to the dilation of certain features of the image by the finite tip size (namely geometrical dilation) [24]. Figure 9.7a, b presents AFM topographies of single-walled carbon nanotubes on silicon oxide. These images were obtained using tips of different radii, thinner in Fig. 9.7a than in Fig. 9.7b. A carbon nanotube is a rolled-up graphene in the form of a cylinder of a few nanometers in diameter.

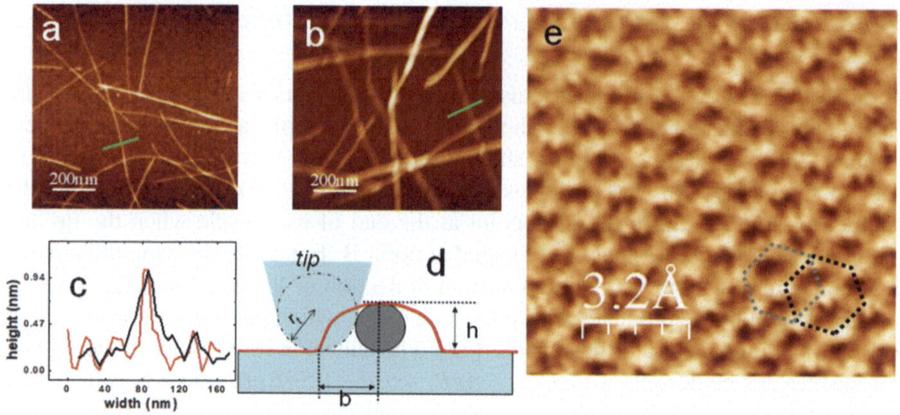

Fig. 9.7 Tip-sample dilation and atomic corrugation in HOPG. (**a**) and (**b**) show AFM images of the same samples of carbon nanotubes adsorbed on silicon oxide. The topographic profiles obtained along the green lines in (a) and (b) show different widths due to dilation in (**c**). (**d**) illustrates the geometric parameters in the dilation process. The red line depicts the dilated section of the carbon nanotube section (gray circle). (**e**) presents the atomic corrugation of HOPG, where the two carbon layers result in two hexagonal networks

Figure 9.7c shows the geometric dilation effects with the different tip radii: a single topography scan line across a carbon nanotube (green lines in Fig. 9.7a, b) results wider for Fig. 9.7b (black) than for Fig. 9.7a (red). This effect arises from the geometrical considerations shown in Fig. 9.7d, and the tip radius r_t can be calculated as $r_t = b^2/2h$, where b and h are the width and the height of the sample, respectively. By using this formula, it can be estimated that the topographies of Fig. 9.7a and b have been obtained using tips with a radius of 15 nm and 70 nm, respectively. We can conclude that the sharper the tip, the better the image, which will become less laterally expanded and more defined. A popular and impressive experiment frequently used to teach AFM is to image HOPG in air [25], where the elastic deformation of the tip-sample contact plus the dilation effects results in atomic corrugation (Fig. 9.7e), although atomic defects are not visualized [26]. During imaging by AFM of virus particles and other biomolecular complexes, the height of the object at each defined point can be determined with subnanometer precision. However, the tip-sample dilation effects are important when considering lateral image dimensions and the object's fine details.

9.4 Imaging Viruses and Other Biological Objects

9.4.1 Nucleic Acids, Proteins, and Biomolecular Complexes

During the virus life cycle, many interactions between viral components and between viral and cellular components occur (see Part II in this book). Thus, it is relevant to revise here some applications of AFM for imaging biological molecules, biomolecular complexes, and their interactions.

AFM imaging in biology can be carried out in air or in liquid. Some relevant examples of AFM imaging of biomolecules in air are described first. For example, DM is frequently used for imaging DNA on a mica surface. As both DNA and mica are negatively charged, $MgCl_2$ is added to the DNA solution; the Mg^{2+} ions are sandwiched between DNA and mica, resulting in the adsorption of DNA molecules on the surface. Subsequently, the sample is dried out and DM is used to image the DNA molecules on the surface. Some structural and physical aspects of the DNA molecule have been the focus of much research using AFM [27]. In addition, an important biological application of AFM is to investigate the binding of proteins to DNA [28]. In this kind of single molecule experiments, the researcher is not probing the average result of the bulk reaction, but rather the action of single protein molecules on a single DNA molecule. Once the protein solution is pipetted into the DNA solution, the reaction starts, and the DNA–protein complex is adsorbed onto the mica at the appropriate time and air dried. Therefore, on the surface, there is a snapshot of the process taking place between proteins and DNA, and the topography provides single molecule information of the protein–DNA complex [29].

However, in biology we are particularly interested in imaging in a liquid milieu, as close as possible to physiological conditions. This is an important point since biomolecules are normally functional only in aqueous solutions. In such a case, several aspects should be considered. First, a strong enough attachment of the biomolecule to the supporting surface is required to achieve good resolution during the AFM imaging process in liquid. Although biomolecules can be covalently linked to a chemically modified surface [30], covalent modifications could potentially damage the biomolecules, and therefore this treatment tends to be avoided. Fortunately, physisorption is usually sufficient and thus, the specimens can be directly adsorbed to a solid surface in a physiological buffer without chemical bonds. The relevant forces that drive the physisorption process are the van der Waals force, the electrostatic double layer (EDL) force, and the hydrophobic effect [31]. Unlike the van der Waals interaction, the EDL force depends strongly on the concentration and valence of charged solutes, as well as the surface charge density of both surface and specimen. The EDL force between two equally charged surfaces is repulsive and hence opposite to the van der Waals attraction [32]. One of the most extended AFM applications is to image 2-dimensional protein crystals in liquid, as well as lipid membranes, by using contact mode. Biological macromolecules become attached to the surface (mica, silicon, gold, glass, etc.) when there is a net attractive force between them and the surface pulling their surfaces into contact. This force per unit of surface (pressure) can be estimated to the light of the DLVO theory [11] like the sum of the electrostatic force between surface and molecule F_{el}, and the van der Waals interaction F_{vdW}.

$$F_{DLVO} = F_{el}(z) + F_{vdW}(z) = \frac{2\sigma_{surf}\sigma_{sample}}{\varepsilon_e \varepsilon_0} e^{-z/\lambda_D} - \frac{H_a}{6\pi z^3}$$

where z is the distance between the surface and specimen; σ_{surf} and σ_{sample} are the charge densities of surface and specimen, respectively; ε_e is the dielectric constant of the electrolyte; ε_0 is the permittivity of vacuum; λ_D is the Debye length, that depends on the electrolyte valence [31]; and H_a the Hamaker constant. The adsorption of a sample onto freshly cleaved mica (atomically flat) can be manipulated by adjusting both the ion content and the pH of the buffer solution.

AFM in liquid is being increasingly used in molecular and cell biology [33–38]. A major application is to produce topographic images of single biomolecules and, especially, larger biomolecular complexes, lipid membranes, and their embedded proteins, and whole cells in close to physiological conditions. Furthermore, by introducing variations in the conditions (i.e., pH variation and addition of other biological molecules) the resulting changes in the general surface structure of the biomolecule or biological object of interest can be identified. Interestingly, in this kind of set-up the cantilever can be used also to apply forces that trigger conformational changes in single protein molecules such as in the chaperonin GroEL; the changes thus induced may mimic those that occur under the action of physiological agents and can be revealed in real time by AFM imaging [39].

AFM is being used also to visualize single proteins doing mechano-chemical work. DM in liquid is generally used for this purpose since it is faster [40]. Maximum peak forces of a few nN are applied [18], in relation to the stiffness of the sample [41] and, although these forces could in principle be damaging to the sample, they are applied for very short periods of time such as 10% of an oscillating period (i.e., for cantilever with a resonance frequency of 10 kHz in liquid, the forces are applied during 10 μs every 0.1 ms). Using this method, it has been possible to visualize, for example, the activity of RNA polymerase on DNA [42] or the conformational changes in a DNA-repair complex upon binding DNA [43]. Further, high-speed AFM developments have enabled to monitor the activity of single molecular machines in real time [44] (see also Sect. 8.5).

In addition to imaging, AFM is also being applied in biology to measure the forces that participate in interactions between and within single biomolecules. AFM and optical tweezers (see Chap. 10) are major single molecule techniques contributing to a better physics-based understanding of protein folding and unfolding, protein–ligand recognition, macromolecular assembly and disassembly, cell adhesion, etc. (e.g., see references [27, 28, 33]).

9.4.2 *Virus Particles*

Structural and chemico-physical characterization has been key to understand the biology of viruses. X-ray crystallography (Chap. 4) and EM (cryogenic electron microscopy, cryo-EM) (Chap. 3) techniques are used to provide direct three-dimensional structural information on whole virus particles, allowing both the interior and the surface of the virus to be visualized at high resolution. However, a limitation of these techniques is that they are both averaging ("bulk") techniques and thus they present an averaged time and space model of the entire population of particles in the crystal or of the many particles selected for obtaining a processed cryo-EM image. As such, these techniques generally provide limited information on possible structural differences between individuals that distinguish them from the average structure. For this reason, the beautifully symmetrical and apparently perfect models of larger viruses derived from these techniques may not be entirely representative of the individual viruses. Therefore, imaging of individual virus particles may provide unique insights into the structural biology and should be considered as an important part of virus characterization [1]. Cryo-EM imaging without imposing symmetry and cryogenic electron tomography of single virus particles (see Chap. 3) allow imaging of asymmetric and/or pleiomorphic virus particles in a vitrified state. In addition, AFM allows imaging of single virus particles in liquid state in close to physiological conditions and enables the visualization in real time of structural changes of individual virus particles in response to different conditions or added agents (see also Sect. 9.5.1). This may enable the analysis, for example, of the stages of virus assembly or disassembly [19], showing the ejection of the genome [45]. Virus particles can be disrupted upon indentations above the rupture strength [46], and mechanical fatigue below the rupture threshold can be used to monitor in real time the intermediate stages of human adenovirus [47] and human picobirnavirus [48] disassembly. Moreover, AFM can be used to determine some physical properties of virus particles including mechanical features (see Chap. 18). Finally, AFM enables to image and measure the interactions of single virus particles with other molecules and even with host cells [49].

Since individual virus particles adsorb weakly to typical surfaces, they are prone to undesired modifications by lateral forces in CM or DM operation, as they are not held by neighbor molecules, unlike proteins in a 2- or 3-dimensional crystal. The typical approach for surpassing this limitation is the use of structural fixation agents, such as glutaraldehyde. In such conditions AFM may provide images whose resolution is comparable to that of some EM images [50], although still far less than that of the best structural models of virus particles recently obtained by cryo-EM (see Chap. 3). Nevertheless, since glutaraldehyde crosslinks different parts of a molecular complex and it structurally reinforces the specimens subjected to study [51], it precludes any characterization of processes or properties of intact native virus particles, such as disassembly or mechanical features [52].

Chemical fixation is, however, not required in many instances to obtain good images or probe mechanical properties of viruses. The adsorption of viruses on solid surfaces is based on the same principles as those involved in the adsorption of DNA and proteins (Sect. 9.4.1). Although the typical approach implies the convenient silanization of glass surfaces [53], treated mica [47], and HOPG [48, 54] have also proved as adequate substrates for virus attachment.

JM has been very useful to study unfixed viruses, since the loading forces can be accurately controlled, and lateral forces are minimized. Firstly, let us exemplify the

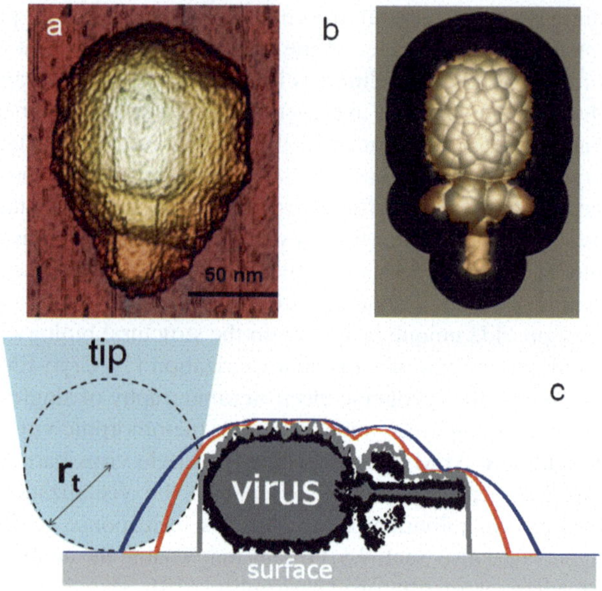

Fig. 9.8 Dilation effects in bacteriophage ϕ29 AFM images. (**a**) shows the topography of a single ϕ29 particle. The bright color image in (**b**) represents a EM structural model of ϕ29, and the black area indicates the dilation effects corresponding to a tip of 10 nm in diameter. The chart in (**c**) indicates the dilation effect as a function of the tip size

tip-virus geometrical dilation with a prolate icosahedral virus. Figure 9.8a shows a topographic image of a bacteriophage ϕ29 virion. Its comparison with the EM model [55] is shown in Fig. 9.8b: AFM enables a reasonable structural delineation of the viral head, the neck region, and the tail. When the tip-sample geometrical algorithm [24] implemented in the WSxM software [56] is applied, one observes (Fig. 9.8b) how the EM model (light contour) is dilated to the black contour in the AFM image. Figure 9.8c demonstrates how the geometrical dilation of a ϕ29 virion image adsorbed on a surface increases with the tip radius. A 0.5-nm radius tip produces the gray color profile, which accurately follows the virus shape. In turn, tips of 5 and 10 nm in radius (red and blue lines, respectively) show an evident dilation and blurring of the structural details in the virion image.

Let us now consider some geometrical aspects of the adsorption of icosahedral viruses (see Chap. 2) to a solid surface. For instance, the roughly spherical minute virus of mice (MVM) has a capsid of icosahedral symmetry with conspicuous protrusions at the threefold symmetry axes. Single MVM particles adsorb to the solid surface in different orientations, which frequently leaves a three-, two-, or fivefold symmetry axes close to the top of the particle (see Fig. 9.9a, b, and c, respectively). The possibility to identify the orientation of single virus particles by resolving protrusions or other structural details allows not only a rather detailed characterization of their topography, but also the determination of local physical properties. For example, by making nanoindentations on single MVM particles adsorbed in different orientations and obtaining *F-z* curves, it has been shown that the viral single-stranded DNA genome is responsible for an anisotropic mechanical stiffening the virion relative to the DNA-free capsid [57], something that is important for preserving viral infectivity. Moreover, it was possible to selectively disrupt these

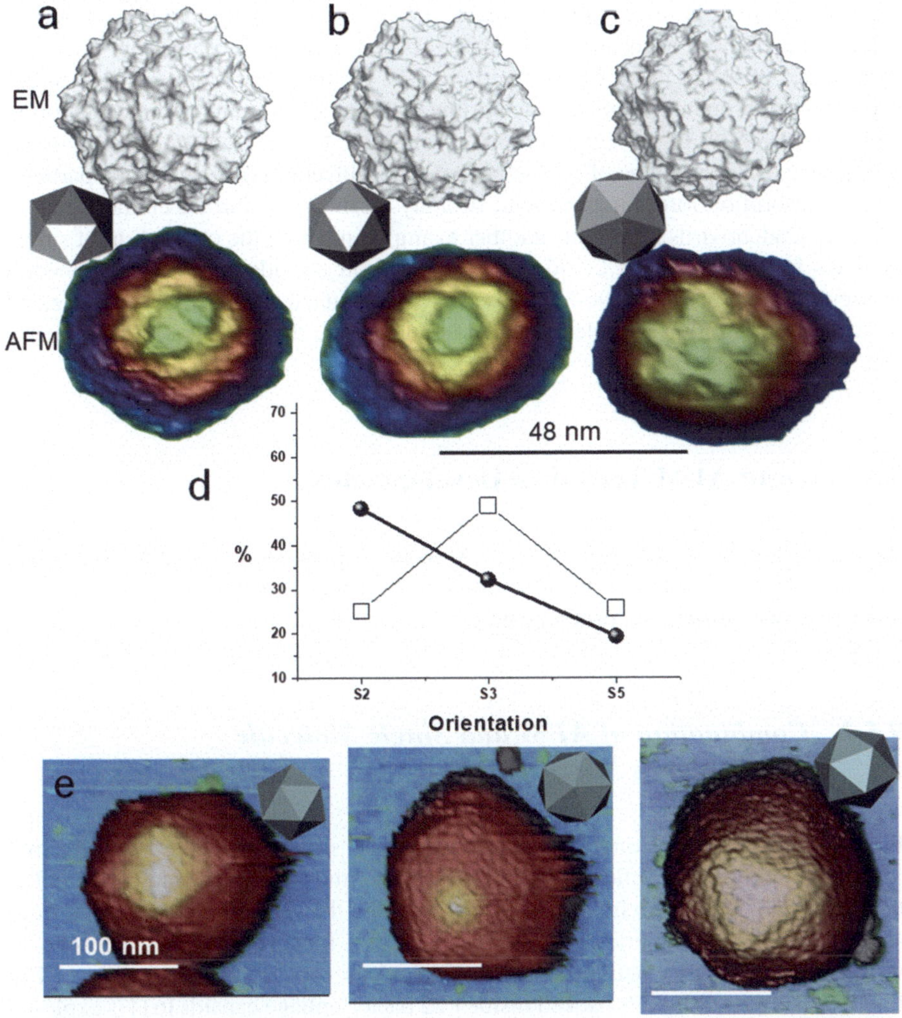

Fig. 9.9 AFM imaging of MVM and human adenovirus. (**a**), (**b**) and (**c**) show the topographies corresponding to single MVM particles oriented with a two-, three-, or fivefold symmetry axis on top, respectively. (**d**) represents the adsorption percentage of each symmetry axis (squares) and the theoretical expectations if the adsorption through edges, faces, and vertices of a perfect icosahedron were equiprobable (circles). In (e) we can observe human adenovirus particles showing different symmetry axes

DNA–protein interactions or replace amino acid residues in selected regions of the MVM capsid and thereby engineer virus particles with altered mechanical properties [58] (see Chap. 18). It may be instructive to quantify the adsorption geometries found in MVM particles (Fig. 9.9d). Black circles represent the orientation likelihood of an icosahedron on a surface provided that the adsorption through vertices (fivefold symmetry), edges (twofold symmetry) and faces (threefold symmetry) is equally probable. If such were the case, a twofold symmetry orientation would exhibit the highest adsorption probability because in an icosahedron there are more edges 30- than faces 20- and vertices 12-. On the other hand, empty squares depict the experimental results, which prime the threefold symmetry axis adsorption with almost a 50% of the adsorbed particles. Since the threefold orientation maximizes the icosahedron area of contact, in contrast with edges or vertices, the particle-surface adsorption forces may depend strongly on the virus-surface contact area.

Human adenovirus represents another example in which the orientation of single particles imaged by AFM could be determined on the adsorption of icosahedral viruses on a surface (Fig. 9.9e). In this case, viruses attach to mica previously treated with $NiCl_2$, exhibiting particles oriented along a two-, three- or fivefold symmetry axes [59].

9.5 Recent AFM Technical Developments

During the last years, the application of AFM to virus research has undergone some technical developments that facilitate the investigation of new processes and properties of virus capsids and protein cages.

9.5.1 Combination of AFM and Single Molecule Fluorescence Microscopy

First, we discuss the combination of AFM and single molecule total internal reflection fluorescence microscopy (TIRFM) to study virus dynamics. Initially, this approach was used in our group to explore the diffusion of dsDNA from semi-disrupted virus particles [46, 60]. Here, we focus on another example in which we studied the activity of individual virus capsids as nanoreactors. The fluorescence of ~220 GFPs packed inside P22 bacteriophage capsids [61] is explored as a function of the alteration of their microenvironment induced by the AFM tip to individual virus particles [60]. The integration of single molecule TIRFM with AFM requires to monitor the fluorescence signal at the surface to avoid not only the background signal of the AFM probe itself, but also the light coming from the bulk solution (Fig. 9.10a). In this approach [53] P22 bacteriophages are immobilized on a glass surface, and a region is scanned with the AFM to visualize virus particles (Fig. 9.10b, left), and imaged with TIRFM to correlate single viral particles with single fluorescence spots (Fig. 9.10b, right). The diffraction-limited resolution of the optical system restricted the size of the smallest spot to $\lambda/2$ ($\approx$250 nm), which is larger than the P22 particles. Consequently, the particles that were too close (arrow, Fig. 9.10b, left) appeared as a single spot (bottom right, Fig. 9.10b, right). The inset of Fig. 9.10b (right) shows the topographic (white) and light (green) profiles of two

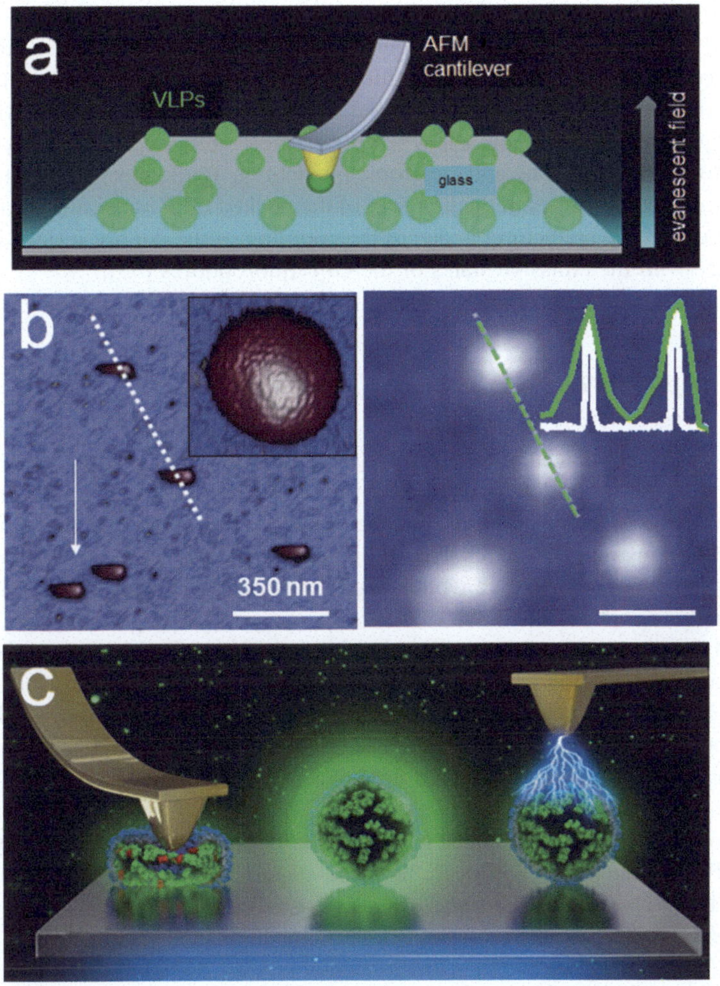

Fig. 9.10 AFM-single molecule fluorescence of viruses. (**a**) presents a diagram of the AFM/TIRF system, showing the P22 particles (light green) and the AFM probed particle in dark green. (**b**) Simultaneous AFM and fluorescence images (right and left, respectively) of individual particles on the glass surface. The white arrow (left) indicates a couple of particles appearing as one spot in fluorescence (right). The inset in a-left shows a high-resolution AFM image of P22 particle resolving the proteinaceous structure of the virus. Inset of the b-right compares the lateral resolution of both microscopies by plotting the signal profile of the AFM topography (white) and the fluorescence (green) obtained at the lines indicated by the dashed white and green lines, respectively. (**c**) Artistic rendition of mechanical quenching (left) and electronic quenching (right). Adapted from [53]

VLPs (Fig. 9.10b, white and green dotted lines) where the particle width in the fluorescence signal is ≈500 nm, approximately five times larger than in the AFM topography. After localizing an intact virus particle that showed a fluorescence signal, a nanoindentation experiment was carried out by using either gold covered or uncovered Si_3N_4 tips. These experiments revealed that a decrease in the P22 particle fluorescence (quenching) occurred in both cases. However, while in the event of gold covered Si_3N_4 tips quenching happened well before the tip established contact with the virus particles, in the case of the non-covered insulating Si_3N_4 tips quenching appeared after contact, when the particle collapsed (Fig. 9.10c). These results unveiled mechanical and electronic quenching. While mechanical quenching takes place due to changes in the integrity of the packed GFP molecules during virus collapse, electronic quenching is due to the coupling of the Au tip Surface Plasmon Resonance with the emissive states of the GFP fluorophores [53].

9.5.2 High-Speed AFM (HS-AFM)

The processes of molecular biology may occur very fast. For instance, kinesin walking on a microtubule moves at 400 nm/s at a few mM of ATP concentrations [62]. One step of 8 nm takes 20 ms, and we would like to capture this movement with the AFM to construct a time-resolved movie of the individual motor's step-by-step activity. Let us assume that we would like to take one image per kinesin step. This would imply to capture 50 frames/second. Assuming a 64 × 64 pixels image, and that the AFM tip has to "touch" the surface once per pixel, we would need to take data at a frequency of $50 \times 64^2 \sim 204$ kHz. If we want to avoid any mechanical vibration noise in our data, we need the mechanical resonance frequency of the AFM system to be above 204 kHz. Although this case is obviously a bit of a stretch, it is useful as a starting point to understand the requirements of the HS-AFM.

There are excellent works published by experts that explain HS-AFM in detail [63]. Here I just pretend to depict a brief glimpse of the topic. The main components of the AFM are the microcantilever, the piezo-stages in *x*-, *y*- and *z*-directions within the AFM head, and the electronics. Thus, it is necessary to improve these three components to increase the scanning speed as much as possible.

Cantilevers. Imaging in HS-AFM is achieved with dynamic force microscopy, where the cantilever oscillates at the resonance frequency. It is obvious this the resonance frequency should be well above the frequency used for measuring the topographical data. It is also important that this cantilever exhibit a low spring constant to avoid damaging the soft biological samples. For a rectangular cantilever, the expressions for the resonance frequency f_r and the spring constant k as a function of the cantilever geometry (Fig. 9.3a, b) are:

$f_r = 0.56\frac{t}{L^2}\sqrt{\frac{E}{12\rho}}$ and $k = E\frac{wt^3}{4L^3}$, where E and ρ are the Young Modulus and density of the cantilever material (typically Si_3N_2), respectively (Section 9.3.1). Considering that we want to have cantilevers with high f_r and low k, it is possible to substitute t/L^2 from k in f_r to derive the expression $\frac{f_r}{k} \sim \frac{L}{t^2 w}$ which reveals that HS-AFM demands very thin cantilevers. For example, in the case of common OMCL-RC800PSA cantilevers, we have $k = 0.05$ N/m, $f_r = 18$ kHz, and the dimensions are 200 μm × 20 μm × 0.4 μm ($L \times w \times t$). Therefore, the coefficient is

~360 kHz m/N. However, in the case of USC-F1.2 cantilevers used for HS-AFM, $k = 0.14$ N/m, $f_r = 1.2$ MHz, and the dimensions are 7 μm × 2 μm × 0.08 μm ($L \times w \times t$), resulting in a coefficient of ~8500 kHz m/N. The comparison between both coefficients of RC800PSA and USC-F1.2 cantilevers results in 0.04, showing a difference of about two orders of magnitude. Another important phenomenology that should be considered is that the companies provide the resonance frequency in air, but in liquid, this operation frequency decreases by a factor ~4.6 due to water viscosity. Therefore, while the RC800PSA cantilever frequency would drop to 4 kHz, the USC-F1.2 would diminish to 240 kHz. The resonance frequency of USC-F1.2 cantilevers (but not that of the RC800PSA cantilevers) is well above the limit we established above for studying kinesin. Although the resonance frequency of the cantilever is the most critical parameter for attaining HS-AFM, there are many more details that can be consulted in the specialized literature [63].

Piezo stages. As commented before, the piezo-tube (Fig. 9.11a) produces the *x*- *y*-, and *z*-movements in AFM. In this case, the three dimensions are coupled in the same tube, whose length and diameter (10 cm × 1 cm, respectively) allow a scanning of 60 μm × 60 μm × 6 μm (*x*-, *y*-, and *z*-directions, respectively). A piezoelectric tube is characterized by the stretching (f_s) and bending (f_b) resonance frequencies, which can be estimated as $f_s \approx c/4L$ and $f_b \approx 0.56 r_g c/L^2$ (c is the speed of sound, L is the piezo length, and r_g is the radius of gyration, related to the unaltered section of the tube [3]. However, even when working with short tubes with less scan capability, the resonance frequency just reaches a few kHz, which does not serve for HS-AFM (Fig. 9.11a). The requirements of the piezo scanners for HS-AFM, among other parameters, include a high resonance frequency, a sufficient scanning amplitude and small crosstalk between the three dimensions. In HS-AFM, the three scanning directions are decoupled by assigning a different piezo for each one (Fig. 9.11b). These three squared piezo stacks are mounted on flexures (blade springs, Fig. 9.11b) that allow to decouple the movement of the sample holder in each direction. The scanning *y*-direction is considered as the slow scan axis because it only moves after the data of each *x*-line scan have been obtained. The *x*-direction is named fast scan because it

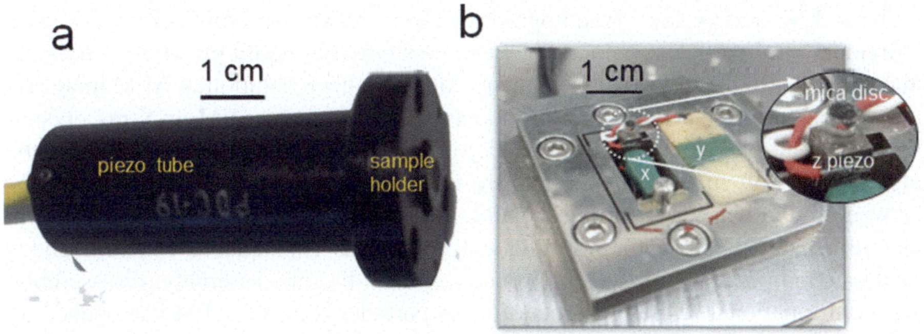

Fig. 9.11 High-Speed AFM (**a**) shows a regular AFM piezo scanner of Nanotec and (**b**) a HS-AFM piezo scanner of WPI-NanoLSI, Kanazawa University [63, 64]

takes the pixel-by-pixel information in a sequential way. The piezo of the *z*-direction must respond with the feedback to the changing topography of the sample and must be the faster system, presenting the lowest dimensions with an approximate volume of less than 18 mm^3 (Fig. 9.11b, inset). The normal sample holder is a mica sheet disk of ~2 mm in diameter (Fig. 9.11b, inset). The HS-AFM design is focused on reducing the dimensions of the mechanical loop in the AFM head as much as possible, including the optical system and the piezo stage. In addition to these considerations, other strategies such as passive damping and counterbalance [63], increase the resonance frequency of the HS-AFM piezo stage until a few hundreds of kHz. A shortage of such designs is the low amplitude scan (~2 μm), but there is always a compromise between scanning size and speed [65].

Electronics and feedback. Last, but not least, it is also necessary to change the electronics and feedback. The methodology here includes to improve the amplitude and force detectors. Active damping is committed to remove unwanted vibrations from the z scanner. The feedback algorithm controlling the tip-sample distance in dynamic force microscopy is key to minimize the tip-sample interaction force. In the first approach, it is convenient to change the feedback loop in dynamic way for uphill and downhill topography regions, instead of having fixed parameters. Thus, a dynamic PID control has been developed, where the gain parameters are automatically changed, depending on the alteration of the cantilever amplitude.

During the last years, HS-AFM has contributed largely to virus research, for instance, in unveiling the dynamics of HIV capsid protein two-dimensional assembly on a flat surface [66]. But it remains a challenge to monitor the three-dimensional assembly processes of individual viruses in real time.

9.6 Understanding Viruses: Some Major Contributions of AFM

AFM enables new possibilities for studying the structure, dynamics, and physical properties of virus particles, complementing long-standing structural approaches such as EM and X-ray crystallography. First, AFM may routinely work with individual viral particles; second, they can be studied in liquid phase, in conditions that are closer to the in vivo environment. Although the resolution of AFM images is not comparable to that of structural models obtained by cryo-EM or X-ray crystallography, and only topographic (surface) images are obtained, AFM facilitates the identification of virus elements that are not symmetrically ordered or that differ between individual virus particles. Third, AFM may eventually allow the study in real time, and in close to physiological conditions (in aqueous medium and a moderate temperature) of dynamic processes including the assembly, disassembly, and conformational rearrangements of virus particles. Fourth, AFM has opened up the possibility to investigate previously unexplored physical properties of virus particles, which is providing novel insights into the relationships between virus structure, physical properties, and biological functions. Along this line, AFM offers the possibility of probing the stiffness of virions and capsids and of selectively disrupting elements in individual virus particles (see Chap. 18). These experiments may contribute to understand virus assembly and disassembly, stability, and dynamics.

9.7 Perspectives

Some of the current developments in AFM imaging aim to increase the data acquisition speed (a typical AFM image takes a few minutes), to directly visualize the dynamics of relatively fast biological processes in real time, including conformational changes and the assembly or disassembly of virus particles. For example, AFM imaging at high (video) rates has already been used to show the 2D assembly of viral proteins on a surface [66].

Mechanical properties of biological objects are generally obtained in indentation experiments [67]. However, there is also the possibility of using mechanical fatigue to extract information about the binding of capsomers, and the role of core proteins in human adenovirus [47, 67, 68]. The manipulation ability of AFM allows to extract of the genome from virus particles in a controlled fashion, allowing to deepen the determinants of nucleic acids delivery during infection [60, 69]. The ability of the AFM for pulling from individual viruses allows to investigate the interaction forces during the initial stages of virus binding on the host [70].

Finally, it is important to stress the ability of AFM to mechanically manipulate and modify individual virus particles. Among other examples, it has been possible to deepen the role of ssRNA in human picobirnavirus disassembly [48] and the physical nature of lemon shape archaeal virus *Sulfolobus* monocaudavirus 1 (SMV1) [54] and coronavirus structures [71, 72]. It is also possible to understand the interaction between shells in multilayered viruses [73, 74]. In the future, AFM could reveal more details on the structural determinants to impair virus infectivity [75] or the effects on virus structure of virus adsorption on surfaces [76]. Last, but not least, the capacity of AFM to drag nanoobjects adsorbed on a surface [77] can be applied to move adsorbed virus capsids, measuring the interaction force between viruses and surfaces for applications in prophylaxis.

In summary, the ongoing developments in AFM promise to improve its applicability in structural virology by: (i) further reducing unwanted sample damage; (ii) increasing the spatial resolution; and (iii) increasing the temporal resolution so that relatively fast structural changes in virus particles can be followed in real time.

Acknowledgments The author acknowledges the support by grants received from the Spanish Ministry of Science and Innovation projects (FIS2017-89549-R, FIS2017-90701-REDT and PID2021-126608OB-I00) and the Human Frontiers Science Program (HFSPO RGP0012/2018). IFIMAC is the Center of Excellence "María de Maeztu." I also want to thank my students for taking care of the hard job: Miguel Cantero, María Jesús Rodríguez, Klara Strobl, Pablo Ibañez, and Alejandro Díez Martínez.

References

1. de Pablo PJ (2019) The application of atomic force microscopy for viruses and protein shells: imaging and spectroscopy. Adv Virus Res 105:161–187
2. Binnig G, Quate CF, Gerber CH (1986) Atomic force microscope. Phys Rev Lett 56:930–933
3. Chen CJ (1993) Introduction to scanning tunneling. Microscopy Oxford University Press
4. de Pablo P, Moreno-Herrero F, Colchero J, Gomez-Herrero J, Herrero P, Baro A, Ordejon P, Soler J, Artacho E (2000) Absence of dc-conductivity in lambda-DNA. Phys Rev Lett 85:4992–4995
5. Baró AM, Miranda R, Alamán J, García N, Binnig G, Rohrer H, Gerber C, Carrascosa JL (1985) Determination of surface topography of biological specimens at high resolution by scanning tunnelling microscopy. Nature 315:253–254
6. Hernando-Perez M, Miranda R, Aznar M, Carrascosa JL, Schaap IAT, Reguera D, de Pablo PJ (2012) Direct measurement of phage phi29 stiffness provides evidence of internal pressure. Small 8:2366–2370
7. Schmaltz G (1929) Über Glätte und Ebenheit als physikalisches und physiologisches Problem. Z Ver Dtsch Ingenieure 12:1461–1467
8. Meyer G, Amer NM (1988) Novel optical approach to atomic force microscopy. Appl Phys Lett 53:1045–1047
9. Sader JE, Chon JWM, Mulvaney P (1999) Calibration of rectangular atomic force microscope cantilevers. Rev Sci Instrum 70:3967–3969
10. Butt H-J, Jaschke M (1995) Calculation of thermal noise in atomic force microscopy. Nanotechnology 6:1
11. Israelachvili JN (2011) Intermolecular and surface forces. In: Intermolecular and surface forces (third edition). Academic Press, San Diego, pp 661–674
12. Johnson KL (1985) Contact mechanics. Cambridge University Press
13. Ohnesorge F, Binnig G (1993) True atomic resolution by atomic force microscopy through repulsive and attractive forces. Science 260:1451–1456
14. Sugimoto Y, Pou P, Abe M, Jelinek P, Pérez R, Morita S, Custance Ó (2007) Chemical identification of individual surface atoms by atomic force microscopy. Nature 446:64–67
15. Carpick RW, Ogletree DF, Salmeron M (1997) Lateral stiffness: a new nanomechanical measurement for the determination of shear strengths with friction force microscopy. Appl Phys Lett 70:1548–1550
16. Martin Y, Williams CC, Wickramasinghe HK (1987) Atomic force microscope–force mapping and profiling on a sub 100-Å scale. J Appl Phys 61:4723–4729
17. Garíca R, Pérez R (2002) Dynamic atomic force microscopy methods. Surf Sci Rep 47:197–301
18. Legleiter J, Park M, Cusick B, Kowalewski T (2006) Scanning probe acceleration microscopy (SPAM) in fluids: mapping mechanical properties of surfaces at the nanoscale. Proc Natl Acad Sci USA 103:4813–4818
19. Kuznetsov YG, Malkin AJ, Lucas RW, Plomp M, McPherson A (2001) Imaging of viruses by atomic force microscopy. J Gen Virol 82:2025–2034
20. Miyatani T, Horii M, Rosa A, Fujihira M, Marti O (1997) Mapping of electrical double-layer force between tip and sample surfaces in water with pulsed-force-mode atomic force microscopy. Appl Phys Lett 71:2632–2634
21. de Pablo P, Colchero J, Gomez-Herrero J, Baro A (1998) Jumping mode scanning force microscopy. Appl Phys Lett 73:3300–3302
22. De Pablo P, Colchero J, Gomez-Herrero J, Baro A, Schaefer D, Howell S, Walsh B, Reifenberger R (1999) Adhesion maps using scanning force microscopy techniques. J Adhes 71:339–356
23. Ortega-Esteban A, Horcas I, Hernando-Pérez M, Ares P, Pérez-Berná AJ, San Martín C, Carrascosa JL, de Pablo PJ, Gómez-Herrero J (2012) Minimizing tip–sample forces in jumping mode atomic force microscopy in liquid. Ultramicroscopy 114:56–61

24. Villarrubia JS (1997) Algorithms for scanned probe microscope image simulation, surface reconstruction, and tip estimation. J Res Natl Inst Stand Technol 102:425–454
25. Marti O, Drake B, Gould S, Hansma PK (1988) Atomic resolution atomic force microscopy of graphite and the '"native oxide"' on silicon. J Vac Sci Technol A 6:287–290
26. Soler JM, Baro AM, García N, Rohrer H (1986) Interatomic forces in scanning tunneling microscopy: Giant corrugations of the graphite surface. Phys Rev Lett 57:444–447
27. Hansma HG, Sinsheimer RL, Li M-Q, Hansma PK (1992) Atomic force microscopy of single- and double-stranded DNA. Nucleic Acids Res 20:3585–3590
28. Lyubchenko YL (2014) Nanoscale nucleosome dynamics assessed with time-lapse AFM. Biophys Rev 6:181–190
29. Dame RT, Wyman C, Goosen N (2003) Insights into the regulation of transcription by scanning force microscopy. J Microsc 212:244–253
30. Wagner P, Hegner M, Guentherodt H-J, Semenza G (1995) Formation and in situ modification of monolayers chemisorbed on Ultraflat template-stripped gold surfaces. Langmuir 11:3867–3875
31. Müller DJ, Amrein M, Engel A (1997) Adsorption of biological molecules to a solid support for scanning probe microscopy. J Struct Biol 119:172–188
32. Müller DJ, Janovjak H, Lehto T, Kuerschner L, Anderson K (2002) Observing structure, function and assembly of single proteins by AFM. Prog Biophys Mol Biol 79:1–43
33. Hansma HG, Pietrasanta L (1998) Atomic force microscopy and other scanning probe microscopies. Curr Opin Chem Biol 2:579–584
34. Hörber JKH, Miles MJ (2003) Scanning probe evolution in biology. Science 302:1002–1005
35. Hinterdorfer P, Dufrêne YF (2006) Detection and localization of single molecular recognition events using atomic force microscopy. Nat Methods 3:347–355
36. Neuman KC, Nagy A (2008) Single-molecule force spectroscopy: optical tweezers, magnetic tweezers and atomic force microscopy. Nat Methods 5:491–505
37. Kirmizis D, Logothetidis S (2010) Atomic force microscopy probing in the measurement of cell mechanics. Int J Nanomedicine 5:137–145
38. Kurland NE, Drira Z, Yadavalli VK (2012) Measurement of nanomechanical properties of biomolecules using atomic force microscopy. Micron 43:116–128
39. Viani MB, Pietrasanta LI, Thompson JB, Chand A, Gebeshuber IC, Kindt JH, Richter M, Hansma HG, Hansma PK (2000) Probing protein-protein interactions in real time. Nat Struct Biol 7:644–647
40. Moreno-Herrero F, Colchero J, Gómez-Herrero J, Baró AM (2004) Atomic force microscopy contact, tapping, and jumping modes for imaging biological samples in liquids. Phys Rev E 69: 031915
41. Xu X, Carrasco C, de Pablo PJ, Gomez-Herrero J, Raman A (2008) Unmasking imaging forces on soft biological samples in liquids when using dynamic atomic force microscopy: a case study on viral capsids. Biophys J 95:2520–2528
42. Kasas S, Thomson NH, Smith BL, Hansma HG, Zhu X, Guthold M, Bustamante C, Kool ET, Kashlev M, Hansma PK (1997) Escherichia coli RNA polymerase activity observed using atomic force microscopy. Biochemistry 36:461–468
43. Moreno-Herrero F, de Jager M, Dekker NH, Kanaar R, Wyman C, Dekker C (2005) Mesoscale conformational changes in the DNA-repair complex Rad50/Mre11/Nbs1 upon binding DNA. Nature 437:440–443
44. Kodera N, Yamamoto D, Ishikawa R, Ando T (2010) Video imaging of walking myosin V by high-speed atomic force microscopy. Nature 468:72–76
45. Strobl K, Mateu MG, de Pablo PJ (2023) Exploring nucleic acid condensation and release from individual parvovirus particles with different physicochemical cues. Virology 581:1–7
46. Valbuena A, Strobl K, Gil-Redondo JC, Valiente L, de Pablo PJ, Mateu MG (2023) Single-molecule analysis of genome Uncoating from individual human rhinovirus particles, and modulation by antiviral drugs. Small n/a 2304722

47. Martin-Gonzalez N, Ibanez-Freire P, Ortega-Esteban A, Laguna-Castro M, San MC, Valbuena A, Delgado-Buscalioni R, de Pablo PJ (2021) Long-range cooperative disassembly and aging during adenovirus Uncoating. Phys Rev X:11
48. Rodríguez-Espinosa MJ, Rodríguez JM, Castón JR, de Pablo PJ (2023) Mechanical disassembly of human picobirnavirus like particles indicates that cargo retention is tuned by the RNA–coat protein interaction. Nanoscale Horiz 8:1665–1676
49. Sieben C, Kappel C, Zhu R, Wozniak A, Rankl C, Hinterdorfer P, Grubmüller H, Herrmann A (2012) Influenza virus binds its host cell using multiple dynamic interactions. Proc Natl Acad Sci USA 109:13626–13631
50. Xiao C, Kuznetsov YG, Sun S, Hafenstein SL, Kostyuchenko VA, Chipman PR, Suzan-Monti M, Raoult D, McPherson A, Rossmann MG (2009) Structural studies of the Giant Mimivirus. PLoS Biol 7:e1000092
51. Vinckier A, Heyvaert I, D'Hoore A, McKittrick T, Van Haesendonck C, Engelborghs Y, Hellemans L (1995) Immobilizing and imaging microtubules by atomic force microscopy. Ultramicroscopy 57:337–343
52. Carrasco C, Luque A, Hernando-Pérez M, Miranda R, Carrascosa JL, Serena PA, De Ridder M, Raman A, Gómez-Herrero J, Schaap IAT et al (2011) Built-in mechanical stress in viral shells. Biophys J 100:1100–1108
53. Strobl K, Selivanovitch E, Ibáñez-Freire P, Moreno-Madrid F, Schaap IAT, Delgado-Buscalioni R, Douglas T, de Pablo PJ (2022) Electromechanical Photophysics of GFP packed inside viral protein cages probed by force-fluorescence hybrid single-molecule microscopy. Small 18:2200059
54. Cantero M, Cvirkaite-Krupovic V, Krupovic M, de Pablo PJ (2023) Mechanical tomography of an archaeal lemon-shaped virus reveals membrane-like fluidity of the capsid and liquid nucleoprotein cargo. Proc Natl Acad Sci USA 120:e2307717120
55. Tang J, Olson N, Jardine PJ, Grimes S, Anderson DL, Baker TS (2008) DNA poised for release in bacteriophage ø29. Structure 16:935–943
56. Horcas I, Fernández R, Gómez-Rodríguez JM, Colchero J, Gómez-Herrero J, Baro AM (2007) WSXM: a software for scanning probe microscopy and a tool for nanotechnology. Rev Sci Instrum 78:013705
57. Carrasco C, Carreira A, Schaap IT, Serena PA, Gómez-Herrero J, Mateu MG, de Pablo PJ (2006) DNA-mediated anisotropic mechanical reinforcement of a virus. Proc Natl Acad Sci USA 103:13706–13711
58. Carrasco C, Castellanos M, de Pablo PJ, Mateu MG (2008) Manipulation of the mechanical properties of a virus by protein engineering. Proc Natl Acad Sci USA 105:4150–4155
59. Ortega-Esteban A, Condezo GN, Perez-Berna AJ, Chillon M, Flint SJ, Reguera D, San Martin C, de Pablo PJ (2015) Mechanics of viral chromatin reveals the pressurization of human adenovirus. ACS Nano 9:10826–10833
60. Ortega-Esteban A, Bodensiek K, San Martin C, Suomalainen M, Greber UF, de Pablo PJ, Schaap IAT (2015) Fluorescence tracking of genome release during mechanical unpacking of single viruses. ACS Nano 9:10571–10579
61. O'Neil A, Prevelige PE, Basu G, Douglas T (2012) Coconfinement of fluorescent proteins: spatially enforced communication of GFP and mCherry encapsulated within the P22 capsid. Biomacromolecules 13:3902–3907
62. Schaap IAT, Carrasco C, de Pablo PJ, Schmidt CF (2011) Kinesin walks the line: single motors observed by atomic force microscopy. Biophys J 100:2450–2456
63. Ando T, Uchihashi T, Fukuma T (2008) High-speed atomic force microscopy for nano-visualization of dynamic biomolecular processes. Prog Surf Sci 83:337–437
64. Uchihashi T, Kodera N, Ando T (2012) Guide to video recording of structure dynamics and dynamic processes of proteins by high-speed atomic force microscopy. Nat Protoc 7:1193–1206

65. Marchesi A, Umeda K, Komekawa T, Matsubara T, Flechsig H, Ando T, Watanabe S, Kodera N, Franz CM (2021) An ultra-wide scanner for large-area high-speed atomic force microscopy with megapixel resolution. Sci Rep 11:13003
66. Valbuena A, Maity S, Mateu MG, Roos WH (2020) Visualization of single molecules building a viral capsid protein lattice through stochastic pathways. ACS Nano 14:8724–8734
67. Martín-González N, Hernando-Pérez M, Condezo GN, Pérez-Illana M, Šiber A, Reguera D, Ostapchuk P, Hearing P, San Martín C, de Pablo PJ (2019) Adenovirus major core protein condenses DNA in clusters and bundles, modulating genome release and capsid internal pressure. Nucleic Acids Res 47:9231–9242
68. Martín-González N, Gómez-González A, Hernando-Pérez M, Bauer M, Greber UF, San Martín C, de Pablo PJ (2023) Adenovirus core protein V reinforces the capsid and enhances genome release from disrupted particles. Sci Adv 9:eade9910
69. Kellermayer MSZ, Vörös Z, Csík G, Herényi L (2018) Forced phage uncorking: viral DNA ejection triggered by a mechanically sensitive switch. Nanoscale 10:1898–1904
70. Alsteens D, Newton R, Schubert R, Martinez-Martin D, Delguste M, Roska B, Muller DJ (2017) Nanomechanical mapping of first binding steps of a virus to animal cells. Nat Nanotechnol 12:177–183
71. Kiss B, Kis Z, Pályi B, Kellermayer MSZ (2021) Topography, spike dynamics, and Nanomechanics of individual native SARS-CoV-2 Virions. Nano Lett 21:2675–2680
72. Cantero M, Carlero D, Chichon FJ, Martin-Benito J, de Pablo PJ (2022) Monitoring SARS-CoV-2 surrogate TGEV individual Virions structure survival under harsh physicochemical environments. Cells 11
73. Jimenez-Zaragoza M, Yubero MPL, Martin-Forero E, Caston JR, Reguera D, Luque D, de Pablo PJ, Rodriguez JM (2018) Biophysical properties of single rotavirus particles account for the functions of protein shells in a multilayered virus. eLife 7:7
74. Azinas S, Bano F, Torca I, Bamford DH, Schwartz GA, Esnaola J, Oksanen HM, Richter RP, Abrescia NG (2018) Membrane-containing virus particles exhibit the mechanics of a composite material for genome protection. Nanoscale 10:7769–7779
75. Domínguez-Zotes S, Valbuena A, Mateu MG (2022) Antiviral compounds modulate elasticity, strength and material fatigue of a virus capsid framework. Biophys J 121:919–931
76. Zeng C, Hernando-Perez M, Dragnea B, Ma X, van der Schoot P, Zandi R (2017) Contact mechanics of a small icosahedral virus. Phys Rev Lett 119:038102
77. Moya A, Aleman J, Gomez-Herrero J, Mas-Balleste R, de Pablo PJ (2022) Nanotribology and electrical properties of carbon nanotubes hybridized with covalent organic frameworks. Carbon 199:80–86

Further Reading

Scanning Force Microscopies Beyond Imaging. Editor Paolo Samorí. Wiley-VCH 2006
Atomic Force Microscopy in Liquid. Editors. A. Baró and R. Reifenberger. Wiley-VCH 2012
Also especially recommended for further reading are references [2, 5, 13, 18, 23, 31, 44, 49, 54, 62, 70] *listed above.*

Chapter 10
Optical Tweezers to Study Viruses

J. Ricardo Arias-Gonzalez

Abstract A virus is a complex molecular machine that propagates by channeling its genetic information from cell to cell. Unlike macroscopic engines, it operates in a nanoscopic world under continuous thermal agitation. Viruses have developed efficient passive and active strategies to pack and release nucleic acids. Some aspects of the dynamic behavior of viruses and their substrates can be studied using structural and biochemical techniques. By the turn of the millennium, physical techniques have been applied to dynamic studies of viruses in which their intrinsic mechanical activity can be measured directly. Optical tweezers are a technology that can be used to measure the force, torque, and strain produced by molecular motors, as a function of time and at the single-molecule level. Thanks to this technique, some bacteriophages are now known to be powerful nanomachines; they exert force in the piconewton range and their motors work in a highly coordinated fashion for packaging the viral nucleic acid genome. Nucleic acids, whose elasticity and condensation behavior are inherently coupled to the viral packaging mechanisms, virion assembly, and virion–cell interactions are also amenable to examination with optical tweezers. In this chapter, we provide a comprehensive analysis of this laser-based tool, its combination with imaging methods, and its application to the study of viruses and viral molecules.

Keywords Biophysics · Virus · Bacteriophage · Capsid · DNA · RNA · Molecular motor · Machine · Single molecule · Mechanochemistry · Optics · Optical tweezers · Magnetic tweezers · Dynamics · Manipulation · Force · Pressure · Elasticity · Condensation

---

J. R. Arias-Gonzalez (✉)
Centro de Tecnologías Físicas, Universitat Politècnica de València, Valencia, Spain
e-mail: ricardo.ariasgonzalez@upv.es

M. G. Mateu (ed.), *Structure and Physics of Viruses*, Subcellular Biochemistry 105,
https://doi.org/10.1007/978-3-031-65187-8_10

Abbreviations

α-Dig	Anti-digoxigenin
AFM	Atomic force microscopy
bp	Base pair
Dig	Digoxigenin
ds	Double-stranded
GC	Guanine-cytosine
HIV-1	Human immunodeficiency virus type 1
IR	Infrared
NA	Numerical aperture
SARS-CoV-2	Severe acute respiratory syndrome coronavirus 2
ss	Single-stranded
3D	Three dimensions

10.1 Introduction: Life Machinery at the Nanoscale

Optical tweezers are an instrument able to measure the forces and motion involved in the dynamics of biomolecular motors. From this viewpoint, they constitute a unique instrument to study the physics underlying some viral processes. In contrast to other structural techniques described in Chaps. 3, 4, 5, 7, 8, and 9 that allow the visualization of virions and capsids, dynamic experiments using optical tweezers are unable to "photograph" the main characters of this book, but can quantitatively track their mechanical activity in real time. This chapter will introduce a new method with which the virus is followed while it is at work, packaging its nucleic acid into the viral capsid. Characteristic results of these experiments are forces and packaging velocities as a function of time or as a function of the amount of packaged genome. Related problems are capsid assembly, virion-cell attachment strength, and the mechanical behavior of nucleic acids; in this regard, force-extension curves of DNA/RNA to determine their elasticity can be also obtained with this technique.

The machinery that makes up life is very small and highly condensed. The dimensions of a globular protein or the thickness of DNA are measured in nanometers. This machinery works in crowded environments (in a cell, the total concentration of macromolecules may reach 300–400 g/l), and is subject to continuous agitation. In these conditions, movement is strongly affected by thermal fluctuations (Brownian motion), and structures are markedly distorted by mechanical stress. The dynamic picture of the motion and activity of a protein or macromolecular complex is that of a boat sailing against furious winds and waves, with rapid changes in direction. Throughout evolution, proteins have developed strategies to rectify and bias thermal fluctuations using the energy from ATP or GTP in a way similar to the orientation of the boat's sail and rudder. These strategies are different from the macroscopic, human-made machinery, whose dynamic behavior more closely resembles a motorboat in a quiet sea, for which fuel energy is used to propel the boat in a chosen direction. Biological machinery is stochastic and soft, unlike macroscopic machines, whose movements are predictable and whose structures are typically rigid. Biological machinery is nonetheless highly efficient in terms of the ratio [developed work]/[spent energy], much more so than human technology to date. A further singularity of biomolecular machinery is its ability to interact with its environment. Biological machines are in essence biochemical entities; they change shape, size, and mass as a result of their function. The action of most proteins and macromolecular assemblies, including viruses, is thus *mechanochemical*, and their complete study must include not only biochemical aspects but also mechanical analyses.

Optical tweezers, described in this chapter, are a technique that allows such mechanochemical analysis. Viruses, like other large biological structures such as ribosomes, ATP synthase, or the bacterial flagellum, are complex machines, and not passive structures. Many complex viruses in particular are coordinated assemblies of motors, structural scaffolds, and information carriers. Many bacteriophages are fascinating from a technological point of view because they are some of the most powerful machines at the nanoscale acting as packaging devices for their own nucleic acid. In turn, the nucleic acids represent soft "hard drives" that store and spread information. Some nucleic acid physical properties, such as their elasticity and denaturation behavior, may have evolved in tune with the machinery that processes them, rendering these processes more efficient over the ages. Understanding DNA and RNA from a mechanochemical point of view may thus shed light on viral performance. This chapter will also be a platform to introduce these novel views of the viral machinery and its peripherals.

10.2 Basic Concepts and General Experimental Design

The forces developed by protein machines including viruses are in the piconewton (pN) range, which compares with one-trillionth (10^{-12}) the weight of a lemon. Measuring these tiny forces was not technologically easy before the invention of the laser in the 1960s; use of this light technology to move and trap micro- and nanostructures began in the 1970s. By the mid-1980s, this application had been developed into a technique commonly known as optical or laser tweezers. Its use in molecular and cell biology became relevant by the 1990s, although experiments manipulating viruses, cells, or organelles within living cells already saw the light in the previous decade [1].

The forces generated by electromagnetic radiation over a physical object are termed *optical forces*. The strength of these forces is generally in the piconewton range, although this depends on the material properties of the object, on the radiation power, and on beam shape. An instrument based on the use of optical forces will therefore be sensitive to the tiny forces exerted by molecular machinery. Since electromagnetic radiation can penetrate the medium without traceable effects, optical

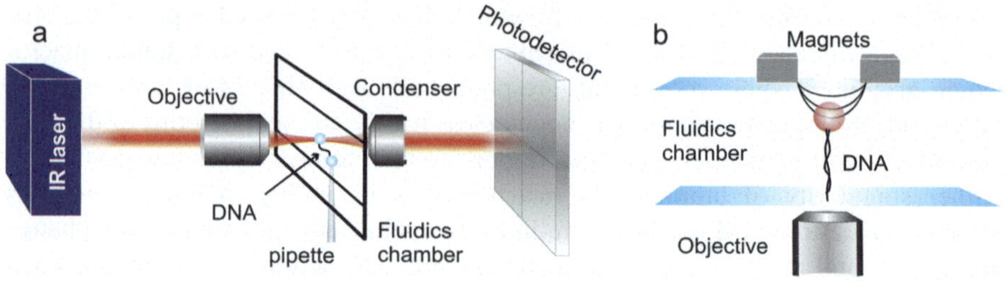

Fig. 10.1 Optical and magnetic tweezers. (**a**) Laser light is highly focused by a microscope objective and collected by a condenser lens. A bead or cell can be trapped in the focal region, which is produced inside a fluidics chamber. A micropipette inserted in the chamber can hold a second bead by suction. Micrometer-sized beads are typically used as mechanical handles to hold a biopolymer by both ends. The collected exiting rays are registered by a position-sensitive photodetector. The scheme shows a single-molecule experiment in which a DNA fiber is stretched between two beads. (**b**) The same experiment is performed using two small magnets that pull a magnetic microsphere inside a fluidics chamber. The DNA is held between the microsphere and the bottom coverslip. A microscope objective is used to monitor the experiment

forces are a non-contact solution to measuring and exerting force on biological particles in their liquid environment and within living cells. Inertial and gravitational effects are insignificant in the long timescales associated with the dynamics of most biological systems.

For reasons that will become clear later, a strongly focused laser beam generates optical forces in all directions toward the focal region, provided that these objects do not absorb light. The result is a three-dimensional (3D) *optical trap*. Most biological structures, natural or synthetic, do not substantially absorb visible or infrared (IR) radiation and can therefore be maneuvered by a strongly focused laser beam within these spectral bandwidths. If the light scattered from the optical trap is analyzed, the forces acting on the trapped object can be deduced. This is the basis of the optical tweezers, an instrument able to trap nanometer and micrometer-sized particles and to measure the external forces that act on them. The resulting displacements can be measured in the sub-nanometer range with millisecond time resolution.

A simplified version of an optical tweezers instrument is shown in Fig. 10.1 (in comparison to its magnetic counterpart, described in Sect. 10.3.7). The basic elements in the optical tweezers can be grouped into those common to an optical microscope and those related to the generation of the optical trap [2]. The *microscope* elements are an objective, a fluidics chamber, and a condenser lens. The *trap* elements are a laser and a photodetector that registers the light at the exit of the condenser. To produce a strongly focused beam, a high-numerical aperture (NA) objective is required (water or oil immersion). Near-IR lasers are optimal for manipulating biological structures because neither water nor biological structures strongly absorb radiation at this wavelength. Laser power in the focal region of several hundred milliwatts provides a good balance that prevents photodamage and produces forces up to a few hundreds of piconewtons. Magnetic tweezers use two small magnets to generate high field gradients in the vicinity of a paramagnetic bead,

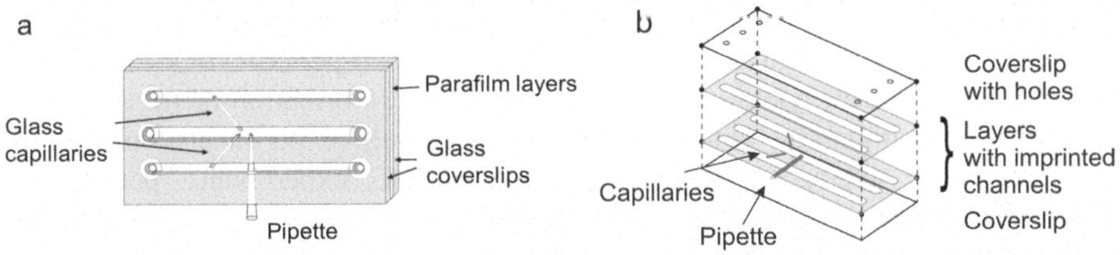

Fig. 10.2 Fluidics chamber. (**a**) The specimens (cells or beads) are passed sequentially through a system of channels. In this design, the experiment takes place in the central channel, which is connected to the top and the bottom channels by glass capillaries. The scheme also shows the disposition of a micropipette, useful in a variety of experiments. (**b**) Mounting a custom-made fluidics chamber for optical tweezers. After assembling the tubes and parafilm layers, the components are heat-sealed

as shown in Fig. 10.1(b). An objective behind the fluidics chamber is used to follow the experiment. This instrument is a good alternative for low forces and to generate torque, as we will see in Sect. 10.3.7.

The fluidics chamber is the test tube for mechanochemical assays. It is usually built using two microscope coverslips sandwiching parafilm layers with imprinted channels. The design shown in Fig. 10.2 allows specimens to be flowed through a top and a bottom channel connected with the central channel by glass capillaries. The optical trap is attained in the central channel. Buffers can be easily exchanged and flowed at a controlled rate, thus adding the ability to exert hydrodynamic pressure. A micropipette can be placed inside the chamber to suck beads and additional tubes can be used to dispense chemicals (for example, ATP) directly to the vicinity of the optical trap. These alternatives are very convenient in a variety of experiments that will be described later in this chapter. In contrast to most structural techniques, optical tweezers allow experiments in physiological conditions that can be changed easily and rapidly (for example, chemical concentration and pH) without affecting the performance of the instrument, which allow the detection in real time of processes triggered by such changes in conditions.

10.2.1 Single-Molecule Approach

As indicated by the above description, this method allows one-by-one trapping and manipulation of specimens (individual molecules and macromolecular assemblies, including virus particles). This ability is the basis of the *single-molecule* approach [3]. *Bulk* experiments, in which a large number of identical (but unavoidably unsynchronized) reactions take place, hinder access to signals from intermediate states of the specimens. Ensemble methods such as electron microscopy (Chap. 3), X-ray crystallography (Chap. 4), or nuclear magnetic resonance (NMR) spectroscopy (Chap. 5) allow "taking snapshots" of specific instants in the temporal flow of events for a certain type of molecule, but only the average state is eventually reconstructed from the data. In addition, data and reconstructed states might be influenced by the cooperative disposition (periodic or non-periodic) of the ensemble before the experiment, yielding misleading structural information or masking rare but potentially important conformations. In fact, bulk experiments only allow observation of average dynamics. Single-molecule experiments permit the unraveling of real-time dynamics and conformations, because the signals are not confused by multiple, asynchronous events.

Another drawback of bulk experiments is that they tend to smear out the information from *thermal fluctuations*. The idealized dynamic picture envisaged from these experiments is a "movie" at long timescales, in which molecular trajectories are smooth and well defined. At short timescales, this movie is very different: dynamics are fast and random (stochastic). As explained in the introduction, proteins and viruses dwell in a background of thermal agitation that affects their dynamic performance. This inherent dynamics can only be captured by real-time monitoring of individual processes. Fluctuations represent an integral part of protein function, and single-molecule information must therefore be included in the overall description of biomolecular processes. The average energy of a thermal fluctuation is $k_B T$, where k_B is the Boltzmann constant and T, the temperature. At 25 °C, this energy is ~4.1×10^{-21} J, a notable number if we consider the free energy released in the hydrolysis of one ATP molecule at physiological conditions, ~20–24 $k_B T$.

10.2.2 Molecular Machinery

Living in a thermal bath of violent agitations might seem to render any machinery inefficient. As pointed out in the introduction, however, biological machines have adapted to this apparently uncomfortable situation by developing a mechanism to extract energy from the thermal bath. This mechanism, the *Brownian ratchet*, consists roughly of rectifying random pushes (like a sailboat) using energy from ATP/GTP. In cases in which thermal pushes are not sufficient (for example, in large structures like the bacterial flagellum), protein machinery might bias diffusional steps by burning ATP as a *power stroke* (like a motorboat), a mechanism similar to that of macroscopic engines. In general, both mechanisms are combined by proteins in an intricate fashion to generate mechanical work [4, 5].

As explained before, the relevant units for quantitating force and size in the cellular context are piconewtons and nanometers, respectively. The discrete steps taken by a biomolecular machine as a consequence of applying a force range from sub-nanometers (e.g., 0.34 nm for DNA polymerases, which is the rise per base pair in B-DNA) to a few nanometers (e.g., 8 nm for kinesins, which is the length of the tubulin dimer). Therefore, the convenient units to express mechanical work are piconewton times nanometers. In fact, the energy level of the thermal fluctuations can be better remembered as $k_B T \approx 4.1$ pN·nm at room temperature.

10.2.3 Limitations of Optical Tweezers

Optical tweezers do not perform real-time microscopy, i.e., structures cannot be resolved during the assay. Optical tweezers provide measurements of force, distance, and time, but cannot ascertain the disposition of all protein domains at a given moment or the structure of nucleic acids, for example, under a stretching force. The specimens are grossly considered as point- or fiber-like structures. Combining fluorescence with optical tweezers (Sect. 10.4.5) is the leading solution: different methods inherited from optical "live" microscopy—including super-resolution fluorescence microscopy (see Chap. 8), which breaks the diffraction limit—can help determine the position and movement of a certain molecule or domain while it is being manipulated optically. This combined technique is still at a puberty stage, subjected to continuous developments, most of them adding great complexity to the setups that blur applications in biophysics. In most cases, the experimenter is "blind": only the position of micrometer-sized structures, such as cells or the beads shown in Fig. 10.1, can be observed during the assay. In these conditions, structural and biochemical information must be known in advance if experiments are to be designed precisely and correctly interpreted.

A second limitation is the low throughput of the experiments with optical tweezers. This technique provides a high degree of detail and rigor when studying the dynamics of a single specimen but at the cost of making assays tedious, because of both low reproducibility and difficulty in making several equivalent assays in parallel. Performing a large series of similar assays is especially important when the signals coming from the specimen are highly heterogenous, given the need to achieve a meaningful statistic, and they must be managed at reasonable times to maintain similar experimental conditions (not mentioning the experimenter's patience!).

10.3 Optical Tweezers

10.3.1 Optical Forces

Light can be viewed for many purposes as a flux of small particles called photons, which can interact with other objects and exchange energy as well as linear and angular momentum, thus generating forces and torques. This simple reasoning is the physical basis underlying optical forces: due to light–matter interaction, light incident on an object can push or modify the object's trajectory in complex ways. These forces are very small, however, and we therefore do not feel the so-called radiation pressure from the Sun, for example, or from the light of a bulb. The fact that light interacts with matter nonetheless explains photodamage and mutations by ultraviolet or X-ray radiation at long exposure times and powers. These radiations are made up of very high-energy photons which, rather than pushing biological matter, alter its structure at the atomic level. At visible and IR wavelengths, radiation is made up of photons with much less energy; in this case, the stream of coherent photons pushes small objects in a similar fashion as water exiting a hose at high pressure. This radiation pressure explains, for instance, why a comet's dust tail points away from the Sun. Radiation pressure was long ago suggested as a propulsion method for spacecraft, by using thin mirrors as "solar sails," although such sails would require areas of square kilometers.

Optical forces are therefore too tiny to be practical for the manipulation of macroscopic objects, but their effect is patent on atoms and on micrometer- and nanometer-sized objects. By adequately shaping the photon bundle from a laser source, it is possible not only to push tiny structures but also to trap them in 3D. The first optical trap was designed to manipulate atoms, which paved the way for the study of many interesting physical phenomena [6]. Trapping of small particles that scale to a few nanometers came soon afterward, an interest that remains active for nanotechnological applications. Trapping of cells was then tested. Whereas light manipulation of small objects and atoms is useful for understanding light–matter interactions, however, the interest of optical traps in biochemical research resides in their use as force sensors, and as a method to mediate mechanochemical reactions.

10.3.2 An Optical Trap

Here we will analyze how to engineer the light from a laser source to confine an object in 3D. The internal structure of a complex biological particle cannot be resolved by visible light wavelengths; consequently, structure is not essential to understand the trapping analysis. We will assume that the particle is dielectric (i.e., non-absorbing) and has a low refractive index (~1.5). These assumptions are common to biological structures including viruses and many synthetic particles. The main effect of light on this kind of object is refraction. The physics of trapping absorbing objects (metallic or semiconductor particles) or non-absorbing particles under resonant conditions are more complex, but to date they have not been exploited in biological experiments. In addition, absorbing and resonant light effects in biological structures are not desired if one wants to observe biological function without perturbation. These effects will therefore not be discussed here.

Figure 10.3a shows a spherical particle under the action of a laser beam whose wavelength is smaller than the size of the sphere. A ray of light can be conceived as a stream of photons, each carrying an associated momentum. The photons following the trajectories labeled "*a*" and "*b*" will be refracted (neglecting relatively minor surface reflections), as depicted, due to their interaction with the particle; as a result, their momentum will change. The momentum change experienced by each photon is transferred to the particle, thus generating the forces F_a and F_b. As more photons follow path "*b*" than "*a*"—because beam intensity is higher along the former path—force F_b is stronger than F_a. After decomposition of vectors F_a and F_b in two axes, one in the propagation direction of the laser and the other perpendicular to it, and the

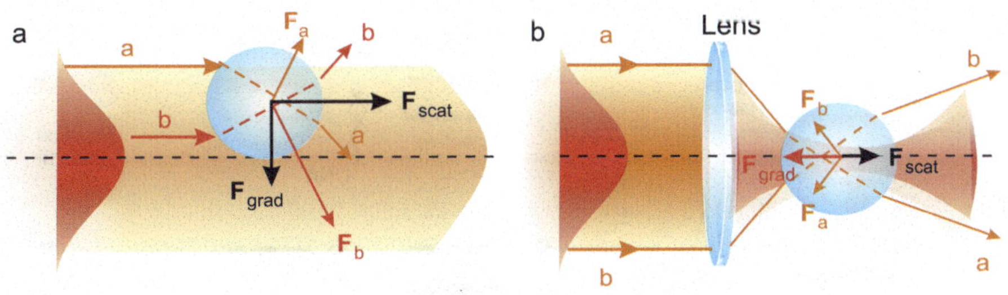

Fig. 10.3 Optical forces acting on a non-absorbing sphere. (**a**) Qualitative view of the forces exerted by a Gaussian laser beam on a sphere. The diagram illustrates the origin of the scattering (F_{scat}) and gradient (F_{grad}) components of the optical force. (**b**) Qualitative view of the forces exerted by a focused Gaussian beam on a sphere on-axis. The diagram shows the axial stability established by the gradient force along the centerline of the beam, toward the focal region. See text for details

extension of this ray analysis to all symmetrical pairs of rays, the net effect of laser light can be understood in terms of two forces: the *scattering force*, which pushes the particle along the incident beam, and the *gradient force*, which points toward the centerline of the beam. From the diagram, it is clear that there is transverse stability for a particle on-axis, since here $F_a = F_b$. To establish a 3D trap, axial stability must be achieved, that is, the scattering force must be compensated as well. Arthur Ashkin first made this possible using two equal laser beams facing each other; a trap generated in this fashion is known as a *dual-beam trap*. Some years later, by focusing the laser beam with an objective lens, he showed that a single-beam trap is also possible (find advanced literature in [1]). Figure 10.3b shows the mechanical stability of a particle in such a trap: the focal region is now a global maximum of optical intensity and in these conditions, an axial gradient force pointing toward the focus is generated before and after the light passes through that region. If the focus is sufficiently strong, the gradient effect also dominates the axial stability thus resulting in the all-optical capture of the particle.

10.3.3 Optical Trapping and Manipulation of Micro- and Nanoparticles

The photon flux from a laser source describes approximately parallel, linear trajectories at the output. Using simple optical elements like lenses and mirrors, the photon flux can be guided to a microscope objective (Fig. 10.4a). The objective can be viewed as a lens that causes the parallel trajectories to converge on an ideal point called the focus. In this approximate description, we do not consider that photons traveling far from the center of the bundle interact with the edges of the objective. These photons will deviate from the linear, perfectly converging trajectories, hence will also slightly affect the trajectories of the remainder of the photons in the bundle. These effects are known more rigorously as *diffraction*; the result is that the photon

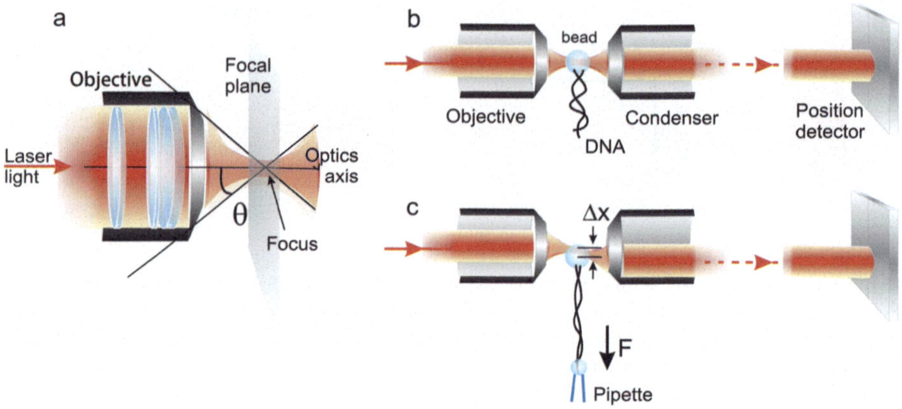

Fig. 10.4 Trapping and detecting force and displacements. (**a**) Representation of a microscope objective and the aperture angle, θ, of the cone of rays which are made to converge to a focus. In this configuration, laser light overfills the objective. (**b**) Diagram illustrating the configuration of the objective in front of a condenser lens, used to collect the light at the exit of the trap, and a photodetector to analyze the scattered light. Between the condenser lens and the photodetector, there are other optical elements not shown here for simplicity (see Fig. 10.5). (**c**) Diagram showing how an external force can be measured by registering light flux offsets in the photodetector. In (**b**) and (**c**), laser light underfills the objectives. Diagrams not to scale

bundle does not converge to a point, but to a focal volume with a size on the order of the light wavelength.

Laser light in fact behaves not only as a set of traveling particles but also like a wave (*wave-particle duality*). It is actually an electromagnetic wave, whose electric and magnetic fields oscillate in both space and time. Laser light is characterized by its wavelength, which is a measure of the distance along which the electromagnetic wave describes a complete oscillation, and by its power in units of energy per transversal area. The intensity of the electromagnetic flux at the output of the laser is highest for the central photon trajectory and decreases rapidly for trajectories away from the center. This type of "light bundle" is called a Gaussian beam (Fig. 10.3a) and is common to most commercial lasers. Other names for these light bundles are TEM_{00} *beams*, and another term for the laser sources that produce them are *single-mode lasers*. Single-mode lasers, in contrast to multi-mode lasers that support more wavelengths, are the ones required for optical trapping. It is, in addition, common to use continuous wave (*CW*) lasers rather than pulsed lasers; in simple terms, they are switched on/off at high frequencies. The laser powers required for optical tweezers are of some hundreds of mW.

The microscope objective is actually a set of many lenses arranged to produce a high-focusing effect (Fig. 10.4a). The focusing power is characterized by the numerical aperture (NA); for reasons that will be explained later, the higher the NA, the higher the trapping strength. Objectives are fabricated to work in air, water, or oil, and the higher the refractive index of the immersion medium, the higher the NA of the objective. For optical tweezers applications, it is thus common to use water and oil immersion objectives. Let n be the refractive index of the immersion medium and θ, the angle of convergence of the rays after the objective (Fig. 10.4a). The NA is then defined by the equation $NA = n \sin \theta$. The maximum numerical aperture, $NA = n$, is achieved when the rays are ideally made to converge on the focus at 90° angles (in practice, objectives cannot curve the rays that way). Water has a refractive index of 1.333 and usual oils, 1.51. Water immersion objectives normally have $NA = 1.2$ and oil immersion objectives can achieve $NA = 1.4$. As explained briefly in Sect. 10.2, the optical trap is attained in a fluidics chamber (Fig. 10.2), a test tube that will be described later to explain how to perform single-molecule experiments.

10.3.4 Force Measurement

If we only need to use the optical trap to manipulate objects, the key instrumentation described so far is sufficient. If we want to measure forces, however, we need to analyze the light scattered from the optical trap. The light exiting the trap is a diverging light ray bundle that can be collected by another objective or by a cost-effective condenser lens. After the condenser lens, the light is redirected to a position-sensitive photodetector. This type of photodetector measures power and registers light flux offsets, as depicted in Fig. 10.4b and c. An external force acting on the optically trapped particle deflects the exiting light cone and subsequently changes its projection after the condenser lens on the photodetector plane. Another method widely used in optical tweezers to measure bead displacement is back-focal plane interferometry, in which displacements are deduced from the interference pattern between the trapping laser and the scattered light (find advanced literature in [8]).

The use of modified position-sensitive photodetectors to register laser light intensity distribution is the basis for determining the externally applied force through the momentum conservation principle [7]. This force calibration method has several advantages (it is independent of particle size, shape, or refractive index, of buffer viscosity or refractive index, and of variations in laser power), although a simpler scheme is used in most optical tweezers designs. This scheme is based on the fact that the external force, F, acting on the optically trapped objects is approximately proportional to the generated displacement, Δx, as measured by the position-sensitive photodetector (see Fig. 10.4c). In other words, $F = \kappa \Delta x$, where κ is the trap stiffness, a parameter that characterizes the "invisible" spring that binds the particle to the optical trap. This parameter depends on the laser power and on the properties of both the particle and the immersion medium. Methods like the use of the Stokes law, or the Brownian motion of the bead in the trap can be used to accurately determine the trap stiffness, as described in-depth in other technical reviews and articles [2].

10.3.5 Optical Tweezers Instrument

The fact that the main elements of an optical microscope are involved in the design of a laser trap made the use of conventional microscopes the first and easiest strategy to set up an optical tweezers instrument. In Sect. 10.2, we divided the main components of the optical tweezers into those corresponding to the "microscope" and to the "trap" parts. The different elements are nonetheless optimally tuned by setting all components on an optical table with vibration isolation, especially when a specific application has to be developed. Many optical tweezers instruments are thus usually homemade and designed in research laboratories, although the use of compact commercial apparatus in the field of molecular and cell biophysics are breaking into the market in recent years.

Experiments in the optical tweezers use microsphere position as the visual reference. Videomicroscopy can easily be integrated in optical trap design because the objective that focuses the laser beam can also be used to generate real-time images of the events in the focal plane (Fig. 10.4a). A white-light lamp positioned on the condenser lens side is often used to illuminate the fluidics chamber interior in Köhler conditions, as in an optical microscope. A CCD camera at the back-focal plane of the objective is used to visualize the experiment. Images from the camera and signals from the photodetectors are processed by a computer.

Figure 10.5 represents the component layout in an optical tweezers instrument. This scheme is not standardized but integrates the general concepts and components of the instrument. In-depth design considerations are discussed in advanced reviews and technical articles [2, 7, 8]. In this scheme, two counter-propagating lasers converge on the fluidics chamber, for which two objectives are used instead of a pair of objective-condenser lenses. If the two lasers converge on the same focus, the instrument is a *dual-beam optical tweezers*. In this design, a micropipette is used to grab a second bead, as in the single-beam optical tweezers described above. If the two lasers are brought to separate focuses, each could be used to grab an independent bead without the need for a micropipette; the instrument so designed is a *dual-trap optical tweezers*. The equipment can also be engineered to split the light from a single laser source into two optical paths, a strategy that increases resolution [9].

In contrast to single-beam optical tweezers, the lasers in dual-beam instruments do not need to overfill the objectives to generate an efficient trap (see Fig. 10.4). Underfilling reduces energy concentration in the focal region, which in turn decreases radiative effects on the trapped specimen. This situation is also exploited to calibrate the optical tweezers by direct measurement of light momentum, as indicated earlier.

When two beams share the same trajectory, it is of paramount importance to avoid interferences. Electromagnetic waves can interfere, generating spurious effects in the optical trap that could lead to undesired consequences in manipulation and measurement. To prevent this, beams from distinct sources should encounter each other with orthogonal polarizations. Polarization is a property of light waves that describes the direction of oscillation of the electric and magnetic fields. Light from a laser

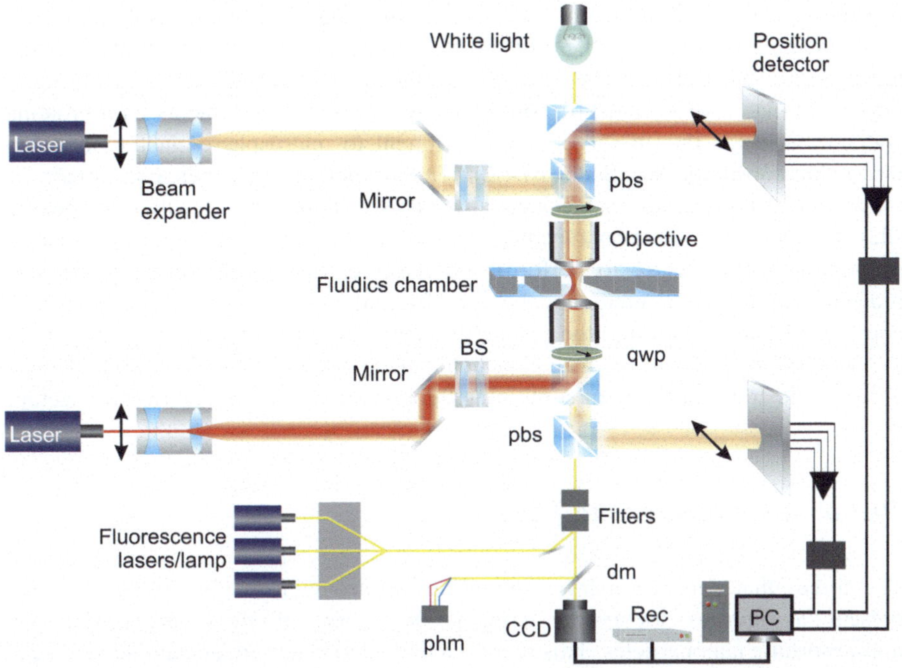

Fig. 10.5 Optical tweezers setup. Schematic summary of the most common designs (not to scale). The arrows show polarization of the beams at the exit of the laser sources and before the photodetectors. In this design, the polarizations of the beams in the trap are made circular and mutually orthogonal [7]. Additional light sources can be introduced in the setup to implement confocal microscopy and fluorescence. The laser wavelengths of the confocal module must be different from the infrared wavelengths used to generate the optical trap(s). BS, beam steering; pbs, polarizing beam splitter; qwp, quarter-wave plate; dm, dichroic mirror; Rec, video recorder; phm, photomultiplier

source is often linearly polarized, as shown in Fig. 10.5. Quarter-wave plates can rotate the polarization of each laser beam immediately before it enters the microscope objectives, so that they do not interfere with each other in the focal region. Polarizing beam splitters are optical elements that transmit or reflect the light depending on its polarization. These elements can be used to guide the two beams before and after the objectives (Fig. 10.5).

The two beams can be steered using motorized elements, such as mirrors and lenses, which can be computer-controlled. These elements allow moving one trap relative to the other in dual-trap optical tweezers. Beam expanders are also used to adjust the area of the laser spot at the exit of the laser sources to the entrance aperture of the objectives. They are used to over- or underfill the objectives, depending on the optical tweezers design. Beam expanders are normally made up of two lenses of different focal lengths. A pinhole can be positioned between them to filter the beam spatially. Astigmatism lenses can be introduced in the laser paths to correct this aberration (not shown in Fig. 10.5).

Finally, it is important to say a word about imaging possibilities. Unless fluorescence imaging is used, only the position of microscopic objects will allow monitoring experiments in the optical tweezers. Cells can be tracked easily, as can large organelles like centrosomes [10], but molecules cannot. Experiments with proteins and nucleic acids thus involve their attachment to microspheres in the so-called molecular construct (see Sect. 10.4.2). These microspheres, generally made of polystyrene or silica, act as mechanical handles to tether the specimen at specific sites. Correct design of the molecular construct and as much structural information as possible is necessary to perform and interpret the assays correctly. Single-molecule fluorescence, including the use of quantum dots,[1] and confocal microscopy[2] can be integrated into the optical tweezers; these are of great help to provide dynamic information on the mechanochemical experiments (see Sect. 10.6.3). Figure 10.5 illustrates one way of combining fluorescence in the optical tweezers setup.

10.3.6 Electromagnetic Theory

The forces that generate a laser source are electromagnetic. In this section, we present a rigorous theory to explain the origin of these interactions on non-absorbing nanoparticles. This section will help to better understand the concepts presented on optical forces in Sect. 10.3.2. The following is not essential for performing experiments or understanding the instrumental description above, and can therefore be skipped.

A particle is small when its characteristic diameter is much smaller than the illumination wavelength. Laser wavelengths for optical tweezers purposes are in the near-IR range, usually from 800 to 1100 nm. Longer wavelengths are strongly absorbed by water and generate undesired heating and convection effects. A particle is then small (also known as a Rayleigh or dipolar particle) when its diameter is below ~100 nm. The Lorentz force in the presence of an electromagnetic field is [11]:

$$\vec{f} = (\vec{\wp} \cdot \nabla)\vec{\mathcal{E}} + \frac{\partial \vec{\wp}}{\partial t} \times \vec{\mathcal{B}} \tag{10.1}$$

where $\vec{\wp}$ is the induced electric dipole moment and $\vec{\mathcal{E}}$, $\vec{\mathcal{B}}$ are the electric and magnetic vectors, respectively (SI units). Experiments in optical tweezers are

[1] *Quantum dots* are luminescent semiconductor nanocrystals which are used as a new class of fluorescent probes. In comparison with organic dyes and fluorescent proteins, they are brighter and exhibit both greater photostability and longer lifetimes. They can also be tuned to specific wavelengths with narrower bandwidths than fluorescent molecules.

[2] *Confocal microscopy* is a high-resolution imaging technique based on a point-by-point illumination of the sample (see Chap. 8). The light source is often a focused laser beam. Lasers of several wavelengths and filters can be used to combine this technique with fluorescent imaging.

performed in liquid media and particle motion is therefore greatly slowed by viscous forces. In these conditions, the particles cannot follow the rapid oscillations of the electromagnetic fields; consequently, the time-average force is the observed value. The electromagnetic field from a single-mode laser is harmonic and can thus be expressed in complex representation as $\vec{\mathcal{E}}(\mathbf{r},t) = \mathcal{Re}\{\mathbf{E}(\mathbf{r})\exp(-i\omega t)\}$, $\vec{\mathcal{B}}(\mathbf{r},t) = \mathcal{Re}\{\mathbf{B}(\mathbf{r})\exp(-i\omega t)\}$, $\vec{\wp}(\mathbf{r},t) = \mathcal{Re}\{\mathbf{p}(\mathbf{r})\exp(-i\omega t)\}$; **E**, **B**, and **p** are in this formalism complex vector functions of position, $\mathbf{r} = (x, y, z)$, and time, t. $\mathcal{Re}$ denotes the real part and ω is the angular frequency of the field. The i-th component of the time-average force is thus (see [12] and references therein):

$$\langle f_i(\mathbf{r},t)\rangle = F_i(\mathbf{r}) = \frac{1}{2}\mathcal{Re}\left\{\alpha \sum_{j=x,y,z} E_j(\mathbf{r})\frac{\partial E_j^*(\mathbf{r})}{\partial x_i}\right\} \tag{10.2}$$

where * denotes the complex conjugate. Eq. (10.2) is the expression of the observed optical force, $\vec{F} = (F_x, F_y, F_z)$, over a small particle with complex electric polarizability α in the liquid medium.

The electric field can be expressed in paraxial representation, that is, a beam or evanescent field with a main propagating direction along **k**, as $\mathbf{E}(\mathbf{r}) = \mathbf{E}_0(\mathbf{r})\exp(i\mathbf{k}\cdot\mathbf{r})$. The modulus of vector **k** is given by $k = n(\omega/c)$, being n the refractive index of the liquid medium and c the speed of light in a vacuum. Substituting this expression into Eq. (10.2), one obtains [12]:

$$\vec{F} = \frac{1}{2}\mathbf{k}\mathcal{Im}\{\alpha\}|\mathbf{E}_0|^2 + \frac{1}{4}\mathcal{Re}\{\alpha\}\nabla|\mathbf{E}_0|^2 - \frac{1}{2}\mathcal{Im}\{\alpha\}\mathcal{Im}\{\mathbf{E}_0\cdot\nabla\mathbf{E}_0^*\} \tag{10.3}$$

where $\mathcal{Im}$ denotes the imaginary part. The first term in Eq. (10.3) is proportional to the field intensity, $|\mathbf{E}_0|^2$, and points along the propagating direction, **k**. This term therefore represents the contribution of the radiation pressure to the scattering force. Since $\mathcal{Im}\{\alpha\} > 0$, this force pushes the particle in the laser-propagating direction. The second term is proportional to the gradient of the field intensity, $\nabla|\mathbf{E}_0|^2$, and it is thus known as the gradient force. Since $\mathcal{Re}\{\alpha\} > 0$ for non-absorbing particles, this force attracts the particle toward the maximum of the field intensity. This term is responsible for optical trapping, as was also explained qualitatively in Sect. 10.3.2. The last term is not significant in typical experiments but may induce rotation of the trapped particle due to spin angular momentum exchange between the field and the particle (see Sect. 10.4.4).

Optical forces on many viral particles can be calculated by using Eq. (10.3). Beads used as mechanical handles are usually in the micrometer range and therefore beyond this approximation. The above description nonetheless allows an understanding of the physics of the problem, which was previously treated with a ray optics description in Sect. 10.3.2. The force on these larger particles can be calculated using the following expression [11, 12]:

$$\vec{F} = \begin{array}{l} \frac{1}{2}\mathcal{R}e\left\{ \int_{\Sigma} d^2 r[(\epsilon\mathbf{E}(\mathbf{r},\omega)\cdot\mathbf{n})\mathbf{E}^*(\mathbf{r},\omega) + (\epsilon\mathbf{H}(\mathbf{r},\omega)\cdot\mathbf{n})\mathbf{H}^*(\mathbf{r},\omega)] \right. \\ \left. - \frac{1}{2}(\epsilon|\mathbf{E}(\mathbf{r},\omega)|^2 + |\mathbf{H}(\mathbf{r},\omega)|^2)\mathbf{n} \right\} \end{array} \tag{10.4}$$

where $\sum$ is a surface enclosing the particle and $\mathbf{n}$ is its local outward normal unit vector. ϵ and μ are the electric and magnetic relative permittivities of the medium, respectively. The complex representation of the H-field, $\vec{\mathcal{H}}(\mathbf{r},t) = \mathcal{R}e\{\mathbf{H}(\mathbf{r})\exp(-i\omega t)\}$, is related to the complex B-field by $\mathbf{B}(\mathbf{r}) = \mu\mathbf{H}(\mathbf{r})$. Eq. (10.4) uses the time-harmonic expression of Maxwell's stress tensor. Computation of forces from this expression involves numerical calculations in most cases. This might be useful to predict the behavior of special particles in the trap but for most biological experiments, it is often only required to characterize empirically the field–particle interaction by a spring constant (see Sect. 10.3.4).

10.3.7 Magnetic Tweezers

Magnetic tweezers are a cost-effective instrument to perform single-molecule experiments with applications similar to those of optical tweezers [13]. This instrument can be used to measure and exert forces in mechanochemical reactions, although to date it has been little used to study viruses and viral structural elements, with the exception of their nucleic acids. To provide a broad palette of possibilities for virus study, we present this technique briefly.

Typical magnetic tweezers setups (Fig. 10.1b) use permanent magnets to generate a field gradient inside a fluidics chamber. By moving these magnets along the field gradient, forces on magnetic particles in a fluidics chamber can be increased/ decreased; by rotating them, torque can be induced. Magnet geometry, including tip poles of micrometer radius, can be determinant in generating field gradients convenient for certain experiments [14]. Electromagnets can also be used to generate stretching forces. In this case, adjusting the current running through the magnetic coils allows a change in the force strength. Superparamagnetic beads are used as mechanical handles for biological molecules in these experiments. Due to space constraints, magnets must be placed outside the fluidics chamber, which prevents the highest field gradients from approaching the magnetic beads, thus establishing an upper boundary to the forces generated. Maximum forces on micrometer-sized superparamagnetic particles in these conditions are usually below ~10 pN, but developments using microscopic magnetic poles inside the fluidics chamber can greatly increase the forces, to a range similar to that of optical tweezers [15, 16]. Minimum generated forces are in the range of 0.01 pN.

The force generated by two magnets on a superparamagnetic bead is:

$$\vec{F} = \frac{1}{2}\nabla\left(\vec{m}\cdot\vec{\mathcal{B}}\right) \tag{10.5}$$

where $\vec{m}$ is the induced magnetic moment on the microsphere in the presence of the magnetic field, $\vec{\mathcal{B}}$, arising from the magnets. Force calibration can be performed against the viscous drag or using Brownian fluctuations. A microscope objective and a camera below the liquid chamber are used to follow the experiments, specifically, to deduce the forces and displacements of the bead. Illumination comes from above the fluidics chamber in this design (Fig. 10.1b). This apparatus can also be built on an inverted microscope.

Although this instrument constitutes the magnetic analogue of optical tweezers, it is not based on the generation of a trap. In other words, particles are not grabbed but dragged. To be manipulated, single particles must be tethered. For example, the DNA molecule in the experiment in Fig. 10.1b is always subject to a pulling force, a reason that makes the constant-force mode (see Sect. 10.4.3) appropriate under near-equilibrium conditions in this instrument. In addition, the capacity of magnetic tweezers to manipulate a particle in 3D is limited. The use of more than two magnetic poles increases maneuverability [15], but 3D control of the particle can remain a problem for many experiments. Mainly through the use of fast cameras, high-resolution magnetic tweezers can cope nowadays with dynamics in the nanoscale with time precision sometimes reaching the submillisecond range [16].

Particles that do not show a magnetic response cannot be manipulated directly with magnetic fields. Cells and organelles, for example, cannot be dragged with magnetic tweezers. While this is a limitation for in vitro experiments, this fact is being exploited in the manipulation of magnetic nanoparticles inside a cell with high specificity [15].

Magnetic and optical tweezers, together with atomic force microscopy (AFM) (Chap. 9), cover a range of forces from 10^{-2} to 10^{5} pN for single-molecule manipulation. These techniques thus provide modern means to look into biochemical reactions in new ways. In addition to the research scope and experimental conditions, spatial and temporal resolution issues should be considered to aid in making the most appropriate choices [17].

10.4 Operation

Forces on biological specimens of a few nanometers are very small, and it is thus difficult to generate a stable trap in liquid media. Thermal fluctuations and instrument noise make trapping efficiency very low and residence times very short. It is also difficult to control specimen orientation and to assure that a single specimen is in the trap. The methodology for manipulating nucleic acids, viruses, and proteins therefore normally involves the use of beads as mechanical handles. We will now explain general strategies for performing experiments with the optical tweezers.

10.4.1 Methodology

Most experiments using optical tweezers require that the molecular system under study be tethered to at least two specific sites. For this purpose, a second bead must be introduced in the experiments, either by suction by a micropipette (shown in Fig. 10.4c, bottom) or using a second optical trap. As stated above, these two beads are mechanical handles to biochemically link, for example, DNA molecules by two ends, or more complex systems such as a virus particle and the viral DNA to be packaged (see below). To exert force, one of the beads must move relative to the other (Fig. 10.4c). In optical designs in which the second bead is held by suction, the micropipette is moved by a computer-controlled stage. In designs in which this micropipette is replaced by another optical trap, the second trap can be displaced by steering one laser beam relative to the other with external mirrors and lenses (Fig. 10.5).

In DNA stretching experiments, the molecule is often modified at one end to insert biotin labels and at the other to insert digoxigenin (Dig) labels. The molecule is then tethered between two beads in the fluidics chamber, one coated with streptavidin to provide biotin-streptavidin linkage and the other with anti-digoxigenin (α-Dig) to provide a Dig-α-Dig linkage. The experiment is followed by videomicroscopy, observing the position of the two beads. DNA extension is measured by recording the relative positions of the two beads in real time. In setups in which a micropipette is used, a low-power laser can be introduced to track the position of the second bead; light scattered from the bead can be detected in another photodetector. Alternatively, if the micropipette rigidly attached to the microchamber is moved relative to the optical trap (Fig. 10.6a), a so-called light-lever device can be used [7]. This device comprises a single-mode laser and a lens attached to the pipette/chamber frame as in Fig. 10.6a. The lens collimates light from the laser and directs it onto a photodetector. A similar light-lever configuration is used in AFM to detect the position of the cantilever in reflection (see Chap. 9).

The procedure for preparing a DNA stretching experiment in the fluidics chamber is shown in Fig. 10.6b. The optical trap is first used to grab a single streptavidin-coated microsphere from one of the channels and place it on the micropipette. A α-Dig-coated bead with attached DNA fibers is then trapped from the other channel and brought toward the bead on the micropipette until a stable biotin-streptavidin bond is formed. This procedure is often known as "fishing."

10.4.2 Molecular Construct

The design of the molecular construct can be a bottleneck in single-molecule optical tweezers experiments. Good design consists of biochemically engineering the specimen for attachment to mechanical handles. Force is a vector magnitude and therefore marks a precise coordinate in the experiments, unlike other parameters

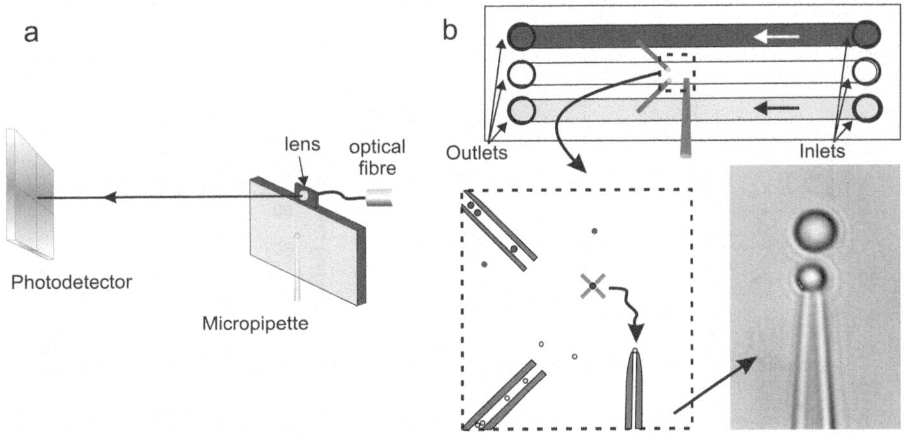

Fig. 10.6 Optical tweezers methodology. (**a**) Use of a light-lever device to track the position of the chamber with the rigidly attached micropipette. (**b**) Top: a three-channel microfluidics chamber is filled with buffer. Two suspensions of different types of microspheres are then flowed into the top and bottom channels through inlet holes; the fluid exits the channels through outlet holes. Bottom, left: microspheres from the top or bottom channel are optically trapped at the exit of the dispenser tubes and deposited on the micropipette in the middle of the central channel. Bottom, right: micrograph as seen by videomicroscopy of a ~ 3-μm diameter bead in the optical trap and a ~ 2-μm diameter bead held by the micropipette

such as temperature or pH. It is important to tether the specimen correctly from the domains that will produce mechanical work or will extend. To study the elasticity of a polymer, for example, DNA, the molecular construct is very simple and consists of tethering the molecule from opposite ends, as explained above. If one wants to unzip double-stranded (ds) DNA to study mechanically induced melting, the molecule must be tethered on opposite strands of the same end. When proteins are involved, structural data are needed to determine the sites expected to unfold. In experiments with molecular motors, structural data, and functional models are desirable to establish the domains involved in motion and the direction the molecule is expected to translate or rotate.

The design of the molecular construct can be seen as the analogue of the crystallization procedure in X-ray diffraction (Chap. 4): proteins need different approaches for their structural analysis with this technique and established strategies help, but do not constitute a standard for new experiments. The interested reader will find chemical approaches to tethering proteins, complexes, and bacteriophages for optical tweezers elsewhere [18, 19]. Much of the success in single-molecule experiments relies on correct construct design. Figure 10.7 shows a molecular construct devised to study bacteriophage packaging by *Bacillus subtilis* phage Φ29. Phage capsids with prepackaged 6.6-μm-long dsDNA were bound directly to a bead on a micropipette, tethering the free DNA end to an optically trapped bead [22]. We will analyze later in this chapter how this construct is used to study the Φ29 packaging mechanism.

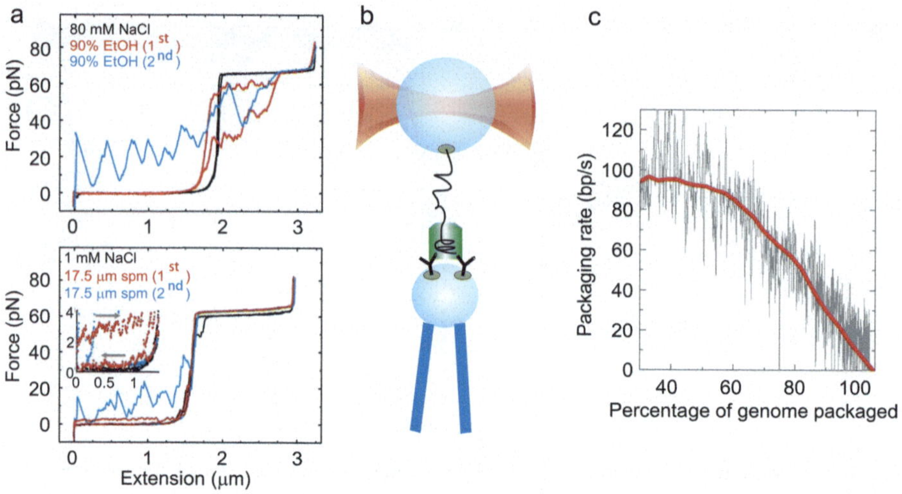

Fig. 10.7 Some applications of optical tweezers. (**a**) DNA stretching experiments. In each graph, a single molecule of dsDNA is first stretched and relaxed in physiological buffer (black curve), followed by two successive stretch/relax cycles (red and blue curves) in a buffer-90% ethanol mixture (top) or in buffer with a low concentration of spermine (bottom). (Reprinted from [20] with permission). (**b**) Molecular construct to study Φ29 packaging activity. The bead in the optical trap is streptavidin-coated for attachment to the biotin-labeled free end of the DNA. The virus is attached to a protein G-coated bead held by a micropipette. The DNA–capsid complex is captured in the tweezers in a stalled, partially prepackaged preparation before the start of the assay (current strategies are not restricted to using prepackaged molecular constructs [21]). (**c**) Packaging rate versus the amount of DNA packaged, relative to the original 19.3-kbp Φ29 genome. Gray line, trace for a single complex; red line, average of eight such measurements. (Reprinted from [22] with by permission)

10.4.3 Operation Modes

There are four basic operation methods: *force-extension*, *constant-force*, *passive,* and *force-jump* modes. The first is used to stretch polymers for analysis of their elastic behavior; Fig. 10.7 shows the force-extension curve of a DNA molecule in this mode. Polymer extension is observed as the force increases continuously; for this reason, this mode is also known as *force-ramp*.

The constant-force (also referred to as force-clamp or force-feedback mode) is used to analyze the activity of a protein machine as a function of a constant load. In this mode, the position of the optical trap relative to the second attachment point is adjusted to maintain constant tension on the bead. The protein kinesin has been tested using this strategy, as has packaging activity of Φ29 (Fig. 10.7) (find particular examples in [1]). The application of distinct constant forces allows observation of varying responses of the specimen.

The passive mode consists of fixing the position of the micropipette and/or traps and letting the specimen work. In this mode, the distance between the beads is not necessarily constant, as the position of the trapped bead undergoes small displacements relative to its resting position, resulting from the activity of the tethered specimen. This mode can be used in the study of DNA polymerization, kinesin transport and viral packaging, and allows measurement of maximum forces developed by protein motors, also known as stall forces.

In force-jump mode, the applied force is changed rapidly in discrete force steps rather than continuously (characteristic of the ramp-mode). This mode is useful for studying kinetics between several molecular states that are stable at different forces. For instance, protein folding or RNA hairpin open/close conformations can be monitored through changes in the molecular extension with this operation mode [23].

10.4.4 Adding Translation Coordinates and Rotation

Measurement in more degrees of freedom is sometimes desired. The result of several coordinated molecular motors is not only linear displacement but also a general 3D motion that might involve several translations and rotations. Multiple optical traps permit measurement of motion in more than one linear coordinate, as well as control over more than one molecule [24]. Moreover, the use of structured beams can create arbitrary trajectories, including vortices in which manipulated beads perform circular trajectories [25]. To date, molecular specimens have been biochemically linked by two specific sites to beads in one or two traps, but the use of other configurations with more than two traps was shown, for example, to allow simultaneous manipulation of several molecules in multiple directions [26].

Another development in manipulation that could help to understand viral packaging processes is the combination of torque measurements in the optical tweezers. Torque is easily produced in magnetic tweezers (see Sect. 10.3.7) and has been generated in optical tweezers using a rotating micropipette [27]; however, torque measurements are technically more difficult. Recent advances point toward the use of the intrinsic properties of light [28]. Light can exchange angular momentum—spin or orbital angular momentum, or both—and can thus exert rotation on particles. Conversely, angular momentum changes in the light scattered from a particle can be analyzed as a method to measure torque [29].

Angular optical tweezers, also termed the optical torque wrench, is a current hybrid, all-optical instrument capable of exerting tension and torsion to individually trapped specimens, as well as of measuring both torque-angle and force-extension quantities as a function of time. The angular momentum of laser polarization is herein used to apply torque on single nucleic acid molecules tethered to anisotropic and aspheric trapped particles [30]. Similar operational modes as those described in Sect. 10.4.3 can additionally equip experiments that involve rotational motion in this apparatus within a resolution of 1 pN·nm. Other optical strategies have been implemented to produce rotational manipulation, including structured beams and phase masks, although they have not been exploited in single-molecule studies [31].

The helical structure of DNA suggests that its binding machinery follows its helical pitch to generate rotation. DNA may move by ratcheting and rotation in the Φ29 phage portal motor in its translocation into the viral capsid, a mechanism that could be examined by a combination of fluorescence microscopy and optical tweezers (see below). Torque-measuring optical tweezers could be then applied to characterize torsion-induced mechanochemical packaging pathways by over- and underwinding the DNA double helix.

10.4.5 Combination with Fluorescence Imaging

Imaging of molecules while they are being manipulated is a demand under continuous progress. There is currently no method with which to perform structural studies with atomic resolution of a molecule in motion or under mechanical stress. The conditions involved in methods such as X-ray diffraction or NMR are incompatible with those used in liquid chambers. It is nonetheless possible to introduce methods developed for optical microscopes in the tweezers setup.

Combining fluorescence with optical tweezers into a hybrid instrument, popularized as "fleezers," not only provides control over the experiment but also allows observation of new conformations induced by force or torque. Single-molecule fluorescence and confocal imaging have already been incorporated successfully to optical tweezers, and their applicability is shown by a number of studies (see [7, 32–34] and references therein). A general strategy to integrate fluorescence detection in optical tweezers was outlined briefly in Fig. 10.5 [35–37]. Fluorescence can be obtained using fluorophores or with quantum dots, which add stability to signal emission. The use of a pair of dyes allows detection of nanoscale distance changes of a protein or of a complex of coordinated proteins using Förster resonant energy transfer (FRET) photophysical interactions [38]. Super-resolution fluorescence microscopy based on stimulated emission depletion (STED), with a resolution down to some tens of nanometers, has also been implemented [39]. The technical hurdles in hybrid instrumentation of this kind arise in maintaining the individual competences of the two technologies when they are combined. While optical traps are compatible with fluorescence imaging, they individually represent sharp challenges when applied at the single-molecule level, often displaying compromised capabilities and resolutions when integrated into the same apparatus, as well as handling difficulties for non-trained users.

Fluorescence has been used to understand DNA structure under tension, ribosome binding events during translocation by one codon [40], the portal motor dynamics of bacteriophages [32, 41, 42], and viral assembly (Sect. 10.5.4), but is underexploited in virology. These studies suggest a fruitful combination of methods to visualize bacteriophage motor dynamics under the control of force and torque.

The coming years will witness a mostly incremental effort to release technical complications from the experimenter, so that the hybrid instruments become user-friendly for biologists, giving continuous birth to detailed dynamics of biomolecules.

In virology, for example, DNA could be fluorescently labeled to follow its twisting dynamics relative to the connector in the packaging experiment or condensation dynamics of the genome under force relative to the capsid assembly.

10.5 Understanding Viruses: Some Major Contributions of Optical Tweezers

In contrast to the techniques studied in previous chapters (Chaps. 3–9), viruses have not been long studied with optical tweezers so far. The reason is the novelty of the single-molecule approaches that involve mechanical manipulation of biological specimens. There is currently much interest not only in characterizing viral mechanochemical functions in real time but also in using this machinery for nanotechnological applications. In this section, we do not intend providing an exhaustive review on advances in virology but rather illustrating the use of optical tweezers in this field through the years.

The replicative protein machinery of viruses, which involves a variety of force-developing motors such as polymerases and helicases, is the object of a growing number of mechanistic studies at the single-molecule level using optical tweezers. These studies exceed the scope of this book; however, interested readers will find illustrative contributions in the literature [43–48].

10.5.1 Elasticity of Nucleic Acids and Relevance for Packaging in Viruses

With the advent of optical tweezers in molecular biology, DNA and RNA in their double- or single-stranded configurations were promptly studied [27]. The first assays in the 1990s elucidated the elasticity of these information carriers. These studies are important for understanding the energetic barriers and the function of proteins and molecular complexes that act on nucleic acids in processes such as replication, transcription, unwinding, or packaging. From the point of view of structural and physical virology, conformational dynamics of nucleic acids is important, for example, for understanding nucleic acid packaging and storage inside viral capsids, from passive encapsulation into preassembled viral capsids to the active packaging process in bacteriophages (see also Chap. 13).

A molecular construct for mechanical analysis of DNA is shown in Fig. 10.4c. Figure 10.7a shows the results of a standard experiment in which individual dsDNA molecules are stretched and then relaxed in different conditions. For these experiments, DNA was obtained from a linear fragment of plasmid piJ702 from *Streptomyces*, whose makeup consists of 70% guanine-cytosine (GC) base pairs (bp). Each molecule was first stretched and relaxed in physiological conditions (aqueous buffer with 80 mM NaCl, 10 mM Tris-HCl, 1 mM EDTA, pH 7.5) and subsequently stretched and relaxed twice in a buffer-90% ethanol mixture (Fig. 10.7a, top) or in the same buffer with 1 mM NaCl and 17.5 μM spermine (Fig. 10.7a, bottom). The ability of optical tweezers to deal with different conditions in the same assay is demonstrated here. Ethanol was used to mimic the behavior of DNA at low humidity, an environment similar to that inside viral capsids; spermine, a polycation involved in genome condensation, was used to study DNA organization in free form, not constrained by, for example, encapsulation in viral proheads.

The force-extension curves in Fig. 10.7a in the physiological buffer (black trace) show the characteristic elasticity regimes of DNA: the entropic regime, in which the molecule behaves as an entropic spring, and the intrinsic regime, in which it is stretched beyond its contour length. At 65 pN, the molecule overstretches as a result of the almost total unwinding it experiences at high stress [27]. Force-extension curves of DNA in the ethanol mixture or with spermine show either a plateau at entropic forces of ~3 pN or a stick-release pattern. These effects indicate the unraveling of collapsed structures, such as coiled DNA or toroidal supercoiling, in the form of a smooth decondensation process or a violent unfolding of the polymer (see [20] and references therein).

The mechanical properties of DNA, as characterized by its persistence length and stretch modulus of the molecules, have been also studied as a function of the GC composition or of external conditions such as pH, ionic strength, temperature, and humidity level (find advanced material in [1, 49–52]). The mechanical properties of the other nucleic acid duplexes, namely, dsRNA and hybrid DNA-RNA molecules have been characterized by using optical and magnetic tweezers in similar configurations [53–55]. These techniques have been used in more varied configurations to study DNA unzipping (the mechanical separation of the two strands), DNA and RNA twisting, the opening of RNA hairpins and the elasticity of single-stranded (ss) DNA [23, 27, 56, 57]. The influence of several DNA-binding proteins on the mechanical properties of the complex has also been analyzed [58]. Finally, non-canonical nucleic acid structures have also been unfolded mechanically by means of these techniques [59–63].

In conclusion, from the point of view of viral nucleic acid storage and packaging in viral capsids, these and other single-molecule manipulation studies have contributed to the understanding of the elastic, electrostatic, and entropic penalties of confining and organizing genetic material at high pressure in a small, closed space. These studies have also shown that nucleic acid structure is dependent on mechanical tension; as indicated, a few piconewtons are sufficient to straighten a dsDNA molecule; some tens of piconewtons are sufficient to promote base-pair unstacking and twist changes in physiological conditions, both of which are mechanical denaturation effects that lead to geometric changes in the DNA double helix. Forces in these ranges are developed by the bacteriophage portal machinery, which will be reviewed in the next section. A dynamic description of viral packaging in the presence of mechanical stress must therefore include the tension-induced conformational alterations of the viral nucleic acid that the portal machinery must process.

10.5.2 Packaging Activity of Tailed Bacteriophages

Bacteriophages are the most abundant organisms on earth (see Chap. 17). Nonetheless, very few virologists have analyzed their machinery at the single-particle level [64–66]. The pioneering work of Smith and Bustamante's team in 2001 reported the direct activity of this impressive, powerful machinery [22]. These results have complemented structural and biochemical studies (see Chap. 13) and also opened entirely new perspectives on the working of viral nanomachines. An artistic view of a typical bacteriophage manipulation assay using optical tweezers is represented in Fig. 10.8.

Measurement of the portal motor of bacteriophage Φ29 in the passive mode showed that its packaging machinery is highly processive (i.e., that can undergo many catalytic cycles without dissociation) and one of the strongest known in nature. Forces against an external load (like that exerted by a micropipette, Fig. 10.7b) or by another optical trap (Fig. 10.9a) can be above 60 pN. These forces are near those that distort the dsDNA in the overstretching transition (see force-extension curves in Fig. 10.7a). Constant-force mode studies indicated that packaging rates, which can reach one hundred base pairs per second (100 bp/s), are punctuated by pauses and slips, and that these rates are dependent on the percentage of total genome packaged

Fig. 10.8 A bacteriophage packaging assay in the optical tweezers. This artist's conception depicts a bacteriophage Φ29 packaging its genome against a force applied from a laser-trapped bead. The viral head is attached to the upper bead by chemical bonds represented by chains. The in vitro experiment is performed in aqueous buffer. A fibreless Φ29 virion map (cross section) at 7.8-Å resolution, obtained from EMdatabank.org (ID: emd-1420 [67]), has been modified for illustrative purposes

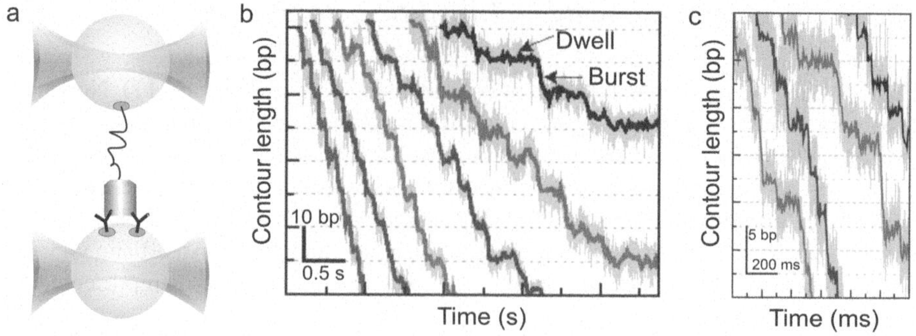

Fig. 10.9 High-resolution measurements of packaging activity. Detecting a structure in motion on increasingly smaller scales involves a simultaneous improvement in spatial and temporal resolutions. (**a**) A dual-trap optical tweezers in which two separate traps are built from the same laser beam allow better sensitivity, as detailed elsewhere [8, 9]. (**b**) Step size in Φ29. The graph shows representative packaging traces collected under low external load ~8 pN, and different [ATP] (from left to right): 250, 100, 50, 25, 10, and 5 μM, all boxcar-filtered and decimated to 50 Hz. Data at 1.25 kHz are plotted in light gray. Packaging is observed to occur in large 10-bp "bursts" separated by flat "dwells." (**c**) Fine structure of the 10-bp bursts: each step consists of four 2.5-bp substeps. The graph shows representative packaging traces collected with external loads of ~40 pN and 250 μM [ATP], all boxcar-filtered and decimated to 100 Hz. Data in light gray are plotted at 1.25 kHz. (Reprinted from [42] with permission)

(see Fig. 10.7c). Since this pioneering work, the Smith and Bustamante labs have contributed to understanding the mechanochemical behavior of this machinery, as we discuss next in the context of relevant results from other groups.

Smith and co-workers extended the use of optical tweezers to other phages in an assay similar to that in Fig. 10.7b. Studies of T4 [68] and λ [69] phages showed that high force generation is a common property of viral packaging motors. Among other results, this group found that T4 works at a strikingly high translocation rate, on average 700 bp/s, with maximum rates of 2000 bp/s [68]. In λ phage, procapsid expansion (see Chap. 13) was detected at 30% of total packaging [69]. The Smith group also analyzed ionic effect on packaging rates and on portal motor function [70], as well as pressure within the capsids at different stages of the packaging process. They developed a new strategy to analyze the packaging process at near-zero force, which also involves initiation of the process without using a prepackaged configuration [21].

The dependence of the internal force on capsid filling is also a significant analysis for its implication in the capsid pressure. The Smith team reported internal forces exceeding 100 pN, which are also indicative of the resistance of the capsid shells; these forces can be characterized in indentation experiments using AFM (see Chaps. 9 and 18). One of these studies showed that the capsid can withstand indentation forces up to ~1000 pN [71], a result compatible with internal forces of 100 pN after complete capsid filling [21]. In addition, Smith and co-workers analyzed the effect of mutations in the phage λ motor on its packaging dynamics [72, 73]. Continuous work over the years with Φ29 showed that packaging initiation after a stall state is a non-equilibrium process during which the motor resumes its operation at a transient high velocity [74]. Polycations were used to show that packaging is facilitated when DNA-DNA repulsive interactions are mildly shielded at low concentrations, but may be overall impeding due to non-equilibrium jamming transition at concentrations that yield attractive DNA-DNA interactions [75, 76].

Specific work with T4 by Chemla, Rao, and co-workers unveiled an off-translocation pathway by which this phage may pause and unpackage DNA presumably to circumvent structural barriers during normal activity [77]. Based on structural analysis, the Smith group showed that an early proposed electrostatic force model [78] was compatible with mechanochemical results from optical tweezers, in which DNA packages by structural switching between a compact and an extended state guided by electrostatic interactions [79]. T4 experiences continuous gripping with ATP binding and possesses an end-clamp mechanism that prevents the complete exit of DNA due to transient slipping in the absence of ATP [80]. The activity of this phage was shown to be insensitive to DNA sequence, thus concluding that although motor gripping may vary with sequence, fluctuations in packaging rate, pausing, and slipping events are stochastic in nature, thus not constraining virus evolution [81].

The Bustamante group has focused on the Φ29 bacteriophage. They determined the kinetic parameters of the packaging motor and their dependence on external load. These researchers also showed that the motor subunits act in a coordinated, successive fashion with high processivity [82], in contrast to more flexible coordination found in T4 bacteriophage [83]. Their study suggests that the packaging rate decrease shown in Fig. 10.7c at high capsid fillings results from build-up of internal pressure due to DNA confinement, which exerts an opposing force and slows the motor. The Φ29 bacteriophage was recognized as equipped with a power stroke mechanism of motor function, much in the same way as kinesin [84] and in contrast to the Brownian ratchet mechanism displayed by DNA and RNA polymerases and the ribosome [85–87]. Bustamante and co-workers also characterized the intersubunit coordination of the homomeric ring ATPase, which powers the DNA packaging process in Φ29. ATPase characterizations of this kind are important because this machinery is also involved in other varied tasks in the cell, from chromosome segregation to protein degradation [42]. For these studies, they used high-resolution optical tweezers (see Sect. 10.6). They found that packaging occurs in increments of 10 bp, in turn, composed of four 2.5-bp steps, as shown in Fig. 10.9. An important conclusion is that a non-integer number of base-pair steps can take place in protein mechanics, in contrast to earlier models that assumed an integer number. In another assay using several modified DNAs, these researchers dissected phosphate interactions in the DNA-motor engagement and showed that promiscuous, nonspecific contacts with the substrates are common in translocase-substrate interactions [88].

Promiscuity by viral DNA packaging machinery was also shown by Rao, Chemla, and colleagues [89], who reported similar packaging activity rates for proheads and previously filled heads of T4 bacteriophage. Full heads and those emptied of most of their packaged DNA can reassemble the packaging machinery and reuse it. This group also found that several exogenous DNA molecules can be used as a substrate to fill the T4 capsid in several initiation steps [89].

The Φ29 pentameric portal motor has been studied profoundly to identify alternating dwell phases, in which the motor does not move, and burst phases, in which the motor translocates in the above-mentioned 2.5-bp steps, Fig. 10.9b. In addition, ATP-binding, ATP-hydrolysis, and ADP-release events, which take place during the dwell phase, and the phosphate-release event, which coincides with the force-generating step in the burst phase, have already been integrated in the mechano-chemical cycle.

The use of magnetic tweezers to provide a tethering force over a single DNA–phage complex was used in conjunction with single-molecule fluorescence. In this experiment, Bustamante and colleagues tested the possibility of connector rotation in Φ29 [41]. Using an impressive variety of instrumentation and techniques, they showed that absolute rotations do not take place; only transient rotation of the connector relative to the capsid shell, followed by a return to the original position, is likely to be involved in the packaging process. The possibility that DNA rotates during packaging, a process compatible with that in which the connector does not rotate, can be directly tested (a translocation mechanism driven by a (non-rotating) compression/extension mechanism is not supported by experimental evidence). DNA twist-measuring assays can be implemented by angular optical tweezers to determine quantitatively torques exerted by the Φ29 ring ATPase, although, so far, a simpler rotor-bead assay has been used just to detect DNA rotation during packaging. In this assay, in which the complex is under tension as in Fig. 10.8, torsion in the DNA is detected by tracking revolutions of an auxiliary bead (rotor bead) attached to one of the strands of the tethered DNA during its packaging [90].

The mechanical properties and structural transitions of nucleic acids in the presence of force are useful information to devise and test DNA confinement models, as prompted in the previous subsection. Several theoretical models have been proposed to understand DNA organization inside the capsids (see [65] and references therein). These models together with the above experimental analysis on single viruses are needed to understand several aspects of the relationship between a virus and its substrate; for example, the connection between internal pressure, non-equilibrium confining dynamics, and DNA ejection forces [91].

Single-molecule manipulation studies have similarly provided new insights into the theoretical modeling of bacteriophage packaging mechanisms, which would be limited if based only on information from structural studies, solution biochemistry, and genetics [92, 93]. For Φ29-phage, single-molecule information has been integrated into a model in which the five ATPases undertake coordinated steps in a cycle in which one ATPase, the so-called "special subunit" of the pentameric ring, takes a regulatory role to trigger the coordinated cascade of 2.5-bp translocating steps [94]. Namely, even though this special subunit binds and hydrolyzes one ATP molecule per cycle, it does not perform mechanical work [95]. The complete Φ29-phage packaging model, compatible with structural information on ring-shaped ATPases, is termed "helical inchworm" translocation mechanism, and it was proposed after new single-molecule experiments using dsRNA and DNA–RNA hybrids were used as nucleic acids substrates during the assays. The model proposes that the ring ATPase portal motor increasingly deviates from planarity to adopt a lock-washer architecture during the ATP loading dwell; phosphate release then triggers the recovery of the planar conformation, a process that generates the translocation of one helical dsDNA turn in 2.5-bp successive steps, while producing a 14° rotation. The motor is then ready to initiate another enzymatic cycle [96, 97].

The connection between DNA structure and the viral packaging mechanism at high force can also be enlightened. Although the portion of DNA engaged in the ATPases is closely in B-form due to structural constraints in the connector, the DNA fragment outside the connector and capsid should strongly deviate from a double helix at forces above 65 pN. It is thus reasonable that ATPases have trouble in doing good contact with the DNA polymer above 65 pN thus decreasing the packaging efficiency. This fact might be one of the reasons why average maximum packaging forces are around 50–60 pN in Φ29. In order to test this scenario, constant forces above 65 pN should be directly applied against the motor.

Single-molecule experiments of RNA packaging have been started with *Pseudomonas* phage Φ6, a dsRNA virus that packages ssRNA [98]. Due to the sticky nature of this substrate, clear differences have been found with respect to dsDNA phages described above. Single-molecule experiments with optical tweezers on dsDNA phages showed maximum velocities of 200 bp/s in Φ29, 800 bp/s in λ, and 2000 bp/s T4 [99], always increasing with viral genome size to complete packaging in roughly 5 min and infection in some 20–30 min. In contrast, Φ6 promotes a helicase activity at the portal motor to unfold ssRNA structural motifs while packaging a rather short genome, thus showing slow and fast packaging rates, in any case, lower than estimated bulk rates of 33 nt/s compatible with an infection cycle of 80 min. In accord with dsDNA phages, Φ6 velocity decreases with the amount of packaged genome, although the explanation may be linked not only to complete capsid filling but, more complexly, to the capsid swelling dynamics of these viruses.

10.5.3 Manipulation of Individual Viruses: Virometry and Infectivity

Direct trapping of individual virions, without using mechanical handles, is possible and was shown using the tobacco mosaic virus in one of the pioneering manipulation experiments by Ashkin. The insights into the field of virology provided by these studies were limited, however, and this method had not found broad applications by the time [100]. Large viruses of hundreds of nanometers are amenable to individual examination in an optical trap, and this setup can be used to test direct responses to external physical stimuli (electric fields, local heat, or a combination of these with chemical changes such as those of pH and ionic strength), as shown for other biological structures [10]. Motion against fluid force can be used to characterize capsid friction and hydrodynamic size. Smaller viruses are more difficult to manipulate because optical forces approach those generated randomly by the thermal fluctuations, which make them escape the trap. In addition, it is difficult to identify single-particle trapping events. However, recently individual human immunodeficiency virus type 1 (HIV-1) virions were manipulated and examined by the effects of the thermal fluctuations and by using fluorescence [101]. The methodology has been labeled as "virometry," with which size, shape, protein composition, or refractive index measurements can be implemented in native conditions [102].

Infection studies can be assayed in this kind of virometry by correlating particular protein densities estimated in the optical tweezers from fluorescence signals with the infectivity of the virions measured in cell culture [103]. Other configurations to manipulate individual viruses include the tethering of one virion to a DNA molecule to analyze virus–cell interactions with single-virion resolution [104].

Virus–cell attachments have already been characterized by force-measuring optical tweezers. In these experiments, a laser-trapped bead coated by influenza virions was approached to a host cell to measure the interactions between an individual virus and the cell receptors. Rupture forces in the range of 10–25 pN and sequential unbinding events were detected and confirmed by similar AFM assays. Molecular dynamics simulations (see Chap. 19) were complementarily used to explain unbinding pathways and interaction dynamics [105]. Infection has also been promoted by trapping directly single virions and transporting them onto single cells in microchips, demonstrating that influenza preferentially attaches to cells in G1 phase, rather than in S, G2, or M2 phase [106, 107]. In another configuration, cell–cell transmission of HIV-1 virus through a synapse was demonstrated by manipulation of single cells with optical tweezers [108].

The physical characterization of virions is important not only to understand function but also for nanotechnological applications of virus particles (see Chap. 22).

10.5.4 Genome Compaction and Viral Assembly

The dynamics of the viral genome and capsid assembly can be real time followed by optical tweezers. In this regard, destabilization of the HIV-1 transactivation response RNA hairpin has been observed by flowing retroviral nucleocapsid proteins near single hairpins tethered between two beads, one held on top of a micropipette and the other optically trapped [109, 110]. The condensation effect of these proteins on dsDNA—which is synthesized in vivo inside the HIV-1 after reverse transcription of the viral RNA (vRNA)—can be monitored by incorporating fluorescence. The nucleocapsid proteins bind non-specifically to the nucleic acid substrate triggering an effect similar to that of polycations, namely, they generate DNA-DNA self-attraction, which compacts the nucleic acid into tight globules in multiple modes [111]. Similar studies were used to follow the coordinated compaction effect of nucleocapsid proteins from severe acute respiratory syndrome coronavirus 2 (SARS-CoV-2)—having two ordered domains associated to either RNA binding or oligomerization—on ssDNA nucleic acid substrates [112].

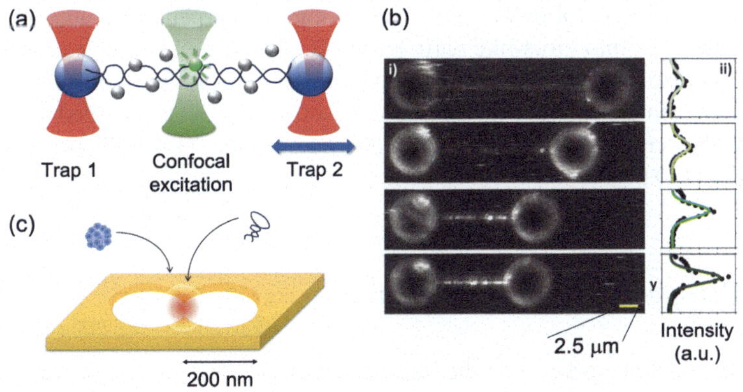

Fig. 10.10 Hybrid and Novel configurations. (**a**) Dual-trap optical tweezers (red focused beams) combined with confocal fluorescence (green beam). Fluorescently tagged proteins (represented by gray dots) light up (light green dot with rays) when they are scanned by the confocal beam. (**b**) Fluorescence images from packaging of DNA by artificial capsid proteins in a hybrid instrument of this kind. Left panels show progressive DNA shortening (from top to bottom, t= 0, 5, 15, and 25 min), which is accompanied by an increase in the fluorescence signal. Right panels show the fluorescence intensity in arbitrary units, which is proportional to the number of bound peptides, along the vertical direction. (Reprinted from [113], Creative Commons License CC-BY-NC-ND 4.0). (**c**) Nanophotonic tweezers. The cartoon shows a nanoaperture structure consisting of two intersecting holes, which generate two opposing sharp tips. A thin metal film (~100 nm-thick) is usually milled to generate such structures, which, under light illumination, yield high electromagnetic field intensity gradients (red region). Such gradients can be used to confine single virions and molecules, as indicated in the cartoon (not to scale), and monitor conformational changes

Optical tweezers and fluorescence with confocal resolution make it possible to follow the condensation dynamics of virus-like particles around a single nucleic acid substrate in real time [113]. These experiments demonstrate the power of combining both techniques. A typical assay is shown in Fig. 10.10a, where a DNA molecule is held stretched between two beads in dual-trap optical tweezers. Fluorescently labeled proteins flowing in the microfluidics chamber are excited when the confocal laser illuminates them. One of the traps (trap 2 in Fig. 10.10a), can be displaced with respect to the other (trap 1) to extend the molecule or a constant force may be applied to follow changes in extension when the proteins bind and condense the DNA, as monitored by the fluorescence. Fluorescent snapshots and quantitative signals during an experiment are shown in Fig. 10.10b; as observed, they become priceless to discuss rigorously the force spectroscopy results and to merge a dynamic model of the packaging process.

In this regard, the combined optical setup allowed tracking the one-dimensional diffusing pathways of small oligomers along stretched DNA demonstrating that large oligomers remain immobile to nucleate rod-shaped virus-like particles [113]. It was also used to reveal real-time multistep self-assembly of simian virus 40-like particles through disordered pathways with non-necessarily sequential events; more in depth, the authors showed that VP1 capsid pentamers bind the optically stretched dsDNA and reduce its contour length presumably by establishing kinks and then loops, followed by the formation of interpentamer interactions that compact the DNA into virus-like particles [114]. The assembly pathway of hepatitis B virus has also been quantitated in real time, demonstrating that its capsid core protein homodimers interact in the presence of a negatively charged substrate to assemble the virion particles through energy minima (1.4 k_BT per condensed nucleotide) [115].

10.6 Perspectives and Conclusions

There is little debate today as to the importance of viewing processes that take place inside cells as a product of nanoscale machinery. Proteins, nucleic acids, and biomolecular complexes are not only chemical entities, they are known to perform mechanical work, an activity inherently related to their function. Manipulation techniques have become key instruments for identifying the mechanochemical activity of this biological machinery. These techniques are thus used not only as confirmatory tests for chemical and structural models but also to address novel mechanisms that cannot otherwise be determined [3].

Optical tweezers are one of the most important tools in this endeavor, and have furnished continuous advances since Arthur Ashkin's pioneering work in 1970 [116]. Not surprisingly, Ashkin was awarded the Nobel Prize in Physics in 2018 for this invention and its application to biological systems. In the previous section, some perspectives on the study of viruses themselves are discussed. Here, we focus on optical advances in the instrument.

Commercial apparatuses are absorbing most of the progress in optical trapping in the last decades, a business effort that is being especially successful in the field of molecular and cell biophysics. The technological pathway involves a modular design, in which a return to optical microscopes as the building platforms is being followed. This market strategy in turn seeks benefiting from advances in super-resolution live microscopy. Custom-built instruments are still the method of choice when commercial systems cannot cope with newly introduced biomolecular systems for analysis at the single-specimen level. They also become unavoidable if one wants to improve the resolution of the measurements and introduce manipulation of single specimens on chip technologies, where multiplexed experiments at controlled temperatures or under crowded conditions may be used to simulate cellular environments.

10.6.1 Improvements in Resolution

One of the most important challenges in the field of biomolecular manipulation is to increase the resolution of the measurements. In bacteriophages, motors that translocate or move nucleic acids are inextricably coupled to their geometric structure. They use these motors to encapsulate nucleic acids to near crystalline densities inside a nanoscopic container. DNA is a double-helical sequence with a distance between base pairs of 0.34 nm in B-form. Resolving discrete steps in DNA packaging can involve spatial resolution of single base pairs, which is limited by drift and noise. Noise can be intrinsic to the experiments, caused by fundamental Brownian fluctuations, and extrinsic, caused by instrumental and environmental disturbances (power fluctuations of the lasers, electronic noise, acoustic and mechanical room vibrations, etc.). A drift in micropipette position or changes in optical components (refractive index variations or thermal expansion/contraction of lenses and mirrors) due to temperature gradients, for example, can alter the calibration. Noise and drift can thus lead to false signals or reduce the accuracy of measurements.

Temporal resolution is another issue that is continually under development. Unlike structural studies in which the specimen is static, dynamic studies involve detection of conformational changes in structures in motion on their most fundamental temporal scale. Resolution is actually a combined function of space and time in optical tweezers and thus, improvements in resolution entail a simultaneous advance on spatial and temporal detection. In addition, there exists a trade-off between the space/force resolution and their temporal counterpart for a given setup, which makes their combined optimization even trickier [31].

Managing a good resolution over the triad of mechanical quantities directly measurable by optical tweezers, namely, force, space, and time, is not only important for these quantities themselves, but to derive any other mechanical quantities, such as velocity or work, which are key to describe the action of biomolecular motors. Improved sensitivity can uncover the discrete stepping structure of a coordinated machine, as shown in Fig. 10.9 for the Φ29 packaging mechanism [42]; in this example, the 10-bp steps of each translocation burst (Fig. 10.9b) were shown to be composed of four 2.5-bp substeps (Fig. 10.9c) by a high-resolution instrument [9].

Recent advances have pushed spatial and temporal resolution to the Angstrom and millisecond scales, respectively, with measurement stability within the second range. The use of two traps from the same laser beam [9] has specifically allowed the direct observation of the intersubunit coordination and of the step size in the packaging motor of Φ29 [42]. Future progress is open to new ideas for specific treatment of each experimental noise source. For example, the use of smaller beads as mechanical handles decreases the effects of Brownian noise and may help increase the measurement bandwidth (due to viscous drag reduction), although beads of a few nanometers in size are technically more difficult to track and manipulate because they present lower trapping forces [8]. Trapping forces can be increased in turn by using higher laser powers, but this strategy may damage the sample. A better solution to compensate for small bead size appears by using high-refractive index materials (like Germanium) to nanofabricate the beads, which increases the stiffness of the traps [117]. In experimental configurations that require the use of molecular tethers, they should also be made as stiff as possible to reduce noise in data acquisition, which is managed by making the tethers as short as possible or by using DNA "origami" beams, which are promising in this challenge [118].

10.6.2 Increasing Throughput and Novel Configurations

Multiplexed trapping is a solution to produce single-molecule experiments in parallel. Spatial light modulators, structured diffractive elements, and acousto-optical deflectors are used to configure several assays both on the focal plane and along the optical axis [8, 119]. These methods, however, suffer from technical difficulties in calibrating the individual traps by the traditional methods (see Sect. 10.3.4), from added noise or from high power requirements. Video-tracking with high-speed cameras has relaxed the issue of calibrating multiplexed traps. Trapping concepts from nanotechnological progress in other fields are bringing ideas for multiplexing experiments with reduced noise and laser powers.

The force responsible for trapping is the gradient component, which is proportional to the gradient of the field intensity (second term in Eq. (10.3)). The larger the field intensity gradient the stiffer the trap. Given that optical forces decrease with the size of the routine dielectric beads used as mechanical handles, a strategy to manipulate nanoparticles or biostructures with sizes below some hundreds of nanometers is to increase the intensity field gradient. Traditional optical tweezers, as described in this chapter, are limited by the numerical aperture of the objectives, but photonics research has made it possible in recent years to create strong field gradients in the proximity of nanofabricated structures. In this radically new strategy, termed "nanotweezers," bulky optical elements (such as lenses, mirrors, waveplates, beam splitter cubes, and modulators) used to prepare and focus a laser beam on a diffraction-limited spot of micrometer dimensions are substituted by light-guiding surface structures that confine the photons within nanoscale regions in geometries such as nanoscale holes, slots, and tips. These so-called near fields are evanescent in the sense that they represent either creeping or standing waves that attenuate very rapidly away from surfaces (intensities are negligible at distances longer than some 200 nm at optical wavelengths).

A specific configuration is depicted in Fig. 10.10c, where two sharp metallic tips irradiated by light generate a strong field gradient at the center with capacity to trap nanoscale objects. Biological nanostructures, such as viruses or small molecules (streptavidin proteins and 10-bp DNAs) have already been robustly trapped and acted against noise this way at low powers (a few milliwatts) [120, 121], what represents proof-of-concepts to model innovative configurations, from sensing to precision (sub-nanometer) manipulation at the single-molecule level. Nanophotonic devices based on both plasmonic and whispering-gallery mode structures, waveguides, and photonic crystal resonators can be implemented on chips for parallel single-specimen analysis. This way, arrays of optically trapped targets can be measured, manipulated, or made chemically react in a controlled manner, expanding the analytical throughput of conventional optical tweezers. Moreover, they can be integrated with other on-chip functionalities, such as microfluidics. Miniaturization reduces reagent requirements, integrates both sample preparation and delivery, and reduces diffusion lengths in multiplexed assays; in addition, it enables system portability, improves stability, and reduces fabrication costs. This technology, today limited by local heating, photothermal damage, and unwanted thermal forces, is expected to mature in the next decades not only to exceed traditional optical tweezers experiments but mainly to broaden the way of interrogating single viruses.

The level of information detail provided by optical tweezers unavoidable involves tedious protocols to fish and operate at the single-molecule level, which requires patient experimenters. Increased resolution yields long data files, often contaminated by artifactual noise, that must be processed mathematically to distill meaningful information. The coming era is witnessing the irruption of artificial intelligence and machine learning, which bring successful solutions both for non-exploratory tasks, cumbersome to program, and for the extraction of significance in the resulting data. The use of these computer science approaches merged with parallel tweezing experiments, either traditional or based on lab-on-a-chip nanophotonic assays, will bring both incremental and innovative answers to the questions we will pose to individual viruses.

10.6.3 Conclusions

We have described a technique with which the mechanical action of viruses as well as the elastic properties and condensation dynamics of nucleic acids can be analyzed. Optical tweezers are a laser-based tool with which to trap and manipulate individual particles with piconewton forces; they are used in molecular biology to measure pairs of associated magnitudes such as force-extension and torque-rotation as a function of time. The resulting dynamic information from these experiments is yielding not only demonstrative experiments of previously conjectured mechanistic models but also unparalleled characterization of the viral machinery. Since the experiments are performed in a liquid chamber, optical tweezers can be adapted to a large variety of physiological conditions. They can also be merged with optical microscopies, allowing information from manipulation experiments and real-time imaging to be integrated into the same experiment.

The dynamic information provided by optical tweezers and its combination with structural and biochemical analysis provide rigorous ways to model viral function and stability. These models in turn permit more precise thermodynamic characterization of the viral machinery, and form a basis for the development of nanotechnological applications.

Acknowledgments It is a pleasure to acknowledge J.R. Moffitt and J.L. Carrascosa for technical insights into different aspects of the chapter, R. Bocanegra and L. Quintana for careful reading of the manuscript, C. Mark and S. Hormeño for editorial and illustration assistance, respectively, and M. de la Guía for graphic design of Fig. 8. This work was funded by the Spanish Ministry of Science and Innovation under the "Ramon y Cajal" program (grant No. RYC-2007-01765) and by grant No. PID2019-107391RB-I00.

References

1. Hormeno S, Arias-Gonzalez JR (2006) Exploring mechanochemical processes in the cell with optical tweezers. Biol Cell 98:679–695
2. Svoboda K, Block SM (1994) Biological applications of optical forces. Annu Rev Biophys Biomol Struct 23:247–285
3. Bustamante C (2008) In singulo biochemistry: when less is more. Annu Rev Biochem 77:45–50
4. Oster G, Wang H (2003) How protein motors convert chemical energy into mechanical work. In: Schliwa M (ed) Molecular Motors. Wiley-VCH, pp 207–227
5. Hwang W, Karplus M (2019) Structural basis for power stroke vs. Brownian ratchet mechanisms of motor proteins. Proc Natl Acad Sci USA 116:19777–19785
6. Ketterle W (1999) Experimental studies of Bose-Einstein condensation. Phys Today 52:30–35
7. Smith SB, Cui Y, Bustamante C (2003) Optical-trap force transducer that operates by direct measurement of light momentum. Methods Enzymol 361:134–162
8. Moffitt JR, Chemla YR, Smith SB, Bustamante C (2008) Recent advances in optical tweezers. Annu Rev Biochem 77:205–228
9. Moffitt JR, Chemla YR, Izhaky D, Bustamante C (2006) Differential detection of dual traps improves the spatial resolution of optical tweezers. Proc Natl Acad Sci USA 103:9006–9011
10. Hormeno S, Ibarra B, Chichon FJ, Habermann K, Lange BMH, Valpuesta JM, Carrascosa JL, Arias-Gonzalez JR (2009) Single centrosome manipulation reveals its electric charge and associated dynamic structure. Biophys J 97:1022–1030
11. Jackson JD (1999) Classical electrodynamics. John Wiley & Sons
12. Arias-Gonzalez JR, Nieto-Vesperinas M (2003) Optical forces on small particles: attractive and repulsive nature and plasmon-resonance conditions. J Opt Soc Am A Opt Image Sci Vis 20:1201–1209
13. Tanase M, Biais N, Sheetz M (2007) Magnetic tweezers in cell biology. In: Yu-Li W, Dennis ED (eds) Methods in cell biology, vol 83. Academic Press, pp 473–493
14. Kilinc D, Lee GU (2014) Advances in magnetic tweezers for single molecule and cell biophysics. Integr Biol (Camb) 6:27–34
15. Neuman KC, Nagy A (2008) Single-molecule force spectroscopy: optical tweezers, magnetic tweezers and atomic force microscopy. Nat Methods 5:491–505
16. Choi HK, Kim HG, Shon MJ, Yoon TY (2022) High-resolution single-molecule magnetic tweezers. Annu Rev Biochem 91:33–59
17. Bustamante C, Macosko JC, Wuite GJ (2000) Grabbing the cat by the tail: manipulating molecules one by one. Nat Rev Mol Cell Biol 1:130–136
18. Maciuba K, Kaiser CM (2022) Tethering complex proteins and protein complexes for optical tweezers experiments. Methods Mol Biol 2478:427–460
19. Keller N, delToro DJ, Smith DE (2018) Single-molecule measurements of motor-driven viral DNA packaging in bacteriophages Phi29, lambda, and T4 with optical tweezers. Methods Mol Biol 1805:393–422
20. Hormeno S, Moreno-Herrero F, Ibarra B, Carrascosa JL, Valpuesta JM, Arias-Gonzalez JR (2011) Condensation prevails over B-A transition in the structure of DNA at low humidity. Biophys J 100:2006–2015
21. Rickgauer JP, Fuller DN, Grimes S, Jardine PJ, Anderson DL, Smith DE (2008) Portal motor velocity and internal force resisting viral DNA packaging in bacteriophage phi29. Biophys J 94:159–167
22. Smith DE, Tans SJ, Smith SB, Grimes S, Anderson DL, Bustamante C (2001) The bacteriophage straight phi29 portal motor can package DNA against a large internal force. Nature 413: 748–752
23. Li PT, Collin D, Smith SB, Bustamante C, Tinoco I Jr (2006) Probing the mechanical folding kinetics of TAR RNA by hopping, force-jump, and force-ramp methods. Biophys J 90:250–260

24. Grier DG (2003) A revolution in optical manipulation. Nature 424:810–816
25. Shen Y, Wang X, Xie Z, Min C, Fu X, Liu Q, Gong M, Yuan X (2019) Optical vortices 30 years on: OAM manipulation from topological charge to multiple singularities. Light Sci Appl 8:90
26. Dame RT, Noom MC, Wuite GJ (2006) Bacterial chromatin organization by H-NS protein unravelled using dual DNA manipulation. Nature 444:387–390
27. Bustamante C, Bryant Z, Smith SB (2003) Ten years of tension: single-molecule DNA mechanics. Nature 421:423–427
28. Gutierrez-Medina B, Andreasson JO, Greenleaf WJ, Laporta A, Block SM (2010) An optical apparatus for rotation and trapping. Methods Enzymol 475:377–404
29. Parkin S, Knoner G, Singer W, Nieminen TA, Heckenberg NR, Rubinsztein-Dunlop H (2007) Optical torque on microscopic objects. Methods Cell Biol 82:525–561
30. Forth S, Sheinin MY, Inman J, Wang MD (2013) Torque measurement at the single-molecule level. Annu Rev Biophys 42:583–604
31. Bustamante CJ, Chemla YR, Liu S, Wang MD (2021) Optical tweezers in single-molecule biophysics. Nat Rev Methods Primers 1
32. van Mameren J, Peterman EJ, Wuite GJ (2008) See me, feel me: methods to concurrently visualize and manipulate single DNA molecules and associated proteins. Nucleic Acids Res 36:4381–4389
33. Whitley KD, Comstock MJ, Chemla YR (2017) High-resolution optical tweezers combined with single-molecule confocal microscopy. Methods Enzymol 582:137–169
34. Yadav R, Senanayake KB, Comstock MJ (2022) High-resolution optical tweezers combined with multicolor single-molecule microscopy. Methods Mol Biol 2478:141–240
35. Lang MJ, Fordyce PM, Engh AM, Neuman KC, Block SM (2004) Simultaneous, coincident optical trapping and single-molecule fluorescence. Nat Methods 1:133–139
36. Richardson AC, Reihani N, Oddershede LB (2006) In: Dholakia K, Spalding GC (eds) Combining confocal microscopy with precise force-scope optical tweezers, vol 6326. SPIE, San Diego, CA, USA, pp 632628–632610
37. Vossen DLJ, van der Horst A, Dogterom M, van Blaaderen A (2004) Optical tweezers and confocal microscopy for simultaneous three-dimensional manipulation and imaging in concentrated colloidal dispersions. Rev Sci Instrum 75:2960–2970
38. Toprak E, Selvin PR (2007) New fluorescent tools for watching nanometer-scale conformational changes of single molecules. Annu Rev Biophys Biomol Struct 36:349–369
39. Heller I, Sitters G, Broekmans OD, Farge G, Menges C, Wende W, Hell SW, Peterman EJ, Wuite GJ (2013) STED nanoscopy combined with optical tweezers reveals protein dynamics on densely covered DNA. Nat Methods 10:910–916
40. Desai VP, Frank F, Lee A, Righini M, Lancaster L, Noller HF, Tinoco I Jr, Bustamante C (2019) Co-temporal force and fluorescence measurements reveal a ribosomal gear shift mechanism of translation regulation by structured mRNAs. Mol Cell 75:1007–1019
41. Hugel T, Michaelis J, Hetherington CL, Jardine PJ, Grimes S, Walter JM, Falk W, Anderson DL, Bustamante C (2007) Experimental test of connector rotation during DNA packaging into bacteriophage phi29 capsids. PLoS Biol 5:e59
42. Moffitt JR, Chemla YR, Aathavan K, Grimes S, Jardine PJ, Anderson DL, Bustamante C (2009) Intersubunit coordination in a homomeric ring ATPase. Nature 457:446–450
43. Herbert KM, Greenleaf WJ, Block SM (2008) Single-molecule studies of RNA polymerase: motoring along. Annu Rev Biochem 77:149–176
44. Pyle AM (2008) Translocation and unwinding mechanisms of RNA and DNA helicases. Annu Rev Biophys 37:317–336
45. van Oijen AM, Loparo JJ (2010) Single-molecule studies of the replisome. Annu Rev Biophys 39:429–448
46. Pitre E, Te Velthuis AJW (2021) Understanding viral replication and transcription using single-molecule techniques. Enzymes 49:83–113

47. Spinks RR, Spenkelink LM, Dixon NE, van Oijen AM (2021) Single-molecule insights into the dynamics of replicative helicases. Front Mol Biosci 8:741718
48. Bianco PR (2022) Insight into the biochemical mechanism of DNA helicases provided by bulk-phase and single-molecule assays. Methods 204:348–360
49. Williams MC, Rouzina I, Bloomfield VA (2002) Thermodynamics of DNA interactions from single molecule stretching experiments. Acc Chem Res 35:159–166
50. Mao H, Arias-Gonzalez JR, Smith SB, Tinoco I Jr, Bustamante C (2005) Temperature control methods in a laser tweezers system. Biophys J 89:1308–1316
51. Hormeno S, Ibarra B, Valpuesta JM, Carrascosa JL, Arias-Gonzalez JR (2012) Mechanical stability of low-humidity single DNA molecules. Biopolymers 97:199–208
52. de Lorenzo S, Ribezzi-Crivellari M, Arias-Gonzalez JR, Smith SB, Ritort F (2015) A temperature-jump optical trap for single-molecule manipulation. Biophys J 108:2854–2864
53. Arias-Gonzalez JR (2014) Single-molecule portrait of DNA and RNA double helices. Integr Biol (Camb) 6:904–925
54. Herrero-Galan E, Fuentes-Perez ME, Carrasco C, Valpuesta JM, Carrascosa JL, Moreno-Herrero F, Arias-Gonzalez JR (2013) Mechanical identities of RNA and DNA double helices unveiled at the single-molecule level. J Am Chem Soc 135:122–131
55. Zhang C, Fu H, Yang Y, Zhou E, Tan Z, You H, Zhang X (2019) The mechanical properties of RNA-DNA hybrid duplex stretched by magnetic tweezers. Biophys J 116:196–204
56. Lipfert J, Skinner GM, Keegstra JM, Hensgens T, Jager T, Dulin D, Kober M, Yu Z, Donkers SP, Chou FC, Das R, Dekker NH (2014) Double-stranded RNA under force and torque: similarities to and striking differences from double-stranded DNA. Proc Natl Acad Sci USA 111:15408–15413
57. Bosco A, Camunas-Soler J, Ritort F (2014) Elastic properties and secondary structure formation of single-stranded DNA at monovalent and divalent salt conditions. Nucleic Acids Res 42: 2064–2074
58. Chaurasiya KR, Paramanathan T, McCauley MJ, Williams MC (2010) Biophysical characterization of DNA binding from single molecule force measurements. Phys Life Rev 7:299–341
59. Garavis M, Bocanegra R, Herrero-Galan E, Gonzalez C, Villasante A, Arias-Gonzalez JR (2013) Mechanical unfolding of long human telomeric RNA (TERRA). Chem Commun (Camb) 49:6397–6399
60. Gutierrez I, Garavis M, de Lorenzo S, Villasante A, Gonzalez C, Arias-Gonzalez JR (2018) Single-stranded condensation stochastically blocks G-Quadruplex assembly in human Telomeric RNA. J Phys Chem Lett 9:2498–2503
61. Yu Z, Schonhoft JD, Dhakal S, Bajracharya R, Hegde R, Basu S, Mao H (2009) ILPR G-quadruplexes formed in seconds demonstrate high mechanical stabilities. J Am Chem Soc 131:1876–1882
62. Mandal S, Hoque ME, Mao H (2019) Single-molecule investigations of G-Quadruplex. Methods Mol Biol 2035:275–298
63. Dhakal S, Schonhoft JD, Koirala D, Yu Z, Basu S, Mao H (2010) Coexistence of an ILPR i-motif and a partially folded structure with comparable mechanical stability revealed at the single-molecule level. J Am Chem Soc 132:8991–8997
64. Casjens SR (2011) The DNA-packaging nanomotor of tailed bacteriophages. Nat Rev Microbiol 9:647–657
65. Chemla YR, Smith DE (2012) Single-molecule studies of viral DNA packaging. Adv Exp Med Biol 726:549–584
66. Feiss M, Rao VB (2012) The bacteriophage DNA packaging machine. Adv Exp Med Biol 726:489–509
67. Tang J, Olson N, Jardine PJ, Grimes S, Anderson DL, Baker TS (2008) DNA poised for release in bacteriophage phi29. Structure 16:935–943

68. Fuller DN, Raymer DM, Kottadiel VI, Rao VB, Smith DE (2007) Single phage T4 DNA packaging motors exhibit large force generation, high velocity, and dynamic variability. Proc Natl Acad Sci USA 104:16868–16873
69. Fuller DN, Raymer DM, Rickgauer JP, Robertson RM, Catalano CE, Anderson DL, Grimes S, Smith DE (2007) Measurements of single DNA molecule packaging dynamics in bacteriophage lambda reveal high forces, high motor processivity, and capsid transformations. J Mol Biol 373:1113–1122
70. Fuller DN, Rickgauer JP, Jardine PJ, Grimes S, Anderson DL, Smith DE (2007) Ionic effects on viral DNA packaging and portal motor function in bacteriophage phi 29. Proc Natl Acad Sci USA 104:11245–11250
71. Ivanovska IL, de Pablo PJ, Ibarra B, Sgalari G, MacKintosh FC, Carrascosa JL, Schmidt CF, Wuite GJ (2004) Bacteriophage capsids: tough nanoshells with complex elastic properties. Proc Natl Acad Sci USA 101:7600–7605
72. Tsay JM, Sippy J, DelToro D, Andrews BT, Draper B, Rao V, Catalano CE, Feiss M, Smith DE (2010) Mutations altering a structurally conserved loop-helix-loop region of a viral packaging motor change DNA translocation velocity and processivity. J Biol Chem 285: 24282–24289
73. Tsay JM, Sippy J, Feiss M, Smith DE (2009) The Q motif of a viral packaging motor governs its force generation and communicates ATP recognition to DNA interaction. Proc Natl Acad Sci USA 106:14355–14360
74. Berndsen ZT, Keller N, Grimes S, Jardine PJ, Smith DE (2014) Nonequilibrium dynamics and ultraslow relaxation of confined DNA during viral packaging. Proc Natl Acad Sci USA 111: 8345–8350
75. Keller N, delToro D, Grimes S, Jardine PJ, Smith DE (2014) Repulsive DNA-DNA interactions accelerate viral DNA packaging in phage Phi29. Phys Rev Lett 112:248101
76. Keller N, Grimes S, Jardine PJ, Smith DE (2016) Single DNA molecule jamming and history-dependent dynamics during motor-driven viral packaging. Nat Phys 12:757–761
77. Kottadiel VI, Rao VB, Chemla YR (2012) The dynamic pause-unpackaging state, an off-translocation recovery state of a DNA packaging motor from bacteriophage T4. Proc Natl Acad Sci USA 109:20000–20005
78. Sun S, Kondabagil K, Draper B, Alam TI, Bowman VD, Zhang Z, Hegde S, Fokine A, Rossmann MG, Rao VB (2008) The structure of the phage T4 DNA packaging motor suggests a mechanism dependent on electrostatic forces. Cell 135:1251–1262
79. Migliori AD, Keller N, Alam TI, Mahalingam M, Rao VB, Arya G, Smith DE (2014) Evidence for an electrostatic mechanism of force generation by the bacteriophage T4 DNA packaging motor. Nat Commun 5:4173
80. Ordyan M, Alam I, Mahalingam M, Rao VB, Smith DE (2018) Nucleotide-dependent DNA gripping and an end-clamp mechanism regulate the bacteriophage T4 viral packaging motor. Nat Commun 9:5434
81. Mo Y, Keller N, delToro D, Ananthaswamy N, Harvey SC, Rao VB, Smith DE (2020) Function of a viral genome packaging motor from bacteriophage T4 is insensitive to DNA sequence. Nucleic Acids Res 48:11602–11614
82. Chemla YR, Aathavan K, Michaelis J, Grimes S, Jardine PJ, Anderson DL, Bustamante C (2005) Mechanism of force generation of a viral DNA packaging motor. Cell 122:683–692
83. Dai L, Singh D, Lu S, Kottadiel VI, Vafabakhsh R, Mahalingam M, Chemla YR, Ha T, Rao VB (2021) A viral genome packaging ring-ATPase is a flexibly coordinated pentamer. Nat Commun 12:6548
84. Block SM, Asbury CL, Shaevitz JW, Lang MJ (2003) Probing the kinesin reaction cycle with a 2D optical force clamp. Proc Natl Acad Sci USA 100:2351–2356
85. Arias-Gonzalez JR (2017) A DNA-centered explanation of the DNA polymerase translocation mechanism. Sci Rep 7:7566

86. Liu T, Kaplan A, Alexander L, Yan S, Wen JD, Lancaster L, Wickersham CE, Fredrick K, Noller H, Tinoco I, Bustamante CJ (2014) Direct measurement of the mechanical work during translocation by the ribosome. eLife 3:e03406
87. Abbondanzieri EA, Greenleaf WJ, Shaevitz JW, Landick R, Block SM (2005) Direct observation of base-pair stepping by RNA polymerase. Nature 438:460–465
88. Aathavan K, Politzer AT, Kaplan A, Moffitt JR, Chemla YR, Grimes S, Jardine PJ, Anderson DL, Bustamante C (2009) Substrate interactions and promiscuity in a viral DNA packaging motor. Nature 461:669–673
89. Zhang Z, Kottadiel VI, Vafabakhsh R, Dai L, Chemla YR, Ha T, Rao VB (2011) A promiscuous DNA packaging machine from bacteriophage T4. PLoS Biol 9:e1000592
90. Liu S, Chistol G, Hetherington CL, Tafoya S, Aathavan K, Schnitzbauer J, Grimes S, Jardine PJ, Bustamante C (2014) A viral packaging motor varies its DNA rotation and step size to preserve subunit coordination as the capsid fills. Cell 157:702–713
91. Keller N, Berndsen ZT, Jardine PJ, Smith DE (2017) Experimental comparison of forces resisting viral DNA packaging and driving DNA ejection. Phys Rev E 95:052408
92. Purohit PK, Kondev J, Phillips R (2003) Mechanics of DNA packaging in viruses. Proc Natl Acad Sci USA 100:3173–3178
93. Yu J, Moffitt J, Hetherington CL, Bustamante C, Oster G (2010) Mechanochemistry of a viral DNA packaging motor. J Mol Biol 400:186–203
94. Tafoya S, Liu S, Castillo JP, Atz R, Morais MC, Grimes S, Jardine PJ, Bustamante C (2018) Molecular switch-like regulation enables global subunit coordination in a viral ring ATPase. Proc Natl Acad Sci USA 115:7961–7966
95. Chistol G, Liu S, Hetherington CL, Moffitt JR, Grimes S, Jardine PJ, Bustamante C (2012) High degree of coordination and division of labor among subunits in a homomeric ring ATPase. Cell 151:1017–1028
96. Castillo JP, Tong AB, Tafoya S, Jardine PJ, Bustamante C (2021) A DNA packaging motor inchworms along one strand allowing it to adapt to alternative double-helical structures. Nat Commun 12:3439
97. Tong AB, Bustamante C (2021) Helical inchworming: a novel translocation mechanism for a ring ATPase. Biophys Rev 13:885–888
98. Hanhijarvi KJ, Ziedaite G, Bamford DH, Haeggstrom E, Poranen MM (2017) Single-molecule measurements of viral ssRNA packaging. RNA 23:119–129
99. Pajak J, Arya G, Smith DE (2021) Biophysics of DNA packaging. In: Bamford DH, Zuckerman M (eds) Encyclopedia of virology, 4th edn. Academic Press, Oxford, pp 160–166
100. Ashkin A, Dziedzic JM (1987) Optical trapping and manipulation of viruses and bacteria. Science 235:1517–1520
101. Pang Y, Song H, Kim JH, Hou X, Cheng W (2014) Optical trapping of individual human immunodeficiency viruses in culture fluid reveals heterogeneity with single-molecule resolution. Nat Nanotechnol 9:624–630
102. Pang Y, Song H, Cheng W (2016) Using optical trap to measure the refractive index of a single animal virus in culture fluid with high precision. Biomed Opt Express 7:1672–1689
103. DeSantis MC, Kim JH, Song H, Klasse PJ, Cheng W (2016) Quantitative correlation between infectivity and Gp120 density on HIV-1 Virions revealed by optical trapping Virometry. J Biol Chem 291:13088–13097
104. Schimert KI, Cheng W (2018) A method for tethering single viral particles for virus-cell interaction studies with optical tweezers. Proc SPIE Int Soc Opt Eng 10723
105. Sieben C, Kappel C, Zhu R, Wozniak A, Rankl C, Hinterdorfer P, Grubmuller H, Herrmann A (2012) Influenza virus binds its host cell using multiple dynamic interactions. Proc Natl Acad Sci USA 109:13626–13631
106. Masuda T, Maruyama H, Honda A, Arai F (2014) Virus enrichment for single virus infection by using 3D insulator based dielectrophoresis. PLoS One 9:e94083
107. Ueda R, Sugiura T, Kume S, Ichikawa A, Larsen S, Miyoshi H, Hiramatsu H, Nagatsuka Y, Arai F, Suzuki Y, Hirabayashi Y, Fukuda T, Honda A (2013) A novel single virus infection system reveals that influenza virus preferentially infects cells in g1 phase. PLoS One 8:e67011

108. McNerney GP, Hubner W, Chen BK, Huser T (2010) Manipulating CD4+ T cells by optical tweezers for the initiation of cell-cell transfer of HIV-1. J Biophotonics 3:216–223
109. McCauley MJ, Rouzina I, Li J, Nunez ME, Williams MC (2020) Significant differences in RNA structure destabilization by HIV-1 GagDp6 and NCp7 proteins. Viruses 12
110. McCauley MJ, Rouzina I, Manthei KA, Gorelick RJ, Musier-Forsyth K, Williams MC (2015) Targeted binding of nucleocapsid protein transforms the folding landscape of HIV-1 TAR RNA. Proc Natl Acad Sci USA 112:13555–13560
111. Gien H, Morse M, McCauley MJ, Kitzrow JP, Musier-Forsyth K, Gorelick RJ, Rouzina I, Williams MC (2022) HIV-1 Nucleocapsid protein binds double-stranded DNA in multiple modes to regulate compaction and capsid Uncoating. Viruses 14
112. Morse M, Sefcikova J, Rouzina I, Beuning PJ, Williams MC (2023) Structural domains of SARS-CoV-2 nucleocapsid protein coordinate to compact long nucleic acid substrates. Nucleic Acids Res 51:290–303
113. Marchetti M, Kamsma D, Cazares Vargas E, Hernandez Garcia A, van der Schoot P, de Vries R, Wuite GJL, Roos WH (2019) Real-time assembly of Viruslike Nucleocapsids elucidated at the single-particle level. Nano Lett 19:5746–5753
114. van Rosmalen MGM, Kamsma D, Biebricher AS, Li C, Zlotnick A, Roos WH, Wuite GJL (2020) Revealing in real-time a multistep assembly mechanism for SV40 virus-like particles. Sci Adv 6:eaaz1639
115. Buzon P, Maity S, Christodoulis P, Wiertsema MJ, Dunkelbarger S, Kim C, Wuite GJL, Zlotnick A, Roos WH (2021) Virus self-assembly proceeds through contact-rich energy minima. Sci Adv 7:eabg0811
116. Ashkin A (1970) Acceleration and trapping of particles by radiation pressure. Phys Rev Lett 24:156–159
117. Sudhakar S, Abdosamadi MK, Jachowski TJ, Bugiel M, Jannasch A, Schaffer E (2021) Germanium nanospheres for ultraresolution picotensiometry of kinesin motors. Science 371
118. Pfitzner E, Wachauf C, Kilchherr F, Pelz B, Shih WM, Rief M, Dietz H (2013) Rigid DNA beams for high-resolution single-molecule mechanics. Angew Chem Int Ed Engl 52:7766–7771
119. Muñoz-Pérez FM, Ferrando V, Furlan WD, Monsoriu JA, Arias-Gonzalez JR (2022) Optical multi-trapping by Kinoform m-Bonacci lenses. Opt Express 30:34378–34384
120. Badman RP, Ye F, Wang MD (2019) Towards biological applications of nanophotonic tweezers. Curr Opin Chem Biol 53:158–166
121. Corsetti S, Dholakia K (2021) Optical manipulation: advances for biophotonics in the 21st century. J Biomed Opt 26

Further Reading

Sheetz MP (1998) Laser tweezers in cell biology. Methods Cell Biol 55
Berns MW, Greulich KO (2007) Laser manipulation of cells and tissues. Methods Cell Biol 82
Schliwa M (ed) (2003) Molecular Motors. Wiley-VCH
Gennerich A (ed) (2017) Optical tweezers. Methods and protocols Methods in Molecular Biology 1486
Peterman, EJG (ed) (2018) Single molecule analysis. Methods and protocols, second edition. Methods in Molecular Biology 1665
Lavelle C (ed) (2018) Molecular motors. Methods and protocols, 2nd edition. Methods in Molecular Biology 1805
Gennerich A (ed) (2022) Optical tweezers. Methods and protocols, 2nd edition. Methods in Molecular Biology 2478

Also especially recommended for further reading are references [1, 8, 15, 31, 53, 64, 65] listed above.

Part III
Structural Foundations of Virus Properties and Functions

Chapter 11
Assembly of Structurally Simple Icosahedral Viruses

Alberto López-Bueno, Jon Gil-Ranedo, and José M. Almendral

Abstract Icosahedral viruses exhibit elegant pathways of capsid assembly and maturation regulated by symmetry principles. Assembly is a dynamic process driven by consecutive and genetically programmed morphogenetic interactions between protein subunits. The non-symmetric capsid subunits are gathered by non-covalent contacts and interactions in assembly intermediates, which serve as blocks to build a symmetric capsid. In some virus examples, the assembly of the protein shell further requires non-symmetric interactions among intermediates to fold into specific conformations. In this chapter, the morphogenesis of some small and structurally simple icosahedral viruses, including representative members of the parvoviruses, picornaviruses, and polyomaviruses as paradigms, is described in some detail. Despite their small size, the assembly of these icosahedral viruses may follow rather complex pathways, as they may occur in different subcellular compartments, involve a panoply of cellular and viral factors, and regulatory protein post-translational modifications that challenge its comprehensive understanding. Mechanisms of viral genome encapsidation may imply direct interactions between the genome and the assembly intermediates, or active packaging into a preformed empty capsid. Further, membranes and factors at specific subcellular compartments may also be critically required for virus maturation. The high stability of intermediates and the process of viral maturation contribute to the overall irreversible character of the assembly process. These and other small, structurally less complex icosahedral viruses were pioneer models to understand basic principles of virus assembly, continue to be leading subjects of morphogenetic analyses, and have inspired ongoing studies on the assembly of larger, structurally more complex viruses as well as cellular and synthetic macromolecular complexes.

A. López-Bueno · J. M. Almendral (✉)
Centro de Biología Molecular "Severo Ochoa" (CSIC-UAM) and Department of Molecular Biology, Universidad Autónoma de Madrid, Madrid, Spain
e-mail: jmalmendral@cbm.csic.es; josemaria.almendral@uam.es

J. Gil-Ranedo
Faculty of Health: Medicine, Dentistry and Human Sciences, University of Plymouth, Plymouth, UK

M. G. Mateu (ed.), *Structure and Physics of Viruses*, Subcellular Biochemistry 105,
https://doi.org/10.1007/978-3-031-65187-8_11

Keywords Icosahedral capsid · Triangulation number · Assembly intermediate · Protein–protein interactions · Protein phosphorylation · Nuclear translocation · Protein folding · Nucleation · Genome encapsidation · Virus factory · Maturation cleavage

Abbreviations

AAP	Assembly-activating protein
AAV	Adeno-associated virus
CBB	Capsid building block
CP	Capsid protein
CPV	Canine parvovirus
DBD	DNA-binding domains
H1-PV	Parvovirus H1
hr-t	Host range-transforming
FPV	Feline parvovirus
MEV	Mink enteritis virus
MVM	Minute virus of mice
NLM	Nuclear localization motif
NLS	Nuclear localization sequence
NPC	Nuclear pore complex
PPV	Porcine parvovirus
SV40	Simian virus 40
VLP	Virus-like particle
VP	Viral protein
2x	Twofold axis
3x	Threefold axis
5x	Fivefold axis

11.1 Introduction

All viral entities have a capsid, which in structurally simple viruses are built up from one or a few types of protein subunits. At late stages of the intracellular phase of their life cycle (see Chap. 1) viruses perform capsid assembly, a process by which the structural capsid protein (CP) subunits are joined by the hydrophobic effect and/or non-covalent electrostatic interactions (and occasionally covalent bonds) to construct the viral particle. This process is essential for viruses to mature (become infectious) and release progeny, and in the simple icosahedral viruses proceeds by strict principles of genetic economy and symmetry (see Chap. 2). This chapter reviews the main stages in the assembly and genome encapsidation of small or medium-sized viruses with a relatively simple icosahedral architecture, exemplified in three distinct virus models: (i) Adeno-associated virus (AAV) and minute virus of mice (MVM; recently renamed *Rodent protoparvovirus 1*) [1], as respective representative models of the *Dependovirus* and *Protoparvovirus* genera of the *Parvoviridae*, which are single-stranded (ss) DNA viruses with a T = 1 capsid (25 nm in diameter) assembling in the nucleus; (ii) Poliovirus, genus *Enterovirus* of the *Picornaviridae*, ssRNA(+) viruses with a pseudoT = 3 capsid (30 nm in diameter) assembling in the cytoplasm; and (iii) Polyomavirus and Simian Virus (SV40; recently renamed as *Macaca mulatta polyomavirus 1*), members of the *Betapolyomavirus* genus of the *Polyomaviridae*, double-stranded (ds) DNA viruses with an all-pentamers T = 7d capsid (45 nm in diameter) assembling in the nucleus. In these so-called "simple" icosahedral viruses, assembly occurs through an orchestrated pattern of interactions of the capsid subunits to form complexes or assembly intermediates, whose composition and conformation usually change along the process. The structural dynamics undergone by the assembly intermediates must also fulfill another important function, which is to traffic the capsid components within the infected cell toward the compartment where the genome is being replicated. It is the accumulation of assembly intermediates at a specific compartment that triggers genome encapsidation and maturation, allowing the virus to finally propagate in nature.

11.2 Icosahedral Capsids: Symmetry and Genetic Regulation

11.2.1 Structural Principles in Icosahedral Capsid Assembly

In structurally simple icosahedral viruses, the capsid is formed by many copies of one or a few protein subunits that assemble by making multiple contacts to build a hollow shell of proper size and symmetry. The regular icosahedron is formed by 60 copies of a capsid building block (CBB), which can be built up by a single CP subunit, by several identical CPs, or by non-identical CPs. The CBBs are related by twofold (2x), threefold (3x), and fivefold (5x) symmetry axes, and CPs within the CBBs establish regular interactions with their neighbors depending on their position in relation to these icosahedral axes. The Caspar and Klug (1962) theory [2] explained how some multiples of 60 identical subunits could be arranged with similar (quasi-equivalent) interactions, according to the rule $T = h^2 + hk + k^2$, where h and k are integers and T is called the triangulation number (see Chap. 2 for a detailed explanation of capsid icosahedral symmetry and quasi-equivalence).

In the members of the *Parvoviridae*, the capsid is a T = 1 perfect icosahedron, formed by a total of 60 CPs (termed viral proteins, VPs, in this and some other virus families), which include two variant types (VP1 and VP2) with identical amino acid sequence and fold except for short stretches of sequences at the C- or N-termini which are intrinsically disordered and are not observed in the X-ray structures of the virus particles (Fig. 11.1a). The topography of the capsid surface differs among the

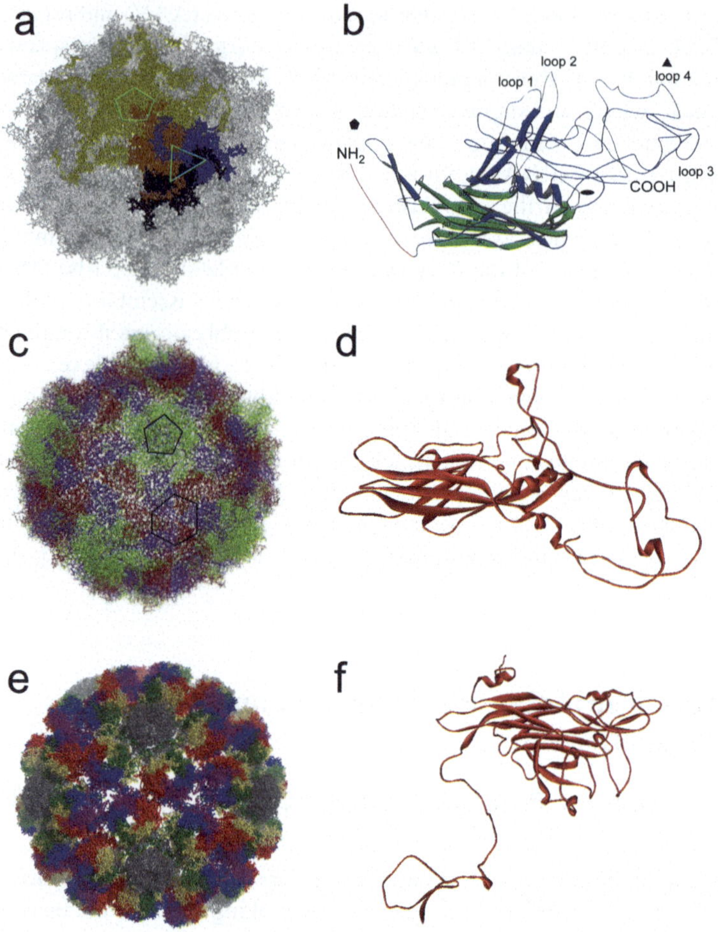

Fig. 11.1 Atomic structure of icosahedral virus particles and capsid subunits whose assembly is described in this chapter. (**a**) Structure of Parvovirus MVM (p and i strains; [3, 4]). Different colors distinguish the subunits surrounding the 5x axes (pentagon) from those interdigitated at the 3x axes (triangle). (**b**) Folding of the VP1 and VP2 subunits in the MVM structure. Note the prominent loops projecting away from the capsid surface. (**c**) Structure of Poliovirus, a picornavirus [5, 6]. The VP1 subunits around the 5x axes (pentagon), and the alternating VP2 and VP3 subunits around the 3x axes (hexagon) are respectively shown in green, red, and violet. (**d**) Folding of the VP1 subunit in the capsid of Poliovirus. The β-strands (arrows) form two antiparallel sheets juxtaposed in a wedgelike structure. (**e**) Structure of the SV40 virus, a polyomavirus, shows the organization of the VP1 pentamers [7]. The 5x-coordinated pentamers are represented in gray and the VP1 subunits of the pentamers in hexameric arrays are represented in different colors (**f**) folding of the VP1 subunit as observed in the SV40 capsid. The β-barrel (jellyroll) is radial to the capsid surface. The figure was prepared using the Pymol program (http://www.pymol.org/) and the VIPER resources [8], using the atomic coordinates of MVM (1MVM, 1Z1C), Poliovirus (2PLV), and SV40 (1SVA), deposited in the Protein Data Bank (PDB)

parvoviruses due to the characteristic prominence of peptide loops and spikes at the 3x symmetry axes, and the depth and contour of the depression surrounding the 5x axes [3, 4, 9–12]. In Poliovirus and related picornaviruses, the protein shell is built up of 60 copies of a fundamental subunit (the protomer) composed of three different proteins (termed VP1, VP2, and VP3) which are not related in amino acid sequence but have a similar fold, and a small extended polypeptide (VP4). VP4 is located at the inner surface of the protein shell and remains covalently linked to the N-terminus of VP2 until the final stages of virus assembly and maturation. The interactions between the VP1 subunits around the 5x axes are not equivalent to those engaging the alternating VP2 and VP3 subunits in the 3x axes, accounting for the features of the surface of the capsid (Fig. 11.1c), including the conspicuous protrusion of the VP1 subunits at the 5x axes [5, 6]. In polyomaviruses such as SV40, the icosahedral capsid is formed by 360 identical CP subunits of a single protein (termed VP1) arrangement does not follow the quasi-equivalence rules of Caspar and Klug, as the basic structural elements (capsomers) are 72 pentamers displayed in a $T = 7d$ surface lattice ([7]; Fig. 11.1e). Each pentamer contains in addition one copy of either of two other proteins (termed VP2 and VP3), which share most of their amino acid sequences. The capsomers located around each 5x axis are referred to as the pentavalent pentamers, and those capsomers arranged around each of the 3x axis are referred to as the hexavalent pentamers.

Although the parvovirus, picornavirus, and polyomavirus capsid proteins are not related in amino acid sequences, the core of all these proteins are folded in their capsids as a β-sheet structure termed β-barrel jelly roll (or Swiss-roll β-barrel, or an eight-stranded antiparallel β-barrel), a wedge-shaped structure comprising two antiparallel β-sheets. The topology of all these capsid subunits is, thus, similar (Fig. 11.1b, d, f). The major structural differences between them are in the loops that connect the strands, which are particularly prominent in parvoviruses (Fig. 11.1b), and the N- and C-terminal segments that extend from the central β-barrel domains. The remarkable similarity of these β-barrel jelly rolls among viral proteins that do not share primary amino acid sequences, and belong to different families of viruses with unrelated biological properties, suggests that it may represent one of the few structural solutions allowing proteins to be packaged in icosahedral capsids, or be a testimony of a common ancestral evolutionary history (see also Chaps. 2 and 7).

11.2.2 Synthesis of the Capsid Subunits: Setting the Assembly Scenario

For successful assembly, synthesis of the capsid proteins must be tightly regulated during the virus life cycle, in order to satisfy at least three key requirements: (i) quantity, viruses must induce the accumulation of high amounts of capsid proteins in the host cells, in order to compete with the vast amount and diversity of preexisting cellular proteins; (ii) timing, capsid proteins synthesis must reach maximum levels by the time the viral genome is being replicated, so it can be efficiently packaged; and (iii) stoichiometry, capsid subunits must be synthesized at the proper ratio to ensure assembly of an infectious viral particle. Genetic regulation of viral gene expression is crucial for a successful assembly process. Mechanisms controlling viral gene expression involve multiple networks, which are out of the scope of this book, so only a brief outline of those involved in assembly is mentioned below.

In Poliovirus (see [13] for a review of its life cycle), the genomic RNA is released in the cytoplasm by the incoming virion, and translated in a single open reading frame to produce a very large polyprotein, which undergoes cotranslational *cis*-cleavage by the viral protease 2A. This cleavage releases from the N-terminus the precursor polyprotein myristoyl-P1, which contains the CP sequences. In this and related picornaviruses, CPs are synthesized to high levels by the combination of the translational competence of the genomic RNA with an effective host protein shut-off induced by viral proteases, whereas the proper protein stoichiometry results from the cleavage of a common precursor.

In parvoviruses, polyomaviruses, and other DNA viruses, the stoichiometry of capsid proteins is regulated mainly at the level of splicing of messenger RNAs. In the parvovirus MVM, for instance, site-directed mutagenesis at the minor splicing sites, or independent cDNA cloning, allows to obtain genomic clones expressing either VP protein [14]. VP2 alone can form a capsid which can encapsidate the viral genome, but VP1 is necessary for the infectivity of the particles due to specific domains residing at its N-terminal segment [15, 16]. A 1:5 ratio of VP1:VP2 is found for the soluble synthesized proteins as well as in the assembled capsid, and preserving this ratio is critically important for an ordered assembly avoiding protein aggregates [17]. In polyomaviruses VP1, VP2, and VP3 levels in the infected cells are regulated by alternative splicing from a common transcript, occurring soon after viral DNA replicative intermediates accumulate. In these small DNA viruses, protein shut-off is not a major mechanism to counteract host protein synthesis, but the nuclear accumulation of protein products prior to assembly (see below) facilitates their interactions at certain nuclearly confined environments.

11.3 Capsid Building Blocks and Assembly Intermediates

The capacity of the viral structural proteins to form capsids ultimately results from their folding and self-assembly properties, which are conferred by the encoded amino acid sequences. Virus capsids are assembled from CBBs which may be CP monomers or, in many cases, CP oligomers. These stable oligomeric CBBs may be considered as the first stable intermediates of the capsid assembly process. The use of in vitro assembly systems in combination with diverse theoretical approaches has allowed the investigation of fundamental principles of virus capsid self-assembly starting from CBBs. The theory of capsid self-assembly is beyond the aims of this chapter, which is dedicated to the description of assembly processes in different viruses from in vitro and in vivo experimental evidences supported by structural and functional studies. The reader is referred to Chap. 1 for a brief overview, and to Chap. 19 for a detailed description of the thermodynamic and kinetic aspects of the assembly of structurally simple virus capsids. These studies are generally consistent with experimental observations on the assembly of very simple virus capsids, frequently carried out in controlled in vitro experiments. The models support assembly as a nucleated cooperative process in which CP or CBB concentration is critical, with a lag phase reflecting the time required to build up an assembly line of intermediate structures. The intermediates are expected to be at very low concentration, but this steady state of intermediates is required for efficient assembly in any stepwise reaction.

11.3.1 Structure of CBBs and Assembly Intermediates

The stepwise assembly pathway has been well characterized in Poliovirus (and other picornaviruses including human rhinovirus and foot-and-mouth disease virus), as some discrete intermediates are stable enough to be isolated. A common strategy that many viruses adopt to build blocks for capsid assembly is to initiate assembly while the structural units are linked into a polyprotein precursor. In Poliovirus, the first intermediate of the assembly pathway is an immature structural unit (unprocessed protomer) formed by a folded polyprotein (termed P1). P1 contains three structural domains. These domains are cleaved from each other upon cleavage at specific sites in the linker sequences by the viral $3CD^{pro}$ protease. The result is a processed protomer (5S protomer) which sediments as a 5S particle and is formed by one copy each of VP0, VP3, and VP1, which correspond to the cleaved P1 domains (Fig. 11.2a).

It is unclear when the β-barrel of these proteins' core fold, but unprocessed P1 of picornaviruses is recognized by panels of virus-neutralizing antibodies elicited against discontinuous epitopes of mature virions, strongly suggesting that the domains in unprocessed P1 are already folded like the mature CPs in the assembled capsid. The structures of the isolated unprocessed or processed P1 protomers have not been solved for any picornavirus yet; however, in the virus capsid structure VP1, VP2, and VP3 form an intricate network of intermolecular interactions among the surfaces of their β-barrel domains that must greatly stabilize the protomer and contribute to the early stages of Picornavirus assembly. VP4 remains covalently linked to VP2 in the VP0 protein until virus assembly is completed and maturation occurs. The 5S precursor is followed in the assembly line by the 14S pentamer, which is formed by oligomerization of five 5S protomers, and is stabilized by extensive protein–protein interactions and by others mediated by myristate chains incorporated in the five VP0 N-termini (see Sect. 11.4.1). These multiple interactions determine a molecular interlocking of the five protomers in the pentamer, conferring high stability and directionality to the whole assembly pathway.

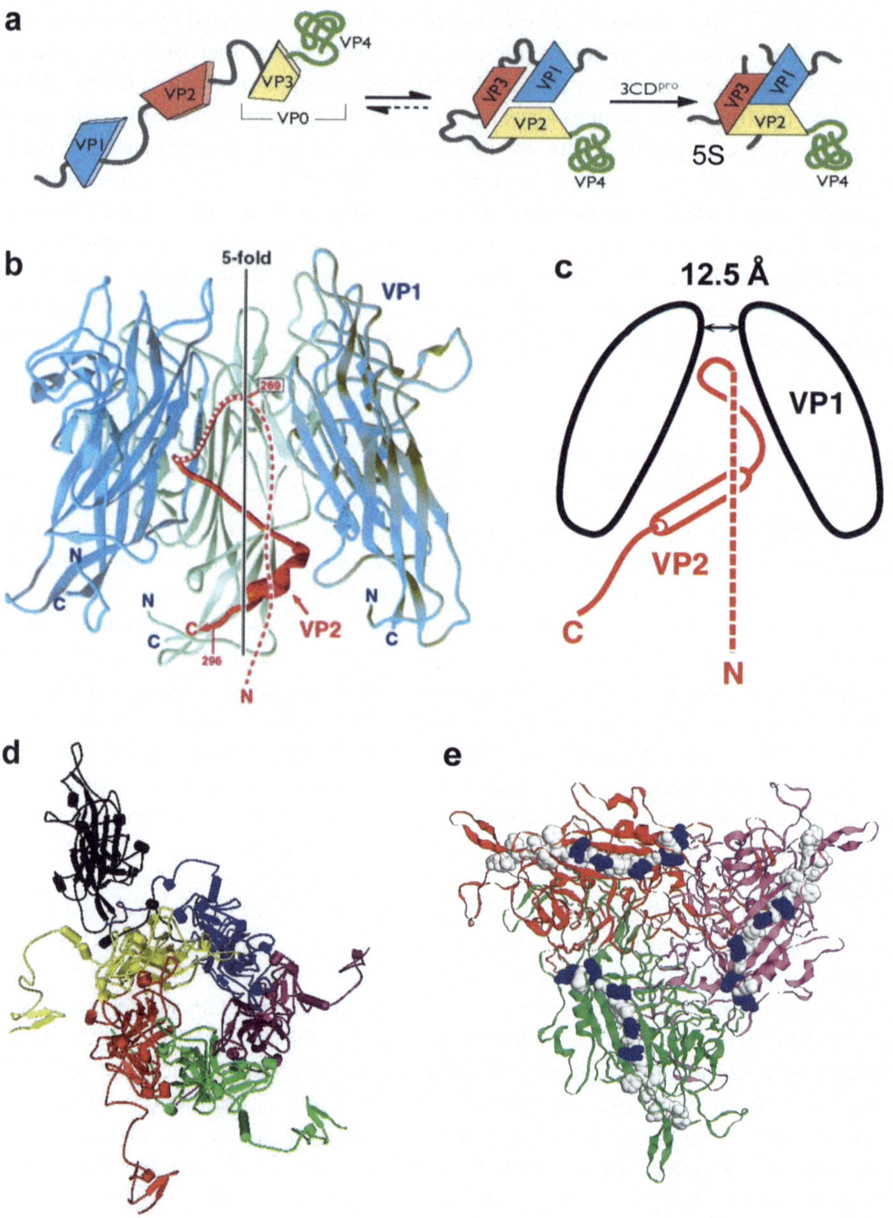

Fig. 11.2 Structural models of assembly intermediates in icosahedral viruses. (**a**) Poliovirus P1 polyprotein precursor showing the four proteins forming the heteromeric structural unit (protomer). Cleavage by $3CD^{pro}$ protease yields the 5S assembly intermediate (Adapted from Flint et al. (2009) Principles of Virology, ASM Press, with permission). (**b**) Structure of the VP1-VP2 assembly intermediate of polyomavirus. VP2 is shown in red and the three VP1 monomers that form contacts with VP2 in green (middle) and blue (left and right). (**c**) Schematic representation of the VP1/VP2 interaction in this complex. VP2 (red) enters the VP1 pentamer (black) from the base, continuing to

In polyomaviruses as well as in parvoviruses, the CP subunits interact in the cytoplasm before nuclear capsid formation. The composition of the assembly intermediates is characteristic of each virus system and plays a role of paramount importance in the correct timing and spatial ordering during assembly. A relatively stable cytoplasmic assembly intermediate seems to be a common need for these viruses achieving capsid assembly and maturing within the nucleus. In Polyomavirus and SV40, stable pentamers of VP1 are formed in the cytoplasm. Each VP1 pentamer binds either one VP2 or one VP3 protein that becomes allocated in the axial cavity of the pentamer (Fig. 11.2b), and the VP1-VP2/3 complex is stabilized by strong hydrophobic interactions [18]. The contacts between subunits alter the configuration of the VP1 pentamer, as demonstrated by changes in epitope accessibility and by direct structural insights obtained from the crystal structure (Fig. 11.2b, c). In SV40 [19], and other polyomaviruses [20], transient disulfide bridges are established intramolecularly, and subsequently intermolecularly, as the monomers assemble into pentamers, which facilitate the folding and interdigitation of structural elements [19]. The non-covalent interactions and covalent bonds collectively conform to a stable cytoplasmic CP pentameric complex (Fig. 11.2d), which is the major assembly intermediate in these viruses.

In parvoviruses, taking the murine MVM as a reference model, trimers of VP subunits assemble in the cytoplasm. The VP1 (82 kDa) and VP2 (63 kDa) proteins synthesized at a VP1:VP2 1:5 molar ratio assemble into two types of trimers (Fig. 11.2d), which are produced in stoichiometric amounts. The larger trimer (200 kDa) is a heterotrimer formed by one VP1 and two VP2 subunits, whereas the smaller (180 kDa) is a VP2-only homotrimer [17]. In the formation of the cytoplasmic trimer, the VP2 protein may act as a scaffolding factor assisting VP1 to acquire a proper folding, as deletion mutants of the VP1-specific region undergo extensive ubiquitination degradative reaction that can be significantly prevented by co-expression of VP2 [16]. The assembly of VP cytoplasmic trimers for these viruses was structurally supported first by the higher stability of the trimer (measured by the buried surface area on oligomer formation) as compared to putative dimers or pentamers of CP subunits, due to the multiple contacts in the intertwined loops of the subunits around the 3x symmetry axes (Fig. 11.1a). Furthermore, crosslinking experiments and mutations disrupting the intertrimer interfaces in the MVM capsid allowed the isolation of stable trimers (see below).

Fig. 11.2 (continued) the upper part of the conical depression. It then loops back to interact specifically with the inner face of VP1, forming a hairpin-like structure. (b,c: reprinted from [18], with permission). (**d**) Structure of the VP1 pentamer in the SV40 capsid. (**e**) Structure of the VP trimer in the MVM capsid. Residues involved in NLM function (see below) are highlighted. (d,e: prepared using the Pymol program (http://www.pymol.org/) and the 1SVA and 1MVM respective atomic coordinates deposited in the PDB)

11.3.2 Intracellular Traffic of CBBs and Assembly Intermediates

For capsid formation, CBBs or other assembly intermediates must accumulate in the subcellular compartment where the viral genome replicates, and at the right time. In some viral systems, the capsid subunits must traffic from the site of synthesis in the cytoplasm to the assembly compartment. This traffic is directed by protein signals in CBBs or intermediates that are accessible to the transport machineries of the cell. The nature of the signals and the configuration of the intermediate exposing them, which will determine the transport route accessed by the intermediate and its fate in the cell, are thus key elements in the viral assembly process.

In polyomaviruses and parvoviruses, the viral genome is replicated in the nucleus, and the CPs of these viruses are therefore karyophilic polypeptides that traverse the nuclear membrane. Translocation of proteins into the nucleus imposes two restrictions, firstly by the presence of nuclear localization sequences (NLS) to access the cellular transport machinery, and secondly by the size of the complex to be transported, which cannot exceed 25–30 nm, the functional diameter of the aperture in the nuclear pore complex (NPC) [21], a supramolecular structure embedded in the nuclear envelope (Fig. 11.3a). The exchange of macromolecules across the NPC is mostly mediated by proteins of the importin β (karyopherin β) superfamily, which comprises importins and exportins. The NLS are, in the so-called conventional configuration [22], single or bipartite stretches of basic amino acids that access, generally upon direct binding to a protein adaptor, the importin α/β transport pathway [23]. However, many karyophilic proteins harbor non-classical NLS that may bind karyopherin β1 directly, or access the alternative karyopherin β2/transportin import pathway. A current active area of research is the identification of the transport routes accessed by the viral proteins in specific cell hosts. Functional conventional NLS are found in the major VP1 as well as in the minor VP2/3 subunits of Polyomavirus [24, 25] and SV40 [26] (Fig. 11.3b). It is remarkable the sequence conservation among the structural proteins of these viruses at the NLS domains, compared to the high sequence divergence in the rest of these proteins. In polyomaviruses, the VP1-VP2/3 cytoplasmic complex is translocated through the NPC by the functional cooperation of the multiple NLS displayed by the VP subunits [27].

In MVM, used as a representative molecular model of the *Parvoviridae*, the two types of VP trimers are major CBBs, or stable assembly intermediates, translocating across the nuclear membrane. The protein subunits within the trimer cooperate for nuclear transport, as both VP1 and VP2 proteins genetically depleted of functional nuclear transport sequences can be co-transported into the nucleus by expressed intact subunits. Indeed, VP1 and VP2 carry independent NLS and efficiently target the nucleus of transfected cells when singly expressed [14]. VP1 harbors two conventional NLS at its N-terminal specific region (Fig. 11.3c), which function independently. These NLS are required for nuclear translocation of the expressed VP1 subunits in MVM [16] and in canine parvovirus (CPV) [28], and also for the

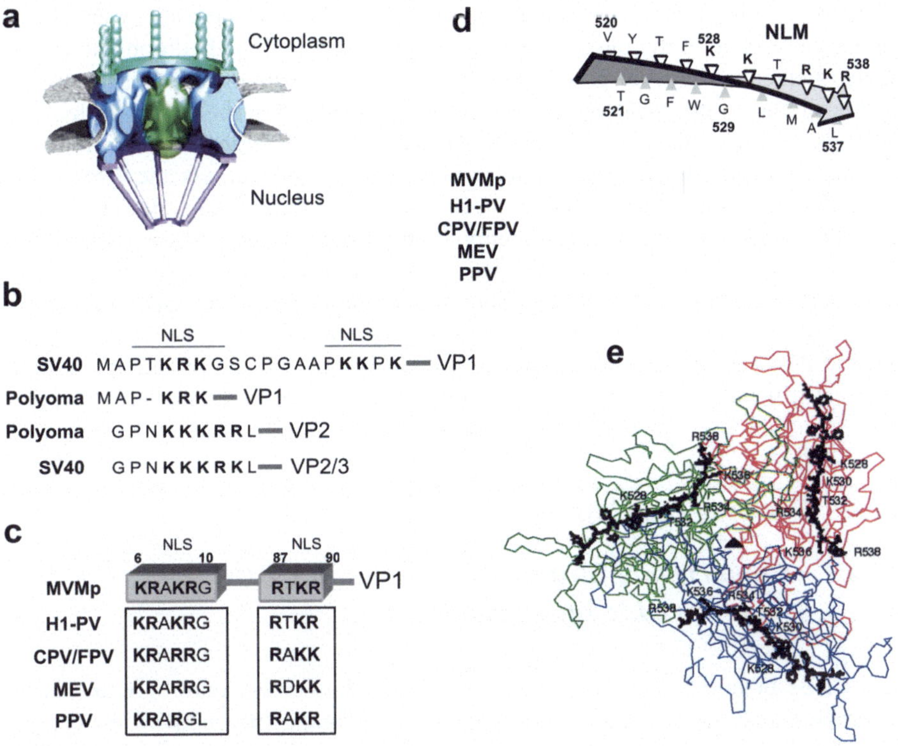

Fig. 11.3 Nuclear localization sequences in polyomavirus and parvovirus capsid proteins. (**a**) Basic architecture of the nuclear pore complex (NPC). The central ring is illustrated with the filament facing the cytoplasm and the basket protruding into the nucleus. (**b**) Examples of identified NLS in the VP proteins of Polyomavirus and SV40. (**c**) Two conserved domains with NLS activity were identified in the VP1 N-terminal sequence of several parvoviruses: MVM, parvovirus H1 (H1-PV), Canine parvovirus (CPV), feline parvovirus (FPV), mink enteritis virus (MEV), porcine parvovirus (PPV). (**d**) Configuration of the NLM in the capsid subunits of MVM and conservation of the NLM sequence in related parvoviruses. Basic residues contributing to nuclear targeting of assembly intermediates are shown in bold. (**e**) The structure of a VP-2 trimer of MVM is viewed down an icosahedral threefold axis (solid triangle) from the interior of the capsid. Residues in the exposed face of βI involved in the NLM function are labeled (Adapted from [30] with permission)

MVM virion to initiate infection [16], suggesting that they are exposed out of the virus shell during the cell entry process. Similar separate regions with basic amino acids essential for assembly and infectivity were identified in the capsid proteins of AAV [29]. In addition, both VP1 and VP2 (the major capsid-forming polypeptide) contain, in their common folded sequence a structured domain with nuclear targeting capacity named nuclear localization motive (NLM) [30]. In the assembled capsid the NLM is localized in the amphipathic ß-strand I at the inner capsid surface (Fig. 11.3d, *upper*). The NLM is the only functional nuclear targeting sequence identified in VP2. The structured NLM, highly conserved in many members of the *Parvoviridae* (Fig. 11.3d, *lower*), shows, when displayed in the capsid structure

(Fig. 11.3e), all its charged basic amino acids placed within the side of the strand facing the solvent at the interior surface of the capsid, and the hydrophobic amino acids oriented toward the protein core. Other non-conventional nuclear transport motifs similar to the NLM of MVM have been identified in the capsid proteins of other parvoviruses like parvovirus B19 [31] and PPV [32], supporting VP trimers as the major parvovirus assembly intermediate translocating across the NPC. This structuration of the NLM probably occurs only upon folding and trimerization of the VPs, which may occur concomitantly. Folding and oligomerization of the capsid subunits occur in the cytoplasm, leading to trimeric assembly intermediates acquiring nuclear transport competence. The structuration of the two types of trimers translocating across the NPC may exert a quality control role in the virus morphogenetic flow, as misfolded subunits (not exposing the NLM) or oligomers with aberrant structures and protein composition would not access the nucleus [16].

11.4 Forming the Capsid

11.4.1 Contacts and Structural Changes in CBBs and Assembly Intermediates

The CBBs or initial stable intermediates of capsid assembly may be found in an assembly-incompetent state, incapable of establishing productive interactions between them to accommodate the final configuration of the capsid [33]. Thus, the assembling oligomers must frequently undergo conformational changes during the late stages of the assembly pathway. In Poliovirus and other picornaviruses, the similar β-barrel folding of the VP1, VP2, and VP3 proteins that facilitate their interactions to form the 60 structural subunits (5S protomers) and the 14S pentameric CBBs (Fig. 11.4a), also favors the subsequent assembly of these latter intermediates into a complete viral particle [34]. For this, the extensive interactions among the β-barrels of adjacent proteins help to form a rather rigid protein shell with a dense network of intersubunit interactions. Inside the capsid, a network of additional protein contacts stabilizes the mature particle. These contacts are largely contributed by the long (40–80 residues) N-terminal arms of the three VP subunits, which have a similar path inside the capsid in different picornaviruses, although their primary amino acid sequence is drastically different. These interactions are most extensive at the 5x axes, where the N-termini of five VP3 molecules are arranged in a tube-like parallel β-sheet. The conformation of these arms is ordered only after capsid assembly, so it is difficult to evaluate the contribution of the interactions involving these arms in the final particle stability based on structural data only.

In members of the *Parvoviridae*, capsid formation in the natural infection starts in the nucleus. Indeed, the nucleolus was identified as the subcellular compartment where AAV assembly initiates, colocalizing with the non-structural replicative viral Rep proteins; subsequently, capsid accumulation spreads across the entire nucleus

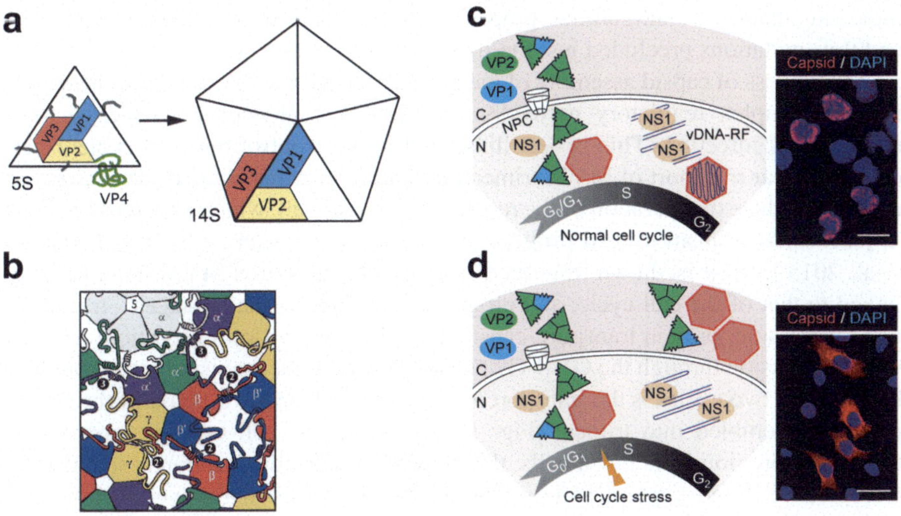

Fig. 11.4 Interactions among assembly intermediates to form the capsid. (**a**) Assembly of the 5S protomer into a 14S pentamer in Poliovirus. The protomer is not identical to the icosahedral asymmetric unit in the capsid. (**b**) Contacts between capsomers in SV40. Four pentamers assembled in the SV40 capsid are shown. Each subunit of the pentamers projects an arm that makes different contacts with the jellyroll of the subunit in another pentamer (Adapted from [7], with permission). (**c, d**) Capsid formation in Parvovirus. The figure shows a model for the temporal and functional coupling of the MVM parvovirus life cycle with the cell cycle. (c) MVM life cycle stages in a cell without cell cycle constraints. Right: assembled virus capsids (red) within the nucleus (blue) visualized by immunofluorescence. (d) Cells under cell cycle stress show misassembly of capsids (red) in the cytoplasm with nuclear exclusion (blue). Acronyms: C: cytoplasm; N: nucleus; NPC: nuclear pore complex; NS1, non-structural viral protein 1; VP1, VP2, structural capsid proteins; vDNA-RF, replicative intermediates of the virus genome (Adapted from [35])

[36]. In MVM, as the assembly intermediates (mainly trimers) accumulate in the nucleus, the nucleation reaction is triggered and multiple non-covalent interactions are established between amino acids localized at the edges of the binding trimers. A few evolutionary conserved residues involved in presumably strong intertrimer contacts were found to be necessary for capsid assembly. These residues buried a large hydrophobic surface upon trimer association or formed buried intertrimer hydrogen bonds or salt bridges [37]. Assembly intermediates other than the CBBs are difficult to isolate in parvoviruses, presumably because of the high efficiency of the assembly reaction in which intermediates are highly transient and accumulate at very low levels, in agreement with theoretical studies on the assembly of structurally simple virus capsids (see Chap. 19). A genetic approach to trap unstable assembly intermediates based on the introduction of disruptive mutations at some of the residues that in the capsid are involved in contacts between MVM trimers yielded high amounts of trimers, without evidence of any other larger intermediate [17]. These trimers could be isolated by sedimentation in sucrose gradients and were competent in a nuclear transport assay performed in permeabilized cells [38],

indicating that their configuration properly exposed the nuclear transport sequences, and thus mutations precluded just the final capsid formation stage.

The process of capsid assembly of many small icosahedral viruses may, however, involve complex regulatory networks with cell host physiology, which in turn restrict virus infection. This is best illustrated by the control that cell cycle exerts on the nuclear transport of MVM trimeric assembly intermediates. Parvoviruses are unable to induce proliferation of the resting (quiescent) cell, which explains—at least in part—their tropism toward proliferative tissues and cancer cells. In Gil-Ranedo et al. 2015 [35], it is shown how the life cycle of parvovirus MVM is exquisitely linked to that of the cell cycle, and progresses through its major stages—transcription, expression, protein transport, capsid assembly, and virion maturation—in step with progression through the G1-S-G2 phases. A central mechanism for the coupling between the two cycles is the transport through the NPC in the S phase of the capsid proteins assembled into trimers (Fig. 10.4c). When the cell cycle progression is blocked in previously infected cells, the trimers nuclear translocation is specifically disrupted, leading to cytoplasmic accumulation of capsids that prevents virus maturation and results in abortive infection (Fig. 10.4d). This tight regulation exerted by the cell cycle on trimer nuclear transport and thus MVM capsid assembly may be a key process in the tropism of parvoviruses to proliferative and tumor cells.

Inside the nucleus, MVM trimers must interact to form larger subviral intermediates through an assembly pathway that is poorly characterized. Evidences of a pattern of supra-trimers assembly was supported by in vitro studies on the reassembly pathway of dissociated VLPs that yielded, besides incomplete capsids, large off-pathway products of unclear biological significance [39], and the formation of non-productive supraviral assemble in transfected cells expressing capsid subunits in the context of the whole MVM genome that was genetically manipulated with exogenous peptides [40]. Final stages of MVM capsid assembly involve further conformational changes of the trimers that have been indirectly detected. An MVM-induced monoclonal antibody recognizing a discontinuous epitope located at each 3x axis of the capsid (which corresponds to the center of each trimer) failed to react with non-assembled trimers accumulated in the cell nucleus or isolated in vitro [17, 38]. Trimers must therefore change their conformation during nuclear capsid assembly, and the process may be triggered by some external factors and/or phosphorylation of the subunits (see below). These induced conformational rearrangements would reorganize some residues located at the vertex of the threefold axes, creating a new epitope.

In Polyomavirus, the VP1 protein can self-assemble into pentamers in heterologous cell expression systems [41], and the purified pentamers can associate to form capsid-like assemblies (VLPs) that become stabilized at low ionic strength by calcium. Interestingly, these VLPs assemble without the minor virion protein components, VP2 and VP3, suggesting that the non-equivalently related subunits of the capsomeres may spontaneously switch their bonding specificity during assembly. VP1-only capsids, however, do not represent the physiological assembly pathway observed in natural virus infections, in which the assembly of the VP1 pentamers is driven upon regular interactions involving VP2 and VP3 with the viral genome DNA (see below), and empty capsids are thought to be minor abortive assembly by-products. In the viral particle of Polyomavirus and SV40, all the capsomers are pentamers, even though the capsid shell is built by subunits arranged in an icosahedral T = 7 lattice. Therefore, the architecture of this virus does not fulfill the arrangement of pentameric and hexameric capsomers predicted by the quasi-equivalent theory for capsids with T > 1 value ([2]; see Sect. 11.2.1). However, the SV40 capsid is stable enough, and this is accomplished in part by virtue of unique intercapsomer bonds established during assembly by the C-terminal domain of the VP1 protein (Fig. 11.4b). This domain acts as a connecting arm which, under six different configurations, makes contact between the pentavalent pentamers (around each 5x axis) and the hexavalent pentamers (around each 3x axis). The different types of contacts in which the arms are involved act as multiple clamps holding the subunits together, and thus ensuring the stability of the capsid shell [7].

11.4.2 Compartments, Factors, and Protein Modifications Influencing Capsid Formation

Icosahedral capsid assembly is influenced by environmental conditions in the cell and also by multiple molecular factors (in addition to the VP protein themselves) that may be encoded by the cells or by the virus genomes. These factors may act at specific subcellular compartments, and their contribution may critically determine in many cases the efficiency of the assembly process in vivo. Some of these factors have been identified in several viral systems, although multiple experimental evidences suggest that many others remain to be identified. In Poliovirus for instance, the rate of assembly of the structural proteins in vitro is reduced by at least two orders of magnitude compared to the rate observed in infected cells, and the empty capsids formed showed altered conformation, unless the reaction is primed by 14S pentamers isolated from infected cells. This experiment implies that proper folding, interactions, and/or modifications of the proteins forming the pentameric 14S intermediate, are essential for assembly to proceed successfully. A candidate for this function is the cellular Hsp70 chaperone, which associates with the P1 polyprotein precursor during its folding to form the 5S structural unit.

The cellular modulations of virus assembly may involve complex signaling pathways. For parvoviruses, capsid assembly during natural infections occurs with high efficiency in the nuclear compartment of infected host cells. However, virus-like particles (VLPs) devoid of nucleic acid may be formed in the cytoplasm of heterologous expression systems (e.g., recombinant baculovirus expressing VP2 of MVM in insect cells) at low efficiency [42, 43]. This distinct assembly efficiency may be accounted, at least in part, by post-translational modifications of the capsid subunits. In MVM-infected cells, VPs and native capsids become extensively post-translationally modified by phosphorylation [44], whereas VLPs purified from heterologous insect cells were not phosphorylated [38]. In spite of the absence of

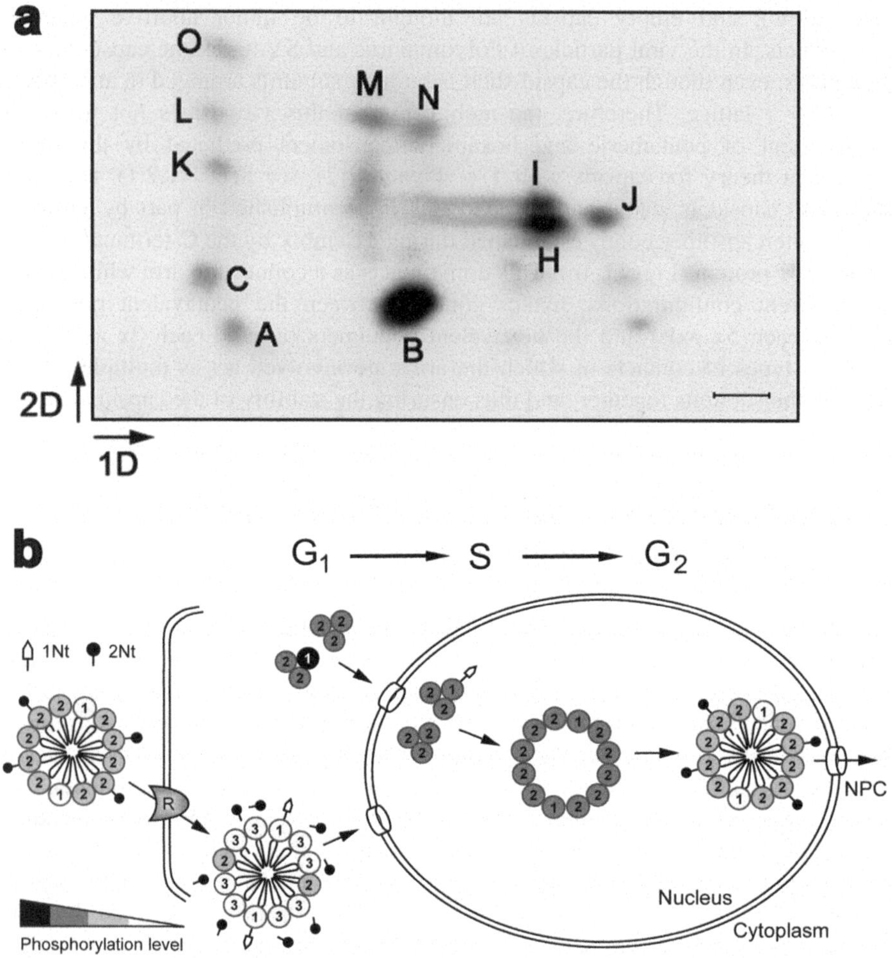

Fig. 11.5 Capsid protein phosphorylation in the parvovirus MVM life cycle. (**a**) Two-dimensional tryptic phosphopeptide maps of the VP2 subunits isolated from in vivo 32P-labeled purified empty capsids. Main phosphopeptides were designed as previously reported [44]. 1D, first dimension; 2D, second dimension (Reproduced from [38], with permission) (b). Phosphorylation levels and exposure of the Nt-domains of the capsid proteins in the MVM life cycle are shown. The G1-S-G2 cell cycle steps at which virus assembly and maturation preferably occur are indicated above. The VP capsid subunits in the viral particles and assembly intermediates are depicted at their estimated stoichiometry. The VP1-Nt and VP2-Nt domains are only illustrated in exposed configuration(s) accessible to antibodies (Reproduced from [35], with permission)

modifications, these VLPs showed a 3D structure identical to the native capsid and virus [4, 42], indicating that phosphorylation is not important for the icosahedral T = 1 ordering. The 2D-tryptic phosphopeptides analysis of native MVM capsid subunits resulted in a complex pattern of phosphoserine and phosphothreonine residues (Fig. 11.5a) which was different for VP1 compared to VP2

[44]. Interestingly, phosphorylation may control the exposure of trafficking domain in assembly intermediates and viral particles (Fig. 11.5b). The MVM cytoplasmic trimeric intermediates harbored a VP1 phosphorylation level fivefold higher than that of VP2, and Nt-sequences non-accessible to antibodies, whereas upon nuclear translocation VP1-Nt became exposed in the subviral assembly intermediates. However, nuclear empty capsids, which are traffic neutral, showed the Nt concealed in their interior and a phosphorylation level restored to the VP1:VP2 stoichiometry. Finally, the newly formed ssDNA-filled virus maturing lacked VP1 phosphorylation and one major VP2 phosphopeptide but exposed the highly phosphorylated VP2-Nt. As in the subsequent infection the VP2-Nt is cleaved in the endosome, the incoming viral particles harbors a much lower phosphorylation status than the exit virion. Furthermore, VP phosphorylation may couple parvovirus nuclear assembly to the cell cycle. The phosphorylation of the MVM capsid subunits was mainly catalyzed by the cytoplasmic activity of the Raf-1 kinase of the MAPK signaling pathway [38], and this modification was crucial for the acquisition of nuclear transport competence by the trimers. Phosphorylation by cellular kinases may be a general strategy to connect capsid assembly with host cell physiology, ensuring a spatially and timely regulated process maximizing virus yield.

As in parvoviruses, in Polyomavirus the efficiency of nuclear assembly is regulated by phosphorylation of the capsid subunits. Polyomavirus is highly tumorigenic in mouse, but host range transforming (hr-t) mutants of this virus defective in tumor induction are blocked in virion assembly when infecting non-permissive cells [45], although viral DNA and capsid proteins are synthesized to wild-type levels. In purified Polyomavirus particles, the several VP1 species identified by 2-D electrophoresis are generated by acetylation and phosphorylation of threonine and serine residues of the initial translation product. The hr-t mutants failed to assemble the complete (240S) viral particle, correlating with the lack of acidic forms of VP1 [46]. These acidic forms resulted from phosphorylation of threonine residues, and at least one of the phospho-threonines was shown to be essential for the encapsidation of the viral minichromosome [47]. In SV40, phosphate groups are added in natural infections on serine and threonine residues flanking the NLS of the VP proteins. Phosphorylation was shown to act on the activity of these NLS indirectly regulating the nuclear accumulation of VPs and virus assembly.

At least two examples of factors encoded by the viral genome favoring assembly of small icosahedral capsids have been characterized. The genome of the polyomaviruses SV40 and JC virus (JCV), the latter causing a fatal demyelinating disease named progressive multifocal leukoencephalopathy (PML), encodes six major proteins including a small auxiliary protein known as agnoprotein, which contributes to viral propagation at various stages in the replication cycle. Best characterized agnoprotein functions are interaction with VP1 pentamers facilitating assembly and ordering morphogenesis of the virion particles [48], and a viroporin-like property [49] that increases membrane permeability leading to an enhancement of virus release [50]. The second example concerns the capsid assembly of the parvovirus AAV, which begins in the nucleolus and spreads throughout the nucleus at later stages of infection. In addition to the three capsid proteins, VP1, VP2, and

VP3, the *cap* gene also encodes, by using an alternative open reading frame, the assembly-activating protein (AAP) that is essential for capsid assembly [36] of many AAV serotypes [51]. The AAP harbors nuclear (NLS) and nucleolar (NoLS) localization signals near the carboxy-terminal region [52], and targets newly synthesized capsid proteins to the nucleolus, where becomes stabilized upon co-expression of the capsid protein VP3 of the same virus serotype, or from an AAV serotype of the same assembly group. The assembly-promoting activity of AAP is mediated by interaction between two hydrophobic domains in the N-terminal region of the molecule with the C-terminus of the VP proteins, which forms the capsid protein interface at the 2x symmetry axes [53]. AAP acts as a scaffolding factor with a modular architecture [54] able to change the conformation of non-assembled VP molecules.

11.5 Genome Encapsidation and Virus Maturation

The pathways of genome packaging adopted by some icosahedral viruses have been, and continue to be, a matter of debate, as it is quite difficult in many cases to distinguish between "concerted assembly," in which the capsid is formed as a result of the ordered association of the protein subunits with the genome, from the "sequential assembly" that occurs when the viral genome is encapsidated into a preformed capsid, (see Chap. 13 for detailed descriptions of mechanisms of encapsidation referred mostly to structurally complex viruses).

11.5.1 Poliovirus Cytoplasmic Maturation

In the *Picornaviridae*, as illustrated in Fig. 10.6a, most evidences support a "concerted" assembly of the 14S pentamer condensing around the RNA genome. Alternatively, an empty procapsid containing 60 copies of the VP0-VP3-VP1 structural unit could be transiently required at low concentrations for packaging [13], but recent evidence substract support to this hypothetical mechanism. Rather, in *Enterovirus*, a *Picornavirus* genus including important human pathogens, many conserved RNA sites along the genome binding viral capsid proteins were identified by bioinformatics and validated by Cryo-electron microscopy and mutagenesis [55]. The data support these sites being RNA Packaging Signals triggering assembly, strongly supporting a concerted capsid assembly model for the entire *Picornaviridae* family of viruses. However, the assembly proceeds through a complex remodeling of cytoplasmic membranes, which are required for viral genome replication, and then the virion assemblies are loaded into autophagic membranes for release through secretory autophagy [56]. Cryogenic electron tomography of infected cells showed membrane-bound half-capsid as a prominent virion assembly intermediate, and that the virus-induced autophagy selects RNA-containing virions over empty capsids, and segregates them from other types of autophagosome contents (Fig. 11.6b).

Furthermore, as characterized by Poliovirus, the maturing virion must undergo a number of modifications prior to becoming infectious [13]. Maturation involves lipid modification and proteolytic cleavage at a specific site, making assembly an irreversible process. Along the final stages of assembly, a molecule of the fatty acid myristate is added post-translationally to the N-terminus of each VP4 subunit. This lipid mediates the interaction of the β-sheet formed by VP3 N-termini with a second β-sheet structure containing strands contributed by both VP4 and VP1 molecules. This feature of the capsid does not form until final stages of the maturation, when proteolytic processing liberates VP2 and VP4 from their precursor VP0, and this reaction is associated with a significant increase in the stability of the viral particle, priming it for entry into a new host cell.

11.5.2 Polymorphic Nuclear Maturation of Polyomaviruses

To form a Polyomavirus virion, the VP1-pentamers become associated with single copies of a minor capsid protein (either VP2 or VP3) and, once imported into the nucleus, interact with the replicating DNA. Although the virus structure is known to atomic resolution, how the virions mature in the nucleus during productive infections is only partly understood. In a stepwise model, the capsid proteins would be sequentially added to and arranged on the viral DNA, resulting in its condensation and packaging to form the virion. The process implies multiple molecular recognition events of the viral genome by the CP subunits, in which DNA sequences near the viral origin of replication and of the T-antigen (major replicating non-structural Polyomavirus protein) play important functions. Several DNA-binding domains (DBD) localized in all three VP proteins contribute to the packaging process. Major DBD of VP2 and VP3 were found in their C-terminal segments, whereas the DBD of VP1 was localized overlapping with the N-terminal bipartite NLS [59]. This VP1 N-terminal sequence of 15 amino acids is not visible in the virion crystal due to structural disorder, but it likely extends into the virion core to interact with the viral DNA.

The dynamics of maturation of the Polyomavirus and SV40 virions is not precisely known yet, but most evidences suggest a scaffolding role of the viral genome for the polymerization of CP subunits. In vitro assembly systems showed intact supercoiled DNA molecules being packaged with disassembled VLPs (Fig. 11.6c) into particles that become infectious, supporting a model of concerted assembly in which multiple VP1 pentamers simultaneously bind to the DNA, which becomes compacted and serves as a scaffold for assembly [57]. Also important for assembly is the capacity of the bound pentamers to increase their effective size due to the length and flexibility of the C-terminal ligands, which can connect to other VP1 pentamers and recruit them to a partially completed capsid [60]. Unfortunately, intermediates of the assembly reaction could not be isolated from infected cells,

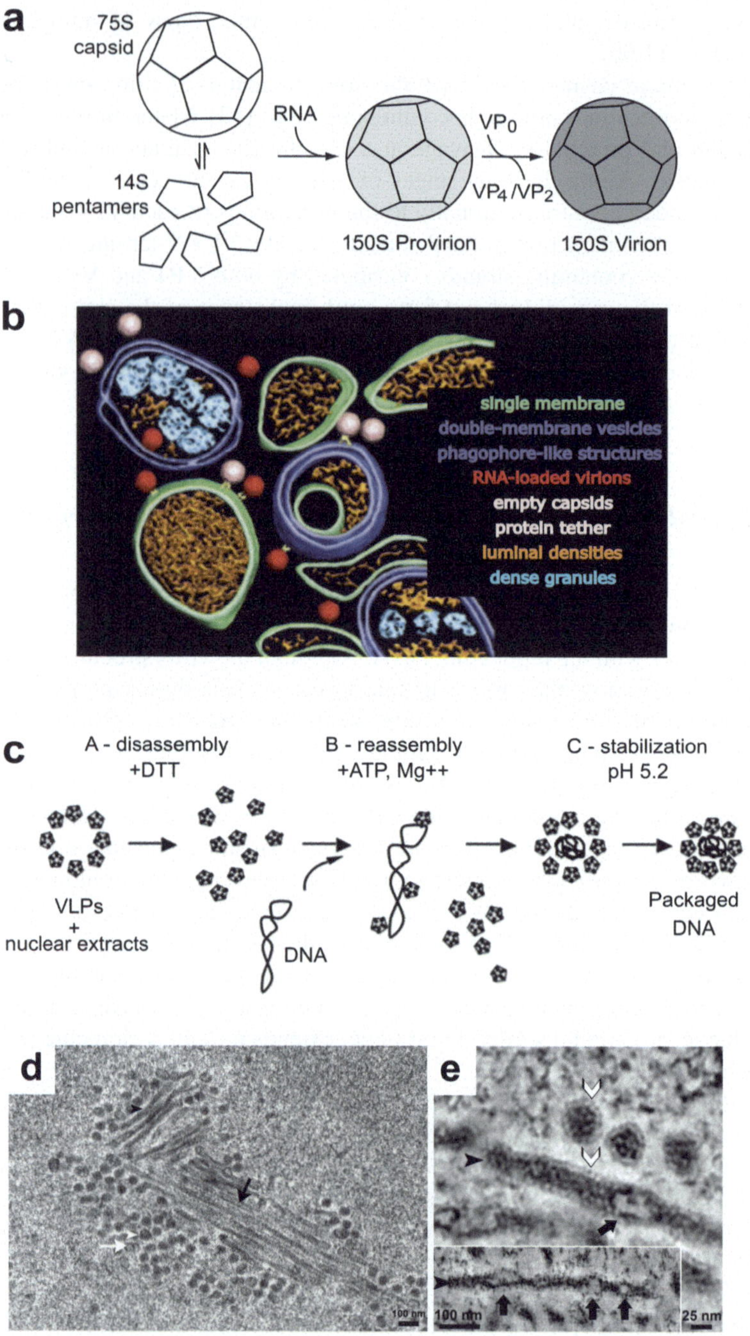

Fig. 11.6 Concerted virus assembly and maturation. (**a**) Late stages of Poliovirus assembly and virion maturation cleavage [34]. (**b**) Segmentation of a cryogenic electron tomography of poliovirus replication and assembly sites in situ showing virus-induced membrane proliferation. Empty capsids (pink) and RNA-loaded virions (red) are represented by their subtomogram averages (Reproduced from [56], with permission). (**c**). Model of SV40 assembly and maturation in vitro.

but the final nuclear assembly process seems to proceed through large, polymorphic structures that may serve as virus factories (Fig. 11.6d, e) (see Chap. 8 for a description of virus factories). Electron microscopy reveals tubular structures in the nucleus adjacent to clusters of virions. They share an organization consisting of a protein shell surrounding an electron-dense DNA core suggesting that they are the main factories able to yield mature virions budding from their ends, although the resolution mechanism is unknown.

11.5.3 Viral Genome Encapsidation into a Preformed Parvovirus Capsid

The final stages of parvovirus virions assembly and maturation are a subject of current active research, but in the best-studied system, the human AAV parvovirus, they proceed through the formation of large amounts of empty capsids in the nucleus, which serve as preformed substrate for active genome encapsidation. In MVM, empty capsids are first detected accumulated in the nucleus of synchronized cells about 2 hours preceding the accumulation of DNA-filled virions. A consensus model for parvovirus maturation, based on the available current data obtained in the AAV and MVM systems [61, 62] is outlined in Fig. 11.7, although many details of the ssDNA packaging mechanism are still poorly understood.

The reaction proceeds through a "packaging stage" in which the major non-structural proteins, NS1 in the *Protoparvovirus* genus and Rep 78/52 proteins in the *Dependoparvovirus* AAV, mediate the association of the empty capsid with DNA replicative intermediates. These multifunctional proteins harbor helicase, DNA binding, and endonuclease activities that are critical for genome replication, and form oligomers in vitro in the presence of ATP. A Rep/NS1 oligomeric packaging motor is probably formed at the fivefold symmetry axis of the capsid, injecting the ssDNA genome through the pore (Fig. 11.7a), which acts as the portal for genome encapsidation. For the packaging reaction, the helicase activity of the non-structural proteins is essential. High-resolution structures for the Rep68 oligomers conforming the helicase domain and the DNA origin-binding domain are available [64], which explains the functioning of this molecular motor

Fig. 11.6 (continued) The addition of supercoiled DNA to disassembled Virus-Like Particles (VLPs) allows pentamers binding to viral dsDNA facilitating concerted assembly and DNA condensation facilitated by presence of chaperones in the nuclear extracts (Reproduced from [57], with permission). (**d**) Organization of the Polyomavirus factory. Electron microscopy of plastic-embedded thin sections of the nucleus of mouse fibroblasts showing tubular structures adjacent to a virus cluster (black arrowhead, "full" tubular structure; white arrowhead, full virion; black arrow, "empty" tubular structure; white arrow, empty virion). (**e**) Spherical virions and tubular structures seen in the nucleus of Polyomavirus-infected cells show a lighter and well-arranged density at their periphery (white arrowhead) corresponding to capsid protein density, and a dense core (black arrowhead) suggesting DNA (**d,e**: adapted from [58])

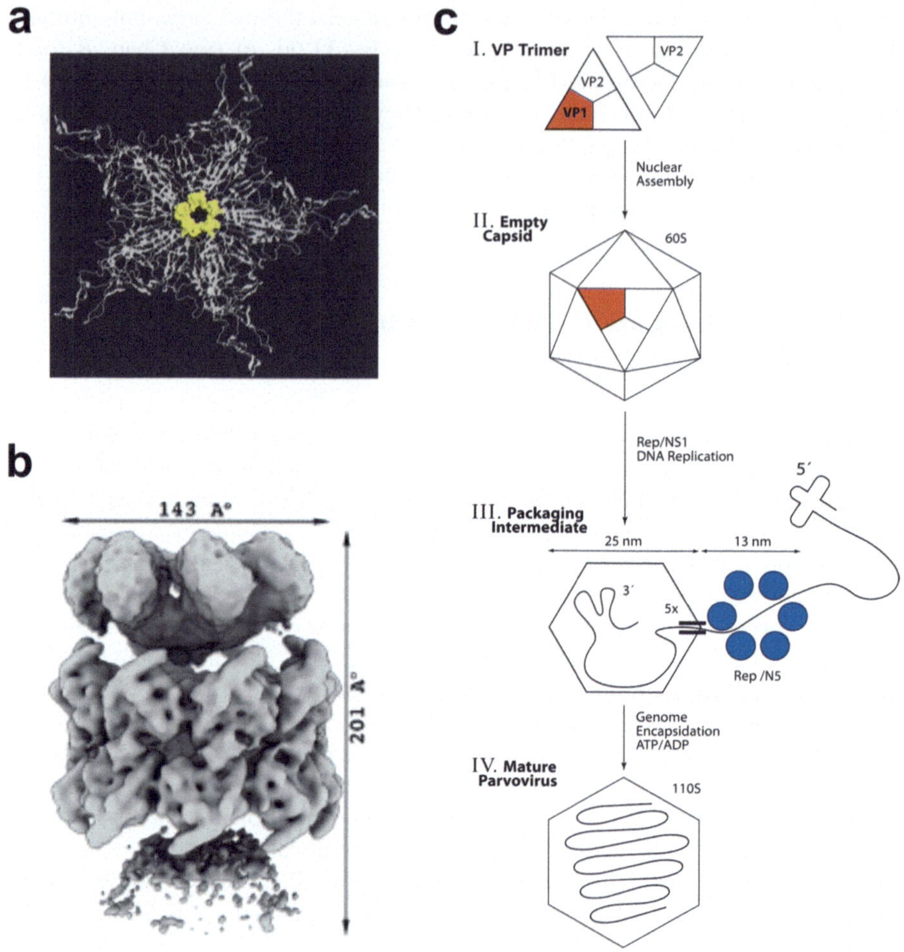

Fig. 11.7 Genome encapsidation into a preformed capsid. (**a**). A backbone model of a VP pentamer of parvovirus AAV showing the pore at the fivefold axis from the outside of the capsid (reproduced with permission from [63]) (**b**). Figure of a 3D Cryo-EM map of the Rep68 protein of AAV in complex with ssDNA showing the indicated dimensions. Upper oligomer corresponds to the helicase domain, and the lower oligomer to the origin binding domain of Rep68 (Reproduced with permission from [64]) (**c**). Integrative model of parvovirus assembly and maturation: I. VP capsid subunits assemble into trimers in the cytoplasm (top image). II. Immature empty capsids (60S) form in the nucleus (top center). III. Genome packaging reaction (bottom center). IV. Mature viral particle (110S) (bottom image)

(Fig. 11.7b). Genome encapsidation operates in the 3′ to 5’ direction, mediated by specific contacts with the resolving hairpins and ATP hydrolysis (Fig. 11.7c). Encapsidation is coupled to DNA replication, ending as the capsid is filled with the complete ssDNA genome, which establishes regular contacts with some internal capsid residues and becomes icosahedrally ordered along certain sequences. The

Rep/NS1 subunits may remain bound to the genome outside of the viral particle. The reader is referred to Chap. 13 for comparisons with equivalent mechanisms of dsDNA and dsRNA packaging into preformed bacteriophage capsids.

11.6 Perspectives and Conclusions

The assembly of small icosahedral viruses proceeds through elegant genetically encoded and virus-specific pathways, which direct the self-association of the synthesized asymmetric protein subunits into a symmetric viral capsid shell. The general mechanic of capsid assembly involves first the oligomerization of CP subunits into early stable intermediates, usually of one major type per virus, which constitute the building blocks (CBBs) from which the capsid is later assembled. Capsid assembly then proceeds along a pathway regulated by ordered interactions between activated CBBs, most likely through a cascade of second-order reactions (see Chap. 19). The general architecture and the organization of the capsid subunits in the intermediates resemble those in the final capsid. However, the assembly process implies further structural rearrangements within and between the capsid subunits, including the establishment of multiple inter-protein non-covalent interactions, protein cleavages, and covalent bonds (in some cases), and intracellular traffic of the assembly intermediates, prior to their final condensation into a closed container. Assembly is a dynamic process, implying that many of these interactions among subunits may have a transient morphogenetic role, and will not be preserved in the assembled virion. Functional capsid assembly cannot be understood without concomitant or later packaging of the genome, a complex process that viruses solve through molecular recognition patterns between the replicating genomes and either the assembly intermediates or a preformed empty capsid (see Chap. 13). During the final stages of morphogenesis some viruses may need a maturation step through limited cleavage of some viral particle subunits that leads to an infectious virion (see Chap. 14 for a description of maturation strategies in different viruses, including structurally complex bacteriophages). Virion morphogenesis in the cell, even for small viruses, is frequently mediated by additional viral and or cellular factors, and may involve particular subcellular factories (see Chap. 8). The efficiency of virus assembly critically determines some important features of viral fitness, such as the virus yield per host cell, or the acquisition of enough stability to prevail in natural environments.

The intensive research on small, structurally simple icosahedral viruses over the past decades have provided detailed information on the atomic structure of viral particles and on the overall organization and mechanisms of the assembly process in vitro and in the host cells. Despite these advances, our knowledge on some crucial steps of the assembly of even structurally simple viruses is still poor or unclear in many respects, as exemplified by the following unsolved problems: the detailed structure of assembly intermediates in picornaviruses and parvoviruses; the factors and transporters regulating the traffic of assembly intermediates in the cell; the enzymology and precise composition of the Rep/NS–Capsid–dsDNA complex mediating parvovirus genome encapsidation; the dynamic processes of protein–protein and protein–nucleic acid recognition leading to the maturation of Poliovirus and Polyomavirus; or the mechanistic resolution of the mature virions from the Polyomavirus factories. Ongoing research on these and other subjects related to the morphogenesis of small viruses promises exciting insights into essential aspects of the structural biology of viruses in the near future.

Acknowledgments ALB and JGR equally contributed to this chapter. We gratefully acknowledge the past dear collaborative work of Mavis Agbandje-Mckenna (University of Florida) on parvovirus MVMp capsid structure, and of Michael Kann (Gothenburg, Sweden) and Mauricio G. Mateu (CBMSO, Universidad Autónoma de Madrid) on assembly. The many inspiring conversations along the years with regular participants to the "Virus Assembly" FASEB meeting (Vermont Academy, USA), in particular with Peter Tattersall (Yale University, Connecticut), are also deeply acknowledged. This review stands from the enthusiastic work in our laboratory of the following students and researchers: Eva Hernando, Eleuterio Lombardo, Beatriz Maroto, Juan C. Ramírez, Laura Riolobos, Noelia Valle, Esther Grueso, Cristina Sánchez-Martínez, Nooshin Bayat, Carlos Gallego, and Tania Calvo-López. Research in JM Almendral's laboratory is currently supported by grants PID2019-111146RB-I00 and PID2022-141799OB-I00 from the Spanish Ministerio de Ciencia e Innovación.

References

1. Cotmore SF, Agbandje-McKenna M, Chiorini JA, Mukha DV, Pintel DJ, Qiu J, Soderlund-Venermo M, Tattersall P, Tijssen P, Gatherer D, Davison AJ (2014) The family Parvoviridae. Arch Virol 159:1239–1247
2. Caspar DLD, Klug A (1962) Physical principles in the construction of regular viruses. Cold Spring Harb Symp Quant Biol 27:1–24
3. Agbandje-McKenna M, Llamas-Saiz AL, Wang F, Tattersall P, Rossmann MG (1998) Functional implications of the structure of the murine parvovirus minute virus of mice. Structure 6: 1369–1381
4. Kontou M, Govindasamy L, Nam H-J, Bryant N, Llamas-Saiz AL, Foces-Foces C, Hernando E, Rubio M-P, McKenna R, Almendral JM, Agbandje-McKenna M (2005) Structural determinants of tissue tropism and in vivo pathogenicity for the parvovirus minute virus of mice. J Virol 79:10931–10943
5. Rossmann MG, Arnold E, Erickson JW, Frankenberger EA, Griffith JP, Hecht H-J, Johnson JE, Kamer G, Luo M, Mosser AG, Rueckert RR, Sherry B, Vriend G (1985) Structure of a human common cold virus and functional relationship to other picornaviruses. Nature 317:145–153
6. Hogle JM, Chow M, Filman DJ (1985) Three-dimensional structure of poliovirus at 2.9 Å resolution. Science 229:1358–1365
7. Liddington RC, Yan Y, Moulai J, Sahli R, Benjamin TL, Harrison SC (1991) Structure of simian virus 40 at 3.8-Å resolution. Nature 354:278–284
8. Natarajan P, Lander GC, Shepherd CM, Reddy VS, Brooks CL, Johnson JE (2005) Exploring icosahedral virus structures with VIPER. Nat Rev Microbiol 3:809–817
9. Tsao J, Chapman MS, Agbandje M, Keller W, Smith K, Wu H, Luo M, Smith TJ, Rossmann MG, Compans RW, Parrish CR (1991) The three-dimensional structure of canine parvovirus and its functional implications. Science 251:1456–1464

10. Xie Q, Bu W, Bhatia S, Hare J, Somasundaram T, Azzi A, Chapman MS (2002) The atomic structure of adeno-associated virus (AAV-2), a vector for human gene therapy. Proc Natl Acad Sci USA 99:10405–10410
11. Gurda BL, Parent KN, Bladek H, Sinkovits RS, DiMattia MA, Rence C, Castro A, McKenna R, Olson N, Brown K, Baker TS, Agbandje-McKenna M (2010) Human Bocavirus capsid structure: insights into the structural repertoire of the Parvoviridae. J Virol 84:5880–5889
12. Kaufmann B, Simpson AA, Rossmann MG (2004) The structure of human parvovirus B19. Proc Natl Acad Sci USA 101:11628–11633
13. Racaniello VR (2013) Picornaviridae: the viruses and their replication. In: Knipe DM, Howley PM (eds) Fields Virology, 6th edn. Lipincott Williams and Wilkins, Philadelphia, USA, pp 453–489
14. Tullis GE, Burger LR, Pintel DJ (1993) The minor capsid protein VP1 of the autonomous parvovirus minute virus of mice is dispensable for encapsidation of progeny single-stranded DNA but is required for infectivity. J Virol 67:131–141
15. Zádori Z, Szelei J, Lacoste M-C, Li Y, Gariépy S, Raymond P, Allaire M, Nabi IR, Tijssen P (2001) A viral phospholipase A2 is required for parvovirus infectivity. Dev Cell 1:291–302
16. Lombardo E, Ramírez JC, Garcia J, Almendral JM (2002) Complementary roles of multiple nuclear targeting signals in the capsid proteins of the parvovirus minute virus of mice during assembly and onset of infection. J Virol 76:7049–7059
17. Riolobos L, Reguera J, Mateu MG, Almendral JM (2006) Nuclear transport of trimeric assembly intermediates exerts a morphogenetic control on the icosahedral parvovirus capsid. J Mol Biol 357:1026–1038
18. Chen XS (1998) Interaction of polyomavirus internal protein VP2 with the major capsid protein VP1 and implications for participation of VP2 in viral entry. EMBO J 17:3233–3240
19. Li PP, Nakanishi A, Clark SW, Kasamatsu H (2002) Formation of transitory intrachain and interchain disulfide bonds accompanies the folding and oligomerization of simian virus 40 Vp1 in the cytoplasm. Proc Natl Acad Sci USA 99:1353–1358
20. Kobayashi S, Suzuki T, Igarashi M, Orba Y, Ohtake N, Nagakawa K, Niikura K, Kimura T, Kasamatsu H, Sawa H (2013) Cysteine residues in the major capsid protein, Vp1, of the JC virus are important for protein stability and oligomer formation. PLoS One 8:e76668
21. Raices M, D'Angelo MA (2012) Nuclear pore complex composition: a new regulator of tissue-specific and developmental functions. Nat Rev Mol Cell Biol 13:687–699
22. Robbins J, Dilwortht SM, Laskey RA, Dingwall C (1991) Two interdependent basic domains in nucleoplasmin nuclear targeting sequence: identification of a class of bipartite nuclear targeting sequence. Cell 64:615–623
23. Mattaj IW, Englmeier L (1998) Nucleocytoplasmic transport: the soluble phase. Annu Rev Biochem 67:265–306
24. Chang D, Haynes JI, Brady JN, Consigli RA (1992) Identification of a nuclear localization sequence in the polyomavirus capsid protein VP2. Virology 191:978–983
25. Moreland RB, Garcea RL (1991) Characterization of a nuclear localization sequence in the polyomavirus capsid protein VP1. Virology 185:513–518
26. Ishii N, Minami N, Chen EY, Medina AL, Chico MM, Kasamatsu H (1996) Analysis of a nuclear localization signal of simian virus 40 major capsid protein Vp1. J Virol 70:1317–1322
27. Ishii N, Nakanishi A, Yamada M, Macalalad MH, Kasamatsu H (1994) Functional complementation of nuclear targeting-defective mutants of simian virus 40 structural proteins. J Virol 68:8209–8216
28. Vihinen-Ranta M, Wang D, Weichert WS, Parrish CR (2002) The VP1 N-terminal sequence of canine parvovirus affects nuclear transport of capsids and efficient cell infection. J Virol 76: 1884–1891
29. Grieger JC, Snowdy S, Samulski RJ (2006) Separate basic region motifs within the adeno-associated virus capsid proteins are essential for infectivity and assembly. J Virol 80:5199–5210

30. Lombardo E, Ramírez JC, Agbandje-McKenna M, Almendral JM (2000) A beta-stranded motif drives capsid protein oligomers of the parvovirus minute virus of mice into the nucleus for viral assembly. J Virol 74:3804–3814
31. Pillet S, Annan Z, Fichelson S, Morinet F (2003) Identification of a nonconventional motif necessary for the nuclear import of the human parvovirus B19 major capsid protein (VP2). Virology 306:25–32
32. Boisvert M, Bouchard-Lévesque V, Fernandes S, Tijssen P (2014) Classic nuclear localization signals and a novel nuclear localization motif are required for nuclear transport of porcine parvovirus capsid proteins. J Virol 88:11748–11759
33. Johnson JE (1996) Functional implications of protein-protein interactions in icosahedral viruses. Proc Natl Acad Sci USA 93:27–33
34. Hogle JM (2002) Poliovirus cell entry: common structural themes in viral cell entry pathways. Ann Rev Microbiol 56:677–702
35. Gil-Ranedo J, Hernando E, Riolobos L, Domínguez C, Kann M, Almendral JM (2015) The mammalian cell cycle regulates parvovirus nuclear capsid assembly. PLoS Pathog 11:e1004920
36. Sonntag F, Schmidt K, Kleinschmidt JA (2010) A viral assembly factor promotes AAV2 capsid formation in the nucleolus. Proc Natl Acad Sci USA 107:10220–10225
37. Reguera J, Carreira A, Riolobos L, Almendral JM, Mateu MG (2004) Role of interfacial amino acid residues in assembly, stability, and conformation of a spherical virus capsid. Proc Natl Acad Sci USA 101:2724–2729
38. Riolobos L, Valle N, Hernando E, Maroto B, Kann M, Almendral JM (2010) Viral oncolysis that targets Raf-1 signaling control of nuclear transport. J Virol 84:2090–2099
39. Medrano M, Fuertes MÁ, Valbuena A, Carrillo PJP, Rodríguez-Huete A, Mateu MG (2016) Imaging and quantitation of a succession of transient intermediates reveal the reversible self-assembly pathway of a simple icosahedral virus capsid. J Am Chem Soc 138:15385–15396
40. Grueso E, Sánchez-Martínez C, Calvo-López T, de Miguel FJ, Blanco-Menéndez N, Fernandez-Estevez M, Elizalde M, Sanchez J, Kourani O, Martin D, Tato A, Guerra M, Andrés G, Almendral JM (2019) Antiangiogenic vascular endothelial growth factor-blocking peptides displayed on the capsid of an infectious oncolytic parvovirus: assembly and immune interactions. J Virol 93:00798–00719
41. Salunke DM, Caspar DLD, Garcea RL (1986) Self-assembly of purified polyomavirus capsid protein VP1. Cell 46:895–904
42. Hernando E, Llamas-Saiz AL, Foces-Foces C, McKenna R, Portman I, Agbandje-McKenna M, Almendral JM (2000) Biochemical and physical characterization of parvovirus minute virus of mice virus-like particles. Virology 267:299–309
43. Yuan W, Parrish CR (2001) Canine parvovirus capsid assembly and differences in mammalian and insect cells. Virology 279:546–557
44. Maroto B, Ramírez JC, Almendral JM (2000) Phosphorylation status of the parvovirus minute virus of mice particle: mapping and biological relevance of the major phosphorylation sites. J Virol 74:10892–10902
45. Garcea RL, Benjamin TL (1983) Host range transforming gene of polyoma virus plays a role in virus assembly. Proc Natl Acad Sci 80:3613–3617
46. Garcea RL, Ballmer-Hofer K, Benjamin TL (1985) Virion assembly defect of polyomavirus hr-t mutants: underphosphorylation of major capsid protein VP1 before viral DNA encapsidation. J Virol 54:311–316
47. Li M, Garcea RL (1994) Identification of the threonine phosphorylation sites on the polyomavirus major capsid protein VP1: relationship to the activity of middle T antigen. J Virol 68:320–327
48. Suzuki T, Semba S, Sunden Y, Orba Y, Kobayashi S, Nagashima K, Kimura T, Hasegawa H, Sawa H (2012) Role of JC virus agnoprotein in virion formation. Microbiol Immunol 56:639–646
49. Carrasco L (1995) Modification of membrane permeability by animal viruses. Adv Virus Res 45:61–112

50. Suzuki T, Orba Y, Okada Y, Sunden Y, Kimura T, Tanaka S, Nagashima K, Hall WW, Sawa H (2010) The human Polyoma JC virus agnoprotein acts as a viroporin. PLoS Pathog 6:e1000801
51. Grosse S, Penaud-Budloo M, Herrmann A-K, Börner K, Fakhiri J, Laketa V, Krämer C, Wiedtke E, Gunkel M, Ménard L, Ayuso E, Grimm D (2017) Relevance of assembly-activating protein for adeno-associated virus vector production and capsid protein stability mammalian and insect cells. J Virol 91:e01198–e01117
52. Earley LF, Kawano Y, Adachi K, Sun X-X, Dai M-S, Nakai H (2015) Identification and characterization of nuclear and nucleolar localization signals in the adeno-associated virus serotype 2 assembly-activating protein. J Virol 89:3038–3048
53. Naumer M, Sonntag F, Schmidt K, Nieto K, Panke C, Davey NE, Popa-Wagner R, Kleinschmidt JA (2012) Properties of the adeno-associated virus assembly-activating protein. J Virol 86:13038–13048
54. Tse LV, Moller-Tank S, Meganck RM, Asokan A (2018) Mapping and engineering functional domains of the assembly-activating protein of adeno-associated viruses. J Virol 92:e00393–e00318
55. Chandler-Bostock R, Mata CP, Bingham RJ, Dykeman EC, Meng B, Tuthill TJ, Rowlands DJ, Ranson NA, Twarock R, Stockley PG (2020) Assembly of infectious enteroviruses depends on multiple, conserved genomic RNA-coat protein contacts. PLoS Pathog 16:e1009146
56. Dahmane S, Kerviel A, Morado DR, Shankar K, Ahlman B, Lazarou M, Altan-Bonnet N, Carlson L-A (2022) Membrane-assisted assembly and selective secretory autophagy of enteroviruses. Nat Commun 13:5986
57. Mukherjee S, Abd-El-Latif M, Bronstein M, Ben-nun-Shaul O, Kler S, Oppenheim A (2007) High cooperativity of the SV40 major capsid protein VP1 in virus assembly. PLoS One 2:e765
58. Erickson KD, Bouchet-Marquis C, Heiser K, Szomolanyi-Tsuda E, Mishra R, Lamothe B, Hoenger A, Garcea RL (2012) Virion assembly factories in the nucleus of polyomavirus-infected cells. PLoS Pathog 8:e1002630
59. Li PP, Nakanishi A, Shum D, Sun PC-K, Salazar AM, Fernandez CF, Chan S-W, Kasamatsu H (2001) Simian virus 40 Vp1 DNA-binding domain is functionally separable from the overlapping nuclear localization signal and is required for effective virion formation and full viability. J Virol 75:7321–7329
60. Waltmann C, Asor R, Raviv U, Olvera de la Cruz M (2020) Assembly and stability of simian virus 40 polymorphs. ACS Nano 14:4430–4443
61. King JA (2001) DNA helicase-mediated packaging of adeno-associated virus type 2 genomes into preformed capsids. EMBO J 20:3282–3291
62. Plevka P, Hafenstein S, Li L, D'Abrgamo A, Cotmore SF, Rossmann MG, Tattersall P (2011) Structure of a packaging-defective mutant of minute virus of mice indicates that the genome is packaged via a pore at a 5-fold axis. J Virol 85:4822–4827
63. Bleker S, Pawlita M, Kleinschmidt JA (2006) Impact of capsid conformation and rep-capsid interactions on adeno-associated virus type 2 genome packaging. J Virol 80:810–820
64. Santosh V, Musayev FN, Jaiswal R, Zárate-Pérez F, Vandewinkel B, Dierckx C, Endicott M, Sharifi K, Dryden K, Henckaerts E, Escalante CR (2020) The Cryo-EM structure of AAV2 Rep68 in complex with ssDNA reveals a malleable AAA+ machine that can switch between oligomeric states. Nucleic Acids Res 48:12983–12999

Further Reading

Carrasco L, Almendral JM (eds) (2006) Virus patógenos. Hélice and Fundación BBVA, Madrid, Spain
Flint SJ, Racaniello VR, Rall GF, Hatziloannou T, Skalka AM (2020) Principles of virology, 5th edn. ASM Press, Washington DC, USA

Kerr JK, Cotmore SF, Bloom ME, Linden RM, Parrish CR (eds) (2006) Parvoviruses. Hodder Arnold, London, UK

Harrison SC (2013) Principles of virus structure. In: Knipe DM, Howley PM (eds) Fields Virology, 6th edn. Lipincott Williams and Wilkins, Philadelphia, USA, pp 52–86

Hunter E (2013) Virus assembly. In: Knipe DM, Howley PM (eds) Fields Virology, 6th edn. Lipincott Williams and Wilkins, Philadelphia, USA, pp 127–152

Reddy VS, Johnson JE (2005) Structure-derived insights into virus assembly. Adv Virus Res 64: 45–68

Zlotnick A, Fane BA (2011) Mechanisms of icosahedral virus assembly. In: Agbandje-McKenna M, McKenna R (eds) Structural Virology. RSC Publishing, Cambridge, UK, pp 180–202

Also especially recommended for further reading is reference [13] listed above

Chapter 12
Architecture and Assembly of Structurally Complex Viruses

Carmen San Martín

Abstract Viral particles consist essentially of a proteinaceous capsid that protects the genome and is also involved in many functions during the virus life cycle. In structurally simple viruses, the capsid consists of a number of copies of the same, or a few different proteins organized into a symmetric oligomer. Structurally complex viruses present a larger variety of components in their capsids than simple viruses. They may contain accessory proteins with specific architectural or functional roles, or incorporate non-proteic elements such as lipids. They present a range of geometrical variability, from slight deviations from the icosahedral symmetry to complete asymmetry or even pleomorphism. Putting together the many different elements in the virion requires an extra effort to achieve correct assembly, and thus complex viruses require sophisticated mechanisms to regulate morphogenesis. This chapter provides a general view of the structure and assembly of complex viruses.

Keywords Virus structure · Virus assembly · Symmetry · Capsid · Cementing proteins · Envelope · Symmetry mismatch · Scaffold · Maturation · Virus evolution

Abbreviations

ABV	*Acidianus* bottle-shaped virus
ASFV	African swine fever virus
ATV	*Acidianus* two-tailed virus
cryo-EM	Cryogenic electron microscopy
cryo-ET	Cryogenic electron tomography
dsDNA	Double-stranded DNA
dsRNA	Double-stranded RNA
EMDB	Electron Microscopy Data Bank

C. San Martín (✉)
Department of Macromolecular Structure, Centro Nacional de Biotecnología (CNB-CSIC), Madrid, Spain
e-mail: carmen@cnb.csic.es

M. G. Mateu (ed.), *Structure and Physics of Viruses*, Subcellular Biochemistry 105,
https://doi.org/10.1007/978-3-031-65187-8_12

ESCRT	Endosomal sorting complexes required for transport
GON	Group of nine
GOS	Group of six
HIV	Human immunodeficiency virus
LUCA	Last universal common ancestor
PBCV-1	*Paramecium bursaria Chlorella* virus-1
PDB	Protein Data Bank
SARS	Severe acute respiratory syndrome
Sid	Size determination protein
SNDV	*Sulfolobus neozelandicus* droplet-shaped virus
ssDNA	Single-stranded DNA
ssRNA	Single-stranded RNA
STIV	*Sulfolobus* turreted icosahedral virus

12.1 Introduction

A viral particle consists essentially of a proteinaceous capsid with multiple roles in protection of the viral genome, cell recognition and entry, intracellular trafficking, and controlled uncoating. Evolutionary forces have caused viruses to adopt different strategies to achieve these goals. Structurally simple viruses (Chap. 11) generally build their capsids from a number of copies of the same, or a few different proteins, organized into a symmetric oligomer. In the case of structurally more complex viruses, capsid assembly requires further elaborations.

Structural complexity on a virus often, but not necessarily, derives from the need to house a large genome, in which case a larger capsid is required. However, capsid or genome sizes by themselves are not determinants of complexity. For example, flexible filamentous viruses can reach lengths in the order of microns, but most of their capsid mass is built by a single capsid protein arranged in a helical pattern [1]. On the other hand, architecturally complex viruses such as the human immunodeficiency virus (HIV) have moderate sized genomes (7–10 kb of single-stranded (ss) RNA) [2]. Structurally complex viruses incorporate a larger variety of components into their capsids than simple viruses. They may contain accessory proteins with specific architectural or functional roles, or incorporate non-proteic elements such as lipids.

The elaborated composition of complex virus particles often involves a rupture of the basic symmetry rules (Chap. 2), from a range of symmetry mismatches in icosahedral shells to completely asymmetric or pleomorphic capsids. The more subtle departure from symmetry is the case when identical subunits occupy similar but slightly different environments, as in the case of quasi-equivalence in icosahedral shells (see Chap. 2). In other cases, virion components with different symmetry may interact with each other, forming a *symmetry mismatch* at the interface. In the extreme case, identical components may form morphological units with no symmetry, or even assemble in a completely different manner for each realization of the virion—this property is called *pleomorphism*. Cryogenic electron microscopy (cryo-EM) (Chap. 3) has lately become the technique of choice for visualizing the organization of complex viruses in great detail, including features that depart from strict icosahedral symmetry. Understanding the architectural details of asymmetric capsids is the most challenging problem, since structural biology techniques heavily rely on the use of symmetries to reach high-resolution detail. Cryogenic electron tomography (cryo-ET) (Chap. 3) is helping to advance our understanding of these viruses.

The presence of many different elements in the virion entails an extra effort to achieve correct assembly. Accordingly, complex virus morphogenesis requires sophisticated mechanisms, tightly regulated in space and time. Here we provide a general view and representative examples of all these variations in complexity, finishing with a consideration on the evolutionary insights provided by structural studies on complex viruses.

12.2 Molecular Composition of Complex Viruses

12.2.1 Different Proteins with Specific Roles

A characteristic feature of complex viruses is the presence of multiple proteins in the virion, playing specific architectural or functional roles during the viral cycle. For example, different proteins may occupy the sixfold and fivefold coordinated positions in the icosahedral net (see Chap. 2). The specific architectural role of proteins at the fivefold vertices is often combined with a specific functional role, as will be described in Sect. 12.3. In icosahedral viruses with triangulation numbers $T > 1$, mobile terminal regions of the capsid proteins may adopt different conformations depending on their position in the capsid. In this way they act as molecular switches, enabling the same protein to occupy the different quasi-equivalent environments (see Chap. 2). In complex viruses, these mobile arms may still exist, but they often appear combined with a variety of minor capsid proteins, required for correct assembly of the virion. These *cementing proteins* can be considered as detached molecular switches, required to modulate the variety of interactions needed for assembly and stability of a complex capsid. One case where the intricate capsid organization includes: (1) biochemically different hexameric and pentameric capsomers; (2) a network of mobile arms; and (3) cementing polypeptides, has been described in great detail: adenovirus [3, 4].

Adenoviruses infect vertebrates. They enclose their dsDNA genome within a *pseudo* $T = 25$ icosahedral shell with a diameter of 950 Å between vertices. Trimers of the major capsid protein, hexon, constitute all the sixfold capsomers, while the vertices are occupied by pentamers of a protein named penton base. Although there is no sequence similarity between hexon and penton base, they have a similar fold based on the eight-stranded β-barrel motif, or "jelly roll," which allows appropriate interdigitation of the different capsomers to form the closed shell. Penton base folds as a single jelly roll. The hexon monomer folds as a double jelly roll perpendicular to the capsid surface (Fig. 12.1a). Thanks to this fold, trimers have a pseudo-hexagonal shape and can occupy sixfold symmetry positions in the icosahedron. Each capsid facet is formed by 12 trimers of hexon. The general icosahedral architecture of adenovirus can be described as two different systems of tiles. Nine hexon trimers form the central plate of each facet, known as GON (Group of Nine). The five peripentonal hexon trimers, together with the penton base, form the second tile system, known as GOS (Group of Six) (Fig. 12.1b). Numerous double-stranded (ds) DNA viruses share the basic jelly roll fold in their capsid proteins and assemble capsids organized as combinations of tiles based on pentagonal and pseudo-hexagonal building blocks. In larger viruses, the tiles composed of pentameric vertex and surrounding capsomers are known as *pentasymmetrons*, while the triangular tiles centered at the threefold icosahedral axis are called *trisymmetrons*. This efficient architectural solution facilitates construction of very large icosahedral capsids; triangulation numbers as large as $T = 499$ have been described, and larger ones are likely to exist [5, 6].

In the prototype adenovirus (human adenovirus type 5), apart from hexon and penton base, there are at least four other proteins making up the icosahedral shell. Minor capsid proteins IIIa, VI, VIII, and IX are required for correct capsid assembly and occupy specific positions in the capsid (Fig. 12.1b, c, d), forming specialized networks that stabilize the two systems of tiles. Polypeptide IX is the only cementing protein located on the outer part of the adenovirus capsid. It has an extended structure and forms a sort of hairnet keeping together the hexon trimers in each GON, and binding GONs to GONs across the icosahedral edges. The N-terminal domains of three IX monomers join via hydrophobic interactions at the icosahedral and local threefold axes in the GONs forming triskelion structures. A long, unstructured domain of each monomer runs in a different direction toward the facet edges, where the C-terminal α-helix joins with the C-terminal helices of another three copies of IX, different from those forming the N-terminal triskelion, to create a leucine zipper 4-helix bundle. On the interior of the shell, each GON is further stabilized by copies of polypeptide VIII located around the icosahedral threefold symmetry axis. Also on the inner capsid surface, polypeptide IIIa mediates the interaction between penton base and the peripentonal hexons, to keep each GOS together. Finally, IIIa and VIII cooperate to bind each GOS to its five surrounding GONs. The remaining minor capsid protein, polypeptide VI, does not follow icosahedral symmetry, but part of it has been assigned to density within an internal cavity present in each hexon trimer [3].

Mobile regions of hexon and penton base also play a role in the extensive interaction networks in the adenovirus capsid. These flexible regions are disordered in the isolated proteins, but they adopt ordered conformations within the capsid context. The N- and C- termini of the hexon monomer, located at the innermost part of the capsomer, adopt a total of 5 (N-) and 6 (C-) different conformations to establish interactions between neighboring hexons, or between hexons and minor capsid proteins (Fig. 12.1e). Similarly, for each penton base monomer an N-terminal

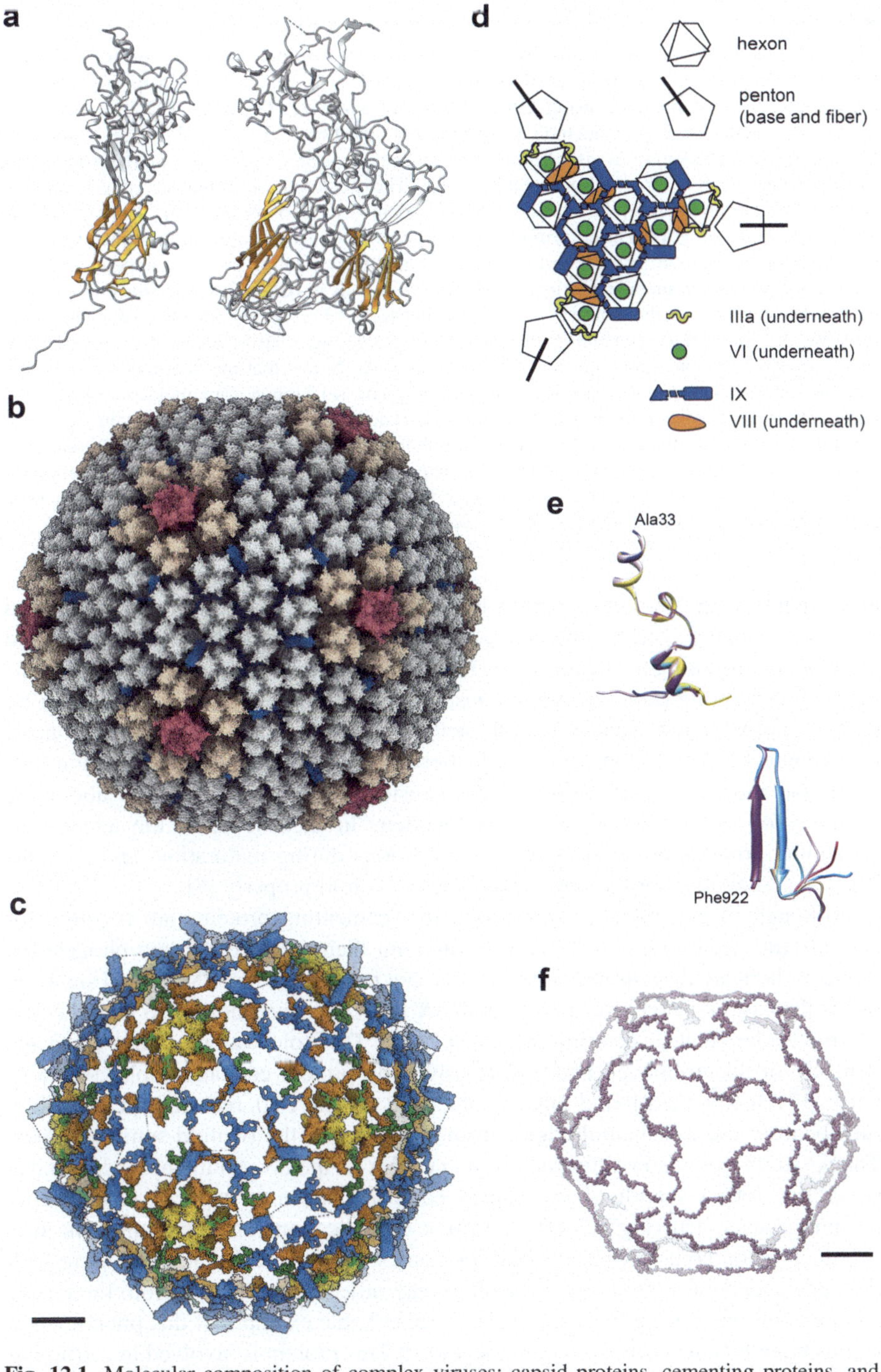

Fig. 12.1 Molecular composition of complex viruses: capsid proteins, cementing proteins, and mobile arms. (**a**) Structure of adenovirus penton base (left) and hexon (right) monomers, with the β-barrel jelly roll motifs highlighted in gold. The molecules are oriented so that the external capsid

Fig. 12.1 (continued) surface is up. Notice the extended N-terminal arm in penton base directed toward the interior of the virion (Unless otherwise indicated, all cryo-EM map and structure representations were prepared using USCF ChimeraX software. Goddard, T.D., Huang, C.C., Meng, E.C., Pettersen, E.F., Couch, G.S., Morris, J.H. and Ferrin, T.E. (2018) UCSF ChimeraX: Meeting modern challenges in visualization and analysis. *Protein Sci*, **27**, 14–25). (**b**) Adenovirus capsid, seen along a threefold icosahedral axis. The penton base pentamer (pink) and the peripentonal hexons (beige) form the GOS, the adenovirus pentasymmetron. The rest of the hexon trimers are in gray. These form the central plate of the icosahedral facet, termed GON, or trisymmetron in the more general terminology. In blue, the cementing protein polypeptide IX. (**c**) Cementing proteins in the adenovirus capsid. The view is as in (**b**), with the pentons and hexons removed to reveal both the external and internal networks of accessory proteins. (**d**) Schematics showing the location of cementing proteins in an adenovirus capsid facet. Reproduced from [4]. Cementing protein colors are the same in (**c**) and (**d**). Notice that the fibers, depicted in this cartoon, are not visible in the capsid model shown in (**b**). (**e**) Mobile arms in adenovirus hexon. Different conformations of the hexon N-terminal (left) and C-terminal (right) regions in the capsid are shown. Panels (**a**), (**b**), (**c**), and (**e**) made from atomic coordinates deposited at the Protein Data Bank (PDB) (Protein Data Bank is at http://www.pdb.org) with entry ID 6B1T. (**f**) Minor capsid proteins as size determinants. The bacteriophage PRD1 "tape measure" protein P30 forms a cage beneath the capsid surface (PDB ID 1W8X). Scale bars correspond to 100 Å

arm extends away from the β-barrels that form the main body of the protein toward the viral core (Fig. 12.1a), interacting with two IIIa monomers along the way, and therefore contributing to anchor the penton within the GOS. Some of the interactions between cementing proteins and hexons, and among cementing proteins, occur by β-sheet augmentation. That is, the interaction is mediated by a β-strand from one of the proteins binding to the edge of a β-sheet in the other [7]. This observation tells about the intricate organization of the capsid and makes us wonder about the difficulty of assembling all elements together. In fact, some of the adenovirus cementing proteins undergo large reorganizations during maturation, and it seems likely that they may require the virion context to fold properly [8].

Although in general it is understood that cementing proteins are required for correct viral assembly, it is difficult to pinpoint their exact role in morphogenesis. Some of them are dispensable for assembly, but required to reach structural stability; this is the case of adenovirus polypeptide IX [9]. Others are thought to play the role of "molecular rulers," determining capsid size. This role was proposed when the structure of bacteriophage PRD1 was solved by protein crystallography (see also Chap. 17). It was then found that minor capsid protein P30, required for assembly, runs beneath the icosahedral edges, from the vertex to the twofold symmetry axis. Thanks to its extended conformation, two copies of the 83-residue polypeptide can cover the 300 Å length of the capsid edge and act as a tape measure during morphogenesis [10] (Fig. 12.1f). A further complication for determining the role of minor virion components in assembly comes from the fact that, in keeping with the genetic economy principle, they often play other roles different from the purely architectural one during the viral cycle. A remarkable example of this phenomenon is illustrated by adenovirus polypeptide VI [4]. This protein is involved in disrupting the endosomal membrane, so that the virus can escape into the cytosol after internalization. It also has a role in facilitating virion traffic to the nucleus along the microtubular network; acts as an activator of the adenoviral gene expression; and promotes transport of newly synthesized hexon to the nucleus. Moreover, a C-terminal peptide of polypeptide VI activates the viral protease for maturation.

Other additional proteins may be incorporated to the viral particles and play fundamental roles for viability. Elucidating the organization of these additional components within the virion is not straightforward, since they usually do not follow defined symmetry rules, and their disposition may even change between particles of the same virus. Notable examples are viral proteases, such as the maturation protease VP24 in herpesvirus [11], or the adenoviral protease AVP [12]; and molecular motors involved in nucleic acid translocation, such as dsDNA packaging machinery in bacteriophage and herpesviruses [13, 14], or the dsRNA packaging ATPase in cystoviruses (bacteriophage Φ6) [15]. Viruses with RNA genomes must carry their own replication and transcription enzymes, to supply RNA metabolism functions absent in the cell [16]. Viruses that carry out their replication in the cytosol (e.g., vaccinia) must also supply DNA and RNA processing enzymes whose cellular counterparts are only present in the nucleus [17]. Some dsDNA viruses encapsidate basic proteins that help screen the nucleic acid charge repulsion, to facilitate compaction of the genome within the reduced capsid space. These basic proteins can be of cellular origin, such as in Simian Virus 40, which uses histones to pack its minichromosome [18]. Baculovirus [19], adenovirus [3], and some giant viruses such as mimivirus encode their own DNA condensing proteins [20]. The genomes of negative strand ssRNA viruses usually appear in the form of ribonucleoproteic structures [21]. More information on the packaging motors and on the organization of nucleic acids within viral capsids can be found in Chap. 13.

12.2.2 Membranes

Apart from the genome and structural proteins, a large number of viruses incorporate lipidic layers into their architecture. Lipid bilayers (membranes), widely present in cells and organelles, are ideally suited for enclosing a defined volume and separating it from neighboring compartments or the surrounding environment. This is the same function they play in viral capsids. For viruses, membranes are particularly advantageous, since they can readily be taken from the cell, are highly scalable in size, and do not consume coding space in the genome.

Viruses with external membranes are called *enveloped viruses*. Envelopes form a protective layer, blocking entry of aggressive chemicals or enzymes into the viral particle. They are composed not only of lipids but also of protein and sugars (glycoproteins). An implication of this arrangement is that in enveloped viruses infectivity is linked to membrane integrity, because the viral attachment proteins that recognize the host cell receptors and trigger internalization are in the viral envelope. Such is the case of influenza virus hemagglutinins [22], or Env proteins in retroviruses [2]. In the simpler enveloped viruses, such as Alphaviruses, the glycoproteins

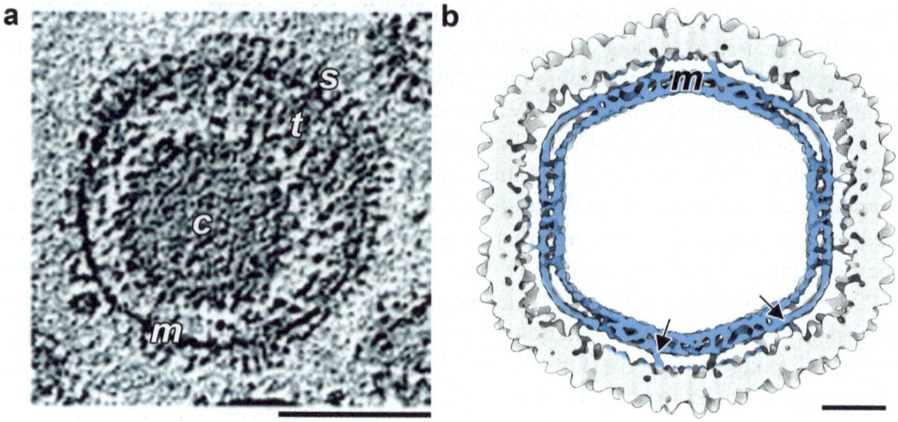

Fig. 12.2 Viruses with lipid bilayers. (**a**) Slice of a cryo-ET reconstruction of Herpes Simplex Virus-1 showing the icosahedral capsid (*c*), tegument (*t*), membrane (*m*), and glycoprotein spikes (*s*). Scale bar, 1000 Å. Modified from [24]. (**b**) Bacteriophage PRD1 has an internal membrane (*m*) that adopts an icosahedral profile due to interactions with capsid proteins (*arrows*). Shown is a slab of the cryo-EM map deposited at the Electron Microscopy Data Bank (EMDB) (Electron Microscopy Data Bank is at http://www.ebi.ac.uk/pdbe/emdb/) with ID EMD_1013, viewed along a twofold icosahedral axis. The scale bar corresponds to 100 Å

completely cover the lipid surface and follow the same icosahedral organization as the inner shell ($T = 4$) [23]. In other cases, such as retroviruses or herpesviruses (Fig. 12.2a), the structure directly in contact with the genome (nucleocapsid or capsid) is not in contact with the membrane, and the organization of glycoprotein spikes is irregular and does not reflect the organization of the virion inner contents [24, 25].

The membrane can also be located beneath the icosahedral shell, such as in tectiviruses (PRD1) and structurally related viruses [10, 23]. In these cases, the membrane itself can adopt an icosahedral layout forced by interactions with the capsid proteins (Fig. 12.2b). In PRD1, some of the mobile arms in its major capsid protein are embedded in the outer leaflet of the membrane, contributing to enforce the icosahedral shape [10]. Icosahedral viruses with internal membranes have a large complement of virion-encoded membrane proteins: approximately half of the 18 proteins present in the PRD1 virion are membrane proteins [26]. The membrane in PRD1 can undergo a large conformational change and protrude forming a tube from one of the vertices. It has been proposed that this tube has a function in injecting the viral genome into the host cell, similar to that played by tail structures in other bacteriophages (see Chap. 17).

Some of the most architecturally complex viruses have membranes. For example, herpesviruses have a $T = 16$ icosahedral capsid (1500 Å in diameter), formed by one major capsid protein and several accessory proteins. This capsid is surrounded by a thick tegument layer, containing at least 13 different viral proteins and also some cellular components. Capsid and tegument are enveloped by a membrane with more than 12 different types of viral glycoproteins [11, 24] (Fig. 12.2a). Giant dsDNA viruses such as *Acanthamoeba polyphaga* Mimivirus (capsid diameter ~500 nm) have internal membranes like bacteriophage PRD1. Asfarviruses (African swine fever virus, capsid diameter ~240 nm) have both an internal membrane surrounded by an icosahedral shell and a loose external envelope [6, 27]. Other examples of complex, lipid-containing viruses will be described in Sect. 12.4. Mechanisms of membrane acquisition are discussed in Sect. 12.5.

12.3 Departures from Symmetry in Quasi-Icosahedral Capsids

12.3.1 Layers with Different T Numbers

In Sect. 12.2.2, it was pointed out that viruses could be organized in multiple layers, intercalating protein (ordered or not) and lipids. In other cases such as adenovirus, multiple cementing proteins combine to form a single icosahedral capsid, while additional components (e.g., dsDNA condensing proteins) do not show any symmetrical organization. In yet another instance, concentric icosahedral protein shells are formed. Remarkably, these shells may have different triangulation numbers, including some not predicted by the theory of quasi-equivalence. This type of organization is most prominently present in the dsRNA viruses.

Rotaviruses and mammalian orthoreoviruses are the best characterized members of the *Reovirales* order [16, 28, 29]. The mature rotavirus virion has a diameter of approximately 100 nm and is organized in three concentric layers composed of four different proteins (Fig. 12.3). The innermost layer, or core shell, is formed by 120 molecules of protein VP2. Because of the number of protein monomers in this shell, it has been described as having a "$T = 2$" triangulation number, a conformation not allowed in the Caspar and Klug formalism [30] (Chap. 2). In fact, the VP2 monomers adopt two different conformations to yield a $T = 1$ icosahedron of asymmetric dimers. VP2 works as a platform for assembly of the next layer, formed by 260 trimers of protein VP6 in a $T = 13$ lattice. VP6 must be able to adopt not only the different conformations required for the $T = 13$ quasi-equivalence requirements, but also those extra ones required to compensate for the symmetry mismatch with the VP2 layer. The third, and outermost layer, is also a $T = 13$ icosahedron composed of 260 trimers of glycoprotein VP7, plus 60 spikes formed by trimers of VP4 that protrude from peripentonal positions. The VP4 spikes must be proteolytically cleaved for the virus to be infectious. Remarkably, after cleavage the fragments remain non-covalently associated on the virion surface, but undergo an intriguing conformational modification, changing from a trimeric arrangement to a mixture of trimeric, dimeric, and monomeric associations [16]. The VP1 and VP3 transcription enzymes are also part of the virion and are located beneath the core shell surface.

The orthoreovirus capsid is also triple layered and shares architectural similarities with rotavirus, namely the mismatch between $T = 1$ and $T = 13$ symmetries.

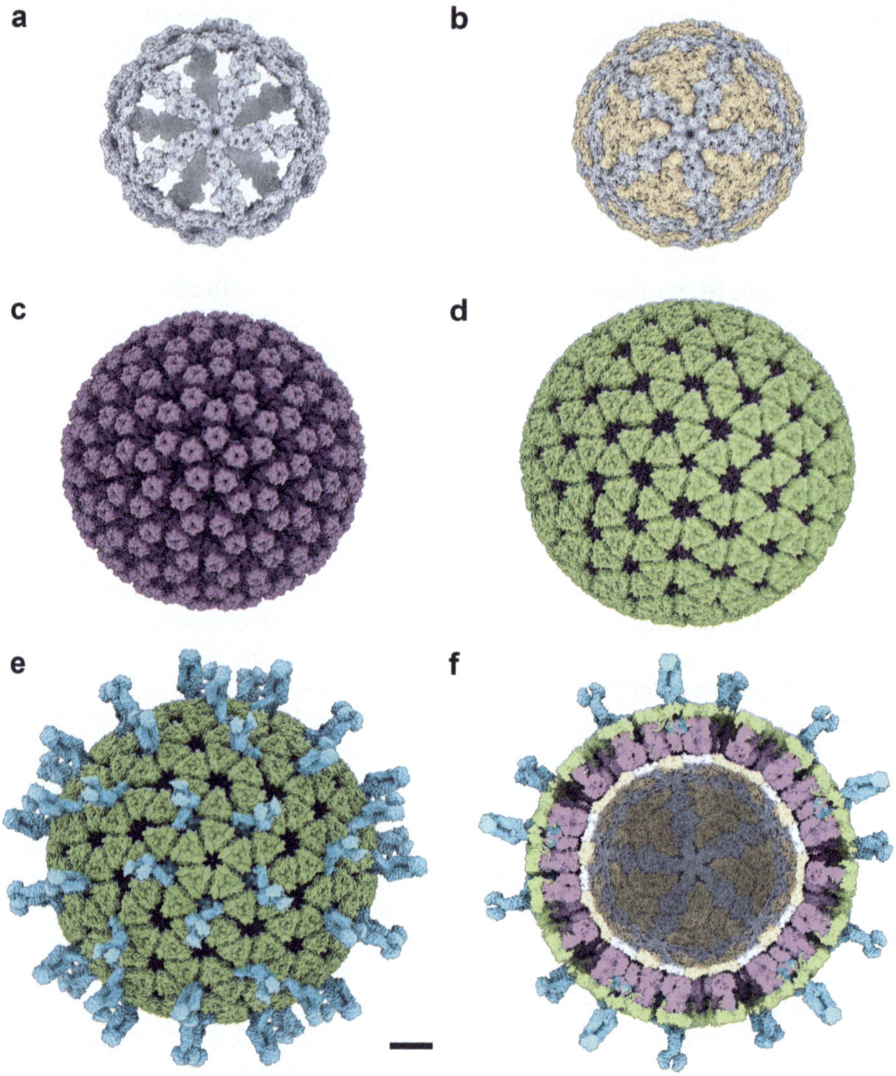

Fig. 12.3 Multiple layered viruses: rotavirus capsid architecture. Panels (**a**) to (**e**) show the consecutive building layers of the mature rotavirus virion. (**a**) Only one type of VP2 monomer in the core shell is represented. (**b**) The VP2 core shell, a $T = 1$ particle formed by asymmetric dimers. One monomer in each dimer is colored beige. (**c**) The intermediate $T = 13$ VP6 layer. (**d**) VP7 glycoprotein layer. (**e**) The complete virion with VP4 spikes. (**f**) Cut-out representation of the virion depicting the layered organization. The view is along a fivefold icosahedral symmetry axis. The scale bar represents 100 Å. Figure prepared using atomic coordinates from PDB IDs 3N09 and 3IYU

However, the composition is more complex, with six different proteins instead of four. The major differences appear in the outermost layers, probably reflecting differences in the viral entry mechanism. Instead of having 60 short spikes distributed in the icosahedral facet, orthoreovirus displays large turreted structures combined with a long, flexible fiber in each of the 12 vertices.

Some of the giant viruses also present nested icosahedral capsids with different T numbers. In Faustovirus, a $T = 16$ shell has been found beneath the $T = 277$ capsid [31]. African swine fever virus (ASFV) also has an external $T = 277$ capsid and an internal core shell, in this case with a $T = 19$ arrangement [32, 33]. The internal membrane in ASFV is intercalated between the two capsids, which do not seem to interact with each other. Faustovirus does not seem to contain any lipid membranes.

Why do viruses have different layers? As in the case of membranes, protein layers help to separate different compartments, and most likely also different functions along the infectious cycle. dsRNA viruses need to keep their genome confined within the core shell at all times during infection, to protect it from aggressions by cellular nucleases and to prevent antiviral reactions triggered by accumulation of dsRNA. In rotavirus, the double-layered particle formed by VP2 and VP6 is the transcriptionally competent form of the virus. The external layers carry the viral components in charge of initial interaction with the host: recognition, attachment, and entry. These are shed once entry into a new host cell has been accomplished [34]. In reovirus however, the turrets are not lost upon entry like the other external layers, but form part of the double-layered, transcriptionally active form of the virus [35]. Little is known about the role of the internal capsid in giant viruses. In ASFV, it has been proposed that the inner shell may help protect the viral genome from host nucleases or dsDNA sensors, delaying activation of the innate immune response during the first infection stages [33].

12.3.2 Symmetry Mismatches

In the previous section, we have seen that in reoviruses there is a symmetry mismatch between two concentric shells with different triangulation numbers. Nevertheless, the two layers still follow icosahedral symmetry, and therefore it has been possible to study their organization at a very detailed level. Symmetry mismatches (two elements with different symmetries in direct interaction) are frequent in icosahedral viral structures, particularly at the vertices, where proteins involved in genome translocation or host attachment reside. Solving the organization of mismatched features represents a remarkable challenge for structural biology techniques, due to the predominance of icosahedral symmetry in the complete virion that obscures them. In the *Cystoviridae* representative bacteriophage Φ6, a hexameric ssRNA packaging ATPase occupies multiple fivefold vertices of the empty procapsid [15]. In dsDNA bacteriophages, 12-fold portal structures occupy one of the vertices (Sect. 12.3.4 and Chaps. 13 and 17). Host recognition elements often take the shape of elongated fibers protruding from the fivefold capsomers. The

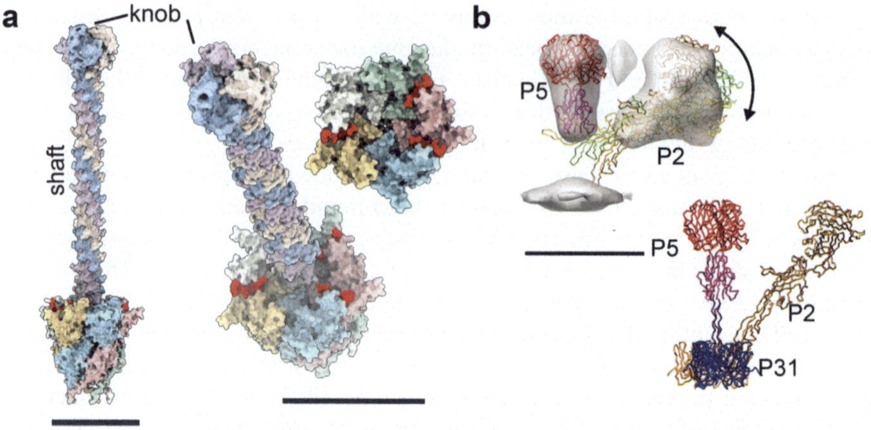

Fig. 12.4 Symmetry mismatches. (**a**) Adenovirus vertex complex, composed of pentameric penton base (compare with Fig. 1a and b) and the trimeric fiber (knob and distal shaft from PDB ID 1QIU, rest of the shaft modeled). In red, three N-terminal fiber peptides bound to penton base (PDB ID 3IZO). Left: side view; center, oblique view; right, top view of the penton base with the fiber shaft and knob removed to highlight the position of the N-terminal fiber peptides. (**b**) A model for the double fiber complex in bacteriophage PRD1 vertex. Top: high-resolution structures of trimeric P5 and monomeric P2 fitted to a cryo-EM reconstruction of the vertex region (side view). The arrow indicates variability in the position of P2. Bottom: a model of the complex interactions between the three vertex components. P31 forms the fivefold capsomers (pentons) in PRD1. This study showed that the two fibers interact with each other at the icosahedral capsid level and that P2 can move relatively to P5. Reproduced from [36], with permission (Copyright (2007) National Academy of Sciences, USA). Scale bars represent 100 Å

oligomerization state and number of fibers per vertex varies and is usually at odds with the pentameric architecture of the capsomers.

In human adenoviruses, a trimer of the fiber forms a non-covalent complex with a pentamer of penton base (Fig. 12.4a). The C-terminal domain of fiber is the main player in trimerization and forms the distal knob, responsible for binding to the adenovirus primary receptor. The knob is followed by a shaft of variable length and flexibility, depending on the virus type. The knob domain has an eight-stranded β-sandwich fold similar to that of the major capsid protein, and the shaft forms a triple β-spiral [37]. Finally, the N-terminal region is responsible for binding to penton base. At the proximal part the shaft becomes frayed and three flexible N-terminal peptides spread out to attach to the penton base pentamer. The crystal structure of penton base in complex with an N-terminal fiber peptide [38] showed the peptide bound to a groove on the outer surface of the pentamer formed by the interface between two penton base monomers, reaching radially from the penton center to the outer rim. Five peptides were observed with equivalent density, implying that all fiber binding sites are equivalent. Therefore, in the virus the three N-terminal tails may adopt two different arrangements: either they occupy three consecutive grooves or two of them are in neighboring binding sites and the third one is flanked by two empty grooves (as depicted in Fig. 12.4a). In the cryo-EM maps of the complete virion, density for the proximal part of the fiber shaft is usually observed protruding from the center of the penton base pentamer. Density for the fiber shaft is blurred by the enforced fivefold symmetry, but it could be observed that its base interacts with a hydrophobic ring at the rim of a narrow channel in the center of the penton [39]. This hydrophobic interaction may allow relative rotation of fiber on penton to accommodate the symmetry mismatch, while the N-terminal tails secure the binding to the penton groves. Fiber binding to the adenovirus receptor in the cell results in its release from the capsid, which in turn induces a cooperative conformational change in the penton base pentamer. This change is thought to play a role in preparing pentons for release at a later stage of adenovirus entry and uncoating [40, 41].

In orthoreovirus, the receptor binding protein σ1 also forms a flexible trimeric spike attached to the fivefold turret [16, 42]. Intriguingly, some viruses can hold two fibers attached to the same vertex: this is the case of fowl adenovirus type-1, where two fibers of different length can be observed bound to the same penton base [43]. Bacteriophage PRD1 also has two different spikes (proteins P2 and P5) attached to a single pentamer of the vertex protein P31 (Fig. 12.4b). Moreover, in PRD1 each spike has a different oligomerization state. P5 is a trimer resembling the structure of the adenovirus fiber; while the other, P2, is a monomer with a pseudo-β propeller head. The role of each spike in PRD1 host recognition and attachment is not fully clarified [26, 36] (see also Chap. 17).

The biological significance of these symmetry mismatches has long intrigued virologists. For the dodecameric nucleic acid packaging motors, the mismatch may allow conformational changes required for the translocation function (see Chap. 13). In the case of host recognition fibers, it is possible that the mismatch facilitates flexibility to scan for and attach to the viral receptor, as well as fiber removal upon binding, a step required to initiate the cascade of signals in both cell and virion for appropriate entry and/or genome delivery.

12.3.3 Special Vertices

Special (or singular) vertices have been found in many icosahedral dsDNA viruses. Singular vertices play key roles in genome packaging and ejection; they may also represent initial or final points in the assembly pathway of the shell. They induce a rupture of icosahedral symmetry (one vertex different from the other 11) and often include a symmetry mismatch (protein with non-fivefold symmetry occupying a fivefold coordinated position in the icosahedral net). Dodecameric proteins involved in genome packaging are found in a single vertex in tailed bacteriophages (see Sect. 12.3.4, and Chap. 13), as well as in herpesviruses [13, 14]. The best characterized case of special vertex is the portal in tailed bacteriophages, which connects the icosahedral head with the conspicuous tail that is characteristic of this viral family.

Giant dsDNA viruses infecting eukaryotic microorganisms also have singular vertices. *Paramecium bursaria Chlorella* virus-1 (PBCV-1) has a 190 nm diameter icosahedral capsid surrounding a lipid bilayer and dsDNA genome. A 250 Å long spike protrudes from one of the capsid vertices [44] (Fig. 12.5a, b). The penton and the peripentonal capsomers beneath the spike are different from the rest. The density for the spike is not well defined, even in cryo-EM maps where icosahedral symmetry has not been enforced. This observation suggests that there may be a symmetry mismatch between the spike and the pentameric capsomer [46]. The spike is too thin to be used as a DNA ejection tube; besides PBCV-1 is thought to deliver its genome into the host by fusion of the internal membrane with the host one. It has been proposed that the function of the PBCV-1 spike is to puncture the cell wall to initiate the fusion process. The capsid side holding the spike is disassembled upon attachment to the host [47]. It has also been proposed that the unique vertex serves as initiation seed for capsid assembly [46].

The giant Mimivirus has a 500 nm large icosahedral capsid structurally related to those of adenovirus, bacteriophage PRD1, and PBCV-1, covered by 125 nm long fibers. Early images of Mimivirus showed a starfish-shape feature with five arms reaching from one of the vertices to the five neighboring ones (Fig. 12.5c). The arms of the starfish are inserted between adjacent facets, opening a gap between them. The starfish is an independent macromolecular assembly that remains together when detached from the virion [48] and is the only part of the capsid not covered by fibers. When Mimivirus enters the cell by phagocytosis, a remarkable structural change occurs, whereby the five icosahedral facets in contact with the starfish feature open, leading to the structure called "stargate" [45]. The internal viral membrane is extruded through the stargate, to fuse with the phagosome membrane and release the viral DNA into the cytosol (Fig. 12.5d). Tailed phages use their special vertex both for genome packaging and delivery (Sect. 12.3.4; Chaps. 13 and 17). In Mimivirus however, the stargate vertex is used for genome delivery, but not for packaging, which occurs instead via an aperture located in the icosahedral facet.

Remarkably, the asymmetry originated by the singular vertex in both PBCV-1 and Mimivirus reflects in an asymmetry of the internal virion contents [44, 48]. The viral genome and surrounding membrane do not occupy the full internal volume of the capsid. Rather, a gap exists between the DNA core and the side of the capsid containing the special vertex. This gap forms a pocket where viral enzymes required for cell membrane penetration may be contained. It may also contain structural elements required to precisely determine the asymmetric location and shape of the genome within the virion. In PBCV-1, the gap contains a membrane vesicle firmly anchored to the unique vertex by a set of five different minor coat proteins [46].

Departures from icosahedral symmetry are not exclusive from complex virus particles, as increasingly shown by advanced cryo-EM image processing methods that enable relaxation of the symmetry enforcement (Chap. 3) [49]. For example, the minor coat protein VP2 in feline calicivirus ($T = 3$, ssRNA) can form a 12-fold portal-like structure not at a vertex, but at a unique threefold symmetry axis [50]. This portal opens upon binding to the virus receptor, the cell adhesion molecule JAM-A. The outermost region of the portal is rich in hydrophobic residues, and it may help open a channel through the endosome membrane for delivery of the virus genome into the cytosol.

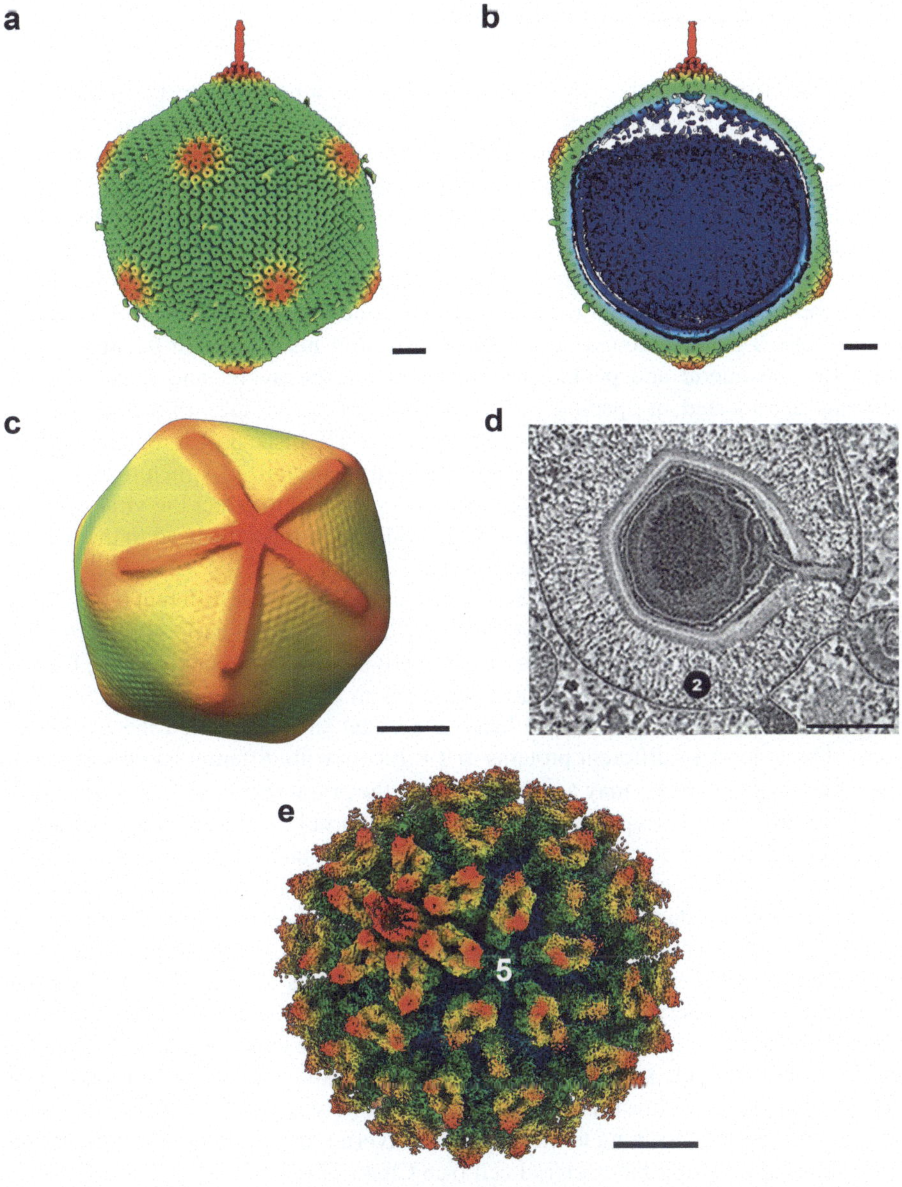

Fig. 12.5 Singular vertices. (**a**) The PBCV-1 capsid (EMDB ID EMD_1597) showing the singular vertex with a spike, and (**b**), a central slab where the asymmetry in internal contents can be appreciated. (**c**) 3D map of the Mimivirus capsid (EMDB ID EMD_5039) showing the starfish feature (red). (**d**) Slice of a tomographic reconstruction showing a Mimivirus particle within a phagosome. The viral membrane is extruded through the open stargate. Modified from [45]. (**e**) Structure of feline calicivirus capsid bound to its receptor, feline JAM-A (EMD-0056), showing the portal-like structure protruding from a singular threefold axis. The neighboring fivefold symmetry axis is indicated in white. Scale bars represent 200 Å in panels (**a**) and (**b**); 1000 Å in (**c**) and (**d**), and 100 Å in (**e**). Maps are colored by distance from the center

12.3.4 Heads, Tails, and Baseplates

Tailed bacteriophages (class *Caudoviricetes*) [51] are among the best described and more complex of the non-enveloped viruses. Their virions are composed of several functionally specialized morphological units, arranged according to different symmetries and connected via multiple symmetry mismatches (see also Chap. 17).

Bacteriophage capsids (heads) contain the dsDNA viral genome. They are icosahedrally ordered, but details vary among the different viruses. For example, HK97 or T7 phages have a strictly icosahedral head with a $T = 7$ net and a single major capsid protein occupying both the sixfold and fivefold coordinated positions. Others, like T4, have elongated, prolate icosahedral heads; two different proteins form the hexameric and pentameric capsomers, respectively, and several minor proteins are located at specific positions on the head surface [52]. One of the 12 vertices in the head is different from the other 11. Instead of a regular penton, it contains the portal structure or connector. This specialized structure is critical during assembly, since it contains the machinery needed to select the viral genome and pump it into the head (see Chap. 13). In addition, it links the head to the feature that differentiates these viral families: the tail.

Bacteriophages in the *Caudoviricetes* class can present three different types of tail morphology: long, contractile (myoviruses); long, non-contractile (siphoviruses); and short, non-contractile (podoviruses). In myoviruses (representative: T4), the tail is composed of two layers of protein, one of them (the outer one) contractile [53] (Fig. 12.6a). The tail connects the head to a distal structure called the baseplate, formed by at least 16 different proteins in T4. Fibers with different lengths protrude from the baseplate; fibers may also be present at the portal region and the icosahedral head. In siphoviruses (representative: phage λ), the tail lacks the outer contractile sheath. Research on the structure of non-contractile tails has unveiled a crucial structural element: the tape measure protein, whose length determines that of the tail by limiting the stacking of tail protein rings to a defined number. The baseplate composition is variable among the different viruses in this group, with some having only a simpler element called the tail tip complex. Fibers can project laterally from the baseplate periphery or longitudinally from the extremity of the tail tip complex [56]. Podoviruses (representatives: T7, P22) have short tails, with a fiber complement that may include long, thin fibers (T7) or thick spikes (P22) [57]. Fibers, tails and baseplates, or tail tips form the complex machinery required to initiate infection by recognizing and attaching to the host and delivering the viral genome through the many layers protecting the bacterial cell (see Chap. 17).

Tailed bacteriophages deviate from the icosahedral symmetry due to their conspicuous genome delivery apparatus; additionally, they are a compendium of symmetry mismatches. Icosahedral (prolate or not) heads have a singular vertex where a fivefold symmetric capsomer is replaced by a 12-fold ring of the portal protein [14]. The portal complex is connected to the tail, which in general follows sixfold symmetry along the tube and baseplate. In the case of myoviruses however, a further mismatch exists, since the contractile sheath presents helical symmetry, while the

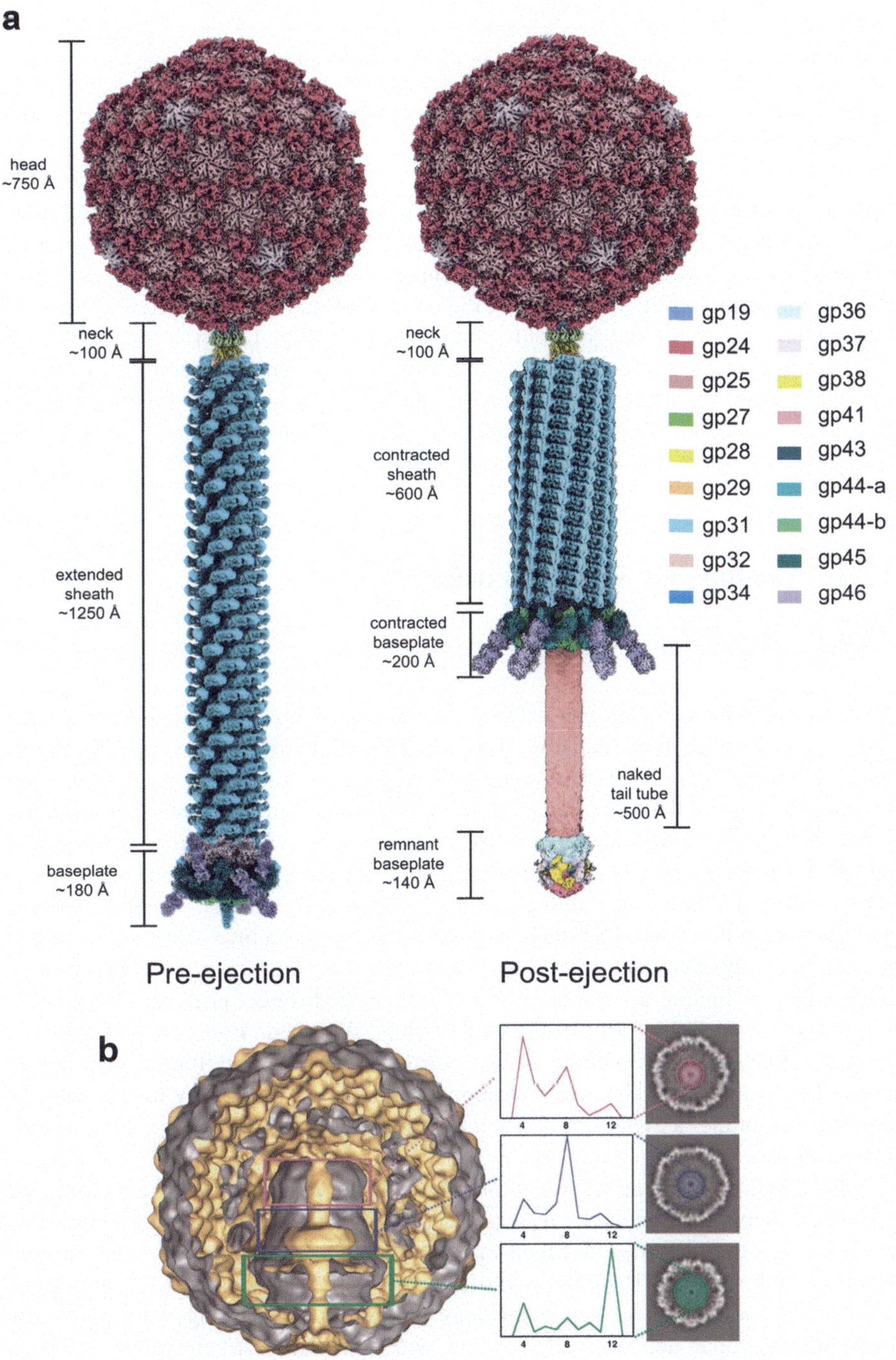

Fig. 12.6 Structure of tailed bacteriophages. (**a**) Structure of *Pseudomonas* bacteriophage E217 (a myovirus), showing the tail in its extended (left) and contracted (right) conformation. The color

Fig. 12.6 (continued) legend conveys the large number of different proteins forming this structure. Modified from [54]. (**b**) 3D map of the bacteriophage T7 procapsid, showing the internal core complex. The plots show the dominant symmetry for each core region. Reproduced from [55], with permission. Diameter of the T7 procapsid = 510 Å

inner tube is formed by stacked hexamers [54]. Additionally, some podoviruses such as T7 incorporate an internal proteic structure referred to as the core. This structure grows from the portal vertex toward the capsid center and is thought to serve as a spindle for wrapping the DNA. In T7, the core is formed by three different proteins and presents eightfold and fourfold symmetries [55] (Fig. 12.6b). Remarkably, the T7 core undergoes a complete reorganization during genome ejection, to form a sixfold channel for DNA delivery across the host cell wall [58]. Finally, another symmetry mismatch may appear in tailed phages when a pentameric packaging motor binds to the portal vertex in the prohead during encapsidation [14].

12.4 Asymmetric Virus Particles

12.4.1 Brick-Shaped Viruses

Poxviruses are large, enveloped dsDNA viruses apparently lacking any kind of high order symmetry in their capsids. The representative of the poxvirus family is vaccinia virus (Fig. 12.7a). The complex composition, large size, and asymmetric organization, plus the sensitivity of these viruses to the different preservation methodologies for EM, have caused the architecture of vaccinia to be a subject of debate for a long time. To complicate things even more, vaccinia virions exist in forms carrying a different number of envelopes: mature virions, wrapped virions, and extracellular virions. The mature virion of vaccinia is a brick-shaped enveloped particle with approximate dimensions 350 × 250 × 150 nm (varying depending on the imaging technique used), composed of at least 75 different proteins [17]. Mature virions are intracellular forms of the viral particle. Wrapped virions are also found in the cells and consist of mature virions surrounded by two additional membranes derived from the Golgi cisternae. Wrapped virions leave the cell by fusing with the plasma membrane, leaving one of their envelopes behind, to produce the extracellular virion.

The 200 kbp genome of vaccinia is contained in a core with an elongated dumbbell shape, surrounded by a protein capsule (core wall). The core also contains a variety of viral enzymes involved in RNA metabolism, required for the virus to replicate in the cytosol. The outer part of the core wall, termed palisade because of its striated appearance, is formed by trimers of protein A10 arranged in a partially ordered hexagonal network, interspersed with portal-like structures. It has been proposed that these portals are the conduits for release of viral mRNA from the core early in infection [61].

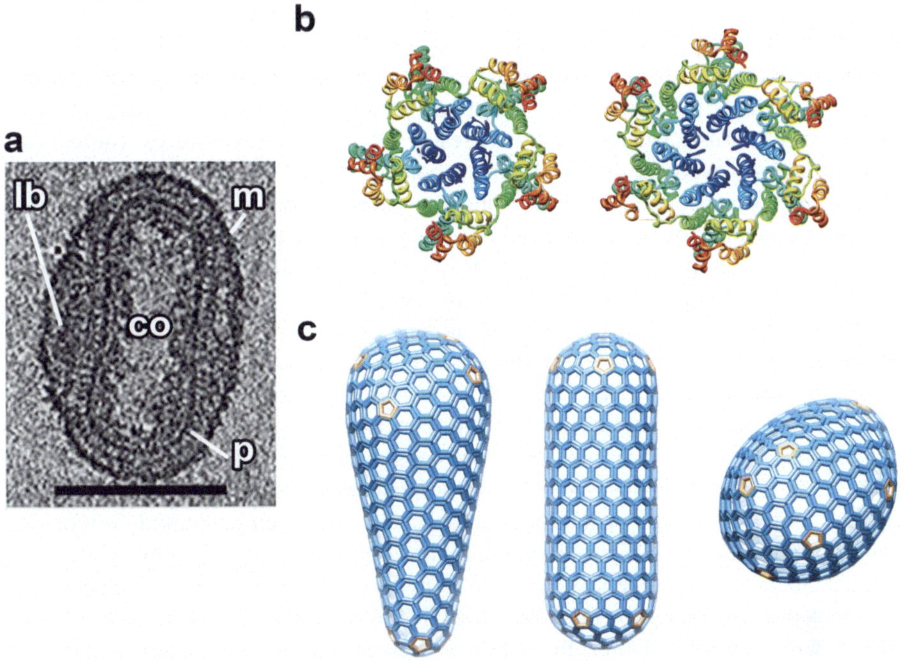

Fig. 12.7 Asymmetric and pleomorphic viruses. (**a**) Cryo-ET reconstruction of mature vaccinia virions. Central section of a virion where the dumbbell shape of the core (**co**), the membrane (**m**), palisade (**p**), and lateral bodies (**lb**) can be appreciated. The bar represents 2000 Å. Modified from [59], with permission (Copyright (2005) National Academy of Sciences, USA). (**b**) Structure of HIV CA protein assembled as a pentamer (left; PDB ID 3P05) or a hexamer (right; PDB ID 3H4E). (**c**) Example of fullerene-like objects generated from 12 pentamers and a variable number of hexamers (Panels (**b**) and (**c**) prepared using USCF Chimera software [60] Pettersen EF, Goddard TD, Huang CC, Couch GS, Greenblatt DM, Meng EC, Ferrin TE (2004) UCSF Chimera—a visualization system for exploratory research and analysis. J Comput Chem 25:1605–1612)

The extremes of the dumbbell rest adjacent to the envelope, while the central part is surrounded by electron-dense material (lateral bodies) of unknown function. In vitro disruption studies suggest that the dsDNA in the core is in complex with condensing proteins [62]. However, the condensing proteins have not been identified yet.

12.4.2 Pleomorphic Viruses

Pleomorphic viruses not only do not follow high symmetry rules when forming the infectious particle, but may even adopt a wide range of sizes, shapes, and composition from particle to particle, making each virion unique. Because of their intrinsic variability, the structural organization of pleomorphic viruses cannot be deduced from structural biology techniques based on averaging data from many identical virions, such as X-ray crystallography or cryo-EM analyses. Cryo-ET (see Chap. 3) has started to reveal the architectural details of this kind of macromolecular machines, which includes many important pathogens for humans. Examples of pleomorphic viruses include retroviruses (HIV); orthomyxoviruses (influenza); coronaviruses (severe acute respiratory syndrome (SARS)-coronavirus); and paramyxoviruses (measles) [2, 63–65]. In addition, atomic force microscopy can be used for surface visualization of any kind of single virus particles in liquid, in close to physiological conditions (see Chap. 8). Possible deformations by adhesion to a solid base should be, however, considered in this case.

Pleomorphism is most pronounced among enveloped viruses, since the lipid envelope readily adapts different shapes and sizes. However, proteins with a tendency to form symmetric aggregations can also give rise to pleomorphic capsids. The capsid protein of retroviruses (CA) can assemble into either hexamers or pentamers, in much the same way as capsid proteins of icosahedral viruses (Fig. 12.7b). Recombinant CA forms only hexamers in certain conditions, giving rise to tubular oligomers or flat, ordered sheets; while when pentameric oligomerization is enforced, $T = 1$ icosahedral particles are formed [66, 67]. However, when CA hexamers and pentamers associate to form the closed mature capsid that contains the nucleocapsid complex including the ssRNA genome, the pentamers are not distributed regularly within the hexamer lattice. Even if a fixed number of 12 pentamers is incorporated into each capsid, the asymmetry of their distribution results in asymmetrical structures that can adopt shapes ranging from roughly spherical to roughly conical and can be modeled using the geometrical principles governing fullerene cones (Fig. 12.7c).

12.4.3 A Glimpse at the Peculiar Architectures of Viruses Infecting Protists and Archaea

In the last two decades, numerous new viruses have been isolated from eukaryotic single-celled organisms and from archaea. Studies on dsDNA viruses infecting free-living amoeba have revealed a wide variety of the so-called giant viruses, with genome and capsid sizes (and therefore complexity) rivaling those of the smallest bacterial cells [68]. Many of these viruses have apparently icosahedral capsids built from proteins with a double jelly roll fold similar to that of the adenovirus hexon (Sect. 12.2.1). The triangulation number of the mimivirus capsid has not been unequivocally determined, but it is expected to lay in the $T \sim 1000$ range. The largest T number described so far ($T = 499$) corresponds to a virus infecting *Cafeteria roenbergensis*, a marine flagellate microbe. This is closely followed by two amoeba viruses, Pacmanvirus and Melbournevirus, with $T = 309$ [5]. Large sets of minor capsid proteins are required to build these colossal shells, and departures from symmetry such as the mimivirus stargate are common (Sect. 12.3.3). A most remarkable deviation in the icosahedral symmetry is found in Tupanvirus (Fig. 12.8a). Tupanviruses, also infecting amoeba, have an icosahedral capsid similar to that of mimivirus (~500 nm diameter), decorated with a long (~550 nm) cylindrical tail, so that the size of the complete viral particle is larger than 1 μm. Both the capsid and cylindrical tail are covered in fibrils [68, 69]. Even larger can be the capsids of the ovoidal-shaped pandoravirus, cedratvirus, or pithoviruses (Fig. 12.8b). Some of these ovoidal particles present specific structures at one or both poles termed "the cork," thought to seal the genome delivery portal (Fig. 12.8c).

The most abundant repertoire of archaeal viruses reported so far is that of dsDNA viruses [73]. Only a few ssDNA archaeal viruses have been described, enclosing their genome in pleomorphic enveloped particles [74]. It is still not clear if RNA archaeal viruses exist [75]. Some dsDNA viruses infecting archaea follow the general architectural types previously known for bacterial and eukaryotic viruses. For example, *Sulfolobus* turreted icosahedral virus (STIV) or *Haloarcula hispanica* SH1 virus are icosahedral, tail-less viruses with an internal membrane, structurally similar to bacteriophage PRD1 [10] or the algae virus PBCV-1 [44]. Archaeal viruses with tailed bacteriophage morphologies have also been reported [76].

Intriguingly, other viral families discovered in archaea are unique to this branch of life, with morphologies not observed anywhere else. According to their overall organization, they have been classified into fusiform, droplet or bottle-shaped, and filamentous viruses. Fusiform viruses are very abundant in habitats dominated by archaeal microorganisms. They have spindle-shaped virions with tails of variable length protruding from the spindle poles. One fusiform virus, *Acidianus* two-tailed virus (ATV) not only has an unusual shape, but is able to assemble new structural features after leaving the host cell. When propagated at temperatures slightly suboptimal for its host (75 °C), isolated virions look like ~0.2 μm long lemons. However, when temperature is raised in the absence of the host cell, these viruses grow two filamentous tails of variable lengths, one from each pole (Fig. 12.8d). This is the only known example of a virus with extracellular assembly, but it is likely that others exist [71]. The tails end in an anchor-like structure, thought to be involved in attachment to the host.

Acidianus bottle-shaped virus (ABV) and *Sulfolobus neozelandicus* droplet-shaped virus (SNDV) are among the few known members of the two viral families termed *Ampullaviridae* and *Guttaviridae* [72] (Fig. 12.8e–f). The enveloped ABV virion contains a conical core formed by a supercoiled nucleoprotein filament. A brush of short filaments protrudes from the bottom of the bottle, but host attachment seems to occur at the opposite side of the virion. Little is known about the architecture of the SNDV virion, except for its droplet shape and the presence of a tuft of long fibers at its narrower pole. Finally, filamentous archaeal viruses can form stiff rods (*Rudiviridae*) or flexible filaments (*Lipothrixviridae*). Rudiviruses are relatively simple in composition, with no envelope and only a few proteins arranged in particles of variable length, usually related to that of the genome. Lipothrixviruses are enveloped, and the ends of their filamentous capsids are capped with structures of varied shapes (spider legs, pincers, bottle brushes), probably involved in attachment to the host (Fig. 12.8g). In spite of their different composition and morphology, rudi-

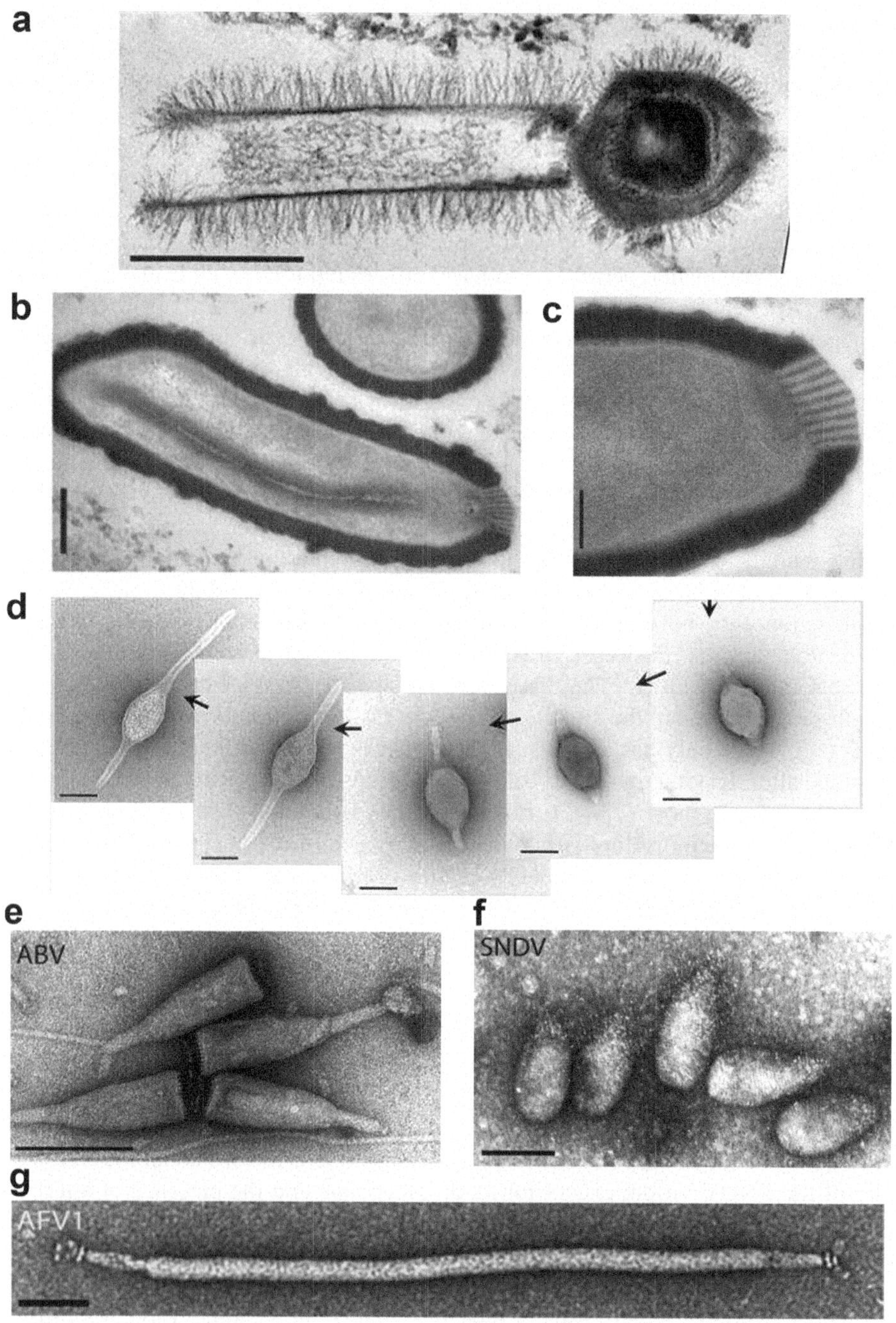

Fig. 12.8 Examples of protist and archaea viruses with unusual shapes. (**a**) A tupanvirus particle showing the icosahedral capsid at the right-hand side and the large, cylindrical tail at the left. (**b**) Longitudinal section of an *Acanthamoeba castellanii* pithovirus particle. Notice the dark, wall-like structure covering the particle and the opening (cork) at the right-hand side. (**c**) Detail of a

Fig. 12.8 (continued) pithovirus cork showing its striated pattern. (**d**) From right to left: extracellular elongation of the two tails in the fusiform ATV virion. (**e**) Bottle-shaped and (**f**) droplet-shaped viruses. (**g**) A lipothrixvirus representative (AFV-1) with claw-shaped structures at both ends. Panel (**a**) modified from [69]; (**b**, **c**) modified from [70]; (**d**) reproduced from [71]; (**e**–**g**) reproduced from [72], with permission. Scale bars represent 500 nm in (**a**), 200 nm in (**b**), and 100 nm in (**c**–**g**)

and lipothrixvirus capsids are built from similar structural principles, with capsid proteins arranged in a filament with helical symmetry, tightly covering the dsDNA genome. Remarkably, the dsDNA molecule is not in the more common B-form, but adopts the more compact A-form double helix architecture. This is another notable characteristic found only in archaeal viruses so far [76].

12.5 Sophisticated Regulation of Assembly and Maturation

In simple viruses, assembly can occur either in a single step where the newly replicated nucleic acid associates with capsid protein subunits during co-assembly, or in a two-step process where an empty capsid is assembled first and the viral nucleic acid is packaged afterward (see Chaps. 11 and 13). For complex viruses, putting together the many different pieces in their proper places at the appropriate time requires elaborated regulation of the morphogenesis process. In the following sections we discuss some of the strategies used by viruses to achieve assembly of complex capsids.

12.5.1 Separate Assembly Lines

In viruses with complex chemical composition, the different morphological components are often built separately, forming subassemblies that will be later put together along carefully regulated pathways. Some of these subassembly reactions may also require chaperones, either of cellular or viral origin, as is also the case for some structurally simpler viruses (Chap. 11). For example, in adenovirus, capsid protein oligomers are formed in the cytosol before being transported to the nucleus, where viral assembly takes place. This preassembly step includes hexon trimerization, which requires a viral chaperone (L4 100K) [77]; and piecing together the vertex complex, composed of a pentamer of penton base bound to a trimer of the fiber protein [78]. In adenovirus, however, the precise temporal order of incorporation of major and minor capsid proteins is not yet understood. Similarly, in tailed phages such as HK97, oligomerization of the major capsid protein into hexamers and pentamers precedes particle assembly. However, this is not the case for all tailed phages, even if they have very similar structures. For example, phage P22 capsid

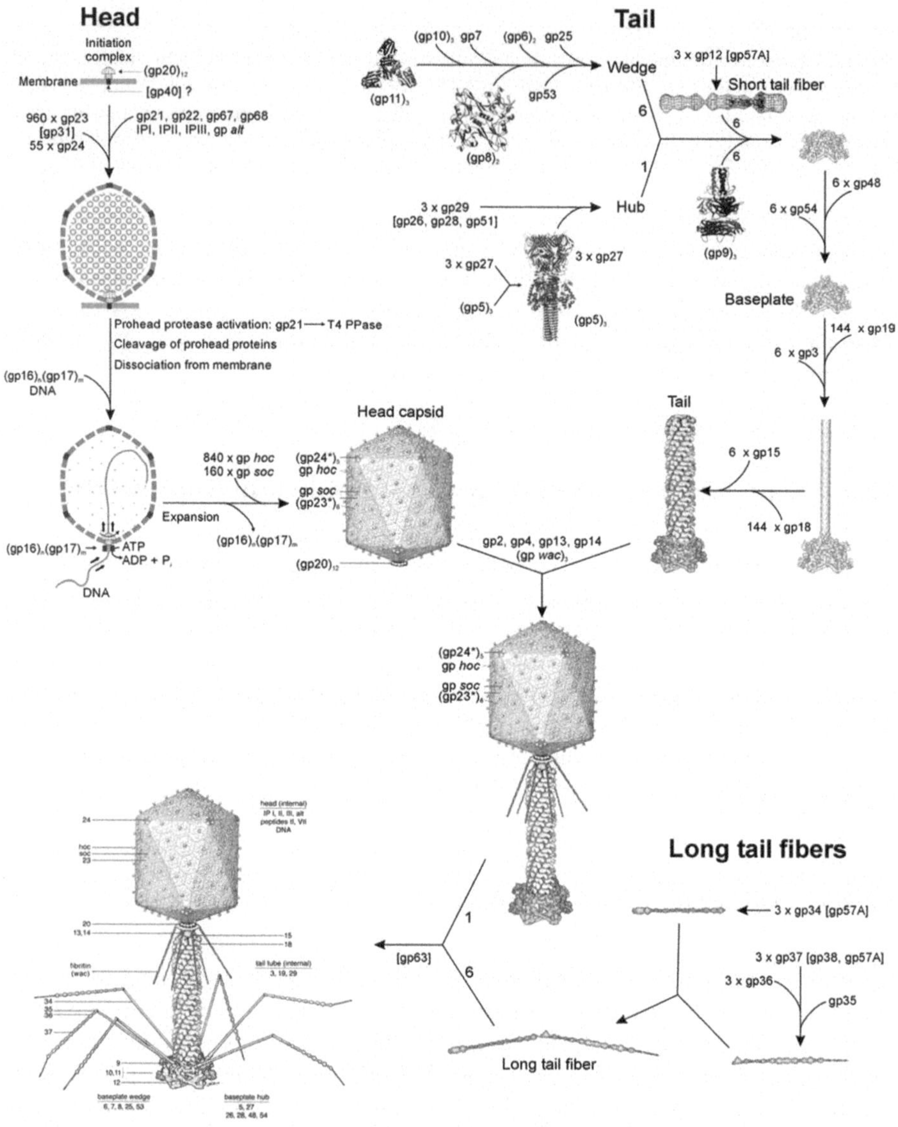

Fig. 12.9 Separate assembly lines. Schematic depiction of the complex bacteriophage T4 morphogenesis pathway. Reproduced from [52], with permission

assembly involves the addition of monomeric coat proteins to the growing edge of the capsid [79].

Tailed phages, and more specifically the myovirus prototype bacteriophage T4, are the best described examples of subassembly formation and integration into a virion [52]. In these long tailed particles, the head, fibers, and tail form separately (Fig. 12.9). Bacteriophage T4 head assembly starts from an initiation complex containing the portal protein gp20 bound to a cellular membrane. This complex recruits the components of a scaffolding core composed of eight different types of viral proteins, among them the main scaffolding protein gp22 and the viral protease (gp21). The core will later be coated by the hexameric and pentameric capsomers (gp23 and gp24) to form the procapsid, or prohead. Once coating is complete, gp21 is activated and cleaves all components of the procapsid except the portal protein. The prohead is detached from the membrane and most of the cleaved peptides exit the particle, opening the way for entry of the viral genome. Genome packaging leads to the large structural rearrangements involved in capsid maturation (Sect. 12.5.3 and Chap. 14), to produce the head in its final form. Finally, the head completion (or neck) proteins gp13, gp14, gp2, and gp4 attach to the portal to form the interface between the head and the tail.

The tail is in turn formed from several preassembled pieces. For contractile tails, the baseplate is assembled first and used as a seeding point for assembly of the inner tube and contractile sheath. In bacteriophage T4, tail assembly involves 19 different proteins and seven viral chaperones. To form the T4 baseplate, proteins gp6, gp7, gp8, gp10, gp11, gp25, and gp53 assemble in the form of hetero-oligomeric wedges. Six wedges bind around a central hub containing gp5 and gp27. Proteins gp9 and gp12 (the short tail fiber) are then inserted at the gaps between wedges, and the interface between wedges and hub is sealed by proteins gp48 and gp54. This seal is the starting point from which the gp19 inner tail tube will grow. The length of the tube is controlled by a tape measure protein gp29, which extends from the hub to the tube end where the tail capping protein gp3 will bind. The tail sheath gp18 assembles around the inner tube, and finally the tail terminator protein gp15 binds to gp3 and the last row of gp18 subunits, making the tail ready to bind to the neck proteins in the head.

The final stage of tail assembly is incorporation of the long tail fibers to the base plate. The fibers also assemble independently, even starting from separate sub-assemblies. The long tail fiber of bacteriophage T4 is kinked. The proximal part of the fiber before the kink is formed by a single protein (gp34), while the distal part contains three different proteins: gp35, gp36, and gp37. The proximal and distal parts assemble separately and then join before attaching to the baseplate. All long tail fiber proteins form trimers, except monomeric gp35 (notice the symmetry mismatch) that sits at the interface between the distal and proximal half fibers. Interestingly, the long fibers cannot join the tail until it is bound to the DNA-filled head. A similar assembly pathway has been described for non-contractile long tails. In podoviruses however, the short tail is not assembled as a separate entity, but it grows outward from the portal vertex on the virion capsid. The careful temporal regulation of these assembly processes is proved by the fact that, when any of the structural proteins is absent, viral morphogenesis is interrupted and the assembly intermediates previous to the disrupted step accumulate in the cell.

12.5.2 Scaffolding

Scaffolding elements are crucial for accurate assembly of large viral capsids. They are present in assembly intermediates (e.g., procapsids), but absent in the final, infectious product. Their role is to facilitate interactions between capsid elements at the early stages of assembly, by promoting nucleation—that is, putting together the viral proteins that may be highly diluted within the crowded cellular context. Scaffolds are also thought to stabilize weak interactions at the initial stages of assembly [80], while simultaneously allowing flexibility for mistake corrections. This last function is most important in large capsids, where the number of interactions to be checked for errors is correspondingly large. Finally, scaffolds have a role in determining the size and shape of viral capsids.

The most studied scaffold proteins are those present in tailed bacteriophages. For example, bacteriophage P22 scaffold is a 33 kDa protein. In the early stages of P22 morphogenesis, a procapsid is formed by 415 copies of the capsid protein, with approximately 300 molecules of scaffold inside. Unlike the capsid protein, the scaffold does not follow icosahedral symmetry; therefore, little is known about its organization in the assembly intermediate. Scaffold proteins in tailed phages have been quite refractory to structural studies. Nuclear magnetic resonance and crystallographic studies on the scaffold of P22 and Φ29 indicate that they have a helical fold. Biophysical analyses indicate that many of them share an extended, rod-like shape and a tendency to dimerize in solution [81]. However, an equilibrium between different oligomeric forms seems to be required to achieve correct capsid assembly. Kinetic studies have revealed that in phage P22, scaffold is predominantly a dimer during assembly, but the presence of free monomers is absolutely required to complete the head. Kinetically trapped intermediates are observed when monomers are depleted by decreasing the ionic strength, while restoring it eliminates the trap and allows elongation to proceed. Phage scaffolds are usually ejected from the procapsid immediately before genome packaging. In P22 and Φ29, the intact protein exits the shell and can be recycled in a new round of assembly. In other cases, the scaffold is removed via cleavage by a viral-encoded protease.

In spite of their apparently simple organization, some small bacteriophages, such as the representative of the *Microviridae* family ΦX174 ($T = 1$), encode both internal and external scaffold proteins [81]. The ΦX174 internal scaffold protein (protein B) helps in the early stages of assembly by preventing aggregation of the capsid protein F into aberrant oligomers and ensuring the recruitment of the vertex spike protein G. The C-terminal region of protein B (24 amino acids) interacts with the capsid and can be observed in the crystal structure of the procapsid, while the rest is disordered and appears to be largely tolerant to mutations. On the contrary, the external scaffold protein D is highly ordered and sensitive to mutations. Protein D is absolutely required for elongation (to assemble capsid pentamers into a spherical particle), while protein B helps to make assembly efficient but is not strictly required. In the absence of B, viral particles can be formed, but the process requires overexpression of protein D and takes as much as 10 times longer than in the

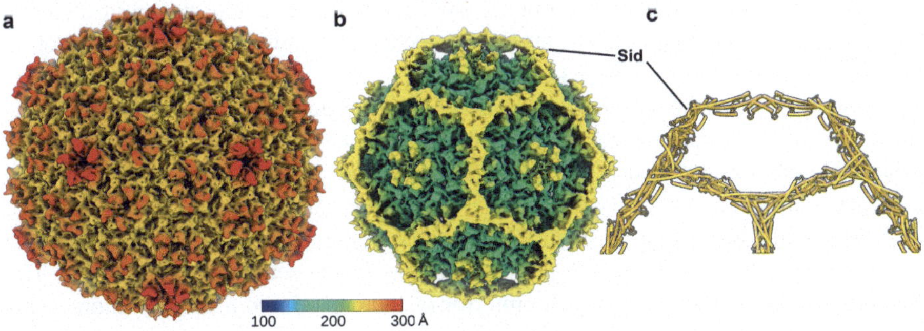

Fig. 12.10 Capsid size determination by scaffolding proteins: the P2/P4 phage system. (**a**) Bacteriophage P2 $T = 7$ procapsid (EMDB ID EMD_5406). (**b**) The parasitic P4 $T = 4$ procapsid (EMDB ID EMD_5405). Note the Sid external scaffold restricting the shell size. The view is along a twofold icosahedral axis. Color key indicates color changes with map radius. (**c**) Detail of the high-resolution structure showing the Sid network formed by interweaving α-helices (represented as tubes; PDB ID 7JW1)

presence of both scaffolds. It is believed that scaffold redundancy confers an evolutionary advantage to ΦX174 by facilitating extremely rapid replication cycles.

In the absence of scaffold, phage capsid proteins self-assemble into aberrant oligomers (tubes, elongated shells, icosahedrons with various T numbers) [79]. It follows from these observations that scaffolding proteins are involved in determining the correct curvature in the interactions between capsomers, so that they can form a closed shell of the appropriate size to hold the viral genome. A remarkable proof of this size determination role comes from the P2/P4 phage system. Phage P2 is a myovirus, with an assembly pathway similar to P22 that includes a $T = 7$ procapsid formed with the help of an internal scaffolding protein, gpO. Remarkably, P2 gpO combines the function of scaffold and protease, able to cleave itself and other shell components at later maturation stages. In the presence of P4, however, the whole assembly pathway is altered. P4 is a mobile genetic element, a replicon that can exist either as a prophage or a free plasmid and does not code for any major structural proteins. Instead, P4 sequesters players of the P2 assembly line to form small $T = 4$ capsids where only its smaller genome can fit, excluding that of P2 (Fig. 12.10). This parasitic process is achieved by synthesis of a P4 encoded scaffold protein, Sid (SIze Determination protein). Sid forms an external scaffold on P2 assembly intermediates, forcing a narrower curvature and therefore a smaller icosahedral net. Unlike the internal scaffolds, P4 Sid forms a well-ordered dodecahedral cage, allowing high-resolution determination of its structure by cryo-EM and single particle averaging [82, 83].

Very complex particles such as those of poxviruses also use scaffolding proteins for assembly. Protein D13 directs the curvature of cell membranes into growing crescents and finally spherical immature particles during vaccinia virus assembly (Chap. 14). In the immature vaccinia virus particle, D13 forms trimers arranged in a hexagonal network on the outer side of the membrane. Similarly to bacteriophage scaffolds, D13 is removed by proteolytic cleavage and is absent in the mature virion. It has been proposed that both the D13 and the palisade networks contribute to define the size of the poxvirus particle [84].

Although one can generally speak about scaffolding proteins, scaffolding functions can also be performed by flexible regions of the capsid proteins, which establish interactions during assembly that are later removed via conformational changes or cleavage by viral proteases. For example, bacteriophage HK97 does not encode a scaffold protein. Instead, a 103-residue stretch at the N-terminus of the capsid protein, known as the delta domain, performs the scaffold function [85]. The delta domain is oriented toward the interior of the capsid and mediates interactions between capsomers during assembly, including the symmetry mismatch at the portal vertex [86]. Once the procapsid is completed, and before the DNA is packaged, the delta domain is cleaved out by the viral protease, allowing the transition to the next assembly intermediate [81] (Chap. 14).

12.5.3 Maturation

In complex viruses, piecing together a number of proteins into a capsule to host the genome is far from producing the final, infectious form. Instead, the newly formed particles (procapsids) need to undergo a series of morphological and/or stability changes to acquire their full infectious potential. This process is known as maturation. There is a double goal for maturation on the viral cycle: first, to produce virions stable enough to protect the genome from aggressive conditions in the extracellular milieu; and second, to prepare the viral particle for correct delivery of the genome into the new host cell.

In dsDNA bacteriophages, maturation encompasses large structural changes and capsid protein rearrangements, concomitant with scaffold removal and genome packaging. The capsid changes from a weak, labile object to a highly stable shell, ready to withstand the high internal pressure imposed by the tightly packed DNA inside [87, 88] (see Chaps. 10, 13, 18 and 19). In other viruses, such as adenovirus, maturation does not end with an extremely stable shell, but with a metastable one [41]. This difference with respect to the bacteriophage case is likely related to the mode of infection specific of each virus. Bacteriophages eject their genome across the bacteria cell wall and membrane leaving the capsid behind, while adenovirus is internalized in the eukaryotic cell, and must be disassembled within in a concerted fashion to ensure exposure of the genome at the appropriate place and time for successful replication. Maturation prepares these viruses to start the programmed uncoating sequence upon reception of the appropriate signal, for example attachment to the receptor, or pH changes along the endocytosis pathway. In enveloped viruses such as dengue virus, surface glycoproteins mature into metastable arrangements needed to trigger fusion of the virus and cell membranes for genome delivery [89].

Maturation processes are not restricted to icosahedral capsids: pleomorphic enveloped viruses such as poxviruses or retroviruses also undergo extensive structural rearrangements to become fully infectious. A more extensive discussion on maturation for several different viruses can be found in Chap. 14.

12.5.4 Membrane Acquisition in Viruses with Lipidic Layers

Morphogenesis of viruses incorporating lipidic layers requires additional steps to coordinate assembly of proteic elements with recruitment of membranes from the cell. Enveloped viruses acquire their external lipid bilayer by specific interactions between viral proteins and cellular membranes, in a process called *budding* [90]. Budding can happen at the cell membrane, in which case it is directly coupled with virus egress. This is the case of retroviruses or influenza viruses. In other cases, budding happens at internal membranes, and the particle needs to follow cell transport routes (for example, the exocytic route) for its final release. This is the case of flaviviruses or coronaviruses.

Budding starts by recruitment of viral proteins to specific membrane regions. Viral proteins (for example Gag in retroviruses) alter the curvature of the bilayer and recruit the rest of viral components at the assembly site. Depending on the virus, interactions between envelope proteins and internal components (for example, capsid in alphaviruses or matrix protein in influenza) may be required for curvature alteration. Once the bud has grown to enclose the complete particle, it remains attached to the original cellular membrane by a thin stalk, which must be pinched off to release the new virion. Many viruses such as HIV-1 or herpesvirus rely on cell resources for this task. In particular, they depend on the ESCRT (endosomal sorting complexes required for transport) machinery [91]. In uninfected cells, ESCRT complexes are involved in many processes requiring membrane constriction and fission, such as formation of multivesicular bodies or separation of daughter cells during mitosis [92]. It is unclear if viruses independent from the ESCRT system (for example, alphaviruses) rely solely on viral proteins or if they recruit other—so far unidentified—cell factors for budding.

Viruses with internal membranes also recruit their lipids from the host. In this case, membranes seem to function as scaffolding platforms for recruitment of proteins forming the icosahedral shell [93]. Although viral membranes are acquired from those already existing in the cell, viruses can select a specific lipidic composition. The membrane of bacteriophage PRD1 contains more phosphatidylglycerol and less phosphatidylethanolamine than the bacterial membrane. In addition, lipids are asymmetrically distributed between the two membrane leaflets. The outer leaflet is enriched in negatively charged lipids (phosphatidylglycerol and cardiolipin), while the inner one is enriched in phosphatidylethanolamine. Specific interactions between lipids and the viral membrane proteins may be the cause of this asymmetry. In turn, asymmetry of lipid charges may be relevant during assembly, helping to condense the genome on one side of the membrane and to recruit capsomers via electrostatic interactions on the other [26].

In the giant viruses such as mimivirus or poxviruses, open membrane fragments located at the periphery of the replication factory seem to serve as nucleation points for recruitment of scaffold (poxvirus) (see also Chap. 8) or capsid (mimivirus) proteins [94]. ASFV recruits its internal membrane by deformation of the endoplasmic reticulum. ASFV capsomers assemble on the convex side of the open membrane fragments, forcing them into an icosahedral shape [27, 95]. At the end of morphogenesis, ASFV particles exit the cell by budding through the plasma membrane, thereby acquiring their second, external membrane [27]. Because it may happen at a late stage of morphogenesis after assembly of the rest of components, membrane acquisition may be considered part of the maturation process. More details on membrane acquisition for complex viruses such as herpes or vaccinia virus, as well as the intriguing envelopment mechanism of the archaea lipothrixviruses, are presented in Chap. 14.

12.6 Overview and Conclusions

Structurally complex viruses incorporate a wide range of molecules into their capsids, including specialized host interaction, genome packaging, and cementing proteins; and in some cases membranes, either internal or external. Accommodation of the different components often requires deviations from high order symmetry, from mismatches to pleomorphism, and involves complex regulation of the assembly dynamics. Key elements in this regulation are: separate assembly lines, scaffold elements, and maturation processes. Host and virus evolution probably act hand in hand to optimize viral particle structure and morphogenesis. In fact, virus structure and evolution are inextricably linked.

Structural studies have shown that many viruses infecting hosts far apart in evolution share similar architectural features, prompting a virus classification in the so-called structural lineages [96] (see also Chap. 7). Three of these lineages include many of the complex viruses described in this chapter. Herpesviruses, which infect animals, share many structural characteristics with tailed bacteriophages. They follow a similar assembly pathway, starting from an empty procapsid formed with the help of scaffold, and maturing to a more angular shell via large structural rearrangements upon DNA packaging. As in tailed phages, one of the vertices in the herpesvirus nucleocapsid is different from the rest and contains a 12-fold symmetric portal structure involved in genome encapsidation. The herpesvirus coat protein folds with a topology very similar to that of the HK97 phage family; and like tailed phages, the genome is tightly packed causing a large internal pressure in the mature virion (Chap. 13). Herpesvirus and tailed phages are grouped in the *Duplodnaviria* realm in the latest taxonomic classification of viruses [97]. Structural parallels also exist between cystoviruses (tail-less dsRNA bacteriophages, representative Φ6) and reoviruses that infect eukaryotic hosts, including the multilayered capsid (Sect. 12.3.1), the core protein fold, and probably the structural transitions undergone during genome encapsidation (Chap. 14). These viruses constitute the phylum *Duplornaviricota* within the vast *Riboviria* realm.

Another lineage comprises dsDNA viruses infecting organisms across the evolutionary tree: bacteria (tail-less phages such as tectiviruses and corticoviruses); archaea (turriviruses); protozoa (mimivirus) and algae (phycodnaviruses); insects, fish, amphibians, and reptiles (iridoviruses); pigs and arthropods (asfarviruses); and vertebrates in general including humans (adenoviruses) [5]. Most major capsid proteins in this lineage adopt the double jelly roll fold. All these viruses belong to the *Varidnaviria* realm. With genome lengths between 10 and 1500 kbp, capsid diameters from 60 to 500 nm, and icosahedral geometries starting at $T = 21$ and going all the way to the $T \sim 1000$ range, this lineage contains the largest and most complex icosahedral viruses discovered so far (Sect. 12.4.3). Virion composition is correspondingly complex: the largest virus solved at sub-4 Å resolution, the phycodnavirus PBCV-1 ($T = 169$d), has at least 13 different types of cementing proteins and six different variants of the pseudo-hexameric capsomer, located in a specific pattern in the capsid [46, 98]. Most of these viruses (except adenoviruses, virophages and possibly Faustovirus) have an internal membrane beneath the icosahedral shell. Some of them pack their genomes bound to virus-encoded proteins. In mimivirus, oxidoreductases form both fibrils on the capsid surface and a helical nucleocapsid inside, and marseilleviruses encode histone doublets [20, 99]. Faustovirus and ASFV have two capsid shells with different triangulation numbers (Sect. 12.4.3). Several morphogenetic pathways have been proposed for these very large icosahedral capsids with pseudo-hexameric, double jelly roll capsomers [6]. For example, the capsid could be assembled from preformed pentasymmetrons and trisymmetrons, with minor coat proteins riveting the seams ("zipper" proteins) and controlling the symmetron size (tape measure proteins, Sect. 12.2.1). Alternatively, the capsid could grow in a spiral fashion by recruiting pseudo-hexameric capsomers one by one around a pentameric capsomer (perhaps the one at the singular vertex) that serves as an assembly initiation point [46]. Capsomer recruitment could be guided by electrostatic interactions between the double jelly roll surfaces, or by minor coat proteins working as "feelers" from a position anchored to the initial vertex assembly.

Genes coding for major capsid proteins are part of the set of hallmark genes that define virus taxonomic classification [97]. The fact that many complex viruses with different hosts share a common structural solution has evolutionary implications. First, since the design has been conserved throughout time, even after most traces of sequence similarity have disappeared, it must be a highly efficient arrangement. Second, it suggests that the architecture was established in the early stages of evolution, before the branches of the evolutionary tree diverged into the three kingdoms known today (archaea, bacteria, and eukarya). Sophisticated sequence/structure comparison analyses indicate that many viral proteins, including the major coat proteins of duplodnaviruses and varidnaviruses, derive from cell proteins in a process called *exaptation* (the repurposing of a host protein for unique virus functions). The cell proteins identified as ancestors of these capsid proteins are present in all domains of life, suggesting that indeed the exaptation may already have happened before the era of the last universal common ancestor (LUCA) [100].

Metagenomics studies indicate that many more of these complex viruses inhabit all environments of the biosphere [5, 99]. Their structural characterization is bound to reveal further architectural and functional diversity and improve our insights on how these large macromolecular complexes assemble. However, going from genome sequences to the actual virus isolation and propagation in laboratory conditions is not straightforward, as in many cases not even the host is known. Feeding of newly solved structures into the databases used for training artificial intelligence tools such as AlphaFold [101] will also improve our ability to predict the fold of proteins available only as their coding nucleotide sequences.

Acknowledgments The C.S.M. group is supported by grants from the Spanish State Research Agency, with co-funding from the European Regional Development Fund (PID2019-104098GB-I00/AEI/10.13039/501100011033 and PID2022-136456NB-I00/AEI/10.13039/501100011033), CSIC (2023AEP082), the European Innovation Council (Horizon Europe) under grant agreement No 101098647 (project iAds), and Marie Skłodowska-Curie Actions (grant agreement 101129778, project INVECTA). The CNB-CSIC is further supported by a Severo Ochoa Excellence grant (CEX2023-001386-S/AEI/10.13039/501100011033). The group is a member of the Spanish Adenovirus Network (AdenoNet, RED2022-134221-T/AEI/10.13039/501100011033), CSIC LifeHub, and CSIC BCBHub.

References

1. Lawrence CM (2021) Structure of helical viruses. In: Bamford DH, Zuckerman M (eds) Encyclopedia of virology, 4th edn. Academic Press, Oxford, pp 362–371
2. Goff SP (2013) Retroviridae. In: Knipe DM, Howley PM (eds) Fields virology, Wolters Kluwer/Lippincott Williams & Wilkins Health, vol 2. Philadelphia, PA, pp 1424–1473
3. Gallardo J, Pérez-Illana M, Martín-González N, San Martín C (2021) Adenovirus structure: what is new? Int J Mol Sci 22
4. San Martín C (2012) Latest insights on adenovirus structure and assembly. Viruses 4:847–877
5. San Martín C, van Raaij MJ (2018) The so far farthest reaches of the double jelly roll capsid protein fold. Virol J 15:181
6. Xian Y, Xiao C (2020) Current capsid assembly models of icosahedral nucleocytoviricota viruses. Adv Virus Res 108:275–313
7. Liu H, Jin L, Koh SB, Atanasov I, Schein S, Wu L, Zhou ZH (2010) Atomic structure of human adenovirus by cryo-EM reveals interactions among protein networks. Science 329: 1038–1043
8. Yu X, Mullen T-M, Abrishami V, Huiskonen JT, Nemerow GR, Reddy VS (2022) Structure of a cell entry defective human adenovirus provides insights into precursor proteins and capsid maturation. J Mol Biol 434:167350
9. Colby WW, Shenk T (1981) Adenovirus type 5 virions can be assembled in vivo in the absence of detectable polypeptide IX. J Virol 39:977–980
10. Abrescia NG, Cockburn JJ, Grimes JM, Sutton GC, Diprose JM, Butcher SJ, Fuller SD, San Martín C, Burnett RM, Stuart DI, Bamford DH, Bamford JK (2004) Insights into assembly from structural analysis of bacteriophage PRD1. Nature 432:68–74

11. Roizman B, Knipe DM, Whitley RJ (2007) Herpes simplex viruses. In: Knipe DM, Howley PM, Griffin DE, Lamb RA, Martin MA (eds) Fields virology, vol II. Lippincott Williams & Wilkins, Philadelphia, pp 2501–2601
12. Mangel WF, San Martín C (2014) Structure, function and dynamics in adenovirus maturation. Viruses 6:4536–4570
13. Liu YT, Jih J, Dai X, Bi GQ, Zhou ZH (2019) Cryo-EM structures of herpes simplex virus type 1 portal vertex and packaged genome. Nature 570:257–261
14. Rao VB, Feiss M (2015) Mechanisms of DNA packaging by large double-stranded DNA viruses. Annu Rev Virol 2:351–378
15. Heymann JB (2023) Structural studies of bacteriophage Phi6 and its transformations during its life cycle. Viruses 15
16. Trask SD, McDonald SM, Patton JT (2012) Structural insights into the coupling of virion assembly and rotavirus replication. Nat Rev Microbiol 10:165–177
17. Condit RC, Moussatche N, Traktman P (2006) In a nutshell: structure and assembly of the vaccinia virion. Adv Virus Res 66:31–124
18. Germond JE, Hirt B, Oudet P, Gross-Bellark M, Chambon P (1975) Folding of the DNA double helix in chromatin-like structures from simian virus 40. Proc Natl Acad Sci USA 72: 1843–1847
19. Tweeten KA, Bulla LA, Consigli RA (1980) Characterization of an extremely basic protein derived from granulosis virus nucleocapsids. J Virol 33:866–876
20. Talbert PB, Henikoff S, Armache KJ (2023) Giant variations in giant virus genome packaging. Trends Biochem Sci 48:1071–1082
21. Ruigrok RW, Crepin T, Kolakofsky D (2011) Nucleoproteins and nucleocapsids of negative-strand RNA viruses. Curr Opin Microbiol 14:504–510
22. Palese P, Shaw ML (2007) Orthomyxoviridae: the viruses and their replication. In: Knipe DM, Howley PM, Griffin DE, Lamb RA, Martin MA (eds) Fields virology, vol II. Lippincott Williams & Wilkins, Philadelphia, pp 1647–1689
23. Huiskonen JT, Butcher SJ (2007) Membrane-containing viruses with icosahedrally symmetric capsids. Curr Opin Struct Biol 17:229–236
24. Grünewald K, Desai P, Winkler DC, Heymann JB, Belnap DM, Baumeister W, Steven AC (2003) Three-dimensional structure of herpes simplex virus from cryo-electron tomography. Science 302:1396–1398
25. Forster F, Medalia O, Zauberman N, Baumeister W, Fass D (2005) Retrovirus envelope protein complex structure in situ studied by cryo-electron tomography. Proc Natl Acad Sci USA 102:4729–4734
26. Butcher SJ, Manole V, Karhu NJ (2012) Lipid-containing viruses: bacteriophage PRD1 assembly. Adv Exp Med Biol 726:365–377
27. Salas ML, Andrés G (2013) African swine fever virus morphogenesis. Virus Res 173:29–41
28. Matthijnssens J, Attoui H, Banyai K, Brussaard CPD, Danthi P, Del Vas M, Dermody TS, Duncan R, Fang Q, Johne R, Mertens PPC, Mohd Jaafar F, Patton JT, Sasaya T, Suzuki N, Wei T (2022) ICTV virus taxonomy profile: Spinareoviridae 2022. J Gen Virol 103
29. Matthijnssens J, Attoui H, Banyai K, Brussaard CPD, Danthi P, Del Vas M, Dermody TS, Duncan R, Fang Q, Johne R, Mertens PPC, Mohd Jaafar F, Patton JT, Sasaya T, Suzuki N, Wei T (2022) ICTV virus taxonomy profile: Sedoreoviridae 2022. J Gen Virol 103
30. Caspar DLD, Klug A (1962) Physical principles in the construction of regular viruses. Cold Spring Harb Symp Quant Biol 27:1–24
31. Klose T, Reteno DG, Benamar S, Hollerbach A, Colson P, La Scola B, Rossmann MG (2016) Structure of faustovirus, a large dsDNA virus. Proc Natl Acad Sci USA 113:6206–6211
32. Wang N, Zhao D, Wang J, Zhang Y, Wang M, Gao Y, Li F, Wang J, Bu Z, Rao Z, Wang X (2019) Architecture of African swine fever virus and implications for viral assembly. Science 366:640–644

33. Andrés G, Charro D, Matamoros T, Dillard RS, Abrescia NGA (2020) The cryo-EM structure of African swine fever virus unravels a unique architecture comprising two icosahedral protein capsids and two lipoprotein membranes. J Biol Chem 295:1–12
34. Baker M, Prasad BV (2010) Rotavirus cell entry. Curr Top Microbiol Immunol 343:121–148
35. Danthi P, Guglielmi KM, Kirchner E, Mainou B, Stehle T, Dermody TS (2010) From touchdown to transcription: the reovirus cell entry pathway. Curr Top Microbiol Immunol 343:91–119
36. Huiskonen J, Manole V, Butcher S (2007) Tale of two spikes in bacteriophage PRD1. Proc Natl Acad Sci USA 104:6666–6671
37. van Raaij MJ, Mitraki A, Lavigne G, Cusack S (1999) A triple β-spiral in the adenovirus fibre shaft reveals a new structural motif for a fibrous protein. Nature 401:935–938
38. Zubieta C, Schoehn G, Chroboczek J, Cusack S (2005) The structure of the human adenovirus 2 penton. Mol Cell 17:121–135
39. Liu H, Wu L, Zhou ZH (2011) Model of the trimeric fiber and its interactions with the pentameric penton base of human adenovirus by cryo-electron microscopy. J Mol Biol 406: 764–774
40. Schoehn G, Fender P, Chroboczek J, Hewat EA (1996) Adenovirus 3 penton dodecahedron exhibits structural changes of the base on fibre binding. EMBO J 15:6841–6846
41. de Pablo PJ, San Martín C (2022) Seeing and touching adenovirus: complementary approaches for understanding assembly and disassembly of a complex virion. Curr Opin Virol 52:112–122
42. Chappell JD, Prota AE, Dermody TS, Stehle T (2002) Crystal structure of reovirus attachment protein σ1 reveals evolutionary relationship to adenovirus fiber. EMBO J 21:1–11
43. Hess M, Cuzange A, Ruigrok RWH, Chroboczek J, Jacrot B (1995) The avian adenovirus penton: two fibres and one base. J Mol Biol 252:379–385
44. Cherrier MV, Kostyuchenko VA, Xiao C, Bowman VD, Battisti AJ, Yan X, Chipman PR, Baker TS, Van Etten JL, Rossmann MG (2009) An icosahedral algal virus has a complex unique vertex decorated by a spike. Proc Natl Acad Sci USA 106:11085–11089
45. Zauberman N, Mutsafi Y, Halevy D, Shimoni E, Klein E, Xiao C, Sun S, Minsky A (2008) Distinct DNA exit and packaging portals in the virus Acanthamoeba polyphaga mimivirus. PLoS Biol 13:e114
46. Shao Q, Agarkova IV, Noel EA, Dunigan DD, Liu Y, Wang A, Guo M, Xie L, Zhao X, Rossmann MG, Van Etten JL, Klose T, Fang Q (2022) Near-atomic, non-icosahedrally averaged structure of giant virus Paramecium bursaria chlorella virus 1. Nat Commun 13:6476
47. Zhang X, Xiang Y, Dunigan DD, Klose T, Chipman PR, Van Etten JL, Rossmann MG (2011) Three-dimensional structure and function of the Paramecium bursaria chlorella virus capsid. Proc Natl Acad Sci USA 108:14837–14842
48. Xiao C, Kuznetsov YG, Sun S, Hafenstein SL, Kostyuchenko VA, Chipman PR, Suzan-Monti M, Raoult D, McPherson A, Rossmann MG (2009) Structural studies of the giant mimivirus. PLoS Biol 7:e92
49. Goetschius DJ, Lee H, Hafenstein S (2019) CryoEM reconstruction approaches to resolve asymmetric features. Adv Virus Res 105:73–91
50. Conley MJ, McElwee M, Azmi L, Gabrielsen M, Byron O, Goodfellow IG, Bhella D (2019) Calicivirus VP2 forms a portal-like assembly following receptor engagement. Nature 565: 377–381
51. Turner D, Shkoporov AN, Lood C, Millard AD, Dutilh BE, Alfenas-Zerbini P, van Zyl LJ, Aziz RK, Oksanen HM, Poranen MM, Kropinski AM, Barylski J, Brister JR, Chanisvili N, Edwards RA, Enault F, Gillis A, Knezevic P, Krupovic M, Kurtboke I, Kushkina A, Lavigne R, Lehman S, Lobocka M, Moraru C, Moreno Switt A, Morozova V, Nakavuma J, Reyes Munoz A, Rumnieks J, Sarkar BL, Sullivan MB, Uchiyama J, Wittmann J, Yigang T, Adriaenssens EM (2023) Abolishment of morphology-based taxa and change to binomial species names: 2022 taxonomy update of the ICTV bacterial viruses subcommittee. Arch Virol 168:74

52. Leiman PG, Kanamaru S, Mesyanzhinov VV, Arisaka F, Rossmann MG (2003) Structure and morphogenesis of bacteriophage T4. Cell Mol Life Sci 60:2356–2370
53. Leiman PG, Shneider MM (2012) Contractile tail machines of bacteriophages. Adv Exp Med Biol 726:93–114
54. Li F, Hou CD, Lokareddy RK, Yang R, Forti F, Briani F, Cingolani G (2023) High-resolution cryo-EM structure of the Pseudomonas bacteriophage E217. Nat Commun 14:4052
55. Agirrezabala X, Martín-Benito J, Castón JR, Miranda R, Valpuesta JM, Carrascosa JL (2005) Maturation of phage T7 involves structural modification of both shell and inner core components. EMBO J 24:3820–3829
56. Davidson AR, Cardarelli L, Pell LG, Radford DR, Maxwell KL (2012) Long noncontractile tail machines of bacteriophages. Adv Exp Med Biol 726:115–142
57. Casjens SR, Molineux IJ (2012) Short noncontractile tail machines: adsorption and DNA delivery by podoviruses. Adv Exp Med Biol 726:143–179
58. Pérez-Ruiz M, Pulido-Cid M, Luque-Ortega JR, Valpuesta JM, Cuervo A, Carrascosa JL (2021) Assisted assembly of bacteriophage T7 core components for genome translocation across the bacterial envelope. Proc Natl Acad Sci USA 118
59. Cyrklaff M, Risco C, Fernández JJ, Jiménez MV, Estéban M, Baumeister W, Carrascosa JL (2005) Cryo-electron tomography of vaccinia virus. Proc Natl Acad Sci USA 102:2772–2777
60. Pettersen EF, Goddard TD, Huang CC, Couch GS, Greenblatt DM, Meng EC, Ferrin TE (2004) UCSF Chimera – a visualization system for exploratory research and analysis. J Comput Chem 25:1605–1612
61. Hernández-González M, Calcraft T, Nans A, Rosenthal PB, Way M (2024) Palisade structure in intact vaccinia virions. mBio:e0313423
62. Kuznetsov Y, Gershon PD, McPherson A (2008) Atomic force microscopy investigation of vaccinia virus structure. J Virol 82:7551–7566
63. Harris A, Cardone G, Winkler DC, Heymann JB, Brecher M, White JM, Steven AC (2006) Influenza virus pleiomorphy characterized by cryoelectron tomography. Proc Natl Acad Sci USA 103:19123–19127
64. Liljeroos L, Huiskonen JT, Ora A, Susi P, Butcher SJ (2011) Electron cryotomography of measles virus reveals how matrix protein coats the ribonucleocapsid within intact virions. Proc Natl Acad Sci USA 108:18085–18090
65. Ke Z, Oton J, Qu K, Cortese M, Zila V, McKeane L, Nakane T, Zivanov J, Neufeldt CJ, Cerikan B, Lu JM, Peukes J, Xiong X, Krausslich HG, Scheres SHW, Bartenschlager R, Briggs JAG (2020) Structures and distributions of SARS-CoV-2 spike proteins on intact virions. Nature 588:498–502
66. Pornillos O, Ganser-Pornillos BK, Yeager M (2011) Atomic-level modelling of the HIV capsid. Nature 469:424–427
67. Cardone G, Purdy JG, Cheng N, Craven RC, Steven AC (2009) Visualization of a missing link in retrovirus capsid assembly. Nature 457:694–698
68. Queiroz VF, Rodrigues RAL, de Miranda Boratto PV, La Scola B, Andreani J, Abrahao JS (2022) Amoebae: hiding in plain sight: unappreciated hosts for the very large viruses. Annu Rev Virol 9:79–98
69. Abrahao J, Silva L, Silva LS, Khalil JYB, Rodrigues R, Arantes T, Assis F, Boratto P, Andrade M, Kroon EG, Ribeiro B, Bergier I, Seligmann H, Ghigo E, Colson P, Levasseur A, Kroemer G, Raoult D, La Scola B (2018) Tailed giant Tupanvirus possesses the most complete translational apparatus of the known virosphere. Nat Commun 9:749
70. Legendre M, Bartoli J, Shmakova L, Jeudy S, Labadie K, Adrait A, Lescot M, Poirot O, Bertaux L, Bruley C, Coute Y, Rivkina E, Abergel C, Claverie JM (2014) Thirty-thousand-year-old distant relative of giant icosahedral DNA viruses with a pandoravirus morphology. Proc Natl Acad Sci USA 111:4274–4279
71. Haring M, Vestergaard G, Rachel R, Chen L, Garrett RA, Prangishvili D (2005) Virology: independent virus development outside a host. Nature 436:1101–1102

72. Pina M, Bize A, Forterre P, Prangishvili D (2011) The archeoviruses. FEMS Microbiol Rev 35:1035–1054
73. Prangishvili D, Bamford DH, Forterre P, Iranzo J, Koonin EV, Krupovic M (2017) The enigmatic archaeal virosphere. Nat Rev Microbiol 15:724–739
74. Pietila MK, Roine E, Sencilo A, Bamford DH, Oksanen HM (2016) Pleolipoviridae, a newly proposed family comprising archaeal pleomorphic viruses with single-stranded or double-stranded DNA genomes. Arch Virol 161:249–256
75. Bolduc B, Shaughnessy DP, Wolf YI, Koonin EV, Roberto FF, Young M (2012) Identification of novel positive-strand RNA viruses by metagenomic analysis of archaea-dominated Yellowstone hot springs. J Virol 86:5562–5573
76. Baquero DP, Liu Y, Wang F, Egelman EH, Prangishvili D, Krupovic M (2020) Structure and assembly of archaeal viruses. Adv Virus Res 108:127–164
77. Hong SS, Szolajska E, Schoehn G, Franqueville L, Myhre S, Lindholm L, Ruigrok RW, Boulanger P, Chroboczek J (2005) The 100K-chaperone protein from adenovirus serotype 2 (subgroup C) assists in trimerization and nuclear localization of hexons from subgroups C and B adenoviruses. J Mol Biol 352:125–138
78. Karayan L, Gay B, Gerfaux J, Boulanger PA (1994) Oligomerization of recombinant penton base of adenovirus type 2 and its assembly with fiber in baculovirus-infected cells. Virology 202:782–795
79. Suhanovsky MM, Teschke CM (2015) Nature's favorite building block: deciphering folding and capsid assembly of proteins with the HK97-fold. Virology 479-480:487–497
80. Johnson JE (2010) Virus particle maturation: insights into elegantly programmed nanomachines. Curr Opin Struct Biol 20:210–216
81. Prevelige PE, Fane BA (2012) Building the machines: scaffolding protein functions during bacteriophage morphogenesis. Adv Exp Med Biol 726:325–350
82. Dearborn AD, Laurinmaki P, Chandramouli P, Rodenburg CM, Wang S, Butcher SJ, Dokland T (2012) Structure and size determination of bacteriophage P2 and P4 procapsids: function of size responsiveness mutations. J Struct Biol 178:215–224
83. Kizziah JL, Rodenburg CM, Dokland T (2020) Structure of the capsid size-determining scaffold of "satellite" Bacteriophage P4. Viruses 12
84. Hernández-González M, Calcraft T, Nans A, Rosenthal PB, Way M (2023) A succession of two viral lattices drives vaccinia virus assembly. PLoS Biol 21:e3002005
85. Huang RK, Khayat R, Lee KK, Gertsman I, Duda RL, Hendrix RW, Johnson JE (2011) The Prohead-I structure of bacteriophage HK97: implications for scaffold-mediated control of particle assembly and maturation. J Mol Biol 408:541–554
86. Huet A, Oh B, Maurer J, Duda RL, Conway JF (2023) A symmetry mismatch unraveled: how phage HK97 scaffold flexibly accommodates a 12-fold pore at a 5-fold viral capsid vertex. Sci Adv 9:eadg8868
87. San Martín C (2019) Virus maturation. Adv Exp Med Biol 1215:129–158
88. Veesler D, Johnson JE (2012) Virus maturation. Annu Rev Biophys 41:473–496
89. Vaney MC, Dellarole M, Duquerroy S, Medits I, Tsouchnikas G, Rouvinski A, England P, Stiasny K, Heinz FX, Rey FA (2022) Evolution and activation mechanism of the flavivirus class II membrane-fusion machinery. Nat Commun 13:3718
90. Rheinemann L, Sundquist WI (2021) Virus budding. Encyclopedia of virology 519–528
91. Votteler J, Sundquist WI (2013) Virus budding and the ESCRT pathway. Cell Host Microbe 14:232–241
92. Scourfield EJ, Martin-Serrano J (2017) Growing functions of the ESCRT machinery in cell biology and viral replication. Biochem Soc Trans 45:613–634
93. Abrescia NG, Grimes JM, Kivela HM, Assenberg R, Sutton GC, Butcher SJ, Bamford JK, Bamford DH, Stuart DI (2008) Insights into virus evolution and membrane biogenesis from the structure of the marine lipid-containing bacteriophage PM2. Mol Cell 31:749–761

94. Suarez C, Welsch S, Chlanda P, Hagen W, Hoppe S, Kolovou A, Pagnier I, Raoult D, Krijnse Locker J (2013) Open membranes are the precursors for assembly of large DNA viruses. Cell Microbiol 15:1883–1895
95. Suarez C, Andres G, Kolovou A, Hoppe S, Salas ML, Walther P, Krijnse Locker J (2015) African swine fever virus assembles a single membrane derived from rupture of the endoplasmic reticulum. Cell Microbiol 17:1683–1698
96. Abrescia NG, Bamford DH, Grimes JM, Stuart DI (2012) Structure unifies the viral universe. Annu Rev Biochem 81:795–822
97. Koonin EV, Dolja VV, Krupovic M, Varsani A, Wolf YI, Yutin N, Zerbini FM, Kuhn JH (2020) Global organization and proposed megataxonomy of the virus world. Microbiol Mol Biol Rev 84
98. Fang Q, Zhu D, Agarkova I, Adhikari J, Klose T, Liu Y, Chen Z, Sun Y, Gross ML, Van Etten JL, Zhang X, Rossmann MG (2019) Near-atomic structure of a giant virus. Nat Commun 10:388
99. Schulz F, Abergel C, Woyke T (2022) Giant virus biology and diversity in the era of genome-resolved metagenomics. Nat Rev Microbiol 20:721
100. Koonin EV, Dolja VV, Krupovic M (2022) The logic of virus evolution. Cell Host Microbe 30:917–929
101. Jumper J, Evans R, Pritzel A, Green T, Figurnov M, Ronneberger O, Tunyasuvunakool K, Bates R, Zidek A, Potapenko A, Bridgland A, Meyer C, Kohl SAA, Ballard AJ, Cowie A, Romera-Paredes B, Nikolov S, Jain R, Adler J, Back T, Petersen S, Reiman D, Clancy E, Zielinski M, Steinegger M, Pacholska M, Berghammer T, Bodenstein S, Silver D, Vinyals O, Senior AW, Kavukcuoglu K, Kohli P, Hassabis D (2021) Highly accurate protein structure prediction with AlphaFold. Nature 596:583–589

Further Reading

Bamford DH, Zuckerman M (eds) (2021) Encyclopedia of virology, vol 1, 4th edn. Academic Press, Oxford
Flint J, Racaniello VR, Rall GF, Hatziioannou T, Skalka AM (2020) Principles of virology, vol 1, chapters 4 and 13, 5th edn. ASM Press, Washington, DC
Rossmann MG, Rao VB (2012) Viral molecular machines. Springer, New York
Also especially recommended for further reading are references [68, 73, 96, 100] listed above.

Chapter 13
Nucleic Acid Packaging in Viruses

María I. Daudén, Mar Pérez-Ruiz, José L. Carrascosa, and Ana Cuervo

Abstract Viruses shield their genetic information by enclosing the viral nucleic acid inside a protein shell (capsid), in a process known as genome packaging. Viruses follow essentially two main strategies to package their genome: Either they co-assemble their genetic material together with the capsid protein or an empty shell (procapsid) is first assembled and then the genome is pumped inside the capsid by a molecular motor that uses the energy released by ATP hydrolysis. During packaging the viral nucleic acid is highly condensed through a meticulous arrangement in concentric layers inside the capsid. In this chapter we will first give an overview of the different strategies used for genome packaging to discuss later some specific virus models where the structures of the main proteins involved are presented and the biophysics underlying the packaging mechanism are discussed.

Keywords Virus · Packaging · Capsid · Nucleocapsid · Encapsidation · Maturation · Molecular motor · Ejection · Bacteriophage · Helical symmetry · Icosahedral symmetry · Portal · DNA · RNA · Shell · Terminase · Cryogenic electron microscopy · X-ray diffraction

M. I. Daudén
Structural Biology Programme, Spanish National Cancer Research Centre, (CNIO), Madrid, Spain

M. Pérez-Ruiz
Faculty of Health and Medical Sciences, Structural Biology of Molecular Machines Group, Protein Structure and Function Program, Novo Nordisk Foundation Centre for Protein Research, University of Copenhagen, Copenhagen, Denmark

J. L. Carrascosa · A. Cuervo (✉)
Department of Macromolecular Structure, Centro Nacional de Biotecnología (CNB-CSIC), Madrid, Spain
e-mail: acuervo@cnb.csic.es

M. G. Mateu (ed.), *Structure and Physics of Viruses*, Subcellular Biochemistry 105,
https://doi.org/10.1007/978-3-031-65187-8_13

Abbreviations

CPV	Cytoplasmic polyhedrosis virus
cryo-EM	Cryogenic electron microscopy
ds	Double stranded
HSV1	Herpes simplex type 1
MRV	Mammalian reovirus
NPC	Nucleoprotein complex
NTP	Nucleoside triphosphate
NTPase	Nucleoside triphosphatase
PaV	Pariacoto virus
SLP	Single layered capsid
pRNA	Packaging RNA
ss	Single stranded
TEC	Transcriptional enzyme complex
TMV	Tobacco mosaic virus

13.1 Introduction

Viruses are composed of a genetic information carrier (either DNA or RNA), a protective multifunctional container usually made of proteins and, eventually, lipidic components [1] (see Chap. 2). Several main aspects must be considered in virus construction. One is that it must use the minimum possible genetic information, making use of geometric principles to build a large container using a limited set of proteins. In addition, the virus particle has to be easy to assemble to facilitate the highest possible progeny production but, at the same time, this assembly process has to incorporate mechanisms to encapsidate the viral genome and reject cellular components. Then, the virus particle must be able to actively participate in its release from the infected cell, transfer to other possible hosts with proper and accurate recognition and, finally, deliver their genetic information to start a new infection cycle.

It is clear that to carry out all these processes in so many different environments the virus particle must be a quite flexible vehicle able to trigger and undergo structural changes to provide at the appropriate moment the required functionality: the resistance needed to survive the harsh extracellular environment may be superfluous and even counter-productive in the well-controlled intracellular conditions. The propagation strategy is also reflected in the virus structure. In certain plants, the viral propagation allows each particle to carry partial genomic information (i.e., the genome is found multipartite in different viral particles), as many viruses will co-infect the same cell. On the other hand, viruses infecting most animal cells present the whole genome either on a single molecule or in several segments within the same viral particle to ensure efficient infection. Also, the requirements to penetrate the eukaryotic cell membrane (see Chaps. 15 and 16) are very different to those required to traverse the thick-walled bacterial envelope (see Chap. 17). Due to these and many other variables, different virus particles have evolved different survival strategies. Thus, viruses provide excellent model systems for the study of biotechnological solutions at the nanoscale.

Viruses offer a unique gradient of structural solutions to encapsidate and protect their genetic material, from the simplest ones to those revealing unexpected levels of sophistication. The co-assembly of the viral RNA and one (or a few) proteins to build a helical particle is one of the simplest arrangements, widely used by single-stranded (ss) RNA and ssDNA viruses (Fig. 13.1a). While easy to build, these viral assemblies may be too rigid to adapt to different environments, and they have to be fully disassembled to allow viral genome replication and expression.

A more complex structural solution is offered by the enveloped RNA viruses, where a flexible nucleocapsid is formed by the interaction of the RNA and multiple copies of a protein (in a way resembling the simple solution mentioned above); but in this case, the nucleoprotein assembly is based on very different types of RNA-protein contacts and may adopt different structural conformations to carry out the transcription and replication of the genetic message. Proper protection to transfer this complex from cell to cell is conferred by the inclusion of the nucleocapsid inside a lipidic envelope derived from the cellular membranes after its modification with viral proteins. The extreme in this type of strategy is offered by the more complex viruses (human immunodeficiency virus, adenoviruses, herpesviruses, vaccinia virus), where the nucleocapsid is enclosed by several layers or envelopes of different composition. Examples of these viruses are given in Chap. 12 and in several other chapters of this book.

An alternative strategy, used by many viruses, is to incorporate the nucleic acid while the capsid is being formed in a co-assembly process. This solution is adopted by a large number of non-enveloped ssRNA viruses with icosahedral capsids (Fig. 13.1a). Icosahedral symmetry is a preferred geometrical solution to build a closed viral container (see Chap. 2). Although the assembly details vary from one virus type to another (see Chap. 11), it seems that the interaction of small segments of the RNA molecule and certain domains of the shell protein facilitates their mutual recognition and their action as chaperones that assist each other to drive the co-assembly process.

The extreme rigidity and other structural properties of double-stranded (ds) nucleic acids impose severe restrictions to the possible ways to enclose these molecules within a protein container. Nevertheless, the evident evolutionary advantages of dsDNA as a genomic substrate have driven viruses to explore efficient solutions to encapsidate this type of DNA molecule, the best-known example being many bacteriophages. These viruses assemble first a protein container (called prohead), in which dsDNA is later packaged using a complex machinery made of different viral proteins (Fig. 13.1a). This process requires an exquisite selection procedure to encapsidate the right (viral) DNA, followed by an energy-consuming process that translocates the DNA up to quasi-crystalline densities inside the virus shell. These viruses can be considered as a paradigm of efficient DNA aggregation:

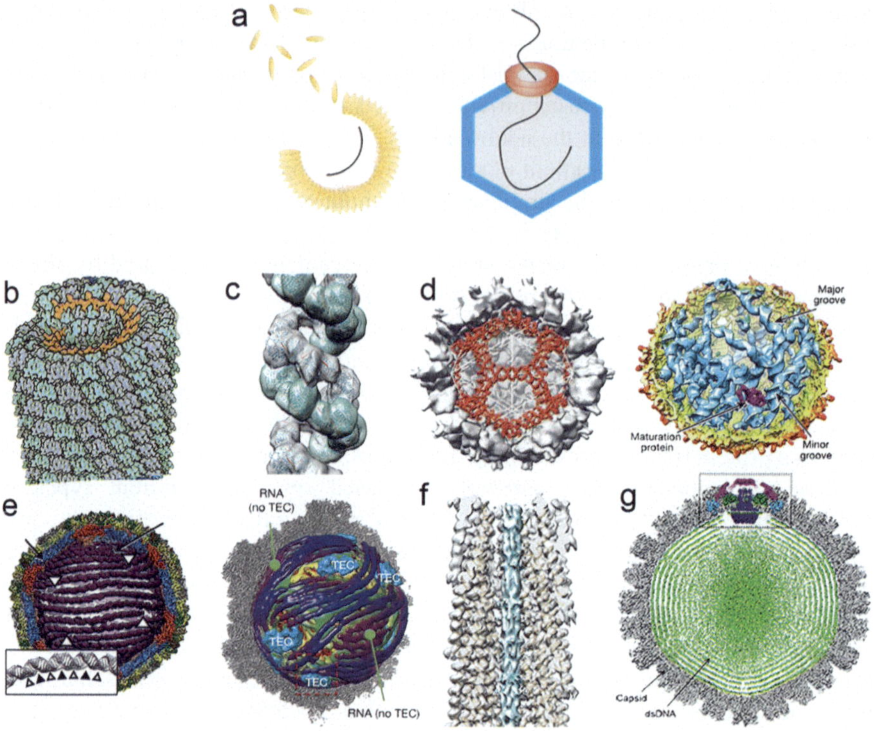

Fig. 13.1 Packaging alternatives and gallery of macromolecular complexes involved in nucleic acid encapsidation in different viruses. (**a**) Schematic of packaging models: left, co-assembly; right, active packaging into a preformed capsid. Reproduced from [2], with permission. (**b**) Structure of TMV, an ssRNA helical virus, showing the RNA coloured orange and the capsid proteins in purple and turquoise. Adapted from the RCSB PDB, with permission. (**c**) Detail of the central region of the structure of the viral ribonucleoprotein complex of influenza virus [3]. Two opposite-polarity nucleoprotein strands are shown in turquoise and blue (EMD-2205). (**d**) Structures of several ssRNA icosahedral viruses. Left, sectioned structure of a Pariacoto virus (PaV) particle; the partial organization of the viral RNA molecule is shown in red and the capsid protein in grey. Reproduced from [4], with permission. Right, sectioned structure of the MS2 bacteriophage, showing the internal RNA assembly (blue). Modified from [5], with permission. (**e**) Structure of the genome in dsRNA viruses. Left, structure of Φ6 bacteriophage; spooled dsRNA (purple) and the inner capsid protein are highlighted with black arrows. The inset shows a model of dsRNA built into the density. Modified from [6], with permission. Right, structure of the cytoplasmic polyhedrosis virus (CPV) reovirus, showing the RNA in purple, the TEC in blue, and capsid proteins in grey. Modified from [7], with permission. (**f**) Cut-open structure of IKe bacteriophage, an ssDNA filamentous virus. The DNA is coloured blue (EMD-6993, [8]). (**g**) Structure of herpesvirus KSHV, a dsDNA icosahedral virus. The dsDNA is coloured green, the viral capsid is shown in grey, and the portal vertex is highlighted with a square. Modified from [9], with permission

there is no other biological example for such an ordered and condensed DNA conformation, an order that plays a critical role in the release of the DNA during infection by these phages. Nevertheless, this packaging solution is expensive in terms of information (several viral proteins are involved), as well as in energy consumption (one ATP molecule is hydrolysed by each two base pairs packaged), but the reward is worth the expense: These viruses exhibit the largest infection efficiency by far, as each viral particle is potentially able to infect host bacteria.

In the following sections we will briefly review examples of viruses using different viral strategies for nucleic acid packaging, making emphasis in common underlying mechanisms and solutions selected by viral evolution.

13.2 Structural Features of the Packaged Nucleic Acids

The extended use of X-ray fibre and crystal diffraction and electron microscopy has provided a great amount of information on the structure of viral particles (see Chaps. 3, 4, and 7). Due to their geometrical shapes and the existence of symmetries (see Chap. 2), these and other structural methods have provided a comprehensive insight into the way the protein components of viral capsids are organized at the molecular or even atomic level. Nevertheless, the same cannot be said for the nucleic acid component of virions, as it is usually far less ordered than the protein counterpart, and in most cases the way the nucleic acid is organized in each viral particle is not identical to those in others. This section reviews the main features of the way the nucleic acids interact with the protein capsids for protection, mobility, and functionality.

13.2.1 RNA Viruses

Tobacco mosaic virus (TMV), a representative positive ssRNA virus with helical symmetry has provided a most successful model for biophysical studies for more than 100 years. The structure of this rigid rod-shaped virion was solved at near-atomic resolution using X-ray fibre diffraction and later by cryogenic electron microscopy (cryo-EM) [10, 11]. The virus is organized as a stack of identical subunits of a single capsid protein that follows a helical path with 17 subunits per ring, leaving an internal cavity (Fig. 13.1b). The ssRNA molecule is located near this inner region, with three nucleotides binding to each of the capsid subunits.

The type of protection for the RNA provided by TMV-like viral particles is limited in functional terms due to their rigid nature. A more flexible solution is adopted by negative ssRNA viruses. The Mononegavirales [12] are characterized by the assembly of a helical symmetry based nucleoprotein complex (NPC) essentially made by the ssRNA and a specific protein (the nucleocapsid protein). Unlike the helical TMV particle, NPCs have a remarkable intrinsic flexibility [3, 13]. Instead of resembling a rigid rod, NPCs have a herring-bone aspect by electron microscopy, revealing different conformations with helical symmetry (Fig. 13.1c). These nucleoproteins are enclosed into a near-spherical or elongated lipidic envelope (integrating viral specific proteins) to yield the infectious viral particle. Examples of these viruses include the *Rhabdoviridae* (like the bullet-shaped vesicular stomatitis virus, or the rabies virus), the *Paramyxoviridae* (respiratory syncytial virus, Sendai virus or measles virus), and the *Orthomyxoviridae* (influenza virus). A general characteristic of all these viruses is that the NPCs (either one or, in the case of viruses with segmented genomes as in influenza virus, several NPCs containing one ssRNA molecule each) are released inside the cell and used as substrates for RNA replication and transcription. This means that the RNA-protein assembly must be flexible enough to allow the RNA polymerase (which also forms part of the NPC) to act on the viral RNA substrate without complete disassembly of the nucleoprotein [14]. Although the viral nucleocapsid proteins of different viruses (N-proteins) show a rather low sequence identity, electron microscopy and X-ray diffraction studies have revealed that NPCs share a basic structural organization that is relatively conserved. The NPC helix has a pitch of around 6–7 nm and 10–13 subunits per turn. Each protein subunit is associated with nine nucleotides, yielding a threading of the proteins into a super-helical complex. The RNA moiety is located in a conserved RNA-binding cavity formed between the carboxyl and amino terminal domains of the protein. This region presents a number of positively charged residues that can interact with the RNA phosphate groups, resulting in the exposure of certain bases of the RNA in the outer face of the NPC, fully available for functional recognition of the RNA polymerase. Three-dimensional reconstructions of different NPCs based on electron microscopy studies showed differences in the overall superhelicity and different parameters of the basic helical assembly and revealed that these complexes have an intrinsic flexibility that accommodates the conformations required in each of the different stages of the viral life cycle, either inside the cell or when packaged inside the viral envelope (reviewed in [15, 16]).

A different solution for the successful storage of the viral genetic information based on ssRNA molecules is found in many icosahedral viruses [4]. Icosahedral symmetry-based capsids are found in about half the virus families. Among them, there is an important group containing positive ssRNA that includes human pathogens such as poliovirus, rhinovirus, and hepatitis A virus of the *Picornaviridae*, as well as many insect and plant viruses (e.g., flock house virus and satellite tobacco mosaic virus). In other ssRNA viruses the icosahedral capsid is enclosed by an outer lipidic envelope, as in the *Togaviridae* and *Flaviviridae*. In many of these viruses, the RNA, besides carrying the genetic information, plays an important role in the assembly of the viral particle and in the definition of the shape and size of the capsid. It has also been observed that the packaged RNA shows in these viruses an increase of secondary structure compared to the cytoplasmic free RNA, including the formation of intrachain duplex RNA stems and non-covalent base-pairing between RNA segments in multipartite RNA viral genomes. However, the knowledge of the RNA structure within viral particles has been largely prevented by the fact that all or a high percentage of the viral RNA is not sufficiently ordered inside the virion. Both X-ray crystallography and three-dimensional reconstructions from cryo-EM images are based on the presence of order and symmetry in identical particles and, thus, only those parts of the RNA sufficiently ordered could be solved with a resolution comparable to that obtained for the capsid. The combination of X-ray diffraction results and electron microscopy reconstructions (see Chap. 7) provided the first structural insights on the organization of the RNA inside the icosahedral capsids of these viruses. One of the best examples is Pariacoto virus (PaV), where the encapsidated ssRNA was observed organized as a dodecahedral cage of double-stranded dsRNA intrachain stems closely following the capsid twofold axes of the icosahedral shell (Fig. 13.1d, left). The location of these RNA moieties close to the interfaces between neighbouring capsid subunits allows them to act as part of molecular switches to generate two types of quasi-equivalent inter-subunit contacts required for the formation of either planar or more wedge-shaped interfaces along the twofold icosahedral axes (see Chap. 2). There is evidence that, in some ssRNA viruses, the RNA plays a specific role in the promotion and direction of viral assembly [17, 18]. The fact that the RNA is folded in precise structures near the capsid and interacts with the capsid proteins supports a condensing role of the RNA in the nucleation and first assembly steps. There is also evidence that the capsid proteins could work as chaperones to facilitate the folding of the RNA in a subset of structures compatible with capsid geometry. Most probably, the mutual interactions between capsid proteins and viral RNA interplay to yield the virion definitive structure [4].

The advances in cryo-EM techniques have allowed obtaining high-resolution asymmetric reconstruction of the MS2 bacteriophage, allowing the tracing of 80% of its genome [5] (Fig. 13.1d, right). These structural studies, together with biophysical analysis and mathematical modelling [19], showed that multiple sequence-structure motifs in MS2 RNA formed stem-loops that act as packaging signals binding cooperatively to the capsid protein. These specific secondary structures in the RNA seem to play a fundamental role by controlling the encapsidation of the viral RNA and the assembly efficiency.

Another type of packaging solution is found in dsRNA viruses, which includes mammalian orthoreoviruses, certain fungal viruses, and some bacteriophages. All of them share many functional and structural characteristics, like the presence of concentric icosahedral capsids [20]. The study of the structure of icosahedral dsRNA viruses is an excellent example of the advantages of the use of hybrid methods: Combining X-ray diffraction of structural components and cryo-EM reconstructions of complete virions allows obtaining quasi-atomic maps of the different virus assemblies (see Chap. 7). The best studied representatives of this type of viruses are the *Sedoreoviridae* family [21] and the *Cystoviridae* bacteriophage Φ6. Both present a multipartite genome (three segments in Φ6 and 10–13 segments in reoviruses), 1 to 3 concentric icosahedral capsids (normally 2) in the virus particle, and an outer lipidic envelope in the case of the *Cystoviridae*. A common trend in all these viruses, except for the *Birnaviridae*, is the presence of an inner icosahedral capsid built by 120 protein subunits [22]; this capsid might be considered as a functional counterpart of the helical NPCs described above, but in this case the capsid plays an active role in the viral RNA metabolism, namely in RNA replication, transcription, and release from the capsid. The inner capsid in these viruses presents, in most cases, specific structures at the fivefold vertices

(Fig. 13.1e). These assemblies, which contain the viral RNA polymerase (transcription enzymatic complexes, TEC), are involved both in the replication of the RNA during assembly and in the generation and extrusion of the viral messenger RNA to the cell during infection [23]. In fact, this capsid maintains the dsRNA enclosed inside during the whole viral cycle, thus preventing a cellular response to the presence of the dsRNA molecules. All these fascinating functions carried out by the inner core have to be performed under quite extreme conditions. The dsRNA segments are densely packaged inside the capsid building concentric layers with spacings of around 3.0 nm [23, 24]. Structural characterization of genome organization in dsRNA viruses showed that in the case of bacteriophage Φ6 tri-segmented RNA adopts a single-spooled genome organization inside its capsid (Fig. 13.1e, left) [6], although dsRNA can be found in multiple conformations and thus RNA polymerases do not seem to direct genome ordering. This high RNA density organization inside the capsid, together with the proposed RNA packaging mechanism, in which the RNA is incorporated into an already performed inner core capsid (discussed below), resembles the case of the dsDNA bacteriophages.

Although reoviruses share some structural and functional similarities with the cystovirus bacteriophage family, they do not seem to have the same packaging mechanism. In these viruses, the packaging process likely initiates by arranging viral RNA and other components into a structured complex, which would then act as the core around which the outer capsid shell assembles prior to RNA replication [25]. It has been observed that these viruses exhibit a notable pattern: they possess an equivalent count of TEC complexes as the segmented dsRNA fragments they are required to package. Thus in some viruses it has been observed that certain fivefold vertices lack polymerase complexes (Fig. 13.1e, right). According to these findings, in rotavirus it has been proved that the polymerase is the one in charge of RNA recruitment inside the procapsid [26]. The arrangement of the dsRNA genome in reovirus has been well characterized for mammalian reovirus (MRV) and cytoplasmic polyhedrosis virus (CPV). In CPV dsRNA fragments extend outward from the TEC complexes gradually expanding within the capsid and forming non-concentric layers following the organization of the liquid crystalline array (Fig. 13.1e, right) [7, 27].

13.2.2 DNA Viruses

The *Inoviridae* family is formed by ssDNA viruses that present some of the simplest helical capsids. They infect bacteria and some of them, including M13 and related phages (such as IKe, fd, and Pf1) have been the subject of detailed studies. M13 has been widely used as a substrate for biotechnological approaches [28] (see also Chap. 22). The capsid protein has a cylindrical shape with the negatively charged amino terminal segment on the outside, and the positively charged carboxyl terminal segment inside, lining a channel where the DNA is enclosed (Fig. 13.1f) [8]. Structural studies in Pf1 and IKe bacteriophages have shown that the ssDNA also adopts a helical structure and it is stabilized by interactions with the capsid shell. In Pf1 it has been observed that lysine and arginine residues inserted between the DNA bases stabilize the phosphate charges [29]. Although the details of the protein-nucleic acid are specific in each case, the overall assembly strategy of these viruses closely resembles that of the ssRNA helical viruses mentioned above.

As in the case of the ssRNA viruses, the most widely preferred geometry of ssDNA viruses is the spherical shape using icosahedral symmetry. Most of these viruses infect mammals and other vertebrates (parvoviruses), while there are several examples infecting plants (geminiviruses) and bacteria (phage ΦX174). The structure of canine parvovirus showed that 13% of the total viral DNA is organized in defined structural stretches. These stretches are 11 nucleotides long and interact with capsid protein pockets at 60 icosahedral symmetry-related positions [30]. There is a low level of sequence specificity in these DNA-protein contacts, and there are no hints of DNA-binding motifs in the capsid subunits. In fact, DNA-binding domains are generated after capsid assembly, thus representing a means to stabilize correct assemblies. A singular aspect in parvoviruses is the fact that the interaction of the DNA with the capsid protein is based on the existence of a series of well-conserved polar residues in the inner face of the capsid, instead of the more common interaction of the nucleic acid with basic residues found in many other viruses.

The dsDNA viruses provide some of the best-known examples of sophisticated strategies to package the nucleic acid in a most efficient way. Most of these viruses have a single linear dsDNA molecule enclosed within an icosahedral capsid, although there are examples of filamentous dsDNA viruses (*Lipothrixviridae*), and others show more complex architectures that include lipidic envelopes (herpesviruses and poxviruses). dsDNA viruses as a group can infect every cell type, from bacteria (some bacteriophage families), invertebrates (*Baculoviridae*, *Iridoviridae*), and vertebrates (*Papovaviridae*, *Adenoviridae*). The best-known examples of this type of viruses include the caudovirales, a widely extended bacteriophage order characterized by their icosahedral capsid attached to a tail which is instrumental for the phage–host interaction [31].

Caudovirales and human herpes virus present a common maturation strategy, building first a proteinaceous prohead, which is later filled up with the dsDNA (for reviews, see refs. [32, 33]. The physical properties of the dsDNA molecule clearly impose severe constraints on the way it can be packaged inside a spherical (icosahedral) container. The persistence length of the dsDNA (around 50 nm) is related to a minimum diameter in the capsid of these viruses: most of them have diameters from 50 to 400 nm, and they present a complex DNA packaging machinery (see Sect. 13.4), which sits in a unique icosahedral vertex of the viral capsid (the portal vertex). This machinery has to deal with the formidable task of filling up the viral capsid with dsDNA up to quasi-crystalline densities (more than 500 mg/ml). Cryo-EM has revealed that the DNA is organized in layers following a traverse spool [34]. This spooled DNA arrangement is also observed in human herpes virus (Fig. 13.1g). The distance between adjacent layers is 2.5–2.7 nm, suggesting that the DNA is tightly arranged in a close hexagonal packaging. Depending on the capsid geometry, the DNA seems to adopt a different topology: fully icosahedral capsids (T7, ε15, P22) show the coaxial spool geometry, while in elongated icosahedral variants (prolate shells in T4, Φ29) the DNA strands adopt an orientation more parallel to the longitudinal axis of the viral capsid [35]. In all these topologies the structure of the outer layers is always better defined than the inner layers: it seems clear that the inner DNA region must present sharp geometrical discontinuities (bends) to allow its accommodation in the very restricted available space. In this context, the presence of inner capsid proteins must play an important role. These components can be individual small proteins (probably instrumental in shielding the electrostatic repulsion between phosphates of neighbouring DNA segments); alternatively, they can form a complex inner core connected to the portal vertex (as in T7, ε15 and myovirus). The presence of this structure along the DNA translocating path is related to both the way the DNA is accommodated in the inner shell surface and the ejection process [36, 37].

An important factor to be considered in these viruses is that this highly condensed DNA inside the viral head has to be also efficiently injected into the host bacterium. This imposes important constraints to the possible topological arrangements of the DNA, especially taking into account that although the packaging and release of DNA are performed along the same capsid vertex (the portal), the respective molecular mechanisms in which they are based are different. While the packaging reaction consumes cellular ATP to develop up to 50 pN forces (see Sect. 13.4.3 and Chap. 10), the ejection of DNA initially depends on structural changes in the tail apparatus that triggers the release of the potential energy stored in the packaged viral head.

Different studies by cryo-EM have shown that the DNA packaging pathway in these viruses follows an initial stage where the DNA is pushed inside the capsid in a poorly ordered way. Only when the percentage of DNA is around 60–70% of the total to be packaged in phages as T3 or Φ29, the layers of DNA start to organize, probably from the layer in contact with the inner face of the capsid shell [38, 39]. At higher DNA concentrations the multiple layer topology is generated until the acquisition of the final coiled spool structure.

13.3 Reorganization of the Viral Capsid During Nucleic Acid Packaging

Viruses that actively package their genetic material (in an energetically demanding process) start their morphogenetic pathway by forming an empty shell (prohead/procapsid) that will be subsequently filled up with the nucleic acid by a molecular motor. This packaging machinery translocates the viral genome to a very high concentration into the head (see Sect. 13.2). Nucleic acid encapsulation in the viral capsid is linked to changes in capsid morphology, a process known as virus maturation (see Chap. 14). These conformational changes have been best characterized for dsRNA and dsDNA bacteriophages. Although associated in many cases to nucleic acid packaging, maturation may occur before, during, and/or after packaging and is described in detail in Chap. 14.

13.3.1 Conformational Changes During RNA Packaging

The advances in structural biology techniques have expanded the knowledge of conformational changes in dsRNA viruses during packaging from bacteriophages to mammalian reoviruses. Remarkably, despite their differences in the initial steps of capsid assembly both seem to share common intermediate structures [26, 40–42]. As discussed before (Sect. 13.2.1), these viruses share functional and structural characteristics (common capsid fold and head symmetries), however they differ in the assembly strategy. While in reoviruses both the RNA genome and capsid components are co-assembled simultaneously, cystovirus actively enclose their genetic material within a preformed capsid [25].

The *Cystoviridae* bacteriophage family (comprising Φ6, Φ8, and Φ12 viruses) includes some of the best understood dsRNA viral assembly systems [40]. These viruses are composed of three concentric capsid layers that enclose a segmented dsRNA genome and a membrane envelope decorated with proteins (Fig. 13.2a) [2]. During viral entry, the membrane proteins are in charge of the fusion of the viral membrane with the host outer membrane. The multi-layered capsid is then disassembled, losing the two outer layers while the innermost shell is delivered inside the cytoplasm activating viral RNA synthesis (Fig. 13.2a) [43]. The assembly of cystoviruses starts with the formation of an empty procapsid, whose main component is protein P1 [40]. The procapsid shows an icosahedral shape composed of cup-like structures assembled on the fivefold vertex [43]. The fivefold vertex of the capsid is occupied by the RNA-dependent polymerase (P2), the packaging nucleoside triphosphatase (NTPase) (P4) and the assembly cofactor (P7). The structure of the procapsid differs from the RNA-filled head (Fig. 13.2b, c). During RNA packaging the prohead structure expands sequentially to a spherical shaped structure with turrets projecting at the fivefold vertex. This conformational change has been suggested to play a biological role by orchestrating RNA segment packaging order by sequentially exposing different genome interaction regions [43]. During this capsid rearrangement the internal volume of the capsid is increased by 2.4-fold [44].

The innermost capsid of most dsRNA viruses is organized in a $T = 1$ symmetry (see Chap. 2) where the asymmetric unit is formed by a dimer of two monomers A and B [45]. In cystoviruses, the P1 class A monomers surround the fivefold axes, while the class B monomers assemble the pentamers together (Fig. 13.2b, c). Capsid expansion can be explained in these viruses by local conformational changes at the P1 monomer interfaces. The most dramatic changes are produced at the A-P1 monomers interface that changes from an angular conformation to a nearly flat position (Fig. 13.2c) [42, 44].

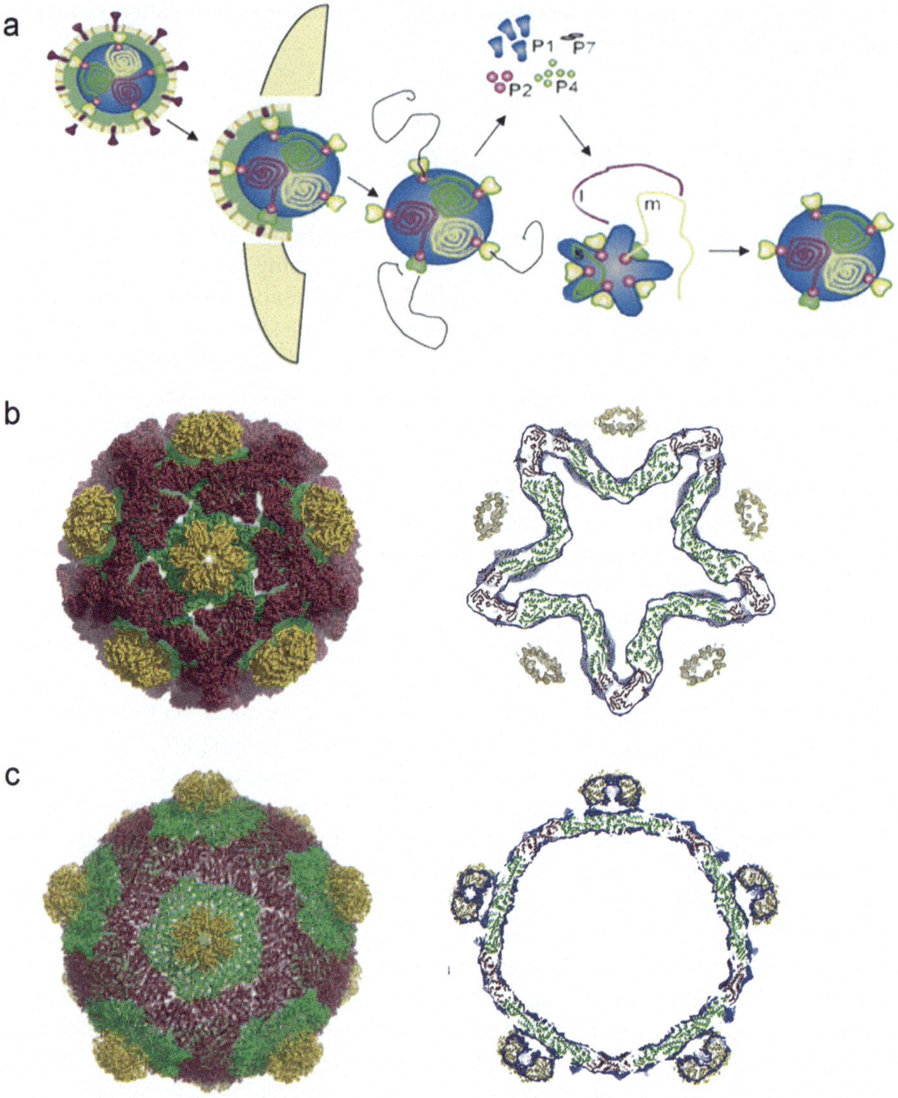

Fig. 13.2 Conformational changes in dsRNA viruses during packaging. (**a**) Diagram representation of Φ6 bacteriophage morphogenetic pathway, presented from left to right: viral entry into the host, mRNA template extrusion, viral protein synthesis, procapsid assembly, and RNA packaging. (**b**) Left, ribbon representation of Φ8 bacteriophage procapsid assembly showing P1-A and P1-B (green and red) and P4 (yellow). Right, cross-section structure from panel left fitted on the Φ6 procapsid (**c**) Left, ribbon representation of Φ8 bacteriophage capsid after expansion. Right, cross-section structure from panel left fitted on the Φ6 expanded capsid model. Figures in panels (**b**) and (**c**) are reproduced from [42], with permission

Implementation of new structural techniques (cryo-EM) able to visualize in situ the assembly of viral particles inside the cells allowed capturing assembly intermediates of single layered capsids (SLP) form rotavirus and orthoreovirus, belonging to the *Sedoreoviridae* family [26, 41]. The observation of these structures, devoid of dsRNA genome, shows an indented shell, reminiscent of the pre-packaging intermediates described above for bacterial viruses (Fig. 13.2b). Although the diameter of these particles is very similar to the one of the mature viruses, in the SLP the capsid is collapsed inwards at the icosahedral fivefold showing a star shape. The conformational changes required for capsid expansion have been studied in detail in orthoreovirus, showing that in these viruses A and B monomers have to follow a flex-hinge movement outwards to acquire a planar conformation in the mature capsid [41]. The contacts of the monomers also have to rotate and slide in a camera iris-like movement. Conservation of the capsid shell fold among the members of the *Sedoreoviridae* family suggests that rotavirus could also follow a similar expansion mechanism.

13.3.2 Expansion and Reinforcement of Caudoviral Procapsids During DNA Packaging

As we have mentioned above, assembly of dsDNA bacteriophages (order Caudovirales) has been extensively studied. In this process viruses package their dsDNA into a preformed viral capsid precursor called prohead or procapsid. This process is a fundamental step in their morphogenesis (Fig. 13.3a). The proheads (Fig. 13.3a, b left and c) are composed of an outer shell, an inner scaffold, and in some cases a core-like structure, as discussed in Sect. 13.2 [31]. The packaging of dsDNA into proheads is a coordinated process, orchestrated by a complex machinery located at the portal vertex (see Sect. 13.4) that utilizes the energy derived from ATP hydrolysis to translocate the DNA inside the preformed protein container [49]. The initiation of DNA packaging is correlated with several structural changes in the prohead (Fig. 13.3c and d): there is a massive rearrangement of the shell subunits, leading to the expansion of the head diameter by about 10%, the scaffold is disassembled, and their subunits are either proteolysed or recycled in new prohead assemblies. Additionally, structural transformations occur in the core components, contributing to the overall dynamics of the packaging process [50]. In several viruses new proteins are incorporated to the viral shell and/or chemical modification of shell proteins take place by cleavage and/or crosslinking. It is worth emphasizing that all these rearrangements must be carried out in situ, maintaining the structural integrity of the viral particle while the DNA molecule is actively being inserted in the protein shell. This reflects the exquisite architectural design of these protein containers. The consequences of the structural rearrangements are fundamental for the packaging reaction: The viral capsid volume becomes almost 50% bigger and a new intersubunit interaction pattern is created (Fig. 13.3c and d). This reorganization

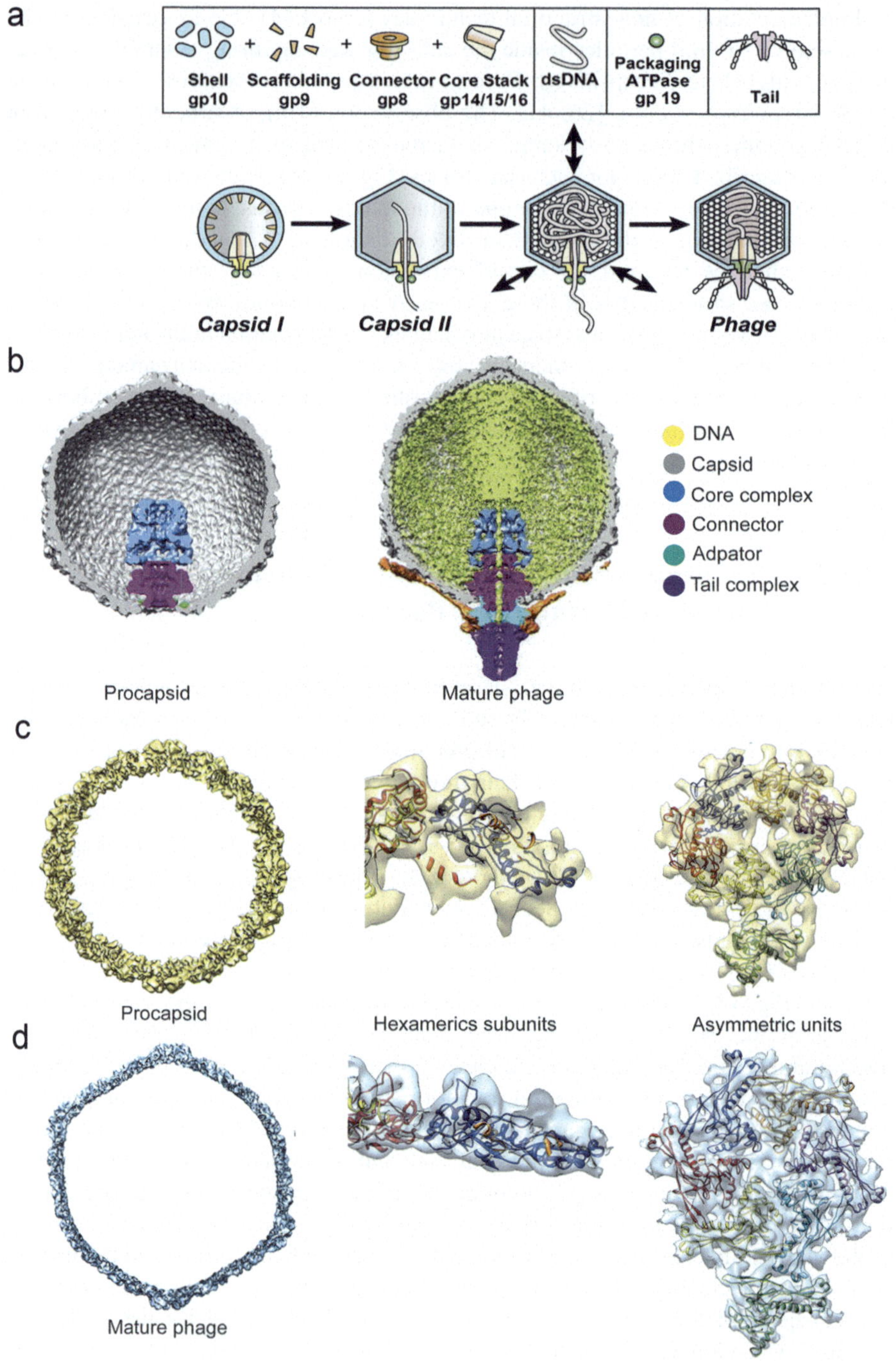

Fig. 13.3 Reorganization of Caudovirales capsids during DNA packaging and maturation. (**a**) Assembly pathway of T3 and T7 bacteriophages. Top panel: structural elements involved. Bottom images, from left to right: procapsid (capsid I) is assembled; then DNA packaging starts (capsid II),

Fig. 13.3 (continued) associated with conformational rearrangements (capsid dynamism is indicated by arrows). Capsid maturation ends when the tail is attached to the portal vertex to yield a mature phage. Reproduced from [46], with permission. (**b**) T7 bacteriophage clipped structure: Procapsid (left) and mature virus (right). Structural elements are shown in different colours as indicated in the figure. Reproduced from [47], with permission. (**c**–**d**) Left, cross-section of the procapsid (**c**, yellow) and the mature capsid (**d**, blue). Middle, detailed structure of the hexameric subunit for the procapsid (**c**) and the mature capsid (**d**). Right, structure of the asymmetric unit (composed of one hexamer and one pentamer subunit) of the procapsid (**c**) and the mature capsid (**d**). In the middle and right panels in **c** (and **d**) T7 proteins pseudo-atomic models are fitted inside the corresponding cryo-EM densities. Reproduced from [48], with permission

stabilizes the viral shell. In addition, the mature head inner surface exposes negatively charged domains that may be important for the interaction of the first layers of DNA to start the organization of the packaged DNA inside the capsid (Fig. 13.3b, right) [48].

All these changes are closely related to the incorporation of the DNA, as the release of the scaffolding proteins and the increment of the capsid volume facilitate the entrance of larger DNA molecules up to high concentrations. At the same time, the shell subunit rearrangement, based on local tilting, rotation, and radial translations of the subunit domains (Fig. 13.3c and d), generates an increment in the contact areas between subunits, thus increasing the stability of the particle [48]. This increment might be further reinforced in different viral systems by either the incorporation of new, accessory proteins (as in T4, λ and ε15) or by covalent inter-subunit crosslinking (HK97) (reviewed in [51]). This rearrangement not only facilitates and likely triggers the release of the scaffolding protein, but also prevents the eventual competition between the DNA molecule and the scaffolding protein subunits for the interaction with the inner surface of the capsid shell. The final result is a more stable capsid, securely enclosing the DNA, and ready to incorporate the tail complex to produce the final infective viral particle.

13.4 Components of the Packaging Machinery

The packaging machinery is a molecular motor that transforms nucleoside triphosphate (NTP) hydrolysis energy into mechanical work, leading to nucleic acid translocation inside the viral head. These motors have evolved from the simplest ones present in dsRNA viruses to very complicated transient multi-complexes present in dsDNA bacteriophages.

13.4.1 dsRNA Packaging

dsRNA viruses present a unique genome encapsidation approach. The existence of RNA as genomic material forces them to use a specialized viral RNA polymerase to replicate and generate its mRNA transcripts [2, 52]. To avoid degradation by host RNAses the viral polymerase has to operate inside the intact viral capsid. Although dsRNA bacteriophages (cystoviruses) and reoviruses share structural and functional features, the initial steps of their assembly pathway differ: while cystoviruses have to translocate RNA segments onto an already performed capsid by a packaging motor, reoviruses seem to assemble their capsid with their RNA fragments already inside.

RNA-dependent RNA polymerases present in dsRNA viruses are found inside the capsid (Fig. 13.4) and are able to produce a dsRNA molecule using ssRNA as a template. As referenced before, in some reovirus not all the vertexes are occupied by TECs, and it has been observed a correlation between the number of RNA genomic segments and the number of polymerase complexes (Fig. 13.1c) [23, 53]. A very complete study of the TEC assembly has been carried out for MRV. In this virus, 10 dsRNA are replicated by 10 TEC complexes. The complex found under the

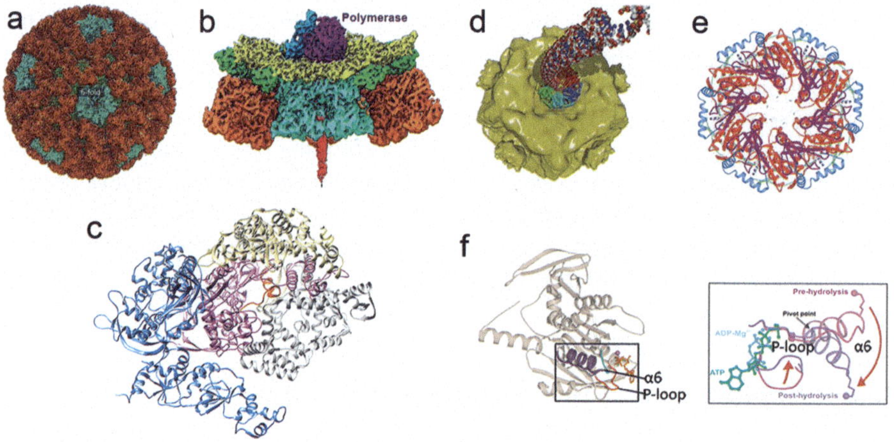

Fig. 13.4 TEC and packaging complex in dsRNA viruses. (**a**–**c**) TEC of MRV. (**a**) MRV structure; the pentons at the fivefold vertices are coloured green. (**b**) Section of one of the MRV pentons depicted in a. The TEC complex is shown in purple (RNA-dependent RNA polymerase) and blue (protein cofactor). Figures in panels (**a**) and (**b**) are adapted from [53], with permission. (**c**) Structural model of the TEC complex: the protein cofactor is shown in blue and the RNA polymerase domains are shown in purple (fingers), grey (palm and C-terminal bracelet), yellow (thumb and N-terminal), or orange (active site) (PDB 7ELL). (**d**–**e**) P4, the packaging motor of Φ12 bacteriophage. (**d**) Model of the Φ12 polymerase complex showing P4 at a fivefold vertex of the capsid, and an RNA molecule being translocated. (**e**) Model of the atomic structure of the P4 hexamer. Reproduced from [54], with permission. (**f**) Left, representation of the atomic model of one P4 monomer before ATP hydrolysis. Right, close-up of the region highlighted in the left panel showing the conformational change that takes place after ATP hydrolysis. Figures in panels (**d**) and (**f**) are adapted from [2], with permission

fivefold vertexes (Fig. 13.4a and b) is composed of a protein cofactor and the RNA-dependent RNA polymerase. Viral RNA polymerases present a conserved core shared also with DNA polymerases, showing the characteristic hand–shape domains: palm, fingers, and thumb (Fig. 13.4c) [23, 53]. It is not completely understood how viral RNA polymerases orchestrate their functions and how they change their activity between dsRNA compaction during assembly and mRNA extrusion during infection. Although the viral polymerase core is conserved the TEC structure differs from one virus to another: some of them present modifications in the N and C-terminal domains, additional co-proteins (as in the case of MRV), or interact with the viral capsid or transcriptional cellular factors. These variations might play an essential role in the regulation and transition between different TEC functions during the viral cycle [23].

RNA packaging has been extensively characterized for cystoviruses [2, 40]. The strategy of these bacteriophages consists in replicating their RNA as it is translocated inside the viral head, and it requires the coordination of two viral complexes, the polymerase and the RNA translocase, both sitting at the fivefold vertex (Figs. 13.2 and 13.4d and Sect. 13.3.1). Viral RNA translocases are structural proteins that are also present in the mature virus. They build the channel for ssRNA entry during packaging and the exit of mRNA transcript during the first steps of viral infection (Figs. 13.2a and 13.4d). RNA translocating systems are simple packaging machines, as they are only constituted by one protein. P4, the packaging motor of bacteriophage Φ12, is an NTPase (it does not have ATP specificity), and it is the only one whose structure is known up to date [2]. Nevertheless, this bacteriophage presents homology with other viruses belonging to the *Cystoviridae* family (Φ6–Φ13). P4 shares important similarities with other helicases: the sixfold symmetry and the characteristic Rossmann fold, consisting of a twisted eight-stranded β-sheet surrounded by five helices responsible for nucleotide binding (Fig. 13.4e and f) [2]. The central channel is only wide enough to accommodate ssRNA. This channel is flanked by an α-helix (helix α6) and two loops (L1 and L2 loops, Fig. 13.4e and f). Biochemical and mutational studies indicated that lysines present in these loops are implicated in RNA interaction and translocation [2]. Structural determination of two different nucleotide interacting protein intermediates suggested a mechanism for RNA translocation: The phosphate binding region (P-loop) changes from a down to up conformation after NTP hydrolysis, indicating that this modification could act as a molecular lever pumping the RNA inside the procapsid (Fig. 13.4f) [54].

The packaging mechanism can be subdivided into three steps: RNA recognition, RNA loading into the hexamer, and RNA translocation. The current model suggests that RNA recognition is mediated by conformational changes into the P1 capsid protein (see Sect. 13.3.1). RNA loading has been postulated to happen by ring opening, as is the case of the Rho protein. The interaction of the RNA molecule triggers NTPase activity leading to the translocation of the RNA molecule [54]. Once the molecule is inside the capsid, the viral polymerase generates the dsRNA molecules present in the mature viral particle.

13.4.2 dsDNA Packaging

The DNA packaging mechanism is well conserved in dsDNA bacteriophages and in some animal viruses such as herpesviruses, showing common essential components. In this section we present a brief review of the most important of them to gather additional functional knowledge about this process.

The Packaging Complex

The genome packaging complex is located at a unique vertex of the prohead, called the portal vertex, and it comprises mainly two components: the portal protein, also called connector, that builds a channel for DNA entry into and exit from the capsid [49, 55]; and the terminase, that is involved not only in the selection and processing of the DNA to be packaged, but also in the ATP-driven DNA translocation. Besides these two proteins, in some viruses the packaging machinery may also include additional components such as the small terminase subunit and the packaging RNA (pRNA; see below). Although the icosahedral head has 12 pentameric vertices, the position of the portal protein and the transient interaction with the terminase is confined to a specific one, the portal vertex. In some viruses, such as T7 and ε15, there is a proteinaceous internal structure, called the core, also interacting with that vertex. This core might facilitate the topological ordering of the dsDNA genome during packaging (Sect. 13.2.2) and it is essential for infectivity [37, 56].

The DNA substrate is also an important component of the packaging complex. The viral dsDNA genome varies in length for each viral system: from 19.3 kpb in the case of Φ29 to 166 kpb in T4 (Table 13.1). Most of them (except Φ29 and some related phages) produce head-to-tail multimers, or concatemers, of DNA as a substrate for packaging. Concatemeric DNA is formed by genomic units linked by tandem repeats generated during the replication process. The recognition of the terminal sequences by the viral proteins ensures the specific packaging of the viral genome from that of the host. Moreover, the terminal repeats indicate the cutting region between genomic units, avoiding the loss of base pairs in one end, and ensuring the packaging of a single molecule. The specific recognition sequences, named cos or pac sequences, differ depending on the viral system. They can be

Table 13.1 Viral components involved in dsDNA packaging in different viruses

Virus	Connector		Large terminase		Small terminase		dsDNA
	Gene product	Mass	Gene product	Mass	Gene Product	Mass	Length
λ	gpB	59 kDa	gpA	73 kDa	gpNu	20 kDa	48.5 kb
SPP1	gp6	57 kDa	gp2	49 kDa	gp1	21 kDa	45.9 kb
T4	gp20	61 kDa	gp17	70 kDa	gp16	18 kDa	166 kb
T7	gp8	59 kDa	gp19	67 kDa	gp18	20 kDa	40 kb
P22	gp1	83 kDa	gp2	58 kDa	gp3	19 kDa	43.4 kb
Φ29	gp10	35 kDa	gp16/pRNA	39/58 kDa	gp3	31 kDa	19.3 kb
HSV1	UL6	74 kDa	UL15	81 kDa	UL28	85 kDa	152 kb

identical in both extremes as in T7, T3, and λ; non-unique and repeated at the ends as in P22 and T4; or miscellaneous, with a protein covalently attached, as in Φ29. Two cleavages are required for the unit length packaging: the initiation cut that generates the free end from which packaging starts; and the terminal cut that delimitates the genome unit. Those cleavages can be either sequence specific, as in λ and T7, or sequence independent as in T4, P22, and SPP1. In the latter case, the connector acts as a termination sensor that produces the 'headful-signal' when the genome is already packaged [57, 58]. The terminase term derives from the phage λ proteins that were first shown to be required for the formation of the termini of the packaged DNA.

The packaging proteins and their size, together with the genome length of several of the most abundant of those viruses (tailed bacteriophages) are presented in Table 13.1. As it happens for the genome [57, 59], portal proteins and terminases also present wide variations in size from one virus to another.

The Portal Protein

Although portal proteins do not share sequence homology and present large variations in size, they all show common structural features. Portals are dodecameric assemblies showing a toroidal morphology with an internal channel wide enough to fit the dsDNA molecule (Fig. 13.5). This ring shape is common to protein complexes involved in DNA translocation such as helicases, sliding clamps, and certain exonucleases and topoisomerases. The structure of portal proteins can be divided into several domains (Fig. 13.5a and b): crown (apical part), wing (central part), stem (tight region in the base of the structure), and clip (at the capsid bottom). While the stem domain is the most conserved region, the wing and the crown present variations in length and complexity among different viruses [55, 61]. Besides these domains, portals can present internal loops, also called channel valve, which generate constrictions in the internal channel. Recent structural studies have shown that the channel valve might present various conformational states (Fig. 13.5c and d) [62, 63], suggesting its ability to open and close the portal pore through movements of its domains and pointing to the portal protein as a dynamic pore. Previously, it was also postulated that positive charges in channel valve loops could interact with the viral genome [55]. These interactions may regulate the DNA flow through the channel and suggest a valve-like mechanism for the translocation. Moreover, structure determination of the herpes simplex type 1 (HSV1) portal inside the virus showed clip and channel loop contacts with the DNA. These interactions might disrupt the dsDNA morphology promoting the formation of a disc-like structure outside the portal (Fig. 13.5c) [60].

Mutational analysis of the portal protein has been essential to describe its functions [55, 64]. The portal protein has been proposed to nucleate the prohead assembly by the interaction with the scaffolding proteins. It also serves as a docking point for other components of the packaging motor (ATPases or pRNA in Φ29, see below in this section) modulating their activity. As mentioned before, the portal has been proposed to be the sensor of the headful mechanism, defining the quantity of DNA to be packaged. Additionally, after DNA packaging the portal protein interacts

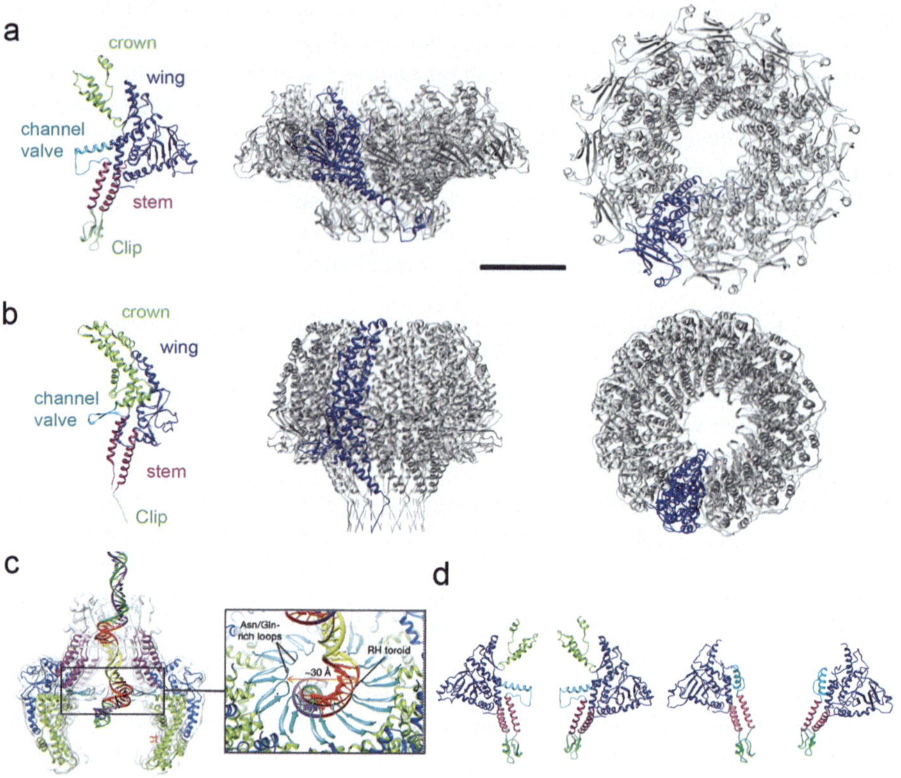

Fig. 13.5 Portal protein structure in dsDNA viruses. Structural models of portal proteins from T7 phage (**a**) and HSV1 (**b**) showing the monomeric (left) and oligomeric assemblies (right). In the monomers the different domains are depicted in different colours. Dodecameric structures (right) are shown in top and side views with a single subunit coloured blue. Scale bar corresponds to 50 Å. (**c**) HSV1 Portal–DNA interactions showing the contacts between the channel valve and the DNA (inlet) (Adapted from [60], with permission). (**d**) Comparison of the two opposed portal monomers showing them in the closed conformation (left) or in the open conformation (right). The crown domain is not visible in the open conformation and has not been shown for the sake of clarity. PDB codes: T7 monomer (6QX5) and oligomer (6QXM), and HSV1 (6OD7)

with the tail proteins and other proteins involved in the closure of the channel, being an essential component for the formation of the mature virus.

The Terminase Complex

The second component of the packaging machinery is the terminase. Terminases are generally hetero-oligomers built by a large protein containing the ATPase and nuclease activities and a small protein involved in viral genome recognition. The composition of the complex might vary between virus and the small terminase can be substituted by a small pRNA in the case of Φ29 or by two alternative proteins in the case of herpes virus. Terminases are the main element of the packaging motor as they are able to convert chemical energy from ATP hydrolysis into mechanical movement of DNA during phage morphogenesis. Besides the ATPase activity, most terminases also carry an endonuclease activity essential to cut concatemeric DNA into genome lengths. During packaging, the large terminase interacts with the portal protein (inserted in the prohead), with the DNA (first to translocate it and finally to cleave it), and with other components of the packaging machinery that might vary between virus (small terminase, pRNA in the case of Φ29).

Phage Φ29, one of the best characterized packaging systems (described below), is an unusual case: its DNA replicates as a unit length genome, thus its large terminase protein (gp16) does not have nuclease activity. Moreover, the terminase functions in coordination with the pRNA molecule essential for docking to the portal vertex. Although this pRNA may be considered as part of the large terminase, it will be described separately in this section.

Even though there is no significant overall sequence similarity, terminase proteins from different phages contain well-conserved patches of amino acid sequences, or structural motifs that are required for packaging [57]. The structural information available about terminases facilitates correlating their topology and interactions to their function for DNA packaging as described in the following sections.

The large terminase subunit. The limited structural information on large terminases due to their flexible, conformationally heterogeneous nature has been partly overcome in the last years by the improvement in structure determination techniques (Fig. 13.6) [65, 66]. Sequence alignments show that the functional signatures of the ATPase domain of viral terminases are conserved and they are similar to those of the translocating monomeric SF2 helicases, restriction endonucleases, and protein translocases. Large terminases consist of two domains: an N-terminal ATPase domain that powers DNA translocation and a C-terminal nuclease domain that generates the termini of the viral genome (Fig. 13.6a). The ATPase domain consists of the classic nucleotide binding Rossmann fold, similar to RNA translocase of dsRNA cystoviruses (see Sect. 13.4.1). It contains the Walker A and B and the catalytic carboxylate, often found in proteins that bind and cleave ATP. Genetic and biochemical studies show that this domain is also essential for the interaction with the small terminase subunit [32, 67]. The nuclease domain presents an also conserved RNase H-like fold formed by seven β-sheets sandwiched between two clusters of α-helices (Fig 13.6a). As mentioned, although these basic folds are conserved in large terminase proteins from various phages, those of herpes virus show almost no amino acid sequence identity and present variations in length [57, 68].

The overexpressed and purified large terminases from many phages exhibit different oligomeric states, being either monomers or oligomers in solution. Nevertheless, during the packaging machinery assembly, the terminase stoichiometry is remodelled. After binding to the proheads through the C-terminal domain and assembling the packaging motor, the viral terminase monomers oligomerize to form a pentameric ring with an internal channel. The channel diameter is ~40 Å, being large enough to allow the passage of the dsDNA molecule (Fig. 13.6b right). In addition, the terminase stoichiometry fits well with the tenfold symmetry of the

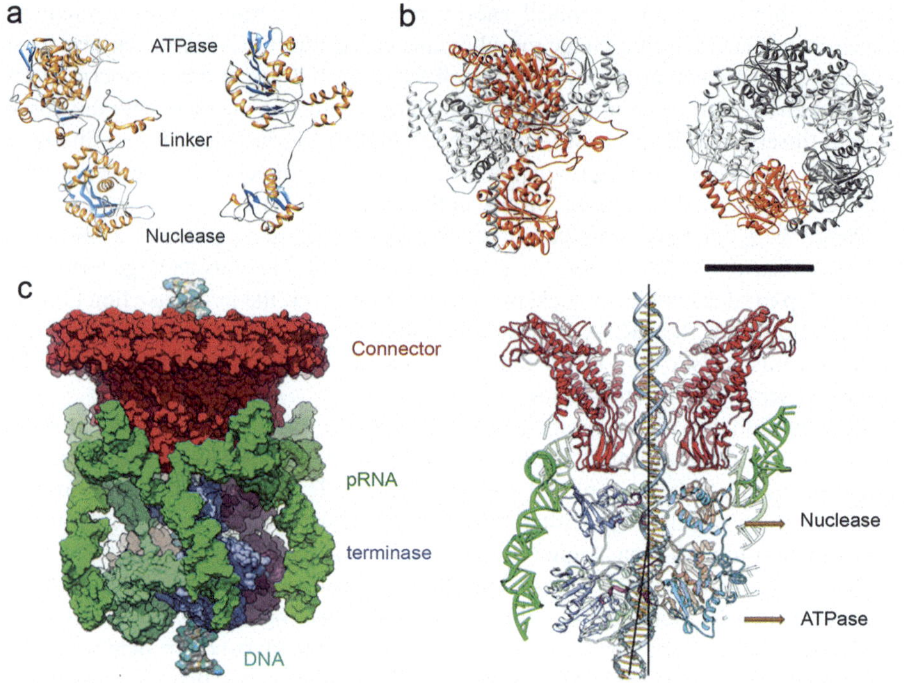

Fig. 13.6 Molecular structures of the large terminases and the full Φ29 packaging motor. (**a**) Large terminase monomers showing the domain structure conservation in herpes virus (left) and Φ29 (right). (**b**) Oligomeric structures from herpes virus (trimer, left) and Φ29 (pentamer, right) with one monomer depicted in red. The scale bar corresponds to 50 Å. (**c**) Structure of Φ29 bacteriophage dsDNA packaging motor stalled during packaging. Left, molecular surface; right, cross-section of a ribbon model. Adapted from [65]. PDB codes: HSV1, 6M5S; Φ29, 7JQQ

dsDNA [32]. The herpes virus large terminase is an unusual one, as it forms a hetero-trimer with two other proteins that show no homology with bacteriophage small terminases (Fig. 13.6b left).

The structure of the Φ29 terminase, stalled during packaging, has provided the best structural description of this machinery in action. The structure showed the terminase pentamer interacting with the portal and the pRNA (Fig. 13.6c). Interestingly, this study revealed for the first time a helical-to-planar dynamic arrangement in the terminase (Fig. 13.6c). This conformational change of the terminase from a planar oligomeric ring-like to the helical form might be tightly related to the translocation process (see next section). Modelling of the DNA inside the helical arrangement of the Φ29 terminase showed that the ATPase domain interacts with the DNA deviating its trajectory in the channel (Fig. 13.6c).

The small terminase subunit. The main function of the small terminases is the recognition of the viral DNA. Besides this task they also regulate the enzymatic activities of large terminases during DNA translocation. This subunit, as it was

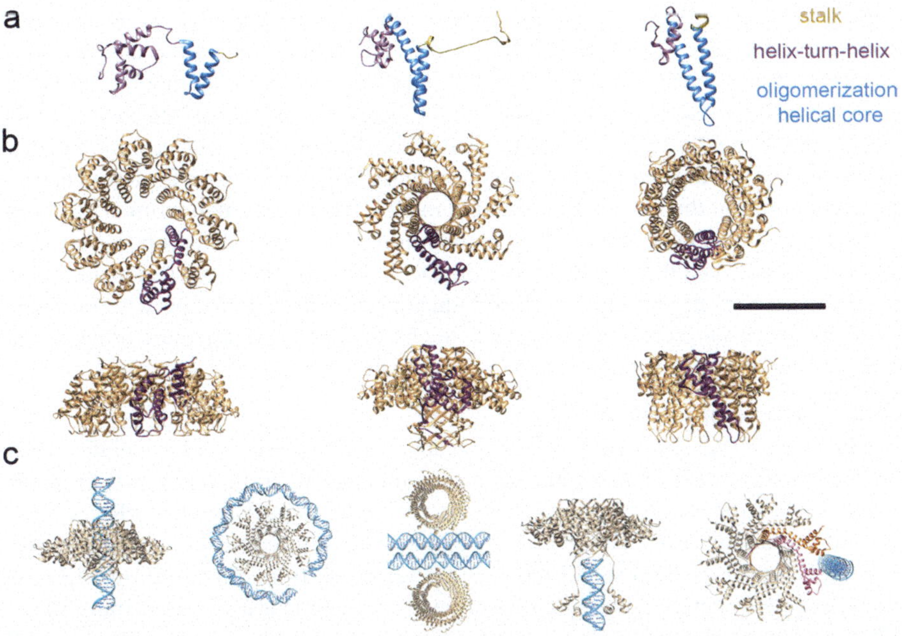

Fig. 13.7 Molecular structures of small terminases and DNA interaction models. (**a**) Monomers of small terminases from phages Sf6 (Podovirus, left), HK97 (Siphovirus, middle) and E217 (Myovirus, right) showing the three conserved structural regions. (**b**) Molecular structures of the corresponding oligomers (top and side views) with one monomer depicted in purple. Scale bar corresponds to 50 Å (**c**) Models for DNA recognition by the small terminase. Reproduced from [69], with permission. Access numbers: phage Sf6 4DYQ, HK97 6Z6E and E217 7UXE

mentioned before, is not present in all packaging systems and these viruses have found another solution to bring specific viral DNA to the packaging vertex. In Φ29 phage, for example, another protein (gp3) primes its DNA developing an analogous function.

The small terminase subunit amino acid sequence is even more variable than the other two motor proteins. Over the last twenty years several structures of these proteins have been obtained using different structural techniques (Fig. 13.7). Despite this heterogeneity, biochemical and structural data revealed some characteristics in domain organization that are highly conserved (Fig. 13.7a). The N-terminal domain formed by a helix-turn-helix is the DNA-binding domain; the central domain formed by two conserved α-helices acts as the oligomerization scaffold and, finally, the C-terminal stalk constituted by two helices of different lengths interacts with the large terminase (Fig. 13.7a) [69]. The oligomerization of the small terminase produces multimer rings (Fig. 13.7b), that vary in size from octamers to decamers (or larger), they also present uneven channel diameters, ranging from 10 to 24 Å in their narrow region. Together, these data suggest that although oligomerization is important, the stoichiometry or a defined inner diameter does not appear to be strictly essential for its function. Alternatively, interactions with other components of the packaging machinery may select for a precise stoichiometry in vivo. How the small terminase recognizes DNA is still an unsolved question, due to the absence of a structure of the small terminase bound to DNA [69]. Several models of DNA recognition have been proposed (Fig. 13.7c), such as: the 'threading model' with DNA interacting through the small terminase channel, the 'nucleosome model' where the DNA wraps around the small terminase ring, the 'two rings pack synapse' as an extension of the former, the 'thimble model' where the small terminase acts as a cap trapping the DNA, and the 'lateral interdigitation model' that proposed specific interactions between the terminase and the DNA. Finally, no conserved residues are located on the small terminase surface indicating that the interaction with the large terminase or portal protein is rather mediated by the overall architecture of the portal complex and by the distinctive shape of the small terminase [32, 69, 70].

Other Components: The pRNA
Bacteriophage Φ29 is a special case because of the presence of a unique RNA molecule that comprises its packaging motor together with the portal and the large terminase. Its role is transitory and likely limited to DNA packaging, as pRNA is not present in mature virus. The pRNA secondary structure consists of two domains: domain I comprises the first 117 bases and is separated from domain II by a 13 base single-stranded region. Although the full-length wild type is a 174-base transcript, a 120-base construct encompassing domain I and lacking 54 nucleotides 3′-domain II is fully competent for packaging assembly in vitro. Despite that the function of domain II is not currently understood, its conservation in all Φ29 relatives suggests that it may have an essential role in vivo. The domain I shows two functional modules: one necessary for prohead binding and the other required for DNA packaging activity, which provides the attachment site for the large terminase [71]. Formation of intermolecular pseudo-knots between adjacent pRNAs would promote circular interactions resulting in an oligomeric ring-like structure. In fact, structural characterization of the in situ packaging machinery showed that five pRNA helices are bound to the prohead and extend as tentacles surrounding the terminase, forming a concentric ring together with the portal protein and the terminase (Fig. 13.6c) [65].

13.4.3 Biophysical Properties of the Packaging Motor

The accurate measurement of viral genome packaging kinetics is difficult without the synchronization of different populations. This process is even harder for dsDNA motors where the motor only assembles transitory but it is not a structural part of the mature virus (see Sect. 13.4.2). While biochemical bulk assays can only determine the packaging efficiency by means of the total time to package the nucleic acid, single-molecule experiments allow collecting statistics of single events, and they have allowed observing for the first time pauses where the motor eventually disassembles from the packaged substrate (see Chap. 10) [72].

The optical tweezers are the main single-molecule technique that has been used to study the biophysical properties of viral packaging motors (Chap. 10). This technique has been mainly employed to study the packaging machinery of the dsDNA viruses. The setting usually consists in a viral prohead assembled with the packaging ATPase in the presence of an ATP analogue, then a microsphere carrying DNA is introduced and the variations in the nucleic acid extension values are carefully followed while different forces are applied to the system [32] (see Chap. 10 for technical details). This technique has allowed characterizing the packaging machinery as a nanomotor able to transform the chemical energy of ATP hydrolysis into mechanical work leading to nucleic acid translocation. Optical tweezers revealed that packaging motors are able to generate forces up to 50–60 pN at low capsid filling (see Chap. 10). These forces, 20–25 times higher than those exerted by the myosine II implicated in muscle contraction, reveal that these motors are among the most powerful identified in nature. Such high values may be essential to allow enough DNA bending and to overcome electrostatic repulsive forces in the DNA molecule, working against the increasing internal capsid pressure, that reaches ~20 atm at the later stages of encapsidation (see [32, 73] and Chaps. 10 and 19).

These experiments have also revealed the speed of the packaging motor for different dsDNA bacteriophages [32, 72, 74]. These values turned out to vary on average from 180 bp/s for Φ29 to 1800 bp/s for T4 which has the longest genome, suggesting that the speed of the motor could be closely related to the length of the phage genome (see Table 13.1 and Sect. 13.4.2). Thus, phages with larger genomes have adapted to faster velocities in order to complete DNA encapsulation within 2–3 min from the total viral cycle of 20–30 min. It has also been observed that the speed decreased at late stages of packaging, when the internal pressure of the capsid increases. For example, λ packaging speed drops threefold when 90% of the genome is inside the capsid.

Another biophysical value determined is the step size, or the translocation length of the motor per ATP molecule hydrolysis. In bulk assays, this value was determined to be 2 bp/ATP. Single-molecule experiments have shown, however, that the step size cannot be taken as a fixed value, as it can vary according to DNA symmetry and filling of the capsid. High-resolution optical tweezers recently showed that the motor translocates in four 2.5 bp steps. This fact strongly suggests a high coordination of the subunits, as the terminase is composed of five subunits and only four of them bind ATP before the hydrolysis and translocation are triggered [72]. Later experiments showed that the motor rotates the DNA during packaging and that the rotation per base pair increases with the filling of the capsid, accompanied by a reduction in the translocation of the bp (step size) [73]. In fact, it seems that one subunit of the terminase would be essential to make periodic, specific, and regulatory contacts with the DNA adapting the step size to the capsid filling to preserve motor coordination. Thus, the packaging motor not only translocates the DNA but it also rotates the viral genome molecule, probably in order to help accommodate the DNA inside the capsid. The current model of genome translocation and the detailed function of the dsDNA packaging motor will be discussed below (Sect. 13.4.5).

13.4.4 dsRNA Packaging Models

Considering the large differences described between DNA and RNA packaging motors (see Sect. 13.4) it would be reasonable to think that RNA packaging machineries would show different characteristics from the ones described for DNA packaging machines. Although no single-molecule experiments have been reported up to date, the atomic resolution of two different RNA translocase conformations (P4 protein, see Sect. 13.4.1) and accurate biochemical studies have given strong structural insights into the molecular mechanism of the dsRNA motors [2]. The main outcome from these data shows several interesting similarities with the DNA packaging motors. Structural snapshots showed that the P-loop of P4 changes its conformation from up to down in the presence of two different ATP analogues suggesting that this movement could help to pump the RNA inside the prohead (see Sect. 13.4 and Fig. 13.4). It has been proposed that P4 would follow a sequential translocation mechanism. In this mechanism the coordination between different subunits would be essential. Communication between adjacent subunits could happen through 'arginine fingers' commonly found in ATPase motors. In this model ATP hydrolysis will trigger a conformational change that places the arginine finger of the adjacent subunit into the ATPase active site starting the next ATP hydrolysis [54]. It is supposed that the ssRNA structure could be similar to the A-dsRNA form presenting 11 bases per turn. It has been suggested that this symmetry mismatch would be solved by the L1 loop that would act as a grommet correcting the RNA position [58].

13.4.5 dsDNA Packaging Models

Composition of dsDNA packaging motors is more complex than dsRNA packaging motors. Usually their packaging machinery is composed of: the portal protein that forms a unique channel for DNA exit and entry in the prohead; the terminase complex that comprises the ATPase that actively pumps the DNA; and other additional components that vary depending on the virus. In the last years a tremendous effort has been made in characterizing the structure (see Sect. 13.4.2) and biophysics (see Sect. 13.4.3) of the dsDNA packaging components, allowing deciphering the detailed mechanism of viral packaging. This knowledge has been put together to accurately describe the packaging mechanism.

A long-standing debate in the field in the past years was whether it was the portal or the large terminase that changed conformation to translocate the DNA inside the prohead. Previously proposed models could be classified into two categories. (1) Portal protein-driven models, where conformational changes induced by ATP hydrolysis in the portal structure lead to DNA packaging. (2) Terminase models, in which the portal was considered as a passive actor in the translocation mechanism, while conformational changes in the ATPase directly translocate DNA. Although

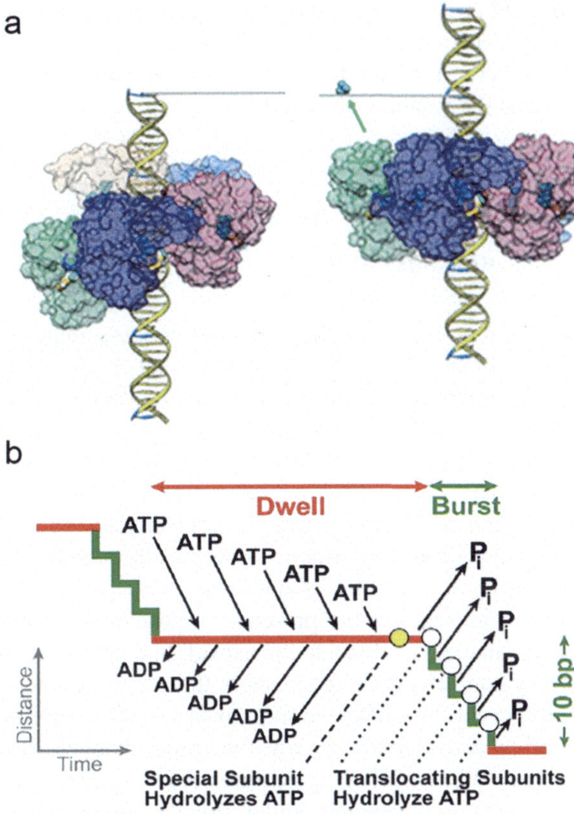

Fig. 13.8 Model of dsDNA packaging in bacteriophages. (**a**) Molecular structure of the Φ29 terminase in its helical (left) and planar (right) conformations. The terminase subunits are shown in different colours, and the DNA inside the terminase channel is shown in yellow. Adapted from [65]. (**b**) Mechanochemical model of the dwell-burst packaging cycle. Adapted from [72], with permission

both models considered the terminase as the ATP-consuming element of the motor, the difference laid in the element that actively pushed the DNA inside the capsid. Nowadays it is clear that the terminase is the component that actively pushes the DNA, and the currently proposed mechanism suggests that ATPase cycles between helical and planar symmetry in the terminase drive the DNA into the procapsid as the subunits adopt the planar configuration (Fig. 13.8a) [65]. However, insights from the structural characterization of the motor components have reconciled terminase and portal driven models, as both conformational changes of the portal protein, consisting in the opening and closing of the portal channel, and the helical-to-planar transition of the terminase facilitate the physical movement of the DNA [63, 65]. While conformational changes of the terminase pump the DNA, those in the portal protein may be essential to retain the viral genome inside the capsid during packaging.

One of the most interesting features to understand the mechanism of nucleic acid packaging is the coupling between ATP hydrolysis and DNA translocation. Biophysical studies revealed that the motor engages DNA after ATP binding and that the genome is translocated right before ADP release [72]. The cycle can be divided into two phases: the dwell phase and the burst phase (Fig 13.8b). In the dwell phase the five ATPase subunits load ATP and interact with the DNA, adopting the helical arrangement. In the burst phase, the subunits sequentially hydrolyse ATP, adopting the planar conformation that results in the translocation of 10 DNA bp (in four 2.5 bp steps) and the rotation of the DNA.

13.4.6 Differences Between Nucleic Acid Packaging and Ejection

In most animal viruses the infection process requires capsid disassembly, pore formation in the capsid after conformational changes to genome exit, or the presence of specialized protein channels. In viruses following an active packaging mechanism (dsDNA and dsRNA bacteriophages) the nucleic acid exits the capsid through the portal/translocase channel located at a fivefold vertex (see Chap. 17), the same passage that is used during genome packaging (see Sect. 13.4). These viruses package their genome in ordered layers (see Sect. 13.1) inside the capsid condensing its nucleic acid at a high density and accumulating an important potential energy (reflected in an internal pressure of up to 60 atm; see Chaps. 10, 18, and 19). This high pressure requires the presence of protein complexes in the portal axis able to retain the nucleic acid inside the capsid. In dsDNA viruses this task is accomplished either by conformational changes in the portal channel or by protein plugs, like the tail proteins or the gatekeeper proteins [75] (see Chap. 17). Even though nucleic acid packaging and ejection follow the same channel pathway, they are not inverse processes. While packaging requires ATP consumption, genome delivery uses the energy stored in the nucleic acid during the condensation process [76]. Nevertheless, this internal energy decreases as the genome is liberated and, thus, it is believed that other mechanisms may participate, as it is the case of proteins inside the host cell that pull the nucleic acid when the replication process is initiated (see [77] and Chap. 19). Differences are also evidenced by the fact that both processes have very different speeds, at least as measured in in vitro experiments: while DNA packaging is carried out at an average speed of 700 bp/s, DNA release can attain 60,000–70,000 bp/s.

In dsDNA bacteriophages there is strong evidence that DNA entry and exit involve different protein complexes. While DNA packaging is done by the packaging motor, DNA delivery uses the tail machine [78] (see Chap. 17). As some protein loops may as well be retaining the exit of the DNA, genome delivery requires conformational changes to modify the ejection channel to allow DNA passage [63, 75]. Some viruses with dsRNA genomes present similar characteristics, as in the first stages of the viral cycle they extrude their mRNAs through the channels at the fivefold vertex (Fig. 13.2 and Sect. 13.3). During this process, the positive strand from dsRNA is displaced to generate the mRNA substrate. Once more, this process is not energy-consuming, as many viruses will co-infect the cells and P4 protein serves as a passive channel for ssRNA.

13.5 Perspectives and Conclusions

Viruses protect their genetic material against aggressions in the extracellular media or by cellular degrading enzymes by enclosing it inside a protein capsid that also serves other biological functions. Viruses have evolved different packaging strategies according to their lifestyles, type of nucleic acid (RNA or DNA, single or double-stranded), amount of genetic material to be packaged, and other variables. A widespread strategy consists in the co-assembly of capsid and nucleic acid; in these cases, the nucleic acid may act as a scaffold to promote capsid assembly, and the capsid proteins may help in the reorganization of the nucleic acid inside the virion. In the simplest cases, this co-assembly strategy results in a helical structure that specifically encloses the viral nucleic acid. Several bacterial, plant or animal ssRNA or ssDNA viruses (but not ssDNA parvoviruses) use packaging pathways involving condensation of capsid and nucleic acid.

Many other icosahedral viruses use dsRNA or dsDNA as genetic material. Packaging of long, stiff double-helical nucleic acid molecules pose additional problems; these viruses evolved sophisticated mechanisms for the packaging of these nucleic acids into preformed capsids. Viral particles that actively package their double-helical nucleic acid genomes into spherical containers usually organize the nucleic acid in concentric layers inside the capsid. This arrangement facilitates genome uncoating by ejection without capsid disassembly.

The packaging machinery in those dsDNA and dsRNA viruses is a molecular motor that transforms the chemical energy of NTP hydrolysis into mechanical work to pump the nucleic acid inside the capsid. During the packaging process the initial immature capsid or prohead suffers a series of conformational changes to become a mature capsid. The nucleic acid is transported inside the capsid through a channel formed by a protein complex present at the fivefold vertex. These proteins present similarities with other nucleic acid translocases, showing ring-like structures with a central channel that serves for genome transport. In dsRNA bacteriophages, this protein is named RNA translocase and constitutes the simplest packaging motor described up to date. Structural snapshots obtained by X-ray crystallography suggest a mechanical mechanism consisting in the movement of a loop that would sequentially change conformation to translocate the genome. dsDNA bacteriophages present a more complicated machinery, formed by a multi-protein complex that assembles transiently to the capsid. In these viruses the translocation pore is built by a protein, the portal, that sits at one single vertex of the capsid. The portal serves as a docking point for a viral ATPase complex, the terminase, that attaches transiently to the capsid during packaging. The terminase is essentially composed of an ATPase and a secondary component, either a protein or RNA, which enhances the activity of the ATPase and specifically recognizes the viral genome. The properties of this multicomponent motor have been characterized using single-molecule techniques (optical tweezers), which have revealed the biophysical characteristics of what constitutes one of the most powerful motors described to date. Several models have been proposed to couple the chemical energy of ATP hydrolysis with the mechanical action of DNA pumping; nevertheless, important aspects of the molecular mechanism are still uncertain. Current evidence suggests that the conformational changes of both the portal protein and the large terminase facilitate the physical movement of the DNA inside the capsid. In some cases, the portal acts as a sensor to indicate the degree of capsid filling to terminate the packing process.

Together, the results obtained so far show that genome packaging is a very complex mechanism. The importance of this process for virus survival has forced some viruses to incorporate sophisticated machineries, sharing structural and mechanistic properties with other protein complexes that perform essential functions during the metabolism of genetic material in living cells.

Acknowledgements This work was supported by Grant P2022/BMD-7232 TomoXliver2 to AC. MID is a recipient of an 'Atracción de Talento Investigador' Fellowship (reference 2018T2_BMD10584) from the Autonomous Region of Madrid. MID wishes to acknowledge the support of O. Llorca during the preparation of this manuscript.

References

1. Hunter E (2013) Virus assembly. In: Knipe DN, Howley PM (eds) Fields virology, vol 1, 6th edn. Lippincott Williams & Wilkins, Philadelphia, pp 127–152
2. Kainov DE, Tuma R, Mancini EJ (2006) Hexameric molecular motors: P4 packaging ATPase unravels the mechanism. Cell Mol Life Sci 63:1095–1105
3. Arranz R, Coloma R, Chichon FJ, Conesa JJ, Carrascosa JL, Valpuesta JM, Ortin J, Martin-Benito J (2012) The structure of native influenza virion ribonucleoproteins. Science 338:1634–1637
4. Schneemann A (2006) The structural and functional role of RNA in icosahedral virus assembly. Ann Rev Microbiol 60:51–67
5. Dai X, Li Z, Lai M, Shu S, Du Y, Zhou ZH, Sun R (2017) In situ structures of the genome and genome-delivery apparatus in a single-stranded RNA virus. Nature 541:112–116
6. Ilca SL, Sun X, El Omari K, Kotecha A, de Haas F, DiMaio F, Grimes JM, Stuart DI, Poranen MM, Huiskonen JT (2019) Multiple liquid crystalline geometries of highly compacted nucleic acid in a dsRNA virus. Nature 570:252–256
7. Zhang X, Ding K, Yu X, Chang W, Sun J, Zhou ZH (2015) In situ structures of the segmented genome and RNA polymerase complex inside a dsRNA virus. Nature 527:531–534
8. Xu J, Dayan N, Goldbourt A, Xiang Y (2019) Cryo-electron microscopy structure of the filamentous bacteriophage IKe. Proc Natl Acad Sci USA 116:5493–5498
9. Gong D, Dai X, Jih J, Liu YT, Bi GQ, Sun R, Zhou ZH (2019) DNA-packing portal and capsid-associated tegument complexes in the tumor herpesvirus KSHV. Cell 178:1329–1343
10. Bhyravbhatla B, Watowich SJ, Caspar DL (1998) Refined atomic model of the four-layer aggregate of the tobacco mosaic virus coat protein at 2.4-Å resolution. Biophys J 74:604–615
11. Sachse C, Chen JZ, Coureux PD, Stroupe ME, Fandrich M, Grigorieff N (2007) High-resolution electron microscopy of helical specimens: a fresh look at tobacco mosaic virus. J Mol Biol 371:812–835
12. Lamb RA (2013) Mononegavirales. In: Knipe DN, Howley PM (eds) Fields virology, vol 1, 6th edn. Lippincott Williams & Wilkins, Philadelphia, pp 880–884
13. Green TJ, Zhang X, Wertz GW, Luo M (2006) Structure of the vesicular stomatitis virus nucleoprotein-RNA complex. Science 313:357–360

14. Coloma R, Arranz R, de la Rosa-Trevin JM, Sorzano COS, Munier S, Carlero D, Naffakh N, Ortin J, Martin-Benito J (2020) Structural insights into influenza A virus ribonucleoproteins reveal a processive helical track as transcription mechanism. Nat Microbiol 5:727–734
15. Coloma R, Valpuesta JM, Arranz R, Carrascosa JL, Ortin J, Martin-Benito J (2009) The structure of a biologically active influenza virus ribonucleoprotein complex. PLoS Pathog 5: e1000491
16. Modrego A, Carlero D, Arranz R, Martin-Benito J (2023) CryoEM of viral ribonucleoproteins and nucleocapsids of single-stranded RNA viruses. Viruses 15:653
17. Bunka DH, Lane SW, Lane CL, Dykeman EC, Ford RJ, Barker AM, Twarock R, Phillips SE, Stockley PG (2011) Degenerate RNA packaging signals in the genome of satellite tobacco necrosis virus: implications for the assembly of a T=1 capsid. J Mol Biol 413:51–65
18. Dykeman EC, Grayson NE, Toropova K, Ranson NA, Stockley PG, Twarock R (2011) Simple rules for efficient assembly predict the layout of a packaged viral RNA. J Mol Biol 408:399–407
19. Twarock R, Bingham RJ, Dykeman EC, Stockley PG (2018) A modelling paradigm for RNA virus assembly. Curr Opin Virol 31:74–81
20. Prasad BV, Prevelige PE Jr (2003) Viral genome organization. Adv Protein Chem 64:219–258
21. Coombs KM (2006) Reovirus structure and morphogenesis. Curr Top Microbiol Immunol 309: 117–167
22. Reinisch KM, Nibert ML, Harrison SC (2000) Structure of the reovirus core at 3.6 Å resolution. Nature 404:960–967
23. Ding K, Celma CC, Zhang X, Chang T, Shen W, Atanasov I, Roy P, Zhou ZH (2019) In situ structures of rotavirus polymerase in action and mechanism of mRNA transcription and release. Nat Commun 10:2216
24. Caston JR, Ghabrial SA, Jiang D, Rivas G, Alfonso C, Roca R, Luque D, Carrascosa JL (2003) Three-dimensional structure of penicillium chrysogenum virus: a double-stranded RNA virus with a genuine T=1 capsid. J Mol Biol 331:417–431
25. Borodavka A, Desselberger U, Patton JT (2018) Genome packaging in multi-segmented dsRNA viruses: distinct mechanisms with similar outcomes. Curr Opin Virol 33:106–112
26. Shah PNM, Gilchrist JB, Forsberg BO, Burt A, Howe A, Mosalaganti S, Wan W, Radecke J, Chaban Y, Sutton G et al (2023) Characterization of the rotavirus assembly pathway in situ using cryoelectron tomography. Cell Host Microbe 31:604–615
27. Liu H, Cheng L (2015) Cryo-EM shows the polymerase structures and a nonspooled genome within a dsRNA virus. Science 349:1347–1350
28. Sidhu SS (2001) Engineering M13 for phage display. Biomol Eng 18:57–63
29. Liu DJ, Day LA (1994) Pf1 virus structure: helical coat protein and DNA with paraxial phosphates. Science 265:671–674
30. Chapman MS, Rossmann MG (1995) Single-stranded DNA-protein interactions in canine parvovirus. Structure 3:151–162
31. Johnson JE, Chiu W (2007) DNA packaging and delivery machines in tailed bacteriophages. Curr Opin Struct Biol 17:237–243
32. Casjens SR (2011) The DNA-packaging nanomotor of tailed bacteriophages. Nat Rev Microbiol 9:647–657
33. Aksyuk AA, Rossmann MG (2011) Bacteriophage assembly. Viruses 3:172–203
34. Cerritelli ME, Cheng N, Rosenberg AH, McPherson CE, Booy FP, Steven AC (1997) Encapsidated conformation of bacteriophage T7 DNA. Cell 91:271–280
35. Petrov AS, Boz MB, Harvey SC (2007) The conformation of double-stranded DNA inside bacteriophages depends on capsid size and shape. J Struct Biol 160:241–248
36. Agirrezabala X, Martin-Benito J, Caston JR, Miranda R, Valpuesta JM, Carrascosa JL (2005) Maturation of phage T7 involves structural modification of both shell and inner core components. EMBO J 24:3820–3829
37. Perez-Ruiz M, Pulido-Cid M, Luque-Ortega JR, Valpuesta JM, Cuervo A, Carrascosa JL (2021) Assisted assembly of bacteriophage T7 core components for genome translocation across the bacterial envelope. Proc Natl Acad Sci USA 118:e2026719118

38. Fang PA, Wright ET, Weintraub ST, Hakala K, Wu W, Serwer P, Jiang W (2008) Visualization of bacteriophage T3 capsids with DNA incompletely packaged in vivo. J Mol Biol 384:1384–1399
39. Comolli LR, Baker BJ, Downing KH, Siegerist CE, Banfield JF (2009) Three-dimensional analysis of the structure and ecology of a novel, ultra-small archaeon. ISME J 3:159–167
40. Poranen MM, Bamford DH (2012) Assembly of large icosahedral double-stranded RNA viruses. Adv Exp Med Biol 726:379–402
41. Sutton G, Sun D, Fu X, Kotecha A, Hecksel CW, Clare DK, Zhang P, Stuart DI, Boyce M (2020) Assembly intermediates of orthoreovirus captured in the cell. Nat Commun 11:4445
42. El Omari K, Sutton G, Ravantti JJ, Zhang H, Walter TS, Grimes JM, Bamford DH, Stuart DI, Mancini EJ (2013) Plate tectonics of virus shell assembly and reorganization in phage phi8, a distant relative of mammalian reoviruses. Structure 21:1384–1395
43. Mindich L (2012) Packaging in dsRNA viruses. Adv Exp Med Biol 726:601–608
44. Huiskonen JT, de Haas F, Bubeck D, Bamford DH, Fuller SD, Butcher SJ (2006) Structure of the bacteriophage phi6 nucleocapsid suggests a mechanism for sequential RNA packaging. Structure 14:1039–1048
45. Luque D, Caston JR (2020) Cryo-electron microscopy for the study of virus assembly. Nat Chem Biol 16:231–239
46. Serwer P, Hunter B, Wright ET (2020) Electron microscopy of in-plaque phage T3 assembly: proposed analogs of neurodegenerative disease triggers. Pharmaceuticals (Basel) 13:18
47. Chen W, Xiao H, Wang X, Song S, Han Z, Li X, Yang F, Wang L, Song J, Liu H et al (2020) Structural changes of a bacteriophage upon DNA packaging and maturation. Protein Cell 11: 374–379
48. Ionel A, Velazquez-Muriel JA, Luque D, Cuervo A, Caston JR, Valpuesta JM, Martin-Benito J, Carrascosa JL (2011) Molecular rearrangements involved in the capsid shell maturation of bacteriophage T7. J Biol Chem 286:234–242
49. Valpuesta JM, Carrascosa JL (1994) Structure of viral connectors and their function in bacteriophage assembly and DNA packaging. Q Rev Biophys 27:107–155
50. Steven AC, Heymann JB, Cheng N, Trus BL, Conway JF (2005) Virus maturation: dynamics and mechanism of a stabilizing structural transition that leads to infectivity. Curr Opin Struct Biol 15:227–236
51. Gertsman I, Gan L, Guttman M, Lee K, Speir JA, Duda RL, Hendrix RW, Komives EA, Johnson JE (2009) An unexpected twist in viral capsid maturation. Nature 458:646–650
52. Lawton JA, Estes MK, Prasad BV (2000) Mechanism of genome transcription in segmented dsRNA viruses. Adv Virus Res 55:185–229
53. Pan M, Alvarez-Cabrera AL, Kang JS, Wang L, Fan C, Zhou ZH (2021) Asymmetric reconstruction of mammalian reovirus reveals interactions among RNA, transcriptional factor micro2 and capsid proteins. Nat Commun 12:4176
54. Mancini EJ, Tuma R (2012) Mechanism of RNA packaging motor. Adv Exp Med Biol 726: 609–629
55. Cuervo A, Carrascosa JL (2012) Viral connectors for DNA encapsulation. Curr Opin Biotechnol 23:529–536
56. Swanson NA, Hou CD, Cingolani G (2022) Viral ejection proteins: mosaically conserved, conformational gymnasts. Microorganisms 10:504
57. Rao VB, Feiss M (2008) The bacteriophage DNA packaging motor. Annu Rev Genet 42:647–681
58. Sun S, Rao VB, Rossmann MG (2010) Genome packaging in viruses. Curr Opin Struct Biol 20: 114–120
59. Mettenleiter TC, Klupp BG, Granzow H (2006) Herpesvirus assembly: a tale of two membranes. Curr Opin Microbiol 9:423–429
60. Liu YT, Jih J, Dai X, Bi GQ, Zhou ZH (2019) Cryo-EM structures of herpes simplex virus type 1 portal vertex and packaged genome. Nature 570:257–261

61. Carrascosa JL, Valpuesta JM (1999) Bacteriophage connectors: structural features of a DNA translocating motors. In: Recent research developments in virology. Transworld Reseach Network, Trivadrum, pp 449–465
62. Rao VB, Fokine A, Fang Q (2021) The remarkable viral portal vertex: structure and a plausible model for mechanism. Curr Opin Virol 51:65–73
63. Cuervo A, Fabrega-Ferrer M, Machon C, Conesa JJ, Fernandez FJ, Perez-Luque R, Perez-Ruiz M, Pous J, Vega MC, Carrascosa JL et al (2019) Structures of T7 bacteriophage portal and tail suggest a viral DNA retention and ejection mechanism. Nat Commun 10:3746
64. Feiss M, Rao VB (2012) The bacteriophage DNA packaging machine. Adv Exp Med Biol 726: 489–509
65. Woodson M, Pajak J, Mahler BP, Zhao W, Zhang W, Arya G, White MA, Jardine PJ, Morais MC (2021) A viral genome packaging motor transitions between cyclic and helical symmetry to translocate dsDNA. Sci Adv 7:eabc1955
66. Mao H, Saha M, Reyes-Aldrete E, Sherman MB, Woodson M, Atz R, Grimes S, Jardine PJ, Morais MC (2016) Structural and molecular basis for coordination in a viral DNA packaging motor. Cell Rep 14:2017–2029
67. Rao VB, Black LW (2010) Structure and assembly of bacteriophage T4 head. Virol J 7:356
68. Yang Y, Yang P, Wang N, Chen Z, Su D, Zhou ZH, Rao Z, Wang X (2020) Architecture of the herpesvirus genome-packaging complex and implications for DNA translocation. Protein Cell 11:339–351
69. Lokareddy RK, Hou CD, Li F, Yang R, Cingolani G (2022) Viral small terminase: a divergent structural framework for a conserved biological function. Viruses 14:2215
70. Buttner CR, Chechik M, Ortiz-Lombardia M, Smits C, Ebong IO, Chechik V, Jeschke G, Dykeman E, Benini S, Robinson CV et al (2012) Structural basis for DNA recognition and loading into a viral packaging motor. Proc Natl Acad Sci USA 109:811–816
71. Morais MC (2012) The dsDNA packaging motor in bacteriophage o29. Adv Exp Med Biol 726: 511–547
72. Chemla YR, Smith DE (2012) Single-molecule studies of viral DNA packaging. Adv Exp Med Biol 726:549–584
73. Liu S, Chistol G, Hetherington CL, Tafoya S, Aathavan K, Schnitzbauer J, Grimes S, Jardine PJ, Bustamante C (2014) A viral packaging motor varies its DNA rotation and step size to preserve subunit coordination as the capsid fills. Cell 157:702–713
74. Keller N, delToro DJ, Smith DE (2018) Single-molecule measurements of motor-driven viral DNA packaging in bacteriophages Phi29, lambda, and T4 with optical tweezers. Methods Mol Biol 1805:393–422
75. Tavares P, Zinn-Justin S, Orlova EV (2012) Genome gating in tailed bacteriophage capsids. Adv Exp Med Biol 726:585–600
76. Purohit PK, Inamdar MM, Grayson PD, Squires TM, Kondev J, Phillips R (2005) Forces during bacteriophage DNA packaging and ejection. Biophys J 88:851–866
77. Roos WH, Ivanovska IL, Evilevitch A, Wuite GJ (2007) Viral capsids: mechanical characteristics, genome packaging and delivery mechanisms. Cell Mol Life Sci 64:1484–1497
78. Vinga I, Sao-José C, Tavares P, Santos M (2006) Bacteriophage entry in the host cell. In: Wegrzyn WG (ed) Modern bacteriophage biology and biotechnology. Research Signpost, Kerala, pp 165–205

Further Reading

Flint SJ, Enquist LW, Racaniello VR, Skalka AM (eds) (2009) Principles of virology, 3rd edn. ASM Press, Washington

Knipe DM, Howley PM (eds) (2013) Fields virology, 6th edn. Lippincott Williams & Wilkins, Philadelphia

Patton JT (ed) (2008) Segmented double-stranded RNA viruses. Structure and molecular biology. Caister Academic Press, Norfolk

Rossmann MG, Rao VB (eds) (2012) Viral molecular machines. Adv Exp Med Biol, vol 726. Springer, New York

Also especially recommended for further reading are references [2, 4, 31, 32] listed above.

Chapter 14
Maturation of Viruses

Gabriela N. Condezo and Carmen San Martín

Abstract Viral genomes are transported between cells using various structural solutions such as spherical or filamentous protein cages, alone or in combination with lipid envelopes, in assemblies of varying complexity. Morphogenesis of the new infectious particles (virions) encompasses capsid assembly from individual components (proteins, and membranes when required), genome packaging, and maturation. This final step is crucial for full infectivity. During maturation, structural and physical changes prepare the viral particles for delivering their genome into cells at the right time and location. The virion must be stabilized for travel across harsh extracellular conditions, while enabling disassembly for genome exposure to replication and translation machineries. That is, maturation has to produce metastable particles. Common maturation strategies include structural reordering, controlled proteolysis, or posttranslational modifications. Here we outline the maturation process in representative members of the six realms proposed by the latest virus taxonomic classification.

Keywords Virus structure · Virus assembly · Capsid · Scaffold · Stability · Uncoating · Envelope · Virus protease · Virus realms

Abbreviations

AdV	Adenovirus
AVP	Adenovirus protease
CA	Capsid
CP	Coat protein
DJR	Double jelly roll
dsDNA	Double-stranded DNA

G. N. Condezo (✉) · C. San Martín (✉)
Department of Macromolecular Structure, Centro Nacional de Biotecnología (CNB-CSIC), Madrid, Spain
e-mail: gncondezo@cnb.csic.es; carmen@cnb.csic.es

M. G. Mateu (ed.), *Structure and Physics of Viruses*, Subcellular Biochemistry 105,
https://doi.org/10.1007/978-3-031-65187-8_14

dsRNA	Double-stranded RNA
EC	Empty capsids
EI	Expansion intermediate
ER	Endoplasmic reticulum
EV	Enveloped virions
FC	Full capsids
FHV	Flock House virus
HDV	Hepatitis delta virus
HIV	Human immunodeficiency virus
HPV16	Human papilloma virus 16
IBDV	Infectious bursal disease virus
ICTV	International Committee on Taxonomy of Viruses
INT	Integrase
IP6	Inositol hexakisphosphate
IV	Immature virion
L-HDAg	Large hepatitis D antigen
MA	Matrix
MCP	Major capsid protein
MV	Mature virus
NC	Nucleocapsid
NωV	*Nudaurelia capensis* omega virus
PC	Polymerase complex
PR	Protease
RNP	Ribonucleoprotein
RT	Reverse transcriptase
S-HDAg	Small hepatitis D antigen
SIFV	*Sulfolobus islandicus* filamentous virus
SJR	Single jelly roll
ssDNA	Single-stranded DNA
ssRNA	Single-stranded RNA
STED	Stimulated emission depletion microscopy
VACV	Vaccinia virus
WV	Wrapped virions
ZmBWV	*Zophobas morio* black wasting virus

14.1 Introduction

In the host cell, viral genomes direct the production of new infectious virus particles, also known as virions. As seen in Chaps. 2, 11, and 12, there are different structural solutions to protect the viral genomes during their transport from one cell to another. These include spherical protein capsids, filaments, or condensates of genetic material. Lipid envelopes studded with viral proteins are also used. Some virus particles can be very complex, with alternating layers of protein and lipids. The process needed to build a new virion is known as morphogenesis and is usually described as divided into three stages: capsid assembly from its individual molecular components (proteins and membranes in some cases), genome packaging, and maturation. In many viruses, after the viral genome is encapsidated, the particle is still not ready for the next infection event. Maturation, the last morphogenetic step, is required for the virus to achieve its full infectious potential. Maturation induces structural and physical changes in the viral particle. These changes prepare the virion to enter the cell, uncoat, and deliver its genome at the proper time and location to start a new transcription and replication round. A crucial part of maturation relates to stability modulation: the virion has to endure deleterious environmental conditions (extremes of temperature, pH, osmotic shock, or dehydration) while traveling outside the host cell, but must properly dismantle to allow access of the transcription and replication machinery to the genome. To fulfill these conflicting stability requirements, maturation sets the viral particle in a metastable state. Maturation strategies include structural reordering, proteolysis, or posttranslational modifications. These strategies may be shared by evolutionarily unrelated viruses.

Virus discovery and characterization throughout the years has revealed a notorious diversity in sequences and architectures, defying attempts at classification. The Baltimore classification, based on the type of genetic material, was the most accepted until recently. However, structural and phylogenetic analyses have identified evolutionary connections among viruses belonging to different Baltimore classes, prompting the establishment of a new classification scheme by the International Committee on Taxonomy of Virus (ICTV). In this scheme, viruses are grouped into six realms: *Duplodnaviria, Varidnaviria, Adnaviria, Monodnaviria, Riboviria,* and *Ribozyviria* (Fig. 14.1) (see also Chap. 7). Viruses in each realm share a small set of hallmark genes involved in genome replication and virion morphogenesis [2]. Here we provide an overview of the maturation process for some of the best-characterized viruses in each realm.

14.2 Maturation in the *Duplodnaviria* Realm

The *Duplodnaviria* realm includes the enormously abundant tailed bacteriophages and tailed archaeal viruses, together with their related animal viruses, herpesviruses [2]. They all have dsDNA genomes, as the realm name indicates, packed inside an icosahedral capsid. Unlike the rest of the realm members, herpesviruses do not have tails, and acquire a lipid envelope from the host cell. All duplodnaviruses share the same major capsid protein (MCP) fold. This fold, also known as the HK97 fold (from the HK97 phage where it was first solved), consists in a mixture of α and β secondary structure elements. The same MCP forms both the hexameric and the pentameric capsomers in the icosahedral shell. The monomer is L-shaped and arranged in two domains, each one corresponding to one arm of the L (Fig. 14.1) [3]. In this realm, the maturation process encompasses large structural

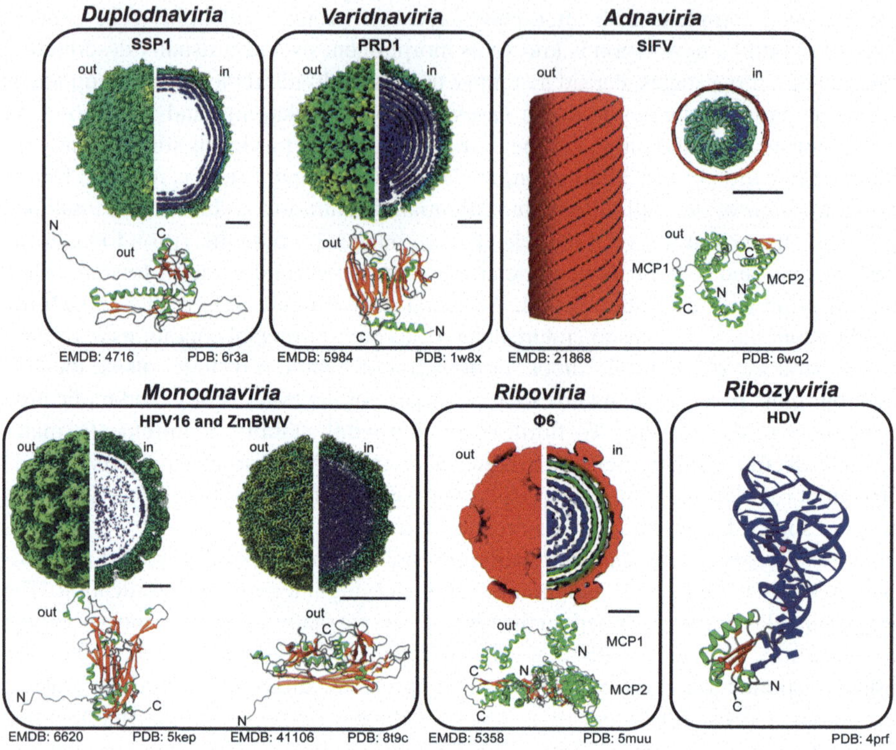

Fig. 14.1 Examples of viruses in each one of the six realms discussed in this chapter. Viral particles are depicted showing their outer surface (out) and in a cross section to show the internal components (in). There is no 3D structure for the complete ribozyvirus virion. The ribbon diagrams show the fold of the major coat proteins (MCPs). *Duplodnaviria*: SSP1 phage; *Varidnaviria*: phage PRD1; *Adnaviria*: *Sulfolobus islandicus* filamentous virus (SIFV); *Monodnaviria*: Human papilloma virus 16 (HPV16) and *Zophobas morio* black wasting virus (ZmBWV) (*Parvoviridae*); *Riboviria*: phage ϕ6; *Ribozyviria*: Hepatitis delta virus (HDV). The capsid surfaces and sections are colored as follows: blue, genome; green, protein capsid; red, membrane and glycoproteins. The MCP structures are colored according to secondary structure elements, with α-helices in green and β-strands in orange. Except for HDV, they are shown in an orientation such that the outer capsid surface (out) is at the top. SIFV and ϕ6 have two capsid proteins (MCP1 and MCP2), and the ϕ6 phage has two capsid layers. Database codes for the maps and structures depicted are shown below each panel (Electron Microscopy Data Bank is at http://www.ebi.ac.uk/pdbe/emdb/; Protein Data Bank is at http://www.pdb.org). Scale bars: 100 Å. All the ribbon and surface structure representations in this chapter were prepared using UCSF ChimeraX [1]

rearrangements, concomitant with scaffold removal and genome packaging. The capsid transforms from a weak, labile particle to a highly stable shell, able to withstand the high internal pressure caused by the tightly confined genome (see Chaps. 10, 18, and 19).

14.2.1 Tailed Bacteriophages and Archaeal Viruses

In tailed bacteriophages, morphogenesis starts with hexameric and pentameric capsomers assembling a spherical procapsid, with the help of the portal and scaffolding proteins which act as nucleating seeds and modulators of capsid assembly elongation. The scaffold proteins can be independent elements, as in phages P22 and ϕ29, or be part of the MCP as in HK97 [3, 4]. In this case, cleavage by a virus-encoded protease is required to remove the scaffold elements during maturation. In P22, the intact scaffold protein is released through large pores in the capsid shell and recycled for a new round of assembly [4].

In HK97, where maturation has been studied in very high detail, the procapsid is also called Prohead-I. With 56 nm diameter and $T = 7$ icosahedral architecture, Prohead-I is a spherical particle containing 60 hexamers and 11 pentamers formed by the MCP in its precursor form, which includes the scaffold domain fused at its N-terminal end (residues 2–103). The remaining vertex in the icosahedral shell is formed by the packaging motor (Chap. 13). The interactions established by the scaffold domain of HK97 induce bending and constraints in some secondary structure elements of the MCP, most notably the so-called spine helix. This bending results in hexameric capsomers with a pronounced skewed shape: they only display local twofold instead of sixfold symmetry (Fig. 14.2, left). Another region of the protein, called the P-loop, is involved in protein–protein interactions at the icosahedral threefold axis. However, because of the presence of the scaffold domain, in the procapsid the P-loop can establish only a limited set of interactions, therefore producing a fragile particle. Upon Prohead-I assembly, the viral protease becomes active, cleaves the scaffolding domains, and cleaves itself, producing small peptides that diffuse out of the particle. The result of these cleavages is another morphogenetic intermediate called Prohead-II. Assembly of Prohead-I is a reversible reaction, but formation of Prohead-II is irreversible because it involves the proteolysis of scaffolding elements. In the absence of the scaffolding domains, Prohead-II is metastable, that is, ready to move to a lower energy state (i.e., a more stable structure) in the next stage of assembly. This is what happens when the dsDNA is packaged [3, 5].

The viral genome is translocated into Prohead-II by the action of the portal, a protein complex that works as a molecular motor converting the chemical energy of ATP hydrolysis into mechanical work (Chap. 13). DNA packaging causes a capsid expansion and triggers a cross-linking reaction between neighboring MCP monomers. These expansion and cross-linking processes have been studied in vitro, subjecting Prohead-II particles to different physicochemical treatments. In a first transition, Prohead-II diameter increases by 15%, producing an expansion intermediate (EI-I) which has a shell thinner than the prohead and a half-life of ~3 min. In EI-I, the spine helix is no longer bent and the hexameric capsomers are not skewed. Then (EI-II), the cross-linking sequence starts, with isopeptide bonds forming between the P-loop in one MCP monomer and another flexible region (E-loop, Fig. 14.2) in a neighboring one. As cross-linking propagates across the hexameric

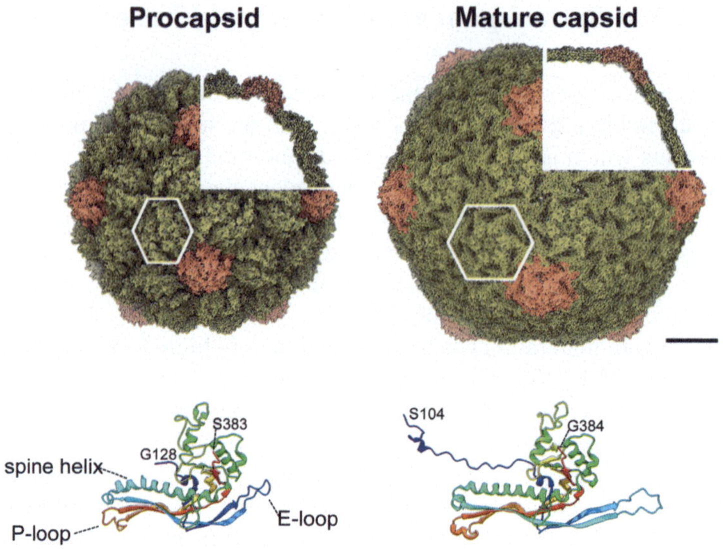

Fig. 14.2 Maturation in *Duplodnaviria:* bacteriophage HK97. Top panels: the immature (left) and mature (right) capsids are shown with the pentamers in pink, and the hexamers in yellow. One quarter of each capsid has been removed and substituted by a slice across the particle, to illustrate the thickness change in the protein shell. The view is along a twofold icosahedral symmetry axis. White hexagons outline one MCP hexamer. Scale bar: 100 Å. Bottom panels: structure of the HK97 MCP monomer in the immature (Prohead-II, PDB: 3E8K, left) and mature (PDB: 1OHG, right) capsids. The structure is rendered as a ribbon with a rainbow color scheme, from blue (N-terminus) to red (C-terminus). The first and last residues traced in each structure, as well as the position of structural elements involved in the conformational and stability changes occurring during maturation, are indicated

capsomers, the particle expands by another 9% producing the so-called Head-I or Balloon (66 nm diameter). Finally, pentamers are also cross-linked to the adjacent hexamers, to produce the mature particle, Head-II (66 nm), a clearly faceted icosahedron (Fig. 14.2, right). With a shell thickness of only 1.8 nm, Head-II is a highly stable particle because all protein subunits are covalently cross-linked to other subunits, resembling the chainmail worn by medieval knights. Although the large conformational rearrangements and capsid stabilization are a common feature in duplodnavirus maturation, not all of them use the cross-linking mechanism observed in HK97. Alternative ways to build stable capsids include cementing by additional proteins (e.g., gpD in phage λ) or MCP domains (phage P22) which interlock several capsomers from the outer capsid surface [6, 7].

The tailed viruses infecting archaeal cells follow a morphogenetic pathway very similar to that of tailed phages. In *Haloarcula sinaiiensis* tailed virus 1 (HSTV-1), the N-terminal domain of the MCP is proteolytically removed and the capsid undergoes expansion and stabilization as the DNA is packaged, mimicking the HK97 assembly pathway. However, the amino acids involved in MCP cross-linking are not conserved in HSTV-1. Instead, it has been proposed that a positively charged residue (Arg311) that forms rings with its counterparts in other MCP monomers around both the quasi-sixfold and fivefold axes of symmetry could coordinate anions for stabilization [8]. Another tailed virus in archaea, *Halorubrum sodomense* tailed virus 2 (HSTV-2), uses an external cementing protein like phage λ [9].

14.2.2 *Herpesviruses*

Members of the *Herpesviridae* family share many structural characteristics with tailed phages, but they present larger complexity. The herpesvirus icosahedral nucleocapsid is assembled from MCP capsomers sharing the basic HK97 fold [10], plus a packaging portal at a single vertex [11]. This nucleocapsid is surrounded by two additional layers: the tegument and the glycoprotein-containing envelope. Herpesviruses resemble bacteriophages in their mode of nucleocapsid assembly: they use a scaffolding protein later removed by the viral protease, and undergo large conformational changes where skewed, twofold hexamers acquire a sixfold symmetric shape. However, some features are unique to herpesviruses. In herpes simplex virus 1, the $T = 16$ procapsid has the same diameter as the mature nucleocapsid (125 nm), that is, there is no expansion, although the nucleocapsid changes from a spherical to a polyhedral shape. Unlike HK97, herpesviruses do not undergo MCP cross-linking. Instead, additional proteins stabilize the nucleocapsid. Some of them (VP23 and VP19, forming the so-called triplexes) are present from the prohead state and change their interaction network during maturation. In contrast, VP26, a capsomer stabilizing protein that connects adjacent MCP upper domains in each hexameric capsomer, is not present in the prohead [12].

The herpesvirus nucleocapsid assembles and matures in the nucleus of the host cell, but additional morphogenetic steps are required to produce the infectious virions. Still in the nucleus, the nucleocapsid starts recruiting tegument proteins. The nucleocapsid exits the nucleus by budding into the inner nuclear membrane, from which it acquires its primary envelope. This envelope is lost by fusion with the outer nuclear membrane, allowing the nucleocapsid to enter the cytoplasm. Then, interactions between the tegument proteins and cytoskeleton direct the viral particle to the Golgi, where it acquires the final tegument proteins and envelope (together with glycoproteins—secondary envelope) by budding; this step completes the virion maturation [10].

14.3 Maturation in the *Varidnaviria* Realm

The *Varidnaviria* realm includes icosahedral, nontailed phages and viruses of archaea, along with diverse viruses of eukaryotes, including the giant viruses found in amoeba and eukaryotic phytoplankton. Most varidnaviruses have dsDNA genomes, except for the *Finnlakeviridae* family formed by viruses with ssDNA genomes [2]. MCPs in most members of this realm are built from a double eight-stranded β-barrel (also called double jelly roll, DJR) orthogonal to the capsid surface (Fig. 14.1), forming pseudo-hexagonal trimeric capsomers. Another common feature among varidnaviruses is having a single jelly roll (SJR) protein (different from the MCP but also with the β-barrel orthogonal to the capsid surface) forming the vertex capsomers. Most varidnaviruses possess an internal lipid membrane intercalated between the genome and the icosahedral capsid. Exceptions are faustovirus (Chap. 12), adenovirus (AdV), and the virophages (satellites of the giant viruses) [13].

The assembly pathway of varidnaviruses is known at a much lower level of detail than in duplodnaviruses. Icosahedral capsids in this realm do not seem to undergo extensive conformational rearrangements upon genome packaging [14, 15]. Maturation proteases are present in varidnaviruses infecting eukaryotic organisms, but no proteolysis seems to be involved in morphogenesis of the realm members infecting prokaryotes.

14.3.1 Maturation in a Non-enveloped, Icosahedral Varidnavirus: Adenovirus

Adenoviruses (AdVs) are varidnaviruses devoid of internal membrane that infect vertebrates. They are pathogens in humans, cattle, and fowl [16]. Recombinant AdVs are used as therapeutic vectors for vaccines, oncolysis, or gene therapies. The *pseudoT* = 25 icosahedral virion (95 nm diameter) is composed of more than 10 different proteins. The MCP (hexon) forms the icosahedral facets, while the vertices are formed by pentamers of a single jelly roll protein (penton base) in complex with trimeric receptor-binding fibers. Minor capsid proteins IIIa, VI, and VIII buttress the inner capsid surface, and some AdVs have genus specific cementing proteins on the outer capsid surface. Inside the capsid, large amounts (up to 25 MDa) of positively charged, virus-encoded core proteins (proteins VII and μ, plus protein V in the *Mastadenovirus* genus) modulate condensation of the dsDNA genome [17–19].

The AdV protease (AVP) cleaves capsid proteins IIIa, VI, and VIII, and core proteins VII and μ, as well as L1 52/55k, one of the four AdV packaging proteins. AVP is packaged together with the viral genome and binds covalently to a peptide (pVIc) cleaved from the C-terminus of protein VI. This peptide enables AVP to slide on the dsDNA molecule and reach all its other targets, in a unique one-dimensional chemistry mechanism [20]. Impaired AVP activity results in non-infectious particles, and the lack of infectivity has been linked to uncoating defects. AdV uncoating happens in a sequential manner, starting at the cell membrane and ending at the nuclear pore, where the capsid completely dismantles and the genome translocates into the nucleus. A crucial step is the exposure of an internal capsid component, protein VI, in the endosome. Protein VI harbors an amphipathic helix that disrupts

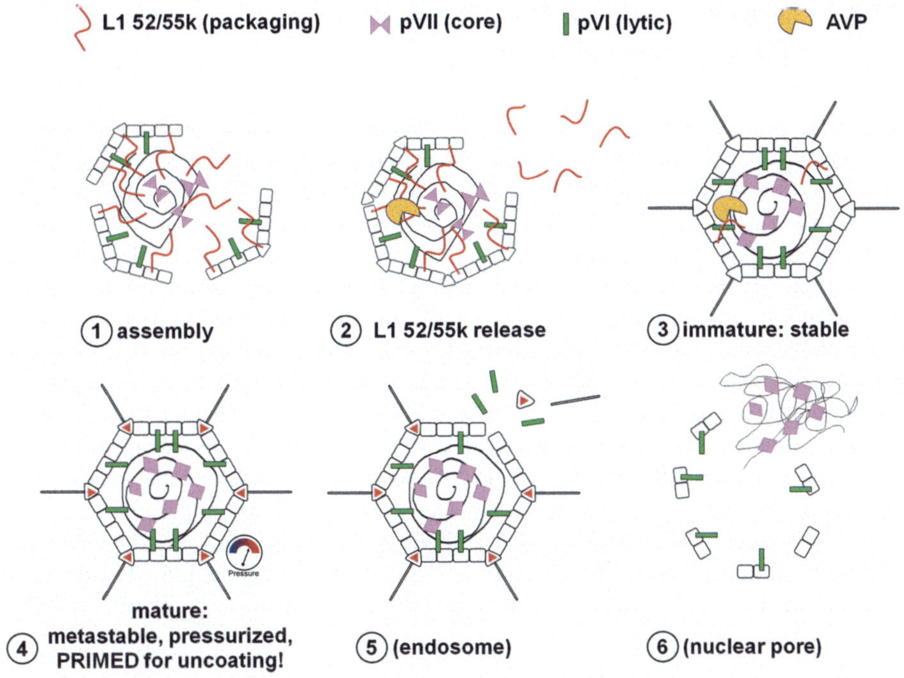

Fig. 14.3 Maturation in *Varidnaviria*: adenovirus. Cartoon summarizing the AdV maturation pathway and its effects in priming for uncoating. See main text for details. Figure reproduced from [19], with permission

the endosome membrane, so that AdV can escape into the cytosol. In immature particles protein VI is not exposed, and as a consequence they become trapped in the endocytic pathway, thereby aborting infection [20–22].

During the initial stages of assembly, protein L1 52/55k interacts with both capsid and core components, helping to recruit capsid fragments to condensing genomes for concurrent assembly and packaging (Fig. 14.3, step 1) [23]. While the capsid grows around the core, AVP starts cleaving L1 52/55k. This process facilitates its removal (similarly to scaffolds in duplodnaviruses) (Fig. 14.3, step 2), as well as separation of the genome from capsid fragments in the final stages of uncoating (Fig. 14.3, step 6) [15, 24]. Additionally, AVP cleavages decrease the genome condensing power of major core protein VII (Fig. 14.3, step 3). This decrease results in internal pressurization of the virion caused by electrostatic repulsion between the numerous DNA negative charges confined in a reduced space. Pressurization destabilizes the weakest part of the icosahedral capsid, providing a physical mechanism to facilitate penton loss in the initial stages of uncoating, and therefore opening an exit channel for protein VI to access the endosome membrane (Fig. 14.3, step 4–5). Thus, maturation makes the AdV virion metastable, priming the particle for stepwise uncoating in the cell. Priming by maturation consists of facilitating penton release (for exposure of membrane lytic protein VI) and genome detachment from the capsid shell [19].

14.3.2 Maturation in an Enveloped, Pleomorphic Varidnavirus: Poxvirus

Vaccinia virus (VACV, *Poxviridae* family) is one of the most intensively studied varidnaviruses, primarily because of its success as a live viral vaccine for smallpox, which led to eradication of the disease. VACV has a complex asymmetric, enveloped virion whose structure and morphogenesis are far from being fully understood. VACV replication and assembly occur in the cytosol. Viral proteins rupture the ER membrane to produce crescent-shaped, open membrane fragments that are the first identifiable VACV assembly intermediate (Fig. 14.4) [25]. Membrane crescents are covered by an external honeycomb lattice composed of trimers of protein D13. D13 is a scaffold protein which allows the crescent to grow until the membrane closes to form a complete sphere, called immature virion (IV). Although it is not present in the mature virion and VACV particles are not icosahedral, D13 folds as a trimeric DJR like the MCPs of icosahedral varidnavirus capsids [26], reflecting common ancestry between symmetric and pleomorphic particles.

As the crescent membranes develop and before they close, they are filled with the viroplasm, which contains viral core proteins and the viral genome [25]. The detailed mechanism for this process remains unclear. The next step of the VACV

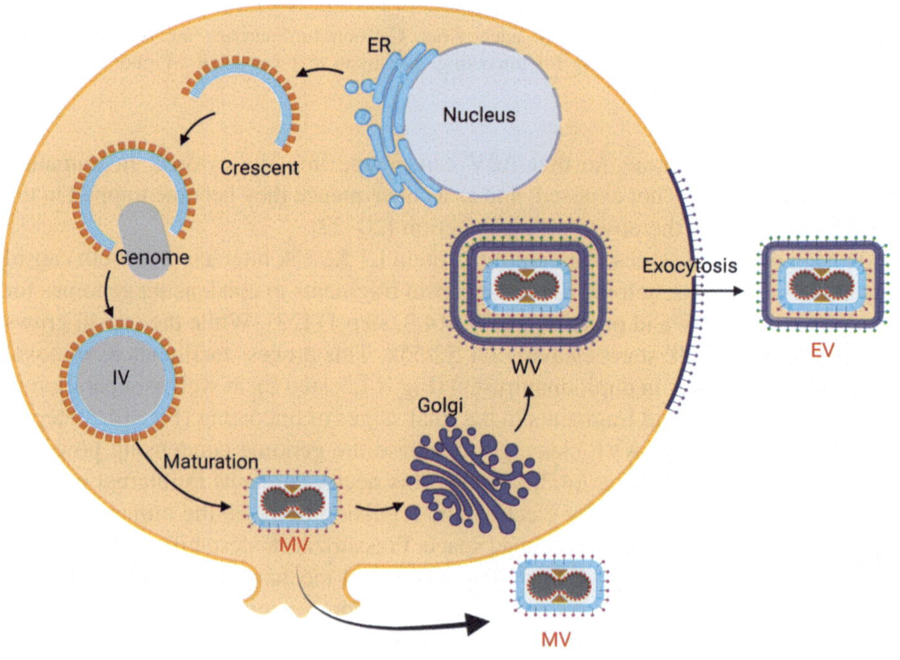

Fig. 14.4 Maturation in *Varidnaviria*: poxvirus. Cartoon summarizing the production of the different viral structures observed in the VACV cycle. Red text labels refer to infectious particle forms (MV and EV). See main text for details. Figure created with BioRender

morphogenesis is transformation of spherical IV to mature virus (MV), which has a brick shape and a volume smaller by 50% [27]. The MV (Fig. 14.4) contains a biconcave core formed by the viral genome and core proteins, including the transcription apparatus. Dense lateral bodies separate the core from the external membrane in the particle equatorial region.

VACV I7 protein, an AdV-like protease, cleaves A17, a membrane protein that serves to anchor D13 to the IV surface, therefore causing removal of the scaffold during maturation [28]. Protein I7 cleaves also the precursor forms of core proteins [25, 29]. The presence of core proteins, positively charged polypeptides bound to the packed dsDNA, is a common trait in eukaryotic varidnaviruses [30]. The evidence collected for AdV suggests that the major role of core proteins is modulating genome condensation and internal capsid pressure [18, 19]. However, it is not clear whether the VACV core proteins are forming a nucleoproteic complex with the genome. Details on physical changes in poxviruses are scarce, and it is not known if pressure changes associated with maturation might also play a role in uncoating regulation. Sequence similarity with metalloendopeptidases suggests VACV may have a second maturation protease, protein G1, although this activity has not been confirmed. It has been suggested that I7 and G1 function sequentially, because I7 mutants arrest morphogenesis prior to core condensation, while the absence of G1 activity arrests the maturation after core condensation has commenced but prior to its completion [25]. Apart from proteolytic maturation, VACV requires an N-myristoylation in at least three viral proteins. In the presence of N-myristoyl transferase inhibitors, viral particles are formed but they are less infectious than those from untreated infected cells. N-myristoylation-defective particles exhibited reduced membrane fusion and core release in the cytoplasm [31], suggesting that the myristate group could have a role in initiating virus-host membrane fusion [32].

An additional maturation step in VACV is the recruitment of surface proteins to the MV membrane, possibly involved in cell attachment, recognition, or other virus–host interaction processes [25]. Particles with different membrane arrangements have been observed, depending on the fate of VACV particles after maturation. A subset of MVs is transported to the ER-Golgi intermediate compartment where the particle acquires two membranes in one step, producing the so-called wrapped virions (WV). WV then fuse their outer envelope with the cell membrane, originating enveloped virions (EV) which contain two membranes [33] (Fig. 14.4). Both MVs and EVs are infectious.

14.4 Maturation in the *Adnaviria* Realm

Adnaviruses infect hyperthermophilic archaea and form filamentous virions packing A-form dsDNA. This dsDNA organization is thought to be one of the adaptations to high temperature environments and is the basis of the realm name [2]. Structural studies have only recently started to provide information on adnavirus structure and assembly, mainly due to the difficulties inherent to identification and cultivation of the hosts in laboratory conditions.

Among the adnaviruses, the lipothrixvirus and ungulavirus linear dsDNA genomes are covered by two homologous MCPs arranged into a helical nucleocapsid, enveloped with a lipid membrane (Fig. 14.1). Both ends of the virion are capped with a terminal structure, thought to play a role in host recognition. The lipid envelope surrounding the nucleocapsid is two times thinner than the membrane of the host cell. This remarkable feature can be explained by selective recruitment of host lipid species that can be bent into a U-shaped horseshoe conformation, or by enrichment in archaeol, a short lipid molecule representing ~1% of lipids in the host membrane [34]. Membrane acquisition is the most salient feature described so far in adnavirus maturation and is remarkably intriguing for *Sulfolobus islandicus* filamentous virus (SIFV). SIFV virions are flexible filamentous particles measuring ~2 μm in length and 24 nm in width. These extremely long particles are assembled inside cells with 1–1.2 μm diameter and are released at the end of the infection cycle through large, six-sided pyramidal structures, dubbed virus-associated pyramids (VAPs). Electron tomography images have shown that viral particles inside the cell form twisted bundles organized in an imperfect hexagonal lattice. Envelopment of the long helical nucleocapsids takes place inside the cell, rather than by budding as in the case of most other known enveloped viruses. This observation was unexpected, because internal membrane-bound compartments have never been observed in archaeal cells, rendering the possibility that SIFV virions are enveloped by budding through intracellular membranes highly unlikely. Consequently, the envelopment of SIFV might occur by a novel mechanism, involving either de novo membrane formation or trafficking of lipids from the cell membrane to the virion assembly centers [35].

14.5 Maturation in the *Monodnaviria* Realm

The *Monodnaviria* realm includes both ssDNA and dsDNA viruses. Most monodnaviruses code for a rolling circle replication endonuclease and a SJR capsid protein. Viruses grouped under the *Monodnaviria* label seem to have arisen from multiple origins via recombinational merging of endonuclease genes from different bacterial plasmids with capsid protein genes from different +RNA viruses [2]. For this reason, some members of this realm differ not only in genome type, but also in capsid architecture (Fig. 14.1). Here we discuss the maturation of two well-studied *Monodnaviria* families: papillomaviruses and parvoviruses.

Papillomaviruses have dsDNA genomes contained in non-enveloped icosahedral particles and infect the skin or mucosa of vertebrate hosts. The MCP, called L1, has a SJR fold [36] and forms $T = 7$ capsids assembled from 72 pentamers (Fig. 14.1). Twelve pentamers are located on fivefold axes and the other sixty occupy 6-coordinated positions [37]. The capsid resulting from the assembly process is infectious, but physically fragile. A maturation process is required to stabilize it and increase its resistance to tryptic proteolysis [38]. This maturation is achieved by the creation of disulfide bonds between MCP monomers in one pentamer with those of the neighboring pentamer [37–39]. Disulfide bonds are a consequence of exposure to oxidizing conditions when the infected cell progresses across different layers of the skin during the desquamation process. Unlike other maturation processes which occur on a time scale of seconds or minutes, papillomavirus capsids undergo an extremely slow maturation process. This slowness could be a reflection of the long overall time scale of the viral cycle in vivo, since cell migration from the deep layers to the skin surface can take several days [38]. Intermediates in the transition from immature to fully mature capsid show a regular progression of capsid shrinking from about 63 to 60 nm [39].

The *Parvoviridae* family includes viruses infecting a large variety of hosts, from *Cnidaria* to vertebrates. Parvoviruses have linear ssDNA genomes (4–6 kpb) and coat proteins (CP) folding as a SJR parallel to the capsid surface. Their small (20–28 nm), non-enveloped capsids are assembled from 60 copies of CP in a $T = 1$ architecture (Fig. 14.1). However, parvovirus CPs may undergo different degrees of proteolytic processing, yielding apparently simple capsids that are actually formed by a mixture of CP forms. These forms have a common C-terminal region but vary in total length [40]. For example, the capsid of Minute Virus of Mice (MVM) contains three forms of the CP (VP1, VP2, and VP3) [41], with the entire sequence of VP3 contained in VP2, which is in turn contained within VP1 [42]. A mixture of VP1 and VP2 is present in empty capsids, while genome-containing capsids have a mixture of the three forms VP1-VP2-VP3 [43].

Using anion-exchange chromatography, which separates complexes by their net surface charge, three types of virus particles have been detected in the nucleus of MVM infected cells: empty capsids (EC), and two populations of full capsids (FC-P1 and FC-P2). It has been proposed that viral genome packaging into EC produces FC-P1 particles, which are infectious but, similarly to EC, cannot be actively exported from the nucleus. Maturation of FC-P1 into FC-P2 consists in capsid surface phosphorylation, which renders the particle suitable for active nuclear export. However, the exact mechanism of nuclear export facilitated by capsid phosphorylation is still unclear, as mutational analyses revealed that the phosphoserine-rich region in VP2 is dispensable for active exportation [44]. During entry, VP2 is cleaved by endosomal proteases to produce VP3 [45]. This cleavage is required to allow exposure of the VP1 N-terminus, which harbors important functional motifs essential for infection [46], particularly for escape from the endosome [47] and nuclear targeting [48].

14.6 Maturation in the *Riboviria* Realm

All viruses with RNA genomes (either single or double stranded), together with the reverse-transcribing viruses with dsDNA genomes (such as Hepatitis B virus), share a common ancestry and form the *Riboviria* realm [2]. The only exception is hepatitis delta virus (HDV) and its relatives, which form a separate realm in the current taxonomic classification (Sect. 14.7). Both enveloped and non-enveloped particles can be found among the riboviruses.

14.6.1 Maturation in Non-enveloped, Single-Stranded RNA Viruses

Noda-, picorna-, and tetraviruses have small naked, icosahedral virions and their morphogenetic pathways have been characterized in considerable detail, taking advantage of their simple capsid composition and availability of in vitro assembly procedures [5].

Flock House virus (FHV), the nodavirus prototype, has a 32 nm diameter icosahedral capsid ($T = 3$) which harbors a bipartite ssRNA genome [49]. The FHV capsid protein is synthesized as a 407 amino acid long polypeptide designated as α and has a SJR fold. After assembly of the provirion (non-infectious particle), autocatalytic cleavage of α generates the β (residues 1–363) and γ (364–407) peptides [50]. Autoproteolysis of the α peptide does not result in any detectable morphological change in the particles, but it makes them more stable and is required for FHV to become infectious. The cleavage happens by a mechanism of acid hydrolysis orchestrated by Asp-75. Mutating either the catalytic aspartate or the residue at position 363 abrogates the maturation cleavage and therefore infectivity. Autoproteolysis does not happen in isolated monomers of α peptide. Only after provirion assembly has placed the catalytic residue and its targets in the appropriate conformation and chemical environment can the cleavage take place. In this way, untimely reactions that would be deleterious for the virus are avoided [51]. The γ peptide remains inside the mature particle, forming an amphipathic helix beneath each β monomer (Fig. 14.5). Because it has membrane lytic activity, the γ peptide is thought to play a role in FHV escape from endosomes. Calcium ions have been observed bound to specific regions in the FHV capsid, possibly helping to stabilize the quaternary structure by strengthening the interaction between subunits. Particles prepared in the presence of metal chelators are purified as immature provirions, indicating a possible role for calcium in the assembly and maturation of the capsid [50].

An autocatalytic cleavage mechanism very similar to that of nodaviruses is responsible for the maturation of the *Alphatetraviridae*, as revealed mainly by studies on *Nudaurelia capensis* omega virus (NωV) [52]. NωV has also a bipartite RNA genome and a naked icosahedral shell assembled from a unique protein, α, but in this case the particle triangulation is $T = 4$. Unlike nodaviruses, alphatetravirus capsids undergo remarkable conformational rearrangements associated with maturation. A spherical, unstable procapsid of 48 nm diameter is assembled at neutral pH and matures by acidification during apoptosis to result in the infectious, faceted 41 nm mature particle [53]. This shell shrinking is driven by the protonation at low

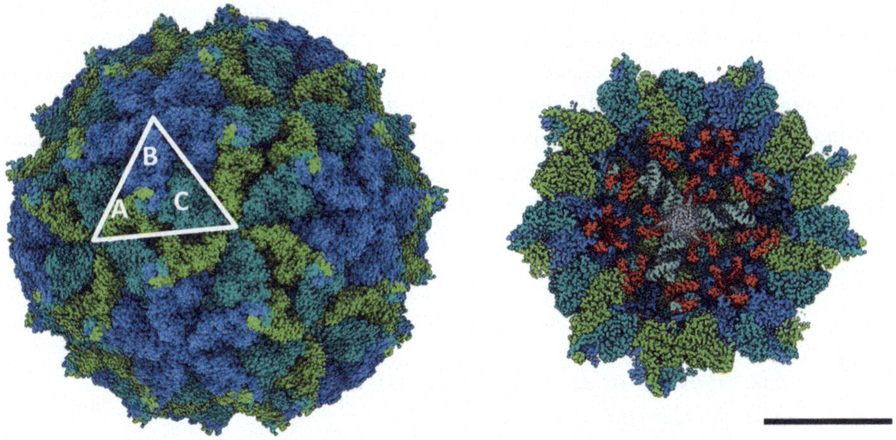

Fig. 14.5 Maturation in *Riboviria*: Flock House virus. A view of the FHV capsid (PDB: 4FTB) from outside (left), and a facet from inside (right) to show the position of the lytic γ peptide (red). An asymmetric unit is delineated by a white triangle. The three independent copies of the MCP are colored in lime, blue, and sea green and labeled A–C. The view is along a threefold symmetry axis. Notice how part of the ssRNA genome forms ordered dsRNA patches (depicted in light aquamarine). Scale bar: 100 Å

pH of acidic residues that caused repulsion at the subunit interfaces in the immature particle. Only when the particle reaches its smaller conformation does proteolytic cleavage of the capsid protein take place, indicating that in the swollen form the catalytic residues are not correctly positioned for cleavage. High resolution studies using cryo-EM have shown that internal regions of the capsid subunits are remodeled during capsid shrinkage, creating the active site required for autocatalytic cleavage [54]. Once the autocatalytic process has started, the particle can no longer swell, even if the pH is raised to neutral values. In NωV, the γ peptides generated by autoproteolysis of α act as a molecular switch between subunits, adopting different conformations depending on their position in the four quasi-equivalent positions in the asymmetric unit. As a result of the proteolytic cleavage, these γ peptides go from a mobile to a more ordered state, therefore stabilizing the particle [55, 56].

Picornaviruses are also small, icosahedral viruses (30–32 nm) with positive sense ssRNA. However, they have a slightly more complex architecture, because their capsids are built from three different polypeptides (*pseudoT* = 3). Important human and animal pathogens such as rhinovirus, poliovirus, foot-and-mouth disease virus, encephalomyocarditis virus, or hepatitis A virus are members of the *Picornaviridae* family. In the picornavirus prototype, poliovirus, the capsid is assembled from SJR proteins VP0, VP1, and VP3. These proteins originate by cleavage from the single, long polypeptide translated from the viral mRNA. After assembly of the provirion and in the presence of the viral genome, an autocatalytic reaction cleaves the N-terminal region of VP0 to produce VP4 and VP2 [57]. The specific mechanism is unclear, although it has been suggested that it is catalyzed by some aspect of the virion structure itself, like the generation mechanism of γ peptides in nodavirus and alphatetravirus. Poliovirus maturation also involves a posttranslational modification of VP4, namely addition of a fatty acid (myristate). Myristoylation of VP4 appears to play a double role: during assembly, it mediates interactions between VP4 and the rest of the icosahedral shell; during entry, it helps to open pores in the host membranes when it is exposed as a result of conformational changes undergone by the particle upon receptor binding [58, 59]. It has been observed that poliovirus maturation is promoted by acidic compartments such as autophagosomes, suggesting that trafficking through this kind of vesicles may be a part of the assembly process [60].

14.6.2 Maturation in Icosahedral, Enveloped Single-Stranded RNA Viruses

Viruses in the *Flaviviridae* and *Togaviridae* families include many human and animal pathogens such as dengue, hepatitis C, Chikungunya, West Nile, Venezuelan equine encephalitis, Zika or Semliki forest viruses [61, 62]. In these families, maturation involves proteolytic processing of the external glycoproteins orchestrated with virus egress along the secretory pathway, to expose the fusion peptide that will enable the virus to enter a new host cell [5, 63, 64].

Flavivirus particles are formed by three structural proteins: protein C located inside the membrane envelope together with the RNA genome, and proteins M and E anchored to the membrane and covering its outer surface. Protein C does not seem to acquire a well-ordered organization, probably due to lack of direct, strong interactions with M and E [65–67]. The glycoproteins E and M undergo a complex reorganization during maturation of dengue virus, one of the best-studied flaviviruses (Fig. 14.6a). In the immature particle, 180 heterodimers composed of glycoprotein E and the precursor version of M (prM) form 60 trimeric spikes protruding from the particle surface. The C-terminal domain of prM is inserted in the membrane. The N-terminal ("pr") domain, which will later be separated by furin cleavage, is anchored to E on the outermost part of the spike. In this conformation, pr hides the fusion peptide present in protein E, while E hides the furin recognition site in pr [66, 68–71].

When the dengue virus immature particle is exposed to low pH in the trans Golgi network, the trimers dissociate, and protein E forms 90 homodimers that lie flat on the membrane surface, giving the particle a smooth aspect and decreasing its total diameter from 60 to 50 nm (Fig. 14.6b) [72, 73]. In the mature dengue virus capsid, protein E follows an atypical icosahedral organization that does not conform to the Caspar and Klug theory, with three monomers per asymmetric unit that do not have quasiequivalent symmetric environments. In addition to the large reorganization undergone by the E protein ectodomains, two amphipathic helices also rearrange during maturation, opening a binding pocket for a lipid headgroup. It has been proposed that this lipid stabilizes the conformation of protein E membrane-

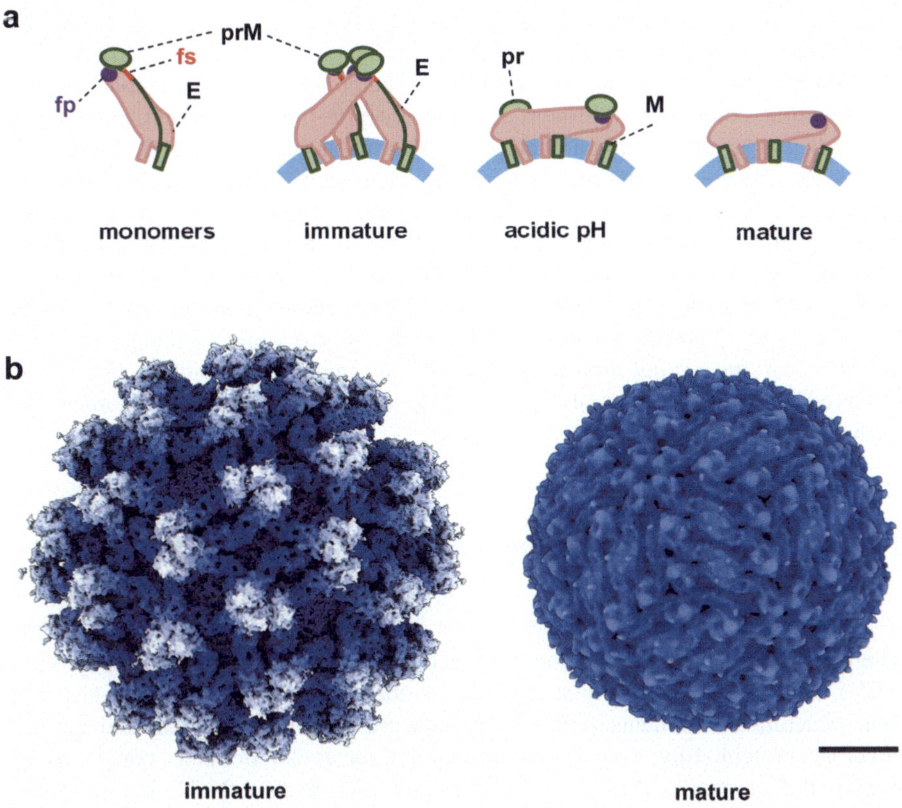

Fig. 14.6 Maturation in *Riboviria*: dengue virus. (**a**) Schematics of the dengue virus maturation pathway. Proteins E and M are represented in pink and green respectively, with the fusion peptide (fp) highlighted in purple in E, and the furin target site (fs) in red in protein M. The prM domain is also indicated. (**b**) Surface rendering representations of the immature (left, EMD:2141) and mature (right, EMD:5520) capsids, colored according to radius from blue to white. The view is along a twofold icosahedral axis. Scale bar: 100 Å

associated helices [71]. The furin target motif in the long linker region between the pr and transmembrane domains of prM becomes accessible to the protease during spike dissociation, facilitating its cleavage. The cleaved peptide (pr) that covers the fusion peptide in protein E remains associated with the particle at acidic pH, but is released when the viral particle encounters neutral pH upon cell exit. Upon pr release, a loop in protein E (150-loop) changes conformation to cover the fusion peptide [74]. The particle is now in a metastable state, prepared for successful entry into the next host, where acidic endosomal pH will trigger a new change and organize the E monomers into the homotrimers required for membrane fusion.

The togaviruses Sindbis virus and Semliki forest virus have a $T = 4$ icosahedral core (40 nm diameter) formed by protein C. This core is enveloped by a membrane covered by glycoproteins (E1 and E2), also arranged in a $T = 4$ icosahedral lattice, to form the ~70 nm diameter particle. Proteins E1 and E2 are arranged in trimeric heterodimers, resulting in 80 three-petal spikes on the virus surface [75–79]. Autoproteolysis separates protein C from the polyprotein product that contains all structural components. Protein E2, the receptor-binding protein, is oriented perpendicular to the membrane and interacts with the core protein C. Protein E1 carries the fusion peptide and lays on the membrane surface. In the immature particle, the N-terminal region of the E2 precursor (preE2) hides the E1 fusion peptide while the virus transits across the Golgi. Later on, preE2 is cleaved by furin-like cell proteases, and the E1–E2 heterodimer changes conformation to expose the fusion peptide which will allow the genome-containing core to escape the endosome. Cryo-EM studies on Semliki forest virus and Chikungunya virus indicate that these conformational changes are much more subtle than those observed in flaviviruses, involving a compaction of the spikes and decompaction of the core [75, 80].

14.6.3 Maturation in Pleomorphic, Single-Stranded RNA Viruses

The pleomorphic retroviruses, and particularly human immunodeficiency virus (HIV), have been extensively studied due to their clinical relevance [81–84]. The main structural components of the HIV particle are the polyprotein Gag, and the viral envelope protein, Env. Gag is composed of three major domains connected by flexible linkers: matrix (MA), capsid (CA), and nucleocapsid (NC). Gag molecules bound to the viral genome via the NC domain are recruited to patches of cell membrane carrying the virus glycoproteins gp120 and gp41, which are cleavage products of Env. Gag recruitment to the membrane induces budding and release of the immature, non-infectious HIV particle. Several thousand copies (~5000) of Gag arrange radially to form a roughly spherical, incomplete shell, with the MA domain in the outermost region, anchored to the viral envelope, and the NC domain in the innermost, in contact with two copies of the ssRNA genome (Fig. 14.7). In the immature HIV particle, Gag monomers oligomerize via CA–CA and MA-MA interactions to form a discontinuous, poorly ordered hexameric lattice. Positively charged holes formed by MA oligomers could serve as pockets for binding the cytoplasmic tails of Env. Immature HIV particles also contain the GagPol polyprotein, which comprises three enzymes critical for the infectious cycle: integrase (IN), reverse transcriptase (RT), and protease (PR).

PR cleaves Gag and GagPol into the individual polypeptides (MA, CA, NC, IN, RT) following a sequential, tightly regulated pathway, and triggering a cascade of large conformational changes required to produce the infectious particle (Fig. 14.7). MA remains anchored to the membrane, and NC remains bound to the RNA inside a shell formed by CA, together with IN and RT. CA can assemble into either hexamers or pentamers, similarly to capsid proteins of icosahedral viruses. A TVGG sequence

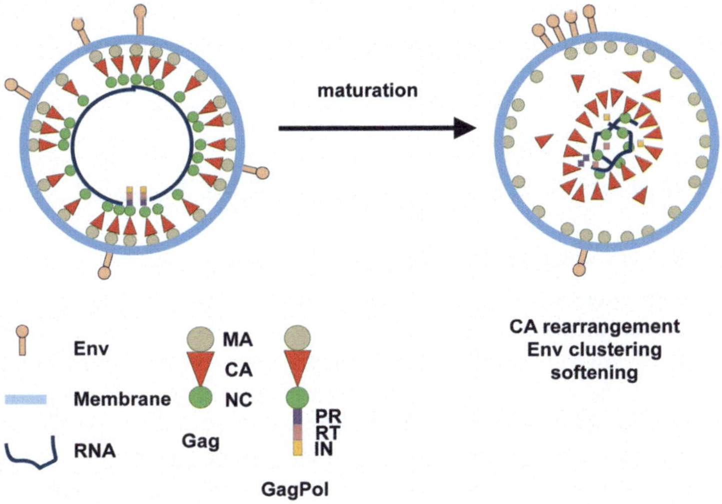

Fig. 14.7 Maturation in *Riboviria* realm: human immunodeficiency virus. Cartoon summarizing the HIV maturation pathway. See main text for details. Figure modified from [85]

motif in CA serves as a molecular switch to modulate the formation of either oligomeric species [86]. Inositol hexakisphosphate (IP6) is a key co-factor for HIV-1 assembly and maturation [87] and may also help in modulation of the pentamer–hexamer TVGG switch [86]. Upon maturation, CA–CA interactions change, resulting in appearance of the pentamers that induce higher curvature points in the hexameric lattice. As a result, the shell is no longer spherical, but instead adopts a fullerene cone shape (see Chap. 12). Cones contain ~1100–1330 copies of CA, with a variable number of hexamers and usually 12 pentamers in HIV-1. Only ~1/3 of the CA molecules present in the immature particle form the mature capsid [88]. Because the position of pentamers in the mature capsid does not correlate with the gaps in the immature Gag lattice, it is thought that the immature CA hexameric lattice needs to disassemble completely and reform de novo to produce the mature capsid [84].

In addition to the remarkable conformational changes undergone by Gag, the physical properties of the HIV particle and the localization of Env proteins on its outer surface also change during maturation (Fig. 14.7). AFM nanoindentation studies (see Chaps. 9 and 18) have shown that immature HIV particles are 14 times stiffer than mature virions. This stiffness change is mediated by the cytoplasmic tail of Env. When this tail is deleted, immature particles become as soft as the mature ones and competent for cell entry, suggesting that the particle mechanical properties play an important role in the viral cycle [89, 90]. HIV-1 particles carry only a few copies (7–14) of Env proteins, responsible for cell binding and membrane fusion during entry. Stimulated emission depletion microscopy (STED) analyses have shown that maturation induces clustering of Env. This clustering is facilitated by an increase in Env mobility that depends on the Env tail (possibly via its interactions with Gag across the membrane). The entry efficiency of HIV-1 is directly correlated to its capacity to form Env clusters, possibly to mediate efficient receptor binding [81, 91, 92].

14.6.4 Maturation in Double-Stranded RNA Viruses

Viruses with dsRNA genomes are found in a wide variety of eukaryotic hosts, including fungi, plants, and animals. Important pathogens such as bluetongue virus or rotavirus (in the *Sedoreoviridae* family) or infectious bursal disease virus (IBDV, in the *Birnaviridae* family) have dsRNA genomes. So far, only one dsRNA virus family has been found in prokaryotic hosts: the *Cystoviridae*. The best-characterized member of this family is the *Pseudomonas* bacteriophage Φ6. Common characteristics of dsRNA viruses include the encapsidation of segmented genomes inside naked, multilayered icosahedral capsids (with some exceptions described below).

Cystoviruses are exceptional among dsRNA viruses because they have a membrane, decorated with viral proteins. This membrane is the outermost part of a triple-layered particle (Fig. 14.1), and it envelops a $T = 13$ shell (58 nm diameter) which in turn contains a 46 nm diameter core. The core is composed of 120 copies of the structural protein P1, with two copies (P1A and P1B) forming the asymmetric unit of a $T = 1$ shell. To protect their genomes from host defense mechanisms, dsRNA viruses package ssRNA genomic precursors, and the complementary strand is synthesized inside the particle. In Φ6, a core shell containing also the packaging NTPase and the viral RNA polymerase is first assembled. This particle, called polymerase complex (PC), will encapsidate the three viral (+) ssRNA segments, synthesize the complementary strands, and transcribe new (+) ssRNA strands. The PC is assembled as a dodecahedral cage with the center of each pentagonal facet deeply recessed toward the particle center. Sequential packaging of the three (+) ssRNA molecules causes three successive capsid expansion transitions, finally yielding a spherical, mature particle (Fig. 14.8). Only after the three segments have been packaged is synthesis of the complementary strands initiated. In this sense, expansion of the PC is considered a maturation process required to activate the particle functionality for virus propagation. During Φ6 PC expansion, the fivefold vertices are pushed out; the fivefold radius doubles (from ~12 to ~22 nm) and the inner capsid volume increases by a factor of 2.4 [93–95]. This large conformational change is driven by rotations of the P1 monomers, and hinging domain movements that alter the intersubunit contacts. As a consequence, the angle between neighboring P1B subunits changes from 98° to 148°, and the buried surfaces increase by 50% for P1B and 16% for P1A [96].

The *Sedoreoviridae* genomes can be composed of as many as 12 segments, housed inside triple-layered icosahedral particles with an architecture similar to that of Φ6, except for the membrane (see Chap. 12). Particle assembly is coupled with replication of the segmented dsRNA genome, by the action of the viral

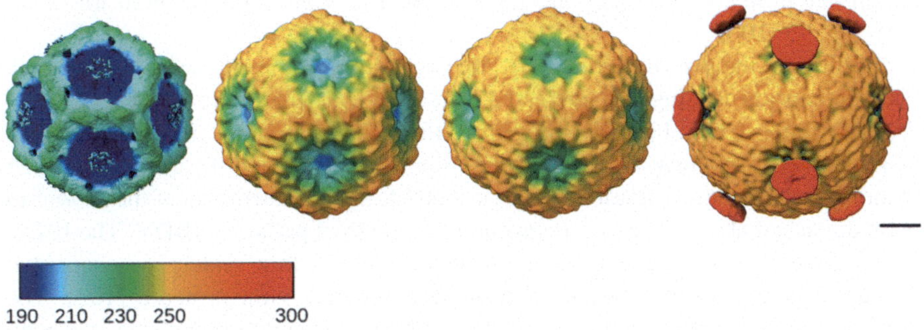

Fig. 14.8 Maturation in *Riboviria* realm: bacteriophage ϕ6. 3D maps showing, from left to right: the genome-less PC (EMDB:1500), two expansion intermediates (EMDB:5355 and 5356), and the fully packaged particle (EMDB:5358). The two expansion intermediates are thought to correspond to the sequential stages occurring upon packaging the first and the second RNA genomic segments [93]. The maps are colored by radius (Å), as indicated by the color bar at the bottom. Scale bar: 100 Å. Figure modified from [85]

polymerase that synthesizes dsRNA molecules from the (+) ssRNA transcripts stored in cytoplasmic structures called viroplasms. Polymerase-(+)RNA complexes are activated by interaction with the core shell protein, producing early subviral particles where synthesis of the complementary RNA strand will happen. Subviral particles competent for dsRNA synthesis extracted from rotavirus infected cells are heterogeneous in size (between 30 and 70 nm) [97], suggesting the possibility of core shell expansion associated with genome packaging, similarly to cystoviruses. However, their fragility made these particles unsuitable for structural studies. Thanks to the development of in situ cryogenic electron tomography to observe assembly intermediates inside the infected cells, detailed structural data have finally been obtained for early subviral particles of a mammalian reovirus encapsidating 10 dsRNA segments [98]. This study proved the existence of single-layered particles composed of 120 copies of the reovirus core shell protein. The single-layered particles are collapsed dodecahedrons, with the fivefold center of the facets pushed inward by ~10 nm compared to the mature virions, as in bacteriophage Φ6. However, it is still unclear if a sequential expansion similar to that of cystoviruses takes place during packaging of the 10 genomic segments. Only completely collapsed or completely expanded particles, but no expansion intermediates, were detected in the cells, suggesting that either the expansion happens rapidly or the collapsed cores are abortive assembly products that failed to complete packaging. Additionally to the potential core expansion upon genome packaging, sedoreovirus particles must recruit their two outer layers. A double-layered particle is first assembled in the viroplasm and may traffic to a different location (for example, endoplasmic reticulum for rotavirus) where the third layer will be acquired to complete the ~80 nm diameter virion. Since the double-layered particle is not infectious, acquisition of this third layer is usually considered a maturation step in reovirus assembly. In rotaviruses, an additional maturation step consisting in cleavage of surface spikes by trypsin-like proteases from the intestinal lumen after cell exit is required for infectivity [97, 99, 100].

Members of the *Birnaviridae* family are exceptional among the dsRNA viruses in that their capsids are single-layered, and instead of consisting only of the $T = 1$ core like the simplest members of this group, they assemble a 65 nm $T = 13$ shell [101]. Birnaviruses have an MCP fold similar to that of picornaviruses, and their maturation also holds similarities with the ssRNA, smaller members of the *Riboviria* realm described above. The family prototype is the fowl pathogen IBDV. The IBDV capsid protein is synthesized as a precursor (pVP2) that undergoes sequential cleavages, both autocatalytic and driven by a separate viral protease [102]. The pVP2 C-terminal peptide is a key player in IBDV assembly, acting as a transient, flexible molecular switch to allow the different interactions between identical MCP monomers required to form the large capsid. Different conformations, or degrees of proteolytic processing, in this region determine whether a hexamer or a pentamer of the MCP is formed. After assembly, the switch is removed, eliminating this flexibility to render the assembly reaction irreversible. Part of the conformational switching role of the pVP2 C-terminal region is based on its ability to interact with a multifunctional viral protein, VP3, which may act as a scaffold, indicating that the birnavirus capsid follows a more complex morphogenetic pathway than the smaller members of this realm [103, 104]. Proteolytic processing of pVP2 also generates a 46 residue peptide with membrane destabilization activity, possibly involved in the disruption of cell membranes during entry, similarly to the γ-peptides of tetra- and nodaviruses [105]. Low pH environments may play a role in birnavirus maturation, as suggested by the observation that cell treatment with chloroquine, which inhibits lysosome acidification, hinders production of VP2 from its precursor pVP2 [106].

14.7 Maturation in the *Ribozyviria* Realm

Viruses related to hepatitis delta virus (HDV, genus *Deltavirus*) comprise this small realm [2]. Deltaviruses [107] are satellite viruses with a small (~1.7 kb), circular negative-sense genomic RNA. The HDV genome codes for a single protein, small hepatitis D antigen (S-HDAg, 195 amino acids), which is required for replication. Genome editing by host factors extends the S-HDAg ORF, coding then for the large hepatitis D antigen (L-HDAg, 214 amino acids). L-HDAg inhibits replication of HDV and is required for viral assembly [108]. The virions are pleomorphic, enveloped particles of approximately 35 nm in diameter [109]. The HDV nucleocapsid (Fig. 14.1) is a ribonucleoprotein (RNP) complex formed by variable amounts of S- and L-HDAg proteins bound to the genomic RNA in an approximately 70:1 protein:RNA ratio. To become fully infectious, HDV requires enclosing its RNP inside a membrane containing the envelope proteins of the helper virus, human hepatitis B virus (HBV) (family *Hepadnaviridae*). In this sense, envelope hijacking can be considered as a maturation step in HDV. Farnesylation, a posttranslational modification by which a hydrophobic molecule is attached to a Cys residue at the C-terminus of L-HDAg, drives the interaction between the HDV nucleocapsid and the HBV envelope for membrane recruitment [108].

14.8 Perspectives and Conclusions

The concept of maturation encompasses modifications in the virus particles following, or concomitant with, genome packaging. These modifications may consist in large conformational rearrangements of capsid proteins, or more in general, in changes in the interactions that keep together the particle components. Architectural rearrangements produce changes in stability or exposure of functional elements (receptor binding, membrane fusion, or membrane lytic peptides) required to prepare the particle for genome delivery in the host cell. Maturation is often driven by virus-encoded proteases, whose activity must be tightly regulated to avoid untimely cleavages that would result in failure to produce infectious progeny. Continuous improvements in structural, physical, and computational biology techniques are helping to detect labile maturation intermediates in infected cells and to define the role of small molecules (lipids, polyanions, cations) or environmental changes (pH) in virus maturation.

Acknowledgments The C.S.M. group is supported by grants from the Spanish State Research Agency, with co-funding from the European Regional Development Fund (PID2019-104098GB-I00/AEI/10.13039/501100011033 and PID2022-136456NB-I00/AEI/10.13039/501100011033), CSIC (2023AEP082), the European Innovation Council (Horizon Europe) under grant agreement No 101098647 (project iAds), and Marie Skłodowska-Curie Actions (grant agreement 101129778, project INVECTA). The CNB-CSIC is further supported by a Severo Ochoa Excellence grant (CEX2023-001386-S/AEI/10.13039/501100011033). The group is a member of the Spanish Adenovirus Network (AdenoNet, RED2022-134221-T/AEI/10.13039/501100011033), CSIC LifeHub, and CSIC BCBHub.

References

1. Goddard TD, Huang CC, Meng EC, Pettersen EF, Couch GS, Morris JH, Ferrin TE (2018) UCSF ChimeraX: meeting modern challenges in visualization and analysis. Protein Sci 27:14–25
2. Koonin EV, Krupovic M, Agol VI (2021) The Baltimore classification of viruses 50 years later: how does it stand in the light of virus evolution? Microbiol Mol Biol Rev 85:e0005321
3. Hendrix RW, Johnson JE (2012) Bacteriophage HK97 capsid assembly and maturation. Adv Exp Med Biol 726:351–363
4. Prevelige PE, Fane BA (2012) Building the machines: scaffolding protein functions during bacteriophage morphogenesis. Adv Exp Med Biol 726:325–350
5. Veesler D, Johnson JE (2012) Virus maturation. Annu Rev Biophys 41:473–496

6. Yang F, Forrer P, Dauter Z, Conway JF, Cheng N, Cerritelli ME, Steven AC, Pluckthun A, Wlodawer A (2000) Novel fold and capsid-binding properties of the lambda-phage display platform protein gpD. Nat Struct Biol 7:230–237
7. Parent KN, Khayat R, Tu LH, Suhanovsky MM, Cortines JR, Teschke CM, Johnson JE, Baker TS (2010) P22 coat protein structures reveal a novel mechanism for capsid maturation: stability without auxiliary proteins or chemical crosslinks. Structure 18:390–401
8. Pietila MK, Laurinmaki P, Russell DA, Ko CC, Jacobs-Sera D, Hendrix RW, Bamford DH, Butcher SJ (2013) Structure of the archaeal head-tailed virus HSTV-1 completes the HK97 fold story. Proc Natl Acad Sci USA 110:10604–10609
9. Pietila MK, Laurinmaki P, Russell DA, Ko CC, Jacobs-Sera D, Butcher SJ, Bamford DH, Hendrix RW (2013) Insights into head-tailed viruses infecting extremely halophilic archaea. J Virol 87:3248–3260
10. Cardone G, Heymann JB, Cheng N, Trus BL, Steven AC (2012) Procapsid assembly, maturation, nuclear exit: dynamic steps in the production of infectious herpesvirions. Adv Exp Med Biol 726:423–439
11. Liu YT, Jih J, Dai X, Bi GQ, Zhou ZH (2019) Cryo-EM structures of herpes simplex virus type 1 portal vertex and packaged genome. Nature 570:257–261
12. Dai X, Zhou ZH (2018) Structure of the herpes simplex virus 1 capsid with associated tegument protein complexes. Science 360
13. Zhang X, Sun S, Xiang Y, Wong J, Klose T, Raoult D, Rossmann MG (2012) Structure of Sputnik, a virophage, at 3.5-Å resolution. Proc Natl Acad Sci USA 109:18431–18436
14. San Martín C, Huiskonen JT, Bamford JK, Butcher SJ, Fuller SD, Bamford DH, Burnett RM (2002) Minor proteins, mobile arms and membrane-capsid interactions in the bacteriophage PRD1 capsid. Nat Struct Biol 9:756–763
15. Condezo GN, Marabini R, Ayora S, Carazo JM, Alba R, Chillón M, San Martín C (2015) Structures of adenovirus incomplete particles clarify capsid architecture and show maturation changes of packaging protein L1 52/55k. J Virol 89:9653–9664
16. Benkö M, Aoki K, Arnberg N, Davison AJ, Echavarria M, Hess M, Jones MS, Kajan GL, Kajon AE, Mittal SK, Podgorski II, San Martín C, Wadell G, Watanabe H, Harrach B, Report Consortium ICTV (2022) ICTV virus taxonomy profile: Adenoviridae 2022. J Gen Virol 103
17. Gallardo J, Pérez-Illana M, Martín-González N, San Martín C (2021) Adenovirus structure: what is new? Int J Mol Sci 22
18. Martín-González N, Gómez-González A, Hernando-Pérez M, Bauer M, Greber UF, San Martín C, De Pablo PJ (2023) Adenovirus core protein V reinforces the capsid and enhances genome release from disrupted particles. Sci Adv 9:ade9910
19. de Pablo PJ, San Martín C (2022) Seeing and touching adenovirus: complementary approaches for understanding assembly and disassembly of a complex virion. Curr Opin Virol 52:112–122
20. Mangel WF, San Martín C (2014) Structure, function and dynamics in adenovirus maturation. Viruses 6:4536–4570
21. Greber UF, Flatt JW (2019) Adenovirus entry: from infection to immunity. Annu Rev Virol 6: 177–197
22. Greber UF, Suomalainen M (2022) Adenovirus entry: stability, uncoating, and nuclear import. Mol Microbiol 118:309–320
23. Condezo GN, San Martín C (2017) Localization of adenovirus morphogenesis players, together with visualization of assembly intermediates and failed products, favor a model where assembly and packaging occur concurrently at the periphery of the replication center. PLoS Pathog 13:e1006320
24. Peréz-Berná AJ, Mangel WF, McGrath WJ, Graziano V, Flint J, San Martin C (2014) Processing of the L1 52/55k protein by the adenovirus protease: a new substrate and new insights into virion maturation. J Virol 88:1513–1524
25. Liu L, Cooper T, Howley PM, Hayball JD (2014) From crescent to mature virion: vaccinia virus assembly and maturation. Viruses 6:3787–3808

26. Bahar MW, Graham SC, Stuart DI, Grimes JM (2011) Insights into the evolution of a complex virus from the crystal structure of vaccinia virus D13. Structure 19:1011–1020
27. Hernández-González M, Calcraft T, Nans A, Rosenthal PB, Way M (2023) A succession of two viral lattices drives vaccinia virus assembly. PLoS Biol 21:e3002005
28. Bisht H, Weisberg AS, Szajner P, Moss B (2009) Assembly and disassembly of the capsid-like external scaffold of immature virions during vaccinia virus morphogenesis. J Virol 83:9140–9150
29. Byrd CM, Bolken TC, Hruby DE (2002) The vaccinia virus I7L gene product is the core protein proteinase. J Virol 76:8973–8976
30. Marion S, San Martín C, Šiber A (2017) Role of condensing particles in polymer confinement: a model for virus-packed "minichromosomes". Biophys J 113:1643–1653
31. Priyamvada L, Kallemeijn WW, Faronato M, Wilkins K, Goldsmith CS, Cotter CA, Ojeda S, Solari R, Moss B, Tate EW, Satheshkumar PS (2022) Inhibition of vaccinia virus L1 N-myristoylation by the host N-myristoyltransferase inhibitor IMP-1088 generates non-infectious virions defective in cell entry. PLoS Pathog 18:e1010662
32. Foo CH, Whitbeck JC, Ponce-de-Leon M, Saw WT, Cohen GH, Eisenberg RJ (2012) The myristate moiety and amino terminus of vaccinia virus l1 constitute a bipartite functional region needed for entry. J Virol 86:5437–5451
33. Sodeik B, Doms RW, Ericsson M, Hiller G, Machamer CE, van 't Hof W, van Meer G, Moss B, Griffiths G (1993) Assembly of vaccinia virus: role of the intermediate compartment between the endoplasmic reticulum and the Golgi stacks. J Cell Biol 121:521–541
34. Baquero DP, Liu Y, Wang F, Egelman EH, Prangishvili D, Krupovic M (2020) Structure and assembly of archaeal viruses. Adv Virus Res 108:127–164
35. Baquero DP, Gazi AD, Sachse M, Liu J, Schmitt C, Moya-Nilges M, Schouten S, Prangishvili D, Krupovic M (2021) A filamentous archaeal virus is enveloped inside the cell and released through pyramidal portals. Proc Natl Acad Sci USA 118
36. Chen XS, Garcea RL, Goldberg I, Casini G, Harrison SC (2000) Structure of small virus-like particles assembled from the L1 protein of human papillomavirus 16. Mol Cell 5:557–567
37. Wolf M, Garcea RL, Grigorieff N, Harrison SC (2010) Subunit interactions in bovine papillomavirus. Proc Natl Acad Sci USA 107:6298–6303
38. Buck CB, Thompson CD, Pang YY, Lowy DR, Schiller JT (2005) Maturation of papillomavirus capsids. J Virol 79:2839–2846
39. Cardone G, Moyer AL, Cheng N, Thompson CD, Dvoretzky I, Lowy DR, Schiller JT, Steven AC, Buck CB, Trus BL (2014) Maturation of the human papillomavirus 16 capsid. MBio 5: e01104–e01114
40. Mietzsch M, Penzes JJ, Agbandje-McKenna M (2019) Twenty-five years of structural Parvovirology. Viruses 11
41. Llamas-Saiz AL, Agbandje-McKenna M, Wikoff WR, Bratton J, Tattersall P, Rossmann MG (1997) Structure determination of minute virus of mice. Acta Crystallogr D Biol Crystallogr 53:93–102
42. Tattersall P, Shatkin AJ, Ward DC (1977) Sequence homology between the structural polypeptides of minute virus of mice. J Mol Biol 111:375–394
43. Tattersall P, Cawte PJ, Shatkin AJ, Ward DC (1976) Three structural polypeptides coded for by minute virus of mice, a parvovirus. J Virol 20:273–289
44. Wolfisberg R, Kempf C, Ros C (2016) Late maturation steps preceding selective nuclear export and egress of progeny parvovirus. J Virol 90:5462–5474
45. Farr GA, Cotmore SF, Tattersall P (2006) VP2 cleavage and the leucine ring at the base of the fivefold cylinder control pH-dependent externalization of both the VP1 N terminus and the genome of minute virus of mice. J Virol 80:161–171
46. Tullis GE, Burger LR, Pintel DJ (1993) The minor capsid protein VP1 of the autonomous parvovirus minute virus of mice is dispensable for encapsidation of progeny single-stranded DNA but is required for infectivity. J Virol 67:131–141

47. Farr GA, Zhang LG, Tattersall P (2005) Parvoviral virions deploy a capsid-tethered lipolytic enzyme to breach the endosomal membrane during cell entry. Proc Natl Acad Sci USA 102: 17148–17153
48. Lombardo E, Ramírez JC, García J, Almendral JM (2002) Complementary roles of multiple nuclear targeting signals in the capsid proteins of the parvovirus minute virus of mice during assembly and onset of infection. J Virol 76:7049–7059
49. Fisher AJ, Johnson JE (1993) Ordered duplex RNA controls capsid architecture in an icosahedral animal virus. Nature 361:176–179
50. Odegard A, Banerjee M, Johnson JE (2010) Flock house virus: a model system for understanding non-enveloped virus entry and membrane penetration. Curr Top Microbiol Immunol 343:1–22
51. Gallagher TM, Rueckert RR (1988) Assembly-dependent maturation cleavage in provirions of a small icosahedral insect ribovirus. J Virol 62:3399–3406
52. Kearney BM, Johnson JE (2014) Assembly and maturation of a T = 4 quasi-equivalent virus is guided by electrostatic and mechanical forces. Viruses 6:3348–3362
53. Canady MA, Tihova M, Hanzlik TN, Johnson JE, Yeager M (2000) Large conformational changes in the maturation of a simple RNA virus, Nudaurelia capensis omega virus (NωV). J Mol Biol 299:573–584
54. Castells-Graells R, Ribeiro JRS, Domitrovic T, Hesketh EL, Scarff CA, Johnson JE, Ranson NA, Lawson DM, Lomonossoff GP (2021) Plant-expressed virus-like particles reveal the intricate maturation process of a eukaryotic virus. Commun Biol 4:619
55. Helgstrand C, Munshi S, Johnson JE, Liljas L (2004) The refined structure of Nudaurelia capensis omega virus reveals control elements for a T = 4 capsid maturation. Virology 318: 192–203
56. Tang J, Lee KK, Bothner B, Baker TS, Yeager M, Johnson JE (2009) Dynamics and stability in maturation of a T=4 virus. J Mol Biol 392:803–812
57. Hellen CU, Wimmer E (1992) Maturation of poliovirus capsid proteins. Virology 187:391–397
58. Moscufo N, Chow M (1992) Myristate-protein interactions in poliovirus: interactions of VP4 threonine 28 contribute to the structural conformation of assembly intermediates and the stability of assembled virions. J Virol 66:6849–6857
59. Strauss M, Filman DJ, Belnap DM, Cheng N, Noel RT, Hogle JM (2015) Nectin-like interactions between poliovirus and its receptor trigger conformational changes associated with cell entry. J Virol 89:4143–4157
60. Richards AL, Jackson WT (2012) Intracellular vesicle acidification promotes maturation of infectious poliovirus particles. PLoS Pathog 8:e1003046
61. Griffin DE (2013) Alphaviruses. In: Knipe DM, Howley PM (eds) Fields virology, vol 1. Wolters Kluwer/Lippincott Williams & Wilkins Health, Philadelphia, PA, pp 651–686
62. Lindenbach BD, Murray CL, Thiel H-J, Rice CM (2013) Flaviviridae. In: Knipe DM, Howley PM (eds) Fields virology, vol 1. Wolters Kluwer/Lippincott Williams & Wilkins Health, Philadelphia, PA, pp 712–746
63. Jose J, Snyder JE, Kuhn RJ (2009) A structural and functional perspective of alphavirus replication and assembly. Future Microbiol 4:837–856
64. Pierson TC, Diamond MS (2012) Degrees of maturity: the complex structure and biology of flaviviruses. Curr Opin Virol 2:168–175
65. Zhang W, Chipman PR, Corver J, Johnson PR, Zhang Y, Mukhopadhyay S, Baker TS, Strauss JH, Rossmann MG, Kuhn RJ (2003) Visualization of membrane protein domains by cryo-electron microscopy of dengue virus. Nat Struct Biol 10:907–912
66. Zhang Y, Corver J, Chipman PR, Zhang W, Pletnev SV, Sedlak D, Baker TS, Strauss JH, Kuhn RJ, Rossmann MG (2003) Structures of immature flavivirus particles. EMBO J 22: 2604–2613

67. Tan TY, Fibriansah G, Kostyuchenko VA, Ng TS, Lim XX, Zhang S, Lim XN, Wang J, Shi J, Morais MC, Corti D, Lok SM (2020) Capsid protein structure in Zika virus reveals the flavivirus assembly process. Nat Commun 11:895
68. Kostyuchenko VA, Zhang Q, Tan JL, Ng TS, Lok SM (2013) Immature and mature dengue serotype 1 virus structures provide insight into the maturation process. J Virol 87:7700–7707
69. Li L, Lok SM, Yu IM, Zhang Y, Kuhn RJ, Chen J, Rossmann MG (2008) The flavivirus precursor membrane-envelope protein complex: structure and maturation. Science 319:1830–1834
70. Yu IM, Zhang W, Holdaway HA, Li L, Kostyuchenko VA, Chipman PR, Kuhn RJ, Rossmann MG, Chen J (2008) Structure of the immature dengue virus at low pH primes proteolytic maturation. Science 319:1834–1837
71. Renner M, Dejnirattisai W, Carrique L, Martin IS, Karia D, Ilca SL, Ho SF, Kotecha A, Keown JR, Mongkolsapaya J, Screaton GR, Grimes JM (2021) Flavivirus maturation leads to the formation of an occupied lipid pocket in the surface glycoproteins. Nat Commun 12:1238
72. Kuhn RJ, Zhang W, Rossmann MG, Pletnev SV, Corver J, Lenches E, Jones CT, Mukhopadhyay S, Chipman PR, Strauss EG, Baker TS, Strauss JH (2002) Structure of dengue virus: implications for flavivirus organization, maturation, and fusion. Cell 108:717–725
73. Zhang X, Ge P, Yu X, Brannan JM, Bi G, Zhang Q, Schein S, Zhou ZH (2013) Cryo-EM structure of the mature dengue virus at 3.5-Å resolution. Nat Struct Mol Biol 20:105–110
74. Vaney MC, Dellarole M, Duquerroy S, Medits I, Tsouchnikas G, Rouvinski A, England P, Stiasny K, Heinz FX, Rey FA (2022) Evolution and activation mechanism of the flavivirus class II membrane-fusion machinery. Nat Commun 13:3718
75. Ferlenghi I, Gowen B, de Haas F, Mancini EJ, Garoff H, Sjoberg M, Fuller SD (1998) The first step: activation of the Semliki Forest virus spike protein precursor causes a localized conformational change in the trimeric spike. J Mol Biol 283:71–81
76. Mancini EJ, Clarke M, Gowen BE, Rutten T, Fuller SD (2000) Cryo-electron microscopy reveals the functional organization of an enveloped virus, Semliki Forest virus. Mol Cell 5: 255–266
77. Mukhopadhyay S, Zhang W, Gabler S, Chipman PR, Strauss EG, Strauss JH, Baker TS, Kuhn RJ, Rossmann MG (2006) Mapping the structure and function of the E1 and E2 glycoproteins in alphaviruses. Structure 14:63–73
78. Pletnev SV, Zhang W, Mukhopadhyay S, Fisher BR, Hernandez R, Brown DT, Baker TS, Rossmann MG, Kuhn RJ (2001) Locations of carbohydrate sites on alphavirus glycoproteins show that E1 forms an icosahedral scaffold. Cell 105:127–136
79. Zhang W, Mukhopadhyay S, Pletnev SV, Baker TS, Kuhn RJ, Rossmann MG (2002) Placement of the structural proteins in Sindbis virus. J Virol 76:11645–11658
80. Yap ML, Klose T, Urakami A, Hasan SS, Akahata W, Rossmann MG (2017) Structural studies of Chikungunya virus maturation. Proc Natl Acad Sci USA 114:13703–13707
81. Krebs AS, Mendonca LM, Zhang P (2022) Structural analysis of retrovirus assembly and maturation. Viruses 14(1):54
82. Lerner G, Weaver N, Anokhin B, Spearman P (2022) Advances in HIV-1 assembly. Viruses 14(3):478
83. Pornillos O, Ganser-Pornillos BK (2019) Maturation of retroviruses. Curr Opin Virol 36:47–55
84. Mattei S, Schur FK, Briggs JA (2016) Retrovirus maturation-an extraordinary structural transformation. Curr Opin Virol 18:27–35
85. San Martín C (2019) Virus maturation. Adv Exp Med Biol 1215:129–158
86. Schirra RT, Dos Santos NFB, Zadrozny KK, Kucharska I, Ganser-Pornillos BK, Pornillos O (2023) A molecular switch modulates assembly and host factor binding of the HIV-1 capsid. Nat Struct Mol Biol 30:383–390
87. Wu C, Xiong Y (2023) Enrich and switch: IP6 and maturation of HIV-1 capsid. Nat Struct Mol Biol 30:239–241

88. Briggs JA, Simon MN, Gross I, Krausslich HG, Fuller SD, Vogt VM, Johnson MC (2004) The stoichiometry of Gag protein in HIV-1. Nat Struct Mol Biol 11:672–675
89. Kol N, Shi Y, Tsvitov M, Barlam D, Shneck RZ, Kay MS, Rousso I (2007) A stiffness switch in human immunodeficiency virus. Biophys J 92:1777–1783
90. Pang HB, Hevroni L, Kol N, Eckert DM, Tsvitov M, Kay MS, Rousso I (2013) Virion stiffness regulates immature HIV-1 entry. Retrovirology 10:4
91. Chojnacki J, Staudt T, Glass B, Bingen P, Engelhardt J, Anders M, Schneider J, Muller B, Hell SW, Krausslich HG (2012) Maturation-dependent HIV-1 surface protein redistribution revealed by fluorescence nanoscopy. Science 338:524–528
92. Chojnacki J, Waithe D, Carravilla P, Huarte N, Galiani S, Enderlein J, Eggeling C (2017) Envelope glycoprotein mobility on HIV-1 particles depends on the virus maturation state. Nat Commun 8:545
93. Nemecek D, Cheng N, Qiao J, Mindich L, Steven AC, Heymann JB (2011) Stepwise expansion of the bacteriophage Φ6 procapsid: possible packaging intermediates. J Mol Biol 414:260–271
94. Butcher SJ, Dokland T, Ojala PM, Bamford DH, Fuller SD (1997) Intermediates in the assembly pathway of the double-stranded RNA virus phi6. EMBO J 16:4477–4487
95. Huiskonen JT, de Haas F, Bubeck D, Bamford DH, Fuller SD, Butcher SJ (2006) Structure of the bacteriophage phi6 nucleocapsid suggests a mechanism for sequential RNA packaging. Structure 14:1039–1048
96. Nemecek D, Boura E, Wu W, Cheng N, Plevka P, Qiao J, Mindich L, Heymann JB, Hurley JH, Steven AC (2013) Subunit folds and maturation pathway of a dsRNA virus capsid. Structure 21:1374–1383
97. Long CP, McDonald SM (2017) Rotavirus genome replication: some assembly required. PLoS Pathog 13:e1006242
98. Sutton G, Sun D, Fu X, Kotecha A, Hecksel CW, Clare DK, Zhang P, Stuart DI, Boyce M (2020) Assembly intermediates of orthoreovirus captured in the cell. Nat Commun 11:4445
99. Roy P (2017) Bluetongue virus structure and assembly. Curr Opin Virol 24:115–123
100. Trask SD, McDonald SM, Patton JT (2012) Structural insights into the coupling of virion assembly and rotavirus replication. Nat Rev Microbiol 10:165–177
101. Coulibaly F, Chevalier C, Gutsche I, Pous J, Navaza J, Bressanelli S, Delmas B, Rey FA (2005) The birnavirus crystal structure reveals structural relationships among icosahedral viruses. Cell 120:761–772
102. Irigoyen N, Garriga D, Navarro A, Verdaguer N, Rodríguez JF, Castón JR (2009) Autoproteolytic activity derived from the infectious bursal disease virus capsid protein. J Biol Chem 284:8064–8072
103. Luque D, Saugar I, Rodríguez JF, Verdaguer N, Garriga D, San Martín C, Velázquez-Muriel JA, Trus BL, Carrascosa JL, Castón JR (2007) Infectious bursal disease virus capsid assembly and maturation by structural rearrangements of a transient molecular switch. J Virol 81:6869–6878
104. Saugar I, Luque D, Oña A, Rodríguez JF, Carrascosa JL, Trus BL, Castón JR (2005) Structural polymorphism of the major capsid protein of a double-stranded RNA virus: an amphipathic alpha helix as a molecular switch. Structure 13:1007–1017
105. Galloux M, Libersou S, Morellet N, Bouaziz S, Da Costa B, Ouldali M, Lepault J, Delmas B (2007) Infectious bursal disease virus, a non-enveloped virus, possesses a capsid-associated peptide that deforms and perforates biological membranes. J Biol Chem 282:20774–20784
106. Wang Y, Duan Y, Han C, Yao S, Qi X, Gao Y, Maier HJ, Britton P, Chen L, Zhang L, Gao L, Gao H, Shen N, Wang J, Wang X (2017) Infectious bursal disease virus subverts autophagic vacuoles to promote viral maturation and release. J Virol 91 (5):e01883–16

107. Magnius L, Taylor J, Mason WS, Sureau C, Deny P, Norder H, Ictv Report C (2018) ICTV virus taxonomy profile: deltavirus. J Gen Virol 99:1565–1566
108. Sureau C, Negro F (2016) The hepatitis delta virus: replication and pathogenesis. J Hepatol 64: S102–S116
109. Asselah T, Rizzetto M (2023) Hepatitis D virus infection. N Engl J Med 389:58–70

Further Reading

Flint J, Racaniello VR, Rall GF, Hatziioannou T, Skalka AM (2020) Principles of virology, vol 1, chapter 13, 5th edn. ASM Press, Washington DC
Koonin EV, Dolja VV, Krupovic M, Varsani A, Wolf YI, Yutin N, Zerbini FM, Kuhn JH (2020) Global organization and proposed megataxonomy of the virus world. Microbiol Mol Biol Rev 84(2):e00061–e00019. https://doi.org/10.1128/MMBR.00061-19
Rossmann MG, Rao VB (2012) Viral molecular machines. Springer, New York
Also especially recommended for further reading are references [5, 85] listed above.

Chapter 15
Virus–Receptor Interactions and Receptor-Mediated Virus Entry into Host Cells

José M. Casasnovas

Abstract The virus particles described in the previous chapters of this book are vehicles that transmit the viral genome and the infection from cell to cell. To initiate the infective cycle, the viral genome must therefore translocate from the viral particle to the cell cytoplasm. Via distinct proteins or motifs in their outermost shell, the particles of animal viruses or bacteriophages attach initially to specific receptors on the host cell surface. These viral receptors thus mediate penetration of the viral genome inside the cell, where the intracellular infective cycle starts. The presence of these receptors on the cell surface is a principal determinant of virus-host tropism. Viruses can use diverse types of molecules to attach to and enter into cells. In addition, virus-receptor recognition can evolve over the course of an infection, and viral variants with distinct receptor-binding specificities and tropism can appear. The identification of viral receptors and the characterization of virus–receptor interactions have been major research goals in virology. In this chapter, we will describe, from a structural perspective, several virus–receptor interactions and the active role of receptor molecules in virus cell entry.

Keywords Virus-host · Virus tropism · Virus attachment · Virus-receptor · Virus structure · Virus entry · Virus neutralization · Capsid dynamics · Uncoating · Membrane penetration · Endocytosis · Cell surface · Cell adhesion · Cell surface molecules · Membrane proteins · Glycoproteins · Carbohydrates · Protein interactions · Molecular recognition · Crystallography · Cryogenic electron microscopy

---

J. M. Casasnovas (✉)
Department of Macromolecular Structure, Centro Nacional de Biotecnología (CNB-CSIC), Madrid, Spain
e-mail: jcasasnovas@cnb.csic.es

M. G. Mateu (ed.), *Structure and Physics of Viruses*, Subcellular Biochemistry 105,
https://doi.org/10.1007/978-3-031-65187-8_15

Abbreviations

AdV	Adenovirus
APN	Aminopeptidase N
CEACAM	Carcinoembryonic antigen cell adhesion molecules
CoV	Coronavirus
CAR	Coxsackievirus-adenovirus receptor
CV	Coxsackievirus
DAF	Decay-accelerating factor
cryo-EM	Cryogenic electron microscopy
DC	Dendritic cells
DC-SIGN	(DC-specific ICAM-3-grabbing nonintegrin)
DSG-2	Desmoglein-2
EchoV	Echovirus
EV	Enterovirus
FMDV	Foot-and-mouth disease virus
g/gp	Glycoprotein
H	Haemagglutinin
HAVcr-1	Hepatitis A virus cellular receptor-1
HIV-1	Human immunodeficiency virus-type 1
hFcRn	Human neonatal Fc receptor
HRV	Human rhinovirus
HS	Heparan sulphate
HSV	Herpes simplex virus
ICAM-1	Intercellular adhesion molecule-1
IgSF	Immunoglobulin superfamily
KREMEN-1	Kringle containing transmembrane protein 1
LDLR	Low-density lipoprotein receptor
MERS-CoV	Middle East respiratory syndrome coronavirus
NDV	Newcastle disease virus
N	Neuraminidase
MCP	Membrane cofactor protein
nAb	Neutralizing antibody
NAG	N-acetyl-glucosamine
PM	Plasma membrane
PV	Poliovirus
PVR	Poliovirus receptor
PRCV	Porcine respiratory coronavirus
PDCoV	Porcine delta coronavirus
SARS-CoV	Severe acute respiratory syndrome coronavirus
SCR	Short consensus repeat
Sia	Sialic acid
Siglec-1	Sialic-acid-binding immunoglobulin-like lectin-1
S	Spike

TGEV	Transmissible gastroenteritis virus
TIM-1	Transmembrane, immunoglobulin, and mucin domain-1
VCAM-1	Vascular cell adhesion molecule-1
VP	Viral capsid protein

15.1 Introduction: Virus Entry into Host Cells, the Recognition of Cell Surface Molecules

The viral particles formed in infected host cells (see Chaps. 11, 12, 13 and 14) are metastable structures that transmit the viral genome and the infection from cell to cell. Viruses must therefore penetrate host cells to initiate the replicative infective cycle by exploiting the cell machinery. In the extracellular transit stage of the viral cycle, animal viruses and bacteriophages attach to specific cell surface molecules (viral receptors) suited for host cell entry following virus-specific entry pathways. Viral receptors must be distinguished from attachment factors, surface molecules to which some viruses can bind but that do not themselves promote virus entry into host cells [1, 2]. Attachment factors concentrate viral particles onto the cell surface and facilitate viruses to encounter specific receptors in the cell surface or in endosomes; they can even mediate conformational changes in viral proteins that increase their affinity for entry receptors [3]. The viral receptor molecules are not just required for initial virus binding to host cells, but also for the transfer of the viral genome through cellular membranes [1]. Virus–receptor interactions can trigger changes in the virus particles that initiate genome translocation, or alterations in the cell, such as signalling events that facilitate virus entry [4]. Viruses enter host cells at the cell *surface* or after endocytosis (Fig. 15.1). Multivalent binding of the virus particles to receptors on the cell surface can mediate uncoating or release of the viral genome in non-enveloped viruses and/or fusion of the virus and cell membranes (Fig. 15.1). Moreover, viruses bound to cell surface molecules can be internalized by following different endocytic pathways [5], where exposure to low pH, proteases, or other cellular factors leads to the delivery of the genome into the cell cytoplasm (Fig. 15.1).

Viruses evolve to recognize specific cell surface receptor molecules appropriate for productive entry and infection of host cells, which frequently determine the host tropism or the cell type a virus can infect. Selection of cell entry receptors by viruses appears to be determined by subtle interactions that regulate the specificity and affinity necessary for efficient cell attachment. Virus–receptor interaction can nonetheless be a highly dynamic process. A single virus can recognize one or several cell entry molecules, which can also differ among viral variants or during the course of an infection [6, 7]. Virus recognition of receptors is under evolutionary pressure to increase their infection efficiency, which can result in the emergence of viruses with altered infectivity or tissue tropism.

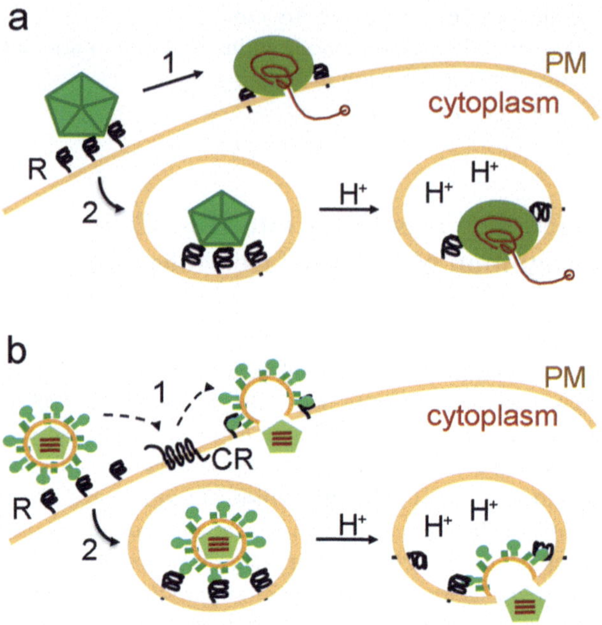

Fig. 15.1 Virus entry into host cells. Scheme illustrating two cell entry pathways by non-enveloped (**a**) or enveloped viruses (**b**). pH-independent entry at the cell surface (pathway 1): Genome uncoating and penetration into cytoplasm (**a**) or virus–cell membrane fusion (**b**) at the plasma membrane (PM). Virus binding to a single receptor (R) or to additional co-receptor molecules (CR) at neutral pH triggers genome translocation. pH-dependent entry in endosomes (pathway 2): Receptor-mediated endocytosis of viruses prior to penetration into the cytoplasm. In the endosomes, viruses firmly bound to multiple receptor molecules are exposed to progressively lower pH, which triggers genome uncoating and penetration (**a**) or fusion (**b**). The entry reaction is mediated by acidification of endosomal compartments, with or without receptor priming

In this chapter, several examples of animal virus–receptor interactions will be presented, together with a description of known models of receptor-switching viruses. The chapter presents a structural view of some virus–receptor interactions that have been characterized by structures of complexes. We will also describe how cell surface receptor molecules participate in the entry process. Membrane penetration events in enveloped viruses are discussed in Chap. 16. Specific aspects of receptor recognition and injection-mediated genome uncoating by some bacteriophages are described in Chap. 17.

15.2 Virus–Receptor Interactions and Receptor Specificity Switch

Viruses recognize a large variety of cell surface molecules, including glycolipids, carbohydrates, and proteins with very distinct folding structures. There are viruses specific for a single receptor molecule, whereas others bind to several structurally distinct receptors for host cell entry [4]. The viral particles use certain proteins in their outermost shell to attach to the cell surface molecules. The multivalent nature of the particles allows viruses to bind with very high avidity to the host cell surface, even though monomeric virus–receptor interactions are usually of moderate affinity (0.1–1 μM) [8].

Some non-enveloped viruses have proteins specialized in attachment to cell surface receptors; however, in other cases unique motifs on the naked capsids are engaged in receptor recognition. Here we discuss both receptor recognition modes, providing a description of adenovirus and picornavirus binding to cell surface molecules. The enveloped viruses carry membrane glycoproteins that bind to the receptor and trigger virus–cell membrane fusion during cell entry. We present the cell surface molecules recognized by human immunodeficiency virus and coronavirus that have been characterized. In the last part of this section, we illustrate the importance of cell surface carbohydrates in virus attachment and infection of host cells.

15.2.1 Virus–Receptor Interactions in Non-enveloped Viruses

In this subsection, we describe receptor recognition by two families of non-enveloped viruses (picornaviruses and adenoviruses) for which virus–receptor interactions have been characterized in detail by structural studies. Picornaviruses use their naked capsid to attach to receptors, whereas adenoviruses have fibres that protrude from the capsid for attachment to several cell surface molecules. Cell entry of other non-enveloped viruses is described elsewhere [9].

Picornaviruses and Their Receptors

The picornaviruses are a large family of non-enveloped viruses responsible for numerous human and animal diseases. Picornavirus particles are formed by an icosahedral protein capsid built by three external viral capsid proteins (VP1 to VP3) and the internal protein VP4 packed inside with a single-stranded RNA genome (Fig. 15.2a). The capsid is composed of 60 basic subunits or protomers arranged as 12 pentamers (see Chap. 11). The members of this virus family bind to distinct types of receptor molecules suited for entry into host cells.

The poliovirus receptor (PVR) was one of the first picornavirus receptors to be characterized [12]. PVR is a type I membrane protein and a member of the immunoglobulin superfamily (IgSF), which has three Ig-like domains (D1 to D3) at the

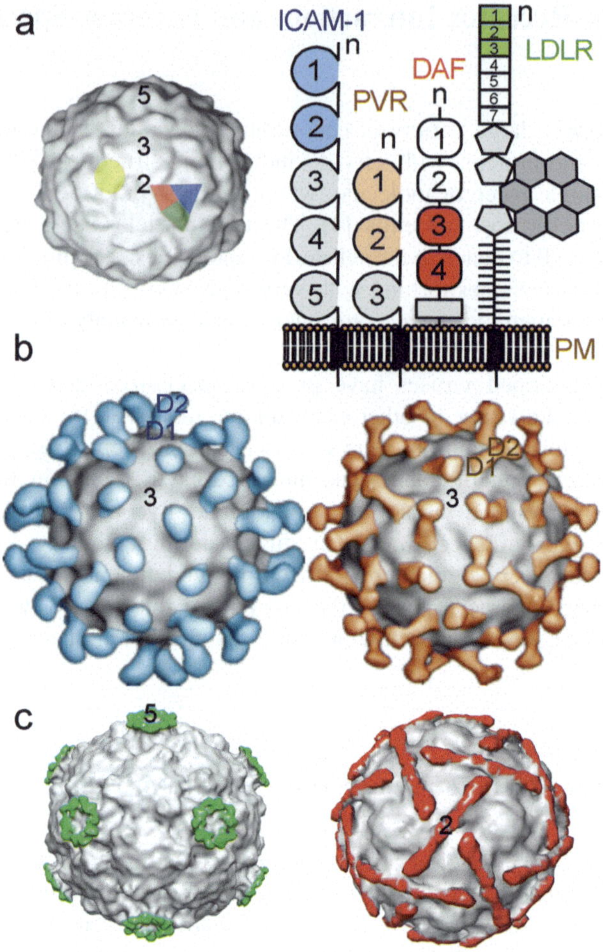

Fig. 15.2 Virus–receptor interactions in picornaviruses. (**a**) Left: Surface representation of the poliovirus (PV) particle computed from the crystal structure of PV1/M (PDB ID 1ASJ). Numbers indicate location of the icosahedral two-, three-, and five-fold axes. The triangle illustrates the approximate location of a capsid building block or protomer, assembled from the external VP1 (blue), VP2 (green), and VP3 (red). The canyon region is marked with a yellow circle. Right: Scheme of picornavirus receptors ICAM-1, PVR, DAF, and LDLR. The N-terminus (n) and extracellular domains are labelled. Receptor domains used to determine structures of virus-receptor complexes shown below are coloured. (**b**) Picornaviruses that use the canyon for receptor recognition. Cryo-EM structure of human rhinovirus HRV3-ICAM-1 (left) and PV1-PVR (right) complexes prepared in solution with purified virus particles and two-domain (D1 and D2) fragments of the receptor molecules. Location of the domains is shown. D1 penetrates the canyon, whereas D2 does not contact the virus and protrudes from the capsid. Images provided by Holland Cheng and Li Xing, adapted from references [10, 11]. (**c**) Cryo-EM structure of HRV2-LDLR (EMD-1049) and EchoV12-DAF (EMD-1057) complexes, representative of picornaviruses that do not use the canyon for receptor binding. Ligand binding repeats 1–3 of LDLR or short consensus repeats (SCR) 3–4 of DAF were used to prepare complexes. Surfaces of the bound receptor fragments are coloured as in (**a**). Images prepared with Chimera (cgl.ucsf.edu/chimera) from cryo-EM maps

extracellular region (Fig. 15.2a). PVR is a member of the nectin or nectin-like subfamily of IgSF receptor, which also function as entry receptors of alphaherpesviruses and measles virus [13–15]. The receptor for most (90%) identified human rhinovirus (HRV) serotypes, the major group of HRV, is intercellular adhesion molecule-1 (ICAM-1) [16, 17], another IgSF member with five Ig-like extracellular domains (Fig. 15.2a). ICAM-1 is also an entry receptor of group A coxsackieviruses (CV-A), whereas other IgSF proteins that function as picornavirus receptors are ICAM-5 [18], the coxsackievirus and adenovirus receptor (CAR) [19], the Hepatitis A virus cellular receptor-1(HAVcr-1 or TIM-1) [20], and vascular cell adhesion molecule-1 (VCAM-1) [21].

PV and HRV bind similarly to the N-terminal membrane-distal domains of their receptor molecules (Fig. 15.2b) [2]. Both viruses use a depressed surface, or canyon, formed by two neighbouring protomers around the five-fold icosahedral vertices of the capsid (Fig. 15.2a and b). The use of the canyon for binding to cell surface receptors was also described for other picornaviruses such as CV-A21, CV-A24, CV-A10, or CV-B3, echoviruses (EchoV), or group B enterovirus (EV-B) [22–26]. CV-A21 or CV-A24 binds to ICAM-1, some CV-A to KREMEN-1, CV-B3 to CAR, and EchoV1 to the $\alpha_2\beta_1$ integrin, whereas other EchoV as well as EV-B attach to the human neonatal Fc receptor (hFcRn). Thus, the canyon is a well-defined and broadly used receptor-binding region in many picornaviruses [22]. The concave nature of the canyon was thought to be suited for hiding conserved receptor-binding residues from neutralizing antibodies (nAbs), which cannot penetrate the depressed surface as efficiently as relatively narrower receptor molecules. This hypothesis was nonetheless challenged by studies showing that nAbs penetrate the canyon [27]. Although the nAbs interact mostly with the walls rather than the bottom of the canyon, the antibody binding epitopes overlapped with receptor-binding sites in both HRV and PV. Receptor binding to the canyon region commonly triggers the capsid uncoating required for cell entry, such as it will be discussed in another section. Cryogenic electron microscopy (cryo-EM, see Chap. 3) studies showed how diverse receptor molecules bind to the canyon. Structural and binding studies with PV and HRV identified differences in the way the viruses engage their IgSF receptors, and they demonstrate that PV uses more exposed residues in the canyon wall than HRV [10, 22]. Recently, progress in the cryo-EM methods allowed determination of high-resolution structures of virus-receptor complexes, which provided a detailed view of the interactions, identified key virus-receptor binding motifs, and revealed new insights on picornavirus cell entry [24, 25].

Other picornaviruses such as foot-and-mouth disease virus (FMDV), some EV or HRV, do not use the canyon for attachment to cell entry receptors, but they recognize receptor molecules with other more exposed capsid regions. FMDV and EV exploit an RGD motif in the VP1 GH flexible loop to bind to integrin receptors [28, 29]. This receptor-binding motif is accessible and is a major antigenic site; however, these viruses escape from immune neutralization by mutating some residues adjacent to the RGD, which prevents detection by antibodies while preserving cell attachment activity and viability. Approximately 10% of HRV serotypes, the minor group, do not bind to ICAM-1 and use members of the low-density lipoprotein receptor

(LDLR) family to enter host cells (Fig. 15.2a) [30]. Even though minor and major group HRV are closely related, the mode by which they bind their respective receptors is strikingly different (Fig. 15.2b and c). Minor group HRV do not use the canyon and contact the receptor through a protruding region close to the capsid five-fold axis (Fig. 15.2c, left) [31]. This group of HRV bear a conserved Lys residue in VP1 that contacts an acidic cluster and a Trp residue in the N-terminal ligand binding domains of the LDLR proteins. This lysine is absent in major group HRV that bind to ICAM-1 and is therefore likely to be a major determinant of the distinct receptor-binding specificity described for rhinoviruses.

A picornavirus–receptor interaction distinct from those described above was reported for EchoV that attach to decay-accelerating factor (DAF, CD55), a complement control protein [32]. These viruses bind to DAF through a capsid protrusion at the southern rim of the canyon, distant from the five-fold axis and near the capsid two-fold axis (Fig. 15.2c, right) [33, 34]. The receptor-binding region in EchoV11 protrudes and is more exposed than the canyon. The kinetic association rate for the monomeric EchoV11-CD55 interaction is thus one to two orders of magnitude higher than those shown for picornaviruses that use the canyon for receptor recognition, which probably reflects the distinct nature of the virus–receptor interactions [35]. EchoV binding to DAF does not trigger capsid uncoating, but these viruses use the canyon to engage the hFcRn, which releases the genome upon cell entry [25, 36]. Nonetheless, DAF plays a key role for EV transport through epithelium, as virus binding to DAF on the apical cell surface triggers signalling events that facilitate virus interaction with CAR at the tight junctions, which initiate cell infection [37].

The structural studies of picornavirus-receptor complexes discussed here show that viruses of this family can use diverse capsid regions for efficient recognition of cell surface molecules and for productive entry into host cells.

Adenoviruses, Non-enveloped Viruses that Bind to Several Cell Surface Molecules

Adenoviruses (AdV) are non-enveloped viruses with icosahedral capsids built by hexon and penton capsomers (see Chap. 2 for definition of capsomers and Chap. 12 for additional information and references). The pentons locate at each of the 12 icosahedral vertices of the capsid and form the base of protruding trimeric fibres (Fig. 15.3a). The penton base and the associated fibres form a complex that recognizes cell surface receptor molecules [38, 39]. The most capsid-distal, globular region of the fibre, the knob (Fig. 15.3a), mediates initial AdV attachment to the cell, whereas the penton base is used for subsequent tight attachment of the virus to a secondary receptor that mediates virus internalization and host cell entry [38]. Receptor-bound AdV are transported to endosomes, where acidification triggers capsid disassembly and virion penetration into the cytoplasm.

Cell tropism and receptor recognition is relatively diverse among AdV. This group of non-enveloped viruses can use at least four distinct types of receptor molecules for initial attachment to the cell surface through the fibre proteins [39], which is mediated by the most distal trimeric knob structure (Fig. 15.3).

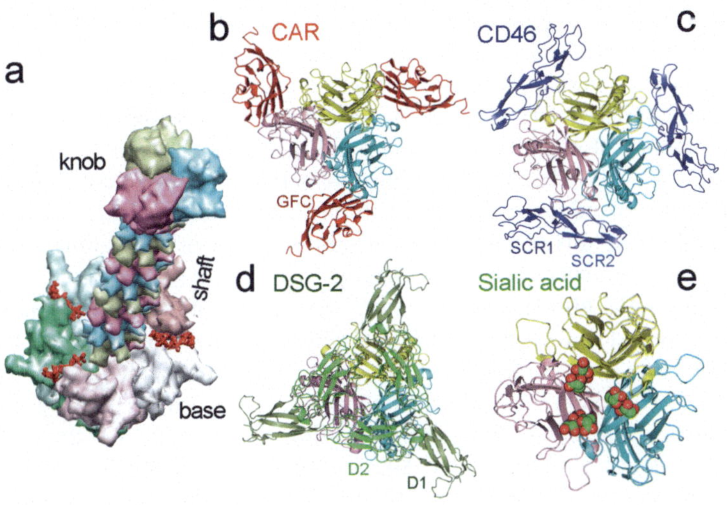

Fig. 15.3 Virus–receptor interactions in adenoviruses. Representation of the receptor-binding adenovirus fibre and penton base (**a**) (see also Chap. 12), and crystal structures of the trimeric AdV knob in complex with CAR (**b**), CD46 (**c**), DSG-2 (**d**), or sialic acid (**e**). Ribbon drawings are shown for the complexes prepared with the N-terminal domain (D1) of CAR (red) and AdV12 (PDB ID 1KAC), with the two N-terminal SCR1-2 of CD46 (blue) and AdV11 (PDB ID 2O39) or with D1 and D2 of DSG-2 and AdV7 (PDB ID 7AGF). The AdV-binding GFC β-sheet of the CAR domain is indicated. Sialic acids in complex with the AdV37 knob (PDB ID 1UXA) are shown with carbons (green) and oxygens (red) as spheres. Figure (**a**) was provided by Carmen San Martín and figures (**b**–**d**) were prepared with PyMOL (pymol.org)

Subgroups A, C, D, E, and F attach to the CAR receptor, a member of the IgSF found in epithelial tight junctions. CAR has two Ig-like domains at the extracellular region and mediates homotypic cell–cell interactions through the same face (GFC β-sheet) of the N-terminal domain engaged by the AdV fibres (Fig. 15.3b) [40, 41]. Fibre binding to CAR interferes with homotypic cell–cell adhesion, which destabilizes epithelial cell layers and facilitates virus release to the airway lumen and spreading to a new host [42]. Most subgroup B AdV use the ubiquitous membrane cofactor protein (MCP, CD46) for initial attachment to the host cell [38]. The AdV knob contacts the two N-terminal short consensus repeats (SCR) of the CD46 molecule (Fig. 15.3c) [43]. Although all AdV use the fibre knob for receptor binding, the mode of contact of the knobs with the CAR or the CD46 proteins is distinct (Fig. 15.3). The switch in receptor-binding specificity between CAR and CD46-binding AdV appears to be related to the distinct conformation of exposed loops in the periphery of the knob. CD46-binding AdV have an extended knob AB-loop that prevents binding to the CAR receptor, as well as specific structural features in the loops that contact the CD46 molecule [7]. Recently, it was shown that subgroup B human AdV can use the hexon trimer instead of knob to attach to CD46 and penetrate into host cells, which revealed a new cell entry mechanism in AdV [44]. Certain subgroup B Ad (3, 7, and 14) attach to desmoglein-2 (DSG-2) [45], a member of the cadherin family. Two Ad knob monomers attach to consecutive DSG-2 domains (Fig. 15.3d) [46]. Certain AdV can also attach to cell surface carbohydrates, such as heparin or sialic acid [38, 47, 48]. Subgroup D adenoviruses 8, 19, and 37, which recognize sialic acid, are responsible for epidemic keratoconjunctivitis, a very contagious ocular disease. The sialic acid binding site has been mapped also at the fibre knob, and it locates at the centre of the trimer (Fig. 15.3e) [47, 48]. The AdV–receptor interaction described here illustrates how different AdV evolved to use distinct knob surfaces to attach to different cell surface receptor molecules; this certainly translates in cell tropism and pathogenesis diversity among the members of this virus family.

Following initial fibre attachment to cell surface receptors, the AdV particles are internalized by endocytosis into endosomal compartments. This process is facilitated by the interaction of the AdV penton base with integrin molecules (αvβ3, αvβ5, and αvβ1) [49]. Each penton base protein contains a loop that carries an integrin-binding RGD sequence and allows multimeric binding to integrins, which mediates its clustering, triggering intracellular signals, actin rearrangement, and clathrin-mediated endocytosis of the virions [38, 39]. During endocytosis, the virion disassembles and provokes the lysis of the endosomal membrane for the genome transfer to the cytoplasm and its subsequent translocation to the nucleus. AdV are thus an example of non-enveloped viruses that use two distinct capsid structures to bind different cell surface receptor molecules for attachment to or penetration of host cells.

15.2.2 Virus–Receptor Interactions in Enveloped Viruses

Enveloped viruses bear membrane-bound glycoproteins specialized in the recognition of cell surface molecules and in subsequent fusion of the viral and cell membranes. The receptor-binding and fusion proteins can be one or two distinct polypeptides, in many cases resulting from the cleavage of a single precursor protein. Both of these proteins associate in the viral envelope, forming the envelope spikes. Among enveloped viruses, the fusion proteins are less diverse than receptor-binding proteins and will be presented in Chap. 16. Here we will discuss the interactions of some viral envelope proteins with cell surface receptor molecules, described by structures of virus-receptor complexes.

Human Immunodeficiency Virus and Its Receptors

Attachment of the human immunodeficiency virus-type 1 (HIV-1) to cell entry receptors and subsequent fusion have been described in great detail. The HIV-1 envelope bears two non-covalently bound viral proteins, gp120 and gp41, which associate to form the trimeric envelope glycoprotein (Env) (Fig. 15.4) [50, 51]. There are around 15 Env anchored to the HIV-1 membrane [52]. The gp120 glycoprotein mediates attachment of the HIV-1 particles to cell surface receptors, whereas the membrane-proximal gp41 completes the fusion of the virus with the cell membrane, resulting in the release of the nucleoprotein into the cell cytoplasm [53, 54] (see

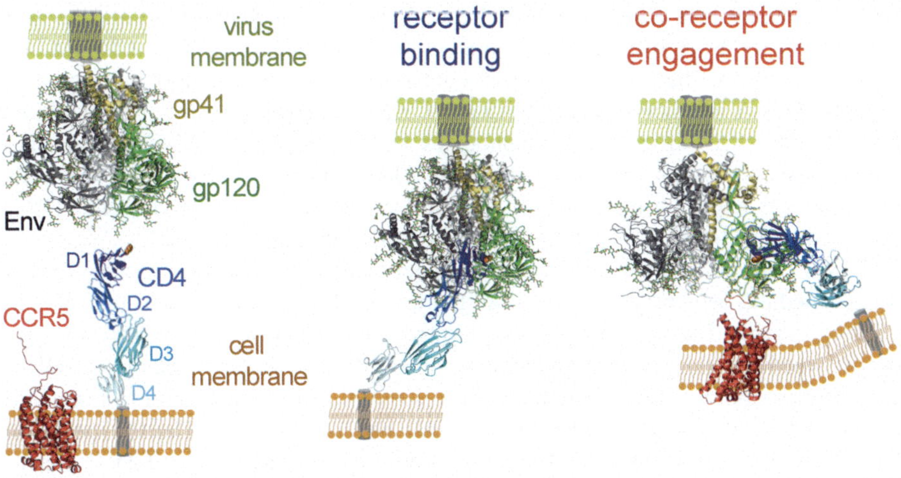

Fig. 15.4 The pathway of HIV-1 receptor recognition during cell entry. Left. Representations of the prefusion HIV-1 Env trimer (PDB ID 5FUU) on the viral membrane, the extracellular portion (D1 to D4, blue or cyan) of the primary CD4 receptor (PDB ID 1WIO) and the CCR5 co-receptor (PDB ID 5UIW) (red) on the target cell membrane. One gp120-gp41 heteromonomer in the trimer is shown in green and yellow, respectively, the other two in grey. The side chain of the key gp120-binding Phe43 in the CD4 D1 tip is shown as orange spheres. Binding of gp120 to the primary CD4 receptor (based on PDB ID 4R2G) (middle) primes gp120 engagement of the CCR5 co-receptor (PDB ID 6MET) (right), which stabilizes an open Env trimer (PDB ID 5VN3) and facilitates virus–cell membrane fusion. Structure representations prepared with PyMOL (pymol.org)

Chap. 16). The HIV-1 entry process is relatively complex [55]. The Env requires engagement of two distinct cell surface molecules to trigger membrane fusion, which occurs at the cell plasma membrane [56]. gp120 initially engages the lymphocyte-specific CD4 (Fig. 15.4), an IgSF cell surface protein. CD4 is a type I membrane protein composed of four Ig-like domains in the extracellular region that forms part of the T cell receptor complex in a subset of T lymphocytes (Fig. 15.4). A depressed groove or pocket in gp120 engages the CD4 N-terminal Ig-like domain (D1) with high affinity [57]. A Phe residue protruding in the most membrane-distal CD4 D1 penetrates into a gp120 narrow pocket (Fig. 15.4). The CD4-binding site is surrounded by long hypervariable loops and glycans, which further hide the depressed site from immune surveillance [58]. Nonetheless, the CD4 receptor-binding site is a major antigenic site in the HIV-1 env and several broadly neutralizing antibodies have been identified from individuals termed elite neutralizing, and some of them entered in clinical trials [59].

CD4 binding triggers local conformational changes in the HIV-1gp120 protein that expose a conserved structure (bridging sheet) that binds to a second molecule termed co-receptor (Fig. 15.4) [57]. The gp120 V3 loop also interacts with the co-receptor [55]. It is proposed that engagement of the co-receptor molecule stabilizes a CD4-bound open conformation of the Env (Fig. 15.4), which primes the virus–cell membrane fusion for the release of the viral nucleoprotein into the host cell cytoplasm. Two distinct HIV-1 co-receptors have been identified, CXCR4 and CCR5, which are chemokine receptors containing seven transmembrane segments [55, 60]; HIV-1 interaction with CCR5 and CXCR4 is strain-specific. HIV-1 strains using CCR5 (R5 viruses) infect monocytes and macrophages (M-tropic) and are mainly associated with sexual transmission. They can also evolve to infect T cells (T-tropic) by acquiring the ability to use the CXCR4 receptor (X4 viruses). HIV-1 viruses that use both co-receptors are termed R5X4 viruses.

HIV-1 particles are also able to attach specifically to cell surface molecules that do not mediate cell entry and infection, but are used instead to present the virus to CD4-expressing T cells. This process, termed trans-infection, is very efficient and occurs when antigen-presenting dendritic cells (DC) encounter T cells [61]. HIV-1 can therefore bind to DC in the periphery, which transports the virus to lymphoid organs, where it is transferred to T cells that become infected. The C-type lectin DC-SIGN (DC-specific ICAM-3-grabbing nonintegrin) was the first HIV-1 attachment factor in DC to be characterized as a mediator of trans-infection [62]. DC-SIGN is specific for simple high-mannose glycans and fucose-containing glycosylations [63, 64]. The Env is heavily glycosylated, and its glycan composition is largely of the high-mannose type, which is recognized by DC-SIGN. The glycan shield of the HIV-1 spike is thus used for attachment to DC that trans-infects T cells. In addition to the spike glycosylation, HIV-1 also uses charged glycans in its envelope lipidome to attach to DC [65, 66]. The HIV-1 envelope membrane is enriched in certain glycosphingolipids called gangliosides. A specific sialic acid-containing subset of gangliosides is recognized by sialic-acid-binding immunoglobulin-like lectin-1 (Siglec-1) on the DC surface and mediates HIV-1 trans-infection [67, 68]. HIV-1 thus exploits different envelope components to attach to cell surface molecules that facilitate the spread of the infection to lymphoid tissues.

Coronaviruses and Their Receptors
The coronaviruses (CoVs) are a large family of enveloped, positive single-stranded RNA viruses divided into four genera, alpha, beta, gamma, and delta, which are involved in respiratory, enteric, hepatic, and neuronal infectious diseases in animals and in man that cause important economic losses and high mortality rates in human infections [69–72]. The recent beta-CoV severe acute respiratory syndrome coronavirus 2 (SARS-CoV-2) outbreak resulted in the COVID-19 pandemic [73], which has caused about seven million deaths. This century, in addition to SARS-CoV-2, we have supported two zoonotic disease outbreaks caused by beta-CoVs, the SARS-CoV in 2003 and the Middle East respiratory syndrome coronavirus (MERS-CoV) a decade later [72], which were associated with cross-species transmission of CoV due to its adaptation to orthologous-receptor molecules for cell infection [74–76].

CoV have a major envelope glycoprotein, the spike (S), which is responsible for CoV cell entry and interspecies transmission [77]. Mutations in the S increase the affinity for orthologous-receptor molecules in other species, facilitating human infection and leading to recurrent zoonotic events. The S is displayed as a peplomer in the CoV envelope [78] and mediates viral attachment to cell surface receptor molecules and viral cell entry. It assembles into homotrimers whose architecture has been revealed by cryo-EM at high resolution [79–83]. The S protein is heavily glycosylated, and it contains a membrane-distal N-terminal S1 portion followed by the S2 region and a single transmembrane segment (Fig. 15.5a). Numerous N-linked glycans shield the S protein from Ab binding and facilitate immune escape [85]. The S1 contains the receptor-binding determinants, whereas the S2 mediates virus–cell fusion and membrane penetration. Like the class I fusion proteins, the S2 region consists mainly of α-helices and is followed by the transmembrane domain; it carries the fusion peptide, a central helix in the trimer, and two conserved heptad-repeat regions, HR1 (N-terminal) and HR2 (C-terminal) (Fig. 15.5a), which form a long three-stranded coiled-coil that projects the fusion peptides to the target cell membrane during the fusion reaction [86]. As in other enveloped viruses, the initiation of the fusion requires partial disassembly of the trimeric spikes and the exposure of the fusion peptide for binding to the host cell membrane [87]. Proteases that cleave the S prime the fusion after attachment to cell surface receptors, either on the cell surface or in endosomes [88, 89]. In addition, some CoV, such as MERS-CoV or SARS-CoV-2, contain a furin recognition site at the N-terminal section of S2 (Fig. 15.5a) where furin cleaves after particle assembly and generates a metastable virion with increased infectivity and extended viral tropism [90, 91].

The S1 region is variable in sequence and length and is specialized for recognition of cell surface receptors [92, 93]. It adopts a V shape that facilitates linkage between the monomers of the trimer and it is composed of several discrete modules or domains such as the N-terminal domain (NTD) and the canonical receptor-binding domain (RBD) in the C-terminal portion of the S1 (Fig. 15.5a). The NTD adopts a galectin-like fold and may function as a receptor-binding domain in some CoV [93]: Mouse Hepatitis Virus (MHV), bovine CoV (bCoV), or human CoV (hCoV-OC43 or hCoV-HKU1) [94]. MHV uses the NTD to bind to the carcinoembryonic antigen cell adhesion molecules (CEACAM) for cell entry [95], whereas bCoV or hCoV-OC43 NTDs bind to sialic acids (9-*O*-Ac-Sia) on the cell surface [94]. A duplicated NTD is located at the N-terminal part of the S in some CoV (hCoV-NL63, hCoV-2018, and TGEV) [96], and it can broaden the tissue tropism of the virus [97].

In most CoV, the major cell tropism determinants map to the RBD in the C-terminal portion of the S1 region [92]. The RBD is an S fragment of about 150 residues that can usually fold independently of the rest of the S, and it can be expressed as a single domain with all receptor-binding determinants [98–102]. The CoV RBD can adopt a closed and an open conformation in the S (Fig. 15.5a), and its open form is primed to attach to cell surface ectoenzymes such as aminopeptidase N (APN), human angiotensin-converting enzyme 2 (ACE2), or dipeptidyl peptidase 4 (DPP4, CD26), which are known CoV cell entry receptors [103–106]. APN, ACE2, and DPP4 are dimeric membrane-bound ectoenzymes that catalyse the hydrolysis of short peptides and have multiple functions such as angiogenesis, cell adhesion, and blood pressure regulation [107–109]. The reason why CoVs use ectoenzymes as entry receptors is unclear; their peptidase function is not required for CoV cell entry and virus-binding regions are distant from their catalytic sites [99, 103–105, 110–112]. Nonetheless, APN dynamics required for peptide

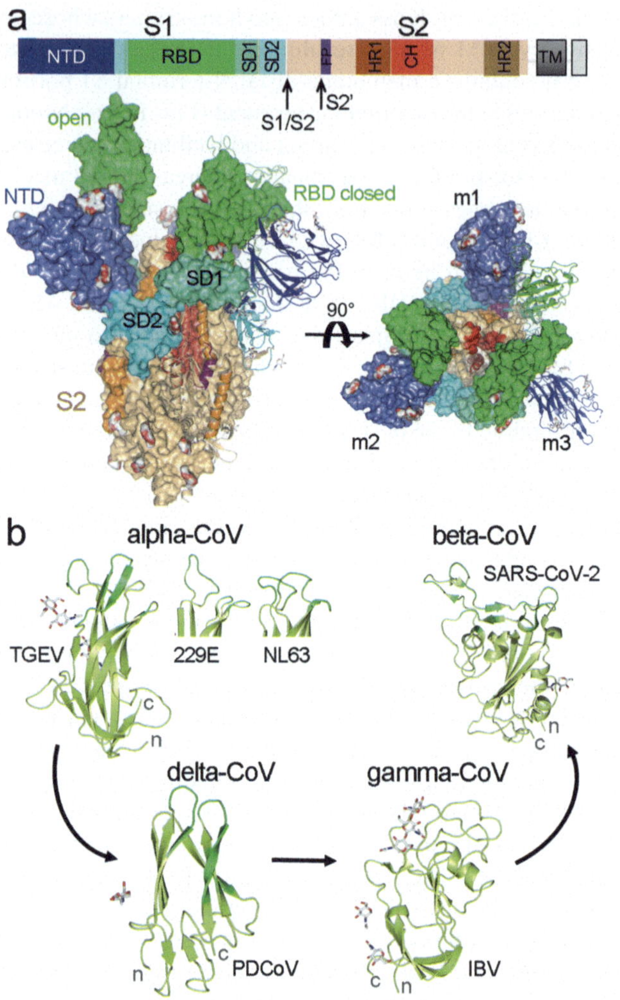

Fig. 15.5 The CoV envelope spike and receptor-binding domains. (**a**) Scheme of the S, with the domains in the S1 region and main S2 motifs coloured: N-terminal domain (NTD), receptor-binding domain (RBD), subdomains 1 and 2 (SD1, SD2), fusion peptide (FP), heptad-repeat 1 and 2 (HR1 and HR2) and the central helix (CH). Transmembrane domain (TM) and cytoplasmic tail are shown at the C-terminal portion of the S as grey boxes. The furin cleavage site at the S1 and S2 junction (S1/S2) in certain CoV such as SARS-CoV-2 and the S2 protease cutting site upstream the FP are indicated with arrows. Lateral and frontal views of the cryo-EM structure of a soluble trimeric S in the prefusion form and with one RBD closed and two RBDs in the open conformation are shown below the scheme (PDB ID 7R4I) [84]. One of the S monomers with the open RBD (m1) and the monomer with the closed RBD (m2) are represented as surfaces with the domains coloured as in (**a**), whereas m3 is shown as ribbons. S1 domains are labelled in m2. The N-linked glycans are shown with carbons in grey, nitrogens blue, and oxygens red. (**b**) Evolution of CoV RBDs and receptor-binding motifs. Ribbon drawing of the indicated CoV RBDs, representative of the four CoV genera: Alpha-CoV, transmissible gastroenteritis virus TGEV (PDB ID 4F2M), hCoV-229E (PDB ID 6U7G) and hCoV-NL63 (PDB ID 3KBH); beta-CoV SARS-CoV-2 (PDB ID 6LZG);

hydrolysis are important for certain CoV binding, and they specifically recognize the open ectodomain form [113].

The sequence and structure of the RBD varies considerably among CoV, and they recognize different receptors. In alpha-CoV, the RBD adopts a β-barrel fold with two highly twisted β-sheets, three disulphide bonds and N-linked glycans on one side of the β-barrel (Fig. 15.5b). The N- and C-terminal ends of the RBD are located on the same side of the domain; on the opposite side, the tip of the barrel has distinct conformations among alpha-CoV due to the variability of the loops forming the receptor-binding motif (RBM) (Fig. 15.5b), which determines different receptor-binding modes. In transmissible gastroenteritis virus (TGEV) or porcine respiratory CoV (PRCV), the RBD tip contains two protruding β-turns (β1-β2 and β3-β4) (Fig. 15.5b), each bearing a solvent-exposed aromatic residue (tyrosine or tryptophan) that penetrates into small cavities of its APN receptor (Fig. 15.6a) [104], whereas in hCoV-229E the RBM adopts a different conformation that engages a different APN site on its DII [114]. The recessed hCoV-NL63 RBM likely determined a switch in the receptor-binding specificity of this alpha-CoV, which recognizes ACE2 instead of APN. The RBM is the major antigenic site in alpha-CoV [104, 114], and this immune pressure likely caused the evolution of the RBM conformation and the different receptor recognition modes identified among alpha genus CoV. Delta and gamma-CoV retained the alpha-CoV RBD fold, although the RBD of infectious bronchitis coronavirus (IBV), gamma-CoV, contains short β-strands and long loops (Fig. 15.5b) [115]. Like alpha-CoV, porcine delta-CoV (PDCoV) also recognizes APN, but binds to a distinct, although overlapping site with that of PRCV and hCoV-229E [116].

The beta-CoV RBD lost the β-barrel fold and they are composed of two subdomains [115], the core, with a central five-stranded β-sheet, and the external subdomain or RBM, which is connected to two adjacent β-strand of the core (Fig. 15.5b). The RBM are structurally variable among beta-CoV and they can accommodate multiple mutations that cause changes in their receptor-binding affinities as well as escape from Ab neutralization [117]. The SARS-CoV-2 RBD, particularly in the omicron variants, accommodates multiple mutations in the RBM and the core, some of which enhance binding to ACE2, but prevent Ab recognition [118]. Substitutions in the RBD of SARS-CoV are also known to facilitate transmission of the virus from bats to palm civets and subsequently to humans, associated with a few RBD residue substitutions (Lys479/Asn and Ser487/Thr) that increase its affinity for human ACE2 [119]. The RBD of SARS-CoV and SARS-CoV-2 similarly attach to the ACE2 ectodomain for cell entry, binding to the

Fig. 15.5 (continued) delta-CoV PDCoV (PDB ID 7VPQ); gamma-CoV infectious bronchitis virus (IBV) (PDB ID 6CV0). The receptor-binding interfaces identified by the PISA server are shown in green and the rest of the RBD in light green, with the N- and C-terminal ends labelled. Only the RBD tips of the hCoV-229E and hCoV-NL63 domains, which bind to APN and ACE2, respectively, are visible. The N-linked glycans are shown as sticks. Presumed RBD evolution is indicated with arrows. Prepared with PyMOL (pymol.org)

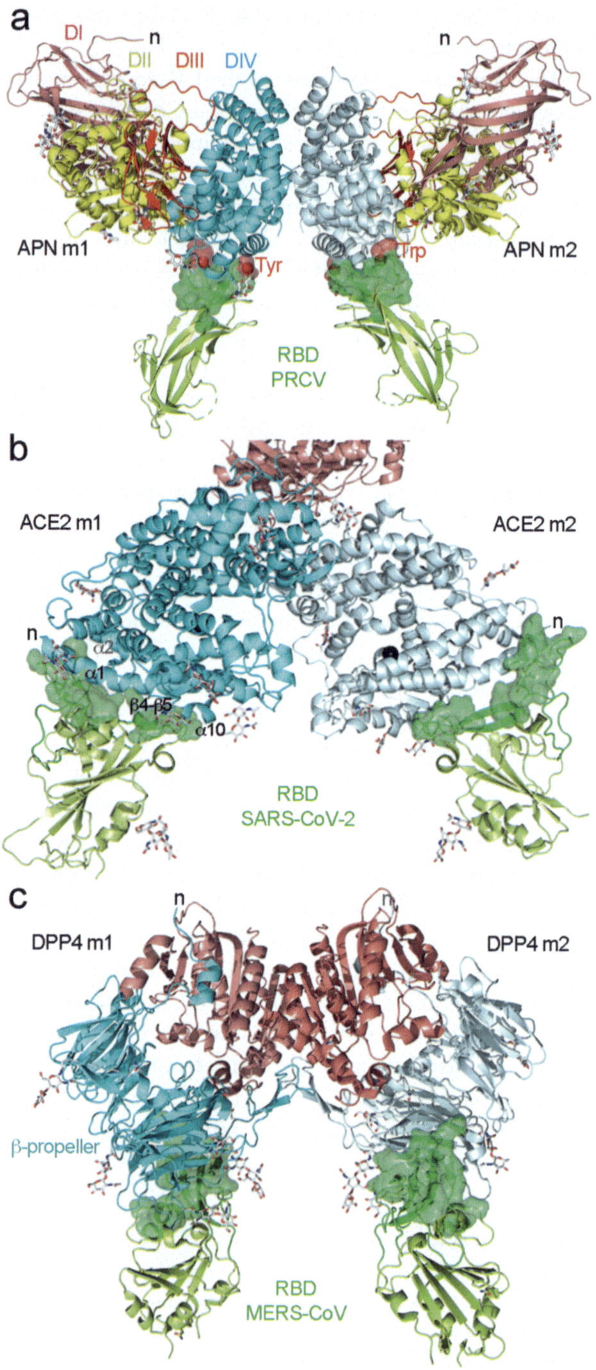

Fig. 15.6 Receptor recognition in coronavirus. Structures of CoV RBD bound to the dimeric APN (PDB ID 4F5C) (**a**), ACE2 (PDB ID 6 M17) (**b**), or DPP4 (PDB ID 4KRO) (**c**). The PRCV, SARS-CoV-2, and MERS-CoV RBDs are shown with their receptor-binding surfaces in green and the rest of the domain in light green. The APN monomers (m) are shown with domains in salmon (N-terminal DI), yellow (DII), red (DIII), and cyan or light cyan (C-terminal DIV), as well as the N-terminal ends near the cell membrane. In the PRCV RBD, two key residues (Tyr and Trp) binding to APN are shown in red. In panel (**b**), the virus binding, N-terminal peptidase domain (cyan), and a fragment of the C-terminal collectrin-like domain (salmon) of the full length dimeric ACE2 receptor are visible, with the main regions recognized by the RBD labelled. The DPP4 monomers in (**c**) are shown with the N-terminal virus binding β-propeller domain in cyan and the C-terminal hydrolase domain in salmon. The N-linked glycans are shown as sticks, with carbons in grey, oxygens red, and nitrogens blue in all the panels. Receptors N-terminus are labelled and the metal ions in their catalytic sites are black sphere. Prepared with PyMOL (pymol.org)

most membrane-distal region of the ectodomain with the RBM (Fig. 15.6b) [102]. The concave surface of their RBM valleys holds the ACE2 N-terminal α-helix, the RBM ridge binds to α2 and the terminal side of the subdomain interacts with the ACE2 β4-β5 turn and α10. There are certain virus-receptor contacts that are conserved between the two SARS-CoVs, but the SARS-CoV-2 bind to ACE2 with higher affinity due to a more elaborate docking. The position of N-linked glycans in the ACE2 is also an important determinant of the receptor-binding species specificity of SARS-CoVs.

The MERS-CoV RBD is structurally similar to the SARS-CoV RBDs, but it uses a different cell entry receptor, DPP4 [105]. Nonetheless, the MERS-CoV external subdomain or RBM is relatively more distinct from that of SARS-CoV than the RBD core, and it contains a four-stranded β-sheet (Fig. 15.6c). The RBM attaches to the most membrane-distal region of the eight-bladed β-propeller of the DPP4 ectodomain, and one RBD binds to each of the DPP4 monomers in the dimer (Fig. 15.6c) [111, 112]. The bound RBD does not appear to interfere with the catalytic activity of DPP4, binding only to the β-propeller domain and away from the regions where the substrate accesses the active site. The RBM contacts β-propeller blades IV and V, including N-linked carbohydrates at blade IV and a helix on the linker between the two blades, which access a concave surface or small "canyon" in the RBM [120]. Interestingly, MERS-CoV recognizes several glycans N-linked to its receptor, such as in the binding of PRCV to APN, where a glycan residue has a key contribution to the virus–receptor interaction [104, 114].

The RBD is the major antigenic site in the CoV S, and the target of nAbs [104, 121]. Numerous nAbs that bind the RBD have been described for alpha and beta-CoV. In alpha-CoV, the loops that configure the RBM are the target of nAbs and this immune pressure induced the conformational evolution of the RBM (Fig. 15.5b), leading to different receptor-binding modes as well as to the switch in receptor-binding specificity in hCoV-NL63 [104, 114]. Numerous Abs against the SARS-CoV-2 RBD were identified during the COVID-19 pandemic, many of them derived from convalescents and others from animals immunized with the S or RBD [122, 123]. The most potent nAbs recognize the RBD and few of them bind to an NTD site close to the RBD [124]. S2-specific Abs show low neutralizing potency [125].

Structures of the SARS-CoV-2 RBD in complex with nAbs have clustered them into at least five distinct subsites [126, 127]. Similar to the way the ACE2 receptor interacts with the RBD, class I and class II Abs lay on top of the RBM. Nonetheless, class I Abs bind exclusively to the RBD in the open conformation, whereas the class II binds to both the open and closed RBD in the S. Class III and IV Abs recognize the outer and inner faces of the RBD, respectively. Recently, several nAbs that recognize numerous SARS-CoV-2 variants or even several members of the sarbecovirus group bind to the RBD front region or site V, which is hidden in the closed conformation of the RBD [127]. Although this site is distant from the ACE2 binding site, some of these Abs can neutralize multiple SARS-CoVs. Abs targeting the RBD core such as class III, IV, and V exhibit a wider neutralization range among sarbecoviruses, albeit some with low neutralizing potency, whereas class I and II, which recognize receptor-binding motifs, have high potency and narrow breadth. The identification of nAb epitopes is important for the design of improved vaccine antigens that elicit focused immune responses against conserved viral epitopes and can be used in future CoV outbreaks [128, 129].

15.2.3 Carbohydrates as Viral Receptors

The cell surface displays a large variety of oligosaccharide determinants linked to glycoproteins, proteoglycans, and glycolipids, some of which are used by viruses to attach to host cells [2, 130]. Most viral carbohydrate receptors are negatively charged and terminate the glycan moiety, features important for virus recognition. Nonetheless, some viruses recognize neutral glycans such as histo-blood group antigens (HBGA), glycoconjugates present in mucosal secretions, epithelia and on red blood cells. Monomeric virus–carbohydrate interactions are usually of low affinity (mM); however, viral particles have many recognition sites that easily engage multiple carbohydrate molecules, and thus attach to the cell surface with high avidity (Fig. 15.7a). Virus recognition of carbohydrates is associated with functions other than cell attachment. Orthomyxovirus, paramyxovirus, and CoV express envelope sialic acid-destroying glycoproteins (sialidases and esterases) that are essential for in vivo host infection. These enzymes can prevent reattachment of newly released viruses from infected cells, can remove cell-bound viruses that fail to enter host cells, and can inhibit virus aggregation during budding. Sialidase inhibitors have proven to be useful antiviral drugs for the treatment of some viral infections.

Sialic acid (Sia) residues linked to glycoproteins and glycolipids act as receptors for many viruses. Sias are derived from N-acetyl-neuraminic acid and mainly occupy the terminal position of a glycan chain, attached to a penultimate galactose by a (2–3) or (2–6) linkage [132], making them easily accessible. Moreover, Sias have a larger number of functional groups than other monosaccharides and can thus participate in a network of polar and non-polar interactions with viral proteins (Fig. 15.7b) [131]. They are receptors for several viruses, such as influenza, corona, paramyxo, toro, adeno, noro, rota, picorna, parvo, polyoma, and reovirus, some of which are important human pathogens [2]. Crystal structures of virus-Sia complexes showed that viruses use relatively recessed binding surfaces that interact mainly with the face bearing a negatively charged carboxylate group (Fig. 15.7b, c and d) [131]. The mode of Sia recognition by different viruses can be similar, although viruses use structurally diverse proteins for receptor recognition [132]. In some cases, however, related viruses such as murine polyomavirus and SV40 use similar capsid surfaces to recognize distinct motifs in the Sia molecules [7]. The virus-glycan contacts are therefore not necessarily conserved (Fig. 15.7c and d).

Viral Sia receptors feature numerous cell-specific modifications that determine viral tropism, cell-to-cell transmission and pathogenicity. There are several examples of viruses responsible for serious diseases, such as adenovirus (AdV8, AdV19, AdV37) and enterovirus, which specifically bind to the α(2–3)-linked Sia. The use of

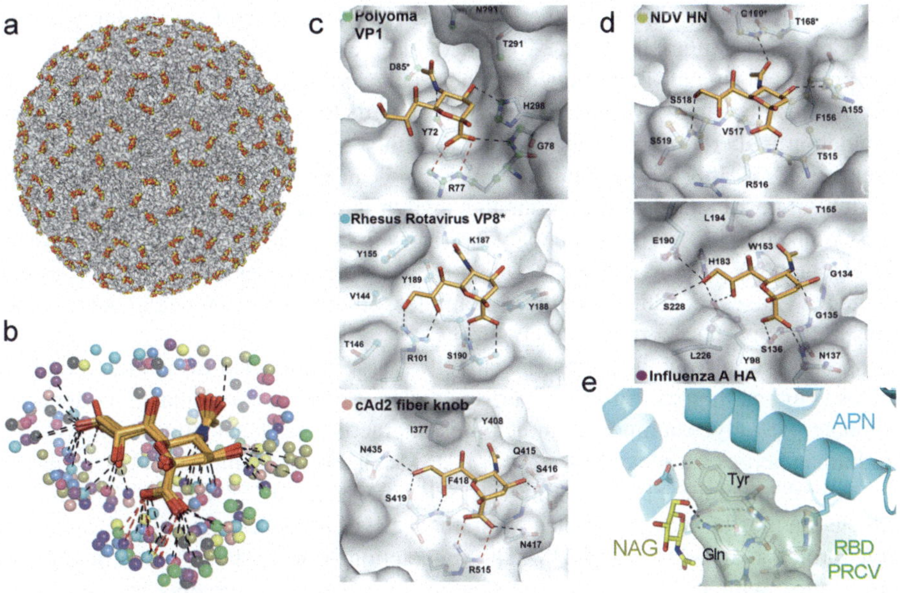

Fig. 15.7 Carbohydrates as viral receptors. (**a**) Multimeric binding of polyomavirus to its sialic acid receptor. Surface representation of the crystal structure (PDB ID 1SIE) of a polyomavirus particle (surface in grey) in complex with a sialic acid receptor fragment (carbons in yellow). (**b**) Possible contacts of viral proteins with terminal sialic acid, based on crystal structures [131]. Sialic acid is shown as sticks with carbons in orange. Viral atoms within 4.0 Å of the carbohydrate receptors in crystal structures of virus-sialic acid complexes are shown as spheres. Hydrogen bonds and salt bridges are indicated by black and red dashed lines, respectively. (**c**) Non-enveloped viruses binding to sialic acid receptors. Detail of sialic acid binding interactions with polyoma VP1 (PDB ID 1VPS), rhesus rotavirus VP8* (PDB ID 1KQR) and canine AdV2 fibre knob (PDB ID 2WBV). (**d**) Enveloped viruses binding to sialic acid receptors. Network of interactions for the Newcastle disease virus (NDV)-HN (PDB ID 1USR) and influenza HA protein (PDB ID 1HGG). (**e**) The PRCV CoV binding to a glycan attached to its APN receptor. Detail of the interaction revealed by the crystal structure of the PRCV RBD in complex with the pig APN shown in Fig. 15.6a. The N-acetyl-glucosamine (NAG) glycan shown as sticks and carbons in yellow. The viral Tyr and Gln residues that contact the glycan in the CoV S protein are labelled. Hydrogen bonds are dashed black lines. Oxygens are red and nitrogens blue in all panels. Figures (**b**–**d**) were provided by U. Neu and T. Stehel, adapted with permission from [131]

this Sia variant, which is part of a branched glycan linked to the CD1a ganglioside, is responsible for the ocular tropism of these viruses, whose infection causes severe ocular disease [48, 130]. Influenza A virus transmissibility and pathogenicity in humans also correlates with recognition of specific Sia molecules by the envelope proteins haemagglutinin (H) and neuraminidase (N), which mediate binding to and release from Sia on airway epithelial cells, respectively [133]. The H (H1 to H15) and N (N1 to N9) protein variants identified in different species differ in their affinities for the α(2-3)- or α(2-6)-linked Sia, which is a key determinant of viral transmissibility [130, 134]. Aquatic birds are the natural reservoir of influenza A, but the avian viruses are periodically transmitted to mammals, causing flu pandemics with a significant number of deaths [134]. The H protein of avian influenza A (HA) binds preferentially to α(2-3)-linked Sia receptors in waterbirds, which are not present in human non-ciliated tracheal cells initially infected by influenza A [135]; transmission of avian influenza A from birds to humans or within the human population is therefore quite inefficient. Nonetheless, variants of some influenza A strains (H1N1, H5N1…) can infect humans and cause sporadic outbreaks with high mortality rates due to HA mutations that switch its receptor-binding specificity for α(2-6)-linked Sia in the human upper respiratory tract [133, 135]. A balance between the HA and NA proteins is also important for the cross-species transmission of avian influenza viruses. In addition to these examples, other viruses can recognize specific substitutions in the Sia molecule [130, 132].

Viruses can also bind to other negatively charged sulphated or neutral carbohydrates. Human herpesviruses are an example of viruses that attach to heparan sulphate (HS) glycosaminoglycan, and this interaction has been studied extensively for human herpes simplex viruses (HSV). HSV cell entry is complex and is mediated by several envelope glycoproteins (gC, gB, gD, gH, and gL) [136]. The initial interaction of the virus particles with the cell is mediated by the gC and gB glycoproteins, which bind to HS. This interaction is nevertheless insufficient for virus entry, which also requires gD binding to cell surface proteins for virus–cell membrane fusion, catalysed by gB and the gH/gL complex. HS can thus be considered an HSV attachment factor that facilitates subsequent interactions with other receptors for cell entry. Viruses such as FMDV can evolve in cell culture and increase its binding affinity for HS [137].

Noroviruses exemplify a group of viruses that use neutral cell surface carbohydrate receptors, known as histo-blood group antigens (HBGA) [138]. Recognition is strain-specific and different strains of virus recognize different types of antigens. These variations affect the tropism and pathogenesis of norovirus infections, as susceptibility to infection depends on the HBGA receptors, which differ among individuals, who may therefore be resistant to infection by certain norovirus strains and susceptible to others. Certain human rotavirus strains also bind to an HBGA abundant in intestinal epithelial cells [139], using a VP8*site that interacts with Sia in other rotaviruses. Subtle modifications in the viral protein allow receptor switching in rotaviruses. As we described above, CoV also interact with N-linked glycans attached to their receptors (Fig. 15.7e) [104], which are a major determinant of the host range of some CoV.

Cell surface carbohydrates are ubiquitous, accessible, and relatively variable "hooks" for cell attachment by highly diverse viral families. Carbohydrate-derived compounds can thus be used to inhibit many types of viral infections. Carbohydrates are main cell entry receptors for some viruses and may be alternative or secondary entry receptors for others. Viruses that bind to protein receptors can evolve to use carbohydrates that expand their host cell tropism [6, 93]. Enveloped viruses can also exploit the glycans linked to their envelope proteins to attach to cell surface lectins. Some of these lectins are specific for particular glycosylation patterns associated with viral membrane glycoproteins, as discussed above for HIV-1. Many enveloped viruses attach to host cells via DC-SIGN [140], a cell surface receptor shared by many pathogens. Glycans are thus intimately involved in virus–cell interactions.

15.3 Receptor-Mediated Virus Cell Entry

15.3.1 Receptor-Mediated Uncoating and Membrane Penetration of Picornaviruses

After binding to cell surface receptors, the viral genome must translocate from the particle to the cell cytoplasm (Fig. 15.1). In enveloped viruses, this transfer may be rather simple and occurs after fusion of the virus and cell membranes, as will be illustrated in Chap. 16. In non-enveloped viruses, mechanisms for genome penetration into the cytoplasm have been studied, particularly in picornavirus; several pieces of the process have been described, and they will be discussed here.

The initial event in picornaviruses membrane penetration is the opening of the viral capsid and egress of the viral genome, a process known as uncoating that is observed during cell infection [141]. Full and empty capsids have marked differences in their sedimentation coefficients, so that the uncoating process in cells and in solution can be followed by ultracentrifugation, which has identified several entry intermediates [141]. Whereas naked picornavirus capsids migrate with coefficients around 160S, receptor binding reduces this [142], externalizes the VP4 and VP1 N-terminus in uncoating intermediates (135S or A particles) [143], and generates empty capsids without RNA (80S particles) at neutral pH and physiological temperatures (Fig. 15.8a, top) [141, 144, 145]. These results demonstrated that the receptor can function as an "unzipper" triggering conformational changes in the picornavirus particles and the exit of the genome from the capsid for membrane penetration [146]. Multivalent receptor binding to the virus capsid is required for a productive uncoating and may be facilitated during receptor-mediated endocytosis of the virus particles [147].

Susceptibility to receptor-mediated uncoating depends on the stability of the viral capsid and the mode of receptor recognition. Receptor-mediated uncoating is temperature dependent and generally accelerates above 30 °C for viruses that use the canyon region for receptor engagement (Fig. 15.2). Nonetheless, there is some diversity, particularly among the major group HRVs, which has been linked to their inherent capsid stability and dynamics. HRV carrying small molecules or fatty acids into a pocket in the VP1 (pocket factor) are less susceptible to receptor-mediated uncoating than serotypes without pocket factor [11, 148], which is known to limit capsid dynamics or breathing (see Chap. 6) [149, 150]. A group of molecules known as WIN compounds also bind to the VP1 pocket, restricting capsid breathing and preventing genome release. The temperature and pocket factor dependence of uncoating suggests that capsid breathing is necessary for receptor-mediated capsid opening at physiological temperatures [11, 151].

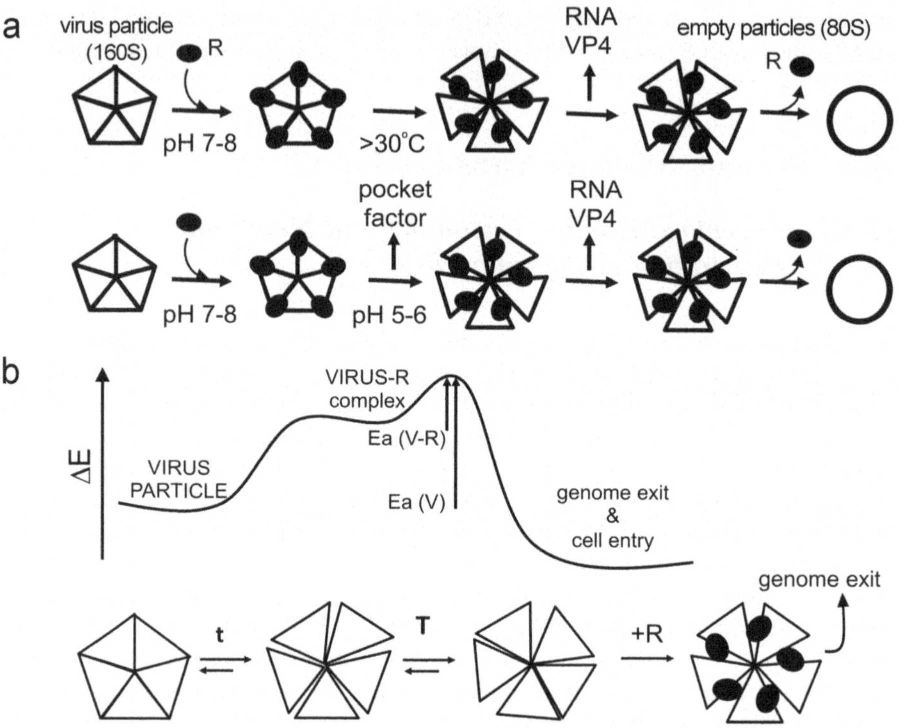

Fig. 15.8 Receptor-mediated virus cell entry. (**a**) Pathways of receptor-mediated uncoating in picornaviruses. Top. In some picornavirus, binding to receptors at neutral pH generates metastable complexes and the exit of the VP4 protein and the RNA genome at physiological temperatures, which results in empty virus particles from which the receptor releases. Bottom. Receptor binding to certain picornavirus at neutral pH generates stable complexes, but its exposure to the mild low pH in endosomes triggers the release of the pocket factor molecules bound to VP1, which primes the exit of VP4 and RNA. These one- or two-step uncoating processes are exclusive of picornavirus using the canyon region to bind to receptors. (**b**) Cell entry receptors function as catalysts. Top. Kinetics and thermodynamics studies indicated that some virus-receptor complexes have higher energy than virus particles. Therefore, receptor engagement can generate metastable virus particles primed for cell entry by reducing the activation energy (*Ea*) for genome release, thus acting as a catalyst. Bottom. Viral capsids or viral envelope proteins are dynamic entities with their building blocks moving randomly at physiological temperatures. Receptor binding maintains the capsid or virus proteins in an expanded or open state, which kinetically facilitates the genome exit from the capsids or the fusion reaction

In addition to receptor binding, low pH at which receptor-bound viruses are exposed during endocytosis is an important factor mediating virus uncoating (Fig. 15.1) [5]. The mildly acidic pH (5.5–6.0) of endosomal compartments is optimal for uncoating and infection of certain picornaviruses such as the minor group HRV [30, 152, 153]. Binding to LDLR does not mediate uncoating, and minor group HRV are thus dependent on endocytosis and endosome acidification for uncoating and host cell entry [30, 152]. Exposure to low pH is also necessary for efficient uncoating and cell entry by some major group HRV [30, 148], although certain serotypes only require ICAM-1 receptor binding, after which they can infect cells, even in the presence of agents that prevent endosome acidification. In the case of relatively stable HRV serotypes with pocket factor molecules, such as HRV16, both receptor binding and low endosome pH are required for entry into host cells [148]. After binding to ICAM-1, HRV16 particles become sensitive to low pH, which triggers rapid RNA release from the capsid, probably due to pH-mediated release of the pocket factor molecule (Fig. 15.8a, bottom). This two-step uncoating mechanism was recently confirmed by cryo-EM structures of EchoV6 or CV-A10 in complex with their receptors, which showed that low pH is the main driving force of pocket factor release in picornaviruses [25, 26]. These results indicate that receptor binding to the picornavirus canyon primes the viral capsid for low pH-mediated uncoating during endocytosis [148]. In addition, for picornaviruses that bind to receptors outside the canyon region (Fig. 15.2), it is likely that low pH alone mediates pocket factor release and uncoating. The cooperative effect of the receptor and low pH in some picornavirus uncoating will help polarize RNA exit at the receptor-bound region of the virus particle in endosomes, which must be important for efficient cell membrane penetration and infection.

The externalized genome of non-enveloped viruses, either on the cell surface or in endosomes, must translocate into the cell cytoplasm to initiate the intracellular phase of the life cycle. The process of genome transfer from the virus particle to the cytoplasm is less well understood in non-enveloped viruses than in enveloped viruses. Structures of isolated envelope fusion proteins have delineated the process of membrane penetration by fusion of virus–cell membranes, described in Chap. 16. Research to date suggests two possible routes by which the genome of non-enveloped viruses can enter the cell, either by membrane rupture or by pores through the membrane [30]. Elegant studies with nanodiscs support the transfer of the picornavirus genome to the cell cytoplasm through a membrane pore after receptor and/or low pH-mediated uncoating [154]. The RNA-associated VP4 is able to form hexameric membrane pores, which may assist in genome membrane penetration. In addition, the N-terminus of VP1, which is externalized during capsid uncoating, interacts with the cell membrane and can attach the particle to the membrane and facilitate RNA movement across the membrane with the help of VP4 [155, 156]. Capsid proteins can thus permeate the membrane for the transfer of the uncoated genome to the cytoplasm.

15.3.2 Structure and Biophysics of Receptor-Catalysed Virus Uncoating

Several structural studies have examined uncoating intermediates by cryo-EM, providing insights into the process of genome exit from the capsid in picornaviruses. An original study analysed the 135S PV particles, which lack VP4 but retain the

RNA inside, and compared this structure with 160S native virions and 80S empty capsids formed after RNA exit [157]. The 135S particle showed VP rearrangements relative to the native particle and a 4% capsid expansion. Major rearrangements occur in the two-fold axes of the capsid and in the canyon region; however, the conformation of the five-fold axis, thought to be the RNA exit port, was similar to that of the native 160S PV capsid. Subsequently, HRV3 complexed with ICAM-1 and in the process of uncoating was analysed by cryo-EM [11]; the structure also showed HRV3 capsid expansion not seen in stable HRV16-ICAM-1 complexes, indicating that capsid expansion must precede uncoating. Even though the structures were determined at low resolution, holes appeared in the two-fold axes of expanded HRV3-ICAM-1 complexes, whereas the five-fold axis regions were almost identical to those of native viruses [11]. These structural findings challenged the hypothesis of RNA exit through the capsid five-fold axis and suggested other exit sites. More recent studies with PV showed that RNA exits the capsid near the two-fold axis [158]. Additional high-resolution structures of picornavirus uncoating intermediates confirmed that capsid expansion creates holes at the two-fold axis of the capsid for RNA uncoating and transfer to the cell cytoplasm [159–161]. As mentioned above, high-resolution cryo-EM has also elucidated how receptors and low pH work together to drive pocket factor release and prime the capsid for genome uncoating [25, 26].

The uncoating process requires multivalent receptor interaction and biophysical studies have shown that receptor binding generates a metastable complex and reduces the activation energy required for uncoating (Fig. 15.8b, top) [162, 163]. Nonetheless, uncoating is temperature dependent and it is inhibited by pocket factor compounds that restrict breathing, suggesting that it must depend on capsid dynamics [151]. At physiological temperatures, capsids are dynamic entities whose protomers can move randomly, opening and closing holes at protomer junctions (Fig. 15.8b, bottom). This rapid, random movement does not allow RNA release, which is kinetically unfavourable due to the transient nature of the holes. After receptor binding at physiological temperatures, the capsid remains expanded and stable holes in the two-fold axis open. Picornaviruses recognize receptors via the canyon, a fragmented region at the junction of two protomers. Canyon-binding receptors can act as wedges at interprotomeric junctions to keep the capsid expanded [10, 11] (Fig. 15.8b, bottom), thus catalysing the uncoating process by locking the capsid in an intermediate open state that allows genome to exit. This mechanism implies that certain receptor molecules catalyse temperature-driven uncoating of dynamic capsids.

15.3.3 Receptor-Mediated Virus–Cell Fusion

As described for non-enveloped capsids, enveloped glycoproteins are dynamic metastable entities that can undergo conformational rearrangements from a prefusion to a postfusion form that fuses cell and virus membranes for the transfer of the viral genome to the cell cytoplasm (see also Chap. 16). This process is triggered by cell surface receptor or low endosomal pH, and it is promoted by other cell factors such as cell proteases. Receptor-mediated conformational switch in virus enveloped proteins has been particularly studied with the HIV-1 Env by different means [58]. The HIV-1 Env exhibits an intrinsic conformational dynamics or breathing, defined by single-molecule fluorescence resonance energy transfer (FRET), which determined the native Env transitions between at least three distinct prefusion conformations [164]. A preeminent closed state represented by low FRET switches to a high FRET state that accesses an intermediate FRET. Ligands are proposed to specifically capture different Env conformations. The CD4 receptor binds to the high FRET state, whereas most broadly nAbs stabilize the ground low FRET form, preventing Env dynamics and virus–cell fusion and entry [164, 165].

Structural studies of the native (membrane-bound) and soluble HIV Env demonstrated the protein plasticity and identified protein conformations associated with binding to receptors and Abs. Low resolution structures of the native Env trimers on viruses identify a conformational switch in the trimer upon CD4 receptor and/or Ab binding, resulting in an outward displacement of gp120 from the centre of the trimer and the opening of the Env [166]. The transition from a closed to an open Env trimer was confirmed by high-resolution cryo-EM structures of soluble proteins alone and in complex with ligands [167, 168], which revealed additional structural details of the receptor-bound Env leading to the fusion reaction. Structures of the Env in complex with broadly neutralizing Abs confirmed that they stabilize the closed form, restrict Env activation, and prevent fusion [169].

Structural and biophysical studies demonstrated the flexible nature of the HIV-1 Env, which can fluctuate or "breathe" between different states at physiological temperatures. Nonetheless, as described for viral capsids in Fig. 15.8b, temperature-induced random breathing of the viral proteins cannot mediate the fusion reaction. Receptor and co-receptor binding to the Env is required to generate a metastable open complex (Fig. 15.4) with a lower activation energy for the transition to a fusion-active form of the Env. This mechanism suggests that the receptors stabilize viral intermediates primed for fusion or uncoating and unifies the basis for receptor-catalysed viral cell entry (Fig. 15.8b).

15.4 Perspectives and Conclusions

Viruses subvert cell surface molecules for cell entry and dissemination of infection. Receptor accessibility and cell expression patterns are important factors that underlie virus selection of specific surface receptors. Most of these receptors promote translocation of the viral genome into the cell. In some cases, however, viruses attach to cellular factors that concentrate infectious particles on the plasma membrane and facilitate virus recognition of entry receptors. In other cases, viruses use cell surface molecules to spread the infection throughout the body; they can attach to migratory cells or induce signalling events that loosen cell–cell contacts and facilitate virus transmission. Cell surface molecules can thus have diverse functions in the dissemination of viral infections; the study of virus–receptor interactions is therefore of great relevance for understanding virus evolution, host tropism, and pathogenesis.

The virus entry receptors have been the main focus of this chapter. These molecules are not just a "hook" for virus attachment, but can also participate in the translocation of the viral genome into host cells, on the cell surface, or within endosomal compartments. Some viral receptors catalyse the entry process alone or in combination with other cell factors, such as the mildly acidic pH found in endosomes. Interaction with certain receptors transforms virus particles into metastable entities, priming delivery of their genomes to the cytoplasm. Alternatively, genome translocation can be mediated by low pH, but this requires endocytosis of receptor-bound viruses. In some picornaviruses, discussed here, the energy absorbed in the virus–receptor interaction translates into virus particle expansion and genome exit. In other non-enveloped viruses, capsid expansion triggered by cell factors might be a necessary intermediate step for genome uncoating. Receptor binding also primes virus–cell membrane fusion in non-enveloped viruses, mediated by conformational changes in virus fusion proteins (Chap. 16). Cell surface molecules are thus key players in virus entry into host cells, and the process of receptor-mediated virus entry continues to be studied. High-resolution cryo-EM structures of virus-receptor complexes are providing further insights into the molecular processes that translocate the viral genome from virus particles to the cell cytoplasm.

Virus–receptor interactions appear to be highly specific. There are many examples of virus selection of only a single member of closely-related cell surface molecules. This specificity is based on the recognition of certain unique structural features of a cell surface molecule. Virus-receptor surface complementarity must be important for virus recognition of certain receptor molecules. Viral proteins are shaped to dock into structural features of their receptors. Many viruses use concave surfaces for binding to cell surface receptors, which optimizes receptor contact area and hides receptor-binding residues from antibodies. Nevertheless, viruses can also have protruding receptor-binding surfaces for receptor recognition. In both cases, viruses can escape from antibody neutralization by mutating non-essential residues in these receptor-binding surfaces. Immune system pressure on viruses is a major determinant for the switch in receptor recognition observed in many virus groups, although this diversity can also be linked to opportunities to spread infection using ubiquitous or more accessible cell surface molecules. In addition to its high degree of specificity, virus-receptor recognition can be dynamic, and viruses evolve to use alternative or distinct receptors.

Preventing virus binding to receptors can efficiently block virus infection and cell damage. Receptor-binding regions in virus proteins are relatively invariant and could potentially be targeted by immune responses to prevent infection. A large number of structural studies have characterized viral antigenic sites targeted by nAbs. Many of these sites overlap receptor-binding regions, showing that blockade of virus binding to cell surface receptors is a major neutralization mechanism. In viruses for which efficient vaccines have been developed, the receptor-binding residues are relatively well-exposed and accessible to antibodies. Viruses have nonetheless developed ways to evade neutralization by hiding receptor-binding regions from antibodies; the use of concave and poorly accessible receptor-binding surfaces is one well-defined strategy of this type. In some viruses, these relatively inaccessible surfaces are also surrounded by variable loops and glycan shields, which further prevent antibody binding and neutralization. Blocking infection of these viruses requires the design of improved vaccines that elicit immune responses focused on relatively inaccessible sites. These therapies require a deep understanding of virus–receptor and virus–antibody interactions through the determination of complex structures. Advances in antibody selection technologies led to the identification of numerous antibody molecules, such as the recent COVID-19 pandemic, which define key viral vulnerabilities. These studies are also opening avenues for the design of molecules that prevent virus entry into cells, as well as antigens that trigger immune responses that neutralize viral infections. Small molecule drugs have also been developed, such as sialic acid analogues to treat influenza virus infection, or molecules that prevent HIV gp120 from binding to its receptor, CD4. The characterization of virus–receptor interactions is thus of considerable interest for the development of antiviral therapies.

Acknowledgments I thank U. Neu, T. Stehle, L. Xing, and H.R. Cheng for providing figures included in this chapter and C. Mark for editorial assistance. The support of PTI + Salud Global and NextGenerationEU through the Plan de Recuperación, Transformación y Resiliencia of the Spanish Ministry of Science is acknowledged.

References

1. Marsh M, Helenius A (2006) Virus entry: open sesame. Cell 124:729–740
2. Casasnovas JM, Stehle T (2021) Viral receptors. In: Bamford DH, Zuckerman M (eds) Encyclopedia of virology, vol 1, 4th edn. Elsevier, Oxford, pp 388–401
3. dos Santos NR et al (2023) Deciphering molecular mechanisms stabilizing the reovirus-binding complex. Proc Natl Acad Sci USA 120:e2220741120
4. Grove J, Marsh M (2011) The cell biology of receptor-mediated virus entry. J Cell Biol 195: 1071–1082
5. Mercer J, Schelhaas M, Helenius A (2010) Virus entry by endocytosis. Annu Rev Biochem 79: 803–833
6. Baranowski E, Ruiz-Jarabo CM, Domingo E (2001) Evolution of cell recognition by viruses. Science 292:1102–1105
7. Stehle T, Casasnovas JM (2009) Specificity switching in virus-receptor complexes. Curr Opin Struct Biol 19:181–188
8. Wang J-H (2002) Protein recognition by cell surface receptors: physiological receptors versus virus interactions. Trends Biochem Sci 27:122–126
9. Johnson JE (2010) Cell entry by non-enveloped viruses. Curr Top Microbiol Immunol Vol. 343. Springer, Berlin
10. Xing L et al (2000) Distinct cellular receptor interactions in poliovirus and rhinoviruses. EMBO J 19:1207–1216
11. Xing L, Casasnovas JM, Cheng RH (2003) Structural analysis of human rhinovirus complexed with ICAM-1 reveals the dynamics of receptor-mediated virus uncoating. J Virol 77:6101–6107

12. Mendelsohn CL, Wimmer E, Racaniello R (1989) Cellular receptor for poliovirus: molecular cloning, nucleotide sequence, and expression of a new member of the immunoglobulin superfamily. Cell 56:855–865
13. Campadelli-Fiume G et al (2000) The novel receptors that mediate the entry of herpes simplex viruses and animal alphaherpesviruses into cells. Rev Med Virol 10:305–319
14. Noyce RS et al (2011) Tumor cell marker PVRL4 (Nectin 4) is an epithelial cell receptor for measles virus. PLoS Pathog 7:e1002240
15. Muhlebach MD et al (2011) Adherens junction protein nectin-4 is the epithelial receptor for measles virus. Nature 480:530–533
16. Greve JM et al (1989) The major human rhinovirus receptor is ICAM-1. Cell 56:839–847
17. Staunton DE et al (1989) A cell adhesion molecule, ICAM-1, is the major surface receptor for rhinoviruses. Cell 56:849–853
18. Wei W et al (2016) ICAM-5/Telencephalin is a functional entry receptor for enterovirus D68. Cell Host Microbe 20:631–641
19. Bergelson JM et al (1997) Isolation of a common receptor for coxsackie B viruses and adenoviruses 2 and 5. Science 275:1320–1323
20. Kaplan G et al (1996) Identification of a surface glycoprotein on African green monkey kidney cells as a receptor for hepatitis A virus. EMBO J 15:4282–4296
21. Huber SA (1994) VCAM-1 is a receptor for encephalomyocarditis virus on murine vascular endothelial cells. J Virol 68:3453–3458
22. Rossmann MG, He Y, Kuhn RJ (2002) Picornavirus-receptor interactions. Trends Microbiol 10:324–331
23. Xing L et al (2004) Structural and functional analysis of integrin alpha2I domain interaction with echovirus 1. J Biol Chem 279:11632–11638
24. Baggen J et al (2018) Role of enhanced receptor engagement in the evolution of a pandemic acute hemorrhagic conjunctivitis virus. Proc Natl Acad Sci USA 115:397–402
25. Zhao X et al (2019) Human neonatal fc receptor is the cellular uncoating receptor for enterovirus B. Cell 177:1553–1565.e16
26. Cui Y et al (2020) Molecular basis of coxsackievirus A10 entry using the two-in-one attachment and uncoating receptor KRM1. Proc Natl Acad Sci USA 117:18711–18718
27. Smith TJ et al (1996) Neutralizing antibody to human rhinovirus 14 penetrates the receptor-binding canyon. Nature 383:350–354
28. Hewat EA et al (1997) Structure of the complex of an Fab fragment of a neutralizing antibody with foot-and-mouth disease virus: positioning of a highly mobile antigenic loop. EMBO J 16: 1492–1500
29. Tuthill TJ et al (2010) Picornaviruses. In: Johnson JE (ed) Cell entry by non-enveloped viruses. Springer, Berlin, pp 43–89
30. Fuchs R, Blaas D (2010) Uncoating of human rhinoviruses. Rev Med Virol 20:281–297
31. Verdaguer N et al (2004) X-ray structure of a minor group human rhinovirus bound to a fragment of its cellular receptor protein. Nat Struct Mol Biol 11:429–434
32. Bergelson JM et al (1994) Decay-accelerating factor (CD55), a glycosylphosphatidylinositol-anchored complement regulatory protein, is a receptor for several echoviruses. Proc Natl Acad Sci USA 91:6245–6248
33. He Y et al (2002) Structure of decay-accelerating factor bound to echovirus 7: a virus-receptor complex. Proc Natl Acad Sci USA 99:10325–10329
34. Pettigrew DM et al (2006) Structural and functional insights into the interaction of echoviruses and decay-accelerating factor. J Biol Chem 281:5169–5177
35. Casasnovas JM, Markarian S, Hammar L (2004) Sensor surface interactions in the study of macromolecular assemblies. In: Cheng R, Hammar L (eds) Conformational proteomics of macromolecular architecture. World Scientific, Singapore, pp 379–390
36. Morosky S et al (2019) The neonatal fc receptor is a pan-echovirus receptor. Proc Natl Acad Sci USA 116:3758–3763

37. Coyne CB, Bergelson JM (2006) Virus-induced Abl and Fyn kinase signals permit coxsackievirus entry through epithelial tight junctions. Cell 124:119–131
38. Zhang Y, Bergelson JM (2005) Adenovirus receptors. J Virol 79:12125–12131
39. Berk AJ (2007) Adenoviridae: the viruses and their replication. In: Knipe DM et al (eds) Fields virology. Lippincott, Williams & Wilkins, Philadelphia, pp 2355–2394
40. Bewley MC et al (1999) Structural analysis of the mechanism of adenovirus binding to its human cellular receptor, CAR. Science 286:1579–1583
41. van Raaij MJ et al (2000) Dimeric structure of the coxsackievirus and adenovirus receptor D1 domain at 1.7 a resolution. Structure 8:1147–1155
42. Walters RW et al (2002) Adenovirus fiber disrupts CAR-mediated intercellular adhesion allowing virus escape. Cell 110:789–799
43. Persson BD et al (2007) Adenovirus type 11 binding alters the conformation of its receptor CD46. Nat Struct Mol Biol 14:164–166
44. Persson BD et al (2021) Human species D adenovirus hexon capsid protein mediates cell entry through a direct interaction with CD46. Proc Natl Acad Sci USA 118:e2020732118
45. Wang H et al (2011) Desmoglein 2 is a receptor for adenovirus serotypes 3, 7, 11 and 14. Nat Med 17:96–104
46. Vassal-Stermann E et al (2019) CryoEM structure of adenovirus type 3 fibre with desmoglein 2 shows an unusual mode of receptor engagement. Nat Commun 10:1181
47. Burmeister WP et al (2004) Crystal structure of species D adenovirus fiber knobs and their sialic acid binding sites. J Virol 78:7727–7736
48. Nilsson EC et al (2010) The GD1a glycan is a cellular receptor for adenoviruses causing epidemic keratoconjunctivitis. Nat Med 17:105–109
49. Smith JG et al (2010) Adenovirus. In: Johnson JE (ed) Cell entry by non-enveloped viruses. Springer, Berlin, pp 195–224
50. Julien J-P et al (2013) Crystal structure of a soluble cleaved HIV-1 envelope trimer. Science 342:1477–1483
51. Lyumkis D et al (2013) Cryo-EM structure of a fully glycosylated soluble cleaved HIV-1 envelope trimer. Science 342:1484–1490
52. Zhu P et al (2006) Distribution and three-dimensional structure of AIDS virus envelope spikes. Nature 441:847–852
53. Freed RO, Martin MA (2007) HIVs and their replication. In: Knipe DM et al (eds) Fields virology. Lippincott, Williams & Wilkins. Philadelphia, pp 2107–2185
54. Wyatt R, Sodroski J (1998) The HIV-1 envelope glycoproteins: Fusogens, antigens, and immunogens. Science 280:1884–1888
55. Chen B (2019) Molecular mechanism of HIV-1 entry. Trends Microbiol 27:878–891
56. Berger EA (1998) HIV entry and tropism. When one receptor is not enough. Adv Exp Med Biol 452:151–157
57. Kwong D et al (1998) Structure of an HIV gp120 envelope glycoprotein in complex with the CD4 receptor and a neutralizing human antibody. Nature 393:648–659
58. Ward AB, Wilson IA (2017) The HIV-1 envelope glycoprotein structure: nailing down a moving target. Immunol Rev 275:21–32
59. Sok D, Burton DR (2018) Recent progress in broadly neutralizing antibodies to HIV. Nat Immunol 19:1179–1188
60. Berger EA, Murphy PM, Farber JM (1999) Chemokine receptors as HIV-1 coreceptors: roles in viral entry, tropism, and disease. Annu Rev Immunol 17:657–700
61. Baribaud F, Pöhlmann S, Doms RW (2001) The role of DC-SIGN an DC-SIGNR in HIV and SIV attachment, infection, and transmission. Virology 286:1–6
62. Geijtenbeek TBH et al (2000) DC-SIGN, a dendritic cell-specific HIV-1 binding protein that enhances trans-infection of T cells. Cell 100:587–597
63. Feinberg H et al (2001) Structural basis for selective recognition of oligosaccharides by DC-SIGN and DC-SIGNR. Science 294:2163–2166

64. Guo Y et al (2004) Structural basis for distinct ligand-binding and targeting properties of the receptors DC-SIGN and DC-SIGNR. Nat Struct Biol 11:591–598
65. Puryear WB et al (2012) HIV-1 incorporation of host-cell-derived glycosphingolipid GM3 allows for capture by mature dendritic cells. Proc Natl Acad Sci USA 109:7475–7480
66. Izquierdo-Useros N et al (2012) Sialyllactose in viral membrane gangliosides is a novel molecular recognition pattern for mature dendritic cell capture of HIV-1. PLoS Biol 10: e1001315
67. Izquierdo-Useros N et al (2012) Siglec-1 is a novel dendritic cell receptor that mediates HIV-1 trans-infection through recognition of viral membrane gangliosides. PLoS Biol 10:e1001448
68. Izquierdo-Useros N et al (2014) HIV-1 capture and transmission by dendritic cells: the role of viral glycolipids and the cellular receptor Siglec-1. PLoS Pathog 10:e1004146
69. Enjuanes L et al (2008) Nidovirales. In: Mahy BWJ, van Regenmortel MHV (eds) Encyclopedia of virology. Elsevier, Oxford, pp 419–430
70. de Groot RJ et al (2011) Coronaviridae. In: King AMQ et al (eds) Virus taxonomy: ninth report of the international committee on taxonomy of viruses. Elsevier Academic Press, San Diego, pp 774–796
71. de Groot RJ et al (2013) Middle East respiratory syndrome coronavirus (MERS-CoV): announcement of the coronavirus study group. J Virol 87:7790–7792
72. Perlman S (2020) Another decade, another coronavirus. New Eng J Med 382:760–762
73. Zhou P et al (2020) A pneumonia outbreak associated with a new coronavirus of probable bat origin. Nature 579:270–273
74. Holmes KV (2005) Structural biology: adaptation of SARS coronavirus to humans. Science 309:1822–1823
75. Graham RL, Baric RS (2010) Recombination, reservoirs, and the modular spike: mechanisms of coronavirus cross-species transmission. J Virol 84:3134–3146
76. Raj VS et al (2014) MERS: emergence of a novel human coronavirus. Curr Opin Virol 5:58–62
77. Perlman S, Netland J (2009) Coronaviruses post-SARS: update on replication and pathogenesis. Nat Rev Microbiol 7:439–450
78. Beniac DR et al (2006) Architecture of the SARS coronavirus prefusion spike. Nat Struct Mol Biol 13:751–752
79. Kirchdoerfer RN et al (2016) Pre-fusion structure of a human coronavirus spike protein. Nature 531:118–121
80. Walls AC et al (2016) Cryo-electron microscopy structure of a coronavirus spike glycoprotein trimer. Nature 531:114–117
81. Yuan Y et al (2017) Cryo-EM structures of MERS-CoV and SARS-CoV spike glycoproteins reveal the dynamic receptor binding domains. Nat Commun 8:15092
82. Wrapp D et al (2020) Cryo-EM structure of the 2019-nCoV spike in the prefusion conformation. Science 367:1260–1263
83. Walls AC et al (2020) Structure, function, and antigenicity of the SARS-CoV-2 spike glycoprotein. Cell 181:281–292.e6
84. Casasnovas JM et al (2022) Nanobodies protecting from lethal SARS-CoV-2 infection target receptor binding epitopes preserved in virus variants other than omicron. Front Immunol 13: 863831
85. Walls AC et al (2016) Glycan shield and epitope masking of a coronavirus spike protein observed by cryo-electron microscopy. Nat Struct Mol Biol 23:899–905
86. Shi W et al (2023) Cryo-EM structure of SARS-CoV-2 postfusion spike in membrane. Nature 619:403–409
87. Walls AC et al (2017) Tectonic conformational changes of a coronavirus spike glycoprotein promote membrane fusion. Proc Natl Acad Sci USA 114:11157–11162
88. Millet JK, Whittaker GR (2015) Host cell proteases: critical determinants of coronavirus tropism and pathogenesis. Virus Res 202:120–134

89. Hoffmann M et al (2020) SARS-CoV-2 cell entry depends on ACE2 and TMPRSS2 and is blocked by a clinically proven protease inhibitor. Cell 181:271–280.e8
90. Shang J et al (2020) Cell entry mechanisms of SARS-CoV-2. Proc Natl Acad Sci USA 117: 11727–11734
91. Park J-E et al (2016) Proteolytic processing of Middle East respiratory syndrome coronavirus spikes expands virus tropism. Proc Natl Acad Sci USA 113:12262–12267
92. Masters PS (2006) The molecular biology of coronaviruses. Adv Virus Res 66:193–292
93. Li F (2012) Evidence for a common evolutionary origin of coronavirus spike protein receptor-binding subunits. J Virol 86:2856–2858
94. Hulswit RJG et al (2019) Human coronaviruses OC43 and HKU1 bind to 9-O-acetylated sialic acids via a conserved receptor-binding site in spike protein domain A. Proc Natl Acad Sci USA 116:2681–2690
95. Williams RK, Jiang GS, Holmes KV (1991) Receptor for mouse hepatitis virus is a member of the carcinoembryonic antigen family of glycoproteins. Proc Natl Acad Sci USA 88:5533–5536
96. Tortorici MA et al (2022) Structure, receptor recognition, and antigenicity of the human coronavirus CCoV-HuPn-2018 spike glycoprotein. Cell 185:2279–2291.e17
97. Schultze B et al (1996) Transmissible gastroenteritis coronavirus, but not the related porcine respiratory coronavirus, has a sialic acid (N-glycolylneuraminic acid) binding activity. J Virol 70:5634–5637
98. Wong SK et al (2004) A 193-amino acid fragment of the SARS coronavirus S protein efficiently binds angiotensin-converting enzyme 2. J Biol Chem 279:3197–3201
99. Wu K et al (2009) Crystal structure of NL63 respiratory coronavirus receptor-binding domain complexed with its human receptor. Proc Natl Acad Sci USA 106:19970–19974
100. Reguera J et al (2011) Antigenic modules in the N-terminal S1 region of the transmissible gastroenteritis virus spike protein. J Gen Virol 92:1117–1126
101. Du L et al (2013) Identification of a receptor-binding domain in the S protein of the novel human coronavirus Middle East respiratory syndrome coronavirus as an essential target for vaccine development. J Virol 87:9939–9942
102. Shang J et al (2020) Structural basis of receptor recognition by SARS-CoV-2. Nature 581: 221–224
103. Li W et al (2003) Angiotensin-converting enzyme 2 is a functional receptor for the SARS coronavirus. Nature 426:450–454
104. Reguera J et al (2012) Structural bases of coronavirus attachment to host aminopeptidase N and its inhibition by neutralizing antibodies. PLoS Pathog 8:e1002859
105. Raj VS et al (2013) Dipeptidyl peptidase 4 is a functional receptor for the emerging human coronavirus-EMC. Nature 495:251–254
106. Yan R et al (2020) Structural basis for the recognition of the SARS-CoV-2 by full-length human ACE2. Science 367:1444–1448
107. Mina-Osorio P (2008) The moonlighting enzyme CD13: old and new functions to target. Trends Mol Med 14:361–371
108. Crackower MA et al (2002) Angiotensin-converting enzyme 2 is an essential regulator of heart function. Nature 417:822–828
109. Boonacker E, Van Noorden CJF (2003) The multifunctional or moonlighting protein CD26/ DPPIV. Eur J Cell Biol 82:53–73
110. Li F et al (2005) Structure of SARS coronavirus spike receptor-binding domain complexed with receptor. Science 309:1864–1868
111. Wang N et al (2013) Structure of MERS-CoV spike receptor-binding domain complexed with human receptor DPP4. Cell Res 23:986–993
112. Lu G et al (2013) Molecular basis of binding between novel human coronavirus MERS-CoV and its receptor CD26. Nature 500:227–231
113. Santiago C et al (2017) Allosteric inhibition of aminopeptidase N functions related to tumor growth and virus infection. Sci Rep 7:46045

114. Tomlinson ACA et al (2017) Receptor-binding loops in alphacoronavirus adaptation and evolution. Nat Commun 8:1735
115. Shang J et al (2018) Cryo-EM structure of infectious bronchitis coronavirus spike protein reveals structural and functional evolution of coronavirus spike proteins. PLoS Pathog 14: e1007009
116. Ji W et al (2022) Structures of a deltacoronavirus spike protein bound to porcine and human receptors. Nat Commun 13:1467
117. Harvey WT et al (2021) SARS-CoV-2 variants, spike mutations and immune escape. Nat Rev Microbiol 19:409–424
118. Behrens GMN, Cossmann A, Hoffmann M (2022) Omicron spike protein: a clue for viral entry and immune evasion. Signal Transduct Target Ther 7:1–2
119. Li W et al (2005) Receptor and viral determinants of SARS-coronavirus adaptation to human ACE2. EMBO J 24:1634–1643
120. Reguera J et al (2014) A structural view of coronavirus-receptor interactions. Virus Res 194:3–15
121. Piccoli L et al (2020) Mapping neutralizing and immunodominant sites on the SARS-CoV-2 spike receptor-binding domain by structure-guided high-resolution serology. Cell 183:1024–1042
122. Hastie KM et al (2021) Defining variant-resistant epitopes targeted by SARS-CoV-2 antibodies: a global consortium study. Science 374:472–478
123. Jiansheng J et al (2023) SARS-CoV-2 antibodies recognize 23 distinct epitopic sites on the receptor binding domain. Commun Biol 6:953
124. Cerutti G et al (2021) Potent SARS-CoV-2 neutralizing antibodies directed against spike N-terminal domain target a single supersite. Cell Host Microbe 29:819–833.e7
125. Gao X et al (2022) Binding and neutralizing abilities of antibodies towards SARS-CoV-2 S2 domain. Hum Vaccin Immunother 0:1–11
126. Barnes CO et al (2020) SARS-CoV-2 neutralizing antibody structures inform therapeutic strategies. Nature 588:682–687
127. Starr TN et al (2021) SARS-CoV-2 RBD antibodies that maximize breadth and resistance to escape. Nature 597:97–102
128. Pantophlet R, Burton DR (2003) Immunofocusing: antigen engineering to promote the induction of HIV-neutralizing antibodies. Trends Mol Med 9:468–473
129. Sesterhenn F et al (2020) De novo protein design enables the precise induction of RSV-neutralizing antibodies. Science 368:eaay5051
130. Olofsson S, Bergström T (2005) Glycoconjugate glycans as viral receptors. Ann Med 37:154–172
131. Neu U, Bauer J, Stehle T (2011) Viruses and sialic acids: rules of engagement. Curr Opin Struct Biol 21:610–618
132. Blaum BS, Stehle T (2019) Sialic acids in nonenveloped virus infections. In: Baker DC (ed) Advances in carbohydrate chemistry and biochemistry. Academic Press, Oxford, pp 65–111
133. Thompson AJ, Paulson JC (2021) Adaptation of influenza viruses to human airway receptors. J Biol Chem 296:100017
134. Liu M et al (2023) Gradual adaptation of animal influenza A viruses to human-type sialic acid receptors. Curr Opin Virol 60:101314
135. Matrosovich MN et al (2004) Human and avian influenza viruses target different cell types in cultures of human airway epithelium. Proc Natl Acad Sci USA 101:4620–4624
136. Navaratnarajah CK et al (2012) Targeted entry of enveloped viruses: measles and herpes simplex virus I. Curr Opin Virol 2:43–49
137. Sa-Carvalho D et al (1997) Tissue culture adaptation of foot-and-mouth disease virus selects viruses that bind to heparin and are attenuated in cattle. J Virol 71:5115–5123
138. Huang P et al (2003) Noroviruses bind to human ABO, Lewis, and secretor histo-blood group antigens: identification of 4 distinct strain-specific patterns. J Infect Dis 188:19–31

139. Hu L et al (2012) Cell attachment protein VP8* of a human rotavirus specifically interacts with A-type histo-blood group antigen. Nature 485:256–259
140. Backovic M, Rey FA (2012) Virus entry: old viruses, new receptors. Curr Opin Virol 2:4–13
141. Hogle JM (2002) Poliovirus cell entry: common structural themes in viral cell entry pathways. Ann Rev Microbiol 56:677–702
142. Casasnovas JM, Springer TA (1994) The pathway of rhinovirus disruption by soluble intercellular adhesion molecule 1 (ICAM-1): an intermediate in which ICAM-1 is bound and RNA is released. J Virol 68:5882–5889
143. Shah PNM et al (2020) Cryo-EM structures reveal two distinct conformational states in a picornavirus cell entry intermediate. PLoS Pathog 16:e1008920
144. Kaplan G, Freistadt MS, Racaniello VR (1990) Neutralization of poliovirus by cell receptors expressed in insect cells. J Virol 64:4697–4702
145. Hoover-Litty H, Greve JM (1993) Formation of rhinovirus-soluble ICAM-1 complexes and conformational changes in the virion. J Virol 67:390–397
146. Racaniello VR (1996) The poliovirus receptor: a hook, or an unzipper? Structure 4:769–773
147. Brandenburg B et al (2007) Imaging poliovirus entry in live cells. PLoS Biol 5:e183
148. Nurani G, Lindqvist B, Casasnovas JM (2003) Receptor priming of major group human rhinoviruses for uncoating and entry at mild low-pH environments. J Virol 77:11985–11991
149. Lewis JK et al (1998) Antiviral agent blocks breathing of the common cold virus. Proc Natl Acad Sci USA 95:6774–6778
150. Roy A, Post CB (2012) Long-distance correlations of rhinovirus capsid dynamics contribute to uncoating and antiviral activity. Proc Natl Acad Sci USA 109:5271–5276
151. Casasnovas JM (2000) The dynamics of receptor recognition by human rhinoviruses. Trends Microbiol 8:251–254
152. Prchla E et al (1994) Uncoating of human rhinovirus serotype 2 from late endosomes. J Virol 68:3713–3723
153. Schober D et al (1998) Major and minor receptor group human rhinoviruses penetrate from endosomes by different mechanisms. J Virol 72:1354–1364
154. Nasr ML et al (2017) Covalently circularized nanodiscs for studying membrane proteins and viral entry. Nat Methods 14:49–52
155. Strauss M et al (2013) RNA transfer from poliovirus 135S particles across membranes is mediated by long umbilical connectors. J Virol 87:3903–3914
156. Butan C, Filman DJ, Hogle JM (2014) Cryo-electron microscopy reconstruction shows poliovirus 135S particles poised for membrane interaction and RNA release. J Virol 88: 1758–1770
157. Belnap DM et al (2000) Molecular tectonic model of virus structural transitions: the putative cell entry states of poliovirus. J Virol 74:1342–1354
158. Bostina M et al (2011) Poliovirus RNA is released from the capsid near a twofold symmetry axis. J Virol 85:776–783
159. Shingler KL et al (2013) The enterovirus 71 A-particle forms a gateway to allow genome release: a cryoEM study of picornavirus uncoating. PLoS Pathog 9:e1003240
160. Ren J et al (2013) Picornavirus uncoating intermediate captured in atomic detail. Nat Commun 4:1929
161. Zhu L et al (2018) Structures of coxsackievirus A10 unveil the molecular mechanisms of receptor binding and viral uncoating. Nat Commun 9:4985
162. Casasnovas JM, Springer TA (1995) Kinetics and thermodynamics of virus binding to receptor: studies with rhinovirus, intercellular adhesion molecule-1 (ICAM-1), and surface plasmon resonance. J Biol Chem 270:13216–13224
163. Tsang SK et al (2001) Kinetic analysis of the effect of poliovirus receptor on viral uncoating: the receptor as a catalyst. J Virol 75:4984–4989
164. Munro JB et al (2014) Conformational dynamics of single HIV-1 envelope trimers on the surface of native virions. Science 346:759–763

165. Lu M et al (2019) Associating HIV-1 envelope glycoprotein structures with states on the virus observed by smFRET. Nature 568:415–419
166. Liu J et al (2008) Molecular architecture of native HIV-1 gp120 trimers. Nature 455:109–113
167. Harris A et al (2011) Trimeric HIV-1 glycoprotein gp140 immunogens and native HIV-1 envelope glycoproteins display the same closed and open quaternary molecular architectures. Proc Natl Acad Sci USA 108:11440–11445
168. Ozorowski G et al (2017) Open and closed structures reveal allostery and pliability in the HIV-1 envelope spike. Nature 547:360–363
169. Tran EEH et al (2012) Structural mechanism of trimeric HIV-1 envelope glycoprotein activation. PLoS Pathog 8:e1002797

Further Reading

Especially recommended for further reading are the following references listed above: References [1, 2, 4, 5] present a general overview of virus-receptor recognition and virus entry into host cells, references [6, 7] describe evolution of receptor-binding specificity in viruses and references [3, 61, 66, 68] are about attachment factors and their role in virus transmission. I also recommend the following references related to specific chapter sections: 15.2.1, [22, 29] (picornavirus), and [38, 49] (adenovirus); 15.2.2, [54, 56–58, 127] (HIV-1), and [71, 76, 78, 79, 87, 92, 105, 119, 121] (coronavirus); 15.2.3, [130, 132, 134]; 15.3.1, [25, 26, 30, 141, 146, 148]; 15.3.2, [11, 150, 154, 158, 163]; 15.3.3, [58, 164, 166, 168]

Chapter 16
Entry of Enveloped Viruses into Host Cells: Membrane Fusion

Vicente Mas and Jose Antonio Melero

† *Prof J.A. Melero prematurely passed away. This chapter is dedicated to his memory.*

Abstract Viruses are intracellular parasites that hijack the cellular machinery for their own replication. Therefore, an obligatory step in the virus life cycle is the delivery of the viral genome inside the cell. Enveloped viruses (i.e., viruses with a lipid envelope) use a two-step procedure to release their genetic material into the cell: (1) they first bind to specific surface receptors of the target cell membrane and then (2) they fuse the viral and cell membranes. This last step may occur at the cell surface or after internalization of the virus particle by endocytosis or by some other route (e.g., macropinocytosis). Remarkably, the virus–cell membrane fusion process goes essentially along the same intermediate steps than other membrane fusions that occur, for instance, in vesicular fusion at the nerve synapsis or cell–cell fusion in yeast mating. Specialized viral proteins, fusogens, promote virus–cell membrane fusion. The viral fusogens experience drastic structural rearrangements during fusion, releasing the energy required to overcome the repulsive forces that prevent spontaneous fusion of the two membranes. This chapter provides an overview of the different types of viral fusogens and their mode of action, as they are currently known. Furthermore, it outlines novel strategies for vaccine development related to stabilized viral fusogens.

Keywords Enveloped viruses · Viral fusogens · Class I, II, and III fusion proteins · Membrane fusion intermediates · Virus entry

V. Mas (✉)
Unidad de Biología Viral, Centro Nacional de Microbiología, Instituto de Salud Carlos III, Madrid, Spain
e-mail: vmas@isciii.es

M. G. Mateu (ed.), *Structure and Physics of Viruses*, Subcellular Biochemistry 105,
https://doi.org/10.1007/978-3-031-65187-8_16

Abbreviations

ACE	Angiotensin-converting enzyme 2
Cryo-EM	Cryogenic electron microscopy
EFC	Entry fusion complex
FAST	Fusion-associated small transmembrane
FP	Fusion peptide
G	Glycoprotein
GPI	Glycosylphosphatidylinositol
HA	Hemagglutinin
HIV	Human immunodeficiency virus
hMPV	Human metapneumovirus
HN	Hemagglutinin-neuraminidase
HRA, HRB, HR1, and HR2	Heptad repeat sequences A, B, 1, and 2, respectively
HSV	Herpes simplex virus
PC	Phosphatidylcholine
PE	Phosphatidylethanolamine
RSV	Respiratory syncytial virus
SARS-CoV-2	Severe acute respiratory syndrome coronavirus 2
SFV	Semliki forest virus
SNARE	Soluble N-ethylamine sensitive factor attachment receptor protein
TBEV	Tick-borne encephalitis virus
TM	Transmembrane
VSV	Vesicular stomatitis virus

16.1 Introduction

Enveloped viruses are characterized by having a lipid bilayer (envelope) surrounding the virus particle (virion). One or several virus-encoded glycoproteins are inserted into the envelope. These proteins are exposed at the virion surface and are responsible for the initial interactions of the virus with the target cell, leading to "virus entry." In fact, it is only the cargo inside the envelope layer and not the virus itself what is discharged inside the cell. The virus glycoproteins are also the main targets of the neutralizing antibody response produced by the host in its defense against the virus, playing a crucial role as antigens in the development of vaccines.

To infect a new cell, the virus particle should first attach to the cell surface through non-covalent interactions of one or more of the viral glycoproteins anchored into the lipid bilayer with specific cell surface receptors. These interactions are described in detail in Chap. 15. Suffice to say here that virus-receptor binding is one of the factors that can influence virus tropism; i.e., which cell types are actually infected by the virus.

For enveloped viruses, fusion of the viral and cell membranes is an obligatory step that follows virus binding to cells. Virus–cell fusion is therefore the step at which the virus particle loses its individuality. Membrane fusion may proceed at the cell surface or alternatively after internalization of the virus particle, generally by endocytosis. In either case, fusion is driven by specialized viral glycoproteins (fusogens) which are activated (triggered) by specific events occurring either at the cell surface or inside the endosome. The viral fusogens are in metastable conformations in the virus particle. Once triggered, they initiate a series of conformational changes (in most cases irreversibly) that facilitate approximation of the two membranes, followed by fusion. At the end of the fusion process, the viral fusogens adopt highly stable conformations. The free energy liberated during the transition from the metastable pre-fusion to the highly stable post-fusion conformation drives the fusion process.

16.2 General Principles of Membrane Fusion

16.2.1 Protein-Free Membrane Fusion

Lipid mixing occurs spontaneously in monolayers, but several forces prevent the spontaneous mixing of lipids between bilayer membranes [1]. Most important among them are: (1) hydrophobic effects that seek to minimize solvent-exposed surfaces, (2) elastic forces that prevent monolayer deformation, and (3) electrostatic repulsions between negatively charged phospholipids. Nevertheless, fusion between protein-free lipid bilayers (e.g., liposomes) can be induced under certain conditions. For instance, certain phospholipids (e.g., phosphatidylcholine, PC) induce positive curvature of the lipid monolayer whereas others (e.g., phosphatidylethanolamine, PE) induce negative curvature. The distribution of PC and PE between the two leaflets of the lipid bilayer can either promote or inhibit spontaneous protein-free membrane fusion. Also, direct dehydration between bilayers promotes fusion by bringing the two membranes into very close contact. However, under most physiological conditions, specialized proteins are needed to overcome the repulsive forces that prevent membrane fusion.

Independently of the driving machinery, fusion of two lipid bilayers occurs in a stepwise manner that includes the formation of an hourglass-like structure, known as the lipid stalk (Fig. 16.1a) [2]. This stalk is then expanded forming what is called the hemifusion diaphragm, in which lipids of the two distal leaflets of the bilayers are now in direct contact. Finally, rupture of the hemifusion diaphragm leads to formation of the fusion pore that is then expanded to complete membrane fusion and content mixing of the two compartments. However, opening of the fusion pore may be a reversible step and does not always lead to full fusion [3]. Energy is therefore required through all steps of the fusion process, including expansion of the fusion

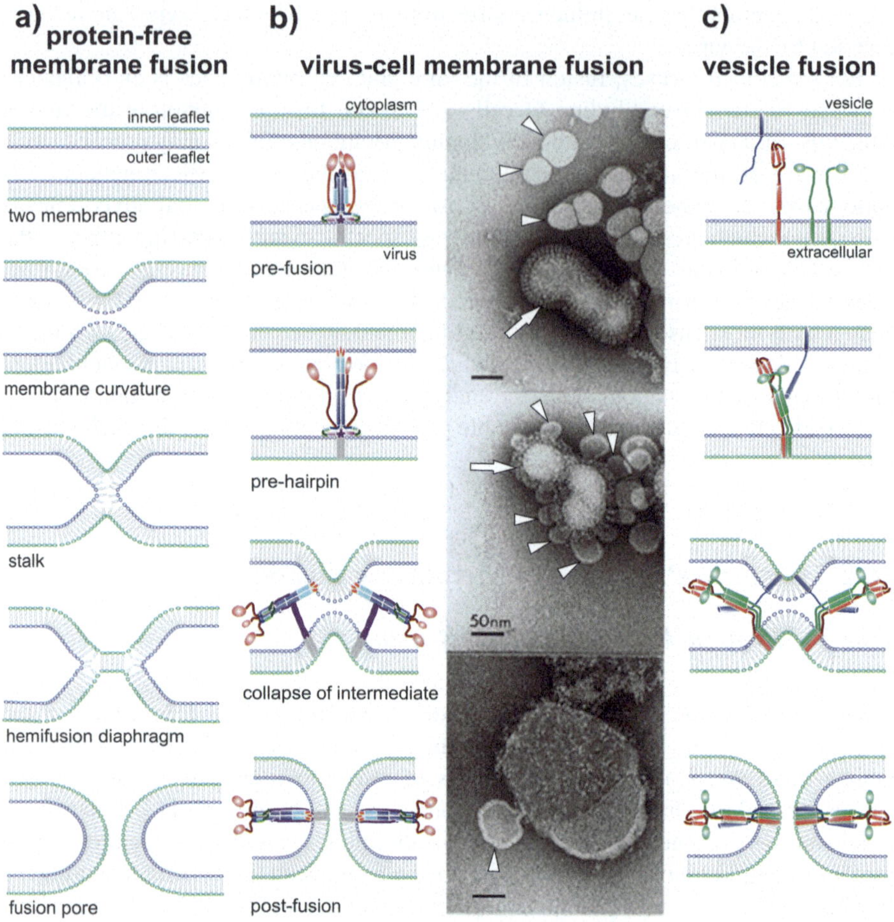

Fig. 16.1 Steps of the membrane fusion process. (**a**) Diagram of the fusion steps between two protein-free lipid bilayers. From top to bottom: the lipids (represented by heads and tails) of the two bilayers are initially curved into nipple-like structures that approach the two membranes. This is followed by formation of the stalk in which the two proximal leaflets are fused. This stalk is then expanded forming a hemifusion diaphragm in which lipids of the distal leaflets of the bilayers are now in direct contact. Finally, rupture of the hemifusion diaphragm leads to formation of the fusion pore. (**b**) Diagram of the virus–cell fusion process: as an example, fusion mediated by the influenza HA is illustrated. Left panels: HA is a homotrimer that initially binds to the cell surface (not shown) by interactions of each monomer head with sialic acid. Then, the virus is internalized by endocytosis. For simplicity, only one HA trimer in the pre-fusion conformation is shown, anchored into the viral membrane. After endosome acidification, the HA globular head falls apart, allowing refolding of the molecule to produce three long α-helices. The fusion peptides, placed at the N-terminal end of each α-helix, insert into the target membrane. This intermediate, dubbed pre-hairpin, collapses to bring the two membranes into proximity leading to formation of the lipid stalk followed by formation of the hemifusion diaphragm (not shown). Finally, the fusion pore is formed by the concerted action of several HA molecules that adopt a very stable post-fusion conformation. Right panels: Electron micrographs of negatively stained mixtures of liposomes (some are indicated by arrowheads) and influenza virus (arrows). Upper panel, the liposome and virus mixture were incubated at neutral pH. Middle panel, the liposome and virus mixture were exposed at pH 5.0

pore. Hence, proteins when present must operate from the initial stages of membrane deformation to final merging of the two membranes.

16.2.2 Virus-Induced Membrane Fusion

Enveloped viruses contain specialized surface glycoproteins that mediate: (1) initial binding of virus to the cell surface and (2) fusion of the virus and cell membranes, either at the cell surface or after endocytosis [4]. The initial attachment of the virus to the cell surface may involve only one type of viral glycoprotein (e.g., influenza virus) or it may require the combined action of several viral glycoproteins (e.g., herpesvirus). However, the actual process of membrane fusion is driven by a specialized type of viral glycoproteins (the viral fusogens) which may or may not have participated additionally in attachment.

Although the sequences of viral fusion proteins vary considerably, they all share certain structural characteristics and are subject to analogous structural rearrangements during the fusion process. As an example, Fig. 16.1b (left panels) illustrates the structural changes that the influenza virus hemagglutinin (HA) undergoes during fusion of the viral and the endosomal membrane. In this case, influenza HA also mediates the initial interaction of the virus with sialic acid of glycoproteins or glycolipids at the cell surface (see Chap. 15). After this initial binding, the virus is internalized by endocytosis. Acidification of the endosome (probably by fusion with lysosomes) triggers HA to start the conformational changes depicted in Fig. 16.1b (left panels) and described next.

Independently of the triggering event, all viral fusion proteins undergo conformational changes upon activation that lead to the formation of an extended unstable intermediate, dubbed pre-hairpin (Fig. 16.1b, "pre-hairpin"). Formation of the pre-hairpin intermediate involves very large-scale structural rearrangements in the viral fusogens with exposure of hydrophobic segments or loops (the fusion peptide, FP). Since the hydrophobic fusion peptide cannot be exposed to a hydrophilic

Fig. 16.1 (continued) for 5 s and then neutralized (note several liposomes bound to a virus particle). Lower panel, the liposome virus mixture were incubated for 30 min at pH 5.0. Note a large vesicle derived from fusion of virus with several liposomes, one of them indicated by the arrowhead in the late stages of fusion. The bar scale is 50 nm. (Courtesy of L.J. Calder and S.A Wharton, Division of Virology, NIMR, MRC, London, UK). (**c**) Diagram of vesicle fusion at the synaptic junction: initially one of the SNARE proteins (synaptobrevin, blue) is inserted into the vesicle membrane, while three other SNAREs (two SNAP 25, green and one syntaxin, red) are inserted in the plasma membrane. After an initial interaction, refolding of the SNAREs leads to formation of a bundle of four parallel α-helices that drives approximation of the two membranes and formation of the stalk and hemifusion intermediates (not shown). Completion of the SNARE complex results in formation of the fusion pore. In the two lower panels of parts (**b**) and (**c**), two HAs and two SNARE complexes are shown surrounding the fusion pore, although the actual number of molecules involved in fusion pore formation is likely to be higher environment it inserts into the target membrane. At this point the viral and target membranes are bridged by two separate segments of the same polypeptide; one is the fusion peptide bound to the target membrane and the other is the transmembrane (TM) region of the viral fusogen inserted into the viral membrane. The pre-hairpin intermediate may have a relatively long half-life; for the human immunodeficiency virus (HIV) gp-41 protein, the half-life seems to be several minutes [5], but in other cases such as influenza HA or West Nile virus E, it may only be a few seconds [6, 7]. The pre-hairpin bridge then collapses bringing into close apposition the viral and target membranes, which are distorted probably into protein-free nipple-like configurations (Fig. 16.1b, "collapse of intermediate"). This is followed by the formation of a lipid stalk and a hemifusion diaphragm (in analogy with the protein-free membrane fusion path), which allows lipid mixing between the two proximal leaflets of the viral and target membranes. Finally, the hemifusion diaphragm opens to form a transient fusion pore that may flicker open and closed until it expands [8], leaving the viral fusogen in a highly stable post-fusion hairpin conformation inserted into the target membrane (Fig. 16.1b, "post-fusion").

Influenza HA-dependent membrane fusion can be reproduced in the test tube with purified virus and liposomes and observed by electron microscopy, as illustrated in Fig. 16.1b (right panels). After mixing of virions (arrows) and liposomes (arrowheads) they remain separated (upper panel), since the former lack the influenza virus receptor (sialic acid). However, a brief pulse at low pH (middle panel) exposes the fusion peptide of the influenza HA which is then inserted into the membrane of multiple liposomes. Longer pulses of low pH result in fusion of the virus with multiple liposomes leading to formation of large vesicles (lower panel; note two vesicles, one small and one large, caught in the process of fusion).

The fusion peptides probably insert only into the outer leaflet of the target membrane. Due to the large number of fusogen molecules present at the viral surface, multiple fusion peptides may interact with the external leaflet of the target membrane upon formation of the pre-hairpin intermediate, potentially initiating membrane deformation. This suggests that cooperativity between several viral fusogens may be required for membrane fusion. Thus, fusion mediated by the influenza HA is positively affected by protein density. It is estimated that 3–4 HA molecules are required for fusion [9], forming a protein ring at the periphery of the fusion pore. Also, electron microscopy and X-ray crystallography results indicate that the E1 glycoprotein of Semliki Forest virus interacts cooperatively during membrane insertion and fusion [10].

Despite the above arguments in favor of cooperativity, calculations indicate that the energy barrier that must be overcome en route to a hemifusion diaphragm is about 40–50 kcal.mol^{-1}. A free energy of roughly this magnitude could be recovered from the collapse of one or two pre-hairpin intermediates, depending on the interactions driving such collapse. In fact, experiments with HIV suggest that only one or two active envelope glycoproteins are sufficient for fusion [11]. However, subsequent estimates have indicated a higher range, which varies depending on the viral strain, exhibiting a direct correlation with virus infectivity and entry kinetics [12, 13]. It may be that the fusion proteins of HIV and other retroviruses have evolved to manage with a single fusion protein, as the number of envelope glycoproteins in the virus particle (estimated 15–20, in contrast to hundreds in other viruses) is rather sparse.

Formation of the pre-hairpin structure and refolding of this intermediate entails some of the most drastic protein rearrangements ever found in biology. Pre-hairpin collapse involves folding back of the membrane proximal domain of the viral fusogen onto a trimeric core whose distal end from the viral membrane is inserted into the target membrane (Fig. 16.1b). Zippering together of these two domains brings the membranes into close proximity. Dehydration of the initial contact site induces monolayer rupture resulting in lipid stalk formation and hemifusion. However, formation of the fusion pore requires further structural rearrangements, including interactions between regions adjacent to the fusion peptide and the transmembrane region [14, 15] and, probably, additional contacts between these two hydrophobic regions that are now inserted into the same membrane. For instance, membrane fusion by the influenza HA with a glycosylphosphatidylinositol (GPI) anchor replacing the TM region halts at the hemifusion stage [16].

Finally, enlargement of the initial fusion pore is the most energy demanding step and requires the coordinated action of several fusogen molecules that surround the early nipple-like fusion intermediate [17, 18].

16.2.3 Vesicle and Cell–Cell Fusion

This topic is brought here only to emphasize the analogies and differences between membrane fusions promoted by unrelated proteins. Vesicle fusion is required for essential biological processes, such as exocytosis and synaptic transmission. Cell–cell fusion is involved in hypodermal cell fusion in *C. elegans*, sperm-egg fusion, yeast mating, placenta formation in mammals, and muscle and bone formation.

In all cases, membrane fusion follows the same steps already described in previous sections; i.e., deformation and approximation of the two membranes, formation of the stalk and hemifusion intermediates, and finally formation and enlargement of the fusion pore (Fig. 16.1c). In analogy with virus–cell fusion, vesicle and cell–cell fusion requires formation of highly stable protein assemblies that provide the energy necessary to overcome the repulsive forces of membranes in close proximity [19]. Also, vesicle and cell–cell fusion, as viral fusion, requires higher order multimerization of the fusogens that delineate the hemifusion diaphragms and the fusion pores [20].

The main difference between virus–cell fusion and vesicle or cell–cell fusion is that in the former process the protein fusogen is present only in the viral membrane. In contrast, the proteins involved in vesicle fusion and cell–cell fusion are initially inserted in the two membranes predestined to fuse (Fig. 16.1c). In synaptic vesicles, the main proteins responsible for membrane fusion are the so-called SNARE (soluble N-ethylamine sensitive factor attachment receptor protein) proteins [21] which share a conserved 60–70 amino acid motif. These proteins, when they find each other refold into a highly stable four-helix parallel coiled-coil bundle that resembles the six-helix bundle formed by the heptad repeats of certain viral fusogens (see below). Formation of the four-helix bundle leads to membrane apposition and hemifusion, as with the collapse of the pre-hairpin intermediate of viral fusogens. However, a unique characteristic of vesicle fusion is that the protein machinery involved in the process is disassembled, once fusion is finished to be reused in subsequent fusion events. This is accomplished by the A + ATPase (adenosine triphosphate) N-ethylmaleimide sensitive factor (NSF) [22].

In contrast to vesicle fusion, cell–cell fusion entails the same set of fusion proteins in the two membranes. For instance, the exceptional process of hypodermal cell fusion in *C. elegans* to form a large multinucleated syncytium of all skin cells is driven by the EFF-1 protein [23]. Unlike SNAREs and viral fusogens, EFF-1 has a homotypic fusion machinery in the opposite membrane. In other words, both membranes must have EFF-1 for fusion to occur. Nevertheless, cell–cell fusion is a multistep process that goes along the same lipid intermediates than viral and vesicle fusion.

16.3 Viral Fusion Proteins

Based on biosynthetic and structural characteristics, viral fusogens have been classified into three categories (Table 16.1). Class I fusion glycoproteins are characterized by being synthesized as inactive precursors that require proteolytic processing to become fusion-competent. They are all homotrimers that upon fusion refold into hairpins containing a long trimeric central coiled-coil core. Class II fusion glycoproteins are derived from longer polyprotein precursors that are proteolytically processed during biosynthesis. The class II fusion proteins form icosahedral scaffolds of protein dimers at the viral surface. During fusion, these proteins undergo an oligomeric rearrangement, converting the metastable pre-fusion dimer into a stable hairpin homotrimer composed of β-sheet structures. Finally, class III glycoproteins are not proteolytically processed. Their post-fusion hairpin trimer displays a central α-helical coiled-coil, as class I glycoproteins, but the fusion domain exposes two fusion loops located at the tip of an elongated β-sheet, revealing a striking convergence with class II fusion proteins.

16.3.1 Class I Viral Fusion Proteins

The first atomic structure of any viral or cellular glycoprotein was determined by X-ray crystallography and reported in 1981 by the laboratories of Wiley and Skehel [24]. It was the structure of the influenza hemagglutinin (HA) trimeric ectodomain, as released from the virus particles by bromelain treatment, which cleaves the HA polypeptides near the TM region.

Table 16.1 Classification of viral fusion proteins

Class	Virus family	Representative	Viral Fusion Protein	Fusion Subunit	Involved in attachment
I	*Orthomyxoviridae*	Influenza virus	Hemagglutinin (HA)	HA2	Yes
	Retroviridae	Human immunodeficiency virus (HIV)	Envelope glycoprotein gp160	gp41	Yes
	Filoviridae	Ebola virus	GP glycoprotein	GP2	Yes
	Coronaviridae	Severe acute respiratory syndrome coronavirus 2 (SARS-CoV-2)	S glycoprotein	S2	Yes
	Arenaviridae	Lassa virus	GPC glycoprotein	GPC2	Yes
	Paramyxoviridae	Sendai virus	F glycoprotein	F1	No
	Pneumoviridae	Respiratory syncytial virus (RSV)	F glycoprotein	F1	No
II	*Togaviridae (Alphavirus)*	Semliki Forest virus	E1 glycoprotein		No
	Flaviviridae	Dengue virus	E glycoprotein		Yes
	Bunyaviridae (Phlebovirus)	Rift Valley fever virus (RVFV)	G_C glycoprotein		Yes
III	*Rhabdoviridae*	Vesicular stomatitis virus	G glycoprotein		Yes
	Baculoviridae	Baculovirus	gp64 glycoprotein		Yes
	Herpesviridae	Herpes simplex virus	gB glycoprotein		No

Influenza HA is synthesized in the infected cell as a polypeptide precursor (HA0) of about 550 amino acids that is cleaved proteolytically to generate the HA1 (roughly the N-terminal two thirds) and HA2 (the C-terminal third) chains that remain covalently linked by a disulfide bond. At the newly created HA2 N-terminus there is a stretch of hydrophobic amino acids, called the fusion peptide, which is inserted into the target membrane during fusion. The overall structure of the influenza HA is that of an elongated spike sticking out of the membrane. The membrane-distal head, formed exclusively by HA1 sequences, bears the receptor (sialic acid) binding site, formed by a shallow pocket exposed on its outward-forming surface (Fig. 16.2a). The stem, primarily composed of HA2 amino acids, contains a sequence with a high propensity to form alpha-helix, known as heptad repeat (HR1), resulting in the formation of a trimeric α-helical coiled-coil.

Recent studies on influenza HA, using diverse techniques such as cryogenic electron microscopy (cryo-EM), X-ray crystallography or single-molecule fluorescence resonance energy transfer (smFRET), have provided the most complete picture of the membrane fusion process driven by class I fusion glycoproteins

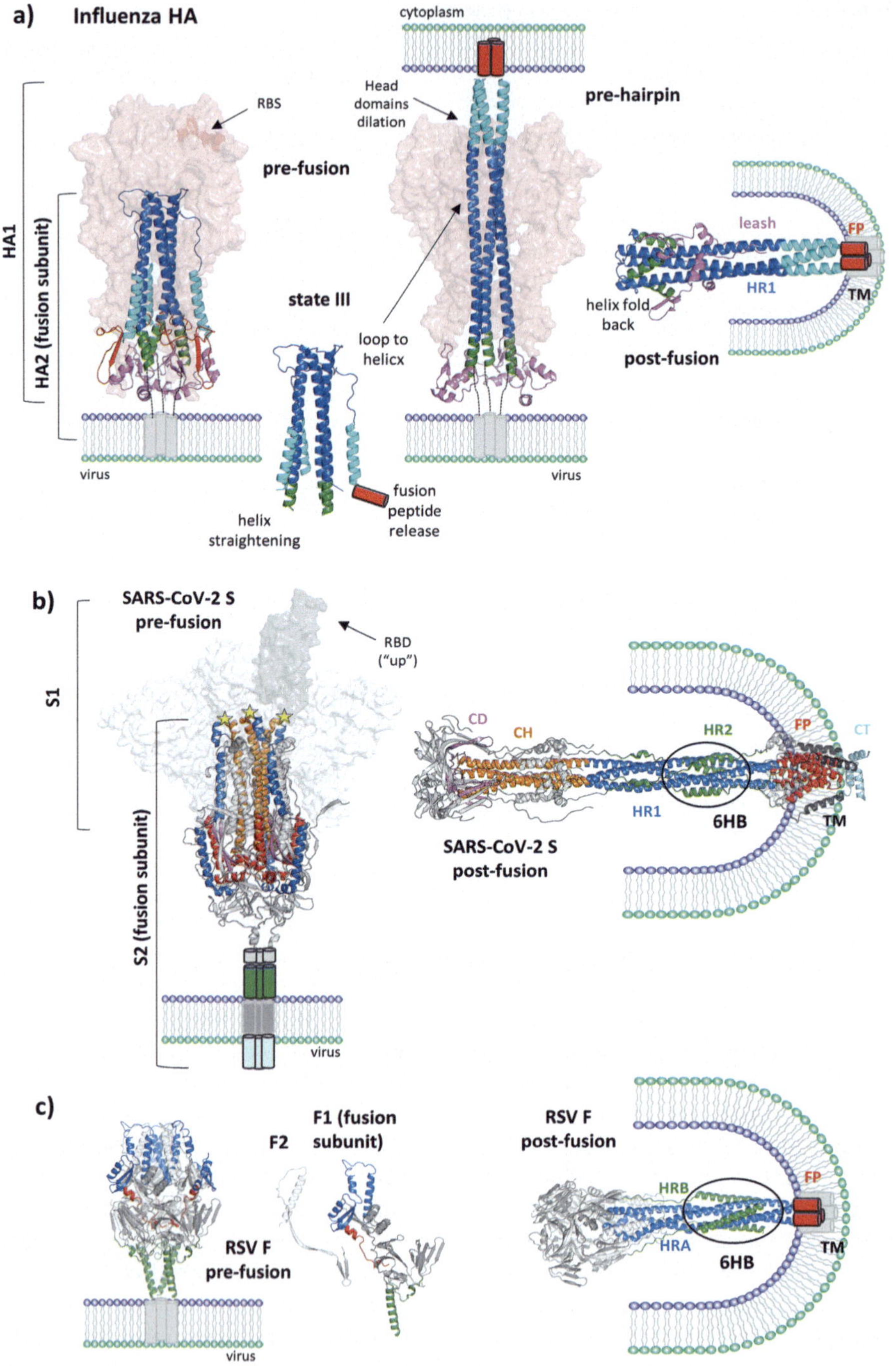

Fig. 16.2 Structure of class I membrane fusion proteins. (**a**) From left to right structures of pre-fusion, intermediate state III, pre-hairpin, and post-fusion conformations of influenza HA [25]. HA trimers are displayed in ribbons colored as in the schematical model of membrane fusion

[25, 30] (Fig. 16.2a). Upon activation of the protein at acidic pH, the head domains initiate a continuous process of gradual dilation. This, coupled with the straightening of the long helix from the C-terminal end, the release of the fusion peptide, and the transition of loops into helices, facilitates the projection of a long trimeric central coiled-coil core structure with the fusion peptides at the N-terminal end, leading to the formation of the extended pre-hairpin form (Fig. 16.2a). In the final hairpin conformation, the upstream polypeptide chains of TM domain, referred to as the leash, extend in an antiparallel manner along the grooves of the central coiled-coil (HR1), while the C-terminal region of the alpha-helix folds backward [25].

Like influenza, other viruses also contain class I fusion glycoproteins that have both receptor and membrane fusion activities (Table 16.1). For instance, the envelope glycoprotein of HIV that is also proteolytically processed and that binds to protein receptors (CD4) and chemokine co-receptors before engaging in membrane fusion at the cell surface. Similarly, the spike glycoprotein of the severe acute respiratory syndrome coronavirus 2 (SARS-CoV-2) undergoes initial cleavage by furin like proteases at the S1/S2 cleavage site within virus-producing cells, leading to the generation of receptor-binding (S1) and fusion (S2) subunits [31]. Structure determination by cryo-EM of a SARS-CoV-2 S protein (spike) in its pre-fusion metastable conformation [32] and in its post-fusion state [27] are shown in Fig. 16.2b.

Once the spike is incorporated into the viral envelope in the pre-fusion conformation, the spike receptor-binding domain in the "up" conformation interacts with the cellular receptor, the angiotensin-converting enzyme 2 (ACE-2). This interaction destabilizes the pre-fusion conformation leading to the shedding of the S1 subunit and the transition of the S2 subunit to the stable hairpin post-fusion conformation. However, for efficient membrane fusion to occur, a second cleavage in S2 (at the S2′

Fig. 16.2 (continued) displayed in Fig. 16.1b. The HA1 (receptor binding) and HA2 (fusion) subunits are delimited in pre-fusion form. For clarity, HA1 is not displayed in intermediate state III and post-fusion conformation. The main protein domains and structural rearrangements involved in membrane fusion are indicated. (**b**) Ribbon representation of the atomic structures of the SARS-CoV-2 S protein in the pre-fusion [26] (left) and post-fusion [27] conformations (right). The S1 (receptor binding) and S2 (fusion) subunits are delineated in the pre-fusion form, while the main protein domains involved in membrane fusion, highlighted in distinct colors, are indicated in the post-fusion conformation. The six-helix bundle structure (6HB) in the post-fusion formation is emphasized with an oval. Sites where proline substitutions (S-2P, see Sect. 16.4) have been engineered for pre-fusion stabilization are marked by yellow stars in each protomer. (**c**) Ribbon representation of the atomic structures of the RSV F protein trimer in the pre-fusion [28] (left) and post-fusion conformation [29]. The panel adjacent to the pre-fusion trimer shows separated protein subunits. Membrane fusion elements, similar to those observed in the SARS-CoV-2 S protein, are illustrated. In all the structures (**a**, **b**, **c**) the location of fusion peptides, transmembrane domains spike that are not resolved in structures are shown schematically. This includes the cytoplasmic tail in the pre-fusion conformation of the SARS-CoV-2 spike. FP: fusion peptide. TM: transmembrane domain. CT: cytoplasmic tail. HR1: heptad repeat 1. HR2: heptad repeat 2. CD: connector domain. CH: central helix. HRA: heptad repeat A. HRB: heptad repeat B. RBS: receptor-binding site. RBD: receptor-binding domain site) is necessary following S1 shedding. This cleavage is mediated either by the host serine protease TMPRSS2 or by the endosomal cysteine protease cathepsin L, enabling membrane fusion at the cell surface or within endosomes, respectively [18, 31, 33].

Unlike HA, the hairpin post-fusion form of the SARS-CoV-2 spike is characterized by the formation of a six-helix bundle structure (6HB). This 6HB involves two heptad repeat sequences HR1 and HR2 within the protein ectodomain. HR1 is located adjacent to the fusion peptide, while HR2 is positioned proximal to the transmembrane (TM) region, which is situated near the C-terminus. The HR2 sequences wrap around the HR1 coiled-coil, forming the stable 6HB and providing most of the energy required to overcome membrane repulsion. Although the length and positioning may vary depending on the viral fusogen, the formation of the 6HB structure is a common characteristic shared by most class I fusion glycoproteins, including the gp41 chain of the HIV envelope protein or the GP2 fusion subunit of the Ebola glycoprotein.

In the case of paramyxoviruses or pneumoviruses the attachment and fusion activities reside in two different surface glycoproteins. The attachment protein (named HN, H, or G) is required for the initial interaction of the virus with the cell surface. Once the virus is bound to the cell, the other major viral glycoprotein (called F, for fusion) is triggered to promote fusion of the viral and cell membranes. Structure determination by X-ray crystallography of a prototypic F protein of respiratory syncytial virus (RSV F) in its pre-fusion conformation [28] and in its post-fusion state [29] are depicted in Fig. 16.2c. The RSV F, as other class I glycoproteins, is synthesized as an inactive precursor (F0) that is translocated co-translationally to the lumen of the endoplasmic reticulum where it assembles into a trimer. Each F protein monomer is proteolytically cleaved during transport to the cell surface, generating two chains, F2 N-terminal and F1 C-terminal that remain linked by two disulfide bonds. The F1 subunit considers the fusion machinery and contains the fusion peptide and the heptad repeat sequences (N-terminal HRA and C-terminal HRB). In analogy with processes describe before, the extended trimeric coiled-coil of pre-hairpin conformation inserts the fusion peptide to the cellular membrane. This step is followed by zipping of the C-terminal part of the molecule along the core coiled-coil to bring together in the post hairpin form the fusion and TM domains, allowing the formation of a large and extremely stable 6HB structure.

Structural and functional data from representative members of the *Paramyxoviridae* family, such as parainfluenza virus 5 or the emerging zoonotic Nipah virus, reveal that attachment glycoproteins shield activating residues within their N-terminal stalk domains. These residues become exposed upon receptor binding, triggering the activation of the fusion protein through a "provocateur" mechanism [34]. Intriguingly, the F protein of pneumovirus (e.g., RSV and human metapneumovirus, hMPV) does not require co-expression of the attachment protein (G) for cell–cell fusion [35]. Furthermore, deletion mutant viruses have been obtained in which the entire G gene is obliterated. These mutants still infect cells in vitro, although less efficiently than the wild-type virus, and are attenuated in animal models of infection [36]. Activation of the F protein of those deletion mutants cannot rely on interactions with the G protein and therefore alternative regulatory mechanisms should control membrane fusion. Of note, a unique characteristic of the RSV F protein is the presence of two proteolytic cleavage sites (instead of one, as in all other paramyxovirus and pneumoviruses) in the F0 protein precursor [35]. The presence of a double cleavage site in F has been found to influence membrane fusion activation by a still poorly understood G independent mechanism [37].

16.3.2 Class II Viral Fusion Proteins

In contrast to class I fusion proteins, the so-called class II fusion proteins (Table 16.1) in their pre-fusion form do not form independent trimers on the viral membrane. Instead, they form a multimeric assembly that encloses the entire viral membrane. In addition, the class II fusion proteins are derived from a polyprotein precursor that is cleaved during biosynthesis to generate the E1 protein of alphaviruses (e.g., Semliki Forest virus and Chikungunya virus), the E protein of flaviviruses (e.g., dengue virus, Zika virus and tick-borne encephalitis virus, TBEV), or the G_C of bunyaviruses (e.g., Rift Valley fever virus). The three proteins fold co-translationally with a companion or regulatory protein, termed p62 for alphaviruses, prM for flaviviruses, and G_N for bunyaviruses [38].

In alphaviruses, the p62-E1 complex is transported to the plasma membrane where they are incorporated into new budding icosahedral virus particles as dimers of p62-E1. p62 is then proteolytically processed into the mature proteins E3/E2. E3 is released in the extracellular environment, while E2 remains bound to the virus where it covers most of the fusion protein E1 and specially its fusion loop. E2 mediates binding to the cell surface receptor.

In contrast, Flavivirus particles bud into the endoplasmic reticulum as immature virions containing prM-E protein complexes. The immature viruses are then transported to the exterior through the exocytic pathway where prM is processed and the peripheral "pr" portion separated from E. The latter protein is then arranged in E-E homodimers at the virion surface with icosahedral symmetry. The flavivirus fusion E protein is additionally responsible for receptor binding.

Bunyavirus particles bud internally within the Golgi apparatus and, upon maturation, once they reach the external environment, they adopt an icosahedral organization composed of heterodimers of G_N and G_C glycoproteins. The G_N or high mannose-type N-linked glycans displayed on one or both of these glycoproteins have been proposed to interact with cellular receptors [39, 40].

The first structure of any class II glycoprotein, solved by X-ray crystallography, was that of the TBE flavivirus E protein ectodomain [41], solubilized from virions by limited trypsin digestion. Similar structures have now been solved for the E ectodomain of dengue virus types 2 and 3 [10]. The polypeptide chain of the E protein follows a complex path, resulting in three globular domains, essentially constituted by β-sheets (Fig. 16.3). The first domain is a β-barrel with up-and-down topology (red). Two adjacent strands in domain I are extended, forming

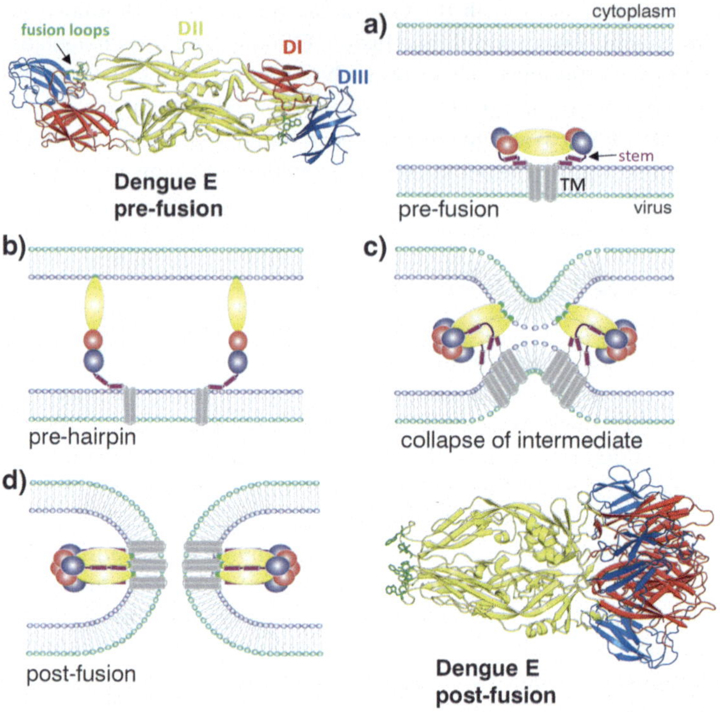

Fig. 16.3 Membrane fusion mediated by a class II fusion protein (Flavivirus). Ribbon representation of the atomic structures of the dengue virus E protein dimer in the pre-fusion conformation (upper left) and the E protein trimer in the post-fusion conformation [42] (lower right). Domains I, II, III, stem, and fusion loops of the E glycoprotein are indicated and colored red, yellow, blue, violet, and green, respectively. (**a**–**d**) Diagram of the fusion process denoting: (**a**) the structure of the flavivirus E glycoprotein dimer already in the endosome before activation, (**b**) dissociation of the E protein monomers, refolding of the fusion domain (yellow) and insertion of the fusion loop (green) into the endosomal membrane, (**c**) formation and refolding of the E protein trimer to approach the two membranes, and (**d**) formation of the fusion pore and stabilization of the E trimer in the post-fusion conformation. In the last two steps, two E protein molecules are represented to indicate the cooperation needed to drive the fusion process

domain II (yellow) which is a long "finger-like" structure that runs parallel to the viral membrane. At the tip of domain II is the hydrophobic fusion loop which remains buried in the virion from the hydrophilic environment by interaction with domain III (blue) of the adjacent monomer in the E-E dimer. Domain I is also connected to domain III which bridges the E ectodomain with the so-called stem region that extends to the TM region of the protein.

Unlike the class I fusion proteins, which are trimeric in their pre- and post-fusion conformations, class II fusion glycoproteins undergo major oligomeric transformations during fusion. As in the case of influenza virus, the flavivirus E protein first binds to a cell surface receptor which induces endocytosis of the virion. Once in the acidic endosome, the E-E homodimer dissociates, resulting in disassembly of the icosahedral scaffold. The individual subunits swing outward through a reorientation of the hinge region that connects domains I and II, and the fusion loops insert into the target membrane. Lateral interactions between monomers facilitate re-clustering into trimers [43]. These rearrangements lead to the formation of an extended trimeric structure, analogous to the pre-hairpin intermediate of class I fusion glycoproteins, in which two different regions of each E polypeptide are inserted into the two membranes to be fused. Collapse of the extended intermediate can proceed by rotation of domain III in each subunit about the segment that links it to domain I and zipping up of the stem alongside the clustered domains II. This refolding brings the two membranes together to initiate formation of the lipid stalk, the hemifusion diaphragm, and the fusion pore.

The structure of the fusion E1 glycoprotein of alphaviruses and G_C of bunyaviruses was found unexpectedly very similar to that of the flavivirus E protein, despite the lack of detectable sequence conservation. E1 and G_C have also three discernible domains, equivalent to those of flavivirus E. In the acidic endosome, companion proteins (E2 and G_N) separate from membrane fusion glycoproteins. These viral fusogens undergo similar conformational changes to those of the flavivirus E protein, leading to fusion of the viral and endosomal membranes. Electron microscopy and X-ray crystallography results provide support for interactions between adjacent E1 trimers when the fusion loops are inserted in the target membrane to produce rings of five or six trimers. It has been postulated that these fivefold interactions would act at the fusion site to induce the formation of a nipple-like curvature in the viral and target membranes, favoring membrane fusion [44]. It is worth noting that, while the companion proteins of viruses such as alphaviruses or bunyaviruses exhibit structural variability, they share a common core composed of a β-ribbon and a C-terminal immunoglobulin-like region [45]. This fact not only highlights the presence of a distant common ancestor for these viruses but also suggests potential implications for improving future antigen designs aimed at vaccine development.

16.3.3 Class III Viral Fusion Proteins

The most well-characterized members of the class III fusion viral glycoproteins include the rhabdovirus G glycoprotein (e.g., vesicular stomatitis virus, VSV), the herpesvirus gB glycoprotein, and the baculovirus gp64 glycoprotein. The pre-fusion [46] and post-fusion [47] structures of the VSV G glycoprotein ectodomain have been elucidated through X-ray crystallography. Initially, only post-fusion conformations for gB and gp64 were determined [48, 49]. However, recent advances, including the introduction of a helix-breaking point mutation into an helix hinge (H516P), a method successfully employed to stabilize class I fusion proteins (as discussed in Sect. 16.4), allowed Vollmer et al. to solve by cryo-EM the structure of the membrane-embedded herpes simplex virus type I (HSV-1) gB in its pre-fusion conformation [50].

Class III fusion glycoproteins are expressed from individual mRNAs and do not require proteolytic processing of either a protein precursor (as in class I proteins) or an accompanying protein (as in class II proteins) for activity. Class III proteins are trimeric before and after fusion and share structural characteristics with both class I and class II fusion glycoproteins, as described below.

The rhabdovirus G protein possesses both receptor-binding and fusion-promoting activities. As in the case of influenza virus, binding of rhabdovirus G to a poorly characterized receptor at the cell surface induces endocytosis of the virus particle. Acidification of the endosome triggers G for membrane fusion. However, and in contrast with all other fusion proteins, the low pH inactivation of rhabdovirus G is reversible. Thus, virions inactivated by prolonged incubation at pH <6 can be reactivated by raising the pH to neutral [51]. This reversibility may be required to allow G to be transported through the acidic Golgi apparatus and to recover its native fusion-competent state when incorporated to new virions [52]. Given this reversibility, it is believed that the energy released during the structural transition of a single trimer from the pre-fusion to the post-fusion conformation is probably small, compared with the energetic barrier of the fusion reaction. In agreement with this hypothesis, the estimated number of rhabdovirus spikes required for fusion is higher (at least 15 trimers) than for other enveloped viruses.

A soluble ectodomain of the VSV_G glycoprotein, released from purified virions treated with thermolysin, was used to solve the structures of the pre-fusion and post-fusion conformations, after exposure to high and low pH, respectively [46, 47]. Several domains could be observed in both structures that are rearranged in their relative orientations during transit from the pre- to the post-fusion structure (Fig. 16.4). In the pre-fusion conformation, the fusion domain contains two fusion loops reminiscent of class II proteins that are oriented downward toward the viral membrane. After low pH exposure, the fusion domain moves upward by flipping relative to the central core of the trimer. Thus, an intermediate equivalent to the pre-hairpin structure of class I proteins is formed. This is followed by the reversal of the molecule around a central rigid block formed by lengthening of the central helix and refolding of the three C-terminal segments into helices that position themselves in the grooves of the central core in an antiparallel manner. This six-helix bundle (6HB) has obvious resemblance with that of the class I proteins.

It is likely that the transition of VSV_G from the pre-fusion to the post-fusion conformation involves disassembly of the trimer into monomers and reassembly into trimers upon interaction of the fusion loops with the target membrane [53]. It is also likely that cooperativity between G glycoproteins is needed to overcome the energy barrier, as mentioned above. As for class II glycoproteins, lattices of G proteins have been observed in virions, particularly in the planar base of the rhabdovirus bullet-shaped particle, which may act to induce nipple-like deformations in the viral and target membranes.

Although it has not been reported, it is likely that low pH exposure also leads to reversible inactivation of the baculovirus gp64 glycoprotein. This protein, as VSV_G, is involved in both receptor binding and membrane fusion after endosome acidification, since baculoviruses also use an endocytic route of entry [54]. In

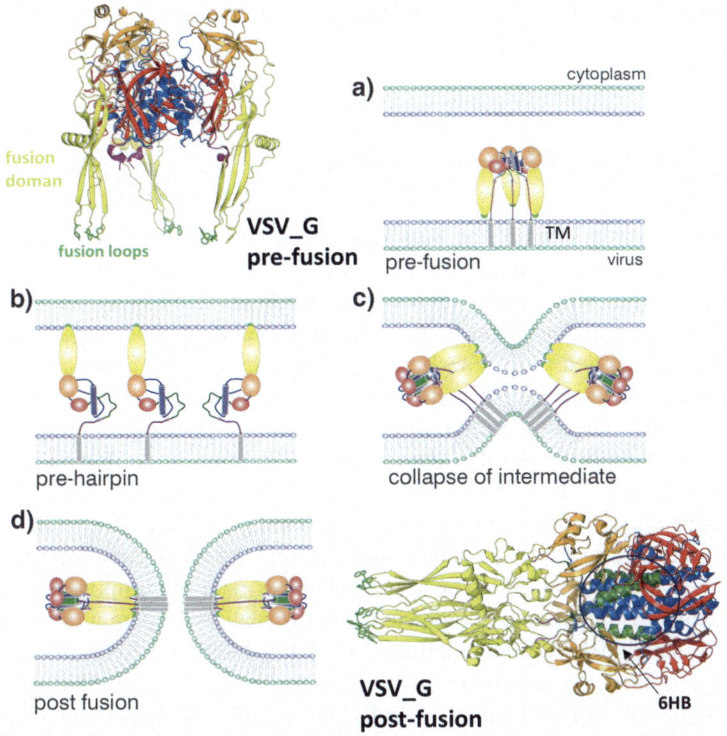

Fig. 16.4 Membrane fusion mediated by a class III fusion protein (Vesicular Stomatitis Virus, VSV). Ribbon representations of the VSV glycoprotein (G), in the pre-fusion conformation [46] (upper left) and post-fusion [47] (lower right) conformation. Domains are colored similarly in all images. The fusion domain is colored in yellow and the fusion loops in green. (**a**–**d**) Diagram of the fusion process denoting: (**a**) the structure of the VSV G glycoprotein trimer already in the endosome before activation, (**b**) dissociation of the G protein subunits, refolding of the fusion domain (yellow) and insertion of the fusion loop (green) into the endosomal membrane, (**c**) formation and refolding of the G protein trimer to approach the two membranes, and (**d**) formation of the fusion pore and stabilization of the G primer in the post-fusion conformation. In the last two steps, two G protein molecules are represented to indicate the cooperation needed to drive the fusion process. 6HB: six-helix bundle

contrast, membrane fusion mediated by herpesvirus gB can occur either at the plasma membrane or in endosomes, depending on the virus and the target cell type. In either case, attachment of herpesvirus to host cells follows a complex mechanism in which several viral glycoproteins interact with cell surface molecules. Some of these interactions trigger fusion, whereas others simply serve to tether the virus to the cell and are dispensable for fusion. The existing structural data on HSV-1 gB in its pre-fusion state demonstrates a notable preservation of conformational domain rearrangements during fusion events involving glycoproteins from both herpesviruses and rhabdoviruses, despite their considerable phylogenetic differences [50].

16.3.4 Other Viral Fusion Proteins

Poxviruses (vaccinia virus is the best known member) represent an extreme case among enveloped viruses, regarding the number of viral glycoproteins required for entry. As for herpes virus, entry can occur by fusion at the plasma membrane or in a low pH-dependent manner from within an intracellular particle, depending on the virus strain and the cell type. Vaccinia virus internalization is believed to occur by macropinocytosis. Four vaccinia virus proteins are involved in attachment to cell surface proteoglycans or laminin [55]. Eleven or twelve other relatively small glycoproteins, ranging in size between 35 and 377 amino acids, form the so-called entry fusion complex (EFC) that mediates membrane fusion [56]. By using conditional lethal mutants of 11 proteins that make the EFC, it was found that eight of them were required to reach the hemifusion step and the other three were needed for completion of virus entry [56]. The structural assessment of components within the poxvirus entry fusion complex conducted so far reveals no apparent homology with other type I, II, and III viral fusion proteins [57]. It is likely that hydrophobic regions of several proteins may assemble in the EFC to form a hydrophobic surface that could bind to the target membrane and drive membrane fusion by some novel mechanism.

Finally, the fusion-associated small transmembrane (FAST) proteins of reoviruses are brought here—despite not being involved in virus entry and reovirus being a non-enveloped virus—because they induce cell–cell fusion and therefore facilitates dissemination of virus to neighboring cells. The FAST proteins are small nonstructural proteins (98–148 amino acids, depending on the viral strain) that are expressed on the surfaces of virus-infected cells, where they induce cell–cell fusion and syncytia formation. Purified FAST proteins, when reconstituted into liposome membranes, induce fusion indicating that they are bona fide fusogens [58, 59]. The orientation of the FAST polypeptides in the cell membrane is also unique among viral fusogens, with a relatively short N-terminal ectodomain followed by a transmembrane region and a long C-terminal cytoplasmic tail. Although they lack a fusion peptide, a relatively hydrophobic region near the N-terminus which is additionally myristoylated seems to insert into the target membrane to drive membrane fusion, at least for certain FAST proteins [60]. Based on structural and functional studies, a FAST protein mediated cell–cell fusion has been proposed. Initially, FAST proteins cluster together in membrane microdomains in the pre-fusion state. Subsequently, cellular cadherins facilitate close membrane apposition, while the cooperative action of FAST protein ectodomains and transmembrane domains induces membrane curvature modifying lipid hydration and lipid packing. Due to their dynamic structure, these ectodomains interact reversibly with both membranes, ultimately leading to hemifusion, lipid mixing, and pore formation [59].

16.4 Novel Viral Vaccines: Targeting Viral Fusogens Through Structure-Based Vaccine Design

Classical vaccines against diseases caused by some enveloped viruses induce antibody production primarily by targeting viral envelope glycoproteins to prevent pathogen entry into host cells. It is crucial to emphasize that, for a protective immune response, vaccine antigens must closely resemble a portion of the pathogen's structure. To address limitations in vaccine development, an innovative approach referred to as "structure-based vaccine design" aims to stabilize antigens in their more vulnerable conformations. Recent advances in protein structure and immunology have underscored the metastable pre-fusion conformation as the main target for vaccine development, being the most vulnerable state adopted by viral fusogens.

The progression of the "structure-based vaccine design" owes much to the development of HIV-1 vaccine approaches, serving as a catalyst. However, the first clinical proof-of-concept for this innovative approach has been effectively proved in the case of RSV [61, 62]. This success has, in turn, sparked the adoption of analogous strategies for the development of vaccines targeting SARS-CoV-2. For both respiratory viruses and owing to the accumulated structural knowledge of their respective viral fusogens over the years, the vaccine antigen design has been directed toward the isolation of pre-fusion stabilized glycoproteins. These strategies have demonstrated the ability to evoke a robust protective immune response against severe respiratory illnesses.

The key principles illustrated in Fig. 16.5 for these innovative strategies involve simple or combined sequence modifications. One extensively characterized example of such strategies is the DSCav-1 subunit of the RSV F protein (Fig. 16.5a) [28]. In this strategy to facilitate the formation of a trimeric soluble subunit from a type I membrane fusion protein, the transmembrane and cytoplasmic tail domains are substituted with the trimerization domain of T4 fibritin (foldon). Additionally, a pair of cysteine substitutions is introduced to create a disulfide bond (DS), and surface cavities are filled by replacing specific residues with amino acids containing larger hydrophobic side chains (Cav-1). This approach effectively maintains the pre-fusion conformation by inhibiting the refolding of secondary structure elements that typically occur during viral membrane fusion, as described in preceding sections.

Four of the most effective approved vaccines for SARS-CoV-2 use a different strategy to stabilize the molecule's stem (S2). This approach involves a singular principle: proline substitutions. The lack of an amine hydrogen in proline residues prevents their involvement in the α-helical hydrogen-bonding network. As a result, two prolines' substitutions (S-2P, marked with stars in Fig. 16.2b) at the loop between HR1 and central helix restrict the formation of the trimeric extended coiled-coil of pre-hairpin structure of S protein [32].

It is noteworthy that, unlike the antibody response elicited against the S protein of SARS-CoV-2, which primarily focuses on blocking receptor binding, RSV vaccines induce an effective response exclusively directed to the membrane fusion process.

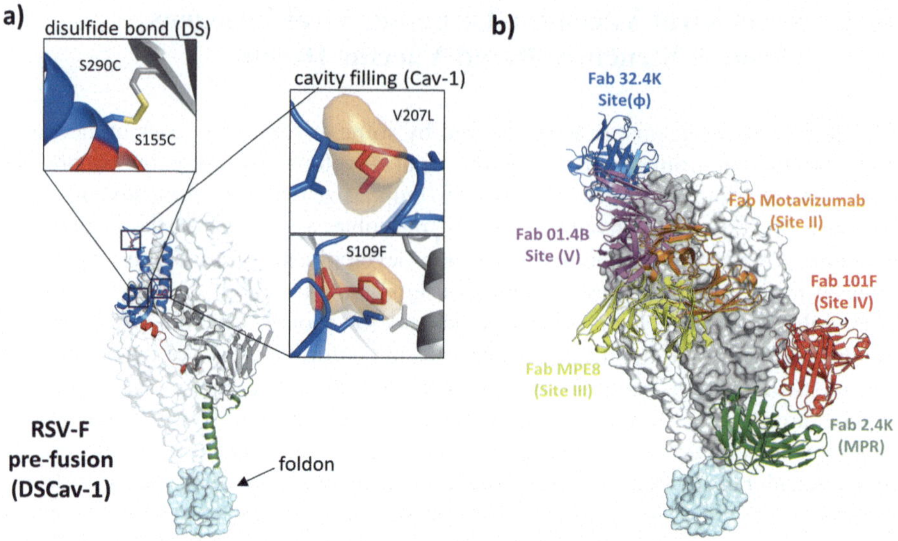

Fig. 16.5 RSV vaccine design by stabilization of RSV F protein in its pre-fusion conformation. (**a**) Structure of construct DSCav-1 with modeled C-terminal foldon. The trimer is depicted with two of the three protomers shown in molecular surface representation, and the third protomer displayed in ribbon representation, color-coded as in Fig. 16.2. Insets illustrate sequence modifications introduced for F protein stabilization: a disulfide bond (DS) and the filling of cavities (Cav-1). (**b**) Antibody responses induced by RSV-F. The binding of neutralizing antibodies to the RSV F protein is modeled based on resolved structures of complexes between the viral glycoprotein and antibody Fabs (antigen-binding fragments). The generated model incorporates representative antibodies targeting the major antigenic sites identified on the F protein, covering nearly the entire F molecular surface. These sites include site ϕ (Fab 32.4k, PDB: 7LUC); site II (Fab Motavizumab, PDB: 4YZP); site III (Fab MPE8, PDB: 5UL8); site IV (Fab 101F, PDB: 3O45); site V (Fab 01.4B, PDB: 7LUC); and the MPR (Fab 2.4K, PDB: 8T7A). Fabs are depicted in ribbons using distinct colors, overlaying just one protomer of the F protein (colored in gray). MPR: membrane proximal region

The neutralizing response against the RSV F pre-fusion antigen targets various antigenic sites, encompassing virtually the entire molecular surface (Fig. 16.5b). The most vulnerable sites for neutralization are those located at the membrane-distal apex of the molecule (site ϕ) [63, 64]. Upon binding to the F protein, antibodies directed to this region block viral infectivity by hijacking the pre-fusion conformation, preventing its transition to the post-fusion conformation, and thereby inhibiting membrane fusion. Importantly, antibodies targeting the fusion subunit (S2) of the S protein of SARS-CoV-2 targeting the post-fusion conformation have also been isolated [31]. However, as mentioned earlier, their contribution to neutralization is less pronounced than those targeting receptor binding.

Other modification types developed for various vaccine approaches include the deletion of non-immunogenic domains, as well as the manipulation of electrostatic interactions or hydrogen bonds [65]. The latter contributes modestly to overall protein stability but plays a prominent role in vaccine design. One example of such substitutions has been included in combination with others in the antigen of a recently approved vaccine for RSV. This bivalent vaccine, likely the one developed with the DSCav-1 construct, incorporates pre-fusion stabilized F subunits [66].

16.5 Early Post-Entry Events

Once membrane fusion has been completed, the viral genome—generally in complex with other proteins or inside a viral nucleocapsid—is found for a second time in a cytoplasmic environment. The first time is when the genome assembles in the cytosol of the infected cell or when it is trafficking from the nucleus to the cell exterior, depending on the virus. However, the fate of the incoming genome is now very different and characteristic for each virus.

Most RNA viruses replicate in the cell cytoplasm, although there are exceptions like influenza virus or Borna virus that do so in the nucleus. If the RNA is of positive polarity, like in flavivirus, the genome may act as mRNA to be translated by the cell protein synthesis machinery. In most cases the primary translational product is a polyprotein that matures into the different viral gene products by proteolytic processing [67]. In the case of negative-stranded RNA viruses, like paramyxoviruses, the first step after entry is the transcription of the viral genome to yield the different mRNAs that are translated into the distinct viral gene products [68].

For RNA viruses that replicate in the nucleus, the nucleoprotein complex of the viral genome and associated proteins have to be transported to the cell nucleus for transcription. Most of the viral proteins required in the nucleus have their own nuclear localization signal (NLS), which is a cluster of basic amino acids. However, actual import of the viral ribonucleoprotein into the cell nucleus may require additional interactions with certain host factors. For instance, the NLS of the influenza nucleoprotein interacts with karyopherin α and this is in turn with karyopherin β which mediates interactions with the proteins of the nuclear pore to promote nuclear import of the viral ribonucleoproteins by an energy-dependent process [69].

Retroviruses represent a special case in which an RNA viral genome (diploid), packed in a capsid inside the virus envelope has to be transcribed to DNA before integration into the cell host genome. In this case, fusion of the virus and cell membrane delivers the capsid into the cell cytoplasm, which interacts with cytoskeleton components for transport to the vicinity of the nucleus. Recent findings corroborate the concept that the intact viral core enters the nucleus mediated by CPSF6 (cleavage and polyadenylation specific 6) recruitment to SC35 nuclear speckles. Then, reverse transcription accompanied HIV-1 uncoating occurs within the nucleus near their integration sites just prior to integration [70].

Most DNA viruses replicate in the nucleus, with exceptions like poxviruses (e.g., vaccinia virus). In the case of herpesviruses, the capsids that contain the viral genome are transported from the site of release to the nuclear pore where importin β promotes nuclear import of the viral DNA by an energy-dependent mechanism

[71]. In contrast, for vaccinia virus the virus core is released into the cytoplasm. The virus core has all the machinery for transcription of the early genes that ensues further progress on DNA replication and transcription of the remaining genome [72].

16.6 Perspectives and Conclusions

Given the success of stabilizing viral fusogens in their pre-fusion conformation for developing vaccines against RSV and SARS-CoV-2, there is an expectation that this technology will be extended to other viruses. In this context, diverse approaches for pre-fusion stabilization have started to be explored for viral fusogens from different viruses, including those containing type I (Nipah, Ebola, Metapneumovirus, or Influenza) or type II (Zika virus) membrane fusion proteins [38, 65]. Similarly, these strategies encompass the introduction of disulfide bridges, domain deletions, filling of cavities, or proline substitutions. Moreover, the use of extended oligomerization motifs is proving to be pivotal in geometric antigen presentation on engineered nanoparticles. Recent studies provide support for the notion that viral fusogens, when coupled in ferritin or lumazine synthase nanoparticles, elicit neutralizing antibody responses that surpass those induced by major vaccines [73, 74].

The viral–cell membrane fusion is the final extracellular step for enveloped viruses, offering susceptibility to inhibition by chemical or biological agents without the drawback of the membrane permeability barrier. Beyond vaccine development, viral fusogens have garnered significant attention as ideal targets for designing effective antivirals. Some of these antivirals are already in clinical use, such as the T20 peptide, a fusion inhibitor for HIV, and monoclonal antibodies like Nirsevimab for RSV, or Sotrovimab and the dual antibody treatment Cilgavimab/Tixagemivab for SARS-CoV-2. However, the rapid evolution of viruses may lead to the emergence of resistant viruses, which requires monitoring and adaptation of treatments and vaccines. Recent advances on high-throughput technologies for screening of large libraries of chemical or biological molecules, the isolation of human monoclonal antibodies, and next-generation sequencing hold promising potential. In the near future, these developments are expected to provide a plethora of drugs to fight a wide range of important diseases caused by enveloped virus.

A critical step in viral-induced membrane fusion is the activation of the viral fusogen. While viruses using low pH endocytosis depend on the protonation of specific residues (mostly histidines) to trigger fusion, the activating mechanism in the case of other viruses is still poorly defined. Better understanding of fusion triggering events may therefore bring new possibilities for the manipulation of virus entry.

Understanding the process of virus–cell membrane fusion holds relevance beyond virology and may extend its impact to other fields in biology. For example, the insights gained from studying viral fusogens could significantly contribute to a better understanding of vesicle fusion in areas such as neurobiology.

Acknowledgments Current research activities in the "Biología Viral" (*Viral Biology*) laboratory are funded by grant PI22/00054 from Instituto de Salud Carlos III. This chapter constitutes a thorough update of the equivalent chapter in the first edition of this book. Prof. José Antonio Melero unfortunately passed away, but he has left us his profoundly impactful contribution to the field of viral biology.

References

1. Chernomordik LV, Kozlov MM (2008) Mechanics of membrane fusion. Nat Struct Mol Biol 15:675–683
2. Yang L, Huang HW (2002) Observation of a membrane fusion intermediate structure. Science 297:1877–1879
3. Chanturiya A, Chernomordik LV, Zimmerberg J (1997) Flickering fusion pores comparable with initial exocytotic pores occur in protein-free phospholipid bilayers. Proc Natl Acad Sci USA 94:14423–14428
4. Harrison SC (2008) Viral membrane fusion. Nat Struct Mol Biol 15:690–698
5. Munoz-Barroso I, Durell S, Sakaguchi K, Appella E, Blumenthal R (1998) Dilation of the human immunodeficiency virus-1 envelope glycoprotein fusion pore revealed by the inhibitory action of a synthetic peptide from gp41. J Cell Biol 140:315–323
6. Floyd DL, Ragains JR, Skehel JJ, Harrison SC, van Oijen AM (2008) Single-particle kinetics of influenza virus membrane fusion. Proc Natl Acad Sci USA 105:15382–15387
7. Harrison SC (2015) Viral membrane fusion. Virology 479-480:498–507
8. Zimmerberg J, Blumenthal R, Sarkar DP, Curran M, Morris SJ (1994) Restricted movement of lipid and aqueous dyes through pores formed by influenza hemagglutinin during cell fusion. J Cell Biol 127:1885–1894
9. Ivanovic T, Choi JL, Whelan SP, van Oijen AM, Harrison SC (2013) Influenza-virus membrane fusion by cooperative fold-back of stochastically induced hemagglutinin intermediates. elife 2: e00333
10. Kielian M, Rey FA (2006) Virus membrane-fusion proteins: more than one way to make a hairpin. Nat Rev Microbiol 4:67–76
11. Yang X, Kurteva S, Ren X, Lee S, Sodroski J (2006) Subunit stoichiometry of human immunodeficiency virus type 1 envelope glycoprotein trimers during virus entry into host cells. J Virol 80:4388–4395
12. Magnus C, Rusert P, Bonhoeffer S, Trkola A, Regoes RR (2009) Estimating the stoichiometry of human immunodeficiency virus entry. J Virol 83:1523–1531
13. Brandenberg OF, Magnus C, Regoes RR, Trkola A (2015) The HIV-1 entry process: a stoichiometric view. Trends Microbiol 23:763–774
14. Buzon V, Natrajan G, Schibli D, Campelo F, Kozlov MM, Weissenhorn W (2010) Crystal structure of HIV-1 gp41 including both fusion peptide and membrane proximal external regions. PLoS Pathog 6:e1000880
15. Chen J, Skehel JJ, Wiley DC (1999) N- and C-terminal residues combine in the fusion-pH influenza hemagglutinin HA(2) subunit to form an N cap that terminates the triple-stranded coiled coil. Proc Natl Acad Sci USA 96:8967–8972
16. Kemble GW, Danieli T, White JM (1994) Lipid-anchored influenza hemagglutinin promotes hemifusion, not complete fusion. Cell 76:383–391
17. Chernomordik LV, Kozlov MM (2003) Protein-lipid interplay in fusion and fission of biological membranes. Annu Rev Biochem 72:175–207
18. White JM, Ward AE, Odongo L, Tamm LK (2023) Viral membrane fusion: a dance between proteins and lipids. Annu Rev Virol 10:139–161

19. Fasshauer D, Otto H, Eliason WK, Jahn R, Brunger AT (1997) Structural changes are associated with soluble N-ethylmaleimide-sensitive fusion protein attachment protein receptor complex formation. J Biol Chem 272:28036–28041
20. Rickman C, Hu K, Carroll J, Davletov B (2005) Self-assembly of SNARE fusion proteins into star-shaped oligomers. Biochem J 388:75–79
21. Rizo J, Rosenmund C (2008) Synaptic vesicle fusion. Nat Struct Mol Biol 15:665–674
22. Sollner T, Bennett MK, Whiteheart SW, Scheller RH, Rothman JE (1993) A protein assembly-disassembly pathway in vitro that may correspond to sequential steps of synaptic vesicle docking, activation, and fusion. Cell 75:409–418
23. Mohler WA, Shemer G, del Campo JJ, Valansi C, Opoku-Serebuoh E, Scranton V et al (2002) The type I membrane protein EFF-1 is essential for developmental cell fusion. Dev Cell 2:355–362
24. Wilson IA, Skehel JJ, Wiley DC (1981) Structure of the haemagglutinin membrane glycoprotein of influenza virus at 3 A resolution. Nature 289:366–373
25. Benton DJ, Gamblin SJ, Rosenthal PB, Skehel JJ (2020) Structural transitions in influenza haemagglutinin at membrane fusion pH. Nature 583:150–153
26. Benton DJ, Wrobel AG, Xu P, Roustan C, Martin SR, Rosenthal PB et al (2020) Receptor binding and priming of the spike protein of SARS-CoV-2 for membrane fusion. Nature 588: 327–330
27. Shi W, Cai Y, Zhu H, Peng H, Voyer J, Rits-Volloch S et al (2023) Cryo-EM structure of SARS-CoV-2 postfusion spike in membrane. Nature 619:403–409
28. McLellan JS, Chen M, Joyce MG, Sastry M, Stewart-Jones GB, Yang Y et al (2013) Structure-based design of a fusion glycoprotein vaccine for respiratory syncytial virus. Science 342:592–598
29. McLellan JS, Yang Y, Graham BS, Kwong PD (2011) Structure of respiratory syncytial virus fusion glycoprotein in the postfusion conformation reveals preservation of neutralizing epitopes. J Virol 85:7788–7796
30. Benhaim MA, Mangala Prasad V, Garcia NK, Guttman M, Lee KK (2020) Structural monitoring of a transient intermediate in the hemagglutinin fusion machinery on influenza virions. Sci Adv 6:eaaz8822
31. Le K, Kannappan S, Kim T, Lee JH, Lee HR, Kim KK (2023) Structural understanding of SARS-CoV-2 virus entry to host cells. Front Mol Biosci 10:1288686
32. Wrapp D, Wang N, Corbett KS, Goldsmith JA, Hsieh CL, Abiona O et al (2020) Cryo-EM structure of the 2019-nCoV spike in the prefusion conformation. Science 367:1260–1263
33. Jackson CB, Farzan M, Chen B, Choe H (2022) Mechanisms of SARS-CoV-2 entry into cells. Nat Rev Mol Cell Biol 23:3–20
34. Jardetzky TS, Lamb RA (2014) Activation of paramyxovirus membrane fusion and virus entry. Curr Opin Virol 5:24–33
35. Gonzalez-Reyes L, Ruiz-Arguello MB, Garcia-Barreno B, Calder L, Lopez JA, Albar JP et al (2001) Cleavage of the human respiratory syncytial virus fusion protein at two distinct sites is required for activation of membrane fusion. Proc Natl Acad Sci USA 98:9859–9864
36. Biacchesi S, Skiadopoulos MH, Yang L, Lamirande EW, Tran KC, Murphy BR et al (2004) Recombinant human Metapneumovirus lacking the small hydrophobic SH and/or attachment G glycoprotein: deletion of G yields a promising vaccine candidate. J Virol 78:12877–12887
37. Rawling J, Cano O, Garcin D, Kolakofsky D, Melero JA (2011) Recombinant Sendai viruses expressing fusion proteins with two furin cleavage sites mimic the syncytial and receptor-independent infection properties of respiratory syncytial virus. J Virol 85:2771–2780
38. Rey FA, Lok SM (2018) Common features of enveloped viruses and implications for immunogen design for next-generation vaccines. Cell 172:1319–1334
39. Halldorsson S, Li S, Li M, Harlos K, Bowden TA, Huiskonen JT (2018) Shielding and activation of a viral membrane fusion protein. Nat Commun 9:349
40. Lozach PY, Kuhbacher A, Meier R, Mancini R, Bitto D, Bouloy M et al (2011) DC-SIGN as a receptor for phleboviruses. Cell Host Microbe 10:75–88

41. Rey FA, Heinz FX, Mandl C, Kunz C, Harrison SC (1995) The envelope glycoprotein from tick-borne encephalitis virus at 2 A resolution. Nature 375:291–298
42. Modis Y, Ogata S, Clements D, Harrison SC (2004) Structure of the dengue virus envelope protein after membrane fusion. Nature 427:313–319
43. Bressanelli S, Stiasny K, Allison SL, Stura EA, Duquerroy S, Lescar J et al (2004) Structure of a flavivirus envelope glycoprotein in its low-pH-induced membrane fusion conformation. EMBO J 23:728–738
44. Gibbons DL, Vaney MC, Roussel A, Vigouroux A, Reilly B, Lepault J et al (2004) Conformational change and protein-protein interactions of the fusion protein of Semliki Forest virus. Nature 427:320–325
45. Guardado-Calvo P, Rey FA (2021) The viral class II membrane fusion machinery: divergent evolution from an ancestral heterodimer. Viruses 13:2368
46. Roche S, Rey FA, Gaudin Y, Bressanelli S (2007) Structure of the prefusion form of the vesicular stomatitis virus glycoprotein G. Science 315:843–848
47. Roche S, Bressanelli S, Rey FA, Gaudin Y (2006) Crystal structure of the low-pH form of the vesicular stomatitis virus glycoprotein G. Science 313:187–191
48. Backovic M, Longnecker R, Jardetzky TS (2009) Structure of a trimeric variant of the Epstein-Barr virus glycoprotein B. Proc Natl Acad Sci USA 106:2880–2885
49. Kadlec J, Loureiro S, Abrescia NG, Stuart DI, Jones IM (2008) The postfusion structure of baculovirus gp64 supports a unified view of viral fusion machines. Nat Struct Mol Biol 15: 1024–1030
50. Vollmer B, Prazak V, Vasishtan D, Jefferys EE, Hernandez-Duran A, Vallbracht M et al (2020) The prefusion structure of herpes simplex virus glycoprotein B. Sci Adv 6:eabc1726
51. Roche S, Gaudin Y (2002) Characterization of the equilibrium between the native and fusion-inactive conformation of rabies virus glycoprotein indicates that the fusion complex is made of several trimers. Virology 297:128–135
52. Albertini AA, Baquero E, Ferlin A, Gaudin Y (2012) Molecular and cellular aspects of rhabdovirus entry. Viruses 4:117–139
53. Albertini AA, Merigoux C, Libersou S, Madiona K, Bressanelli S, Roche S et al (2012) Characterization of monomeric intermediates during VSV glycoprotein structural transition. PLoS Pathog 8:e1002556
54. Hefferon KL, Oomens AG, Monsma SA, Finnerty CM, Blissard GW (1999) Host cell receptor binding by baculovirus GP64 and kinetics of virion entry. Virology 258:455–468
55. Ho Y, Hsiao JC, Yang MH, Chung CS, Peng YC, Lin TH et al (2005) The oligomeric structure of vaccinia viral envelope protein A27L is essential for binding to heparin and heparan sulfates on cell surfaces: a structural and functional approach using site-specific mutagenesis. J Mol Biol 349:1060–1071
56. Laliberte JP, Weisberg AS, Moss B (2011) The membrane fusion step of vaccinia virus entry is cooperatively mediated by multiple viral proteins and host cell components. PLoS Pathog 7: e1002446
57. Kao CF, Tsai MH, Carillo KJ, Tzou DL, Chang W (2023) Structural and functional analysis of vaccinia viral fusion complex component protein A28 through NMR and molecular dynamic simulations. PLoS Pathog 19:e1011500
58. Top D, de Antueno R, Salsman J, Corcoran J, Mader J, Hoskin D et al (2005) Liposome reconstitution of a minimal protein-mediated membrane fusion machine. EMBO J 24:2980–2988
59. Duncan R (2019) Fusogenic reoviruses and their fusion-associated small transmembrane (FAST) proteins. Annu Rev Virol 6:341–363
60. Top D, Read JA, Dawe SJ, Syvitski RT, Duncan R (2012) Cell-cell membrane fusion induced by p15 fusion-associated small transmembrane (FAST) protein requires a novel fusion peptide motif containing a myristoylated polyproline type II helix. J Biol Chem 287:3403–3414
61. Crank MC, Ruckwardt TJ, Chen M, Morabito KM, Phung E, Costner PJ et al (2019) A proof of concept for structure-based vaccine design targeting RSV in humans. Science 365:505–509

62. Graham BS, Gilman MSA, McLellan JS (2019) Structure-based vaccine antigen design. Annu Rev Med 70:91–104
63. Ngwuta JO, Chen M, Modjarrad K, Joyce MG, Kanekiyo M, Kumar A et al (2015) Prefusion F-specific antibodies determine the magnitude of RSV neutralizing activity in human sera. Sci Transl Med 7:309ra162
64. Melero JA, Mas V (2015) The Pneumovirinae fusion (F) protein: a common target for vaccines and antivirals. Virus Res 209:128–135
65. Byrne PO, McLellan JS (2022) Principles and practical applications of structure-based vaccine design. Curr Opin Immunol 77:102209
66. Venkatesan P (2023) First RSV vaccine approvals. Lancet Microbe 4:e577
67. Lindenbach B, Randall G, Bartenschlager R, Rice C (2020) Flaviviridae: the viruses and their replication. In: Howley PM, Knipe DM, Whelan S (eds) Fields virology, vol 1, 7th edn. Wolters Kluwer, Netherlands, pp 246–301
68. Plemper RK, Lamb RA (2020) Paramyxoviridae: the viruses and their replication. In: Howley PM, Knipe DM, Whelan S (eds) Fields virology, vol 1, 7th edn. Wolters Kluwer, Netherlands, pp 504–558
69. Melen K, Fagerlund R, Franke J, Kohler M, Kinnunen L, Julkunen I (2003) Importin alpha nuclear localization signal binding sites for STAT1, STAT2, and influenza A virus nucleoprotein. J Biol Chem 278:28193–28200
70. Selyutina A, Persaud M, Lee K, KewalRamani V, Diaz-Griffero F (2020) Nuclear import of the HIV-1 core precedes reverse transcription and uncoating. Cell Rep 32:108201
71. Krug LT, Pellet PE (2021) The family Herpesviridae: a brief introduction. In: Howley PM, Knipe DM, Cohen JL, Damania BA (eds) Fields virology, vol 2, 7th edn. Wolters Kluwer, Netherlands, pp 212–234
72. Broyles SS (2003) Vaccinia virus transcription. J Gen Virol 84:2293–2303
73. Houser KV, Chen GL, Carter C, Crank MC, Nguyen TA, Burgos Florez MC et al (2022) Safety and immunogenicity of a ferritin nanoparticle H2 influenza vaccine in healthy adults: a phase 1 trial. Nat Med 28:383–391
74. Geng Q, Tai W, Baxter VK, Shi J, Wan Y, Zhang X et al (2021) Novel virus-like nanoparticle vaccine effectively protects animal model from SARS-CoV-2 infection. PLoS Pathog 17: e1009897

Further Reading

Especially recommended for further reading are references [1, 4, 7, 10, 18, 28, 31, 38, 62, 65] from those listed above.

Chapter 17
Bacteriophage Receptor Recognition and Nucleic Acid Transfer

Mark J. van Raaij

Abstract Correct host cell recognition is important in the replication cycle for any virus, including bacterial viruses. This essential step should occur before the bacteriophage commits to transferring its genomic material into the target bacterium. In this chapter, we will discuss the mechanisms and proteins bacteriophages use for receptor recognition (just before full commitment to infection) and nucleic acid injection, which occurs just after commitment. Some bacteriophages use proteins of the capsid proper for host cell recognition, others use specialised spikes or fibres. Usually, several identical recognition events take place, and the information that a suitable host cell has been encountered is somehow transferred to the part of the bacteriophage capsid involved in nucleic acid transfer. The main part of the capsids of bacteriophages stays on the cell surface after transferring their genome, although a few specialised proteins move with the DNA, either forming a conduit, protecting the nucleic acids after transfer and/or functioning in the process of transcription and translation.

Keywords Adsorption · Attachment · Bacterium · Bacteriophage · Baseplate · Capsid · Cell wall · Flagellum · Nucleic acid conduit · Contractile tail · Gram-negative · Gram-positive · Lipopolysaccharide · Membrane · Peptidoglycan · Pilus · Receptor · Tail fibre · Tailspike · Teichoic acid

Abbreviations

cryo-EM	Cryogenic electron microscopy
DNA	Deoxyribonucleic acid
ds	Double-stranded
EMDB	Electron microscopy data bank

M. J. van Raaij (✉)
Department of Macromolecular Structure, Centro Nacional de Biotecnología (CNB-CSIC), Madrid, Spain
e-mail: mjvanraaij@cnb.csic.es

M. G. Mateu (ed.), *Structure and Physics of Viruses*, Subcellular Biochemistry 105,
https://doi.org/10.1007/978-3-031-65187-8_17

gp	Gene product
Hoc	Highly antigenic outer capsid protein
LPS	Lipo-polysaccharide
nm	Nanometer
Mat	Maturation protein (of phage MS2)
ORF	Open reading frame
Omp	Outer membrane porin
PDB	Protein structure data bank
RNA	Ribonucleic acid
Soc	Small outer capsid protein
ss	Single-stranded
stf	Side tail fibre
tfa	Tail fibre assembly

17.1 Introduction

Bacteriophages, or phages, are viruses that infect bacteria and were independently discovered by both Félix d'Herelle and Frederick Twort just over a century ago. Bacteriophages are the most numerous replicating biological entities on earth; for every bacterium there are thought to be on average ten phages [1]. The observation that different bacteriophages specifically recognise and kill different bacterial species and strains has led to multiple applications [2]. Félix d'Herelle himself pioneered phage therapy, although with the advent of modern antibiotics it was largely abandoned. Phage therapy is now undergoing a revival because more and more bacterial strains develop resistance against many or all useful antibiotics. For a discussion of phage therapy, see Chap. 22. Bacteriophages and the lytic enzymes they produce are also investigated for the control of microbes in the food industry, of water-borne pathogens and pathogenic bacteria in hospital settings.

Phages may also have negative consequences. For example, they may interfere with industrial fermentation processes such as those in dairy plants. Bacteriophages, especially lysogenic ones, may also transfer DNA encoding virulence factors to bacteria, making them more pathogenic. Therefore, bacteriophages to be used in the food industry and medical applications should be well characterised and proven to be free of sequences encoding virulence factors. They should also be non-lysogenic or be genetically converted to be non-lysogenic (see Chap. 22).

In the laboratory, phages have been used in many fundamental experiments in microbiology, genetics, biochemistry and molecular biology since Max Delbrück and Salvador Luria started investigating them in the middle of the twentieth century. Nowadays, bacteriophages are used as cloning and phage-display vectors, DNA encoding phage T7 RNA polymerase is part of protein expression vectors and phage enzymes such as DNA polymerases and ligases are used in routine DNA manipulations. The ease with which many bacteriophages can be grown, their stability and relatively facile genetic modification have led to the development of advanced applications such as their use as labelled ground tracers to follow the flow of underground water, as innocuous test particles for equipment designed to remove dangerous viruses and phage-based diagnostics to detect low amounts of dangerous bacteria. Some phages are being used as nanoparticles for nanotechnological developments (see Chap. 22).

The ecological importance of bacteriophages is illustrated by the wide range of defence systems that bacteria have developed against them [3]. Examples of these defence systems are restriction enzymes, which stood at the basis of gene manipulation and cloning in molecular biology, and more recently the CRISPR-CAS system, which has applications in precise genetic manipulations of all types of organisms, potentially including humans.

Bacteriophages can be divided into virulent, or obligatory lytic, phages on one hand and temperate, or lysogenic, phages on the other. Upon infection of their host cell, temperate phages may take one of two paths, lytic or lysogenic. In the lytic pathway, the phage takes over the host cell and redirects its biosynthetic pathways towards producing many daughter phages and ultimately, lysis of the host cell to release its progeny. Alternatively, a temperate phage may integrate its DNA into the host genome—this phage form is called a prophage. In this way, its genome replicates with that of the bacterium in conditions favourable for bacterial growth (some phages even encode virulence factors that favour bacterial growth inside a eukaryotic host). When conditions cease to be favourable for bacterial growth and the bacterial SOS system is activated, the phage may switch to the lytic pathway, produce daughter phages, lyse the bacterium and "escape". Virulent phages do not have the ability to integrate into the host cell genome and can only follow the lytic pathway. Filamentous phages (see below) are an exception and follow an intermediate pathway. When they infect bacteria, they redirect the bacterial biosynthetic systems to produce phages, but phage egress does not lead to lysis. Instead, the bacterium turns into a phage factory, continuously extruding phages. The bacterium continues to grow and divide, albeit at a slower rate.

Bacteriophages can be divided into several different types based on their size and nature of their genetic material: large DNA phages, small DNA phages, and RNA phages [4]. Large DNA phages have a double-stranded (ds) DNA genome ranging from 10 to 20 kilobases to more than 700 kb and can infect both Gram-negative and Gram-positive bacteria. All or nearly all large DNA phages belong to the *Caudoviricetes* class (tailed phages), which contains over 95% of all observed phage species. Their characteristic tails function in efficient host cell recognition and infection and can be long, straight and contractile (myoviruses), long, flexible and non-contractile (siphoviruses), or short and stubby (podoviruses). Please note that the families *Myoviridae, Siphoviridae* and *Podoviridae* have been abolished in the most recent virus classification (2022) by the ICTV (International Committee on Taxonomy of Viruses) because they represent polyphyletic groups and do not reflect shared evolutionary histories. However, the value of grouping the tailed viruses as myovirus, siphovirus or podovirus according to morphology is still recognised, and this morphological grouping will be used in this chapter.

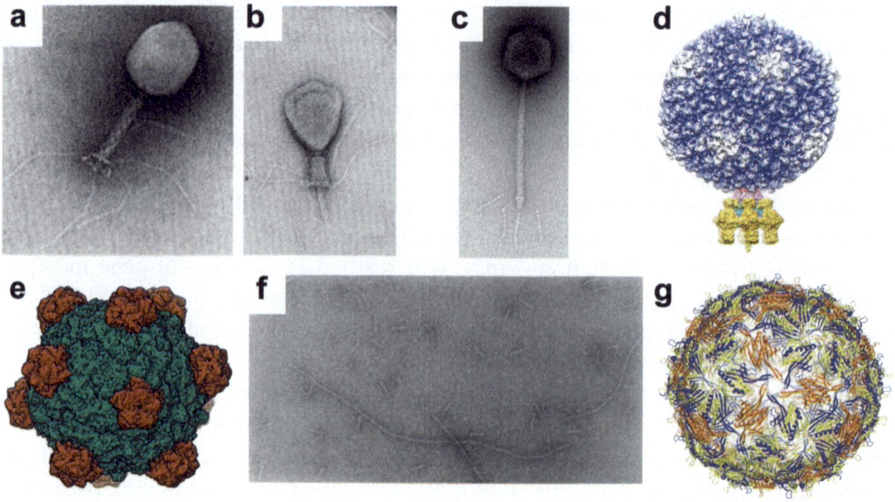

Fig. 17.1 Diverse members of the bacteriophage family. (**a**, **b**) Electron micrographs of extended (**a**) and contracted (**b**) T4 phage; the capsid is about 86 nm wide (reprinted from [5] with permission from Elsevier). (**c**) Electron micrograph of bacteriophage T5; the capsid width is 550 nm (reprinted from [6] with permission from Elsevier). (**d**) Cryo-EM density of phage P22 [7]; here the capsid has a width of 70 nm (reprinted from [7] with permission from Elsevier). (**e**) Crystal structure of phage ϕX174; the capsid diameter is 33 nm (PDB entry 2BPA). (**f**) Electron micrograph of the filamentous phage M13; the filament is about 8 nm wide (reprinted from [8] with permission from Elsevier). (**g**) Structure of bacteriophage MS2 viewed down the fivefold symmetry axis (PDB entry 2MS2)

Unlike most other viruses, empty particles of tailed phages stay on the cell surface after they have transferred their nucleic acid genome into the host cell. Well-known examples of phages with a contractile tail are the virulent *Escherichia coli* phage T4 (Fig. 17.1a, b) and the temperate *Escherichia coli* phage μ, whilst examples of *Escherichia coli* phages with a flexible tail are the virulent phage T5 (Fig. 17.1c) and the temperate phage λ. Examples of bacteriophages with a short tail are the virulent *Escherichia coli* phage T7, the temperate *Salmonella* phage P22 (Fig. 17.1d) and the virulent *Bacillus subtilis* phage ϕ29. Phages with a long tail contain a tape measure protein, around which the tail is assembled and which thus determines the length of the tail [4].

A family of intermediate-size dsDNA phages are the tectiviruses. These bacterial viruses have the peculiarity of an internal membrane located below the outer protein shell and can be divided into two subgroups—those infecting Gram-positive and those infecting Gram-negative bacteria. Bacteriophage PRD1 is the most well-known example [9]. It is a non-tailed icosahedral virus with a double-stranded DNA genome of 15 kilobases. PRD1 has a broad host range and infects several species of Gram-negative bacteria including *Escherichia coli* and *Salmonella*. Susceptibility of bacteria to PRD1 depends on the presence in the bacteria of a conjugative IncP, IncN or IncW plasmid, which codes for the receptor

[10]. Corticoviruses, which are also dsDNA phages, have an internal membrane, whilst dsRNA-containing cystoviruses have an external membrane [4].

Small DNA phages or isometric phages have a covalently closed circular single-stranded (ss) DNA of up to 10 kilobases. They have an icosahedral capsid of around 30 nm. The prototype is ϕX174 (Fig. 17.1e), for which the different steps of its replication cycle (infection, DNA replication, capsid assembly and host lysis) have been studied [11, 12].

Filamentous phages also contain a circular ssDNA molecule that is folded back onto itself in the elongated phage particle [13, 14]. Prototypes of this family of phages are the Ff phages (e.g. f1, fd and M13) (Fig. 17.1f). They are useful cloning vectors for producing ssDNA due to the lack of a packaging limit—larger circular DNA molecules simply lead to longer phage particles.

RNA phages are small, have icosahedral capsids and contain a single linear ssRNA of about 4 kilobases, coding for only 4 genes (maturation protein, capsid protein, lysis protein and replicase protein). A well-studied RNA phage is MS2 (Fig. 17.1g) [15]. Like the filamentous phages, they attach to the F-coded pilus and can thus only infect bacteria containing the F-plasmid. However, they enter bacterial cells by attachment to the side of the F-pilus, rather than the end as filamentous phages do.

Phages may change their host range by mutation of their primary receptor-binding proteins or adoption of new receptor-binding proteins via horizontal gene transfer, and bacterial mutants that are resistant to infection by a particular bacteriophage have generally lost or modified the primary phage receptor. In this chapter, we will discuss the specialised proteins used by bacteriophages for the very first step of infection, receptor recognition. We mention what is known about the mechanism different prototype bacteriophages belonging to different groups use to decide that a correct host cell has been encountered and that the phage should commit itself and proceed to the following step, DNA transfer. We will also treat what is known about the DNA transfer step for different prototype phages.

17.2 Proteins Used in Receptor Recognition and Nucleic Acid Transfer

Phages take advantage of host physiology for their infection process and replication. Like eukaryotic viruses, some bacteriophages recognise their host cell receptors with proteins of the capsid itself. Others have specialised proteins for this function. In this section, we discuss the receptor-binding mechanism of one or several examples of important phage groups. We will also discuss what is known about nucleic acid transfer for different types of phages. Naming of bacteriophage proteins may be confusing to readers unfamiliar with phages. In many cases gp (gene product) followed by letters (capital or not) or numbers (Arabic or Roman) is used, i.e. gp37 for phage T4 or gpF in case of ϕX174. The single letter p or P (protein) may also be used, i.e. pIII of bacteriophage M13. In the case of bacteriophage T5, pb (protein band) is used, i.e. pb1. Proteins may be numbered or lettered according to the order of discovery, size, location on the genome or other schemes the discoverers fancied. Other names or the designation ORF (open reading frame) followed by a number may also be used.

17.2.1 ssRNA Phages

ssRNA phages contain 90 dimers of the capsid protein, together forming a $T = 3$ icosahedral capsid [15]. They bind to the side of bacterial pili, which are made up of multiple copies of the pilin protein, in the first step of infection. For MS2, it is known that one copy of the pilin-binding maturation protein (Mat) is located at one of the fivefold vertices. Cryo-EM reconstruction at 2 nm resolution on pilus-bound MS2 phage showed bound phages are attached at a slight angle, consistent with the symmetry-mismatch of a single copy of the maturation protein bound to one of the fivefold vertices [16]. Recently, single-particle cryogenic electron microscopy (cryo-EM, see Chap. 3) was used to determine the asymmetric structure of the MS2 virion bound to a bacterial F-pilus [17]. Binding to the pilus results in a small reorientation of Mat relative to the capsid, which could facilitate the release of the Mat-connected viral RNA. Mat contacts four pilin subunits. It is not known if specific secondary receptor interaction of the phage with membrane components takes place.

Apart from binding pilin on the outside of the capsid, Mat binds the 5′ and 3′ ends of the viral RNA on the inside. At the vertex contacting the pilus, density that presumably includes contributions from both RNA and maturation protein is poised near the channel that goes through the fivefold vertex to the outside. This suggests that the RNA-maturation protein complex leaves the capsid as the first step of the nucleic acid transfer process [16], presumably through the hole observed at the fivefold vertex. Retraction of the pilus may bring the phage to the bacterial membrane (Fig. 17.2a). Combined cryo-EM and cryogenic electron tomography (cryo-ET) analyses suggest that arrival of the phage at the membrane may crack open the viral capsid, facilitating the unpacking of the RNA and the transfer of the Mat-RNA complex into the cell [17, 18]. The mechanism of the transfer process is still unknown.

17.2.2 dsRNA Phages

The only known dsRNA bacteriophages, the cystoviruses, are enveloped viruses with a diameter of around 85 nm. Embedded in the membrane are spikes. The cystovirus internal capsid is made up of 60 dimers of the P1 protein, in a $T = 1$ geometry [19]. The outer shell is formed by 200 trimers of the P8 protein arranged as a $T = 13\ l$ shell. The nucleocapsid is enclosed by a lipid bilayer, which interacts with P4 and P8. The membrane contains four viral integral membrane proteins: P6, P9,

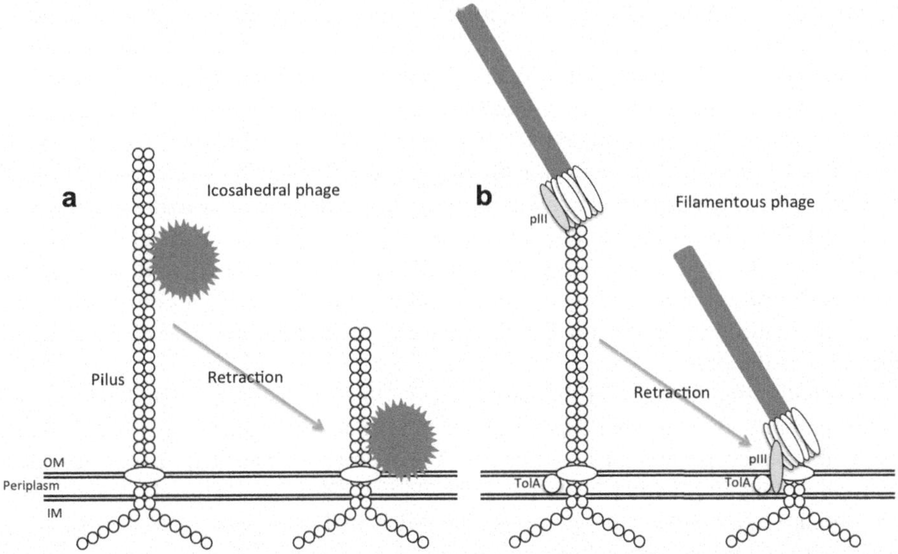

Fig. 17.2 Receptor binding of pilus-binding bacteriophages. (**a**) Icosahedral RNA phage (such as MS2 or Qβ) bound to the side of the pilus. After retraction of the pilus, the phage is close to the membrane and transfer of the RNA into the host takes place. (**b**) Filamentous RNA phage (such as M13 or fd) bound to the end of the pilus. After retraction of the pilus, pIII interacts with the secondary receptor TolA and transfer of the DNA into the host takes place

P10, and P13. In addition, the spike protein P3 is attached to P6 and protrudes 2 nm from the membrane surface. The genome consists of three double-stranded RNA segments, each encoding several proteins. All known cystoviruses infect *Pseudomonas* species, although φ8 also infects *E. coli* and other hosts [20].

Cystoviruses start infection by the trimeric P3 protein spikes attaching to host cell pili or to rough lipopolysaccharides [21, 22]. P6 mediates fusion of viral membrane with the bacterial outer membrane, releasing the capsid into the periplasmic space. The P5 capsid protein then digests the peptidoglycan layer of the host, leading to endocytosis of the capsid into the cytoplasm, now covered by the host's inner membrane. This membrane and the P8 protein shell are subsequently lost, but the rest of the capsid remains intact, hiding the viral genome from antiviral host factors. mRNAs are transcribed by the viral RNA-dependent RNA polymerase (P2) inside the core particle and released into the cell cytoplasm.

17.2.3 Filamentous ssDNA Phages

Filamentous bacteriophages are thin and long particles (naturally about 8 by 900 nm) that are assembled around a circular single-stranded DNA. They can be made significantly longer depending on the length of the encapsulated genome and of its packaging density [13]. Filamentous phages include the comparatively well-studied

Ff phages (of which M13 is an example), which are intensively used in phage-display technology (Chap. 22, Sect. 22.5.1). The end of the virus that assembles first is formed by proteins pVII and pIX, this end is also called the distal end. The major capsid protein is protein pVIII, a small α-helical protein that assembles in a helical fashion around the encapsulated DNA molecule. The proximal end is formed by proteins pVI and pIII. The receptor-binding protein pIII has three domains, N1, N2 and a C-terminal domain that anchors the protein to the phage through its interaction with pVI. Infection is initiated by absorption of protein pIII to the distal terminus of F-pilus (Fig. 17.2b), via the N2 domain of pIII [13]. As both the tip and sides of the conjugative pilus are formed by the protein pilin, filamentous phages must recognise a different epitope than the small RNA phages—an epitope that is not available on the sides of the pilus.

The C-terminal domain of the *Escherichia coli* periplasmic protein TolA acts as coreceptor for filamentous phages [23]. Upon retraction of the pilus by the bacterium, the phage protein pIII approaches the membrane and a normally shielded epitope of domain N1 of pIII binds to TolA (Fig. 17.2b). Details of subsequent steps are obscure, although it has been postulated that pIII may participate in forming a channel through the bacterial membrane [13, 23].

17.2.4 Small Icosahedral ssDNA Phages

Small ssDNA phages (microviruses) are isometric icosahedral phages that recognise their receptor by the capsid proper. In the case of the well-studied phage ϕX174 (Fig. 17.3) [12], primary, reversible receptor interaction with the terminal N-acetyl-

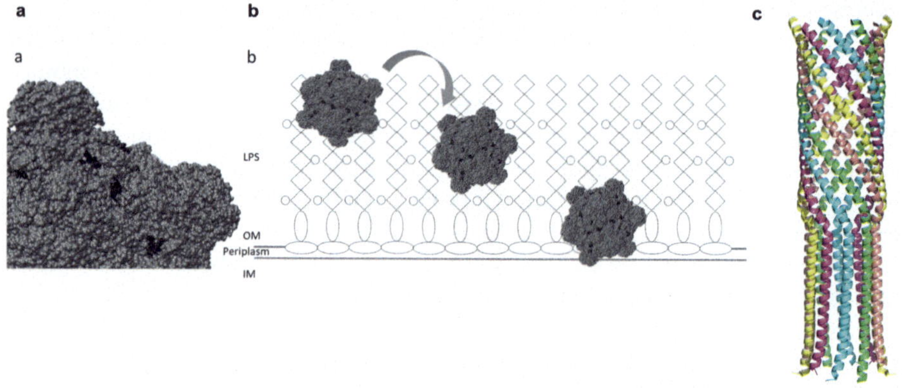

Fig. 17.3 Receptor binding of bacteriophage ϕX174. (**a**) Location of the primary, reversible, receptor binding site (in black) on the phage capsid (grey) formed by protein gpF. (**b**) Phage ϕX174 binds reversibly to the non-reducing ends (grey circles) of the core polysaccharide of bacterial lipo-polysaccharide. ϕX174 may roll over the cell surface until a suitable, unknown, secondary receptor is encountered. (**c**) Structure of the gpH decamer tube viewed from the side (PDB entry 4JPP [24])

glucosamine at the non-reducing end of the core polysaccharide of bacterial lipopolysaccharide (LPS; [11]) is mediated by the capsid protein gpF. The binding site has been localised to the side of depression on gpF [25]. Spike proteins gpG and gpH are necessary for irreversible cell attachment to a second factor [11], which may be encoded by the *phxB* gene of *Escherichia coli*. Usually, two to three spikes become embedded in the cell wall and the phage submerges to about one-half of its diameter. Tailed bacteriophages are thought to diffuse laterally along the cell surface after reversible attachment until they encounter their secondary receptor and bind irreversibly (see Sect. 17.2.6). Bacteriophage ϕX174 cell attachment may proceed similarly. However, instead of walking along the surface, ϕX174 may roll (Fig. 17.3b).

Adsorbed ϕX174 particles are often located at regions of adhesion between the cell wall and the inner membrane. This suggests that the secondary receptor is also located here and indicates that the DNA genome may be transferred directly into the cytoplasm. DNA eclipse from the phage has a high activation energy barrier and requires the LPS lipid A moiety of the host, and DNA transfer into the cell requires protein gpH of the phage (also called the DNA pilot protein) and an active host cell metabolism [11]. The gpG proteins at one of the icosahedral fivefold symmetry axes of the viral capsid recognise an LPS receptor to initiate DNA ejection through a tube made by the gpH protein [24–26]. The gpG spike is jettisoned before DNA transfer. Ten monomers of gpH come together to form a long alpha-helical barrel, with an inside diameter wide enough to allow the passage of dsDNA (Fig. 17.3c).

The gpH protein also enters the cell and directs stage I DNA synthesis. It has been proposed that DNA synthesis is necessary for complete DNA transfer, in the same way that active transcription is necessary for the transfer of phage T7 DNA (see Sect. 17.2.6). After entering into the bacterium with the DNA, gpH promotes phage replication, perhaps by stimulating transcription, being involved in DNA replication or just stabilising the DNA and protecting it against degradation by host cell enzymes.

17.2.5 Lipid-Containing Icosahedral dsDNA Phages

The best studied and prototype tectivirus is the virulent bacteriophage PRD1 (Fig. 17.4) [9] The PRD1 virion has an external pseudo $T = 25$ icosahedral capsid shell. Beneath the capsid lies a membrane which, in the mature virion, follows the shape of the capsid due to the pressure of the packaged genome. PRD1 first reversibly binds to its receptor, followed by an irreversible step (Fig. 17.4a, b, c). The vertex of bacteriophage PRD1 shows two different types of spikes associated with the pentameric P31 capsid protein, composed of the proteins P2 (Fig. 17.4d) and P5 (Fig. 17.4e) [9, 27]. Specific binding to the cell surface receptor depends on the protein P2. The cell surface receptor is a component of the conjugative system coded for by a plasmid, as bacteria without these plasmids are resistant to PRD1 infection. The entire conjugative apparatus coded by the plasmid is necessary for

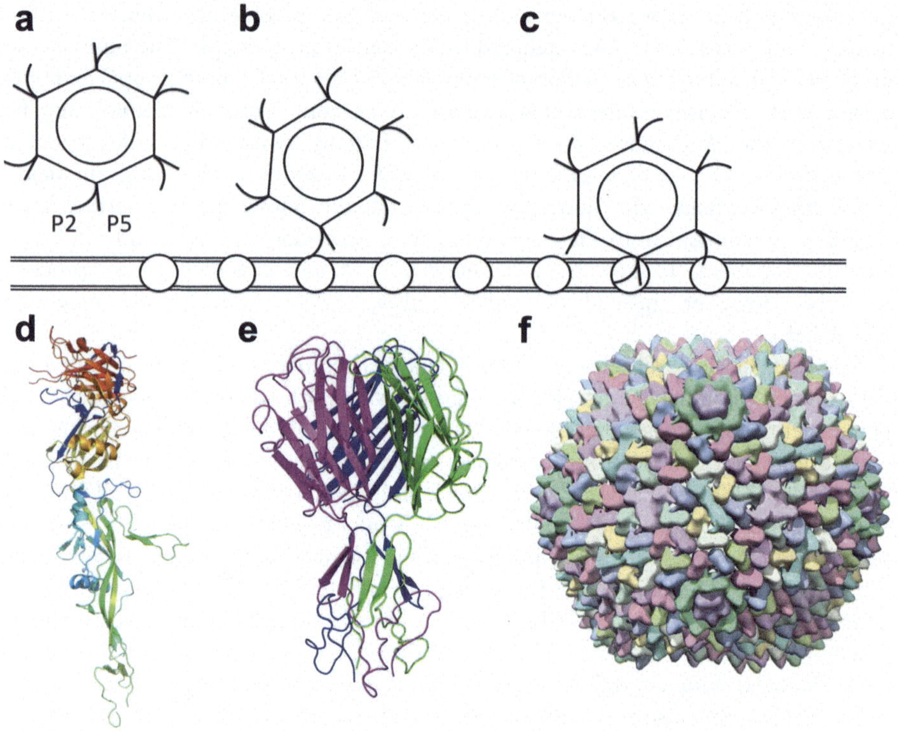

Fig. 17.4 Receptor binding of bacteriophage PRD1. Schematic drawings of (**a**) phage PRD1 and its spike proteins P2 and P5, (**b**) phage PRD1 binding to its receptor using the P2 protein and (**c**) opening of the vertex for membrane penetration and DNA ejection. (**d**) Crystal structure of the monomeric receptor-binding protein P2 (PDB entry 1N7V). (**e**) Crystal structure of the carboxy-terminal distal part of the trimeric vertex protein P5 (PDB entry 1YQ8). (**f**) Structure of bacteriophage PRD1 (based on PDB entry 1W8X)

infection, and the TraF protein is considered to be the phage receptor protein. Receptor binding of the P5 protein has not been demonstrated, but it might give the virus a selective advantage in the wild by binding to an unknown, alternative receptor.

After binding to the receptor, P2 is thought to trigger conformational changes within the spike complex that are transmitted to the internal membrane. Information about receptor binding is somehow transferred to the DNA delivery apparatus, which contains at least the proteins P7, P11, P14, P18 and P32. This leads to considerable conformational changes in the vertex structure. The removal of the spike complex creates an opening in the vertex, which enables an appendage to protrude that penetrates the outer membrane of the host. The protruding tube probably contains protein P11, but it also contains an extension of the phage lipid membrane [28]. The lytic transglycosylase (P7) is thought to assist in genome entry by locally degrading the peptidoglycan layer. The appendage formed extends some 35 nm (the thickness of the cell envelope) penetrating the cytoplasmic membrane

[28]. DNA transfer is dependent on active membrane tube formation and reduction of the membrane vesicle volume assisted by at least the proteins P14, P18 and P32. The force required for DNA transfer may result in part from the pressure release of the tightly packed DNA in the capsid, but contraction of the lipid membrane of the phage may also play a role [29]. Consistent with this proposal is the fact that in the phage the membrane is forced into a spherical form, whilst in isolation it prefers to assume a tubular conformation [9]. After DNA injection, transcription and protein-primed genome replication take place. The phage protein shell stays associated with the host cell surface.

Corticoviruses are icosahedral and, like tectiviruses, contain an internal lipid membrane. Only one species is known: *Pseudoalteromonas* phage PM2. PM2 particles are formed by an icosahedral capsid of 60 nm in diameter and with a triangulation number $T = 21d$, containing spikes at the vertices. This capsid surrounds an inner lipid bilayer (the lipid core) [30]. In turn, the lipid core encloses a 10 kb, circular, supercoiled molecule of double-stranded DNA. The corticovirus phage PM2 recognises its Gram-negative host using its P1 spike protein [31]. Binding of P1 to the host triggers the uncoating of the virion, allowing the lipid core to interact and fuse with the host's outer membrane [32]. Then, protein P7 degrades the periplasmic peptidoglycan layer [33], allowing the highly supercoiled DNA to reach and pass through the cytoplasmic membrane.

17.2.6 Large dsDNA Phages

Large dsDNA phages of the *Caudoviricetes* class possess a tail for efficient host cell recognition and DNA transfer [34–36]. In general, in a first step reversible attachment is achieved, with a subsequent irreversible attachment, after which the phage is fully committed to DNA transfer into the selected host cell. Receptor structures in the host cells recognised by tailed phages as a group are highly diverse and, depending on phage and host bacterium, include proteins or protein complexes (e.g. flagella, pilins, porins and other transport proteins), and oligosaccharides and derivatives (including LPS and teichoic acid). Proteins on the outside of the DNA-containing capsid head may be involved in a first, highly reversible attachment to the host, augmenting the local phage concentration near the bacterium and favouring efficient infection. Candidates for these reversible capsid head–cell interactions include head fibres and immunoglobulin-like domains of phage structural proteins. Binding of the phage tail to host receptors often involves two steps. In the first step, a reversible interaction with a primary receptor is established; in a second step, irreversible binding to a secondary receptor may occur. Phage tail proteins that bind cell receptors include fibres and tailspikes [35, 37]. Phage receptor-binding proteins are typically present in multiple copies and are in general homotrimers, which results in enhanced avidity.

Phage tails are also directly involved in DNA transfer into the bacterial cytoplasm [38–40]. Podoviruses expel internal capsid proteins that form a tube that traverses the cell wall and contacts the cell membrane [39, 41]. Myoviruses contract their outer tail sheath, driving the inner tail tube into the cell wall [42, 43]. Infection by Siphoviruses is less well known [44], but they may recognise integral protein complexes in the cell membrane and use them for genome transfer. In the following three sections, we will discuss different aspects of receptor binding and DNA transfer for some of the better-known podoviruses (including P22, T7 and ϕ29), siphoviruses (T5, λ, SPP1 and *Lactobacillus* phages) and myoviruses (T4, P2, A511 and spore-binding phage 8a).

Podoviruses

Podoviruses have a tail that is not long enough to span the periplasmic space of Gram-negative bacteria or the cell wall of Gram-positive bacteria. Here, we will discuss the host recognition and DNA transfer steps of three podoviruses as examples, two that infect the Gram-negative bacteria *Salmonella enterica* (phage P22) y *Escherichia coli* (phage T7), and one that infects the Gram-positive bacterium *Bacillus subtilis* (phage ϕ29).

Bacteriophage P22. P22 infects the host cell using a three-step mechanism (Fig. 17.5a, b, c): (1) the virus binds through its tailspikes to LPS; (2) the tail needle contacts with the secondary receptor and/or pierces the membrane; (3) the tail proteins change their conformation and the DNA is ejected. The complete architecture of the P22 tail apparatus has been determined [45]. As in many other podoviruses, the P22 tail (Fig. 17.5d) contains six tailspikes (Fig. 17.5e). Each one is a stable homotrimer of gp9 [46] and can be divided into three parts: an amino-terminal phage-binding domain, a central parallel β-helix domain and a carboxy-terminal, highly interdigitated part. The amino-terminal phage-binding domain is more evolutionary conserved between related phages, whilst the sequences of the carboxy-terminal domains have evolved more, apparently depending on the bacteria to infect. In phage P22, a hinge is present between the amino-terminal domain and central domain, which allows flexibility between the two. Folding and trimeric assembly of the tailspikes are dependent on their interdigitated carboxy-terminal domains. The tailspike possesses receptor-destroying endorhamnosidase enzymatic activity. It cleaves the α(1,3)-O-glycosidic bond between rhamnose and galactose of the bacterial LPS O-antigen repeats (endorhamnosidase activity). This results in cleavage of the LPS O-antigen receptor during the adsorption of the phage to the cell surface, rendering the binding reversible. More importantly, this activity may be used by the phage to clear a path through the bacterial outer LPS region (the capsule) to gain access to the secondary receptor on the cell surface. The enzymatic activity may also serve to release progeny phage particles from the cellular debris at the end of the lytic cycle.

From the centre of the tail, the needle protein projects, which is a thin trimer of gp26 [47] (Fig. 17.5f). The tip of the needle projects beyond the tailspikes (Fig. 17.5d) and is thus likely the first protein to contact the cell surface. At the same time, gp26 is the plug that keeps the DNA in the P22 particle; which suggests that this interaction is somehow connected to DNA release. Penetration of the outer membrane of a Gram-negative bacterium by a podovirus must involve forming a

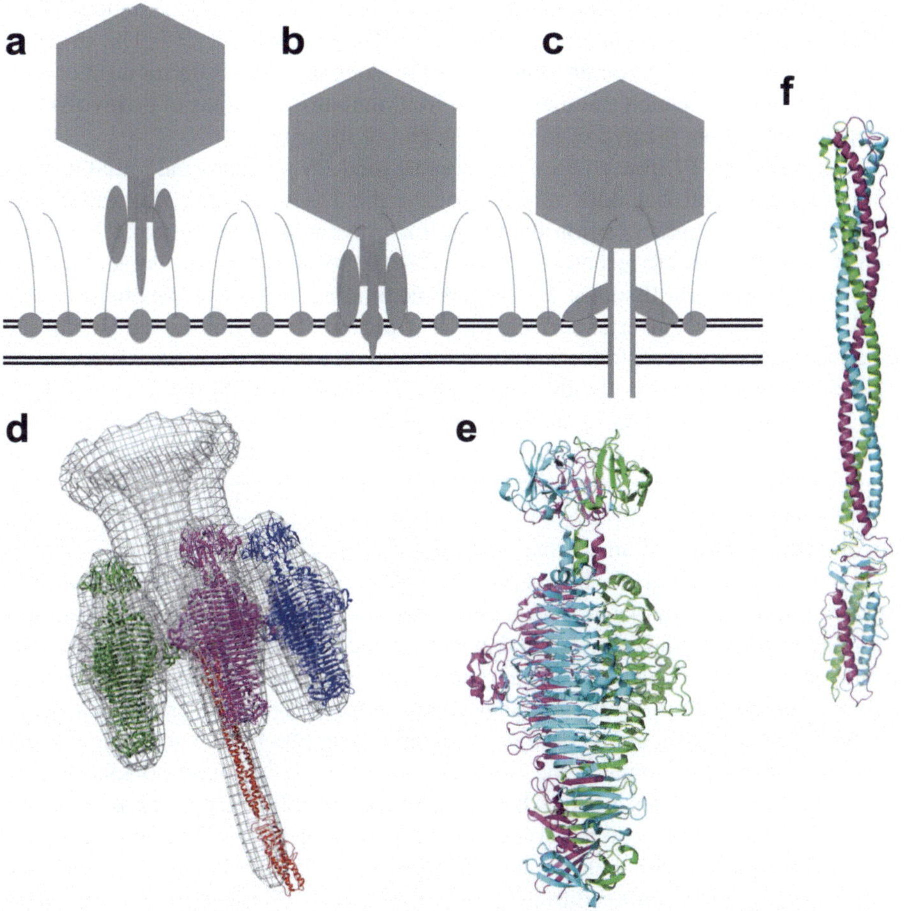

Fig. 17.5 Bacteriophage P22. (**a**) The bacteriophage binds through its tailspikes to LPS. (**b**) The tail needle contacts with the secondary receptor and/or pierces the membrane. (**c**) The tail proteins change their conformation and the DNA is ejected. (**d**) Cryo-EM reconstruction (EMDB: 1119) of the tail with crystal structures of tailspikes and tail needle fitted. (**e**) Crystal structure of the tailspike (gp9, PDB: 2XC1). (**f**) Crystal structure of the tail needle (gp26, PDB entry 2POH)

channel for DNA transport across the cytoplasmatic membrane [48]. Bacteriophage P22 contains three so-called ejection proteins: gp7, gp16 and gp20 [48]. These proteins are involved in DNA transfer from the phage into the host bacterium. They form a conduit through the cell envelope for genome ejection into the cell cytoplasm [49].

Bacteriophage T7. Not all podoviruses contain enzymatically active receptor-binding tailspikes. Instead, T7 contains six thin, kinked, tail fibres. The fibres, homotrimers of the protein gp17, are responsible for the first specific, but still reversible, attachment to *Escherichia coli* LPS. The kinked fibres are comprised of an amino-terminal tail attachment domain, a slender shaft, a flexible kink, and a carboxy-terminal domain composed of several nodules [50]. The structure of the carboxy-terminal region comprising residues 371–554 is known [50]. The structure revealed a β-stranded pyramid domain and a globular carboxy-terminal receptor-binding domain, in which residues known from mutational studies to be involved in receptor binding are located in loops at the end of the trimer.

Binding of the T7 fibre to the inner core of the LPS mediates the initial contact with the target cell but does not yet trigger the DNA release. This initial and reversible binding may position the phage for irreversible binding and DNA ejection. Whether a specific receptor is involved in the secondary, irreversible, binding, is currently unknown. If so, the first, reversible binding may allow the phage to stay associated with the bacterial surface whilst it diffuses two-dimensionally until the putative secondary receptor is encountered. For T7, these secondary interactions may be mediated by its outer tail proteins gp11 and/or gp12. As in the case of P22, the T7 tail is too short to span the periplasmic space of its Gram-negative host (*Escherichia coli* for T7), and a channel needs to be made to allow the phage genome to travel from the virion to the cytoplasm. A cryo-ET study captured T7 in the act of ejecting its genome inside *E. coli* minicells [51]. The tail tip inserts a DNA ejection conduit (the ejectosome) inside the host cell envelope. It is thought that T7 extends its tail by ejecting the internal head proteins (gp14, gp15, gp16) into the cell prior to DNA transfer into cytoplasm. These three proteins are known to form a cylindrical structure inside the phage head, and they are ejected from the infected virion into the cell envelope before the phage genome. They should disaggregate from their structure in the mature virion and also almost completely unfold in order to leave the head and pass through the narrow head–tail connector channel and form a trans-envelope conduit that connects the virion tail tip to the cell cytoplasm. Bacteriophage T7 has therefore been described as having an extensible tail, like phage phiX174 (see Sect. 17.2.4). A recent comparison of the structures of DNA-filled and empty T7 particles in complex with their receptor [52] have revealed the structural changes that occur in the tail complex during genome delivery. T7 ejection proteins were observed in two states: an initial state previous to ejection, forming rings, and a final state after ejection in which they are assembled as a trans-envelope channel. For a recent review on genome ejection in phage T7 see [41].

After phage adsorption and protein ejection, approximately 1 kilobase of the 40-kilobase genome is normally internalised by the cell [53], perhaps as a result of the release of the pressure of the DNA encapsulated in the capsid (see Chaps. 13, 18 and 19). Subsequent to internalisation of the leading first kilobase, translocation is coupled to transcription. The leading three kilobases include mainly promoters for *Escherichia coli* RNA polymerase. The RNA polymerase moves along the T7 DNA until it encounters the cytoplasmatic membrane but continues to transcribe, thereby pulling DNA from the capsid into the cell. Transcription by *Escherichia coli* RNA polymerase produces messenger RNA for the T7 proteins located at the leading end of the genome, including the single-subunit T7 RNA polymerase and an inhibitor of bacterial restriction enzymes that might otherwise degrade the T7 DNA. The leading end of the genome does not contain restriction sites, whilst the rest of the genome does. At 30 °C, the T7 genome is internalised at a constant 40 base pairs per second across the genome, the same rate as transcription of ribosomal RNA operons. *Escherichia coli* RNA polymerase is one of the strongest motor proteins known and it is thus not surprising that it can pull a DNA molecule into the cell cytoplasm.

Many encapsulated bacteria are infected by specialised bacteriophages that carry host-capsule degrading hydrolases. These usually tailspike-associated enzymatic activities enable the phages to penetrate the capsular layer and are important determinants of the bacteriophage host range. Podovirus K1F infects *Escherichia coli* K1, a polysialic acid encapsulated pathogen. K1F has degrading tailspikes known as endosialidases that break the α2,8-glycosidically linked sialic acid oligomers or polymers [54]. The tailspike endoNF permits the phage to adsorb to the polysialic capsule, depolymerise the capsule to tunnel towards the outer bacterial membrane and to mediate the adhesion to the membrane. Infection may then proceed as for other podoviruses. There are only some isolated phages capable of degrading biofilms. The T7-related bacteriophage ϕ15 is able to degrade biofilms of two *Pseudomonas putida* strains indicating exopolysaccharide depolymerase activity. The ϕ15 tailspike protein is responsible for exopolysaccharide degrading activity and this exopolysaccharide is probably a primary and essential receptor for ϕ15.

Bacteriophage ϕ29. Phage ϕ29 [55] infects *Bacillus subtilis,* a Gram-positive host. Thus, it is perhaps not surprising that this phage has some characteristics that are different from the above-mentioned podoviruses that infect Gram-negative hosts. The phage capsid is decorated with 55 head fibres attached to the quasi-threefold symmetry positions (Fig. 17.6). The head fibres are trimers of the protein p8.5. Although phage particles without head fibres are infective, the head fibres might enhance attachment of the virions onto the host cell wall, interacting with the cell teichoic acids [56].

The tail of bacteriophage ϕ29 consists of the dodecameric portal protein p10, 12 radially arranged trimeric appendages formed by p12 and an extension formed by proteins p11, p9 and p13 (Fig. 17.6). Protein p12 can cleave the teichoic acid of the bacterial cell wall and is thus the functional equivalent of the P22 tailspike in ϕ29 in mediating the first specific, but reversible binding of the phage to the bacterial host. The p12 appendages are kinked; an amino-terminal cylindrical outward-pointing arm links them to the lower collar region of the ϕ29 tail and a carboxy-terminal spindle-shaped domain contains the polyglycerol phosphate degrading active site [57]. Protein p12 monomers are β-helical and assemble into a homotrimer in a sideways manner (Fig. 17.6). An intramolecular chaperone at the very carboxy-terminus helps the p12 monomers to trimerise and is then cleaved off by an autoproteolytic reaction to yield the mature protein. Proteins p11, p9 and p13 make up the central rod of the tail, with p13 being the most distal protein and thus the one likely to make the first irreversible contact with the bacterial cell surface. Its crystal structure revealed both a transglycolase domain and a metalloendoprotease domain. Both likely function in degrading the peptidoglycan layer before DNA transfer can occur [58].

Most of the carboxy-terminal enzymatic domains point downwards in a cryo-EM reconstructed structure of the whole phage, but two point horizontally outwards. The movement of the appendages was suggested to allow the phage to move downwards

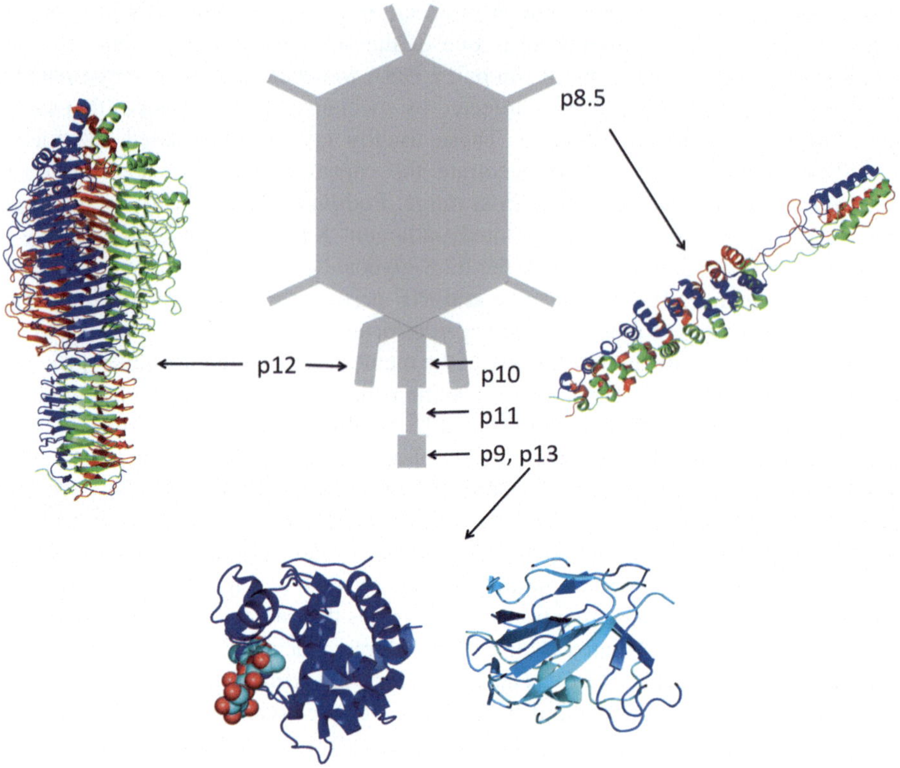

Fig. 17.6 Bacteriophage ϕ29 and its constituent proteins are important in the first steps of infection. A schematic drawing of phage ϕ29 is presented with p8.5, p10, p12, p11, p9 and p13 indicated. The phage is surrounded by crystal structures of the relevant proteins. Crystal structures are shown of the head fibre p8.5 (PDB entry 3QC7), the carboxy-terminal part of the appendage p12 (PDB entry 3GQ7) and p13. The amino-terminal lysozyme-like domain and the carboxy-terminal endopeptidase-like domain of p13 are shown (left and right, respectively), the amino-terminal domain with the co-crystallised substrate tri-N-acetyl-D-glucosamine (PDB entries 3CSZ and 3CSQ, respectively). P13 is located at the end of the phage tail

through the teichoic acid layer to the bacterial membrane [59]. More recently, the structural changes in the ϕ29 particle during adsorption and genome transfer to *Bacillus subtilis* minicells were visualised by cryo-ET [60]. Together with previous observations, the results led to a model in which the transient interactions of the head fibres with the host cell may facilitate the reversible binding of the 12 appendages (tailspikes) to teichoic acids and the hydrolysis of the latter. As the teichoic acids are degraded, the appendages pull the capsid and the tail knob closer to the peptidoglycans of the cell wall. Digestion of the peptidoglycan allows the tail knob to penetrate deeper in the cell wall, until the virion becomes embedded in the cell wall. Further appendages now bind teichoic acids and the virion approaches the cytoplasmic membrane.

The next step in the infection process involves the opening of a pore for DNA ejection. Several observations (reviewed in [40]) suggest that six tail knob gp9 subunits form a tube that is blocked by a long loop of this protein. Upon genome release, these loops form a cone-shaped structure at the tail distal end. A reconstituted in vitro system using liposomes and ϕ29 virions was used to reveal that the cone-shaped structure formed by the released long gp9 loops inserts into the cell membrane and serves as a pore for DNA release.

Bacteriophage ϕ29 DNA transfer proceeds via a push-pull mechanism ([61] and references therein). The right end of the genome enters the bacterium first and the leading two-thirds enter the cell in the absence of phage protein synthesis, presumably by release of the high pressure exerted by the packaged DNA in the capsid (the push phase) (see Chaps. 10 and 18). Transfer of the remaining third of the genome depends on synthesis of the specific phage proteins p16.7 and p17 encoded by the part of the genome transferred in the push phase. Energy is also required for this pull phase, but the exact mechanism is unknown.

Siphoviruses

Siphoviruses have an apparent communication problem between the two ends of their long, flexible tail tube [44]. Somehow, the fact that a suitable bacterium to infect has been encountered with the distal end of the tail has to be transmitted to the end most proximal to the head domain, where the packaged DNA must be released. A conformational change taking place in a domino effect, opening the inner diameter of the tail, has been proposed for this purpose [62], although a simpler mechanism may be that opening the bottom of the tail allows the tape measure protein to leave and DNA simply follows [63, 64].

Bacteriophage T5. Like many phages that infect Gram-negative bacteria, bacteriophage T5 (Fig. 17.7a) has two different types of fibres: the outward-pointing L-shaped tail fibres and the central straight tail fibres [65]. The three L-shaped tail fibres are trimers of the protein pb1, which each monomer containing 1396 amino acids. They bind reversibly to the O-antigen of *Escherichia coli* LPS and accelerate the absorption [66]. However, the loss of this first receptor does not lead to T5 resistance. The interaction of the L-shaped tail fibres with O-antigen accelerates the adsorption of the phage to the cell and keeps the fibred phage at the surface of the host for a considerably longer time than T5 phage lacking fibres, making binding to the secondary receptor more likely the longer the phage can move along the cell surface. Efficient surface attachment occurs when more than one of the three tail fibres binds to an O-antigen at the same time [67]. The L-shaped tail fibres have a thin outward-pointing part followed by a kink and a thicker downward-pointing part consisting of consecutive bead-shaped domains. The first 200 amino acids are rich in alanines and serines and are predicted to form a coiled coil, which may correspond to the thin domain of the L-shaped tail fibres. The remaining part is more varied in sequence and contains large hydrophobic and aromatic residues, so likely forms the consecutive globular domains. The very carboxy-terminal part of the protein is an intramolecular chaperone and probably functions in correct trimerisation of the protein as it does for p12 of phage ϕ29, the carboxy-terminal 132 amino acids get

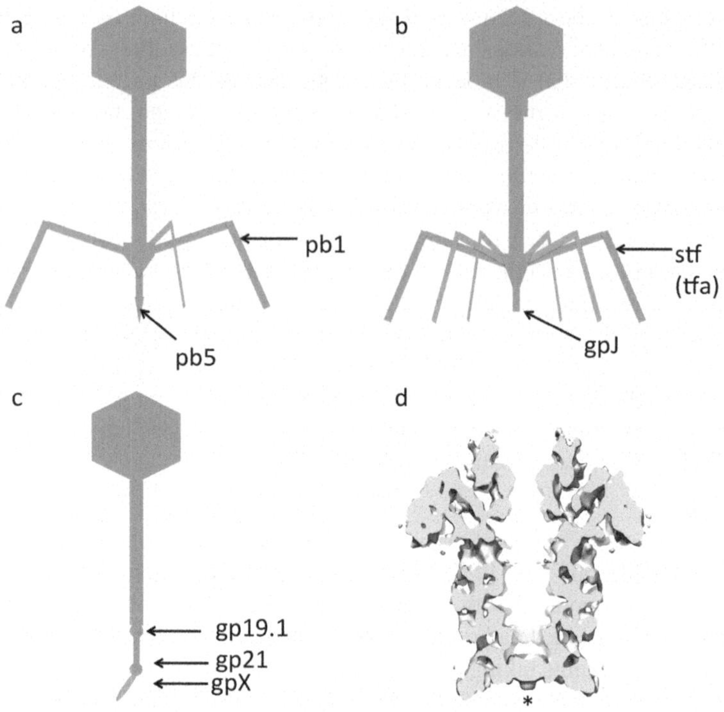

Fig. 17.7 Schematic drawing of some siphoviruses. (**a**) Bacteriophage T5 with the L-shaped tail fibres (pb1) and the FhuA-binding protein pb5 indicated. (**b**) Bacteriophage λ with the side tail fibres (stf, tfa) and the LamB-binding protein gpJ indicated. (**c**) Bacteriophage SPP1. The positions of the hub protein gp19-1, the tail protein gp21, which may have a glycan hydrolysing function, and the unknown receptor-binding tail tip, gpX, are indicated. (**d**) Cut-through of the neck region of phage SPP1 consisting of the portal protein gp6, gp15 and gp16. Electron microscopy density (EMDB 1021) is shown in grey. An asterisk indicates the stopper of gp16 that prevents DNA from exiting the phage head before infection

autoproteolysed after trimerisation. The atomic structure of the C-terminal domain of pb1 contains a β-structured receptor-binding domain and an α/β-structured intramolecular chaperone domain [68]. Whilst present, the chaperone domain appears to prevent pb1 from binding its receptor, suggesting that apart from functioning as a chaperone protein, it may function as a transient cap to prevent newly hatched phages from binding to the membrane debris of the bacterium they have just emerged from [69].

In a second adhesion step, pb5, the central tail fibre, binds irreversibly to the outer membrane ferrichrome transporter FhuA. Comparison of the structures of the T5 tail tip before and after interaction with FhuA revealed important conformational changes that are undergone by this protein complex upon infection, including bending of the central fibre on the side of the tail tip, tail anchoring to the membrane, tail tube opening, and formation of a transmembrane pore next to FhuA, through which the DNA is transferred [70] (Fig. 17.8). In the mature phage particle, the DNA

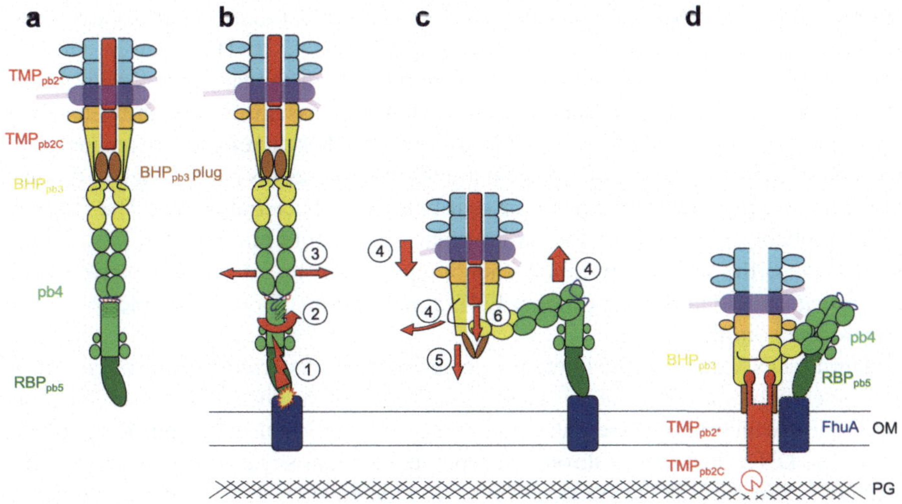

Fig. 17.8 Schematic representation of the conformational changes in the bacteriophage T5 tail tip upon receptor binding and in preparation for DNA transfer. (**a**) The bacteriophage T5 tail tip. (**b**) When the receptor-binding protein pb5 contacts the FhuA receptor (dark blue), a signal (1) is transferred to pb4, which rotates (2) and leads to re-organisation of domain higher up in the tail tip (3). (**c**) The re-organisation causes pb4 to bend (4) and brings the baseplate hub protein pb3 (yellow and brown) closer to the outer membrane (4). This liberates the pb3 plug (brown), opening the tube (5) and expelling the tape measure protein C-terminal fragment TMPpb2C (6). (**d**) The baseplate hub protein pb3 plugs refold as a β-hairpin leg and insert in the outer membrane (OM). The rest of the tape measure protein TMPpb2* is also expelled and forms and outer membrane channel connected to pb3. TMPpb2C (Pacman shape), is released into the periplasm, digests the peptidoglycan (PG). This figure was adapted from [70] Fig. 5, as permitted by the CC-BY license it was published under

is connected to the proximal end of the tail. Upon binding to the receptor, a signal is transmitted from the receptor-binding protein, along the tail, to the head-tail connection, which induces DNA penetration through the tail tube. Signal transmission may occur as a progressive conformational change of the major tail protein starting at the distal end of the tail tube, perhaps widening the inner volume to allow for DNA passage.

Bacteriophage T5 transfers its double-stranded DNA in two steps. In the first step, 8% of the total T5 chromosome is transferred, perhaps by simple pressure release of the tightly packed DNA in the capsid. After a pause of a few minutes during which proteins coded by this fragment are synthesised, the remaining part of the DNA is transferred [38, 71]. The section of the T5 genome that is injected in the first step encodes proteins that degrade host cell DNA and shut down the synthesis of host proteins. How the remaining section of DNA is transferred into the bacterium is unknown.

Bacteriophage λ. The lysogenic bacteriophage λ (Fig. 17.7b) has traditionally been an important model system for molecular biology. Laboratory strains of λ appear to have lost their side tail fibres and can still infect their Gram-negative host *Escherichia coli* efficiently in laboratory conditions [72]. However, λ Ur (the original λ) does have fibres that extend from the side of the tail tip (side tail fibres, stf; Fig. 17.7b). The side tail fibres need the tail fibre assembly protein tfa for their correct trimerisation and assembly. It is not sure whether tfa remains associated with the side tail fibres in the mature phage particles, but this seems likely—again, tfa may also function as a transient cap to prevent binding to bacterial debris. The side tail fibres probably recognise OmpC in a primary, reversible binding step [73]. In the second step, the carboxy-terminal end of gpJ at the tip of the tail binds to the *Escherichia coli* lamB gene product [74], a maltoporin. These two proteins form a very stable complex.

During DNA transfer, the linear phage genome is transferred past the *Escherichia coli* outer membrane. During this process, transmembrane channels are formed in the liposomes, which permit the entry and escape of small molecules, but not proteins [75]. The DNA then passes through a separate bacterial sugar transport protein, the mannose permease of the bacterial phosphotransferase system, in the cytoplasmic membrane [76]. The two hydrophobic subunits II-PMan and II-MMan alone are sufficient for penetration of λ DNA. It seems thus that bacteriophage λ has subverted the bacterial maltose entry system into a transfer system for its DNA. Once in the cytoplasm, the linear DNA circularises using the *cos* sites, which are the 12-base G-C-rich cohesive sticky ends of the genome.

Bacteriophage SPP1. The tail of the Gram-positive *Bacillus subtilis* phage SPP1 (Fig. 17.7c) has a single central fibre for host interaction [77]. The tail tip, gp21, of SPP1 binds to YueB, a membrane protein whose extracellular region crosses the thick cell wall to expose a receptor region at the bacterial surface. The tail tip also mediates degradation of the bacterial cell wall. The tip is 31 nm long and 10 nm wide at its largest diameter. It can be divided into three domains: a sphere-like region, a broad flattened domain and a terminal rod (Fig. 17.7c). Gp21 has predicted β-helical structure and may resemble the P22 tailspike in structure and glycan hydrolysing function.

The tip does not have a DNA channel and therefore probably moves out of the way before DNA transfer. The binding of the tail fibre to YueB leads to the irreversible commitment of the virus particle to eject its DNA. The structure of the siphovirus SPP1 tail has been determined before and after DNA ejection [62]. The results revealed extensive structural rearrangements in the internal wall of the tail tube. A proposal was made that the binding of the tail tip to YueB triggers a conformational switch that is propagated as a cascade of conformational changes along the 160-nm long helical tail structure to reach the head-to-tail connector. This would lead to the connector opening and allowing passage of DNA into the host cell through the tail tube. The structure of the connector region between the tail and the head of DNA-containing phage has been determined by electron microscopy and fitting of atomic structures. It consists of three dodecameric proteins: the portal protein gp6, the adaptor protein gp15 and a "stopper" protein, which prevents premature DNA release [78, 79] (Fig. 17.7d).

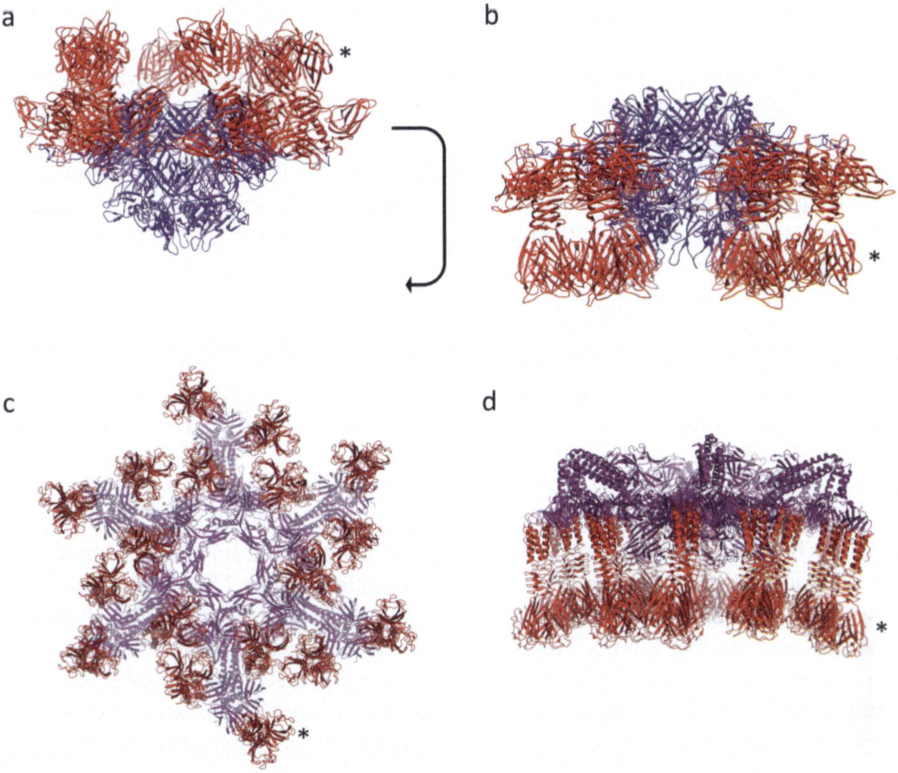

Fig. 17.9 Crystal structures of *Lactococcus lactis* bacteriophage baseplates. (**a**) The baseplate of phage p2 before activation, with the receptor-binding domains of the receptor-binding proteins (in red) pointing upwards, away from the bacterium (PDB entry 2WZP). The movement of 200° by the receptor-binding proteins that completely reverse their orientation is indicated by an arrow. (**b**) The baseplate of phage p2 after activation by calcium ions, with the receptor-binding domains of the receptor-binding proteins (in red) pointing downwards, towards the bacterium (PDB entry 2X53). (**c**) The baseplate of phage TP901-1 (PDB entries 4DIV and 4DIW) is viewed from below (i.e. from the side of the bacterium) and viewed from the side (**d**). The 18 receptor-binding protein trimers are in red. One of the 18 receptor-binding domains in parts (**a**) and (**b**), as well as one of the 54 receptor-binding domains in parts (**c**) and (**d**), are indicated by asterisks

Lactococcus lactis Bacteriophages. Siphoviruses that infect the Gram-positive *Lactococcus lactis* bacterium contain rather compact receptor-binding proteins. The crystal and electron microscopy structures of the baseplates of two *Lactococcus lactis* phage p2 [80] and TP901-1 [81] have been determined (Fig. 17.9). The baseplate of phage p2 is composed of three protein species [82]. The central part of the baseplate is formed by a circular hexamer of ORF15 with a central hole. A trimer of ORF16 is located at the bottom of the baseplate, forming a closed dome that does not allow DNA passage. Six ORF18 trimers are attached to the central ring, each trimer interacting with a carboxy-terminal extension of an ORF15 monomer. ORF18 is the receptor-binding protein and its structure shows an N-terminal β-sandwich shoulder domain, which binds to the baseplate. The remainder of the structure is a short triple β-helical neck and a carboxy-terminal head domain. In the unactivated baseplate, the six trimeric receptor-binding proteins point upwards, i.e. away from the bacterial membrane. Activation by calcium ions causes large conformational changes in p2 baseplate, leading to rotation of receptor-binding domains by 200° to point downwards towards the host cell. The head domains of ORF18 may recognise lipoteichoic acids, which are phospho-glycerol polymers. The conformational change of the baseplate also leads the three ORF15 monomers to separate, opening up a hole in the centre of the baseplate, presumably allowing passage of DNA for transfer into the host.

The baseplate of phage TP901-1 is composed of multiple copies of four different proteins [83]. The centre is a hexameric circular core formed by the Dit protein. The hole in this core is filled by a central tail fibre, Tal. From the core, six arms emanate, each arm being composed of a trimer of the BppU protein. The arms are α-helical up to the elbow. The rest of the arm points downwards and is an adaptor domain for the receptor-binding proteins. To each adaptor, three trimeric receptor-binding proteins bind, leading to a total of 54 sites for the host cell envelope saccharides. The receptor-binding proteins point downwards, i.e. towards the host bacterium, ready for adhesion. No calcium ions are necessary for activation of TP901-1, so there are likely no conformational changes that have to take place before receptor binding. How receptor binding is related to DNA transfer in these phages is less clear, perhaps the strong binding with up to 54 receptor molecules pushes the central tail fibre against the cell wall and the force of the cell wall pushing against the tail fibre is sensed by the other end of the tail fibre, which opens a hole at the end of the tail tube.

Myoviruses

With their double-shafted tail, myoviruses pierce the bacterial outer membrane and traverse the periplasmic space or peptidoglycan layer with the inner tail tube upon contraction of the outer tail tube (Fig. 17.10a) [42, 43]. The tips of myovirus inner tail tubes even have a spike for efficient membrane penetration [86] (Fig. 17.10d). In this way, even large genomes of up to several hundred kilobases can be transferred efficiently into the bacterial cell. Here, we discuss what is known about receptor attachment and DNA transfer of some well-known myoviruses.

Bacteriophage T4. Phage T4 [87] is considered the archetypal myovirus. Receptor binding and DNA injection have been studied in detail for this virus. Like the head fibres of phage ϕ29, immunoglobulin-like domains on the surface of T4 (see also Chap. 12) could interact with the cell surface and help to maintain the phage in the vicinity of the cell surface by weak and non-specific interactions until the correct receptor is contacted. The bacteriophage T4 capsid is decorated with two non-essential outer capsid proteins: Hoc (highly antigenic outer capsid protein) and Soc (small outer capsid protein). Soc provides stability to T4 to survive under hostile environments; Hoc has immunoglobulin-like domains and may interact with the bacterial surface providing certain advantages [88].

Phage T4, and presumably other myoviruses as well, infect their host bacteria very efficiently, probably due to the sophistication of its contractile tail [43, 89, 90]

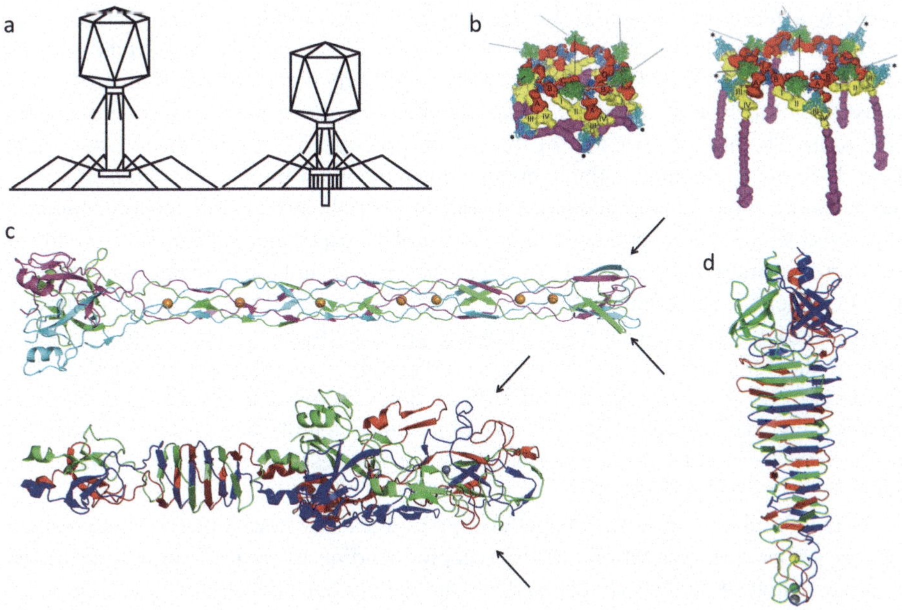

Fig. 17.10 Receptor recognition and DNA transfer by *Escherichia coli* myoviruses. (**a**) Schematic drawing of bacteriophage T4 before and after tail contraction (adapted from [84], copyright by the National Academy of Sciences). (**b**) Structure of the baseplate in the hexagon and star conformations with the different proteins coloured differently. The short tail fibres are shown in magenta; folded away in the hexagon-state (extended tail sheath)—and extended in the star-state (contracted tail sheath); reproduced from [85]. (**c**) Structures of the long tail fibre protein gp37 tip domain and the carboxy-terminal domain of gp12. Arrows indicate the receptor-binding regions. (**d**) The central spike of bacteriophage P2. The structure contains three ions at the tip: a calcium ion (yellow), a chloride ion (green) and an iron ion (grey)

(Fig. 17.10a). Infection of phage T4 starts with the interaction of the long tail fibres [35, 91, 92] with cell surface receptors (see below). The long tail fibres of bacteriophage T4 are kinked structures of about 145 nm long with a variable width of up to 5 nm [85]. They can be divided into proximal and distal half-fibres, attached to each other at an angle of about 20°. Upon phage assembly in the previous host cell and in adverse conditions for phage multiplication, the long tail fibres are in a retracted conformation, lying against the tail sheath and head of the bacteriophage. Because the neck whisker protein (called gp*wac* or fibritin) is necessary for incorporation of the long tail fibres into the phage [93], we can suppose that this interaction is also responsible for maintaining the long tail fibres in a retracted state. In the extended conformation, only the proximal end of the fibre is attached to the baseplate and the fibre tips point outwards.

The long tail fibres are composed of four different proteins: gp34, gp35, gp36 and gp37 [85]. The proximal half-fibre (the thigh) is formed by a homotrimer of protein gp34. The amino-terminal part of gp34 is attached to the trimeric protein gp9 of the baseplate whilst the carboxy-terminal part interacts with gp35. Protein gp35 (the knee) is a monomer and may be responsible for creating the angle between the proximal and distal half-fibres. The distal half-fibre is composed of a trimer of gp36 and a trimer of gp37. Protein gp36 forms the upper part of the shin and gp37 constitutes the rest of the shin, including the very distal receptor recognising tip (the foot). The atomic structure of the receptor recognising tip has been determined [84]. It forms a globular collar domain, followed by a thin needle region in which seven iron atoms are coordinated and end in a small intertwined receptor-binding domain (Fig. 17.10). A phage-encoded molecular chaperone, gp57A, is required for the correct trimerisation of short tail fibre protein gp12 and the long tail fibre proteins gp34 and gp37. A specific chaperone, gp38, is required for the correct trimeric assembly of gp37 but does not stay bound to gp37. In the extended conformation, the tips of the long tail fibres can interact reversibly with receptor molecules. The glucosyl-α-1,3-glucose terminus of LPS or outer membrane porin C (OmpC) can function as alternative receptors and both outer membrane protein C and the interaction region of LPS have to be absent or inaccessible for *Escherichia coli* to be resistant to infection by T4.

In the T4-like *Salmonella* phage S16, gp38 does stay bound to gp37 and instead of just a chaperone, functions as the receptor-binding tip [94]. A monomer of the glycine-rich gp38 binds strongly to the trimeric beta-helical C-terminal end of gp37. The glycines contribute to a new protein fold made up of a sandwich of polyglycine type II helices.

Several studies have explored different aspects of the structural rearrangements that occur in the T4 virion once attached to the cell surface [42, 43]. In one study, cryo-ET was used to visualise details of the structural intermediates during T4 infection of *E. coli* minicells [95]. When at least three long tail fibres have bound a receptor molecule, a signal is transferred to the baseplate of the phage, which then changes conformation. The binding information transfer is likely related to the angle of attachment of the long tail fibre to the baseplate. In the free phage, this angle is variable and the fibres are flexible up to certain limits. When several fibres are attached to their receptor, external forces on the phage may force these angles to less or more than these allowed angles, pushing proteins near the long fibre attachment site in different conformations and triggering the conformational change of base plate, from the hexagon conformation to the star conformation [85]. This conformational change is very extensive, involving changes in location and interaction partners for several baseplate proteins. Amongst others, it leads to extension of the short tail fibres, which are normally folded away in the baseplate (Fig. 17.10b). The short tail fibres are trimers of the protein gp12. When the baseplate is in its hexagon conformation, the short tail fibres are incorporated into the baseplate by interactions along their lengths. However, when the baseplate switches to its star conformation, only the proximal (amino-terminal) end of gp12 remains bound to the baseplate from which the rest of the molecule extends. The distal carboxy-terminal part of the short tail fibres comprising residues 317–517 is responsible for a tight, irreversible interaction with the core region of the LPS (Fig. 17.10). The short tail fibres (gp12) have been seen in thin sections of phage-infected bacteria, forming rigid connections of the baseplate to the host cell surface.

In a subsequent step, contraction of the tail sheath (a helical polymer of protein gp18) is initiated [96]. The sheath of T4 is 24 nm wide and 93 nm long and is composed of 138 gp18 subunits, which are arranged in a six-start helix. The pitch and twist of the helix are 4 nm and 17°, respectively. The contracted T4 sheath is 33 nm wide and 42 nm long. The six-start helix changes its pitch and twist to 1.6 nm and 33°, respectively. In the process of sheath contraction, gp18 subunits move as rigid bodies without refolding or significant domain shifts. Upon contraction, the contact area between gp18 molecules increases by about 4 times. The events that lead to sheath contraction have been explained in atomic detail [96].

As a result of tail sheath contraction, the rigid inner tail tube (a helical polymer of gp19) is driven through the outer cell membrane. The trimeric protein gp5 at the tip of the tail tube contains three lysozyme domains that are activated when they enter in contact with the periplasmic peptidoglycan layer. They create a hole, which allows reaching the cytoplasmic membrane. When the tail tube interacts with the cytoplasmic membrane, the phage DNA is released through the tail tube. Interaction of the tip of the central tail tube with the periplasmatic chaperone PpiD has been proposed. PpiD is anchored to the cytoplasmatic membrane and may help the phage DNA enter the cytoplasm [97]. As for other phages, the pressure of the tightly packed DNA in the head domain probably accounts for the force that initiates DNA transfer, but its completion may need further mechanisms. It appears a proton-motive force over the inner membrane is necessary for successful DNA transfer [98]. Under favourable conditions, the 172-kilobase genome of phage T4 DNA enters the bacterium in less than 1 min, making it a very efficient process.

Bacteriophage P2. Phage P2 (not to be confused with the above-mentioned lactococcal phage p2) [99] is a myovirus with a smaller genome than that of T4, namely 33 kilobases [50]. It also has a simpler tail, for which the synthesis of only 11 gene products is necessary. Phage P2 has a baseplate with a single central spike, which is a trimer of gpV, and six tail fibres, which are formed by gpG. The structure of the central spike protein has been solved at high resolution [43]. The P2 fibres have a specialised chaperone to aid them in their folding, like those of the λ side tail fibre and the bacteriophage T4 distal tail fibre protein gp37. In the first step of infection, the fibres attach to the core region of the LPS of. Calcium ions have been shown to greatly improve adsorption. The baseplate then presumably changes conformation to allow the central tail tube to pass through it. Contraction of the tail sheath drives the tail tube through the baseplate and the bacterial outer membrane and periplasm, delivering the DNA exit site to the inner membrane. The tip of the tail tube consists of a triple-helical conical spike, of which the very tip is stabilised by three ions: iron, chloride and calcium (Fig. 17.10d).

Myoviruses infecting Gram-positive hosts. Myoviruses infecting Gram-positive hosts tend not to have long side tail fibres. Phage A511 infects the Gram-positive bacteria *Listeria monocytogenes* and *ivanovii.* The structure of phage A511 was solved in both pre- and post-host attachments, i.e. uncontracted and contracted, respectively [100]. The structure of the baseplate is simpler than that of phage T4.

The structures of the pre- and post-attachment states of the *Staphylococcus* phage φ812 have also been determined [101], and this baseplate contains the same basic framework. In both A511 and φ812 phages, in the pre-attachment state, the receptor-binding proteins are pointing upward, towards the head of the phage. After DNA ejection, the baseplates transform into a double-layered structure, and the receptor-binding proteins change to a downward orientation, i.e. towards the bacterium, in a similar process as observed for the *Lactococcus* siphovirus p2.

The tail tube and central part of the baseplate have structures similar to those of phage T4, but phage A511 has different outer part of the baseplate, notably with different receptor-binding proteins attached to it. Contraction of the phage tail sheath starts at the baseplate and propagates through the sheath towards the neck in a domino-like motion, a mechanism that is likely true for all myoviruses. During this transformation, the symmetry changes from three- to sixfold. Interestingly, although the receptor-binding proteins of A511 are different from T4, the tip of the tail tube is very similar, despite the differences between the cell envelopes of Gram-positive and Gram-negative bacteria.

Bacteriophage A511 contains the conserved tube-baseplate core complex, consisting of the tube proteins (gp19 and gp54 in T4), the baseplate hub proteins (gp48 and gp27 in T4), the tailspike protein (gp5 in T4), and the four wedge proteins (gp6, gp7, gp25 and gp56 in T4). These proteins are likely to be also conserved in simpler myoviruses that infect Gram-negative bacteria, like the phage P2 and Mu. Simpler baseplates contain only one trimeric protein (the host receptor-binding fibre) that is attached to gp7 [89]. More complicated phages like T4 developed gp9 and the long tail fibres for reversible host selection on the one hand and irreversible binding via gp10, gp11 and gp12 on the other hand.

Bacteriophage 8a. The structure of the *Bacillus anthracis* spore-binding phage 8a was determined by cryo-electron tomography (see Chap. 3) [102] (Fig. 17.11). This phage infects the Gram-positive *Bacillus anthracis* and, as the name suggests, it also binds *Bacillus anthracis* spores. Like other myoviruses, the tail consists of a six-start helical sheath surrounding a central tail tube, but it has a baseplate different from other phages. The baseplate recognises and attaches to host cells. Cryo-electron tomographic analysis was performed on four different conformations that likely correspond to four states of the infection process: DNA-filled particles with extended tails, DNA-filled particles with contracted tails, partially DNA-filled particles with contracted tails and empty particles with contracted tails.

The baseplate has a claw-like structure, with six extensions in a closed conformation when the tail is extended, and in an open conformation when the tail is contracted. Primary receptor interaction probably takes place by the closed baseplate, although it is unknown which part of the baseplate is responsible, as phage 8a lacks long tail fibres or other obvious primary receptor-binding modules. However, densities reminiscent of the bacteriophage T4 short tail fibres [85] are present at the periphery of the baseplate, in a folded-away state when the tail sheath is extended. Perhaps the folded-away short tail fibre-like densities present a sufficient receptor-binding site for interaction. In any case, the baseplate undergoes a conformational change from a closed to an open conformation upon binding to the host cell surface, because in the contracted state, these densities point downwards. The downward extension of the densities may bind host cell receptors more efficiently to irreversibly

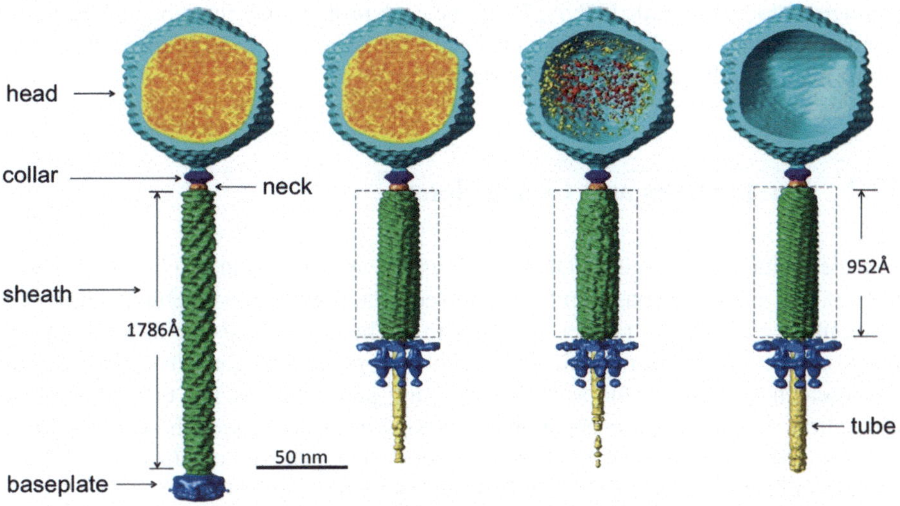

Fig. 17.11 Myoviruses of Gram-positive bacteria. *Bacillus anthracis* spore-binding phage 8a. Electron cryotomographic images of four stages on the pathway of DNA ejection are shown: the phage with extended sheath, the phage with contracted sheath but still containing DNA, the phage with contracted sheath in the process of ejecting its DNA and the phage with contracted sheath that has fully ejected all DNA (reprinted from [102] with permission from Elsevier). The conformational change of the baseplate from a disk-shaped structure to a structure with sideways- and downward-pointing extensions is very clear

immobilise the baseplate on the host cell surface and cause its structural rearrangement. As the baseplate and tail sheath are physically connected, the structural rearrangement of the baseplate likely initiates contraction of the tail sheath.

The extended tail sheath is assembled around the 8-nm diameter tail tube as a right-handed, six-stranded helical structure measuring 20 nm in diameter and 180 nm in length. Each helical strand consists of 45 subunits related to each other by a 4-nm helical pitch and a 22° twist. The 45 subunits in each helical strand of the contracted tail are related by a 2-nm pitch and a 31° twist; the contracted tail sheath is thus about half the length of the extended one. Like for bacteriophage T4, the structure of individual tail sheath subunits is similar in both conformations, indicating that tail contraction results from rigid body rotations and translations of entire subunits rather than by conformational changes within subunits. Contraction of the tail sheath forces the central tail tube downwards into the cell surface. A central spike structure is present at the distal end of the tail tube in extended tails, which may function as a cell-puncturing device as proposed for T4. No equivalent structure is seen in contracted tails, so it is likely the tail spike falls off during or after contraction to allow passage of the DNA.

A comparison of the neck regions in extended and contracted tails indicates that there are two conformations. In the extended state, the neck is assembled into a compact shape plugging the connection between the head and the tail tube. In the contracted state, the neck forms a central pore that is large enough to allow the passage of double-stranded DNA. Thus, the neck protein may function as a gate that regulates the passage of DNA from the head into the tail, by undergoing a substantial conformational change upon tail contraction.

17.3 Perspectives and Conclusions

A bacteriophage that ejects its DNA spuriously or into an unsuitable host cell ceases to exist [63]. Evolution has therefore endowed bacteriophages with efficient host cell recognition and DNA transfer mechanisms. These mechanisms differ between phages, and it is therefore difficult to propose a general mechanism. Also, due to the research interests of the scientists studying particular phages, they have often concentrated on certain aspects of phage biology. However, given these restrictions, here we give a brief overview, attempt to draw some general conclusions and draw attention to some interesting phage facts and potential applications.

In general, bacteriophages attach to host cells using a two-step mechanism. In the first step, attachment is specific but reversible. This allows the phage to diffuse two-dimensionally over the bacterial surface until the second receptor is encountered. This second receptor is then bound irreversibly and at this point, the phage is immobilised and committed to DNA transfer. The sequences of receptor-binding proteins of phages are highly variable; and varying the receptor-binding properties is probably the main mechanism a phage has to adapt to hosts that have varied their cell wall structure or to adapt to a new bacterial host strain or species. Horizontal gene transfer between phages plays an important role in this variation, even between different phage types (horizontal gene transfer is probably most likely when a phage infects a bacterial cell containing a prophage). An example of relatively recent horizontal gene transfer is between the tip of the temperate siphovirus λ side tail fibre and the virulent myovirus T4 long tail fibre, in which residues that are structurally important are conserved, but the distal domain has evolved to bind a different receptor [84]. Another example of evolutionary recent horizontal gene transfer is the sequence and structural homology between the receptor-binding domains of the tailspikes of the podovirus P22 and the myovirus Det7 [103]. There are of course many phages still to be discovered, and their biological characterisation, the sequencing of their genomes and structure determination by electron microscopic (Chap. 3) and crystallographic techniques (Chap. 4) will likely reveal new receptor-recognition mechanisms.

Some tailed phages encode several alternate tail fibre genes and may express them either alternately or simultaneously in the same particle [37]. An example is the myovirus ϕ92, which encodes a Swiss army knife-like set of fibres to infect most laboratory strains of *Escherichia coli* (including those with a sialic acid capsule) and many *Salmonella* strains [104]. In the temperate myoviruses μ and P1 the specificity for the host receptor is altered upon inversion of a genome segment. The inversion results in a tail fibre with an alternative carboxy-terminal receptor-binding part. The amino-terminal part, which binds to the phage, is coded outside the inversion region and thus stays constant. In this way, in one form bacteriophage μ carries tail fibres that allow it to infect *Escherichia* and *Salmonella* bacteria and in a second form tail fibres that allow it to infect *Shigella*, *Enterobacter*, *Erwinia* and *Citrobacter* [105]. *Bordetella* phage BPP-1 has been shown to use an error-prone reverse transcriptase to generate variation in the distal knob of its tail fibre proteins [106].

As Gram-positive bacteria lack an outer membrane and instead have a thicker layer of peptidoglycan, one would expect their receptor-binding mechanisms to be different. Phages infecting Gram-negative hosts tend to use porins, transport proteins, enzymes or LPS available on the outer membrane as primary or secondary receptors. Phages infecting Gram-positive hosts tend to use peptidoglycan, teichoic acids or proteins that span the cell wall as receptors. One characteristic of tailed phages infecting Gram-positive hosts is that they appear to lack the side tail fibres many tailed phages infecting Gram-negative hosts possess. Phages may also bind to pili, flagella and capsules of both kinds of bacteria.

In most bacteriophages, much less is known about the mechanisms of nucleic acid transfer. Because the nucleic acid in many phages is packaged under considerable pressure (see Chaps. 13, 18 and 19), it is likely initial transfer is at least helped by the partial release of this pressure when an appropriate transfer channel has been opened. Transfer of the remaining nucleic acid may then be promoted by phage proteins co-transferred into the cell, produced by transcription from the first region of nucleic acid transferred or by host factors. For a phage that always starts transfer with a specific end of the genome, like T7, proteins coded for at this end may be involved in successful genome transfer. For circularly permuted phages like T4, the entire nucleic acid is likely transferred by pressure and host factors, as the phage cannot count on a specific protein being expressed from the part of the genome first transferred. Phages have also developed different mechanisms to protect the nucleic acid that is transferred into the cells. Nucleic acids may be protected from nucleases in the periplasm by a physical barrier: the extensible tail tube of podoviruses and microviruses, the inner tail tube in myoviruses, or by adopting a transport system already native to the bacterium that is about to be infected. Once in the cytoplasm, restriction enzymes are a major danger. Some phages modify the nucleotides of their DNA so they cannot be recognised by restriction enzymes. Other phages (T7, T5) only transfer a small part of their genome in the first step, and this part encodes proteins that inactivate restriction enzymes or otherwise protect DNA against destruction.

For nucleic acid transfer, phages infecting Gram-positive bacteria only have to cross the inner membrane. Therefore, if they can somehow get close to it by digesting the peptidoglycan cell wall, they may not need to extend a tube-like tailed phages infecting Gram-negative bacteria need to do to allow their genomes to safely traverse the nuclease-ridden periplasmic space. This may be an explanation for the fact that there appear to be relatively fewer myoviruses infecting Gram-positive bacteria [63]. Instead, podoviruses and siphoviruses can, at least in theory, digest their way straight to the cytoplasmic membrane of Gram-positive bacteria if their baseplates contain the appropriate enzymes.

Future structural studies of bacteriophage proteins, phage protein assemblies and whole bacteriophages by crystallography (Chap. 4) and cryo-EM (Chap. 3) will certainly resolve more details of phage receptor-binding and nucleic acid transfer. The recent "resolution revolution" in cryo-EM (Chap. 3) is leading to the determination of entire phage tail complexes. Cryo-ET (Chap. 3) using minicells of bacteria is also already playing an important role in understanding the mechanisms of infection by phages. Apart from structural studies, future studies will aim to measure individual interactions of phages and their proteins with their host cells. Single molecule techniques may be used to quantify the force of binding of bacteriophages or their receptor-binding proteins to the receptor. In atomic force microscopy (see Chap. 9) a phage fibre may be bound to the tip and used to probe a surface containing its receptor, whilst optical tweezers (see Chap. 10) could also be used to measure these interactions.

The remarkable properties of bacteriophages have led to many ideas for applications (see Chap. 22). Now that information in atomic detail is becoming available on receptor-binding proteins of different bacteriophages, this information can be exploited to perform directed mutation and modulate their receptor-binding properties. Future knowledge of the nucleic transfer mechanisms may also lead to applications in designing nucleic acid delivery vehicles.

Acknowledgements I thank Carmela Garcia-Doval for co-authoring a previously published version of this chapter and M.G. Mateu for critical revision of the first version of this manuscript and for valuable help in revising literature. Structure figures were made with PyMol (The PyMOL Molecular Graphics System, Schrödinger, LLC) or the UCSF Chimera package (developed by the Resource for Biocomputing, Visualization and Informatics at the University of California, San Francisco and supported by NIH P41 RR-01081).

References

1. Wommack KE, Colwell RR (2000) Virioplankton: viruses in aquatic ecosystems. Microbiol Mol Biol Rev 64:69–114
2. Harper DR, Abedin ST, Burrowes BH, McConville ML (eds) (2021) Bacteriophages. Springer Nature, Switzerland, p 2
3. Millman A et al (2022) An expanded arsenal of immune systems that protect bacteria from phages. Cell Host Microbe 30:1556–1569.e5
4. Sanz-Gaitero M, Seoane-Blanco M, van Raaij MJ (2019) Structure and function of bacteriophages. In: Harper DR, Abedin ST, Burrowes BH, McConville ML (eds) Bacteriophages. Springer Nature, Switzerland, pp 1–73
5. Mesyanzhinov VV (2004) Bacteriophage T4: structure, assembly, and initiation infection studied in three dimensions. Adv Virus Res 63:287–352
6. Noirclerc-Savoye M, Flayhan A, Pereira C, Gallet B, Gans P, Ebel C, Breyton C (2015) Tail proteins of phage T5: investigation of the effect of the His6-tag position, from expression to crystallisation. Protein Expres Purif 109:70–78
7. Tang J, Lander GC, Olia A, Li R, Casjens S, Prevelige P Jr, Cingolani G, Baker TS, Johnson JE (2011) Peering down the barrel of a bacteriophage portal: the genome packaging and release valve in P22. Structure 19:496–502

8. Specthrie L, Bullitt E, Horiuchi K, Model P, Russel M, Makowski L (1992) Construction of a macrophage variant of filamentous bacteriophage. J Mol Biol 228:720–724
9. Oksanen HM, Abrescia NGA (2019) Membrane-containing icosahedral becteriophage PRD1: the dawn of viral lineages. In: Greber UF (ed) Physical virology. Springer Nature, Switzerland, pp 85–109
10. Butcher SJ, Manole V, Karhu NJ (2012) Lipid-containing viruses: bacteriophage PRD1 assembly. In: Rossmann MG, Rao V (eds) Viral Molecular Machines. Springer, New York
11. Fane BA, Brentlinger KL, Burch AD, Hafenstein SL, Moore E, Novak CR, Uchiyama A (2005) PhiX174 et al., the Microviridae. In: Calendar RL (ed) The bacteriophages, 2rd edn. Oxford University Press, Oxford
12. Kirchberger PC, Ochman H (2023) Microviruses: a world beyond ΦX174. Annu Rev Virol 10:99–118
13. Hay ID, Lithgow T (2019) Filamentous phages: masters of a microbial sharing economy. EMBO Rep 20:e47427
14. Conners R, León-Quezada RI, McLaren M, Bennett NJ, Daim B, Rakonjac J, Gold VAM (2023) Electron microscopy of the f1 filamentous phage reveals insights into viral infection and assembly. Nat Commun 14:2724
15. Golmohammadi R, Valegard K, Fridborg K, Liljas L (1993) The refined structure of bacteriophage MS2 at 2.8 Å resolution. J Mol Biol 234:620–639
16. Toropova K, Stockley PG, Ranson NA (2011) Visualising a viral RNA genome poised for release from its receptor complex. J Mol Biol 408:408–416
17. Meng R, Jiang M, Cui Z, Chang JY, Yang K, Jakana J, Yu X, Wang Z, Hu B et al (2019) Structural basis for the adsorption of a single-stranded RNA bacteriophage. Nat Commun 10: 3130
18. Gorzelnik KV, Zhang J (2021) Cryo-EM reveals infection steps of single-stranded RNA bacteriophages. Prog Biophys Mol Biol 160:76–83
19. Sun Z, El Omari K, Sun X, Ilca SL, Kotecha A, Stuart DI, Poranen MM, Huiskonen JT (2017) Double-stranded RNA virus outer shell assembly by bona fide domain-swapping. Nat Commun 8:14814
20. Mäntynen S, Sundberg LR, Poranen MM (2018) Recognition of six additional cystoviruses: Pseudomonas virus phi6 is no longer the sole species of the family Cystoviridae. Arch Virol 163:1117–1124
21. Bamford DH, Palva ET, Lounatmaa K (1976) Ultrastructure and life cycle of the lipid-containing bacteriophage φ6. J Gen Virol 32:249–259
22. Hu GB, Wei H, Rice WJ, Stokes DL, Paul Gottlieb P (2008) Electron cryo-tomographic structure of cystovirus φ12. Virology 372:1–9
23. Riechmann L, Holliger P (1997) The C-terminal domain of TolA is the co-receptor for filamentous phage infection of *E. coli*. Cell 90:351–360
24. Sun L, Young LN, Zhang X, Boudko SP, Fokine A, Zbornik E, Roznowski AP, Molineux IJ, Rossmann MG, Fane BA (2014) Icosahedral bacteriophage ΦX174 forms a tail for DNA transport during infection. Nature 505:432–435
25. McKenna R, Ilag LL, Rossmann MG (1994) Analysis of the single-stranded DNA bacteriophage PhiX174, refined at a resolution of 3.0 Å. J Mol Biol 237:517–543
26. Sun Y, Roznowski AP, Tokuda JM, Klose T, Mauney A, Pollack L, Fane BA, Rossmann MG (2017) Structural changes of tailless bacteriophage ΦX174 during penetration of bacterial cell walls. Proc Natl Acad Sci USA 114:13708–13713
27. Hong C, Oksanen HM, Liu X, Jakana J, Bamford DH, Chiu W (2014) A structural model of the genome packaging process in a membrane-containing double stranded DNA virus. PLoS Biol 12:e1002024
28. Peralta B, Gil-Carton D, Castaño-Díez D, Bertin A, Boulogne C, Oksanen HM, Bamford DH, Abrescia NG (2013) Mechanism of membranous tunnelling nanotube formation in viral genome delivery. PLoS Biol 11:e1001667

29. Santos-Pérez I, Oksanen HM, Bamford DH, Goni FM, Reguera D, Abrescia NGA (2017) Membrane-assisted viral DNA ejection. BBA-Gen Subjects 1861:664–672
30. Kivelä HM, Abrescia NGA, Bamford JKH, Grimes JM, Stuart DI, Bamford DH (2008) Selenomethionine labeling of large biological macromolecular complexes: probing the structure of marine bacterial virus PM2. J Struct Biol 161:204–210
31. Kivelä HM, Kalkkinen N, Bamford DH (2002) Bacteriophage PM2 has a protein capsid surrounding a spherical proteinaceous lipid core. J Virol 76:8169–8178
32. Huiskonen JT, Kivelä HM, Bamford DH, Butcher SJ (2004) The PM2 virion has a novel organization with an internal membrane and pentameric receptor binding spikes. Nat Struct Mol Biol 11:850–856
33. Kivelä HM, Daugelavičius R, Hankkio RH, Bamford JKH, Bamford DH (2004) Penetration of membrane-containing double-stranded-DNA bacteriophage PM2 into Pseudoalteromonas hosts. J Bacteriol 186:5342–5354
34. Nobrega FL, Vlot M, de Jonge PA, Dreesens LL, Beaumont HJE, Lavigne R, Dutilh BE, Brouns SJJ (2018) Targeting mechanisms of tailed bacteriophages. Nat Revs 16:760–773
35. Seoane-Blanco M, van Raaij MJ, Granell M (2021) Bacteriophage tail fibres, tailspikes, and bacterial receptor interaction. In: Reference module in life sciences, Encyclopedia of Virology, vol 4, 4th edn. Academic Press, pp 194–205. https://doi.org/10.1016/B978-0-12-814515-9.00152-1
36. Zinke M, Schröder GF, Lange A (2022) Major tail proteins of bacteriophages of the order Caudovirales. J Biol Chem 298:101472
37. Ouyang R, Ongenae V, Muok A, Claessen D, Briegel A (2024) Phage fibers and spikes: a nanoscale Swiss army knife for host infection. Curr Opin Microbiol 77:102429
38. Molineux IJ, Panja D (2013) Popping the cork: mechanisms of phage genome ejection. Nat Rev 11:194–204
39. Bhardwaj A, Olia AS, Cingolani G (2014) Architecture of viral genome-delivery molecular machines. Curr Opin Struct Biol 25:1–8
40. Xu J, Xiang Y (2017) Membrane penetration by bacterial viruses. J Virol 91:e00162–e00117
41. Swanson NA, Hou C-FD, Cingolani G (2022) Viral ejection proteins: mosaically conserved, conformational gymnasts. Microorganisms 10:504
42. Brackmann M, Nazarov S, Wang J, Basler M (2017) Using force to punch holes: mechanics of contractile nanomachines. Trends Cell Biol 27:623–632
43. Taylor NMI, van Raaij MJ, Leiman PG (2018) Contractile injection systems of bacteriophages and related systems. Mol Microbiol 108:6–15
44. Linares R, Arnaud C-A, Degroux S, Schoehn G, Breyton C (2020) Structure, function and assembly of the long, flexible tail of siphophages. Curr Opin Virol 45:34–42
45. Iglesias SM, Lokareddy RK, Yang R, Li F, Yeggoni DP, Hou C-FD, Leroux MN, Cortines JR, Leavitt JC, Bird M, Casjens SR, White S, Teschke CM, Cingolani G (2023) Molecular architecture of Salmonella Typhimurium P22 genome ejection machinery. J Mol Biol 435:168365
46. Betts S, King J (1999) There's a right way and a wrong way: in vivo and in vitro folding, misfolding and subunit assembly of the P22 tailspike. Structure 7:R131–R139
47. Olia AS, Casjens S, Cingolani G (2007) Structure of phage P22 cell envelope–penetrating needle. Nat Struct Mol Biol 14:1221–1226
48. Casjens SR, Molineux IJ (2012) Short noncontractile tail machines: adsorption and DNA delivery by Podoviruses. In: Rossmann MG, Rao V (eds) Viral molecular machines. Springer, New York
49. Wang C, Tu J, Liu J, Molineux IJ (2019) Structural dynamics of bacteriophage P22 virions during the initiation of infection. Nat Microbiol 4:1049–1056
50. Garcia-Doval C, van Raaij (2012) Structure of the receptor-binding carboxy-terminal domain of bacteriophage T7 tail fibers. Proc Natl Acad Sci USA 109:9390–9395
51. Hu B, Margolin W, Molineux IJ, Liu J (2013) The bacteriophage T7 virion undergoes extensive structural remodeling during infection. Science 339:576–579

52. Chen W, Xiao H, Wang L, Wang X, Tan Z, Han Z, Li X, Yang F, Liu Z, Song J, Liu H, Cheng L (2021) Structural changes in bacteriophage T7 upon receptor-induced genome ejection. Proc Natl Acad Sci USA 118:e2102003118
53. Chang CY, Kemp P, Molineux IJ (2010a) Gp15 and gp16 cooperate in translocating bacteriophage T7 DNA into the infected cell. Virology 398:176–186
54. Schulz EC, Schwarzer D, Frank M, Stummeyer K, Mühlenhoff M, Dickmanns A, Gerardy-Schahn R, Ficner R (2010) Structural basis for the recognition and cleavage of polysialic acid by the bacteriophage K1F tailspike protein EndoNF. J Mol Biol 397:341–351
55. Xu J, Wang D, Gui M, Xiang Y (2019) Structural assembly of the tailed bacteriophage Phi29. Nat Commun 10:2336
56. Xiang Y, Rossmann MG (2011) Structure of bacteriophage Phi29 head fibers has a supercoiled triple repeating helix-turn-helix motif. Proc Natl Acad Sci USA 108:4806–4810
57. Xiang Y, Leiman PG, Li L, Grimes S, Anderson DL, Rossmann MG (2009) Crystallographic insights into the autocatalytic assembly mechanism of a bacteriophage tail spike. Mol Cell 34: 375–386
58. Xiang Y, Morais MC, Cohen DN, Bowman VD, Anderson DL, Rossmann MG (2008) Crystal and cryoEM structural studies of a cell wall degrading enzyme in the bacteriophage Phi29 tail. Proc Natl Acad Sci USA 105:9552–9557
59. Xiang Y, Morais MC, Battisti AJ, Grimes S, Jardine PJ, Anderson DL, Rossmann MG (2006) Structural changes of bacteriophage Phi29 upon DNA packaging and release. EMBO J 25: 5229–5239
60. Farley MM, Tu J, Kearns DB, Molineux IJ, Liu J (2017) Ultrastructural analysis of bacteriophage F29 during infection of *Bacillus subtilis*. J Struc Biol 197:163–171
61. Alcorlo M, Gonzalez-Huici V, Hermoso JM, Meijer WJ, Salas M (2007) The phage Phi29 membrane protein p16.7, involved in DNA replication, is required for efficient ejection of the viral genome. J Bacteriol 189:5542–5549
62. Plisson C, White HE, Auzat I, Zafarani A, Sao-Jose C, Lhuillier S, Tavares P, Orlova EV (2007) Structure of bacteriophage SPP1 tail reveals trigger for DNA ejection. EMBO J 26: 3720–3728
63. Davidson AR, Cardarelli L, Pell LG, Radford DR, Maxwell KL (2012) Long noncontractile tail machines of bacteriophages. In: Rossmann MG, Rao V (eds) Viral molecular machines. Springer, New York
64. Arnaud C-A, Effantin G, Vivès C, Engilberge S, Bacia M, Boulanger P, Girard E, Schoehn G, Breyton C (2017) Bacteriophage T5 tail tube structure suggests a trigger mechanism for Siphoviridae DNA ejection. Nat Commun 8:1953
65. Letellier L, Boulanger P, Plancon L, Jacquot P, Santamaria M (2004) Main features on tailed phage, host recognition and DNA uptake. Front Biosci 9:1228–1339
66. Heller K, Braun V (1979) Accelerated adsorption of bacteriophage T5 to *Escherichia coli* F, resulting from reversible tail fiber-lipopolysaccharide binding. J Bacteriol 139:32–38
67. Heller K, Braun V (1982) Polymannose O-antigens of *Escherichia coli*, the binding sites for the reversible adsorption of bacteriophage T5 via the L-shaped tail fibers. J Virol 41:222–227
68. Garcia-Doval C, Castón JR, Luque D, Granell M, Otero JM, Llamas-Saiz AL, Renouard M, Boulanger P, van Raaij MJ (2015) Structure of the receptor-binding carboxy-terminal domain of the bacteriophage T5 L-shaped tail fibre with and without its intra-molecular chaperone. Viruses 7:6424–6440
69. Gonzalez-Serrano R, Rosselli R, Roda-Garcia JJ, Martin-Cuadrado AB, Rodriguez-Valera F, Dunne M (2023) Distantly related Alteromonas bacteriophages share tail fibers exhibiting properties of transient chaperone caps. Nat Commun 14:6517
70. Linares R, Arnaud C-A, Effantin G, Darnault C, Epalle NH, Boeri Erba E, Schoehn G, Breyton C (2023) Structural basis of bacteriophage T5 infection trigger and *E. coli* cell wall perforation. Sci Adv 9:eade9674

71. Guihard G, Boulanger P, Letellier L (1992) Involvement of phage T5 tail proteins and contact sites between the outer and inner membrane of *Escherichia coli* in phage T5 DNA injection. J Biol Chem 267:3173–3178
72. Hendrix RW, Duda RL (1992) Bacteriophage Lambda PaPa: not the mother of all Lambda phages. Science 258:1145–1148
73. Montag D, Schwarz H, Henning U (1989) A component of the side tail fiber of Escherichia coli bacteriophage Lambda can functionally replace the receptor-recognizing part of a long tail fiber protein of the unrelated bacteriophage T4. J Bacteriol 171:4378–4384
74. Berkane E, Orlik F, Stegmeier JF, Charbit A, Winterhalter M, Benz R (2006) Interaction of bacteriophage Lambda with its cell surface receptor: an *in vitro* study of binding of the viral tail protein gpJ to LamB (maltoporin). Biochemistry 45:2708–2720
75. Roessner CA, Ihler GM (1986) Formation of transmembrane channels in liposomes during injection of Lambda DNA. J Biol Chem 261:386–390
76. Erni B, Zanolari B, Kocher HP (1987) The mannose permease of *Escherichia coli* consists of three different proteins. Amino acid sequence and function in sugar transport, sugar phosphorylation, and penetration of phage Lambda DNA. J Biol Chem 262:5238–5247
77. Seul A, Brasilès S, Petitpas I, Lurz R, Campanacci V, Cambillau C, Weise F, Zairi M, Tavares P, Auzat I (2021) Biogenesis of a bacteriophage long non-contractile tail. J Mol Biol 433(18):167112
78. Lhuillier S, Gallopin M, Gilquin B, Brasiles S, Lancelot N, Letellier G, Gilles M, Dethan G, Orlova EV, Couprie J, Tavares P, Zinn-Justin S (2009) Structure of bacteriophage SPP1 head-to-tail connection reveals mechanism for viral DNA gating. Proc Natl Acad Sci USA 106: 8507–8512
79. Orlov I, Roche S, Brasilès S, Lukoyanova N, Vaney MC, Tavares P, Orlova EV (2022) CryoEM structure and assembly mechanism of a bacterial virus genome gatekeeper. Nat Commun 13:7283
80. Bebeacua C, Tremblay D, Farenc C, Chapot-Chartier M-P, Sadovskaya I, van Heel M, Veesler D, Moineau S, Cambillau C (2013a) Structure, adsorption to host, and infection mechanism of virulent Lactococcal phage p2. J Virol 87:12302–12312
81. Bebeacua C, Lai L, Vegge CS, Brondsted L, van Heel M, Veesler D, Cambillau C (2013b) Visualizing a complete Siphoviridae member by single-particle electron microscopy: the structure of lactococcal phage TP901-1. J Virol 87:1061–1068
82. Sciara G, Bebeacua C, Bron P, Tremblay D, Ortiz-Lombardia M, Lichière J, van Heel M, Campanacci V, Moineau S, Cambillau C (2010) Structure of lactococcal phage p2 baseplate and its mechanism of activation. Proc Natl Acad Sci USA 107:6852–6857
83. Veesler D, Spinelli S, Mahony J, Lichiere J, Blangy S, Bricogne G, Legrand P, Ortiz-Lombardia M, Campanacci V, van Sinderen D, Cambillau C (2012) Structure of the phage TP901-1 1.8 MDa baseplate suggests an alternative host adhesion mechanism. Proc Natl Acad Sci USA 109:8954–8958
84. Bartual SG, Otero JM, Garcia-Doval C, Llamas-Saiz AL, Kahn R, Fox GC, van Raaij MJ (2010) Structure of the bacteriophage T4 long tail fiber receptor-binding tip. Proc Natl Acad Sci USA 107:20287–20292
85. Leiman PG, Arisaka F, van Raaij MJ, Kostyuchenko VA, Aksyuk AA, Kanamaru S, Rossmann MG (2010) Morphogenesis of the T4 tail and tail fibers. Virol J 7:355
86. Browning C, Shneider MM, Bowman VD, Schwarzer D, Leiman PG (2012) Phage pierces the host cell membrane with the iron-loaded spike. Structure 20:326–339
87. Yap ML, Rossmann MG (2014) Structure and function of bacteriophage T4. Future Microbiol 9:1319–1327
88. Chen Z, Sun L, Zhang Z, Fokine A, Padilla-Sanchez V, Hanein D, Jiang W, Rossmann MG, Rao VB (2017) Cryo-EM structure of the bacteriophage T4 isometric head at 3.3-Å resolution and its relevance to the assembly of icosahedral viruses. Proc Natl Acad Sci USA 114:E8184–E8193

89. Leiman PG, Shneider MM (2012) Contractile tail machines of bacteriophages. In: Rossmann MG, Rao V (eds) Viral molecular machines. Springer, New York
90. Arisaka F, Yap ML, Kanamaru S, Rossmann MG (2016) Molecular assembly and structure of the bacteriophage T4 tail. Biophys Rev 8:385–396
91. Hyman P, van Raaij MJ (2018) Bacteriophage T4 long tail fiber domains. Biophys Rev 10: 463–471
92. Mourosi JT, Awe A, Guo W, Batra H, Ganesh H, Wu X, Zhu J (2022) Understanding bacteriophage tail fiber interaction with host surface receptor: the key "blueprint" for reprogramming phage host range. Int J Mol Sci 23:12146
93. Wood WB, Eiserling FA, Crowther RA (1994) Long tail fibers: genes, proteins, structure, and assembly. In: Karam JD (ed) Molecular biology of bacteriophage T4. ASM Press, Washington, DC
94. Dunne M, Denyes JM, Arndt H, Loessner MJ, Leiman PG, Klumpp J (2018) Salmonella phage S16 tail fiber adhesin features a rare polyglycine rich domain for host recognition. Structure 26:1573–1582
95. Hu B, Margolin IJ, Molineux IJ, Liu J (2015) Structural remodeling of bacteriophage T4 and host membranes during infection initiation
96. Taylor NMI, Prokhorov NS, Guerrero-Ferreira RC, Shneider MM, Browning C, Goldie KN, Stahlberg H, Leiman PG (2016) Structure of the T4 baseplate and its function in triggering sheath contraction. Nature 533:346–352
97. Wenzel S, Shneider MM, Leiman PG, Kuhn A, Kiefer D (2020) The central spike complex of bacteriophage T4 contacts PpiD in the periplasm of Escherichia coli. Viruses 12:1135
98. Goldberg EB, Grinius L, Letellier L (1994) Recognition, attachment and injection. In: Karam JD (ed) Molecular biology of bacteriophage T4. ASM Press, Washington, DC
99. Christie GE, Calendar R (2016) Bacteriophage P2. Bacteriophage 6:e1145782
100. Guerrero-Ferreira RC, Hupfeld M, Nazarov S, Taylor NMI, Shneider MM, Obbineni JM, Loessner MJ, Ishikawa T, Klumpp J, Leiman PG (2019) Structure and transformation of bacteriophage A511 baseplate and tail upon infection of Listeria cells. EMBO J 38:e99455
101. Novacek J, Siborova M, Benesik M, Pantucek R, Doskar J, Plevka P (2016) Structure and genome release of Twort-like Myoviridae phage with a double-layered baseplate. Proc Natl Acad Sci USA 113(33):9351–9356
102. Fu X, Walter MH, Paredes A, Morais MC, Liu J (2011) The mechanism of DNA ejection in the *Bacillus anthracis* spore-binding phage 8a revealed by cryo-electron tomography. Virology 421:141–148
103. Walter M, Fiedler C, Grassl R, Biebl M, Rachel R, Hermo-Parrado XL, Llamas-Saiz AL, Seckler R, Miller S, van Raaij MJ (2008) Structure of the receptor-binding protein of bacteriophage Det7: a podoviral tail spike in a Myovirus. J Virol 82:2265–2273
104. Schwarzer D, Buettner FF, Browning C, Nazarov S, Rabsch W, Bethe A, Oberbeck A, Bowman VD, Stummeyer K, Muhlenhoff M, Leiman PG, Gerardy-Schahn R (2012) A multivalent adsorption apparatus explains the broad host range of phage Phi92: a comprehensive genomic and structural analysis. J Virol 86:10384–10398
105. Paolozzi L, Ghelardini P (2006) The bacteriophage Mu. In: Calendar RL (ed) The bacteriophages, 2rd edn. Oxford University Press, Oxford
106. Liu M, Deora R, Doulatov SR, Gingery M, Eiserling FA, Preston A, Maskell DJ, Simons RW, Cotter PA, Parkhill J, Miller JF (2002) Reverse-transcriptase mediated tropism switching in *Bordetella* bacteriophage. Science 295:2091–2094

Further Reading

Harper DR, Abedin ST, Burrowes BH, McConville ML (eds) (2021) Bacteriophages. Springer Nature, Cham, p 1

Hatfull GF (2021) Bacteriophages (chapter 21). In: Howley PM, Knipe DM (eds) Fields virology, Fundamentals, vol 4, 7th edn. Wolters Kluwer, Dordrecht, pp 590–623

Rossmann MG, Rao V (2012) Viral molecular machines. Springer, New York

Youle M (2017) Thinking like a phage. Wholon, San Diego

Also especially recommended for further reading are references [2, 9, 11, 12, 34, 35, 38, 41–44, 65, 85, 87, 90–93, 97, 99, 105, 106] *listed above.*

Chapter 18
Mechanical Properties of Viruses

Pedro J. de Pablo and Mauricio G. Mateu

Abstract Structural biology techniques have greatly contributed to unveiling the interplay between molecular structure, physico-chemical properties, and biological function of viruses. In recent years, classic structural approaches are being complemented by single-molecule techniques such as atomic force microscopy and optical tweezers to study physical features of viral particles that are not accessible to classic structural techniques. Among these features are mechanical properties such as stiffness, intrinsic elasticity, tensile strength, and material fatigue. The field of virus mechanics is contributing to materials science by investigating some physical parameters of "soft" biological matter and biological nano-objects. Virus mechanics studies are also starting to unveil the biological implications of some physical properties of viruses and their contribution to virus function. Virus particles are subjected to internal and external forces and they may have adapted to withstand, and even use those forces. This chapter focuses on the mechanical properties of virus particles, their structural determinants, their use to study virus function, and some possible biological implications, of which several examples are provided.

Keywords Virus · Capsid · Atomic force microscopy · Mechanical properties · Elasticity · Stiffness · Spring constant · Tensile strength · Material fatigue · Molecular structure · Stability · Dynamics · Biological materials · Virus engineering · Biotechnology · Nanotechnology Physical Virology

---

P. J. de Pablo (✉)
Department of Physics of the Condensed Matter, C03, and IFIMAC (Instituto de Física de la Materia Condensada), Facultad de Ciencias, Universidad Autónoma de Madrid, Madrid, Spain
e-mail: p.j.depablo@uam.es

M. G. Mateu (✉)
Centro de Biología Molecular "Severo Ochoa" (CSIC-UAM), and Department of Molecular Biology, Universidad Autónoma de Madrid, Madrid, Spain
e-mail: mgarcia@cbm.csic.es

M. G. Mateu (ed.), *Structure and Physics of Viruses*, Subcellular Biochemistry 105,
https://doi.org/10.1007/978-3-031-65187-8_18

Abbreviations

AAV-2	Adeno-associated virus 2
AFM	Atomic force microscopy
AdV	Adenovirus
BMV	Brome mosaic virus
CCMV	Cowpea chlorotic mottle virus
ds	Double-stranded
cryo-EM	Cryogenic electron microscopy
DLP	Double-layer particle
DLVO	Derjaguin-Landau-Verwey-Overbeek
ERT	Endogenous reverse transcription
FEA	Finite-element analysis
F-z	Force vs. z-piezo displacement
HBV	Hepatitis B virus
hPBV	Human picobirnavirus
HDX-MS	Hydrogen-deuterium exchange mass spectrometry
HIV-1	Human immunodeficiency virus type 1
HOPG	Highly oriented pyrolytic graphite
HRV	Human rhinovirus
HSV-1	Herpes simplex virus type 1
IBDV	Infectious bursal disease virus
IAV	Influenza A virus
MD	Molecular dynamics
MLV	Murine leukemia virus
MVM	Minute virus of mice
NMR	Nuclear magnetic resonance
Nt	N-terminal
NV	Norovirus (Norwalk virus)
RNP	Ribonucleoprotein particle
RV	Rotavirus
ss	Single-stranded
SARS-CoV-2	Severe acute respiratory syndrome coronavirus 2
SLP	Single-layer particle
SMV1	*Sulfolobus* monocaudavirus 1
SV40	Simian virus 40
TBSV	Tomato bushy stunt virus
TLP	Triple-layer particle
TMV	Tobacco mosaic virus
vdW	Van der Waals

18.1 Introduction

The mechanical properties of some biomolecular complexes are essential for their function. Indeed, forces at the nanoscale play a central role in biochemistry, from myosin-actin systems [1], involved in cargo transport in the cell or as the ultimate responsible of muscular action, to cellular or viral motor protein assemblies ([2], see Chaps. 10, 13).

Viruses are striking examples of biomolecular complexes endorsed with specific material properties that may provide a basis not only for their biophysical function, but also for their ability to endure a variety of environmental aggressions. During the passive extracellular stage of the infection cycle the viral capsid can be considered mainly as a container ([3], see Chap. 2) that protects the viral genome against physicochemical assaults [4]. This critical function of many viral capsids has imposed strong selection pressures leading to their structural stabilization. On the other hand, the need to uncoat the viral genome during infection of host cells (Chaps. 15–17) has favored the evolution of many capsids as metastable, conformationally dynamic biological complexes. Thus, nature may have modulated some of the capsid material characteristics for both structural stability and dynamics (see Chap. 1). Different examples of the delicate balance between the conformational stability and dynamics of virus particles and their control during the viral cycle are provided in Chaps. 1, 11, 12, 13, 14, 15, and 19.

Mechanical properties such as elasticity or tensile strength are among the material properties of any solid object, including virus particles. The development of atomic force microscopy (AFM) (see [5] and Chap. 9) has enabled the experimental study of the mechanical properties of virions and virus capsids at the single-particle level [6]. In this chapter, we describe current approaches to determine the mechanical properties of virus particles that are inherent to materials science analyses. These properties include stiffness, intrinsic elasticity of the particle material, mechanical strength, brittleness, and material fatigue. We then discuss the relationships between these mechanical properties and the molecular structure of viruses, possible biological implications, and bio/nanotechnological applications of virus mechanics. Some theoretical studies on virus mechanics are described in Chap. 19.

18.2 Mechanical Stiffness and Breaking Force of Virus Particles Determined by AFM in Indentation Assays

The advent of single-molecule techniques, such as AFM (Chap. 9), optical tweezers (Chap. 10), and similar methods [7] opened the possibility of measuring the tiny forces (nN or pN) that take place in biological molecules and processes at the nanometer scale, such as conformational changes [8], protein folding [9], or the mechanochemical action of motor proteins [10] (see Chaps. 10 and 13). AFM is ideally suited to probe the mechanical elasticity of biological specimens in liquid, under near physiological conditions (see also Chap. 9. AFM was applied early to cells [11] and microtubules [12], and it was just a matter of time that it was also applied to measure virus elasticity. It may be illustrative to briefly refer here to the origin of the first experiments on the mechanical properties of small, spherical virus particles in liquid using phage ϕ29 as a model [6]. Optical tweezers had been used to study the mechanochemical action of the ϕ29 DNA packaging motor [13] (see Chap. 10), and other experiments using optical tweezers on this same phage were later started by colleagues of one of the authors of this chapter (P.J.P.). Since these experiments were not working properly, we decided to check the virus integrity by AFM imaging. From that moment on, it was obvious that the elasticity experiments we and others had carried out by indenting cells or rod-like microtubules with the AFM tip could be readily adapted to quantitatively probe the mechanical features of even the smallest spherical viruses in a physiologically relevant liquid medium.

Chapter 9 describes the technique of AFM and some experimental setups for its use for virus imaging. In this section, we will briefly describe some particular conditions required for measuring the mechanical stiffness and breaking resistance of viral particles using AFM. The procedure requires first the strong adsorption of the specimens onto a suitable surface, in order to minimize unwanted displacements during relatively deep indentations. In the case of viruses, a typical recipe is to silanize glass surfaces to enhance the hydrophobic interactions with the viral particles [6]. The virus suspension is diluted to get a few particles (<10) per square micron on the surface and guarantee their convenient localization (Fig. 18.1a). Jumping mode [14] has shown to be extremely successful for imaging individual viruses on surfaces (see Chap. 9). It is convenient to prewet the tip with ~20 μl of buffer to avoid undesired cantilever bending. A great amount of the work published on virus mechanics has been performed with AFM systems provided by *Nanotec Electrónica* [15]. Rectangular cantilevers RC800PSA, and Biolevers (BL-AC40TS) (*Olympus*) with nominal spring constants of 0.05 N/m and 0.03 N/m, respectively, are typically used. Cantilever spring constants are typically calibrated using Sader's method [16]. Stiffness determination is based on the monitored indentation of individual viral particles with the AFM tip. Therefore, in order to perform nanoindentations, single force *vs.* z-piezo displacement (*F-z*) experiments are performed by *pushing* on the top of a selected single virus particle (Fig. 18.1b). The particle is continuously zoomed in by reducing the *x-y* scanning size until the bump of the very top is under the whole piezo scan (~50 nm × ~50 nm). Afterward a *F-z* curve is obtained by indenting the top of the particle (Fig. 18.1c, red), likely within a few nm of uncertainty (mainly provoked by thermal drift and the intrinsic non-linearity and creep of the piezo). Still, this method has proven to be robust enough to establish electrical contact with carbon nanotubes [17], which are much smaller in diameter than viral particles.

When the tip is in contact with the sample, the tip-sample force F can be described by:

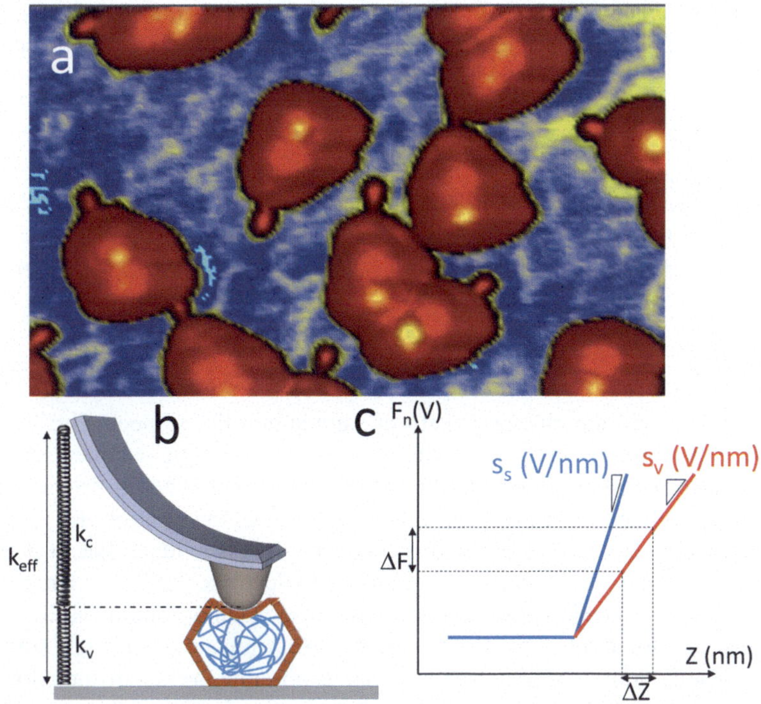

Fig. 18.1 Single indentation assay. (**a**) AFM topography of ϕ29 phage particles adsorbed on highly oriented pyrolytic graphite (HOPG). (**b**) A sketch of an ideal nanoindentation of a viral particle and the interpretation of the mechanical result by considering the AFM cantilever and the virus particle as a system of two springs in series. (**c**) *F-z* plot showing the expected bending of the cantilever when indenting a hard substrate (blue) or a virus particle (red)

$$F = F_{\text{contact}} + F_{DLV0}$$

where F_{contact} is determined by the contact mechanics. In general, the expression of this force is a complicated solution of the elasticity equations, but there are two special and useful situations where it adopts a simple form:

(i) *Hertzian contact* for the gentle indentation of a spherical tip on an elastic solid half-space.

$$F_{\text{contact}} = 4/3E^{*}\sqrt{R^{*}}d^{3/2}$$

where *, d is the indentation depth, $R^{*-1} = R_t^{-1} + R_v^{-1}$ (R_t and R_v tip and virus radii, respectively) and E^* is given by:

$$1/E^{*} = \left(1 - \nu_t^2\right)/E_t + \left(1 - \nu_s^2\right)/E_s$$

where ν_t, E_t, ν_s, and E_s are the Poisson's ratio ν and elastic modulus E of the tip (t) and the sample (s) (defined below).

(ii) *Linear contact*—when the effective response is like that of a spring with a constant k_{eff}. This situation occurs for small indentations of a thin shell-like structure [18].

$$\boldsymbol{F}_{\mathbf{contact}} = -\boldsymbol{k}_{\mathbf{eff}}\boldsymbol{d}$$

F_{DLVO} follows the Derjaguin–Landau–Verwey–Overbeek model (DLVO) [19] and accounts for the double Debye double-layer electrostatic force (F_{el}) and van der Waals (vdW) force (F_{vdW}). In the simplest case of two planes the forces are given by:

$$\mathbf{F}_{\mathrm{DLVO}} = \mathbf{F}_{\mathrm{el}} + \mathbf{F}_{\mathrm{vdW}} = \frac{2\sigma_s\sigma_t}{\varepsilon_0\varepsilon} e^{\frac{-z}{\lambda_D}} - \frac{H_a}{6\pi z^3}$$

where σ_s and σ_t are the charge density of sample and tip, respectively; $\varepsilon_0\varepsilon$ is the dielectric constant times the permittivity of vacuum; z is the tip-sample distance, H_a is the Hamaker constant describing the vdW interactions between tip and sample, and λ_D is the Debye length that determines the range of the electrostatic forces.

We must consider two issues at this point. First, it is evident that to prime the measurement of sample elasticity it is necessary to decrease the Debye length. Since the Debye length decreases with the concentration c of salt in solution as $\lambda_D \sim 1/\sqrt{c}$ [20], when investigating viruses relatively high (but physiological) salt concentrations are usually employed to screen as much as possible the electrostatic forces. In those circumstances, the tip-sample interactions are dominated by the contact forces and sample elasticity (Fig. 18.1b). Second, since in a first approximation most icosahedral virus capsids can be considered as hollow, relatively thin shells, it is tempting to directly use the linear approximation for modeling the tip-virus contact. Nevertheless, some caution is necessary because of two main reasons. On one hand, the linear model can only be applied if the total indentation of the virus capsid is at most the shell thickness (see [18] and Fig. 18.1c), which in most protein viruses is about a couple of nanometers. On the other hand, capsids or virions may be filled with, for example, either scaffolding proteins or nucleic acid, thus deviating from an ideal shell-like model. Even so, since these fillings cannot be considered as homogeneous solids to use the Hertzian contact approach, the linear model is still approximately valid and provides an easy and direct framework to interpret the experimental results.

In most AFM nanoindentation experiments on viruses it has been found that, for small deformations, the force is linear with the indentation. In such cases, the spring constant of the virus k_v can be easily obtained by simply considering the virus like a spring in series with the cantilever with a total spring constant of k_{eff} (Fig. 18.1b):

$$\boldsymbol{k}_v^{-1} = \boldsymbol{k}_{\mathbf{eff}}^{-1} - \boldsymbol{k}_c^{-1}$$

where k_c is the spring constant of the cantilever (here the substrate is considered as non-deformable). For practical use, this equation is customarily rewritten in terms of the slope of the cantilever deflection obtained on the solid surface, which is assumed to be non-deformable, s_s (V/nm), and the slope of the cantilever deflection obtained on the virus s_v (V/nm). For this purpose, in Fig. 18.1c a cartoon of the normal force obtained on the substrate (blue) and the virus (red) is depicted, in such way that for the same ΔF the slope on the hard surface (s_s) is larger than the slope on the soft virus (s_v). The bending force ΔF (nN) corresponds to a cantilever bending of Δy (nm), thus assuming a Hookean response $\boldsymbol{\Delta F = k_c \Delta y = k_c S_s^{-1} \Delta V}$. On the other hand, in agreement with Fig. 18.1c $\boldsymbol{\Delta V = \Delta Z S_v}$. Therefore, the effective spring constant obtained on the virus can be expressed as $\boldsymbol{k_{eff} = \Delta F / \Delta Z = k_c \frac{S_v}{S_s}}$. Hence, the spring constant of the virus can be conveyed as $\boldsymbol{k_v = k_c \frac{S_v}{S_s - S_v}}$.

The standard protocol of an AFM nanoindentation experiment (Fig. 18.2a) consists of the location of an intact particle, where a *F-z* is carried out showing three steps (Fig. 18.2b). First, the tip approaches the virus particle until contact is established, and then the particle starts to be deformed (Fig. 18.2a–b). When the breaking force is reached, at ~2.0 nN (Fig. 18.2b), the bending force F_n decreases suddenly in a series of steps (Fig. 18.2a–b) and the virus particle is disrupted. In the case of an intact rotavirus (RV) particle (Fig. 18.2c, left) a crack appeared at the very center and propagated outwards (Fig. 18.2c, right) [21]. The *F-z* curve can be converted in indentation data (Fig. 18.2d) [23], where the initial slope reports the spring constant of the virus (~ 0.35 N/m) and the first peak downwards informs about the breaking force (~1.9 nN). It is convenient to adjust the *F-z* indentation speed to about ~60 nm/s to allow the water molecules to exit the virus when it is squeezed [24]. Even if the shell integrity is maintained, only those viral particles showing stable values for the spring constant along successive *F-z* measurements should be considered. Thus, artifacts such as particle mobility effects that often occur when the particle is loosely bound to the surface can be mostly avoided. However, effects such as tip sliding or particle tilting are difficult to quantify, and techniques to rule them out remain to be developed. Some viruses do not exhibit a neat linear deformation and breaking force. For instance, indentation data obtained on the lemon-shaped Sulfolobus monocaudavirus 1 (SMV1) particles present a non-linear contour (Fig. 18.2e) without any signature of breakage during several deformations [22]. In fact, the initial and final topographies (Fig. 18.2f) obtained before (#1) and after (#5) mechanical probing did not reveal any breakage of the SMV1 structure.

18.3 Intrinsic Elasticity of Virus Capsids: Young's Modulus

In materials science, the Young's modulus of elasticity E is a measure of the *intrinsic* stiffness of an elastic material. Let us first consider a solid piece of some material, say a cylinder of initial length L_O and a cross-section with a surface A (Fig. 18.3a). If we apply a force F along the main axis, the cylinder elongates from its original length L_O to $L_O + \Delta\ell$. For small deformations, the stress $\sigma = F/A$, i.e., the force per unit area A, is proportional to the relative deformation or strain ε, given by $\Delta\ell/L_0$

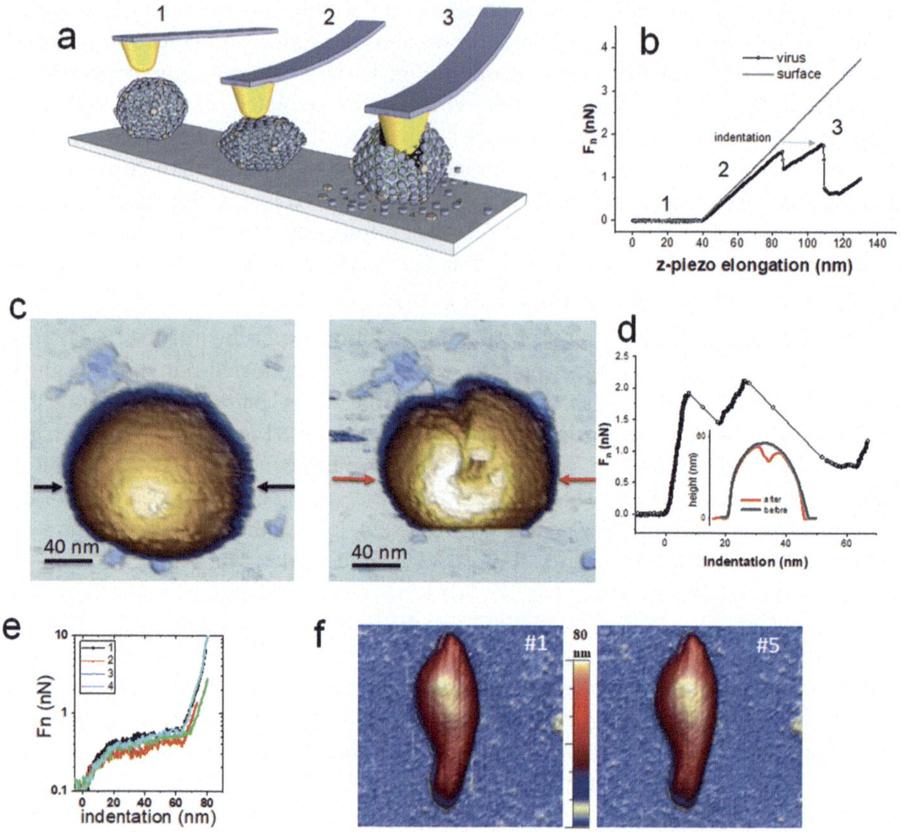

Fig. 18.2 Nanoindentation and virus fracture. (**a**) Cartoon showing the three stages during a typical nanoindentation experiment on a virus particle: (1) before contact, (2) deformation, and (3) breakage. (**b**) Evolution of the normal force (F_n) as a function of z-piezo elongation for indentation on a hard substrate (grey line) or a virus particle (dots and line, whose contour exhibits the three stages described in panel **a**). (**c**) AFM topographies of a rotavirus particle before and after nanoindentation, showing a crack on the viral particle (adapted from [21], with permission). (**d**) Nanoindentation data extracted from panel b, showing the linear deformation and the breaking force close to 2 nN. Inset compares topographical profiles of panel (**c**) before (blue) and after (red) de experiment. (**e**) Four consecutive indentations carried out on an SMV1 particle. (**f**) AFM topography of the SMV1 particle before (left) and after (right) the nanoindentations charted in panel (**e**) (adapted from [22])

(Fig. 18.3a). Thus, $\boldsymbol{\sigma} = \boldsymbol{E\varepsilon}$, where the proportionality constant E is the Young's modulus. From the latter, we can find an expression for the force [25]:

$$\boldsymbol{F} = \frac{\boldsymbol{AE}}{\boldsymbol{L_0}}\boldsymbol{\Delta\ell}$$

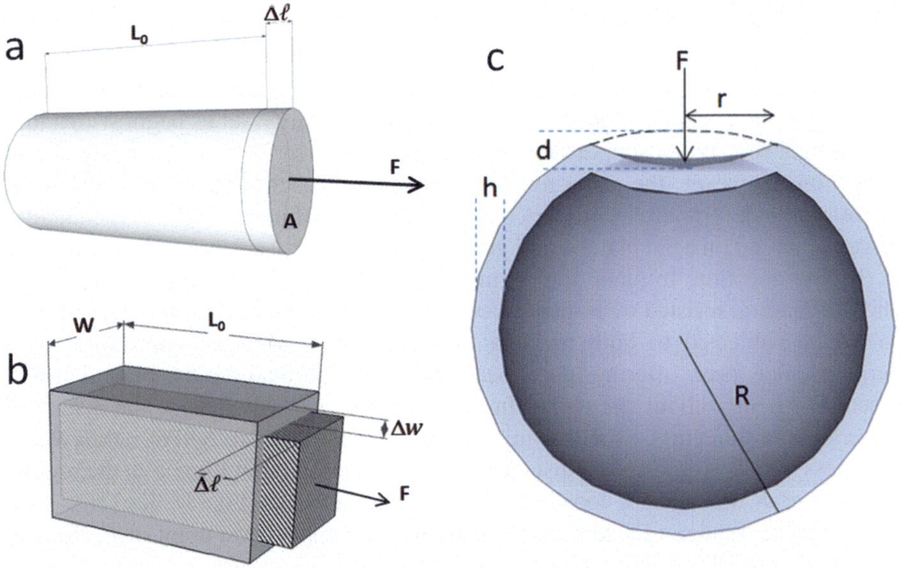

Fig. 18.3 Physical magnitudes of materials mechanics. (**a**, **b**) Geometrical parameters required for estimating the Young's Modulus (**a**) or the Poisson Ratio (**b**). (**c**) Main geometrical definitions for the application of thin shell theory

The coefficient of $\Delta\ell$ has units of N/m, being the effective longitudinal spring constant *k* of the object (see Fig. 18.3a). Another important mechanical magnitude is the non-dimensional Poisson ratio parameter μ, which accounts for the ratio of transverse and longitudinal strains:

$$\mu = -\frac{\varepsilon_{\text{trans}}}{\varepsilon_{\text{long}}}$$

Transversal strain $\boldsymbol{\varepsilon}_{\text{trans}}$ is typically negative, since when a piece of material is pulled by the force *F* the transversal dimension decreases by Δw (Fig. 18.3b). There are exceptions like auxetic materials that, when stretched, become thicker perpendicularly to the applied force, and thus have negative values of the Poisson ratio. In general, the Poisson ratio μ value is constrained between −1 and 0.5, and steel and rigid polymers show values around 0.3 [26]. In studies of virus mechanics μ is assumed to be 0.3.

In the following, we will go through the typical approaches followed in most published works to extract the Young's modulus of viral shells, i.e., thin shell theory and finite element analysis (see also Chap. 19). Certainly, from a physical point of view, one outcome of most works published on virus mechanics so far is the estimation of the Young's modulus of a virus shell.

18.3.1 Shell Theory

It is well known that shells, i.e., thin plates that are curved in their undeformed state, exhibit special mechanical properties. For the sake of clarity, let us begin with a thin and straight rod. When supported at both ends, transverse forces cause bending and deflection. However, the same rod can withstand a much greater axial load with unnoticeable deformation. Similar concepts apply to a thin flat shell. Perpendicular forces actuating on a flat plate would generate bending, resisted by bending moments in the cross-section. In-plane forces, in turn, would cause stretching, compression, and/or shearing, resisted by in-plane forces on a cross-section. A closed curved shell would be subjected to both bending and stretching under the application of a perpendicular force, offering more mechanical strength than the flat plate that will essentially bent without stretching. In the following, we will analyze the details of these ideas in thin curved shells (Fig. 18.3c). When a spherical shell of radius R and thickness $h \ll R$ is deformed a distance d by an indenting point force F in the regime where $d \ll R$, the shell is locally both stretched and bent (Fig. 18.3c) [18]. As a consequence, a bulge of size r develops right at the application point of F. We can express the stretching and bending energies as [27]:

$$\mathbf{E_{str}} \sim \mathbf{K}\left(\frac{\mathbf{d}}{\mathbf{R}}\right)^2 \mathbf{r}^2, \text{ and } \mathbf{E_b} \sim \boldsymbol{\kappa}\left(\frac{\mathbf{d}}{\mathbf{r}^2}\right)^2 \mathbf{r}^2$$

where $K = \frac{Eh}{2(1-\mu)}$ and $\kappa = \frac{Eh^3}{12(1-\mu^2)}$ are the 2D stretching and bending modulus, respectively. Note that these refer to the three-dimensional properties of the material making up the shell. We can write the sum of bending and stretching energies as:

$$E_{\text{total}} \sim E\frac{h^2}{R}\left[\frac{hR}{r^2} + \frac{r^2}{hR}\right]d^2$$

By imposing the equilibrium condition $\frac{dE_{\text{total}}}{dr} = \mathbf{0}$, we can find that the size of the bent region scales as $r \sim \sqrt{hR}$. This means, for instance, that for two spherical shells with identical thickness h, doubling the radius R only increases the lateral deformation by a factor $\sqrt{2}$. By making the derivative of E_{total} with respect to the indentation d and using the obtained value of r, we find that the force is linear in the indentation with a spring constant of the shell that scales like $\sim E\frac{h^2}{R}$. To calculate the proportionality constant accurately requires expanding the shape of the deformed shell in spherical harmonics [12, 28], and for a thin shell it turns out to be $\frac{2}{\sqrt{3(1-\nu^2)}}$.

18.3.2 Finite Element Analysis

Another popular approximation for interpreting the results of experiments on virus mechanics is to use Finite Element Analysis (FEA). In simple terms, FEA attempts to obtain a numerical solution of the complicated differential equations describing a physical problem concerning a physical object considered as a continuous medium [29]. The key point is to divide the object into many small non-intersecting domains (finite elements) where the physical equations are discretized and solved. This numerical methodology can be applied to a variety of physical phenomena, including mechanical deformations, circulation of fluids, or electrostatic problems.

Here, we consider a case study of FEA application for human RV, Fig. 18.4a) [21]. The RV infectious particle is a 100 nm non-enveloped triple-layered particle (TLP) composed of three concentric protein shells enclosing the dsRNA genome and the viral RNA polymerase and capping enzyme (Fig. 18.4b) [30]. Finite elements simulations mimicking the AFM nanoindentation of the different RV particles were performed using the program COMSOL Multiphysics 4.3 (Comsol, Stockholm, Sweden). In the simulations, each layer was modeled as a homogenous spherical shell made of a material with Young's modulus E and Poisson ratio $\nu = 0.3$

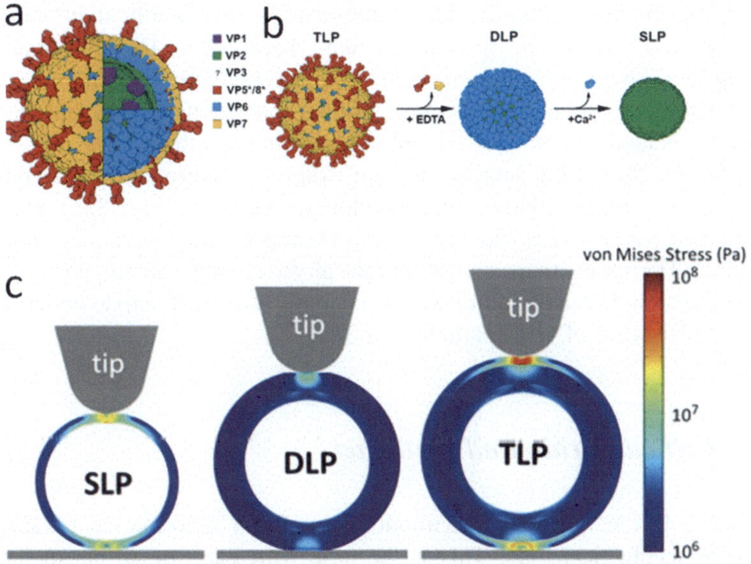

Fig. 18.4 Finite elements modeling of RV particles. (**a**) Schematic representation of the mature RV triple layer particle (TLP). The different viral proteins are color coded. (**b**) Double-layer particle (DLP) or single-layer particle (SLP) obtained from a TLP. VP7 and VP5*/VP8* are disassembled from the TLP in the presence of EDTA. High concentration of Ca^{2+} ions takes apart VP6 trimers from the DLP to yield a SLP. (**c**) Stress distribution at 4 nm of indentation in the models used in the finite elements simulations of SLP (left), DLP (middle), and TLP (right). Inset color scale represents the value of the von Mises stress in Pa

(a standard value for protein-like materials). This model shell was placed on a hard flat substrate and indented by a hard spherical object with radius $R_{in} = 15$ nm, mimicking the nominal radius of the AFM tip. This model shell was placed on a hard flat substrate and indented by a hard spherical object with radius $R_{in} = 15$ nm, mimicking the nominal radius of the AFM tip (Fig. 18.4c). The system was simulated using a two-dimensional axisymmetric model that was meshed with over 1400–6000 triangular elements. The contacts between the shell and the tip as well as the supporting surface during indentation were implemented with a contact normal penalty factor (more details can be found in [21]). In this way, it was possible to derive the Young's Moduli $E_1 = 0.53 \pm 0.20$ GPa, $E_2 = 0.082 \pm 0.070$ GPa, and $E_3 = 1.0 \pm 0.9$ GPa for the external (TLP), medium (DLP) and internal (SLP) capsid protein layers, respectively.

18.4 Beyond Stiffness and Breaking Force: Brittleness, Material Fatigue, and Reversible Damage of Virus Particles

It may be tempting to directly relate virus particle stiffness with structural stability against disruption into subunits. This interpretation may imply that "soft" virus particles are less resistant to dissociation than "hard" ones. However, very often elastic objects are mechanically more resilient than rigid ones. We can find examples in everyday life, e.g., rubber versus glass. Rubber is softer than glass, but glass does not support mechanical deformation without breaking as rubber does. The example represented in Fig. 18.2 applies to the fragile fracture observed for most non-enveloped viruses. However, enveloped viruses [31, 32] and some non-enveloped viruses [22] did not exhibit breakage under punctual mechanical deformation. Therefore, there are additional physical properties of virus particles beyond stiffness and breaking force explained in Fig. 18.2 that help to understand the mechanical resistance of virus particles.

18.4.1 Critical Strain and Brittleness

A material is brittle if it breaks without a significant deformation (strain) when subjected to mechanical load. This is the case with glass or an eggshell. Brittle materials present elastic deformation before breakage as it happens, for example, with the rotavirus particle (Fig. 18.2c). We can define the critical strain ε_c, which is obtained as the coefficient of the height of the deformed virus particle at the breaking point, divided by the original height of the virus. In the case of RV (Fig. 18.2d), we obtained $\varepsilon_c \approx 7.4$ nm/55.6 nm $= 0.13$. This means that the particle deforms elastically by ≈13% before breaking. In the case of human adenovirus (AdV), the absence of protein VII in the core induces a decrease of the brittleness (or increase of the elasticity) from 17% (wild type) to 22% [33]. This parameter is also adequate to characterize RNA viruses, such as human picobirnavirus (hPBV) [34]). The estimation of ε_c requires a neat identification of the breaking force, which is clearly resolved in many protein capsids, but difficult to recognize in soft and non-fragile viruses such as SMV1 (Fig. 18.2e).

18.4.2 Material Fatigue of Viruses

In many materials, repeated application of limited force leads to their eventual disruption because of material fatigue. Mechanical fatigue studies have been largely carried out on macroscopic metal objects, but this property has been rarely studied for protein-based bionanomaterials such as viral capsids and other biomolecular complexes [35].

Mechanical fatigue studies apply cycling loading forces below rupture strength of the object under examination [36], revealing not only the evolution of its resistance to breakage over time but also the location of the object's weakest parts. A fundamental requirement of these studies is monitoring the structural integrity of the sample along the experiment. This investigation requires a precise control of the force applied on the particle [37] to induce the sequential disruption of capsid subunits. These novel mechanical studies on viral particles provide biologically relevant information about virus disassembly [38]. In addition, mechanical fatigue studies help to discern the physical differences between distinct classes of a virus particle when the studies of stiffness and breaking force do not provide insightful differences. This is the case in the following example.

Human picobirnavirus (hPBV) is an excellent candidate nanocontainer for heterologous cargos [34]. In particular, the conditions of the cargo delivery entail tailoring the escape of the molecular payload when the hPBV particle is disassembled. To this end, it is necessary to control the cargo–container interaction which, in turn, would tune the retention of cargo when the disassembly of the nanocage takes place. Two hPBV mutants were designed where the RNA–coat protein interaction is modified by changing the length of the N-terminal end of the coat protein [34]. The difference in stiffness between these mutants was not determinant [34]. Mechanical fatigue experiments were carried out on each mutant, that crack-opened individual particles in real time to monitor the cargo release by taking successive images of individual particles at ~100 pN of force in jumping mode [14]. In this way, in a 128 × 128 pixels frame, the virus particle is hit by the tip around 6000 times [39]. While the hPBV particle is initially intact at frame #0 (Fig. 18.5a), it is continuously eroded by the AFM tip, showing evident signals of degradation in successive frames. The height variation along the fatigue experiment charts for the hPBV variants with short and long N-terminal (Fig. 18.5b, c, respectively) indicate the mechanical unpacking rate for each virus cage. Once the particles start the disassembly process, right after the variation of height Δh stops being ~0, it is possible to observe different variation

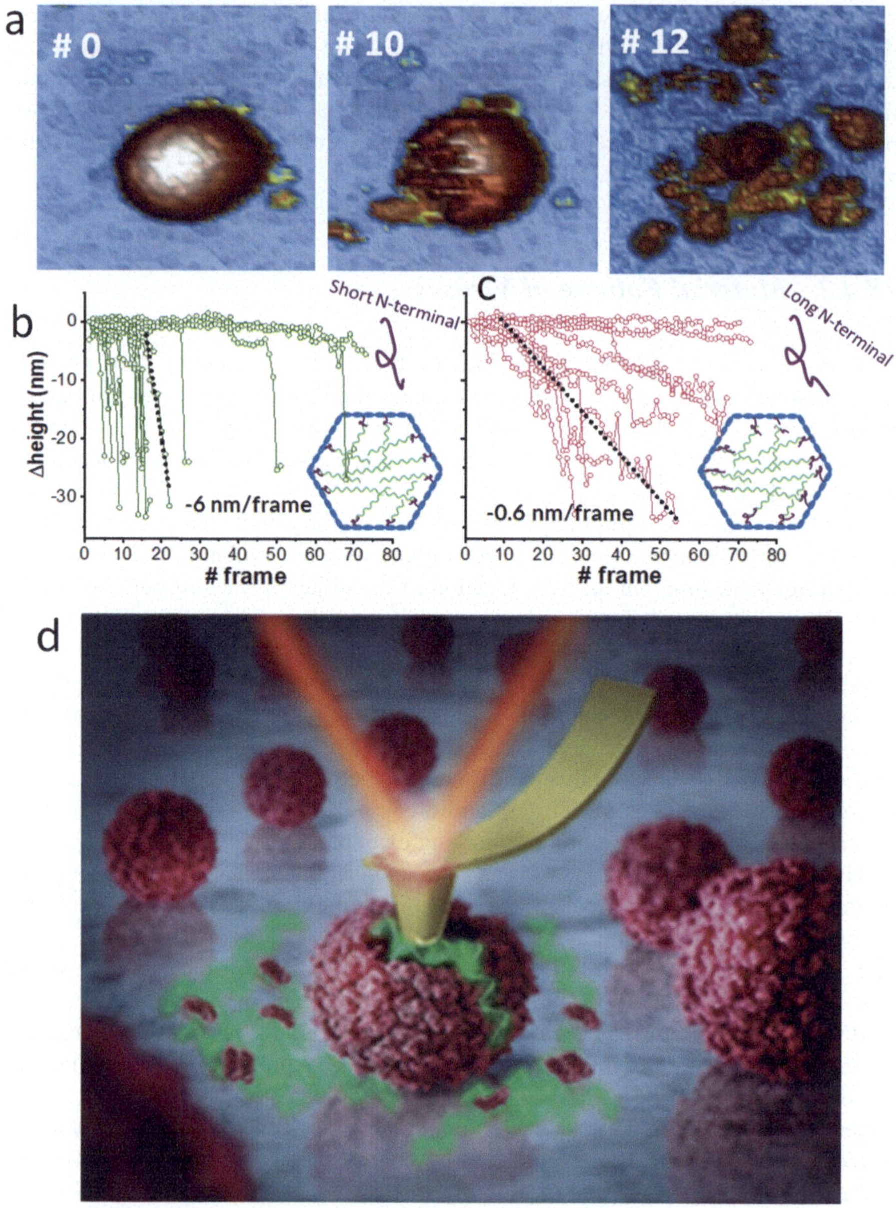

Fig. 18.5 Probing the cargo retention of hPBV with mechanical fatigue. (**a**) AFM topographies corresponding to frames #0, #10, and #12, show the gradual degradation of a hPBV particle with long N-termini during a mechanical fatigue experiment at 100 pN. (**b**) Charts of the height variation recorded for 10 hPBV particles with short N-termini. (**c**) Charts of the height variation recorded for 10 hPBV particles with long N-termini. In (**b**) and (**c**), dotted dark lines indicate Δh evolution during unpacking. (**d**) Artistic rendition of the cargo retention experiments. Adapted from [39]

rates of Δh (charts of two particles are marked with dotted black lines in Fig. 18.5b–c). Specifically, the hPBV mutants with a short N-terminal exhibit ~6 nm/frame rate, while the mutants endorsed with an N-terminal just 8% longer, show a rate of ~0.6 nm/frame. Therefore, it is possible to conclude that the virus particle material escapes 10 times slower from the long N-terminal hPBV particles (Fig. 18.5c) than from the short N-terminal structures (Fig. 18.5b).

18.4.3 Assessing the Structural Recovery of Viruses After Extreme Deformation

Disruptive events of some protein assemblies, including protein cages and virus particles, have been found to be reversible. Three-dimensional cellular protein complexes such as microtubules and vaults have demonstrated self-healing capabilities [40, 41]. Such self-healing ability has been demonstrated also in monolayers of the capsid protein of human immunodeficiency virus type 1 (HIV-1) [42]. The bacteriophage T7 capsid has also shown this recovery ability in very rare occasions [43]. Reversible stepwise rupture events in T7 bacteriophage have been attributed to the reversible buckling of individual capsomers during the nanoindentation process [44].

The above-mentioned observations of self-healing after mechanically induced structural damage refer to protein shells that are relatively fragile and easily fractured. In contrast, enveloped viruses such as coronaviruses showed massive recovery abilities [31, 32]. Another interesting example is the reversible deformation of the non-enveloped SMV1, which shows a helical structure [45]. In this case, the imaging force was varied from frame to frame (Fig. 18.6a–c) alternating low values (500 pN) with high values (up to 1100 pN, Fig. 18.6d) during five cycles. The experiments evinced the recovery of the virus structure, even after being subjected to deformations of 80% relative to the height of the particle (Fig. 18.6b). This fact suggests that under the pressure exerted by the AFM tip, the virus contents flow and can migrate to other parts of the capsid lumen (Fig. 18.6e). This high elasticity demonstrates the wall-to-wall collapse of the virus without damage, suggesting a protein shell with membrane-like fluidity and a liquid genomic cargo that only fills ~11% of the virion's internal volume [22].

18.5 A Case Study: The Mechanics of Bacteriophage ϕ29

In this section, we will exemplify how AFM experiments can help to understand some biophysical features of viruses using bacteriophage ϕ29 (Fig. 18.7a) as a model. First, we will discuss the values obtained for the spring constant of the ϕ29 prohead, or immature capsid (as a measure of capsid mechanical stiffness), and its

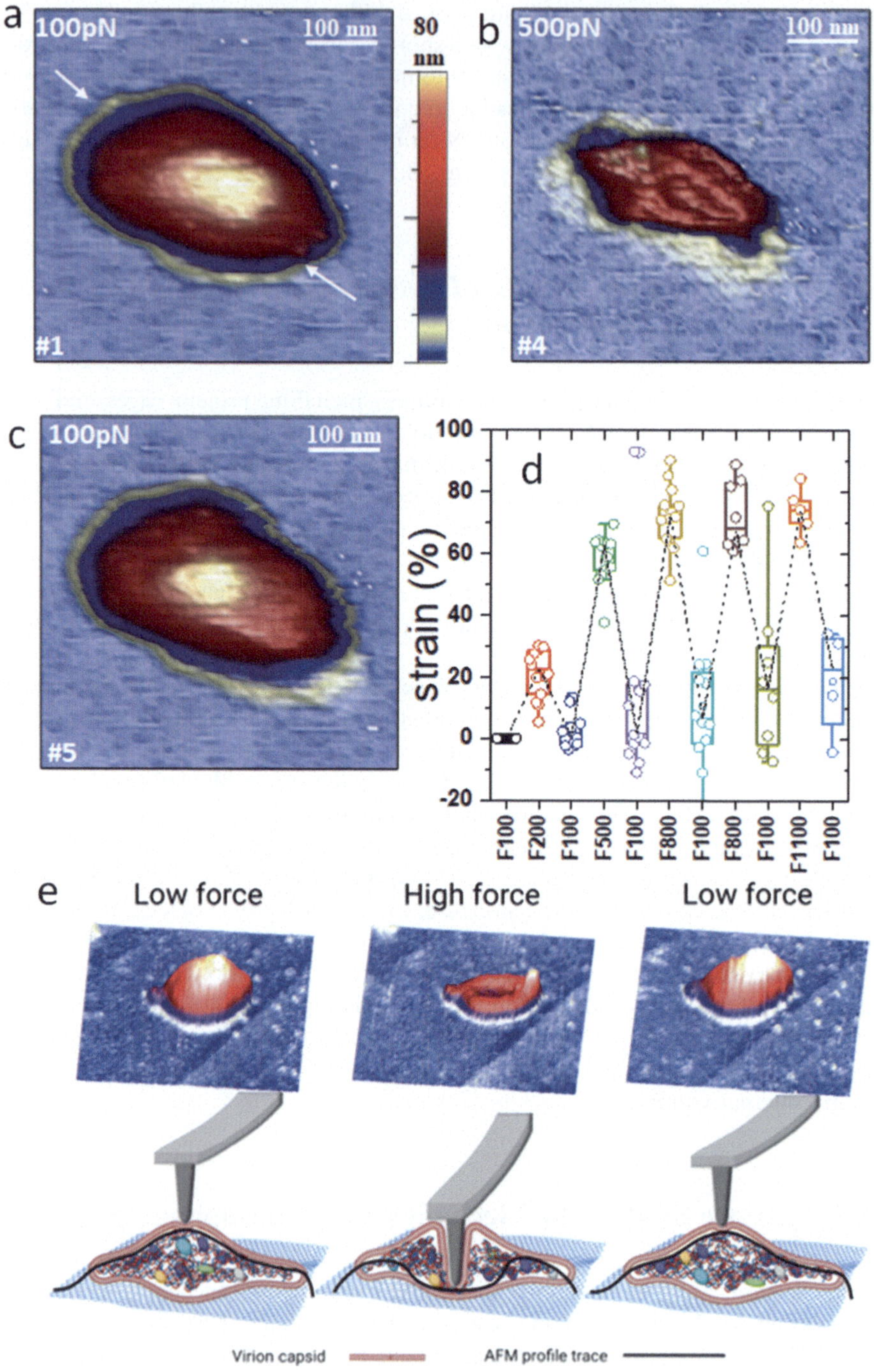

Fig. 18.6 Reversible deformation of SMV1 virus in response to alternate force values. (**a**, **b**, **c**) Topography of a single SMV1 particle obtained using a force of 100 pN (**a**), raising the force to 500 pN (**b**), or lowering the force to 100 pN again (**c**). (**d**) Strain evolution for several SMV1

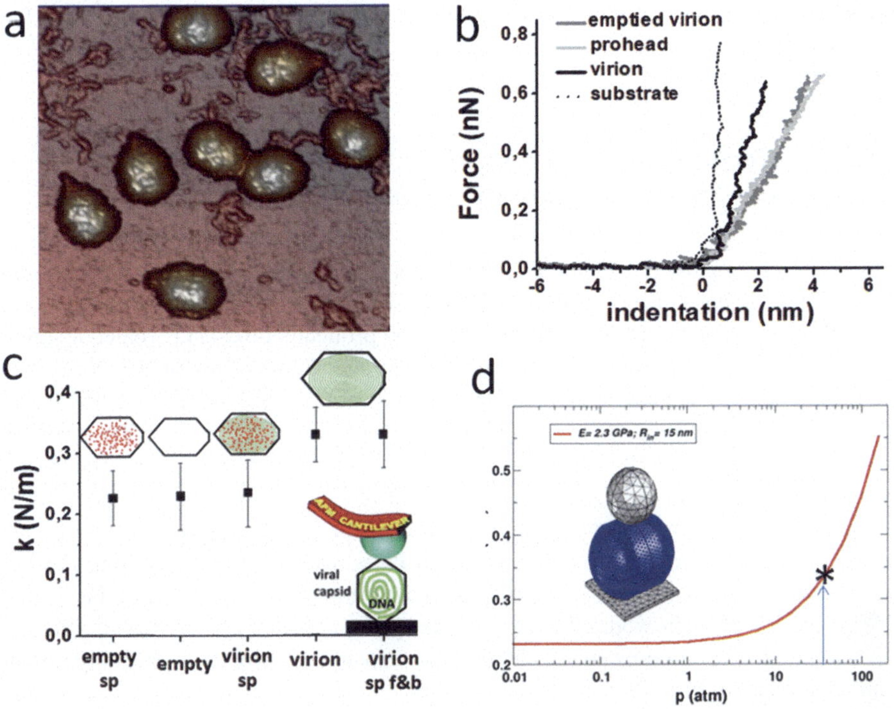

Fig. 18.7 Phage ϕ29 internal pressure. (**a**) AFM topography of mature ϕ29 particles where the tail is clearly visible. (**b**) Individual indentation curves on the substrate (dotted line), DNA-filled ϕ29 virion (black line), empty ϕ29 virion (dark gray line), or ϕ29 prohead (gray line). (**c**) Values of the spring constants obtained for the different types of ϕ29 particles analyzed in (**b**). (**d**) FEA results of the spring constant as a function of the pressure for a phage ϕ29-like structure. Adapted from [46]

relationship with the Young's modulus (as a measure of the intrinsic elasticity of the capsid material) by using Finite Element modeling. Second, we will explain the detection and evaluation of internal pressure in bacteriophage ϕ29. Third, we will describe the identification of built-in mechanical stress in viral particles and its influence on capsid stiffness. Finally, we will describe how the electrostatic forces

Fig. 18.6 (continued) virions. The strain was calculated as the coefficient of the maximum height at each force divided by the initial maximum height at 100 pN. (**e**) An SMV1 particle imaged by mechanical tomography (top images) and graphical interpretation of the results obtained (bottom schemes). Left, the virus particle was imaged at low force (no deformation). The black line on the bottom scheme represents the apparent height profile at every pixel, which coincides with the actual virus topography. Center, the same virus particle imaged at high force shows a depressed apparent plateau (black line on the bottom scheme) that forms only at the pixel where the tip is touching the virus. Right, the same virus particle imaged again at low force (like in panel A; no apparent damage. Again, the height profile at every pixel (black line at the bottom scheme) coincides with the actual virus topography. Adapted from [22] measured in this phage are influenced by the presence/absence of genome inside the virus cavity.

18.5.1 Mechanical Properties of the ϕ29 Capsid: Stiffness and Structural Stability Against Breakage

Indentation assays of individual ϕ29 proheads revealed the existence of a bimodal distribution of the spring constant [6], and yielded average k values of 0.16 and 0.31 N/m. However, further studies of the same proheads [46, 47] revealed a single distribution of the spring constant with an average value of about 0.3 N/m. It is possible to calculate analytically the deformation of a homogeneous, spherical, and elastic shell that is subjected to equal and opposite forces applied at the poles by expanding the shape of the deformed sphere in spherical harmonics [28]. The spring constant for the total indentation becomes $\boldsymbol{k} = \mathbf{2.25}\frac{\boldsymbol{Eh^2}}{\boldsymbol{R}}$, and the Young's modulus of the shell can be estimated, by using the thickness h and the radius R of the phage particle (obtained from EM studies [48]), to be between 1.2 and 1.6 GPa. Further modeling with FEA (Sect. 18.3.2) results in a Young's modulus of 1.8 GPa. Phage ϕ29 proheads also showed a decrease in the spring constant after repeated indentations. Further experiments suggested that this capsid softening was probably due to capsid disruption [49, 50].

18.5.2 Internal Pressure in the ϕ29 Virion

It has been proposed that translocation of the double-stranded (ds) DNA genome of phage ϕ29 through the virus tail into the bacterium host is initiated by a *push* mechanism, followed by a *pull* mechanism mediated by bacterium proteins [51]. The push mechanism would be driven by the elastic energy provided by the internal pressure accumulated during the DNA packaging process (see also Chaps. 17 and 19). However, the very existence of pressure in ϕ29 has remained elusive, and single-molecule techniques have added valuable information. Specifically, AFM nanoindentation experiments on individual viral particles reveal that the ϕ29 virion is stiffer than the ϕ29 prohead (Fig. 18.7b, c). Since no major mechanically relevant structural differences between the prohead and the empty mature head particles are apparent, it is reasonable to consider that the packaged dsDNA is the main responsible for the observed stiffening.

To distinguish between structural or pressure-induced reinforcement, further experiments were performed in the presence of spermidine^{3+}, a positively charged counter ion that can induce DNA condensation, even inside viruses [52], because it neutralizes the negative charges of the nucleic acid phosphates. Real-time indentation experiments probing the same individual particles before and during treatment with spermidine^{3+} showed that this compound induces a mechanical softening of the full virion that is made as elastic as the prohead, as determined by measuring spring constant values. Experiments carried out by adding and later removing spermidine^{3+} showed that this is a reversible process (Fig. 18.7c). Theoretical calculations using the model described in [53] yielded an internal pressure in the ϕ29 virion of about 30 ± 10 atm in the absence of spermidine^{3+}. The experimental evaluation of the internal pressure from the determination of spring constant values in the absence or presence of spermidine^{3+} and the use of FEA (see Sect. 18.3.2 and Chap. 19) yielded a pressure of 40 ± 10 atm, in good agreement with the theoretical predictions (Fig. 18.7d) [46].

18.5.3 Built-in Mechanical Stress in the ϕ29 Capsid

We have already commented on the mechanical advantage that curved shells present when subjected to deformation (e.g., by indentation), as they undergo simultaneous stretching and bending. Nevertheless, there are other mechanisms that viral shells may take advantage of to improve their mechanical performance. Residual mechanical stress is the stress remaining after its cause has been removed. If we take a straight rod and simultaneously push from both ends (Fig. 18.8a), it bulges up in a stressed bump. This deformation quickly disappears when forces are removed, and the rod recovers the original straight shape. However, this stress can be conserved if both ends of the bulged rod are pinned.

The shell of bacteriophage ϕ29 prohead exhibits an intriguing anisotropic stiffness: nanoindentations show that the prolate ϕ29 prohead is about twofold stiffer along the short axis (particle laying on its side) than along the long axis (upright particle) (Fig. 18.8b). The experimental and theoretical evidence indicates that this anisotropy is due to residual mechanical stress, as discussed next.

If the ϕ29 prohead is considered as a shell, continuum elasticity theory (see Sect. 18.3.1 and Chap. 19) predicts spring constant opposite trends along the short and long axes of the empty virus particle to those revealed by experiment. The ϕ29 capsid can be considered as a shell made of a cylindrical body closed by two spherical caps. By considering independently, the elastic response of the cylindrical and spherical parts of the virus particle one would expect that the laid-down virus would be mechanically softer than the upright one, since a cylindrical shell is easier to deform than a spherical shell of the same dimensions [18]. This is because a cylindrical shell, unlike a spherical one, can be bent without much stretching. We have already seen that, for a spherical shell, the effective spring constant is $\boldsymbol{k_{\mathrm{eff}}^{\mathrm{sph}} \sim Eh^2/R}$. However, when indentation of a cylinder is considered [12], the spring constant equals $\boldsymbol{k_{\mathrm{eff}}^{\mathrm{cyl}} \sim Eh^{5/2}/R^{3/2}}$. Therefore, the ratio of spring constants is $\boldsymbol{k_{\mathrm{eff}}^{\mathrm{sph}}/k_{\mathrm{eff}}^{\mathrm{cyl}} \sim \sqrt{R/h}}$, which for the geometry of ϕ29 ($R = 21$ nm and $h = 1.6$ nm [48]) becomes approximately $\boldsymbol{k_{\mathrm{eff}}^{\mathrm{sph}}/k_{\mathrm{eff}}^{\mathrm{cyl}} \sim 3.6}$. Hence, in the framework of continuum elasticity theory of shells, the cylindrical body of the virus is expected to be more

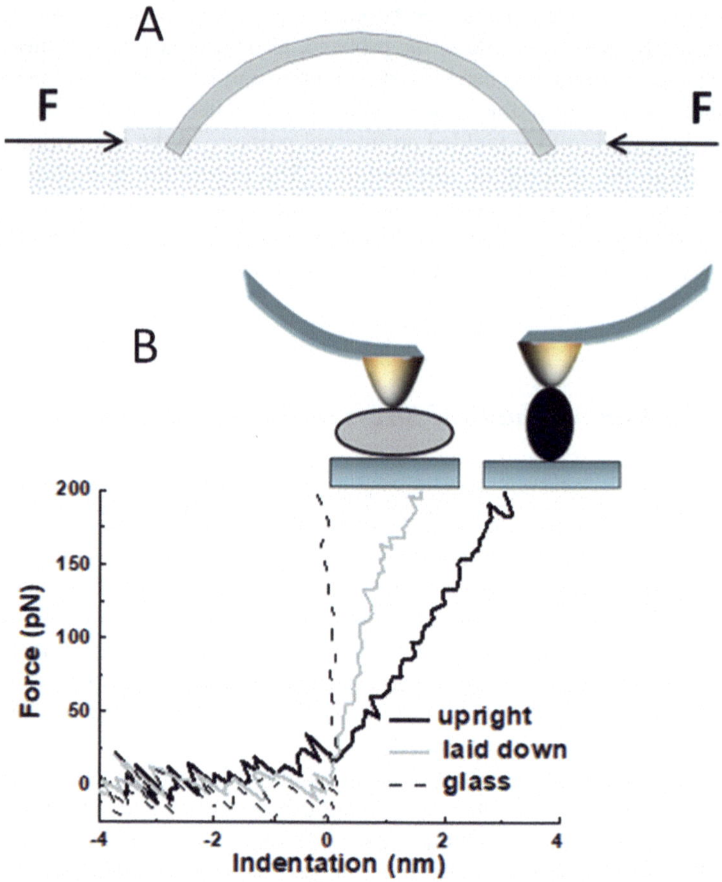

Fig. 18.8 Built-in mechanical stress. (**a**) The concept of residual stress. (**b**) Experimental results of indentations on ϕ29 particles laying on its side (gray) or upright (dark). Interpretation of these results indicates the presence of built-in mechanical stress in these particles (see text)

than three times softer than the spherical caps. This disagreement with the experimental results can be attributed to the existence of residual stress on the equatorial zone of the shell imposed by the protein curvature, a hypothesis that was confirmed by coarse-grained simulations [47].

This built-in stress in the virus prohead could be an evolutionary strategy to provide extra mechanical strength. It is interesting to discuss prestress generation during the formation of the capsid. In the assembly process, there is a competition between the tendency of proteins to aggregate at the preferred curvature and the need to minimize the rim area exposed in a partially assembled capsid by completing a closed structure [54]. If the curvature of the closed capsid is different from the spontaneous one, lateral stress will develop. Interestingly, not only ϕ29 [55], but other complex dsDNA bacteriophages have a scaffolding protein that co-assembles with the major capsid protein to produce a prohead with the correct shape and size

[56] (see Chap. 12). Once the shell is built, the scaffolding proteins are released concomitantly to DNA packaging. The possibility that some scaffolding protein subunits might be released from the proheads used in the experiments during previous storage cannot be excluded, but no change in prohead structure was in fact detected. Since protein-protein binding interactions in viral capsids are relatively weak, on the order of a few $\boldsymbol{k_B T}$ [57] (see Chap. 19), it is likely that scaffolding protein help impose a curvature in the capsid very different from the spontaneous one, by assisting the bending of the proteins at the junctions between subunits. This will generate a much larger in-plane stress, which will mechanically reinforce the capsid. In the absence of a scaffold, the stress generated during assembly could help to better tolerate DNA packaging, avoiding capsid disruption.

18.5.4 Electrostatics of Bacteriophage ϕ29 Capsid

The recognition events between viruses and host cells are dominated by both specific and non-specific interactions. Electrostatic force is a non-specific interaction that is determined by the charge of viral particles. Detection of electrostatics in liquid environments with AFM was already developed for surfaces in the early 1990s [58]. The electrostatics of single biomolecules, such as dsDNA [59], and the avidin-streptavidin system [60] have also been explored. We will focus here on the simplest methodology based on single indentations carried out on a ϕ29 virus capsid (Fig. 18.9a). A careful look at the contact point between the tip and the sample in *F-z* curves (Fig. 18.9b) reveals a sharp kink when the solid substrate is contacted (black curves) but a smooth curvature when the virus is contacted (red curves). The origin of this curvature is the repulsive electrostatic force between the protein cage and the AFM tip, both negatively charged. The charge of a virus particle can be estimated by fitting the nanoindentation curves to two-sphere DLVO models [61]. Interestingly, the presence of dsDNA affects the charge density of bacteriophage ϕ29 (Fig. 18.9c), although the underlying mechanism of this influence remains unknown [61].

18.6 Mechanical Properties of Adenovirus (AdV): Single Indentation, Material Fatigue, Genome Release, and Aging

Adenovirus (AdV), an animal virus, is structurally and functionally quite different from the tailed bacteriophage ϕ29. Both, however, are naked (no lipid envelope), structurally complex viruses based on a fundamentally icosahedral capsid that contains a dsDNA molecule. The family prototype, human AdV type 5 (Ad5), has a 35-kbp dsDNA genome housed inside a 95-nm, pseudo $T = 25$ icosahedral capsid.

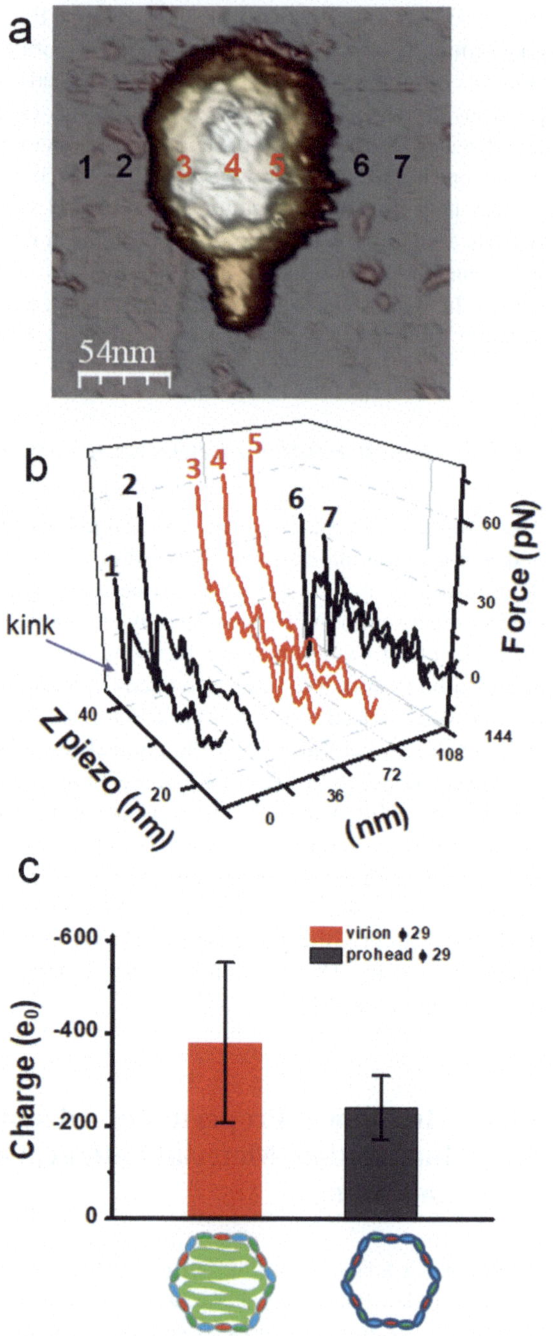

Fig. 18.9 Measuring the electrostatics of bacteriophage ϕ29. (**a**) AFM topography of a phage particle on HOPG. The numbers (1–7) indicate the location of the indentations performed on the substrate or virus. (**b**) *F-z* curves corresponding to indentations at locations 1–7 in panel (**a**). Black and red traces, respectively, correspond to indentations on the substrate or virus particle. (**c**) Influence of the viral DNA on the charge of virus particles. Modified from [61], with permission

The icosahedral facets are formed by trimers of the major coat protein, hexon. A pentamer formed by five copies of the penton base protein is located at each vertex, in complex with a trimer of the receptor-binding fiber. In addition, correct assembly requires four different minor coat proteins: IIIa, VI, and VIII on the inner capsid surface, and IX outside [62] (Fig. 18.10a). In the non-icosahedral core, the genome is packed together with ~25 MDa of positively charged, virus-encoded proteins V, VII, and μ. The AdV protease (AVP) is also packaged together with the genome.

18.6.1 The Role of Core Proteins VII and V

The AFM investigation of AdV has provided several insights into the interplay between mechanical properties and virus functionality [64]. During virus maturation, a viral protease cleaves several capsid and core proteins, making the virions metastable and primed for sequential uncoating. A mutant called ts1 does not undergo proteolytic cleavages and is stalled at the immature state. The presence of uncleaved precursor proteins makes ts1 virions more stable than the wild-type (wt) virus, impairing proper uncoating and aborting infection. AFM revealed that the extra stability of ts1 does not equate with greater stiffness [65]. However, mechanical fatigue experiments [37], the exploration of the core mechanics [66], and genome release [63] revealed that maturation increases the internal pressure of the virus capsid. More recently, we focused on exploring the role of core proteins VII and V in the virus stability and genomic core externalization. In the case of protein VII, successive indentations (~5) performed in the same particle crack-opened the virus capsid (Fig. 18.10b). After three indentations, the VII-particle showed more externalized DNA (yellow) than the wild-type particle. This occurrence was supported by measuring the height inside the crack (Fig. 18.10c). While the height of AdV-VII particles decreased faster and faster after each indentation (blue), the heights of wild-type particles reached a plateau (red). These data indicate that the role of protein VII (Fig. 18.10a, pink diamond) is to condense the core [67]. In the case of protein V, we compared the final stages of wild-type Ad5 and a mutant particle lacking protein V (Ad5-V) after mechanical fatigue at 100 pN. The comparison of the AFM topographies (Fig. 18.10d) indicated that the DNA of the virus core (yellow) was more externalized in the wild-type particle than in the Ad5-V case. Therefore, protein V helps in the genome release process during the virus particle disassembly because it binds the coat protein to the core (Fig. 18.10a, brown ellipse) pulling away the genome when single capsomers are removed during disassembly [33].

Fig. 18.10 The role of core proteins V and VII in human AdV genome uncoating. (**a**) Cartoon of human AdV structure. The different viral proteins are color coded. Adapted from [62], with permission. (**b**) Images showing the evolution of structural changes of Ad5-wild type and Ad5-VII—particles during three successive indentations. Images are colored by height, as indicated by the color bar on the right. Externalized viral DNA is visible as yellow strands. (**c**) Progression of viral core height obtained at the center of every crack for 8 Ad5-VII- and 6 Ad5-wild-type particles. Adapted from [63] (**d**) Final stages of genome uncoating in mechanical fatigue experiments using a force of 100 pN for AdV-wild type and AdV–V particles. Adapted from [33]

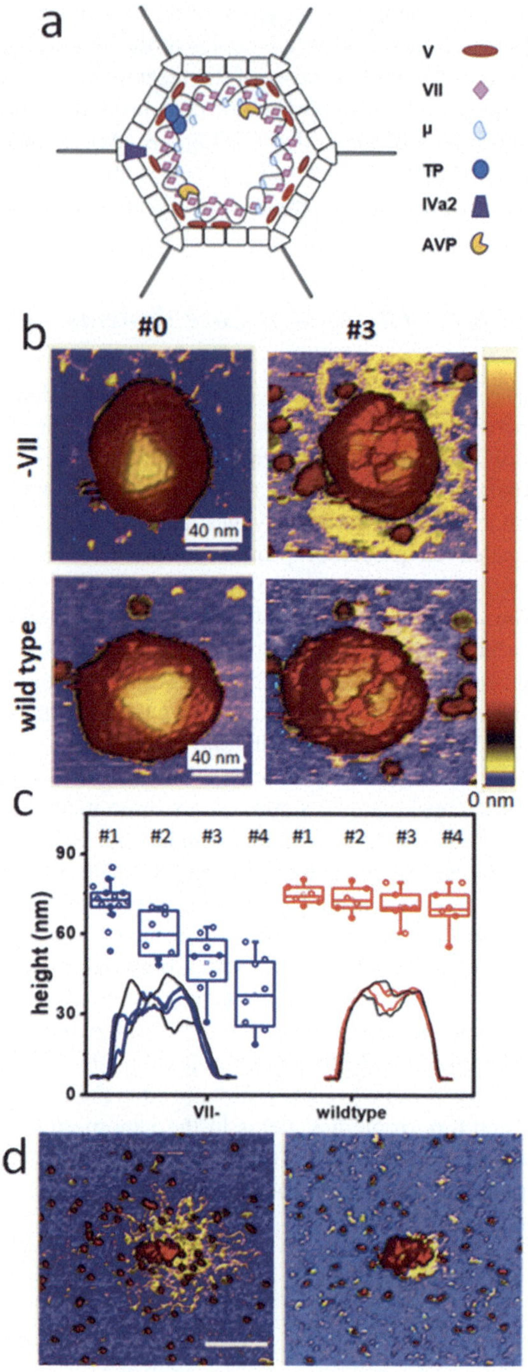

18.6.2 Aging in Human Adenovirus

The particles of human adenovirus travel through the cytoplasm to the nuclear pore to deliver its genome inside the cell nucleus. Inside the host cell, the virus finely tunes the sequential loss of pentons (the proteins located at the icosahedron corners) to render a semi-disrupted capsid at the nuclear pore [68]. Virus particles with too many pentons would not be able to release their DNA through the cell's nuclear pore, whereas those with too few pentons would externalize their genome before reaching the nucleus. Mechanical fatigue induces penton failure [37] (Fig. 18.11a), which mimics the stresses the virus sustains during its journey to the nucleus [70]. The mechanical fatigue approach allows to study of the transition kinetics of penton release as a two-state process, from where it is possible to derive the spontaneous escape rate and the free energy barrier of release of a single penton [69]. Moreover, it is possible to fit the survival probability P of penton N (Fig. 18.11b) under different fatigue forces to Weibull analysis: $\boldsymbol{P(N) = 1 - \exp\left[-\left(\frac{N+1}{\lambda}\right)^{\beta}\right]}$, where λ is the half-life and β determines the nature of the penton dynamics release. In particular, $\beta < 1$ would imply that failure rate decreases over time (*Lindy effect*). This effect can be found in ancient monuments, since the older something is, the more likely it will survive. $\beta = 1$ would imply a Poisson process with a constant probability of penton failure over time. However, $\beta > 1$ (Fig. 18.11b) involves the phenomenon termed as aging in materials science, which means that the failure rate increases over time. This indicates that the probability of the virus losing a penton increases if another penton has been lost before. This aging process ($\beta > 1$) accelerates the overall penton escape rate by about 50% (Fig. 18.11c) with respect to a sequence of independent escape events ($\beta = 1$). Specifically, while the first penton resists for about 60 min, the second and third would last 30 and 20 min more, respectively (Fig. 18.11c). In this way, aging guarantees to have the virus particle just dismantled enough to release the genome through the nuclear pore. Without aging, virus particles would be still quite intact and the genome could not escape (Fig. 18.11c), with negative consequences for infection [69].

18.7 Mechanical Differences Between Virus Particles

The studies on bacteriophage ϕ29 or AdV reviewed in Sects. 18.5 and 18.6 by the group led by one of the authors of this chapter (P.J.P.) provide two relevant examples of the researches on virus mechanics that have been carried out by several groups using about two dozen model viruses so far. During the last 15–20 years, mechanical properties have been analyzed for virions and/or capsids of bacterial viruses, plant viruses, or animal viruses that have different compositions, sizes, shapes, and architectures (Fig. 18.12). Together, these studies have quantified the mechanical properties of virus particles; unveiled some structural determinants of virus

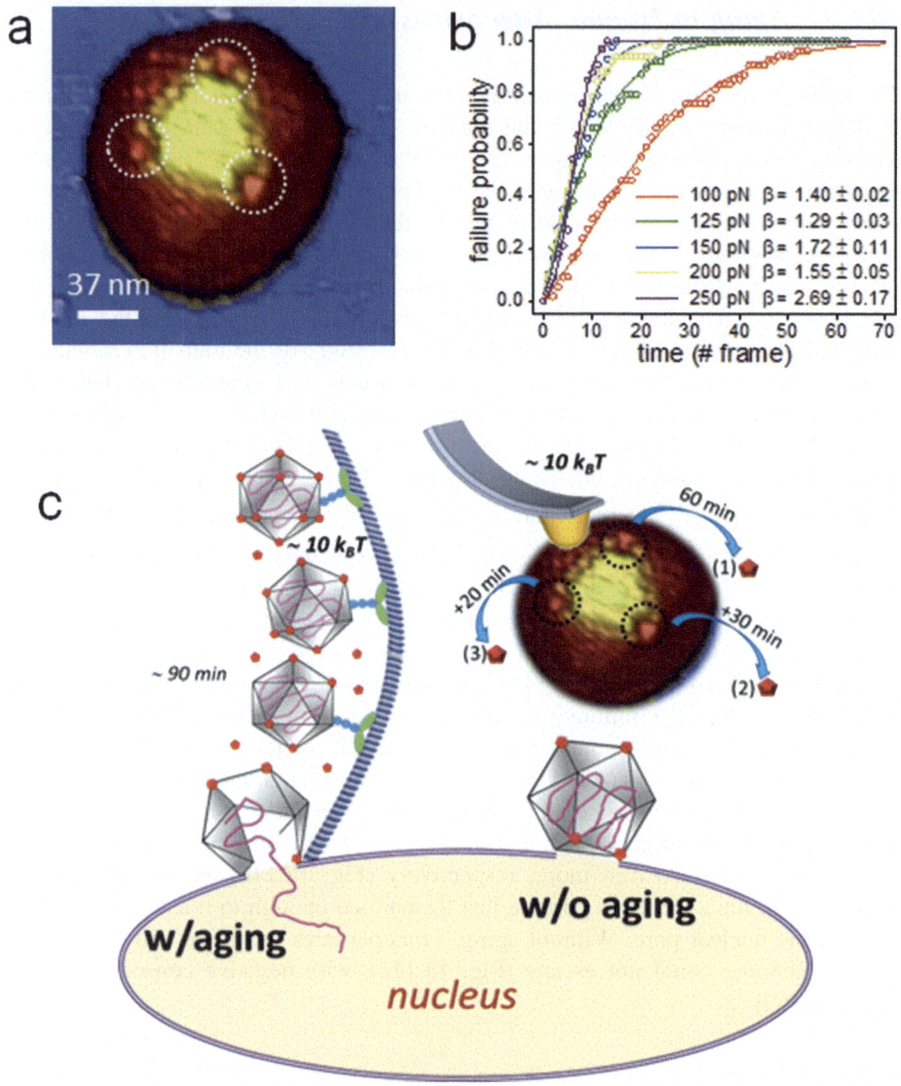

Fig. 18.11 Aging during human AdV uncoating. (**a**) AFM topography of a human AdV particle showing three vacancies of pentons on a triangular facet (encircled). (**b**) Change in the probability of penton failure probability over time for different fatigue forces. The inset shows the fitting parameters of Weibull statistics (see text). (**c**) Diagram showing the implications of aging on virus infectivity (see text). Adapted from [69]

mechanics; provided further mechanistic insights into different stages of the viral infectious cycle; revealed the direct biological relevance of the mechanical behavior of some viruses; and contributed to the development of biomedical or technological applications of virus-based nanoparticles (reviewed in [64, 74–83]).

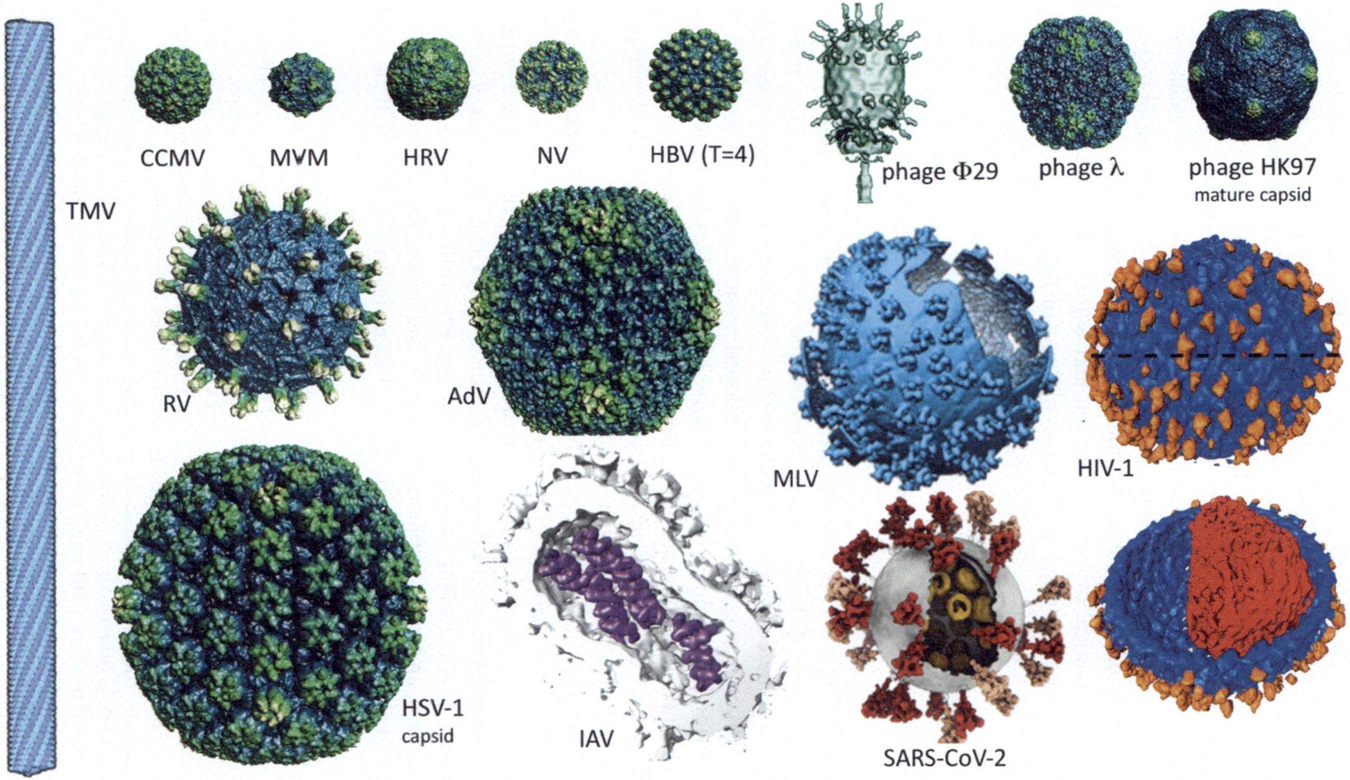

Fig. 18.12 Some virus particles whose mechanical properties have been studied to date. Virus particles shown include: (i) Helical viruses (TMV, tobacco mosaic virus). (ii) Structurally simple icosahedral plant viruses (CCMV, cowpea chlorotic mottle virus) or animal viruses (MVM, minute virus of mice; HRV, human rhinovirus; NV, norovirus; HBV, hepatitis B virus T = 4 particle). (iii) Structurally complex non-enveloped icosahedral viruses (phages Φ29 (complete),

The results obtained to date indicate that some fundamental mechanical concepts may apply to virus particles in general (see Chap. 19). However, they have also revealed that virus particles, even those that are evolutionarily related and structurally similar, may be mechanically quite different. In part, mechanical differences between viruses may be the result of different material properties and/or architecture. However, several findings, described and discussed in later sections (Sects. 18.8–18.10), indicate that some important differences in the mechanical behavior of viruses may have arisen as a consequence of their biological evolution in response to specific selective pressures for adaptation to different niches. In this section, some differences in mechanical properties between and within viral particles are briefly summarized. For more detailed descriptions of those differences, and for citation of the original research articles, see [77, 78].

18.7.1 Differences in Stiffness and Intrinsic Elasticity

Very large (over 100-fold) differences in stiffness (k values) were found between some structurally quite different viruses. For example, for the enveloped severe acute respiratory syndrome coronavirus 2 (SARS-CoV-2; Fig. 18.12) virion the k value was 0.013 N/m, whereas for the non-enveloped minute virus of mice (MVM; Fig. 18.12) virion the k value reached 1.4 N/m. In principle, such large differences could be explained as a consequence of the different materials that form the outer layer of the virion (e.g., a "soapy" lipidic envelope in SARS-CoV-2 versus a compact protein capsid in MVM), and/or the inner components.

However, equally large differences in stiffness were found between simple, spherical virus capsids made up of same material (protein). How could we explain such differences? A quick look at the molecular structures of spherical viruses determined by cryogenic electron microscopy (cryo-EM; Chap. 3) or X-ray crystallography (Chap. 4) reveals that their capsids can be very different from each other in size and thickness, and in the number, size, shape, and geometric arrangement of the protein tiles they are made of (Chap. 2). A thicker hollow sphere will be more

Fig. 18.12 (continued) λ (without its tail), and HK97 (mature empty capsid, without the DNA and with no tail); RV rotavirus; AdV, adenovirus). (iv) Structurally complex enveloped viruses (HSV-1, herpesvirus type 1 (only the naked capsid is shown); IAV, influenza A virus (cross-section); MLV, murine leukemia virus (cross-section); HIV-1, human immunodeficiency virus type 1 (a cross-section showing the mature capsid (in red) is illustrated below); SARS-CoV-2, severe acute respiratory syndrome coronavirus 2 (cross-section)). All viruses are represented at approximately the same scale. TMV is 300 nm long × 18 nm wide, and provides a reference for scale. The TMV image is reproduced from the PDB molecule of the month shown in https://doi.org/10.2210/rcsb_pdb/mom_2009_1. All icosahedral virus models except that of Φ29 have been obtained from VIPERdb (https://viperdb.org) [71]. The Φ29 model is reproduced from [72], with permission. Enveloped virus images except that of HIV-1 were taken from Fig. 3.13 in Chap. 13. The HIV-1 image is reproduced from [73], with permission difficult to deform by pushing with an AFM tip than a thinner one made of the same material. A smaller hollow sphere will be more difficult to deform than a larger one having the same thickness and composition. The stiffness of a virus capsid, or any other solid object, as given by the *k* value, depends not only on the material it is made of, but also on its dimensions and geometry (in the widest sense of the term).

Remarkably, equally large differences in stiffness were detected even when protein capsids of similar size, shape, and architecture were compared. The *k* values ranged from 0.01 N/m (as low as that of a lipidic viral envelope) for the norovirus (NV; Fig. 18.12) capsid under certain conditions, to 1.3 N/m for some mutant MVM capsids. Relatively large differences in stiffness were also found between mutant capsids that were structurally nearly identical, even at atomic resolution (e.g., some MVM capsids that differed in only one amino acid residue per capsid subunit).

When comparing elasticity between capsids of different virus species with different architectures, it could be more meaningful to consider not the stiffness, as given by the spring constant *k* value (Sect. 18.2), but the intrinsic elasticity of the capsid material (the protein crust), as given by the Young's modulus *E* value (Sect. 18.3). As already discussed, some simplifications are needed to calculate the *E* value for a virus capsid. Although many icosahedral virus capsids can be approximated to thin-shelled perfect spheres or icosahedrons, closer inspection reveals that virus capsids are ridden with protrusions and depressions, sometimes very conspicuous. Capsid thickness is not uniform, either. However, for some virus capsids, the approaches described in Sect. 18.3 or other approaches yielded reasonably similar *E* values. It must be also realized that the *E* value calculated for a viral particle provides an averaged elasticity value for what can be actually regarded as an inhomogeneous or composite material. However, the estimated *E* value may still provide a fair indicator to compare the average intrinsic elasticity of virus particles, especially those that are structurally simpler and similar to each other.

Interestingly, the average *E* values obtained for different spherical virus capsids span as much as two orders of magnitude. For example, from 0.03 GPa for the NV capsid at acidic pH to 1.8 GPa for the phage Φ29 (Fig. 18.12) prohead (see Sect. 18.5.1) or 2.8 GPa for an MVM capsid mutant. From these values, some of the hollow quasi-spherical protein shells of viruses could, at the nanoscale, mechanically resemble soft rubber balls, while others could be compared to hard acrylic plastic globes.

18.7.2 Differences in Strength and Brittleness

Virions and virus capsids can differ also in their ability to withstand mechanical loads (forces) without being irreversibly deformed or disrupted. Some icosahedral virus particles withstood forces of up to 6–7 nN (e.g., herpes simplex virus type 1 (HSV-1) particles devoid of their lipid envelope (Fig. 18.12), or phage T7 virions), about 10-fold higher than those required to disrupt some other virus particles.

Likewise, some virions (e.g., SARS-CoV-2) and icosahedral virus capsids (e.g., NV at neutral or acidic pH, cowpea chlorotic mottle virus (CCMV; Fig. 18.12) at pH = 6) behaved like hollow rubber balls as they withstood very large, even wall-to-wall deformations before being disrupted (see Sect. 18.4). In contrast, the capsids of other viruses were fractured at moderate deformations (10–30% of the capsid diameter (e.g., MVM, adeno-associated virus type 2 (AAV-2), NV at alkaline pH, Φ29 and the mature phage λ (Fig. 18.12) capsid). Thus, some viral particles are mechanically much stronger and/or much less brittle than others.

18.7.3 Differences in Resistance to Material Fatigue and Self-Healing

Structurally very different virus particles are susceptible to material fatigue (e.g., the Φ29, λ, HK97 or AdV capsids (Fig. 18.12), the HIV-1 (Fig. 18.12) capsid protein lattice, or the T7 or SARS-CoV-2 virions). However, systematic, controlled comparisons of the resistance to material fatigue between different virus particles are still lacking.

Like susceptibility to material fatigue, self-healing may be a general feature of virus capsids. Self-healing under certain conditions has been detected for the NV capsid at acidic or neutral pH, the HIV-1 capsid protein lattice, the phage T7 capsid and virion, and the SARS-CoV-2 virion. However, no controlled comparisons of self-healing potential between different viral particles have been described so far.

18.7.4 Mechanical Anisotropy of Virus Particles

Some viral capsids and/or virions are mechanically anisotropic. When different discrete regions of some virus particles were indented, the k values obtained showed up to several-fold differences [6, 47, 84, 85]. Anisotropic stiffness may, in some cases, be partly due to geometry (as for the phage Φ29 elongated capsid). However, the quasi-spherical capsid of some MVM mutants was also mechanically anisotropic, whereas the structurally nearly identical capsid of the parent MVM strain was isotropic. Thus, a single value of k or of the Young's modulus E may not be appropriate to describe the elasticity of a virus particle. Moreover, the inclusion of certain components may have a dissimilar mechanical effect on different particle regions. For example, the presence of the viral nucleic acid turned a mechanically isotropic MVM capsid into an anisotropic virion.

Mechanical anisotropy of virus particles is not restricted to elasticity. Some virus particles, even quasi-spherical particles (e.g., some mutant MVM capsids), are anisotropic regarding mechanical strength and/or brittleness as well. Mechanical anisotropy of a virus particle, like the absolute values of their mechanical parameters themselves, may partly arise as a physical consequence of a particular material composition or architecture, but it may have also been modulated by biological adaptation [84–89] (Sect. 18.10).

18.7.5 Are the Different Mechanical Properties of Virus Particles Correlated to Each Other?

Stiffness, brittleness, and resistance to material fatigue in viral capsids are not necessarily correlated, and may also vary in complex ways. The same capsid may be quite stiff but, at the same time, relatively brittle and fail under relatively low mechanical loads causing small deformations (Sect. 18.4.1). This lack of correlation was clearly shown by comparing many mutant MVM capsids having the same size and shape and nearly identical atomic structure. The results revealed a good correlation between stiffness and brittleness, but no correlation at all was found between those two properties and mechanical strength [90]. Stiffness and mechanical strength have different structural and physicochemical foundations (Sect. 18.9).

To sum up, comparison of the mechanical properties between and within virus particles have revealed that (1) different virus particles can show quite different mechanical properties, including stiffness, intrinsic elasticity, mechanical strength, brittleness, and resistance to material fatigue; (2) different regions in a same virus particle can, in certain cases, show significantly different mechanical properties (mechanical anisotropy); and (3) mechanical properties of viral capsids such as elasticity and strength against disruption are not necessarily linked to each other.

18.8 Modulation of the Mechanical Properties of Virus Particles

The mechanical behavior of any virus particle is not invariant. It is well known that many physical or (bio)chemical effectors can promote or interfere with virus infectivity by modulating the structure and/or dynamics of viral particles during some stage of the infectious cycle (Chaps. 11–19). Interestingly, a growing number of AFM studies reveal that many of those same effectors can also modulate the intrinsic elasticity, strength, brittleness, and/or fatigue resistance of virus particles under mechanical load. The analysis of changes in the mechanical properties of virions or capsids upon the action of infectivity-determining effectors is being used to obtain new insights into the structural modifications and mechanisms that control specific stages of the viral infection process.

A growing list of effectors that, depending on the virus species, both affected virus infectivity and modulated some mechanical property of a viral particle includes: (1) changes in pH, ionic strength, temperature, macromolecular crowding, and osmotic pressure; (2) heterologous ligands, such as some ions, organic molecules, host cell proteins, and other biomolecules that may bind a virus particle; (3) macromolecular viral components, including the viral nucleic acid, and structural proteins, either recruited by a virus particle (e.g., during morphogenesis or maturation) or removed from the particle (e.g., during maturation, genome uncoating, or disassembly); (4) chemical modifications of viral macromolecular components such as proteolytic processing, introduction of disulfide bridges or other covalent bonds, and other post-translational modifications; or (5) during virus multiplication in the cell, mutations in the viral genome that result in amino acid substitutions in viral capsid proteins. Some relevant examples are briefly summarized next. For more detailed descriptions, and for citation of many original studies that yielded the results summarized in this section, see [78].

18.8.1 Physical or Chemical Conditions

pH, Ionic Strength, and Temperature pH variations modulated the stiffness of widely diverse viruses including CCMV, NV, AdV, and phages C22 and T7. Ionic strength modulated the stiffness of phage C22. Heating had an effect on the stiffness of phage C22 and on both the stiffness and strength of phage T7. These variations in mechanical properties are indicative of diverse effector-induced structural changes in the viral particles that could modulate virus infectivity (see Sect. 18.6 for the effect of pH on the mechanical behavior of AdV, and its relationship to virion disassembly).

Osmotic Pressure and Macromolecular Crowding A higher osmotic pressure increased the mechanical strength of both phage λ and HSV-1. The effect was traced to increased condensation of the viral dsDNA, relieving the mechanical stress on the capsid wall, which also inhibited genome ejection and, thus, impaired virus infectivity. Macromolecular crowding, in turn, modulated the stiffness and strength of brome mosaic virus (BMV) particles. Analysis of the mechanical effect of crowding agent concentration and the presence or absence of the viral single-stranded (ss) RNA led the authors to propose that the RNA molecule exerts some internal pressure that pre-stresses the viral capsid.

18.8.2 Heterologous Ligands

Bound Ions Ca^{2+} ions increased the stiffness and mechanical strength of tomato bushy stunt virus (TBSV) and simian virus 40 (SV40). Mg^{2+} ions and/or polycations such as spermine or spermidine decreased the stiffness of dsDNA viruses including phages λ and Φ29, AdV (see Sect. 18.6), and the Herpes simplex virus 1 (HSV-1) and of a ssRNA virus, triatoma virus. For some of the dsDNA viruses, the softening effect of Mg^{2+} and polycations was traced to increased condensation of the viral nucleic acid molecule, reducing the internal pressure on the capsid walls.

Bound Organic Molecules Virus infectivity-inhibiting compounds that fill hydrophobic pockets in the human rhinovirus (HRV; Fig. 18.12) capsid (pleconaril, pirodavir) increased virion stiffness. Binding of tannin to the helical tobacco mosaic virus (TMV; Fig. 18.12) virion both inhibited its infectivity and increased its strength against disruption by pulling the RNA out of its long helical capsid. Binding of specific antiviral or proviral compounds to different sites on the mature HIV-1 capsid modified either capsid stiffness or mechanical strength, depending on the compound. These effects were, respectively, associated with changes in the conformational dynamism of the capsid at equilibrium or with the establishment of further enthalpic interactions between protein capsid subunits. Binding of the proviral cellular compound IP6 to the HIV-1 core either stiffened or softened the viral particle, depending on the activation of RNA reverse transcription in the core. The above results revealed an association between small-molecule binding to virus particles, complex changes in the mechanical response of the latter due to changes in structure, molecular interactions or conformational dynamics, and modulation of virus infectivity. For HRV, an inverse correlation was detected between increased stiffness and reduced infectivity. These results suggested the possibility of fighting viral diseases through the development of antiviral drugs based on small organic compounds that exert their action by modifying infectivity-determining mechanical properties of virus particles.

Bound Host Cell Proteins Integrin or defensin, respectively, softened or stiffened the vertex capsid regions of the AdV virion. These antagonic effects were, respectively, traced to the facilitation or impairment of the release of individual pentons from the capsid vertices during the induced disassembly of the virus particle. The proviral cellular protein CypA to the mature HIV-1 capsid modulated its stiffness. The specific mechanical effects of different cellular proteins bound to viral particles are providing new insights into biologically relevant structural changes controlled by interactions between host molecules and viruses during the infection process.

18.8.3 Macromolecular Viral Components

The Viral Nucleic Acid Irrespective of the type of viral nucleic acid, or the mechanisms by which it is packaged in the capsid or released during infection, the presence of the viral nucleic acid inside a virus capsid generally (though not always) modulates the stiffness, strength, and/or brittleness of the viral particle. Examples include ssDNA viruses (MVM, but not AAV-2); dsDNA viruses (phages λ, ϕ29 (Sect. 18.5.2), T7 and PRD-1, AdV and the non-enveloped HSV-1 particle); and ssRNA viruses (CCMV, BMV, and ribonucleoprotein-containing viruses such as the infectious bursal disease virus (IBDV) and influenza A virus (IAV) (Fig. 18.12).

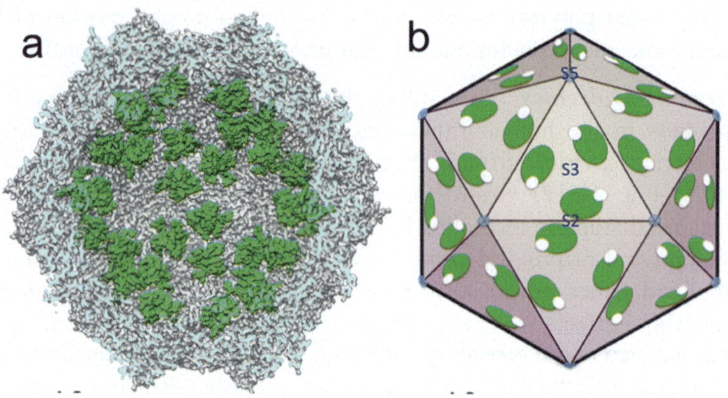

Fig. 18.13 The MVM virion. (**a**) cross-section of the MVM virion atomic structure [91]. ssDNA segments bound to the capsid inner wall are colored green. (**b**) scheme of the icosahedral MVM virion. The approximate location of capsid-bound ssDNA segments or nearby cavities are respectively indicated by green or white ovals. The positions of different icosahedral symmetry axes in the capsid (S2, S3, and S5) are indicated. Figure reproduced from [78]

There is, however, no common mechanism by which a viral nucleic acid molecule may stiffen a virion. Different mechanisms have been unveiled depending on the virus species and nucleic acid type, as summarized next.

The MVM empty capsid (devoid of nucleic acid) and the capsid as a part of the MVM virion have nearly the same atomic structure, but the presence of the viral ssDNA molecule in the virion stiffens certain capsid regions (around the S2 and S3 icosahedral symmetry axes). Wedge-shaped ssDNA segments are bound to pockets at the capsid inner wall close to the S2 axes (Fig. 18.13). These ssDNA wedges act like internal molecular buttresses. Genetic removal of some capsid-ssDNA interactions in the virion reduced the stiffness of the virion which became as soft as the empty capsid, despite it still containing the full ssDNA molecule. The ssDNA-mediated anisotropic stiffening of the virion appears to be a biological adaptation for improving extracellular virion survival without impairing its infectivity (Sect. 18.10.5).

Unlike the ssDNA in MVM, the stiff dsDNA genome of phage λ, Φ29 and other tailed phages, HSV-1 or AdV generate a substantial internal pressure that stresses the capsid. This pressurization results from the long persistence length of the dsDNA molecule, increased by the mutual repulsion of non-neutralized negative charges of phosphates, and its tight packaging inside the capsid. The pressure exerted by the dsDNA is controlled by cations and/or positively charged, DNA-condensing proteins. Virion pressurization requires a high mechanical strength of the capsid to impair its disruption and loss of infectivity. dsDNA viruses provide a clear example of the adaptive value of the mechanical properties of viruses (Sect. 18.10).

In contrast to the above ssDNA or dsDNA viruses, in ssRNA viruses such as CCMV, the ssRNA genome is not packaged in a preformed capsid. Instead, both are assembled together (Chap. 13), and the CCMV virion may not become pressurized.

The mechanisms by which the viral ssRNA molecule, or ssRNA-containing ribonucleoprotein (RNP) complexes, may stiffen a virus particle are not well understood yet.

The studies on virus-mediated stiffening and/or strengthening by the packed viral nucleic acid have provided new knowledge on the mechanisms of virus morphogenesis, maturation, disassembly, and/or genome uncoating. Nucleic acid-mediated mechanical strengthening or stiffening of virions can also be directly relevant for the survival of some viruses (Sect. 18.10).

Capsid Protein Subunits Removal of weakly bound penton subassemblies at the vertices of the AdV virion by different means reduced particle stiffness and strength and facilitated in vitro capsid disassembly and/or genome uncoating for AdV, HSV-1, and phage P22. These and other observations have provided new insights into the mechanisms by which disassembly and/or uncoating may be controlled during infection by dsDNA viruses (for the AdV case see Sect. 18.6).

Accessory Viral Proteins and Covalent Modifications of Viral Proteins Binding of positively charged viral proteins that condense the viral dsDNA made the AdV virion less stiff (Sect. 18.6). In contrast, viral cementing proteins bound to phages λ or P22 generally increased both stiffness and strength of the viral particles. Auxiliary proteins also stiffened the HSV-1 capsid at the vertices where the pentons are weakly bound. Phage HK97 maturation involved the introduction of covalent bonds between capsid subunits that increased the capsid mechanical strength but not its intrinsic elasticity. Likewise, the introduction of disulfide bonds increased the mechanical strength of the SV40 capsid without affecting its elasticity during the first indentation. Recruiting of cementing proteins or introduction of covalent bonds between capsid subunits provide alternative adaptive strategies to strengthen the capsid during virion morphogenesis and maturation.

Capsid Layers Some non-enveloped viruses contain several concentric protein layers (see Chap. 12). The strength and/or stiffness of rotavirus (Fig. 18.12) or phage PRD-1 particles increased with the number of macromolecular layers they contained. In the NV capsid the protruding (P) domain of the capsid protein forms a protein layer on top of the basal capsid protein layer. Removal of the P domain decreased capsid strength and stiffness, which was attributed to the introduction of prestress that would stabilize the capsid. Like the structural modifications introduced during maturation of some viruses, the building of additional protein layers during morphogenesis of some other viruses may strengthen the virion.

The Viral Envelope Removal of proteins embedded in the lipid envelope of IAV reduced its stiffness to that of pure liposomes, and removal of the matrix protein layer under the envelope also softened the viral particle. Analysis of the stiffness variations led to a model of the structural events that occur during virus fusion to the cell membrane and RNP release in the cell. Maturation of retroviruses MLV and HIV-1 (Fig. 18.12) softened the virion. The softening of the mature HIV-1 virion facilitated its entry into the host cell (Sect. 18.10). Those results, taken together, reveal that, for enveloped viruses, the outer lipid envelope, envelope-embedded proteins, and the protein matrix layer underneath can be major determinants of the mechanical properties of the virus particle.

18.8.4 Genetic Modulation of the Mechanical Properties of Virus Particles

Even the smallest structural changes in virions or capsids due to mutations that result in the substitution of one or more amino acid residues in a capsid protein can potentially modify the mechanical properties of the viral particles.

MVM has been used as a structurally simple model virus to dissect the effects on virus particle mechanics of many individual amino acid substitutions and to begin to understand the relationships between structure, mechanical behavior, and virus function. Over 40 different amino acid substitutions have been individually introduced in the MVM capsid so far (Fig. 18.14). The mutant virions or capsids analyzed generally showed anisotropic changes in stiffness, mechanical strength, and/or brittleness. In many cases, changes that modify some mechanical properties of

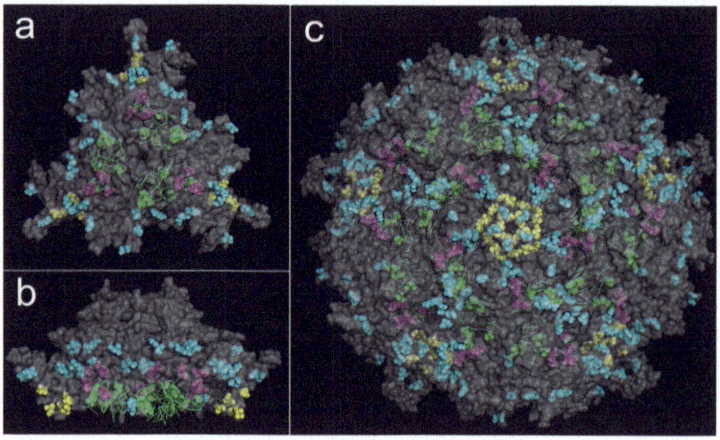

Fig. 18.14 Some amino acid residues in the MVM capsid whose individual role on the mechanical properties of the capsid and/or virion has been analyzed. Left, front view (**a**) or side view (**b**) of the atomic structure of a trimeric capsid building block. Right, (**c**), front view of the atomic structure of an ensemble of five trimers around an S5 axis in the MVM virion [91]. A semi-transparent depiction allows the visualization of buried amino acid residues. Thirty-two residues per capsid subunit that were individually mutated to investigate their role in the mechanical properties of MVM and their biological relevance are color coded: cyan, residues at the interfaces between trimeric capsid building blocks; yellow, residues surrounding the base of the capsid pores around each S5 symmetry axis; magenta, residues delimiting capsid small cavities close to ssDNA-binding sites; green, residues involved in interactions with genomic ssDNA segments in the virion. ssDNA segments that interact with those capsid residues are represented as green ribbons. Figure reproduced from [78]

MVM particles could be linked to different mechanisms by which those mutations impair virus infectivity (Sect. 18.10.5).

The effects of amino acid substitutions on the mechanics of viral particles have been much less studied for viruses other than MVM. However, some tested amino acid substitutions did affect virion and/or capsid stiffness for viruses as diverse as CCMV, HRV, AdV, and HIV-1. A salt-stable natural mutant of CCMV with a single amino acid substitution was stiffer and mechanically stronger than the parent strain. The capsid of a mutant AdV with an amino acid substitution in an auxiliary capsid protein was stiffer than the wild type. Four single amino acid substitutions that impaired disassembly of the mature HIV-1 capsid also increased its stiffness. Mutant HRV virions carrying single amino acid substitutions that increased the volume of residues lining capsid cavities were stiffer than the parent virions. Importantly, an inverse quantitative correlation was found between HRV infectivity and stiffness, which was dependent on the size of the introduced side chain. These results indicated that HRV infectivity depends on the mechanical elasticity conferred to the capsid by the presence of large enough cavities. A reduction in cavity size reduced the elasticity of the virion, which resulted in impaired infectivity (Sect. 18.10).

To sum up, the analysis of changes in the mechanical response of virus particles upon mutation or the action of environmental, cellular, or viral effectors, is being used to explore relationships between virus structure, physical properties, and biological function. From an evolutionary perspective, the high sensitivity of the mechanical behavior of viruses to the introduction of point mutations in the capsid genes may contribute to the rapid biological adaptation of viruses to withstand, or even use, mechanical forces to which they may be subjected during the infectious cycle (Sect. 18.10).

18.9 Mechanical Properties of Virus Particles as Signatures of Changes in Structure, Stability, and/or Dynamics During the Infectious Cycle

It could be argued that viruses in nature may not be subjected to the relatively strong mechanical forces typically applied during AFM indentation experiments. Thus, it could be thought that the mechanical properties determined for any virus, and their modulation in response to the action of different effectors, may be irrelevant from a biological standpoint.

In fact, virus particles in nature are indeed subjected to direct mechanical stress, in or out of infected cells. Section 18.10 reviews considerable experimental support for the view that the mechanical behavior of (at least) some viruses may have been tailored by evolution to withstand, or even take advantage, of the mechanical forces they may be subjected to during their infectious cycle.

But, even if viruses were not subjected to biologically significant mechanical forces in nature, the analysis of the mechanical behavior of a virus particle under load would still be biologically relevant, as it provides important insights into virus structure and function. In many AFM studies to date, the force applied on a virus particle by the cantilever was not intended to mimic any mechanical force that could be acting on the particle in nature. Instead, mechanical force was used as an in vitro effector to influence the thermodynamics or kinetics of the virus particle under specific conditions. Mechanical force in single-particle AFM experiments, like heat, pH extremes, or chemical denaturants frequently used in other in vitro experiments, provides an adequate substitute for in vivo (biological) effectors that, in many occasions, have not been identified yet or are poorly understood. More specifically, mechanical force constitutes a powerful physical probe to identify and investigate effector-mediated changes in molecular structure, conformational stability, or dynamics of the viral particle that can modulate virus infectivity.

The results obtained so far have provided new insights into previously known but incompletely understood processes that control different stages of the viral infectious cycle. They have also contributed to the discovery of previously unknown structural changes and/or mechanisms required for infection. Moreover, AFM indentation and/or imaging of single virus particles have uniquely allowed the detection of sequences of events in position-dependent and component-specific changes in particle structure mediated by different cues. The studies with AdV (Sect. 18.6) provide a particularly significant example; other relevant examples are briefly mentioned in Sect. 18.8. To reiterate, studies on the mechanical response of viruses are important to understand their biology, irrespective of whether virus particles are actually subjected to mechanical forces in vivo.

18.9.1 An Atomic Structure-Based View on the Mechanical Properties of Virus Particles

Why do some effector-mediated changes in the atomic structure of a virus particle, and/or atomic movements within it, modify its mechanical properties? Here a simplified view is outlined based on structural, thermodynamic, and kinetic considerations. For a more detailed description of this view and citation of original references see [78]. This model is an adaptation of concepts previously established for the mechanical unfolding of proteins, or for the dissociation of macromolecular complexes by mechanically stretching them. Fundamental similarities regarding structure, thermodynamics, and kinetics may occur when, instead of "pulling" on a protein or molecular complex, the AFM tip "pushes" on a virus particle to determine its stiffness, brittleness, strength, or fatigue [78]. This model is consistent with theoretical and experimental results on virus mechanics obtained so far.

Structurally simple, quasi-spherical virus capsids or virions (capsid plus viral nucleic acid) are considered here, to avoid dealing with complications introduced by other components present in more complex viral particles. According to this model, effector-mediated structural changes in a virus particle that modulate its equilibrium dynamics and/or its propensity to undergo a conformational transition may lead to changes in its mechanical stiffness.

Let us consider what could occur at the atomic level when a protein-based, quasi-spherical virus capsid is being deformed by pushing with an AFM tip on some part of its surface.

If the Elastic Limit Is Not Reached As the capsid "crust" is gradually deformed, atomic groups (including amino acid side chains) directly under the tip are gradually displaced from their equilibrium positions in the particle. In the compact capsid shell, these atomic groups are non-covalently bonded to other nearby groups belonging to the same or a neighbor protein subunit, so other groups not directly in "contact" with the AFM tip will also be displaced from their equilibrium positions. In some simulations of virus capsid mechanics, each non-covalent bond between atomic groups in the capsid has been considered to be a virtual spring. When the atoms and atomic groups in the capsid are at their equilibrium positions, the springs are relaxed. As the AFM tip pushes deeper, the weaker springs (weak interatomic interactions, e.g., individual van der Waals contacts) are increasingly stretched, and may eventually be broken. Ionic interactions and hydrogen bonds may eventually be also stretched and disrupted. Moreover, as atomic groups are displaced from their equilibrium position, steric clashes with other groups will have to be relieved through additional atomic displacements which will stress, and eventually break, other weak interatomic interactions. As clusters of apolar groups in the capsid become disrupted, the hydrophobic effect that contributes to stabilize each capsid protein subunit and hold the subunits together will be attenuated. As a consequence of all of those events, the free energy of the virus particle will increase.

In terms of an energy diagram (Fig. 18.15), the capsid will move along the mechanical reaction coordinate (along the direction of the applied force) away from its equilibrium position at the bottom of the free energy well of the native state. However, if the deformation does not reach a certain threshold (the elastic limit), the capsid will not reach the free energy peak that corresponds to the transition state of any reaction that could lead to a change of thermodynamic state (e.g., to a conformationally rearranged capsid, a capsid missing some subunits, or a disrupted capsid). After unloading by retraction of the AFM tip, the stretched or broken springs (interatomic interactions) will usually be restored; the capsid atomic groups will be back to their equilibrium positions; and the capsid will return to its lower free energy conformation at equilibrium.

From the standpoint of mechanics, the fully reversible stretching and eventual disruption of multiple non-covalent interactions will be observed as an elastic deformation of the capsid (see Sect. 18.2). The gradual, reversible breaking of many weak individual interactions could lead to multiple non-linearities along the slope of the F-z curve, but they would be too small to be resolved. The stiffness, i.e., the mechanical force required to achieve a certain (elastic) deformation of the capsid (the k value) will be related to the energy needed to move the atoms a certain distance along the mechanical reaction coordinate from their equilibrium positions. In other

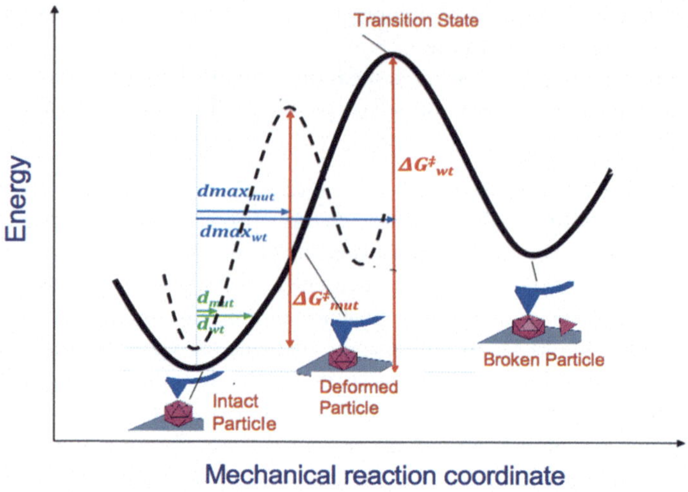

Fig. 18.15 Energy diagram associated to the application of a mechanical force on a virus particle. The energy of the system during the elastic deformation of the virus particle, followed by particle disruption in a two-state reaction, is represented along the mechanical reaction coordinate. *Thick black line*: variation in free energy in the absence of any effector (*wt* subscript). *Thin, dashed black line*: variation in free energy in the presence of an effector (particle-binding compound or capsid amino acid substitution introduced by genetic mutation) that modifies particle elasticity and strength (*mut* subscript). In this example, the effector stiffens the capsid and makes it more brittle and less strong, as respectively indicated by: -*Stiffness* (compare the *green arrows*): distance d_{mut} is shorter than d_{wt}. From a mechanical standpoint, under the elastic regime the same applied force achieves a lower deformation for the *mut* capsid than for the *wt* capsid; the elastic constant k_p will be higher for *mut* than for *wt*. From a thermodynamics/kinetics standpoint, the slope of the free energy well of the initial state (intact capsid) along the mechanical reaction coordinate is steeper for *mut* than for *wt*. -*Brittleness* (compare the blue arrows): distance $dmax_{mut}$ is shorter than $dmax_{wt}$. From a mechanical standpoint, the *mut* capsid is disrupted at a lower deformation than the *wt* capsid; the critical deformation d_{rp} will be lower for *mut* than for *wt*. From a thermodynamics/kinetics standpoint, the position along the mechanical coordinate of the transition state ‡ for capsid disruption is closer to the position of the initial state for *mut* than for *wt*. -*Strength* (compare the red arrows): the $\Delta G^{\ddagger}_{mut}$ value is lower than the $\Delta G^{\ddagger}_{wt}$ value. From a mechanical standpoint, the *mut* capsid requires less applied force than the *wt* capsid to become disrupted; the yield force F_{rp} will be lower for *mut* than for *wt*. From a thermodynamics/kinetics standpoint, the free energy barrier along the mechanical coordinate for capsid disruption is lower for *mut* than for *wt*. See Sect. 18.9 for a detailed description

words, the k value will be related to the steepness of the free energy well of the native state along the reaction coordinate (Fig. 18.15).

If the Elastic Limit Is Surpassed If the force applied on a virus capsid is high enough and the deformation surpasses the elastic limit, the ensemble of interatomic interactions lost may lead to a change of the capsid thermodynamic state. The capsid could transit from its native state to, for example, a locally deformed (buckled) capsid, a capsid with a different global conformation, an incomplete capsid with one or more subunits missing, or a disintegrated capsid. In terms of an energy diagram

(Fig. 18.15), the capsid will move along the mechanical reaction coordinate out of the free energy well of the native state, and up into the free energy peak of the transition state of the reaction. From there, the system will fall into the free energy well that corresponds to the final state of the reaction.

From the standpoint of mechanics, when enough non-covalent interactions are lost and the elastic limit is surpassed, one or more non-linear events may be observed in the F-z curve. If a two-state reaction occurs from an intact capsid of diameter D to a disrupted capsid, the force required to overcome the single free energy barrier of the reaction will correspond to the yield force F_r that defines the capsid mechanical strength (see Sect. 18.4.1). The distance between the native state and the transition state along the mechanical reaction coordinate will correspond to the critical deformation d_r, and will determine the (d_r/D) value that defines capsid brittleness (Fig. 18.15).

If the reaction from the initial state of the viral particle to the final state proceeds through populated intermediate states, each non-linear event in the F-z curve may correspond to the transition between one state and the next state, separated from the previous one by a transition state. In such a case, the considerations mentioned for a two-state reaction could, in principle, be applied to each of the individual transitions between one state and the next along the reaction coordinate.

Material Fatigue During cyclic indentation of a virus particle using a force well below the breaking force F_r, small "cracks" between capsid subunits may be initiated by disruption of a small number of weaker non-covalent interactions at the interfaces between particle subunits (see Sect. 18.4.2). As mentioned above, once the load is removed, and given enough time, each lost interaction may be restored, and an elastic deformation of the viral particle will be observed. However, if the indentation frequency is high enough, not every broken interaction in every crack will have time to be restored before the next indentation disrupts additional interactions. If so, subsequent cyclic application of a comparatively low force can lead to the gradual enlargement and merging of several cracks, until complete interfaces between particle components are disrupted. Some of these components will then be lost and, eventually, the particle will fall apart. In terms of an energy diagram (Fig. 18.15), the mechanism leading to fatigue of a virus particle could be contemplated as a kinetic ratchet. Cyclic indentation using small forces will gradually move the particle, through many small steps, upwards along the mechanical reaction coordinate. It will eventually reach the free energy peak(s) corresponding to the transition state (or the transition states if intermediates are present), until the final state (a disrupted capsid) is reached.

In the following paragraphs, the simple structural/ thermodynamics/ kinetics scenario just outlined is invoked to justify the experimentally observed associations between effector-mediated changes in conformational dynamics or structural stability of a virus particle, and changes in its stiffness or strength.

18.9.2 Stiffness Modulation as a Signature of Changes in the Equilibrium Dynamics of a Virus Particle

Virus particles may sample a relatively large ensemble of similar conformations at equilibrium, and this conformational dynamism can be biologically relevant [92, 93]. The free energy well of a virus particle in its native state (Fig. 18.15) must, thus, be relatively "flat," i.e., very low thermal energy may be enough to allow large displacements of some atomic groups from their equilibrium positions in the capsid minimum free energy conformation (the bottom of the energy well).

Equilibrium dynamics of virus particles in solution has been quantified by a number of biophysical techniques, including nuclear magnetic resonance (NMR) spectroscopy (Chap. 5) and, especially, hydrogen–deuterium exchange determined by mass spectrometry (HDX-MS) (Chap. 6) [92, 93]. Some parameters obtained by X-ray crystallography (B-factors) (Chap. 4), or cryo-EM (local resolution) (Chap. 3), can also provide information on local differences in the equilibrium dynamics of viral particles [94]. In addition, AFM imaging (Chap. 8) has provided information on the large-scale equilibrium dynamics of a virus capsid lattice [42]. Differences in local and/or global equilibrium dynamics between structurally similar virus particles, probed by some of the above techniques, have been compared with differences in mechanical stiffness. The results using different model viruses revealed a remarkable linkage between changes in mechanical stiffness/intrinsic elasticity, and changes in equilibrium dynamics, as summarized next.

Some mutant MVM capsids containing single amino acid substitutions in the capsid protein were structurally nearly identical to the parent capsid, even at atomic resolution. However, minor differences led to substantially reduced equilibrium dynamics as revealed by X-ray crystallography or cryo-EM results supported by HDX-MS analysis [94, 95]. This reduced mobility of atoms, atomic groups, and structural elements indicated that more thermal energy is required to displace them a certain distance from their equilibrium positions (at the bottom of the free energy well) (Fig. 18.15). According to the view expressed in Sect. 18.9.1, it could be expected that those mutant capsids would show a higher stiffness and lower intrinsic elasticity than the parent capsid: more mechanical force exerted by the AFM tip would be required for those atomic groups to be displaced from their equilibrium positions a certain distance along the mechanical reaction coordinate. This expectation was experimentally verified [94, 95].

A similar linkage between changes in equilibrium dynamics and mechanical stiffness was unveiled for the HRV virion. Binding of antiviral drugs to hydrophobic pockets in the capsid, or individual amino acid substitutions that partially filled the cavity, led to both a reduction in the equilibrium dynamics of the virion and an increase in its stiffness [96]. Likewise, betaine or certain infectivity-inhibiting compounds bound to specific sites in the mature HIV-1 capsid protein lattice led to both increased "breathing" (amplitude of the relative movements at equilibrium of capsid protein hexamers) and reduced stiffness [42, 97].

The above results, obtained with very different virus models, provided strong support for an inextricable relationship between changes in the equilibrium dynamics of a virus capsid, and changes in its mechanical stiffness (or rather, intrinsic elasticity).

One must note, however, that such a relationship may, in some cases, be blurred by the structural complexity of the virus particle and/or by certain particularities of the techniques used. As stiffness depends on size and geometry, the capsids compared should be structurally similar, as was the case in the studies mentioned above. It should also be considered that HDX-MS (or other methods in solution) provide a direction-averaged value of equilibrium dynamics. In contrast, stiffness is determined by AFM through an indentation along the perpendicular to the capsid surface (the mechanical coordinate of the reaction). If the deformability of the indented capsid region were significantly anisotropic, the k value obtained could not provide a good estimation of the stiffness along other directions. However, the relatively blunt AFM tip contacts a large region of the capsid, and the deformability of such a large region may be generally expected to be fairly isotropic. Thus, the equilibrium dynamics-stiffness correlation could still be detected, as actually found for the virus models already studied.

To sum up, effector-mediated changes in equilibrium dynamics (determined by HDX-MS or other techniques) or stiffness (determined by AFM) appear to be different manifestations of the same physical phenomenon: A structure-based variation in the energy required to move atomic groups and larger structural elements in a virus particle a certain distance from their equilibrium positions. The detected stiffening of a virus particle in the presence of some effector may signal that, at the atomic level, such effector may be locally or globally impairing the equilibrium dynamics of the virus particle.

18.9.3 Stiffness Modulation as a Signature of Changes in the Propensity for Conformational Rearrangements of a Virus Particle

During the infectious cycle, different physical or (bio)chemical effectors facilitate transitions between different conformations of a virus particle [92, 98–100]. Those different conformations constitute different thermodynamic states separated by a free energy barrier. Maturation of many viruses provides an important and well-studied example. In this process, the assembled virus particle undergoes irreversible structural modifications, resulting frequently in a metastable infectious virion [99] (Chap. 14).

Mechanical force can provide the energy needed to trigger such conformational rearrangements in vitro. For example, indentation of the T7 virion or capsid under the elastic regime revealed stepwise, non-linear events that were associated to relatively large, reversible changes in particle structure and dynamics [44].

A connection between an effector-mediated change in the propensity of a virus particle to undergo a conformational rearrangement, and a change in its mechanical stiffness has been detected in several studies (Sect. 18.8). Relevant examples provided by mutational studies of two different conformational transitions in MVM are summarized next.

A conformational rearrangement of the MVM capsid related to externalization of signaling peptide sequences Exit of virus infectivity-determining peptide signals through pores at the fivefold symmetry axes of the MVM capsid [101] was associated to a subtle, reversible conformational rearrangement of the capsid. This transition was promoted *in vitro* by moderate heating, and led to changes both in the exposure to solvent of some capsid tryptophans and in the capsid equilibrium dynamics [102, 103].

Individual removal of atomic groups (truncation of amino acid side chains by site-directed mutation) around the base of the capsid pores (yellow residues in Fig. 18.14) were highly deleterious for the virus and invariably abolished the pore-associated conformational rearrangement [102]. When tested, they also impaired the equilibrium dynamics of the pore and other regions [94, 95]. In each case, the S5 capsid regions around the pores were specifically stiffened. Pseudo-reversions that restored the pore-associated conformational transition restored the native elasticity of the S5 regions. Covalent crosslinking both eliminated the transition and stiffened the S5 regions. Individual removal of atomic groups farther from the pores (cyan, magenta, or green residues in Fig. 18.14) were not highly deleterious, and neither impaired the transition nor stiffened the S5 regions. Together, these results revealed an inextricable linkage in MVM between the impairment of an infectivity-determining conformational transition associated to signal translocation through capsid pores, and the local stiffening of the pore region [85].

A conformational transition related to heat-induced inactivation of the MVM virion Individual removal of atomic groups that eliminated interactions between the ssDNA molecule and the capsid, or that were located close to the DNA-binding sites (green or magenta residues in Fig. 18.14), increased the rate of a heat-induced, virus-inactivating transition [104, 105] that may be related to unproductive DNA release [106] through one of the capsid pores. For the ensemble of mutant capsids, an exponential decrease in the rate constant of the virus-inactivating transition was observed as a function of the increase in k value (stiffness) for the S2 and S3 capsid regions, which are close to the DNA-binding sites [105].

The correlation between either an increase or decrease in the propensity for a transition between two capsid conformational states separated by a free energy barrier, and the respective decrease or increase in capsid stiffness, can be justified by invoking again the model described in Sect. 18.9.1 (Fig. 18.15). Suppose a mutation introduces additional non-covalent interactions that increase the resistance of atomic groups to move from their equilibrium positions. In terms of reaction kinetics, if there is no change in the distance between the native state and the transition state along the reaction coordinate, more energy (force) will be required to disrupt enough interactions between atomic groups to allow these groups to reach their positions in the transition state. The free energy barrier $\Delta G^{\dagger}$ will increase, and the reaction rate will decrease (Fig. 18.15). In mechanical terms, a higher force will be required to deform the capsid along the mechanical reaction coordinate enough to reach the position of the transition state. From a mechanical standpoint, when the region involved in the transition is indented, the k value for the mutant capsid will be higher than that of the parent capsid. Indeed, a direct linear relationship was experimentally found between the k value of a region in a virus particle, and the free energy barrier ($\Delta G^{\ddagger}$) of a structural transition in which that region is involved ($\Delta G^{\ddagger} = Ck$, where C is an empirical proportionality factor). Substituting Ck for $\Delta G^{\ddagger}$ in the Eyring equation results in an equation that describes an exponential relationship between the reaction rate constant and the k (stiffness) value, as experimentally observed for the heat-induced MVM-inactivating transition [105].

The above results do not mean that if the capsid of a virus species is stiffer than that of another virus species, the former will be less prone to undergo a conformational change. Due to significant differences in architecture and molecular structure, the proportionality factor C between $\Delta G^{\ddagger}$ and k will likely be different for the two species. The relationship found between stiffness and propensity for a conformational transition may apply only when structurally very similar capsids are compared, such as mutants of the same virus species. Also, the linkage between stiffness and propensity for a conformational rearrangement may be blurred if the deformability of the region considered was largely anisotropic. However, as indicated in Sect. 18.9.2, a large anisotropy may not be frequently expected for relatively large regions in the viral particle.

It must be emphasized that the expected effect on virus infectivity of a change in stiffness is context dependent. If a conformational rearrangement promotes infection (e.g., the pore-associated transition in MVM), stiffening the capsid region involved in the transition (S5 in MVM) will *impair* infectivity. If the conformational rearrangement impairs infection (e.g., the heat-induced, virus-inactivating transition in MVM), stiffening the capsid region involved (S2 and S3 in MVM) will *promote* infectivity. Of course, in some cases the degree of local or global stiffening of a virus particle achieved by some effector may not be enough to elicit any infectivity-determining conformational rearrangement. Thus, it should not be expected that any change in particle stiffness for any virus species will necessarily lead to a change in infectivity.

To sum up, effector-mediated structural changes in a virus particle that either impair or promote a conformational transition will also respectively increase or decrease the stiffness of the region(s) involved in the transition. An effector-mediated change in mechanical stiffness of a virus particle may provide a signature for a change in its propensity to undergo a conformational rearrangement that could be biologically relevant.

18.9.4 Changes in Mechanical Strength as a Signature of Changes in the Resistance of a Virus Particle to Irreversible Loss of Components

In several studies with different viruses, including some mentioned in Sect. 18.8, changes in the mechanical strength of a virus particle (as deduced from the loss of molecular components or particle disintegration under load) were accompanied by changes in its resistance to thermal or chemical disruption. As one example, the binding of cementing proteins to a phage capsid increased its thermal, chemical, and mechanical stability [38]. According to what was discussed in Sects. 18.9.1 and 18.9.3, the introduction of additional non-covalent interactions between components of a virus particle, or between the particle and heterologous ligands, will likely increase the free energy barrier of the reaction that leads to particle disruption. Additional energy needed to break all those extra inter-molecular interactions could, in principle, be provided by thermal, chemical, or mechanical means. In AFM experiments, an increase in the force F_r required to separate some component(s) from a virus particle, or to disintegrate the particle, would provide a signature for a higher activation-free energy being required for the separation of some particle components (Fig. 18.15).

In some cases, however, no correspondence was found between changes in mechanical strength of a virus particle, and changes in its thermal or chemical stability against disruption. For example, certain mutations in the mature HIV-1 capsid protein lattice increased its thermal stability but not its mechanical strength or fatigue resistance [107]. Careful consideration should be given to each particular case: technique and type of analysis; reversibility or irreversibility of the dissociation process being observed; whether the activation free energy or the difference in free energy between two states at equilibrium are being determined. In addition, it should be stressed again that mechanical force is applied to a viral particle in one direction only. The free energy barrier along the mechanical reaction coordinate may, in some cases, be significantly different from the free energy barrier for thermal or chemical dissociation. Interestingly, disassembly of some virions may be facilitated by inbuilt internal pressure that will push on the inner capsid wall (Sect. 18.10.1). In such a case, AFM studies on the virion mechanical stability (strength) may be biologically more relevant than analysis of thermal or chemical stability in which no consideration can be given to the vectorial forces actually exerted on the virus particle.

18.9.5 Structural Determinants of the Mechanical Properties of Virus Particles

As viral capsids are all made of protein, what structural differences underlie their differences in intrinsic elasticity, mechanical strength, brittleness, or fatigue resistance? Neither the capsid protein subunits themselves nor the lattice formed by these subunits in the viral capsid constitute a mechanically homogeneous material. The intrinsic elasticity and/or the strength of a viral capsid may depend on the number, distribution, directionality, type, and energy of the covalent or non-covalent interactions between atomic groups and amino acid residues within the same capsid protein subunit or between neighboring subunits.

Covalent interactions in a virus particle include the linkages between chemical groups and residues in each polypeptide chain and, in a few capsids, disulfide bonds or other covalent crosslinks between amino acid residues. Non-covalent interactions are extremely numerous and include very different types, such as van der Waals contacts, hydrogen bonds, and ionic interactions, and the entropically driven hydrophobic effect may also play a major role in particle elasticity and/or strength. These considerations have been invoked to explain, for example, the pH-dependent reduction in stiffness of the CCMV capsid [108, 109]; the large increase in stiffness of the phage HK97 capsid during its initial maturation stage (prohead-I to prohead-II) [110]; or the remarkable (up to 120%) increases in mechanical stiffness caused by a single amino acid substitution (per protein subunit) in the MVM capsid [85].

What specific structural differences at the atomic level are responsible for the observed mechanical differences between virus particles? Theoretical analyses of virus mechanics have been undertaken to propose fundamental physical explanations of different experimentally observed mechanical features of virus particles. In addition, theoretical analyses and modeling are leading to predictions related to the fundamental mechanical properties of viral particles and their materials foundations (e.g., [76]), many of which have not been experimentally addressed yet. Many such analyses are based on elasticity theory using simplified models, finite element or coarse-grained molecular dynamics (MD) simulations (Chap. 19). In addition, AFM-based indentation experiments are addressing the identification at atomic level of some structural determinants of the mechanical properties of virus particles, as briefly summarized next.

An experimental structure–function relationship approach has been followed to investigate, at different levels of resolution, the structural foundations of mechanical features in viral capsids. In this type of study, a difference in mechanical behavior may be traced to a structural difference between two similar capsids. So far, most of the virus structure–mechanics comparisons of viral particles have been made between structurally quite different particles (Sect. 18.8). However, large structural differences make it very difficult to identify at close to atomic resolution the changes in structural elements or specific substitutions of amino acid residues responsible for the mechanical differences observed.

Ideally, the higher the structural similarity between the capsids compared, the better the chance to find the true structural foundation of any change in a physical property or function. Either two evolutionarily very closely related capsids can be compared or, better yet, rationally chosen single amino acid substitutions can be introduced by site-directed mutagenesis in a given viral capsid, and their effect on capsid structure and mechanics analyzed.

The effects of the substitution of single amino acids in the capsid proteins on the mechanical properties of virus particles have begun to be investigated. The results so far have revealed that single amino acid replacements may result in very small changes in the equilibrium (minimum free energy) atomic structure of the viral particle. Such minute structural changes can, however, be translated into substantial changes in the number and/or energy of intersubunit interactions, leading to remarkable changes in the mechanical strength of the particle [90, 111]. Moreover, small changes in many atomic positions and interatomic interaction energies may be translated into very large local or global changes in the conformational dynamics of the viral particle, leading to large changes in stiffness and intrinsic elasticity. Many specific amino acid residues in the MVM capsid (Fig. 18.14) have been found to control either the local or global stiffness or the mechanical strength of the viral particle [85, 86, 94–96, 105].

18.10 Mechanical Properties of Viruses as Phenotypic Traits for Biological Adaptation

The solid-state physicist and the structural biologist/chemist may appreciate the value of the results being obtained on the mechanical properties of viruses for a better understanding of "soft" condensed matter and self-organizing biomolecular assemblies. However, the virologist may question whether the mechanical characterization of virus particles is relevant from a biological standpoint.

Previous sections have reviewed some of the results that demonstrate the relevance of analyzing the mechanical properties of virus particles as probes to investigate structural, thermodynamic or kinetic aspects of processes involved in the viral infectious cycle, even if viruses were not subjected to mechanical forces in vivo. In this section, we review some evidence in support of the hypothesis that the mechanical properties of viruses may have been directly shaped by evolution for survival when subjected to mechanical forces, or even for using those forces during the infection process.

Are viruses actually subjected to mechanical stress that may select for mechanically more robust variants? Do virus particles require a certain mechanical elasticity to complete some step in their life cycle? Indeed, virions and their capsids are subjected to physical forces that may deform or break them either outside or inside cells. Depending on the virus species and the stage of the virus cycle being considered, such forces may include: shear forces while drifting in viscous fluids, or while being extruded through nuclear pores or other tight openings; osmotic effects; pressurization due to tightly packaged dsDNA; hydrostatic pressure; capillary forces on particle desiccation; physical deformation of both non-enveloped virions during binding to multiple cell receptors and internalization, or of enveloped virions during their fusion to cell membranes; forces generated inside a virus particle during in situ genome replication or transcription; etc. Viruses with a tubular, very elongated shape, such as tobacco mosaic virus (Fig. 18.12) may be particularly sensitive to mechanical stress, but any virus may seriously risk mechanical disruption at some point or other in its existence.

In addition to being subjected to relatively strong point forces, many virus particles in nature may face weaker, but repeated mechanical forces that could lead to material fatigue. For example, through frequent collisions with other macromolecules in crowded extracellular or intracellular environments [112], or by repeated force strokes during their intracellular transport by molecular motors [69, 113].

Moreover, some virus particles must change its shape (undergo conformational rearrangements) to fulfill its biological function (see Chaps. 1, 13, 15); thus, their ability to deform or withstand a certain deformation may confer them with a selective advantage.

However, one must be cautious before accepting that the response of any specific virus particle to mechanical force in vitro constitutes a direct biological adaptation to a selective pressure exerted by some mechanical force in vivo. Some forces on virus particles during the infectious cycle may be too weak to exert any significant selective pressure; the particles could be intrinsically strong, stiff, or elastic enough to be infectious without a need for adaptation of their mechanical properties through mutation and selection. Moreover, as mentioned above, changes in the mechanical properties of a virus particle may frequently occur as a consequence of biologically relevant, effector-mediated changes in structure or dynamics, without a need for a mechanical force acting directly on the viral particle. It must be noted that non-natural effectors (such as laboratory-designed mutations or synthetic cargos) did modulate in vitro, in the absence of any biological selection, the mechanical properties of virus capsids, or even of non-viral protein cages [41, 114–117].

In the face of such considerations, several experimental studies have directly addressed the question of whether the mechanical strength or stiffness showed by a virus particle could be the result of biological adaptation in response to mechanical forces acting on that particle in vivo. The results of some of those studies, summarized next, provide strong support for an affirmative answer to that question.

18.10.1 Mechanical Strengthening to Withstand Pressurization for Genome Injection: Tailed Phages and HSV-1

The dsDNA genomes of tailed phages Φ29, λ, HK97, P22, and T7, and that of HSV-1, are actively packaged inside a preformed capsid that is barely large enough to contain the hydrated full-length nucleic acid molecule. As a consequence of the very high packing density, the dsDNA can exert a very high pressure (tens of atmospheres) on the inner capsid wall [46, 47, 118–121]. Such a high internal pressure will tend to disrupt the capsid, and one may ask why those viruses have not alleviated that pressure by evolving, for example, a slightly shortened DNA molecule or a slightly larger mature capsid.

In fact, pressurization in mature-tailed phages has a biological function by driving injection of their DNA genomes into the host bacterium [118, 122–124] (Sect. 18.5.2 and Chaps. 17 and 19). Likewise, pressurization of HSV-1 facilitates injection of its dsDNA into the nucleus through the nuclear pore during infection [120, 125]. The biological requirement for high internal pressure in those viruses imposes, in turn, a selective pressure for evolving a mechanically strong capsid.

During maturation (Chap. 13), the capsids of tailed phages and HSV-1 are mechanically strengthened [38, 44, 50, 110, 112, 126–130]. Evolutionary strategies for strengthening are, however, different. They include the introduction of prestress in the capsid (Φ29; [6]), or external reinforcements through either a chainmail of protein crosslinks (HK97; [110]) or the binding of auxiliary (cementing) proteins (λ, [38]; P22, [127, 129]; HSV-1, [126, 128, 130, 131]).

To sum up, strong evidence indicates that dsDNA-tailed phages and HSV-1 have evolved a pressurized virion to drive genome injection during infection of the host cell; this, in turn, led to the evolution of a mechanically strong enough mature capsid able to withstand the high internal pressure exerted by the dsDNA genome and required for infection.

18.10.2 Balanced Mechanical Strength for Pressure-Mediated Capsid Disassembly and Genome Uncoating: AdV

Studies on the mechanical behavior of AdV during virion assembly and maturation and genome packaging, or during its regulated disassembly and uncoating of its dsDNA genome have been reviewed and referenced in detail in Sect. 18.6. Those studies indicate that AdV has evolved a mechanically carefully balanced capsid, strong enough to maintain virion integrity during the extracellular stage, but weak enough to allow its pressure-induced gradual disassembly to release the viral genome into the cell nucleus.

18.10.3 Mechanical Softening to Facilitate Entry into Cells: HIV-1

Infection by retroviruses such as murine leukemia virus (MLV) and HIV-1 involves the internalization of the viral core (the capsid with the RNP complex inside) through fusion of the virion lipidic envelope and the cell membrane. Membrane fusion must involve a mechanical pulling force that results from a dramatic conformational rearrangement of the Env protein attached to the viral envelope. This mechanism requires the viral and cellular membranes to be brought close together and deformed enough to allow their fusion.

Non-infectious immature MLV and HIV-1 virions enter host cells very inefficiently compared with the mature, infectious virions. Interestingly, maturation makes these viruses much less stiff [132, 133]. Immature HIV-1 virions in which the Ct domain of Env had been deleted were nearly as elastic as mature virions, and were also able to fuse their membrane and enter host cells [133]. Immature HIV-1 virions that still contained the Env Ct domain, but lacked the rest of this protein, were as stiff as the immature virion, and also incompetent for cell entry [134]. Analysis of immature HIV-1 virions that incorporated a fixed amount of Env protein without the Ct domain (to enable cell entry) and that contained increasing amounts of the Env Ct domain revealed an inverse correlation between Env Ct-mediated stiffening of the HIV-1 virion and competence for entry into host cells [134]. These results led the authors to propose that the decreased stiffness of the HIV-1 virion during maturation would lower the free energy required to deform the viral envelope enough to enable its fusion with the cell membrane [83, 134].

To sum up, strong evidence indicates that the reduced stiffness of the mature HIV-1 virion may be a biological adaptation that facilitates the mechanically driven membrane fusion process required for infection.

The high mechanical elasticity of a very different enveloped virus, influenza virus, suggests that enough mechanical elasticity may be required for fusion of other enveloped viruses as well.

NV, a non-enveloped virus, must withstand the acidic conditions encountered during passage through the stomach, but must be able to dissociate to release its genome at weakly basic conditions during infection within the ileum. Remarkably, the NV capsid was found stiffer at neutral and acidic pH and less rigid at alkaline pH [135]. The authors suggested that the pH-dependent mechanical softening of the NV capsid could be a result of a putative, pH-triggered conformational rearrangement needed for infection, or even a direct requirement to facilitate virus entry and RNA release. If this hypothesis is correct, mechanical features could have a biologically critical role not only during entry of enveloped virions into cells through membrane fusion but also during entry of at least some non-enveloped viruses through other mechanisms (see Chap. 15).

18.10.4 Ligand-Mediated Balanced Strength for in-Capsid Genome Replication and Controlled Capsid Disassembly: HIV-1

During infection by HIV-1, its ssRNA genome is reverse transcribed inside the mature capsid (Fig. 18.12, bottom right). Binding of the cellular factor IP6 stabilizes the capsid and increases endogenous reverse transcription (ERT) efficiency. As ERT progresses the capsid is destabilized and the genome is uncoated [136].

AFM analysis revealed a transient stiffening of the HIV-1 core as ERT progressed. Peaks of stiffness of the HIV-1 core bound to IP6 during ERT were associated to specific stages of viral DNA synthesis in the viral core. These stiffness peaks suggested the action of transient point forces that would push on the inner capsid wall. Further incubation led to a reduction in core stiffness, eventual particle disassembly, and genome uncoating [87, 136, 137].

These and other results led the authors to propose a model in which transient mechanical forces generated during ERT induce localized cracks in the mature HIV-1 capsid, which eventually lead to the partial disruption of the capsid [136, 137]. According to this model, IP6 binding to the mature HIV-1 capsid may constitute an adaptive trait that (among other infectivity-promoting effects) modulates the strength of the viral capsid to enhance virus infectivity.

To sum up, the mechanical strength of HIV-1 capsid appears to have been finely tuned by binding the proviral factor IP6 to preserve its integrity during ERT, while still allowing its mechanically induced disruption for uncoating when the reverse transcription process is completed [83, 136].

18.10.5 Anisotropic Stiffening to Prevent Inactivation Without Impairing Infection: MVM

As mentioned in Sect. 18.8.3, the MVM virion is mechanically anisotropic. Segments of the viral ssDNA bound to equivalent sites at the capsid inner wall (Fig. 18.13) act like molecular buttresses that increase the mechanical stiffness of the capsid regions (S2 and S3) close to the DNA-binding sites. In contrast, the regions (S5) surrounding the capsid pores are free from bound DNA, and remain mechanically as elastic in the DNA-filled virion as in the DNA-free capsid. Is this peculiar anisotropic distribution of mechanical stiffness in the virion biologically irrelevant, or has it evolved through mutation and selection because it provides some advantage for virus survival?

During MVM morphogenesis the ssDNA genome is actively packaged in a preassembled capsid through a capsid pore, and the N-terminal (Nt) peptide segments of many VP2 capsid subunits are externalized through other pores. In vitro, the VP2 Nts can be externalized in the absence of the viral DNA by supplying external energy through moderate heating or by chemical action [101]. The results suggest that, during genome encapsidation, the ssDNA molecule fills the capsid and exerts enough mechanical force to push the Nt peptide segments out of the capsid.

Uncoating of the MVM ssDNA genome also occurs through a capsid pore, and can be promoted by moderate heating [106] or mechanical force [138]. If, during propagation between hosts, the virion encounters a warm enough environment, the genome may be prematurely released, and infectivity will be lost.

The MVM virion is, thus, under antagonic selective pressures. It must be conformationally stable enough to impair the transition that leads to the release of its genome before infecting a cell. However, it must also be conformationally unstable enough to allow the transition associated to the externalization of the VP2 Nt peptide signals required for infectivity.

A mechanical model of how MVM may have adapted to those conflicting selective pressures is supported by the results reviewed in Sect. 18.9.3. MVM appears to have evolved by positive selection ssDNA-binding sites at the capsid inner wall, close to S2 and S3 regions (Fig. 18.13). The capsid-bound ssDNA segments mechanically stiffen the S2/S3 regions and impair the transition that could lead to DNA release before infecting a host cell. The probability of the virion surviving the transit between hosts is, thus, increased. In turn, the capsid S5 regions around the pores have been kept free of bound DNA by negative selection, thus maintaining a comparatively low stiffness, and allowing the conformational transition required for externalization of infectivity-determining peptide signals through the capsid pores (which, in vivo, would be mechanically induced during genome packaging by the DNA molecule as it fills the capsid interior).

To sum up, strong evidence indicates that the anisotropic distribution of stiffness in the MVM virion may be the result of biological adaptation. Some capsid regions evolved to become stiff enough to impair a heat-induced transition that may lead to untimely viral nucleic acid release; other capsid regions were kept elastic enough to allow, during genome encapsidation, a mechanically induced transition that leads to externalization of peptide signals required for virus infectivity [77, 105].

18.11 Engineering Mechanical Properties of Virus Particles

Research on the mechanical properties of virus particles could be relevant also from an applied perspective. As mentioned above, some small organic compounds inhibit virus infectivity by modifying the stiffness or mechanical strength of the virus particles [88, 96, 97, 139]. This discovery may lead to the development of novel antiviral drugs based on the interference with the biologically relevant mechanical behavior of specific viruses.

In addition, there is a rapidly growing interest in the use of virus particles and virus-based nanostructured materials and devices for different biomedical, biotechnological, or nanotechnological applications (see Chaps. 21 and 22; [140]). Unfortunately, virus particles and other soft protein-based materials may be disrupted under the mechanical stress they may experience during purification, fabrication, storage, and/or use. In addition, many viral particles are metastable and particularly sensitive to conformational rearrangements (see Chaps. 1 and 14). Conformational instability of a virus particle may be desirable for some applications but not others. For example, virus particles for gene therapy or targeted drug delivery must be capable of undergoing structural rearrangements required to deliver their cargo into target cells. For many other potential applications including contrast agents, diagnostic components, vaccines, building blocks for nanomaterials, and others, conformational flexibility may be detrimental because it may lead to loss of function. In such cases, one may want to consider virus-based particles that, in addition to the highest possible structural strength and resistance to fatigue would have also the highest possible mechanical stiffness.

It may, thus, be important to acquire the knowledge needed to be able to genetically engineer the mechanical strength, resistance to fatigue, or stiffness of virus particles (and other protein-based supramolecular materials; [141]) that would make them more suitable for some intended applications.

The studies undertaken to determine the mechanical properties of virus particles, their modulation by physical, (bio)chemical effectors or genetic mutations, and their biological relevance, reviewed in the previous sections, have yielded as a by-product a number of quite different virus particles with substantially modified stiffness or mechanical strength.

MVM and HRV particles with increased stiffness were engineered in the laboratory through the introduction of rationally chosen non-natural point mutations [84–86, 105]. Because the primary goal of those studies was the high-resolution exploration of the molecular determinants of the mechanical elasticity and conformational flexibility of virus particles and their biological relevance most, though not all, of the tested mutations, led to increased particle stiffness. Some other point mutations, however, have resulted in significantly decreased stiffness ([84]; Escrig et al., unpublished results).

Removal of intersubunit interactions in the MVM capsid through individual amino acid substitutions has generally resulted in reduced mechanical strength [90], as was expected. However, the results pointed to a complex relationship between changes in local interactions and changes in the global strength of a virus capsid. Increased mechanical strength was obtained in nature as a result of a point mutation in a salt-stable CCMV mutant [111]. Mechanical strengthening of viral particles has also been achieved by nature as a consequence of large structural modifications including the binding of auxiliary (cementing) proteins in phage λ [38] and other tailed phages, or the establishment of a chainmail of covalent crosslinks between capsid subunits during maturation of phage HK97 [110]. The discovery that, during maturation, some viruses are mechanically strengthened by intersubunit covalent bonds inspired the rational engineering of a mechanically improved (viral) protein-based material. A complete network of intersubunit disulfide bonds were genetically introduced in the mature HIV-1 capsid protein lattice. The result was a chain mail-like protein-based nanostructured bidimensional material with genetically highly improved thermostability and resistance to mechanical stress and material fatigue, with no unwanted stiffening [107] (see Chap. 22).

18.12 Perspectives and Conclusions

The first study on the mechanical properties of a virus (the rod-like TMV) in air was published 25 years ago [142]. Six years later, the first study on the mechanical properties of a virus (phage Φ29) in a liquid [6] set the now-established procedure for the study of virus mechanics under physiologically relevant conditions. In the last two decades, different mechanical features of virions and/or capsids from about two dozen different virus species (Fig. 18.12) have been experimentally characterized. Stiffness, intrinsic elasticity, brittleness, mechanical strength and resistance to material fatigue have been defined, quantified, and compared. The modulation of the mechanical properties of virus particles by a large variety of virus infectivity-determining physical or (bio)chemical effectors has been analyzed. The structural foundations for the mechanical properties of virus particles, and their variation in response to mutation or external agents, have begun to be explored.

For some bacteriophages including Φ29, λ, HK97, and other tailed phages, and for several animal viruses including HSV-1, AdV, HIV-1, and MVM, several series of connected researches have been pursued over the years. The results obtained with these or with a fair number of other virus species have shown that biologically relevant, effector-mediated changes in the structure or dynamics of virions or capsids can be sensitively probed by AFM through the detection of changes in their mechanical properties. Those results, combined with those obtained using other experimental or theoretical physical, chemical, or biological approaches, are improving our understanding of the physicochemical mechanisms that govern different stages of the infectious cycle: virion assembly, maturation, exit from the host cell, survival in the extracellular environment, entry into a new cell, intracellular trafficking, virion disassembly, and genome release.

Moreover, and most importantly, several studies have provided strong evidence for the view that the mechanical properties of (at least some) virions or their capsids may be the result of biological adaptation in response to mechanical forces directly acting on those particles in nature. Mechanical strengthening achieved through the establishment of built-in stress, binding of auxiliary proteins, or covalent crosslinks between capsid subunits, for example, may help virus particles to withstand mechanical forces acting on them. dsDNA phages and HSV-1 may, respectively, use the internal pressure exerted by the tightly packaged nucleic acid to drive genome injection in the infected bacteria or in the eukaryotic cell nucleus, respectively. In order to withstand the biologically useful, but potentially deleterious internal pressure, they appear to have evolved a mechanically strong mature capsid. AdV seems to have evolved a capsid that is mechanically strong enough to maintain its integrity during the extracellular stage, but weak enough to allow controlled changes in internal pressure that facilitate its gradual disassembly in the cell nucleus during uncoating of the viral genome. Controlled softening of the retroviral particles during virus maturation could help entry into cells by facilitating virus–cell membrane fusion. The DNA-mediated anisotropic mechanical stiffness of the MVM virion reflects the unequal distribution of conformational dynamism in the virus particle, which may have arisen as a biological adaptation to better resist thermal inactivation outside the cell, without impairing infectivity inside the cell. The specific degree of stiffness, intrinsic elasticity, mechanical strength, brittleness, and/or fatigue resistance exhibited by a virus particle may have been tailored by evolution to withstand, or even benefit from the mechanical forces it may encounter during the infectious cycle.

The results obtained on virus mechanics are not only relevant in the field of virology for understanding virus structure, physics, and biology. They have also implications or applications in other scientific fields, including materials science and bio/nanotechnology.

From the point of view of materials science, not much is known so far about the mechanical properties of “soft” biological materials, and especially, natural or artificial protein-based nanostructured assemblies. Virus particles provide excellent models to understand the mechanical behavior of biological (soft) condensed matter. The analyses reveal both differences and limitations regarding mechanical characteristics that are already relatively well understood for other kinds of condensed matter. For example, mechanical characterization includes the estimation of some volumetric physical magnitudes such as the Young’s modulus, based on homogenous models of continuum mechanics. However, proteins and protein-based complexes and, especially, virus particles have very complex shapes, and are made of discrete subunits connected by networks of intersubunit interactions. They even involve the presence and interaction of different biological materials including protein, nucleic acid, and (in some cases) lipids, and can be regarded as composite materials. What specific structural elements, including individual amino acid residues in a capsid, and what specific interatomic interactions are responsible for elasticity, strength, or resistance to fatigue of a viral capsid or a protein-based supramolecular complex? What kind of “cracks,” in atomic terms, are responsible for the observed fatigue of such complex materials? Further experiments are required to gain insights into the relationship between the discontinuous and heterogeneous atomic structure of virus capsids and protein assemblies, and their mechanical properties.

From the point of view of bio/nanotechnology, studies on the molecular determinants of the mechanical stability of viral particles are starting to provide insights into the rational engineering of virus-based nanoparticles and nanostructured materials with increased physical resistance to unwanted structural deformations or disruption during their envisaged applications in medicine, biotechnology, or nanotechnology.

Acknowledgments We gratefully acknowledge J.L. Carrascosa, J.R. Castón, C. Catalano, R. Delgado-Buscalioni, T. Douglas, J. Gómez-Herrero, U. Greber, D. Luque, R. Podgornik, D. Reguera, W. Roos, C. San Martín, F. Sansbury, A. Valbuena, and N. Verdaguer for collaboration on virus mechanics, advice and fruitful discussions; M. Cantero, C. Carrasco, P.J.P. Carrillo, M. Castellanos, A. Díez-Martínez, S. Domínguez-Zotes, J. Escrig, A. García-Arribas M. Hernando-Pérez, P. Ibañez, M. Jiménez-Zaragoza, A. Llauró, N. Martín-González, M. Medrano, A. Ortega-Esteban, R. Pérez, MJ Rodríguez-Espinosa, K. Strobl, and A. Valbuena for excellent work on virus mechanics; and M.A. Fuertes and A. Rodríguez-Huete for great experimental assistance. Work in P.J.P.’s laboratory is currently funded by PID2021-126608OB-I00, the Human Frontiers Science Program (HFSPO RGP0012/2018), and an institutional grant from the Maria de Maeztu Programme for Units of Excellence. Work in M.G.M.’s laboratory is currently funded by MICINN/FEDER EU (grant PID2021-126973OB-I00), and by institutional grants from the Severo Ochoa Programme for Centers of Excellence and Fundación Ramón Areces. M.G.M. is an associate member of the Institute for Biocomputation and Physics of Complex Systems, Zaragoza, Spain.

References

1. Shiroguchi K, Kinosita K (2007) Myosin V walks by lever action and Brownian motion. Science 316:1208–1212
2. Moreno-Herrero F, de Jager M, Dekker NH, Kanaar R, Wyman C, Dekker C (2005) Mesoscale conformational changes in the DNA-repair complex Rad50/Mre11/Nbs1 upon binding DNA. Nature 437:440–443
3. Moody MF (1999) Geometry of phage head construction. J Mol Biol 293:401–433
4. Cordova A, Deserno M, Gelbart WM, Ben-Shaul A (2003) Osmotic shock and the strength of viral capsids. Biophys J 85:70–74
5. Binnig G, Quate CF, Gerber C (1986) Atomic force microscope. Phys Rev Lett 56:930–933
6. Ivanovska IL, de Pablo PJ, Ibarra B, Sgalari G, MacKintosh FC, Carrascosa JL, Schmidt CF, Wuite GJL (2004) Bacteriophage capsids: tough nanoshells with complex elastic properties. Proc Natl Acad Sci USA 101:7600–7605
7. Gittes F, Schmidt CF (1998) Thermal noise limitations on micromechanical experiments. Eur Biophys J Biophys Lett 27:75–81
8. Rief M, Oesterhelt F, Heymann B, Gaub HE (1997) Single molecule force spectroscopy on polysaccharides by atomic force microscopy. Science 275:1295–1297
9. Fisher TE, Carrión-Vázquez M, Oberhauser AF, Li H, Marszalek PE, Fernández JM (2000) Single molecule force spectroscopy of modular proteins in the nervous system. Neuron 27: 435–446
10. Svoboda K, Schmidt CF, Schnapp BJ, Block SM (1993) Direct observation of kinesin stepping by optical trapping interferometry. Nature 365:721–727
11. Alcaraz J, Buscemi L, Grabulosa M, Trepat X, Fabry B, Farre R, Navajas D (2003) Microrheology of human lung epithelial cells measured by atomic force microscopy. Biophys J 84:2071–2079
12. de Pablo PJ, Schaap IAT, MacKintosh FC, Schmidt CF (2003) Deformation and collapse of microtubules on the nanometer scale. Phys Rev Lett 91:98101
13. Smith DE, Tans SJ, Smith SB, Grimes S, Anderson DL, Bustamante C (2001) The bacteriophage ϕ29 portal motor can package DNA against a large internal force. Nature 413:748–752
14. Ortega-Esteban A, Horcas I, Hernando-Pérez M, Ares P, Pérez-Berná AJ, San Martín C, Carrascosa JL, de Pablo PJ, Gómez-Herrero J (2012) Minimizing tip-sample forces in jumping mode atomic force microscopy in liquid. Ultramicroscopy 114:56–61
15. Horcas I, Fernandez R, Gomez-Rodriguez JM, Colchero J, Gomez-Herrero J, Baro AM (2007) WSXM: a software for scanning probe microscopy and a tool for nanotechnology. Rev Sci Instrum 78:013705
16. Sader JE, Chon JWM, Mulvaney P (1999) Calibration of rectangular atomic force microscope cantilevers. Rev Sci Instrum 70:3967–3969
17. de Pablo PJ, Gomez-Navarro C, Martinez MT, Benito AM, Maser WK, Colchero J, Gomez-Herrero J, Baro AM (2002) Performing current versus voltage measurements of single-walled carbon nanotubes using scanning force microscopy. Appl Phys Lett 80:1462–1464
18. Landau LD, Lifshizt E (1986) Theory of elasticity. Pergamon, London
19. Derjaguin B, Landau L (1993) Theory of the stability of strongly charged lyophobic sols and of the adhesion of strongly charged particles in solutions of electrolytes. Prog Surf Sci 43:30–59
20. Israelachvili J (2002) Intermolecular and surface forces. Academic Press, London
21. Jiménez-Zaragoza M, Yubero MPL, Martín-Forero E, Castón JR, Reguera D, Luque D, de Pablo PJ, Rodríguez JM (2018) Biophysical properties of single rotavirus particles account for the functions of protein shells in a multilayered virus. elife 7:e37295
22. Cantero M, Cvirkaite-Kuprovic V, Kruprovic M, de Pablo PJ (2023) Mechanical tomography of an archeal lemon-shaped virus reveals membrane-like fluidity of the capsid and liquid nucleoprotein cargo. Proc Natl Acad Sci USA 120:e2307717120

23. Cantero M, Rodríguez-Espinosa MJ, Strobl K, Ibáñez P, Díez-Martinez A, Martín-González N, Jiménez-Zaragoza M, Ortega-Esteban A, de Pablo PJ (2023) Atomic force microscopy of viruses: stability, disassembly, and genome release. In: Heller I, Dulin D, Peterman EJ (eds) Single molecule analysis, Methods Mol Biol, vol 2694, pp 317–338
24. Zink M, Grubmuller H (2009) Mechanical properties of the icosahedral shell of southern bean mosaic virus: a molecular dynamics study. Biophys J 96:1350–1363
25. Sokolnikoff IS (1983) Mathematical theory of elasticity. Kriger Pub. Co., Malabar, Florida
26. Greaves GN, Greer AL, Lakes RS, Rouxel T (2011) Poisson's ratio and modern materials. Nat Mater 10:823–837
27. Helfer E, Harlepp S, Bourdieu L, Robert J, MacKintosh FC, Chatenay D (2001) Buckling of actin-coated membranes under application of a local force. Phys Rev Lett 87:088103
28. Niordson FI (1985) Shell theory. North Holland, Amsterdam
29. Hutton DV (2004) Fundamentals on finite element analysis. McGraw Hill, New York
30. Settembre et al (2011) 1 E. C. Settembre, J. Z. Chen, P. R. Dormitzer, N. Grigorieff and S. C. Harrison. EMBO J 30:408–416
31. Kiss B, Kis Z, Pályi B, Kellermayer MSZ (2021) Topography, spike dynamics, and nanomechanics of individual native SARS-CoV-2 virions. Nano Lett 21:2675–2680
32. Cantero M, Carlero D, Chichón FJ, Martín-Benito J, de Pablo PJ (2022) Monitoring SARS-CoV-2 surrogate TGEV individual virions structure survival under harsh physicochemical environments. Cells 11:1759
33. Martín-González N, Gómez-González A, Hernando-Pérez M, Bauer M, Greber UF, San Martín C, de Pablo PJ (2023) Adenovirus core protein V reinforces the capsid and enhances genome release from disrupted particles. Sci Adv 9:eade9910
34. Ortega-Esteban A, Mata CP, Rodríguez-Espinosa MJ, Luque D, Irigoyen N, Rodríguez JM, de Pablo PJ, Castón JR (2020) Cryo-electron microscopy structure, assembly, and mechanics show morphogenesis and evolution of human picobirnavirus. J Virol 94:e01542–e01520
35. Kellermayer M, Smith SB, Bustamante C, Granzier HL (2001) Mechanical fatigue in repetitively stretched fibers of titin. Biophys J 80:852–863
36. Schijve J (2009) Fatigue of structures and materials. Kluwer Academic Publishers, Dordrecht
37. Ortega-Esteban A, Pérez-Berná AJ, Menéndez-Conejero R, Flint SJ, San Martín C, de Pablo PJ (2013) Monitoring dynamics of human adenovirus disassembly induced by mechanical fatigue. Sci Rep 3:1434
38. Hernando-Pérez M, Lambert S, Nakatani-Webster E, Catalano CE, de Pablo PJ (2014) Cementing proteins provide extra mechanical stabilization to viral cages. Nat Commun 5:4520
39. Rodríguez-Espinosa MJ, Rodríguez JM, Castón JR, de Pablo PJ (2023) Mechanical disassembly of human picobirnavirus like particles indicates that cargo retention is tuned by the RNA-coat protein interaction. Nanoscale Horizons 8:1665
40. Schaap IAT, Carrasco, de Pablo PJ, MacKintosh FC, Schmidt CF (2006) Elastic response, buckling, and instability of microtubules under radial indentation. Biophys J 91:1521–1531
41. Llauró A, Guerra P, Irigoyen N, Rodríguez JF, Verdaguer N, de Pablo PJ (2014) Mechanical stability and reversible fracture of vault particles. Biophys J 106:687–695
42. Valbuena A, Mateu MG (2015) Quantification and modification of the equilibrium dynamics and mechanics of a viral capsid lattice, self-assembled as a protein nanocoating. Nanoscale 7: 14953–14964
43. de Pablo PJ, Hernando-Pérez M, Carrasco C, Carrascosa JL (2018) Direct visualization of single virus restoration after damage in real time. J Biol Phys 44:225–235
44. Vörös Z, Csík G, Herényi L, Kellermayer MSZ (2017) Stepwise reversible nanomechanical buckling in a viral capsid. Nanoscale 9:1136–1143
45. Wang F, Cvirkaite-Krupovic V, Vos M, Beltran LC, Kreutzberger MAB, Winter J-M, Su Z, Liu J, Schouten S, Krupovic M, Egelman EH (2022) Spindle-shaped archeal viruses evolved from rod-shaped ancestors to package a larger genome. Cell 185:1297–1307

46. Hernando-Pérez M, Miranda R, Aznar M, Carrascosa JL, Schaap IAT, Reguera D, de Pablo PJ (2012) Direct measurement of phage ϕ29 stiffness provides evidence of internal pressure. Small 8:2366–2370
47. Carrasco C, Luque A, Hernando-Pérez M, Miranda R, Carrascosa JL, Serena PA, de Ridder M, Raman A, Gómez-Herrero J, Schaap IAT, Reguera D, de Pablo PJ (2011) Built-in mechanical stress in viral shells. Biophys J 100:1100–1108
48. Tao YZ, Olson NH, Xu W, Anderson DL, Rossmann MG, Baker TS (1998) Assembly of a tailed bacterial virus and its genome release studied in three dimensions. Cell 95:431–437
49. Ivanovska IL, Miranda R, Carrascosa JL, Wuite GJL, Schmidt CF (2011) Discrete fracture patterns of virus shells reveal mechanical building blocks. Proc Natl Acad Sci USA 108: 12611–12616
50. Roos WH, Radtke K, Kniesmeijer E, Geertsema H, Sodeik B, Wuite GJL (2009) Scaffold expulsion and genome packaging trigger stabilization of herpes simplex virus capsids. Proc Natl Acad Sci USA 106:9673–9678
51. Gonzalez-Huici V, Salas M, Hermoso JM (2004) The push-pull mechanism of bacteriophage ϕ29 DNA injection. Mol Microbiol 52:529–540
52. Podgornik R, Leforestier A, Siber A, Livolant F (2011) Protein-DNA interactions determine the shapes of DNA toroids condensed in virus capsids. Biophys J 100:2209–2216
53. Purohit PK, Kondev J, Phillips R (2003) Mechanics of DNA packaging in viruses. Proc Natl Acad Sci USA 100:3173–3178
54. Zandi R, van der Schoot P, Reguera D, Kegel W, Reiss H (2006) Classical nucleation theory of virus capsids. Biophys J 90:1939–1948
55. Choi KH, Morais MC, Anderson DL, Rossman MG (2006) Determinants of bacteriophage ϕ29 head morphology. Structure 14:1723–1727
56. Dokland T (1999) Scaffolding proteins and their role in viral assembly. Cell Mol Life Sci 56: 580–603
57. Zlotnick A (2003) Are weak protein-protein interactions the general rule in capsid assembly? Virology 315:269–274
58. Butt HJ (1991) Electrostatic interaction in atomic force microscopy. Biophys J 60:777–785
59. Sotres J, Baro AM (2010) AFM imaging and analysis of electrostatic double layer forces on single DNA molecules. Biophys J 98:1995–2004
60. Almonte L, Lopez-Elvira E, Baró AM (2014) Surface-charge differentiation of streptavidin and avidin by atomic force microscopy-force spectroscopy. ChemPhysChem 15:2768–2773
61. Hernando-Pérez M, Cartagena-Rivera AX, Bozic AL, Carrillo PJP, San Martín C, Mateu MG, Raman A, Podgornik R, de Pablo PJ (2015) Quantitative nanoscale electrostatic of viruses. Nanoscale 7:17289–17298
62. San Martín C (2012) Latest insights on adenovirus structure and assembly. Viruses 4:847–877
63. Ortega-Esteban A, Bodensiek K, San Martín C, Suomalainen M, Greber UF, de Pablo PJ, Schaap IAT (2015) Fluorescence tracking of genome release during mechanical unpacking of single viruses. ACS Nano 9:10571–10579
64. de Pablo PJ, San Martín C (2022) Seeing and touching adenovirus: complementary approaches for understanding assembly and disassembly of a complex virion. Curr Opin Virol 52:112–122
65. Pérez-Berná AJ, Ortega-Esteban A, Menéndez-Conejero R, Winkler DC, Menéndez M, Steven AC, Flint SJ, de Pablo PJ, San Martín C (2012) The role of capsid maturation on adenovirus priming for sequential uncoating. J Biol Chem 287:31582–31595
66. Ortega-Esteban A, Condezo GN, Pérez-Berná AJ, Chillón M, Flint SJ, Reguera D, San Martín C, de Pablo PJ (2015) Mechanics of viral chromatin reveals the pressurization of human adenovirus. ACS Nano 11:10826–10833
67. Martín-González N, Hernando-Pérez M, Condezo GN, Pérez-Illana M, Siber A, Reguera D, Ostapchuk P, Hearing P, San Martín C, de Pablo PJ (2019) Adenovirus major core protein condenses DNA in clusters and bundles, modulating genome release and capsid internal pressure. Nucleic Acids Res 47:9231–9242

68. Greber UF, Willets M, Webster P, Helenius A (1993) Stepwise dismantling of adenovirus 2 during entry into cells. Cell 75:477–486
69. Martín-González N, Ibáñez-Freire P, Ortega-Esteban A, Laguna-Castro M, San Martín C, Valbuena A, Delgado-Buscalioni R, de Pablo PJ (2021) Long-range cooperative disassembly and aging during adenovirus uncoating. Phys Rev X 11:021025
70. Yamauchi Y, Greber UF (2016) Principles of virus uncoating: clues and the snooker ball. Traffic 17:569–592
71. Montiel-Garcia D, Santoyo-Rivera N, Ho P, Carrillo-Tripp M, Johnson JE, Reddy VS (2021) VIPERdb v3. 0: a structure-based data analytics platform for viral capsids. Nucleic Acids Res 49:D809–D816
72. Wikoff WR, Johnson JE (1999) Virus assembly: imaging a molecular machine. Curr Biol 9: R296–R300
73. Briggs JAG, Grünewald K, Glass B, Förster F, Kräusslich H-G, Fuller SD (2006) The mechanism of HIV-1 core assembly: insights from three-dimensional reconstructions of authentic virions. Structure 14:15–20
74. Roos WH, Ivanovska IL, Evilevitch A, Wuite GJL (2007) Viral capsids: mechanical characteristics, genome packaging and delivery mechanisms. Cell Mol Life Sci 64:1484–1497
75. Roos WH, Wuite GJL (2009) Nanoindentation studies reveal material properties of viruses. Adv Mater 21:1187–1192
76. Roos WH, Bruinsma R, Wuite GJL (2010) Physical virology. Nat Phys 6:733–743
77. Mateu MG (2012) Mechanical properties of viruses analyzed by atomic force microscopy: a virological perspective. Virus Res 168:1–22
78. Mateu MG (2023) Virus mechanics: a structure-based biological perspective. In: Comas-Garcia M, Rosales-Mendoza S (eds) Physical virology. Springer, From the state of the art to the future of applied virology, pp 237–282
79. Marchetti M, Wuite GJL, Roos WH (2016) Atomic force microscopy observation and characterization of single virions and virus-like particles by nano-indentation. Curr Opin Virol 18:82–88
80. de Pablo PJ (2018) Atomic force microscopy of virus shells. Sem Cell Dev Biol 73:199–208
81. de Pablo PJ (2019) The application of atomic force microscopy for viruses and protein shells: imaging and spectroscopy. Adv Virus Res 105:161–187
82. Buzón P, Maity S, Roos WH (2020) Physical virology: from virus self-assembly to particle mechanics. WIREs Nanomed Nanobiotechnol 12:e1613
83. Rousso I, Deshpande A (2022) Applications of atomic force microscopy in HIV-1 research. Viruses 14:648
84. Carrasco C, Castellanos M, de Pablo PJ, Mateu MG (2008) Manipulation of the mechanical properties of a virus by protein engineering. Proc Natl Acad Sci USA 105:4150–4155
85. Castellanos M, Pérez R, Carrasco C, Hernando-Pérez M, Gómez-Herrero J, de Pablo PJ, Mateu MG (2012) Mechanical elasticity as a physical signature of conformational dynamics in a virus particle. Proc Natl Acad Sci USA 109:12028–12033
86. Carrillo PJP, Medrano M, Valbuena A, Rodríguez-Huete A, Castellanos M, Pérez R, Mateu MG (2017) Amino acid side chains buried along intersubunit interfaces in a viral capsid preserve low mechanical stiffness associated with virus infectivity. ACS Nano 11:2194–2208
87. Rankovic S, Varadarajan J, Ramalho R, Aiken C, Rousso I (2017) Reverse transcription mechanically initiates HIV-1 capsid disassembly. J Virol 91:e00289–e00217
88. Rankovic S, Ramalho R, Aiken C, Rousso I (2018) PF74 reinforces the HIV-1 capsid to impair reverse transcription-induced uncoating. J Virol 92:e00845–e00818
89. Xu C, Fischer DK, Rankovic S, Li W, Dick RA, Runge B, Zadorozhnyi R, Ahn J, Aiken C, Polenova T, Engelman AN, Ambrose Z, Rousso I, Perilla JR (2020) Permeability of the HIV-1 capsid to metabolites modulates viral DNA synthesis. PLoS Biol 18:e3001015
90. Medrano M, Valbuena A, Rodríguez-Huete A, Mateu MG (2019) Structural determinants of mechanical resistance against breakage of a virus-based protein nanoparticle at a resolution of single amino acids. Nanoscale 11:9369–9383

91. Agbandje-McKenna M, Llamas-Saiz AL, Wang F, Tattersall P, Rossmann MG (1998) Functional implications of the structure of the murine parvovirus, minute virus of mice. Structure 6:1369–1381
92. Bothner B, Hilmer JK (2011) Probing viral capsids in solution. In: Agbandje-McKenna M, McKenna R (eds) Structural virology. RSC Publishing, Cambridge, pp 41–61
93. Worner TP, Shamorkina TM, Snijder J, Heck AJR (2021) Mass spectrometry-based structural virology. Anal Chem 93:620–640
94. Luque D, Ortega-Esteban A, Valbuena A, Vilas JL, Rodríguez-Huete A, Mateu MG, Castón JR (2023) Equlibrium dynamics of a biomolecular complex analyzed at single amino acid resolution by cryo-electron microscopy. J Mol Biol 435:168024
95. Guerra P, Valbuena A, Querol-Audí J, Silva C, Castellanos M, Rodríguez-Huete A, Garriga D, Mateu MG, Verdaguer N (2017) Structural basis for biologically relevant mechanical stiffening of a virus capsid by cavity-creating or spacefilling mutations. Sci Rep 7:4101
96. Valbuena A, Rodríguez-Huete A, Mateu MG (2018) Mechanical stiffening of human rhinovirus by cavity-filling antiviral drugs. Nanoscale 10:1440–1452
97. Domínguez-Zotes S, Valbuena A, Mateu MG (2022) Antiviral compounds modulate elasticity, strength and material fatigue of a virus capsid framework. Biophys J 121:919–931
98. Johnson JE (2003) Virus particle dynamics. Adv Protein Chem 64:197–218
99. Veesler D, Johnson JE (2012) Virus maturation. Annu Rev Biophys 41(473–496):2012
100. Sherman MB, Smith HQ, Smith TJ (2020) The dynamic life of virus capsids. Viruses 12:618
101. Valle N, Riolobos L, Almendral JM (2006) Synthesis, post-translational modification and trafficking of the parvovirus structural polypeptides. In: Kerr JR, Cotmore SF, Bloom ME, Linden RM, Parrish CR (eds) Parvoviruses. Edward Arnold, London, pp 291–304
102. Reguera J, Carreira A, Riolobos L, Almendral JM, Mateu MG (2004) Role of interfacial amino acid residues in assembly, stability, and conformation of a spherical virus capsid. Proc Natl Acad Sci USA 101:2724–2729
103. van de Waterbeemd M, Llauró A, Snijder J, Valbuena A, Rodríguez-Huete A, Fuertes MA, de Pablo PJ, Mateu MG, Heck AJR (2017) Structural analysis of a temperature-induced transition in a viral capsid probed by HDX-MS. Biophys J 112:1157–1165
104. Reguera J, Grueso E, Carreira A, Sánchez-Martínez C, Almendral JM, Mateu MG (2005) Functional relevance of amino acid residues involved in interactions with ordered nucleic acid in a spherical virus. J Biol Chem 280:17969–17977
105. Castellanos M, Carrillo PJP, Mateu MG (2015) Quantitatively probing propensity for structural transitions in engineered virus nanoparticles by single-molecule mechanical analysis. Nanoscale 7:5654–5664
106. Ros C, Baltzer C, Mani B, Kempf C (2006) Parvovirus uncoating in vitro reveals a mechanism of DNA release without capsid disassembly and striking differences in encapsidated DNA stability. Virology 345:137–147
107. Domínguez-Zotes S, Fuertes MA, Rodríguez-Huete A, Valbuena A, Mateu MG (2022) A genetically engineered, chain mail-like nanostructured protein material with increased fatigue resistance and enhanced self-healing. Small 2105456
108. Klug WS, Bruinsma RF, Michel J-P, Knobler CM, Ivanovska IL, Schmidt CF, Wuite GJL (2006) Failure of viral shells. Phys Rev Lett 97:228101
109. Wilts BD, Schaap IAT, Young M, Douglas T, Knobler CM, Schmidt CF (2010) Swelling and softening of the CCMV plant virus capsid in response to pH shifts. Biophys J 98:656a
110. Roos WH, Gertsman I, May ER, Brooks CL III, Johnson JE, Wuite GJL (2012) Mechanics of bacteriophage maturation. Proc Natl Acad Sci USA 109:2342–2347
111. Michel JP, Ivanovska IL, Gibbons MM, Klug WS, Knobler CM, Wuite GJL, Schmidt CF (2006) Nanoindentation studies of full and empty viral capsids and the effects of capsid protein mutations on elasticity and strength. Proc Natl Acad Sci USA 103:6184–6189
112. Hernando-Pérez M, Pascual E, Aznar M, Ionel A, Castón JR, Luque A, Carrascosa JL, Reguera D, de Pablo PJ (2014) The interplay between mechanics and stability of viral cages. Nanoscale 6:2702–2709

113. Liashkovich I, Hafezi W, Kühn JE, Oberleithner H, Kramer A, Shahin V (2008) Exceptional mechanical and structural stability of HSV-1 unveiled with fluid atomic force microscopy. J Cell Sci 121:2287–2292
114. Llauró A, Guerra P, Kant R, Bothner B, Verdaguer N, de Pablo PJ (2016) Decrease in pH destabilizes individual vault nanocages by weakening the inter-protein lateral interaction. Sci Rep 6:34143
115. Heinze K, Sasaki E, King NP, Baker D, Hilvert D, Wuite GJL, Roos WH (2016) J Phys Chem B 120:5945–5952
116. Snijder J, Kononova O, Barbu IM, Uetrecht C, Ryrup WF, Burnley RJ, Koay MST, Cornelissen JJLM, Roos WH, Barsegov V, Wuite GJL, Heck AJR (2016) Assembly and mechanical properties of the cargo-free and cargo-loaded bacterial nanocompartment encapsulin. Biomacromolecules 17:2522–2529
117. Guerra P, González-Alamos M, Llauró A, Casañas A, Querol-Audí J, de Pablo PJ, Verdaguer N (2022) Symmetry disruption commits vault particles to disassembly. Sci Adv 8:eabj7795
118. Evilevitch A, Lavelle L, Knobler CM, Raspaud E, Gelbart WM (2003) Osmotic pressure inhibition of DNA ejection from phage. Proc Natl Acad Sci USA 100:9292–9295
119. Ivanovska I, Wuite G, Jonsson B, Evilevitch A (2007) Internal DNA pressure modifies stability of WT phage. Proc Natl Acad Sci U SA 104:9603–9608
120. Bauer DW, Huffman JB, Homa FL, Evilevitch A (2013) Herpes virus genome, the pressure is on. J Am Chem Soc 135:11216–11221
121. Keller N, Berndsen ZT, Jardine PJ, Smith DE (2017) Experimental comparison of forces resisting viral DNA packaging and driving DNA ejection. Phys Rev E 95:052408
122. Jeembaeva M, Castelnovo M, Larsson F, Evilevitch A (2008) Osmotic pressure: resisting or promoting DNA ejection from phage? J Mol Biol 381:310–323
123. Köster S, Evilevitch A, Jeembaeva M, Weitz DA (2009) Influence of internal capsid pressure on viral infection by phage λ. Biophys J 97:1525–1529
124. Bauer DW, Li D, Huffman J, Homa FL, Wilson K, Leavitt JC, Casjens SR, Baines J, Evilevitch A (2015) Exploring the balance between DNA pressure and capsid stability in herpes viruses and phages. J Virol 89:9288–9298
125. Brandariz-Nuñez A, Robinson SJ, Evilevitch A (2020) Pressurized state inside herpes capsids-a novel antiviral target. PLoS Pathog 16:e1008604
126. Sae-Ueng U, Liu T, Catalano CE, Huffman JB, Homa FL, Evilevitch A (2016) Major capsid reinforcement by a minor protein in herpesviruses and phage. Nucleic Acids Res 42:9096–9107
127. Llauró A, Schwarz B, Koliyatt R, de Pablo, PJ, Douglas T: Tuning capsid nanoparticle stability with symmetrical morphogenesis. ACS Nano 10: 8465–8473
128. Snijder J, Radtke K, Anderson F, Scholtes L, Corradini E, Baines J, Heck AJR, Wuite GJL, Sodeik B, Roos WH (2017) Vertex-specific proteins pUL17 and pUL25 mechanically reinforce herpes simplex virus capsids. J Virol 91:e00123–e00117
129. Kant R, Llauró A, Rayaprolu V, Qazi S, de Pablo PJ, Douglas T, Bothner B (2018) Changes in the stability and biomechanics of P2 bacteriophage capsid during maturation. Biochem Biophys Acta 1862:1492–1504
130. Evilevitch A, Sae-Ueng U (2022) Mechanical capsid maturation facilitates the resolution of conflicting requirements for herpesvirus assembly. J Virol 96:e01831–e01821
131. Klug WS, Roos WH, Wuite GJL (2012) Unlocking internal prestress from protein nanoshells. Phys Rev Lett 109:168104
132. Kol N, Gladnikoff M, Barlam D, Shneck RZ, Rein A, Rousso I (2006) Mechanical properties of murine leukemia virus particles: effect of maturation. Biophys J 91:767–774
133. Kol N, Shi Y, Tsvitov M, Barlam D, Shneck RZ, Kay MS, Rousso I (2007) A stiffness switch in human immunodeficiency virus. Biophys J 92:1777–1783
134. Pang H-B, Hevroni L, Kol N, Eckert DM, Tsvitov M, Kay MS, Rousso I (2013) Virion stiffness regulates mature HIV-1 entry. Retrovirology 10:4

135. Cuellar JL, Meinhoevel F, Hoehne M, Donath E (2010) Size and mechanical stability of norovirus capsids depend on pH: a nanoindentation study. J Gen Virol 91:2449–2456
136. Aiken C, Rousso I (2021) The HIV-1 capsid and reverse transcription. Retrovirology 18:20
137. Rankovic S, Deshpande A, Harel S, Aiken C, Rousso I (2021) HIV-1 uncoating occurs via a series of rapid biomechanical changes in the core related to individual stages of reverse transcription. J Virol 95:e00166–e00121
138. Strobl K, Mateu MG, de Pablo PJ (2023) Exploring nucleic acid condensation and release from individual parvoviruses with different physicochemical cues. Virology 435:168024
139. Wang H, Chen Y, Zhang W (2019) A single-molecule atomic force microscopy study reveals the antiviral mechanism of tannin and its derivatives. Nanoscale 11:16368–16376
140. Mateu MG (2016) Assembly, engineering and applications of virus-based protein nanoparticles. In: Cortajarena AL, Grove TZ (eds) Protein-based engineered nanostructures. Springer, Switzerland, pp 83–120
141. Cortajarena AL, Grove TZ (eds) (2016) Protein-based engineered nanostructures. Springer, Switzerland
142. Falvo MR, Washburn S, Superfine R, Finch M, Brooks FP Jr, Chi V, Taylor RM II (1997) Manipulation of individual viruses: friction and mechanical properties. Biophys J 72:1396–1403

Further Reading

Comas-Garcia M, Rosales-Mendoza S (eds) (2023) Physical virology: from the state of the art to the future of applied virology. Springer
Kurland NE, Drira Z, Yadavalli VK (2012) Measurement of nanomechanical properties of biomolecules using atomic force microscopy. Micron 43:116–128
Roos WH (2011) How to perform a nanoindentation experiment on a virus. In: Peterman EJG, Wuite GJL (eds) Single molecule analysis: methods and protocols, methods in molecular biology, vol 783. Humana Press, Totowa, NJ, pp 251–264
Stockley PG, Twarock R (eds) (2010) Emerging topics in physical virology. Imperial College Press, London

Chapter 19
Theoretical Studies on Assembly, Physical Stability, and Dynamics of Viruses

Antoni Luque and David Reguera

Abstract All matter must obey the general laws of physics and living matter is not an exception. Viruses have not only learnt how to cope with them but have managed to use them for their own survival. In this chapter, we will review some of the exciting physics that are behind viruses and discuss simple physical models that can shed some light on different aspects of the viral life cycle and viral properties. In particular, we will focus on how the structure and shape of the viral capsid, its assembly and stability, and the entry and exit of viral particles and their genomes can be explained using fundamental physics theories.

Keywords Biophysics · Virus · Bacteriophage · Capsid · DNA · RNA · Budding · Statistical mechanics · Continuum elasticity theory · Nucleation · Self-assembly · Wrapping · Pressure · Electrostatics

Abbreviations

AFM	Atomic force microscopy
BMV	Brome mosaic virus
CBB	Capsid building block
CCMV	Cowpea chlorotic mottle virus
CK	Caspar and Klug
CNT	Classical nucleation theory
cryo-EM	Cryo-electron microscopy
DH	Debye-Hückel

A. Luque
Department of Biology, University of Miami, Coral Gables, FL, USA
e-mail: antoni.luque@miami.edu

D. Reguera (✉)
Department of Physics of the Condensed Matter, Universitat de Barcelona, Barcelona, Spain

Universitat de Barcelona Institute of Complex Systems (UBICS), Barcelona, Spain
e-mail: dreguera@ub.edu

M. G. Mateu (ed.), *Structure and Physics of Viruses*, Subcellular Biochemistry 105,
https://doi.org/10.1007/978-3-031-65187-8_19

dsDNA	Double-stranded DNA
FEA	Finite element analysis
FvK	Föppl-von Kármán number
HBV	Hepatitis B virus
HIV	Human immunodeficiency virus
HK97	Bacteriophage Hong Kong 97
HPV	Human polyoma virus
NMA	Normal mode analysis
ssRNA	Single-stranded RNA
STMV	Satellite tobacco mosaic virus
TMV	Tobacco Mosaic Virus
TST	Thin shell theory

19.1 Introduction

Viruses are an endless source of fascination for biophysicists. Contrary to most biological organisms, viral particles are made of a minimal number of relatively simple components that are not capable of any metabolic activity, except when their genome takes over the metabolism of the infected host to achieve the replication of new particles (see Chap. 1). Despite the lack of sophisticated biological machinery, viruses have found the way to efficiently infect the host, assemble, and egress the cell following, in many cases, a coordinated sequence of passive and spontaneous processes. This strongly suggests that, during their life cycle, viruses must rely on general physical and chemical mechanisms to succeed in their different tasks and to achieve the required resistance against possible extreme environmental conditions.

The assembly of viruses (Chaps. 11–14) provides a good example to illustrate the importance of physical mechanisms in the virus life cycle. In physiological conditions, viral capsids assemble directly from their building blocks, which involve the participation of the capsid proteins, the genetic material and, in some cases, auxiliary proteins. Indeed, it has been shown that many viruses can be reconstituted in a fully infective form by mixing these constitutive elements in vitro in the right conditions, as in the paradigmatic case of tobacco mosaic virus (TMV). This means that the interactions among these elements spontaneously drive the formation of shells in a process that can be described by basic thermodynamic and kinetic principles. Moreover, the physical properties of the viral genome, which differ depending on the type of nucleic acid, have a crucial influence on the viral strategy of assembly and infection. In particular, single-stranded RNA (ssRNA), which is a relatively flexible macromolecule, usually co-assembles with the viral capsid proteins to form infective viruses. Instead, double-stranded DNA (dsDNA) is a more rigid polymer that many viruses pack at high densities inside a spontaneously preformed shell, the procapsid, using the most powerful molecular motors known in nature. In this way, the elastic energy and high electrostatic self-repulsion of the packed dsDNA phosphate backbones generate an internal pressure that is crucial to initiate the subsequent infection of a new host (see Chaps. 10, 13, and 17).

The wealth of biological details described in previous chapters in this book is essential to confer specificity to viruses and make possible their success in infecting a particular host. However, even viruses that infect hosts from different species or even domains of life also share some common physical mechanisms, whose study would pave the way to understand key points in the viral life cycle and to a myriad of potentially useful applications. It is worth noticing that viruses have been subjected to natural selection, some of them for billions of years, perhaps since the origin of life. Thus, they constitute a pool of optimal solutions at the nanoscale of incalculable value, and their efficiency in performing all kinds of functions makes them a perfect model to guide the design of a new generation of nanostructures. In fact, the impressive properties of viral capsids have already grounded the development of promising biomedical and nanotechnological applications, as described in Chaps. 20–22.

In this chapter, we will review the physics behind some of the major steps of the virus infection cycle, and we will introduce the main theoretical tools developed in the emerging field of physical virology. In particular, we will discuss some of the underlying physical principles and likely mechanisms involved in virus architecture, viral particle assembly and mechanical stability, nucleic acid packaging, and virus exit from the infected cell and entry into another host cell; and we will see how simple theoretical models provide very useful insight into many aspects of viruses, despite not incorporating most of the structural and biological elements introduced in other chapters of this book.

More specifically, first, we will see how the exquisite regular shape and structure of viral particles is a natural consequence of simple geometrical principles and free energy minimization. Then, we will discuss how the favorable interactions among capsid proteins lead to the spontaneous formation of capsids in a self-assembly process that can be described using the same models that explain the phase transformations of well-known non-biological systems. The presence and the physical properties of the genome play a central role in this process, hence ssRNA and dsDNA viruses will be discussed separately. Once assembled, the physical stability of viruses is crucial to ensure their infectivity, and this stability is related to the mechanical properties of the capsid. Classical elasticity theory will be the starting framework to study capsid mechanical stability and, complemented with molecular modeling, will provide essential information to understanding the physical resistance of viral shells. Finally, in the last section of this chapter, we will describe some basic physical mechanisms behind genome delivery and virus egress strategies during the infection cycle.

19.2 Architecture of Viral Shells

An astonishing aspect of viruses is the well-defined shape and symmetry of their capsids. As discussed in Chap. 2, viral shells essentially adopt four basic different architectures: quasi-spherical, rod-like, bacilliform, and, more rarely, conical. Thus, completely unrelated viruses can form similar structures which is truly remarkable, especially considering that they generally infect different hosts and are made of capsid proteins whose amino acid sequences, sizes, or conformations can be very diverse. This suggests that, underlying the virus-specific structural details, there must be some common physical and geometrical principles dictating the shape and architecture of capsids.

The aim of this section is to provide a brief overview of the main theoretical ideas and models that explain the architecture of viral capsids. They have been grouped into three categories. First, the geometrical concepts that justify the overall shape and symmetry of viruses are discussed in Sect. 19.2.1. These geometrical principles provide a catalogue of the possible architectures adopted by viral capsids using very precise rules of construction. However, these ideas do not suffice to explain why a specific virus prefers a particular architecture. The selection of concrete structures obeys the physical principle of free energy minimization, and several models of increasing level of complexity have been proposed to understand the details of this process. The simplest models describe viral shells as continuum elastic materials, and their description in terms of thin shell theory (TST) is introduced in Sect. 19.2.2. Nevertheless, the fundamental understanding of architectural details of capsids requires the use of discrete models, and in Sect. 19.2.3 we discuss simulations of coarse-grained and discrete biomolecular models that allow exploring more complex and realistic scenarios. The combination of continuum and simulation approaches is fundamental to establish a solid theoretical framework that can rationalize our understanding of viral structures, helping us to guide further experiments and applications.

19.2.1 Geometrical Theories

One can justify the construction of the four main morphologies of viral capsids using basic geometrical ideas. As mentioned in Chap. 2, the small size of virus capsids, on the order of tens to hundreds of nanometers, restricts the amount of information that can be coded in the viral genome contained inside. Therefore, for the sake of genetic economy, capsids are typically built from multiple copies of a single or a few different small proteins [1]. These identical subunits interact with each other building a regular structure, in a process that shares some similarities with the crystallization of molecules that form regular lattices [1, 2]. However, in the case of viruses, capsid proteins construct a hollow shell rather than a three-dimensional solid. In a planar surface, the organization allowing the highest packing of identical units, which maximizes the number of interactions, is the hexagonal lattice or, equivalently, the triangular lattice. Now, starting from this hexagonal network, it is possible to rationalize the construction of all basic capsid shapes. In particular, open helical tubes typical of rod-like viruses can be obtained by simply wrapping up the lattice. Alternatively, to obtain a closed shell, one needs to introduce 12 pentagonal defects (see Chap. 2). If those are evenly distributed, one gets an icosahedral shell, characteristic of quasi-spherical viruses. In a similar way, prolate or bacilliform capsids can be obtained by wrapping the lattice into a helical tube and closing each of its ends with hemispheres containing six defects. Finally, conical viruses are obtained by making a tube with a conical section closed with a different number of defects at the ends.

The geometrical characterization of viral capsids is particularly important because there is only a limited number of ways to introduce the pentameric defects that lead to perfectly closed regular shells. For spherical viruses, the different possibilities give rise to the triangulation T-number used in the structural classification of icosahedral capsids (see Chap. 2). For tubular viruses, the symmetry and arrangement of the pentameric defects in the caps determine a discrete set of compatible lengths, radii, and helicities that can be adopted. Finally, the different distribution of the 12 pentameric defects in both ends of a conical shell, e.g., 7 and 5, or 9 and 3, explains the discretized values of cone angles found for viruses like human immunodeficiency virus (HIV) [3]. Indeed, these simple geometrical rules provide the basis for the structural classification of spherical viruses introduced back in the 1960s by Caspar and Klug (CK) [2] and recently generalized to accommodate different protein configurations by Twarock and Luque [4]. This classification was extended for some bacilliform viruses by Moody [5] and generalized by Luque and Reguera [6] (see Chap. 2). Conical capsids display a lower symmetry than their spherical and elongated counterparts, challenging their full geometrical characterization, although a few subsets of capsids in these groups have been abstracted geometrically [7].

Graphically, these structures can be considered as the result of covering a sphere, a spherocylinder, or a conical tube using equilateral triangular tiles that are associated to the capsid subunits (see Fig. 19.1a, b). Nevertheless, certain viruses do not fit strictly into this classification. For instance, polyoma virus has a structure like a $T = 7$ capsid where the expected hexagonal and pentagonal positions in the icosahedral lattice are occupied by clusters of five proteins (see also Chap. 2). To tackle this puzzling situation, an alternative tiling theory[1] has been developed by Twarock et al., introducing tiles of other shapes as rhombs or kites rather than the triangular tessellation of the CK theory (see Fig. 19.1b). These ideas have been successful in characterizing the structure of some spherical and tubular all-pentamer viruses [9]. A more recent generalization has reconciled the Caspar and Klug and the tiling theory using Archimedean and Laves lattices, providing a framework for quasi-equivalent and non-quasi-equivalent capsids [4].

The geometrical approaches discussed so far facilitate the characterization of capsids by mapping the convex shell onto a 2D net representation. However, this approximation does not capture the alternative 3D configurations of the amino acids of the capsid proteins through the capsid thickness. This problem has been addressed using icosahedral point arrays [19]. This approach captures structural details on the

[1] Tiling theory is a set of mathematical operations that can be used to cover an arbitrary surface with geometrical shapes (or tiles) with no overlaps and no gaps.

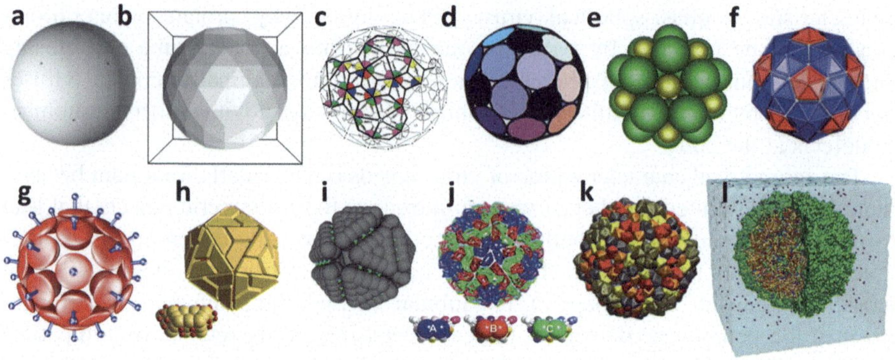

Fig. 19.1 Different levels of modelization of viral capsids. (**a**) Continuum elastic model of a spherical capsid in TST. (**b**) Representation of a shell as a triangulated network [8]. (**c**) Tiling representation of polyoma virus using kite and rhombic tiles [9]. (**d**–**g**) Coarse-grained models where capsomers and proteins are represented as (**d**) disks [10] or (**e**) balls [11] over the surface of a sphere, (**f**) hexagonal and pentagonal pyramids [12], and (**g**) ellipsoidal capsomers with additional attractive and repulsive sites [13]. (**h**–**j**) Coarse-grained models at the level of proteins are described as: (**h**) multiple bead trapezoidal objects with many interaction sites (shown below) [14], (**i**) truncated pyramids [15], (**j**) trapezoidal multi-beads of three different types (shown below) [16]. (**k**) Shape-based coarse-grained model of phage ΦX174 [17]. (**l**) Full atom model, including water and ions, of STMV [18]

tertiary structure of viral proteins and even the organization of the genomic material inside the capsid. Recent refinements of this theory are consolidating the viral *phrenology* approach of generating geometrical molecular blueprints [20].

These geometrical models are very helpful in characterizing the structure of viral capsids. Nonetheless, they do not explain *why* viruses adopt those architectures. In this sense, several simple physical models have tried to justify the prevalence of these structures in terms of their possible energetic or biological advantages. One of these desirable features is storage efficiency, which could be translated into maximizing the volume-to-surface ratio, i.e., maximizing the storage capacity of the capsid with the minimum amount of protein. The first ideas in this direction were based on exploring models that maximized the number of contacts, the packing, or the covering of a sphere by simple geometric objects like disks, representing the capsomers or the structural subunits of the capsid [21]. These models were able to reproduce the symmetries of some all-pentamer viruses but did not recover the sequence of T-numbers described by CK and adopted by most spherical viruses. Interestingly, the mathematical ideas of packing and covering are somehow connected to a more general principle in physics: the free energy minimization. This idea, already suggested by CK [2], has been explored using different continuum and discrete models as described in the next two subsections.

19.2.2 Continuum Theories

A first approximation to the energetic description of different capsid shapes can be made using the well-known continuum elasticity theory [22]. In this framework, the viral capsid is described as a continuum homogeneous material and the focus is placed on the elastic energy required to deform it to obtain the different capsid shapes. At first sight, this continuum description might seem very crude, since viruses are relatively small and made of discrete elements (proteins). However, even the smallest viruses contain hundreds of thousands of atoms that render this approximation not so unrealistic. Moreover, in most viruses, the radius of their protein shell is substantially larger than its thickness, which means that the capsid can be approximated as a two-dimensional surface and described using a simplified limit of the elastic theory known as thin shell theory (TST). This approach has been already used to describe the structural properties of several biological systems, like membranes and vesicles and constitutes a very convenient starting point to understand the energetics of viral capsids.

Thin Shell Theory (TST)
In the context of TST, the elastic energy of a viral shell is decomposed into two main contributions [8]. First, one has a *stretching* term that is the energetic cost associated with elongating or compressing a planar piece of capsid material. This cost is commonly characterized by two elastic constants: the two-dimensional (2D) Young's modulus Y, which describes the stiffness of the material and is given by the ratio between the applied stress and the resulting elongation per unit length; and the Poisson ratio ν, which describes the relative contraction in the transverse directions when a material is stretched in the longitudinal direction (see Chap. 18). For many protein-like materials, this is taken to be around $\nu \sim 0.3$ or 0.4, which is like rubber. The second energetic contribution is the *bending* term, which accounts for the out-of-plane deformation. This is characterized by a bending constant κ that describes the resistance to flexion and a spontaneous curvature C_0, which is the inverse of the preferred radius of curvature of the surface. It is worth noticing that the latter is ultimately defined by the directionality of the interactions between capsid proteins and the steric effect of their surfaces in contact. Additionally, there is a contribution associated with the Gaussian rigidity, which for closed surfaces with fixed topology, like a sphere, a spherocylinder, or a cone, adds a constant that is generally not considered.

Furthermore, as discussed in Chap. 2, spherical, bacilliform, and conical viruses require the insertion of 12 pentameric defects, or *disclinations*, in the original planar hexagonal network to form a closed shell. These disclinations are made of the same material as the rest of the capsid and generally accumulate an important degree of stress, which grows with the size of the shell and can promote the faceting of the structure, as we will see in Sect. 19.4.2. The contribution of these disclinations is implicitly captured in the TST description, although explicit expressions can be also derived [23, 24]. Interestingly, the total elastic energy of a shell obtained by adding up all these contributions depends only on a single non-dimensional parameter known as the Föppl-von Kármán (FvK) number, $\gamma = YR^2/\kappa$, where R is the radius of the shell. The FvK number is essentially the ratio of the stretching and bending energies. It considerably simplifies the study of the capsid properties in the context of TST [8, 25, 26] and, as we will see in Sect. 19.3.1, it is crucial for the assembly.

As has been discussed above, the particular shape and structure of viral capsids is the consequence of the free energy minimization principle. This means that to obtain the possible architectures of viral capsids within TST, one must minimize the energy functional containing the stretching and bending energies, including the contribution of the disclinations. This can be challenging in many cases, but, rather than solving the analytical expressions of TST, one can also minimize the elastic energy of a shell in a more convenient way: discretizing the structure into a triangular network and performing numerical simulations using the so-called *mass-and-spring model*, which approximates the interactions by a set of harmonic springs. In this approach, the stretching energy is given by $E_s = \frac{1}{2}\varepsilon \sum_{ij} (r_{ij} - \sigma)^2$, where the energy parameter ε is related to the 2D Young modulus, $Y = 2\varepsilon/\sqrt{3}$, r_{ij} is the distance between nearest neighbor nodes of the triangular network, and σ is the equilibrium spacing. For the bending energy one assumes $E_B = k\sum_{ij} [1 - \cos(\theta_{ij} - \theta_0)]$, where θ_{ij} is the angle between nearest neighbor triangles, k is the local bending related to the bending modulus by $\kappa = \sqrt{3}k/2$, and $\theta_0 = \sqrt{3}\sigma C_0/2$ the spontaneous curvature [25, 27].

In this way, it has been possible to investigate the phase diagram of the optimal viral shapes as a function of the spontaneous radius of curvature and the FvK number [25]. Spherical capsids are more stable than spherocylindrical shapes at low values of the spontaneous curvature for a fixed surface, i.e., number of capsid proteins. However, conical capsids were never found to be energetically better than spherical or spherocylindrical capsids. This suggests that in the context of TST some alternative physical constraints are necessary to explain the existence of conical architectures, such as those adopted by the mature capsid of HIV-1, which require displacive and disassembly/reassembly kinetic processes as shown by recent models and experiments [28] (see also Chap. 2). In addition, for large FvK, capsids prefer to adopt a relatively faceted shape in order to release the stress accumulated in the disclinations, in a process known as the buckling transition [8], which has been well reported in the maturation of quasi-spherical capsids (see Sect. 19.4.2).

Thus, TST provides a very useful starting point to explain the overall shapes of viral capsids. However, to capture the possible distribution of capsomers or capsid subunits in a shell, it is more convenient to use discrete models, as discussed below.

19.2.3 Coarse-Grained, Discrete Models, and Simulations

The structure adopted by a viral capsid is ultimately a consequence of the interactions between its fundamental building blocks. In the framework of TST, the molecular interactions between capsid proteins are effectively replaced by the stretching and bending elastic terms. However, to explore equilibrium and dynamical properties of capsids in a higher degree of detail, one cannot ignore the discrete nature of viral shells.

In this context, the discrete modeling of capsids has been performed at different levels of resolution, from all-atom approaches to simple coarse-grained physical models (see Fig. 19.1). The appropriate choice of resolution depends on different factors, like the properties to be probed, the existence of experimental data, and the computational limitations. Here we have organized these modeling strategies in two groups. The first one contains *low-resolution* models that are usually grounded in basic physical ideas, aiming at understanding the essential ingredients required to explain common aspects of viral structures; the second group includes *high-resolution* models, which generally use specific molecular reconstructions derived from experiments as a starting point, and then introduce force fields that can vary in complexity in order to properly characterize the finer structural details of particular virus capsids.

Low-Resolution Models

Low-resolution models might seem a rough approximation to describe complex molecular structures like capsids, but they constitute an excellent strategy to improve our understanding of viruses. The absence of molecular details allows the simulation of relatively large systems, and the simplicity of the models facilitates the qualitative characterization of the underlying mechanisms that explain many important properties of viral capsids. In general, these models replace the stretching and bending energies, discussed above in the context of continuum theory, by several types of effective interactions between the capsid building blocks (CBBs). These interactions require at least three main ingredients. First, one needs a short-range repulsion force that mimics the steric effects between capsid proteins, which is intended to avoid their overlapping. Then, a longer-range attraction is also introduced to drive the self-assembly of the capsid and to give stability to the final structure. The third essential ingredient is the existence of an imposed curvature or preferential angle of interaction between the building blocks, which is ultimately responsible for the selection of a particular size and curvature in the shell.

The seminal work by Zandi et al. [10] proved that the icosahedral and other symmetrical structures adopted by spherical viruses are the natural consequence of free energy minimization of a very generic interaction, which essentially contains the contributions discussed above. In their low-resolution model, the description was made in terms of capsomers, represented as disks or spheres of two different radii (see Fig. 19.1d), corresponding to pentamers and hexamers, which were confined to the surface of a sphere. Then, using a computational technique known as Monte Carlo simulation, it was possible to find the optimal arrangement of capsomers. For the case where hexamers and pentamers coexist in the capsid, the distribution that minimized the total energy corresponded precisely to the T-number architectures of the CK classification (see Fig. 19.2). Additionally, when only one type of morphological unit was allowed, the optimal free energy minima corresponded to the icosahedral and non-icosahedral architectures observed in all-pentamer viruses both in vivo and in vitro.

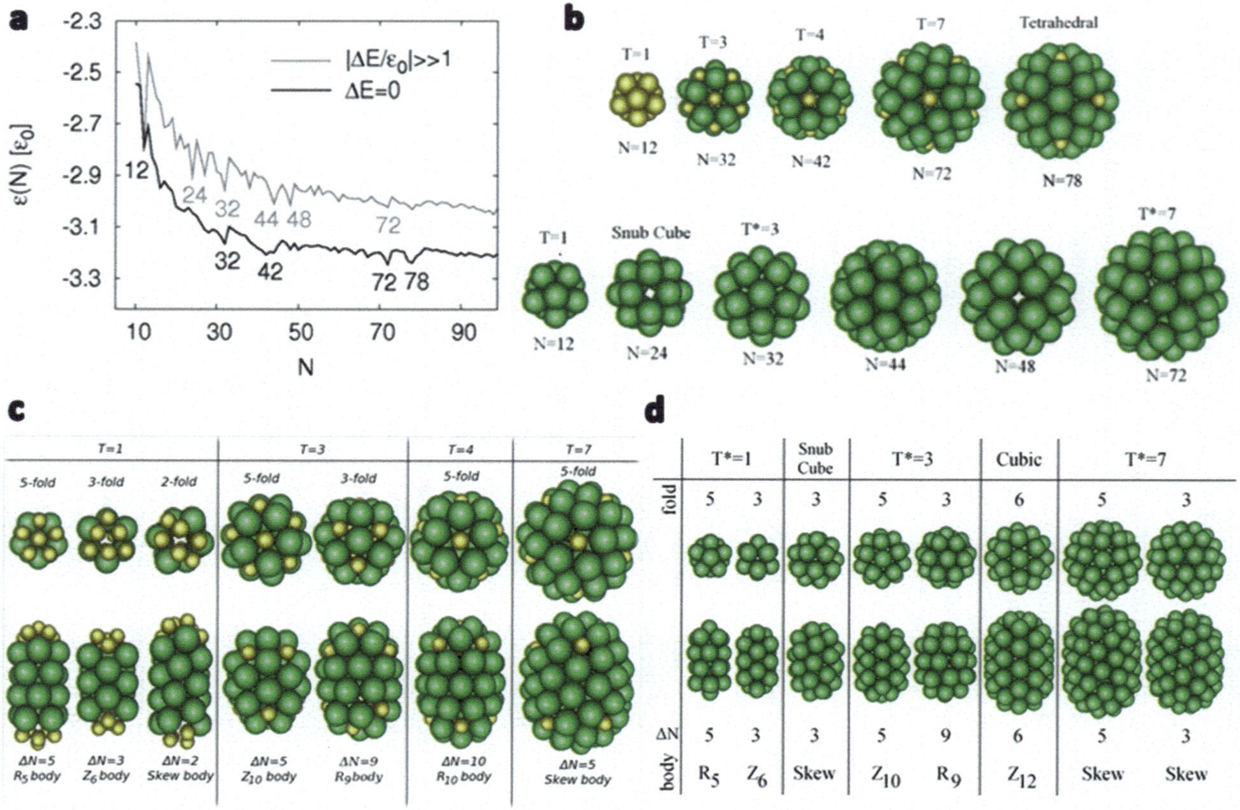

Fig. 19.2 Gallery of optimal structures obtained from coarse-grained simulations. (**a**) Energy landscape of the optimal structures obtained in a coarse-grained model of a quasi-spherical capsid with two types of capsomers (solid line), i.e., pentamers and hexamers, or just one type (gray line). The figure plots the average energy per capsomer as a function of the total number of capsomers of the shell, and the deepest minima correspond to the optimal structures, that for the two

Fig. 19.2 (continued) capsomer types coincide with the CK T-number structures. Notice that the apparently small energy differences *per capsomer* plotted correspond to quite large energy differences *per capsid.* (**b**) Gallery of optimal structures for quasi-spherical viruses made of two types of capsomers, i.e., pentamers (gold) and hexamers (green) (top), or just of one type (bottom). (**c**) and (**d**) Gallery of optimal structures for spherocylindrical viruses with two (**c**) and just one (**d**) capsomer types. The top images show a zenithal view of the caps, and the lower images show a lateral view of the complete bacilliform virus. In each case, the T-number of the cap, the minimum number of capsomers in the body ΔN required to elongate the structure, and the arrangements of capsomers in the body (forming rings R_n or zig-zag patterns Z_n of n capsomers) are also indicated. Figures adapted from [11, 29]

Luque et al. [11] extended this model to study the optimal structures of bacilliform capsids (see Fig. 19.2c, d) showing that icosahedral symmetry is also predominant on the ending caps of these shells. Interestingly, the optimal structures followed the set of geometrical rules derived in [6] as extension of CK for elongated viruses. In addition, this study justified the existence of non-icosahedral-capped capsids for all-pentamer viruses (e.g., polyoma virus), and demonstrated that the bacilliform geometrical construction also captures most of the properties of icosahedrally-capped all-pentamer viruses. The incorporation of a preferred angle and a bending energy cost to deform it in the interaction potential between CBBs leads to simple models that do not require the presence of a spherical template and explain how viruses select a particular size and structure [30, 31].

The important message of these simple coarse-grained models is that the impressive regular and symmetric architectures of viral capsids emerge naturally from free energy minimization of very generic interactions between their building blocks. In other words, the details of these interactions turn out to be not so crucial in determining the possible catalogue of viral shell architectures, although they can be essential to determine the physiological conditions of capsid formation. In fact, various models that use different types of interaction potentials and more realistic representations lead to similar conclusions [13, 32]. The exhaustive thermodynamic exploration of these models has shown that the set of basic parameters introduced in these simple physical approaches are able to spontaneously reproduce spheroidal shells, tubular and helical shapes, and even head-tail morphologies [13]. In this way, these low-resolution strategies have played a very important role in unveiling the key factors that determine the architecture of viral capsids. Furthermore, they are contributing to the determination of fundamental mechanisms controlling the mechanical properties, and the assembly and disassembly of viral capsids either in the presence or absence of genetic material, as will be discussed in Sects. 19.3 and 19.4.

High-Resolution Models

High-resolution models are the most appropriate computational strategy to study in detail the properties of specific viruses. Contrary to the models discussed above, these are based on structures reconstructed from X-ray crystallography (see Chap. 4) and/or cryogenic electron microscopy (cryo-EM; see Chap. 3) studies. The interactions between atoms are modeled through effective force fields, and techniques like molecular dynamics or Monte Carlo are applied to sample the structural dynamics of the viruses.

All-atom simulations are gaining popularity in the field of structural and chemical biology and are of particular interest for biomedical applications. The constant improvement of computational power and sampling techniques has made this approach more feasible in the last years, though there is still ongoing research to develop more accurate force fields and to reach longer time scales in simulations. The first all-atom simulation with explicit solvent of a complete virus was developed for the relatively small satellite tobacco mosaic virus (STMV), involving a million atoms (135,960 protein atoms, 30,330 nucleic acid atoms, plus water and ions) and achieving timescales of nanoseconds. In that case, it was shown that at physiological conditions the virus becomes unstable when the viral RNA is removed [18]. To study larger systems or processes that involve longer time scales it is required to use multiscale, or mesoscale models, and some degree of coarse-graining [33]. Nonetheless, in the last decade there has been significant progress [34, 35] showing that millisecond scales are feasible even with relatively large capsids like those of HIV [36].

These simulations are starting to provide very useful insights into the detailed structural mechanisms of different viruses. For instance, in the case of HIV, it has been possible to study the interactions between the main capsid protein and the lipidic membrane in the immature virus [37], the role of the C-terminal binding interface between the capsid proteins controlling the curvature and shape of the capsid [37], and the phase diagram for the assembly of the capsid proteins [38]. Another well-studied case has been the plant virus cowpea chlorotic mottle virus (CCMV), where all-atom simulations have characterized its swelling process that is triggered by changes in the pH [39]. In particular, the combination of elastic networks with coarse-grained models has allowed simulating the structure of this virus in the scale of microseconds at a low computational cost. Besides, general multiscale models have been applied to simulate the *Nudaurelia capensis* ω virus and human papilloma virus. More recently, the combination of elastic network models and spectral analysis have facilitated the identification of physical quasi-rigid domains in capsids [40, 41], successfully identifying disassembly units on multiple structural viral lineages [42]. Alternatively, coarse-grained simulations of the bacteriophage MS2 indicate that the organization of the genetic material within the capsids of RNA viruses depends on the non-specific self-repulsion of the RNA in the inner layers, but the organization of the outer layers strongly depends on the interactions with the inner part of the capsid. Finally, simulations of the structure of Pariacoto virus suggest assembly mechanisms of RNA viruses in agreement with alternative physical models of assembly [43] (see Sect. 19.3).

19.3 Assembly of Viruses

The assembly of viral particles is a crucial step in the replication cycle of viruses. Depending on the virus, the capsid can successfully assemble from the capsid proteins alone, or in combination with scaffolding proteins, or the genetic material, following specific assembly pathways reviewed in Chaps. 11–14. However, the assembly of viruses shows also some general trends that seem susceptible to be modeled using basic physical ideas. As discussed in Sect. 19.2, a very important fact is the spontaneity of the self-assembly process of viruses with single-stranded genomes, which is driven by free energy minimization, meaning that no external energy in the form of ATP is required. This makes the proper assembly of some viral shells possible even in the absence of genetic material or in the presence of other types of cargoes. In fact, many viruses can be reconstituted in vitro from their basic components, yielding particles that are fully infective and indistinguishable from native virions (see Chaps. 21 and 22). The best-known example is TMV, which was the first virus to be fully reconstituted in vitro by mixing the capsid proteins and the viral RNA. In addition, by properly tuning the assembly conditions it has been possible to produce empty capsids of viruses that in vivo require the genetic material to assemble, as has been shown for plant viruses like CCMV and brome mosaic virus (BMV), or animal ones like human polyoma virus (HPV) and hepatitis B virus (HBV). The modification of assembly conditions, like capsid protein concentration, pH, or ionic strength, can also lead viral capsid proteins to assemble into capsid structures different from the native one [44, 45], and to encapsulate foreign genetic materials and even non-organic cargoes [46, 47]. Obviously, this has opened the possibility to engineer viral capsid proteins for different types of applications in nanoscience, e.g., encapsulation of materials or templating of nanostructures (see Chap. 22).

Diverse viruses can follow different assembly pathways (see Chaps. 1, 11–14). In particular, their building blocks can be made of different numbers of capsid protein subunits and are typically defined as being the most stable oligomer in solution. For instance, the CBBs correspond to single proteins in the case of *Penicillium chrysogenum* virus, dimers for CCMV, pentamers in HPV, or pentamers and hexamers in solution for bacteriophage Hong Kong 97 (HK97). Starting from these building blocks in solution, the assembly process for empty capsids proceeds first by the formation of some aggregates of CBBs which stochastically grow incorporating more CBBs, forming what will be called *partial capsids* or *intermediates*, until a complete shell is assembled (see Fig. 19.3). It is worth noticing that understanding the steps during viral assembly is a key question to identify, for instance, possible targets for antivirals (see Chap. 20).

From a theoretical point of view, most modeling efforts have focused on the simplest case, i.e., the in vitro self-assembly of *empty* viral capsids from capsid proteins that do not require any auxiliary proteins. The abundance of in vitro experiments has facilitated the investigation of this scenario, and it is the first topic to be described in this section. Subsequently, we will discuss separately the peculiarities of the assembly of ssRNA and dsDNA viruses, since the mechanism

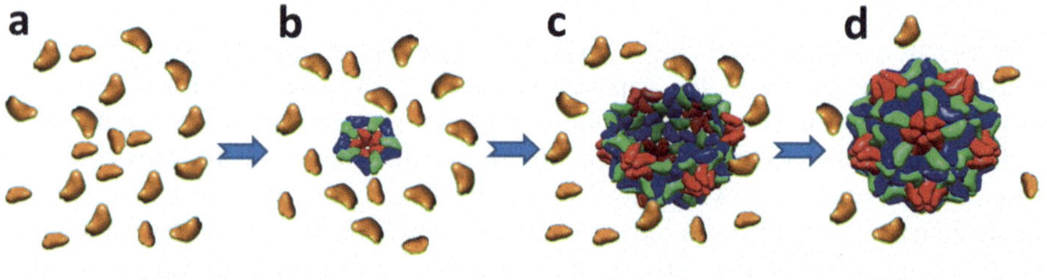

Fig. 19.3 Sketch of the different steps in capsid assembly. (**a**) The process is initiated by the presence of a sufficiently high concentration of free capsid building blocks, CBBs (in gold). (**b**) These building blocks aggregate and disaggregate by thermal fluctuations until a sufficiently large cluster of subunits (in color) forms that act as the *nucleus* or embryo of the capsid; (**c**) the embryo grows by the addition of building blocks and forms a partial capsid or intermediate of *n* building blocks; (**d**) finally a complete capsid is formed

followed in the assembly of nucleocapsids depends crucially on the type of nucleic acid of the viral genome, which strongly influences also the infection pathway.

19.3.1 Empty Capsids

The fact that many viruses can assemble in the absence of genetic material indicates that the interactions between viral capsid proteins generally contain all the information to form a capsid. The range and specificities of these interactions can vary among viruses, but they are usually dominated by two non-specific contributions: hydrophobic and electrostatic [48].

Hydrophobic and Electrostatic Interactions

Many (though not all) viruses whose structure has been solved at close to atomic resolution, have interfaces between subunits with a strong component of hydrophobic amino acid side chains. In addition, structural and biochemical studies with some model viral particles have shown that the strength of viral capsid protein–protein interactions tends to increase when the temperature is raised, indicating that hydrophobic interactions are the main driving force in the assembly of those particles. The hydrophobic effect is ubiquitous in biological systems, although it is still a phenomenon under intense investigation. However, some effective characterizations have been already tested in non-biological soft matter systems, which conveniently serve as a starting point for the study of viruses. In particular, the total hydrophobic interaction for a complete capsid can be modeled as $G_H = - A_H\gamma$, where A_H is the total hydrophobic area buried in the formation of the shell and γ denotes an effective surface tension between the hydrophobic protein surface and water.

In addition, electrostatics plays an important role in many viral capsids. Capsid proteins are in many cases rich in basic residues leading to an excess of positive charge in the interior of viral shells. This allows the subunits to interact with the negatively charged phosphate backbone of the viral nucleic acids, promoting the assembly and stability of the capsid. In the absence of the genetic material, this overcharge introduces an effective repulsion between capsid proteins. As a consequence, the assembly of empty capsids may require higher salt concentrations (to screen the electrostatic repulsions) and/or lower pHs (to reduce the charge of protein residues), compared with the co-assembly with genetic material. In addition, in several capsids, electrostatic repulsions between either positively or negatively charged residues that may occur under certain conditions in vivo are used to regulate capsid conformational stability, disassembly, or genome uncoating (e.g., in TMV, CCMV, and foot-and-mouth disease virus).

In a first approximation, the electrostatic contribution in a viral capsid can be described using Debye-Hückel (DH) theory [49]. This takes into account that the ions of the salt in the medium will screen the electrostatic interaction between the charged residues of a capsid. For two unit charges e separated by a distance r this description leads to the potential $V_c = k_B T(\lambda_B/r)\exp(-\kappa r)$, where k_B is the

Boltzmann constant, T the absolute temperature, $\lambda_B = e^2/4\pi\varepsilon_0\varepsilon_r k_B T$ the Bjerrum length, and $\kappa^{-1} = \left(4\pi\lambda_B \sum_i z_i^2 c_0^i\right)^{-1/2}$ the Debye screening length. Here $\varepsilon_0\varepsilon_r$ is the permittivity of the medium (water in this case), c_0^i is the number density of charged species i in the electrolyte (in units of m^{-3}), and z_i its corresponding valence. In the crudest approximation, a quasi-spherical capsid can be considered as a uniformly charged thin sphere of radius R and surface charge density σ, surrounded by an (ideal) ionic solution (1:1 electrolyte) of bulk concentration c_0. The electrostatic contribution to the free energy of the capsid is then obtained by integrating the DH potential over the surface of the sphere, yielding $G_{\text{elec}} = \frac{\pi\sigma^2 R^2}{\varepsilon_0\varepsilon_r\kappa}$, provided that the Debye length is short enough, i.e., $\kappa R >> 1$, which is usually satisfied even at low salt concentrations, e.g., on the order of mM (for a more detailed derivation of both terms, the reader is referred to [48, 50]). This simple approximation can be refined by incorporating the effects of a finite capsid thickness, a non-uniform distribution of charges in the capsid, lower ionic strengths, or even solving the more rigorous description of this electrostatic problem given by the Poisson-Boltzmann equation [51]. Interestingly, all these approaches lead to qualitatively similar results [50].

In this context, the total free energy of a complete spherical capsid is obtained by adding the hydrophobic and electrostatic contributions, yielding $G \approx -A_H\gamma + \frac{\pi\sigma^2 R^2}{\varepsilon_0\varepsilon_r\kappa}$. Rough estimates of the order of magnitude of both contributions can be made using HBV as an example. For HBV, the total hydrophobic buried surface is $A_H \sim 10^3$ nm^2, and the energy per unit area of exposed protein (accounting also for van der Waals interactions) is roughly $\gamma \sim 6$ mJ/m^2, yielding an estimate of $\sim 10^4$ $k_B T$ as a typical attractive contribution. The order of magnitude of its surface charge is $\sigma \sim 0.7$ e/nm^2, and its radius is $R \sim 30$ nm, giving also an electrostatic contribution of $\sim 10^4$ $k_B T$ at physiological conditions ($c_0 = 150$ mM) where $\kappa^{-1} \sim 1$ nm. Both contributions have thus similar strengths, illustrating the delicate balance between the two required for a successful assembly and its sensitivity to environmental conditions.

The main parameter controlling the strength of the hydrophobic interaction is the temperature, whose increase favors a stronger attraction. In addition, the electrostatic contribution can be tuned by two accessible physicochemical conditions: the pH and the salt concentration of the solution. The pH controls the fraction of charged residues in the capsid protein, and the salt concentration varies the ionic strength, which screens the electrostatic repulsion between capsid proteins. The interplay between electrostatic repulsion and steric forces between capsid proteins also regulates the optimal angle of interaction between them. Thus, these three factors, i.e., temperature, pH, and salt concentration, have major roles in controlling the feasibility of capsid assembly both in vivo and in vitro.

Capsid Assembly as a Thermodynamic Process

The fourth important factor in the self-assembly of capsids is the concentration of subunits, and its impact on the assembly of empty capsids has been probed in vitro in the last years. In those assembly experiments, one starts from a given concentration

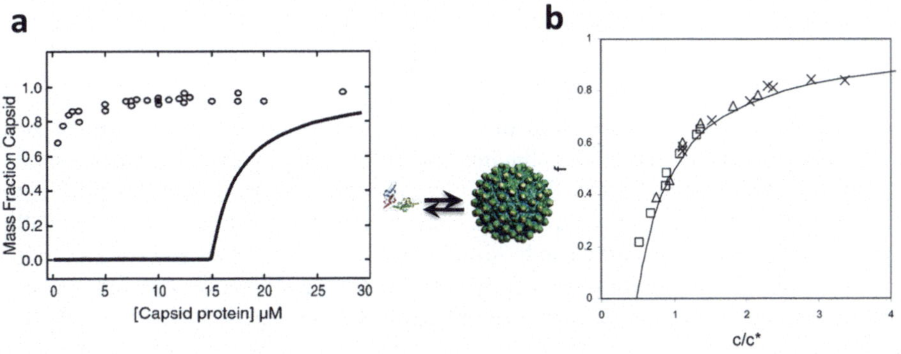

Fig. 19.4 In vitro assembly of HBV capsids. (**a**) Mass fraction of proteins in capsids as a function of the total capsid protein concentration in assembly (solid line) and disassembly (open circles) experiments. (**b**) Fraction of subunits in complete capsids, f, versus the total concentration of capsid protein scaled by the critical micelle concentration $c_{\text{cmc}} = 2c^{*}$. The symbols are data from [52] obtained from experiments of HBV assembly at 25 °C and different buffer concentrations of NaCl (crosses $c_{\text{salt}} = 0.7$ M; triangles $c_{\text{salt}} = 0.5$ M and squares $c_{\text{salt}} = 0.3$ M). The line is the universal aggregation curve predicted by Eq. (19.2). Adapted from [48] and [53]

of capsid proteins in solution and monitors the formation of capsids by scattering techniques, electron microscopy, or size-exclusion chromatography. These studies have revealed a steep dependence of the assembly yield on protein concentration as shown in Fig. 19.4a for the paradigmatic case of HBV. For low protein concentrations capsids are not formed; then the production starts above a certain threshold concentration, and grows steeply with concentration, until eventually reaching saturation. Another distinguishing feature observed experimentally is the lack of intermediates, i.e., most subunits are found either free in solution or in fully formed capsids.

This behavior can be described theoretically by using standard ideas from equilibrium thermodynamics. The process of assembly can be thought of as a phase transition between two states: the capsid building blocks in solution and in a fully formed capsid. The equilibrium state is reached when the chemical potentials of both states are equal, leading to the well-known law of mass action:

$$K^{\text{eq}}_{\text{capsid}} = \frac{c_{\text{capsid}} c_s^{q-1}}{c_1^q} = e^{-\frac{G}{kT}} \tag{19.1}$$

where q is the total number of subunits in a complete capsid, c_{capsid} and c_1 are the equilibrium concentrations of capsids and subunits, respectively, c_s is the molarity of pure water taken as a reference concentration, and G is related to the free energy of capsid formation described above. The law of mass action can be conveniently rewritten in terms of the capsid mass fraction $f \equiv qc_q/c$ as

$$1-f=f^{1/q}\left(\frac{c^*}{c}\right)^{1-1/q}\approx\begin{cases}0 & c<c^*\\ 1-c/c^* & c>c^*\end{cases}\tag{19.2}$$

where c is the total concentration of proteins in the solution, and $c^*=c_s e^{\frac{\Delta g}{kT}}$ is a threshold concentration related to the binding energy per protein $\Delta g \equiv G/q$. In terms of this threshold concentration, the final yield of capsid production can be described by a universal curve given by Eq. (19.2). Interestingly, since for viruses one typically has $q \gg 1$ this universal curve can be simplified to $f = 0$ for $c < c^*$ and $f = 1\text{-}c^*/c$ for $c > c^*$. This simple model successfully described the temperature and salt concentration dependence of the in vitro assembly of HBV [48], as shown in Fig. 19.4b.

Kinetic Aspects of Virus Assembly

Different studies using light scattering or turbidimetry techniques have been able to monitor also the kinetics of the assembly process for empty capsids like HBV or HPV. In recent years, native mass spectrometry [54, 55], time-resolved small-angle X-ray scattering [56–58], interferometric scattering microscopy [59, 60], and high-speed atomic force microscopy [61–63] are opening the door to monitoring the real-time kinetics of viral assembly with unprecedented resolution. These experiments suggest again the existence of common features in the kinetics of capsid assembly. It turns out that initially there is a lag time period where no capsids are formed, and then the production rapidly increases following a sigmoidal curve. Another characteristic is that the disassembly of the capsid takes place at different concentration conditions than the assembly, giving place to a hysteresis phenomenon, which is observed even in viruses that do not go through any apparent post-assembly maturation (see Fig. 19.4a). Indeed, once capsids are formed, they do not dissociate easily even if the concentration of free subunits is drastically reduced. This has a clear biological relevance, since viruses must resist different environments to reach a new host, where no viral proteins are present.

Again, these features match the expected behavior of a standard first-order phase transition. Accordingly, different models have borrowed physical concepts involved in the kinetics of phase transitions to quantify and characterize the assembly of viral capsids [64].

The first kinetic model for viral assembly was introduced by Zlotnick and collaborators [65]. There, the assembly of a capsid is conceived as a series of concurrent molecular reactions between individual subunits and intermediates of all possible sizes, which are schematically represented as subunits ⇔ nuclei + subunits ⇔ capsids (see Fig. 19.3). The population of intermediates containing n subunits is given by the set of equations:

$$\frac{\mathrm{d}[n]}{\mathrm{d}t}=k_{\mathrm{elong},n-1}[n-1][\mathrm{subunit}]+k_{\mathrm{dis},n+1}[n+1]-k_{\mathrm{elong},n}[n][\mathrm{subunit}] - k_{\mathrm{dis},n}[n]\tag{19.3}$$

The kinetics is split into a nucleation step, which involves the formation of the capsid embryos, and an elongation process, which corresponds to the growth of the nuclei by the sequential addition of subunits. In this context, nuclei are treated as a special kind of intermediates that cannot disassemble and form at a slower rate k_{nuc} than the elongation stage, in order to justify the initial lag time of viral assembly. Numerical simulations have shown that this set of reactions quickly reaches a steady-state concentration of intermediates that asymptotically approach an apparent equilibrium characterized by a law of mass action Eq. (19.1), from which one can obtain the association free energy of the capsid.

Using this model, it has been possible to describe the kinetics of in vitro assembly experiments of empty capsids of viruses like HBV [52] and human papillomavirus [66]. Further refinements were proposed by Dragnea et al. [67] to describe the assembly of BMV capsids, which include certain simplifications in the rate equations and an additional activation step for the monomers to fit the short time initial takeoff of the assembly. Remarkably, these studies conclude that the interactions driving the assembly are on the order of a few k_BT per protein, typically 5–6 k_BT per contact, which is much weaker than a covalent bond [52]. This has two clear biological motivations: first, weak interactions facilitate the correction of possible mistakes during the assembly; second, disassembly must be feasible for many viruses to release the genetic material and infect a new host. Interestingly, the capsid of viruses like bacteriophages that do not require disassembling undergo a maturation process that in many cases reinforces the shell with additional or alternative subunit-subunit interactions (see Sect. 19.4.2 and Chap. 14).

However, despite the utility of these kinetic models, they present several drawbacks. In particular, they involve a large set of reaction equations that need to be solved numerically, which can be very demanding computationally; the reaction rates must be fitted to experimental data, which limits the predictive power of the theory; and the size of the nucleus is usually guessed a priori, which means that different cases should be explored for each experiment. Moreover, these models assume aggregative equilibrium. Hence to justify that disassembly takes place at much lower protein concentrations than expected, they have to assume that the binding energies for assembly and disassembly are different.

Classical Nucleation Theory of Capsids

Alternatively, an approach, based on classical nucleation theory (CNT), has overcome some of these limitations, explaining in a simpler way the common trends found experimentally in the assembly of capsids, and providing also a more direct connection between theory and experiments [29, 68]. The key point in CNT is the development of a physical model for the free energy of formation of partial capsids that, contrary to the equilibrium aggregation theory discussed above, contains explicitly the dependences in both the interaction between subunits and the concentration of free subunits. As we show below, this energy can be related to the rate of capsid production, reducing considerably the number of parameters to be fit, if any. Let us describe the main elements of CNT applied to viral capsids.

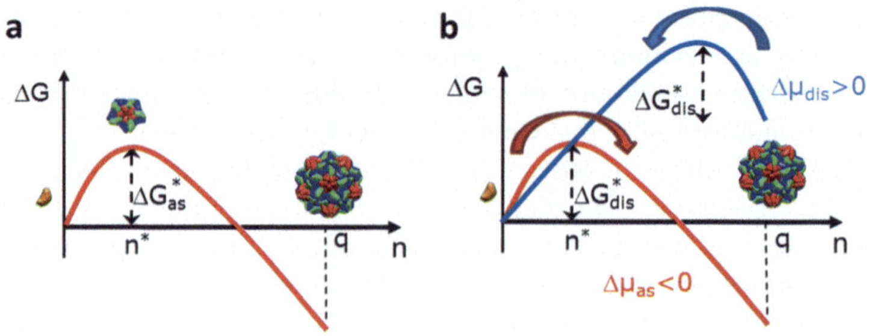

Fig. 19.5 Free-energy barriers of capsid formation in the context of CNT. (**a**) Total free energy of formation of a capsid as a function of the number of capsid building blocks (CBBs) n at favorable assembly conditions ($\Delta\mu_{\mathrm{as}} < 0$). The complete capsid contains q building blocks. The nucleation barrier ΔG^*_{as} and the critical cluster size n^* are indicated. (**b**) Free energy of capsid formation for favorable assembly ($\Delta\mu_{\mathrm{as}} < 0$, red line) and disassembly ($\Delta\mu_{\mathrm{dis}} > 0$, blue line) conditions. For disassembly, the barrier must be jumped in the opposite direction (i.e., from right to left) than for the assembly. To have similar barrier heights for assembly and disassembly, different chemical potentials, and accordingly concentrations of capsid proteins, are required. That is ultimately the reason for the hysteresis between assembly and disassembly conditions found experimentally (see Fig. 19.4a)

As we discussed above, the aggregation of a capsid becomes thermodynamically favorable when a building block in a fully formed capsid has a lower chemical potential than in solution, i.e., when the interaction energy of a building block in the capsid overcomes the entropic penalty of removing it from the solution. Nevertheless, if one takes a closer look at the assembly process, one realizes that, in the successive addition of building blocks to intermediate structures, those CBBs placed at the rim of a partial shell will be exposed and miss few energy contacts, i.e., the formation of intermediates intrinsically entails an energetic cost that originates an energy barrier. It is precisely this barrier that justifies the existence of a lag time at the beginning of the assembly process and the scarcity of intermediates observed experimentally. This is a well-known scenario in physics and is analogous to a vapor–liquid phase transition, where the formation of a liquid drop in the vapor phase has an energetic penalty associated with its surface tension. Following this analogy, the simplest way to model the energetic cost of an intermediate shell is by introducing a rim energy penalty associated to a line tension, equivalent to the surface tension in a drop.

In the framework of CNT applied to quasi-spherical capsids, a partial shell can be approximated as a spherical cap made of n subunits, whose free energy of formation is given by $\Delta G(n) = n\Delta\mu + a\sqrt{n(q-n)}$. Here, $\Delta\mu = -k_B T \ln\left(\frac{c}{c^*}\right)$ is the difference in chemical potentials between a subunit in a complete capsid and in solution, and a is a constant related to the line tension. Even at favorable assembly conditions, the competition between these two terms gives rise to an activation barrier (see Fig. 19.5a). The maximum of this barrier defines the critical nucleus size:

$$n^* = \frac{q}{2}\left(1 + \frac{\Delta\mu}{\sqrt{\Delta\mu^2 + a^2}}\right) \tag{19.4}$$

and its height

$$\Delta G^*_{\text{as}} = \frac{q}{2}\left(\Delta\mu + \sqrt{\Delta\mu^2 + a^2}\right) \tag{19.5}$$

defines the activation (or nucleation) barrier. As we see in Fig. 19.5a, for $n < n^*$ the free energy of formation of an intermediate increases with the number of CBBs, so these intermediates will tend to disassemble back into free subunits. Contrarily, partial capsids of size $n > n^*$ reduce their free energy upon the addition of CBBs, and they will tend to grow spontaneously until complete a capsid. Thus, n^* is the critical size that partial capsids must reach to trigger the formation of shells and can be considered as the *nucleus* or embryo of the assembly process. It is worth emphasizing that, contrary to the assumption in the previous kinetic model, the critical nucleus is the most unstable intermediate because it corresponds to a maximum of the energy, and both its size and the height of the nucleation barrier depend on the concentration of subunits.

The rate of capsid formation J, i.e., the number of capsids that form per unit volume and unit time, is controlled by the presence of this barrier. Like in other activation processes, it follows an Arrhenius behavior given by:

$$J = c_s\beta^* Z e^{-\frac{\Delta G^*_{\text{as}}}{k_B T}} \tag{19.6}$$

Hence, it depends exponentially on the barrier height and linearly on β^*, which is the rate of attachment of CBBs to the critical cluster size. Here, Z is a correction factor associated with the local curvature at the top of the barrier and accounts for the possibility that nearly critical-sized clusters dissociate.

The activated mechanism described by CNT has several biological advantages. First, the requirement of a threshold concentration and the existence of a critical size that must be reached to trigger capsid formation favor the production of enough proteins to guarantee the completion of capsids. In other words, if there were no barrier there would be multiple intermediates growing at the same time, which would rapidly deplete the protein by generating many incomplete shells, a remarkably inefficient scenario. Additionally, the presence of a barrier is the origin of the hysteresis, which prevents the disassembly of capsids under conditions where the concentration of subunits is reduced or nonexistent, favoring the stability of the virus during the extracellular process before a new infection (see Sect. 19.5.2). A more detailed derivation of CNT for viral capsids can be found in [29, 68].

The CNT model tacitly assumed that the capsid grows at a fixed radius defined by the preferred angle of interaction between CBBs. In the last years, this CNT model has been extended by accounting for the elastic energy contributions to the free energy of capsid formation due to the energetic cost of bending the shell out of its

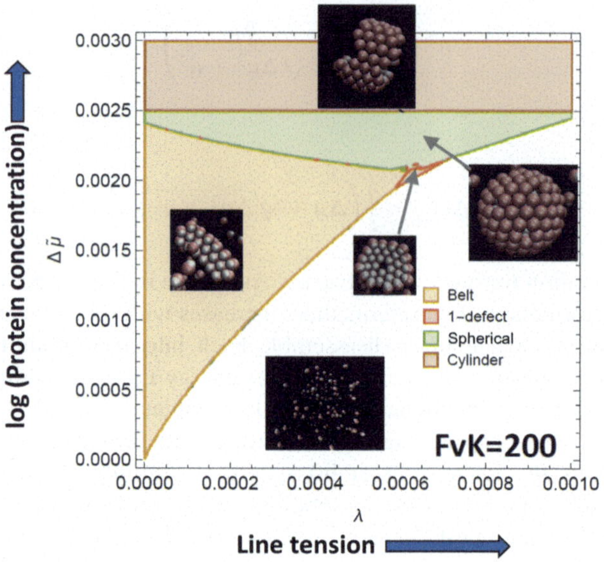

Fig. 19.6 Phase diagram of the outcome of viral assembly predicted by CNT with elastic contributions. Phase diagram of the most stable shape of a viral shell predicted by CNT incorporating bending and stretching energy contributions, as a function of the protein concentration (expressed in terms of a scaled chemical potential) and the scaled line tension for a value of the FvK number of 200. Four possible outcomes are shown, corresponding to the formation of tubes (dark orange), spherical shells (green), ribbons (light orange), and geometrically frustrated shells with one defect (red). The inset shows the structures observed in coarse-grained simulations, confirming the predictions of the model. Adapted from [70]

preferred curvature and the accumulation of stress due to the Gaussian curvature [23, 69, 70]. The accumulation of stress can prevent the completion of the capsid by a mechanism known as geometric frustration [23, 69], which can be relieved by the incorporation of pentamers or by growing anisotropically. The competition between binding, line tension, bending, and stress determines the shape of the resulting viral capsid, yielding a very interesting phenomenology. The outcome of the assembly can be recast into a universal phase diagram depending on three parameters (Fig. 19.6): the chemical potential difference $\Delta\mu$ and the line tension λ (both scaled by the bending energy), and the FvK number [70]. As shown in Fig. 19.6, by changing the assembly conditions, a virus whose native capsid is spherical can get trapped into a partial shell or ribbon-like structure, or end up forming tubes instead of spheres, which will compromise its viability. This competition between different energetic contributions also explains why some viruses need a maturation step since capsids with high mechanical resistance (i.e., high FvK number) cannot be self-assembled directly as spherical structures. Altogether, the results of these studies suggest promising routes to hinder viral replication by inducing mis-assembly.

Simulations of the Assembly of Empty Capsids

Despite the advances in computational power, all-atom simulations of the assembly of even the smallest viruses are still out of reach. Accordingly, current simulations necessarily involve coarse-graining at different levels, which, apart from computational advantage, represent also a great strategy to isolate the relevant details involved in the assembly process at a microscopic level.

At the coarser level, several models have explored the geometrical particularities required for the formation of T-number icosahedral shells. In particular, models based on local rules [71], which warrant the formation of the proper contacts between subunits, have shown that, due to the similarity between quasi-equivalent positions (see Chap. 2), the set of rules to successfully assemble a T-capsid can be less than T. For instance, four local rules suffice to produce $T = 7$ shells [72]. In a similar spirit, tiling approaches have studied the assembly of capsids focusing on the combinatorial aspects involved in the addition of subunits to intermediates of all possible sizes, concluding that only a small set of intermediates are relevant for the assembly pathway of empty shells [73, 74]. Various models have introduced a more physical implementation of these ideas by proposing effective interaction potentials with local constraints or strict orientational forces designed to produce a particular T-number target. This embraces patchy particle potentials [75] and bond-based potentials [76] (see Fig. 19.1). Remarkably, minimal models based on incorporating the angular anisotropy of the interaction between CBBs have been very successful in mimicking the assembly of empty spherical capsids [77, 78]. These models suggest that the conditions required to maximize assembly efficiency may depart from those at which the resulting capsid is energetically more stable, and that the preferred curvature is the main parameter selecting the size and T-number of the capsid, while the bending stiffness determines the rounded or faceted aspect.

The next level of complexity for coarse-grained strategies includes models that aspire to partially represent the shape of the assembly building blocks, like protein monomers, dimers, or pentamers and hexamers with a different degree of accuracy (see Fig. 19.1). In this way, subunits have been represented as pyramids [12], ellipsoids [13], triangular, and trapezoidal elements [14, 16, 79], which are agglomerates of relatively small spherical beads that help to place the interactions sites that are handcrafted to succeed in the assembly of a specific target structure [64].

Models based on growing a triangulated network minimizing the elastic energy, have also been very successful in unveiling important features of viral assembly. They have been used to describe for instance the growth of spherical and polyhedral shells [80, 81], the (irregular) shapes of conical virus capsids like HIV [82, 83], the mechanisms selecting the size and structure of spherical viruses [84] or the need of scaffolding proteins in the assembly of large icosahedral viruses [85].

Finally, higher-resolution coarse-grained strategies have been more challenging to investigate [86]. However, some groups have studied, for instance, the assembly of HIV capsid proteins [38].

These simulation studies have confirmed many of the assumptions discussed in the aggregation and CNT theories and have provided a better understanding of the difficulties and ingredients required for the successful assembly of empty viral capsids. The combination of these theoretical and simulation studies is opening the door to a higher understanding about the assembly of viral capsids, and hopefully will facilitate the development of novel antiviral strategies.

19.3.2 RNA Co-assembly

Although in vitro experiments have demonstrated that in certain conditions the viral capsid proteins of ssRNA viruses alone are able to reconstruct the native capsid structure, in vivo their assembly generally (though not always) requires the presence of the genetic material [87] (see Chap. 13). Consequently, the self-repulsion and entropy of the nucleic acid, and its interaction with the inner surface of the capsid are new ingredients that must be considered to characterize the free energy of the complete system.

In physiological conditions, ssRNA behaves like a flexible polymer with a persistence length on the range of the nucleotide size. In this scenario, the main driving force in the co-assembly process is the electrostatic interaction between the ssRNA and the positively charged residues of the capsid, which often concentrate in disordered terminal segments of the capsid proteins. By virtue of electroneutrality one would expect a 1:1 correspondence between the length of the viral RNA and the charge of the inner part of the capsid in contact with the genetic material. However, experiments have determined that a variable part of the nucleic acid negative charge is neutralized by metal ions and/or polycations (e.g., spermine or spermidine) in different viruses. Moreover, the distribution of charges in the capsid, the excluded volume effects, and the Donnan potential (i.e., the difference in electrostatic potential between inside and outside the capsid due to the unbalance of charged macromolecules) have been shown to play a crucial role in the possible overcharging of viruses. Accordingly, there seems to be no universal genome-to-capsid charge ratio [51, 88].

In a first approach, several models have investigated the thermodynamics of co-assembly combining basic electrostatic and polymer physics theories, as discussed below. As a starting point, the complexation energy between ssRNA and the capsid can be evaluated again using the DH mean field theory [50]. For spherical thin shells this yields $\Delta F_c = \frac{\pi\sigma^2 R^2}{\varepsilon_0 \varepsilon_r}(fd - \kappa^{-1})$, where d is the absorbed layer thickness of the ssRNA, f is a numerical factor, and κ^{-1} is again the Debye screening length. The co-assembly is thermodynamically favored when this complexation energy becomes negative, which depends on the competition between the characteristic scales d and κ^{-1}. Contrary to the case of dsDNA bacteriophages discussed below, here the nucleocapsid reaches an energy minimum that can be evaluated from the previous energy of formation, indicating that ssRNA viruses may not be pressurized [50, 89]. More refined thermodynamic models that include the distribution of charges in the capsid, the possible presence of charged peptide arms, and the contribution of charged macromolecules, have been developed and can justify the overcharging of some ssRNA viruses in nature and the favorable co-assembly of viruses with no net capsid charge [88].

Remarkably, many RNA viruses store several strands of genetic material rather than a long single sequence. This was thermodynamically characterized by Zandi et al. [90] using a simple Flory mean-field theory, which accounts for the elastic compression of the chain, the self-repulsion of the genetic material, and the interaction of the polyelectrolyte with the inner wall of the capsid. This theory also justified the results from several in vitro experiments showing that the genome length can control the size of the final capsid [91]. Interestingly, the results of these thermodynamic models are valid also for other negatively charged polymers different from RNA, as has been proved experimentally [46].

Regarding the organization of ssRNA, experimental studies have shown that it is distributed inhomogeneously inside the capsid, being very dense close to the inner surface and looser as we move toward the center, which in many cases is completely devoid of genetic material. The RNA in the central region does not adopt any particular structure and its organization is a consequence of the self-repulsion of the genetic material [92]; instead for some viruses the outer part sews the capsid proteins tracing a Hamiltonian path (i.e., a path that visits each vertex of the capsid only once) through the edges of a dodecahedron partially recovering the icosahedral symmetry of the shell, which could have an important role during the assembly of the capsid [43, 87].

Recently, several simulations have started to investigate the co-assembly of RNA viruses. Again, this is a challenging task that requires coarse-graining and strong computational efforts. One of the first models explored the encapsulation of a polymer defined on a cubic lattice, but more realistic off-lattice simulations with flexible polymers encapsulated by coarse-grained capsomers or capsid proteins have been feasible lately [15, 93–95]. These simulations have assessed the role played by the genome length, the relative stoichiometry of genome versus capsid proteins, and the strength of the different interactions in the efficiency of the encapsulation process. It has been found that there is an optimal genome length that maximizes the encapsulation efficiency, which depends on the competition between capsid–capsid and capsid–genome interactions. Remarkably, depending on these interactions, two plausible assembly pathways have been identified (see Fig. 19.7). In the first one, the polymer is encapsulated in concert with the capsid formation, a process that is the natural extension of the assembly of empty capsids. Interestingly, in this case the capsid formation can be also described by classical nucleation theory, where the genome reduces significantly the nucleation barrier [93]. The other pathway takes place for strong subunit-RNA interactions. Initially many subunits (typically in excess of the total needed) attach to the genome and then they cooperatively rearrange to form the closed capsid. Other studies suggest the existence of special sequences of RNA with high-affinity binding to the capsid protein, called packaging signals [96–99], that strongly favor the interaction with the capsid proteins and are crucial for the assembly. They can also be potentially used to deregulate capsid assembly [100]. However, the process of nucleic acid-assisted assembly of ssRNA viruses is still far from well understood; further experimental and theoretical studies

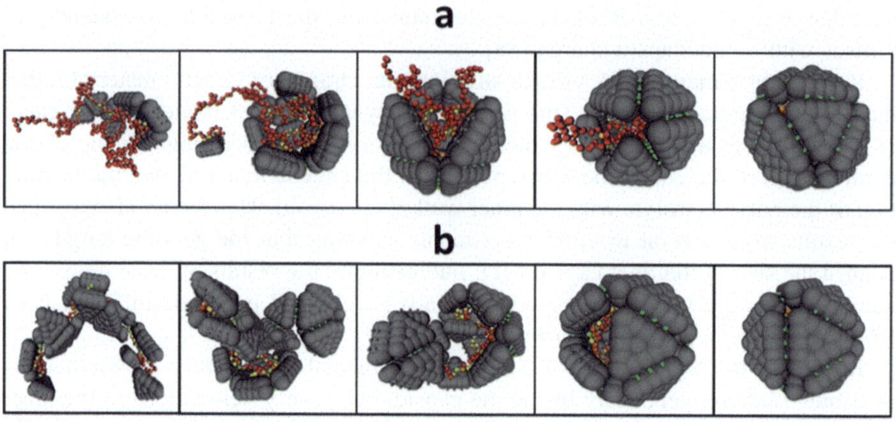

Fig. 19.7 Mechanisms of co-assembly for RNA virus. Different snapshots of a coarse-grained simulation showing two possible pathways in the formation of an RNA virus particle. (**a**) The RNA (red chain) is encapsulated in concert with the formation of the capsid by the addition of capsid proteins (modeled by multiple gray beads with green interaction sites). (**b**) For strong RNA–capsid protein interactions, first, many capsid subunits (in excess above the stoichiometric ratio) attach to the RNA in a disordered way and then cooperatively rearrange to form the complete capsid. Reproduced from [15] with permission

are clearly required, and the interested reader is referred to Chap. 13 for current views on the process based on experimental observations.

The encapsulation of non-genetic materials, such as gold nanoparticles, emulsions, or negatively charged polyelectrolytes by viral proteins, has also been explored in experiments and simulations, mainly driven by their potential technological applications (see Chap. 22).

19.3.3 dsDNA Packaging

There are two physical properties of dsDNA that play a crucial role in the assembly of many viruses: elasticity and charge. Contrary to ssRNA and ssDNA, dsDNA is a relatively stiff polymer. In a first approximation, its rigidity is characterized by its persistence length, i.e., the length over which the DNA keeps straight under thermal fluctuations, which is about 50 nm for typical monovalent salt physiological conditions. This persistence length is similar to the characteristic capsid dimensions. Since dsDNA is usually stored at high densities, this implies that a significant elastic cost is paid during the packaging of the dsDNA genome inside the protective viral shell. Regarding dsDNA charge, every base pair contributes with two negative charges. As for the viral ssRNA and ssDNA genomes, part of the dsDNA charge may be neutralized by metal ions and/or polycations. But, in general, the charge causes the DNA chains to repel each other. The electrostatic term tends to dominate over the elastic one, although the latter seems to determine the conformation adopted by the confined dsDNA [101]. These two factors do not facilitate a spontaneous co-assembly of the capsid, except in the presence of condensing proteins like in polyoma virus or adenovirus, and this is, most likely, one of the main reasons why many dsDNA viruses first assemble an empty capsid, the procapsid, and then package the genetic material using a molecular motor (see Chap. 13). Another reason is that spherical capsids with the mechanical resistance required to hold large internal pressures cannot be assembled directly [70].

Along the years, several models have evaluated the energetics of the packed DNA [102–104]. Here we will describe the so-called inverse spool model. This theoretical approach assumes that the total free energy of confined DNA consists of the two contributions, elastic and electrostatic, discussed above. Then the total energy is given by $G_{\mathrm{tot}}(L,d_s) = G_b + G_{\mathrm{int}}$ and depends on both the total length of the genetic material L and the separation between strands d_s. The elastic contribution, G_b, is calculated assuming an inverse spool arrangement of the DNA, where the strands are packed in a hexagonal array with spacing d_s. Then the elastic energy and the total DNA length become:

$$G_b(L) = 2\pi\xi_p k_B T / \left(\sqrt{3}d_s\right) \int_R^{R_{\mathrm{out}}} (N(R')/R')dR' \tag{19.7}$$

$$L = 4\pi / \left(\sqrt{3}d_s\right) \int_R^{R_{\mathrm{out}}} R'N(R')dR' \tag{19.8}$$

where ξ_p is the persistence length, $N(R')$ is the number of hoops of radius R' in the capsid, and R and R_{out} are the inner and the outer radius of the inverse spool, the latter taken as the radius of the inner surface of the capsid (see Fig. 19.8).

An accurate first principles calculation of the DNA–DNA interactions is a difficult and unachieved task. Thereby, the electrostatic interaction G_{int} has been commonly characterized phenomenologically from osmotic stress experiments at the proper conditions [105]. These experiments have shown that the electrostatic interaction between DNA strands has a different behavior depending on the valence of the added salt. For monovalent and divalent salts, the contribution is purely repulsive and the dependence of the osmotic pressure $\Pi(d_s)$ with respect to the separation between strands is given by:

$$\Pi(d_s) = F_0 e^{-\frac{d_s}{c}} \tag{19.9}$$

where c and F_0 are parameters that characterize the decay length and strength of interactions, which depend on salt conditions. In the case of trivalent and tetravalent salts the effective electrostatic interaction has an optimal distance d_0 between strands, i.e., at smaller separations leads to a repulsive interaction whereas at higher separations promotes attraction between the strands. A convenient empirical expression for the osmotic pressure in this situation is:

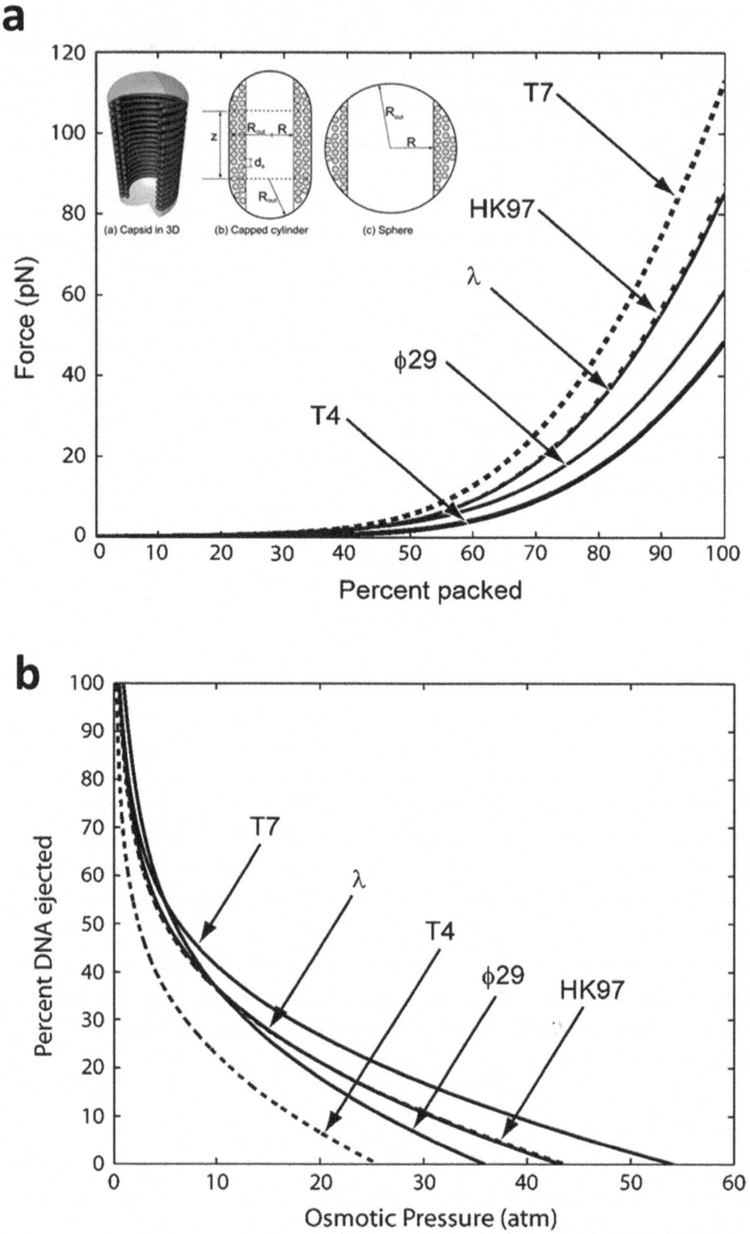

Fig. 19.8 Forces and pressures of confined dsDNA. (**a**) Forces resisting dsDNA packaging (and assisting dsDNA ejection) were estimated using the inverse spool theory for different bacteriophages (T7, λ, T4, Φ29, HK97) in 0.5× PBS buffer. The inset shows the idealized organization of the dsDNA inside different capsids assumed in the model. (**b**) Osmotic pressure inside the capsid (in the horizontal axis) as a function of the percentage of DNA ejected estimated by the inverse spool theory, for different bacteriophages (T7, λ, T4, Φ29, HK97) in a buffer containing10 mM $MgSO_4$. Figures reproduced from [103] with permission

$$\Pi(d_s) = F_0\left(e^{-\frac{(d_s - d_0)}{c}} - 1\right) \tag{19.10}$$

From the osmotic pressure, the interaction energy can be obtained yielding:

$$G_{\text{int}} = \sqrt{3}F_0L(c^2 + cd_s)e^{-\frac{d_s}{c}} \tag{19.11}$$

in the purely repulsive regime and

$$G_{\text{int}} = \sqrt{3}F_0L\left[(c^2 + cd_s)e^{-\frac{(d_s - d_0)}{c}} - (c^2 + cd_0) - \frac{1}{2}(d_0^2 - d_s^2)\right] \tag{19.12}$$

in the presence of tri- and tetra-valent counterions. It is important to emphasize that these interaction free energies G_{int} obtained from osmotic stress experiments not only account for the electrostatics of the DNA and counterions, but also for entropic and hydration contributions [103, 105].

Knowing the contributions of bending and interaction terms, the optimal energy and DNA spacing are then obtained by minimizing the total energy $G_{\text{tot}}(L, d_s)$ with respect to d_s for a given total genome length L. The internal pressure P of the confined DNA or the force F at which it will be ejected can be finally obtained from it by their standard thermodynamic definitions as $P = -\partial G_{\text{tot}}/\partial V$ or $F = -\partial G_{\text{tot}}/\partial L$. This model predicts the forces during the ejection process and the internal pressure in bacteriophages [103, 106], which are in the range of 20–60 atmospheres as determined by AFM, osmotic pressure experiments, and recently using SAXS [107–109] (see Chaps. 10 and 18, and Fig. 19.8). Remarkably, bacteriophages require the strongest molecular motors reported so far to store the dsDNA at such high densities. In many cases, this stored energy is used to initially drive the injection of the dsDNA genome into the bacteria during the infection process. Interestingly, some bacteriophages like PRD1, might use an internal membrane to sustain this pressure during genome delivery thus speeding up injection [110].

Although the fact that the dsDNA is compacted in an inverse spool configuration is still controversial [101], the estimates of the pressures are not very sensitive to the configuration, since they are mostly controlled by the electrostatic contributions [111] and the packing density of the DNA. In fact, using in Eqs. (19.9) and (19.10) a simple estimate of the strand separation:

$$d_s = \sqrt{\frac{2V}{\sqrt{3}}\frac{1}{\sqrt{L}}} \tag{19.13}$$

which is obtained by equating the total volume available inside the capsid V with that of the closed-packed DNA (assuming that it occupies the whole volume), one gets an accurate approximation of the capsid pressure due to the confined DNA.

19.4 Mechanical Stability of Capsids

Several studies indicate that virions can be subjected to substantial mechanical stress, for instance, due to osmotic shock, related to a sudden change of salt concentration in the environment, or to the packaging of the viral genome at high densities (see Chaps. 13 and 18). Interestingly, the last scenario concerns many dsDNA bacteriophages, which must withstand up to tens of atmospheres arising from the confined viral genome, as we have just seen. Additionally, after the self-assembly of the capsid, to become infective many viruses undergo a maturation process that may properly tune their physical and mechanical properties. In bacteriophages, this maturation often involves the modification of the capsid shape through the so-called *buckling transition*, where an initial spherical procapsid undergoes a transformation into a polyhedral-like shell resembling an icosahedron (see Chap. 13). All these complex scenarios illustrate different sophisticated strategies that viruses have developed to protect their genetic material, remaining infective despite the hazards of the environment. Thus, their mechanical characterization becomes of particular interest to understand the stability of viruses.

The characterization of the mechanical properties of different viral capsids have been the subject of intense investigations in the last years. A few studies relied on *bulk* experiments using biochemical and physical techniques, like inelastic Brillouin light scattering. However, the recent application of atomic force microscopy (AFM; Chap. 9) and optical tweezers (Chap. 10) have allowed the experimental determination of the mechanical properties and mechanochemical function of individual viral particles (see Chaps. 9, 10, and 18). This has also boosted the interest in developing and applying theoretical frameworks to characterize the mechanical response of viral capsids.

Thus, in this section we will discuss the main theoretical approaches applied to the mechanical characterization of viral capsids. The studies are grouped into three blocks: first, we introduce some theoretical strategies to investigate the elastic response of viral capsids, which are essentially based on the TST framework and computational simulations; then some physical ideas behind the maturation of viruses are discussed; finally, we briefly review some techniques that, by analyzing thermal vibrations of the capsid, facilitate the extraction of important structural properties and general information about shell transformations.

19.4.1 Elastic Response of Viral Capsids

Elastic Properties from TST

TST constitutes the natural starting framework to study the mechanical properties of viral capsids. This model has been particularly useful to complement AFM indentation experiments (see Chap. 18), facilitating their interpretation and allowing the extraction of the elastic properties of the capsid [112]. To describe the local deformations of the capsids generated in AFM experiments, it is more convenient to rewrite the elastic energy in terms of a free energy functional of the indentation profile $\zeta(\mathbf{r})$ given by $\Delta F = \int dS\left\{\frac{1}{2}\kappa(\Delta\zeta)^2 + \frac{1}{2}Y\left(\frac{2\zeta}{R}\right)^2 + \frac{1}{2}\tau(\nabla\zeta)^2\right\}$. The first term describes the bending energy (with $\Delta\zeta$ being the curvature), the second term the stretching energy, and the third term the extra work required to deform the capsid, assuming that the shell is subjected to a tension τ [112]. The indentation profile can be obtained from the differential equation $\delta\Delta F/\delta\zeta(\mathbf{r}) = f(\mathbf{r})$, where $f(\mathbf{r})$ is the applied radial force per unit area. The previous equation is valid only for small deformations and can be solved analytically only in special situations, for instance, in the case of a point force applied on an empty spherical shell. Let us focus first on the scenario where the tension τ, which can be related, for example, to an internal pressure, vanishes.

For indentations much smaller than the shell thickness, TST predicts a linear relation between the applied force and the resulting indentation, as follows from the simple scaling arguments described in Chap. 18. The corresponding effective spring constants (as a measurement of mechanical stiffness; see Chap. 18) for spherical and cylindrical shells are $k_{\mathrm{sph}} = \frac{2}{\sqrt{3(1-\nu^2)}}\frac{Eh^2}{R}$, and $k_{\mathrm{cyl}} \propto Eh\left(\frac{h}{R}\right)^{3/2}$, respectively [22]. As shown in Chap. 18, using these expressions the Young's modulus of a virus can be derived from the slope of the force indentation curves obtained with the AFM, provided the size and the thickness of the capsid are known.

In addition, as described in Chaps. 10 and 18 and Sect. 19.3.3, the presence of the confined dsDNA genome can build up a substantial pressure in the capsid, which may ultimately modify its mechanical response. In a first approximation, the classical Laplace's law is able to estimate the tension generated in a shell by a pressure difference Δp, which for a spherical shell is $\tau = \Delta pR/2$. Using this value in the previous free energy functional, and solving for a point force, one obtains an effective spring constant [113],

$$k = k_{\mathrm{empty}}\frac{\pi}{2}\frac{\sqrt{\tilde{\tau}^2 - 1}}{\operatorname{arctanh}\sqrt{(1 - 1/\tilde{\tau}^2)}} \tag{19.14}$$

which depends on the ratio $\tilde{\tau} = \Delta pR/k_{\mathrm{empty}}$ between the pressure difference Δp multiplied by the radius R and the spring constant k_{empty} of the empty shell. This expression provides a simple way to estimate the internal pressure of spherical viruses by comparing the spring constants of the emptied mature capsid and the virion, provided experiments reveal that the cause of this difference is indeed the existence of internal pressure (see Chap. 18).

It is worth stressing that experiments and calculations have shown that tension does not necessarily originate from an internal pressure. For instance, in the icosahedral minute virus of mice, viral ssDNA segments bind to equivalent sites at the capsid inner wall and act like molecular buttresses that reinforce anisotropically the viral particle (see Chap. 18); in the case of bacteriophage Φ29, it has been shown that the capsid is also mechanically anisotropic, because of built-in stress [114]. The development of pre-stress is a common strategy in engineered structures and probably reinforces the capsid to better tolerate the pressure developed during the packaging of the viral genome.

An important mechanical property of a capsid, especially for dsDNA viruses, is its tolerance to pressure before bursting. TST provides simple expressions that relate the largest pressure that a capsid can resist with the maximum relative radial expansion and the maximum tangential stresses. The largest stresses can be in turn evaluated in terms of the inflection point of the potential between subunits, i.e., the maximum force that the interaction can hold [115, 116]. In addition, by simply using Laplace's law, one obtains that the maximum pressure Δp^* that a virus can tolerate, at equal interaction strengths, decays with the radius of the shell as $1/R$. In other words, at equivalent values of the molecular interactions, big capsids should be less efficient at tolerating pressures. Notice that this decay also applies to the effective spring constant k, and it can be also shown that this follows for the bulk modulus [29, 117].

Finally, TST and scaling arguments can also be used to characterize large indentations of viral capsids using point-like axial loads. For large deformations, the response of the viral capsid becomes strongly nonlinear, which in many cases is associated to an inversion of the curvature (*buckle*) in a region of the shell around the indentation point [22, 118]. The elastic analysis shows that for spherical shells this inversion typically takes place when the indentation becomes comparable to the shell thickness h, or alternatively, when the load force exceeds a threshold value $F_{inv} \sim \kappa/R$. In the nonlinear deformation regime, the force-indentation curve is expected to show a squared root dependence on the indentation [22, 118]. However, in these situations, the characterization of the indentation using TST becomes very challenging, and other techniques, like finite element analysis (FEA) discussed in Chap. 18, are more suitable. FEA studies have demonstrated that capsids become stiffer and show higher non-linearities (associated to local bulging or buckling) for high FvK numbers [26, 112]. The relevance of the finite thickness and non-uniformity of viral shells has been also explored, for instance, in the softening of CCMV after its pH swelling transition, or in the anisotropic response of HBV when indented at the different axis of icosahedral symmetry [26, 112].

Discrete Models and Simulations

However, FEA is essentially a numerical solution of continuum elasticity equations, and accordingly it shares also certain limitations. For instance, the plasticity or the breaking of viral shells observed after repeated indentations cannot be captured. In addition, the discrete and possible anisotropic interactions between capsid building blocks cannot be easily incorporated in this description. To overcome these limitations, one must resort to the use of discrete models.

Unfortunately, contrary to the case of the structure and assembly of viral capsids (described in Sects. 19.2 and 19.3), there have not been yet many models of this kind studying the deformation of viral capsids. The most popular simulation methods have been based on the discrete triangularization of TST, described in Sect. 19.1

[119], which have been able to reproduce the non-linearities in the indentation of viral capsids, and estimate mechanical differences between some T-numbers. It is worth noticing that this should not be confused with FEA, which discretizes the capsid in tiny elements that do not correlate with the positioning of the capsid proteins in the shell.

Also at the low-resolution level, the model of Zandi et al. [116] has highlighted the importance of the particular arrangement of the capsomers and the triangulation number on the mechanical properties of spherical capsids. The application of virtual-AFM simulations on these models confirmed the impact of capsomer arrangement on the mechanical response of capsids [120]. At the high-resolution level, Schulten's group developed a coarse-grained model of HBV complementing AFM experiments. The simulations were in agreement with the experiments without the necessity of any type of fitting [112, 121]. One of the most interesting features of this coarse-grained approach was the identification of the local rearrangements of capsid proteins and the irreversible conformational changes produced during the indentation of HBV [122], which was not captured by FEA studies. Perilla's group recently refined this technology to capture molecular details while reducing the computational cost of molecular dynamics significantly [123].

19.4.2 Maturation and Buckling

As described in Chap. 14, to become infective many viruses require a maturation step after the self-assembly of the procapsid. This process seems to tune different properties of the viral shell that are manifested in many cases by a noticeable change in the capsid shape, generally associated to a *buckling transition*, which is of relevance among bacteriophages. In this transformation, the initial spherical procapsid with icosahedral symmetry undergoes a transition into a polyhedral shell with flatter triangular faces and an overall shape resembling an icosahedron.

TST provides again an excellent framework to establish the physical basis of the buckling transition [8]. In this context the asphericity of the capsid, i.e., the degree of buckling, is exclusively determined by the FvK number, $\gamma = YR^2/\kappa$, which compares the stretching and bending contributions in the shell. When FvK exceeds a certain threshold value, $\gamma_B \sim 10^2$, the stretching energy of a spherical shell can be reduced if the 12 pentamers buckle out inducing the faceting of the shell into a polyhedral shape (see Fig. 19.1). Remarkably, it can be shown that within TST, the FvK number can be expressed as $\gamma = 12(1 - \nu^2)(R/h)^2$, which means that it essentially depends on the square of the ratio between the capsid radius R and its thickness h. This suggests that big and thin shells should be more likely to adopt polyhedral rather than spherical shapes, in agreement with experimental observations of the faceted nature of large viruses [117].

Alternatively, a low-resolution discrete model has shown that the buckling transition depends on the triangulation number T and the icosahedral class[2] P of the virus structure [117]. In particular, $P = 1$ shells are most likely to produce polyhedral shapes, whereas viruses with $P = 3$ prefer to remain spherical (see Fig. 19.9). The addition of minor proteins can modify these mechanical properties facilitating the evolution of capsids from different classes within the same viral structural lineage [124]. For big capsids, the polyhedral shape becomes systematically the most stable, in agreement with TST. This study showed that in general the buckling transition can be also induced by a capsid expansion, in consonance with virus maturation. Furthermore, in comparison with the spherical shape, the resulting icosahedral shell is mechanically stiffer, tolerates larger expansions, and withstands higher internal pressures before failing. This suggests that the polyhedral shape could have a certain advantage for dsDNA phages that rely on the pressurization of their genetic material to be infective, justifying their buckling transition.

However, despite the very useful insights provided by the theoretical tools described above, one should not forget that the occurrence of the buckling transition during virus maturation is a complex process that often involves different molecular mechanisms including conformational changes, cleavage, and even the formation of covalent bonds. These mechanisms are virus-specific and cannot be easily characterized by most of the models described previously.

19.4.3 Viral Capsid Vibrations

An interesting fundamental property of materials is the relation between the unavoidable thermal vibrations that they suffer at finite temperatures and their structural properties. In the case of capsids, this has been largely explored by applying normal mode analysis (NMA) of elastic networks based on all-atom or cryo-EM structural models of specific viruses [125]. Essentially, in NMA the interaction between atoms or pseudo-atoms is approximated by a harmonic potential, and the individual vibrations of the subunits are characterized in terms of collective harmonic oscillations, called normal modes, which can be obtained by diagonalizing the second derivative of the potential energy (Hessian) [126]. In this description, the low frequency modes are associated with conformational changes in the capsid, and the high frequency modes describe localized motions.

In this way, NMA has allowed the study of the transition during the swelling of CCMV, the maturation of *Nudaurelia capensis* ω virus and phage HK97 [126], successfully identifying disassembly units on different viral structural lineages [42]

---

[2]The icosahedral class P was introduced by Caspar and Klug to classify triangulation numbers and it is related to them by $T = h^2 + k^2 + hk = Pf^2$, where f is the greatest common divisor of the integers h and k. It should not be confused with the pseudo-triangulation number P, used to label icosahedral viruses when the subunits that occupy quasi-equivalent positions are chemically or structurally different (see Chap. 2).

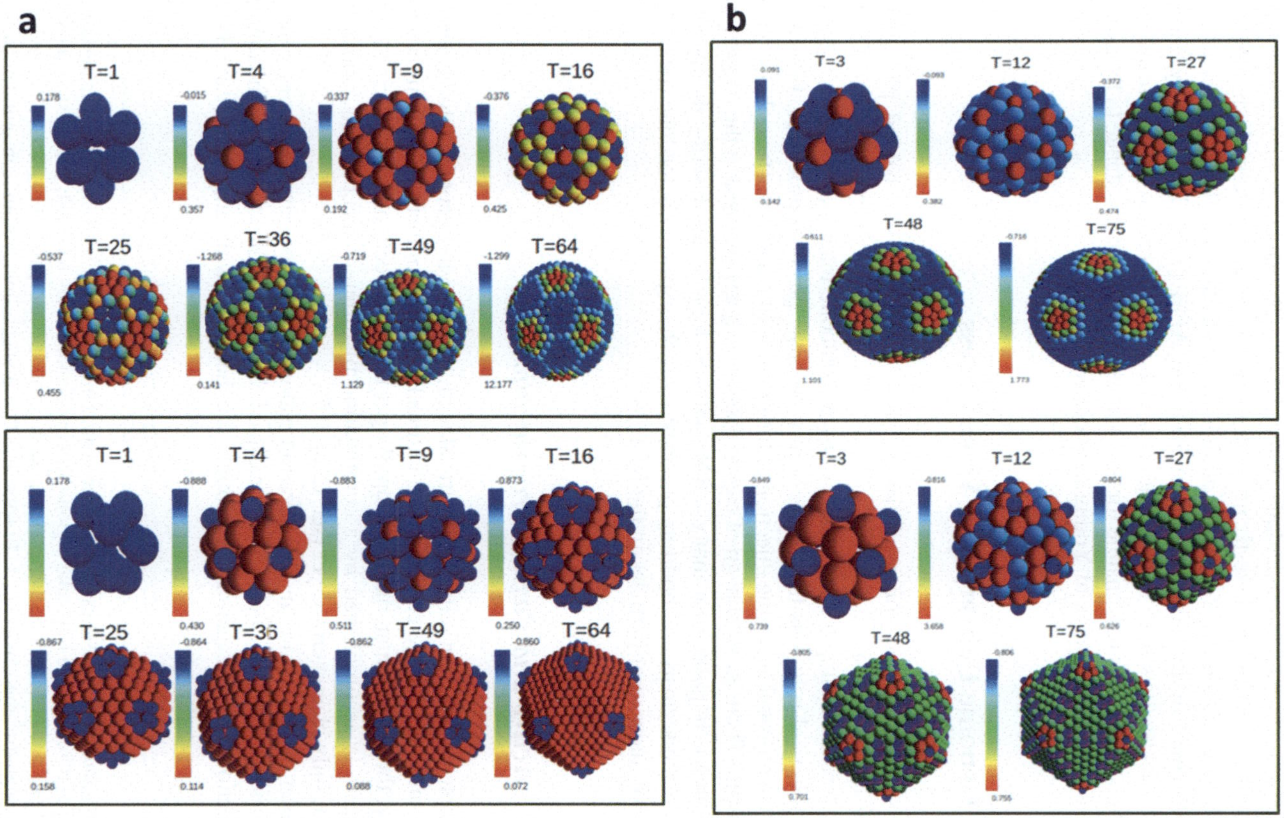

Fig. 19.9 Relevance of capsid structure in the buckling of spherical viruses. **(a)** The distribution of local stresses (indicated by the color scales) for spherical and polyhedral (faceted) T-number shells belonging to the class $P = 1$. Capsomers in blue are stretched, whereas those in red are compressed. In the spherical capsids, positive stress concentrates on the lines connecting pentamers for $T > 4$, delimiting triangular areas with stretched hexamers. The absolute values of

or uncovering latent instabilities [127]. Besides, investigations combining high-resolution simulations and NMA have shown that the maturation of HK97 follows a pathway dominated by icosahedral symmetry [128]. Remarkably, the combination of all-atom simulations and elastic network models constitutes a promising alternative method to obtain the Young's modulus of viruses that should complement the values obtained by the combination of TST and AFM [129]. Interestingly, these multiscale techniques have shown that cross-linking between capsomers during the maturation of HK97 is responsible for the increase of the shell resistance to breaking without noticeable changes in Young's modulus [130].

19.5 Genome Delivery and Virus Egress

In this section, we are going to discuss some of the physics behind particular mechanisms of genome delivery and virus exit from and entry into cells. Once again, the relative simplicity and lack of own metabolic machinery of viruses have led to biological solutions that strongly rely on basic physical processes. First, we describe in physical terms the injection of the genome of dsDNA bacteriophages into the host cell. Then, a mechanism of capsid disassembly is discussed in the context of CNT, which is important for many ssRNA viruses. Finally, we describe the modeling of the endocytosis and budding processes, which respectively mediate the entry and exit of many viral particles in the host.

19.5.1 Active and Passive Translocation

Bacteriophage infection relies on the translocation of the viral genome into bacteria, leaving an intact empty capsid in the extracellular environment. In this infection pathway, the genetic material must rapidly get inside the cell, and the efficiency of the process is accomplished by a combination of different mechanisms, some of them not fully characterized yet (see Chap. 17).

In order to succeed in this translocation pathway, many bacteriophages rely on a proteinaceous tail that is attached to the capsid (head) and helps to initiate and direct the injection of their genomes through the bacterial wall and membrane (see

Fig. 19.9 (continued) these stresses get larger as the T-number of the shell increases. The hexamer stretching is relieved if the triangular regions between pentamers are flattened; thus these structures have a strong preference for buckling. (**b**) Same for spherical and polyhedral (faceted) T-number shells belonging to the class $P = 3$. In spherical $P = 3$ shells, compression concentrates on the pentamers, whereas the hexamers surrounding them are slightly stretched, leading to a dodecahedral pattern instead of the icosahedral one observed in $P = 1$. In this case, nothing indicates any particular relief of the stress on the hexamers upon adopting an icosahedral shape. Reproduced from [117]

Chap. 17). This process can involve complex conformational changes in the tail proteins requiring energy. A typical example is the contractile tail of bacteriophage T4. The contraction of the tail sheath drives the rigid inner tube through the bacterial wall and allows the virus to inject its DNA into the bacterium (see Chap. 17).

This contraction has been described in the context of physics as a *martensitic transition* [131, 132], a well-known phase transition occurring typically in solids. The idea is that the tail proteins can adopt two possible conformational arrangements: one extended and strained, and the other contracted and relaxed. A small activation barrier separates both states, where the initial one (the extended) has a higher energy. When the tail attaches to the membrane it triggers a rapid transition that releases the elastic energy stored by the proteins in the extended conformation adopted during the assembly of the virus. The contraction force derived from these studies is about 100 pN, roughly twice the stalling force of the motor packaging the DNA of Φ29 (see Chap. 10). This type of martensitic mechanism seems to be also behind other large conformational changes in the fusion proteins of different enveloped viruses, such as HIV and influenza virus [131] (see Chap. 16).

After a triggering event the dsDNA starts to enter the cell. As mentioned in Sect. 19.3.3, the capsid internal pressure drives the initial ejection of the genetic material. The initial forces and velocities of this entrance can be properly quantified by the inverse spool model. But after roughly 50% entrance of the DNA the pressure difference vanishes, and the remaining genetic material relies on other mechanisms to get in (see Chap. 13).

Different physical strategies can be also conceived to explain the entrance of the remaining viral genome length inside the cell. The simplest one is by pure diffusion. In this case, thermal fluctuations shake the nucleic acid back and forth through the cell wall pore until the genome completely enters into the host. However, the translocation time of this mechanism scales as the square of the genetic material length L leading to $\tau_{\mathrm{dif}} = L^2/D$, where D is the diffusion coefficient of the chain, and for typical viral parameters, this time becomes prohibitively long. Alternatively, a second mechanism involving the presence of binding proteins has been studied to explain how the remaining viral genome is driven into the cell. Here, as soon as a portion of DNA gets inside the cell, certain proteins bind to it and impede its diffusion backward, which rectifies the diffusion of the chain and speeds up the entrance by a factor proportional to the number of binding sites. Nevertheless, despite this rectification, the entrance time estimated for this mechanism is still too long. These two simple strategies cannot explain the entrance of the viral genome. However, it was shown that the proteins that bind to the DNA not only rectify its diffusion but simultaneously exert an effective translocating force that considerably increases the speed of the DNA entrance [133]. This additional force was characterized by a simple model evaluating the free energy of a stiff portion of DNA entering a spherical cell in the presence of nonspecific binding proteins (see Fig. 19.10). This simple model shows that the total free energy becomes more favorable the more genetic material gets inside the cell translating into a powerful effective force that drives the translocation. The effective force observed in coarse-grained simulations was also quantified using this free energy in a simple

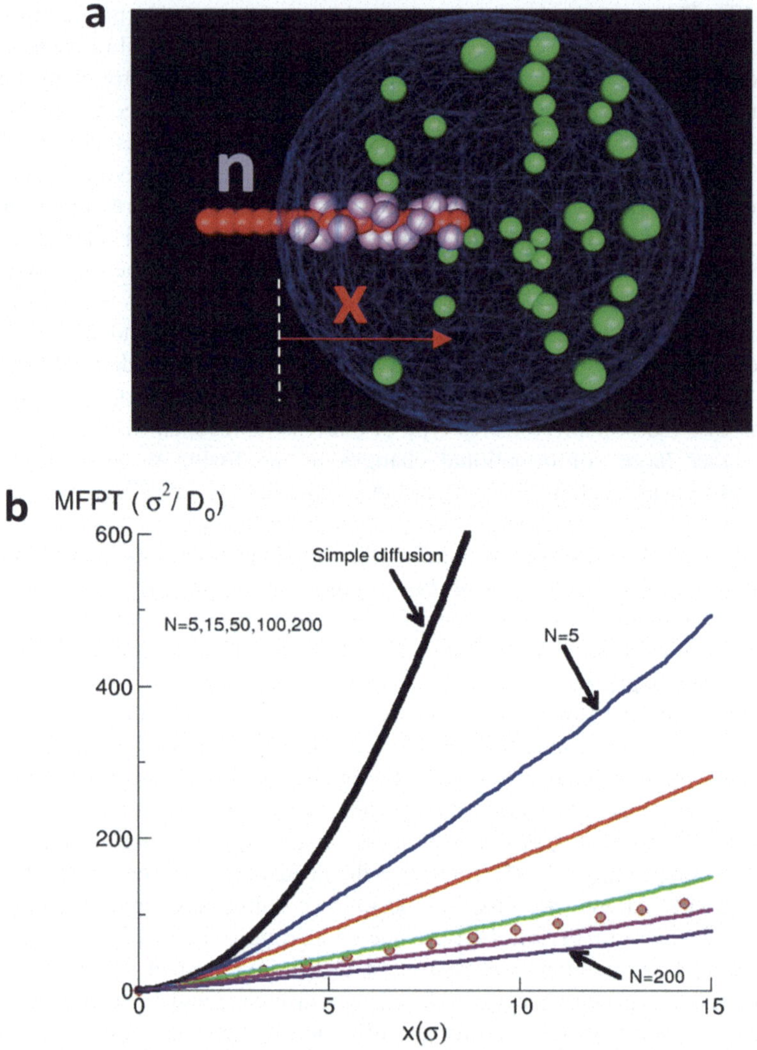

Fig. 19.10 Model for the translocation of a stiff portion of dsDNA in the presence of binding proteins. (**a**) Schematic illustration of the simulation setup, where a stiff portion of dsDNA (in red) enters a cell (in blue) in the presence of binding proteins (green (unbound) and violet (bound)). (**b**) Average time required for the entrance of a given distance x (measured in units of the bead size σ) by simple diffusion (black line) and in the presence of different number N of binding proteins. For a sufficiently large concentration, the entrance becomes even faster than by ratcheted diffusion (red dots). Reproduced from [133]

non-equilibrium kinetic model. Interestingly, this speed up effect associated to the binding energy of proteins has been verified experimentally in in vitro experiments that measure the kinetics of ejection (in that case in the presence of DNAse) [134]. On the other hand, although the model can also be refined by accounting for the entropy of the DNA chain [106], there are still many puzzling features of this translocation process of bacteriophages to be described, as has been shown by the experimental studies of Phillips et al. [135, 136].

DNA polymerases can also play an active role in pulling in the DNA of some viruses, which has been studied experimentally with optical tweezers (see Chap. 10) and analyzed with stochastic models.

19.5.2 Capsid Disassembly

Eukaryotic viruses and some bacteriophages typically follow an infection pathway that requires virions to penetrate the host. Once inside, the capsid of these viruses must change its conformation or, in many cases, disassemble to allow the release of the genetic material in specific regions of the cell, which will ultimately initiate the viral replication factory (see Chap. 8).

In many cases, viruses follow a post-assembly maturation stage that strengthens the initial weak interactions responsible for the assembly (see Chap. 14). Nevertheless, other viruses do not necessarily go through a maturation process, but they have still to avoid disassembly during their trip to infect a new host. If the assembly of viruses were an equilibrium process, viruses would disassemble as soon as the capsid protein concentration drops below the threshold value c^*. In this context the models introduced in Sect. 19.3.1 can be invoked. In particular, CNT offers a good explanation of how viruses may accomplish this delicate balance. It is precisely the existence of an activation barrier that had to be overcome during the assembly that prevents their disassembly. Then the process of viral disassembly can be thought of as the inverse of the assembly and can be mathematically described using the same CNT theory [29].

The disassembly of viruses like HBV has been probed by in vitro experiments (see Fig. 19.4a). This virus does not suffer any maturation or conformational change and the HBV capsid proteins can be reassembled repeatedly. In these experiments, one starts from a given total concentration of proteins in the form of complete capsids and dilutes this concentration monitoring the disappearance of capsids. The results of these experiments have shown that the threshold concentration at which capsids disassemble is much lower than c*, the value at which they start to assemble. This can be easily understood and quantified in the framework of CNT.

To initiate the disassembly, one has now to jump the barrier in the opposite direction to that of the assembly (see Fig. 19.5b). The height of this disassembly barrier is given by $\Delta G^*_{\mathrm{dis}} = \Delta G^*_{\mathrm{as}} - \Delta G(q)$, where ΔG^*_{as} and n^* are given by Eqs. (19.4) and (19.5), and the disassembly rate depends also exponentially on $\Delta G^*_{\mathrm{dis}}$ as in Eq. (19.6). At the conditions at which the assembly took place, this barrier is too high to be surmounted and capsids remain stable. In fact, negative $\Delta\mu$, implying $c \ll c^*$ are needed to lower the disassembly barrier up to a point that disaggregation is feasible. Since it is the height of the assembly and disassembly barriers that dominates the kinetics of both processes, a simple argument can be made to get a reasonable estimate of what would be the concentration required for disassembly. Assuming that the barrier height to be surmounted for disassembly has to be the same than that of the assembly (to observe the processes at similar time scales) one gets the following simple expression:

$$(c^*)^2 = c_1^{\mathrm{as}} c_1^{\mathrm{dis}} \tag{19.15}$$

that relates the threshold protein concentrations required for the assembly, c_1^{as} and disassembly, c_1^{dis} with the equilibrium concentration c^*, and through Eq. (19.1), with the binding energy per subunit. Thus, these simple ideas provide a way to extract useful information from combined assembly and disassembly experiments [29], as confirmed recently in experiments with CCMV [137].

19.5.3 Budding and Endocytosis

Many animal viruses require also crossing in and out of cellular membranes to proceed with the infection. In enveloped viruses, the lipid envelope and fusion proteins help to accomplish this task (see Chap. 16); non-enveloped viruses are generally able to get inside the cell through endocytosis (see Chap. 15). Different physical factors control the feasibility of these processes: the membrane elasticity, the adhesion energy between the capsid and the membrane, the spontaneous thermal fluctuations, and the diffusion of spike proteins. Accordingly, several physical models have been proposed to address these interesting problems.

After the complete assembly of the nucleocapsid, most enveloped viruses acquire their envelope during the exit process from the cell by wrapping and pinching off the host membrane. However, this deformation of the membrane entails an energetic cost. The first mechanism suggested to accomplish that was the spontaneous wrapping of the virus taking advantage of thermal fluctuations of the membrane [138]. However, the energy provided by thermal fluctuations alone is not enough to wrap completely the virus unless the membrane stiffness is low [139]. To overcome this cost, the virus counts on the adhesion energy between the membrane and its spike or capsid proteins. Several models accounting for these factors have been proposed to study the competition between these contributions and the optimal conditions for virus entry.

This process has been characterized by analyzing the wrapping of a spherical particle by an elastic membrane [140]. The elastic deformation of the membrane is described using a continuous elasticity model (see Sect. 19.2.2) where an additional term accounting for the adhesion energy between the virus and the membrane is introduced. This model has shown that a minimum value of the adhesion energy is required to have full wrapping, and that the process is very sensitive to the size of the viral particle. These results suggest the possibility of interfering with the entrance or budding of viruses by modifying the adhesion energy or the membrane tension. Recent simulations show that elasticity theory can account quantitatively for different mechanisms of budding, including the stalling prior to the release of the virus [141–143].

An extension of these ideas has been performed to account for the concurrent assembly and budding processes of HIV [144], ascertaining the conditions in terms of Gag–Gag interaction strength and membrane rigidity required to have complete budding.

During the budding and exit from an infected cell, many enveloped viruses embed in their lipid envelope spike-shaped viral proteins that subsequently play a crucial role to infect another cell. Statistical models of viral budding, which incorporate the entropy and distribution of spikes to the elastic and adhesion effective free energy terms, have also been proposed [145]. These studies allow analyzing the simultaneous formation of many buds and the competence of viruses to sequester the spike proteins. The results show that, for biologically reasonable values of the membrane elastic constant and viruses with relatively strong spike binding strength, all binding sites of the capsid get occupied by spikes upon successful budding.

Interestingly, with slight modifications, models developed to study budding can also be applied to study endocytosis. The endocytosis of the virus is mediated by the binding interactions between ligand molecules on the viral capsid and their receptor molecules on the cell membrane. These receptors are mobile and must diffuse in the membrane to wrap around the virus and facilitate its entrance. In some cases, this can be the rate-limiting step for virus entry, and the corresponding kinetics can be modeled using a diffusion equation, as done in [146]. These studies have shown that particles in the size range of tens to hundreds of nanometers can enter or exit cells via wrapping without requiring the help of clathrin or caveolin. The optimal size to get an optimal wrapping time was characterized and turns out to be in the range 25–30 nm in radius, which is quite similar to the size of many spherical viruses.

19.6 Perspectives and Conclusions

Theoretical models and simulations are starting to make a significant contribution to our understanding of how viruses function. As we have seen in this chapter, the regular architecture of viruses obeys general geometrical rules and basic physical principles. For instance, the icosahedral symmetry of spherical viruses and the structure of prolate viruses are a consequence of free energy minimization of generic interactions between the capsid proteins. In addition, the study of continuous elastic theories and TST has established a reference framework to rationalize the distinguishing morphologies and the faceting of viral capsids, based on the minimization of the effective elastic and bending energies. Complementing TST, coarse-grained models have shown the dependence of buckling on the triangulation number, and high-resolution NMA has described basic aspects of capsid maturation for several viruses. The combination of TST, coarse-grained, FEA, and NMA simulations has been an invaluable source characterizing the remarkable elastic properties of capsids facilitating the interpretation of AFM experiments.

In the context of viral assembly, CNT and kinetic models provide a solid background for explaining the physical mechanisms involved in the assembly and disassembly of capsids. In CNT, the most important element is the existence of an energy barrier controlling the formation of capsids, which justifies the main features observed experimentally, i.e., the scarcity of intermediates, the existence of a lag time at the beginning of the assembly, its sigmoidal kinetics, and the hysteresis in the disassembly. The incorporation of elastic energy contributions to CNT has improved our understanding on how viruses select a particular size and offer interesting perspectives on how to interfere with this process. Coarse-grained simulations have started to complement CNT and other models by exploring the pathways of capsid assembly and capsid-genome co-assembly of some ssRNA viruses, showing that the genetic material is able to reduce the energy barrier of capsid formation. In addition, simulations have shown that the inhomogeneous distribution of RNA inside the capsid is a generic feature due to the self-repulsion of this highly flexible genetic material; several models have also justified the relation of charges between the genetic material and capsid proteins, the relation between the capsid size and the encapsulated cargo, the assembly competition of different ssRNA chains, or the ordered structure of the RNA layer in contact with the interior of the shell.

Alternatively, for some dsDNA viruses, it has been possible to estimate the pressures built up during the packaging of the genetic material and the corresponding ejection forces, which assist the initial stage of the dsDNA entry inside a new host cell. The translocation of the remaining material can be directed by either a nonspecific physical mechanism, like the binding of proteins to the viral genome, or by the pulling action of a polymerase. In the first case, the speed up due to the binding of proteins is associated to the lowering of the free energy with the length inside and can be described by a simple adsorption model. Finally, statistical thermodynamical models are starting to describe the general physical mechanisms involved in the recruiting of viral envelope and membrane viral proteins, as well as the exit and entry of some viral particles by budding and endocytosis, respectively. The results indicate that viruses have the right size to efficiently favor the spontaneous occurrence of these processes.

Theoretical physical virology is a relatively young field, and most of the *smart* physical strategies developed by viruses to accomplish different stages in their replication cycle are only starting to be unveiled. At the structural level, the understanding of irregular, conical, pleomorphic, and exotic shapes like bottle and spindle shapes [147] recently discovered in archaeal and giant viruses are major challenges to be faced. Regarding the mechanical stability of viruses, the main goal is to overcome the limitations of continuum techniques characterizing the elastic properties of viral shells. For this purpose, the development of coarse-grained discrete models, which incorporate the anisotropy of the interactions between the capsid building blocks (CBBs), is very timely and will facilitate a more accurate description of the protective role of the capsid, including the rich mechanical phenomenology obtained in AFM experiments. In the context of assembly, a clear direction is the further development of CNT to reach a quantitative and predictive understanding of assembly experiments on specific viruses, and its extension to explicitly incorporate the effects of the genetic material and auxiliary (scaffolding and/or condensing) proteins. Simulations will play a very important role to test these theories and to gain further insight into the molecular details of this process. Finally, regarding the delivery of the genetic material and the egress of viral particles, simulations, and theory are needed to clarify especially the non-equilibrium and kinetic aspects of these processes.

Obviously, the answer to all these questions will require a close collaboration between physicists and molecular virologists. In this endeavor, the main challenge faced by theory is to properly adapt the main physical ideas into a precise biological context and relate them to accessible experimental magnitudes. One new approach in this direction has been the combination of environmental genomic sequencing with physical models, statistical learning, and bioinformatics to investigate the structural properties of capsids from uncultured viruses, which represent most viruses on the planet [148, 149]. This should allow us to progressively incorporate other important biological information, demonstrating the utility of theories in the prediction, guidance, interpretation of experiments, and, ultimately, revealing how viruses have physically adapted to different conditions. With this effort, physics would provide not only a theoretical understanding of viruses but also a valuable tool in the development of novel antiviral strategies and useful applications.

Acknowledgments We acknowledge useful discussions with María Aznar, Robijn Bruinsma, Mauricio Comas-García, William Gelbart, William Klug, Charles Knobler, Joseph Rudnick, Jan Wedekind, Roya Zandi, Diana Y. Lee, Colin Brown, Aurora Vogel, Reidun Twarock, Rees Garmann, Forest Rohwer, Rob Edwards, Anca Segall, and Cynthia Silveria, among many others. D.R. acknowledges the support from the Spanish State Research Agency through grants No. PID2021-126570NB-I00 and RED2022-134654-T co-financed by European Union's FEDER funds. A.L. acknowledges the support from the National Science Foundation award #1951678 and the Gordon and Betty Moore Foundation award GBMF9871 (https://doi.org/10.37807/GBMF9871).

References

1. Crick F, Watson J (1956) Structure of small viruses. Nature 177:473–475
2. Caspar D, Klug A (1962) Physical principles in the construction of regular viruses. Cold Spring Harb Symp Quant Biol 27:1–24
3. Ganser BK, Li S, Klishko VY et al (1999) Assembly and analysis of conical models for the HIV-1 core. Science 283:80–83
4. Twarock R, Luque A (2019) Structural puzzles in virology solved with an overarching icosahedral design principle. Nat Commun 10:4414
5. Moody MF (1965) The shape of the T-even bacteriophage head. Virology 26:567–576
6. Luque A, Reguera D (2010) The structure of elongated viral capsids. Biophys J 98:2993–3003
7. Sadre-Marandi F, Liu J, Tavener S, Chen C (2014) Generating vectors for the lattice structures of tubular and conical viral capsids. Comput Math Biophys 2:128–140

8. Lidmar J, Mirny L, Nelson DR (2003) Virus shapes and buckling transitions in spherical shells. Phys Rev E 68:051910
9. Twarock R (2006) Mathematical virology: a novel approach to the structure and assembly of viruses. Philos Trans R Soc A Math Phys Eng Sci 364:3357–3373
10. Zandi R, Reguera D, Bruinsma RF et al (2004) Origin of icosahedral symmetry in viruses. Proc Natl Acad Sci USA 101:15556–15560
11. Luque A, Zandi R, Reguera D (2010) Optimal architectures of elongated viruses. Proc Natl Acad Sci 107:5323–5328
12. Fejer S, James T, Hernandez-Rojas J, Wales D (2009) Energy landscapes for shells assembled from pentagonal and hexagonal pyramids. Phys Chem Chem Phys 11:2098–2104
13. Fejer S, Chakrabarti D, Wales D (2010) Emergent complexity from simple anisotropic building blocks: shells, tubes, and spirals. ACS Nano 4:219–228
14. Rapaport DC (2010) Modeling capsid self-assembly: design and analysis. Phys Biol 7:045001
15. Elrad OM, Hagan MF (2010) Encapsulation of a polymer by an icosahedral virus. Phys Biol 7: 045003
16. Nguyen HD, Reddy VS, Brooks CL (2009) Invariant polymorphism in virus capsid assembly. J Am Chem Soc 131:2606–2614
17. Arkhipov A, Freddolino PL, Schulten K (2006) Stability and dynamics of virus capsids described by coarse-grained modeling. Structure 14:1767–1777
18. Freddolino PL, Arkhipov AS, Larson SB et al (2006) Molecular dynamics simulations of the complete satellite tobacco mosaic virus. Structure 14:437–449
19. Keef T, Twarock R (2008) New insights into viral architecture via affine extended symmetry groups. Comput Math Methods Med 9:221–229
20. Wilson DP, Roof DA (2021) Viral phrenology. Viruses 13:2191
21. Bruinsma RF, Gelbart WM, Reguera D et al (2003) Viral self-assembly as a thermodynamic process. Phys Rev Lett 90:248101
22. Landau LD, Lifshitz EM (1975) Theory of elasticity. Pergamon Press, London
23. Grason GM (2016) Perspective: geometrically frustrated assemblies. J Chem Phys 145: 110901
24. Dong Y, Zandi R, Travesset A (2022) Exact solution for elastic networks on curved surfaces. Phys Rev Lett 129:88001
25. Nguyen TT, Bruinsma RF, Gelbart WM (2005) Elasticity theory and shape transitions of viral shells. Phys Rev E 72:051923
26. Klug WS, Bruinsma RF, Michel J-P, Knobler CM (2006) Failure of viral shells. Phys Rev Lett 97:228101
27. Seung HS, Nelson DR (1988) Defects in flexible membranes with crystalline order. Phys Rev A 38:1005–1018
28. Ning J, Erdemci-Tandogan G, Yufenyuy EL et al (2016) In vitro protease cleavage and computer simulations reveal the HIV-1 capsid maturation pathway. Nat Commun 7:13689
29. Luque Santolaria A (2011) Structure, mechanical properties, and self-assembly of viral capsids. Thesis, University of Barcelona
30. Aznar M, Reguera D (2016) Physical ingredients controlling stability and structural selection of empty viral capsids. J Phys Chem B 120:6147–6159
31. Llorente JMG, Hernández-Rojas J, Bretón J (2014) A minimal representation of the self-assembly of virus capsids. Soft Matter 10:3560–3569
32. Chen T, Glotzer S (2007) Simulation studies of a phenomenological model for elongated virus capsid formation. Phys Rev E 75:051504
33. Saunders M, Voth G (2012) Coarse-graining of multiprotein assemblies. Curr Opin Struct Biol 20:144–150
34. Jana AK, May ER (2020) Structural and dynamic asymmetry in icosahedrally symmetric virus capsids. Curr Opin Virol 45:8–16
35. Hadden JA, Perilla JR (2018) All-atom virus simulations. Curr Opin Virol 31:82–91

36. Perilla JR, Schulten K (2017) Physical properties of the HIV-1 capsid from all-atom molecular dynamics simulations. Nat Commun 8:15959
37. Krishna V, Ayton GS, Voth GA (2010) Role of protein interactions in defining HIV-1 viral capsid shape and stability: a coarse-grained analysis. Biophys J 98:18–26
38. Chen B, Tycko R (2011) Simulated self-assembly of the HIV-1 capsid: protein shape and native contacts are sufficient for two-dimensional lattice formation. Biophys J 100:3035–3044
39. Miao Y, Johnson JE, Ortoleva PJ (2010) All-atom multiscale simulation of cowpea chlorotic mottle virus capsid swelling. J Phys Chem B 114:11181–11195
40. Polles G, Indelicato G, Potestio R et al (2013) Mechanical and assembly units of viral capsids identified via quasi-rigid domain decomposition. PLoS Comp Biol 9:e1003331
41. Ponzoni L, Polles G, Carnevale V, Micheletti C (2015) SPECTRUS: a dimensionality reduction approach for identifying dynamical domains in protein complexes from limited structural datasets. Structure 23:1516–1525
42. Brown C, Agarwal A, Luque A (2024) pyCapsid: identifying dominant dynamics and quasi-rigid mechanical units in protein shells. Bioinformatics 40:btad761
43. Devkota B, Petrov AS, Lemieux S et al (2009) Structural and electrostatic characterization of pariacoto virus: implications for viral assembly. Biopolymers 91:530–538
44. Lavelle L, Gingery M, Phillips M et al (2009) Phase diagram of self-assembled viral capsid protein polymorphs. J Phys Chem B 113:3813–3819
45. Kanesashi SN, Ishizu KI, Kawano MA et al (2003) Simian virus 40 VP1 capsid protein forms polymorphic assemblies in vitro. J Gen Virol 84:1899–1905
46. Hu Y, Zandi R, Anavitarte A et al (2008) Packaging of a polymer by a viral capsid: the interplay between polymer length and capsid size. Biophys J 94:1428–1436
47. Dixit SK, Goicochea NL, Daniel M-C et al (2006) Quantum dot encapsulation in viral capsids. Nano Lett 6:1993–1999
48. Kegel WK, van der Schoot P (2004) Competing hydrophobic and screened-coulomb interactions in hepatitis B virus capsid assembly. Biophys J 86:3905–3913
49. Verwey EJW, Overbeek JTG (1999) Theory of stability of lyophobic colloids. Dover Press, Minneola, NY
50. Siber A, Božič AL, Podgornik R (2012) Energies and pressures in viruses: contribution of nonspecific electrostatic interactions. Phys Chem Chem Phys 14:3746–3765
51. Zandi R, Dragnea B, Travesset A, Podgornik R (2020) On virus growth and form. Phys Rep 847:1–102
52. Ceres P, Zlotnick A (2002) Weak protein–protein interactions are sufficient to drive assembly of hepatitis B virus capsids. Biochemistry 41:11525–11531
53. Singh S, Zlotnick A (2003) Observed hysteresis of virus capsid disassembly is implicit in kinetic models of assembly. J Biol Chem 278:18249–18255
54. Snijder J, Uetrecht C, Rose RJ et al (2013) Probing the biophysical interplay between a viral genome and its capsid. Nat Chem 5:502–509
55. Wörner TP, Shamorkina TM, Snijder J, Heck AJR (2021) Mass spectrometry-based structural virology. Anal Chem 93:620–640
56. Kler S, Asor R, Li C et al (2012) RNA encapsidation by SV40-derived nanoparticles follows a rapid two-state mechanism. J Am Chem Soc 134:8823–8830
57. Chevreuil M, Lecoq L, Wang S et al (2020) Nonsymmetrical dynamics of the HBV capsid assembly and disassembly evidenced by their transient species. J Phys Chem B 124:9987–9995
58. Khaykelson D, Raviv U (2020) Studying viruses using solution X-ray scattering. Biophys Rev 12:41–48
59. Garmann RF, Goldfain AM, Manoharan VN (2019) Measurements of the self-assembly kinetics of individual viral capsids around their RNA genome. Proc Natl Acad Sci 116: 22485–22490

60. Garmann RF, Goldfain AM, Tanimoto CR et al (2022) Single-particle studies of the effects of RNA–protein interactions on the self-assembly of RNA virus particles. Proc Natl Acad Sci 119:1–11
61. Valbuena A, Maity S, Mateu MG, Roos WH (2020) Visualization of single molecules building a viral capsid protein lattice through stochastic pathways. ACS Nano 14:8724–8734
62. Marchetti M, Kamsma D, Cazares Vargas E et al (2019) Real-time assembly of viruslike nucleocapsids elucidated at the single-particle level. Nano Lett 19:5746–5753
63. van Rosmalen MGM, Kamsma D, Biebricher AS et al (2020) Revealing in real-time a multistep assembly mechanism for SV40 virus-like particles. Sci Adv 6:1–8
64. Hagan MF (2014) Modeling viral capsid assembly. In: Advances in chemical physics, pp 1–68
65. Zlotnick A (2003) How does your virus grow? Understanding and interfering with virus assembly. Trends Biotechnol 21:536–542
66. Mukherjee S, Thorsteinsson MV, Johnston LB et al (2008) A quantitative description of in vitro assembly of human papillomavirus 16 virus-like particles. J Mol Biol 381:229–237
67. Chen C, Kao CC, Dragnea B (2008) Self-assembly of brome mosaic virus capsids: insights from shorter time-scale experiments. J Phys Chem A 112:9405–9412
68. Zandi R, van der Schoot P, Reguera D et al (2006) Classical nucleation theory of virus capsids. Biophys J 90:1939–1948
69. Hagan MF, Grason GM (2021) Equilibrium mechanisms of self-limiting assembly. Rev Mod Phys 93:25008
70. Mendoza CI, Reguera D (2020) Shape selection and mis-assembly in viral capsid formation by elastic frustration. elife 9:e52525
71. Schwartz R, Shor PW, Prevelige PE, Berger B (1998) Local rules simulation of the kinetics of virus capsid self-assembly. Biophys J 75:2626–6636
72. Berger B, Shor P, Tucker-Kellogg L, King J (1994) Local rule-based theory of virus shell assembly. Proc Natl Acad Sci USA 91:7732–7736
73. Keef T, Micheletti C, Twarock R (2006) Master equation approach to the assembly of viral capsids. J Theor Biol 242:713–721
74. Moisant P, Neeman H, Zlotnick A (2010) Exploring the paths of (virus) assembly. Biophys J 99:1350–1357
75. Wilber AW, Doye JPK, Louis AA et al (2007) Reversible self-assembly of patchy particles into monodisperse icosahedral clusters. J Chem Phys 127:085106
76. Hagan MF, Chandler D (2006) Dynamic pathways for viral capsid assembly. Biophys J 91: 42–54
77. Aznar M (2013) Coarse-grained modeling of the assembly and mechanical properties of viruses. Universitat de Barcelona
78. Reguera D, Hernández-Rojas J, Gomez Llorente JM (2019) Kinetics of empty viral capsid assembly in a minimal model. Soft Matter 15:7166–7172
79. Hagan MF, Elrad OM (2010) Understanding the concentration dependence of viral capsid assembly kinetics – the origin of the lag time and identifying the critical nucleus size. Biophys J 98:1065–1074
80. Morozov A, Rudnick J, Bruinsma R, Klug W (2010) Assembly and disassembly of deltahedral viral shells. In: Stockley P, Twarock R (eds) Emerging topics in physical virology. Imperial College Press, London, pp 159–183
81. Wagner J, Zandi R (2015) The robust assembly of small symmetric nanoshells. Biophys J 109: 956–965
82. Levandovsky A, Zandi R (2009) Nonequilibirum assembly, retroviruses, and conical structures. Phys Rev Lett 102:198102
83. Hicks SD, Henley CL (2006) Irreversible growth model for virus capsid assembly. Phys Rev E 74:031912
84. Panahandeh S, Li S, Zandi R (2018) The equilibrium structure of self-assembled protein nanocages. Nanoscale 10:22802–22809

85. Li S, Roy P, Travesset A, Zandi R (2018) Why large icosahedral viruses need scaffolding proteins. Proc Natl Acad Sci 115:10971–10976
86. Lynch DL, Pavlova A, Fan Z, Gumbart JC (2023) Understanding virus structure and dynamics through molecular simulations. J Chem Theory Comput 19:3025–3036
87. Bruinsma RF (2006) Physics of RNA and viral assembly. Eur Phys J E Soft Matter 19:303–310
88. Ting CL, Wu J, Wang Z-G (2011) Thermodynamic basis for the genome to capsid charge relationship in viral encapsidation. Proc Natl Acad Sci USA 108:16986–16991
89. van der Schoot P, Bruinsma R (2005) Electrostatics and the assembly of an RNA virus. Phys Rev E 71:061928
90. Zandi R, van der Schoot P (2009) Size regulation of ss-RNA viruses. Biophys J 96:9–20
91. Cadena-Nava RD, Comas-Garcia M, Garmann RF et al (2012) Self-assembly of viral capsid protein and RNA molecules of different sizes: requirement for a specific high protein/RNA mass ratio. J Virol 86:3318–3326
92. ElSawy KM, Caves LSD, Twarock R (2011) On the origin of order in the genome organization of ssRNA viruses. Biophys J 101:774–780
93. Mahalik JP, Muthukumar M (2012) Langevin dynamics simulation of polymer-assisted virus-like assembly. J Chem Phys 136:135101
94. Perlmutter JD, Hagan MF (2015) Mechanisms of virus assembly. Annu Rev Phys Chem 66: 217–239
95. Perlmutter JD, Perkett MR, Hagan MF (2014) Pathways for virus assembly around nucleic acids. J Mol Biol 426:3148–3165
96. Patel N, Dykeman EC, Coutts RHA et al (2015) Revealing the density of encoded functions in a viral RNA. Proc Natl Acad Sci 112:2227–2232
97. Twarock R, Bingham RJ, Dykeman EC, Stockley PG (2018) A modelling paradigm for RNA virus assembly. Curr Opin Virol 31:74–81
98. Comas-Garcia M (2019) Packaging of genomic RNA in positive-sense single-stranded RNA viruses: a complex story. Viruses 11:253
99. Perlmutter JD, Hagan MF (2015) The role of packaging sites in efficient and specific virus assembly. J Mol Biol 427:2451–2467
100. Patel N, Abulwerdi F, Fatehi F et al (2022) Dysregulation of hepatitis B virus nucleocapsid assembly in vitro by RNA-binding small ligands. J Mol Biol 167557
101. Harvey S, Petrov A, Devkota B (2009) Viral assembly: a molecular modeling perspective. Phys Chem Chem Phys 11:10541–10542
102. Odijk T (1998) Hexagonally packed DNA within bacteriophage T7 stabilized by curvature stress. Biophys J 75:1223–1227
103. Purohit PK, Inamdar MM, Grayson PD et al (2005) Forces during bacteriophage DNA packaging and ejection. Biophys J 88:851–866
104. Kindt J, Ben-Shaul A, Tzlil S, Gelbart WM (2001) DNA packaging and ejection forces in bacteriophage. Proc Natl Acad Sci USA 98:13671–13674
105. Rau D, Lee B, Parsegian V (1984) Measurement of the repulsive force between polyelectrolyte molecules in ionic solution: hydration forces between parallel DNA double helices. Proc Natl Acad Sci USA 81:2621–2625
106. Inamdar MM, Gelbart WM, Phillips R (2006) Dynamics of DNA ejection from bacteriophage. Biophys J 91:411–420
107. Evilevitch A, Lavelle L, Knobler CM et al (2003) Osmotic pressure inhibition of DNA ejection from phage. Proc Natl Acad Sci USA 100:9292–9295
108. Villanueva Valencia JR, Li D, Casjens SR, Evilevitch A (2023) 'SAXS-osmometer' method provides measurement of DNA pressure in viral capsids and delivers an empirical equation of state. Nucleic Acids Res 51:11415–11427
109. Smith DE, Tans SJ, Smith SB et al (2001) The bacteriophage φ29 portal motor can package DNA against a large internal force. Nature 413:748–752

110. Santos-Pérez I, Oksanen HM, Bamford DH et al (2017) Membrane-assisted viral DNA ejection. Biochim Biophys Acta - Gen Subj 1861:664–672
111. Cordova A, Deserno M, Gelbart WM, Ben-Shaul A (2003) Osmotic shock and the strength of viral capsids. Biophys J 85:70–74
112. Roos WH, Bruinsma R, Wuite GJL (2010) Physical virology. Nat Phys 6:733–743
113. Vella D, Ajdari A, Vaziri A, Boudaoud A (2012) The indentation of pressurized elastic shells: from polymeric capsules to yeast cells. J R Soc Interface 9:448–455
114. Carrasco C, Luque A, Hernando-Perez M et al (2011) Built-in mechanical stress in viral shells. Biophys J 100:1100–1108
115. Purohit PK, Kondev J, Phillips R (2003) Mechanics of DNA packaging in viruses. Proc Natl Acad Sci USA 100:3173–3178
116. Zandi R, Reguera D (2005) Mechanical properties of viral capsids. Phys Rev E 72:021917
117. Aznar M, Luque A, Reguera D (2012) Relevance of capsid structure in the buckling and maturation of spherical viruses. Phys Biol 9:036003
118. Buenemann M, Lenz P (2008) Elastic properties and mechanical stability of chiral and filled viral capsids. Phys Rev E 78:051924
119. Vliegenthart GA, Gompper G (2006) Mechanical deformation of spherical viruses with icosahedral symmetry. Biophys J 91:834–841
120. Aznar M, Roca-Bonet S, Reguera D (2018) Viral nanomechanics with a virtual atomic force microscope. J Phys Condens Matter 30:264001
121. Roos WH, Gibbons MM, Arkhipov A et al (2010) Squeezing protein shells: how continuum elastic models, molecular dynamics simulations, and experiments coalesce at the nanoscale. Biophys J 99:1175–1181
122. Arkhipov A, Roos WH, Wuite GJL, Schulten K (2009) Elucidating the mechanism behind irreversible deformation of viral capsids. Biophys J 97:2061–2069
123. Bryer AJ, Rey JS, Perilla JR (2023) Performance efficient macromolecular mechanics via sub-nanometer shape based coarse graining. Nat Commun 14:2014
124. Indelicato G, Cermelli P, Twarock R (2020) Surface stresses in complex viral capsids and non-quasi-equivalent viral architectures. J R Soc Interface 17:20200455
125. Tama F, Wriggers W, Brooks CL (2002) Exploring global distortions of biological macromolecules and assemblies from low-resolution structural information and elastic network theory. J Mol Biol 321:297–305
126. Gibbons MM, Klug WS (2007) Mechanical modeling of viral capsids. J Mater Sci 42:8995–9004
127. Martín-Bravo M, Gomez Llorente JM, Hernández-Rojas J (2020) A minimal coarse-grained model for the low-frequency normal mode analysis of icosahedral viral capsids. Soft Matter 16:3443–3455
128. May ER, Feng J, Brooks CL (2012) Exploring the symmetry and mechanism of virus capsid maturation via an ensemble of pathways. Biophys J 102:606–612
129. May E, Brooks C (2011) Determination of viral capsid elastic properties from equilibrium thermal fluctuations. Phys Rev Lett 106:188101
130. Roos W, Gertsman I, May E et al (2012) Mechanics of bacteriophage maturation. Proc Natl Acad Sci USA 109:2342–2347
131. Olson G (1999) New directions in martensite theory. Mater Sci Eng A 273–275:11–20
132. Falk W, James R (2006) Elasticity theory for self-assembled protein lattices with application to the martensitic phase transition in bacteriophage T4 tail sheath. Phys Rev E 73:011917
133. Zandi R, Reguera D, Rudnick J, Gelbart WM (2003) What drives the translocation of stiff chains? Proc Natl Acad Sci USA 100:8649–8653
134. Evilevitch A (2006) Effects of condensing agent and nuclease on the extent of ejection from phage lambda. J Phys Chem B 110:22261–22265
135. Grayson P, Han L, Winther T, Phillips R (2007) Real-time observations of single bacteriophage lambda DNA ejections in vitro. Proc Natl Acad Sci USA 104:14652–14657

136. Chen Y, Wu D, Gelbart W et al (2018) Two-stage dynamics of in vivo bacteriophage genome ejection. Phys Rev X 8:021029
137. Timmermans SBPE, Ramezani A, Montalvo T et al (2022) The dynamics of viruslike capsid assembly and disassembly. J Am Chem Soc 144:12608–12612
138. Simons K, Garoff H (1980) The budding mechanisms of enveloped animal viruses. J Gen Virol 50:1–21
139. Lerner DM, Deutsch JM, Oster GF (1993) How does a virus bud? Biophys J 65:73–79
140. Deserno M (2004) Elastic deformation of a fluid membrane upon colloid binding. Phys Rev E 69:031903
141. Ruiz-Herrero T, Velasco E, Hagan MF (2012) Mechanisms of budding of nanoscale particles through lipid bilayers. J Phys Chem B 116:9595–9603
142. Lázaro GR, Mukhopadhyay S, Hagan MF (2018) Why enveloped viruses need cores—the contribution of a nucleocapsid core to viral budding. Biophys J 114:619–630
143. Dharmavaram S, She SB, Lázaro G et al (2019) Gaussian curvature and the budding kinetics of enveloped viruses. PLoS Comput Biol 15:e1006602
144. Zhang R, Nguyen TT (2008) Model of human immunodeficiency virus budding and self-assembly: role of the cell membrane. Phys Rev E 78:051903
145. Tzlil S, Deserno M, Gelbart WM, Ben-Shaul A (2004) A statistical-thermodynamic model of viral budding. Biophys J 86:2037–2048
146. Gao H, Shi W, Freund LB (2005) Mechanics of receptor-mediated endocytosis. Proc Natl Acad Sci USA 102:9469–9474
147. Wang F, Cvirkaite-Krupovic V, Vos M et al (2022) Spindle-shaped archaeal viruses evolved from rod-shaped ancestors to package a larger genome. Cell 185:1297–1307.e11
148. Luque A, Benler S, Lee DY et al (2020) The missing tailed phages: prediction of small capsid candidates. Microorganisms 8:1944
149. Lee DY, Bartels C, McNair K et al (2022) Predicting the capsid architecture of phages from metagenomic data. Comput Struct Biotechnol J 20:721–732

Further Reading

Bruinsma RF, Wuite GJL, Roos WH (2021) Physics of viral dynamics. Nat Rev Phys 3:76–91
Israelachvili J (2011) Intermolecular and surface forces, 3rd edn. Academic Press, London
Katen S, Zlotnick A (2009) The thermodynamics of virus capsid assembly. Methods Enzymol 455: 395–417
Knobler CM, Gelbart WM (2009) Physical chemistry of DNA viruses. Annu Rev. Phys Chem 60: 367–383
Marenduzzo D, Micheletti C, Orlandini E (2010) Biopolymer organization upon confinement. J Phys Condens Matter 22:283102
Nurmemmedov E, Castelnovo M, Catalano CE, Evilevitch A (2007) Biophysics of viral infectivity: matching genome length with capsid size. Q Rev. Biophys 40:327–356
Phillips R, Kondev J, Theriot J, Orme N (2010) Physical biology of the cell. Garland Science, New York
Also especially recommended for further reading are references [42, 45, 56, 57, 62, 63, 92, 101, 110] *listed above*

Part IV
Applied Structural and Physical Virology

Chapter 20
Antiviral Agents: Structural Basis of Action and Rational Design

Luis Menéndez-Arias and Federico Gago

Abstract During the last forty years, significant progress has been made in the development of novel antiviral drugs, mainly crystallizing in the establishment of potent antiretroviral therapies and the approval of drugs eradicating hepatitis C virus infection. Although major targets of antiviral intervention involve intracellular processes required for the synthesis of viral proteins and nucleic acids, a number of inhibitors blocking virus assembly, budding, maturation, entry, or uncoating act on virions or viral capsids. In this review, we focus on the drug discovery process while presenting the currently used methodologies to identify novel antiviral drugs by means of computer-based approaches. We provide examples illustrating structure-based antiviral drug development, specifically neuraminidase inhibitors against influenza virus (e.g., oseltamivir and zanamivir) and human immunodeficiency virus type 1 protease inhibitors (i.e., the development of darunavir from early peptidomimetic compounds such as saquinavir). A number of drugs acting against hepatitis B virus and human immunodeficiency virus and their mechanism of action are presented to show how viral capsids can be exploited as targets of antiviral therapy. The recent approval of the antiretroviral drug lenacapavir illustrates the successful application of this knowledge.

Keywords Antiretroviral drugs · Capsid proteins · DNA polymerases · Drug development · Fusion inhibitors · Hepatitis virus · Herpesviruses · Human immunodeficiency virus · Influenza virus · Ligand docking · Neuraminidase · Nucleoside analogs · Proteases · Viral assembly · Viral entry · Viral replication · Virtual screening

---

L. Menéndez-Arias (✉)
Centro de Biología Molecular "Severo Ochoa" (Consejo Superior de Investigaciones Científicas and Universidad Autónoma de Madrid), Madrid, Spain
e-mail: lmenendez@cbm.csic.es

F. Gago (✉)
Department of Biomedical Sciences, Universidad de Alcalá, Alcalá de Henares, Madrid, Spain
e-mail: federico.gago@uah.es

M. G. Mateu (ed.), *Structure and Physics of Viruses*, Subcellular Biochemistry 105,
https://doi.org/10.1007/978-3-031-65187-8_20

Abbreviations

AIDS	Acquired immune deficiency syndrome
HIV	Human immunodeficiency virus
HSV	Herpes simplex virus
VZV	Varicella-zoster virus
CMV	Cytomegalovirus
MPXV	Monkeypox virus
RSV	Respiratory syncytial virus
SARS-CoV-2	Severe acute respiratory syndrome coronavirus 2
HBV	Hepatitis B virus
HCV	Hepatitis C virus
EBV	Epstein–Barr virus
mRNA	Messenger RNA
RT	Reverse transcriptase
NNRTIs	Nonnucleoside RT inhibitors
PDB	Protein Data Bank
RSV	Respiratory syncytial virus
HR	Heptad repeat
HTS	High-throughput screening
SBVS	Structure-based virtual screening
LBVS	Ligand-based virtual screening
NMR	Nuclear magnetic resonance
CoMFA	Comparative molecular field analysis
Neu5Ac	*N*-acetylneuraminic acid
THF	Tetrahydrofuran
CTD	C-terminal domain
NTD	N-terminal domain

20.1 Introduction

Antiviral drugs are compounds that stop the development and propagation of a virus without causing a relevant damage in the host cell. In the past 50 years, great achievements have been made in the field of antiviral drug discovery. Around 100 antiviral drugs have been approved for the treatment of viral infections caused by human immunodeficiency virus (HIV), herpes simplex virus (HSV), varicella-zoster virus (VZV), cytomegalovirus (CMV), monkeypox virus (MPXV), respiratory syncytial virus (RSV), influenza virus, severe acute respiratory syndrome coronavirus 2 (SARS-CoV-2), and hepatitis B and C viruses (HBV and HCV, respectively), as well as drugs used in the treatment of external anogenital warts caused by human papillomavirus infections [1, 2].

The approval of more than 40 antiretroviral drugs has been one of the greatest accomplishments in recent years. The introduction of single-pill fixed-dose combinations with demonstrated high potency and minimal toxicity has transformed HIV infection into a chronic, but manageable disease [3]. During the last decade, the discovery and approval of direct-acting antiviral medicines to treat HCV infection has been a stunning triumph [4]. Studies show that these medicines are highly effective at clearing the infection in more than 90% of people. However, despite these success stories and the availability of drugs targeting different viruses infecting humans, a definitive cure for HIV, HBV, HCMV, HPV, HSV, RSV, VZV, SARS-CoV-2, or influenza virus infections is yet to be discovered, and there are no drugs approved for treatment against many human pathogenic viruses (e.g., dengue virus, tick-borne encephalitis virus, Ebolavirus and other filoviruses, arenaviruses, etc.. . .).

There are several reasons that account for the difficulties in developing antiviral agents. First, viruses are obligatory intracellular parasites. Because every step in the viral life cycle engages host functions, it is difficult to interfere with virus growth without having a negative impact on the host cell. Side effects are relatively common. Second, clinically important viruses are often dangerous and cannot be propagated or tested in model systems. Thus, viruses causing fatal diseases in humans (e.g., smallpox or several hemorrhagic diseases) have to be handled by well-trained and experienced scientists, and in facilities with strict containment requirements. Not surprisingly, these labs are expensive and difficult to maintain. Apart from the biological safety limitations, sometimes viral infections cannot be properly monitored for antiviral drug development due to the lack of appropriate animal models of human disease (e.g., smallpox or measles) or to difficulties in growing the virus in cell culture (e.g., HBV).

A third factor that limits the efficacy of antiviral drugs is their potency requirements. Ideally, an antiviral agent should be extremely potent. Partial inhibition is not acceptable for an antiviral drug. The reason is that limited viral replication under drug pressure allows for the generation of resistant variants that can be selected under treatment. The emergence of resistance is a major drawback of many antiviral therapies [5]. For example, in the case of HIV, therapies prescribed in the late 80s or early 90s were based on a single drug (mostly zidovudine) or combinations of two drugs (usually two inhibitors of the viral polymerase) [6]. However, those drugs were not potent enough to limit the emergence of drug-resistant variants [7] and therefore these viruses were almost impossible to combat successfully with the available drug armamentarium. In a patient with full-blown AIDS, HIV production has been estimated at about 10^{12} virions per day. The high replication rate of HIV and the low fidelity of its DNA polymerase [8, 9] trigger the appearance of drug resistance under suboptimal therapies.

Another issue that deserves some attention is the short duration of many viral infections (e.g., flu, coronavirus disease, common cold, etc.. . .). Very often, the symptoms appear when the virus is no longer replicating and are due to the patient's immune response. In those cases, antiviral drugs should be administered early in infection or as a prophylactic measure in populations at risk. However, this could be potentially harmful for healthy populations.

20.2 Drug Discovery and Potential Targets of Antiviral Intervention

Early evidence of activity against vaccinia virus was reported for several thiosemicarbazone derivatives [10], and one of them (*N*-methylisatin-β-thiosemicarbazone) entered clinical studies for the prophylaxis of smallpox [11] at the time when vaccination against smallpox virus took over. The first antiviral drug licensed for clinical use was a thymidine analog known as idoxuridine (5′-iodo-2′-deoxyuridine), whose synthesis was described in 1959 [12]. Idoxuridine has been used topically to treat eye and skin infections caused by HSV. The drug acts on viral replication by interfering with the normal function of thymidylate phosphorylase and viral DNA polymerases.

Encouraged by the success with antibiotics in the 60s and 70s, drug companies launched huge blind-screening programs that were not very fruitful. Today, recombinant DNA technology and sophisticated chemistry [13], as well as impressive advances in structural and functional genomics [14], have facilitated the identification and analysis of particular proteins or mechanisms. Essential viral genes can be cloned and expressed in appropriate organisms so that the encoded proteins can be purified and analyzed in molecular and atomic detail.

Drug discovery programs start with the identification of suitable drug targets (Fig. 20.1). A drug target can be defined as a biomolecule (usually a single protein or a protein complex) linked to a disease and containing a suitable binding site that can be exploited to modulate its function. These targets need to be validated to demonstrate that they are critically involved in a disease and that their modulation is likely to have a therapeutic effect. In virology, many drug targets are viral proteins (e.g., enzymes), nucleic acids, or other biological macromolecules required in the virus life cycle. Infection and viral propagation can be blocked by small compounds binding to relevant targets. Once the target has been validated in vitro and/or in animal models, lead identification starts with the design and development of a suitable assay to monitor the biological function under study. Active compounds that demonstrate

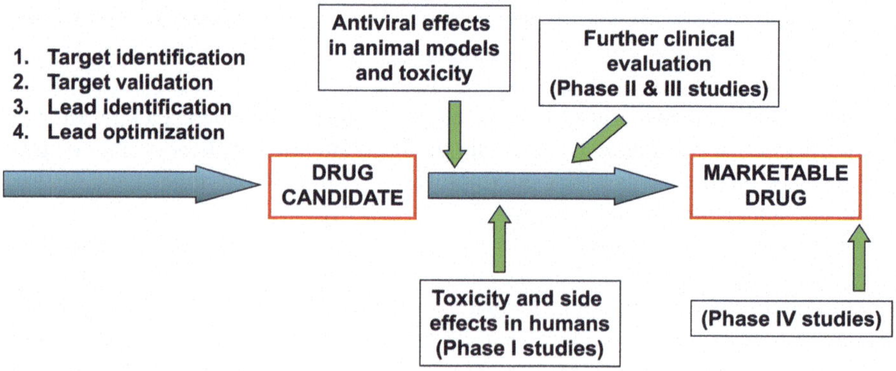

Fig. 20.1 Schematic outline of the drug discovery process

dose-dependent target modulation and some degree of selectivity for the target under study are called lead compounds. These molecules are optimized in terms of potency and selectivity to become drug candidates.

In order to become a marketable drug, the candidate undergoes additional preclinical evaluation, including pharmacokinetic and toxicity studies in animal models. Clinical trials involving new drugs commonly consist of four phases. Approval is usually granted if the drug advances successfully through phases I, II, and III. The main objective of phase I is to assess drug safety and pharmacokinetics in a relatively small number of healthy individuals who receive small doses of the compound. In phase II, the testing protocol is established. These are trials trying to find out appropriate doses, prescription regimens, etc.... and are carried out with patients. Phase III constitutes the final testing before approval and these trials try to show whether the drug has a measurable benefit or advantage over other treatments on a relatively large number of patients (usually >100 patients for antiviral drugs). Phase IV includes all studies carried out after approval of the drug and may address many different issues, from side effects to efficacy in comparison with other drugs or regimens, and usually involve a large number of patients worldwide.

Recent years have witnessed a remarkable progress in the discovery and development of antiviral agents, fueled by advances in the understanding of viral life cycles which, in turn, have provided new opportunities for therapeutic intervention. Although the mechanisms of replication and propagation can show significant variations between different viruses, a prototypic life cycle is presented in Fig. 20.2 (see also Chap. 1). Typically, a virion (i.e., an infectious viral particle) first attaches to the surface of the host cell. This interaction can be specific and involve the participation of one or more different types of proteins. For example, in HIV-1, the main receptor is CD4, but other proteins (e.g., chemokine receptors CCR5, CXCR4, etc....) can facilitate viral entry acting as coreceptors (see Chap. 16).

Viral uptake occurs through different mechanisms such as endocytosis (for example, in the case of influenza virus) or the fusion of the cellular membrane with the viral envelope, as demonstrated in the case of retroviruses (see Chaps. 15 and 16). These events allow the internalization of the capsid containing the viral genome. Uncoating or disassembly is a still poorly understood process that releases the viral genome from its protein shell (see Chaps. 15 and 16). Disassembly may occur rapidly after fusion, as occurs in HIV, or be triggered by pH variations, as observed in viruses entering the cell by an endocytic pathway (e.g., influenza virus). The replication of the viral genome occurs by different mechanisms depending on whether the viral genome is a single- or double-stranded DNA or RNA. Broadly speaking, DNA viruses use cellular pathways to replicate their genomes, but RNA viruses (and in general, those replicating in the cytoplasm) provide their own enzymes to complete virus replication, even though some host proteins, lipids, and membranes are usually co-opted to build the viral replication organelles [15]. Cellular factors together with specific proteins encoded within the viral genome contribute to transcription and post-transcriptional modification of viral messenger RNAs (mRNAs). In some viruses (for example, in retroviruses), viral mRNAs are

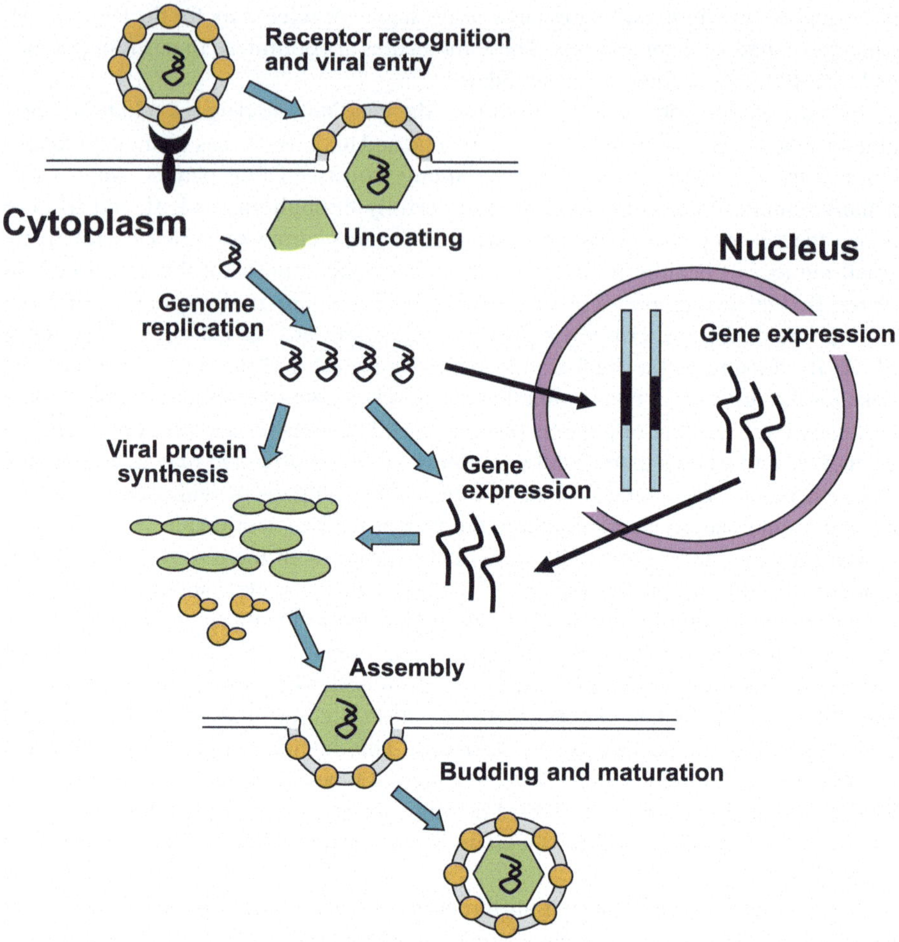

Fig. 20.2 Basic steps of viral replication as potential targets of antiviral therapy

generated in the nucleus by the action of the cellular RNA polymerase. This is due to the integration of the viral genome (as a double-stranded DNA) into the genome of the host cell.

Viral polyproteins are synthesized by the cellular translational machinery. After post-translational modification, viral proteins and their genomes are transported to assembly sites within the host cell (see Chap. 8). Viral factors may also participate in these processes. The sites of assembly are frequently located at the plasma membrane or in intracellular factories (often associated with membranes). Virion assembly is a complex process which involves multiple molecular recognition steps and conformational transitions. The viral capsid is assembled in a multimerization reaction, with or without the help of scaffolding proteins or viral nucleic acids (see Chaps. 11, 12, and 19). The viral nucleic acids are packaged into the capsid during or after its assembly (see Chap. 13). Finally, the viral particle can be transformed into an infectious virion through a maturation process that involves changes in the structure and properties of the capsid (see Chap. 14).

In enveloped viruses, a membrane with embedded viral proteins is incorporated into the virion (see Chap. 12), and the resulting viral particles are released after budding. In contrast, cell lysis mediates the release of non-enveloped viruses (see Chap. 11). In some cases (e.g., in retroviruses), maturation occurs once the immature virion has been released and involves the proteolytic processing of precursors containing the viral proteins, including those that form the capsid.

All of those steps of the viral life cycle constitute potential targets of antiviral intervention. At present, drugs inhibiting viral enzymes involved in the replication or expression of the viral genome are commonly used in antiviral treatments. However, recent developments including the determination of structures, properties, and functions of capsids and virions, as well as the elucidation of events involving interactions between components of the viral particle or between them and host cell molecules (the subjects of this book) have opened novel avenues for the discovery of drugs acting directly on the viral particle. This chapter provides examples of approved or preclinical antiviral strategies directed at inhibiting viral nucleic acid metabolism, as well as others aimed at interfering with cell recognition, entry, uncoating, assembly, or maturation of virus particles.

20.3 Antiviral Drugs and Mechanisms of Action

Licensed compounds used in the treatment of viral infections target HIV, HBV, HCV, RSV, SARS-CoV-2, influenza virus, MPXV, HSV, and other herpesviruses such as VZV and CMV. A number of drugs act on steps that lead to the formation of the viral capsid or the mature virion (i.e., assembly, budding, and maturation) while others, whose target is the assembled capsid or the virion, interfere with processes affecting viral entry and uncoating [1]. Nonetheless, most of the approved drugs block intracellular events affecting the synthesis and dynamics of viral proteins and nucleic acids (Table 20.1). Within this group, viral polymerases are the major targets for many antiviral drugs.

20.3.1 Drugs Blocking Intracellular Processes Required for the Synthesis of Viral Components

Replication Inhibitors Targeting Viral Polymerases

Compounds inhibiting the replication of HSV, VZV, and CMV include prodrugs of nucleoside analogs (e.g., valacyclovir, valganciclovir, and famciclovir) that need to be phosphorylated in order to become substrates of the viral DNA polymerase

Table 20.1 Antiviral drugs approved for clinical use

Target	Step of the viral life cycle or cellular function inhibited	Virus	Drug type and name
Viral	Entry	HIV	*Fusion inhibitor:* Enfuvirtide *CD4 receptor binding inhibitor:* Fostemsavir
		HSV	*Fusion inhibitor:* Docosanol
	Disassembly/ uncoating	Influenza virus	*Drugs binding to the viral protein M2 (an ion channel):* Amantadine and rimantadine
	Genome replication	HIV[a]	*Nucleoside/nucleotide RT inhibitors:* Zidovudine, lamivudine, abacavir, emtricitabine, and tenofovir (tenofovir disoproxil fumarate and tenofovir alafenamide fumarate)[b] *Nonnucleoside RT inhibitors:* Nevirapine, efavirenz, etravirine, rilpivirine, and doravirine
		HBV	*Nucleoside/nucleotide analogs:* Lamivudine, emtricitabine, entecavir, telbivudine, adefovir (adefovir dipivoxil)[b] and tenofovir (tenofovir disoproxil fumarate and tenofovir alafenamide fumarate)[b]
		HCV	*Ribonucleotide analog:* Sofosbuvir *Nonnucleoside RNA polymerase inhibitor:* Dasabuvir *Ribonucleoside analog (mutagen)*: Ribavirin
		Influenza virus	*Endonuclease inhibitor (RNA polymerase subunit)*: Baloxavir marboxil *Ribonucleotide precursors (mutagens)*: Ribavirin and favipiravir
		SARS-CoV-2	*Ribonucleotide analog:* Remdesivir *Ribonucleoside analog (mutagen)*: Molnupiravir
		HSV and VZV	*Nucleoside/nucleotide analogs:* Acyclovir (valaciclovir)[b], penciclovir (famciclovir)[b], idoxuridine, vidarabine, trifluridine, and brivudine
		CMV	*Nucleoside/nucleotide analogs:* Ganciclovir (valganciclovir) and cidofovir *Pyrophosphate analog:* Foscarnet *DNA terminase complex inhibitor*: Letermovir
		Mpox	*Nucleotide analog:* Cidofovir
	Integration into the host genome	HIV	*HIV integrase inhibitors:* Raltegravir, elvitegravir, dolutegravir, bictegravir, and cabotegravir

(continued)

Table 20.1 (continued)

Target	Step of the viral life cycle or cellular function inhibited	Virus	Drug type and name
	Synthesis of viral mRNAs	CMV	*Antisense (phosphorothioate) oligonucleotide:* Fomivirsen
	Regulators of the viral life cycle	HCV	*Nonstructural protein 5A (NS5A) inhibitors:* Ledipasvir, ombitasvir, daclatasvir, elbasvir, velpatasvir, and pibrentasvir
	Cleavage of viral polyproteins	HIV[a]	*HIV protease inhibitors:* Saquinavir, amprenavir and its prodrug fosamprenavir, lopinavir, atazanavir, tipranavir, and darunavir
		HCV[a]	*HCV protease inhibitors:* Paritaprevir, asunaprevir, vaniprevir, grazoprevir, voxilaprevir, and glecaprevir
		SARS-CoV-2	*SARS-CoV-2 M^{pro} inhibitors:* Nirmatrelvir, ensitrelvir, simnotrelvir
	Capsid assembly	HIV	*Capsid binding compound:* Lenacapavir
	Budding	Influenza virus	*Viral neuraminidase inhibitors:* Oseltamivir, zanamivir, laninamivir octanoate, and peramivir
		Orthopoxviruses (smallpox, MPXV)	*VP37 envelope wrapping protein binder:* Tecovirimat
Cellular	Viral entry	HIV	*Viral coreceptor inhibitor:* Maraviroc *CD4-binding antibody:* Ibalizumab
	Innate immunity	HBV, HCV, and HPV	*Interferons:* Pegylated interferons α-2a and α-2b, and interferons α-2a and α-2b
	mRNA capping enzymes	HCV and other RNA viruses	*Ribonucleoside analog:* Ribavirin (also acts in replication as a lethal mutagen)

[a] Other approved anti-HIV drugs include less-frequently used or discontinued inhibitors of the RT (didanosine, zalcitabine, stavudine, and delavirdine) and protease (ritonavir, indinavir, and nelfinavir). First-generation HCV protease inhibitors boceprevir, telaprevir, and simeprevir are no longer marketed due to falling demand for the drugs caused by competition from newer and more effective anti-HCV combination therapies

[b] Compounds approved as prodrugs, whose names are indicated between parentheses

(Fig. 20.3). Viral enzymes (e.g., thymidine kinases in HSV and VZV, and a protein kinase in CMV) are responsible for the transformation of acyclovir, ganciclovir, and penciclovir into their monophosphate derivatives. Further phosphorylation steps are carried out by host cell kinases. The triphosphate derivatives of acyclovir, ganciclovir, and penciclovir, which mimic the natural substrates of the viral DNA polymerase, are incorporated into the growing DNA chain [16]. Acyclovir blocks DNA synthesis due to the lack of a 3′-OH in the ribose ring of the nucleotide, thereby impeding the formation of a phosphodiester bond with the 5′-OH of the next dNTP. On the other hand, ganciclovir and penciclovir possess 3′-OH groups and allow for

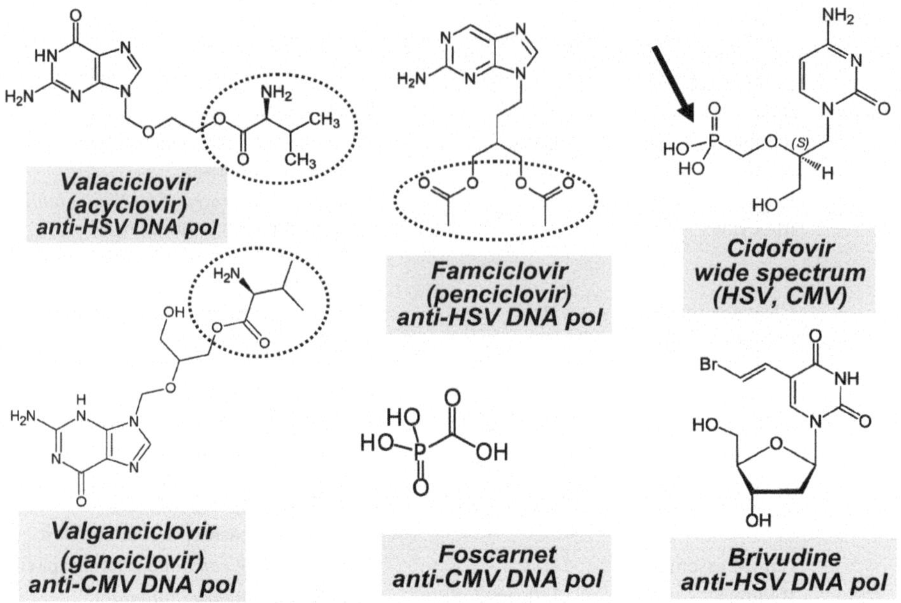

Fig. 20.3 Approved nucleoside, nucleotide, and pyrophosphate analogs with antiviral effect on herpesviruses. Valacyclovir, famciclovir, and valganciclovir are prodrugs of acyclovir, penciclovir and ganciclovir, respectively. Protecting groups that favor oral absorption are shown within dotted ellipses. The arrow shows the location of the phosphate group in the nucleotide analog. HSV, herpes simplex virus; CMV, cytomegalovirus

nucleotide incorporation. However, an unstable structure is formed due to the absence of the deoxyribose ring, and the DNA synthesis reaction stops after addition of several nucleotides.

Cidofovir is a phosphonate-containing acyclic cytosine analog that, unlike the inhibitors described above, does not depend on viral enzymes for its conversion to the triphosphorylated form that competes with the dNTP substrates [17]. Cidofovir-diphosphate is an inhibitor of the viral DNA polymerases of herpesviruses, orthopoxviruses, adenoviruses, polyomaviruses, and papillomaviruses [18]. Another group of DNA synthesis inhibitors that interfere with DNA polymerase function without being incorporated into the DNA is represented by vidarabine (a nucleoside analog) and foscarnet (a pyrophosphate analog). Foscarnet is a broad-spectrum inhibitor active against HSV-1, HSV-2, VZV, HCMV, Epstein–Barr virus (EBV), HIV, and HBV [19], but inactive against RNA polymerases of RNA viruses.

Nucleoside analogs constitute the backbone of current anti-HIV therapies. HIV is a retrovirus that replicates through a proviral double-stranded DNA intermediate. Its polymerase, known as reverse transcriptase (RT), is able to synthesize DNA by using either RNA or DNA as templates. Nucleoside/nucleotide RT inhibitors are prodrugs that need to be converted to active triphosphate analogs in order to be incorporated into the DNA during reverse transcription. They act as chain terminators thereby blocking DNA synthesis [20]. The HIV-1 RT is a heterodimer composed of subunits of 560 and 440 amino acids (known as p66 and p51, respectively) [21]. Both polypeptide chains have the same amino acid sequence, but the large subunit contains an RNase H domain that includes the 120 amino acids of its C-terminal region. The DNA polymerase domain is formed by residues 1–315 of which aspartic acid residues 110, 185, and 186 in the palm subdomain of p66 make up the catalytic triad.

There are many crystal structures available for HIV-1 RT, including relevant complexes such as the ternary complex of HIV-1 RT/double-stranded DNA/dTTP, or RT bound to RNA/DNA, or to DNA/DNA template-primers [22]. Additional structures containing DNA primers blocked with nucleoside analogs have been described as well. A large number of nonnucleoside RT inhibitors (NNRTIs) have also been crystallized in complex with the viral polymerase [22]. NNRTIs bind to a hydrophobic pocket located about 8–10 Å away from the DNA polymerase active site. The first NNRTIs were identified by using an antiviral screening approach, but structure-based drug design has played a prominent role in the discovery and development of more flexible next-generation NNRTIs such as etravirine or rilpivirine [23, 24]. The latest NNRTI approved for clinical use is doravirine. This drug showed limited adverse effects and potent activity in individuals with low CD4 counts and high viral loads and was found to be effective against NNRTI-resistant HIV strains containing K103N, Y181C, or K103N/Y181C amino acid substitutions.

Apart from retroviruses, reverse transcription is also involved in the replication of HBV [21]. These viruses form an immature RNA-containing nucleocapsid that, inside the cell, is converted into a mature nucleocapsid containing a relaxed circular DNA [25]. The DNA found in extracellular virions is synthesized by viral RT using the pregenomic RNA as a template. The HBV RT shows sequence similarity with HIV-1 RT, and several inhibitors of HIV-1 replication (e.g., lamivudine, adefovir, or tenofovir) have been approved for treating HBV infections (Table 20.1) [26]. Unfortunately, there is no available crystal structure for the HBV RT. The existence of this gap of knowledge can be attributed to major difficulties in obtaining a catalytically active polymerase.

RNA polymerases play a key role in the replication of important pathogenic RNA viruses such as HCV, poliovirus, rhinovirus, influenza virus, and others. The HCV replicase (NS5B, an RNA-dependent RNA polymerase) is a 65-kDa protein that contains a C-terminal membrane insertion sequence that traverses the phospholipid bilayer as a transmembrane segment. In 2013, the approval of sofosbuvir as the first antiviral drug targeting the HCV replicase represented a tremendous breakthrough in the fight against HCV infection [27]. Approval made sofosbuvir the first oral HCV-specific nucleotide polymerase inhibitor with potent, broad antiviral activity, and also the first all-oral treatment option for patients who could not receive interferon. Discovered in 2007 by a team led by Michael Sofia [28], sofosbuvir is a ProTide prodrug [29, 30], whose active metabolite is 2′-deoxy-2′-α-fluoro-β-C-methyluridine-5′-triphosphate (Fig. 20.4). This uridine nucleotide analog acts as a chain terminator in RNA synthesis and shows pan-genotypic antiviral activity. ProTide prodrugs are useful for delivering nucleoside analog monophosphates and

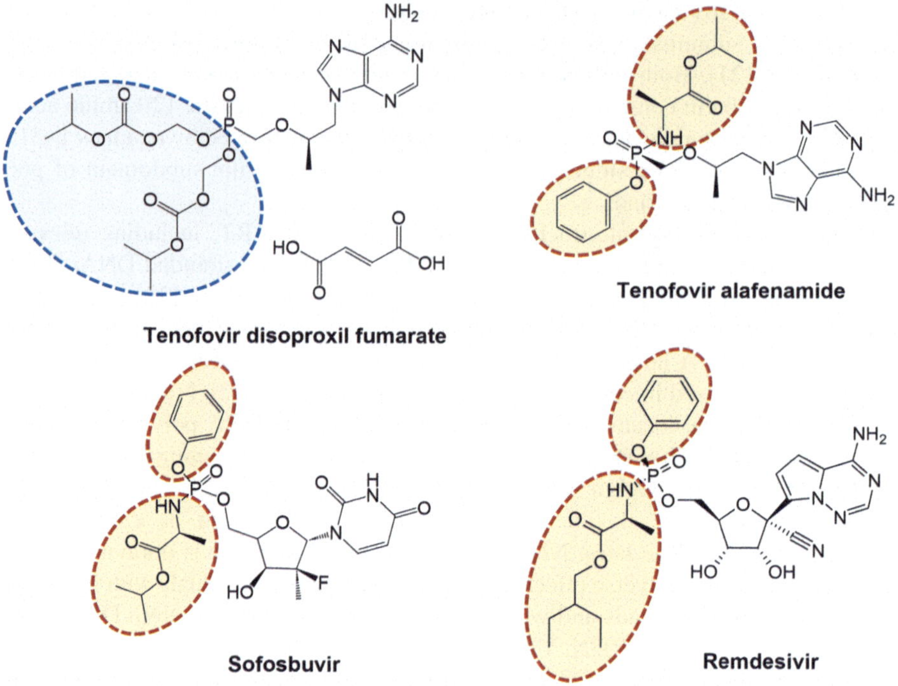

Fig. 20.4 Applications of the ProTide technology in the development of antiviral prodrugs. This approach was developed to deliver nucleotide analogs (as monophosphates or monophosphonates) into the cell and successfully applied to the delivery of tenofovir in HIV-infected cells. Tenofovir disoproxil fumarate has increasingly been replaced by tenofovir alafenamide because of reduced kidney and bone toxicity. Tenofovir alafenamide, sofosbuvir, and remdesivir are examples of antiviral prodrugs using the ProTide technology. Protecting groups are shown within dotted ellipses: light orange for the alafenamide and light blue for the disoproxil moieties

monophosphonates to cell types with high expression of cathepsin A and carboxylesterase 1, such as immune cells [29, 30]. Apart from sofosbuvir, a successful application of the ProTide technology is represented by tenofovir alafenamide (Fig. 20.4), an antiretroviral drug which improved the delivery of tenofovir into cells by facilitating efficient phosphorylation of the drug by endogenous nucleoside kinases [31].

Remdesivir is also a ProTide prodrug able to diffuse into cells, where it is converted to the monophosphate derivative of (2*R*,3*R*,4*S*,5*R*)-2-(4-aminopyrrolo[2,1-*f*][1,2,4]triazin-7-yl)-3,4-dihydroxy-5-(hydroxymethyl)tetrahydrofuran-2-carbonitrile through the actions of esterases and a phosphoamidase. The triphosphate form of the drug is incorporated into RNA at position +1 by the viral RNA polymerases. Enzymatic studies with the SARS-CoV-2 RNA polymerase showed that remdesivir causes delayed termination. Thus, RNA synthesis is halted after subsequent incorporation of three additional nucleotides [32]. Remdesivir has been approved or authorized for emergency use to treat COVID-19 in numerous countries. The drug has inhibitory activity in cell culture against different types of viruses, most notably *Filoviridae* (e.g., Ebola and Marburg viruses), *Paramyxoviridae* (e.g., Nipah and Hendra viruses), *Pneumoviridae* (e.g., RSV and human metapneumovirus), and *Coronaviridae* (e.g., SARS-CoV and Middle East respiratory syndrome coronavirus) [33–35].

As discussed above for HIV-1 RT, viral RNA polymerases can also be inhibited by nonnucleoside inhibitors. Dasabuvir is a selective inhibitor of HCV genotype 1 RNA polymerases that likely binds to an allosteric site in the palm subdomain of the enzyme [36]. Although dasabuvir is not as effective as sofosbuvir, it can be used as part of a multiple drug regimen to protect against selection of resistant HCV variants.

Apart from interfering with RNA or DNA synthesis, some nucleoside analogs also have a mutagenic activity that can lead to viral extinction. This process, known as lethal mutagenesis, has been demonstrated in RNA viruses for approved drugs such as ribavirin, favipiravir, and molnupiravir (Table 20.1) [37].

Drugs Interacting with Other Targets in the Viral Genome Replication Machinery

The influenza virus RNA polymerase consists of three subunits: polymerase basic 1 (PB1), PB2 and polymerase acidic (PA) in influenza A and B virus or polymerase 3 (P3) in influenza C virus [38]. PB1 contains characteristic sequence motifs of RNA-dependent RNA polymerases, and the fingers, palm, and thumb subdomains found in all viral polymerases. The synthesis of viral mRNAs requires a host pre-mRNA to be used as a primer for viral transcription. In the nucleus of infected cells, PB2 binds to the 5′-cap structure of host pre-mRNAs, and PA cleaves the captured RNA 10–13 nucleotides away from the cap structure. The N-terminal region of PA contains highly conserved residues for metal ion binding (His41, Glu80, Asp108, and Glu119) and catalytic activity of type II endonucleases. Approved in 2019, baloxavir marboxil is a prodrug of baloxavir acid. This molecule blocks the nucleolytic activity of PA by interacting with cations found in the active site of the enzyme [39]. Baloxavir marboxil shows potent broad-spectrum activity against influenza viruses, although resistance develops relatively soon due to the emergence of I38T/M variants. These amino acid substitutions involve the removal of a methyl group required for binding to baloxavir acid.

In herpesviruses, DNA maturation and packaging are performed by the terminase complex, which cleaves DNA to package the genome into the capsid [40, 41]. Letermovir is a 3,4-dihydro-quinazoline that targets the pUL56 subunit of the HCMV terminase complex, which is encoded by the viral genes *UL56*, UL51, and *UL89*. Letermovir inhibition blocks the cleavage of DNA concatemers, resulting in long uncleaved DNA and noninfectious HCMV particles. Letermovir is only active against HCMV and has no effect on other herpesviruses [42].

Other Targets of Antiviral Drugs

Gene expression is another possible target for antiviral intervention. A phosphorothioate derivative of the oligonucleotide 5′-GCG TTT GCT CTT CTT CTT GCG-3′ (known as fomivirsen) has been used as a repressor of the synthesis of early gene products in CMV infections. It binds to the complementary sequence in the viral mRNA, blocking its translation. In retroviruses, integration of the double-stranded proviral DNA (the product of reverse transcription) into the host cell DNA is essential for expression of viral mRNA. Retroviral mRNA is generated by transcription of the integrated provirus by cellular RNA polymerases.

Raltegravir, elvitegravir, dolutegravir, bictegravir, and cabotegravir are inhibitors of HIV-1 integrase [3, 43], an enzyme endowed with 3′ endonucleolytic and strand transfer activities. It is composed of three structural domains: an N-terminal domain of unknown function, the central catalytic domain, and a C-terminal domain with DNA binding affinity. Approved drugs bind to the central domain, close to the catalytic residues Asp64, Asp116, and Glu152, and act by blocking the strand transfer activity of the integrase. First-generation integrase inhibitors such as raltegravir and elvitegravir showed a relatively low genetic barrier to resistance. However, dolutegravir, bictegravir, and cabotegravir were effective against both wild-type HIV-1 integrase and many raltegravir- and elvitegravir-resistant mutants [43]. These drugs show a very high genetic barrier to development of drug resistance, and this is important given the rising trend of resistance to NNRTI-based regimens, particularly in low- and middle-income countries. A series of X-ray crystallography and cryogenic electron microscopy (cryo-EM) studies carried out in the last decade have revealed the structural basis of retroviral DNA integration [44]. They have shown that integrase molecules congregate on viral DNA ends to assemble a conserved intasome core machine that facilitates integration. Recently, allosteric integrase inhibitors have been developed as a promising class of small molecules that antagonize viral morphogenesis. However, these compounds are still in clinical trials.

Viral proteins are frequently synthesized as precursor polypeptides that are cleaved by viral proteases. Sometimes, this cleavage occurs before the assembly of the viral capsid, which is formed by the association of mature proteins into capsomers. For example, in poliovirus, rhinovirus, and other picornaviruses, 60 protomers made of processed capsid proteins VP0, VP1, and VP3 assemble to form the icosahedral particle.

In HCV, its serine protease (designated as NS3) interacts with an NS4A peptide co-factor that modulates its proteolytic activity. Promising NS3/4A inhibitors were developed after rational structure-based design using a hexapeptide (Asp-Asp-Ile-Val-Pro-Cys) derived from the natural NS5A/5B cleavage site as a lead compound. After systematic replacement of single amino acid residues, important substituents in the protease binding pocket were identified and compounds were docked in the crystal structure of the unliganded NS3 protease. Further optimization led to the identification of ciluprevir (BILN-2061), one of the first promising drug candidates (Fig. 20.5). Boceprevir and telaprevir were the first selective, potent, and orally bioavailable NS3 serine protease inhibitors approved for treatment of chronic HCV infection [45]. However, both were discontinued due to commercial reasons derived from fierce competition in the pharmaceutical industry. Today, there are six drugs available for HCV therapy that are usually prescribed as components of approved drug combinations (Table 20.1).

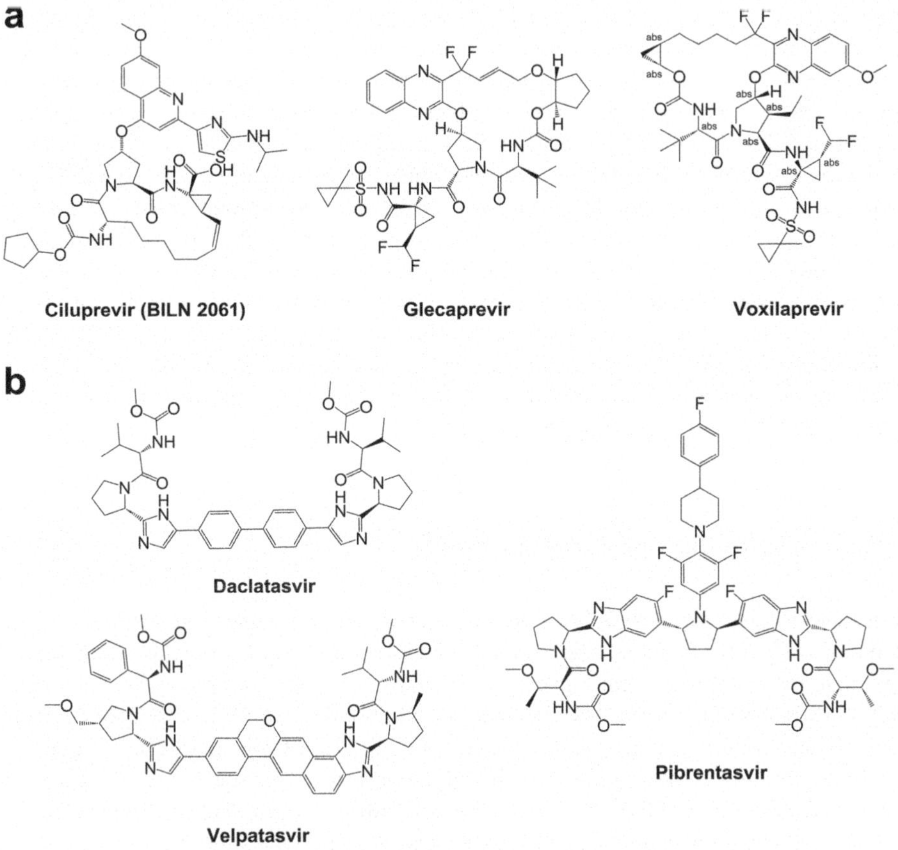

Fig. 20.5 Approved drugs targeting HCV protease (**a**) and NS5A protein (**b**). Depicted molecules are frequently used in combination with sofosbuvir in HCV eradicating therapies

An interesting new drug targeting the main protease (M^{pro}) of SARS-CoV-2 is nirmatrelvir. This is a low-molecular-weight (499.5 Da) covalent inhibitor that mimics the peptide substrate of M^{pro}, the cysteine protease responsible for the hydrolysis of viral replicase polyproteins pp1a and pp1ab to produce functional proteins during replication [46]. It is administered together with ritonavir as a "booster." Ritonavir is an HIV protease inhibitor that slows the metabolism of nirmatrelvir via cytochrome P450 (especially CYP3A4) inhibition. Also targeting M^{pro}, but acting as a noncovalent inhibitor, ensitrelvir [47] has been approved for emergency use in Japan in November 2022, and is currently in phase III clinical trials elsewhere.

The nonstructural 5A (NS5A) protein of HCV has been identified as the target of daclatasvir (BMS-790052), a thiazolidinone-containing small molecule that was identified using phenotypic screening in a cell-based HCV genotype 1b replicon system [48]. The NS5A protein does not have enzymatic activity but plays a critical role in the formation of endoplasmic reticulum-derived double-membrane vesicles, thereby facilitating efficient viral replication and assembly. NS5A inhibitors are usually effective against all HCV genotypes, although velpatasvir and pibrentasvir provide greater pan-genotypic activity than the previously approved drugs of this class [4]. All NS5A-inhibitors are presumed to bind to the domain I of NS5A protein, reducing hyperphosphorylation and blocking its dimerization. These inhibitors possess a relatively low threshold for resistance. Mutations most commonly observed in patients failing therapy appear at positions 24, 28, 30, 31, 58, 92, and 93, while the most relevant from a clinical point of view is Y93H [49]. Daclatasvir has been largely abandoned in Western countries because of high treatment costs. However, effective drug combinations combining NS5A inhibitors with HCV RNA polymerase and/or protease inhibitors (e.g., sofosbuvir/velpatasvir, glecaprevir/pibrentasvir, and voxilaprevir/sofosbuvir/velpatasvir) are currently recommended for curing HCV infection (Fig. 20.5).

20.3.2 Inhibitors of Viral Assembly

For many years, viral assembly stood out as a challenging target in antiviral research. Considerable efforts have been focused on creating small molecules that either stabilize or destabilize viral capsids (see Chaps. 11 and 12). Lenacapavir, a long-acting drug that targets the HIV-1 capsid, showed potent antiviral activity in persons with HIV-1 and was approved for medical use in 2022 [50]. In addition, there are several compounds undergoing preclinical development that are effective inhibitors of capsid assembly in HBV (e.g., heteroaryldihydropyrimidines) [28, 29] and HIV-1 (CAI, CAP-1, bevirimat, etc....) [51] (see Sects. 20.6.1 and 20.6.2).

20.3.3 Antiviral Drugs Targeting Budding and Production of Enveloped Extracellular Virions

At present, antiviral treatment and prophylaxis of human influenza virus infections, including aggressive strains such as avian H_5N_1, is based on the use of oseltamivir, zanamivir, peramivir, and laninamivir, which inhibit the release of infectious viruses [52]. The influenza virus membrane contains a glycoprotein (hemagglutinin) that recognizes sialic acid in cell-surface glycoproteins and glycolipids [53]. Sialic acid is therefore the receptor for virus entry. However, newly formed virions may remain attached to the cell membrane because of the hemagglutinin–sialic acid interaction. The viral neuraminidase (also known as sialidase) facilitates the release of virions from infected cells by cleaving sialic acid residues. It is this hydrolytic activity that is blocked by neuraminidase inhibitors.

Tecovirimat (formerly known as ST-246) targets the major envelope protein of poxviruses and blocks the formation of intracellular enveloped virus coming out of the Golgi apparatus, thereby preventing the production of extracellular virions [54, 55]. Because the viruses cannot leave the infected cells, viral dissemination within the human body is thwarted. Tecovirimat targets p37, a phospholipase that is part of a protein complex required for extracellular virus particle formation. The drug blocks the interaction of p37 with cellular proteins Rab9 GTPase and TIP47, which are needed for the formation of cell-associated enveloped virus and extracellular enveloped virus [56, 57]. Tecovirimat is a small tetracyclic acylhydrazide molecule, recognized in 2002 as an effective drug against orthopoxvirus-induced cytopathic effects, after an extensive high-throughput screening process involving >350,000 compounds [54]. Its discovery was a result of a big effort to identify compounds suitable for stockpiling in response to smallpox-related bioterror concerns. Today, with MPXV being considered a global threat, tecovirimat has become the drug of choice for a safe and effective treatment against the infection, either alone or in combination with cidofovir (see above in Sect. 20.3.1).

20.3.4 Maturation Inhibitors

Maturation is a relatively common process in animal and human viruses (see Chap. 14). In contrast to HCV and other virus families, specific cleavage of the polypeptides encoding the mature viral proteins occurs only on assembled immature particles. HIV-1 protease inhibitors block the proteolytic processing of precursor polypeptides Gag and Gag-Pol. Cleavage of Gag and Gag-Pol leads to the formation of structural proteins (e.g., MA, CA and NC), p6 and the viral enzymes protease, RT and integrase (Fig. 20.6). This process occurs extracellularly and is essential for the formation of an infectious virion. The HIV-1 protease is a homodimeric enzyme composed of subunits of 99 amino acids with a symmetric substrate binding pocket. Inhibitors were initially designed as peptide derivatives mimicking the transition state of the reaction that contained a non-hydrolyzable bond at the position where the cleavage was expected to occur. All of the approved HIV protease inhibitors, except tipranavir, can be considered as peptidomimetics [6].

20.3.5 Drugs Targeting Viral Entry

Drugs acting at this step include compounds that block cell attachment, receptor recognition or the mechanisms leading to the translocation of the viral capsid to the host cell cytoplasm (see Chaps. 15 and 16). Antiretroviral drugs belonging to this category are fostemsavir, maraviroc, and enfuvirtide, as well as ibalizumab, a non-immunosuppressive humanized monoclonal antibody that binds to CD4, the primary receptor for HIV. Fostemsavir interferes with CD4 binding, while maraviroc

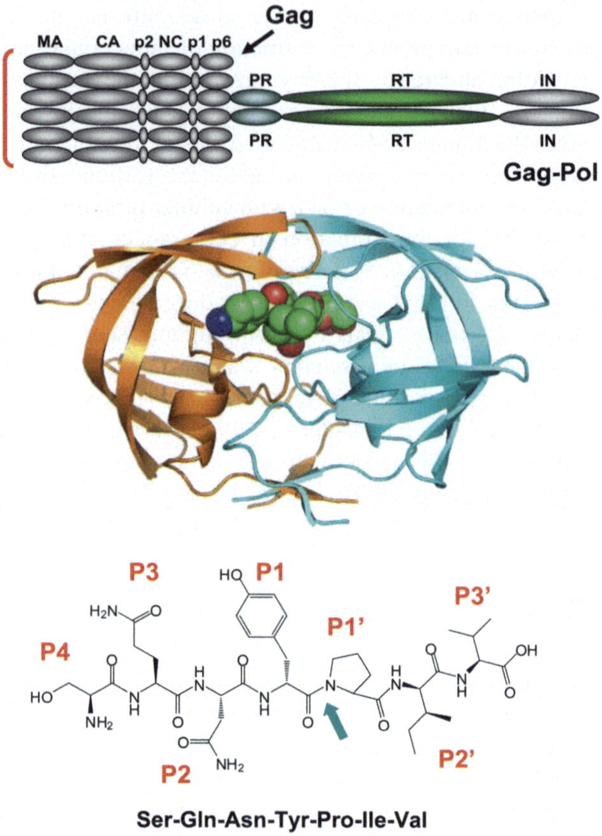

Fig. 20.6 Protein arrangement in Gag and Gag-Pol precursor polyproteins, HIV-1 protease structure and model substrate. Cartoon representation of the protease bound to darunavir (PDB entry 1T3R). Inhibitor C, O, and N atoms are displayed as green, red, and purple spheres, respectively. The minimal peptide substrate SQNY*PIV is shown below the protease structure. The asterisk in the peptide sequence and the arrow in the lower panel indicate the cleavage site

binds to the chemokine coreceptor CCR5 required for viral entry. Enfuvirtide blocks the fusion of the viral envelope with the cell membrane.

Receptor and Coreceptor Recognition in HIV-1 Infection

HIV-1 enters target cells through a multi-step process that requires non-specific interactions with surface heparan sulfates, followed by binding of the viral envelope glycoprotein gp120 to its primary cellular receptor CD4. Attachment inhibitors in HIV-1 include dextran sulphate and other polyanions (e.g., tailor-made dendrons rich in decorated tryptophan residues [58]), as well as proteins binding glycosidic residues (e.g., mannose-specific lectins or cyanovirin N).

CD4 binds into a carbohydrate-free depression in gp120 that is poorly accessible to immunoglobulins. However, crystallographic studies carried out with complexes of gp120, CD4, and Fab fragments of neutralizing antibodies have shown that Phe43 of CD4 and residues 365–371 and 425–430 of gp120 make the largest contributions to interatomic contacts. Phe43 is located on the CDR2-like loop of CD4 and binds within a hydrophobic pocket of gp120, termed the "Phe43 cavity," while Arg59 (in CD4) is located on a neighboring β-strand and forms an ion pair with the gp120 residue Asp368. The disruption of this interaction by small molecule binding has been validated as a promising approach to prevent HIV-1 entry. Inhibitors containing a tetramethylpiperidine ring such as NBD-556 displayed remarkable antiviral activity and altered the gp120 conformation in a similar way as observed upon CD4 binding [59]. Structure-based modification of the critical oxalamide mid region of NBD-556 led to the discovery of NBD-11021, a potent CD4 antagonist [60]. This compound was the precursor of BMS-626529, later named temsavir and currently marketed as a prodrug (fostemsavir) [61]. Approved in 2020, fostemsavir represented an important advance against multidrug resistant HIV-1 due to its novel mechanism of action and lack of cross-resistance to other antiretroviral drugs [62]. Docking studies and molecular dynamics simulations predicted that temsavir binds to the unliganded conformation of HIV-1 gp120 within the structurally conserved outer domain, adjacent to the CD4 binding loop [63].

CCR5 and CXCR4 are the necessary coreceptors for cellular entry of HIV-1. These molecules are members of the seven-transmembrane G protein-coupled receptor superfamily and can be blocked by natural ligands such as MIP-1α, MIP-1β and RANTES for CCR5 and SDF-1 for CXCR4. RANTES derivatives and SDF-1α have been investigated as potential inhibitors of viral entry. However, smaller CCR5 antagonists such as TAK-652 and TAK-779, aplaviroc, vicriviroc, cenicriviroc, and maraviroc were further studied as potential anti-HIV agents. So far, maraviroc remains the only clinically approved coreceptor antagonist. However, resistance to this drug appears mostly as a result of HIV-1 using CXCR4 as an alternate coreceptor for entry [7]. Cenicriviroc, a dual CCR2/CCR5 antagonist, is currently in clinical trials as an anti-inflammatory drug that ameliorates the effects of COVID-19 in both early and late stages of the disease [64].

The most advanced CXCR4 antagonist, AMD3100 (plerixafor), has not been further studied as a therapeutic agent against HIV-1 [1]. However, in combination with granulocyte colony stimulating factor (and sold under the brand name Mozobil®), it has been approved for the mobilization of hematopoietic stem cells to the peripheral blood for collection and subsequent autologous transplantation in patients with non-Hodgkin's lymphoma and multiple myeloma.

Fusion Inhibitors: Targeting the Formation of Coiled-Coil Structures

Viral recognition by cellular receptors triggers a mechanism that leads to the fusion of the viral envelope with the target cell membrane (see Chap. 16). Crystallographic studies of pre- and post-fusion conformations of the influenza virus hemagglutinin showed that the formation of a six-helix bundle played a major role in the fusogenic events. Drugs targeting the helix bundle have been developed to block entry of HIV-1, RSV, SARS-CoV-2, and other viruses having class I fusion proteins.

In the case of HIV-1, enfuvirtide is a 36-amino acid peptide that derives from residues 127–162 of the transmembrane protein gp41. This glycoprotein has a trimeric structure, and each monomer contains a transmembrane region of 21 residues, located between an N-terminal ectodomain (175 residues) and a long cytoplasmic domain of 150 amino acids. The sequence of enfuvirtide overlaps with the heptad repeat 2 (HR2) region (residues 117–154) that interacts with the HR1 region (residues 29–82) of gp41. Enfuvirtide binds HR1 and prevents formation of the trimeric coiled-coil structure that is required for membrane fusion. Despite showing efficacy in vivo, the clinical use of the drug has been limited by its reduced bioavailability and its large production costs. Investigational anti-HIV drugs based on similar principles include tifuvirtide, sifuvirtide, and T-2635. These peptides mimic in part the HR2 structure and exert their effects through a mechanism of action akin to that shown by enfuvirtide. Similar approaches have been considered for viruses of clinically relevant families such as *Arenaviridae*, *Coronaviridae*, *Filoviridae*, *Orthomyxoviridae*, and *Paramyxoviridae* [65].

An antiviral agent that inhibits viral entry by interfering directly with cell-surface phospholipids is *n*-docosanol (also known as behenyl alcohol). This is a saturated fatty alcohol containing 22 carbon atoms that restricts entry in a broad range of enveloped viruses that uncoat at the plasma membrane of target cells. The drug was approved for medical use in the USA to treat cold sores produced by HSV.

20.3.6 Uncoating Inhibitors

Amantadine and rimantadine were used more than 30 years ago to prevent influenza A virus infections. However, they never gained wide acceptance and today virtually all strains are resistant to these agents. These compounds block the release of the viral capsid to the cytoplasm of the infected cell because they interact with M2, an integral protein located in the viral envelope. It consists of four identical units, each one having three distinct domains: (1) an ectodomain made up of 24 residues exposed to the outside environment, (2) a transmembrane region (mostly hydrophobic) composed of 22 amino acids, and (3) 52 amino acids at the C-terminal region exposed to the intraviral milieu. M2 is activated by low pH and functions as an ion channel that facilitates entry of protons into the viral particle. Amantadine and rimantadine bind to the transmembrane region (Fig. 20.7), sterically blocking the channel and inhibiting the release of the viral capsid to the host cell cytoplasm.

20.4 Strategies in the Development of Antiviral Drugs: From Random Screening to Structure-Based Design

Most of the antiviral drugs clinically used today and mentioned in previous and following sections of this chapter were discovered serendipitously or as a result of screening campaigns that made use of large libraries of synthetic compounds or natural products together with biochemical and whole cell replicon assays. In the traditional drug discovery process, it is estimated that screening of around 80,000 molecules can render one hit that needs to be further optimized into a lead compound by improving its activity and reducing its toxicity before entering clinical trials. In

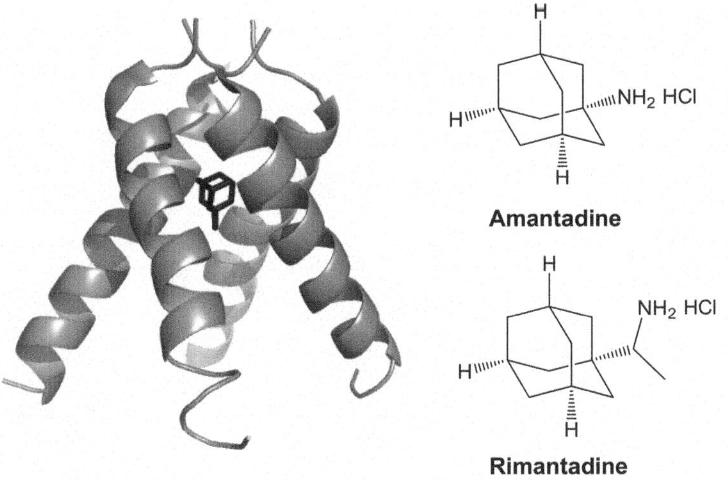

Fig. 20.7 Transmembrane region of influenza virus M2 protein bound to amantadine and chemical structures of drugs blocking the release of the viral capsid into the cytoplasm. Cartoon representation of the transmembrane regions of the M2 tetramer and stick model of the bound inhibitor (coordinates were taken from PDB entry 3C9J)

this scenario, pharmaceutical companies invested in the development of high-throughput screening (HTS) assays and in the generation of large chemical libraries obtained by combinatorial chemistry. Nevertheless, this process is expensive, time-consuming and requires further efforts in target validation.

Random approaches including HTS and combinatorial chemistry are used in the absence of structural information about ligands or target proteins or macromolecular complexes. Over the last three decades, we have been able to witness many profound changes in the drug discovery paradigm due to major advances in such diverse areas as HTS methods, genome sequence projects, macromolecular structure determination, quality and scope of publicly accessible databases, computer performance, and new algorithms for "in silico" approaches [66, 67] (Fig. 20.8). Among the latter, structure-based and ligand-based virtual screening (SBVS and LBVS, respectively) stand out as powerful tools that are being increasingly used in both industry and academia [68]. These techniques rely on the availability of the three-dimensional structure(s) of the macromolecular target (SBVS) and/or one or more known ligands (LBVS). This information is used to search in chemical or fragment libraries for putative hits (i.e., putative target-binding compounds) that can thereafter be tested experimentally for confirmation of affinity/activity. The positive hits are expected to be eventually transformed into lead compounds. Due to the huge number of molecules that can be found in currently available databases (e.g., 1.4 billion in ZINC20 [69]), it is customary to employ a series of computational cost-effective filters to

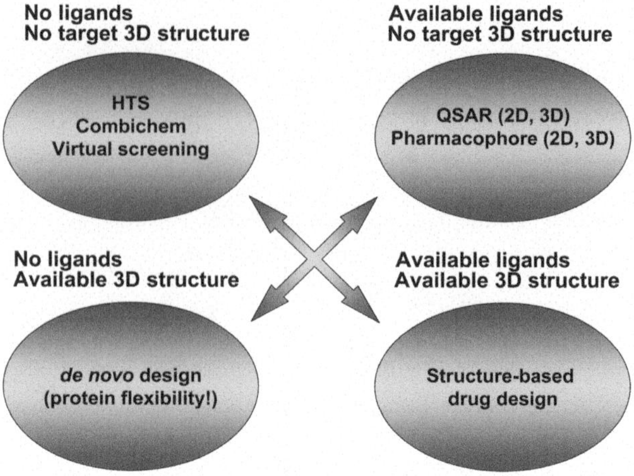

Fig. 20.8 Strategies in drug design depending on the availability of structural information for ligands and biological targets. Abbreviations: 2D, bidimensional; 3D, three-dimensional; HTS, high-throughput screening; and QSAR, quantitative structure–activity relationship

narrow down the number of putative ligands that will be subjected to virtual screening. Lipinski's rule of five[1] [70] and/or other physicochemical-based principles, as well as pharmacophoric hypotheses [71], can be used for filtering out many undesired candidates. One goal of LBVS is to find similarly shaped molecules that belong to novel chemotypes ("scaffold hopping") [72]. A word of caution is necessary, however, as it has been conclusively shown that docking scores and other theoretically derived metrics are often not good enough to fully support the selection of candidates [73]. Likewise, the need for continuous improvement of the scoring functions and validation datasets [74], as well as the importance of a critical visual inspection of the results to better inform the decision-making process, cannot be overemphasized [75].

20.4.1 Structure-Based Drug Design

Structure-based drug design relies on knowledge of the three-dimensional structure of the biological target gained by means of experimental techniques such as X-ray crystallography (see Chap. 4), nuclear magnetic resonance (NMR) spectroscopy (see Chap. 5), or cryo-EM (see Chap. 3), as well as novel highly reliable theoretical methods for predicting protein structure such as RoseTTAFold [76] and AlphaFold [77]. A number of computational tools have also been developed to identify putative ligands that can fit into an active site or functional epitopes in the target protein or

[1] Lipinski's rule states that, in general, an orally active drug should meet at least three of the following criteria: (1) Not more than 5 hydrogen bond donors (nitrogen or oxygen atoms with one or more hydrogen atoms), (2) not more than 10 hydrogen bond acceptors (nitrogen or oxygen atoms), (3) a molecular mass < 500 daltons, and (4) an octanol-water partition coefficient log P not greater than 5.

molecular complex. Success in molecular docking, which may be defined as an optimization problem that describes the "best-fit" orientation of a ligand within a binding site [78], very often depends on the quality of the three-dimensional target protein structure. Structures determined by X-ray crystallography at atomic or near-atomic resolution are the most adequate for docking. More specifically, successful identification of lead compounds usually requires a crystallographic resolution below 2.5 Å and an R factor well below 25% (see Chap. 4).

20.4.2 Ligand Docking and Virtual Screening

Docking methods are calibrated and tested for their ability to predict, at the top of the list of possible solutions, the native conformation of a ligand bound to a protein as found in a representative series of crystallographic complexes. To assess the performance of a docking program in SBVS, its ability to discriminate true binders from a pool of fake ligands ("decoys") is also examined. The goodness of fit between the ligand and the receptor is evaluated by means of a scoring function composed of different terms that attempt to account for the forces driving the binding event. Although the underlying physical laws describing the association process are well understood, accuracy and computational processing (mainly execution time) evolve in opposite directions, and fine tuning the appropriate balance between them is not an easy task. Therefore, accuracy is normally sacrificed for speed, especially in SBVS where very often too simplistic scoring functions are employed which plainly cannot capture the true free energy of binding that is measured experimentally [79]. Ligand conformers can be generated within the binding site itself "on the fly" for each target or created beforehand as a collection of all major theoretically possible conformers following some pre-defined rules. The former method is more computer-intensive but, as an advantage, it can generate strained conformations that adapt better to each specific binding site environment. The latter method allows the stored coordinates to be reused as many times as needed in different SBVS campaigns, but the drawback is that a relevant conformation that deviates slightly from a canonical one can be missed.

Fragments [molecular weight $\leq$ 250 Da, log P $\leq$ 2.5, (where P refers to the octanol-water partition coefficient) and less than six rotatable bonds] are usually endowed with reduced affinities compared to conventional ligands but are better suited for chemical modifications aimed at producing novel drug candidates [80]. They can sample chemical space more effectively than do regular ligands. It is generally accepted that fragment docking outperforms traditional small molecule docking in terms of hit rates even though erroneous binding modes can be obtained for them because their volumes are much smaller than those of typical binding site cavities (Fig. 20.9). Docked fragments can then evolve virtually (by adding different chemical decorations), be linked (to join two or more fragments that occupy different regions of the binding site), self-assemble (through direct bond formation between different reacting fragments), and/or be optimized to better attain drug-like

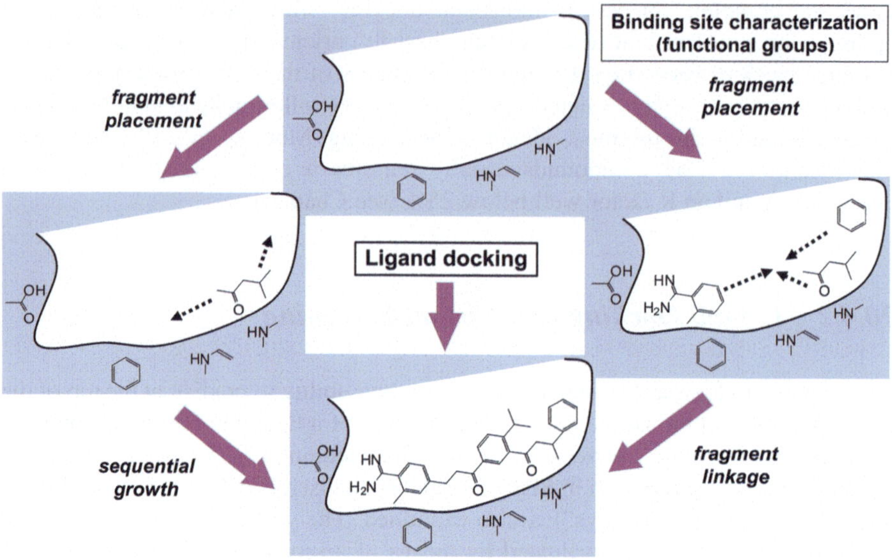

Fig. 20.9 Schematic diagram showing ligand- and fragment-based virtual screening strategies

properties. Since the binding affinity of compounds has been shown to grow approximately linearly with their molecular weight, the affinity of fragments and standard drugs is usually compared in terms of a metric called ligand efficiency index, which is defined as the binding free energy per heavy atom. Given that affinity/activity information related to ligand–target complexes is currently stored in several databases (e.g., BindingDB, ChEMBL, and PDBBind, among others), ligand efficiency indices have been proposed as useful variables to chart the chemico-biological space and visualize it in an atlas-like type of representation [81].

The binding pockets that are explored in most SBVS campaigns are relatively small cavities that are meant for natural agonists, substrates, or cofactors in a limited number of macromolecular targets. These pockets represent what are known as the "orthosteric" binding sites for drugs. However, other drugs (e.g., NNRTIs) exert their action by altering the receptor's conformation ("allosteric modulators") in such a way that the binding and/or functional properties of agonists or substrates is/are negatively affected. Besides, other drugs are known to modulate or prevent some key protein–protein interactions.

Most protein–protein interfaces have been usually regarded as "undruggable" because conventional analysis reveals that they lack well-defined binding pockets and display shallow grooves instead. However, it has been noted that a large percentage of the free energy of interaction normally arises from small regions designated as "hot spots." These facts supported the concept that protein–protein interaction disruptors do not need to mimic the entire protein binding surface but rather a smaller subset of key residues in a relatively compact region of one of the binding partners.

Different computational tools have been developed for exploring the properties of protein–protein interfaces and facilitating the identification of starting points for small-molecule design (e.g., PocketQuery [82], http://pocketquery.csb.pitt.edu/). More generally, "druggable" sites can be located using FTMap [83], which is an efficient fast Fourier transform correlation method that samples multiple positions of up to 16 different probes on a three-dimensional grid centered on a protein of interest. A simple energy expression then allows the identification of consensus sites where probes tend to cluster, giving rise to favorable interactions. These sites correlate with locations where high-affinity ligands are found in protein complexes. The method is sensitive to conformational changes in the protein and can be used to map affinity sites prior to docking. The FTMap algorithm is implemented in the public domain Web server http://ftmap.bu.edu/.

20.4.3 Structure–Activity Relationships

Finally, visual examination of the crystal structure of a single ligand–receptor complex is usually highly informative and can guide the design of new analogs that will display graded affinities towards the target. However, a quantitative assessment of these variations will generally be difficult because the observed free energy differences result from a subtle interplay of binding forces within the receptor binding site that normally take place in the face of competition with water molecules. An approach that has been shown to be useful in these cases is to model the whole set of complexes and decompose the ligand–receptor interaction energies into a number of van der Waals and electrostatic contributions emanating from individual receptor amino acids, together with additional terms representing the desolvation of both binding partners. The resulting matrix of energy terms can then be projected onto a small number of orthogonal "latent variables" or principal components using partial least squares in a way similar to that used in the formerly very popular—but target uninformed—comparative molecular field analysis (CoMFA). At the end of the procedure, those pair-wise interactions between the ligands and individual protein residues that are predictive of binding affinity or activity are selected and weighted according to their importance in the model. Since its inception, this chemometric method, termed comparative binding energy (COMBINE) analysis, has been successfully applied to a variety of biologically relevant targets including the protease and the RT of HIV-1 [84].

20.4.4 Ligand Optimization

Following Nobel Prize winner James Black's famous advice that "the most fruitful basis for the discovery of a new drug is to start with an old drug" [85], a reasonable fast approach to drug discovery in the face of emerging viral pandemics is to identify active molecules from existing currently or formerly approved drugs. A favorable outcome of these "drug repurposing" or "drug repositioning" efforts using best practices [86] is a series of candidate molecules—akin to early prototypes in a classical medicinal chemistry project—that can be used to redesign the weak hits into more potent inhibitors [87], possibly incorporating additional beneficial features such as improved oral bioavailability by purposeful exploitation of prodrug strategies.

20.5 Case Studies in Structure-Based Antiviral Drug Development

During the last three decades, we have witnessed an unprecedented effort to discover new antiviral drugs, most of them directed towards HIV, HCV, and more recently towards SARS-CoV-2. Many of those studies have benefited from the availability of high-resolution crystal structures (see Chap. 4). In recent years, the atomic or quasi-atomic structures of many more viral proteins and virus particles that may constitute new targets for antiviral drug development have been determined by cryo-EM, which has become the method of choice for some purposes (see Chap. 3). We have selected a few case studies to illustrate how structure-based design has contributed to antiviral drug discovery. In this section, we describe two successful examples that include oseltamivir and zanamivir acting on influenza virus neuraminidase, and mutation-resilient inhibitors of HIV-1 protease (e.g., darunavir). In the next section (Sect. 20.6), we describe current experimental efforts in structure-based design, applied to inhibitors acting on the assembly or disassembly of viral capsids.

20.5.1 Neuraminidase Inhibitors Against the Threat of Influenza Pandemics

Although the crystal structure of influenza virus neuraminidase (also known as sialidase) was determined in 1983 [88], its resolution was relatively low (approx. 3.0 Å) and, therefore, of limited value in structure-based drug design. Initial efforts to design neuraminidase inhibitors were based on substrate-like compounds such as *N*-acetylneuraminic acid (Neu5Ac) derivatives (Fig. 20.10). One of them, the 2-deoxy-α-Neu5Ac demonstrated good activity in mouse model experiments. The improvement in resolution of the neuraminidase X-ray structures obtained in

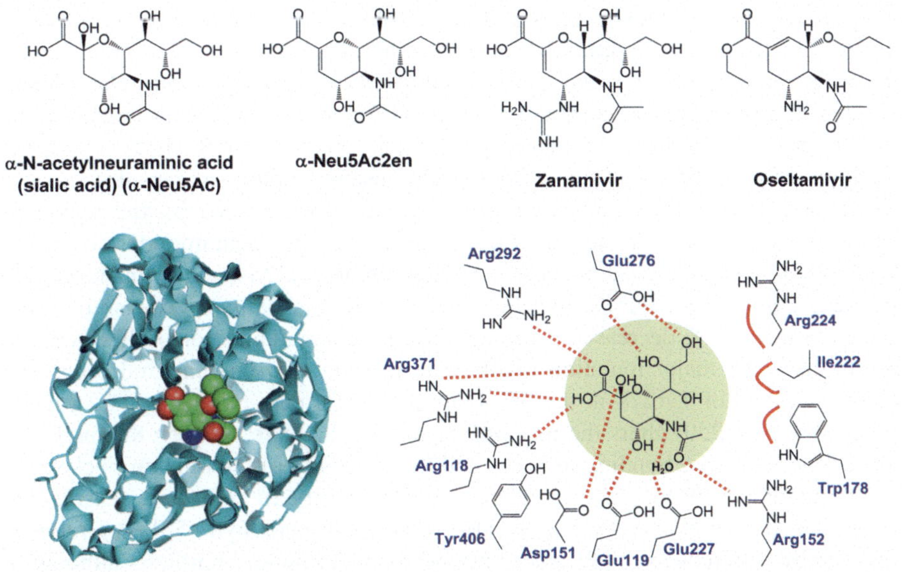

Fig. 20.10 Influenza virus neuraminidase inhibitors and structural basis of their mechanism of action. The lower part shows the structure of a neuraminidase monomer with a molecule of GS 4071 (the oseltamivir prodrug). C, O, and N atoms in the inhibitor are shown in green, red, and purple, respectively (the structure shown was taken from PDB entry 2HT8). The scheme on the right shows the hydrogen bonding network (red dotted lines) and the hydrophobic contacts (continuous red lines) in the sialic acid binding site of influenza virus neuraminidase

complex with Neu5Ac or with the derivative Neu5Ac2en facilitated the design of more potent inhibitors. These studies revealed that sialic acid adopts a quite deformed conformation when bound to neuraminidase, due to strong ionic interactions of the sialic acid carboxyl moiety with three arginine residues of the enzyme (Arg118, Arg192, and Arg371). Starting with Neu5Ac2en, the substitution of the C-4 hydroxyl group with a guanidinyl functionality rendered the 4-deoxy-4-guanidino-Neu5Ac2en (zanamivir) that showed significantly improved affinity promoted by interactions between the guanidinyl moiety and the neuraminidase residues Glu119 and Glu227 [89]. Zanamivir was found to be highly selective for influenza virus sialidase. It was approved in 1999 as the first neuraminidase-targeting anti-influenza drug (Fig. 20.10) even though it can be administered only as a powder for oral inhalation. Laninamivir was derived by methylation of the C-7 hydroxyl group of zanamivir, and it is administered as an octanoyl ester prodrug [52].

Oseltamivir was developed as a result of efforts at Gilead Sciences to develop neuraminidase inhibitors based on non-carbohydrate templates [85]. In this context, a cyclohexene ring was considered as a transition-state mimic that was amenable to chemical modifications to optimize its biological activity. X-ray crystallographic studies of sialic acid and related analogs bound to neuraminidase (Fig. 20.10) demonstrated that the C7 hydroxyl group of the glycerol side chain does not interact with amino acid residues in the sialidase active site. Replacement of the glycerol moiety to improve the lipophilicity of those inhibitors was carried out even though the hydroxyl groups at C8 and C9 were involved in interactions with Glu276 and Arg224. A 3-pentyl ether side chain at position C7 gave optimal results and led to the development of GS 4071. Oseltamivir (originally known as GS 4104) is an ethyl ester prodrug of GS 4071 that can be administered orally as capsules or as a suspension, and is then converted to the active form in vivo by the action of endogenous esterases. Crystal structures of influenza virus neuraminidase bound to GS 4071 showed a rearrangement of the sidechain of Glu276 that now interacts with Arg228, generating a larger hydrophobic area in this domain. The amino acid substitution H274Y decreases neuraminidase binding affinity for oseltamivir and results from the major mutation involved in resistance to this antiviral drug. The repulsive effect of Tyr274 in the 250-loop (a loop formed in the enzyme structure around position 250) alters the neuraminic acid binding site, through the modification of the hydrogen bonding network and the flexibility of the binding pocket.

Oseltamivir-resistant influenza virus A(H1N1) viruses remained sensitive to zanamivir and laninamivir but were also resistant to peramivir [52]. Peramivir is structurally different from the other approved neuraminidase inhibitors although it shares similar functional groups with both zanamivir (a guanidine moiety) and oseltamivir (a hydrophobic side chain). As in the case of oseltamivir, zanamivir, and laninamivir binding is also reduced by amino acid substitutions affecting the 250-loop. One of them is S247P which was detected in influenza A(H3N2) variant virus. This substitution increases the rigidity of the loop, altering the hydrogen bonding network involved in drug binding.

20.5.2 Development of Darunavir to Combat Resistance to HIV-1 Protease Inhibitors

The HIV-1 protease is a homodimer that contains two catalytic aspartic acid residues in the active site (one from each subunit) (Fig. 20.6). The Asp25 residues share an acidic proton and interact with a water molecule in the absence of substrate or inhibitor. The hydrolysis mechanism involves activation of a water molecule by these aspartates and nucleophilic attack of the water oxygen to the amide carbonyl of the bond to be cleaved between P1 and P1′. Breakdown of the resulting tetrahedral intermediate by successive proton transfers leads to the amino and carboxylate products. The scissile bond of the peptide substrate is in close proximity of the active site. Upon binding, flexible flaps (one per monomer) that adopt an open conformation in the apoenzyme fold down over the substrates. The peptide substrates contain at least seven residues extending from P4 to P3′, where the cleavage site lies between P1 and P1′. The side chains of the substrate are lodged in the protease subsites S4 to S3′. Replacement of the hydrolyzable peptide linkage by a

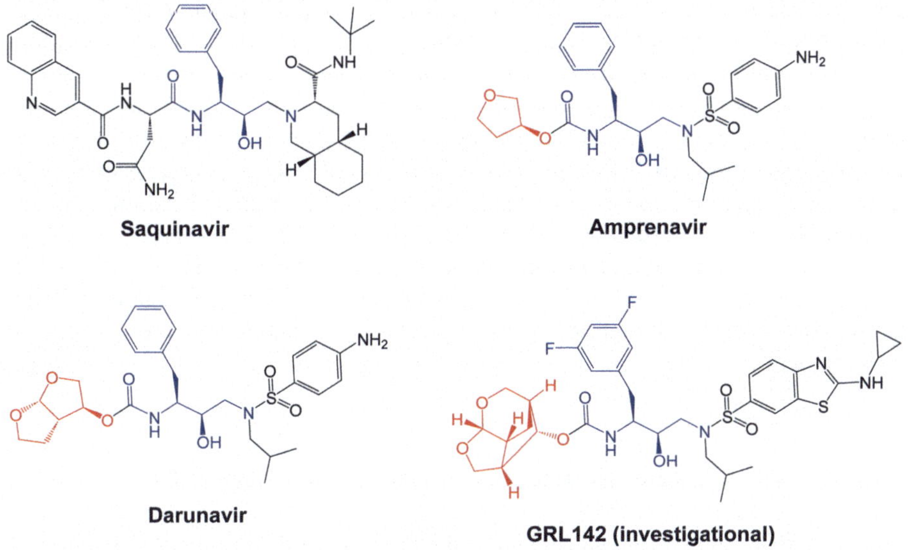

Fig. 20.11 Chemical structures of saquinavir, amprenavir, darunavir and the investigational derivative GRL142

non-cleavable bond was a basic principle for the development of peptidomimetic inhibitors that culminated in the approval of saquinavir [90].

Saquinavir and other related protease inhibitors have serious drawbacks that include severe side effects and toxicities, poor bioavailability (high doses required to maintain their therapeutic effect), costly synthesis, and the rapid emergence of drug resistance. Extensive analysis of protein–ligand X-ray structures of wild-type HIV-1 protease and resistant variants showed that there were minimal alterations in the active site backbone conformation of proteases resistant to inhibitors. Therefore, drugs effective on resistant HIV-1 strains should maintain extensive hydrogen-bonding interactions with the protein backbone of the wild-type enzyme.

The replacement of the asparagine moiety at the P2 position in saquinavir with 3*R*-tetrahydrofuranyl glycine resulted in a more potent inhibitor (Fig. 20.11). Other studies leading to the development of fosamprenavir/amprenavir showed an improvement in antiviral activity by introducing the 3*R*-tetrahydrofuranyl urethane moiety onto a hydroxyethylamine sulfonamide isostere. However, crystal structures revealed that 3*R*-tetrahydrofuranyl urethane-bearing inhibitors had weak hydrogen bonding potential with main chain atoms of residues Asp29 and Asp30. Therefore, a bicyclic tetrahydrofuran (bis-THF) ligand was designed and developed to improve hydrogen bonding with both aspartic acid residues. Inhibitors incorporating a P2 bis-THF and a *p*-methoxysulfonamide as the P2′ ligand (e.g., TMC-126) were remarkably potent in enzyme inhibition assays and displayed strong antiviral activity in cell culture. Darunavir was similar to TMC-126 but contained *p*-aminosulfonamide instead of *p*-methoxysulfonamide at the P2′ position. The P2′-amine group provided more favorable pharmacokinetic properties compared to the P2′-methoxy group. X-ray crystal structures of darunavir-bound HIV-1 protease showed that both P2 and P2′ moieties are involved in extensive hydrogen bonding with the protein backbone. These tight interactions were consistently observed with mutant proteases and might be responsible for the unusually high resistance profile of darunavir. Detailed reviews describing the design and development of darunavir have been published [91]. Recent designs, such as GRL142, incorporate fluorine to improve penetration into the central nervous system [92]. GRL142 exhibits 20-fold better affinity than darunavir for an extremely resistant HIV-1 protease variant harboring 20 amino acid substitutions (Q7K, L10F, I13V, I15V, D30N, V32I, L33F, E35D, M36I, S37N, I47V, I54L, Q58E, I62V, L63P, A71V, I84V, N88D, L89T, and L90M). Therefore, it is a promising inhibitor for further clinical development.

20.6 Viral Capsids as Targets of Antiviral Intervention

The availability of crystal structures of poliovirus and rhinovirus (see Chap. 4) led to the discovery of compounds that could bind to a hydrophobic pocket in the capsid protein VP1. Those molecules, known as WIN compounds (for Winthrop, the firm that developed the drugs), mimic natural factors that bind to the VP1 pocket and stabilize the capsid in a metastable conformation. WIN compounds and derivatives such as pleconaril (WIN-63843) bind into the hydrophobic pocket and stabilize the metastable capsid conformation in a way that results in inhibition of both viral adsorption to cell membranes and RNA uncoating [93]. Although pleconaril shows antiviral activity against rhinovirus and enteroviruses (including the dangerous enterovirus D68), its efficacy has not been demonstrated in clinical trials. On the other hand, there are remarkable examples of structure-based design of preclinical drugs binding to the mature or immature viral capsid or interfering with their assembly or conformational transitions required for genome uncoating directed against HBV and, more extensively, HIV-1. As mentioned in Sect. 20.3.2, these efforts led to the approval of lenacapavir, a first-in-class HIV-1 capsid inhibitor [50]. In the following sections, we will summarize recent progress towards the development of HBV and HIV capsid inhibitors.

20.6.1 Hepatitis B Virus

In HBV, the capsid is composed of 120 copies of a dimeric structural protein (known as CA protein) that assemble around an RNA-reverse transcriptase complex that will transcribe the RNA pregenome to the mature DNA genome. Ten years ago, the fluorescent probe 5,5′-*bis*[8-(phenylamino)-1 naphthalenesulfonate was shown to disturb the geometry of intersubunit contacts and to act both as a noncompetitive inhibitor of capsid assembly in HBV and as an assembly “misdirector” that facilitates the formation of noncapsid polymers or aberrant capsids [94]. Assembly misdirection has also been shown to occur upon exposure to activator heteroaryldihydropyrimidines such as HAP1 [95, 96], BAY 41–4109 ($EC_{50} = 50\,nM$) [63] or GLS-4 ($EC_{50} = 12\,nM$) [97]. These molecules are capsid assembly modulators that bind to core dimers and perturb the proper nucleocapsid assembly [98].

Depending on their chemical structure, core protein allosteric modulators can also form empty capsids, devoid of the pregenomic RNA-polymerase complex. Chemical compounds generating genome-free nucleocapsids are represented by phenylpropenamide derivatives (e.g., AT-61 and AT-130) [99], sulfamoylbenzamides (e.g., DVR-23, NVR-3-778) [100–102], sulfamoylpyrroloamides, and glyoxamoylpyrroloxamides [103]. These molecules act by stabilizing dimer–dimer interactions and thereby increasing the rate of nucleocapsid assembly.

Several core protein allosteric modulators, such as NVR3-778, GLS4JHS, RO7049389, JNJ-0440, JNJ-6379, vebicorvir (ABI-H0731), canocapavir (ZM-H1505R), and AB-423, were well tolerated with favorable pharmacokinetics in healthy volunteers, and several of these compounds have showed antiviral properties in patients with chronic hepatitis B in early clinical trials [98, 104, 105]. However, development of verbicorvir and others has been terminated, based on optimistic prospects for obtaining more effective capsid allosteric modulators in future [106].

20.6.2 Human Immunodeficiency Virus

In the case of HIV-1, we have discussed above the role of the viral protease in the release of the capsid protein (CA) from the precursor Gag polyprotein, and how protease inhibitors can prevent the formation of the mature capsid. Another compound that interferes with the maturation process is bevirimat (3-*O*-(3′,3′-dimethylsuccinyl)betulinic acid, also known as PA-457). This compound is a derivative of betulinic acid and interferes with the processing of the Gag polyprotein at the CA-SP1 cleavage site (i.e., between the C-terminal of the capsid protein and a spacer peptide located between CA and the nucleocapsid protein). This inhibitory action is independent of the protease activity. Tomographic reconstructions suggest that the CA-SP1 of Gag forms a helical bundle that serves to anchor the carboxy-terminal domain (CTD) of CA below the plane of the N-terminal domain (NTD). In the absence of CA-SP1 cleavage, mature cores cannot be formed because the immature particles cannot disassemble (for reviews, see [107, 108]). Virions produced from bevirimat-treated cells display aberrant core morphologies. A similar mechanism of action has been recently described for a pyridine-based compound (PF-46396) whose inhibitory activity results in the accumulation of CA/SP1 (p25) precursor proteins and the blockade of viral core particle maturation [109].

Assembly of the HIV-1 capsid depends on CA–CA interactions involving structural domains NTD and CTD (Fig. 20.12). The peptides CAI and CAC1 and

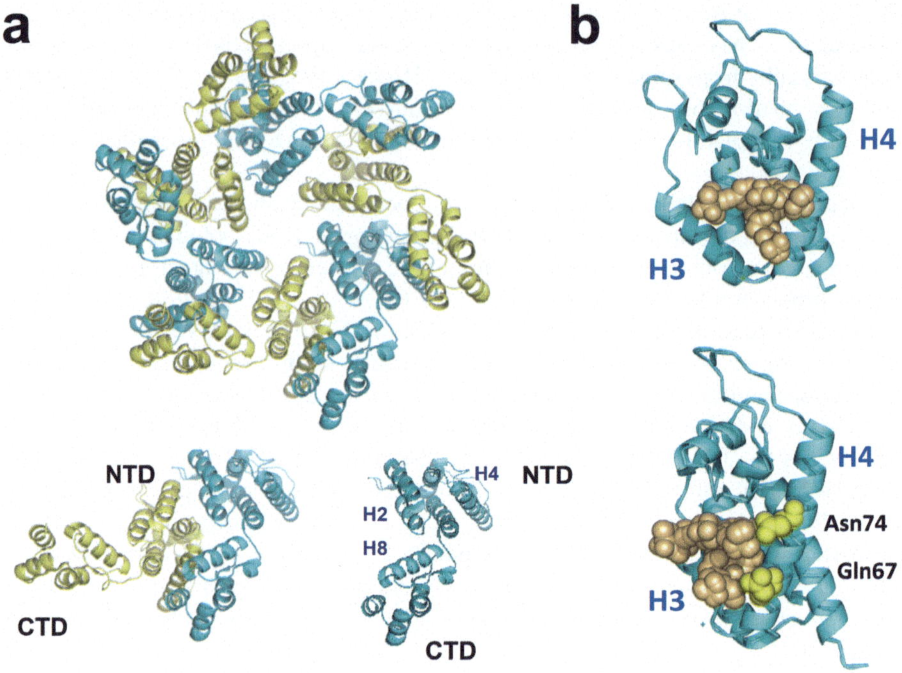

Fig. 20.12 HIV-1 capsid structure as a target of specific inhibitors. (**a**) Structural model of the HIV-1 CA hexamer [107] (coordinates taken from PDB entry 3GV2). The cartoons below show the structures of a CA dimer and a monomer, indicating the location of the structural domains NTD and CTD. Helices H2, H4, and H8 contribute to the NTD-CTD and NTD-NTD interfaces. CTD–CTD interactions are important for the formation of the capsid lattice by connecting hexameric rings. CTD dimerization is largely dependent on interactions between H9 helices (CTD dimers shown on the right are models obtained with data from the PDB entry 1A43). (**b**) Structure of CA NTD bound to lenacapavir (GS-6207) (PDB entry 6VKV), showing the location of α-helices H3 and H4. The drug is shown with brown spheres while critical residues involved in drug resistance are shown in yellow in the lower panel figure. All images were produced using PyMol (http://www.pymol.org/)

derivatives, as well as small molecules such as α-hydroxyglycineamide (a C-amidated tripeptide), benzodiazepine-based compounds, and CAP-1 have been identified as inhibitors of HIV-1 capsid assembly [110, 111]. CAP-1 contains a urea-based scaffold and two substituted aromatic groups at the molecule ends and acts in the late stage of the viral life cycle. When added to preformed HIV-1 particles, the drug has no inhibitory effect on HIV-1 infection. The CAI peptide was identified in a screen using a random peptide library. The CAC1 peptide was rationally designed based on structural and thermodynamic information to include most of the residues critically involved in the CTD-CTD dimerization interface that participates in mature HIV-1 capsid assembly [111, 112]. CAP-1, CAI, CAC1 and their combinations and/or derivatives were able to partially inhibit HIV-1 infectivity in cultured cells.

One of the limitations of CAI and other peptide inhibitors of capsid assembly is their inability to penetrate cells. Chemical modifications consisting of covalent crosslinks between amino acids found in those inhibitors led to the development of conformationally restricted peptides that showed increased affinity for monomeric CTD and improved delivery to target cells. Examples of those "stapled" peptides are NYAD-1 (derivative of CAI) and NYAD-201 [111, 112]. NYAD-1 inhibits the formation of both immature and mature virions. At low micromolar concentrations, this drug interferes with early events in the viral life cycle. Despite their higher solubility and low toxicity, further development of HIV-1 CA-derived natural or stapled peptides was limited by their relatively modest binding affinities. Also, problems related with the general metabolic instability of peptides may need to be addressed by using peptidomimetic compounds.

A high-throughput antiviral screening campaign led to the identification of a cluster of structurally related compounds that inhibited HIV-1 replication at submicromolar concentrations by targeting a post-entry stage prior to reverse transcription. Among them, PF-3450074 (or PF74) triggers premature HIV-1 uncoating in target cells [113]. Interestingly, point mutations in CA that stabilize the HIV-1 core conferred strong resistance without influencing inhibitor binding. Structural studies of the complex of the CA N-terminal domain and PF74 (PDB entry 2XDE) revealed a new binding site, different from those described for CAP-1 (PDB entry 2PXR) or CAI (PDB entry 2BUO) (Fig. 20.12), and several in-silico screened inhibitors [114]. PF74 increased the rate of HIV-1 CA multimerization in vitro, while CAI and CAP-1 decreased this rate under the same assay conditions. In a model of an assembled CA hexamer in complex with PF74, an indole group protruding from the N-terminal domain was found to localize to the interface between capsid monomers suggesting both early- and late-stage effects on viral propagation. The N-terminal domain of CA is also the target of novel benzodiazepines that prevent virus release or benzimidazoles that inhibit the formation of the mature capsid [115]. These compounds bind to a different site as compared with PF74 and have no effects at the early steps of viral replication. Benzimidazole inhibitors are expected to interfere with the formation of CA_{NTD}/CA_{CTD} interfaces in the CA hexamer [115].

By using the structural framework of PF74 as a template, researchers developed two capsid inhibitors, GS-CA1 and its analog GS-6207, that inhibited HIV-1 replication in T-lymphocytes and peripheral blood mononuclear cells (PBMCs) with EC_{50}s in the range of 140–240 pM [116]. This potent inhibition together with the relatively long half-life of GS-6207 (7–11 weeks) made this compound very attractive for the development of long-acting therapies. Lenacapavir (GS-6207) has been approved for use with other antiretrovirals in heavily treatment-experienced people with multidrug resistant HIV [50]. Resistance to lenacapavir is associated with HIV-1 CA amino acid substitutions L56I, M66I, Q67H, K70N, N74D, N74S, and T107N (alone or in different combinations). Q67H and N74D are the most frequently selected mutations in HIV-infected cells, while Q67H has been identified in some patients after one-week exposure to low doses of the drug [117, 118]. Structural analysis of HIV-1 CA/lenacapavir complexes demonstrates that lenacapavir and GS-CA1 bind to the same site as PF74 [119]. Besides its effects in disrupting capsid assembly, lenacapavir interferes with HIV-1 CA binding to the cellular cofactors Nup153 and CPSF6 that mediate viral nuclear import and direct integration into gene-rich regions of chromatin [116].

20.7 Conclusions and Perspectives

Despite the recent success in developing new antiviral drugs, particularly against HIV and HCV, important challenges still lie ahead. Drug resistance is a frequent obstacle that limits the efficacy of antiviral treatments. As learnt from the HIV experience, monotherapies should be considered suboptimal therapies whereas combination therapies are expected to prevail in the future treatment of viral infections. Good combination therapies rely on the exploitation of multiple targets and/or mechanisms of action. A better knowledge of the biological processes involved in specific viral infections, including interaction with specific host factors, is required to develop efficient screening methods and to identify viable targets. A wealth of information derived from massive sequencing techniques applied to humans and pathogens will probably be helpful in this process. Drug efficacy, toxicity, and cost are also issues that need to be considered in the development of future antiviral drugs.

From a technical point of view, high-throughput three-dimensional structural analysis, computational modeling, and better synthetic methods will provide tools to accelerate drug discovery, once the target mechanism has been validated. The rapidly increasing availability of high-resolution structures of viral particles and substantial advances in our understanding of their morphogenesis and conformational dynamics are opening new avenues for the design of novel antiviral agents. Although the odds of finding effective inhibitors of novel targets have probably increased, the outcome still remains unpredictable.

HIV and HCV have been major actors in the drug development field for the last 40 years, while a lot of attention has been devoted to SARS-CoV-2 since the start of the COVID-19 pandemic. Nonetheless, antiviral therapies against other viral pathogens such as HBV, influenza virus, flaviviruses (e.g., dengue virus), and coronaviruses are still needed and will probably gain attention in the coming years due to the general impact of those viruses on human health.

Acknowledgments This work was supported in part by grants of the Ministry of Science and Innovation of Spain PID2022-136725OB-I00/AEI/MCI/10.13039/501100011033 awarded to L. M.-A, and PID2022-136307OB-C22/AEI/10.13039/501100011033 awarded to F.G. An institutional grant to CBMSO from the Fundación Ramón Areces is also acknowledged. The CBMSO has been certified since 2023 as Severo Ochoa Center of Excellence by AEI/MCI/10.13039/501100011033.

References

1. De Clercq E, Li G (2016) Approved antiviral drugs over the past 50 years. Clin Microbiol Rev 29(3):695–747
2. Tompa DR, Immanuel A, Srikanth S, Kadhirvel S (2021) Trends and strategies to combat viral infections: a review on FDA approved antiviral drugs. Int J Biol Macromol 172:524–541
3. Menéndez-Arias L, Delgado R (2022) Update and latest advances in antiretroviral therapy. Trends Pharmacol Sci 43(1):16–29
4. Spengler U (2018) Direct antiviral agents (DAAs) - a new age in the treatment of hepatitis C virus infection. Pharmacol Ther 183:118–126
5. Ma Y, Frutos-Beltrán E, Kang D et al (2021) Medicinal chemistry strategies for discovering antivirals effective against drug-resistant viruses. Chem Soc Rev 50(7):4514–4540
6. De Clercq E (2009) Anti-HIV drugs: 25 compounds approved within 25 years after the discovery of HIV. Int J Antimicrob Agents 33:307–320
7. Menéndez-Arias L (2013) Molecular basis of human immunodeficiency virus type 1 drug resistance: overview and recent developments. Antiviral Res 98(1):93–120
8. Menéndez-Arias L (2009) Mutation rates and intrinsic fidelity of retroviral reverse transcriptases. Viruses 1:1137–1165
9. Sebastián-Martín A, Barrioluengo V, Menéndez-Arias L (2018) Transcriptional inaccuracy threshold attenuates differences in RNA-dependent DNA synthesis fidelity between retroviral reverse transcriptases. Sci Rep 8(1):627
10. Hamre D, Brownlee KA, Donovick R (1951) Studies on the chemotherapy of vaccinia virus. II. The activity of some thiosemicarbazones. J Immunol 67:305–312
11. Bauer DJ, St. Vincent L, Kempe CH, Downie AW (1963) Prophylactic treatment of smallpox contacts with N-methylisatin β-thiosemicarbazone (compound 33T57, Marboran). Lancet 2: 494–496
12. Prusoff WH (1959) Synthesis and biological activities of iododeoxyuridine, an analog of thymidine. Biochim Biophys Acta 32:295–296
13. Jordan AM, Roughley SD (2009) Drug discovery chemistry: a primer for the non-specialist. Drug Discov Today 14:731–744
14. Bassetto M, Massarotti A, Coluccia A, Brancale A (2016) Structural biology in antiviral drug discovery. Curr Opin Pharmacol 30:116–130
15. Nagy PD (2022) Co-opted membranes, lipids, and host proteins: what have we learned from tombusviruses? Curr Opin Virol 56:101258
16. Shiraki K (2018) Antiviral drugs against alphaherpesvirus. Adv Exp Med Biol 1045:103–122
17. De Clercq E, Holý A (2005) Acyclic nucleoside phosphonates: a key class of antiviral drugs. Nat Rev Drug Discov 4:928–940
18. Beadle JR, Hartline C, Aldern KA et al (2002) Alkoxyalkyl esters of cidofovir and cyclic cidofovir exhibit multiple-log enhancement of antiviral activity against cytomegalovirus and herpesvirus replication in vitro. Antimicrob Agents Chemother 46(8):2381–2386
19. Oberg B (1989) Antiviral effects of phosphonoformate (PFA, foscarnet sodium). Pharmacol Ther 40(2):213–285
20. Menéndez-Arias L (2008) Mechanisms of resistance to nucleoside analogue inhibitors of HIV-1 reverse transcriptase. Virus Res 134:124–146
21. Menéndez-Arias L, Sebastián-Martín A, Álvarez M (2017) Viral reverse transcriptases. Virus Res 234:153–176
22. Xavier Ruiz F, Arnold E (2020) Evolving understanding of HIV-1 reverse transcriptase structure, function, inhibition, and resistance. Curr Opin Struct Biol 61:113–123
23. De Corte BL (2005) From 4,5,6,7-tetrahydro-5-methylimidazo[4,5,1-*jk*](1,4)benzodiazepin-2 (1*H*)-one (TIBO) to etravirine (TMC125): Fifteen years of research on non-nucleoside inhibitors of HIV-1 reverse transcriptase. J Med Chem 48:1689–1696

24. Zhuang C, Pannecouque C, De Clercq E, Chen F (2020) Development of non-nucleoside reverse transcriptase inhibitors (NNRTIs): our past twenty years. Acta Pharm Sin B 10(6): 961–978
25. Hu J, Seeger C (2015) Hepadnavirus genome replication and persistence. Cold Spring Harb Perspect Med 5(7):a021386
26. Menéndez-Arias L, Álvarez M, Pacheco B (2014) Nucleoside/nucleotide analog inhibitors of hepatitis B virus polymerase: mechanism of action and resistance. Curr Opin Virol 8:1–9
27. Lawitz E, Jacobson IM, Nelson DR et al (2015) Development of sofosbuvir for the treatment of hepatitis C virus infection. Ann N Y Acad Sci 1358:56–67
28. Sofia MJ, Bao D, Chang W et al (2010) Discovery of a β-d-2'-deoxy-2'-α-fluoro-2'-β-C-methyluridine nucleotide prodrug (PSI-7977) for the treatment of hepatitis C virus. J Med Chem 53(19):7202–7218
29. Mehellou Y, Balzarini J, McGuigan C (2009) Aryloxy phosphoramidate triesters: a technology for delivering monophosphorylated nucleosides and sugars into cells. Chem Med Chem 4(11):1779–1791
30. Mehellou Y, Rattan HS, Balzarini J (2018) The ProTide prodrug technology: from the concept to the clinic. J Med Chem 61(6):2211–2226
31. Ray AS, Fordyce MW, Hitchcock MJ (2016) Tenofovir alafenamide: a novel prodrug of tenofovir for the treatment of human immunodeficiency virus. Antiviral Res 125:63–70
32. Gordon CJ, Tchesnokov EP, Woolner E et al (2020) Remdesivir is a direct-acting antiviral that inhibits RNA-dependent RNA polymerase from severe acute respiratory syndrome coronavirus 2 with high potency. J Biol Chem 295(20):6785–6797
33. Lo MK, Jordan R, Arvey A et al (2017) GS-5734 and its parent nucleoside analog inhibit Filo-, Pneumo-, and Paramyxoviruses. Sci Rep 7:43395
34. Konkolova E, Dejmek M, Hřebabecký H et al (2020) Remdesivir triphosphate can efficiently inhibit the RNA-dependent RNA polymerase from various flaviviruses. Antiviral Res 182: 104899
35. Malin JJ, Suárez I, Priesner V et al (2020) Remdesivir against COVID-19 and other viral diseases. Clin Microbiol Rev 34:e00162–20
36. Kati W, Koev G, Irvin M et al (2015) In vitro activity and resistance profile of dasabuvir, a nonnucleoside hepatitis C virus polymerase inhibitor. Antimicrob Agents Chemother 59(3): 1505–1511
37. Hadj Hassine I, Ben M'hadheb M, Menéndez-Arias L (2022) Lethal mutagenesis of RNA viruses and approved drugs with antiviral mutagenic activity. Viruses 14(4):841
38. Fodor E, Te Velthuis AJW (2020) Structure and function of the influenza virus transcription and replication machinery. Cold Spring Harb Perspect Med 10(9):a038398
39. O'Hanlon R, Shaw ML (2019) Baloxavir marboxil: the new influenza drug on the market. Curr Opin Virol 35:14–18
40. Ligat G, Cazal R, Hantz S, Alain S (2018) The human cytomegalovirus terminase complex as an antiviral target: a close-up view. FEMS Microbiol Rev 42(2):137–145
41. Gentry BG, Bogner E, Drach JC (2019) Targeting the terminase: an important step forward in the treatment and prophylaxis of human cytomegalovirus infections. Antiviral Res 161:116–124
42. Piret J, Boivin G (2019) Clinical development of letermovir and maribavir: overview of human cytomegalovirus drug resistance. Antiviral Res 163:91–105
43. Smith SJ, Zhao XZ, Passos DO et al (2021) Integrase strand transfer inhibitors are effective anti-HIV drugs. Viruses 13(2):205
44. Maertens GN, Engelman AN, Cherepanov P (2022) Structure and function of retroviral integrase. Nat Rev Microbiol 20(1):20–34
45. Hazuda DJ, Burroughs M, Howe AY et al (2013) Development of boceprevir: a first-in-class direct antiviral treatment for chronic hepatitis C infection. Ann N Y Acad Sci 1291:69–76
46. Toussi SS, Hammond JL, Gerstenberger BS, Anderson AS (2023) Therapeutics for COVID-19. Nat Microbiol 8(5):771–786

47. Unoh Y, Uehara S, Nakahara K et al (2022) Discovery of S-217622, a noncovalent oral SARS-CoV-2 3CL protease inhibitor clinical candidate for treating COVID-19. J Med Chem 65(9): 6499–6512
48. Gao M, Nettles RE, Belema M et al (2010) Chemical genetics strategy identifies an HCV NS5A inhibitor with a potent clinical effect. Nature 465:96–100
49. Sorbo MC, Cento V, Di Maio VC et al (2018) Hepatitis C virus drug resistance associated substitutions and their clinical relevance: Update 2018. Drug Resist Updat 37:17–39
50. Dvory-Sobol H, Shaik N, Callebaut C, Rhee MS (2022) Lenacapavir: a first-in-class HIV-1 capsid inhibitor. Curr Opin HIV AIDS 17(1):15–21
51. McFadden WM, Snyder AA, Kirby KA et al (2021) Rotten to the core: antivirals targeting the HIV-1 capsid core. Retrovirology 18(1):41
52. Gubareva L, Mohan T (2022) Antivirals targeting the neuraminidase. Cold Spring Harb Perspect Med 12(1):a038455
53. Gamblin SJ, Skehel JJ (2010) Influenza hemagglutinin and neuraminidase membrane glycoproteins. J Biol Chem 285:28403–28409
54. Almehmadi M, Allahyani M, Alsaiari AA et al (2022, 1870) A glance at the development and patent literature of tecovirimat: the first-in-class therapy for emerging monkeypox outbreak. Viruses 14(9):1870
55. Wang J, Shahed-Ai-Mahmud M, Chen A et al (2023) An overview of antivirals against monkeypox virus and other orthopoxviruses. J Med Chem 66(7):4468–4490
56. Chen Y, Honeychurch KM, Yang G et al (2009) Vaccinia virus p37 interacts with host proteins associated with LE-derived transport vesicle biogenesis. Virol J 6:44
57. Duraffour S, Lorenzo MM, Zöller G et al (2015) ST-246 is a key antiviral to inhibit the viral F13L phospholipase, one of the essential proteins for orthopoxvirus wrapping. J Antimicrob Chemother 70(5):1367–1380
58. Martí-Marí O, Martínez-Gualda B, Fernández-Barahona I et al (2022) Organotropic dendrons with high potency as HIV-1, HIV-2 and EV-A71 cell entry inhibitors. Eur J Med Chem 237: 114414
59. Curreli F, Choudhury S, Pyatkin I et al (2012) Design, synthesis, and antiviral activity of entry inhibitors that target the CD4-binding site of HIV-1. J Med Chem 55:4764–4775
60. Curreli F, Kwon YD, Zhang H et al (2015) Structure-based design of a small molecule CD4-antagonist with broad spectrum anti-HIV-1 activity. J Med Chem 58(17):6909–6927
61. Wang T, Kadow JF, Meanwell NA (2021) Innovation in the discovery of the HIV-1 attachment inhibitor temsavir and its phosphonooxymethyl prodrug fostemsavir. Med Chem Res 30(11):1955–1980
62. Grant PM, Kozal MJ (2022) Fostemsavir: a first-in-class HIV-1 attachment inhibitor. Curr Opin HIV AIDS 17(1):32–35
63. Langley DR, Kimura SR, Sivaprakasam P et al (2015) Homology models of the HIV-1 attachment inhibitor BMS-626529 bound to gp120 suggest a unique mechanism of action. Proteins 83(2):331–350
64. Files DC, Tacke F, O'Sullivan A et al (2022) Rationale of using the dual chemokine receptor CCR2/CCR5 inhibitor cenicriviroc for the treatment of COVID-19. PLoS Pathog 18(6): e1010547
65. Monteiro A, Yu KOA, Hicar MD (2021) Peptide-based fusion inhibitors for preventing the six-helix bundle formation of class I fusion proteins: HIV and beyond. Curr HIV Res 19(6): 465–475
66. Śledź P, Caflisch A (2018) Protein structure-based drug design: from docking to molecular dynamics. Curr Opin Struct Biol 48:93–102
67. Yu W, Weber DJ, MacKerell AD Jr (2023) Computer-aided drug design: an update. Methods Mol Biol 2601:123–152
68. Wingert BM, Camacho CJ (2018) Improving small molecule virtual screening strategies for the next generation of therapeutics. Curr Opin Chem Biol 44:87–92

69. Irwin JJ, Tang KG, Young J et al (2020) ZINC20-A free ultralarge-scale chemical database for ligand discovery. J Chem Inf Model 60(12):6065–6073
70. Lipinski CA, Lombardo F, Dominy BW, Feeney PJ (2001) Experimental and computational approaches to estimate solubility and permeability in drug discovery and development settings. Adv Drug Deliv Rev 46:3–26
71. Koes DR, Camacho CJ (2012) ZINCPharmer: pharmacophore search of the ZINC database. Nucleic Acids Res 40(Web Server issue):W409-414
72. Hu Y, Stumpfe D, Bajorath J (2017) Recent advances in scaffold hopping. J Med Chem 60(4): 1238–1246
73. Macip G, Garcia-Segura P, Mestres-Truyol J et al (2022) Haste makes waste: a critical review of docking-based virtual screening in drug repurposing for SARS-CoV-2 main protease (M-pro) inhibition. Med Res Rev 42(2):744–769
74. Boyles F, Deane CM, Morris GM (2022) Learning from docked ligands: ligand-based features rescue structure-based scoring functions when trained on docked poses. J Chem Inf Model 62(22):5329–5341
75. Fischer A, Smieško M, Sellner M, Lill MA (2021) Decision making in structure-based drug discovery: visual inspection of docking results. J Med Chem 64(5):2489–2500
76. Baek M, DiMaio F, Anishchenko I et al (2021) Accurate prediction of protein structures and interactions using a three-track neural network. Science 373(6557):871–876
77. Jumper J, Evans R, Pritzel A et al (2021) Highly accurate protein structure prediction with AlphaFold. Nature 596(7873):583–589
78. Kitchen DB, Decornez H, Furr JR, Bajorath J (2004) Docking and scoring in virtual screening for drug discovery: methods and applications. Nat Rev Drug Discover 3:935–949
79. Guido RV, Oliva G, Andricopulo AD (2008) Virtual screening and its integration with modern drug design technologies. Curr Med Chem 15:37–46
80. Sun C, Petros AM, Hajduk PJ (2011) Fragment-based lead discovery: challenges and opportunities. J Comput Aided Mol Des 25:607–610
81. Abad-Zapatero C, Perišić O, Wass J et al (2010) Ligand efficiency indices for an effective mapping of chemico-biological space: the concept of an atlas-like representation. Drug Discov Today 15:804–811
82. Koes DR, Camacho CJ (2012) PocketQuery: protein-protein interaction inhibitor starting points from protein-protein interaction structure. Nucleic Acids Res 40:W387–W392
83. Hall DR, Ngan CH, Zerbe BS et al (2012) Hot spot analysis for driving the development of hits into leads in fragment-based drug discovery. J Chem Inf Model 52:199–209
84. Morreale A, Gago F (2012) Chapter 10: COMparative BINding Energy (COMBINE) analysis as a structure-based 3D-QSAR method. In: Luque J, Barril X (eds) Physico-chemical and computational approaches to drug discovery. Royal Society of Chemistry
85. Raju TN (2000) The Nobel chronicles. 1988: James Whyte Black, (b 1924), Gertrude Elion (1918-99), and George H Hitchings (1905-98). Lancet 355(9208):1022
86. Lewis RA (2021) Best practices for repurposing studies. J Comput Aided Mol Des 35(12): 1189–1193
87. Zhang CH, Stone EA, Deshmukh M et al (2021) Potent noncovalent inhibitors of the main protease of SARS-CoV-2 from molecular sculpting of the drug perampanel guided by free energy perturbation calculations. ACS Cent Sci 7(3):467–475
88. Varghese JN, Laver WG, Colman PM (1983) Structure of the influenza virus glycoprotein antigen neuraminidase at 2.9 Å resolution. Nature 303:35–40
89. Von Itzstein M (2007) The war against influenza: discovery and development of sialidase inhibitors. Nat Rev Drug Discov 6:967–974
90. Roberts NA, Martin JA, Kinchington D et al (1990) Rational design of peptide-based HIV proteinase inhibitors. Science 248:358–361
91. Ghosh AK, Anderson DD, Weber IT, Mitsuya H (2012) Enhancing protein backbone binding – a fruitful concept for combating drug-resistant HIV. Angew Chem Int Ed Engl 51:1778–1802

92. Weber IT, Wang YF, Harrison RW (2021) HIV protease: historical perspective and current research. Viruses 13(5):839
93. Patick AK (2006) Rhinovirus chemotherapy. Antiviral Res 71:391–396
94. Zlotnick A, Ceres P, Singh S, Johnson JM (2002) A small molecule inhibits and misdirects assembly of hepatitis B virus capsids. J Virol 76:4848–4854
95. Stray SJ, Bourne CR, Punna S et al (2005) A heteroaryldihydropyrimidine activates and can misdirect hepatitis B virus capsid assembly. Proc Natl Acad Sci USA 102:8138–8143
96. Bourne CR, Finn MG, Zlotnick A (2006) Global structural changes in hepatitis B virus capsids induced by the assembly effector HAP1. J Virol 80:11055–11061
97. Ren Q, Liu X, Luo Z et al (2017) Discovery of hepatitis B virus capsid assembly inhibitors leading to a heteroaryldihydropyrimidine based clinical candidate (GLS4). Bioorg Med Chem 25(3):1042–1056
98. Feng S, Gao L, Han X et al (2018) Discovery of small molecule therapeutics for treatment of chronic HBV infection. ACS Infect Dis 4(3):257–277
99. Katen SP, Tan Z, Chirapu SR et al (2013) Assembly-directed antivirals differentially bind quasiequivalent pockets to modify hepatitis B virus capsid tertiary and quaternary structure. Structure 21(8):1406–1416
100. Campagna MR, Liu F, Mao R et al (2013) Sulfamoylbenzamide derivatives inhibit the assembly of hepatitis B virus nucleocapsids. J Virol 87(12):6931–6942
101. Klumpp K, Shimada T, Allweiss L et al (2018) Efficacy of NVR 3-778, alone and in combination with pegylated interferon, vs entecavir in uPA/SCID mice with humanized livers and HBV infection. Gastroenterology 154(3):652–662.e8
102. Ren Y, Ma Y, Cherukupalli S et al (2020) Discovery and optimization of benzenesulfonamides-based hepatitis B virus capsid modulators via contemporary medicinal chemistry strategies. Eur J Med Chem 206:112714
103. Nijampatnam B, Liotta DC (2019) Recent advances in the development of HBV capsid assembly modulators. Curr Opin Chem Biol 50:73–79
104. Berke JM, Dehertogh P, Vergauwen K et al (2020) Antiviral properties and mechanism of action studies of the hepatitis B virus capsid assembly modulator JNJ-56136379. Antimicrob Agents Chemother 64(5):e02439–19
105. Naggie S, Lok AS (2021) New therapeutics for hepatitis B: the road to cure. Annu Rev Med 72:93–105
106. Jin Y, Wang S, Xu S et al (2023) Targeting hepatitis B virus cccDNA levels: recent progress in seeking small molecule drug candidates. Drug Discov Today 28(7):103617
107. Adamson CS, Freed EO (2008) Recent progress in antiretrovirals – lessons from resistance. Drug Discov Today 13:424–432
108. Prevelige PE Jr (2011) New approaches for antiviral targeting of HIV assembly. J Mol Biol 410:634–640
109. Blair WS, Cao J, Fok-Seang J et al (2009) New small-molecule inhibitor class targeting human immunodeficiency virus type 1 virion maturation. Antimicrob Agents Chemother 53:5080–5087
110. Pornillos O, Ganser-Pornillos BK, Kelly BN et al (2009) X-ray structures of the hexameric building block of the HIV capsid. Cell 137:1282–1292
111. Bocanegra R, Rodríguez-Huete A, Fuertes MÁ et al (2012) Molecular recognition in the human immunodeficiency virus capsid and antiviral design. Virus Res 169(2):388–410
112. Obr M, Kräusslich HG (2018) The secrets of the stability of the HIV-1 capsid. Elife 7:e38895
113. Shi J, Zhou J, Shah VB et al (2011) Small-molecule inhibition of human immunodeficiency virus type 1 infection by virus capsid destabilization. J Virol 85:542–549
114. Blair WS, Pickford C, Irving SL et al (2010) HIV capsid is a tractable target for small molecule therapeutic intervention. PLoS Pathog 6:e1001220
115. Lemke CT, Titolo S, von Schwedler U et al (2012) Distinct effects of two HIV-1 capsid assembly inhibitor families that bind the same site within the N-terminal domain of the viral CA protein. J Virol 86:6643–6655

116. Bester SM, Wei G, Zhao H et al (2020) Structural and mechanistic bases for a potent HIV-1 capsid inhibitor. Science 370(6514):360–364
117. Margot N, Vanderveen L, Naik V et al (2022) Phenotypic resistance to lenacapavir and monotherapy efficacy in a proof-of-concept clinical study. J Antimicrob Chemother 77(4): 989–995
118. Margot NA, Naik V, VanderVeen L et al (2022) Resistance analyses in highly treatment-experienced people with human immunodeficiency virus (HIV) treated with the novel capsid HIV inhibitor lenacapavir. J Infect Dis 226(11):1985–1991
119. Bester SM, Adu-Ampratwum D, Annamalai AS et al (2022) Structural and mechanistic bases of viral resistance to HIV-1 capsid inhibitor lenacapavir. mBio 13(5):e0180422

Further Reading

De Clercq E (2019) Fifty years in search of selective antiviral drugs. J Med Chem 62(16): 7322–7339

Feld JJ, Foster GR (2016) Second generation direct-acting antivirals - do we expect major improvements? J Hepatol 65(1 Suppl):S130–S142. Erratum in: J Hepatol (2017) 66(5):1105

Luczkowiak J, Álvarez M, Sebastián-Martín A, Menéndez-Arias L (2019) DNA-dependent DNA polymerases as drug targets in herpesviruses and poxviruses. In: Gupta SP (ed) Viral polymerases: structures, functions and roles as antiviral drug targets. Elsevier-Academic Press, pp 95–134

Maeda K, Das D, Kobayakawa T et al (2019) Discovery and development of anti-HIV therapeutic agents: Progress towards improved HIV medication. Curr Top Med Chem 19(18):1621–1649

Schlicksup CJ, Zlotnick A (2020) Viral structural proteins as targets for antivirals. Curr Opin Virol 45:43–50

von Delft A, Hall MD, Kwong AD et al (2023) Accelerating antiviral drug discovery: lessons from COVID-19. Nat Rev. Drug Discov 22(7):585–603

Zhan P, Pannecouque C, De Clercq E, Liu X (2016) Anti-HIV drug discovery and development: current innovations and future trends. J Med Chem 59(7):2849–2878

Books

Jiang S, Lu L (eds) (2022) Virus entry inhibitors: stopping the enemy at the gate (Advances in Experimental Medicine and Biology 1366). Springer Nature Singapore Pte Ltd.

Liu X, Zhan P, Menéndez-Arias L, Poongavanam V (eds) (2021) Antiviral drug discovery and development (Advances in Experimental Medicine and Biology Book 1322). Springer Nature Singapore Pte Ltd.

Tang H (ed) (2019) Hepatitis B virus infection: molecular virology to antiviral drugs (Advances in Experimental Medicine and Biology Book 1179). Springer Nature Singapore Pte Ltd.

Chapter 21
Design of Novel Vaccines Based on Virus-Like Particles

Juan Bárcena, María Zamora-Ceballos, and Esther Blanco

Abstract Virus-like particles (VLPs) are formed by viral proteins that, when overexpressed, spontaneously self-assemble into particles that structurally are similar to infectious virus or subviral particles (e.g. the viral capsid). VLPs are appealing as vaccine candidates because their inherent properties (i.e. virus-sized, multimeric antigens, highly organised and repetitive structure, not infectious) are suitable for the induction of safe and efficient humoral and cellular immune responses. VLP-based vaccines have already been licensed for human and veterinary use, and many more vaccine candidates are currently in late stages of evaluation. Moreover, the development of VLPs as platforms for foreign antigen display has further broadened their potential applicability both as prophylactic and therapeutic vaccines. This chapter provides an overview on the design and use of VLPs for the development of new-generation vaccines.

Keywords Virus · Virus-like particles · VLPs · Chimeric VLPs · Viral nanoparticles · Capsid structure · Structural proteins · Vaccines · Immune response · Multimeric presentation · Antigen display · Epitopes · B-cell · T-cell

Abbreviations

ANTXR2	Anthrax toxin receptor 2
APC	Antigen-presenting cell
BTV	Bluetongue virus
COVID-19	Coronavirus disease 2019
CpG-ODN	CpG-containing oligodeoxynucleotide
CPMV	Cowpea mosaic virus
CPV	Canine parvovirus
cryo-EM	Cryogenic electron microscopy

J. Bárcena (✉) · M. Zamora-Ceballos · E. Blanco
Centro de Investigación en Sanidad Animal (CISA-INIA/CSIC), Madrid, Spain
e-mail: barcena@inia.csic.es

M. G. Mateu (ed.), *Structure and Physics of Viruses*, Subcellular Biochemistry 105,
https://doi.org/10.1007/978-3-031-65187-8_21

CTL	Cytotoxic T cell
DC	Dendritic cell
FHV	Flock house virus
GFP	Green fluorescent protein
HBcAg	Hepatitis B core antigen
HBsAg	Hepatitis B small surface antigen
HBV	Hepatitis B virus
HCV	Hepatitis C virus
HIV	Human immunodeficiency virus
HPV	Human papillomavirus
HRV-14	Human rhinovirus 14
IBDV	Infectious bursal disease virus
MEV	Mink enteritis virus
MHC	Major histocompatibility complex
MIR	Major immunodominant region
PA	Protective antigen
PAMP	Pathogen-associated molecular pattern
PCV2	Porcine circovirus type 2
PPV	Porcine parvovirus
PRR	Pattern-recognition receptor
RHDV	Rabbit haemorrhagic disease virus
SARS-CoV	Severe acute respiratory syndrome coronavirus
TLR	Toll-like receptor
VLP	Virus-like particle

21.1 Introduction

Vaccination is one of the most important and cost-effective methods of preventing infectious diseases. No other method in human or veterinary medicine has had such an impact in reducing morbidity and mortality and increasing the overall well-being of humans and animals [1, 2]. Life-threatening and devastating diseases, such as smallpox, have been eradicated by vaccination campaigns and the incidence of many others, including the recent COVID-19 pandemic, has been greatly reduced [3]. Historically, this notable achievement has been brought about by an essentially empirical approach. Most of currently licensed vaccines are still produced by relatively simple methods of reducing the pathogen's virulence. These include biological attenuation of virulence, for example by repeated passage of viruses in cell-culture or embryonated eggs (live-attenuated vaccines), and chemical inactivation of whole microorganisms (inactivated or 'killed' vaccines) [4]. Upon administration, these classical vaccines, especially the live-attenuated ones, frequently induce a spectrum of immune responses which closely resembles that elicited by the actual infectious agent: a potent B-cell mediated (or humoral) immune response plus a T-cell mediated (or cellular) immune response, eventually conferring life-long protection of the vaccinated host against subsequent infection by the corresponding pathogen.

Nevertheless, in spite of the successful development of numerous efficient classical vaccines, their use presents relevant drawbacks regarding safety and other issues. There is a limited but present risk of reversion to a virulent phenotype in vivo, incomplete inactivation, or accidental pathogen release from vaccine manufacturing facilities, any of which can lead to spreading of the disease the vaccine was intended to prevent. Examples of these damaging outcomes are known for classical vaccines such as those against poliovirus or foot-and-mouth disease virus [5, 6]. Live-attenuated vaccines are contraindicated for use in immunocompromised individuals or newborns. Besides, in the veterinary field, the use of classical vaccines does not usually allow for a clear-cut serological discrimination between vaccinated and pathogen-infected animals, a mandatory requirement to gain access to the international market of livestock products. These concerns have led many countries to a non-vaccination policy against important livestock diseases, that relies on slaughtering infected and contact herds (stamping-out), together with strict limitations on animal movements. Consequently, livestock populations are highly susceptible to such diseases, and there is a serious risk of reintroduction, as has happened repeatedly in several European countries, with diseases such as classical swine fever or foot-and-mouth disease. Another critical issue regarding the production of both human and animal classic viral vaccines is the need to develop a cell culture system susceptible of infection by the viral pathogen, that has not been always possible and/or advisable due, for example, to technical or safety reasons. Finally, the traditional strategies have shown their limits for the development of vaccines against organisms with complex pathogen–host interactions, which undermine, evade, or misdirect the host's immune response, such as human immunodeficiency virus (HIV). Hence, the focus of modern vaccinology is the development of alternative vaccine approaches circumventing the limitations of classical vaccines.

The advent of genetic engineering techniques and new biotechnological methods have enabled the development of synthetic or subunit vaccines, based on the use of isolated components of pathogens, such as nucleic acid (DNA or RNA) sequences, recombinant immunogenic proteins or synthetic peptides, often portions of virus capsids [7]. Subunit vaccines are considered an intrinsic safe approach as they are based on non-replicating immunogens and are thus, absolutely non-infectious. Importantly for veterinary use, they allow for discrimination between vaccinated and pathogen-infected animals (the so-called marker vaccines). However, safety comes at a price. Vaccines based on isolated, soluble components of viruses and other pathogens are usually poorly immunogenic per se, requiring the administration of frequent and high doses of antigen. Moreover, they usually need to be formulated with potent immune-stimulating adjuvants, which may cause considerable side effects such as toxicity or pain.

Basic studies on virus structure and assembly (see Chaps. 1, 2, 11, 12, and 22) led to the experimental observation that many viral structural proteins have the intrinsic ability to self-assemble into virus-like particles (VLPs) that are structurally either indistinguishable or very similar to authentic viral particles produced during viral infection. These VLPs have been used in fundamental studies on virus assembly, conformational stability, dynamics, and disassembly [8, 9]. Furthermore, VLPs are receiving increasing attention due to their potential use as tools in a wide array of medical and biotechnological applications, such us cell targeting, drug delivery, imaging or as nanomaterials [10, 11] (see Chap. 22). Of particular relevance is the role of VLP technology as a powerful tool for vaccine development. VLPs resemble the structure of viruses from which they derive, thus representing better immunological mimics of virion particles compared to soluble capsid subunits, resulting in improved effectiveness as vaccines [12]. VLP based-vaccines combine many of the advantages of whole-virus-based and recombinant subunit vaccines, exhibiting a high safety profile. VLPs, produced using recombinant protein expression systems (thus avoiding the need to handle large amounts of pathogenic microorganisms), can stimulate strong B- and T-cell immune responses and have been shown to exhibit self-adjuvanting properties. These characteristics have made VLPs attractive stand-alone vaccine candidates for many viral diseases [13–16]. In addition, VLPs can also be used as platforms for the multimeric display of foreign antigens of interest, derived from viruses or other pathogens. This strategy couples delivery of the chosen antigen with the ability of VLPs to induce strong immune responses, thus acting as efficient adjuvants [17–21].

Until recently, vaccines were exclusively used to prevent infectious diseases (prophylactic vaccines). This situation has evolved and many efforts are currently focused on developing therapeutic vaccines to target diseases such as cancer, allergy, and inflammatory autoimmunity, or even other conditions like nicotine dependence or hypertension [18, 19, 22]. Over the last years, mechanisms underlying the outstanding immunogenicity of viruses have been elucidated, identifying important key parameters, like their size, particulate nature, their highly repetitive and ordered structures (see Chap. 2), or their ability to activate critical subsets of cells from the innate immune response, which in turn triggers appropriate conditioning of the adaptive immune response [21, 23–25]. These features now serve as a paradigm for rational, virus structure-based vaccine design, and numerous new-generation tailor-made vaccines based on VLPs aim to harness these viral properties to target a wide array of human and animal diseases [17, 26]. This chapter provides an overview on the development and use of VLPs as vaccines for prophylactic and therapeutic applications.

21.2 Immunology of Vaccines, an Overview

What makes VLPs good vaccine candidates? In order to answer this question it is important to briefly review how vaccines induce a protective immune response.

The immune system has evolved to distinguish between non-infectious self and infectious nonself. Accordingly, it can mount strong immune responses to microbial infections but usually fails to respond to self-molecules. Unravelling the

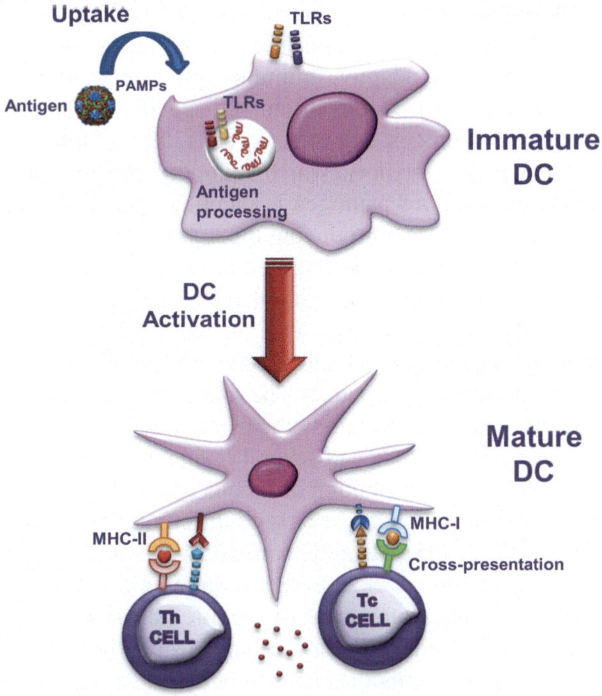

Fig. 21.1 The innate immune system, first line of host defence against microbial infections. PAMPs present at the surface of pathogens stimulate antigen uptake and processing by special subsets of APCs, DCs, resulting in upregulation of DC maturation markers. Subsequently, mature DCs present pathogen-derived antigens to naive Th and Tc cells via MHC class-II and class-I molecules, respectively. Secretion of cytokines by DCs stimulates differentiation into B and T effector cells resulting in antibody release and CTL responses

mechanisms used by the immune system to discriminate self from nonself is critical for our understanding of diseases and our ability to rationally design vaccines.

The immune system is composed of the innate (non-specific) and adaptive (specific) branches [23, 25, 26]. The innate immune system represents the first line of host defence against microbial infections (Fig. 21.1). After a pathogen breaches the host's physical barriers such as mucosal surfaces or skin, the immune system mounts an early, innate immune response within minutes. Cells of the innate immune system, like dendritic cells (DCs) and macrophages, known as antigen-presenting cells (APCs), can recognise general characteristics of the pathogen and initiate a response. Pathogen-associated molecular patterns (PAMPs) are molecules or motifs that are common to many pathogens and generally not present in the host, such as lipopolysaccharide, which can be found on the cell wall of many bacterial species, double-stranded RNA, or unmethylated CpG motifs, which are normally associated with virus infection. PAMPs can be recognised by Toll-like receptors (TLRs) and other pattern-recognition receptors (PRRs) which are present in host cells. The multivalent display and highly ordered structure exhibited by many pathogens, notably icosahedral virus surfaces that are essentially two-dimensional crystals, are also recognised as PAMPs, resulting in increased immunogenicity. PAMPs trigger innate immune sensing mechanisms and stimulate antigen uptake and processing by APCs. Subsequently, activated APCs present the antigens to cells of the adaptive immune response.

Adaptive immunity represents the second line of immunological defence (Fig. 21.2) and takes longer to develop, on the order of days. Once the antigen is delivered to the adaptive immune system and stimulates the proliferation (clonal selection and expansion) of antigen-specific effector cells, these cells begin eliminating the pathogen and infected cells. The adaptive response is also the basis for immunological memory, which is important for ensuring a fast and strong response when an infectious pathogen is encountered again (secondary infections). Vaccines use this process by introducing the immune system to a specific pathogen, stimulating the formation of specific memory cells and antibodies, which are then available to rapidly recognise the natural pathogen upon future exposures, resulting in its elimination and termination of the disease. The primary players of the adaptive immune response are B and T cells (lymphocytes). The main function of B cells is the production of antibodies. The role of T cells is more diverse ranging from lysis of infected cells, carried out by the cytotoxic T cells (CTLs), to secretion of cytokines, stimulation of B cells and several regulatory functions, performed by helper T cells (Th cells). The contribution of B cells versus T cells, and CTLs versus Th cells to the resolution of infection varies with different pathogens and from primary to secondary responses [25]. Usually, antibodies are more important during secondary infections. In contrast, T cells have a prominent role during many primary infections, as well as for the containment of chronic infections. Thus, for prophylactic vaccines, aiming at inducing effector mechanisms that protect from reinfection, the induction of neutralising antibodies is usually the most important factor. However, for therapeutic vaccination against chronic infections, the focus is usually placed on the stimulation of CTL responses. Hence, rational design of vaccines entails knowledge on the type of immune response to be evoked, and this in turn has structural implications for the design of new-generation vaccines.

21.3 The Role of Size, Geometry, and Molecular Patterns in Vaccine Design

The key properties of viruses responsible for eliciting potent immune responses include their size, geometry, and the ability to induce innate immunity, which in turn conditions the subsequent adaptive immune response [24]. These properties should be taken into account for the rational design of new-generation vaccines.

The dimensions of vaccine antigens can vary greatly (Fig. 21.3) [27]. The smallest (<10 nm) are subunit antigen vaccines (i.e. peptide or recombinant protein vaccines). Usually, these types of antigens are formulated with adjuvants to form larger particles or aggregates. Particulate antigens, such as VLPs and nanoparticles, are larger (20–100 nm). Virosomes (VLPs derived from viruses with lipid envelopes) and liposomes are in the range of 100 nm–1 μm. Antigens presented in the context of microparticles, emulsions (like Montanide), mineral salts (alum adjuvants), and whole-cell vaccines are the largest (100 nm–20 μm).

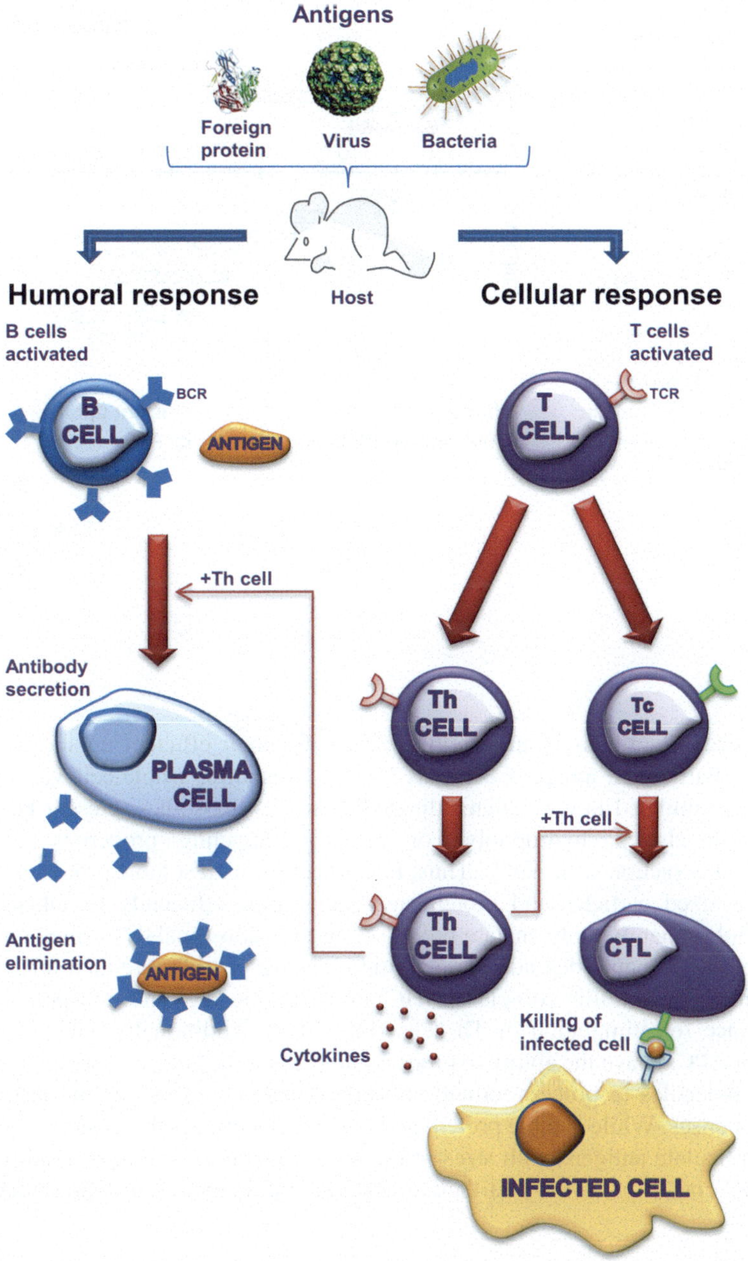

Fig. 21.2 The humoral and cell-mediated branches of the adaptive immune response. The humoral branch comprises lymphocytes of the B-cell lineage. Antibodies are the effector molecules produced by this response. The process begins with the interaction of B-cell receptors (BCRs) with antigens. Binding of the antigen promotes differentiation into antibody-secreting cells (plasma cells). The cell-mediated branch comprises lymphocytes of the T-cell lineage. T-helper cells

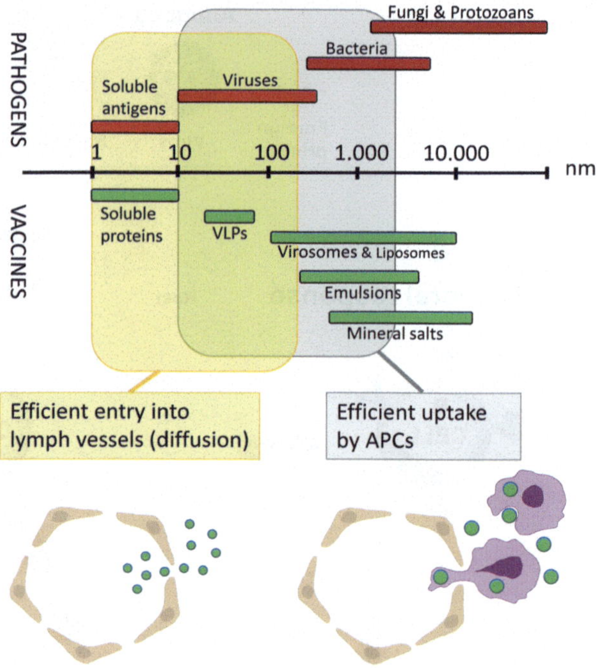

Fig. 21.3 Sizes of pathogens and vaccine components. Size ranges of different types of pathogens and vaccine delivery systems indicated in nm. Down, mechanisms of entry of different sized particles into lymphatic vessels on their way to secondary lymphoid organs. Particles ranging 10–200 nm can directly diffuse through lymphatic endothelial cell junctions, while particles larger than 200–500 nm are carried into the lymphatic system by specialised APCs

The size of antigens is an important factor for their efficient uptake by APCs [27, 28]. Particulate antigens, such as VLPs, virosomes, whole-cell vaccines, or antigens formulated in particulate adjuvants, such as liposomes, present large surfaces with charge, hydrophobic or receptor-interacting properties, enabling enhanced interaction with APCs. Thus, in contrast to small soluble protein antigens, pathogen sized particles and protein aggregates may efficiently be taken up by APCs. Subsequently, antigens are processed in the endosomal–lysosomal compartments of APCs, being degraded into peptides. These peptides are then loaded onto major histocompatibility complex (MHC) class II molecules and transported to the cell surface for stimulation of Th cells (Fig. 21.1). Additionally, DCs (a special subset of APCs) have the ability to process and present the vaccine antigen via MHC class I molecules (a process termed cross-presentation), to induce the priming of CTL responses. While soluble protein antigens are not efficiently cross-presented by DCs, particulate antigens with sizes in the range of viruses, or larger, readily reach the MHC class I pathway, and their peptides are effectively cross-presented (3–4

Fig. 21.2 (continued) (Th cells) and cytotoxic T cells (CTLs) are the key effectors of this response. The process begins when T lymphocytes recognise antigens on the surface of antigen-presenting cells (APCs). The Th cells recognise antigens bound to MHC class II molecules and produce cytokines that promote activation of other lymphocytes (both B- and T-cells). Once activated by Th cells, Tc cells differentiate into CTLs that can kill pathogen-infected cells orders of magnitude higher than peptides from soluble antigens). Processing of particulate antigens that are 20 nm–3 μm in diameter has been reported, suggesting APCs have evolved to efficiently process any antigen with dimensions similar to pathogens, ranging from viruses to bacteria and cells (Fig. 21.3). Therefore, particulate vaccines enhance the uptake, processing, and presentation of the antigens by APCs.

Vaccines are usually injected intramuscularly or subcutaneously. However, adaptive immune responses are mainly induced in the secondary lymphoid organs. So, the transport of antigens from the peripheral tissues to lymphoid organs through the lymphatic system is a relevant issue for vaccine design. Molecules of 10–200 nm efficiently enter the lymphatic system through diffusion and drainage (Fig. 21.3), with an optimal size being ~40 nm, just within the range size of viruses and VLPs. By contrast, particles that are larger than 200–500 nm do not efficiently enter lymph capillaries in a free form. Instead, large particles are carried into the lymphatic system by specialised cells (particular subsets of APCs), which can squeeze through openings between overlapping endothelial cells. Therefore, size determines whether particles freely drain to lymphoid organs or arrive at these sites in association with cells, and this has major implications for vaccine design and immune cell targeting. As antigens with virus-like dimensions (and smaller) most efficiently reach the lymph nodes in a cell-free state, vaccine delivery systems with similar dimensions will enable direct interaction of antigens, in an authentic native configuration, with follicular B cells. This property dramatically increases their ability to induce antibody responses, and it is crucial for the induction of antibodies recognising conformation-dependent epitopes. In addition, for the induction of CTL responses, specific subsets of DCs ($CD8^+$ lymphoid DCs) that are essentially restricted to lymphoid organs must be loaded with antigen. Hence, again particulate antigens with virus-like dimensions are preferable, since antigen size limits the ability of antigens to directly reach the adequate lymph node compartments.

Regarding the structure of antigens, most viral surfaces, as well as bacterial components, consist of one or a few proteins and consequently are repetitive and highly organised. Both the innate and adaptive immune systems recognise such highly repetitive structures as PAMPs. B cells, for example, recognise specifically and respond strongly to the ordered structure characteristic of viruses [29]. Most importantly, highly repetitive surface patterns efficiently crosslink B cell receptors on the surface of B cells, which represents a strong activation signal for these cells, leading to a strong antibody response (Fig. 21.4) [30]. Moreover, the stimulation of B cells by repetitive, highly organised antigens can be strong enough to elicit Th cell-independent induction of antibody responses (T-cell independent B-cell antigens). Therefore, these antigens can activate B cells more efficiently and at much lower concentrations than monomeric antigens (soluble proteins).

Finally, it has become apparent that the innate and adaptive immune systems are intimately connected (Fig. 21.1). No relevant adaptive immune response will be mounted unless certain types of APCs, like DCs, are activated, resulting in a rapid release of inflammatory cytokines, and up-regulation of costimulatory molecules, which all serve to condition and shape the subsequent adaptive response [23]. For

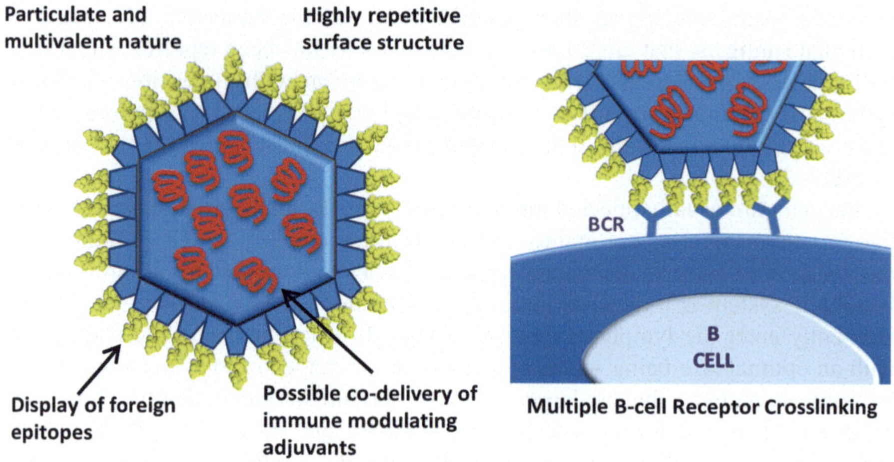

Fig. 21.4 Immunogenic features of VLPs presenting foreign antigens

instance, cross-presentation alone is not sufficient to drive induction of a CTL response, as DCs need to be previously activated (indeed, non-activated DCs presenting antigen to T cells induce T cell tolerance rather than immunity). In this regard, several viruses and VLPs have been shown to induce the phenotypic and functional maturation (activation) of DCs, by triggering TLRs which recognise viral-specific PAMPs [31]. Interestingly, human papillomavirus (HPV) derived-VLPs, but not unassembled recombinant L1 protein (the constituent monomer of HPV VLPs), can activate DCs directly [32], suggesting that the size and organised structure of VLPs are important parameters to induce DC activation. However, it should be noted that different VLPs differ in their ability to induce maturation of DCs, even if the VLPs are similar in size and structure (e.g., VLPs from different polyomavirus species) [33], indicating that other factors that are VLP-specific can also affect the outcome of the interaction between VLPs and DCs.

Alternatively, an interesting approach being investigated is the use of PAMPs, in particular TLR ligands, as immune modulating adjuvants. These molecules have the ability to enhance B- and T-cell responses induced by an antigen, offering advantages over existing conventional vaccine adjuvants. Virus-related PAMPs being developed as immune modulating adjuvants include CpG-containing oligodeoxynucleotides (CpG-ODNs), double-stranded RNA derivatives, and analogs of poly(I:C). Unfortunately, the use of such TLR ligands as vaccine adjuvants is associated with several problematic issues. Because TLR ligands activate their cognate receptors which are present in many types of immune and non-immune cells, they have the potential to be toxic. Thus, co-administration of TLR ligands and antigen by simple mixing may lead to a degree of systemic exposure causing non-specific stimulation of the immune system, with resultant unwanted reactogenicity or immune related pathology. A strategy to solve these problems is to target the TLR ligands directly to DCs (Fig. 21.4). This can be achieved by packaging viral TLR ligands within a particulate delivery system, providing an environment in which the ligand is stable and importantly, preventing its systemic distribution. Moreover, when combined with the antigen, the particulate nature of the delivery system serves to target both the antigen and the TLR ligand to the APCs, conveying them to the appropriate intracellular compartments, thus eliciting highly efficient activation of innate and adaptive immunity. VLPs, virosomes, and liposomes have been used to co-deliver antigens with a variety of TLR ligands [17, 26].

In summary, to induce potent immune responses, vaccines of viral size are preferable, as only particles in the nanometre range can reach lymph nodes and directly interact with B cells. Moreover, the antigenic epitopes should be displayed on the nanoparticles in an ordered and highly repetitive way to optimally activate B cells. Nanoparticle-sized vaccines are also preferable for the induction of T-cell responses, as size limits the ability of particles to reach the appropriate DC subsets in lymphoid organs. Viruses (and some VLPs) have the ability to directly induce the activation of DCs by triggering TLRs. Delivery of TLR ligands by nanoparticles together with the antigen may facilitate activation of DCs, which is essential for the induction of T-cell responses. Hence, vaccine carrier systems, such as VLPs, that mimic the size, geometry, and PAMPs of viruses may be one possible way to optimally harness viral immunogenic properties, without the risks associated with viral infection (Fig. 21.4).

21.4 VLPs as Antiviral Vaccines

21.4.1 Characteristics of VLPs

VLPs are supramolecular assemblies with a well-defined geometry, usually icosahedrons or rod-like helical structures, and diameters in the range of 20–120 nm [34–36] (Fig. 21.5). They are based on the natural intrinsic ability of many viral structural proteins to spontaneously assemble into multimeric structures when expressed using recombinant expression systems. VLPs mimic the organisation and conformation of authentic native viruses (Fig. 21.6), with or without lipid envelope depending on the virus of origin and the viral structural protein expressed (i.e. capsid proteins or viral envelope proteins), or in some cases subviral particles. Thus, VLPs are usually structurally and antigenically indistinguishable from infectious virus particles, generally retaining the ability to bind and penetrate host cells, and are consequently highly immunogenic, but unable to replicate, since they lack the viral genome. VLPs are able to stimulate both innate and adaptive immune responses, elicit protective systemic and mucosal immunity, and have been shown to exhibit self-adjuvanting abilities [23, 25, 37, 38]. In addition, immunisation with VLP-based vaccines does not induce antibody responses to internal or non-structural viral proteins (which do not form part of the VLPs but are present in virus-infected cells, and therefore induce antibody responses in the infected host). This enables serological differentiation between VLP-vaccinated and virus-infected animals (marker vaccines), which

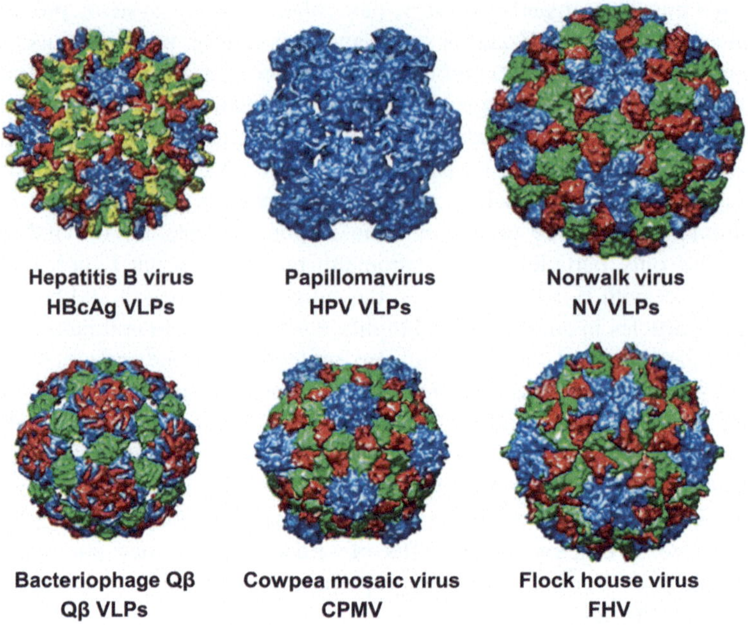

Fig. 21.5 Examples of viruses and VLPs mentioned in this chapter that have been developed as VLP-based vaccines or platforms for foreign antigen display. The images are coloured to distinguish the symmetry-related subunits and are taken from the VIPER database (http://viperdb.scripps.edu). PDB IDS: hepatitis B virus (HBV) capsid, 1QGT; Papillomavirus capsid, 1DZL; Norwalk virus capsid, 1IHM; bacteriophage Qβ, 1QBE; cowpea mosaic virus, 1NY7; flock house virus, 2Z2Q

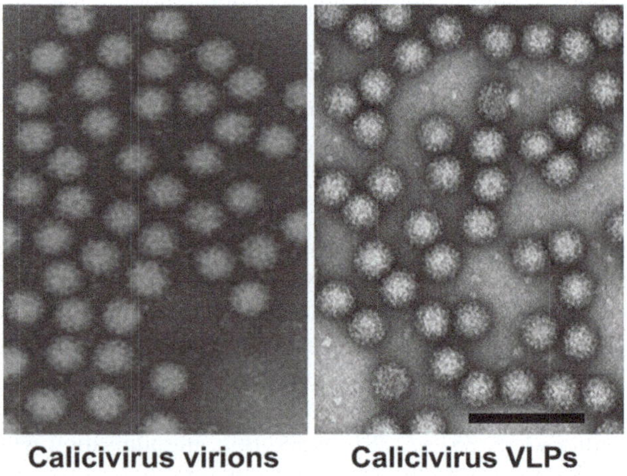

Fig. 21.6 Electron microscopy of negatively stained calicivirus particles: virions (left) and VLPs (right). Scale bar = 100 nm

represents an important feature for vaccines intended for veterinary use against notifiable diseases of livestock. Overall, these advantages have made VLPs attractive vaccine candidates against many viral diseases [13, 14, 16, 39].

The generation of VLPs derived from certain viruses has been strongly motivated by the lack of efficient cell culture propagation systems for those viruses, as was the case for hepatitis B virus (HBV), hepatitis C virus (HCV), HPV, or caliciviruses. An additional important reason for these efforts has been the possibility to substitute infectious viruses requiring high-level biosafety containment laboratories for their handling, like Ebola virus or severe acute respiratory syndrome coronavirus (SARS-CoV) by VLPs. Hence, those VLPs have been used as efficient surrogates for virus diagnostics, basic research applications (like understanding the assembly or architecture of these viruses), and vaccine development. Moreover, an important challenge nowadays consists in the substitution of currently used live vaccines, for example against influenza A virus, by an appropriate highly efficient and safe VLP-based vaccine [40].

In some instances, the VLPs are derived from internal structural viral proteins, like the HBV core antigen (HBcAg), or the Gag capsid proteins from retroviruses (including HIV), which are capable of self-assembling into subviral particles. Such VLPs do not mimic the external surface structure of the corresponding native virions, and antibodies induced by them are not able to interfere with the course of virus infection. Hence, these VLPs are not suited per se for vaccine development against the viruses from which they were derived, but they can be used as scaffolds for foreign antigen presentation (see Sect. 21.5). Of the utmost interest for the development of antiviral vaccines are VLPs composed of viral surface proteins, usually corresponding to the external capsids of non-enveloped virions, particularly when virus-neutralising antibodies recognise those proteins and/or their assemblies. Examples of such VLPs are those derived from HPV, bluetongue virus (BTV), rotavirus, parvovirus, calicivirus, and infectious bursal disease virus (IBDV) [12, 14].

21.4.2 Structural Classification of VLPs

VLPs have been produced for a wide range of taxonomically and structurally distinct viruses that infect humans, other animals, plants, yeasts, or bacteria [12, 14, 34–36]. These comprise viruses that have a single capsid protein, multiple capsid proteins, and those with or without lipid envelopes, indicating that the ability to develop VLPs does not seem to be limited to certain types of virus families or by the complexity or structure of the virus particle. The largest number of VLPs have been obtained from viruses with single-stranded RNA genomes of positive polarity. Many families and genera from this most structurally diverse class of viruses have been involved in the generation of differently sized and structured VLPs. Regarding organisation and symmetry (see Chap. 2), this group of VLPs is represented by structures of two different symmetries: icosahedral (higher number of prototypes,

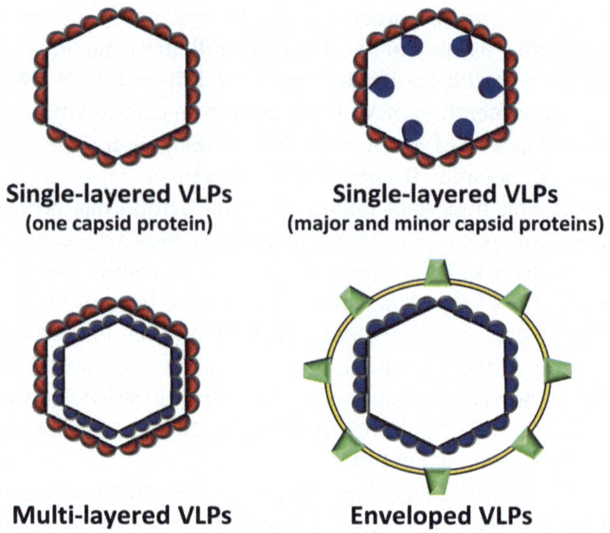

Fig. 21.7 Structural variants of VLPs

18 in total) and rod-like (three prototypes), derived from nonenveloped viruses, as well as by isometric VLPs without any distinct symmetry (three prototypes) derived from enveloped viruses [35].

Structural variants of VLPs are depicted in Fig. 21.7. The simplest variant is represented by the self-assembly of a single viral structural protein into a single-shelled VLP. One of the most studied VLPs of structurally simple viruses is the HPV-derived VLP. HPV virions contain the major and minor capsid proteins (L1 and L2, respectively). The virus capsid (60 nm) contains 72 pentamers of L1 protein, centred on the vertices of a $T = 7$ icosahedral lattice (see Chap. 2), while L2 protein is present at about 1/30 of the abundance of L1. HPV VLPs are usually formed just by L1 protein, with morphological characteristics very similar to those of HPV virus capsids [41]. Parvovirus VLPs are also formed by a single protein (60 copies of VP2, the major viral structural protein) (see Chap. 11). Other examples of VLPs formed by a single viral structural protein are those derived from the HBcAg, hepatitis E virus (HEV), calicivirus, nodavirus, circovirus, or bacteriophage Qβ [12, 14, 35]. Similarly, VLPs can be composed by both, a major and a minor protein forming a single shell, like in the case of HPV-derived VLPs composed by L1 and L2 proteins [42].

More complex multi-layered VLPs are generated by the co-expression of several viral structural proteins, like in the case of the BTV-derived VLPs (*Sedoreoviridae* family). The icosahedral BTV virions are formed from non-equimolar amounts of seven structural proteins (VP1–VP7). However, only the four major structural proteins of the virus (VP2, VP3, VP5, and VP7) are necessary for the formation of double-layered VLPs, which are structurally indistinguishable from authentic virus particles [43]. VLPs derived from rotaviruses, also belonging to the *Sedoreoviridae* family (see Chap. 12), can be generated by co-expressing different subsets of viral structural proteins [44]. Expressing VP2 alone produces single-layered VLPs. Co-expression of proteins VP2 and VP6 that form the innermost and middle capsid layers of rotavirus particles, respectively, produces double-layered 2/6-VLPs. Co-expression of VP4 with VP2 and VP6 form double-layered VLPs with VP4 spikes (2/4/6-VLPs). Co-expression of VP2, VP6, and VP7 produces triple-layered 2/6/7-VLPs, and finally, co-expression of VP2, VP4, VP6, and VP7 results in 2/4/6/7-VLPs that are complete triple-layered VLPs with the VP4 spike protein.

Another prototypic example of complex VLPs are enveloped VLPs containing unilamellar liposomes with embedded viral envelope proteins. The first VLP vaccine to be produced and characterised (in the early 1980s) consisted of isometric, spherical, pleomorphic particles with diameter between 17 and 25 nm, formed by the HBV small surface antigen (HBsAg), co-assembled with host cellular membranes [45]. Several other enveloped VLPs have been successfully developed by the simultaneous synthesis and co-assembly of two different viral protein species, i.e. capsid and envelope proteins. This is the approach used for the generation of VLPs derived from influenza virus, which contain both viral glycoproteins: haemagglutinin and neuraminidase, as well as matrix structural proteins. Other examples of enveloped VLPs are those derived from HCV, formed by the co-expression of core, E1 and E2 proteins, and retrovirus-based VLPs (like those of HIV), containing the Gag capsid protein and the envelope protein (gp120). The VLPs derived from viruses with lipid envelopes are sometimes referred to as virosomes [46].

21.4.3 Expression Systems for VLP Production

VLPs can be produced in many different production platforms, including mammalian cells, insect cells, plants, yeast, bacteria, and recombinant expression systems based on viral vectors: baculovirus, poxviruses, alphavirus replicons, etc. [14, 47]. One of the most popular choices is expression in insect cells using the recombinant baculovirus technology [48–50]. This expression system has many advantages for VLP production. Large amounts of correctly folded recombinant proteins can be produced with mammalian-like post-translational modifications. Although yeast and bacterial cells can achieve similar yields, the complexity of the VLPs produced with the baculovirus expression system is remarkably higher (VLPs formed from the co-expression of up to five proteins). An additional advantage is that baculoviruses have a limited host range (namely for insects) and are hence safe for vertebrates. The design of recombinant baculoviruses is simple and fast, providing a high versatility to this expression system. This is important when producing vaccines for viruses whose surface proteins rapidly mutate (i.e. influenza A virus), a fundamental requirement to contend with potential pandemics in a timely manner. The baculovirus expression system is used to produce approved commercial VLP-based vaccines for human use, such as Cervarix® (HPV), as well as veterinary vaccines like Porcilis®, Ingelvac Circoflex®, and Circumvent® (PCV2)

[49]. Alternatively, virus-free expression in insect cells has been increasingly explored, based either on transient transfection [51], or stable gene expression from a recombinant insect cell line [52].

Although VLPs have been produced for a wide range of viruses, as indicated above, clearly not all are equally suitable for the development of vaccines. Even if a VLP-based product candidate has been shown to elicit a robust immune response under laboratory conditions, it might not be developed as a vaccine for widespread use, if its manufacturing process is not scalable or cost-effective. According to the experiences accumulated over the last years, some predictive considerations about the chances of success in adequate VLP production yields can be enumerated. Structurally simple non-enveloped icosahedral viruses (see Chap. 11) are the most suitable candidates for successful high yield production of VLPs. VLPs made by the assembly of a single protein are usually able to be produced in large amount and high quality (i.e. HBcAg, parvovirus, calicivirus, papillomavirus, or bacteriophage Qβ). An increase in the number of proteins to be expressed and assembled proportionally raises the difficulties of correct assembly [14], usually leading to lower yields of VLP production. The efficient assembly of capsids derived from proteins encoded by multiple discrete mRNAs (i.e. orbivirus, birnavirus), requires the co-expression of the interacting proteins close to each other and within the same cell. On the other hand, capsid proteins that need to be processed from a polyprotein precursor in a timely fashion (monocistronic RNAs, i.e. picornavirus, flavivirus) are usually assembled with low yields, probably due to relatively inefficient processing of the precursors in recombinant expression systems and/or the toxicity of the viral protease required for processing. Finally, due to the inherent properties of the lipid envelope, production of enveloped VLPs is technically complex [14]. However, although large-scale manufacturing processes for VLP production remains challenging [53, 54], current and developing synthetic biology approaches (i.e. integration of multiple scientific disciplines like, molecular biology, genetics, structural biology, biophysics, and bioinformatics) offer prospects to overcome specific challenges in VLP engineering and manufacturing [55, 56].

21.4.4 VLPs as Stand-Alone Vaccines

The most straightforward application of VLPs is to use them for vaccination against the virus from which they were derived. Currently, VLP-based vaccines against human diseases are in various stages of development, spanning from preclinical evaluation to market (Table 21.1) [14, 16].

VLP-based vaccines for HBV have been licensed commercially. Recombivax® (Merck) and Engerix® (GlaxoSmithKline), both licensed in 1986, are subvirion 22-nm spherical lipid-protein particles consisting of self-assemblages of the HBV small surface protein, HBsAg, co-assembled with membrane lipids, expressed in genetically engineered yeast (*Saccharomyces cerevisiae*). Although these vaccines were proved to be highly effective, they suffered from a lack of immunogenicity in a

Table 21.1 Selected VLPs tested as vaccine candidates for different human and animal diseases

Virus	Family	Recombinant proteins	Envelope	Morphology of VLPs	Target species
Hepatitis B virus	*Hepadnaviridae*	Surface antigen	Yes	Isometric	Human
Human papillomavirus	*Papillomaviridae*	L1	No	Icosahedral, $T = 7d$, 1	Human
Hepatitis E virus	*Hepeviridae*	Capsid protein	No	Icosahedral, $T = 1$	Human
Influenza A	*Orthomyxoviridae*	HA, NA, M1, M2	Yes	Pleomorphic	Human
Norwalk virus	*Caliciviridae*	Capsid protein	No	Icosahedral, $T = 3$	Human
Hepatitis C virus	*Flaviviridae*	Core, E1, E2	Yes	Isometric	Human
Rotavirus	*Sedoreoviridae*	VP2, VP4, VP6, VP7	No	Icosahedral, $T = 13$	Human
Rift Valley fever virus	*Bunyaviridae*	N, Gn, Gc	Yes	Pleomorphic	Human
Porcine circovirus type 2	*Circoviridae*	Coat protein	No	Icosahedral, $T = 1$	Pig
Bluetongue virus	*Sedoreoviridae*	VP2, VP3, VP5, VP7	No	Icosahedral, $T = 131$	Sheep
Rabbit haemorrhagic disease virus	*Caliciviridae*	Capsid protein	No	Icosahedral, $T = 3$	Rabbit
Porcine parvovirus, canine parvovirus, mink enteritis virus	*Parvoviridae*	VP2	No	Icosahedral, $T = 1$	Pig, dog, mink
Infectious bursal disease virus	*Birnaviridae*	VP2, VP3, VP4	No	Icosahedral, $T = 13$, 1	Chicken
Newcastle disease virus	*Paramyxoviridae*	NP, M, F, HN	Yes	Pleomorphic	Chicken
Avian influenza virus	*Orthomyxoviridae*	HA, NA, M1	Yes	Pleomorphic	Chicken
Nodavirus	*Nodaviridae*	Coat protein	No	Icosahedral, $T = 3$	Fish

small percentage of patients (~5–10% nonresponders), which was determined to be due to an absence of epitopes derived from PreS proteins (PreS1 and PreS2) on the surface of the VLPs. A more immunogenic VLP vaccine was subsequently developed that contained, in addition to HBsAg, the proteins PreS1 and PreS2, which are present in the native HBV envelope at lower levels than HBsAg. This new HBV VLP-based vaccine (Sci-B-Vac®, SciGene), expressed in the mammalian CHO cell line, was found to elicit a strong antibody response and 100% seroconversion and seroprotection rates. Hence, a better mimic of the authentic virion surface structure resulted in broadened effectiveness of the HBV VLP-based vaccine [45].

The second type of commercially available VLP-based vaccines was developed against HPV, for the prevention of cervical cancer. These vaccines consist of 40 nm

VLPs assembled from the HPV major capsid protein, L1. Gardasil® (Merck) is produced in yeast (*Saccharomyces cerevisiae*) and was licensed in 2006. Cervarix® (GlaxoSmithKline) is produced using the recombinant baculovirus expression system and was licensed in 2009. Both contain mixtures of recombinant VLPs derived from different HPV types associated with genital infection (current Gardasil®9 vaccine is intended to protect against HPV types 6, 11, 16, 18, 31, 33, 45, 52, and 58) [41]. These vaccines have been shown to provide 100% protection against cervical precancerous as well as genital wart development, with an expected protective effect of a minimum of 4.5 years.

There are other VLP-based licensed vaccines for human diseases, like Hecolin®, against HeV, QVLP®, a plant produced influenza vaccine displaying quadrivalent HA antigens, and several more in an advanced stage of clinical development, including those directed against human parvovirus (B19), Norwalk virus, rotavirus, influenza A virus, and SARS-CoV-2 virus [14–16]. In the veterinary field, there are currently several VLP-based vaccines commercially available against porcine circovirus type 2 (PCV2), which were licensed in 2006–2009: Ingelvac Circoflex®, Circumvent® PCV, and Porcilis PCV®. Other veterinary VLP-based vaccines are ReproCyc® ParvoFLEX, licensed in 2019 against porcine parvovirus (PPV), and Yurvac®, launched in 2023 against rabbit haemorrhagic disease virus (RHDV). Further promising candidate vaccines in different stages of development include those against: BTV, IBDV, canine parvovirus, nodavirus, Rift Valley fever virus, avian influenza virus, or Newcastle disease virus [13, 57].

21.5 VLPs as Platforms for Foreign Antigen Display. Structure-Based Engineering of VLPs for Vaccine Development

As indicated in previous sections, VLPs can also be used as platforms for the multimeric display of foreign antigens, that can be incorporated into VLPs either by genetic fusion (chimeric VLPs) or by chemical conjugation (conjugated VLPs). In such cases, VLPs serve both as a presentation scaffold for antigens derived from other pathogens, in a suitable repetitive and highly organised configuration, and as adjuvants to boost the immune response. Ideally, the immunogenic 'viral fingerprint' of the VLP is transmitted to the attached antigen, making it as potent an immunogen as the VLP itself.

The poor immune response of many soluble antigens can be overcome by rendering them highly repetitive through their incorporation onto a carrier viral structural protein [58]. In contrast to monomeric and oligomeric carrier proteins commonly used for antigen presentation, such as bovine serum albumin and keyhole limpet haemocyanin, chimeric VLP structures are able to provide not only a high density of introduced foreign antigens per particle, but also a distinctive three-dimensional conformation, which is especially important for the presentation of conformation-dependent epitopes. Hence, the regular, repetitive pattern and correct conformation of inserted epitopes are factors encouraging the development of VLPs as platforms for inducing immunological response against foreign antigens.

Insertion in specific sites of VLPs of continuous epitopes (those included in a short peptide segment) or continuous parts of discontinuous epitopes (those made up of several peptide segments in defined conformations) may alter the conformational preferences of the peptide segment relative to the native ones (i.e. in their native protein or viral capsid), which may have dramatic effects on their immunogenicity as a part of the chimeric VLP. If complete discontinuous epitopes must be inserted in the VLP, the problem is severely compounded, as different peptide segments must be inserted in such a way that not only their conformational preferences, but also their geometric relationships are preserved in the chimeric VLP, relative to those in the native protein or viral particle. Detailed structural information and computational docking approaches may be critical for the rational design of highly immunogenic chimeric VLPs [55].

VLP chimeras have been extensively explored as vaccine candidates since the mid-1980s [59]. The interest in rational-based manipulations on chimeric VLPs was reinforced by the simultaneous development of recombinant DNA engineering techniques and structural knowledge about virions, capsids, and envelope proteins obtained by the use of high-resolution techniques (X-ray crystallography (see Chap. 4) and cryogenic electron microscopy (cryo-EM) (see Chap. 3) [60]. Concurrently, research in immunology, molecular biology, and structural studies led to the identification, characterisation, and fine mapping of immunogenic epitopes derived from pathogens causing relevant diseases. X-ray crystallographic studies of antigen–antibody complexes as well as peptide-T-cell receptor interactions provided considerable insight into the structural basis of immunological recognition [61].

Currently a wealth of high-resolution viral structural information (see Chaps. 2–7) has facilitated the ability to modify VLPs deliberately so that they function essentially as molecular scaffolds for antigen presentation. VLPs derived from both double-and-single-stranded DNA and RNA viruses encompassing 14 different families of virus have been successfully used for the display of foreign antigens [18, 19, 35] (Table 21.2). Results obtained using VLPs derived from different viruses, like HBV, HPV, parvovirus, calicivirus, or bacteriophages have illustrated the potential of this approach to induce strong immune responses against foreign B- and T-cell epitopes [16, 18, 19, 22, 25].

21.5.1 Chimeric VLPs Inducing Antibody Responses Against Target Molecules

In order to induce high-titre antibody responses effectively, target antigens must be displayed on the surface of VLPs, in immunodominant regions, at a high density. The genetic insertion of target sequences into viral structural proteins to generate

Table 21.2 Examples of VLPs and chimeric virions used as platforms for foreign antigen display

VLP platform or Viral nanoparticle	Foreign antigen	Incorporation method
HBcAg	Influenza A M2 extracellular domain, B- and T-cell epitopes from HCV, epitopes derived from malaria	Genetic fusion
HBsAg	Dengue virus envelope, HIV gp41epitopes	Genetic fusion
Bovine papillomavirus	CTL epitopes from HPV and HIV, L2 HPV epitopes, Aβ peptide	Genetic fusion
HPV	HPV E7 oncoprotein	Genetic fusion
HPV	Aβ peptide	Chemical conjugation
RHDV	Model T-cell epitopes (i.e. Chicken ovoalbumin CTL epitope) Influenza A M2 extracellular domain, B-cell epitopes from feline calicivirus, CPV, foot-and-mouth disease virus T-cell epitopes from foot-and-mouth disease virus, Cancer related CTL epitopes	Genetic fusion and chemical conjugation
PPV	Model T-cell epitopes (i.e. CTL epitope derived from LCMV nucleoprotein)	Genetic fusion
Phage Qβ	Nicotine, angiotensin II, CCR5 extracellular domain	Chemical conjugation
CPMV	B-cell epitope (NIm-1A) of HRV-14, B-cell epitope of canine parvovirus	Chemical conjugation
FHV	ANTXR2 domain, ectodomain of G protein from Rift Valley fever virus	Genetic fusion

chimeric particles is the most common method for displaying foreign epitopes on VLPs. Essentially, DNA fragments encoding continuous or discontinuous immunological epitopes are cloned into the genes encoding the self-assembly-competent polypeptides that form the VLPs. Upon assembly of the chimeric subunit proteins into supramolecular structures, the introduced epitopes are presented at a relatively high density (usually one copy per VLP subunit) and, ideally, in an accessible and conformationally relevant manner. An advantage of this chimeric VLP technique is that successful incorporation of a target epitope guarantees that the antigen will be displayed in the same conformation and at high density throughout the particle surface. This technique also has substantial advantages from a manufacturing standpoint. Chimeric particles can be purified using the same well-established methods used to purify unmodified parental VLPs.

The key to this technology is the identification of adequate permissive insertion sites to incorporate the target epitopes, within the primary sequence of the VLP subunits. That is, sites that do not compromise the correct folding and assembly of the VLP protein subunits and do not alter the structural integrity and immunogenicity of the VLP. Obviously, depending on the type of foreign epitope to be displayed, the structural requirements of the insertion sites would be different. B-cell epitopes should be located at exposed sites on the surface of the particles, preferentially located at immunodominant regions, readily accessible for direct interaction with B-cells. However, T-cell epitopes do not need to be located at exposed sites within the VLP structure, because these epitopes have to be proteolytically processed by APCs before being presented to the target T-cells (as explained in Sect. 21.3), and thus can be inserted at nonaccessible permissive sites of the VLP. Hence, knowledge of the three-dimensional structure of the VLP scaffold is a critical requirement for successful generation of chimeric VLPs.

Since many B-cell epitopes in their native environments are found at surface loops of the pathogen's proteins, this is often a good choice for the insertion of foreign antigens in VLP protein subunits. In some cases, proper folding and presentation can be accomplished by replacing exposed, immunodominant viral epitopes located at loop structures on VLPs, with the desired target epitope. Indeed, neutralising viral B-cell epitopes incorporated at immunodominant exposed loops of chimeric VLPs can, in some instances, elicit higher antibody responses than the corresponding epitope in its native viral capsid (i.e. epitope displayed in its natural context) [62]. However, generating chimeric VLPs is still largely empirical; it is almost impossible to predict whether individual peptides will be compatible with VLP assembly at defined insertion sites or whether the insertions will be immunogenic in the resulting chimeric VLP. For instance, it has been shown that minor displacements in the insertion site (two or three amino acid positions) can cause drastic changes in the immunogenicity of the inserted sequence [63]. Peptide insertions with high hydrophobicity, a high β-strand index, a large volume, or a strong positive charge are prone to cause problems in VLP assembly. The success rate can be improved using combinatorial technologies. This type of approach was used to generate chimeric VLPs derived from the core protein of woodchuck hepadnavirus [64], enabling the identification of 17 different sites in the subunit protein that could be used to insert target sequences.

Another limitation attributed to the chimeric approach is that the size of the antigens that can be inserted into VLPs, in particular into their surface-exposed immunodominant regions, is usually restricted to short peptides. Chimeric protein subunits containing foreign peptide insertions longer than 20–30 amino acids often fail to assemble into VLPs. However, it should be noted that different VLPs exhibit marked differences regarding the range of insertion sizes they can tolerate. Relatively large insertions have been successfully incorporated into exposed sites in VLPs, like three tandem copies of a foreign epitope spanning 62 amino acids in RHDV-derived VLPs [62], or 238 amino acids of the complete green fluorescent protein (GFP), inserted in HBcAg-derived VLPs [45]. Other whole proteins or complete protein domains, with more relevance as immunogens to induce antibody responses, have been successfully inserted in chimeric VLPs, like a 120 amino acid-long polypeptide from the hantavirus nucleocapsid protein in HBcAg-derived VLPs, or a 395 amino acid portion of dengue virus envelope protein type 2, in HBsAg-derived VLPs [59]. Hence, in some cases, it is possible to display large antigens in the surface of chimeric VLPs. This might be important for targeting complex discontinuous B-cell epitopes, or in order to increase the number of foreign epitopes displayed by chimeric VLPs, offering the possibility of producing multi-target vaccines, by incorporating different epitopes in each VLP subunit [65].

The VLP derived from the HBcAg is one of the most extensively investigated chimeric carriers of foreign antigens [66]. HBcAg polypeptide is 183 to 185 amino acids long (21-kDa). The organisation of HBcAg-derived particles is largely α-helical and thus, quite different from previously known viral capsid proteins with β-sheet jellyroll packings. The HBcAg monomer fold is stabilised by a hydrophobic core that is highly conserved among human HBV variants. Association of two amphipathic α-helical hairpins results in the formation of a dimer with a four-helix bundle as the major central feature. The dimers assemble into two types of particles, large and small, that are 34 and 30 nm in diameter and correspond to triangulation numbers $T = 4$ and $T = 3$, containing 240 and 180 monomers, respectively (see Chap. 2). The major immunodominant region (MIR) with the central amino acid positions 76–81 is located at the tips of the α-helical hairpins that form spikes on the capsid surface. In addition to the highly exposed MIR, the region between amino acids 127 and 133 is the next exposed and accessible epitope on the particle surface. This region is located at the end of the C-terminal α-helix and forms small protrusions on the surface of the HBcAg particle. Foreign antigens can be inserted at the N- and C-termini, and at the MIR (i.e. the tips of the VLP spikes), without affecting the ability of the resulting chimeric proteins to assemble into VLPs [66]. The N-terminal end of HBcAg has been widely used as insertion site to display a variety of short foreign epitopes. The capacity for incorporating foreign peptides at this site is around 50 amino acids. The inserted epitopes are accessible to specific antibodies and in general terms, high levels of specific antibody responses are achieved. Regarding the C-terminal insertions, amino acid positions 144, 149, 156, 163, and 167 are the most frequently used sites for the incorporation of foreign epitopes. The capacity of the C-terminal end to accept foreign insertions usually exceeds 100 amino acids residues, depending on the structure of the inserted polypeptide. In some cases, quite large polypeptides have been successfully inserted at this location. A 559 amino acid long insertion, containing three copies of the HCV core protein, did not prevent self-assembly of chimeric proteins. However, the C-terminal end of HBcAg is not exposed at the surface of the VLPs and thus, chimeric constructions harbouring foreign antigens at this site usually induce weak to moderate specific antibody responses. Hence, this site is mainly used for the insertion of T-cell epitopes. The most interesting and promising site for foreign insertions is the MIR, which allows full exposure of the inserted peptides on the VLP surface. In addition, this insertion site has revealed an unexpected high capacity for the incorporation of large foreign polypeptides. The entire 120 amino acids long immunoprotective region of the hantavirus nucleocapsid protein, and the 238 amino acids long GFP (as indicated above), have been successfully incorporated into chimeric HBcAg VLPs at this insertion site. A versatile approach based on HBcAg VLPs is the SplitCore technology [67], in which the HBcAg gene is split into two parts within the MIR region, and both fragments are expressed in *E. coli* cells. Thus, both portions of HBcAg are expressed separately within the same cell but associate through hydrophobic interactions to reconstitute functional dimers and complete VLPs. Interrupting the central part of the protein, at the most surface-exposed region of the particle structure, provides great flexibility in the choice of antigens that can be presented. In this system, inserts are fused at either of the fragments, allowing the display of large antigens (i.e. 300 amino acids), on the VLP surface.

Mosaic VLPs have been developed to aid in the assembly of complex VLP chimeras. These VLPs are typically formed by co-expression of a wild-type VLP subunit along with a chimeric VLP subunit. The resultant VLPs contain copies of both wild-type and chimeric subunits which are incorporated with differing proportions. The use of mosaic VLPs has enabled the formation of HBcAg VLPs containing fragments of up to 213 amino acids from the hantavirus nucleoprotein, or 163 amino acids from the HBV preS domain [66]. Alternatively, co-expression of VLP subunit proteins derived from different virus strains may be used to assemble mosaic VLP vaccines that generate antibodies that protect against multiple pathogen serotypes. This approach has been used for the generation of mosaic VLPs derived from HPV formed by the co-expression of L1 and L2 capsid proteins of both, HPV-6 and HPV-16 types. The resulting VLPs comprised all four co-assembled subunit proteins.

21.5.2 Chimeric VLPs Targeting Self-Antigens

Traditionally, antibody-inducing vaccines were used solely to provide prophylactic protection against pathogens. However, there has been a growing interest in the development of new vaccines with the goal of deliberately inducing antibody responses against self-molecules that are involved in chronic disease processes. These autoantibodies are expected to serve as competitive inhibitors to fight diseases. The ability to induce antibody responses against self-molecules is seemingly limited by the mechanisms of B-cell tolerance, which eliminate, desensitise, or change the specificity of potentially self-reactive B cells, in order to prevent autoimmune disorders. Nevertheless, B-cell tolerance is somehow inefficient; it has been estimated that approximately 20% of long-lived mature B cells are self-reactive. Thus, antibodies against self-molecules can be induced by vaccination.

The main hurdle for eliciting a high-titre antibody response against self-antigens is overcoming T-cell tolerance. Stringent T-cell tolerance mechanisms ensure that autoreactive B cells cannot receive T-cell help and, therefore, cannot proliferate and produce potentially damaging autoantibodies. Thus, one strategy for inducing antiself antibodies has been to immunise with self-antigens chemically conjugated or genetically linked to foreign T-helper epitopes. This approach requires large doses of antigen and the use of powerful adjuvants. In addition, antibody titres induced against the self-antigen are often very low and rapidly diminish, particularly in comparison with the responses induced by vaccines against foreign antigens. By contrast, self-antigens displayed on VLPs are inherently immunogenic at low doses and without exogenous adjuvants. It has been shown that self-antigens incorporated to VLPs can induce antibody titres up to 1000-times higher than those induced by the same self-antigen linked to a foreign T-helper epitope. Moreover, VLP display can make a self-antigen as immunogenic as a foreign antigen presented in the same context. The magnitude of the antiself antibody responses correlates with the density at which the self-antigen is displayed on the VLP surface. In conclusion, B cell unresponsiveness can be overcome by antigens presented in a repetitive and highly organised 'foreign-like' manner. In fact, the immune system is largely unable to distinguish between self and foreign proteins based on antigenic epitopes, but does so based on antigenic structural organisation.

VLP display of self-antigens has been successfully used to target molecules that are involved in the pathogenesis of a variety of chronic diseases, including Alzheimer's disease, rheumatoid arthritis, hypertension, acquired immunodeficiency syndrome, and certain cancers [16, 18, 22]. Some of these vaccines have shown clinical efficacy in animal models and several are currently in clinical trials. However, a vaccine inducing an antibody response against a self-protein raises safety concerns that must be addressed.

Alzheimer's disease is a neurodegenerative disease that results in accumulation of plaques consisting of aggregated amyloid-β (Aβ) peptide (42 amino acids in length), neurofibrillary tangles, and loss of neurons leading to dementia. One therapeutic strategy has been the use of vaccination against Aβ peptide to reduce the accumulation of amyloid plaques in the brain. Papillomavirus VLPs displaying an Aβ peptide induced high levels of specific antibodies, inhibiting effective assembly of peptides into neurotoxic peptides in vitro, as well as reducing Aβ deposits in a mouse model of Alzheimer's disease. However, a clinical trial using synthetic Aβ peptides for immunisation showed minimal efficacy and resulted in development of encephalitis in 6% of the patients. This effect was attributed to a T-cell mediated autoimmune response caused by the vaccine adjuvant. Hence, a safe and effective Alzheimer's disease vaccine might rely on the induction of high-titre anti-Aβ antibodies, delivering only B-cell epitopes located at the N-terminus of the Aβ peptide, in the absence of inflammatory antiself T-cell responses. Different VLP platforms (based on HPV, bacteriophage Qβ, HBcAg) have been used to display a short Aβ peptide which does not contain a T-cell epitope [16, 18]. Thus, the flexibility of the VLP display system allows the tailoring of vaccines, hopefully enabling to conform to both safety and efficacy requirements.

Another example is one potential therapeutic approach to prevent HIV infection, based on the production of autoantibodies against the HIV cellular receptor CCR5 [18]. It is an indirect route to vaccination, avoiding the problem of constant mutation of the viral antigens. A peptide from CCR5 was attached to bovine papilloma virus VLPs and upon administration in mice, these VLPs elicited the production of autoantibodies, which inhibited binding of the ligand and blocked infection of an indicator cell line expressing CCR5. CCR5 peptides have also been conjugated to bacteriophage Qβ VLPs, which have been evaluated in rhesus macaques.

Vaccines directed against self-molecules hold promise for treating a variety of chronic diseases by the induction of antibodies targeting endogenous proteins. In contrast to passive vaccination with monoclonal antibodies, active immunisation with VLP-based vaccines displaying self-peptides induce long-lasting antibody responses [38], potentially providing affordable and convenient therapy for patients.

21.5.3 Chimeric VLPs Inducing Cell-Mediated Immunity

The use of chimeric VLPs as platforms to induce CTL responses against foreign antigens presents different requirements than vaccines aimed at inducing humoral responses. Unlike B-cell epitopes, T-cell epitopes do not need to be exposed on the surface of the particle or presented at high density and are included in single peptide segments (i.e. they are continuous epitopes). This enables diverse approaches for the generation of chimeric VLPs intended for this purpose. Chimeric VLPs can be generated not only by fusing target antigens to viral major capsid proteins (as were the examples cited in the previous sections), but also by making fusions between target antigens and proteins that are minor structural components of VLPs, and are often not necessary for particle formation. In this case, the target antigens would not be displayed in high numbers per particle and will usually be located facing the interior of the VLPs. This is not detrimental in the case of T-cell epitopes, and in fact, in some situations can be advantageous. Thus, targeting large polypeptides might be necessary in situations in which the most relevant target CTL epitopes in a protein have not yet been precisely mapped. Large peptides and even full-length proteins can be added to minor structural proteins without interfering with capsid assembly. For instance, the non-structural HPV encoded oncoprotein, E7, was fused to the HPV minor capsid protein L2. When co-expressed alongside the HPV major capsid protein (L1), L2–E7 was incorporated efficiently into VLPs and because E7 was incorporated on the inside of the VLP, the fusion domain did not interfere with cell binding. Chimeric L1/L2–E7 VLPs induced strong CTL responses that protected mice from challenge with an E7-expressing tumour cell line [68].

As indicated in Sect. 21.3, in order to activate cytotoxic T-cells, DCs must also provide appropriate costimulatory signals that are only upregulated upon DC maturation. Certain VLP types, such as HPV VLPs, can provoke these signals by directly stimulating the phenotypic and functional maturation of DCs, and thus, are highly efficient in inducing CTL responses. In addition, the immunogenicity of many VLPs can be enhanced incorporating molecules with adjuvant activity, such as TLR ligands. Proof of concept was established using bacteriophage Qβ VLPs co-delivering TLR ligands and CTL epitopes to DCs in animal models [17]. It was shown that Qβ-VLPs devoid of any TLR ligands essentially failed to induce a CTL response, while Qβ-VLPs which were internally loaded with CpG-ODNs induced a robust response against the foreign CTL epitope incorporated to the VLPs (the p33 epitope from lymphocytic choriomeningitis virus glycoprotein).

In summary, numerous chimeric VLPs designed to present B and T-cell epitopes in this manner have been tested in preclinical research [15, 16, 22, 47, 57]. The final intended application of such vaccines spans human to veterinary use and prophylactic to therapeutic treatment. Epitopes have been derived from viral, bacterial, eukaryotic parasitic, and disease related self-molecules.

The vaccine RTS, S/AS01 Mosquirix® against malaria is one of such chimeric VLP-based vaccines that has reached the market [15, 16]. It combines a target region eliciting neutralising antibodies with a T-cell epitope derived from *P. falciparum* circumsporozoite protein (CSP), which are fused to HBsAg as a mosaic VLP platform. Particle formation requires four-fold excess scaffold protein (native HBsAg monomers) as compared to target antigen-HBsAg fusion protein. This vaccine was approved for use by the European Medicines Agency (EMA) (outside of EU), and in October 2021 the World Health Organization (WHO) recommended its widespread use among children in Sub-Saharan Africa, as well as in regions with moderate to high *P. falciparum* transmission. Subsequently, a next generation VLP-based malaria vaccine has been developed, R21/Matrix-M, in which particles are formed only with target antigen-HBsAg fusion protein (it is not a mosaic VLP vaccine), therefore displaying larger proportion of malaria target antigens [15, 16]. This vaccine has been shown to induce sterile protection in mice and in October 2023 has received WHO recommendation. Other candidate vaccines based on chimeric VLPs in late stages of clinical evaluation include different influenza A vaccines based on M2-HBcAg VLPs.

21.5.4 Further VLP Modifications for Vaccine Improvement

The design of new vaccine strategies based on VLPs may involve complementary modifications, other than the incorporation of target antigens, aimed at improving the elicited immune response, enhancing the self-adjuvant abilities of VLPs. One possible approach is the incorporation of PAMPs into VLPs (as indicated previously in Sect. 21.3). For example, the introduction of CpG oligonucleotides inside VLPs, which act as ligands for TLRs, may contribute to enhance both, the humoral and cellular immune responses. Other TLR ligands that can be incorporated into VLPs are single-stranded RNA or double-stranded RNA molecules [31]. Some VLPs spontaneously encapsidate nucleic acids derived from the cells in which they are expressed. For instance, VLPs derived from cucumber mosaic virus, expressed in *E. coli*, contain bacterial RNA which enhances the stability of VLPs, as indicated by cryo-EM studies, while improving their immunogenicity [17, 26]. Alternatively, VLPs can be deliberately loaded with nucleic acids (by VLP disassembly/reassembly in the presence nucleic acids), to enhance their immunogenicity.

Other types of modifications that can be applied to VLPs of interest as vaccine platforms are those aimed at increasing their stability, since this can provide several advantages. Engineered VLPs with enhanced stability, able to tolerate variations in temperature and other conditions, would be advantageous during manufacturing, distribution, and storage of vaccines. VLPs able to withstand transport or storage at different temperatures, thus not dependent on a strict cold chain maintenance, would facilitate the logistics of vaccine distribution, particularly in areas of the developing world. Moreover, enhancing VLP stability might reduce vaccine costs, due to longer shelf-life and reduced energy costs associated with cold-storage. An approach used to increase the structural stability of viral capsids (VLPs or virus nanoparticles) is the substitution (through structure-based directed mutagenesis) of amino acids located at the interfaces between subunits, with cysteines or charged amino acids, with the aim of introducing new intersubunit disulfide bonds or electrostatic interactions. This approach has been successfully used with VLPs derived from bacteriophages Qβ and MS2, HBV (HBcAg), or with viral particles and VLPs from foot-and-mouth disease virus [8, 19]. Interestingly, engineering intersubunit disulfide bonds in the norovirus capsid not only stabilised the resulting VLPs preventing disassembly, it also enhanced the immunogenicity elicited by focusing the immune response towards relevant exposed neutralising epitopes [69].

21.5.5 Chemical Conjugation of VLP-Based Assemblies

An alternative approach for displaying antigens on the surface of VLPs is the use of modular systems, in which the native VLP and target antigen are synthesised separately and then chemically conjugated in vitro, covalently or noncovalently, linking the antigen to the surface of preassembled VLPs (Fig. 21.8) (see also Chap. 22) [19, 55, 70]. An advantage of this approach is that the size and structure of the target antigen to be attached are not constrained by the requirements of the correct folding of the VLP monomers and particle assembly. Furthermore, chemical conjugation allows the attachment to the surface of VLPs of diverse kinds of target antigens: short linear peptides, cyclic peptides, full-length proteins, and even non-protein targets, such as glycans or small haptens (organic molecules that can elicit an immune response only when attached to a protein carrier). One example is the antismoking candidate vaccine Nicotine-Qbeta (NicQb), which is produced by covalently coupling nicotine to the surface of VLPs derived from the bacteriophage Qβ [18].

The conjugation techniques (see also Chap. 22) rely on the presence of addressable moieties on the surface of VLPs. For instance, the amino groups of naturally occurring surface-exposed lysines are convenient sites for conjugation to the thiols of cysteine-containing target antigens, using chemical crosslinkers. Otherwise, VLPs can be engineered to contain these convenient sites, by introducing surface-exposed cysteines or lysines, providing accessible reactive groups [71]. For instance, HBcAg VLPs, bacteriophage MS2, and cowpea mosaic virus have been mutated so that they contain a single surface-exposed reactive residue (lysine or cysteine) per capsid subunit, suitable for antigen conjugation [19]. To ensure that the antigen is coupled to the VLP in a directed and oriented fashion, peptide antigens can be engineered to contain either an amino- or carboxy-terminal amino acid linker sequence containing a free cysteine group.

The classic approach to chemical conjugation of VLPs consists in the use of crosslinkers, normally, heterobifunctional conjugation reagents

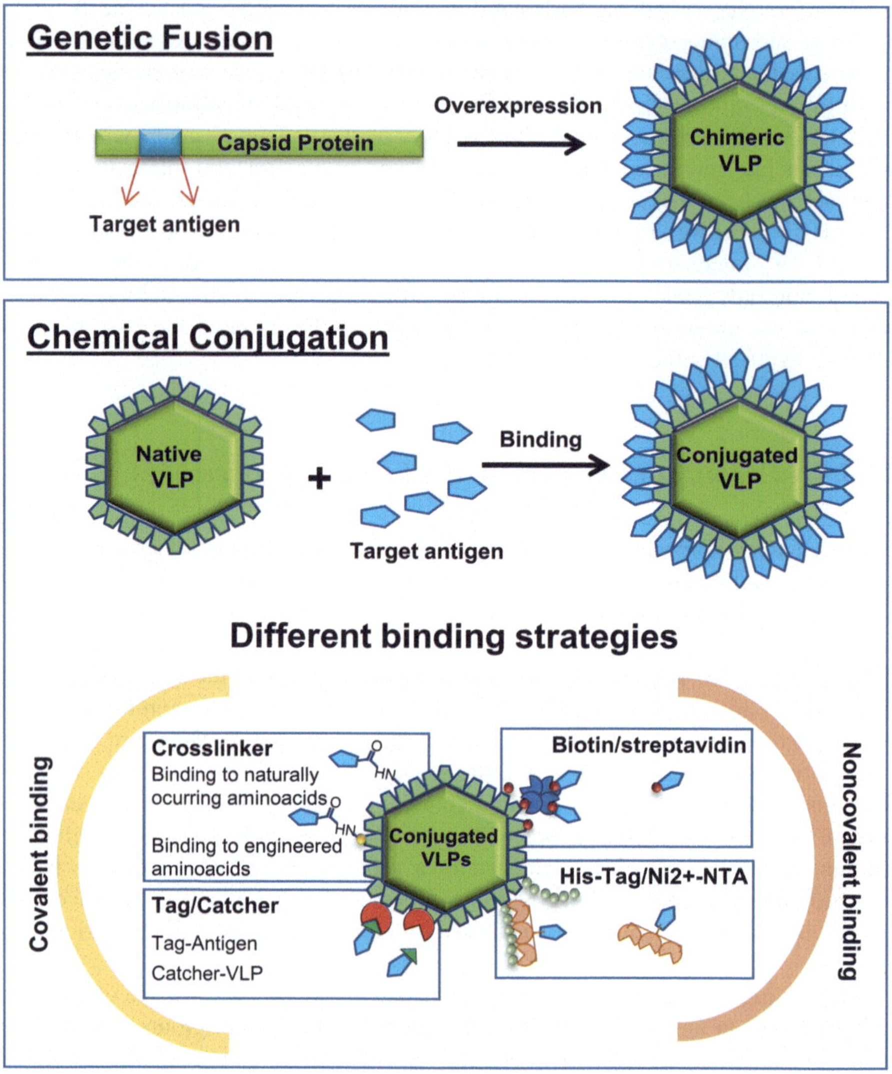

Fig. 21.8 Different approaches for displaying antigens on the surface of VLPs. For the genetic insertion of target sequences, the DNA encoding an antigenic peptide is cloned into the gene encoding the self-assembly-competent polypeptide that forms the VLPs. The resulting construction self-assembles into chimeric VLPs. Chemical conjugation techniques rely on the use of modular systems, in which the native VLP and target antigen are synthesised separately and then conjugated in vitro, covalently or noncovalently, linking the antigen to the surface of preassembled VLPs

(i.e. maleimidobenzoic acid sulfosuccinimidyl ester, Sulfo-MBS), promoting covalent attachment of the antigen to the VLP.

For noncovalent conjugation of antigens to VLPs, exposed lysine residues can be biotinylated and then attached to biotinylated target antigens through the use of a streptavidin linking molecule [70]. Using this approach, VLPs displaying peptides derived from self-antigens (i.e. Aβ, CCR-5, or TNF-α) have been obtained. Similarly, norovirus VLPs have been engineered to display poly-histidine stretches (His-Tag) for noncovalent conjugation to antigens with nickel-loaded nitrilotriacetic acid (Ni^{2+}-NTA) moiety [70].

Recently developed technologies, known as Tag/Catcher 'plug and display' decoration of VLPs, are based on two components, a peptide (the tag) and a protein (the catcher), which react irreversibly (forming covalent isopeptide bonds) when in close proximity to each other [72]. If the catcher is fused to a VLP platform and the tag is fused to a target antigen, these form a two-component conjugated VLP ready to use. Tag/Catcher technology was developed from certain cell-surface proteins of Gram-positive bacteria, which are able to form autocatalytic intramolecular isopeptide bonds. The SpyTag peptide (13 amino acids) forms a spontaneous isopeptide bond upon mixing with the SpyCatcher protein (116 amino acids) [72]. The Tag/Catcher-AP205 platform was developed by fusion of the SpyCatcher into the capsid protein of the Acinetobacter phage AP205, yielding a VLP platform able to covalently couple malaria antigens linked to SpyTag. Since its development, the Tag/Catcher-AP205 platform has been used to display structurally and functionally diverse target antigens, ranging from small peptides to large (>300 kDa) trimeric proteins [72]. The Tag/Catcher approach has been extended to other VLP platforms, such as VLPs based on HBcAg, norovirus, porcine circovirus, or canine parvovirus. Besides SpyTag/SpyCatcher, there are other related systems like SNAP-Tag (a 20-kDa protein which binds irreversibly to alkyl-guanine ligands) or SnoopTag/SnoopCatcher.

Hence, the flexibility of the chemical conjugation approach offers substantial inherent advantages. However, from the manufacturing point of view, it poses some drawbacks. Chemical linkage results in less than 100% of foreign antigen insertion achieved with the genetic fusion method and is not as reproducible. Thus, the production process of conjugated VLPs involves additional challenges and quality control is inevitably more complex.

21.6 Use of Plant and Insect-Derived Chimeric Virions for Foreign Antigen Display

An interesting approach related to the VLP-based vaccine technology is the use of genetically engineered plant viruses as vectors for the presentation of foreign epitopes. In this approach, the heterologous antigens are inserted by genetic fusion at exposed loops within viral capsid proteins, but using infectious viral clones for cell transfection, resulting in the formation of viable chimeric virus particles. In contrast with VLPs (in the strict sense frequently given to this definition), these chimeric whole-virus particles are replication competent and are therefore infectious in plants, their natural hosts. The engineered viruses are propagated in plants and subsequently, the corresponding purified chimeric virions are used as immunogens in target animals. Since the plant viruses are unable to replicate in mammals, these chimeric virions behave similarly to VLPs in terms of safety and the immune response induced. Both filamentous and icosahedral plant viruses have been successfully developed as foreign antigen presentation systems [73, 74]. Some of the most frequently used systems are cowpea mosaic virus (CPMV), tobacco mosaic virus, cucumber mosaic virus, alfalfa mosaic virus, and papaya mosaic virus.

CPMV has been extensively studied for vaccine applications [74, 75]. It is a comovirus in the picornavirus superfamily, with a capsid composed of two protein subunits termed small (S) and large (L). The S subunit is about 23 kD and folds into a jellyroll β-sandwich, while the L subunit folds into two jellyroll β-sandwiches, with a total mass of 41 kD. Sixty copies of both proteins form the virus capsid 30 nm particle with a $T = 3$ icosahedral symmetry. Knowledge of the detailed three-dimensional structure of the particle allowed the identification of suitable sites for the insertion of foreign peptides, and the availability of infectious cDNA clones enabled engineering the corresponding genetic changes in the viral genome. Thus, CPMV became the first plant virus to be successfully developed as a peptide presentation platform [75]. In most cases, the foreign sequence has been inserted into the most exposed loop of the virus surface, the βB-βC loop of the S protein, resulting in the display of 60 copies of inserted peptide on the surface of the particle. Other sites, such as the βC′-βC″ loop of the S protein and the βE-αB loop of the L protein, have also been used successfully. When the inserted peptide is less than 40 amino acids, the yields of chimeric particles are generally similar to those obtained with wild-type CPMV (up to 1 mg of particles per gram of infected leaf tissue).

CPMV particles have been genetically modified to display foreign epitopes derived from mink enteritis virus (MEV) and canine parvovirus (CPV) VP2 proteins, as well as HIV gp41 protein. These chimeric particles were shown to induce neutralising antibody responses when injected into experimental animals, and in the cases of the chimeric particles harbouring epitopes derived from MEV and CPV, the immune responses induced resulted in protection against viral challenges with the corresponding viruses. Likewise chimeric CPMV particles expressing a foreign antigen derived from the outer membrane protein F of *Pseudomonas aeruginosa* were able to protect mice against challenge by two different immunotypes of *P. aeruginosa* in a model of chronic pulmonary infection.

Biochemical analysis of chimeric CPMV virions with foreign antigens inserted at the βB-βC loop or the βC′-βC″ loop identified a proteolytic cleavage event near the carboxy-terminus of the inserted sequence, which appears to be position, rather than sequence, dependent. This cleavage does not result in the loss of the epitope from the surface of the virion, but as a result the epitope is only anchored to the viral surface by its N-terminus.

An interesting feature of CPMV as an antigen presentation system is the fact that the conditions for crystallisation of the virus are well established. Hence, modified particles can readily be crystallised and the structure of the inserted peptide determined by X-ray crystallography, providing a unique opportunity for analysing how the mode of presentation of a peptide affects its immunogenicity. In fact, such a study was conducted by generating a series of chimeric CPMV constructions expressing the 14 amino acid NIm-1A epitope from human rhinovirus 14 (HRV-14), at different positions on the capsid surface [75]. Biochemical and crystallographic analysis of constructions expressing the NIm-1A epitope inserted into the βC′-βC″ loop of the S protein revealed that, although the inserted peptide was free at its C-terminus, it adopted a conformation distinct from that found when a similarly cleaved peptide was expressed in the βB-βC loop of the S protein. Adjustment of the insertion site within the βB-βC loop (by moving it one residue to the left) resulted in the isolation of a chimeric virus in which cleavage at the C-terminus of the foreign epitope was greatly reduced. Crystallographic analysis confirmed that in this case the epitope was presented as a closed loop. Polyclonal antisera raised against this chimeric virus had a significantly enhanced ability to bind to intact HRV-14 particles, compared with antisera raised against other constructs presenting the same epitope as peptides with free C-termini (the antisera raised against these constructions reacted strongly against HRV-14 VP1 in Western blots but bound poorly to intact HRV-14). In conclusion, this study demonstrated that the mode of presentation of an epitope on a heterologous carrier can dramatically affect its immunological properties. For epitopes such as NIm-1A, which adopt a constrained structure in their native context (HRV-14 particles), presentation as a closed loop is probably essential for good mimicry. In contrast, the success of the chimeric CPMV particles displaying foreign antigens derived from parvoviruses, or the bacteria *P. aeruginosa* (described above), in stimulating protective immunity, probably stems from the fact that those antigens act as linear epitopes that are active even in a denatured form, because they fold on recognition by the corresponding antibodies and the entropic cost of their folding is not enough to prevent binding with enough affinity. Thus, the accurate structural mimicry is not an essential requirement in those cases. Overall, the results of this study demonstrated the potential of the CPMV-based display platform for studying the relationship between peptide structure and immunogenicity [75].

Whereas no known plant virus can infect mammalian cells, the ability of the chimeric viruses to replicate in plants and spread in the environment, raises biosafety concerns associated with the transport, distribution, and administration of these plant virus-based vaccines. One option to circumvent these problems is to render the chimeric viruses non-infectious using inactivation methods, such as irradiation with ultraviolet light or chemical treatments. However, these processes have to be carefully monitored as they risk altering the structural properties of the particles and, therefore, their immunogenic characteristics. Other possibility is to use VLPs derived from the plant virus capsid proteins instead of the infectious chimeric virus particles, for the presentation of the foreign epitopes [76]. Furthermore, since the VLPs do not longer need to be competent at packaging RNA or spreading within plant tissues, as the replication competent chimeric viruses do, this enables to extend the range of mutations (i.e. more insertion sites and longer insertions) that it is possible to introduce into the capsid proteins, thereby extending the versatility of the particles as antigen display systems for vaccine development.

Insect viruses can also be used as platforms for antigen display (they do not replicate in mammalian cells). Flock house virus (FHV) particles are widely used for antigen display, attachment, and delivery in animals [77, 78]. FHV is a member of the insect virus family nodavirus. The capsid (30 nm) is composed of 180 subunits of the single coat protein, arranged with a $T = 3$ icosahedral symmetry. Several surface-exposed loops have been successfully used as sites for insertion of foreign antigens. FHV particles can be used either as intact virions (and thus, replication competent in insect cells), or as VLPs derived from the expression of the coat protein in the baculovirus expression system.

An interesting application of the FHV-based display platform was the construction of chimeric FHV particles incorporating the extracellular domain of the anthrax toxin receptor 2 (ANTXR2), in order to use them as both, an antitoxin and vaccine for *Bacillus anthracis*, the causative agent of anthrax [77]. The foreign sequence was inserted into the two most exposed loops, at positions 206 and 264, on the FHV coat protein. The extracellular ANTXR2 domain is 181 amino acids long and adopts a Rossmann-like α/β fold. In the native structure, the termini of this domain are only 4.8 Å apart, thus, easing the insertion into a loop of the VLP capsid protein. Indeed, the corresponding chimeric proteins readily assembled into VLPs. Cryo-EM density maps confirmed successful insertion of the foreign domain at the VLP surface. In the case of insertion at site 206, the ANTXR2 domains were clustered in groups of six at the twofold axes, while insertion at site 264 resulted in more uniform distribution of the foreign domain. In both cases, the inserted domain appeared functional, exhibiting ability to bind anthrax protective antigen (PA, which forms part of anthrax lethal toxin and oedema toxin), inhibiting PA-cellular receptor binding and intoxication in vitro and in vivo. Since they could bind PA, the chimeric VLPs were evaluated as a therapeutic compound to treat anthrax infections, using the FHV-ANTXR2 particles as a multivalent scaffold for displaying the antigenic PA protein. Immunisation of rats with the FHV–ANTXR2–PA complex induced a strong toxin-neutralising antibody response as well as protection against lethal toxin challenge, confirming the dual-action ability for treatment and protection. Notably, higher antibody titres and improved protection levels were observed following immunisation with multivalent FHV-ANTXR2-PA particles as compared to monomeric (soluble) recombinant PA alone. These results further confirm that multivalency of antigen presentation facilitates effective antibody responses.

21.7 Perspectives and Conclusions

VLPs are appealing as vaccine candidates because their inherent properties (i.e. virus-sized, multimeric antigens, highly organised and repetitive structure, not infectious) make them suitable for the induction of safe and efficient humoral and cellular immune responses. Currently, there is a clear trend towards the establishment of VLPs as a powerful tool for vaccine development. VLP-based vaccines have already been licensed for human diseases (caused by HBV, HPV, HEV, or influenza virus) as well as for use in the veterinary field (PCV2, PPV), and many more vaccine candidates are currently in late stages of evaluation. Moreover, the development of VLPs as platforms for foreign antigen display has further broadened their potential applicability both as prophylactic and therapeutic vaccines.

Currently, main efforts in VLP technology are focused on the development of new VLP-based antigen display platforms for vaccine development. Structural characterisation of VLPs is an important requirement for this aim. Structural studies are mainly performed by advanced cryo-EM and X-ray crystallography, focusing on the comparison of VLPs composed of different numbers and combinations of structural proteins. The use of biochemical methods provides details on individual viral structural components, as well as insights into the structural basis of assembly, packaging, and the interactions of VLPs with host components. Many studies are aimed at characterising the minimal requirements for VLP formation or contemplate prospects of modifying the original proteins without hampering the natural ability of these proteins to assemble into highly organised macromolecules. As a consequence, modified structural proteins appropriate for assembly of multipurpose chimeric VLPs can be eventually designed.

The relative ability of diverse VLP types to induce the different branches of the immune response is influenced by a number of factors that are VLP-specific. Therefore, it appears unlikely that a single VLP platform will meet all the desired requirements. However, the continued parallel development of multiple VLP platforms will ensure that individual vaccines can be tailored appropriately to the type of immune response required in each case.

Acknowledgments The authors are indebted to current and former members of their group (Iván Angulo, Carolina Cubillos, Horacio Almanza, Noelia Moreno, Ignacio Mena, Beatriz Guerra, Yolanda Gómez, Elisa Torres, and Carolina Cabezas), and other collaborators: José R. Castón (CNB-CSIC), Johann Mertens (IMDEA Nanoscience), for their work and skills, that made their work in this field possible. Their work is funded by grants from the Spanish Ministry of Science and Innovation (PID2019-107145RB-I00 and PID2022-140925OB-I00); Comunidad de Madrid co-financed with ECFEDER funds (S2018/BAA-4370 PLATESA 2) and European Union's Horizon 2020 Research and Innovation Programme (VetBioNet, grant agreement No. 731014 and LAGMED, PRIMA S2-11-PCI2019-103732).

References

1. Metwally S, Viljoen G, El Idrissi A (2022) Veterinary vaccines: principles and applications. Wiley and FAO, Chichester
2. Pollard AJ, Bijker EM (2021) A guide to vaccinology: from basic principles to new developments. Nat Rev Immunol 21:83–100
3. Rodrigues CMC, Plotkin SA (2020) Impact of vaccines; health, economic and social perspectives. Front Microbiol 11:1526
4. Ghosh I, Gandhi M (2022) Emerging frontiers in vaccine development: a review of changing paradigm. J Biosci Med 10:123–145
5. Murdin AD, Barreto L, Plotkin S (1996) Inactivated poliovirus vaccine: past and present experience. Vaccine 14:735–746

6. Cottam EM, Wadsworth J, Shaw AE, Rowlands RJ, Goatley L, Maan S, Maan NS, Mertens PP, Ebert K, Li Y, Ryan ED, Juleff N, Ferris NP, Wilesmith JW, Haydon DT, King DP, Paton DJ, Knowles NJ (2008) Transmission pathways of foot-and-mouth disease virus in the United Kingdom in 2007. PLoS Pathog 4:e1000050
7. Koff WC, Rappuoli R, Plotkin SA (2023) Historical advances in structural and molecular biology and how they impacted vaccine development. J Mol Biol 435:168113
8. Mateu MG (2016) Assembly, engineering and applications of virus-based protein nanoparticles. Adv Exp Med Biol 940:83–120
9. Sherman MB, Smith HQ, Smith TJ (2020) The dynamic life of virus capsids. Viruses 12
10. Chung YH, Cai H, Steinmetz NF (2020) Viral nanoparticles for drug delivery, imaging, immunotherapy, and theranostic applications. Adv Drug Deliv Rev 156:214–235
11. Wen AM, Steinmetz NF (2016) Design of virus-based nanomaterials for medicine, biotechnology, and energy. Chem Soc Rev 45:4074–4126
12. Kheirvari M, Liu H, Tumban E (2023) Virus-like particle vaccines and platforms for vaccine development. Viruses 15
13. Crisci E, Barcena J, Montoya M (2013) Virus-like particle-based vaccines for animal viral infections. Inmunologia 32:102–116
14. Roldão A, Silva AC, Mellado MCM, Alves PM, Carrondo MJT (2017) Viruses and virus-like particles in biotechnology: fundamentals and applications. In: Comprehensive biotechnology, pp 633–656
15. Morales-Hernandez S, Ugidos-Damboriena N, Lopez-Sagaseta J (2022) Self-assembling protein nanoparticles in the design of vaccines: 2022 update. Vaccines (Basel) 10
16. Mohsen MO, Bachmann MF (2022) Virus-like particle vaccinology, from bench to bedside. Cell Mol Immunol 19:993–1011
17. Mohsen MO, Augusto G, Bachmann MF (2020) The 3Ds in virus-like particle based-vaccines: "design, delivery and dynamics". Immunol Rev 296:155–168
18. Lee KL, Twyman RM, Fiering S, Steinmetz NF (2016) Virus-based nanoparticles as platform technologies for modern vaccines. Wiley Interdiscip Rev Nanomed Nanobiotechnol 8:554–578
19. Frietze KM, Peabody DS, Chackerian B (2016) Engineering virus-like particles as vaccine platforms. Curr Opin Virol 18:44–49
20. Nguyen B, Tolia NH (2021) Protein-based antigen presentation platforms for nanoparticle vaccines. NPJ Vaccines 6:70
21. Lung P, Yang J, Li Q (2020) Nanoparticle formulated vaccines: opportunities and challenges. Nanoscale 12:5746–5763
22. Tornesello AL, Tagliamonte M, Buonaguro FM, Tornesello ML, Buonaguro L (2022) Virus-like particles as preventive and therapeutic cancer vaccines. Vaccines (Basel) 10
23. Mohsen MO, Gomes AC, Vogel M, Bachmann MF (2018) Interaction of viral capsid-derived virus-like particles (VLPs) with the innate immune system. Vaccines (Basel) 6
24. Mohsen MO, Zha L, Cabral-Miranda G, Bachmann MF (2017) Major findings and recent advances in virus-like particle (VLP)-based vaccines. Semin Immunol 34:123–132
25. McFall-Boegeman H, Huang X (2022) Mechanisms of cellular and humoral immunity through the lens of VLP-based vaccines. Expert Rev Vaccines 21:453–469
26. Gomes AC, Mohsen M, Bachmann MF (2017) Harnessing nanoparticles for immunomodulation and vaccines. Vaccines (Basel) 5
27. Bachmann MF, Jennings GT (2010) Vaccine delivery: a matter of size, geometry, kinetics and molecular patterns. Nat Rev Immunol 10:787–796
28. Shah RR, O'Hagan DT, Amiji MM, Brito LA (2014) The impact of size on particulate vaccine adjuvants. Nanomedicine (Lond) 9:2671–2681
29. Zabel F, Kundig TM, Bachmann MF (2013) Virus-induced humoral immunity: on how B cell responses are initiated. Curr Opin Virol 3:357–362
30. Vogel M, Bachmann MF (2020) Immunogenicity and immunodominance in antibody responses. Curr Top Microbiol Immunol 428:89–102

31. Zepeda-Cervantes J, Ramirez-Jarquin JO, Vaca L (2020) Interaction between virus-like particles (VLPs) and pattern recognition receptors (PRRs) from dendritic cells (DCs): toward better engineering of VLPs. Front Immunol 11:1100
32. Lenz P, Day PM, Pang YY, Frye SA, Jensen PN, Lowy DR, Schiller JT (2001) Papillomavirus-like particles induce acute activation of dendritic cells. J Immunol 166:5346–5355
33. Gedvilaite A, Dorn DC, Sasnauskas K, Pecher G, Bulavaite A, Lawatscheck R, Staniulis J, Dalianis T, Ramqvist T, Schonrich G, Raftery MJ, Ulrich R (2006) Virus-like particles derived from major capsid protein VP1 of different polyomaviruses differ in their ability to induce maturation in human dendritic cells. Virology 354:252–260
34. Pumpens P, Pushko P (2022) Virus-like particles: a comprehensive guide, 1st edn. CRC Press, Boca Raton
35. Pushko P, Pumpens P, Grens E (2013) Development of virus-like particle technology from small highly symmetric to large complex virus-like particle structures. Intervirology 56:141–165
36. Zeltins A (2013) Construction and characterization of virus-like particles: a review. Mol Biotechnol 53:92–107
37. Vacher G, Kaeser MD, Moser C, Gurny R, Borchard G (2013) Recent advances in mucosal immunization using virus-like particles. Mol Pharm 10:1596–1609
38. Chackerian B, Peabody DS (2020) Factors that govern the induction of long-lived antibody responses. Viruses 12
39. Brun A, Barcena J, Blanco E, Borrego B, Dory D, Escribano JM, Le Gall-Recule G, Ortego J, Dixon LK (2011) Current strategies for subunit and genetic viral veterinary vaccine development. Virus Res 157:1–12
40. Tregoning JS (2020) First human efficacy study of a plant-derived influenza vaccine. Lancet 396:1464–1465
41. Rosalik K, Tarney C, Han J (2021) Human papilloma virus vaccination. Viruses 13
42. Kirnbauer R, Taub J, Greenstone H, Roden R, Durst M, Gissmann L, Lowy DR, Schiller JT (1993) Efficient self-assembly of human papillomavirus type 16 L1 and L1-L2 into virus-like particles. J Virol 67:6929–6936
43. Gwynn A, Mbewana S, Lubisi BA, Tshabalala HM, Rybicki EP, Meyers AE (2023) Chimaeric plant-produced bluetongue virus particles as potential vaccine candidates. Arch Virol 168:179
44. Ruiz V, Mozgovoj MV, Dus Santos MJ, Wigdorovitz A (2015) Plant-produced viral bovine vaccines: what happened during the last 10 years? Plant Biotechnol J 13:1071–1077
45. Pumpens P, Ulrich RG, Sasnauskas K, Kazaks A, Ose V, Grens E (2008) Construction of novel vaccines on the basis of virus-like particles: Hepatitis B virus proteins as vaccine carriers. In: Khudyakov YE (ed) Medicinal protein engineering. CRC Press, pp 205–247
46. Moser C, Muller M, Kaeser MD, Weydemann U, Amacker M (2013) Influenza virosomes as vaccine adjuvant and carrier system. Expert Rev Vaccines 12:779–791
47. de Pinho Favaro MT, Atienza-Garriga J, Martinez-Torro C, Parlade E, Vazquez E, Corchero JL, Ferrer-Miralles N, Villaverde A (2022) Recombinant vaccines in 2022: a perspective from the cell factory. Microb Cell Factories 21:203
48. Possee RD, Chambers AC, Graves LP, Aksular M, King LA (2020) Recent developments in the use of baculovirus expression vectors. Curr Issues Mol Biol 34:215–230
49. Felberbaum RS (2015) The baculovirus expression vector system: a commercial manufacturing platform for viral vaccines and gene therapy vectors. Biotechnol J 10:702–714
50. Fernandes F, Teixeira AP, Carinhas N, Carrondo MJ, Alves PM (2013) Insect cells as a production platform of complex virus-like particles. Expert Rev Vaccines 12:225–236
51. Puente-Massaguer E, Lecina M, Gòdia F (2018) Nanoscale characterization coupled to multi-parametric optimization of Hi5 cell transient gene expression. Appl Microbiol Biotechnol 102: 10495–10510
52. Dias MM, Vidigal J, Sequeira DP, Alves PM, Teixeira AP, Roldao A (2021) Insect High Five™ cell line development using site-specific flipase recombination technology. G3-Genes Genomes Genetics 11:10

53. Moleirinho MG, Silva RJS, Alves PM, Carrondo MJT, Peixoto C (2020) Current challenges in biotherapeutic particles manufacturing. Expert Opin Biol Ther 20:451–465
54. Gupta R, Arora K, Roy SS, Joseph A, Rastogi R, Arora NM, Kundu PK (2023) Platforms, advances, and technical challenges in virus-like particles-based vaccines. Front Immunol 14: 1123805
55. Charlton Hume HK, Vidigal J, Carrondo MJT, Middelberg APJ, Roldao A, Lua LHL (2019) Synthetic biology for bioengineering virus-like particle vaccines. Biotechnol Bioeng 116:919–935
56. Lua LH, Connors NK, Sainsbury F, Chuan YP, Wibowo N, Middelberg AP (2014) Bioengineering virus-like particles as vaccines. Biotechnol Bioeng 111:425–440
57. Cid R, Bolivar J (2021) Platforms for production of protein-based vaccines: from classical to next-generation strategies. Biomol Ther 11
58. Liu W, Chen YH (2005) High epitope density in a single protein molecule significantly enhances antigenicity as well as immunogenicity: a novel strategy for modern vaccine development and a preliminary investigation about B cell discrimination of monomeric proteins. Eur J Immunol 35:505–514
59. Plummer EM, Manchester M (2011) Viral nanoparticles and virus-like particles: platforms for contemporary vaccine design. Wiley Interdiscip Rev Nanomed Nanobiotechnol 3:174–196
60. Lee KK, Johnson JE (2003) Complementary approaches to structure determination of icosahedral viruses. Curr Opin Struct Biol 13:558–569
61. Arnon R, Van Regenmortel MH (1992) Structural basis of antigenic specificity and design of new vaccines. FASEB J 6:3265–3274
62. Moreno N, Mena I, Angulo I, Gomez Y, Crisci E, Montoya M, Caston JR, Blanco E, Barcena J (2016) Rabbit hemorrhagic disease virus capsid, a versatile platform for foreign B-cell epitope display inducing protective humoral immune responses. Sci Rep 6:31844
63. Rueda P, Hurtado A, del Barrio M, Martinez-Torrecuadrada JL, Kamstrup S, Leclerc C, Casal JI (1999) Minor displacements in the insertion site provoke major differences in the induction of antibody responses by chimeric parvovirus-like particles. Virology 263:89–99
64. Billaud JN, Peterson D, Barr M, Chen A, Sallberg M, Garduno F, Goldstein P, McDowell W, Hughes J, Jones J, Milich D (2005) Combinatorial approach to hepadnavirus-like particle vaccine design. J Virol 79:13656–13666
65. Zamora-Ceballos M, Moreno N, Gil-Cantero D, Caston JR, Blanco E, Barcena J (2022) Immunogenicity of multi-target chimeric RHDV virus-like particles delivering foreign B-cell epitopes. Vaccines (Basel) 10
66. Roose K, De Baets S, Schepens B, Saelens X (2013) Hepatitis B core-based virus-like particles to present heterologous epitopes. Expert Rev Vaccines 12:183–198
67. Kolb P, Nguyen TTA, Walker A, Nassal M (2015) SplitCore: advanced nanoparticulate molecular presentation platform based on the hepatitis B virus capsid. In: Khudyakov YE, Pumpens P (eds) Viral nanotechnology CRC Press, Boca Raton, pp. 187–208
68. Greenstone HL, Nieland JD, de Visser KE, De Bruijn ML, Kirnbauer R, Roden RB, Lowy DR, Kast WM, Schiller JT (1998) Chimeric papillomavirus virus-like particles elicit antitumor immunity against the E7 oncoprotein in an HPV16 tumor model. Proc Natl Acad Sci USA 95:1800–1805
69. Verardi R, Lindesmith LC, Tsybovsky Y, Gorman J, Chuang GY, Edwards CE, Brewer-Jensen PD, Mallory ML, Ou L, Schon A, Shi W, Tully ES, Georgiou G, Baric RS, Kwong PD (2020) Disulfide stabilization of human norovirus GI.1 virus-like particles focuses immune response toward blockade epitopes. NPJ Vaccines 5:110
70. Brune KD, Howarth M (2018) New routes and opportunities for modular construction of particulate vaccines: stick, click, and glue. Front Immunol 9:1432
71. Mateu MG (2011) Virus engineering: functionalization and stabilization. Protein Eng Des Sel 24:53–63
72. Aves KL, Goksoyr L, Sander AF (2020) Advantages and prospects of tag/catcher mediated antigen display on capsid-like particle-based vaccines. Viruses 12

73. Nikitin N, Vasiliev Y, Kovalenko A, Ryabchevskaya E, Kondakova O, Evtushenko E, Karpova O (2023) Plant viruses as adjuvants for next-generation vaccines and immunotherapy. Vaccines (Basel) 11
74. Nkanga CI, Steinmetz NF (2021) The pharmacology of plant virus nanoparticles. Virology 556: 39–61
75. Steinmetz NF, Lin T, Lomonossoff GP, Johnson JE (2009) Structure-based engineering of an icosahedral virus for nanomedicine and nanotechnology. Curr Top Microbiol Immunol 327:23–58
76. Balke I, Zeltins A (2020) Recent advances in the use of plant virus-like particles as vaccines. Viruses 12
77. Venter PA, Schneemann A (2008) Recent insights into the biology and biomedical applications of flock house virus. Cell Mol Life Sci 65:2675–2687
78. Jariyapong P (2015) Nodavirus-based biological container for targeted delivery system. Artif Cells Nanomed Biotechnol 43:355–360

Further Reading

Especially recommended for further reading are references [8, 10, 14, 16–20, 25, 35, 55, 70] *listed above*

Chapter 22
Engineering and Bio/Nanotechnological Applications of Virus Particles

Mauricio G. Mateu and Alejandro Valbuena

Abstract Virus particles (VPs) are naturally evolved nanomachines. Their outstanding molecular structures, physical and chemical properties, and biological activities make them potentially useful for many biomedical or technological applications. Natural VPs such as virions or capsids must, however, be modified by genetic and/or chemical engineering in order to become adequate for many specific uses. We present first a general overview of the methods used for obtaining virions and viral capsids, and of genetic and chemical engineering approaches to suitably modify VPs. In the second part of the chapter, we present an updated overview on current or developing applications of engineered VPs as tools, materials, reagents, or nanodevices in biomedicine, biotechnology, or nanotechnology.

Keywords Virus · Virion · Capsid · Virus particle · Virus-like particle · Nanoparticle · Virus engineering · Chemical functionalization · Genetic modification · Site-directed mutagenesis · Directed evolution · Virotherapy · Gene therapy · Targeted drug delivery · Diagnostic imaging · Nanobiosensors · Confined reactions · Inorganic nanoparticles · Nanoscale materials

Abbreviations

AAV	Adeno-associated viruses
AdV	Adenovirus
AFM	Atomic force microscopy
BMV	Brome mosaic virus
CA	Capsid protein of HIV-1
CCMV	Cowpea chlorotic mottle virus
CPMV	Cowpea mosaic virus

M. G. Mateu (✉) · A. Valbuena
Centro de Biología Molecular "Severo Ochoa" (CSIC-UAM), and Department of Molecular Biology, Universidad Autónoma de Madrid, Madrid, Spain
e-mail: mgarcia@cbm.csic.es

M. G. Mateu (ed.), *Structure and Physics of Viruses*, Subcellular Biochemistry 105,
https://doi.org/10.1007/978-3-031-65187-8_22

cryo-EM	Cryo-electron microscopy
CAR	Coxsackie and adenovirus receptor
CT	Computer tomography
CTL	Cytotoxic T-lymphocyte
1D, 2D, 3D	Mono-, bi-, tri-dimensional
DOX	Doxorubicin
EMDB	Electron Microscopy Data Bank
FDA	Food and Drug Administration
FI	Fluorescence imaging
FMDV	Foot-and-mouth disease virus
FRET	Förster resonance energy transfer
5-FU	5-fluorouracil
GFP	Green fluorescent protein
HBV	Hepatitis B virus
HCRSV	*Hibiscus* chlorotic ringspot virus
HIV-1	Human immunodeficiency virus type 1
HPV	Human papillomavirus
mAb	Monoclonal antibody
MHT	Magnetic hyperthermia therapy
MRI	Magnetic resonance imaging
MVM	Minute virus of mice
NMR	Nuclear magnetic resonance
NP	Nanoparticle
NTA	Nitriloacetic acid
PCR	Polymerase chain reaction
PDB	Protein data bank
PDT	Photodynamic therapy
PEG	Polyethyleneglycol
PET	Positron emission tomography
PhMV	*Physalis* mottle virus
PTT	Photothermal therapy
QD	Quantum dot
RCNMV	Red clover necrotic mottle virus
PVX	Potato virus X
SBMV	Southern bean mosaic virus
SARS-CoV-2	Severe acute respiratory syndrome coronavirus 2
scFv	Single-chain antibody fragment
SV40	Simian virus 40
TMV	Tobacco mosaic virus
VLP	Virus-like particle
VP	Virus particle

22.1 Introduction

Viruses and their protein capsids have acquired, through evolution of their complex molecular structures, remarkable physical and chemical properties and sophisticated biological activities. As most chapters of this book attest, knowledge on the structure, properties, and function of virions and virus capsids is growing at an increasingly rapid rate. This accumulated knowledge has been guiding during the last few decades the development of many novel applications of virions, viral capsids, and virus-based nanoparticles (NPs).

In this chapter, we will use the term *virus particle* (VP) in its broadest sense to refer to any supramolecular assembly made, in full or in part, from viral structural proteins, either natural or modified by molecular engineering, and either containing or lacking the viral nucleic acid.

This chapter attempts to provide a succinct, updated overview of different approaches to engineer VPs, and of current or developing applications of engineered VPs. It is addressed mainly to students and non-specialists, but it may also provide researchers working on a particular subject within this field with brief, updated summaries of related subjects. Specific sections include: (1) an outline of procedures available to produce assembled VPs (Sect. 22.2); (2) an outline of the genetic and chemical strategies used for engineering VPs (Sect. 22.3); (3) a brief, categorized, non-exhaustive review of current or developing applications of natural or engineered VPs (Sects. 22.4 and 22.5).

Two extremely important applications of VPs in biomedicine involve their use as targets for the development of novel antiviral drugs, or as epitope carriers for the development of vaccines. Some aspects concerning these two long-recognized and increasingly exploited uses of VPs have, respectively, been covered in Chaps. 20 and 21. Thus, in the present chapter, these applications are only mentioned for completeness, but are not reviewed again. Here we summarize other current or potential biomedical applications of VPs including phage therapy, virotherapy, gene therapy, targeted drug delivery, and diagnostics (Sect. 22.4). The advent of nanoscience and nanotechnology has opened up a vast number of additional possibilities for the use of engineered VPs in biotechnology or nanotechnology; some of them are also summarized in this chapter (Sect. 22.5).

The width of the contemplated subject, the rapidly increasing number of related studies, the intended main audience, and space limitations have generally precluded mention of original studies. We have purposely limited most citations to reviews that contain references to many relevant original studies. The reader may wish to consult some of those reviews for a deeper insight into specific subjects. We apologize to the researchers whose work, however important, could not be mentioned or directly referenced here.

Virions and viral capsids range in size from ~15 nm up to ~1 μm in diameter for quasi-spherical particles, or from ~5 nm in diameter and up to ~2 μm in length for tubular particles (Chaps. 2 and 12). Thus, they are comparable in dimensions to many other biological, non-biological, or hybrid NPs that are being designed, constructed, studied, and/or developed for biomedical or technological uses [1, 2]. VPs in general show a number of strengths that make them excellent choices for many applications. Some of their desirable features are shared with other protein-based NPs of non-viral origin: (1) a defined chemical composition, size, architecture, and molecular structure; (2) suitability for modification of their structure, properties, and/or functionalities by genetic engineering; (3) suitability for highly diverse functionalization by chemical engineering; (4) suitability for easy and cost-efficient large-scale production; (5) biocompatibility; (6) biodegradability; and (7) low toxicity.

VPs have also several more specific strengths. Some of these are shared with a few other protein-based particles, while others are rather unique. A non-exhaustive list includes: (1) virions and virus capsids can efficiently self-assemble in infected host cells, and conditions have been found for self-assembling many VPs in cells and/or in vitro; (2) self-assembly of virions and viral capsids can be extremely faithful and efficient; (3) virus capsids are made of a number of repetitive protein subunits that form a regular, symmetric, multivalent supramolecular assembly; (4) particular assembly conditions can be found for altering the size and architecture of some VPs; (5) virus capsids (and a few cellular protein assemblies, such as ferritin-like proteins or vaults) contain a relatively large cavity that makes them excellent nanocontainers; (6) virions and virus capsids have been exquisitely engineered by biological evolution; they have been endowed with outstanding features, including the ability to perform structural rearrangements and mechano-chemical actions that can be controlled and adapted for specific uses (e.g., the controlled uptake or delivery of cargo molecules); (7) as a result of different biological niches and evolutionary pressures, virions and capsids from different virus species show a high diversity in size, shape, molecular composition, structural organization, type of nucleic acid genome, mechanisms of action, and/or cell types they can infect; thus, they provide a diverse toolkit of NP platforms (Fig. 22.1); and (8) last but not least, the knowledge being acquired on the structure, properties, and function of viruses and viral structural components is inspiring the design and development of synthetic or hybrid NPs endowed with novel architectures, properties, and activities [1, 2].

VPs also present potential weaknesses for some applications. Several of those weaknesses are shared with non-viral NPs, while others are rather specific. Important issues, depending on the VP considered, may include: (1) insufficient physical and/or chemical stability (including limited thermostability or susceptibility to mechanical disruption by point forces or material fatigue); (2) high dynamism at equilibrium, and susceptibility to conformational rearrangements in response to specific external stimuli, which may be useful for some applications but undesirable for others; (3) high immunogenicity; (4) susceptibility to rapid clearance from the organism; (5) safety concerns. These and other issues are being addressed, in most cases, by engineering adequately modified VPs.

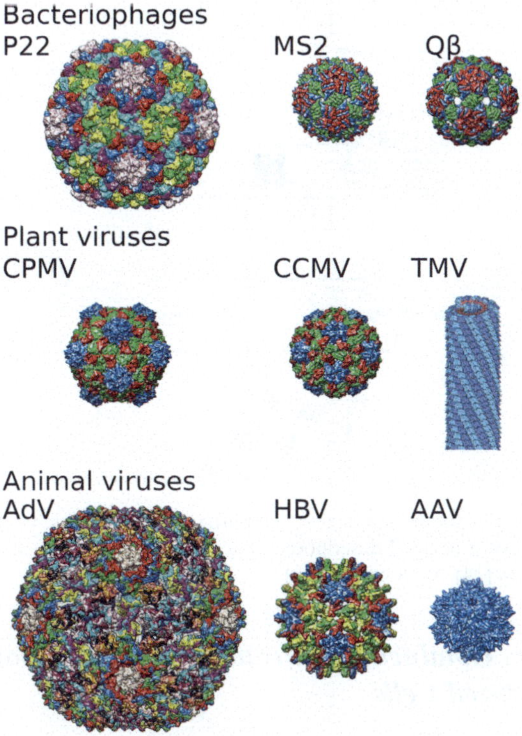

Fig. 22.1 Structural models of some viruses used for the development of engineered VPs for biomedical or technological applications. Viruses are represented at approximately the same scale. Adeno-associated virus (AAV) is 25 nm in diameter. For Tobacco mosaic virus (TMV) only a part of the 300 nm-long rod-like particle is represented. Top: bacteriophages Qβ (PDB:1QBE), MS2 (PDB:2MS2), and P22 (PDB:5UU5). Middle: plant viruses Cowpea mosaic virus (CPMV) (PDB:2BFU), Cowpea chlorotic mottle virus (CCMV) (PDB:1CWP), and TMV (PDB:2TMV; composed from the image of the PDB molecule of the month https://doi.org/10.2210/rcsb_pdb/mom_2009_1). Bottom: mammalian viruses AAV8 (PDB:6PWA), Adenovirus (AdV) (PDB:6CGV), and Hepatitis B virus (HBV) (PDB:1QGT). Rendered virus surfaces at 3 Å resolution have been obtained from VIPERdb (https://viperdb.org) [3]

22.2 Production of Virions, Viral Capsids, and Virus-Like Particles (VLPs)

Infectious virions, viral capsids devoid of nucleic acid, and VLPs derived from viral capsids can be produced and purified using different procedures [4–14] (Fig. 22.2). Some of these procedures are briefly outlined next.

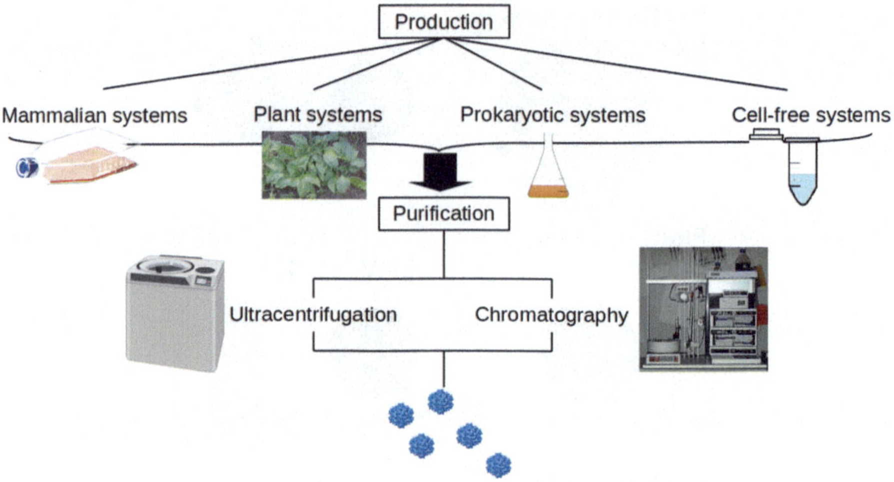

Fig. 22.2 Some strategies for the production and purification of virions, viral capsids, and VLPs. See text (Sect. 22.2) for a detailed description. Diagram inspired in Fig. 1 from [6]. (centrifuge image reproduced from [14]. With permission)

22.2.1 Virions Obtained by Infection or Transfection of Cultured Cells

Suitable cells are either infected with virions or transfected with an infectious viral nucleic acid. Transfection by an infectious viral genomic DNA clone or (in the case of RNA viruses) genomic RNA transcribed from a DNA clone, allows the production of virions, either natural or previously modified in the laboratory by reverse genetics [15, 16]. The progeny virions produced in the infected or transfected cells are usually purified.

For some applications that may require very large amounts of virions, bacteriophages or plant viruses have definite advantages over animal or human viruses regarding economy and simplicity of the production process [9, 12, 17]. Many milligrams of phages or plant viruses can typically be obtained per gram of infected cells. Moreover, the process can be readily escalated by growing huge amounts of host bacteria in large fermenters, or by growing extensive plant cultures. In addition, phages and plant viruses typically require less stringent biosafety measures, as they do not infect humans or animals. However, potential environmental issues should still be considered.

Virions from animal viruses are typically obtained in lower amounts and at higher cost compared to phages or plant viruses. Facilities for the production of animal viruses are relatively expensive, and upscaling production may be difficult and costly. The requirements and expense for more complex biosafety, containment, and validation procedures should also be taken into account. However, for some applications animal viruses may be favored, or even strictly required. For example, adequately modified human or animal viruses are generally used for oncolytic virotherapy or gene therapy (Sects. 22.4.4 and 22.4.5).

22.2.2 *Nucleic Acid-Free VPs Obtained from Virions*

In some cases, viral nucleic acid-free (empty) capsids or VLPs can be obtained as a byproduct of an inefficient infection process. They can be readily separated from the nucleic acid-filled progeny virions using, for example, centrifugation in density (CsCl) gradients. Empty VPs from some virus species may be readily obtained from pure virion preparations by inducing the controlled release of the viral genome through capsid openings (i.e., without particle disassembly) by heating, acidification, or other means. Alternatively, virions may be disassembled into their individual components, and the dissociated capsid protein subunits can be purified and used as building blocks for the reassembly of genome free VPs in vitro [11] (Sect. 22.2.4).

For many technological developments, nucleic acid-free VPs from plant viruses or phages have been favored over nucleic acid-free VPs from animal viruses [9, 12, 17], largely because very large amounts of the former may be obtained from massive virion preparations.

22.2.3 *Recombinant Viral Capsid Proteins and VPs Produced in Transfected Cells*

Capsids or VLPs from animal or human viruses may be desirable or required for certain applications. The structure, properties, and function of a large number of animal or human virions or capsids have been extensively studied. These viruses have been endowed by natural selection with unique functionalities that can be exploited, especially for biomedical uses. Issues regarding animal virus-derived particle production and use, such as adequate facilities, biosafety, and validation procedures, can be satisfactorily dealt with, although they may come at a cost.

Most importantly, nucleic acid-free VPs from animal or human viruses may be obtained in very large amounts that can be compared to those obtained for VPs derived from plant viruses or phages. In general, the procedure involves the expression of the capsid protein(s) and the self-assembly of the VPs in suitable cells [4–14]. The capsid protein gene(s) is (are) cloned in a suitable expression vector, and the recombinant DNA obtained is used to transfect cultured bacteria (e.g., *Escherichia coli*), yeast, plant, insect or mammalian cells, or even whole plants. Depending on the virus species, cell type, and experimental conditions, the expressed capsid protein subunits may associate to form VPs in situ. In some cases, auxiliary proteins are required for capsid assembly (e.g., scaffolding proteins to promote and guide the process, or proteases for the processing of assembly-incompetent precursor capsid proteins).

One of the best systems for large-scale production of viral capsids and VLPs is based on the use of recombinant baculovirus vectors for expression of heterologous proteins in insect cells [5, 6] (see Chap. 21). In many cases, substantial efficiency and high yields of rather pure VPs can be obtained. The baculovirus system also allows the introduction, in VPs derived from animal or human viruses, of many post-translational modifications similar to those that occur in virus-infected host cells.

Production in heterologous expression systems of viral genome-free VPs avoids many limitations of the procedures required for the production of virions. There is no risk of escape of infectious particles, no special containment facilities are required, and the process can be readily scaled up. In addition, expression of recombinant capsid proteins facilitates their straightforward genetic modification using site-directed mutagenesis or directed evolution (Sect. 22.3.1). The use of efficient protein expression systems may lead to an increased use of VPs from animal or human viruses for the development of VLP-based vaccines (Chap. 21) and many other applications (Sects. 22.4 and 22.5).

22.2.4 Virus Capsids and VLPs Assembled In Vitro

Conditions have been found for the self-assembly in the test tube of virus capsids and virions from their protein and nucleic acid components, previously produced in transfected cells and purified [18]. Physical and/or chemical conditions required for VP assembly in vitro (including protein concentration, temperature, pH, ionic strength, presence of macromolecular crowding agents, etc.) are in many instances nonphysiological. However, the assembled particles obtained are frequently identical or very similar to authentic VPs produced by infection of host cells. Assembly within infected cells of some virus capsids, including those of cowpea chlorotic mottle virus (CCMV), brome mosaic virus (BMV), hepatitis B virus (HBV), and human papillomavirus (HPV), requires the presence of the viral nucleic acid (Chap. 12). However the viral nucleic acid is not required for self-assembly of those capsids under certain conditions in vitro.

Many heterologous biomolecules, organic compounds, or inorganic NPs can promote the assembly of some VPs or influence the size and architecture of the assembled particles [8, 19–23]. In this way, the toolkit of available VPs for different applications can be further expanded. For example, the CCMV capsid proteins can self-assemble in vitro into $T = 3$ or $T = 1$ icosahedral particles of different sizes, rods of different widths or lengths, or multilamellar forms, depending on the conditions. Truncation of the C-terminal segment of the capsid proteins of HBV or southern bean mosaic virus (SBMV) favored the assembly of icosahedral $T = 4$ or $T = 1$ (respectively), instead of $T = 3$ capsids. Tubular structures of the wild-type human immunodeficiency virus type 1 (HIV-1) capsid (CA) protein are typically favored in vitro over truncated cones similar to authentic mature capsids; in turn, some CA mutations favored the assembly of thinner tubes or cone-like structures. Protein- or nucleic acid-assisted assembly of some VPs (e.g., tobacco mosaic virus (TMV), phage P22) has also been achieved in vitro by including the appropriate components in the mixture.

The possibility of assembling VPs in vitro has brought about great benefits for both basic research and many applications. From an applied perspective, the strengths of in vitro assembly include biosafety, controlled production, simpler purification procedures, and lower cost. Most importantly, different organic or inorganic cargos can be enclosed in natural or adequately engineered VPs by simply adding them to the assembly mixture. Another advantage is that some genetic modifications (Sect. 22.3.1) or chemical modifications (Sect. 22.3.2) that may be difficult to introduce directly in a labile VP can be introduced in the free capsid proteins before the VPs are assembled from them in vitro.

In addition to recombinant systems for in-cell expression of viral capsid proteins, cell-free translation systems [24, 25] are being used to obtain capsid proteins to be assembled later under adequate conditions in vitro. In some cases, VPs are directly formed in the cell-free mixture.

Despite the advantages of capsid self-assembly in vitro over other procedures to obtain VPs, finding the right conditions for faithful and efficient assembly of a particular VP in vitro is generally not straightforward, and the yields may be rather limited. These and other issues are being addressed for the scaling up of in vitro VP assembly for commercial application [6].

22.2.5 Purification of VPs

Purification of virions typically include ultracentrifugation through a density cushion (e.g., a sucrose solution) followed by further ultracentrifugation in a velocity gradient (e.g., a sucrose concentration gradient) and/or a density gradient (e.g., a CsCl gradient). Due to the relatively large size of assembled virions compared to many cellular contaminants, those procedures tend to be very effective. Other purification steps may include size exclusion, ion exchange, and/or affinity chromatography [4, 6]. Affinity chromatography of viruses using virus-specific protein receptors or antibodies may be particularly effective, although elution of intact virions can be difficult in some cases because of their conformational instability. Procedures used for the purification of capsids or VLPs produced in cells or in vitro are similar to those used for purifying virions.

The *bottom-up* approaches described above for the production and purification of virions, virus capsids, and VLPs can result in the production of very large numbers of monodisperse, structurally identical nanoparticles with remarkable physical properties, multivalent chemical functionalities, and complex biological activities.

22.3 Engineering VPs

The structure, properties, and function of a virion and its protein capsid have evolved in nature for efficient propagation of the virus in host cells and organisms. For some intended applications natural, unmodified VPs may be adequate (e.g., when used as targets for developing new antiviral drugs (Chap. 20) or as homologous viral vaccines (Chap. 21)). However, even in those cases there may be a need for improving some particular VP feature(s). For example, infectious virions and empty capsids of foot-and-mouth disease virus (FMDV) have been genetically engineered for increased thermostability, in order to develop FMD vaccines that are less dependent on a strict cold chain [26, 27]. The spike (S) fusion protein of the severe acute respiratory syndrome coronavirus 2 (SARS-CoV-2) in some current COVID-19 vaccines has been genetically engineered to elicit higher amounts of virus-neutralizing antibodies by impairing a conformational transition (Chap. 16). For most intended applications, natural VPs and structural viral proteins are simply inadequate, unless they are suitably modified. The knowledge acquired on the structure, physical and chemical properties, and biological activities of VPs, summarized in many chapters of this book, is guiding their adequate modification for many different uses.

Two complementary general approaches are being extensively used for the modification of the structure, properties, or function of proteins and protein (or nucleoprotein) assemblies, including VPs: genetic engineering and chemical engineering [11, 28–30] (Fig. 22.3). In many instances, the two approaches are combined for achieving the intended modification. These approaches will be outlined here, with emphasis on their use for the specific engineering of VPs, rather than individual proteins.

22.3.1 *Genetic Engineering of VPs*

The structure, properties, and/or function of virions and virus capsids can be dramatically modified by genetic engineering using recombinant DNA techniques. A general approach to engineer infectious virions involves the site-directed or combinatorial introduction of mutations in an infectious cDNA clone that contains the full-length viral genome. In most cases, one or several amino acid residues of the capsid protein(s) are replaced by mutation of the corresponding codons in the capsid gene(s). The same procedure can be used to introduce mutations in other viral genes, or in regulatory regions of the viral genome that may result in the desired virion phenotype. The mutagenized infectious DNA clone, or a viral RNA transcribed from it, is used to transfect infection-competent cells [15, 16]. Progeny infectious virions carrying the genetic modifications are assembled in the transfected cells. They are further multiplied by infecting fresh cultured cells in serial passages, in order to

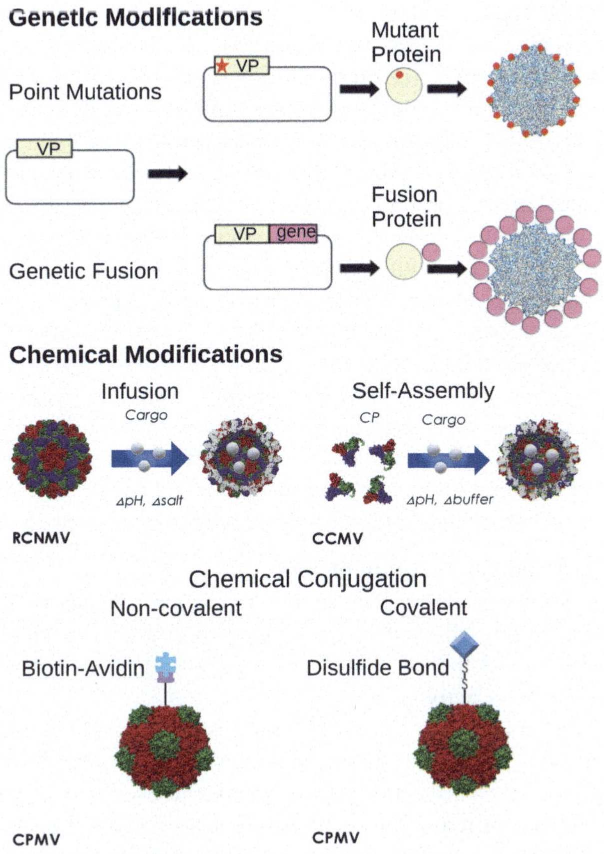

Fig. 22.3 Some genetic or chemical approaches for engineering VPs. See text (Sect. 22.3) for detailed descriptions. (Adapted from [31]. With permission)

obtain high enough amounts of modified virions. The virions obtained are then purified.

Of course, this approach cannot be used if the introduced mutations are lethal for the virus. In addition, non-lethal but still deleterious mutations may compromise the yield or usefulness of the modified virion. Importantly, many introduced mutations will reduce the biological fitness of the virus and may be genetically unstable. The mutant virus will be outcompeted by the wild-type or another, more fit variant in the population, or secondary mutations will be introduced in the viral genome during its replication that will compensate for the deleterious effect of the primary mutation. These considerations may apply to any virus species, but they are most important for RNA viruses, whose error-prone multiplication, combined with other factors, results in genetically highly heterogeneous virus populations termed quasispecies [32].

Genetic engineering is an even more powerful method for the modification of the structure, properties, and/or functions of non-infectious VPs, without the severe limitations that may apply to the genetic modification of infectious virions. Many mutations, even those that would prevent or impair virus multiplication and propagation, can be introduced by site-directed mutagenesis and/or combinatorial methods in the capsid protein gene(s) (previously cloned in a suitable vector). The variant proteins encoded by the mutant capsid genes can be produced either in transfected cells or in a cell-free protein synthesis system in vitro. The VPs can be assembled in the producing cells or protein synthesis mixture, or the capsid protein(s) can be purified in free form and used to assemble the VPs in vitro (Sect. 22.2).

Genetic engineering generally leads to the permanent modification of the VPs, especially in non-replicating empty capsids and VLPs, in which genetic reversion or compensating mutations cannot occur. It also results in particles that can be identical to each other, within and between production batches. This feature can greatly simplify production and quality control. Moreover, it will preclude differences between batches regarding stability, specific activity, or side effects. In addition, it will avoid the eventual loss of the engineered property or activity. The accumulated experience over many years has shown that even modest genetic changes (one or a few amino acids per capsid subunit) can sometimes result in profound changes in structure, dynamics, physical or chemical properties, and/or functions of a VP. Protein engineering by genetic means, thus, constitutes an excellent approach for tailoring VPs and other viral components for many applications.

Rational Engineering of VPs

A fully rational protein engineering approach using site-directed mutagenesis can ideally be followed if enough knowledge is available on the detailed (atomic) structure of the VP and on the relationship between its structure and the targeted property or activity.

Atomic or quasi-atomic structures have been determined for a large, and rapidly growing, number of virions and capsids using either X-ray crystallography (Chap. 4) or, more recently, cryogenic electron microscopy (Cryo-EM) (Chap. 3). The detailed structure of many VP components and other viral proteins has been determined by either X-ray crystallography or nuclear magnetic resonance (NMR) spectroscopy (Chap. 5). These techniques have sometimes been combined with each other, and with other biophysical techniques, for an in-depth understanding of the structure and dynamics of VPs, either free or bound to different ligands (Chaps. 3–9). In very recent times, the "resolution revolution" in the cryo-EM field has led to the choice of this technique for the determination of most VP structures. In turn, the technical advances in X-ray crystallography have allowed the rapid determination of the structure of many smaller viral components in unliganded and/or ligand-bound forms (see Chaps. 3 and 4).

Atomic coordinates for the construction of structural models of virions, capsids, VLPs, and viral components, either free or in complex with different ligands, obtained by X-ray crystallography, cryo-EM, or NMR spectroscopy, are available on the Protein Data Bank (PDB; http://www.rcsb.org). Atomic coordinates and other structural information obtained by EM are also available on the Electron Microscopy Database (EMDB; http://www.ebi.ac.uk/pdbe/emdb). The VIPERdb database (http://viperdb.scripps.edu) contains abundant structural information on many VPs.

Knowing the atomic structure of the VP one wishes to rationally engineer (or, in some instances, that of a close homolog) is a critical requirement for attempting a rational virus engineering approach. In addition, this approach requires the gathering of as much information as possible on the structural basis and mechanisms that determine the property or activity to be modified, and on other aspects of the relationships between structure, properties, and function of the targeted VP. The gathering of all this information is even more critical in the case of nonenveloped viruses. The relatively simple structures of many small, naked viral capsids are packed with multiple, complex biological functions (Chap. 11). Structure–properties–function studies by extensive mutational analysis of the capsids of some viruses (e.g., FMDV, the minute virus of mice (MVM), HIV-1), as well other less systematic mutational studies of many other viral capsids, have provided abundant evidence of a complex structural overlapping of properties and functions, even at the single amino acid residue level [18]. Thus, single substitutions that attempt to modify some targeted feature of a VP may have unwanted consequences if they affect also some other viral feature required for the intended application (e.g., self-assembly or conformational stability). The above issues must be addressed on a case-by case basis, ideally through an exhaustive information-driven approach.

Introducing Point Mutations in VPs by Site-Directed Mutagenesis. Experience has shown that, in quite a few cases, substitution of even a single amino acid residue per protein subunit in a VP may result in a large phenotypic change that can significantly improve the suitability of that particle for the intended application. For example, a dramatic thermostabilization of the FMDV virion without loss of infectivity (required for vaccine production) was achieved through the rationally designed substitution of a single amino acid in a capsid protein [26]. Designed amino acid substitutions are also frequently introduced in virus capsids as a prerequisite for their adequate functionalization by chemical engineering (Sect. 22.3.2).

Site-directed mutagenesis for replacing one or a few amino acid residues in a virus capsid protein is typically done using mutagenic oligonucleotides in a polymerase chain reaction (PCR) to amplify the desired segment of the capsid protein gene inserted in a recombinant DNA molecule. Cultured cells are transfected with the mutagenized viral gene contained in a suitable expression vector, to obtain either the free capsid protein or the assembled VP. The mutagenized viral gene (or gene segment) can alternatively be subcloned in an infectious cDNA clone and the latter, or an infectious RNA transcript, is used to transfect host cells in order to obtain modified infectious virions (Sect. 22.2.1). Introduction of non-natural amino acids at specific positions in a VP by using specialized genetic means has facilitated its site-specific chemical functionalization (Sect. 22.3.2).

Any single amino acid substitution introduced in a structural viral protein will be present in each of the many copies of that protein contained in the VP. Especially in the capsids of nonenveloped viruses, the many identical amino acid substitutions introduced at symmetrically equivalent positions may have synergistic effects. In some cases, it may be desirable to reduce the number of identical substitutions in the capsid. For example, to limit the number of anchor points for chemical functionalization (Sect. 22.3.2). One procedure that does not require genetic engineering for reducing the number of reactive groups in a VP involves the phenotypic mixing of mutant and wild-type capsid proteins. However, the exact number of substitutions or the relative positions of mutated and non-mutated capsid proteins in the VP may be difficult to control with precision if this procedure is used.

The rational site-specific substitution of amino acid residues has repeatedly shown its extreme usefulness to modify VP properties or functions. Individual amino acid substitutions in viral capsids have frequently abolished some feature that may interfere with the intended application (e.g., binding to a natural receptor, in case a change in tropism is desired). One or a few well-chosen amino acid substitutions have repeatedly led to the acquisition of novel features or functions (e.g., high affinity for a different cell receptor or other ligands, including non-biological ones such as organic compounds or metal ions). Physical or chemical properties of different VPs, including thermal, acid or mechanical stability, or conformational propensities, have also been improved by site-directed mutagenesis.

Site-Directed Deletion or Insertion of Heterologous Peptides in VPs and Fusion of Heterologous Proteins with VP Proteins. Standard mutagenic techniques can be used for deleting peptide segments from viral capsid proteins. Likewise, foreign peptides may be inserted by PCR-based procedures by using oligonucleotides that contain the desired sequence. Protein domains or complete proteins can be covalently linked to a VP by fusion of the gene coding for the exogenous protein with a capsid protein gene. The chimeric protein is expressed, and the VPs are assembled either in cells or in vitro. These modifications can dramatically modify the structure, properties, and/or activities of the VP, provided they do not prevent its assembly.

In the capsids of many nonenveloped viruses, the N- or C-terminal segments of the capsid proteins are frequently exposed to solvent. Thus, these segments usually provide excellent loci to insert relatively large foreign peptides or fuse heterologous proteins, typically using a peptide spacer. In some cases, especially when guided by detailed structural information, N- or C-terminal insertions did not severely compromise capsid assembly, stability, conformational propensities, or other desirable features of the VP. Insertion of foreign peptides in exposed loops of some VPs impaired VP assembly or stability in some, but not all cases. In addition, steric constraints may alter the conformational propensities of the inserted peptide and impair its function (see also Chap. 21). Despite these and other potential issues, chimeric VLPs with inserted peptides or proteins are being extensively explored for diverse applications, including vaccines (Chap. 21), virotherapy, and gene therapy (Sect. 22.4).

In many instances, the viral nucleic acid not only carries genetic information, but also exerts some adaptive role on the structure, properties, and/or function of the virion. Thus, deletion, modification, or addition of nucleotide segments to the viral genome by genetic manipulation may also be considered for the modification of virions intended for some applications.

Combinatorial Engineering/Directed Evolution of VPs

In many cases, the information available on the structure, properties, and/or function of a specific VP may be insufficient to undertake its modification using a fully rational protein engineering approach. Moreover, introduction by site-directed mutagenesis of only one or very few amino acid substitutions in a viral protein may not yield a long enough jump in sequence space to achieve the desired phenotype. In these cases, combinatorial engineering/directed evolution approaches may be followed. Combinatorial protein engineering involves two main steps: (1) generation of genetic diversity leading to large libraries (populations) of variants of the target protein or protein complex (e.g., a viral protein or VP) and (2) screening or selection to identify and isolate the specific variant(s) with the desired phenotype [33–35].

Step i: Generation of genetic diversity. The number of possible polypeptide sequences is huge. For example, as many as 20^{200} different sequences are possible for a 200 amino acid-long polypeptide (albeit the vast majority of those sequences will not even be able to fold). The largest possible combinatorial protein libraries will still contain only a tiny fraction of all of the possible variants. Even if much larger libraries could be created, and the best selection methods could be used, it would be increasingly difficult to identify one specific variant among so many others. Thus, directed evolution has the best chances of success when all of the available information on the structure, properties, and/or functions of the target VP are used to restrict, as much as possible, the part of the viral genome that will be combinatorialized. For example, if one wishes to increase the affinity of a VP for a given cell receptor, and the receptor-binding site is known to reside entirely in a particular loop of a capsid protein, only the region of the capsid protein gene that encodes that loop should possibly be randomized. Approaches to generate huge numbers of protein variants include PCR using degenerate oligonucleotides, error-prone PCR, DNA shuffling in different implementations, or, frequently, a combination of some of the above.

Degenerate oligonucleotides are mixtures of oligonucleotides whose sequence differs in one or a few nucleotide positions only. Using degenerate oligonucleotides to amplify a DNA segment by PCR can result in a library of protein variants whose sequences differ in one or a few amino acid residue positions. For example, if two positions are fully randomized with the 20 proteinogenic amino acids, the maximum number of different protein variants will be $20^2 = 400$. Obviously, this site-directed combinatorial approach is limited to very small jumps in sequence space and requires a rather detailed previous knowledge on the specific amino acid residues that may determine the desired property or function.

Error-prone PCR involves the use of conditions that increase the error rate of the DNA polymerase used to amplify a DNA segment (e.g., a complete or partial viral genome, a single viral gene or gene segment). In this way, a substantial number of mutations can be randomly introduced along the amplified segment. Very large libraries of variant VPs or viral components that differ in one or several amino acids can be obtained. While the jumps in sequence space using this procedure are relatively limited, many useful phenotypic changes in VP properties or function have already been achieved, or are within reach.

DNA shuffling involves the in vitro recombination of relatively small fragments of homologous DNA segments (again, a complete or partial viral genome, a single viral gene or gene segment). In this way, relatively large jumps in sequence space (compared to other combinatorial methods) can be achieved. However, the procedure is dependent on the availability of homologous variants of the targeted DNA segment. A collection of natural variants of a virus species containing several serotypes or strains, or even a single, highly heterogeneous RNA virus population (quasispecies) provide excellent starting points for directed evolution by DNA shuffling. Ex vivo methods for the shuffling of DNA fragments to generate increased variation by increasing recombination events have also been developed.

Step ii: screening or selection. After producing a large enough library of protein variants, the variant(s) with the desired phenotype must be identified and isolated, either by screening or through selection. Screening is generally case-specific, may not be readily implemented in each case, and time considerations severely limit the number of tested variants.

A much better alternative is the direct selection of the desired phenotype. Some selection methods are case-specific and, again, they may not be readily implemented. Fortunately, powerful, rather general selection methods have been developed. Display methodologies are based on a physical linkage between genotype and phenotype in the same particle. Phage display was the first of the display methodologies, and since its implementation it has been heavily used for the directed evolution of many peptides and proteins [36–39] (see Sect. 22.5.1). Other display methodologies based on the same principles have been developed using different biological entities for the linkage of genotype and phenotype. These include ribosome display, yeast display, bacterial display, etc.

Viruses can be readily subjected to directed evolution. Many variant virions are already generated during their replication in the host (especially for RNA viruses). Additional variation may be obtained by infection of cultured cells, and increased in the presence of systems to increase the recombination of the replicating viral genomes. Genetic variation may also be generated in vitro by error-prone PCR and/or DNA shuffling of an appropriate viral nucleic acid segment, followed by subcloning in an infectious cDNA (Sect. 22.2.1). Importantly, in any virion population thus generated, the variant genotypes and the corresponding phenotypes are naturally linked. Thus, the desired virion may be readily identified and isolated by direct selection. Directed evolution has been repeatedly used to obtain variant virions with altered cellular tropism, reduced antigenicity, or increased mechanical stability, among other useful modifications [40–42] (Sect. 22.4).

Directed evolution of virions can be used also to obtain viral capsids or VLPs with improved properties or altered functions. The variant capsids or capsid proteins can be obtained from the corresponding variant virions as indicated in Sect. 22.2.2. This approach may be relatively straightforward, provided the nucleic acid inside the virion has no significant influence on the targeted capsid property or function (e.g., stability or cell tropism).

Directed evolution using selection by phage display has been used also to obtain variant peptides that have then been fused to capsid proteins to promote recognition between the engineered VP and certain ligands (e.g., metals or other inorganic compounds; [43]; see Sect. 22.5.3).

Combinatorial engineering on recombinant DNAs that encode viral capsid protein genes can be used to directly obtain libraries of variant viral nucleic acid-free capsids or VLPs, in the same way libraries of non-viral proteins can be obtained. However, the absence of the viral genome inside the variant VPs means that no physical linkage between genotype and phenotype will exist in this case, which may complicate the implementation of an adequate screening/selection procedure.

Directed evolution approaches are frequently combined with fully rational approaches in an iterative and sometimes complex, but extremely powerful protein engineering strategy to modify VPs.

22.3.2 Chemical Engineering of VPs

Chemical engineering is another powerful strategy for the modification and functionalization of VPs [14, 29, 44–47]. In general, an exogenous compound harboring some specific function is bound (conjugated) to an assembled VP through non-covalent or covalent interaction (Sect. 22.2). Sites on the outer or inner surface of a hollow VP are generally chosen [48]. Exogenous compounds that have been enclosed and/or bound to the outer or inner surfaces of different VPs include proteins, protein domains, and peptides; nucleic acids and oligonucleotides; carbohydrates; lipids; organic polymers (e.g., polyethyleneglycol (PEG)) and small organic compounds (e.g., fluorophores or drugs); metals, metal compounds, and other inorganic compounds; and preformed NPs (e.g., quantum dots (QDs) or gold NPs). Depending on the VP and the attached compound(s), the functionalized particle can be made suitable for one or other of a vast number of current, developing or potential uses (see Sects. 22.4 and 22.5).

Infusing or Enclosing Exogenous Compounds into VPs

In some protocols for VP-mediated targeted drug delivery (Sect. 22.4.5), and also for a few other applications of VPs, an exogenous compound (e.g., a toxic drug for targeted cancer therapy) is infused or enclosed in the VP without a need for a strong or specific binding between them. Many VPs have pores and channels that are permanently open, or that are transiently open during conformational changes at equilibrium (VP "breathing"). In such cases, if the exogenous compound is small enough and has the right physicochemical features (e.g., adequate hydrophilicity or hydrophobicity, net charge, etc.), it may suffice to infuse it through the VP pores using a high concentration of the compound under adequate conditions.

A more general approach involves the (re) assembly of the VP in the presence of the exogenous compound. The capsid proteins are expressed in free form or are obtained through the controlled disassembly of a preformed VP (typically an empty capsid or VLP). The capsid proteins and the exogenous compound at a high concentration are mixed under conditions that will allow the reassembly of the VP in vitro. When attempting this approach, it should be taken into account that certain compounds have been shown to interfere with the assembly of the VP, alter the morphology or properties of the assembled VP, and/or reduce the final yield.

Even if an exogenous compound can be infused in a preformed VP or enclosed during VP (re) assembly, there is a definite chance that the compound will prematurely diffuse out of the VP through permanent or transient openings. Thus, devising a way to bind the internalized compound to the VP inner surface is usually desirable. In some cases, non-specific, relatively weak non-covalent binding may suffice. The inner surface of many virus capsids is positively charged (to promote an interaction with the negatively charged viral nucleic acid). This feature has been exploited to facilitate the internalization of positively charged exogenous compounds and to retain them through Coulombic interactions. Enclosing some negatively charged compounds has been facilitated by adding them together with positively charged molecules.

Non-Covalent Attachment of Exogenous Compounds to VPs

Highly specific and relatively strong non-covalent binding of exogenous compounds to defined sites on either the inner or outer surface of a VP most frequently requires the introduction, by genetic or chemical means, of a suitable moiety at the chosen location(s). In certain cases, the exogenous compound must also be genetically or chemically modified, to endow it with the capacity to bind the moiety that was linked to the VP.

One general approach for achieving the non-covalent functionalization of VPs with exogenous compounds involves the use of genetic engineering to fuse of a histidine tag ($(His)_6$ hexapeptide) to a surface-exposed N- or C-terminal segment of a capsid protein. A Tris-nitriloacetic acid (Tris-NTA) is chemically linked to the exogenous compound. The nickel-loaded Tris-NTA-compound will bind the $(His)_6$-tagged VP with high affinity and specificity.

An analogous procedure is based on the very strong and specific non-covalent interaction between biotin and avidin (or streptavidin). The VP is chemically biotinylated using biotin ligase, and (strept)avidin is chemically linked to the exogenous compound.

A different approach for the non-covalent functionalization of a nucleic acid-containing VP with an exogenous compound is based on the hybridization of a nucleic acid segment with a complementary oligonucleotide chemically linked to the exogenous compound.

The reversible non-covalent binding of exogenous compounds to VPs may be inadequate for applications where a permanent modification is desired. For example, a drastic decrease in VP concentration or changes in the physical or chemical conditions may lead to the untimely dissociation and loss of an exogenous compound that had been non-covalently attached to the VP. However, non-covalent binding may greatly facilitate the release of the attached active compound if the latter must exert its action in free form. For example, a toxic drug enclosed in a VP may have to be released from the carrier VP to exert its effect in a targeted tumor cell (see Sect. 22.4.6).

Covalent Chemical Functionalization of VPs with Exogenous Compounds

Covalent linkage of certain non-biological compounds to VPs may be difficult or virtually impossible to achieve using genetic engineering alone. Instead, different chemical reactions have been used to irreversibly bind an enormous variety of exogenous compounds to VPs through the establishment of covalent bonds.

In a limited number of cases, the desired compound can be covalently linked to adequate reactive groups (amino acid side chains) that are already present on the unmodified VP surface. However, in the vast majority of cases, the particle will not contain the desired moieties at appropriate locations and/or will present equally reactive groups at undesired locations. Thus, structure-guided genetic engineering of the VP frequently constitutes an obligatory first step for the removal of unwanted reactive groups and/or for insertion of adequate reactive groups at specific positions.

Procedures for the irreversible functionalization of VPs by chemical modification may use either standard chemical reactions or fast and highly specific "click" chemistry. Target reactive groups include those of cysteines (thiol), lysines (amine), aspartic or glutamic acids (carboxylate), or tyrosines (phenyl) that may be naturally present, or that have been genetically introduced at appropriate sites in the particle. Chemical functionalization is achieved using different reactions, depending on the targeted group and other considerations. For example, thiols in cysteines can be reacted with maleimide, or amines in lysines with N-hydroxysuccinimide esters or derivatives. By choosing the right group and chemistry, many different exogenous compounds have been covalently linked to specific locations, either on the VP outer surface (exposed to the external environment) or inner wall (confined within the capsid cage) (see Sects. 22.4 and 22.5).

The number of targeted reactive groups (e.g., lysines) that are naturally present on the VP outer surface or inner wall may be too high to conveniently remove by site-directed mutagenesis all those that occupy positions one does not want to functionalize. Moreover, some of those reactive groups may have to be preserved, because they have a role in some desirable property or activity of the VP (e.g., stability or cell receptor binding). Special procedures have been developed to achieve some degree of selective functionalization, while still targeting reactive sites naturally present in the particle. For example, asymmetric derivatization of the CCMV capsid was achieved by reversibly binding one face of the CCMV capsid with exposed thiol groups to an activated resin, passivating the exposed thiol in the other face with iodoacetic acid, and then dissociating the particle from the resin [19].

In many cases, exogenous compounds must be covalently linked to only one or a few specific sites in each capsid subunit. One strategy to achieve this goal involves the introduction of a non-natural amino acid at those specific positions, using especially modified bacteria or cell-free protein synthesis systems for production of the capsid proteins [49]. Then, a click-chemistry reaction specific for the reactive group present in the introduced non-natural amino acid residue(s) (e.g., alkynes with azide moieties) may be used for limited, spatially precise functionalization [46, 47].

Other general methods for the site-specific covalent coupling of exogenous compounds to VPs are being increasingly used. The SpyCatcher/SpyTag technology and analogous procedures have been met with great interest. SpyCatcher is a bacterial protein that can form a covalent isopeptide bond with the SpyTag peptide. SpyTag is genetically fused to one end of a viral protein. SpyCatcher is conjugated to the exogenous compound to be attached to the VP. The covalent isopeptide bond formation between SpyTag and SpyCatcher is formed in a spontaneous and highly efficient reaction [29].

Some issues may arise when an exogenous compound is chemically coupled, either non-covalently or covalently, to a pre-assembled VP. They include reduced particle stability, and unwanted effects on the conformational dynamism or function of the viral particle (e.g., cell receptor binding). Moreover, chemical functionalization must be carried out each time a new batch of modified particles must be produced. This requirement may result in unwanted heterogeneity within and between batches, which will make production and quality control more complex, time-consuming, and expensive. In contrast, genetic engineering of a virus capsid must be performed only once, and every batch will be made of identical particles. However, the versatility of chemical engineering approaches combined with genetic engineering for the attachment of non-biological compounds to VPs is far superior to what can currently be achieved by genetic engineering alone.

22.4 VPs for Biomedical Applications

The number of studies on engineered VPs for biomedical or bio/nanotechnological uses is increasing at an outstanding rate [8, 13, 19–21, 23, 30, 31, 45, 48, 50–60] (Fig. 22.4). A detailed review of even a fraction of the individual achievements to date is beyond the scope and space limitation of this chapter. In this and the next section, we have merely attempted to provide a categorized outline of some current, developing, or potential uses of engineered virions, virus capsids, or VLPs.

This section (Sect. 22.4) deals mainly with applications that can be more clearly perceived as related to biomedicine; the next section (Sect. 22.5) deals mainly with other biotechnological or nanotechnological applications. Of course, any such categorization cannot be clear-cut. For example, VP-based bioinorganic NPs can be used for developing biosensors for either industrial uses or diagnosis. Some nanostructured VP arrays could be used for industrial catalysis, while others could be used for tissue regeneration. For each contemplated application, we have chosen one or a few studies to provide representative examples. As justified in the Introduction to this chapter, citations have generally been limited to some recent review articles. Some reviews that cover multiple applications of engineered VPs rather than one or a few specific uses have been cited at the beginning of this paragraph; some more specialized reviews covering one or a few related applications are cited in the corresponding subsections below. The interested reader is directed to the cited reviews for references to many original research articles.

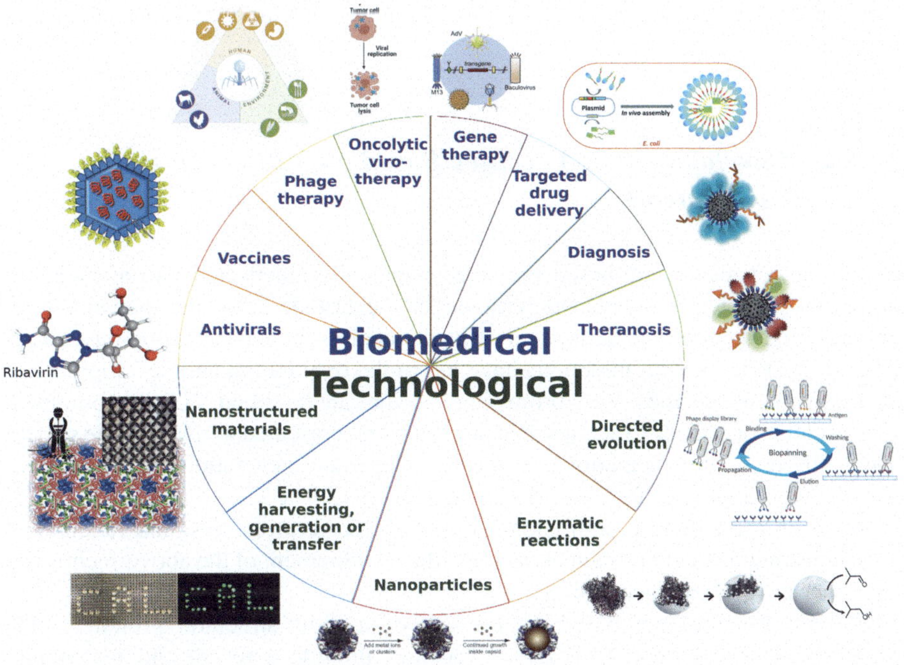

Fig. 22.4 Some current or developing applications of engineered VPs. See text (Sects. 22.4 and 22.5) for detailed presentations of the different applications and of the different images summarized in this Figure. Illustrations are taken from other Figures in this chapter, where their procedence is indicated; the illustration for oncolytic virotherapy is reproduced from [61]. With permission

22.4.1 Design of Novel Antiviral Drugs Targeting VPs or Viral Components

The biomedical importance of developing new effective antiviral drugs and viral vaccines cannot be overemphasized. Finding novel drugs that can inhibit virus infection has traditionally relied, and still partly relies, on the screening of natural or synthetic small organic compounds for adequate antiviral activity. In addition, the knowledge being acquired on the relationships between structure and function of viruses and viral components is guiding the rational design and/or improvement of antiviral drugs. Most known antiviral agents to date are directed at interfering with the action of viral proteins during the replication and expression of viral components in the infected cell. However, an increasing number of studies are devoted to identify, design, or study antiviral compounds that can interfere with steps in the viral cycle in which virus particles are involved (the steps described in Part II in this book). An antiviral drug that interferes with the assembly of the mature HIV-1 capsid has recently been approved for clinical use. Chapter 20 provides a detailed account on the design and development of antiviral drugs in general, with emphasis on antiviral drug-based strategies to interfere with steps of the infection process that involve VPs.

22.4.2 *Development of Vaccines Based on VPs or Viral Components*

Classic viral vaccines are based on either chemically inactivated virions or infectious, but genetically attenuated, virions. In the last decades, the unavailability, unfeasibility, or different limitations of inactivated or attenuated virion-based vaccines to treat many viral infections have led to the exploration and development of alternative viral vaccines. Very promising vaccines are based on VLPs, recombinant viral proteins or peptides vaccines that are made and presented to the immune system using different implementations. Some of these "new generation" vaccines have already been approved for clinical or veterinary use. The knowledge that is being acquired on the structure, immunogenic and antigenic properties, and function of VPs and structural viral proteins is guiding the development of the above-mentioned and other novel viral vaccines.

Subunit vaccines are based on the use of a virus structural protein as the immunogen. The current COVID-19 vaccines provide a recent and resounding example of this approach. Approved COVID-19 vaccines contain recombinant adenovirus, mRNA, or protein, but in every case the immunogen finally presented to the immune system is the spike (S) membrane fusion protein of SARS-CoV-2 or a protein S fragment (see Chap. 16). Some of those vaccines are based on a genetically engineered S protein with improved immunogenicity, a result of the extensive knowledge acquired over many years on the structure and function of homologous proteins during infection by enveloped viruses including respiratory syncytial virus, HIV-1, and coronaviruses. Chap. 16 in this book provides a detailed account of membrane fusion for entry of enveloped viruses into host cells and includes a brief account of the development of COVID-19 vaccines.

Unmodified or engineered VLPs are being developed as novel vaccines against infection by viruses or other pathogens, or even against cancer or other diseases. VLPs are very attractive vaccine candidates because they may carry the desired immunogenic moieties and may present them in a suitable way to the immune system, but are not infectious. Some VLPs have already been licensed for human or veterinary use, and many other VLP-based vaccines are being developed. Chapter 21 provides a detailed account on the design of new vaccines based on VLPs or chimeric virions.

22.4.3 Phage Therapy

Phage therapy involves the administration of bacteriophages for treating bacterial infections [62–66]. More than a century ago, d'Herelle used phage preparations to treat children suffering from bacterial dysentery. In the 1920s and 1930s, before antibiotic drugs were available, phage therapy was received with great interest and became extensively used in some countries. The first phage therapy programs were implemented in Georgia and Poland, and these programs are still active. However, early phage therapy treatments were sometimes poorly implemented or controlled, and met variable success. With the widespread use of penicillin and, later, other antibiotics, phage therapy was abandoned in most countries. The situation has recently changed again because of the rapidly increasing emergence of microbial resistance to antibiotics (to which 1.2 million deaths were directly attributed in 2019 alone). In the last 5 years or so, a renewed interest in phage therapy has been developed, and a number of recent reports have described the limited use of lytic phages to treat severe, multi-antibiotic resistant bacterial infections of individual patients. Modern, though still very limited, phage therapy programs have been established in the last years in Western countries including Belgium, France, Sweden, and the United States, with several clinical trials underway [62, 65, 66].

The use of phages to treat bacterial infections presents several important issues, which are currently being contemplated, and eventually addressed. Some issues are related to the biological production and purification of phages, including the presence of toxins or other harmful contaminants. Other issues come from the targeted bacteria, including genetic variability, presence of prophages (phage DNA genomes integrated in the bacterial genome), and inbuilt phage resistance, or its rapid acquisition during treatment. Most issues concern the phage particles themselves. Candidate lytic therapeutic phages may show eventual lysogenic activity or very restricted host range. They may lack a mechanism to counteract defense systems in bacteria (e.g., restriction-modification, CRISPR), or carry toxin genes. Candidate phages may be physically or chemically not stable enough or susceptible to biological inactivation during preparation or use. They may induce an immune response or may genetically acquire undesirable features during their multiplication.

Addressing some of the issues related to the phage particle itself is being helped by the extensive knowledge on the relationships between the structure and function of phages. The vast majority of phages known belong to the order *Caudovirales* (dsDNA phages with tails), and these viruses have been extensively studied in exquisite molecular detail since the times of the *Phage Group* in the 1940s (see Chaps. 12 and 17 for some important aspects of phage structure, assembly, host recognition, and genome delivery). General genetic and chemical engineering approaches (Sect. 22.3) are being used to modify phages intended for therapy. Special procedures have also been devised for the directed evolution of phages through enhanced recombination in infected cells and improved selection methods or for the in vitro synthesis of tailored phage genomes [63, 65].

Some examples on the engineering of phages for therapeutic use are mentioned next. Therapeutic applications of natural phages would benefit from their engineering for a widened or modified host range. Most natural phages have an extremely narrow host range, which largely depends on the specificity of the recognition between specialized proteins in the phage and receptor molecules in the host bacterium. This specific event ultimately results in the injection of the phage genome into the host cell (Chap. 17). Changes in tropism have been achieved by the genetic modification or replacement of cell receptor-binding proteins on phage particles (e.g., fd, T2, T3, T4, T7), or by displaying on the phage surface antibody fragments that bind heterologous cell receptors (M13).

Likewise, undesirable immunogenicity of phage particles has been reduced by genetic modification of immunogenic regions, or through chemical concealment from the immune system by coating the phage capsid with PEG or other low-immunogenicity compounds.

Resistance of bacterial biofilms to both antibiotics and infection by phages has been reduced by genetically fusing a biofilm-degrading enzyme to a T7 phage capsid protein. The enzyme displayed on the phage surface degraded the extracellular matrix of the biofilm, exposing the bacterial cells to both the action of antibiotic drugs and to infection by phages.

Phages have also been chemically modified with chloramphenicol to drive the antibiotic to the bacterial cell wall, providing a fine example of combined phage-antibiotic therapy.

In 2019, genetically modified phages were first clinically used. Several clinical trials using engineered phages for specific therapies of bacterial infections are currently under way. Other therapeutic or prophylactic treatments using phages in humans are being contemplated. For example, to treat non-infectious diseases where bacteria exert some role, or for controlling the composition of human microbiomes [65].

The use of phages to control harmful bacteria in farm animals, including the prevention or treatment of some animal diseases caused by bacteria, is also being actively explored. In 2005, the first phage-based product for use in agriculture was approved by the corresponding United States regulatory agency to treat a bacterial disease of crops (bacterial spot disease) [67].

Phages are also being used or developed as bio-tools for the control of bacterial growth in other situations. Some of these applications can also be considered as therapeutic, as they will have a clear positive incidence in health (One Health approach, at the intersection of human health, animal health, and environmental health). One such application involves the use of phages for the control of bacterial contamination in foods [67, 68]. In 2006, the Food and Drug Administration (FDA) approved in the United States the first phage-based product to control contamination in meats and poultry products by a bacterium (*Listeria monocytogenes*). More recently, a number of phage-based products to treat a diversity of animal or plant-derived foods for human consumption have been approved in different countries and are commercially available. Most of those products contain natural phages that have not been genetically modified [68].

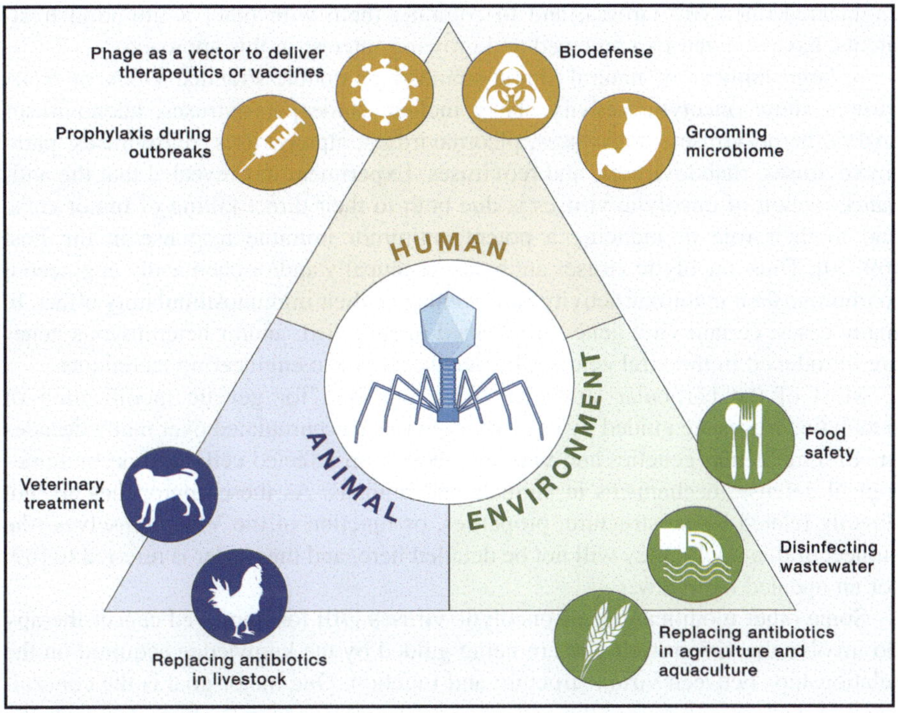

Fig. 22.5 Potential uses of phage therapy. See text (Sect. 22.4.3) for detailed descriptions (Reproduced from [65]. With permission)

Infectious phages are being contemplated also for other health-related applications, including the control of bacteria in the environment, their use as bactericides for decontaminating surfaces or for wastewater treatment, or as sensitive biosensors for bacterial detection [59] (Fig. 22.5). It may be anticipated that, to guide many of those developments, intensified research will be required on the structure, properties, and function of phages (Chap. 17), and on their adequate modification by chemical or genetic engineering.

22.4.4 *Oncolytic Virotherapy*

Oncolytic virotherapy is a developing anticancer treatment that involves the administration to a patient of infectious, but attenuated viruses to selectively infect and eliminate cancer cells [61, 69, 70]. Oncolytic virotherapy was already contemplated 100 years ago, when experiments using mice confirmed that some viruses could eliminate solid tumors. Clinical trials during several decades using natural oncolytic viruses alone showed, however, limited efficacy. Much more recently, the possibilities to improve the efficiency of virus-based anticancer therapies by administering engineered oncolytic viruses, and to combine them with other antitumoral treatments, have resulted in a renewed and growing interest in this approach.

A large number of natural single-stranded or double-stranded DNA or RNA viruses show oncolytic activity. They include some parvoviruses, adenoviruses (AdV), herpesviruses, poxviruses, picornaviruses, alphaviruses, retroviruses, paramyxoviruses, rhabdoviruses, and reoviruses. Experiment has revealed that the anticancer action of oncolytic viruses is due both to their direct killing of tumor cells, and to their role in inducing a potent antitumor immune response in the host [69, 70]. Thus, oncolytic viruses are being genetically and/or chemically engineered to improve their cytotoxic activity and/or enhance their immunostimulatory effect. In many cases, certain viral genes are deleted or modified, and/or heterologous genes are introduced in the viral genome by standard genetic engineering techniques.

Most of the particular strategies being explored for genetic modification of oncolytic viruses are guided by the vast knowledge accumulated over many decades on viral molecular genetics and viral metabolism in infected cells, and on immunological defense mechanisms in animals and humans. As these approaches are not directly related to the structure, properties, or function of the VPs themselves (the subject of this book), they will not be detailed here, and the reader is referred to [69] for an updated overview.

Some other modifications of oncolytic viruses [70] for improved cancer therapy do involve the virion itself and are being guided by the knowledge acquired on the relationships between virion structure and function. One major goal is the concealment of the virion to reduce the antiviral immune response in the host. To evade the immune response herpes simplex virus type 1, AdVs, and other oncolytic viruses have been encapsulated into liposomes, or coated with an organic polymer (e.g., PEG) or with a mineral shell (e.g., iron oxide, manganese, or calcium carbonate). In addition to concealment, some of those coatings improved the oncolytic activity of the virions, or facilitated their detection in the host by using certain imaging techniques (Sect. 22.4.7). Some oncolytic viruses have been coated with tumor cell membranes, or internalized in whole cells (sometimes after adequate functionalization for efficient intake), which resulted both in their concealment from the immune system, and in improved delivery into tumors [70]. Oncolytic viruses are also being engineered for a higher affinity and specificity for cell receptors on the target tumor cells. The approaches used include chemical functionalization with organic or peptide compounds, or the display of target cell receptor-binding peptides or proteins. Many of those modifications are being also introduced in VPs intended as vectors for gene therapy (Sect. 22.4.5) or for targeted drug delivery (Sect. 22.4.6), and the reader is referred to those sections for a more extended coverage of this subject.

Many clinical trials with chemically and/or genetically modified oncolytic viruses for cancer treatment are underway, and some of them are already delivering very promising results [70].

22.4.5 *Virus Vector-Based Gene Therapy*

Gene therapy involves the introduction of genes into target cells in a patient, using a suitable vector, in order to prevent or treat a genetic disorder by correction or replacement or the defective gene(s) [40, 42, 71, 72].

The multiplication and perpetuation of viruses in nature relies on viral particles that efficiently deliver their genetic material into host cells through well-regulated mechanisms (Chaps. 15–17). Over the last several decades, the detailed knowledge acquired on virus genetics and metabolism, as well as on the structure and function of VPs, has guided many studies to convert VPs into suitable vectors for gene therapy. Many viral vectors have already been developed and used in many hundreds of gene therapy protocols and clinical trials to treat many genetic disorders, including cardiovascular, muscular, neurologic, hematological, and ophthalmological diseases, infectious disorders, and different cancers.

In essence, the development of a viral vector for gene therapy involves the manipulation of the viral genetic material to include the therapeutic gene (and/or other genetic elements) and to prevent virus replication in the host. Genetically manipulated retroviruses, AdVs, adeno-associated virus (AAV) and, to a much lesser extent, many other animal viruses belonging to different families, are being used as gene therapy vectors for laboratory studies and clinical trials (Fig. 22.6). As for the development of better oncolytic virions (Sect. 22.4.4), the specific modifications introduced by genetic engineering in the viral genome constitute a key element for developing improved gene therapy vectors.

Many such modifications are introduced to prevent virus replication, allow the inclusion of the exogenous therapeutic gene, optimize its packaging, improve the transduction of the transgene payload, increase its replication in the transduced cells, regulate its expression, etc. For reasons mentioned in Sect. 22.4.4, a description of even a few of those modifications is outside the scope of this chapter. References

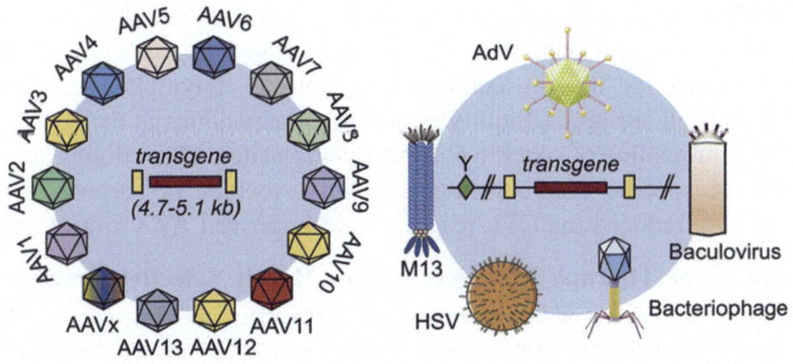

Fig. 22.6 Some viral vectors for gene therapy. Left, different AAV serotypes can be chosen or recombined to engineer improved vectors for gene therapy by DNA shuffling. Right, other current (AdV) or potential viral vectors for gene therapy. See text (Sect. 22.4.5) for descriptions (Reproduced from [73]. With permission)

[40, 42] provide recent reviews of many modifications introduced in AAV or AdV genomes for the development of more efficient gene therapy vectors.

In addition to the genetic modifications mentioned above, a growing number of modifications are being introduced in the VPs themselves (not only in the viral genome), in order to improve their suitability as therapeutic gene carriers. Here we provide several examples of the modifications introduced in gene therapy viral vectors by genetic or chemical means (Sect. 22.3), with special reference to the widely used AAV and AdV vectors [34, 40, 42, 74].

Engineering Gene Therapy Viral Vectors for Concealment from the Immune Response

The antigenicity and immunogenicity of viral vectors constitute a serious issue for efficient gene delivery, as they may contribute to the premature elimination of the vector from the patient. A large fraction of the human population has been infected with natural AAV or AdV viruses and carry antibodies that can recognize epitopes on AdV or AAV viral vectors. In addition, inoculation with the viral vector itself may generate a potent immune response, especially when an extended treatment is required. Several strategies have been explored to reduce recognition by pre-existing antibodies of viral vectors for gene therapy and/or to reduce their immunogenicity. One approach, extensively tested with AAV and AdV, involves the coating of the capsid surface with poorly immunogenic substances. For example, organic polymers (PEG), lipids, or cell-derived extracellular vesicles have been used to conceal neutralizing epitopes on the AAV capsid.

Another tested strategy to reduce the antigenicity and/or immunogenicity of AAV and other viral vectors involves the structure-guided removal or replacement, by site-directed mutation, of a few amino acids that participate in immunodominant epitopes and antigenic sites on the capsid surface. Directed evolution (Sect. 22.3.1) using error-prone PCR or DNA shuffling has been used to select antibody-resistant AAV variants. Variants obtained evaded a pre-existent immune response and, in some cases, also showed improved tropism and enhanced gene transduction as added benefits.

AAV vectors have also been genetically modified to evade the cytotoxic T-cell immune response (see Chap. 21). Ubiquitin-mediated degradation of the AAV capsid is required for the presentation of cytotoxic T-lymphocyte (CTL) viral epitopes on the cell surface. Ubiquitylation of lysine residues in the AAV capsid is facilitated by elimination of specific phosphorylation sites. The rational, site-directed replacement of specific lysine residues, or of tyrosines or serines involved in phosphorylation, reduced the CTL response and improved AAV transduction.

Engineering Gene Therapy Viral Vectors for Highly Selective Tropism

Adequate viral vectors for gene therapy must show efficient and selective transduction of target cells. For example, a viral vector for cancer therapy should efficiently deliver the therapeutic gene in the tumor cells, but not in the patient's normal cells. Efficient and selective transduction involves a strong and highly specific interaction between a receptor molecule on the target cell surface and a cell receptor-binding moiety on the VP. Structure–function studies have revealed that different serotypes and strains of AAV, AdV, and other viruses recognize different receptors. Thus, gene therapy vectors based on different natural virus variants may be used to target different cell types. However, a higher affinity between VP and receptor may be required, or no known natural virus variants may recognize receptors on the targeted cells. In those cases, the viral vector may be suitably engineered for increased affinity or altered cell tropism.

The engineering of a viral vector with a specific cell tropism may require a combination of steps: (1) any moiety on the viral capsid that recognize cells other than the targeted cells should be removed or inactivated, (2) if no suitable receptor-binding site is present, an exogenous compound carrying an adequate receptor-binding site should be attached to the viral particle (preferably through a covalent linkage), and (3) the affinity and/or specificity of a suitable receptor-binding site may have to be increased. Both chemical engineering and genetic engineering have been extensively used to enhance or modify the tropism of AAV- or AdV-based viral vectors for gene therapy [40, 42]. Some examples are mentioned next.

Some AdVs show a broad tropism due to their binding to the coxsackie and adenovirus receptor (CAR), which is expressed in many cell types. A rational protein engineering approach removed the CAR-binding site in the AdV capsid by site-directed mutation of one or a few amino acids in the knob domain of the fiber capsid protein. Further modifications in the other two major capsid proteins (hexon and penton proteins) eliminated the interaction with other receptors. Complete retargeting of the AdV vectors can then be achieved by genetically introducing other, specific receptor-binding sites in the viral capsid. For example, retargeting of AdV vectors to some specific cancer cells was achieved by replacing either the fiber protein or its knob domain by those of other AdV variants, thus combining in a chimeric viral vector the different strengths of different virus variants. In other cases, retargeting of AdV vectors was achieved by replacing the fiber knob by heterologous protein domains, for example the fibritin protein of phage T4. Many other moieties have been covalently attached to AdV VPs for retargeting, including integrin-binding, RGD-containing and other heterologous peptides, single-chain antibody fragments (scFv), T-cell receptors, etc. In a different approach, antibodies against specific cell receptors were non-covalently attached to the AdV capsid through the specific interaction between the antibody Fc domain and protein A molecules displayed on the VP surface.

Directed evolution approaches have been used to select AdV variants with adequate cell tropism and improved transduction of target cells. For example, serial passaging of a mixture of natural AdV serotypes in conditions that enhance recombination was undertaken to select for a chimeric AdV with tropism for a colon cancer cell line. The selected variant showed high selectivity for transduction of colon cancers compared to the parental natural variants.

Many studies have dealt with adequate modifications of AAV gene therapy vectors [40]. Specific knowledge of structure–function relationships in AAV has allowed the rational genetic engineering of different AAV serotypes to obtain improved viral gene therapy vectors. In one example, five rationally selected amino acid residues in the AAV-2 capsid were replaced by the residues at equivalent positions in AAV-1. The chimeric AAV retained desirable AAV-2 features, but showed improved efficiency to transduce muscle cells compared to AAV-2. In another example, specific residues in the AAV-2 capsid were replaced by those present in AAV-9 to engineer the AAV-9 glycan-binding site in the chimeric AAV particle.

Directed evolution has also been used to modify the tropism of AAV vectors. Randomized peptide sequences, including those that bind receptors expressed in different target cells, were genetically inserted in permissible sites on the AAV surface, to create large libraries of virion variants that display the peptide sequences. AAV variants with tropism for specific cell types were selected and isolated from those libraries as described in Sect. 22.3.1. In another example, error-prone PCR was used to combinatorialize the capsid protein gene of AAV-2 or AAV-9, and mutants with enhanced transduction of different target cells were selected. DNA shuffling of several AAV serotypes was used to generate combinatorial libraries of chimeric AAV variants. In different studies, this approach led to the selection of variant AAVs with tropism for different target cells, or with improved properties (i.e., mechanical resistance). In another study, rational engineering using site-directed mutagenesis and informed directed evolution were combined to develop novel AAV vectors with tropism for liver cells.

Virus-based gene therapy is becoming a reality for the clinical treatment of a growing number of monogenic disorders. In 2003, an AdV-based vector encoding tumor suppressor p53 was approved in China for treatment of certain cancers; in 2012, an AAV-based vector encoding a variant lipoprotein lipase for treatment of an inherited disorder was approved in the European Union. In 2016, a retroviral vector to treat severe combined immunodeficiency was also approved in Europe. In the United States, the FDA approved in 2017 two retroviral gene therapy vectors for the treatment of acute lymphoblastic leukemia or large B cell lymphoma, respectively. Since then, the number of approved viral vector-based gene therapy products to treat specific monogenic disorders, including certain types of cancer, has been increasing rapidly [40, 42].

However, many important issues remain to be solved before safe and efficient gene therapies can be developed for the efficient treatment of the many monogenic disorders known in humans and, eventually, more complex genetic disorders as well. The increasing knowledge being acquired on the structure–function relationships in VPs may contribute to guide further studies for the genetic or chemical engineering of gene therapy vectors. The results of those studies are not only leading to improved viral-based vectors, but are also guiding the design of non-viral or synthetic particles as gene therapy vectors.

22.4.6 VPs for Targeted Drug Delivery

To date, most chemotherapeutic drugs for the treatment of cancer or other pathologies are systemically administered using relatively high doses. Systemic administration and high doses, combined with the high cytotoxicity but poor selectivity of the available drugs, frequently result in the death of both cancer cells and some healthy cells in the patient, leading to severe side effects. The advent of nanotechnology has greatly intensified the efforts to direct drugs to cancer cells only, using highly selective NP carriers [13, 30, 31, 45, 50, 58, 75–78]. Drug carrier NPs are also being developed for other applications, including the treatment of cardiovascular diseases, infectious diseases and the selective delivery of pesticides in agriculture [31, 77].

NPs of biological or synthetic origin being tested for targeted drug delivery include liposomes, organic polymers, inorganic particles, and engineered VPs. The same considerations that led to the development of VPs as gene therapy vectors (see Sect. 22.4.5) have also led to the development of adequately modified VPs for targeted drug delivery. The inbuilt features of VPs and the results obtained in targeted drug delivery using engineered VPs are also inspiring the design and/or improvement of synthetic NPs as alternative carriers for selective drug delivery [2].

Molecular engineers confronted with the modification of natural VPs for targeted drug delivery face some of the same problems encountered with VPs for phage therapy (Sect. 22.4.3), oncolytic virotherapy (Sect. 22.4.4), or gene therapy (Sect. 22.4.5). These include the efficient protection and transport of a molecular cargo; evasion from the patient's immune system; adequate pharmacokinetics; absence of toxicity; high selectivity for target cells; and efficient and controlled delivery of their cargo. Engineered VPs for targeted drug delivery must fulfill, in addition, several specific requirements. To cite a few examples, a small drug enclosed in the VP should not diffuse out of the particle before the latter reaches the target cell; the drug carried either inside or outside the VP must not interfere with the integrity, stability, solubility, trafficking in the organism, and cell-targeting capacity of the latter; in many cases, the drug must be dissociated and released from the VP through an adequately controlled mechanism; the dose of the delivered drug should be high enough to achieve the desired therapeutic action, but low enough to avoid side effects.

A widely researched approach for targeted chemotherapy involves the use of engineered VPs loaded with the same cytotoxic drugs that are currently employed for systemic chemotherapy (e.g., doxorubicin (DOX), cisplatin, and 5-fluorouracil (5-FU)). Anticancer strategies based on the targeted delivery of other compounds, such as therapeutic peptides or proteins, are being developed (see below).

Loading VPs with Anticancer Drugs

Many genetic, chemical, or combined strategies (Sect. 22.3) have been used to functionalize VPs for targeted delivery of anticancer drugs. Adequate procedures depend, in part, on the nature of the specific drug and VP considered. In some cases, the outer surface of the VP has been chemically functionalized with the drug. However, confinement of the drug inside the VP cavity usually presents several advantages, including protection from degradation, and prevention of drug-associated toxicity during circulation in the organism. Encapsulation of both small drugs and larger compounds has been achieved during the in vitro assembly of the

VP from free capsid proteins in the presence of a high concentration of the compound. Conditions for the reversible disassembly of the capsid and its reassembly in the presence of a drug have been achieved for different viral systems, including CCMV, BMV, red clover necrotic mosaic virus (RCNMV), *Hibiscus* chlorotic ringspot virus (HCRSV), etc.

Alternatively, small drugs have been infused into VPs through capsid pores. The opening/closing of pores or channels in some viral capsids is modulated by environmental conditions (e.g., pH, ionic strength, temperature) or chemical agents (e.g., metal ions) during some stages of the viral cycle, including genome packaging and uncoating. This inbuilt conformational dynamism of some VPs has been used to advantage for the controlled infusion of small drugs into VPs and to avoid leaching during storage and circulation. For example, some drugs have been infused by opening pores in the RCNMV capsid with calcium ions, or in CCMV by changing the pH and salt concentration.

Specific non-covalent interactions between the drug and the VP (Sect. 22.3.2) have been frequently used to drive drug encapsulation and to prevent leaching of small drugs. For example, DOX, cisplatin, or 5-FU were encapsidated in phage MS2 VPs by conjugating them with MS2-derived RNA segments that non-covalently bind to the capsid inner wall [31].

In addition to small drugs, other anticancer agents have been encapsulated in VPs through specific non-covalent interactions. For example, prodrug-modifying enzymes were encapsulated into phage Qβ VPs via the interaction between a Rev peptide conjugated to the enzyme and specific segments in the viral RNA. In some cases, the cargo itself (small organic compounds, polymers, inorganic NPs) has facilitated its own encapsulation by acting as a template that promoted VP assembly [20, 22, 23].

The irreversible functionalization of the VP with an exogenous compound through covalent binding (Sect. 22.3) can be advantageous for many applications, including the engineering of VPs intended for drug delivery. However, if a drug that must be released from the VP is covalently linked to the VP, a way for the controlled breakage of the covalent linkage should be devised (see below).

Engineering Drug-Carrying VPs for Concealment from the Immune Response
The procedures used to engineer VPs for targeted drug delivery that can evade recognition by the host immune system are similar to those used for VPs intended for virotherapy or gene therapy. For example, coating of CPMV VPs intended for targeted drug delivery with PEG, carbohydrates (heparin, hyaluronic acid, sialic acid polymers), lipids (cell membranes, synthetic membranes), proteins (serum albumin, cell membrane glycoproteins), or peptides (elastin-like peptides, synthetic peptides) showed promising results for immune evasion.

Engineering Drug-Carrying VPs for Selective Cell Tropism
In some cases, the natural tropism of natural VPs has been used to advantage for drug delivery into certain cancer cell types. For example, some AAV, MVM, and other parvoviruses are naturally oncotropic. Remarkably, cowpea mosaic virus (CPMV) is also able to bind receptors that are overexpressed in some cancer cells, and CPMV

VPs carrying DOX molecules enhanced the cytotoxicity of the drug for the cancer cells. However, many VPs intended for targeted drug delivery must be engineered for selective binding to a receptor that is overexpressed in the targeted cells. Again, this goal is generally pursued by using similar approaches to those followed for VPs intended for virotherapy or gene therapy. For example, folate or derivatives have been attached to the surface of different VPs (CPMV, cucumber mosaic virus, HCRSV, AdV, HBV) to target many types of cancer cells in which the folate receptor is overexpressed. The surface of many different VPs (polyomavirus, RCNMV, BMV, phages Qβ, HK97, MS2) has been functionalized for targeting specific cell receptors with highly selective receptor-binding sites contained in heterologous peptides, proteins (including anti-receptor antibodies or scFvs), nucleic acid aptamers, carbohydrates, etc.

Engineering VPs for Controlled Drug Delivery

The inherent conformational dynamism of some VPs is being exploited not only to infuse some drugs and prevent leaching, but also for the controlled delivery of their cargo. For example, the pH dependent disassembly of HBV-derived VPs has been harnessed for the controlled release of DOX in the targeted cells [79]. Some VPs have been chemically modified for triggering their controlled disassembly for the release of cargo under a controllable stimulus that does not rely on environmental conditions [30].

Methods to release a drug that had been covalently linked to the VP have also been devised. For example, in one study, the prodrug AlDOX covalently linked to a P22 VLPs through thiol-maleimide coupling was released by the pH-induced cleavage of the bond between DOX and the maleimide moiety. In another example, two therapeutic peptides were genetically fused to the N-terminus of P22 capsid proteins through a peptide spacer that included a cleavage site recognized by cathepsin B, a lysosomal protease that is overexpressed in cancer cells. Release of the free peptide was achieved by specific cleavage in the target cancer cells of the peptide spacer by cathepsin B [30].

Multiple Functionalization of VPs for Targeted Drug Delivery

In many instances, multiple features of a same VP have been modified for its efficient use as a drug delivery vector. In a representative early study, a high molecular weight polyacid, poly(styrene sulfonate), was encapsulated during assembly of HCRSV VPs, to electrostatically retain DOX molecules inside the viral particle and prevent leaching. In addition, folic acid was covalently linked to the capsid surface for targeting tumor cells that overexpress the folate receptor. The VPs thus engineered improved DOX uptake and cytotoxicity in ovarian cancer cells.

In another, more recent study, P22 VPs were modified in multiple ways for the specific delivery of therapeutic peptides into breast cancer cells [80] (Fig. 22.7). The covalent functionalization with the peptides and their protease-mediated release in the target cells have been already mentioned. In addition, the FLAG peptide was fused to one of the therapeutic peptides to visualize by fluorescence the intracellular traffic of the VP. Moreover, the negatively charged, enhanced green fluorescent protein (GFP) was fused to the P22 scaffold protein involved in P22 VP assembly to

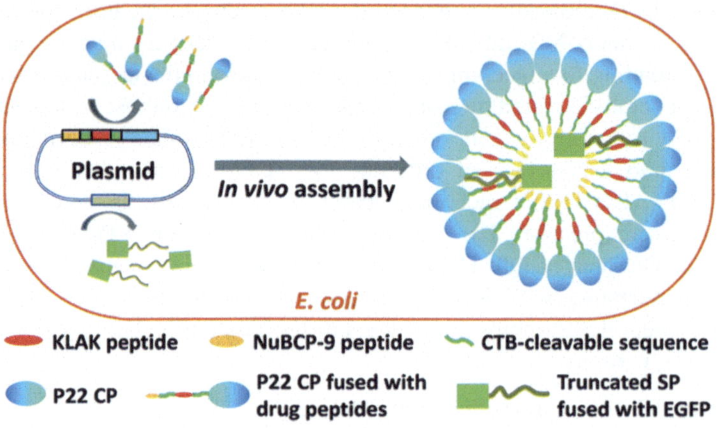

Fig. 22.7 An engineered phage P22-based VP for targeted drug delivery. KLAK and NuBCP-9 therapeutic peptides were covalently linked to the P22 capsid protein CP by genetic fusion, including also intermediate cleavable peptide sites. The FLAG peptide was fused to one of the therapeutic peptides to be used as a fluorescence probe (not shown). Enhanced GFP (EGFP in the diagram) was covalently linked to the P22 scaffold protein SP by genetic fusion. Finally (not shown), a RGD-containing peptide was chemically linked to the thus engineered P22 VP surface (deep blue regions in the VP cartoon at right) to target breast cancer cells. See text (Sect. 22.4.6) for a more detailed description. (Reproduced from [80]. With permission)

counteract the potential electrostatic repulsion between the positively charged peptide moieties displayed on the VP. Finally, a RGD-containing peptide was chemically linked to the outer VP surface to target breast cancer cells in which RGD-binding integrins are overexpressed.

Some Developing Strategies for VP-Mediated Targeted Chemotherapy
Many different VPs have been used for targeted delivery of DOX, cisplatin, 5-FU, and/or other chemotherapeutic drugs or prodrugs to many different cancer cell types. Examples include VPs derived from phages (e.g., MS2, Qβ), plant viruses (e.g., RCNMV, TMV, potato virus X (PVX), *Physalis* mottle virus (PhMV), CPMV), and animal viruses (e.g., influenza virus, FMDV, HPV, HBV, certain polyomaviruses, nodaviruses) [31].

VPs are being developed not only for the targeted delivery of cytotoxic organic compounds, but also for the targeted delivery of immunostimulatory oligonucleotides; small interfering RNAs that inhibit the expression of factors that promote metastasis and inhibit anticancer immune responses; cytotoxic peptides; or prodrug-activating proteins. These anticancer agents are generally enclosed in the VPs to protect them from inactivation in the organism, reduce non-specific toxic effects, and promote their internalization in cancer cells only. In one example, two cytotoxic peptides were encapsulated in VPs from P22 for their targeted delivery to breast cancer cells. In another example, a bacterial cytochrome encapsulated in CCMV efficiently activated the prodrugs tamoxifen and resveratrol.

Other strategies for specifically killing specific cells through the targeted delivery of certain compounds that are activated by light or magnetic fields are being actively explored and are briefly described next.

Photodynamic Therapy (PDT)
Targeted PDT is emerging as a promising therapeutic strategy to treat infectious diseases and cancers. The procedure involves the NP-mediated delivery into target cells (e.g., bacteria or tumor cells) of a photosensitizer (e.g., porphyrin derivatives) that produce reactive oxygen species and can kill the targeted cells [77, 81]. Some examples are mentioned here. CCMV VPs were engineered with a photosensitizer and a cell receptor-binding ligand to specifically kill *Staphylococcus aureus* cells. Engineered MS2 or Qβ VPs were used to deliver photosensitizers and kill cancer cells or cells from the immune system, respectively. TMV VPs targeted to melanoma cells were loaded with a photosensitizer that killed the targeted cells. VPs from CPMV were modified to display dendron polymers to which photosensitizer molecules were attached. These modified VPs were able to kill both melanoma cells and macrophages.

Photothermal Therapy (PTT)
Targeted PTT is another emerging and promising therapeutic strategy. PTT involves the NP-mediated delivery to target cells of a photo-absorbing compound or inorganic NP (e.g., polydopamine or gold NPs). The heat generated by the photo-absorbing compound when exposed to near-infrared light will specifically kill the targeted cells, but not the normal cells that did not internalize the compound [31, 79, 81]. For example, IR-780 iodide or ICG, two photosensitizing and photo-absorbing compounds, were encapsulated into HBV VPs that were targeted to cancer cells by a RGD peptide displayed on the VP outer surface. In both cases, significantly higher antitumor activity was observed relative to that of the free compounds. VPs from Qβ, TMV, or HBV have also been used for experimental PTT.

Magnetic Hyperthermia Therapy (MHT)
Targeted MHT is based on the NP-mediated delivery to target cells of a magnetic NP. The heat generated by the magnetic NPs when exposed to an alternating magnetic field will kill the targeted cells, but not the normal cells that did not internalize the NPs [82]. Iron oxide NPs are typically used as magnetic NPs for experimental MHT. Most studies on MHT have used synthetic nanoparticles. However, in one study, superparamagnetic gold nanoparticle clusters were delivered into tumor cells by an engineered HBV VP that carried a tumor cell recognition peptide [79], and these NPs were used for MHT of tumor-bearing mice. The many strengths of engineered VPs as vectors for targeted drug delivery may soon lead to their increased use in the development of PDT, PTT, or MHT.

To date, no VPs (or any other NPs) have been approved for treatment of cancers or other diseases through targeted drug delivery. Despite the many partial successes, there are still a number of important issues to be solved (see above). However, those issues are being addressed in a rapidly growing number of studies that involve the clever engineering of drug-carrying VPs and other NPs. Some preclinical trials are already underway.

22.4.7 VPs for Diagnosis

Noninvasive imaging technologies to visualize biological structures and processes in vivo are widely used in biological research and for diagnosis in the clinic. VPs have been modified to effectively carry and specifically deliver to target cells and tissues certain compounds that provide signals for noninvasive in vivo imaging [31, 79, 83–85] (Fig. 22.8). In some cases, VPs have been functionalized with two different agents for combined in vivo imaging. Targeted delivery of those compounds is generally achieved by engineering the VP carriers to display cell-targeting moieties, using the same genetic or chemical approaches that are followed for VPs intended for virotherapy, gene therapy, or targeted drug delivery (Sects. 22.3, 22.4, 22.5, and 22.6).

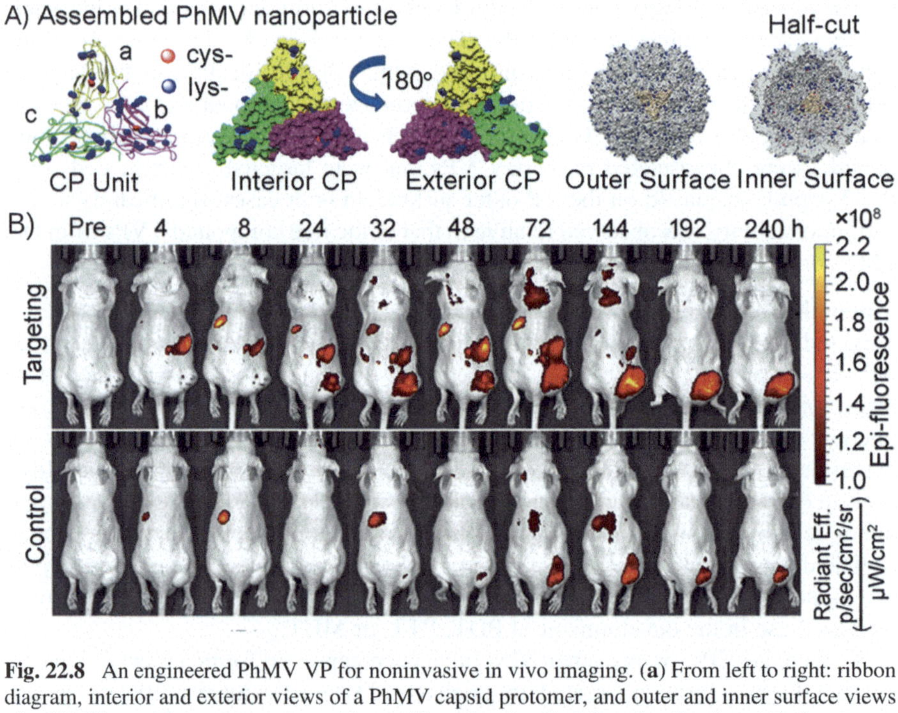

Fig. 22.8 An engineered PhMV VP for noninvasive in vivo imaging. (**a**) From left to right: ribbon diagram, interior and exterior views of a PhMV capsid protomer, and outer and inner surface views of a PhMV capsid protomer; cys and lys residues for chemical functionalization with fluorescent compounds are indicated. (**b**) The functionalized PhMV VPs were used for in vivo near-infrared fluorescence imaging of subcutaneous PC-3 ($\alpha2\beta1$ positive) prostate tumors in athymic nude mice. See also the text on fluorescence imaging (Sect. 22.4.7) (Adapted from [86]. With permission)

VPs for Fluorescence Imaging (FI)

FI requires a suitable fluorescent probe. Cyanines are frequently used in free form for in vivo near-infrared FI, but their application is limited by their low stability in the organism and weak fluorescence. VPs are being explored for carriers of fluorescent probes to increase their stability and for their targeted delivery to specific cells (e.g., cancer cells), and to increase the sensitivity of the process [79, 81].

Some fluorescent probes have been encapsulated by disassembly of different VPs and its reassembly in the presence of the fluorescent compound. Examples of the engineering and use of VPs for FI include the encapsulation of the fluorescent compound CH1 in simian virus 40 (SV40) VPs, or the negatively charged fluorescent compound ICG in BMV. Negatively charged QDs provided both a fluorescent probe and a template for driving the reassembly of CCMV VPs around them. Cyanines were covalently linked to PhMV VPs through a suitable spacer that minimized fluorescence quenching. Fluorescent dyes have been linked to CPMV VPs for vascular imaging. For better control of the spacing between functionalized sites in the VP, fluorescent proteins were fused to HBV capsid proteins through adequate spacers, and the functionalized particles were assembled from the fusion proteins. The results obtained encouraged the use of fluorescent VPs to help in the early diagnosis of cancer using FI.

VPs for Magnetic Resonance Imaging (MRI)

Low tissue penetration is a limitation of FI. Fortunately, other in vivo imaging techniques, including MRI and those mentioned in the next paragraphs, can provide high-resolution images deep within tissues. MRI is widely used in the clinic for noninvasive diagnosis and evaluation of therapeutic treatments. MRI requires the use of contrast agents to increase the difference in the signals obtained by irradiating different tissues (e.g., diseased versus normal tissues). Paramagnetic metal complexes (e.g., chelated Gd^{3+}) are generally used as MRI contrast agents in the clinic, sometimes bound to carrier inorganic NPs. Superparamagnetic iron oxide (e.g., Fe_3O_4) can also be used as a contrast agent, and synthetic iron oxide-modified NPs have already been approved for clinical use. However, those inorganic NPs generally show limited cargo capacity, low biocompatibility, and low selectivity. The low sensitivity of MRI with current contrast agents hampers the use of this technique for early diagnosis.

Different engineered VPs are being used to develop selective MRI contrast agents with improved sensitivity [8, 23, 31, 79, 83, 84]. In several studies by different groups, large numbers of Gd^{3+} ions were attached to VPs from CCMV, CPMV, PhMV, TMV, Qβ, or MS2 following different strategies: direct chelation of the ion to the RNA genome (CPMV) or to binding sites in the VP (CCMV); linking of Gd^{3+}-chelating groups to the VP (PhMV, P22); fusion of a Gd^{3+}-binding peptide to the N-terminus of the capsid proteins for its display on the assembled VP surfaces; or binding of metal chelates to in situ synthesized polymers (P22, Qβ). In other studies, functionalized iron oxide NPs were enclosed in VPs derived from BMV, rotavirus, SV40, HBV, or M13 by driving capsid assembly around the metal core [79, 83]. These and other studies have proven that VPs can be functionalized with large amounts of MRI contrast agents; can protect the contrast agent and reduce its toxicity; and can deliver the contrast agent to specific cells, increasing the contrast between the targeted cells (e.g., tumor cells) and non-targeted (e.g., healthy) cells and facilitating accurate and timely diagnosis by MRI.

VPs for Positron Emission Tomography (PET)
PET is a high-resolution, highly sensitive in vivo imaging technique that can detect pathological changes, and that is widely used for early cancer diagnosis. However, radioisotopes used for PET are not specific and have short half-lives in the organism. VPs are being explored as radioisotope carriers to slow down their clearance from the organism and to drive them to target cells (e.g., cancer cells) for increasing PET sensitivity [31, 79, 83].

Examples of VP engineering for improved PET include the chemical functionalization with some radioisotope-containing compounds (e.g., [^{18}F]-fluorobenzaldehyde) of VPs derived from MS2, Qβ, or NV1203 (an oncotropic herpesvirus).

VPs for Computer Tomography (CT)
CT is another in vivo imaging technique that is widely used for diagnosis. Engineered VPs are being explored also for improved CT. To provide one example, gold-coated CPMV VPs were successfully used to increase CT sensitivity and resolution between tissues [31].

VPs as Nanobiosensors
Nanobiosensors are nanometric-sized sensors for the detection of a substance (the analyte) through its specific interaction with an immobilized biomolecule or material of biological origin (the detector). The detector biomolecule must be coupled to another component (the transducer) that, upon recognition of the analyte by the detector, provides an amplified signal (optical, electronic, or other) that can be quantified. The design of a biosensor involves the identification or design of a biomolecule or biomolecular assembly that can specifically interact with the analyte.

VPs constitute particularly adequate platforms for the construction of highly sensitive and specific biosensors [59, 60, 81, 85, 87]. VPs can be readily functionalized (Sect. 22.3) with many copies of both the detector and the transducer, and their relative positions can be controlled if required for appropriate signal amplification. Several engineered VPs derived from different viruses, including CPMV, TMV , P22, M13, and HBV, are being explored for the design and development of new nanobiosensors. VP-based biosensors have already been devised for the detection of bacteria, viruses, metabolites, disease (e.g., cancer) biomarkers in human samples, drugs, or other analytes of biomedical interest. Nanobiosensors are also being developed for agricultural or industrial applications.

In one representative study on the design and evaluation of VP-based biosensors [88] (Fig. 22.9), a SpyCatcher/SpyTag-based procedure was used to covalently attach to HBV VPs many molecules of an anti-thrombin antibody (the detector) and of fluorescent nanoluciferase (the transducer). This nanobiosensor detected very small concentrations (1 nM) of thrombin, a biomarker for diagnosis of pulmonary metastasis and coagulation disorders. In another example, HBV VPs were genetically engineered for displaying at high density nickel nanohairs and antibodies

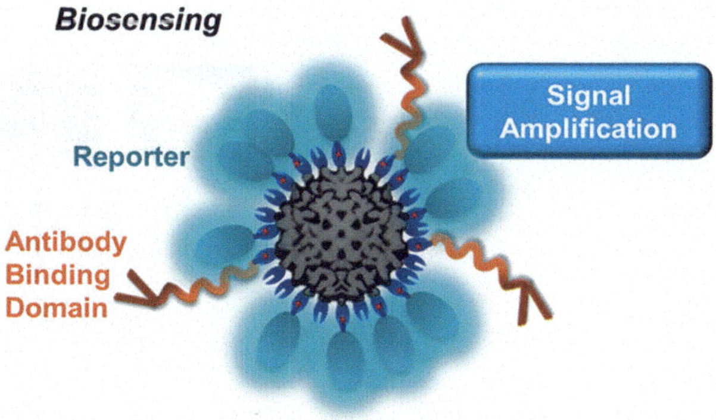

Fig. 22.9 An engineered VP-based biosensor. A SpyCatcher/SpyTag-based procedure was used to covalently attach to HBV VPs many molecules of an anti-thrombin antibody (the detector) and of fluorescent nanoluciferase (the reporter or transducer). See text for details (Sect. 22.4.7) (Reproduced from [88]. With permission)

against troponin I, a marker associated to risk of heart attack. The nanohairs led to the formation of a regular array of functionalized NPs. Troponin binding to the antibody was detected using a secondary antibody and a tertiary antibody labelled with a fluorescent QD. Biosensors based on phage VPs as detectors are being developed for the highly specific and sensitive detection of bacterial species and strains in human samples, food, or the environment [59, 60].

22.4.8 VPs for Theranosis

The term theranosis is applied to procedures that combine diagnosis and therapy. Some attractive theranostic strategies are being developed by combining noninvasive in vivo imaging methods (Sect. 22.4.7) with therapies based on targeted drug delivery (Sect. 22.4.6). Engineered multifunctional VPs may, thus, provide particularly well-suited tools for theranosis [31, 79, 89]. Accordingly, an increasing number of studies are focusing on the engineering of multifunctional VPs that can simultaneously provide a contrast agent for in vivo diagnostic imaging and a drug delivery vector.

In a representative study [90] (Fig. 22.10), Qβ VPs were functionalized with GFP fused to the capsid proteins for fluorescence imaging. A cell-penetrating peptide was conjugated to the surface of the GFP-VP particle for enabling its passage through the blood-brain barrier. ^{68}Ga was bound to the VP outer surface for PET imaging. Finally, epirubicin was attached as a cytotoxic drug to be released in the slightly acidic tumor microenvironment. The theranostic action of the multifunctional engineered Qβ VPs was tested in a glioblastoma mouse model. The biodistribution of the VPs was followed by FI, PET, and CT. The engineered VPs efficiently delivered the cytotoxic drug in a localized manner and with reduced side effects compared to those of the cytotoxic drug used alone. The tumor was eradicated after 8 days with two doses of the modified VPs.

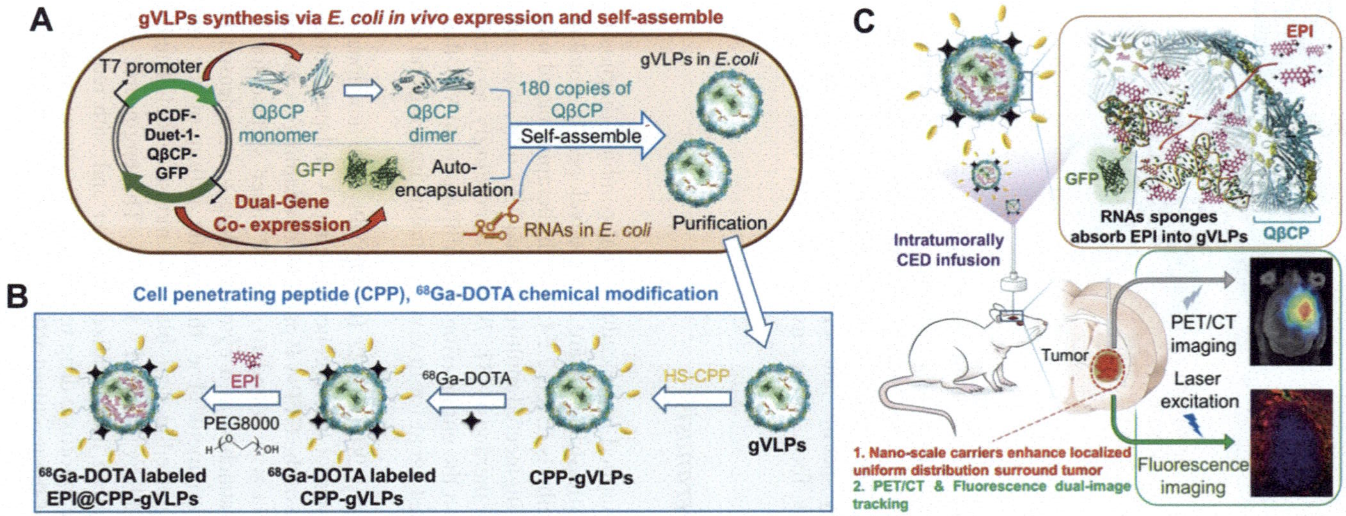

Fig. 22.10 An engineered VP for cancer theranosis. (**a**) procedure for obtaining Qβ VPs containing GFP as a fluorescent marker (gVLPs). (**b**) gVLPs were chemically modified to present a cell-penetrating peptide (CPP), ^{68}Ga-DOTA for PET/CT imaging and epirubicin (EPI) for killing target cells. (**c**) Schematics of the infusion of the functionalized VPs in a brain tumor. See text for details (Sect. 22.4.8). (Reproduced from [90] which is subjected to license CC BY-NC 4.0 DEED, https://creativecommons.org/licenses/by-nc/4.0/)

PDT, PTT, and MHT (Sect. 22.4.7) are particularly suited to develop promising theranostic treatments. In certain cases, both therapy and in vivo imaging can be based on the same active compound.

22.5 VPs for Bio/Nanotechnological Applications

22.5.1 Bacteriophages for the Directed Evolution of Peptides and Proteins

Directed evolution is a powerful genetic approach for engineering proteins with modified structure, properties, and/or functions. This approach is heavily used both for basic protein research and for biomedical or biotechnological applications [33–35]. As described in more detail in Sect. 22.3.1, the sequence of a target protein gene is partially randomized by point mutation and/or recombination, and expressed to obtain a combinatorial library of protein variants. Then, the variant in the library that has the desired feature(s) is identified and isolated by either screening or selection.

Phage display is a general and extremely powerful approach for the selection of the desired variant from a combinatorial library of peptides or proteins [36–39] (Fig. 22.11). The original phage display approach was devised to identify short peptides with high affinity for a target ligand, either a biomolecule (e.g., a cell receptor protein or antibody) or a non-biologic ligand. A synthetic peptide with a randomized amino acid sequence is genetically fused to the major capsid protein pVIII of phage M13, which yields a large library of recombinant phages, each one displaying many copies of a peptide variant on the capsid surface. The phage library is panned using an immobilized target ligand. Under stringent enough conditions, only those phages that carry variant peptides with a high enough affinity for the ligand are retained. They are then eluted, and the panning cycle is repeated until only the phages with the highest affinity for the ligand remain bound. The selected phages are eluted and amplified by infection of an *E. coli* host. The peptide-coding region in the genome of the selected recombinant phages is sequenced to identify the sequence (s) of the high-affinity peptides they carry.

The above procedure was later modified to display on the phage surface one or a few copies of protein domains or full-length proteins fused to the minor capsid protein pIII through an adequate peptide spacer. Based on the same principle, other display methodologies that use other viruses, ribosomes, or cells instead of phages to link phenotype and phenotype have been later devised. Phage display and related methodologies directly select for higher affinity between protein or peptide variants and the chosen ligand. However, these procedures have been modified in different ways to select variant proteins based on other features, such as higher stability or increased catalytic activity.

Phage display and related methodologies have led to the identification of many engineered peptides or proteins with high affinity for different ligands (even for

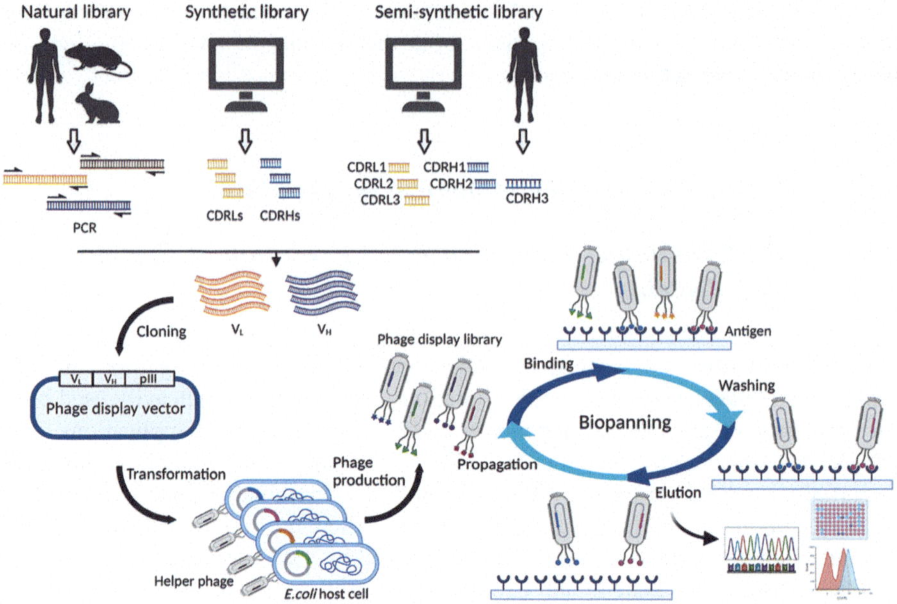

Fig. 22.11 Phage display for the selection of variant proteins from combinatorial libraries. See text for a description of the general procedure (Sect. 22.5.1). The figure illustrates the main steps for obtaining a combinatorial library of scFv antibody fragments displayed on recombinant M13 phages (left), followed by the main steps for selecting from the phage library those scFv variants with a higher affinity for an antigen immobilized on a solid substrate. (Reproduced from [39]. With permission)

non-biological ligands, such as metals, metal compounds, or other inorganic compounds [43]; see Sect. 22.5.4); altered ligand specificity; increased stability at high temperatures, pH extremes or organic solvents; and, in the case of enzymes, increased catalytic activity or modified substrate specificity. Many engineered proteins selected by phage display have been approved for clinical use, or commercialized for industrial applications. One particularly relevant outcome of the phage display technology has been the development of improved therapeutic monoclonal antibodies. By the end of 2022, 17 phage display-selected antibodies had been approved for clinical use, and many others are under clinical trial [39].

22.5.2 *VPs for Improved or Confined Enzymatic Reactions*

Modified VPs (and other NPs) containing heterologous enzymes are being actively explored as nanoreactors for improved biocatalysis [8, 20, 23, 91–93]. Some enzymes have been displayed on the outer surface of engineered VPs for increased activity. However, most studies so far on nanobioreactors for understanding and/or improving enzyme catalysis have focused on the effects of confining enzymes in the relatively large cavities of VPs or other hollow protein cages (e.g., ferritin, vaults, or bacterial microcompartments).

Engineered VPs used as experimental nanoreactors include those derived from CCMV, P22, or MS2. One or multiple copies of one or several different enzymes have been enclosed in those VPs using different procedures (outlined in Sect. 22.3). One simple approach involves the (re)assembly of the VP under adequate conditions in the presence of free enzyme molecules. VPs that encapsulated enzyme molecules at either low or high density were obtained by adjusting the relative concentrations of capsid protein and enzyme. In one study, CCMV VPs containing one enzyme molecule only were obtained to study catalysis at the single-molecule level. Encapsulation was, in some cases, promoted by non-covalent interactions between suitably modified enzyme molecules and capsid proteins. A better control of the number and localization of the enzyme molecules inside or outside the VP was achieved by genetically fusing the enzyme, through an adequate peptide spacer, to viral capsid proteins or scaffolding proteins.

Entry of small substrates or exit of small products through pores in enzyme-containing VP nanoreactors has been modulated by genetic engineering of the pore walls. For example, substitution of some amino acid residues lining the pores in P22 or MS2 VPs altered the local electrostatics and influenced the entry of charged substrates into the VP nanoreactor. Modulation of local electrostatics and/or pore size can be used to enhance the reaction or even modify the selectivity of the nanoreactor.

Multienzyme nanoreactors have been built by encapsulating two or three different enzymes to enhance the coupling of serial enzymatic reactions along a metabolic pathway via substrate channeling. In one study, a nicotinamide adenine dinucleotide enzyme cofactor was chemically tethered to the inner wall of a P22 VP nanoreactor that contained two functionally coupled enzymes. The immobilized cofactor could shuttle between the two encapsulated enzymes, but could not escape from the nanocontainer. Multienzyme catalysis by a regular array of VP-based nanoreactors was demonstrated by using P22 VPs [94] (Fig. 22.12). A range of peptides were displayed on the outer surface of VPs that contained encapsulated enzyme molecules. The peptides modulated the VP surface charge allowing the interaction between the modified VPs and branched organic polymers carrying the opposite charge to form lattices of enzyme-loaded VPs. This strategy allowed the construction of an extended nanostructured material with multienzyme catalytic activity for the production of isobutanol. The VP lattice required less solvation than the equivalent number of free VPs, which allowed for very high enzyme and substrate concentrations that resulted in increased catalytic efficiency.

Studies with nanoreactors have revealed many interesting effects of confinement on enzymatic catalysis. These include complex changes in enzyme kinetics, selectivity, or substrate stability and, when enzymatic reactions were coupled by confining two enzymes, substrate channeling between the active sites was detected. Many of those changes could be traced to macromolecular confinement and crowding; physicochemical differences between the microenvironment inside the particle and

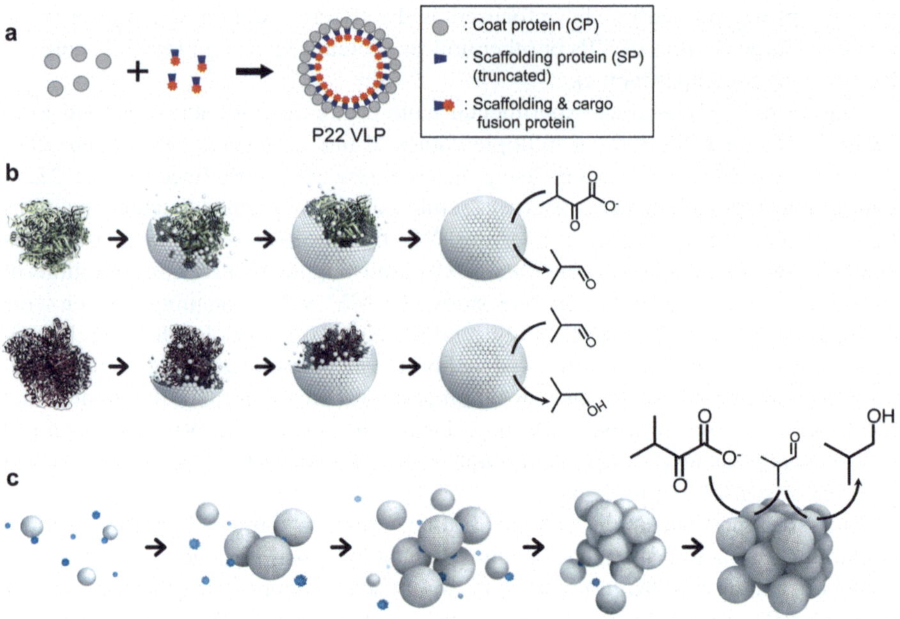

Fig. 22.12 A VP-based nanoreactor. (**a**) Strategy for assembly of a P22 VP with encapsulated cargo from the capsid protein CP and cargo molecules fused to the viral scaffolding protein SP. (**b**) The strategy illustrated in (**a**) was followed to encapsulation in P22 VPs the enzyme KivD (green), which catalyzes the conversion of α-ketoisovalerate to isobutyraldehyde, or the enzyme AdhA (purple), which catalyzes the conversion of isobutyraldehyde to isobutanol. (**c**) The VPs with the encapsulated enzymes assemble into a higher order material mediated by dendrimers (blue). See text (Sect 22.5.2). (Reproduced from [94] with permission)

the solution; and/or limitations imposed by the VP shell barrier to substrate availability or product diffusion through existing pores. Those fundamental studies are guiding the design of efficient VP-based nanoreactors for higher reaction rates and yields, higher selectivity, increased enzyme half-life, and lower cost.

22.5.3 VPs for the Fabrication of Metal NPs and Bioinorganic NPs

Nanostructured NPs containing metals, metal compounds, or other inorganic materials are being fabricated and tested for an increasing number of developing applications. Unfortunately, inorganic NPs of defined size, geometry, and organization are very difficult to obtain using most current procedures. The precise size and geometry of VPs, the many options for their functionalization, and other features (Sect. 22.1) have stimulated the use of VPs as templates or nanoreactors for the

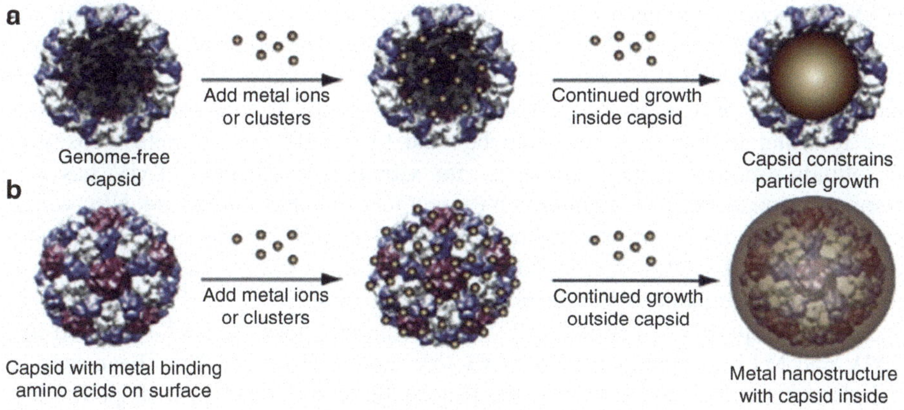

Fig. 22.13 Two general approaches for VLP-templated growth of inorganic (metal) NPs. The inorganic NP is nucleated in the VLP cavity (**a**) or formed on the VLP outer surface (**b**). See text (Sect. 22.5.3). Reproduced from [54]. With permission

synthesis of nanostructured inorganic NPs and the fabrication of bioinorganic hybrid materials. VP-based, metal-containing bioinorganic NPs may find different applications in biomedicine, including improved contrast agents for MRI or CT-based diagnosis, highly sensitive biosensors (Sect. 22.4.7), or cell type-specific magnetic NPs for localized magnetic hyperthermia therapy (Sect. 22.4.6). Developing or potential technological applications of bioinorganic NPs include their use as improved catalysts or as components of novel physical, chemical, or biological sensors, batteries, electronic or optoelectronic devices (e.g., solar cells), or data storage (memory) devices [8, 19–21, 23, 31, 43, 46, 48, 52–56, 81, 95–100].

Many VPs of different sizes and shapes (quasi-spherical, elongated, or rod-like) are being used for the fabrication of bioinorganic NPs. The list includes VPs from plant viruses (e.g., CCMV, CPMV, BMV, RCNMV, TMV), phages (e.g., P22, M13), and animal viruses (e.g., SV40, HIV-1, alphavirus). Inorganic materials can be attached to the VP outer surface or synthesized on it. They can also be encapsulated or built inside the VP cavity. Adequate modification by genetic or chemical means of the VP outer surface or inner wall is required in most cases for the attachment, encapsulation, or synthesis of metallic and other inorganic NPs (Fig. 22.13). Engineering strategies followed in those cases include the introduction of suitable amino acid residues at chosen positions on or in the VP for site-specific covalent functionalization; non-covalent binding using procedures based, for example, on His tag-Ni-NTA or (strept)avidin-biotin interactions; or other strategies (Sect. 22.3). Depending on the foreseen application, the fabricated metal-VP composites are used as such, or the VP proteins are removed to yield a purely metallic NP.

Attachment or Templated Synthesis of Metallic NPs on the VP Outer Surface
In some studies, the VP was used as a template for attaching preformed metallic NPs. For example, CPMV VPs were modified by site-directed mutagenesis to introduce cysteine residues at specific positions on the capsid surface. Multiple iron oxide or

gold NPs were conjugated to those cysteine residues to produce a geometrically organized array of metal NPs on the VP surface. The attached gold NPs were connected through oligophenylenevinylenes for the fabrication of conductive nanonetworks. In another study, TMV was functionalized with cobalt ferrite NPs. The resulting hybrid particles could be used to modify the magnetoviscosity of ferrofluids, improving their properties for vibration dampening. TMV has also been functionalized with platinum NPs or QDs to yield composite NPs whose conductance could be switched on and off, and that could be used to build memory devices. Gold NPs, QDs, or other NPs have also been attached to the outer surface of adequately functionalized VPs from CPMV, P22, or M13.

Many adequately functionalized VPs, including those derived from rod-like TMV and M13, or spherical CPMV, CCMV, and P22 viruses, have been used as templates for the synthesis of metallic nanoshells with defined size, shape (hollow nanospheres or nanorods), and organization. TMV in particular has been intensively used for the templated synthesis of metal or metal compound NPs including iron oxide nanotubes, and CdS or PbS QDs among many others. M13-based phage display was used for the selection from combinatorial peptide libraries of variant peptides with high affinity for metals, metal compounds, or other inorganic compounds [43]. The M13 phages that displayed peptides with high affinity for metal or metal compounds were used as templates to fabricate, for example, semiconductor (ZnS, CdS) or magnetic nanomaterials. M13 has also been used to fabricate gold-platinum nanowires for catalytic oxidation, metal oxides, polymers, and NPs for improved battery electrodes. CPMV VPs have been used to synthesize many different metallic nanoshells. CCMV and P22 VPs have been used to template the synthesis of gold NPs and QDs.

Encapsulation or Fabrication of Inorganic NPs in the VP Cavity

In other studies, inorganic NPs were encapsulated in BMV, CCMV, alphavirus, RCNMV, or SV40 VPs using different strategies to drive or promote encapsulation (Sect. 22.3). For example, in early studies, surface-functionalized gold nanoparticles or QDs drove their own encapsulation in BMV VPs by acting as templates for capsid assembly.

Many VPs have been used as nanoreactors for the confined nucleation and growth of inorganic NPs of defined size and shape (nanospheres or nanowires). In pioneering studies [19, 48], positively charged groups on the inner wall of CCMV VPs were removed by site-directed mutagenesis to promote the mineralization of polyoxometalate salts in the VP cavity, yielding monodisperse crystalline inorganic NPs. In other early studies, the CCMV cavity was used for mineralization of paratungstate or decavanadate to yield monodisperse inorganic NPs. In another study, positively charged groups at the CCMV capsid inner wall were replaced by negatively charged groups, which promoted electrostatic interactions with ferrous ions leading to the formation of crystalline iron oxide NPs in the VP cavity. TMV VPs whose central cavity had been activated with Pd or Pt ions were used as nanoreactors to fabricate thin conducting nanowires made of Ni, Co, Cu, or Co-Fe alloy by electroless deposition.

In a few studies, combined strategies were followed for incorporating metallic nanostructures both on the outer surface and in the cavity of a chosen VP. For example, QDs were encapsulated and gold NPs were attached to the outer surface of adequately functionalized SV40 VPs. The resulting hybrid NPs were used to study surface plasmon resonance coupling of gold NPs and energy transfer between the internal QDs and the external gold NPs. In another study, engineered MS2 VPs with encapsulated gold NPs fluorescent dyes displayed on the outer surface were used to study energy transfer between the dye and the gold NP for research and potential applications in nanophotonics.

22.5.4 VPs for Energy Harvesting, Generation, or Transfer

Light Harvesting

Light harvesting NPs are being fabricated to develop new photovoltaic or photocatalytic devices. In these devices, light is collected by a donor moiety, and the energy is transferred to an acceptor moiety to provide the desired output (electricity or an increased chemical reaction rate, respectively). VPs provide suitable platforms for the precise positioning of light harvesting molecules in order to maximize its efficiency through interchromophore Förster resonance energy transfer (FRET) [54, 81].

Engineered CPMV, TMV, MS2, and M13 VPs have been used as platforms for experimental light harvesting devices. In one example [101], a photocatalytic NP was built by conjugating a donor chromophore to the MS2 capsid inner wall, and an acceptor zinc porphyrin to a non-natural amino acid introduced on the capsid outer surface. FRET between donor and acceptor occurred through the 1-nm thick protein shell. Energy transfer to the acceptor was determined by a colorimetric response upon reduction of methyl viologen. The catalytic cycle was completed by reduction of the resulting porphyrin radical cations by β-mercaptoethanol.

Piezoelectricity, Triboelectricity, and Pyroelectricity

Piezoelectricity, triboelectricity, and pyroelecticity, respectively refer to the generation of an electrical potential upon mechanical pressure, transient contact, or heating. Three studies by the same research group demonstrated the piezoelectric, triboelectric, and pyroelectric behavior of a virus, phage M13, and proved that a virus can be used to generate electrical energy.

In the first study [102], phage M13 films were self-assembled on gold substrates, and the virus piezoelectric behavior was determined by piezoresponse force microscopy, a variant of atomic force microscopy (AFM; Chap. 9). A metal-coated tip was used to apply a periodic mechanical load to a phage film, and the generated electrical potential was quantified. The piezoelectric response was modulated by genetically modifying the phage particle to alter the number of negatively charged amino acid residues in the pVIII major capsid protein N-terminus. Films made of genetically

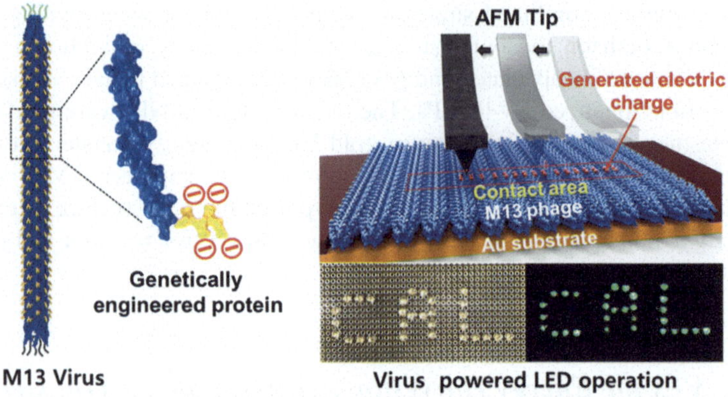

Fig. 22.14 An engineered array of VPs for converting mechanical energy into electrical energy. Left: M13 virions were modified by genetic engineering to introduce more surface charges in order to generate higher triboelectric potentials. Right: arrays of M13 virions when contacted with an AFM tip generated enough electricity to power 30 light-emitting diodes. See text for a detailed description (Sect. 22.5.4). (Reproduced from [103]. With permission)

engineered M13 phages were used to build a piezoelectric generator that provided enough electrical energy to operate a liquid crystal display.

In the second study [103] (Fig. 22.14), a triboelectric generator was built, again using genetically engineered M13 phages. In this case, negatively charged amino acid residues were fused to the N-terminus of the pVIII protein, and electricity was generated upon contact of the engineered phage film with a metal-coated AFM tip. The M13-based triboelectric generator generated enough electrical power to operate 30 light-emitting diodes.

In a recent study by the same group [104], the pyroelectric behavior of phage M13 was demonstrated. A His-tag was fused to the minor capsid protein pIII of M13, and the engineered phages were bound through the His-tag to Ni-NTA-coated gold substrates to create a film of unidirectionally polarized phage particles. The pyroelectric properties were tuned by genetically engineering the major capsid protein pVIII to contain a variable number of negatively charged residues at its N-terminus. The electric potential generated upon heating of the phage films was quantified by Kelvin probe force microscopy using an AFM microscope with a metal-coated tip. The pyroelectric behavior was traced to conformational changes in the pVIII capsid subunits, that resulted in changes in polarization along the longitudinal axis of the phage particle.

These results encourage the use of engineered VPs for the development of devices that can harvest mechanical energy (piezoelectric or triboelectric materials) or thermal energy (pyroelectric materials) to generate electrical energy.

22.5.5 VPs to Develop Mono-dimensional (1D), Bi-dimensional (2D), and Three-dimensional (3D) Nanostructured Materials

Most studies mentioned in Sects. 22.4 and 22.5 so far involved the use of functionalized VPs in free (non-aggregated) form. However, some studies used engineered VPs as building blocks for the hierarchical construction of nanostructured 1D, 2D, or 3D arrays. Engineered VP arrays may present unique physical or chemical properties or functionalities that are not present in the isolated engineered VPs in free form, and that can be exploited for further technological or biomedical uses [23, 51, 52, 55, 56, 76, 81, 97].

Rod-like TMV or M13 phages at high concentrations under defined conditions tend to self-assemble into organized superstructures (liquid crystals). Examples on the potential use of such VP arrays are provided, for example, by the studies on the use of M13 phage films attached to gold substrates for building piezoelectric, triboelectric, or pyroelectric generators (Sect. 22.5.4) (Fig. 22.14).

As an alternative to self-assembled arrays, functionalized VPs derived from rod-like (e.g., M13, TMV) or spherical viruses (e.g., CCMV, CPMV) have been chemically linked to fabricate nanostructured arrays. In addition, ordered 1D, 2D, or 3D hybrid materials have been fabricated by combining functionalized VPs with non-biological components (e.g., CCMV or P22 VPs with organic polymers, HBV VPs with nickel nanohairs, or hydrogels formed by M13 particles and gold NPs). In one study (already described in Sect. 22.5.2), a lattice of P22 VP-based nanoreactors with encapsulated enzymes was made for highly efficient multistep biocatalysis. In another study, thin and flexible electrodes for lithium-ion batteries were fabricated using arrays of metallized VPs. A third example is provided by the engineering of a 3D lattice of adequately functionalized HBV NPs as a biosensor for detection of a biomarker associated to risk of heart attack (Sect. 22.4.7).

There is also great interest in the development of 2D or 3D arrays of engineered VPs as biocompatible scaffolds that can mimic the extracellular matrix, for use in tissue growth, regeneration, and engineering [23, 46]. To promote tissue growth, VPs able to form organized arrays can be engineered to display peptides that recognize cellular receptors, promoting and organizing the attachment of specific cell types into tissue patches. These patches could be implanted into injured tissues and organs for tissue regeneration. In one study, engineered M13 VPs that displayed signaling peptides for neurite extension and neural cell adhesion self-assembled into 3D arrays of aligned nanofibers [105]. These scaffolds supported neural cell proliferation and differentiation and oriented cell growth in three-dimensions, which holds promise for spinal injury repair. In another study, films of M13 phages displaying cell-binding peptides induced, in the presence of calcium and phosphate precursor ions, the biomineralization of bone-like hard tissue composites. TMV, M13, CPMV, turnip yellow mosaic virus VPs have also been used in studies related to tissue growth and engineering.

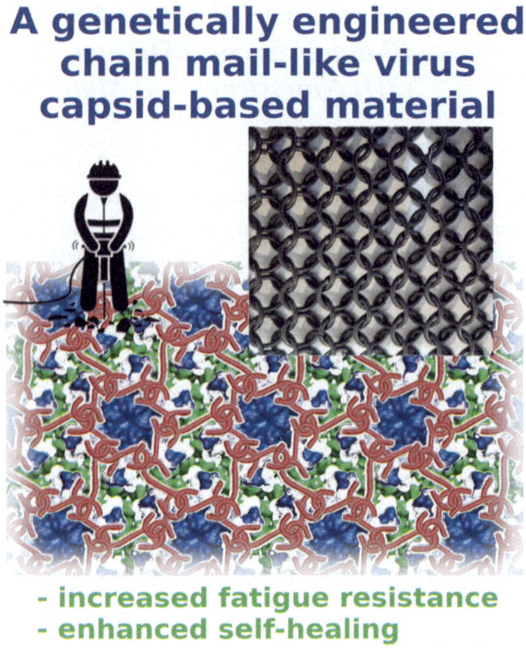

Fig. 22.15 A VP-based material engineered for increased mechanical robustness. Cysteine residues were introduced in appropriate positions of the HIV-1 capsid protein CA. Under non-reducing conditions, a chain mail-like network of interprotein disulfide bonds were formed which dramatically increased the thermal and mechanical resistance of a CA-based 2D nanostructured material [106]. See text for a more detailed description

A different approach was followed to obtain a mechanically robust VP-based 2D material with potential uses as a biocompatible nanostructured coating [106] (Fig. 22.15). A negatively charged inorganic substrate (mica) was used as a flat template to promote the self-assembly of the mature capsid protein of HIV-1 (CA). In this way, an open 2D lattice of CA subunits was obtained instead of a closed 3D particle. The hexagonal lattice architecture was indistinguishable from that of the authentic viral capsid, except for the absence of pentameric defects that lead to lattice curvature to form a closed capsid. To increase the mechanical resistance of the CA monolayer, cysteines were rationally introduced in CA by site-directed mutagenesis. Under normal non-reducing conditions (as found in the extracellular environment), a network of intersubunit disulfide bonds was formed in the CA lattice. The resulting "biomolecular chain mail" preserved the desirable elasticity of the unmodified protein monolayer, but became extremely resistant to disruption by heat, point mechanical forces, or material fatigue.

22.6 Perspectives and Conclusions

Viruses and viral capsids possess remarkable physical and chemical features and can perform sophisticated biological actions. VPs are being modified and functionalized in multiple ways by genetic and/or chemical engineering for many uses. Studies on the fabrication, characterization, and applications of engineered VPs and VP-based nanostructured materials are increasing at an outstanding pace. Some engineered VPs are in current use; many others are being explored for a myriad of future uses. Developing biomedical applications include novel antiviral drugs and vaccines; phage therapy; oncolytic virotherapy; gene therapy; targeted drug delivery; tissue engineering; in vivo diagnostic imaging; and biosensors. Developing technological uses include protein engineering; enzyme nanoreactors and novel catalysts; physical, chemical, or biological sensors; new hybrid materials; batteries; electronic and optoelectronic devices; data memory devices; and devices for energy harvesting, transfer or storage. Many issues must be adequately addressed before most of those applications become an everyday reality in the clinics or the industry. To address many of those issues, the importance of continued fundamental and applied research on the structure and physics of viruses cannot be overemphasized.

Acknowledgements The authors gratefully acknowledge former and current collaborators and members of their group for their invaluable contributions to different studies on structure–properties–function relationships and engineering of VPs. Work in M.G.M.´s laboratory is currently funded by MICINN/FEDER EU (grant PID2021-126973OB-I00), and by institutional grants from the Severo Ochoa Programme for Centers of Excellence and Fundación Ramón Areces. M.G.M. is an associate member of the Institute for Biocomputation and Physics of Complex Systems, Zaragoza, Spain.

References

1. Cortajarena AL, Grove TJ (2016) Protein-based engineered nanostructures. Adv Exp Med Biol 940
2. Liao Z, Tu L, Li X, Liang X-J, Huo S (2021) Virus-inspired nanosystems for drug delivery. Nanoscale 13:18912–18924
3. Montiel-Garcia D, Santoyo-Rivera N, Ho P, Carrillo-Tripp M, Johnson JE, Reddy VS (2021) VIPERdb v3. 0: a structure-based data analytics platform for viral capsids. Nucleic Acids Res 49:D809–D816
4. Pattenden LK, Middelberg APJ, Niebert M, Lipin DI (2005) Towards the preparative and large-scale precision manufacture of virus-like particles. Trends Biotechnol 23:523–529
5. Palomares LA, Ramirez OT (2009) Challenges for the production of virus-like particles in insect cells: the case of rotavirus-like particles. Biochem Eng 45:158–167
6. Gurda BL, Agbandje-McKenna M (2011) Production and purification of viruses for structural studies. In: Agbandje-McKenna M, McKenna R (eds) Structural virology. RSC Publishing, Cambridge, pp 3–21
7. Saunders K, Lomonosoff GP (2013) Exploiting plant virus-derived components to achieve in planta expression and for templates for synthetic biology applications. New Phytol 200:16–26
8. Glasgow J, Tullman-Ercek D (2014) Production and applications of engineered viral capsids. Appl Microbiol Biotechnol 98:5847–5858
9. Culver JN, Brown AD, Zang F, Gnerlich M, Gerasopoulos K, Ghodssi R (2015) Plant virus directed fabrication of nanoscale materials and devices. Virology 479–480:200–212
10. Kim HJ, Kim H-J (2017) Yeast as an expression system for producing virus-like particles: what factors do we need to consider? Lett Appl Microbiol 64:111–123

11. Steele JFC, Peyret H, Saunders K, Castells-Graells R, Marsian J, Meshcheriakova Y, Lomonossoff GP (2017) Synthetic plant virology for nanobiotechnology and nanomedicine. Wiley Interdiscip Rev Nanomed Nanobiotechnol 9:e1447
12. Rybicki EP (2020) Plant molecular farming of virus-like nanoparticles as vaccines and reagents. Wiley Interdiscip Rev Nanomed Nanobiotechnol 12:e1587
13. Nooraei S, Bahrulolum H, Hoseini ZS, Katalani C, Hajizade A, Easton AJ, Ahmadian G (2021) Virus-like particles: preparation, immunogenicity and their roles as nanovaccines and drug nanocarriers. J Nanobiotechnol 19:59
14. Wijesundara YH, Herbert FC, Kumari S, Howlett T, Koirala S, Trashi O, Trashi I, Al-Kharji NM, Gassensmith JJ (2022) Rip it, stitch it, click it: a chemist's guide to VLP manipulation. Virology 577:105–123
15. Evans DJ (1999) Reverse genetics of picornaviruses. Adv Virus Res 53:209–228
16. Nagyova A, Subr Z (2007) Infectious full-length clones of plant viruses and their use for construction of viral vectors. Acta Virol 51:223–237
17. Eiben S, Koch C, Altintoprak K, Southan A, Tovar G, Laschat S, Weiss IM, Wege C (2019) Plant virus-based materials for biomedical applications: trends and prospects. Adv Drug Deliv Rev 145:96–118
18. Mateu MG (2013) Assembly, stability and dynamics of virus capsids. Arch Biochem Biophys 531:65–79
19. Uchida M, Klem MT, Allen M, Suci P, Flenniken M, Gillitzer E, Varpness Z, Liepold LO, Young M, Douglas T (2007) Biological containers: protein cages as multifunctional nanoplatforms. Adv Mater 19:1025–1042
20. De la Escosura A, Nolte R, Cornelissen J (2009) Viruses and protein cages as nanocontainers and nanoreactors. J Mater Chem 19:2274–2278
21. Bronstein LM (2011) Virus-based nanoparticles with inorganic cargo: what does the future hold? Small 12:1069–1618
22. Liu Z, Qiao J, Niu Z, Wang Q (2012) Natural supramolecular building blocks: from virus coat proteins to viral nanoparticles. Chem Soc Rev 41:6178–6194
23. Li F, Wang Q (2014) Fabrication of nanoarchitectures templated by virus-based nanoparticles: strategies and applications. Small 10:230–245
24. Bundy BC, Franciszkowitz MJ, Swartz JR (2008) Escherichia coli-based cell-free synthesis of virus-like particles. Biotechnol Bioeng 100:28–37
25. Hu VT, Kamat NP (2023) Cell-free protein synthesis systems for vaccine design and production. Curr Opin Biotechnol 79:102888
26. Mateo R, Luna E, Rincón V, Mateu MG (2008) Engineering viable foot-and-mouth disease viruses with increased thermostability as a step in the development of improved vaccines. J Virol 82:12232–12240
27. Porta C, Kotecha A, Burman A, Jackson T, Ren J, Loureiro S, Jones IM, Fry EE, Stuart DI, Charleston B (2013) Rational engineering of recombinant picornavirus capsids to produce safe, protective vaccine antigen
28. Vaidya AJ, Solomon KV (2022) Surface functionalization of rod-shaped viral particles for biomedical applications. ACS Appl Bio Mater 5:1980–1989
29. Brune KD, Howarth M (2018) New routes and opportunities for modular construction of particulate vaccines: stick, click, and glue. Front Immunol 9:1432
30. Essus VA, Souza GSE Jr, Nunes GHP, Oliveira JDS, De Faria BM, Romão LF, Cortines JR (2023) Bacteriophage P22 capsid as a pluripotent nanotechnology tool. Viruses 15:516
31. Chung YH, Cai H, Steinmetz NF (2020) Viral nanoparticles for drug delivery, imaging, immunotherapy, and theranostic applications. Adv Drug Deliv Rev 156:214–235
32. Domingo E, Biebricher C, Eigen M, Holland J (2001) Quasispecies and RNA virus evolution: principles and consequences. Landes Bioscience, Austin
33. Hida K, Hanes J, Ostermeier M (2007) Directed evolution for drug and nucleic acid delivery. Adv Drug Deliv Rev 59:1562–1578

34. Bartel MA, Weinstein JR, Schaffer DV (2012) Directed evolution of novel adeno associated viruses for therapeutic gene delivery. Gene Ther 19:694–700
35. Wang Y, Xue P, Cao M, Yu T, Lane ST, Zhao H (2021) Directed evolution: methodologies and applications. Chem Rev 121:12384–12444
36. Kehoe JW, Kay BK (2005) Filamentous phage display in the new millennium. Chem Rev 105: 4056–4072
37. Paschke M (2006) Phage display systems and their applications. Appl Microbiol Biotechnol 70:2–11
38. Carnazza S, Guglielmino S (2010) Phage display as a tool for synthetic biology. Nanotechnology Science and Technology Series. Nova Science Publications, New York
39. Zhang Y (2023) Evolution of phage display libraries for therapeutic antibody discovery. MAbs 15:2213793
40. Li C, Samulski RJ (2020) Engineering adeno-associated virus vectors for gene therapy. Nat Rev Genet 21:255–272
41. Torres-Barceló C, Turner PE, Buckling A (2022) Mitigation of evolved bacterial resistance to phage therapy. Curr Opin Virol 53:101201
42. Swift EA, Pollard SM, Parker AL (2022) Engineering cancer selective virotherapies: are the pieces of the puzzle falling into place? Human Gene Therapy hum.2022.178
43. Cui Y (2016) Engineered phages for electronics. Biosens Bioelectron 85:964–976
44. Strable E, Finn MG (2009) Chemical modification of viruses and virus-like particles. In Manchester M Steinmetz NF (eds) Viruses and Nanotechnology. Curr Top Microbiol Immunol 327:1–21
45. Mateu MG (2011) Virus engineering: functionalization and stabilization. Protein Eng Des Sel 24:53–63
46. Pokorski JK, Steinmetz NF (2011) The art of engineering viral nanoparticles. Mol Pharm 8: 29–43
47. Smith MT, Hawes AK, Bundy BC (2013) Reengineering viruses and virus-like particles through chemical functionalization strategies. Curr Opin Struct Biol 24:620–626
48. Douglas T, Young M (2006) Viruses: making friends with old foes. Science 312:873–875
49. Ryu Y, Schultz PG (2006) Efficient incorporation of unnatural amino acids into proteins in *Escherichia coli*. Nat Methods 3:263–265
50. Garcea RL, Gissmann L (2004) Virus-like particles as vaccines and vessels for the delivery of small molecules. Curr Opin Biotechnol 15:513–517
51. Fischlechner M, Donath E (2007) Viruses as building blocks for materials and devices. Angew Chem Int Ed 46:3184–3193
52. Young M, Willits D, Uchida M, Douglas T (2008) Plant viruses as biotemplates for materials and their use in nanotechnology. Annu Rev Phytopathol 46:361–384
53. Steinmetz NF, Lin T, Lomonosoff GP, Johnson JE (2009) Structure-based engineering of an icosahedral virus for nanomedicine and nanotechnology. In Manchester M Steinmetz NF (eds) Viruses and Nanotechnology. Curr Top Microbiol Immunol 327:23–58
54. Dedeo MT, Finley DT, Francis MB (2011) Viral capsids as self-assembling templates for new materials. Prog Mol Biol Transl Sci 103:353–392
55. Lee S-Y, Lim J-S, Harris MT (2012) Synthesis and applications of virus-based hybrid nanomaterials. Biotechnol Bioeng 109:16–30
56. Bittner AM, Alonso JM, Górzny ML, Wege C (2013) Nanoscale science and technology with plant viruses and bacteriophages. In: Mateu MG (ed) Structure and physics of viruses. Springer, Dordrecht, pp 667–702
57. Alonso JM, Górzny ML, Bittner AM (2013) The physics of tobacco mosaic virus and virus-based devices in biotechnology. Trends Biotechnol 31:530–538
58. van Kan-Davelaar HE, van Hest JCM, Cornelissen JJLM, Koay MST (2014) Using viruses as nanomedicines. Br J Pharmacol 171:4001–4009

59. Harada LK, Silva EC, Campos WF, Del Fiol FS, Vila M, Dąbrowska K, Krylov VN, Balcão VM (2018) Biotechnological applications of bacteriophages: state of the art. Microbiol Res 212–213:38–58
60. Pizarro-Bauerle J, Ando H (2020) Engineered bacteriophages for practical applications. Biol Pharm Bull 43:240–249
61. Ma R, Li Z, Chiocca EA, Caligiuri MA, Yu J (2023) The emerging field of oncolytic virus-based cancer immunotherapy. Trends Cancer 9:122–139
62. Gordillo Altamirano FL, Barr JJ (2019) Phage therapy in the Postantibiotic era. Clin Microbiol Rev 32:e00066–e00018
63. Lenneman BR, Fernbach J, Loessner MJ, Lu TK, Kilcher S (2021) Enhancing phage therapy through synthetic biology and genome engineering. Curr Opin Biotechnol 68:151–159
64. Hatfull GF, Dedrick RM, Schooley RT (2022) Phage therapy for antibiotic-resistant bacterial infections. Annu Rev Med 73:197–211
65. Strathdee SA, Hatfull GF, Mutalik VK, Schooley RT (2023) Phage therapy: from biological mechanisms to future directions. Cell 186:17–31
66. Petrovic Fabijan A, Iredell J, Danis-Wlodarczyk K, Kebriaei R, Abedon ST (2023) Translating phage therapy into the clinic: recent accomplishments but continuing challenges. PLoS Biol 21:e3002119
67. Endersen L, O'Mahony J, Hill C, Ross RP, McAuliffe O, Coffey (2014) Phage therapy in the food industry. Annu Rev Food Sci Technol 5:327–349
68. Moye Z, Woolston J, Sulakvelidze A (2018) Bacteriophage applications for food production and processing. Viruses 10:205
69. Harrington K, Freeman DJ, Kelly B, Harper J, Soria J-C (2019) Optimizing oncolytic virotherapy in cancer treatment. Nat Rev Drug Discov 18:689–706
70. Duan S, Wang S, Qiao L, Yu X, Wang N, Chen L, Zhang X, Zhao X, Liu H, Wang T, Wu Y, Li N, Liu F (2023) Oncolytic virus-driven biotherapies from bench to bedside. Small 19: 2206948
71. Wang D, Gao G (2014a) State-of-the-art human gene therapy. Part I. Gene delivery technologies. Discov Med 18:67–77
72. Wang D, Gao G (2014b) State-of-the-art human gene therapy. Part II. Gene therapy strategies and clinical applications. Discov Med 18:151–161
73. Fakhiri J, Grimm D (2021) Best of most possible worlds: hybrid gene therapy vectors based on parvoviruses and heterologous viruses. Mol Ther 29:3359–3382
74. Kotterman MA, Schaffer DV (2014) Engineering adeno-associated viruses for clinical gene therapy. Nat Rev 15:445–451
75. Wen AM, Rambhia PH, French RH, Steinmetz NF (2013) Design rules for nanomedical engineering: from physical virology to the applications of virus-based materials in medicine. J Biol Phys 39:301–325
76. Farr R, Choi DS, Lee S-W (2014) Phage-based nanomaterials for biomedical applications. Acta Biomater 10:1741–1750
77. Czapar AE, Steinmetz NF (2017) Plant viruses and bacteriophages for drug delivery in medicine and biotechnology. Curr Opin Chem Biol 38:108–116
78. Ma Y, Nolte RJM, Cornelissen JJLM (2012) Virus-based nanocarriers for drug delivery. Adv Drug Deliv Rev 64:811–825
79. Kim KR, Lee AS, Kim SM, Heo HR, Kim CS (2023) Virus-like nanoparticles as a theranostic platform for cancer. Front Bioeng Biotechnol 10:1106767
80. Wang J, Fang T, Li M, Zhang W, Zhang ZP, Zhang XE, Li F (2018) Intracellular delivery of peptide drugs using viral nanoparticles of bacteriophage P22. J Mater Chem B 6:3716–3726
81. Petrescu DS, Blum AS (2018) Viral-based nanomaterials for plasmonic and photonic materials and devices. WIREs Nanomed Nanobiotechnol 10:e1508
82. Kaur P, Aliru ML, Chadna AS, Asea A, Krishnan S (2016) Hyperthermia using nanoparticles-promises and pitfalls. Int J Hyperth 32:76–88

83. Shukla S, Steinmetz NF (2015) Virus-based nanomaterials as positron emission tomography and magnetic resonance contrast agents: from technology development to translational medicine: virus-based nanomaterials as PET and MRI contrast agents. WIREs Nanomed Nanobiotechnol 7:708–721
84. Schwarz B, Douglas T (2015) Development of virus-like particles for diagnostic and prophylactic biomedical applications: development of virus-like particles. WIREs Nanomed Nanobiotechnol 7:722–735
85. Yur D, Lieser RM, Sullivan M, Chen W (2021) Engineering bionanoparticles for improved biosensing and bioimaging. Curr Opin Biotechnol 71:41–48
86. Hu H, Masarapu H, Gu Y, Zhang Y, Yu X, Steinmetz NF (2019) Physalis mottle virus-like nanoparticles for targeted cancer imaging. ACS Appl Mater Interfaces 11:18213–18223
87. Lee JW, Song J, Hwang MP, Lee KH (2013) Nanoscale bacteriophage biosensors beyond phage display. Int J Nanomedicine 8:3917–3925
88. Hartzell EJ, Lieser RM, Sullivan MO, Chen W (2020) Modular hepatitis B virus-like platform for biosensing and drug delivery. ACS Nano 14:12642–12651
89. Beatty PH, Lewis JD (2019) Cowpea mosaic virus nanoparticles for cancer imaging and therapy. Adv Drug Deliv Rev 145:130–144
90. Pang HH, Chen PY, Wei KC, Huang CW, Shiue YL, Huang CY, Yang HW (2019) Convection-enhanced delivery of a virus-like nanotherapeutic agent with dual modal imaging for besiegement and eradication of brain tumors. Theranostics 9:1752–1763
91. Cardinale D, Carette N, Michon T (2012) Virus scaffolds as enzyme nano-carriers. Trends Biotechnol 30:369–376
92. Maity B, Fujita K, Ueno T (2015) Use of the confined spaces of apo-ferritin and virus capsids as nanoreactors for catalytic reactions. Curr Opin Chem Biol 25:88–97
93. Wang Y, Douglas T (2023) Tuning properties of biocatalysis using protein cage architectures. J Mater Chem B 11:3567–3578
94. Uchida M, McCoy K, Fukuto M, Yang L, Yoshimura H, Miettinen HM, LaFrance B, Patterson DP, Schwarz B, Karty JA, Prevelige PE Jr, Lee B, Douglas T (2018) Modular self-assembly of protein cage lattices for multistep catalysis. ACS Nano 12:942–953
95. Love AJ, Makarov V, Yaminsky I, Kalinina NO, Taliansky ME (2014) The use of tobacco mosaic virus and cowpea mosaic virus for the production of novel metal nanomaterials. Virology 449:133–139
96. Mohammadniaei M, Park C, Min J, Sohn H, Lee T (2018) Fabrication of electrochemical-based bielectronic device and biosensor composed of biomaterial-nanomaterial hybrid. Adv Exp Med Biol 1064:263–296
97. Selivanovitch E, Douglas T (2019) Virus capsid assembly across different length scales inspire the development of virus-based biomaterials. Curr Opin Virol 36:38–46
98. Xing X, Chen M, Gong Y, Lv Z, Zhou HS-T (2020) Building memory devices from biocomposite electronic materials. Sci Technol Adv Mat 21:100–121
99. Yang W, Li X, Fei L, Liu W, Liu X, Xu H, Liu Y (2022) A review on sustainable synthetic approaches toward photoluminescent quantum dots. Green Chem 24:675–700
100. Wang W, Gao S, Wang Y, Li Y, Yue W, Niu H, Yin F, Guo Y, Shen G (2022) Advances in emerging photonic memristive and memristive-like devices. Adv Sci 9:2105577
101. Stephanopoulos N, Carrico ZM, Francis MB (2009) Nanoscale integration of sensitizing chromophores and porphyrins with bacteriophage MS2. Angew Chem Int Ed 48:9498–9502
102. Lee BY, Zhang J, Zueger C, Chung W-J, Yoo SY, Wang E, Meyer J, Ramesh R, Lee S-W (2012) Virus-based piezoelectric energy generation. Nat Nanotechnol 7:351–356
103. Kim H, Lee J-H, Lee JH, Lee BY, Lee BD, Okada K, Ji S, Yoon J, Le H, Lee S-W (2021) M13 virus triboelectricity and energy harvesting. Nano Lett 21:6851–6858
104. Kim H, Okada K, Chae I, Lim B, Ji S, Kwon Y, Lee S-W (2023) Virus-based pyroelectricity. Adv Mat 35:230553

105. Merzlyak A, Indrakanti S, Lee SW (2009) Genetically engineered nanofiber-like viruses for tissue regenerating materials. Nano Lett 9:846–852
106. Domínguez-Zotes S, Fuertes MA, Rodriguez-Huete A, Valbuena A, Mateu MG (2022) A genetically engineered, chain mail-like nanostructured protein material with increased fatigue resistance and enhanced self-healing. Small 18:e2105456

Further Reading

Especially recommended for further reading are references [8, 14, 17, 31, 48, 77, 78, 84, 89] *listed above.*

Index of Virus Names

M. G. Mateu (ed.), *Structure and Physics of Viruses*, Subcellular Biochemistry 105,
https://doi.org/10.1007/978-3-031-65187-8

O

P

Q

R

S

T

V

W

Note: Listed virus names generally refer to virus species. In some cases names refer to more than one species, which in the book and the literature are then specified (sometimes) by a number or letter (or some other code).

Index of Virus Families

M. G. Mateu (ed.), *Structure and Physics of Viruses*, Subcellular Biochemistry 105,
https://doi.org/10.1007/978-3-031-65187-8

Morphological grouping of tailed phages cited in the text (no longer recognized by ICTV as virus families)

Note: Some virus genera in the above families, or genera unassigned to any family, are occasionally cited in the text. Most genera cited in this book are prototype genera, and their names include the root of the family they belong to (e.g., *Tymovirus* genus of the *Tymoviridae*). A few other genera are occasionally mentioned (e.g., Alphavirus (*Togaviridae*), Enterovirus (*Picornaviridae*); Aquareovirus, Orbivirus, Rotavirus (*Reoviridae*); Lentivirus (*Retroviridae*), etc.).

General Index[1]

[1]Note: For keywords with more than 100 occurrences in the text (capsid, electron, glycoprotein, layer, maturation, particle, receptor, sequence, spike, structure, trimer, viral), page numbers are indicated for the first time that keyword is cited in each chapter.

M. G. Mateu (ed.), *Structure and Physics of Viruses*, Subcellular Biochemistry 105,
https://doi.org/10.1007/978-3-031-65187-8

B

G

N

The manufacturer's authorised representative in the EU is Springer Nature Customer Service Centre GmbH, Europaplatz 3, 69115 Heidelberg, Germany. If you have any concerns regarding our products, please contact ProductSafety@springernature.com

Printed and bound by CPI Group (UK) Ltd, Croydon, CR0 4YY

01/10/2026

02225914-0019